ADVANCES IN PHOTOGRAMMETRY, REMOTE SENSING AND SPATIAL INFORMATION: 2008 ISPRS CONGRESS BOOK

International Society for Photogrammetry and Remote Sensing (ISPRS) Book Series

Book Series Editor

Paul Aplin
School of Geography
University of Nottingham
Nottingham, UK

information from imagery

Advances in Photogrammetry, Remote Sensing and Spatial Information Sciences:

2008 ISPRS Congress Book

Editors

Zhilin Li
Department of Land Surveying and Geo-Informatics, Hong Kong Polytechnic University, Hong Kong

Jun Chen
National Geomatics Centre of China, Beijing, China

Emmanuel Baltsavias
Institute of Geodesy and Photogrammetry, ETH-Hoenggerberg, Switzerland

CRC Press
Taylor & Francis Group
Boca Raton London New York Leiden

CRC Press is an imprint of the
Taylor & Francis Group, an **informa** business
A BALKEMA BOOK

Note on the front cover image:

The cover image relates to the ISPRS 2008 Congress motto "Silk Road for Information from Imagery".

The ancient Silk Roads were regarded as the information super-highway of their age, serving as the conduit not only for goods but also for the transmission of knowledge and ideas between east and west.

Nowadays, the need for timely, quality, long-term, global spatial information requires building and operating new Silk Roads based on knowledge sharing and international cooperation that can transfer information from an unprecedented amount of imagery to everyone in an emerging, people-centered, inclusive and development-oriented Information Society.

CRC Press/Balkema is an imprint of the Taylor & Francis Group, an informa business

Typeset by Vikatan Publishing Solutions (P) Ltd., Chennai, India
Printed and bound in Great Britain by Cromwell Press Ltd, Trowbridge, Wiltshire.

Published by: CRC Press/Balkema
P.O. Box 447, 2300 AK Leiden, The Netherlands
e-mail: Pub.NL@taylorandfrancis.com
www.crcpress.com – www.taylorandfrancis.co.uk – www.balkema.nl

Library of Congress Cataloging-in-Publication Data is applied for

ISBN: 978-0-415-47805-2 (hbk)
ISBN: 978-0-203-88844-5 (ebook)

Advances in Photogrammetry, Remote Sensing and Spatial Information Sciences:
2008 ISPRS Congress Book – Li, Chen & Baltsavias (eds)
© 2008 Taylor & Francis Group, London, ISBN 978-0-415-47805-2

Table of contents

Advances in Photogrammetry, Remote Sensing and Spatial Information Sciences: 2008 ISPRS Congress Book – Li, Chen & Baltsavias (eds)
© 2008 Taylor & Francis Group, London, ISBN 978-0-415-47805-2

List of contributors

Altan, Orhan
Department of Geodesy and Photogrammetry, Faculty of Civil Engineering, Istanbul Technical University, 34469 Ayazaga-Istanbul, TURKEY. E-mail: oaltan@itu.edu.tr.

Anton, François
Informatics and Mathematical Modelling, Technical University of Denmark, Building 321, 2800 Kgs. Lyngby, DENMARK. E-mail: fa@imm.dtu.dk

Armenakis, Costas
Department of Earth and Space Science and Engineering, Geomatics Engineering, York University, 4700 Keele St., Toronto, Ontario, M3J 1P3, CANADA. E-mail: armenc@yorku.ca

Baltsavias, Emmanuel
Institute of Geodesy and Photogrammetry, ETH-Hoenggerberg, CH-8093 Zurich, SWITZERLAND. E-mail: manos@geod.baug.ethz.ch

Boccardo, Piero
Politecnico di Torino—DITAG, Torino, ITALY. E-mail: piero.boccardo@polito.it

Böhm, Jan
Institute for Photogrammetry, Universität Stuttgart, GERMANY. E-mail: jan.boehm@ifp.uni-stuttgart.de

Budge, Amelia M.
Earth Data Analysis Center, MSC01 1110, Bandlier West RM 111, 1 University of New Mexico, Albuquerque, NM 87131-0001, USA. E-mail: abudge@edac.unm.edu

Chen, Jun
National Geomatics Centre of China, No 1 Baishengcun, Zizhuyuan, Beijing 100044, PR CHINA. Email: chenjun@nsdi.gov.cn or chenjun_isprs@263.net

Cho, Kohei
Department of Network and Computer Engineering, Tokai University, 2-28-4, Tomigaya, Shibuya-ku, 151-0063 Tokyo, JAPAN. E-mail: cho@yoyogi.ycc.u-tokai.ac.jp

Crosetto, Michele
Institute of Geomatics, Av. Canal Olímpic s/n, Castelldefels, Barcelona, SPAIN. E-mail: michele.crosetto@ideg.es

D'Apuzzo, Nicola
Hometrica Consulting—Dr. Nicola D'Apuzzo, Zurich, SWITZERLAND. E-mail: nda@hometrica.ch

de Jong, Steven
Department of Physical Geography, Faculty of Geosciences, University of Utrecht, Utrecht, The NETHERLANDS. E-mail: s.dejong@geog.uu.nl.

Dees, Matthias
Department of Remote Sensing and Landscape Information Systems—FeLis, University of Freiburg, Freiburg, GERMANY. E-mail: matthias.dees@t-online.de

Dowman, Ian
Department of Civil, Environmental and Geomatic Engineering, University College, Gower Street, London, WC1E 6BT, UK. E-mail: idowman@cege.ucl.ac.uk

El-Hakim, Sabry
National Research Council, Ottawa, CANADA. E-mail: Sabry.el-Hakim@nrc.ca

El-Sheimy, Naser
Department of Geomatics Engineering, The University of Calgary, CANADA.
E-mail: naser@geomatics.ucalgary.ca

Everaerts, Jurgen
Centre for Remote Sensing and Earth Observation Processes, Flemish Institute for Technological Research, BELGIUM. E-mail: jurgen.everaerts@vito.be

Fritz, Lawrence W.
Senior Scientist Emeritus, Lockheed Martin Corp. 14833 Lake Terrace, Rockville, MD 20853-3632, USA. E-mail: lwfritz@gmail.com

Gerke, Markus
International Institute for Earth Observation and Geoinformation (ITC),
Department of Earth Observation Science, Enschede, The NETHERLANDS. E-mail: gerke@itc.nl

Giulio Tonolo, Fabio
ITHACA—Information Technology for Humanitarian Assistance Cooperation and Action, Torino, ITALY. E-mail: fabio.giuliotonolo@ithaca.polito.it

Gold, Christopher M.
Department of Computing and Mathematics, University of Glamorgan, Pontypridd, Wales, CF37 1DL, UK. E-mail: cmgold@glam.ac.uk

Gong, Jianya
State Laboratory for Information Engineering in Surveying, Mapping and Remote Sensing, Wuhan University, 129 Luoyu Road, Wuhan, Hubei, 430079, CHINA.
E-mail: jgong@lmars.whu.edu.cn

Gruen, Armin
Institute of Geodesy and Photogrammetry, ETH Zurich, SWITZERLAND.
E-mail: agruen@geod.baug.ethz.ch

Grussenmeyer, Pierre
Photogrammetry and Geomatics Group, INSA Strasbourg, Graduate School of Science and Technology, FRANCE. E-mail: Pierre.Grussenmeyer@insa-strasbourg.fr

Hanke, Klaus
Surveying and Geoinformation Unit, University of Innsbruck, Technikerstrasse 13, A-6020 Innsbruck, AUSTRIA. E-mail: klaus.hanke@uibk.ac.at

Hecker, Chris
Department of Earth Systems Analysis, International Institute for Earth Observation and Geoinformation (ITC), Enschede, The NETHERLANDS. E-mail: hecker@itc.nl

Heipke, Christian
Institute of Photogrammetry and Geoinformation, Leibniz Universität Hannover, Nienburger Str. 1, D30167 Hannover, GERMANY. E-mail: heipke@ipi.uni-hannover.de.

Hinz, Stefan
Remote Sensing Technology, Technische Universität München, Arcisstrasse 21, D-80333 Munich, GERMANY. E-mail: Stefan.Hinz@bv.tum.de

Höhle, Joachim
Department of Development and Planning, Research Group of Geoinformatics, Aalborg University, 11 Fibigerstraede, DK-9220 Aalborg, DENMARK. E-mail: jh@land.aau.dk

Jacobsen, Karsten
Institute of Photogrammetry and Geoinformation, Leibniz Universität Hannover, Nienburger Str. 1, D30167 Hannover, GERMANY. E-mail: jacobsen@ipi.uni-hannover.de

Kainz, Wolfgang
Department of Geography and Regional Research, University of Vienna, Vienna, AUSTRIA. E-mail: wolfgang.kainz@univie.ac.at

Kavouras, Marinos
School of Rural and Surveying Engineering, National Technical University of Athens, GREECE. E-mail: mkav@survey.ntua.gr

Kneubühler, Mathias
Remote Sensing Laboratories, University of Zurich, SWITZERLAND. E-mail: kneub@geo.unizh.ch

Koenig, Gerhard
Geodesy and Geoinformation Science, Berlin University of Technology, Strasse des 17. Juni 135 (H 12), 10623 Berlin, GERMANY. E-mail: gerhard.koenig@tu-berlin.de

Koch, Barbara
Department of Remote Sensing and Landscape Information Systems—FeLis, University of Freiburg, Freiburg, GERMANY. E-mail: barbara.koch@felis.uni-freiburg.de

Kokla, Margarita
School of Rural and Surveying Engineering, National Technical University of Athens, GREECE. E-mail: mkokla@survey.ntua.gr

Li, Songnian
Department of Civil Engineering, Ryerson University, 350 Victoria Street, Toronto, Ontario, M5B 2K3, CANADA. E-mail: snli@ryerson.ca

Li, Zhilin
Department of Land Surveying and Geo-Informatics, Hong Kong Polytechnic University, HONG KONG. E-mail: lszlli@polyu.edu.hk.

Liang, Shunlin
Department of Geography, University of Maryland, College Park, MD 20742, USA. E-mail: sliang@geog.umd.edu

Luhmann, Thomas
Institute for Applied Photogrammetry and Geoinformatics, University of Applied Sciences, D-26121 Oldenburg, GERMANY. E-mail: luhmann@fh-oldenburg.de

Ma, Guorui
State Laboratory for Information Engineering in Surveying, Mapping and Remote Sensing, Wuhan University, 129 Luoyu Road, Wuhan, Hubei, 430079, CHINA. E-mail: maguorui_rs@yahoo.com.cn

Maas, Hans-Gerd
Institute of Photogrammetry and Remote Sensing, Dresden University of Technology, Dresden, GERMANY. E-mail: hans-gerd.maas@tu-dresden.de

Mayer, Helmut
Institute for Photogrammetry and Cartography, Bundeswehr University Munich, D-85577 Neubiberg, GERMANY. E-mail: Helmut.Mayer@unibw.de

Mioc, Darka
Department of Geodesy and Geomatics Engineering, University of New Brunswick, P.O. Box 4400, Fredericton, New Brunswick, E3B 5A3, CANADA. E-mail: dmioc@unb.ca

Mitchell, Harvey
School of Engineering, University of Newcastle, Newcastle, AUSTRALIA.
E-mail: harvey.mitchell@newcastle.edu.au

Morain, Stanley A.
Earth Data Analysis Center, MSC01 1110, Bandlier West RM 111, 1
University of New Mexico, Albuquerque, NM 87131-0001, USA. Email: smorain@edac.unm.edu.

Murai, Shunji
Japan Association of Surveyors, 1-3-4 Koishikawa, Bunkyo-ku, Tokyo 112-0002, JAPAN.
E-Mail: sh1939murai@nifty.com

Pasquali, Paolo
Sarmap S.A., Cascine di Barico, 6989 Purasca, SWITZERLAND. E-mail: paolo.pasquali@sarmap.ch

Patias, Petros
Faculty of Rural & Surveying Engineering, Aristotle University of Thessaloniki, Univ. Box 473, GR-54124 Thessaloniki, GREECE. E-mail: patias@auth.gr

Pfeifer, Norbert
Institute for Photogrammetry and Remote Sensing, Vienna University of Technology, AUSTRIA.
E-mail: np@ipf.tuwien.ac.at

Petrie, Gordon
Department of Geographical and Earth Sciences, University of Glasgow, Glasgow, G12 8QQ, Scotland, UK. E-mail: g.petrie@ges.gla.ac.uk

Robson, Stuart
Department of Civil, Environmental and Geomatic Engineering, University College London, UK.
E-mail: srobson@cege.ucl.ac.uk

Sandau, Rainer
German Aerospace Center (DLR), Rutherfordstr. 2, 12489 Berlin, GERMANY.
E-mail: rainer.sandau@dlr.de

Schaepman, Michael
Centre for Geo-Information, Wageningen University, The NETHERLANDS.
E-mail: Michael.Schaepman@wur.nl

Sester, Monika
Institute of Cartography and Geoinformatics, Leibniz Universität Hannover, GERMANY.
E-mail: monika.sester@ikg.uni-hannover.de

Smillie, Kenneth
Leica Geosystems AG, CH-9435 Heerbrugg, SWITZERLAND.
E-mail: kenneth.smillie@leica-geosystems.com

Stilla, Uwe
Department of Photogrammetry and Remote Sensing, Technische Universität München,
Arcisstrasse 21, 80333 Muenchen, GERMANY. E-mail: Stilla@tum.de

Sui, Haigang
State Laboratory for Information Engineering in Surveying, Mapping and Remote Sensing, Wuhan
University, 129 Luoyu Road, Wuhan, Hubei, 430079, CHINA. E-mail: haigang_sui@263.net

Tang, Xinming
Key Lab for GIS, Chinese Academy of Surveying and Mapping, 16 Beitaiping Road, Haidian District,
Beijing 100039, CHINA. E-mail: tang@casm.ac.cn

Trinder, John
School of Surveying and SIS, University of New South Wales, Sydney, NSW 2052, AUSTRALIA.
E-mail: J.Trinder@unsw.edu.au

van der Meer, Freek
International Institute for Earth Observation and Geoinformation (ITC), Department of Earth
Systems Analysis, Enschede, The NETHERLANDS. E-mail: vdmeer@itc.nl &
University of Utrecht, Faculty of Geosciences, Department of Physical Geography, Utrecht,
The NETHERLANDS. E-mail: f.vandermeer@geo.uu.nl

van der Meijde, Mark
International Institute for Earth Observation and Geoinformation (ITC), Department of Earth
Systems Analysis, Enschede, The NETHERLANDS. E-mail: vandermeijde@itc.nl

van der Werff, Harald
International Institute for Earth Observation and Geoinformation (ITC), Department of Earth
Systems Analysis, Enschede, The NETHERLANDS. E-mail: vdwerff@itc.nl

van Ruitenbeek, Frank
International Institute for Earth Observation and Geoinformation (ITC), Department of Earth
Systems Analysis, Enschede, The NETHERLANDS. E-mail: ruitenbeek@itc.nl

Wehr, Aloysius
Institute of Navigation, Universität Stuttgart, Breitscheidstr. 2, 70174, Stuttgart, GERMANY.
E-mail: wehr@nav.uni-stuttgart.de

Woodsford, Peter A.
1Spatial, Cambridge and University College London, UK. E-mail: woodsford@1spatial.com

Zhang, Yun
Department of Geodesy and Geomatics Engineering, University of New Brunswick, CANADA.
E-mail: yunzhang@unb.ca

Zhou, Qiming
Department of Geography, Hong Kong Baptist University, Kowloon Tong, Kowloon, HONG KONG.
E-mail: qiming@hkbu.edu.hk

Advances in Photogrammetry, Remote Sensing and Spatial Information Sciences:
2008 ISPRS Congress Book – Li, Chen & Baltsavias (eds)
© 2008 Taylor & Francis Group, London, ISBN 978-0-415-47805-2

Foreword

ISPRS holds a Congress every four years and although scientific work goes on all the time, the Congress provides an occasion to review progress and think about what is important for the coming years. The proceedings of the Congress normally fill eight or nine bulky volumes and it is not easy to pick out the papers which review progress. This Congress Book provides the solution: experts have provided a review of developments in their fields of interest, and these reviews have been refereed by other specialists so that the book provides a comprehensive, expert view on the sciences of ISPRS during the previous four years.

The first volume of the ISPRS Book Series was published in 2004 and the sixth hit the bookshops early this year. The volumes have covered a range of topics, but have not been planned to be comprehensive. This volume is different in that it has been carefully designed and covers most of the topics covered by ISPRS and is an excellent companion for the Congress and will be a valuable resource until the next congress.

ISPRS Council believes that this book will be a valuable resource for anybody interested in photogrammetry, remote sensing and the spatial information sciences, and that the book is authoritative and comprehensive. After two introductory chapters, historical and technical, the five sections cover sensors, platforms and data acquisition; data processing and analysis; data modelling, management and visualisation; applications and education. These closely follow the technical commissions of ISPRS, but with photogrammetry, remote sensing and spatial information sciences covered in all sections. The applications section could have taken up a whole book, so the most important applications, in the eyes of the editors have been selected. The authors come from Europe, North America and Asia, reflecting the main regions where research and development takes place; most come from academia, but all are experts in their fields.

The book is the result of a lot of work by the authors and the editors and ISPRS Council is very grateful for their efforts. It is the role of ISPRS to set standards in the promotion of the sciences of photogrammetry and remote sensing and this book sets the standard at a high level. Council very much hopes that this book will be the first of a series of Congress Books published by ISPRS and that the quality and comprehensive coverage will make it attractive to lecturers, students and practitioners.

Ian Dowman & Orhan Altan

Advances in Photogrammetry, Remote Sensing and Spatial Information Sciences: 2008 ISPRS Congress Book – Li, Chen & Baltsavias (eds)
 ISBN 978-0-415-47805-2

Preface

The Congress is the largest ISPRS event, taking place every four years. It is the most important event in the community of photogrammetry, remote sensing and spatial information sciences.

It is always the case that during the period of four years between two successive ISPRS congresses significant scientific and technological progress will have been made in the fields of photogrammetry, remote sensing and spatial information sciences. As a result, thousands of papers are submitted to each congress, and recorded in many thick volumes of proceedings—the International Archives of Photogrammetry, Remote Sensing and Spatial Information Sciences. For example, at the XXth Congress held in Istanbul (Turkey), during 12–23 July 2004, more than 50 poster and 100 oral presentation sessions were scheduled and nearly 2000 papers were presented over the 12 days. At this coming Congress to be held in Beijing during 3–11 July 2008, a total of 2600 papers are to be presented in 143 oral and 45 poster sessions.

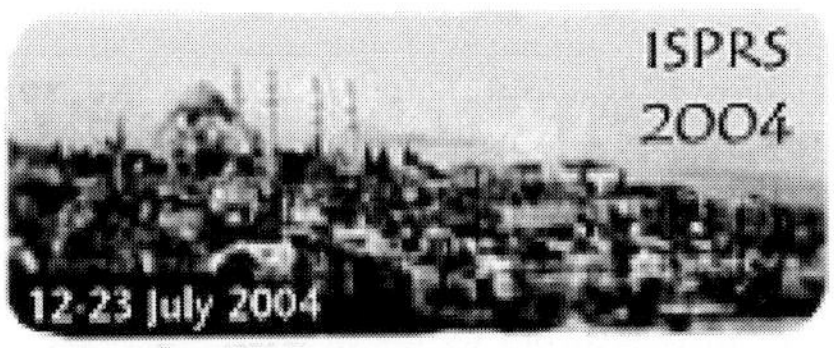

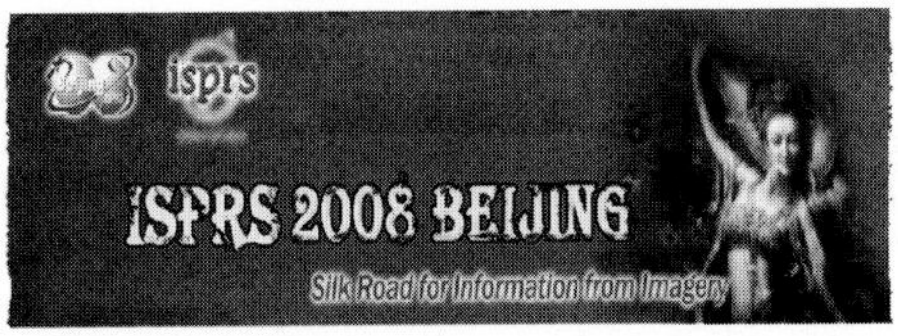

It is always pleasant to see significant progress being made in different fields within ISPRS. On the other hand, this also causes difficulties for researchers and practitioners when attempting to digest such a huge body of literature. Therefore, as one can imagine, it would be very desirable to have a few overview/review papers that describe the major achievements and look to future developments in different fields. This motivated Council in 2003 to initiate the ISPRS congress book series on a permanent basis. It was proposed to produce the first book for the Istanbul Congress but it was not materialized due to time constraints. Thus, this is the first attempt to realize such an endeavour. The Congress Director, Chen Jun, discussed the idea with Zhilin Li in the middle of 2005 and together they produced a proposal in June 2006. This proposal was strongly supported by the ISPRS Council and the Second Vice President, Emmanuel Baltsavias, then joined the team as one of the editors.

The volume is intended to cover the major themes in all eight technical commissions. It was not intended to cover the topics of all existing ISPRS Working Groups, for various reasons. The emphasis is on the substantial developments since the Istanbul Congress. After intensive discussions with council members, a list of 36 chapters was finalized. In this volume, a total of 34 chapters are included. They are divided into six parts as follows:

Parts	Title	No. of Chapters
I	Introduction	2
II	Sensors, Platforms and Data Acquisition Systems	7
III	Data Processing and Analysis	8
IV	Data Modelling, Management and Visualization	6
V	Applications	9
VI	Education and Cooperation	2

Apart from the shorter Parts I and VI, the technical chapters have been grouped naturally into four parts, i.e. data acquisition (sensors, platforms and systems), (b) processing and analysis (methods and algorithms) of image data, (c) modelling and representation of spatial data which are more related to spatial information sciences, and (d) image-based applications (photogrammetry and remote sensing). In Part V Applications, only nine topics were selected although the number of existing applications is large and continuously increasing. This selection was based on the following criteria:

- their importance, i.e. wide applications;
- the availability of contributors, i.e. commitment of authors to contribute chapters; and
- mainstream activities of the ISPRS community.

After approval of the contents by the ISPRS Council, the active members of the Society were invited to contribute. Most of them accepted our invitations and completed their chapters in time, although understandably some of our colleagues were too heavily committed to participate in this project. In the end, we have a list of 68 researchers in the authors' list. From this list, you could also see big names there such as the President Ian Dowman, Past Presidents Shunji Murain (1992–1996), Lawrence W. Fritz (1996–2000) and John Trinder (2000-2004), and the 2008 Brock Gold Medal Award recipient Armin Gruen. All authors completed the drafts according to guidelines, revised their chapters based on reviewers' and editor's comments and proofread edited versions with great efficiency.

This Congress Book is intended for a wide audience, not only for scientists and researchers but also for university students and practitioners. As mentioned previously, it is intended to emphasize the state-of-the-art developments in photogrammetry, remote sensing and spatial information sciences, particularly the major progress in the last four years since the 2004 Congress. However, as this is the first book of its kind, a broad spectrum has been covered to make the book more comprehensive. For example, a chapter on historical development is also included. In the end, this particular ISPRS Congress Book also offers a more general overview instead of just the latest developments.

It is not an easy task to edit such a large volume, written by so many authors and under strict time constraints. Indeed, it is a challenging task. At this stage, we realise that the Book has its imperfections. It can be noted that (a) not all topics that should be covered are included in the book; (b) some chapters don't give the full picture but rather authors' personal views based on their experiences and knowledge; (c) an in-depth review, with multiple revision cycles, was not possible due to the time constraints; (d) better coordination between the different chapters could have been achieved if time permitted this. These points might be taken into consideration in the editing of Congress Books in the future. Anyway, we believe that, in spite of these imperfections, this Book would be a useful product for many persons interested in photogrammetry, remote sensing and spatial information sciences.

Naturally, we now felt relieved after the compilation of the material submitted by authors. At the same time, we feel obliged to express our thanks to:

- All the authors who worked very hard to make this book possible;
- All the reviewers for their swift comments;
- The ISPRS council members and technical commission presidents for their support and help on various topics (e.g. defining book topics and recommending possible contributors);
- The State Bureau of Surveying and Mapping (of China), Chinese Society of Geodesy, Photogrammetry and Cartography, and National Natural Science Foundation of China for their support (including financial support);
- ISPRS book series editor, Paul Aplin, for his assistance;
- Léon Bijnsdorp and his colleagues of Taylor & Francis for making this publication possible; and
- David Tait for editing the English of the chapters.

All in all, we appreciate the efforts made by the various parties on this project and now we are pleased to present this volume to you.

Zhilin Li, Jun Chen & Emmanuel Baltsavias
April 2008

Part I
Introduction

Advances in Photogrammetry, Remote Sensing and Spatial Information Sciences: 2008 ISPRS Congress Book – Li, Chen & Baltsavias (eds)

CHAPTER 1

Historical development of ISPRS

John Trinder & Lawrence W. Fritz

ABSTRACT: ISPRS was created in Vienna, Austria, as the International Society of Photogrammetry (ISP) in 1910. Remote sensing was added to its name and activities in 1980. For almost the past 100 years, the areas of study of ISPRS have been changing markedly according to developments in technologies. This chapter describes chronologically the formation of the Society, the evolution of its structure and operations for the periods 1910–1930, 1930–1980, and 1980 to the present, and briefly reviews the scientific and technological areas of study of the Society.

Keywords: History, development of ISPRS, photogrammetry, remote sensing, spatial information

1.1 INTRODUCTION

The International Society for Photogrammetry and Remote Sensing (ISPRS) is a non-governmental international organization, whose mission is devoted to the development of international cooperation for the advancement of knowledge, research, development, education and training in the photogrammetry, remote sensing and spatial information (P&RS&SI) sciences, their integration and applications, to contribute to the well-being of humanity and the sustainability of the environment. The Society pursues its mission without any discrimination on grounds of race, religion, nationality or political philosophy. Established as ISP in 1910, ISPRS is the oldest international umbrella organization in its fields, which may be summarized as addressing 'Information from Imagery'.

ISPRS has been and continues to be a thriving international community of Societies, Institutes and related member organizations and groups that are allied by having and sharing common interests in the P&RS&SI sciences and technologies. This chapter provides a brief summary of its illustrious history with highlights of the evolution of its structure and of its most significant scientific achievements over the past century.

The primary sciences and technologies of ISPRS are defined in the current Statutes and Bylaws (ST&BL) as:

> *Photogrammetry and Remote Sensing* is the art, science and technology of obtaining reliable information from non-contact imaging and other sensor systems, about the Earth and its environment, and other physical objects and processes through recording, measuring, analysing and representation.
>
> *Spatial Information Science* is the art, science and technology of obtaining reliable spatial, spectral and temporal relationships between physical objects, and of processes for integration with other data for analysis, portrayal and representation, independently of scale.

Information on the current status of ISPRS structure and activities is maintained on its website [www.isprs.org] and published quadrennially in volumes as *"The International Archives of the Photogrammetry, Remote Sensing and Spatial Information Sciences"*.

1.2 EVOLUTION OF THE SOCIETY STRUCTURE

All quotations and information presented in Section 2 "Evolution of the Society Structure" and in Appendices 1–4 are derived from the Part A volumes of "*The International Archives of the Photogrammetry, Remote Sensing and Spatial Information Sciences*".

1.2.1 *ISP. The founding years. 1907–1930*

The ISPRS was founded as the International Society of Photogrammetry on 4 July 1910 in Vienna, Austria, under the leadership of its first President, Prof. Dr. Eduard Doležal (1862–1955), who was Professor for Practical Geometry at the Technical University of

Vienna, and also Rector of that University from 1908 to 1909. In Doležal's words,

> "... in the year 1907, I founded the 'Austrian Society of Photogrammetry' in anticipation of my expectations of the rapid development of photogrammetry in order to gather interested specialists in this field for the common work. By the following year I was able to publish the 'International Archiv für Photogrammetrie' as the first professional review of photogrammetry. Due to the influence of that event, the foundation of the Section Laussodat of the 'Société Française de Photographie' took place the next year and was followed in 1909 by the founding of the 'German Society of Photogrammetry' at Jena. More and more national societies were organized, although their members still belonged to the Austrian Society. In 1910 I reorganized the Austrian Society into the International Society of Photogrammetry. The 'Austrian Society' was the first national section admitted into this International Society, followed by the 'German Society' as its second section. In September 1913, I convened the first International Congress of Photogrammetry in Vienna, at which 400 photogrammetrists and interested scientists took part from almost all European countries. Germany, France and Austria pointedly participated in an exhibition supplementing the Congress.
>
> This first Congress was a very promising beginning and a magnificent base for international cooperation. Unfortunately this achievement was interrupted by the First World War and thereafter all international activities were suspended for many years. Nevertheless I was able to continue the publication of further volumes of the International Archives of Photogrammetry. At my instigation the 2nd International Congress of Photogrammetry was held in 1926 at Berlin with aerial photography and photogrammetry in the foreground of interest. ...As a consequence of this Congress international competition between Germany, France, Italy, Switzerland, and other countries resulted in the inventing of better and improved ingenious photogrammetric cameras, plotting apparatus and other photogrammetric equipment. The Congress again paved the way for international scientific cooperation which contributed to the reestablishment of connection between nations torn apart by the war."
>
> After this Congress many new national photogrammetric societies were organized in swift succession, so that the 3rd International Congress, held at Zurich in 1930, was attended by 13 national societies."

In its earliest years, the national member bodies of ISP were named Sections.

> "The 2nd Congress was held during 22–26 November in Berlin at the Technische Hochschule, where a very valuable exhibition was also arranged, with many new constructions. Once again the impulse to coordinate was felt, within either different countries or groups of such. Sweden, Norway, Denmark, Finland and the Baltic countries agreed to form a 'Northern Section', intended to function until the creation of national societies in these countries. It was decided at the Congress that new Congresses should be held every four years.
>
> Congresses were then held in Zurich in 1930, in Paris in 1934 and in Rome in 1938. This last Congress was held at a time of great international tension—the Munich agreement was made on the first day of the Congress: several nations were not represented; others left Rome during the meeting.
>
> Then war broke out again. No meeting could be held until ten years later, at Scheveningen in 1948, where good results were obtained, although all the consequences of the past years were not overcome. In 1952 the Congress was held in more tranquil circumstances in Washington ..." (Odencrants 1956)

Appendices 1 and 2 provide a chronological compendium of ISPRS Congresses and Memberships, respectively. A Congress has convened every quadrennium (except the World War years). The Congress has served well as the forum for its Member Organizations to administrate their common interests; to present, discuss and coordinate their scientific activities; and to exhibit their most recent individual and joint scientific and technological (S&T) successes in a public arena.

The original Statutes of ISP were prepared in German and published in the International Archives for Photogrammetry in 1911 as *"Organ der Internationalen Gesellschaft für Photogrammetrie"*. Membership of the Society comprised separate National Sections, which were to include a minimum of 10 individuals. Individual membership was possible for persons originating from countries that were not members. The category of Corresponding Member also existed. However, societal administrations such as voting rights of members were not defined in this document.

> "At a meeting in Jena, 28–29 September 1925, it was decided to arrange an International Congress in 1926, and to elect a Society and Congress Board independent of all commercial firms" (Odencrants 1956).

Thus at the 1926 Berlin Congress the national delegations present elected the first ISP Council composed of a President, Honorary President, Secretary General and Congress Director. Doležal, who was President until 1926, was elected Honorary President and continued to play a significant role in the scientific activities of the Society until 1938. The Presidents and Council members who succeeded Doležal are listed in Appendix 3. The 1926 Congress also selected 11 "Technical Committees" (three of which had two parts) covering 14 specific areas of the science and assigned each to a member nation with its named individual to preside. The 1930 Statutes named these as "Scientific Commissions" but by 1948, the title had evolved to become "Technical Commissions". Appendix 4 is a full list of all Commissions with their elected host countries and Presidents.

Delegations from 13 National Societies attended the September 1930 Zurich Congress. National delegate meetings were held at which they decided upon a new set of ISP Statutes to codify societal organization, operations and administration. Proudly, General Perrier, President of the French Society of Photogrammetry, stated:

> "…that our International Society is a perfectly private organization and that with us neither the right of voting nor the imposing of rates (for the rest very little) are subordinated to the control of Governments. It was at Zurich in 1930 that we established the By-laws of our Society ourselves. Completely independent from any duty as regards Governments, we can abstract ourselves from political troubles dealing only with our science." (1938).

The 1930 ISP regulations were briefly outlined in 17 Statutes (ST) in five chapters covering the Society's: I-Object; II-Organization; III- Administration; IV-Scientific Work; V-Additional regulations. Membership was defined in Statute 3. (ST.3) as:

> "ST.3.
>
> a. National societies for photogrammetry whose statutes are not opposed to the principles of the International Society.
> b. Individual members (private persons, public authorities, institutes, business houses, etc.) from countries in which there has hitherto been no affiliation to form a national society, or in which no need is felt for such.
> c. Private persons may be admitted to honorary membership for their eminent services to photogrammetry."

The organs of the Society were defined as: The General Meeting; the Delegate Meeting; the Council; the Auditors. One could roughly extrapolate these into the current Congress Plenary, General Assembly, Council and Financial Commission.

> "ST.8. The ordinary general meeting is to take place every time in connection with an international Congress for Photogrammetry. The general meeting approves the president's statement of accounts, the keeping of the books and the preliminary estimates. It appoints the Council and the Scientific Commissions, and decides on the motions put forward by the delegate meeting, the Council, the National Societies and individual members.
>
> At the general meeting every national society is entitled to one vote for every 50 members, provided that its membership is at least 10. Its minimum number of votes is two."…"The highest number of votes for a single national society is 10. There is one vote at the general meeting for each 50 individual members from a country in which there is no national society "

Officer composition and appointment was defined as:

> "ST.10. The Council consists of 7 members belonging to 5 different countries … appointed by the general meeting."
> "ST.11. The two auditors and two deputies are appointed by the delegate meeting."

Scientific work was guided by:

> "ST.14. The technical and scientific meetings of the ISP are held in the form of international congresses. All members are to be invited to these. … Visitors may be admitted. The Congresses are grouped into plenary meetings and meetings of the Commissions. … Reporters will report briefly on the results of the deliberations of the Commissions … and will prepare these reports for print."

It was recognized then that the 1930 Statutes lacked much specificity but they were designed for additional regulations to be submitted for decision at a general meeting. Most ensuing changes were primarily to enact variations of the Commission topics to be addressed in the succeeding term. At the Rome 1938 Congress, the delegates approved the initiative of Prof. Schermerhorn to found an International Review of Photogrammetry publication at no expense to the Society. This became *"Photogrammetria"* which was first published in 1948. In 1989, it was re-titled *"The ISPRS Journal of Photogrammetry and Remote Sensing"* to reflect the full S&T scope of ISPRS.

1.2.2 *Restructuring & development years. 1930–1980*

The dual membership issue for U.K. representation in 1952 was the first significant issue for which the 1930 Statutes were clearly insufficient. Further issues surfaced at the 1956 Stockholm Congress.

> "It has proved during the last 4-year period that the Statutes need a complete revision. We do not think it possible to make one alteration here and one there as the structure and volume of the ISP has changed a lot since its Statutes were written. Among questions to be taken into consideration are e.g. …scope of ISP … Commission organization … membership rules … cooperation with other international organizations … organization within the ISP … the keeping of documents, transactions, etc. … economy, secure funds, calculate fees, etc. … motions, voting, etc."

The 1956 General Assembly (GA) tasked revision to a Statutes Committee consisting of Brown (Chair, UK), Härry (Switzerland), Janicot (France), Möller/Fagerholm (Sweden), Reading (USA) and Schermerhorn (Netherlands). On 9 July 1968 at the Lausanne Congress, the GA approved their more comprehensive set of 29 Statutes and 34 Bye-Laws. These new regulations basically contained the essence and much of the prose contained in the current (pre–2008) ST&BL. In brief, the notable administrative and operational changes to the ISP structure enacted in 1968 were to:

- State ISP is an international Non-Governmental Organization (NGO) devoted to the development of international cooperation for the advancement of Photogrammetry and its application.
- State ISP's non-discriminatory character and define its societal aims for: holding Congresses; initiating and coordinating research; publishing the *International Archives of Photogrammetry* and an international review (i.e. *Photogrammetria*); stimulating formation of National Societies of Photogrammetry; and encouraging publication of scientific papers and journals.
- Redefine National Membership as:

  "ST.4. A country in which activity in photogrammetry exists may apply to join the Society. A country joining the Society shall do so through a single Member Organization. This Member Organization shall be the Member of the Society and shall be responsible to the Society for the proper discharge of all duties of membership including specifically those of:

  - representing the whole community of Photogrammetrists of the country
  - paying the dues of membership
  - participating in the scientific work of the society.

  The Member Organization of a country shall be its National Society for Photogrammetry, or in the absence of a National Society of Photogrammetry

  - an Association of Societies, each having the advancement of Photogrammetry as one of its principal objects,
  - the principal Scientific Academy of that country or its National Research Council,
  - or failing these and provisionally, any other Institute or Association or institutions whether government or non-government concerned with Photogrammetry'.

- State total of living Honorary Members be limited to seven.
- Define Member Organization (MO) categories and their financial obligations.
- Specify the composition of the General Assembly, Council, Financial Commission, Technical Commissions and their powers, tasks, responsibilities, and operating procedures.
- Simplify voting eligibility (one delegate per MO) and modify GA voting procedures to single vote per delegate on administrative and scientific matters, and weighted voting (using MO category as weight) on financial matters.
- Formalize purpose of Commission Boards, National Correspondents, Working Groups (WGs), Inter-Commission WGs and their meeting, coordination, report responsibilities.
- Name the Technical Commissions and list their scope, i.e. terms of reference.
- Specify Congress and Symposia scientific and operational activities, including Archives.
- Designate seat of Society to be country of the Secretary General; official languages to be English, French (definitive version) and German.

The next significant changes to these ST&BL were made at the 1980 Hamburg Congress. There the General Assembly approved amendments to:

- ST.1—change the name of the Society to International Society for Photogrammetry and Remote Sensing;
- BL.18—include remote sensing in the scope and description of all Technical Commissions;
- ST.4—define and add Sustaining Members and Honorary Members;
- ST.7—provide for nomination of Council and clarify roles of two Vice Presidents;
- ST.9—clarify how Congress Director and ST.8—Congress location is selected;
- ST.20 & ST.22—make all voting on all topics be related to category of Member (weighted voting);
- BL.28—clarified authority of Council to act in exceptional circumstances

Thus, after 70 years of functioning under its original name, the Society had changed its name to the International Society for Photogrammetry and Remote Sensing. It was agreed that while 'Photogrammetry' could be considered as a subset of 'Remote Sensing' and hence should be second in the name, out of respect for the forefathers of ISPRS, the Society decided that 'Photogrammetry' should be named first. The ST&BL introduced remote sensing formally into the terms of reference of all commissions with the intention that all commissions would embrace remote sensing needs and interests. Commission VII was assigned primary responsibility for remote sensing applications. It was recommended that all Members consider extending their name and activities to include remote sensing. However, the changes to the name and terms of reference of the commissions did not immediately lead to ISPRS being considered as the major organization studying remote sensing. There were already a number of organizations promoting remote sensing in their scientific activities and hence it has taken time for ISPRS to establish its reputation in remote sensing. Soon after the Hamburg Congress, the Society adopted its first logo, having a globe-shape overprinted with ISPRS. It was designed by Commission II President, Dr. Jarko Jaksic of NRC, Canada.

1.2.3 *Outreach and positioning years. 1980 to present*

At the 1984 Rio de Janeiro Congress, 2nd Vice President Antipov presented the results of a Statutes committee originated by 1972 President Gamble (†1977) and completed by Council under the leadership of President Doyle. Antipov stated:

> "The main disadvantage of the present Statutes and Bylaws is that there is no punctual linkage between their structures. The Statutes consist of 5 chapters and 29 articles. Bylaws have no chapters and in their contents there are 34 articles. … If there were no marginal notes then it would be rather inconvenient to use Statutes and Bylaws. Practically it seems impossible to improve the matter by means of separate amendments. … In a revised version the Statutes are divided into 20 articles. … Accordingly the Bylaws have also 20 articles and they are closely linked with corresponding ones of the Statutes. … As far as the wording is concerned, about 90% of old sentences are used in the new version without or with very minor modification. At the same time, some principle amendments have been suggested."

The reordered ST&BL were approved by the 1984 GA. This brought a high level of order and clarity for managing the affairs of the Society. The ST&BL promptly became a convenient, ready reference for international understanding of MO responsibilities, as well as for ISPRS officers. After 1984, most GA modifications to the ST&BL were relatively minor. But some had a significant impact on how ISPRS conducted its scientific activities and developed its widening interaction with other international organizations, especially due to a rapidly changing S&T world. Other than the quadrennial changes in the scope of the Commissions, the more notable ST&BL changes were:

- Added cooperation with other international organizations as a Society goal (1984)
- Changed definitive text of ST&BL from French to English (1984)
- Changed Members to Ordinary Members; added Regional Memberships (1988)
- Increased tasks of Technical Commissions (promote S&T, cover international standards, organize tutorials, submit progress reports to Council) (1992).

As computer storage techniques advanced, the ISPRS image acquisition activities began to blend with its cartographic activities. This was first recognized by denoting Commission IV in 1980 as "Cartographic & Data Bank Applications" and then in 1992 as "Cartographic & Data Base". The topic was initially identified as Geographic Information Systems (GIS), but by 1996 it was evident that in reality ISPRS addresses data acquisition, data processing, data storage, data integration, and data presentation to produce information at close—as well as far-range, which in sum reality is "the topic of the spatial information sciences".

At the 2000 Amsterdam Congress, President Fritz presented *ISPRS—A Strategic Plan for the 21st Century*. He stated to the GA:

> "In early 1997, the leadership of ISPRS realized the need to define the future of the Society and enhance its ability to stay abreast of the changing global community that it serves. The sciences and technologies we work with and depend upon are advancing ever faster. These are not times for a strong vibrant society such as ISPRS to sit back and assume our activities, structure, outreach and most importantly, value, will be sufficient to meet the challenges and opportunities of the future. Your ISPRS Council colleagues agreed unanimously that it is our elected responsibility to conduct an ISPRS self-evaluation, openly solicit ideas and to formulate a long-range plan—A Strategic Plan of the 21st Century. This is the tale of this journey and the common vision for the future of ISPRS."

In the Strategic Plan, the 1996–2000 Council redefined the ISPRS Mission and the 11 aims of ISPRS to that listed in the current Statutes. To enact the Plan, Council developed the following objectives with each having a list of specified goals and actions to fulfil them.

- Encourage and Facilitate Research and Development
- Advance Knowledge by Scientific Network Creation
- Promote International Cooperation
- Pursue Inter-Disciplinary Integration
- Facilitate Education and Training
- Enhance and Promote Applications
- Develop Recognition of the P&RS&SI Sciences.

The GA approved the Strategic Plan by acclamation. This resulted in approval of numerous modifications to the ST&BL articles covering mission, definitions, activities, member responsibilities, and the addition of Associate Membership. Other Societal activities were initiated as a result of the plan such as: The ISPRS Foundation; registration of ISPRS as a formal entity; ISPRS Annual Reports; ISPRS website posting of opportunities and data sets; meeting rebates to support ISPRS administration; formalization of international outreach with membership in UN agencies such as COPUOS, and other NGOs and

IGOs such as ICSU and CEOS; creation of an International Policy Advisory Committee (IPAC) and an International Science Advisory Committee (ISAC); adding an ISPRS tagline "Information from Imagery"; and a modernized ISPRS logo (determined by a worldwide competition held in 1999–2000 and won by Mr. Mike Kierstead, a graphics design student from Canada). As a result, the activities of ISPRS are now defined as The Photogrammetry, Remote Sensing and Spatial Information Sciences. The Strategic Plan concluded with:

> "The vision of ISPRS is to be 'the' international focal point for excellence in the photogrammetric, remote sensing and spatial information sciences. Its mission and activities all have altruistic roots and it flourishes on the voluntary efforts of its Members. It works on open democratic principles and it subscribes to an internationally accepted non-discriminatory doctrine. ...Its success is now contingent on the collective efforts and dedication of its Members."

In 2004, the areas of activity of the Technical Commissions were completely revised and an eighth commission was added. Two commissions were assigned to each of the three science areas of photogrammetry (Commissions III and V), remote sensing (Commissions VII and VIII), and spatial information (Commissions II and IV), with one emphazising theory and the other applications. Included were two further overarching commissions (Commissions I and VI) to cover image acquisition and education and outreach that support all commissions. This action was a significant departure from the structure that had been in place for 55 years, but this step was considered necessary to ensure that the future communities of individuals in ISPRS will be adequately identified with ISPRS by the titles and the terms of reference of the Commissions. The 2004 GA instantiated the reallocations of these Commission Terms of Reference and formalized the following in the ST&BL:

- Included role and relationship of the ISPRS Foundation—an independent entity devoted solely to support ISPRS philanthropic activities
- Added IPAC, ISAC and International Committee on Remote Sensing of the Environment (ICORSE) as permanent committees
- Established a quorum to require presence of 50% of eligible votes for convening a GA.

1.3 EVOLUTION OF THE ISPRS SCIENCES AND TECHNOLOGIES

1.3.1 *Development of photogrammetry science and technology. 1850s–1970s*

The invention of the photographic process in the mid 19th century stimulated the formation of small, but dedicated, scientific and industrial groups whose goals were to develop creative applications of photography and apparatus for exploiting its potential. In the late 1850s, Aimé Laussedat carried out the first topographical survey of an area by means of a pair of photographs suitably distanced from each other. Concurrently, Ignazio Porro developed the "photogoniometer" and many other ingenious apparatus. Laussedat named the method "Metrical Photography", which after further development was later named "Photogrammetry by Intersection". Early applications of photogrammetry were primarily for terrestrial purposes, although placing cameras on balloons had been attempted as early as the 1860s. By the end of the 19th century, the development of binocular measuring methods using stereopairs of photographs, led by Carl Pulfrich, resulted in a new field of "stereoscopic photogrammetry". It was realized that for extracting metric information from photographs, instruments were required to overcome the need for significant manual computations. Indeed, Otto von Gruber once stated, "He who computes much does not think." Therefore, emphasis was placed on developing analogue restitution instruments based on stereo observations of overlapping photographs, incorporating the concept of the floating mark for measurements of 3-D coordinates.

With the invention of the airplane in 1903, and subsequently the development of aerial cameras, opportunities for applications of aerial photogrammetry expanded rapidly. Doležal had written in 1911: "85% of the earth's surfaces are topographically unknown, the knowledge depends greatly on vague descriptions by explorers." The potential for photogrammetry to overcome the deficiencies in "topographic knowledge" in the form of maps, had clearly been recognized by Doležal and others, once the airplane and aerial cameras were available. The task was to develop instrumentation that would improve the efficiency and accuracy of photogrammetric methods for mapping worldwide. Early examples of these instruments were demonstrated at the 2nd ISP Congress in Berlin in 1926, where significant interest was in aerial applications and instrumentation for restitution of aerial photographs. During this 2nd Congress, Technical Committees were established to coordinate efforts of nations in the scientific areas of study of ISP, the topics of which were determined from Member resolutions agreed at the Congress. Appendix 4 shows the 14 topics of study for the period 1926–1930.

During the 3rd Congress in Zurich in 1930, interest continued in restitution of aerial photographs as well as aerial triangulation, which was an ongoing activity amongst practitioners. Seven areas of study were selected for the Scientific Commissions to address, based on the resolutions at the Congress (see Appendix 4). In 1934, except for the introduction of a commission on aerial triangulation, the commissions were unchanged from those determined in 1930, thus maintaining a significant emphasis on non-topographic applications. By 1938 the ISP reported that the various fields of application conducted by its members had expanded from aerial and terrestrial activities for mapping to many other fields of application, including astronomy, industrial, architecture, archaeology, farming, planning, aerodynamics and the military sciences. At the 5th Congress in Rome in 1938, the only significant change in the commissions was that medical applications replaced the topic of X-ray photogrammetry. These commissions lasted for the period 1938–1948 because World War II interrupted the activities of the Society.

During the 6th Congress in The Hague in 1948, the commissions were reorganized by President Schermerhorn into the fundamental structure that lasted until the 20th Congress in Istanbul in 2004. That is, seven commissions based on the sequence of processes in photogrammetry, which generalized were Commissions: I-data acquisition, II-instrumentation, III-aerial triangulation, IV-mapping, V-non-topographic applications, VI-profession matters, and VII-photo-interpretation.

The first aerial survey camera was developed in 1915. Initially aerial cameras were important for intelligence and mapping purposes in World War I. The early frame cameras had a narrow field angle, but the field angle was increased and both image and geometric quality were improved throughout the period 1930 to 1980. Preceding and during WW II many unique cameras, some with multiple lens designs (up to 9 lens or more) enabling oblique imaging, were developed. A single Wide-Angle lens was developed by Carl Zeiss in Jena, Germany, in the 1930s with nearly 100° field angle and a format of 18 cm square. Bausch and Lomb Optical Company in USA developed a Wide-Angle lens with approximately 90° field angle in 1938 for the US Corps of Engineers and a similar camera was built by Fairchild Camera and Instrument Company in 1940. Wild Heerbrugg introduced the RC5 Wide-Angle aerial camera in 1944. The Super-Wide-Angle camera was developed by Wild Heerbrugg and exhibited for the first time at the 9th ISP Congress in London in 1960, together with analogue instrumentation for mapping from the Super-Wide-Angle photographs. The standard commercial format for Wide-Angle aerial frame cameras in Great Britain and USA was 9 inches square (230 mm) with a focal length of 6 inches (150 mm) while those in other European countries typically had a format of 180 mm square. Glass plates were also used for high precision photogrammetry (terrestrial and aerial) until about the 1970s. A standard format was adopted for all aerial frame cameras of 230 mm square after World War II. Aerial cameras continued to be improved in design and efficiency into the 1970s, with greater levels of automation made possible with the availability of micro-electronics.

Instrumentation continued to be based on an analogue solution of the restitution of images and aerial triangulation from the 1930s into the 1960s. In 1957, the development of the concept of the analytical stereoplotter by Uki Helava brought about a major shift in the approach to the design of stereoplotters, which, except for Ottica Italiana Mechannica, would not be realized in the inventory of the major instrument companies for about 20 years. The analytical plotter is basically a two-stage comparator controlled in real-time by an on-line computer, which provides a feedback loop to maintain the photographs continuously in the correct positions for stereo viewing and coordinate measurement. In 1957, dedicated computers were necessary to perform the high speed computations required for the real-time operations of analytical stereoplotters. But within 10 years computers had been developed with sufficient speed for commercial analytical plotters to be produced. By the 13th Congress of ISP in Helsinki in 1976, almost all major instrument companies had abandoned development of analogue instrumentation and were marketing analytical stereoplotters. However, for some time after the development of analytical stereoplotters and prior to the introduction of digital stereoplotters, hybrid and semi-analytical systems involving analogue plotters with digital output of 3-D coordinates, and innovative solutions for the conversion of analogue stereoplotters to analytical, were developed to extend the life of many analogue stereoplotters throughout the world.

Attempts to automate height measurement in a photogrammetric stereoplotter by the process of stereo-correlation (now referred to as image matching) was initially demonstrated by Gilbert Hobrough in 1957, and several prototype systems were demonstrated at the 10th ISP Congress in Lisbon, Portugal, in 1964. Because of the limitations of digital image processing at that time, the stereo-correlation was based on analogue signal processing. The Raytheon-Wild B8 Stereomat was developed in 1964 as a prototype for orthophoto production, but was not adopted by mapping organizations primarily due to issues of reliability. Hobrough subsequently developed the Gestalt Photo Mapper (GPM) in 1967 for automated orthophoto production, but it did not gain wide acceptance by mapping organizations and therefore was not developed further. Digital photogrammetric systems based on digital processing were not achieved in commercial stereoplotters until the 1990s.

The potential of aerial triangulation to improve the economy of the mapping process by eliminating the need for large numbers of ground control points was recognized in the 1930s and hence became an ongoing topic of study within ISP for about 50 years. Initially, aerial triangulation was carried out on analogue instruments and adjustments were undertaken by analogue and computational interpolation processes, such as stereo-templates. Attempts were also initiated in the 1930s to record the tilts of the aircraft using such equipment as sun and horizon cameras. As computers became available in the 1950s and their power increased in the 1960s, aerial triangulation adjustments could be undertaken by analytical formulations (so-called bundle block adjustment) for almost unlimited photo block sizes, together with appropriate statistical analyses and self-calibration to correct for lens, film and atmospheric systematic errors. The origin of this approach was developed in the 1950s by Dr. Helmut Schmid and Duane Brown in USA, and implemented by the US Coast & Geodetic Survey in the early 1960s into a suite of operational aerotriangulation programs that eliminated the need for laborious computations and analogue template solutions. By 1967 computer capabilities enabled large blocks of photos (>200) to be coordinate refined and adjusted. Work on enhancing the accuracy potential of the bundle adjustment continued well into the 1970s.

While orthophotography had been discussed as early as the 1930s, development of equipment for their production gained momentum into the 1960s when it was apparent that manual methods of map production were unable to satisfy the demands for up-to-date maps throughout the world. Orthophotos, which include the image content of the photos and geometry of a map, could be produced rapidly using specially designed analogue instrumentation in the 1960s and 1970s. Although the analogue approach was accepted for about 20 years, there were deficiencies in the orthophotos. Subsequent computer supported developments such as the Gestalt and Wild OR1 largely overcame the geometric deficiencies. But the correction of uneven illumination across photos and between adjacent photos by analogue photographic equipment produced inconsistent results. Eventually the development of digital systems replaced the analogue approach.

Since its inception, ISPRS Commissions have addressed a variety of close range applications of photogrammetry. Each close range application has its own characteristics, and hence special processes usually have to be developed for each. In many early applications, the measurements were based on stereo observations of the overlapping photos in specially designed stereoplotters, or adaptations of stereoplotters designed for aerial photos. Analytical solutions were also possible, but not common. Specially designed metric cameras for close range applications were available from the instrument manufacturers, but they were inflexible in respect of object distances over which the images were in focus, and were expensive. The majority of these cameras were phased out in the 1980s.

At the 1956 Stockholm Congress, Commission VII President G. C. Coleman, Jr. stated:

> "Photo-interpretation (PI) has been informally defined as the act of identifying objects imaged on photographs and deducing their significance."

PI had formally become a part of ISP activities in 1948 and was normally confined to black & white (B/W) aerial photographs. In the 1950s, colour and colour-IR films were developed and were adapted for PI applications such as for vegetation analysis. Radar images were also available at that time but they were very low resolution. The development of detector technology for image acquisition of reflected solar radiation from the Earth's surface in the infra-red regions of the electro-magnetic spectrum opened up the potential for obtaining images over a much wider range of wavelengths than had been previously possible. This technology became the foundation of the electro-optical imaging systems that are now used by remote sensing experts for detecting visible and infra-red radiation from the Earth's surface. Airborne electro-optical systems in USA originally were tested by NASA for the interpretation of ground cover in the 1960s and early 1970s. However, it is generally agreed that remote sensing was born in July 1972 (during the 12th ISP Congress in Ottawa) when the first of the Landsat satellites was launched. At the 13th Congress in Helsinki, Finland, in 1976, there were strong arguments by remote sensing specialists in ISP for the Society to embrace this technology more strongly in its activities. This stance was supported by Council, which recommended to the GA at the Society's 14th Congress in Hamburg, Germany, in 1980 that the Society's name be changed to include remote sensing.

1.3.2 *The P&RS&SI sciences and technologies. 1980–present*

The period 1980 to the present has seen the influence of the rapid developments in electronic digital technologies, including: faster computing and virtually unlimited data storage capacities at continually reducing costs; faster computer graphics technologies; advanced digital imaging techniques and digital image processing; very high resolution commercial satellite imaging; satellite positioning based on Global Navigation Satellite Systems (GNSS); terrain laser scanning also referred to as LiDAR (Light Detection And Ranging); Synthetic Aperture Radar (SAR) and Interferometric SAR (IfSAR or InSAR) imaging; small satellite technologies; and growth of the Internet. Many of these developments are described later in this book but a brief summary will be given here.

1.3.2.1 *Data acquisition*

The technologies of aerial frame cameras were advanced in the 1980s with the inclusion of forward motion compensation (FMC) to correct for the blurring effects caused by the forward movement of the aircraft during film exposure. This meant that slower high resolution films could be used for aerial photography, leading to higher quality images being available. With the introduction by USA of the Global Positioning System (GPS) in the 1980s, it became possible to determine the position of the aircraft for each exposure during flight. This further advanced the ability of aerial triangulation to reduce the number of ground control points, as was demonstrated at the 16th ISPRS Congress in Kyoto, Japan, in 1988. Positioning systems integrated with Inertial Navigation Systems (INS) and Inertial Measuring Units (IMU) have enabled the determination of camera positions and tilts with sufficient accuracies, in some cases leading to so-called direct orientation. These approaches have substantially reduced or completely eliminated the need for ground control points. Applications of these technologies have been presented at recent ISPRS Congresses and ISPRS WGs have undertaken studies on accuracies of direct orientation.

Although the principles of automatic photogrammetric systems were demonstrated as early as the 1950s, it was not until 1988 that a prototype commercial system based on digital image processing was demonstrated at the 16th ISPRS Congress in Kyoto, Japan. Digital Photogrammetric Workstations (DPW) (or Softcopy Workstations) for undertaking photogrammetric operations on digital images became commercially available in the early 1990s, based on digital images produced by digital cameras or from digitally scanned photos using high precision scanners. These scanners have now reached maturity and can scan rolls of colour film in a matter of minutes per image. The main products from DPWs are digital elevation models (DEM) and orthophotos, since these operations require only minimum manual input. Automated or semi-automated line mapping is also being performed by some institutions.

With the first commercially available high resolution images from space by Space Imaging Co. in 2000, and the first digital aerial camera demonstrated at the 19th ISPRS Congress in Amsterdam, and available commercially in 2001, came the new era of purely digital photogrammetry. While DPWs were commercially available in the early 1990s, it took another 10 years for the technologies to develop before digital aerial cameras could acquire high resolution images with sufficient coverage to compete economically and technically with analogue film cameras. High resolution, highly redundant images with radiometric resolutions of 11 or 12 bits and ground sampling distances (GSD) smaller than 10 cm, are acquired by this new generation of digital aerial cameras. The introduction of these cameras has been claimed by some experts to represent a paradigm shift in photogrammetry, because it is no longer necessary to limit the number of images acquired, as was the case for film images that had to be processed manually. Digital images are processed automatically, and hence can be acquired with much higher overlaps. This greater redundancy in the image acquisition, as well as the improved image quality, leads to more reliable, robust and economic 3-D coordinate determination. Since these cameras have only been introduced over the past 6–8 years, further improvements on the original designs are expected.

The commercial licensing in 1993 of space imaging technologies, which had been developed in the USA for military applications, led to a race by numerous international companies to produce high resolution cameras and systems for the acquisition and processing of high resolution satellite images for commercial applications. The first successful satellite launched was IKONOS (now operated by GeoEye, Inc.) in 1999. This satellite acquires images with 1 m resolution in B/W panchromatic mode, in stereo if requested, and 4 m resolution multi-spectral images. Several similar satellites have since been launched by other US companies. Geometric positional accuracies of derived features to 2 m or better are now common. In October 2007, DigitalGlobe, Inc. launched a 0.5 m resolution commercial satellite and GeoEye has begun work for a 0.25 m satellite to be launched in 2011. Organizations from other countries have also launched satellites that can acquire digital images with resolutions similar to or approaching those of the current US satellites. The panchromatic images provide a digital source of data for digital mapping that competes with medium scale aerial photography, while the multi-spectral images provide a new source of high resolution data for remote sensing applications.

Airborne laser scanning (ALS) or LiDAR has been growing in importance in ISPRS over the past 10–15 years. LiDAR scans the terrain surface with a laser beam at right angles to the flight direction of an aircraft. The measured distances from the aircraft to visible points on the terrain surface enable the position and elevation of points to be determined. The equipment includes a GPS receiver to determine the location of the aircraft and an IMU to continuously determine the tilts of the aircraft. A dense point cloud is determined at a separation typically of about 1 m that represents a digital surface model (DSM) of the visible terrain surface as well as such objects as buildings and trees, cars, lamp posts, etc., but the laser beam may also penetrate vegetation to measure to the terrain surface. The accuracy of the elevation posts is of the order of 10–20 cm. LiDAR systems are improving rapidly in terms of the pulse repetition frequency, power and hence range of operation, the number of echoes recorded for each pulse, leading also to full-wave recording of the echoes, and recently the emission and recording of echoes of more than one frequency cycle. ISPRS has had WGs studying the applications of LiDAR since 1996. There are many applications of LiDAR data such as DEMs of the bare Earth surface, beach erosion studies, infrastructure analysis, flood risk analysis and many more. Interferometric SAR is based on the reception by two radar antennas of a reflected SAR beam from the terrain, which has been emitted by one of these antennas. Elevations can be derived from the phase differences in the data received by the two antennas. The technique has been adapted for satellite and aircraft borne systems. Interferometric SAR has gained increasing applications for determining DEMs with accuracies of the order of 10 m or better for satellite systems to less than 1 m for airborne systems. Differential interferometric SAR analyses changes in elevations over a certain period with accuracies better than 0.01 m.

Remote sensing is based on the acquisition of satellite images, and to a lesser extent aerial images, in the visible wavelengths ($\lambda = 0.3$ to $0.7\ \mu m$), near infrared ($\lambda = 0.7$ to $1.2\ \mu m$), the reflected infrared ($\lambda = 1.2$ to $3.2\ \mu m$), thermal infrared ($\lambda = 3.0$ to $14\ \mu m$), and the microwave ($\lambda = 1$ mm to 1 m) regions of the electromagnetic spectrum. Earth observation satellites launched since 1972 for remote sensing applications have resolutions (GSD) ranging from 0.4 m to >1 km and detect radiation in the various parts of the electro-magnetic spectrum. The number of planned launches for the next 10 years totals more than 100. Small lower cost satellites have recently been launched enabling many more countries to enter the space industry. Remote sensing images are typically processed through reconstruction, correction (radiometric and geometric), transformation (enhancement and data compression), classification (feature categorization/labelling), and output to databases (usually as GIS layers). The imagery time, season and scale is used in conjunction with feature size, shape, shadow, tone, colour, texture, pattern and context relationships to develop interpretation keys. Nowadays processing is primarily all-digital—from acquisition to products. Remote sensing experts therefore have a vast range of data and processes for addressing innumerable applications. Much of the data is retained in archives for future comparisons and studies.

1.3.2.2 *Data extraction for digital mapping and GIS*

The typical applications of aerial photogrammetry prior to 1980 were for orthophotography and line mapping based originally on manual plotting and later on digitization of features. Line mapping was partly automated using on-line computers in the semi-analytical and the analytical stereoplotters, but the process was still time consuming. However, by the 1980s, spatial information systems, referred to also as GIS, were being developed in many countries. There was a need for production of geo-coded digital spatial data that could be input to a local GIS with an appropriate structure, thus enabling overlaying of this data with other layers of spatial data for display and spatial analysis. This process opened up many new issues in the handling, processing and analysis of the spatial data that have become part of the studies of Commission II and IV since 2000. Networked, distributed and Web based GISs are now areas of study of ISPRS in these commissions.

There have been many attempts to extract features automatically from aerial and high resolution satellite images, including roads and buildings. Successful demonstration of an operational method of extracting such data would enable line maps to be produced more efficiently than can be achieved manually. However, the content of these images is extremely complex and varies greatly from one image to the next. Automation of the processes requires an understanding of the content of the image, the context in which the images are recorded, and should preferably apply to a range of image scales, i.e. GSD size. No commercial system has yet been developed, but research continues. It is likely that methods will be developed that are suitable for specific applications before more versatile commercial methods are available. Systems have been developed based on a semi-automatic approach, which uses the skill of an operator to approximately locate and identify a feature, and the speed of the computer to accurately determine its location and attributes. However, they have not become regular features in DPWs.

1.3.2.3 *Close range photogrammetry*

In the late 1970s and 1980s, a new approach to close range photogrammetry was possible due to the availability of digital imaging and adequate computing power. This led to a much broader range of applications, including high precision industrial and engineering applications, referred to as Vision Metrology. This approach is based on specially designed cameras made from off-the-shelf components, calibrated to achieve high precision measurements. Multiple overlapping convergent photos are recorded of the object with all object points targeted. Coordinate observations can be achieved manually on a monocomparator for film images or more likely automatically, either on digitized film images, or on directly recorded digital images. The design of the camera configuration to achieve a desired accuracy for the specific application can be determined prior to imaging by network analysis. The availability of high quality consumer grade digital cameras and software designed specifically for close range applications now provide for a high degree of automation in close range photogrammetry.

1.3.2.4 *Remote sensing applications*

Although remote sensing is a relatively new field, developments and applications of the technologies have grown rapidly in a very broad range of areas, from vegetation studies, geological applications, surface subsidence, transportation, meteorology, anthropogenic effects, environmental monitoring, sea surface and ocean colour, disaster monitoring and many, many more. Electro-optical multispectral and hyperspectral aerial and satellite images with resolutions from less than 1 m to >1 km continue to be used for the extraction of terrain information and interpretation of features, using new software tools, such as those based on an object-oriented approach. Extraction of terrain information is also based on multi-polarized and multi-frequency radar images, as well as the fusion of multiple data sources. There are 20 WGs in ISPRS Commissions VII and VIII for the period 2004–2008 covering the theory and applications of remote sensing. Some examples of these applications will be covered in later chapters of this book.

1.4 CONCLUSIONS

Over the nearly 100 years of history of ISPRS, its basic goals and structures have been retained, yet expanded. Similarly, there have been major developments in the sciences and technologies that have driven the methods and applications of the P&RS&SI sciences. In the early days of ISP, the processing of images was based on analogue methods to reduce the computations. The development of efficient algorithms, electronic computers and digital technologies has transformed the methods used in ISPRS. New types of images, digital processing techniques and a much broader range of applications are now pursued. It is virtually impossible to project successful developments in the next 10 years, let alone the next 100 years, but one can be confident that the activities of the Society and the disciplines it supports will continue to expand. Recognizing the contributions of the forefathers in establishing ISPRS should ensure that the Society will continue to attract dedicated people to manage ISPRS and successfully pursue its goals.

BIBLIOGRAPHY

Birdseye, C.H. 1940. Stereoscopic Photogrammetric Mapping. *Annals of the* Association of *American Geographers* 30 (1): 1–24.

Hughes, D., Fricker, P., Chaupis, A., Traversari, E., Schreiber, P. & Schapira, F., 2003. The development of Photogrammetry in Switzerland. *ISPRS Highlights* 8, No. 4: 33–39.

ISPRS, 1907–1984. *International Archives of the Photogrammetry, Remote Sensing and Spatial Information Sciences* Part A of Volumes I–XXXV.

Konecny, G. 1996. Paradigm Changes in ISPRS from the 1st to the 18th Congress in Vienna. *International Archives of Photogrammetry & Remote Sensing* 31-A: 62–67.

Livingston, R.G. 1964. A history of Military Mapping Camera Development. *Photogrammetric Engineering* 30: 97–110.

Schenk, T. 1999. *Digital Photogrammetry*. Laurelville, OH: TerraScience.

Schlögl, M. 1996. Extraordinary General Assembly, 1996 Vienna Congress. *International Archives of Photogrammetry & Remote Sensing* 31, Part A: 113–116.

Slama, C. (ed.) 1980. *Manual of Photogrammetry 4th Edition*. Falls Church, VA, USA: American Society of Photogrammetry.

Anonymous. *Carl Zeiss–History of a Most Respected Name in Optics*—http://www.inflenses.com/carl-zeiss-history-2.html accessed 12 November 2007.

APPENDIX 1 ISPRS Congresses—1913 to 2008.

Term	Active at end of Term			Congress		
	OdM	AsM	Term	OdM	AsM	Term
1910–1913	5		1910–1913	5		1910–1913
1913–1926	14		1913–1926	14		1913–1926
1926–1930	15		1926–1930	15		1926–1930
1930–1934	19		1930–1934	19		1930–1934
1934–1938	23		1934–1938	23		1934–1938
1938–1948	8+		1938–1948	8+		1938–1948
1948–1952	17		1948–1952	17		1948–1952
1952–1956	25		1952–1956	25		1952–1956
1956–1960	38		1956–1960	38		1956–1960
1960–1964	43		1960–1964	43		1960–1964
1964–1968	52		1964–1968	52		1964–1968
1968–1972	58		1968–1972	58		1968–1972
1972–1976	61		1972–1976	61		1972–1976
1976–1980	66		1976–1980	66		1976–1980
1980–1984	73		1980–1984	73		1980–1984
1984–1988	81		1984–1988	81		1984–1988
1988–1992	94		1988–1992	94		1988–1992
1992–1996	99	6	1992–1996	99	6	1992–1996
1996–2000	103	12	1996–2000	103	12	1996–2000
2000–2004	89	9	2000–2004	89	9	2000–2004
2004–2008			2004–2008			2004–2008

OdM = Ordinary members; AsM = Associate members; RgM = Regional members; StM = Sustaining members.

APPENDIX 2 ISPRS ordinary memberships chronology—1910 to 2008.

Listed by year of initial admittance (re-admittance) * denotes member is currently inactive

1910
- Austria (1948)
- Germany (1952)

1913
- Denmark
- Sweden
- Norway

1926
- France
- Italy
- Switzerland
- Latvia
- Spain (1956)
- Romania (1964)
- Poland (1960)
- *Czechoslovakia (1960)
- Hungary (1960)

1930
- Belgium

1934
- Finland
- The Netherlands
- U.S.A.
- *Portugal

1938
- Canada (1952)
- *Yugoslavia (1952)
- U.K. (1952)
- Chile
- *Senegal

1952
- Thailand (1968)
- Israel
- Japan

1953
- Pakistan

1955
- Egypt
- India
- Iran

1960
- China-Taipei
- Myanmar (Burma)
- South Africa
- Turkey
- Argentina
- *Tunisia
- Morocco
- Iraq
- Australia
- *Sudan
- *German D.R.

1964
- Brazil
- *Luxembourg
- Malaysia
- Philippines

1968
- Algeria
- Bulgaria
- Cuba
- Libya
- Syria
- Russia (U.S.S.R.)
- Mexico
- Peru

1972
- * Papua-New Guinea
- Sri Lanka (Ceylon)
- Cyprus
- * Nigeria
- Ireland
- Greece

1976
- Indonesia
- Jordan
- Kuwait
- Surinam

1980
- China-Beijing
- Korea
- Hong Kong
- New Zealand
- * Madagascar
- * PR Congo

1984
- * Bolivia
- Ethiopia
- Tanzania
- Uruguay
- * Zaire
- Colombia
- Burkina Faso

1988
- Nepal
- Venezuela
- Kenya
- Cote d'Ivoire
- Malawi
- Mongolia
- Qatar
- United Arab Emirates

1992
- Vietnam
- Zimbabwe
- Brunei
- * Zambia
- * Albania
- Azerbaijan
- * Belarus
- * Bophuthatswana
- Estonia
- Lithuania
- Saudi Arabia
- Slovenia
- Ukraine

1996
- Czech Republic
- Slovakia
- Croatia
- Ghana
- Lebanon
- Namibia

2000
- Botswana
- * Eritrea
- El Salvador
- Bangladesh
- * Benin
- Cameroon

2004 –

2008 –

APPENDIX 3 ISPRS council—1910 TO 2008.

1910–1913
President — E. Doležal, Austria

1913–1926
President — E. Doležal, Austria

1926–1930
President — O. Eggert, Germany
Honorary President — E. Doležal, Austria
Secretary General — Kärner, Germany
Congress Director — F. Bäschlin, Switzerland

1930–1934
President — Gen. Perrier, France
Honorary President — E. Doležal, Austria
Secretary General — H. Roussilhe, France
Treasurer — M. Labussiere, France
Council Members — Torroja, Spain; Buchholtz, Latvia; H. von Langendorff, Germany; F. Bäschlin, Switzerland

1934–1938
President — G. Cassinis, Italy
Hon. President — E. Doležal, Austria
Secretary General — M. Tucci, Italy
Treasurer — P. Dori, Italy
Council Members — H.v. Langendorff, Germany; J. Maury, Belgium; G. Perrier, France; K. Weigel, Poland

1938–1948
President — W. Schermerhorn, The Netherlands
Secretary General — B. Scherpbier, Netherlands
Treasurer — v. Freytag Drabbe, The Netherlands
Council Members — G. Perrier, France; H.v. Langendorff, Germany; O.S. Reading, USA; G. Cassinis, Italy

1948–1952
President — O. S. Reading, USA
Secretary General — E. S. Massie, USA
Treasurer — W. C. Cude, USA
Council Members — F. Bäschlin, Switzerland; G. Cassinis, Italy; R. Janicot, France; W. Schermerhorn, Netherlands

1952–1956
President — P. Mogensen, Sweden
Secretary General — P.O. Fagerholm, Sweden
Treasurer — S.G. Möller, Sweden
Council Members — R. Ll. Brown, UK; R. Janicot, France; O.S. Reading, USA; W. Schermerhorn, Netherlands

1956–1960

President	R. Ll. Brown, UK
Secretary General	R.T.L. Rogers, UK
Treasurer	J.E. Odle, UK
Council Members	P.O. Fagerholm, Sweden
	R. Janicot, France
	O.S. Reading, USA
	W. Schermerhorn, Netherlands

1960–1964

President	A. Paes Clemente, Portugal
Secretary General	A.D. Calvário, (Portugal 1960–61)
	F. Alexandre, Portugal (1961–64)
Treasurer	A. Santos Silva, Portugal
Vice President	R.Ll. Brown, UK
Council Members	P.O. Fagerholm, Sweden
	H. Härry, Switzerland
	G.D. Whitmore, USA

1964–1968

President	H. Härry, Switzerland
Secretary General	W. Bachmann, Switzerland
Treasurer	E. Huber, Switzerland
V. President	P.O. Fagerholm, Sweden
Council Members	A. Paes,Clemente, Portugal
	S.G. Gamble, Canada
	L. Solaini. Italy

1968–1972

President	L. Solaini. Italy
Secretary General	G.C. Tewinkel, USA
Congress Director	S.G. Gamble, Canada
1st V. President	W. Bachmann, Switzerland
2nd V. President	L. Skládal, Czechoslovakia
Treasurer	R.S. Halonen, Finland

1972–1976

President	S.G. Gamble, Canada
Secretary General	J. Cruset, France
Congress Director	R.S. Halonen, Finland †(1972–75)
	K.G. Löfström, Finland (1975–76)
1st V. President	G.C. Tewinkel, USA
2nd President	T. Maruyasu, Japan
Treasurer	A.J. van der Weele, Netherlands

1976–1980

President	J. Cruset, France
Secretary General	F. J. Doyle, USA
Congress Director	G. Konecny, F. R. Germany
1st V. President	S.G. Gamble, Canada †(1976–77)
	E. O. Dahle, Norway
2nd V. President	P. Fagundes, Brazil
Treasurer	Mrs. A. Savolainen, Finland

1980–1984

President	F.J. Doyle, USA
Secretary General	G. Konecny, F. R. Germany
Congress Director	P. Fagundes, Brazil
1st V. President	G. Zarzycki, Canada
2nd V. President	I. Antipov, Soviet Union
Treasurer	H. Jerie, Netherlands

1984–1988

President	G. Konecny, F. R. Germany
Secretary General	K. Torlegård, Sweden
Congress Director	S. Murai, Japan
1st V. President	G. Zarzycki, Canada
2nd V.President	J. C. Trinder, Australia
Treasurer	Mrs. G. Togliatti, Italy

1988–1992

President	K. Torlegård, Sweden
Secretary General	S. Murai, Japan
Congress Director	L.W. Fritz, USA
1st V. President	G. Konecny, F.R. Germany
2nd V. President	I. Katzarsky, Bulgaria
Treasurer	K. Atkinson, UK

1992–1996

President	S. Murai, Japan
Secretary General	L.W. Fritz, USA
Congress Director	K. Kraus, Austria
1st V. President	K. Torlegård, Sweden
2nd V. President	A. Grün, Switzerland
Treasurer	J.C. Trinder, Australia

1996–2000

President	L.W. Fritz, USA
Secretary General	J.C. Trinder, Australia
Congress Director	K.J. Beek, The Netherlands
1st V. President	S. Murai, Japan
2nd V. President	M. Barbosa, Brazil
Treasurer	H. Rüther, South Africa

2000–2004

President	J.C. Trinder, Australia
Secretary General	I.J. Dowman, UK
Congress Director	O. Altan, Turkey
1st V. President	L. W. Fritz, USA
2nd V. President	G. Begni, France
Treasurer	A. Peled, Israel

2004–2008

President	I.J. Dowman, UK
Secretary General	O. Altan, Turkey
Congress Director	Chen Jun, China
1st V. President	J.C. Trinder, Australia
2nd V. President	E.Baltsavias, Switzerland
Treasurer	S. Morain, USA (1977–80)

APPENDIX 4 ISPRS technical commissions—1926 TO 2008.

1926–1930

I.	Terrestrial photogrammetry	H. Dock	Austria
II.	Rectification	H. Roussilhe	France
III.	Stereo-aerial photogrammetry	O. Eggert	Germany
IV.	Aerial triangulation	F. Bäschlin	Switzerland
V.	X-ray measurements	A. Hasselwander	Germany
VIa.	Architectural and engineering photogrammetry	J. Torroja	Spain
VIb.	Photogrammetry for flying objects	Th. Ween	Norway
VII.	Economy	A. Kruttschnitt	Hungary
VIII.	Instruments, optics, norms	G. Cassinis	Italy
IX	Plates and films	A. von Odencrants	Sweden
Xa.	Education at universities and research institutes	A. Buchholz	Latvia
Xb.	Training of technical personnel	A. Ivanceanu	Romania
XIa.	Photographic airplanes	K. Weigel	Poland
XIb.	Navigation	J. Petrik	Czechoslovakia

1930–1934

I.	Terrestrial photogrammetry	F. Bäschlin	Switzerland
II.	Aerial photogrammetry	M. Labussiere	France
III.	Mapping	H. von Langendorff	Germany
IVa.	Various applications	E. Doležal	Austria
IVb.	X-ray photogrammetry	A. Hasselwander	Germany
V.	Industrial photogrammetry and economy	K. Weigel and E. Warchalowski	Poland
VI.	Education, bibliography, terminology	Medvey & K.v. Oltay	Hungary

1934–1938

I.	Ground photogrammetry	F. Bäschlin	Switzerland
II.	Air photography	H.H. Blee	USA
III.	Obtaining control by terrestrial methods or by aerial triangulation for rectification or for stereoscopic plotting	W. Schermerhorn	Netherlands
IV.	Plotting of air photographs	H. von Langendorff	Germany
V.	Various applications of photogrammetry	E. Doležal	Austria
VI.	X-ray photogrammetry & close-up photogrammetry	C. Sannié	France
VII.	Industrial organization of photogrammetry and statistics of works	G. Cassinis	Italy
VIII.	Teaching, terminology, bibliography	K.v. Oltay	Hungary

1938–1948

I.	Ground photogrammetry and its applications	O.S. Reading	USA
II.	Air photography	M. Zeller	Switzerland
III.	Preliminary operations on the ground for aerial photogrammetry	F. Bäschlin	Switzerland
IV.	Plotting of air photographs	P. Than	Sweden
V.	Geodetical applications of photogrammetry	G. Poivilliers	France
VI.	Applications of photogrammetry to biology and medicine	J. Didier & Coliez	France
VII.	Industrial organization of photogrammetry and statistics	G. Cassinis	Italy
VIII.	Teaching & bibliography	G. Harding	USA

1948–1952

I.	Photography and navigation	L.E. Howlett	Canada
II.	Plotting machines & instruments	G. Poivilliers	France
III.	Geodetic or control operations	P. Wiser	Belgium
IV.	Mapping from photographs	G. Cassinis	Italy
V.	Special applications & measurements	B. Hallert	Sweden
VI.	Education, terminology, bibliography, history, polyglot dictionary	K. Lego	Austria
VII.	Photo interpretation	R.N. Colwell	USA

1952–1956

I.	Photography and navigation	J. Cruset	France
II.	Plotting, theory and instruments	W.K. Bachmann	Switzerland
III.	Aerial triangulation	P. Wiser	Benelux
IV.	Mapping from photographs	G.S. Andrews	Canada
V.	Non-topographic photogrammetry	G. Boaga	Italy
VI.	Education, terminology, bibliography	K. Neumaier	Austria
VII.	Photo interpretation	G.C. Coleman	USA

1956–1960

I.	Photography and navigation	J. Cruset	France
II.	Plotting, theory and instruments	F. Vanderheyden	Belgium
III.	Aerial triangulation	G. Cassinis	Italy
IV.	Mapping from photographs	H. Härry	Switzerland
V.	Special applications of photogrammetry	R. Burkhardt	F.R. Germany
VI.	Education, terminology and bibliography	A. Barvir	Austria
VII.	Photo interpretation	G.C. Coleman	USA

1960–1964

I.	Photography and navigation	G.C. Brock	UK
II.	Plotting, theory and instruments	A.L. Nowicki	USA
III.	Aerial triangulation	G. de Maison d'Autume	France
IV.	Mapping from photographs	E.F. Gigas	F.R. Germany
V.	Special applications of photogrammetry	K. Hubeny	Austria
VI.	Education, terminology and bibliography	R.S. Halonen	Finland
VII.	Photo interpretation	C.H. Edelman	Netherlands

1964–1968

I.	Photography and navigation	R.W. Fish	UK
II.	Theory, methods, instruments of restitution	K. Schwidefsky	F. R. Germany
III.	Aerial triangulation	G.C. Tewinkel	USA
IV.	Mapping from photographs	L. Skládal	Czechoslovakia
V.	Non-topographic photogrammetry	T. Maruyasu	Japan
VI.	Education, terminology and bibliography	W. Sztompke	Poland
VII.	Photo interpretation	R. Chavallier	France

1968–1972

I.	Photography and navigation	M.B. Scher	USA
II.	Theory, methods and instruments of restitution	H. Deker	F. R. Germany
III.	Aerial triangulation	E.H. Thompson	UK
IV.	Mapping from photographs	A.J. van der Weele	Netherlands
V.	Non-cartographic applications of photogrammetry	M. Carbonnel	France
VI.	Bibliography, education and terminology	P. Gal	Czechoslovakia
VII.	Photo-interpretation	A. Reinhold	German D. R.

1972–1976

I.	Primary data acquisition	E. Welander	Sweden
II.	Instrumentation for data reduction	G. Inghilleri	Italy
III.	Mathematical analysis of data	F. Ackermann	F. R. Germany
IV.	Topographic and cartographic applications	G. Ducher	France
V.	Non-topographic photogrammetry	H.M. Karara	USA
VI.	Economic, professional and educational aspects of photogrammetry	W. Sztompke	Poland
VII.	Interpretation of data	L. Sayn-Wittgenstein	Canada

1976–1980

I.	Primary data acquisition	I. Nakajima	Japan
II.	Instrumentation for data reduction	M. Baussart	France
III.	Mathematical analysis of data	I. Antipov	USSR
IV.	Topographic and cartographic applications	J.M. Zarzycki	Canada
V.	Non-topographic photogrammetry	K. Torlegård	Sweden

VI.	Economic, professional and educational aspects of photogrammetry	H.Z. Sitek	Poland
VII.	Interpretation of data	G. Hildebrandt	F. R. Germany

1980–1984

I.	Primary data acquisition	J.C. Trinder	Australia
II.	Instrumentation for data reduction and analysis	Z. Jaksic	Canada
III.	Mathematical analysis of data	E. Kilpelä	Finland
IV.	Cartographic and data bank applications of photogrammetry and remote sensing	R. Mullen	USA
V.	Other applications of photogrammetry and remote sensing	J.W. Gates	UK
VI.	Economic, professional and educational aspects of photogrammetry and remote sensing	J. Hothmer	F. R. Germany
VII.	Interpretation of photographic and remote sensing data	L. Laidet	France

1984–1988

I.	Primary data acquisition	Ph. Hartl	F. R. Germany
II.	Instrumentation for data reduction and analysis	L.W. Fritz	USA
III.	Mathematical analysis of data	E. Kilpelä	Finland
IV.	Cartographic and data bank applications of photogrammetry & remote sensing	A.S. Macdonald	UK
V.	Other applications of photogrammetry and remote sensing	V. Kratky	Canada
VI.	Economic, professional and educational aspects of photogrammetry and remote sensing	Mrs. O. Adekoya	Nigeria
VII.	Interpretation of photographic and remote sensing data	K.J. Beek	Netherlands

1988–1992

I.	Primary data acquisition	M. Barbosa	Brazil
II.	Systems for data processing and analysis	K. Szangolies	German D. R.
III.	Mathematical analysis of data	D. Li	China
IV.	Cartographic and data base applications of photogrammetry and remote sensing	T. Hirai	Japan
V.	Close-range photogrammetry and machine vision	A. Grün	Switzerland
VI.	Economic, professional and educational aspects of photogrammetry and remote sensing	J. Badekas	Greece
VII.	Interpretation of photographic and remote sensing data	F. Hegyi	Canada

1992–1996

I.	Sensors and platforms and imagery	L. Mussio	Italy
II.	Systems for data processing, analysis and representation	M. Allam	Canada
III.	Theory and algorithms	H. Ebner	Germany
IV.	Mapping and geographic information systems	R. Welch	USA
V.	Close-range techniques and machine vision	J.G. Fryer	Australia
VI.	Economics, professional matters and education	D. Li	China
VII.	Resource and environmental monitoring	R.P da Cunha	Brazil

1996–2000

I.	Sensors, platforms and imagery	G. Joseph	India
II.	Systems for data processing, analysis and representation	I.J. Dowman	UK
III.	Theory and algorithms	T. Schenk	USA
IV.	Mapping and geographic information systems	D. Fritsch	Germany
V.	Close-range techniques and machine vision	H. Chikatsu	Japan
VI.	Education and communications	K. Villanueva & L. Aziz	Indonesia
VII.	Resource and environmental monitoring	G. Remetey-Fülöpp	Hungary

2000–2004

I.	Sensors, platforms and imagery	S. Morain	USA
II.	Systems for data processing, analysis and representation	J. Chen	China
III.	Theory and algorithms	F. Leberl	Austria
IV.	Spatial information systems and digital mapping	C. Amenakis	Canada
V.	Close-range vision techniques	P. Patias	Greece
VI.	Education and communications	T.M. Sausen	Brazil
VII.	Resource and environmental monitoring	R. Navagund	India

2004–2008

I.	Image data acquisition—sensors & platforms	A. Baudoin	France
II.	Theory and concepts of spatio-temporal data handling and information	W. Kainz	Austria
III.	Photogrammetric computer vision and image analysis	W. Förstner	Germany
IV.	Geodatabases and digital mapping	S. Nayak	India
V.	Close-range sensing—analysis and applications	H.-G. Maas	Germany
VI.	Education and outreach	K. Cho	Japan
VII.	Thematic processing, modeling and analyses of remotely sensed data	J. van Genderen	Netherlands
VIII.	Remote sensing applications and policies	A. Peled	Israel

Advances in Photogrammetry, Remote Sensing and Spatial Information Sciences: 2008 ISPRS Congress Book – Li, Chen & Baltsavias (eds)
 ISBN 978-0-415-47805-2

CHAPTER 2

Scientific-technological developments in photogrammetry and remote sensing between 2004 and 2008

Armin Gruen

ABSTRACT: This chapter gives an overview of the developments in sensor hardware and processing methodology within ISPRS over the past four years. The focus is on photogrammetry and optical remote sensing. For details we refer to the more specific contributions in this book and also to other developments that are not within the mainstream of ISPRS research. We distinguish three major data acquisition platforms: satellites, aerial and terrestrial. We show that progress is incremental and stepwise and depends largely on innovations on the sensor side and in the computer hardware. We work out the major trends and also discuss briefly what effects these may have on the profession.

Keywords: Sensors, algorithms, photogrammetry, remote sensing, spatial information

2.1 INTRODUCTION

Scientific-technical developments in the last four years—this is a very ambitious topic. While a period of four years seems to be reasonably short to be described and analysed, we nevertheless find in high-tech areas such as photogrammetry, remote sensing and spatial information systems a large amount of sensor and system developments, algorithmic research, investigations, improvements (even if only gradually), and new applications, such that it is difficult to keep the overview. This situation gets even worse because many interesting developments are taking place outside our communities—in electrical engineering, computer science, computer vision, machine vision, robotics, visualization, animation, artificial intelligence, multimedia, geo-sciences, etc. Just looking at the topics of the many computer vision conferences and SPIE Symposia makes us wonder how one can digest all this information. The published literature is tremendous, especially if one would consider also those publications that are found outside ISPRS-related media. We therefore will not use any references—they can be found in the specialized chapters of this book.

Progress in most of these areas is characterized by small steps forward, rather than by one-at-a-time inventions and revolutionary findings. It definitely cannot be the purpose of this chapter to report on all efforts of system and process improvements. We can only work out the main lines of technological and scientific advancements. Detailed assessments of these developments can be found in most of the other contributions to this book. This applies in particular to integration and fusion of data and multiple information sources, DSM generation and deformation measurements from SAR data, early stages of LiDAR processing, object extraction and attribution from hyper-spectral images, automated extraction of roads, buildings and vegetation from multi-source data, change detection from multi-temporal images, advanced classification techniques, texturing, visualization and VR/AR, and the GIS-related topics of data representation, spatio-temporal modelling, dynamic GIS, semantic integration of heterogeneous spatial information and data quality and uncertainty propagation.

Concerning the organization of this article, we will pay attention to the fact that we are using three major platforms for our sensors: satellites, airborne and terrestrial vehicles. Therefore, we will arrange our material accordingly.

Progress in our fields is predominantly technology-driven. New sensors or processing devices will allow us to develop new methods and systems at a higher level of performance. Therefore, in this report, sensors will play an essential role in the analysis of the situation. New sensors create new capabilities and opportunities, and therefore new applications or better performance in existing ones.

2.2 CURRENT ENVIRONMENT FOR DATA ACQUISITION AND PROCESSING

Most of traditional photogrammetry was characterized by the use of photographic cameras as sensors, but a great variety of processing devices (analogue stereo instruments of many types, mono- and stereo-comparators, rectifiers, triangulators, orthophoto projectors, analytical plotters and the like).

Today the situation is completely reversed. For data acquisition, we use a great number of different sensors: film cameras, special CCD and CMOS cameras, still video cameras, camcorders, linear array cameras of various types, among them the latest Three-Line-Scanners and digital panoramic cameras, laser scanners, microwave and ultrasound sensors, X-ray devices and electronic imaging devices, and all sorts of combined and hybrid systems. On the data processing side, however, only digital stations (and maybe image scanners) are required, if equipped with the needed software, which also may include GIS, CAD and specific visualization and simulation packages. All photogrammetric and remote sensing functions that had been executed in the past on different instruments can now be integrated under such a single system. In the past, we could distinguish three lines of system development for digital stations, according to the three major sensor platforms and application areas: satellite remote sensing, aerial photogrammetry and close range photogrammetry, leading to different kinds of system with greatly varying functionality. Today, we see a tendency towards integrating all functions under one unique system. Given the flexible nature of such platforms, radiometric manipulations, special remote sensing software, 3-D modelling functions, database functionality, data analysis procedures, visualization and animation functions and all kind of third party software can be also integrated or connected to. This makes a digital station a truly universal system for data processing, administration, analysis and representation. Needless to say, in reality not all desired functionality is available yet and there is much room for future improvements and developments.

2.3 SATELLITE IMAGING

During the past few years an amazing development has taken place. Within less than 10 years, the spatial resolution of satellite images has increased by more than a factor 10. After the launch and successful operation of SPOT-5, IKONOS, QuickBird and EROS, Asia made a strong move by deploying Beijing-1, Cartosat-1 and -2, ALOS/PRISM, ROCSAT-2 and KOMPSAT-2, all within the last 4 years. With the stereo capabilities of most of these sensors we are entering a new domain of data processing. The need to consider the third dimension in modelling and data processing more accurately, brings photogrammetric methodology to the processing of satellite images. Issues like sensor and trajectory modelling, network and system analysis, self-calibration, image matching for DSM generation and 3-D object extraction are topics that are well-known in the photogrammetric domain. And indeed, during the past four years substantial progress has been made in sensor modelling, geo-referencing, self-calibration and DSM generation from high-resolution satellite imagery.

There is no end in sight. Worldview-1 has already shown its first amazing 41 cm GSD images and GeoEye is almost ready to launch their new sensors. This brings satellite remote sensing "down to earth" and into the realm of aerial medium scale imagery (1:30,000) with all the consequences involved. Others, like Cartosat 2A and THEOS, are due to be launched this February/March. China has also big plans concerning near future earth observation missions.

The available software for data processing, the appetite of global databases (Google Earth and others) for high-resolution 3-D data and the increased worldwide availability of these types of images will give the field another boost and will lead to new applications.

This development on the optical side of satellite sensors is matched by the latest achievements in the microwave domain. ALOS/PALSAR, TerraSAR-X and RADARSAT-2 are further success stories, which broaden our capabilities in sensing and value adding considerably and open the case for many R&D projects.

2.4 AERIAL SENSING

It is only eight years since the first large format aerial camera came onto the market (Leica ADS40 at the XIXth ISPRS Congress in Amsterdam). Today we see the replacement of aerial photographic cameras worldwide at a remarkable pace. Very often this is accompanied by the integration of GPS and INS systems. This fact brings new topics for research. What is the accuracy of the new digital cameras? Are there new kinds of systematic errors creeping into the systems? What is the status of calibration? How about the stability of calibration over time? Do the traditional network design conditions hold in the case of GPS/INS integration? What are the accuracy properties of Linear Array Camera networks? Which sensor and trajectory models are the best? How can pushbroom technology be used advantageously over single frame technology? How can we best make use of the on-line and real-time capabilities? There are plenty of interesting research issues, which were all somehow addressed in the past years—some more, others less. Yet, it is startling to see how fast the digital camera systems have been accepted in professional practice and used extensively in projects, although really fundamental and extended studies on these previously mentioned issues are still missing. Here is another case where professional practice is jumping ahead of even the system developers' experiences and state of knowledge. On top of it is the big issue of automation in information extraction (DSM/DTM generation, modelling of cities, road networks, vegetation and the like, together with the extraction of dynamic processes), which will stay with us for the next 50 years.

The development of large format aerial cameras will continue. There are already new products on the horizon; existing ones will be improved. The revitalization of old sensing concepts, now in the digital domain, can be observed. Oblique imaging, combined with a multi-camera approach is extensively used by companies like Pictometry. Huge amounts of aerial image data have already been collected over the USA and Europe.

Some people have started lately to look more closely into the real-time processing issue—an important aspect when it comes to natural and man-made hazards monitoring, homeland security and other new application fields.

Another sensor having a great impact is LiDAR. The ease of use in generating millions of object points in 3-D in a very short time has led to a tremendous surge in system acquisitions and applications. We are facing here a similar situation as discussed before: the practical applications are over-rolling the serious scientific investigations and the development of algorithms for automated object extraction. Therefore, people are surprised here and there if in special cases the systems do not perform according to expectations. Exaggerated expectations are then met by frustrations. On the other hand, these sensors have been further developed in order to record multiple pulses or even full waveforms. This gives additional capabilities to LiDAR.

The biggest issue in research has been and will continue to be the automation of information extraction. Here progress in image and/or point cloud understanding is extremely slow. Therefore, people are trying to get around this problem by applying a multi-sensor approach, combining many and different cues for object reconstruction. While hybrid systems or data processing concepts are often mentioned, we have seen only relatively few being actually used and investigated. This will and has to change in the near future, if progress is to be achieved.

Another basic approach is semi-automated processing. This is understood as a methodology, which either has the human operator in the processing chain by design from the very beginning or uses first a fully automated procedure, whose results are subsequently manually edited. At present, semi-automated approaches are extensively used, e.g. in 3-D city modelling, or in the second mode, in DSM/DTM generation.

The use of spectral and radiometric properties in the processing of hyper-spectral aerial pushbroom scanner data is rather marginal in ISPRS. This is a field that deserves more attention in the future.

A particular topic of vividly growing interest is that of UAVs (Unmanned Aerial Vehicles). They can operate at different flying heights (up to stratospheric platforms, operating in geo-stationary mode), but in most civilian applications we see them flying at heights of 50 to 400 m above ground. Low altitude UAV systems are small, of low to moderate cost, very flexible in terms of use and image acquisition (vertical, oblique and quasi-terrestrial imaging modes can be easily realized).

The research issues involved are similar to the standard aerial cases, but here there are additional topics of relevance, as for instance image-based navigation, because usually the integrated GPS and INS systems are low cost components, which do not deliver a high accuracy. Currently, the systems are operating in a pre-programmed waypoint-following mode. In the future, a perception-action based approach would be desirable as well, making the platform able to react intelligently to specific object space-related events. Also, on-line and real-time performance is required in many of the new applications. In the near future, we will also see laser scanners integrated with digital medium format cameras.

2.5 TERRESTRIAL SENSING

Close range applications have always seen (other than aerial photogrammetry and satellite remote sensing) a large number of diverse sensors and systems. This has not changed in the past years. A variety of active and passive sensing devices is employed and also the application fields vary greatly, from industrial design and quality control and robotics to cultural heritage, biomedical imaging, generation of virtual environments, animation and so forth. The list of successful applications is almost unlimited.

Sensors come in many different forms: single area array chip cameras (SLR-type, industrial, high-speed), multiple head cameras, panoramic cameras, structured light systems of different types, laser scanners, and lately also range cameras.

While customer-type still video cameras already come with 12 Megapixels at very low prices, digital back cameras, as high-end systems, are now offering up to 39 Megapixels. This trend will continue. At the same time, even mobile phone cameras already have up to 5 Megapixels.

High speed cameras fill a niche market. They are available with an image size of 1000 × 1000 pixels at a frame rate of up to 5000 frames per second.

Combined systems are also gaining much attention. GPS-integrated still video cameras are already available as off-the-shelf systems, now even reaching into the mobile phone domain. Laser scanners with integrated cameras have already shown their potential and are under development in different places. As in other fields, the algorithmic development of hybrid data processing cannot follow the pace in hardware advancements.

Multiple camera arrangements are nowadays pretty much standard. The motion analysis industry is offering systems, used in biometrics and the movie industry, with virtually unlimited numbers of cameras (applications with up to 245 cameras have been reported). Those systems are based on the measurement and tracking of retro-reflective targets. The processing of these images with very simplistic content is done in the camera and only pixel coordinates are passed on to the host computer.

Range cameras are currently being developed in different places. They obtain grey-value images and at the same time record the depth information with a single sensor. Distance measurement is achieved via modulation-based time-of-flight techniques. A serious restriction is still the limited in-plane and depth resolution, but there is justified hope that this situation will improve in the near future. To what extent—this remains to be seen.

Panoramic imagers are becoming very popular in art and multi-media applications. They have also been investigated with respect to their suitability for measurement purposes. The first results are quite encouraging. Very large formats of panoramic images of up to one Gigapixel can be obtained and, with concepts of multiple imaging at different exposure settings, images with very high radiometric depth can be produced, yielding a number of novel applications.

In terrestrial applications we see an increasing number of recordings of dynamic events and/or mobile platform applications. One example is the land-based mobile mapping systems. The main developments took place in the 1990s, but many are working now in an operational mode and there are more than a dozen commercial systems on the market. They are all using manned platforms (cars, trains), but there are also substantial efforts being put into the development of un-manned platforms. But most of this is going on in the robotics community.

A central issue is often the integration of sensor technology with reliable data processing schemes to generate highly automated on-line or real-time image-based measurement systems.

Another trend observed is the development of systems that can be applied by non-expert users. This takes photogrammetry "out-of-the-box" and makes it available and affordable to a wide range of users.

2.6 R&D ISSUES IN PHOTOGRAMMETRY AND REMOTE SENSING

With the great changes in technology come many challenges for research and development. Here we list a number of topics that are of great relevance for the current R&D scenery.

2.6.1 *New sensors*

The development and use of new sensors requires the study and testing of innovative sensor models, and the investigation of the related network structures and accuracy performance. Of particular interest here are high-resolution satellite and aerial cameras, especially of the linear array type, terrestrial panoramic cameras and laser scanners.

2.6.2 *Sensor and data integration*

The combination of different sensors and their related datasets requires new approaches in sensor modelling and in the combined processing of different kinds of data. Combinations of cameras and GPS/INS, cameras of various types and laser scanners optical images and radar data and the like are of particular interest. Also, the appropriate use of *a priori* data information, e.g. from GIS, for image data processing is of relevance.

2.6.3 *On-line and real-time processing*

The need for very fast processing requires algorithmic redesigns in many areas. Sequential estimation methods present a suitable tool to tackle some of those problems.

2.6.4 *3-D modelling*

Our environment is essentially three-dimensional. With digital technology the traditional approach of 2.5-D modelling should be overcome in favour of consequent 3-D modelling. Here many new problems are emerging in measurement, surface modelling, topology generation and data model definition.

2.6.5 *Image understanding*

The overwhelming research issue today is the automation of all processing functions, from the orientation processes to image matching and feature and object extraction. While we have seen some progress lately in those procedures where mainly geometric issues are dealt with, the great problem to be solved is automated image interpretation. Methods of artificial intelligence have not delivered the promised performance and we observe a certain stagnation in the development of image understanding algorithms. Therefore many researchers have turned towards semi-automated approaches, where the human capacity in image content interpretation is paired with the speed and precision of computer-driven measurement and modelling algorithms.

2.7 CONCLUSIONS AND OUTLOOK

In the past years, digital photogrammetry with its online and real-time processing capabilities has already opened many new areas of application. In industrial inspection and quality control many photogrammetric techniques and systems have found their way into daily practice. A similar development is currently noted in cultural heritage applications, although, due to the great variation in requirements in this area, more research efforts are urgently required there.

With the recent expansion of photogrammetry's data acquisition tools (sensors) and processing techniques, we see many more novel applications emerging in the close range area. The generation of reality-based data for virtual environments, animation, video gaming and the like constitutes a huge potential for photogrammetry. In fact, there are many recent movies (The Lord of the Rings, Matrix, etc.), where photogrammetry has been used extensively, especially for the recording of 3-D movements of bodies (motion analysis) and the tracking of face expressions.

The pressing need for modelling of our 3-D environment (3-D city and terrain modelling) from aerial and high-resolution satellite images and laser scanners will have a tremendous impact in natural hazard damage monitoring, risk analysis, car navigation with 3-D models, location-based services, virtual tourism, and in many more applications.

Global databases with free access (e.g. Google Earth), with their need to improve the completeness and quality of spatially-related data will continue to have a severe and positive impact on image-based techniques of 3-D modelling.

With the new generation of high-resolution satellite sensors (SPOT, ALOS/PRISM, IKONOS, QuickBird), the issue of 3-D modelling is gaining much more prominence. Therefore photogrammetric techniques are also becoming more important in satellite image applications. On the other hand, radiometric analyses are also attaining more attention in photogrammetry. We observe that the originally different techniques of remote sensing and photogrammetry are partly converging today.

Within the image-related sciences, computer vision, robot vision, remote sensing, visualization, simulation, animation and spatial information science, modern photogrammetry and remote sensing have managed to position themselves as indispensable members, whose specific procedures and techniques are required for problem solving.

The impacts of these developments and requirements on the profession are manifold, out of which questions arise such as:

- How do we handle the inflation of data, in particular images and point clouds? Our processing capabilities today are already trailing way behind the data generation rate.
- How does the increased system complexity affect our daily production work? It would be a great mistake to assume that photogrammetry can be handled by a black-box approach. Most of the procedural aspects of photogrammetry are much too complex to be left to untrained personal. Only a good education of the operators can ensure that the procedures are executed properly and that the results are of high quality and reliability.
- How do we cope with the competition from neighbouring disciplines? Depending on our own capabilities, flexibilities and attitudes, our discipline will either disappear or emerge with greater strength than ever before.

We have shown that photogrammetry and remote sensing have expanded their techniques very much in recent years. This has opened many new fields of application. There is no good reason why this process should not continue in the years to come. Already now, but even more in the near future, we are and will be flooded by huge amounts of data (images, point clouds) emerging from satellite, aerial and terrestrial platforms. A good deal of those images and point clouds will have to be processed using quantitative techniques. This is why we see a very bright future for photogrammetry and remote sensing, in research, development and with respect to practical applications.

Part II
Sensors, platforms and data acquisition systems

Advances in Photogrammetry, Remote Sensing and Spatial Information Sciences: 2008 ISPRS Congress Book – Li, Chen & Baltsavias (eds)

CHAPTER 3

Spaceborne digital imaging sensors and systems

Gordon Petrie

ABSTRACT: This chapter presents an account of the developments that have taken place in the area of spaceborne digital imagers and systems during the period since the last ISPRS Congress was held in 2004. First of all, the developments in the high- and medium-resolution optical systems mounted on satellites are covered. After which, the corresponding developments in spaceborne radar systems are outlined and discussed.

Keywords: Spaceborne optical imagers, spaceborne radar imagers, pushbroom scanners, micro-satellites, mini-satellites

3.1 INTRODUCTION

The developments in spaceborne digital imaging systems that have taken place since the last ISPRS Congress was held in 2004 will be reviewed under two main headings -

i. optical imaging systems; and
ii. radar imaging systems.

Those systems that generate imagery from space with a low spatial resolution—those with ground spatial distance (GSD) values larger than 40 m; for example, the imagery produced by meteorological or weather satellites and that generated by oceanographic satellites—will not be discussed in this review, which is concerned specifically with those systems that generate imagery that is mainly used for mapping and GIS purposes. Furthermore, purely military spaceborne imaging systems will also be excluded from this review, although, as will be seen later, some systems have been included that are intended for joint military and civilian use.

Before going on to discuss developments in spaceborne imaging systems and imagery in some detail, it may be useful to make a few observations of a general nature about the trends that have become apparent during the inter-Congress period. Starting with spaceborne optical imaging systems, at the highest resolution level, the two American commercial companies (GeoEye & DigitalGlobe) and one Israeli company (ImageSat International) that operate systems providing imagery at GSD values smaller than 1 m, are all privately funded, but they are very heavily focused on satisfying the requirements of military intelligence and mapping agencies. Closely behind these are a group of high-resolution spaceborne scanner systems that generate imagery with GSD values in the range 1 to 5 m. A substantial number of these originate from countries in South and East Asia—India (Cartosat-1&-2), Taiwan (Formosat-2), China (Beijing-1), Korea (Kompsat-2) and Japan (ALOS/Prism) (Petrie 2007a). All of these satellites have been funded and are supported by the respective national governments. Again, in most cases, defence and security concerns appear to underlie much of this support. When one comes to the area of systems delivering still lower (medium-resolution) optical imagery—mainly for land cover, land use and environmental purposes—a point of particular note has been the growth in the number of micro-satellites that have been orbited, much of this being the result of the activities of SSTL (Surrey) and its partnerships with a number of developing countries through technology transfer projects. Moreover, if the German RapidEye series of five satellites (for which the satellite platforms have also been built by SSTL) is launched and comes into operation successfully during 2008, this will have a major impact on this area. Stimulated by the success of SSTL, several other system suppliers have moved into the area of micro- and mini-satellites (Stoney 2006).

With spaceborne radar imaging systems, the most obvious point to be made at the outset is the overall revival of activity in this area. Previously it has had a rather low profile with a comparatively low level of activity, mostly based on the C-band imagery provided by the SAR imagers mounted on the Canadian Radarsat and ESA ERS and Envisat platforms that have been focused mainly on scientific and environmental applications of SAR imagery. Now, however, there is strong development in this area with higher resolution X-band SAR imagers, especially from two European countries; Germany, with its TerraSAR-X and TanDEM-X projects and its (military-funded) SAR-Lupe programme and Italy with its COSMO-SkyMed series, which is partly funded by the Italian Space Agency and partly by the country's Defence Ministry. Canada has also reinforced the current upsurge of interest in spaceborne radar imagery with the launch of the long-delayed Radarsat-2 satellite, as has Japan with its PALSAR system mounted on the ALOS satellite. Both Israel and India are also constructing satellites equipped with SAR imagers and these can be expected to feature before long, though whether the imagery will be made available to civilian users is still a matter of debate in both countries.

3.2 SPACEBORNE OPTICAL IMAGING SYSTEMS

Starting with a general comment, it should be noted that almost all current spaceborne optical imaging systems are based on the use of pushbroom scanners equipped with CCD linear arrays. The use of whiskbroom scanners employing optical-mechanical scanning of the ground as the basis of their imaging systems, as used in the Landsat series, has almost disappeared in those satellites that have come into operation in recent years. It is also worth noting the use of asynchronous imaging modes with certain satellites such as the EROS and QuickBird, by which the attitude of the imaging scanner can be continually altered during the image acquisition process. Another distinct change that has taken place over the last few years has been the methodology that is used to acquire stereo-coverage from spaceborne platforms. Previously the use of overlapping cross-track coverage acquired from multiple orbits occurring at different times was very common, e.g. that acquired by the SPOT-1 to SPOT-4 satellites. With the recent introduction of dedicated imagers designed specifically to acquire stereo-imagery from the SPOT-5, Terra/ASTER, Cartosat-1 and ALOS satellites, the use of along-track stereo-imagers using two forward- and backward-pointing linear arrays or three arrays (with an additional nadir-pointing array) has come into widespread operation, as pioneered with the German MOMS-02 scanner that was mounted both on the Space Shuttle and the Russian Mir space station during the 1990s. This arrangement allows the overlapping images to be acquired with only a very short time gap between them, thus eliminating the difficulties that are encountered with overlapping cross-track imagery when the images may have been acquired months apart during different seasons.

Turning next to the actual sensor technologies that are employed in spaceborne optical imagers, as noted above, these relate almost entirely to pushbroom scanners and to the linear arrays of charged-coupled devices (CCDs) that form the imaging elements of these scanners. A limitation is the maximum length of an individual array, which currently stands at 12,000 to 14,000 CCD detectors. If a wide swath needs to be acquired by a particular satellite, then the use of multiple arrays is unavoidable. Thus, for example, the ground swath covered by the new WorldView-1 (WV-1) satellite is 15.2 km, while the GSD is 0.41 m, which means that over 37,000 individual detectors need to be available in the cross-track direction to achieve this coverage. Since the individual linear arrays cannot be butted together, they have to be offset from one another with consequences for the geometric processing of the resulting imagery. On the optical side, the use of folded optical trains to allow telescopes with very long focal lengths to be accommodated in very high-resolution satellites has always been a necessity in such satellites. With the rapid growth in the number of mini-satellites that can acquire high-resolution imagery—which has been such a feature of the inter-Congress period—the problem of fitting such telescopes into the small volume that is available in mini-satellites has become even more acute. This has led to the development of innovative three-mirror anastigmatic (TMA) telescopes of different designs as fitted, for example, to the TopSat and RapidEye satellites built by SSTL.

Spaceborne optical imaging systems can be conveniently reviewed in some more detail on the basis of the spatial resolution of the imagery that they generate, sub-divided into five main groups, as follows:

i. those spaceborne platforms and imaging systems generating very high-resolution imagery at ground sampled distances (GSD) of 1 m or below, i.e. at a sub-metre level (<1 m);
ii. those generating high-resolution imagery at GSD values between 1 m and 4 m;
iii. those generating higher medium-resolution imagery at GSD values between 4 and 8 m;
iv. those generating medium-resolution imagery at ground sampled distance (GSD) values between 8 and 20 m; and
v. those generating lower medium-resolution imagery at a lower level with GSD values between 20 and 40 m.

3.2.1 *Very high-resolution imaging systems*

Within this group, attention will be focused first of all on the background against which the spaceborne imaging systems are currently being operated on a commercial basis by the relevant companies based

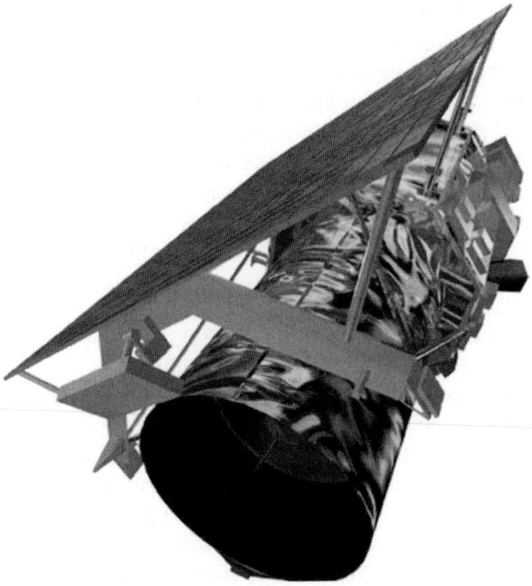

Figure 3.1. An artist's impression of the forthcoming GeoEye-1 satellite (formerly called OrbView-5) that is currently being built by General Dynamics (Source: GeoEye).

in the USA. On the one hand, they have had a very favourable home market place within which they have been able to pursue their business. This has been created largely through the failure of the Boeing company to produce the military reconnaissance satellites with an optical imaging capability that it was contracted to deliver under the Future Imagery Architecture (FIA) programme of the National Reconnaissance Office (NRO). This failure led to the Boeing company's contract for the supply of these satellites being cancelled in 2005. In turn, this led to the National Geo-spatial Intelligence Agency (NGA) contracting with the commercial spaceborne imaging companies to supply the very high-resolution space imagery that is required for its mapping and intelligence gathering purposes under its ClearView and NextView programmes (Petrie 2004a). These huge contracts have ensured the guaranteed funding that has enabled the commercial operating companies to obtain the massive financial investment that was needed to undertake the development of the next generation of very high-resolution satellites. The main beneficiaries of these two NGA-funded programmes have been the ORBIMAGE and DigitalGlobe companies. As a result, the Space Imaging company, which was the unsuccessful bidder in the NextView programme, was taken over by ORBIMAGE in 2005, the merger creating the new GeoEye company.

Notwithstanding the success of the two American companies in obtaining these contracts, the fact remained that the very high-resolution imagery that was required was still being gathered by the same group of ageing satellites, with the Ikonos (launched in 1999) and OrbView-3 (launched in 2003) satellites being operated by the GeoEye company and QuickBird (launched in 2001) by DigitalGlobe. This small fleet of three satellites was reduced still further to two when the OrbView-3 satellite ceased to function in March 2007. The situation was made still more difficult with the decision by the GeoEye company to postpone the launch of the new OrbView-5 satellite (now renamed GeoEye-1) from the late summer of 2007 to the middle of 2008. If and when the new GeoEye-1 is launched and comes into operation successfully, it will generate the highest resolution imagery of any commercially operated satellite: 0.41 m GSD for its panchromatic imagery and 1.64 m for its multi-spectral imagery with a swath width of 15.2 km. However, this rather gloomy story about these commercial spaceborne systems that generate very high resolution imagery has been relieved somewhat with the successful launch and entry into operational service of the WorldView-I (WV-I) satellite of the DigitalGlobe company in September 2007. This produces panchromatic imagery only with a GSD of 0.5 m and a swath width of 17 km. A second satellite, WorldView-II (WV-II), is scheduled for launch in 2008. Besides its panchromatic imaging channel, this second satellite in the WV series will collect multi-spectral image data in eight different spectral channels in the VNIR part of the spectrum with a GSD of 1.8 m.

Besides the two American companies, the only other commercial operator of a spaceborne system producing sub-metre imagery is ImageSat International from Israel. In 2004, it announced that, due to lack of demand for very high-resolution imagery from the non-military sector, it would no longer offer the imagery acquired

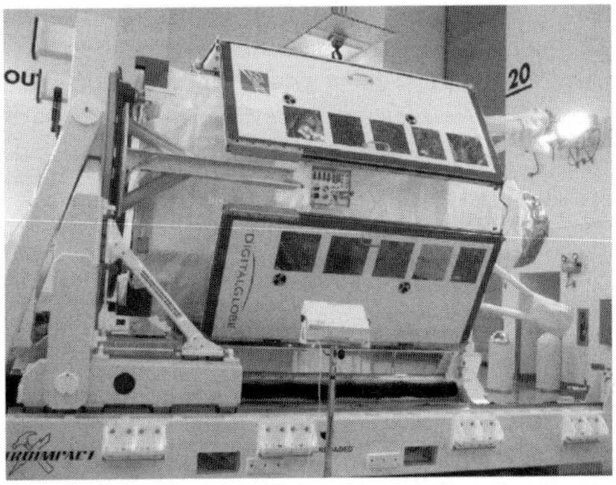

Figure 3.2. The WorldView-1 satellite of the DigitalGlobe company was built by Ball Aerospace. Here it is shown mounted on a platform for transportation prior to launch (Source: Ball Aerospace).

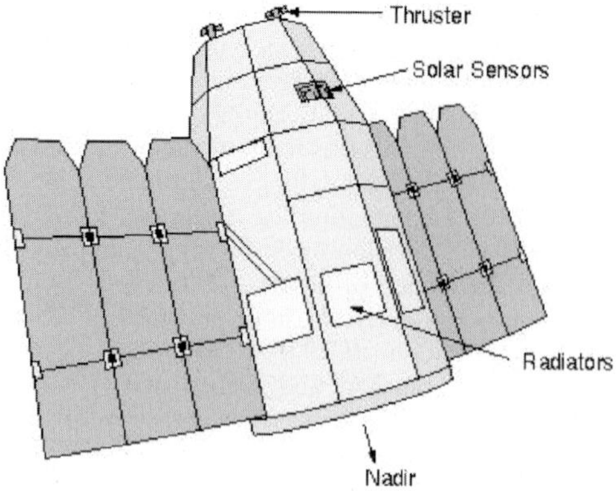

Figure 3.3. A drawing showing the main features of the EROS-B satellite that acquires very high-resolution panchromatic linescan imagery (Source: ImageSat International).

by its EROS satellites for sale to the public. Instead it would concentrate on supplying the EROS imagery to government and national security agencies in Israel and elsewhere and to those partner countries (principally in Asia) that cooperate in the operation and downloading of imagery from its EROS satellites via its Satellite Operating Partner (SOP) programme. To serve these customers, the company launched and brought its EROS-B satellite into successful operation in April 2006. This produces panchromatic imagery only with a GSD of 0.7 m and a swath width of 7 km.

Also falling within this group of very high-resolution spaceborne imaging systems are the two new government funded and operated systems from other Asiatic countries that are producing panchromatic imagery with a GSD of 1 m, similar to that of Ikonos (Petrie 2004b, 2007a). The first of these is the Kompsat-2 satellite from South Korea, which was launched in July 2006. This not only collects pan imagery but it also generates multi-spectral imagery with a GSD of 4 m and a swath width of 15 km. The satellite was constructed in close collaboration with EADS Astrium from France, while the very high-resolution pushbroom scanner was built by the El-Op company from Israel. The distribution of Kompsat-2 imagery outside South Korea, the USA and the Middle East is being undertaken by the French SPOT Image company. The second satellite from Asia is the Indian Cartosat-2, which was launched in January 2007. This system generates panchromatic imagery only with a swath width of 10 km. It appears to be oriented primarily at supplying the needs of India's mapping and intelligence agencies in much the same way as the country's previous TES reconnaissance satellite that was launched in 2001.

Finally coming into this group is the very interesting Russian Resurs DK-1 spacecraft, which was launched in June 2006. Several Russian sources have stated that it is a modified version of the Yantar-4KS1 series of large and highly-manoeuvrable military reconnaissance satellites. The Resurs DK-1 produces both panchromatic and three-band (green, red, NIR) multi-spectral imagery with GSD values of 1 and 3 m, respectively, and can cover a very wide swath of 28 km over the ground from an orbital altitude of 360 km. Like most of the other satellites in this class, the satellite can be body pointed ±30° in the cross-track direction to obtain images of areas on the ground located to the side of the satellite ground track. However, compared with the other satellites in this class, it has some quite unusual features. In particular, it has been placed in an elliptical orbit, which can have a perigee (lowest altitude) value as low as 200 km and a corresponding apogee of 360 km. Furthermore, the satellite's orbital inclination can be varied over the range 64° to 71°. These orbital characteristics are very different to the near-polar Sun-synchronous circular orbits of the other satellites in this group. Obviously if the DK-1 satellite is operated from its

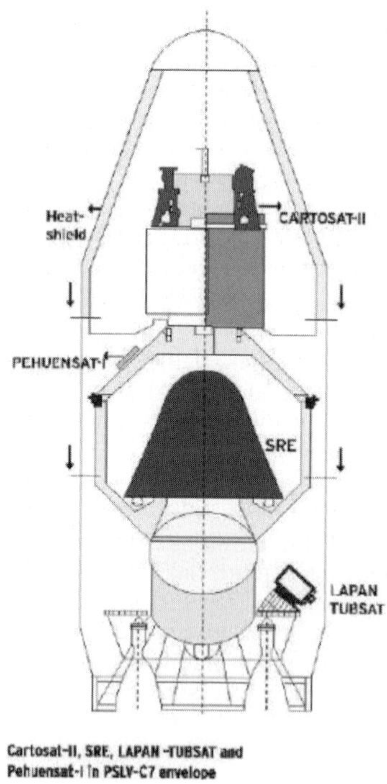

Figure 3.4. The drawing shows the arrangement of the multiple satellites that were launched simultaneously by an Indian PSLV rocket in January 2007. At the top of the stack is the Indian Cartosat-2 very high-resolution imaging satellite; at the foot of the stack is the tiny Lapan-Tubsat remote sensing micro-satellite (Source: ISRO).

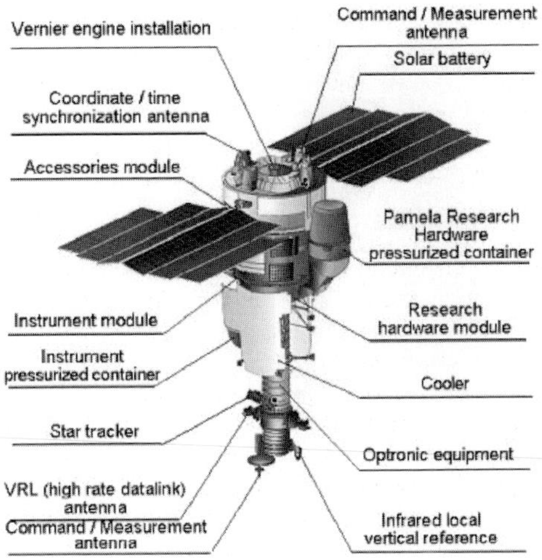

Figure 3.5. A drawing showing the main features of the Resurs DK-1 satellite that generates very high-resolution scanner imagery (Source: NTs OMZ).

lowest perigee value, it will generate its panchromatic imagery with a GSD of 0.5 m. However, frequent use of such a low altitude would result in atmospheric drag starting to be significant. This would need to be followed by corrections to the orbit using the satellite's rocket engine to maintain altitude and prevent re-entry into the atmosphere. In turn, this would cause a heavy drain on the satellite's fuel reserves. So the use of higher perigee altitudes such as the 360 km value mentioned above would be more likely in actual operational use. The Resurs DK-1 satellite is operated by the Research Centre for Earth Operative Monitoring of the Russian Federal Space Agency (Roskosmos), which receives the imagery at two ground stations located in Moscow and in Western Siberia. The DK-1 imagery is available to commercial customers through certain authorized Russian companies such as Sovzond. Comparisons show that the quality of the DK-1 images is quite comparable to those for the same areas that have been acquired by the American Ikonos and QuickBird satellites.

In summary, it is obvious that the replacement of the well known American commercial satellites in this class has been a very difficult task and is only now beginning to be resolved with the coming into service of the first WorldView satellite. Besides which, a very significant development during the inter-Congress period has been the entry of the three Asiatic countries, Israel, India and South Korea, into this area of activity with the primary task of providing very high-resolution imagery for national geo-spatial mapping and intelligence gathering purposes. The advent of the Russian DK-1 satellite with its unusual orbital and imaging characteristics is another interesting development, although, looking to the future, it is still not clear if this is the first in a series of such systems.

3.2.2 *High-resolution imaging systems*

Within this category, there has been a parallel development during the period 2004 to 2008 to that described above with many of the main developments occurring in Asia and with a strong orientation towards satisfying the needs of national security organizations. The several new satellites within this category that have entered service since 2004 have included Formosat-2 (launched in 2004), Cartosat-1, Beijing-1 and TopSat (all launched in 2005), ALOS (launched in 2006) and THEOS (currently scheduled for launch early in 2008). This particular area was previously rather sparsely occupied by the EROS-A and SPOT-5 satellites that were launched in 2000 and 2002, respectively, and producing panchromatic imagery with GSD values of 2 and 2.5 m, respectively. Now it is very well populated and served by these various new operational systems.

Starting with Cartosat-1, this large Indian satellite carries twin pushbroom scanners equipped with CCD linear arrays having 12,000 detectors, each 7 μm in size (Lutes 2006a, 2006b). These produce overlapping images with a 2.5 m GSD and a 30 km swath width. The two scanners are tilted in the along-track direction at angles of +26° (forward) and -5° (backward) to produce stereo coverage with a view to DEM production. In some ways, it resembles the dedicated HRS imager of SPOT-5 with its twin scanner imagers, each having linear arrays that are 12,000 pixels long. However, the HRS imager has a much lower spatial resolution with a 5 × 10 m GSD and a much wider swath of 120 km. In the case of the Taiwanese Formosat-2, there is a strong French connection, since the satellite and its scanner were largely constructed in France by EADS Astrium. Besides which, SPOT Image is acting as the worldwide distributor of the imagery that is generated by this satellite, except for the area of Taiwan and continental China. The Formosat-2 images have GSD values of 2 m (pan) and 8 m (multi-spectral) and cover a swath of 24 km. The forthcoming THEOS (THailand Earth Observation System), which is also being constructed by EADS Astrium for Thailand, has an almost identical specification to that of Formosat-2 (Piboon et al. 2005). Furthermore, there is a similar proposal to that implemented with the Formosat-2 imagery that the world-wide reception and distribution of the THEOS imagery will be carried out through the SPOT Image network.

The ALOS satellite, which was placed in orbit by the Japanese JAXA organization in January 2006, is a very large and heavy (4 ton) satellite with three very

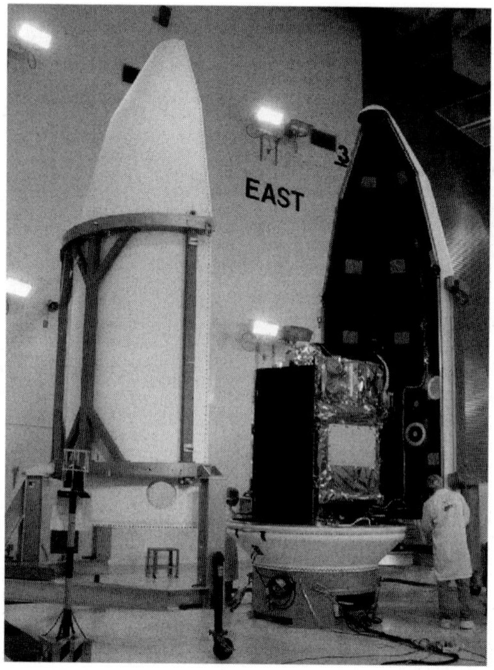

Figure 3.6. The Formosat-2 satellite is about to be enclosed within its protective heat shield prior to its launch at Vandenburg AFB, California (Source: NSPO, Taiwan).

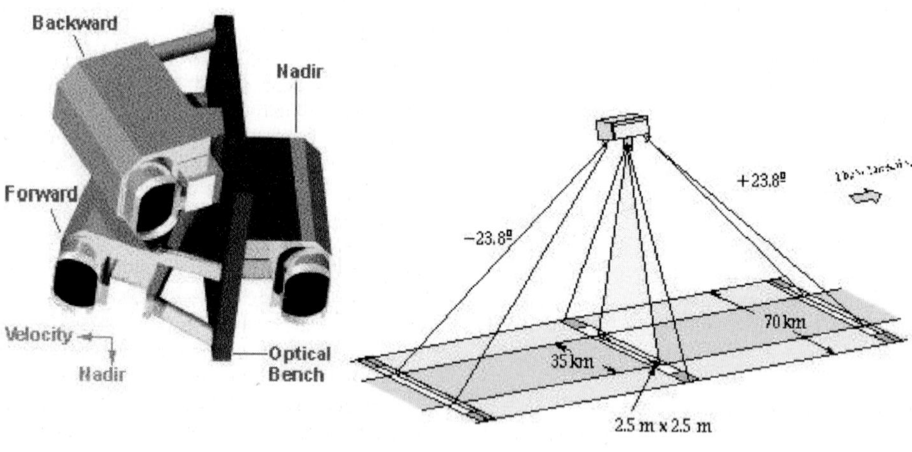

Figure 3.7. On the left is the PRISM three-line pushbroom scanner mounted on the Japanese ALOS satellite: on the right is a diagram showing the ground coverage for each of forward, nadir and backward-pointing CCD linear arrays of the PRISM imager (Source: JAXA).

different imagers (PRISM, AVNIR-2 and PALSAR) mounted on it (Osawa 2004). Its PRISM (Panchromatic Remote-sensing Instrument for Stereo Mapping) imager is a full blown three-line pushbroom scanner with the backward- and forward-pointing telescopes being set at angles of ±24° to the vertical (Tadono et al. 2004). These two scanners and the third (nadir-pointing) scanner produce overlapping panchromatic images having a 2.5 m GSD and a 35 km swath width. This stereo-imagery is designed specifically for the production of DEMs, in a similar manner to that of the Cartosat-1 and SPOT-5 HRS imagers,

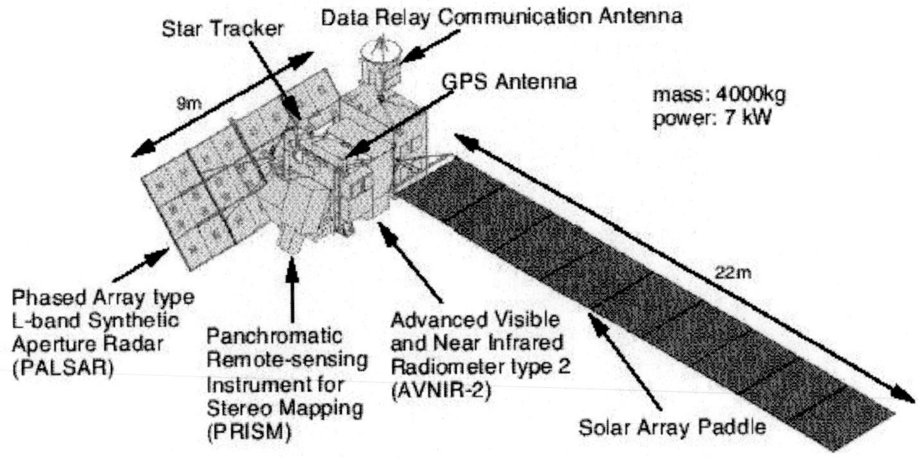

Figure 3.8. The large and heavy Japanese ALOS satellite carries three imaging devices—the high resolution PRISM stereo-imager; the medium-resolution AVNIR-2 optical imager; and the PALSAR L-band synthetic aperture radar (Source: JAXA).

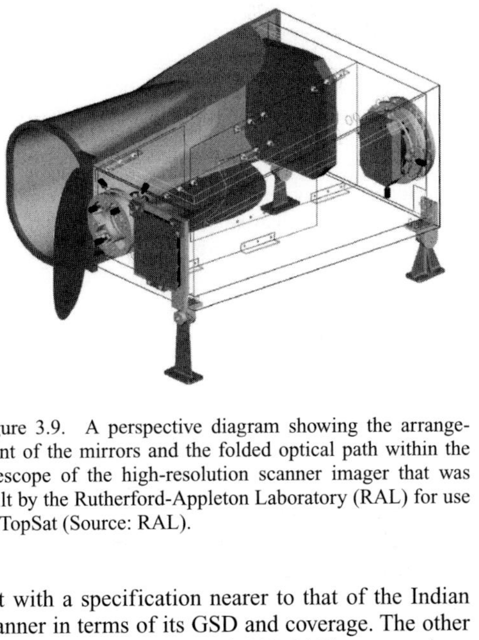

Figure 3.9. A perspective diagram showing the arrangement of the mirrors and the folded optical path within the telescope of the high-resolution scanner imager that was built by the Rutherford-Appleton Laboratory (RAL) for use on TopSat (Source: RAL).

but with a specification nearer to that of the Indian scanner in terms of its GSD and coverage. The other two imagers (AVNIR and PALSAR) mounted on ALOS fall into rather different categories and will be discussed later in this account.

The remaining satellites in this category are all micro- or mini-satellites. Two of them—TopSat and Beijing-1—have been constructed by SSTL (Surrey) in the UK and were launched together in 2005 using a Russian rocket. In the case of TopSat, the funding was provided by the UK Ministry of Defence and the British National Space Centre, while the operation of the satellite and the reception of the image data using both fixed and mobile ground stations is carried out by the QinetiQ defence research company (Crawley 2005). The TopSat imagery has GSD values of 2.8 m (pan) and 5.6 m (multi-spectral) over swaths of 10 and 15 km, respectively. The sale and distribution of the TopSat imagery to non-military users is carried out by the Infoterra and DMCII companies, the latter being an offshoot of SSTL, the company that built the satellite (Stephens 2007). With regard to Beijing-1, this satellite combines its high-resolution pushbroom scanner giving panchromatic imagery with a GSD of 4 m over a swath width of 24 km with a much lower (32 m GSD) resolution scanner that produces multi-spectral linescan imagery over a swath of 600 km. A development that combines some of the features of both TopSat and Beijing-1 will be the NigeriaSat-2 satellite that is currently under construction by SSTL and is planned for launch in 2009. This 300 kg mini-satellite will combine a high-resolution pushbroom scanner producing 2.5 m GSD pan and 5 m GSD multi-spectral imagery over a 20 km swath with a medium-resolution imager generating 32 m GSD imagery over a 300 km swath. The compact and lightweight construction of these micro- and mini-satellites has been achieved partly through the use of innovative folded telescopes that form the basis of their pushbroom scanner imagers.

Stimulated by the striking success of the SSTL-built satellites, several other suppliers have moved into this area. In particular, EADS Astrium has secured the contract to supply two satellites, Alsat-2A and AlSat-2B, to Algeria, which follow on from the previous AlSat-1 built by SSTL. The first of these new satellites is scheduled for launch in 2008; the second, in 2009. Like TopSat and NigeriaSat-2, each of the

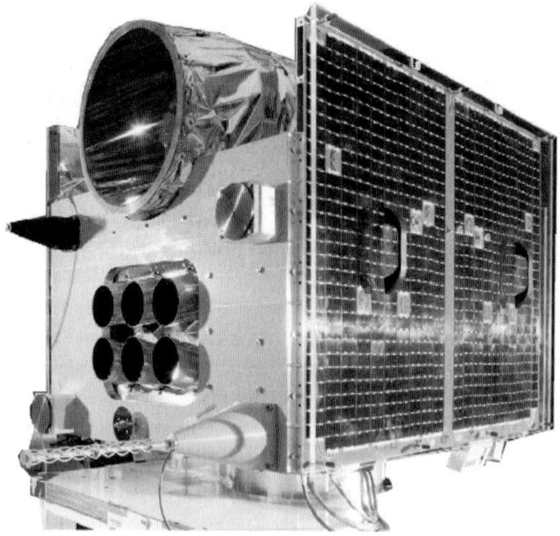

Figure 3.10. The telescope of the high-resolution pushbroom scanner is mounted on the top of the Beijing-1 micro-satellite built by SSTL. The multiple objective lenses of the telescopes of the lower-resolution multi-spectral scanner can be seen on the front panel of the satellite below that of the high-resolution pushbroom scanner (Source: SSTL).

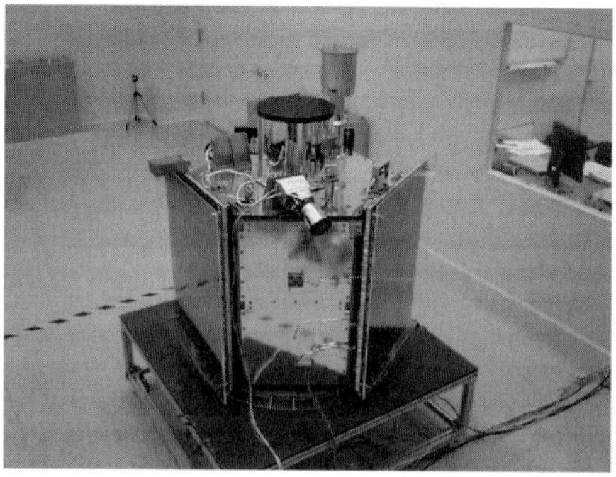

Figure 3.11. The RazakSAT satellite that is being built for the Malaysian government by the Satrec Initiative company in South Korea (Source: Satrec Initiative).

two AlSat-2 satellites will have a high-resolution pushbroom scanner generating panchromatic line-scan imagery with 2.5 m GSD and a 17 km swath over the ground. Both satellites will also acquire multi-spectral imagery with a GSD of 10 m and having the same 17 km swath width.

Another new supplier is the Satrec Initiative company from South Korea, which is building the RazakSAT micro-satellite for an agency of the Malaysian government. The specification is not too dissimilar to that of AlSat-2 in that it will produce both panchromatic and multi-spectral imagery with 2.5 and 10 m GSD values over a swath width of 20 km. A unique aspect of RazakSAT is, however, that it will be placed in a near-equatorial orbit inclined at 9° to the Equator. This will ensure that the territory of Malaysia, which is located between 1 and 9 degrees latitude north, will be overflown on several occasions each day. The satellite is scheduled to be launched late in 2008 using the American Space-X company's low-cost Falcon launcher. Besides RazakSAT, the Satrec Initiative company is constructing another satellite, DubaiSat, that is designed to produce 2.5 m GSD imagery for a United Arab Emirates (U.A.E.) government agency. This is also scheduled to be launched in 2008, possibly in combination with two of the satellites being built by SSTL, using a Russian rocket.

In summary, this particular area has seen much activity during the inter-Congress period. It is interesting to note the two quite different approaches that have been followed in the provision of high-resolution imagery in the range 1 to 4 m. The first group uses conventional large and comparatively heavy satellite platforms that need fairly powerful launchers to place them in orbit. By contrast, the second group comprises low-cost, lightweight micro- or mini-satellites that can be placed in orbit using much less powerful and less expensive launchers. To illustrate the point, the TopSat and Beijing-1 mini-satellites were launched in tandem using a single Russian rocket and together weighed well under 300 kg. By contrast, the Formosat-2 and THEOS satellites each weigh 750 kg and Cartosat-1 (with its dual scanners) weighed 1,560 kg at launch. ALOS with its three imagers weighed over 4.5 tons!

3.2.3 *Higher medium-resolution imaging systems*

Previously, many spaceborne imaging systems had been launched that fell into this category. Indeed quite a number of them are still in operation, including three of the satellites in the Indian IRS series—IRS-1C, IRS-1D and IRS-P6 (ResourceSat-1)—that were launched in 1995, 1997 and 2003, respectively. However, since the 2004 Congress, these have only been joined by a couple of new satellites. The first of these is the Russian Monitor-E-1, which was launched into a near-polar Sun-synchronous orbit in August 2005. This has provided panchromatic and three-band multi-spectral imagery with GSD values of 8 m and 20 m and swath widths of 90 and 160 km, respectively. From presentations made by Russian contributors at a recent conference in Bulgaria, it appears that the satellite has been experiencing technical problems in communicating with its ground stations. Therefore only a limited amount of imagery has been acquired up till now. Indeed some sources have reported that the satellite has ceased to function. At the present time, with the help of SSTL, a new

Earth Observation (EO) satellite, called Kanopus-2, is under construction by the same agency (NPP VNIEM) that built the older Russian series of Meteor remote sensing satellites. This new satellite is scheduled for launch in 2009.

A second operational satellite falling within this category is the LAPAN-Tubsat micro-satellite that has been developed jointly by the Technical University of Berlin and the Indonesian National Institute of Aeronautics & Space. The satellite was launched in January 2007 using an Indian rocket (Fig. 3.4). The satellite is equipped with a pair of Sony colour video cameras fitted with f = 50 mm and 1,000 mm lenses, respectively. These cameras generate images in the European PAL TV format that are 752 × 582 pixels in size with GSD values of 6 m and 200 m and swath widths of 3.5 km and 85 km, respectively. Apparently only images of those areas lying within the field of view of the ground station in West Java can be received from this satellite. Another micro-satellite that falls into this category is the Sumbandila that is being constructed by the University of Stellenbosch in South Africa. This will acquire six-band multi-spectral imagery with a 6.5 m GSD and a 45 km swath width. The launch, using a Russian rocket, is scheduled for 2008.

However, there is in prospect a very significant addition to this category of higher medium-resolution satellites in the form of the constellation of five mini-satellites of the RapidEye company based in Brandenburg, Germany. All five satellites have been built by SSTL in the U.K., while the multi-spectral pushbroom scanner imagers have all been supplied by Jena-Optronik. The imager is based on the use of a very innovative three-mirror anastigmatic (TMA) telescope having an all-aluminium construction with

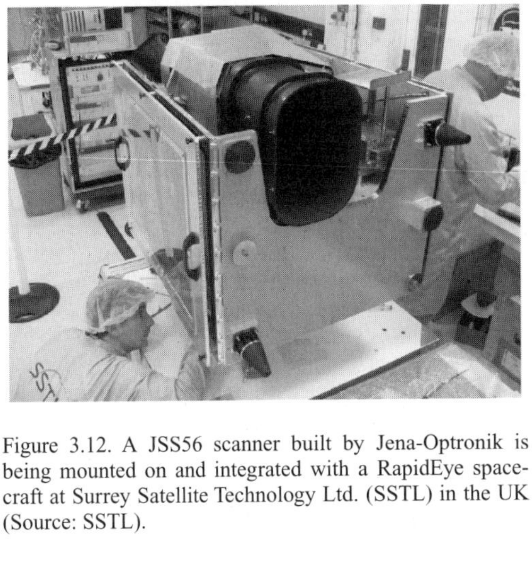

Figure 3.12. A JSS56 scanner built by Jena-Optronik is being mounted on and integrated with a RapidEye spacecraft at Surrey Satellite Technology Ltd. (SSTL) in the UK (Source: SSTL).

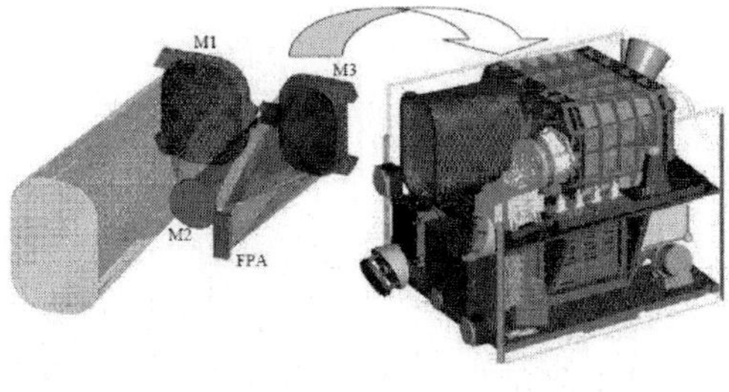

Figure 3.13. The three-mirror anastigmatic (TMA) telescope showing the mirrors (M1, M2 & M3) and the focal plane array (FPA) and their relationship to the overall construction of the JSS-56 spaceborne scanner on which it is mounted (Source: Jena-Optronik).

a reflective silver coating instead of using optical glass (Petrie 2007b). All five satellites are scheduled to be launched on a single Russian Dnepr rocket early in 2008. The pushbroom scanner that is mounted on each satellite has five linear arrays, each with 12,000 detectors and the appropriate spectral filter. The resulting five images will cover the blue, green, red, red edge and near infra-red spectral bands. Each image will have a GSD of 6.5 m and a swath width of 78 km from the orbital height of 600 km. The RapidEye company is aiming its imagery specifically at the agriculture, forestry and environmental sectors with frequent re-visits for monitoring purposes. If the system can be launched and brought into operation successfully, it promises to have a great impact on this particular area.

3.2.4 *Medium-resolution imaging systems*

The situation regarding spaceborne systems generating medium-resolution imagery is much like that of the previous group producing higher medium-resolution imagery in that nearly all the satellites that are currently in operation in this category were brought into service prior to the last Congress held in 2004. They include the following:

i. The well-known SPOT-2 and SPOT-4 satellites (launched in 1990 and 1998) generating pan and multi-spectral linescan imagery at 10 and 20 m GSD, respectively;
ii. Landsat-7 (launched in 1999) producing 15 m GSD pan and 30 m GSD multi-spectral imagery and now crippled due to its malfunctioning image corrector mechanism;
iii. The Terra satellite (launched in 1999) with its ASTER scanner generating stereo and multi-spectral imagery with a 15 m GSD;
iv. The Chinese-Brazilian CBERS-2 satellite with its combination of pan and multi-spectral imagery with 20 m GSD; and

v. The ESA PROBA satellite equipped with its CHRIS scanner and the NASA EO-1 satellite with its Hyperion scanner are both generating hyper-spectral imagery with 20 m and 30 m GSD values, respectively. Both were launched in 2001.

There are only a limited number of dedicated medium-resolution satellites to be added during the period covered by this review. One is the CBERS-2B satellite, which was launched in September 2007 and generates imagery with the same characteristics as the previous CBERS-2. These CBERS satellites appear mainly to serve the local markets for remotely sensed imagery within Brazil and China. Besides these, the AVNIR-2 imager that is mounted on board the Japanese ALOS satellite also generates medium-resolution imagery that, in this case, takes the form of four-band multi-spectral imagery having a GSD of 10 m and a swath width of 70 km.

In this general context, it may also be appropriate to mention the Egyptsat-1 micro-satellite, which was placed successfully into orbit in April 2007. This micro-satellite was built for the Egyptian National Authority for Remote-sensing & Space Sciences (NARSS) by a group of Ukrainian companies. However, no details of the imaging sensors have been released, other than the brief statement from Ukrainian sources that the satellite carries two of them—a "high-resolution multi-spectral imager" and "an infra-red imaging sensor". No examples of the images acquired by Egyptsat-1 have been released so far, nor is there any reference to the Egyptsat-1 satellite or its imagery on the NARSS Web site. However, unofficial Web sites in Egypt state that the images from Egyptsat-1 have a GSD of 8 m and a swath width of 45 km. Egyptsat-1 was launched together with the Saudisat-3 mini-satellite. Again little information is available on this satellite, though a Ukrainian source mentions that Saudisat-3 operates a pushbroom scanner with a 15 m GSD.

It should be noted that the imagery generated by all of the older satellites in this class has been used extensively in the past, especially for agricultural and landuse monitoring and for scientific studies into land cover and environmental change. However, this extensive use has been based on the fact that the imagery has, to a large extent, been available at a comparatively low cost that does not recover the expense involved in constructing, launching and operating the satellites and their ground segments. The many difficulties and failures that have been associated with the Landsat Data Continuity Mission (LDCM) over the years only confirm and highlight the situation that, while the user community in this area is large, and indeed it extends world-wide, it is not able to sustain commercially viable spaceborne imaging operations in the manner that has been successful with high-resolution satellites where national security and defence agencies form a very large part of the user community.

However, there is some new activity in this area, mainly of a local character, which should bear fruit shortly. This takes the form of micro-satellites that are being built for local use by research institutes in Asian countries. These include the X-Sat that is being constructed by Nanyang Technical University in Singapore, and RASAT, which is being built by the Turkish Tubitak research agency. In both cases, the optical imager is being supplied by the Satrec Initiative company from South Korea. In the case of X-Sat, it will produce multi-spectral RGB imagery with a 10 m GSD and 50 km swath width. With RASAT, both panchromatic and multi-spectral imagery will be generated with 7.5 m and 15 m, respectively. Again the target date for the launch of both of these satellites is 2008.

3.2.5 *Lower medium-resolution imaging systems*

Apart from the rather ancient but still operational Landsat-5 satellite (launched in 1984), this area has become the domain of the micro-satellite systems that have been pioneered to a large extent by SSTL. However, even here, there has been a lull in the launching of new satellites over the period 2004 to 2008. Those satellites of the Disaster Monitoring Constellation (DMC)—AlSat-1, UK-DMC and NigeriaSat-1 —that are still operational were all launched during the period 2002-2004. So was the Turkish DMC satellite, Bilsat, which is no longer in operation, "having completed its mission in 2006". However the future is somewhat brighter since a second generation of DMC satellites is currently under construction by SSTL, including the UK-DMC-2 and Spanish Deimos-1 satellites, both of which are scheduled to be launched during 2008. These two new satellites will generate multi-spectral imagery with a GSD of 22 m, which represents a substantial improvement over the 32 m GSD of the current generation of DMC satellites. They will be joined by the NigeriaSat-2 satellite, which will contribute 32 m GSD multi-spectral imagery—as already discussed above.

3.2.6 *Overview of spaceborne optical imaging systems*

Taking an overall view of the situation regarding spaceborne optical imaging during the period since the last Congress in 2004, a number of points can be made as follows:

a. One point that really stands out has been the remarkable development and adoption of micro- and mini-satellites in all areas except those systems that produce the very highest resolution imagery. However, even here, the EROS-B satellite can readily be classed as falling into the category of a mini-satellite. A consequence of this development has been that spaceborne imaging systems are now being operated by many countries other than the traditional major space powers. It appears fairly certain that this trend will continue.

b. Another very noticeable development has been that of dedicated spaceborne systems, such as SPOT-5/HRS, Cartosat-1 and ALOS/PRISM, that produce stereo-coverage of the ground with a good base:height ratio that permits the generation of DEMs over large areas of the Earth's terrain.
c. At the top end of the scale, the imagery acquired by the very high-resolution spaceborne systems continues to find a ready market. However, it does seem that by far the biggest part of this market comprises national security and intelligence gathering agencies. Without this demand, it is doubtful that the commercial companies supplying very high-resolution imagery would be viable. What is also noticeable is that currently there are no spaceborne imaging systems being operated by commercial companies outside the very high-resolution sector. The only addition to this short list will be the Rapid Eye company when it launches its constellation of satellites that will generate medium-resolution imagery.
d. There has been a noticeable lack of development of the medium-resolution satellites that are needed to replace the three well-known series of Landsat, SPOT and IRS satellites that have formed the backbone of spaceborne remote sensing for Earth resources applications for so many civilian agencies and academic departments world-wide over the past 20 or 30 years. This has caused much controversy as evidenced by the debate over the Landsat Data Continuity Mission.

3.3 SPACEBORNE RADAR IMAGING SYSTEMS

For a considerable time, there has been comparatively little development to report in the field of spaceborne radar imaging, which has been almost wholly dependent on the SAR imagery produced by the Radarsat-1 and ERS-2 satellites, both of which came into operation in 1995, and the Envisat ASAR radar that came into operation in 2002. All of their SAR imagers operate in the C-band of the microwave spectrum. Now however a complete transformation of this sector is taking place with the launch of a substantial number of new satellites equipped with SAR imagers. Most of these operate in the X-band giving the promise of much higher-resolution radar imagery than before, while others are ensuring continuity in the supply of C-band imagery. There is also a single system providing imagery in the L-band part of the spectrum that was used originally by the first civilian radar satellite, Seasat.

3.3.1 *X-band SAR imaging systems*

This particular area is now being exploited by two European countries, Germany and Italy, both of which are pursuing their activities in a very substantial and dynamic manner, strongly supported by their military defence establishments (Petrie 2004c). In the case of Germany, it is in the middle of launching and bringing into operation a fleet of five satellites under its SAR-Lupe programme (Merkle 2003).

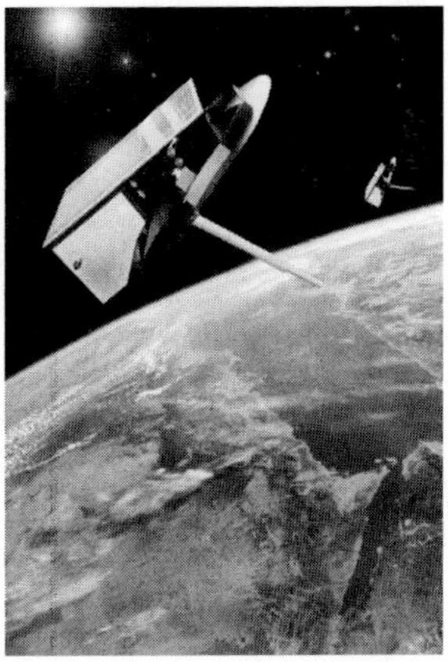

Figure 3.14. (Left) An artist's impression of a pair of SAR-Lupe radar satellites in orbit (Source: OHB-System).

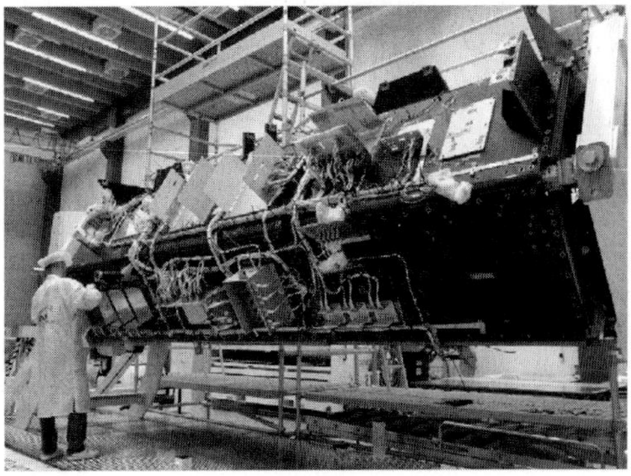

Figure 3.15. (Right) The TerraSAR-X satellite under construction at EADS Astrium (Source: EADS Astrium).

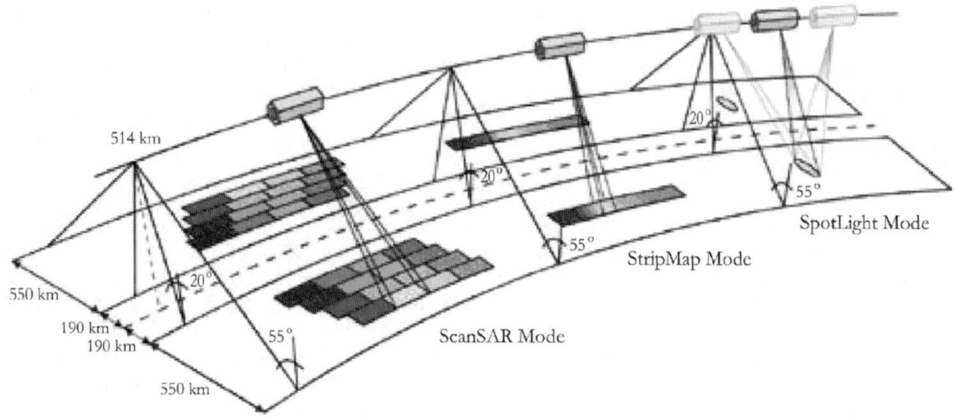

Figure 3.16. This diagram shows the respective ground coverages of the Spotlight, StripMap and ScanSAR modes of the TerraSAR-X satellite (Source: DLR).

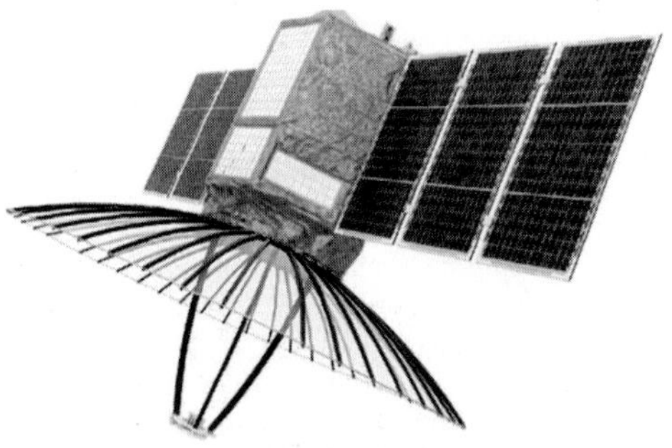

Figure 3.17. (Left) The TecSAR satellite with its X-band radar antenna deployed (Source: IAI).

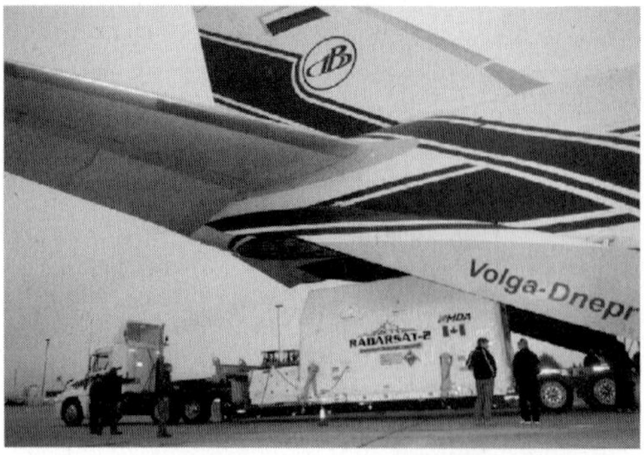

Figure 3.18. (Right) The Radarsat-2 satellite in its transportation box being loaded on to an Antonov An-224 freight aircraft for its flight to Baikonur, Kazakhstan where it was launched using a Russian Soyuz rocket in December 2007 (Source: Canadian Space Agency).

Three of these satellites have already been launched in December 2006, July 2007 and November 2007, respectively. The remaining two satellites are scheduled to be launched in March and August 2008. The whole fleet is being launched using Russian rockets. The five satellites are being placed in three different orbital planes, spaced out to ensure world-wide coverage. Plane 1 will contain two satellites; Plane 2 will have a single satellite; while the remaining two satellites will occupy Plane 3. The five satellites are being built by the OHB System company based in Bremen, Germany, with various components being supplied by other European and American companies. The GSD values of the resulting SAR imagery will be better than 1 m over an area of 5 × 5 km when the X-band SAR is operated in its highest resolution mode.

The other major German spaceborne SAR project is the TerraSAR project, which is a public/private partnership between the German Space Agency (DLR) and the German arm of the EADS Astrium aerospace company. The TerraSAR-X satellite that has resulted from this cooperative programme was launched in June 2007. Once again, its X-band imager will generate imagery with a best spatial resolution (GSD) of 1 m. The commercial marketing of the TerraSAR-X imagery is being carried out by Infoterra GmbH, a subsidiary company of EADS Astrium. A further development of the programme is the TanDEM-X satellite, which is a near identical twin to the TerraSAR-X satellite and will operate in tandem with it in a close formation orbit (Krieger et al. 2007). The objective of this extension to the programme is to generate a global digital elevation model (DEM) using SAR interferometric techniques. The product should prove to be attractive to many users world-wide. Currently the TanDEM-X satellite is scheduled to be launched in 2009.

In the case of Italy, its government has funded a fleet of four COSMO-SkyMed satellites that will provide X-band SAR imagery for both civilian and defence applications (Candela et al. 2003), though the defence authorities will have priority in the mission planning and operation of the satellites. On the civilian side, the SAR image data acquired by the four satellites will form part of the European Global Monitoring for Environment & Security (GMES) programme. The four satellites are being built by the Alenia Spazio company (now called Thales Alenia Space). Each of the four satellites will be placed in a Sun-synchronous orbit placed 90° apart in longitude to ensure world-wide coverage. The X-band SAR mounted in each of the satellites will generate imagery with a GSD of 1 m over a 10 km wide swath when operated in its highest resolution (spotlight) mode. The first two COSMO-SkyMed satellites were placed in orbit in June and December 2007 using Boeing Delta-II launchers. The remaining two satellites in the constellation are scheduled to be launched during the summer of 2008 and early in 2009.

The remaining satellite that has been equipped with an X-band SAR is the TecSAR, which has been constructed by Israel Aerospace Industries (IAI) with the radar being supplied by its subsidiary company, Elta Systems. Like the other Ofek and EROS satellites launched by Israeli organizations, it is a comparatively lightweight mini-satellite, weighing in the region of 400 kg. TecSAR has just been launched on 21 January 2008 by an Indian PSLV rocket, reportedly into elliptical orbit with an orbital inclination of 41°. The TecSAR imagery will be acquired for use by Israel's defence and intelligence agencies in any one of its scan-SAR, strip-SAR or spotlight modes, with a GSD of 1 m using the last of these three modes.

3.3.2 *C-band SAR imaging systems*

As noted above, C-band is the part of the microwave spectrum that has been used by several civilian spaceborne SAR systems, including the European ERS-1, ERS-2 and Envisat satellites and the Canadian Radarsat-1 satellite. The last of these is still in operational use after more than twelve years' service and it has just been joined by its sister Radarsat-2 that was launched in December 2007. This satellite has been produced through another public/private partnership between the Canadian Space Agency (CSA) and the MacDonald Dettwiler & Associates (MDA) company. Radarsat-2 was originally scheduled to be launched in 2001, but the programme underwent a series of difficulties with American government regulatory authorities that resulted in the original American contractors who were undertaking the satellite construction and the launch provision having to be changed. In the end, the satellite bus was built by Alenia Spazio in Italy, while the launch was carried out using a Russian rocket. The acquisition of the Radarsat-2 imagery can be carried out in an almost bewildering number of modes providing different coverages, incident angles and ground resolution values. The ultra-fine mode can provide a GSD of 3 m over a swath width of 20 km.

A second spaceborne SAR that will operate in the C-band is that being mounted on the Indian RISAT satellite, which is scheduled to be deployed in 2008. It will utilize the now standard operating modes—scan-SAR, strip-SAR and spotlight—producing images having a 3 m GSD in the finest of these modes. Along with Radarsat-2, it will ensure the continuity of the experience that has been built up world-wide with the several previous C-band SAR missions.

3.3.3 *L-band SAR imaging systems*

An L-band SAR system was mounted on the very first civilian radar satellite, Seasat, which was operated by NASA during 1978. The Japanese JERS-1 SAR also generated its imagery in the L-band and operated very successfully from 1992 till 1998. The resulting imagery was quite different to and complemented that produced by the contemporary ERS and Radarsat satellites with their C-band SAR imagery. In particular, the JERS-1 SAR imagery was used extensively in the mapping of huge areas of tropical rain forest and boreal forest world-wide. The new PALSAR imager that has been mounted on the Japanese ALOS satellite follows on from this and also operates in the L-band region (Osawa 2004). However, it is fully polarimetric instead of operating solely in the single HH mode that was used with the JERS-1 SAR. The ground resolution of the PALSAR imagery ranges from a GSD of 7 m (in fine mode) to 100 m in scan-SAR mode with respective swath widths ranging between 40 km out to 350 km in scan-SAR mode. The PALSAR imagery has already come into widespread use since the launch of the ALOS satellite in January 2006.

3.3.4 *Overview of spaceborne radar imaging systems*

As will be apparent from the account set out above, the whole area of spaceborne radar imaging is being completely transformed by the developments that have taken place over the last two years. What is of special importance has been the introduction of civilian satellites operating in the X-band and generating high resolution SAR imagery on a consistent basis. Previously the only spaceborne X-band SAR imagery that has been available to non-military users has been that generated by the short-term experimental flights of the SIR-C and SRTM projects on board the Space

Shuttle. It will be very interesting to see if the X-band imagery generated by the several new satellites that are being launched will result in SAR imagery being used more widely now that it will be available on a more regular basis than before. For the rest, the launch of the Radarsat-2 satellite has ensured that users will also have continuity of access to C-band SAR imagery as they have enjoyed for many years through that generated by the existing ESA and Radarsat-1 satellites. The launch of the Palsar SAR has revived the possibility of using longer wavelength L-band imager for forest mapping applications as was done with the JERS-1 SAR imagery.

3.4 OVERALL SUMMARY

It is apparent that, notwithstanding the many problems associated with the funding, construction, launch and operation of satellites equipped with optical imaging systems, the current situation in 2008 is reasonably healthy—in large measure due to the successful introduction of mini-satellites that have allowed many new countries to enter the field of spaceborne remote sensing. In the case of the spaceborne radar imaging systems, they are experiencing a remarkable revival, largely due to military interest in the capabilities of X-band SAR systems to generate high-resolution imagery from space, irrespective of the weather and illumination conditions that are present over the Earth.

REFERENCES & FURTHER READING

Candela, L. & Caltagirone, F. 2003. COSMO-SkyMed: Mission Definition & Main Application and Products. *Proceedings, POLinSAR Workshop on Applications of SAR Polarimetry and Polarimetric Interferometry, ESA-ESRIN, Frascati, Italy, 14–16 January, 2003.*

Crawley, S. 2005. TopSat: Low Cost High Resolution Imagery from Space. *Acta Astronautica* 56(1–2): 147–152.

Krieger, G., Moreira, A., Fiedler, H., Hajnsek, I., Werner, M., Younis, M. & Zink, M. 2007. TanDEM-X: A Satellite Formation for High Resolution SAR Interferometry. *IEEE Transactions on Geoscience & Remote Sensing* 45(11): 3317–3341.

Lutes, J. 2006a. Photogrammetric Processing of Cartosat-1 Stereo Imagery. *GeoEye Technical White Paper*. [See http://www.geoeye.com/techpapers/default.htm]

Lutes, J. 2006b. First Impressions of Cartosat-1. *JACIE 2006 Civil Commercial Imagery Evaluation Workshop, Laurel, Maryland, March 14–16, 2006.* [See http://www.geoeye.com/techpapers/default.htm]

Merkls, S. 2003. The SAR-Lupe Program: An Industrial View. *Conference on Security & Defence Aspects of Space—The Challenge for EU, Athens, Greece, May 8–9, 2003.*

Osawa, Y. 2004. Optical and Microwave Sensors on Japanese Mapping Satellite—ALOS. *ISPRS Congress, Istanbul, Commission I, Paper no. 53.*

Petrie, G. 2004a. High-Resolution Imaging from Space—A World-Wide Survey (Part I—North America). *GeoInformatics* 7(1): 22–27.

Petrie, G. 2004b. High-Resolution Imaging from Space—A World-Wide Survey (Part II—Asia). *GeoInformatics* 7(2): 22–27.

Petrie, G. 2004c. High-Resolution Imaging from Space—A World-Wide Survey (Part III—Europe & Part IV—ISS). *GeoInformatics* 7(3): 38–43.

Petrie, G. 2007a. Space Image Acquisition for Geospatial Intelligence: Developments in South & East Asia. *GeoInformatics* 10(3): 38–45.

Petrie, G. 2007b. Jena-Optronik's Imaging Scanners: Earth Observation from Space & the Air. *GeoInformatics* 10(6): 42–47.

Piboon, M., Peanvijarnpong, C., Dowreang, D., Aphicholati, N. & Kodchabudthada, W. 2005. Thailand Earth Observation System (THEOS): A New Dimension of Thailand Remote Sensing. *Proceedings, Asian Conference on Remote Sensing (ACRS), Paper D3-P23.*

Stephens, P. 2007. Small Satellite Constellations: Opening a New Era in Sustainable Earth Observation. *GeoInformatics* 10(3): 46–48.

Stoney, W.E. 2006. ASPRS Guide to Land Imaging Satellites. *NOAA Commercial Remote Sensing Symposium "Key Trends and Challenges in the Global Marketplace".*

Tadono, T., Shimada, M., Watanabe, M., Hashimoto, T. & Iwata, T. 2004. Calibration and Validation of PRISM Onboard ALOS. *International Archives of Photogrammetry & Remote Sensing* XXXV Part B1: 13–18.

WEB SITES OF SATELLITE CONSTRUCTORS AND OPERATORS

DigitalGlobe, U.S.A. WorldView-1 http://www.digitalglobe.com/index.php/86/WorldView-1.

GeoEye, U.S.A. GeoEye-1 http://www.geoeye.com/products/imagery/geoeye1/default.htm.

German Aerospace Center (DLR). TerraSAR-X http://www.dlr.de/tsx/start_en.htm.

ImageSat International, Israel. EROS-B http://www.imagesatintl.com/.

Indian Space Research Organisation (ISRO). Cartosat-1 http://www.isro.org/Cartosat/Page3.htm.

Israel Aircraft Industries. TecSAR http://www.iai.co.il/.

Italian Space Agency. COSMO-SkyMed http://www.asi.it/SiteEN/ContentSite.aspx?Area=Osservare+la+Terra.

Japan Aerospace Exploration Agency (JAXA). ALOS http://www.jaxa.jp/projects/sat/alos/index_e.html.

Korea Aerospace Research Institute (KARI), Korea. Kompsat-2. https://www.kari.re.kr/index1.asp.

MacDonald, Dettwiler & Associates (MDA), Canada. Radarsat-2. http://www.radarsat2.info/.

National Institute for Space Research (INPE), Brazil. CBERS. http://www.cbers.inpe.br/en/index_en.htm.

National Institute of Aeronautics & Space (LAPAN), Indonesia LAPAN-Tubsat http://lapantubsat.org/eng/index.php?option=com_mjfrontpage&Itemid=35.

National Space Program Office (NSPO), Taiwan. Formosat-2 http://www.nspo.org.tw/2005e/projects/project2/intro.htm.

OHB-System, Germany. SAR-Lupe http://www.ohb-system.de/gb/Satellites/satellites.html.

QinetiQ, U.K. TopSat http://www.qinetiq.com/home/commercial/space/topsat.html.

RapidEye AG, Germany. RapidEye 1 to 5 http://www.rapideye.de/.

Research Centre for Earth Operative Monitoring (NTs OMZ), Russia. Resurs DK-1 http://eng.ntsomz.ru/spacecraft/resurs_dk.

Satrec Initiative, Korea. RazakSAT, DubaiSat, RASAT. http://www.satreci.com/eng/progra(m/program_01.htm.

SPOT Image, France. SPOT-2, -4 and -5; Formosat-2, Kompsat-2 http://www.spot.com/web/SICORP/1070—formosat-2-images.php.

Surrey Satellite Technology Ltd. (SSTL), U.K. TopSat, Beijing-1, RapidEye, UK-DMC-2, Deimos-1, NigeriaSat-2 http://www.sstl.co.uk/index.php?loc=5.

Advances in Photogrammetry, Remote Sensing and Spatial Information Sc
2008 ISPRS Congress Book – Li, Chen & Baltsavia

CHAPTER 4

Airborne digital imaging sensors and systems

Gordon Petrie & Kenneth Smillie

ABSTRACT: This chapter presents an account of the developments that have taken place in the area of airborne digital imaging sensors and systems during the period since the last ISPRS Congress was held in 2004. First of all, the many developments in digital frame cameras that have taken place during this period are covered, including both single cameras and multiple camera systems. After which, the developments in digital line scanners over the same period are reviewed, including both single-line and three-line scanners. Finally, the developments that have taken place in airborne radar systems are outlined and discussed.

Keywords: Digital frame cameras, digital line scanners, airborne radar systems, instrument development

4.1 INTRODUCTION

This chapter gives an overview of the airborne digital imaging sensors and systems that have been developed and have come into operational use during the period since the last ISPRS Congress was held in Istanbul in 2004. It expands on and updates the review of the systems that was presented at the ISPRS Commission I Symposium held in Marne-la-Vallee, France, in July 2006 (Petrie & Walker 2007). Overall the chapter is divided into two parts covering optical and radar imaging systems, respectively. In the case of the optical systems, they can be further divided into two main groups—digital frame cameras and digital line scanners. Thus, the developments in airborne digital imaging systems will be reviewed and discussed under three main headings:

(i) optical digital frame cameras;
(ii) optical digital line scanners; and
(iii) radar digital imaging systems.

4.2 AIRBORNE DIGITAL FRAME CAMERAS

Since the last ISPRS Congress in 2004, airborne digital frame camera technology has continued to develop apace (Cramer 2005). Underlying much of this development have been the advances that have been taking place in the two main digital imaging technologies—CCD and CMOS—on which the airborne optical digital imagers are based. In particular, the individual area arrays that are used in airborne digital frame cameras are now much larger in size in terms of the number of pixels that can be imaged in a single exposure. Indeed, this particular development has resulted in a re-definition of the format sizes that are commonly used in these cameras. Thus the 6 to 8 Megapixel image size of the small-format cameras that was regarded as normal in 2004 has now doubled to 12 to 16 Megapixels in 2008. A similar development has occurred with medium-format frame cameras. In 2004, these cameras produced images that typically were 16 Megapixels in size. Now, very often, they generate images that are 39 Megapixels in size.

However, in contrast to the situation with the two smaller format sizes, with large-format digital frame cameras—which are those generating individual frame images that are over 50 Megapixels in size—the situation is very different. Development has to a large extent stagnated, although Fairchild Imaging has shown a prototype 100 Megapixel array at the ASPRS 2007 Annual Conference. However, the upper limit in terms of the arrays that are available commercially in production quantities is still the very expensive 85 Megapixel CCD area array produced by Fairchild Imaging, whose use is confined to certain U.S. military cameras. To overcome this limitation, the development of systems that can generate large-format digital frame images has gone in a totally different

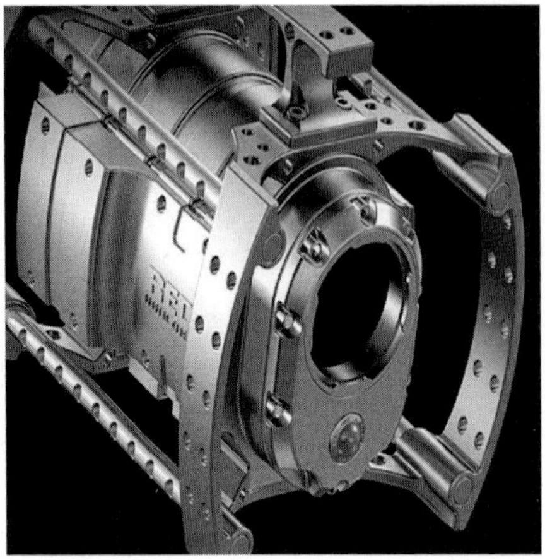

Figure 4.1. The Red One ultra HDTV video camera generates its 11 Megapixel images at a rate of 50 Hz with a high potential for airborne surveillance and monitoring operations (Source: Red One).

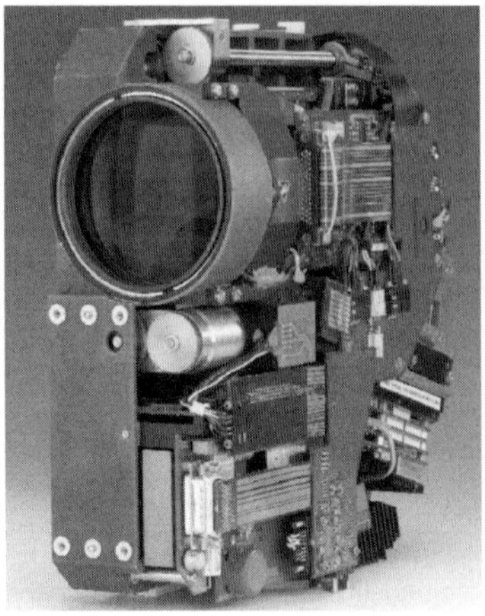

Figure 4.2. The Carl Zeiss VOS40 camera with its folded optics, zoom lens and fast framing rates is purpose-built for aerial reconnaissance operations (Source: Carl Zeiss).

direction. Currently it is based on the use of multiple cameras, lenses and arrays. Some of these systems, such as the Intergraph DMC and Vexcel/Microsoft UltraCam, were shown in the Istanbul Congress in 2004 (Petrie 2003 & 2004). Both of these cameras are highly integrated systems combining eight or nine individual small- and medium-format cameras, each with its own lens, shutter and CCD area array that were purpose-built for photogrammetric applications and are closely coupled together within a single unit. However, the systems that have been developed since then mostly make use of existing medium- and small-format cameras that are available off-the-shelf and are much more loosely coupled together. A number of these cameras are then grouped together on a single mount, often in an oblique pointing configuration to provide a greater angular coverage of the terrain. Regardless of the specific configuration of an individual system or the degree of integration that has been adopted, all of these multiple camera systems generate their final large-format images through the stitching together of the component images either in-flight or in a post-flight data processing operation.

Having regard to the current situation that has been outlined in the two paragraphs above, it seems appropriate to review these developments in the area of airborne digital frame cameras in some more detail under two main headings:

(i) single frame cameras equipped with a single lens and a single area array; and
(ii) multiple frame camera systems making use of multiple lenses and multiple area arrays.

4.2.1 *Single frame cameras*

The developments in the area of single cameras will be reviewed under the three main groupings based on format size, namely those producing small-, medium- and large-format frame images, respectively.

4.2.1.1 *Small-format frame cameras*

Within this particular category, a number of distinctive trends can be discerned. At the lower end in terms of image size—in the range 2 to 4 Megapixels—there has been a very strong development of digital cameras that can acquire images at high framing rates. This results principally from the requirements of military and other security services for the continuous surveillance of relatively small areas of terrain from low altitudes from slow-flying UAVs, helicopters, blimps and STOL aircraft. Previously this requirement was met using video cameras producing images in the standard video formats and rates of PAL (767 × 575 pixels) @ 50 Hz and NTSC (640 × 480 pixels) @ 60 Hz. However, arising from the demands from users for much larger formats with an increased number of pixels, many new cameras have been developed that produce digital images to an HDTV standard, typically 1920 × 1020 pixels = 2 Megapixels (2 Mpix) at the standard video rates of 50 or 60 Hz. Current developments in this particular segment of digital video cameras are headed in the direction of still larger formats. For example, the new Red One camera provides images that are 4520 × 2540 pixels = 11 Megapixels (11 Mpix) in size at an imaging rate of 50 to 60 Hz. Essentially this type of camera provides coloured images having the same size and resolution of those provided by a high quality small-format still camera, but at the acquisition rate of a video camera. The resulting data volumes are a matter that requires special attention. At the time of writing (at the end of 2007), over 100 of these Red One cameras had been delivered to customers mainly in the media and entertainment industries. Given the attractive specification, it appears inevitable that cameras of this type will make their way on to numerous different types of airborne platform for surveillance and monitoring purposes. Since the resulting imagery will almost always need to be geo-referenced for surveillance purposes, the suppliers of photogrammetric software need to prepare for the advent of this type of imagery.

Besides the surveillance and monitoring applications, another quite different set of requirements from military users that has led to the need for fast acquisition rates arises from the need to equip the high-speed military aircraft used for reconnaissance purposes from low altitudes. An example of an airborne digital frame camera built to satisfy these demanding requirements is the Carl Zeiss VOS40, which is in service with certain NATO air forces. This camera features folded optics (to reduce the overall size of the camera); fast framing rates; and a small format (2k × 2k = 4 Mpix). Thus the VOS40/270 model in the series, which is equipped with an f = 270 mm lens, can acquire its monochrome frame images at the rate of 15 frames per second, while the VOS40/500 model, which is equipped with a long-focus f = 500 mm lens, can acquire its data at the rate of 3 frames per second. Again there is a need for the resulting imagery to be geo-referenced for targeting and other military purposes.

In particular, cameras from the Kodak DCS 460/560/660/760 series have been used widely, producing colour images up to 6 Megapixels in size using mosaic filters and Bayer interpolation. With the addition of an anti-vibration mount, a suitable control box, a GPS receiver, and sometimes a lower grade IMU, these cameras could readily be fitted to light aircraft. This has allowed them to be used for local projects involving repeated flights for monitoring purposes and

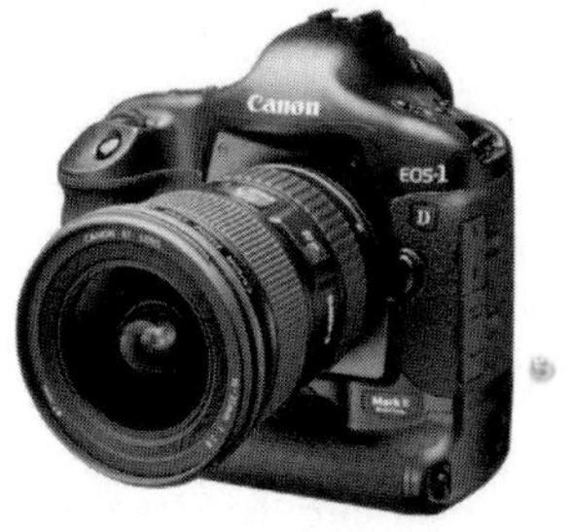

Figure 4.3. The Canon EOS-1Ds Mark II high-quality SLR camera generating 16.7 Megapixel colour images is now in widespread use for airborne imaging (Source: Canon).

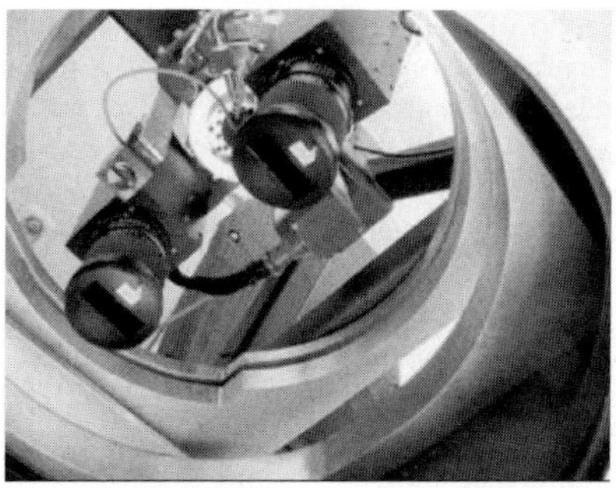

Figure 4.5. The New Metric Cameras that have been built for the GeoPIE project by the VTT research institute in Finland generate images with a rectangular format. This example comprises twin convergent cameras that are used to generate overlapping stereo frame images (Source: VTT).

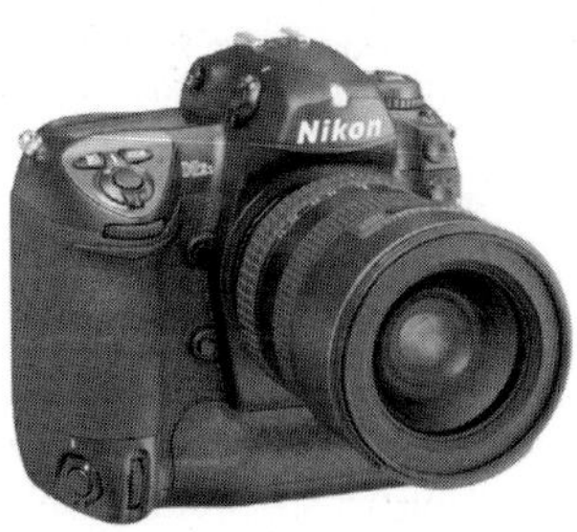

Figure 4.4. The Nikon D2X SLR camera generating images that are 12.4 Megapixels in size is also being used in airborne imaging operations (Source: Nikon).

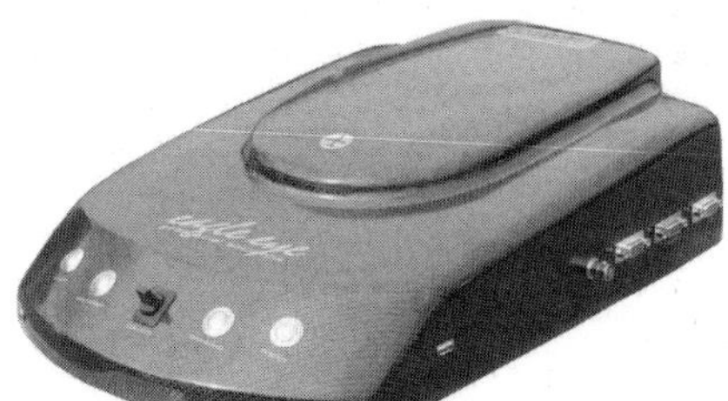

Figure 4.6. The G2010 "eagleeye" camera from Global Imaging Technologies also generates rectangular format frame images (Source: GIT).

the provision of aerial images in response to environmental disasters such as floods, fires, landslides, etc. However, since the last ISPRS Congress in 2004, the performance of these high-quality SLR still cameras has increased greatly, especially in terms of the image format sizes that can be generated. In particular, two cameras—the Nikon D2X model producing colour frame images that are 12.4 Mpix in size and the Canon EOS-1Ds Mk. II model that generates images that are 16.7 Mpix in size—have proven to be popular with aerial photography operators for digital image data acquisition. These cameras are being operated both from fixed mounts pointing vertically downwards but also in an oblique configuration, either from a fixed mount or hand-held. Both cameras use CMOS area arrays that have largely replaced the previously dominant CCD arrays that were used, for example, in the older Kodak DCS series.

Another interesting development in the area of small-format digital frame cameras has been the recent introduction of purpose-built airborne cameras equipped with highly rectangular formats. The first of these was the New Metric Camera (NMC) built by the VTT research institute in Finland on behalf of a group of six European commercial companies and research institutes under the GeoPIE project supported by funding from the European Union (Mayr & Ohlhof 2004). The NMC employs nine inexpensive CCD area arrays, the resulting nine individual images being optically butted within the camera to form individual rectangular frame images that are approximately 10,000 pixels (cross-track) × 1600 pixels (along-track) in size. A second group of cameras that also falls into this category is the G2010 series developed by Global Imaging Technologies based in Baltimore in the USA. These cameras also make use of multiple (in this case, five) inexpensive CCD area arrays to produce images that, when stitched together, are 7880 × 1020 pixels in size. The company also produces its ARIES II camera, which generates images that are 7020 × 1030 pixels in size. Again this uses five CCD arrays whose images are then stitched together to form a single image. The third camera in this group is the Medusa camera that is being developed by the VITO research institute in Belgium for use on board the Mercator-1 HALE (High Altitude, Long Endurance) UAV (van Achteren et al. 2006). Unlike the previous two examples that utilize multiple CCD arrays, the Mercator camera will use a single custom-made CMOS chip producing an image that is 10k × 1.2k pixels in size. When used in conjunction with a long focal length (f = 330 mm) lens from an altitude of 18 km, this will generate an image with a ground sampled distance (GSD) of 0.3 m.

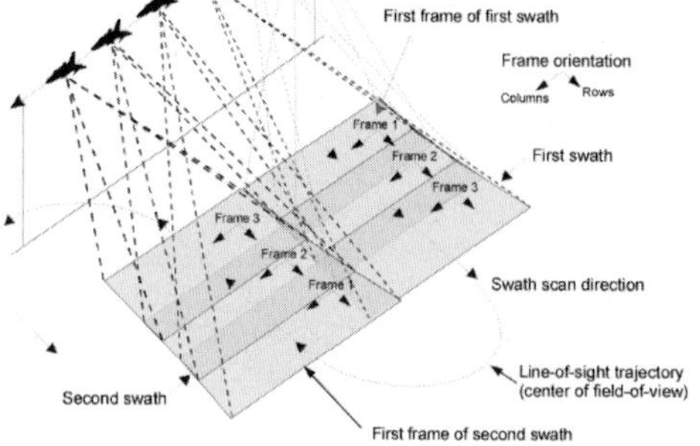

Figure 4.7. Stepping frame cameras generate a sequence of frame images at very high speed in the cross-track direction to provide a very wide angular coverage of the ground (Source: Goodrich).

A fourth example is the family of digital frame cameras equipped with multiple area arrays and lenses that have been introduced by the Russian manufacturer, NPO KSK. These include (i) a miniature low-altitude model with four lenses and four area arrays that produces images that, when stitched together, have a swath width of 6000 pixels cross-track; and (ii) another low-altitude model with 8 arrays and 4 lenses that gives images with 12,200 pixels in the cross-track direction. These small-format cameras are supplemented by larger-format frame cameras that follow the same principle. They include (iii) a medium-altitude model with 18 arrays and 6 lenses producing images that are 27,000 pixels wide and (iv) a high-altitude model with 32 arrays and 8 lenses resulting in images that are 49,000 pixels wide in the cross-track direction.

Yet another innovative development has been that of stepping frame cameras. Using this configuration, a series of frame images are exposed at a very high rate in the cross-track direction to provide a very wide coverage of the ground in that direction (Iyengar & Lange 2006). To achieve this, the cameras are rotated rapidly in a series of angular steps using a motorized mount. This concept has been implemented for both military and civilian applications. Examples of the former case are the CA-270 (medium-altitude) and CA-295 (high-altitude) cameras from Recon/Optical. These feature two cameras mounted side-by-side, the first of these being a 5k × 5k (25 Megapixel) camera operating in the visible part of the electromagnetic spectrum. The second camera generates a 2k × 2k (4 Megapixel) image in the MWIR (λ = 3 to 5 μm) part of the spectrum. Both cameras can operate at speeds up to four frames per second, providing horizon-to-horizon coverage. In the civilian mapping domain, an example of this new technology is the VisionMap system from Israel. This also features twin cameras, in this case two 11 Megapixel frame cameras that are mounted side-by-side and are equipped with synchronized shutters to acquire their images simultaneously during each individual cross-track scan.

In summary, the use of small-format digital frame cameras from airborne platforms is now widespread with a marked increase in the format size of the images arising from current developments in CMOS area array technology. As noted above, the more conventional types of small-format cameras are frequently deployed on light aircraft for monitoring purposes. There is also a great deal of innovation taking place in this area arising from the very specific requirements of military and security agencies for the surveillance and monitoring of areas of terrain being carried out from UAVs and other slow-flying airborne platforms over long periods of time and from the requirements of high-speed reconnaissance aircraft flying at low altitudes.

4.2.1.2 *Medium-format frame cameras*

Unlike the area of small-format frame cameras where CMOS arrays have taken a dominant position, CCD arrays continue to be the principal digital imaging technology that is used as the basis of airborne medium-format frame cameras, albeit with a similar doubling in the size of the format in terms of the number of pixels that can be generated within a single frame image. Many of the cameras that fall within this category originated as high-quality film cameras designed for professional use in studios and outdoors. They included cameras from Contax, Rollei and Hasselblad that were modified by replacing their film magazines with digital backs from MegaVision (USA.), Phase One (Denmark) and Imacon (Sweden), respectively. Initially most of these digital backs utilized the Kodak 4k × 4k (= 16 Megapixel) CCD area array, but soon this was replaced by the 22 Megapixel array. Currently the arrays that are commonly being used are 31.6 and 39 Megapixels in size. Besides these Kodak arrays, a similar range is available from

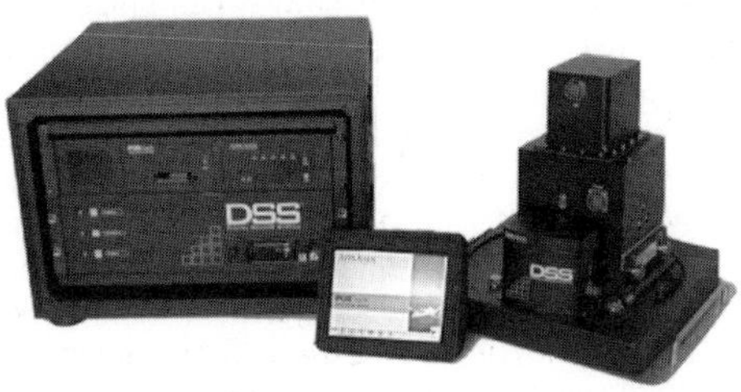

Figure 4.8. The Applanix DSS medium-format digital camera is sitting on its mount (at right), together with its control electronics (at left) and touch screen display (at centre) (Source: Applanix).

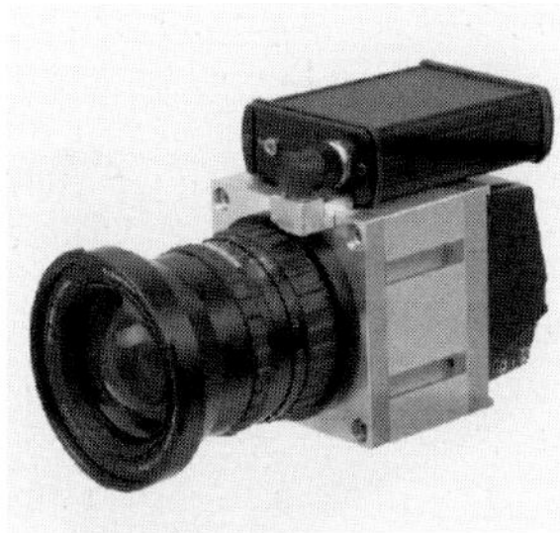

Figure 4.9. The Rolleimetric AIC medium-format airborne digital camera with its electronics interface module mounted rigidly on top of the camera (Source: Rolleimetric).

Dalsa. All of the digital backs that are utilizing these arrays generate colour or false-colour images through the use of mosaic filters and Bayer interpolation. The digital backs from all of these suppliers are (or should be) constructed in a manner that ensures they remain rigidly stable in their position within the image plane of the camera during the time of exposure, as required for photogrammetric measurements. Needless to say, the resulting cameras will also be calibrated for this purpose.

This particular category of airborne imager has grown in quite spectacular fashion since 2004. The two biggest suppliers of medium-format airborne digital cameras are the Applanix and Rolleimetric companies, which have sold over 90 and 70 of their cameras, respectively. Both companies offer a comparable range of models with 16, 22 and 39 Megapixel backs and a range of high-quality lenses with different focal lengths and angular coverages, the latter being a valuable feature that is not available on larger format imagers. In the case of Applanix, its DSS (Digital Sensor System) camera is often supplied as a fully integrated system including a special camera mount to correct for aircraft crabbing; and its own in-house produced POS/AV system comprising a GPS/IMU or GNSS/IMU unit that will be used for direct geo-referencing purposes (Artes 2004). The overall DSS system will also include a computer sub-system that stores both the image data and the measured auxiliary GPS/IMU data. The computer is also used to implement the control, navigation, flight management and direct geo-referencing functions of the overall system both in-flight and post-flight, in this case, using Applanix's own POSTrack and POSPAC Air software. In the case of Rolleimetric, similar functionality is provided through the connection of its AIC (Aerial Industrial Camera) to a laptop computer and to commonly available GPS/IMU units and flight control and management systems that are provided by third party suppliers such as IGI (Toelg 2007). In fact,

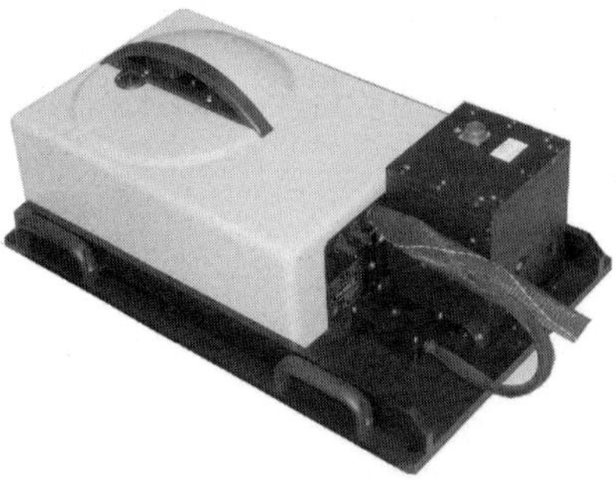

Figure 4.10. This Applanix medium-format digital frame camera is mounted rigidly together with a Leica ALS-50 airborne laser scanner on a common base plate (Source: Leica Geosystems).

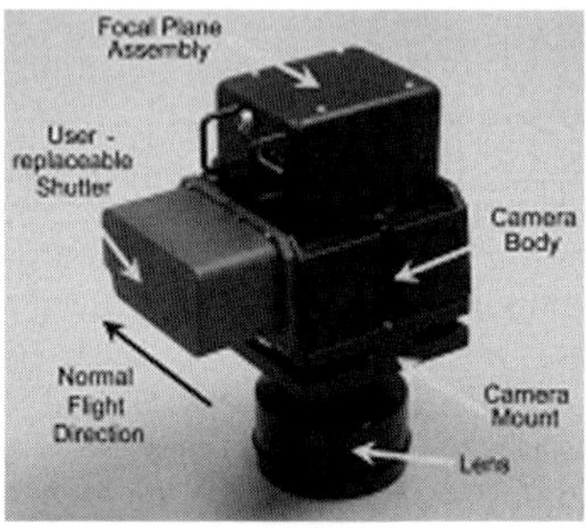

Figure 4.11. The new Leica RCD105 medium-format airborne digital frame camera (Source: Leica Geosystems).

IGI also provides a fully integrated system on its own account with its DigiCAM medium-format digital camera unit, which is based on the Hasselblad camera equipped with digital backs from Imacon (Petrie 2006b).

Like the DSS, the DigiCAM is available with a range of lenses having different focal lengths and angular coverages. The camera can be used in conjunction with IGI's own CCNS flight control system and AEROcontrol GPS/IMU unit, either stand-alone or with the company's LiteMapper laser scanner.

Indeed, while many of these medium-format cameras are sold and operated as stand-alone units, quite a substantial part of the market for these medium-format cameras has been to supply them for integration into airborne laser scanning systems. Within this particular context, the leading supplier, Optech, has supplied many of its ALTM airborne LIDAR systems in combination with either Applanix DSS or Rolleimetric AIC cameras. Leica Geosystems, which is the other major commercial supplier of airborne laser scanners, has in the past also supplied these two cameras (and the Spectrum NexVue camera) integrated with many of its ALS airborne LIDARs. So much so that, in the summer of 2007, Leica started to offer its own medium-format digital frame camera in the form of its RCD105 product, which is, in fact, manufactured on its behalf by an OEM supplier based in north-eastern USA. The RCD105 camera is available either integrated with Leica's own ALS50-II and ALS-CM laser scanners or as a stand-alone product. Like the other cameras in this particular class, the RCD105 generates 39 Megapixel colour (RGB or CIR) frame images using the Bayer interpolation procedure. The camera has its own CC105 camera controller that is mounted in a separate stand-alone box and can be used to control the operation of two cameras simultaneously. The RCD105 camera is also available with a range of lenses having different focal lengths and a user-replaceable shutter that allows rapid exposure times of up to 1/4000th second to be used. Much the same remarks about the integration of medium-format cameras with airborne laser scanners apply to IGI, whose LiteMapper scanner systems are mostly sold complete with its own DigiCAM camera that has been described above. Similarly, the TopoSys company has supplied its Harrier airborne LIDARs with Applanix DSS and Rolleimetric AIC cameras.

Yet another recent development within this area has been the lightweight BuckEye airborne camera system developed by the US Army Topographic Engineer Center (TEC) for rapid change detection to identify the locations of improvised explosive devices (IEDs) in Iraq and Afghanistan. A number of Buckeye systems have been built by the Flight Landata company based in Massachusetts. They comprise either a 22 Megapixel or a 39 Megapixel camera and a GPS/IMU unit that are mounted together on a base plate that allows the system to be fitted to a wide variety of airborne platforms. The resulting geo-referenced imagery that has been generated from different flights flown at different times is then processed using the SOCET SET software from BAE Systems to detect the changes that have taken place on the ground. Newer versions of the BuckEye system have had airborne laser scanners added to them, the data being used to generate 3-D models of urban areas where military operations are taking place.

Another consequence of the explosive growth in the market for small- and medium-format frame cameras has been the development of gyro-stabilized mounts that are designed specifically to accommodate these small-sized cameras. The existing widely used and commonly available camera mounts such as the Leica PAV30, Intergraph T-AS and Somag GSM3000 models were designed originally for use with Leica and Zeiss film cameras. As such, they are too large, heavy and power hungry, especially for use in the small aircraft that are now being used extensively as platforms for the small- and medium-format

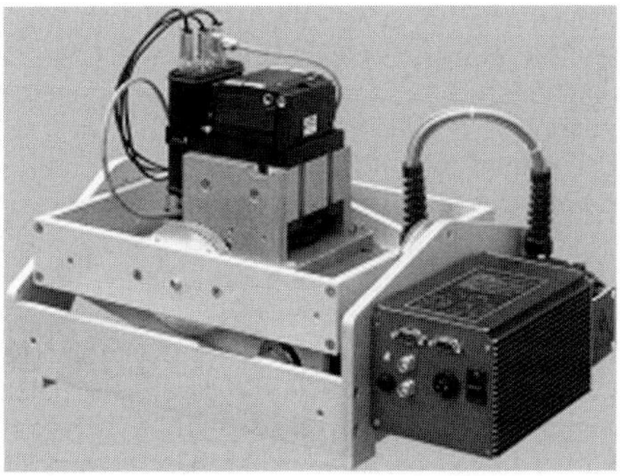

Figure 4.12. This Rolleimetric AIC medium-format airborne digital frame camera is mounted on a GGS Speyer gyro-stabilized platform (Source: GGS Speyer).

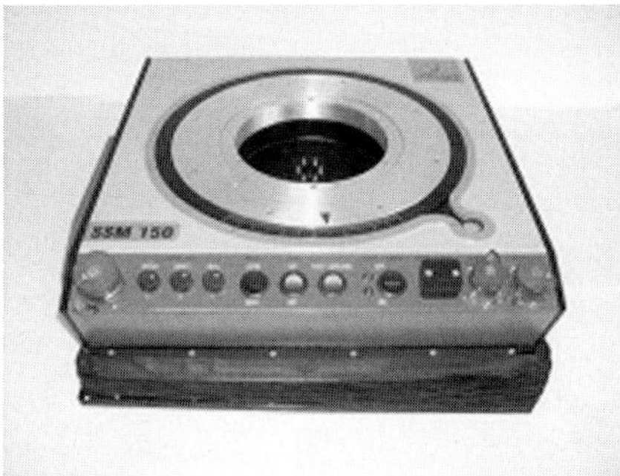

Figure 4.13. The Somag SSM 150 gyro-stabilized mount is designed for use with small-format airborne cameras (Source: Somag).

digital frame cameras, which are very much smaller in size and weight than conventional metric film cameras. The new smaller and much less expensive mounts include the AeroStab-2 units constructed by GGS Speyer and the SSM150 and SSM250 models from Somag that are designed specifically for use with small- and medium-format digital cameras, respectively.

4.2.1.3 *Large-format frame cameras*

As noted above, single cameras producing images that are greater than 50 megapixels in terms of their frame size are still rather rare. Consequently there is little development to report in this area. Both the Recon/Optical CA-260/50 reconnaissance camera with its 50 Megapixel image size and the various models from BAE Systems North America producing 85 Megapixel frame images were developed some years ago. The use of these cameras is confined to US military agencies. In 2006, the Dalsa company reported the successful development of a CC array that is 10.5k × 10.5k in size (giving a 10 apixel image), which it produced on behalf of t Navy. However, so far, there is no public knowledge of this being used in an operational airborne camera. Until the production of large-format area arrays becomes a less esoteric process and the individual arrays are reduced greatly in cost, their use appears to be limited to military reconnaissance cameras and the generation of monochrome frame images.

4.2.2 *Multiple frame camera systems*

The numerous developments that have taken place in this field will be reviewed under two main headings:

(i) multiple small-format cameras, and
(ii) multiple medium-format cameras.

4.2.2.1 *Multiple small-format frame camera systems*

This particular area has been one of ceaseless development over the four years of the inter-Congress period. One especially lively area of activity has been that of multiple camera systems generating combinations of vertical and oblique images simultaneously. On the one hand, these systems have been produced in large numbers for civilian use, most notably by the Pictometry company based in the Rochester area in the United States (Smith & Merrill 2004). Over 70 of these multiple camera systems are reported to be in use by Pictometry in the United States with a further dozen systems in use by its European licensee, Blom Aerofilms (Artes 2006). Similar systems are being used by other competing commercial companies in both regions, for example, by Geospan and Aerial Cartographics of America (ACA) in the USA and by COWI (in Denmark) and GetMapping (in the UK). All of these systems have been used extensively to provide systematic coverage of urban areas. Apparently the oblique images are more easily interpreted by non-expert users. Besides which, users in the emergency (police, fire, ambulance) services can also utilize the information regarding building use and structure that is provided by the oblique images, which are also being made available on Microsoft's Virtual Earth Web site. A typical system, called MIDAS, is that produced by Track'Air in the Netherlands comprising five of the Canon EOS-1Ds Mk. II cameras mentioned above (Petrie 2006a). These are configured to produce one vertical and four oblique colour images providing over 80 Megapixels of data in a single combined exposure. It seems only a matter of time before these systems are developed further to fill the gaps in the present "Maltese Cross" arrangement of the ground coverage. If so, then we might expect to

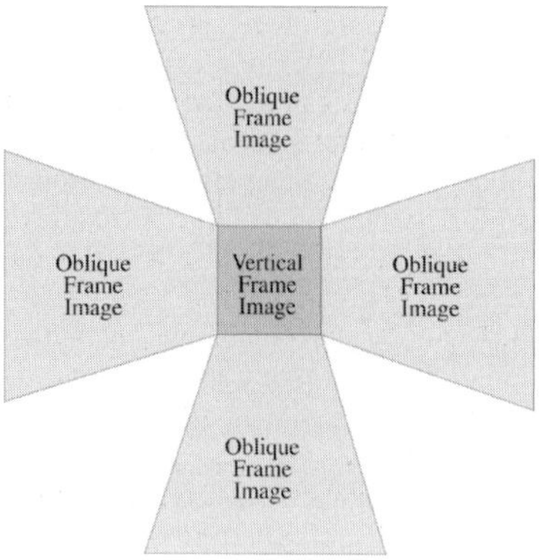

Figure 4.14. The ground coverage of a five camera airborne digital imaging system that generates one vertical and four oblique frame images simultaneously (Drawn by M. Shand).

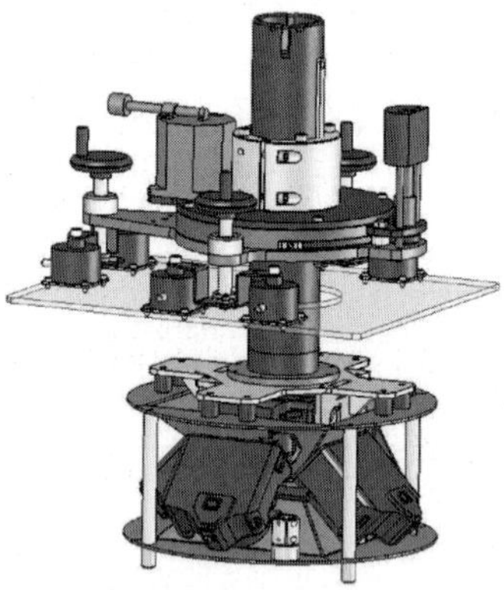

Figure 4.15. A drawing showing the configuration of the Track'Air MIDAS multiple camera system, with the vertical column supporting the cylinder containing the five cameras that are fitted inside an external pod hanging below the aircraft, while the adjustable mount sits on the aircraft floor (Source: Track'Air).

see a nine camera system comprising one vertical and eight oblique pointing cameras, in a configuration similar to the nine-lens film cameras from the 1930s.

Several other combinations and configurations of multiple vertical and oblique small-format cameras have been developed specifically for use by military and homeland security agencies. For example, various prototypes, called the Sonoma Persistent Surveillance System, have been developed by a team from the Lawrence Livermore National Laboratory (LLNL) in California. One of these is a 66 Megapixel system comprising six individual 11 Megapixel cameras coupled together on a simple mount together with a GPS/IMU unit. This configuration has been developed further into a production system—the so-called Airborne Persistent Surveillance System—by ITT Space Systems. It combines a similar set of six 11 megapixel cameras supplied by Geospatial Systems Inc., a rotatable gimbal mount from Vinten, and an IPAS GPS/IMU system that has been supplied by Leica Geosystems. The processing of the image and geo-referencing data takes place on-board the airborne platform and is integrated with a radio system that broadcasts the merged 66 Megapixel frame image data set to users on the ground in near real-time over low-frequency radio links. Still larger versions of these systems, including one producing a 176 Megapixel image data

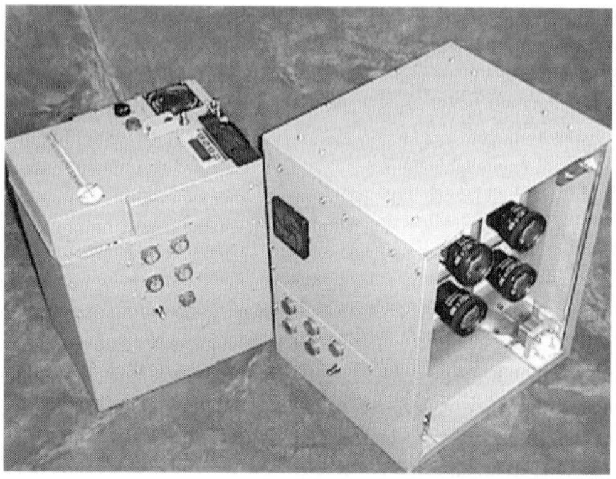

Figure 4.16. The SpectralView multiple camera system from Airborne Data Systems that generates multi-spectral colour and false-colour composite images. Each of the four small-format cameras is fitted with a different spectral filter (Source: Airborne Data Systems).

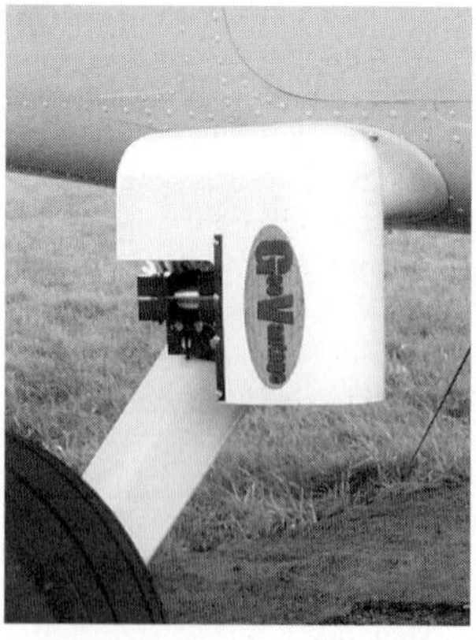

Figure 4.17. This GeoVantage multiple camera system comprising four small-format cameras has been fitted to the undercarriage of a light aircraft (Source: GeoVantage).

set, have been implemented by LLNL. Still larger ones are planned

Also falling into this group of multiple small-format systems are the conventional multi-spectral digital frame camera systems producing colour or false-colour images with a full radiometric resolution (i.e. without Bayer interpolation). Usually three or four small-format cameras are coupled together with their optical axes set precisely parallel to one another (i.e. bore-sighted) and their shutters synchronized to ensure the exact coverage of the same piece of ground by each camera simultaneously. Each of the cameras is fitted with the specific spectral filter (RGB and/or NIR) that is needed to capture the appropriate band image from which the final colour or false-colour composite image will be generated by image fusion. Large numbers of these multi-spectral systems have come into service, especially in the USA. Examples are the SpectraView systems built in quantity by Airborne Data Systems (ADS) in Minnesota and the numerous GeoVantage systems operated by the John Deere company for agricultural applications (Petrie 2006a).

4.2.2.2 *Multiple medium-format frame camera systems*

Following on directly from the previous group of small-format multi-spectral systems are the similar multi-spectral systems based on the use of medium-format digital frame cameras. There are rather few of these, the most prominent being the Quadricom system developed in-house by the French national mapping agency, IGN. This is based on four medium-format cameras utilizing Kodak 4k × 4k area arrays, each with its appropriate filter. The resulting colour-separated monochrome images are then used to generate the final colour and false-colour images by image fusion.

Much more common has been the development of systems employing pairs of medium-format frame cameras in a twin low oblique configuration pointing in opposite directions cross-track—essentially the classical "split-vertical" configuration used in military reconnaissance aircraft to improve the ground coverage that could be achieved during a single flight. The revival of this old idea was first implemented by DIMAC Systems based in Luxembourg who have developed their DiMAC camera system comprising two camera modules, each equipped with the 39 Megapixel digital back from Phase One that generates colour images via Bayer interpolation (Petrie 2006a). The two cameras sit side-by-side within a cylindrical drum case and are tilted so that the areas on the ground to the left and right of the flight line are covered with a tiny overlap between them. The two images are rectified and mosaiced together using in-house developed software resulting in a final colour image that is 10.5k × 7.2k (= 75.6 Megapixels) in size. So, as with several of the other multiple camera arrangements discussed above, the final merged image falls into the category of a large-format image. This simple arrangement has also been implemented more recently by other system suppliers such as Rolleimetric and IGI, using pairs of their AIC and DigiCAM medium-format cameras, respectively.

4.2.2.3 *Multiple medium- & small-format frame camera systems*

The final group comprise the highly integrated multiple camera systems that utilize combinations of medium- and small-format arrays and cameras. In these systems, the medium-format frame cameras

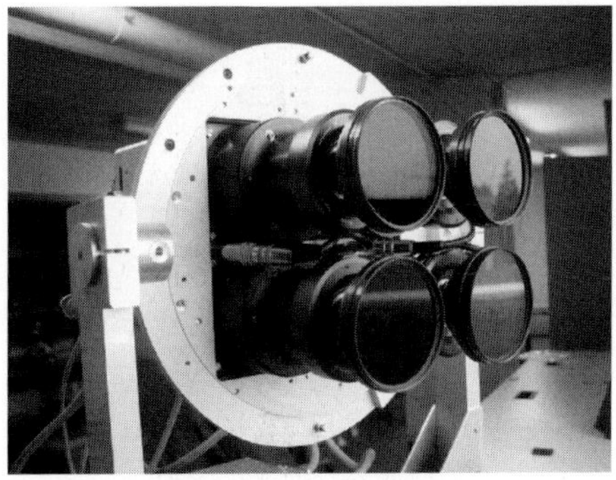

Figure 4.18. The IGN Quadricam uses four medium-format cameras, each equipped with a different spectral filter. The sets of four multi-band images acquired by the camera are merged to form the final colour or false-colour composite images without interpolation (Source: IGN).

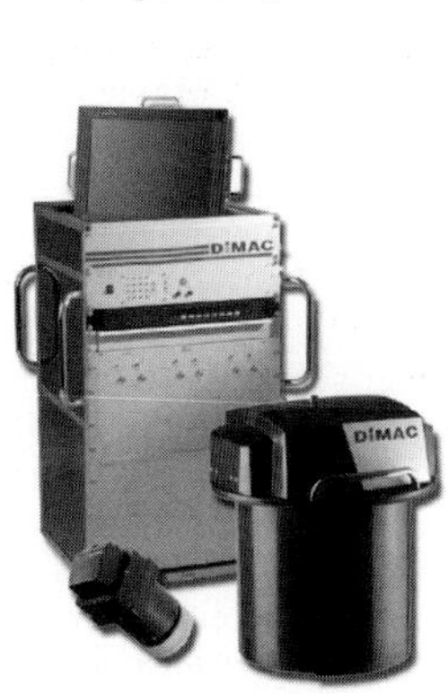

Figure 4.19. The DiMAC 2.0 camera system comprises a cylindrical case inside which the camera modules are mounted; the system's electronics cabinet is shown at the rear (Source: DIMAC Systems).

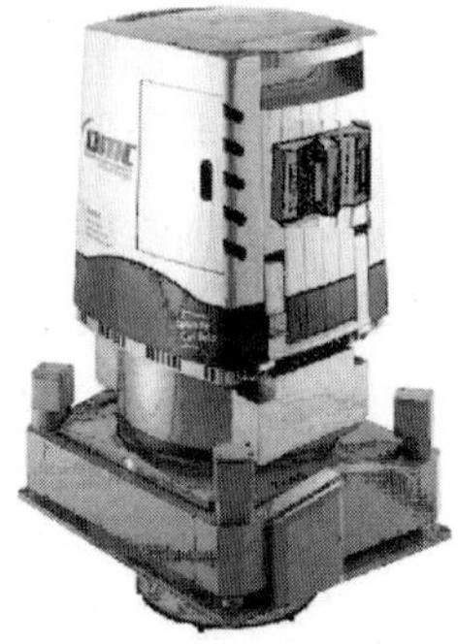

Figure 4.20. The latest version of the Intergraph DMC camera with its solid state disk (SSD) storage facility located on the side of the camera unit. The camera is mounted on a T-AS gyro-stabilized platform (Source: Intergraph).

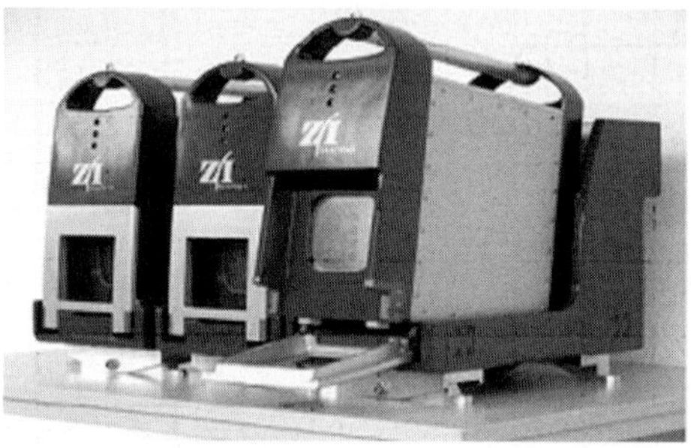

Figure 4.21. The external Flight Data Storage (FDS) unit of the DMC, which is used if very large numbers of images need to be recorded and stored (Source: Intergraph).

are used to generate higher resolution panchromatic images that are merged to form a single large-format image, while the small-format cameras provide multi-band images that can be used to colourize the panchromatic image to generate colour or false-colour frame images.

The first of these systems is the Intergraph DMC (Digital Mapping Camera), the prototype of which was first shown at the ISPRS Congress held in Amsterdam in 2000. Since the Istanbul Congress held in 2004, various incremental improvements have been made to the DMC. The latest of these have been targeted at reducing the weight and cost of the DMC with a view to the system being operated from smaller aircraft than before. A major component in this effort was the adoption of a solid-state disk (SSD) that can be slotted into the side of the main body of the camera. Each SSD cartridge can store 1000 images and can be exchanged during flight. The use of the SSD unit not only reduces the overall weight drastically, but results in a marked reduction in the power requirements as compared with the previous very heavy flight data storage (FDS) units. A second change was to offer the IGI AEROcontrol GPS/IMU unit as an option instead of only the Applanix POS/AV unit that was standard before. The IGI unit is less expensive and has fewer export restrictions, allowing the DMC to be sold for example in China. Eighty of these DMC systems have been sold to date.

The other system falling into this category is the UltraCam, now marketed by Microsoft Photogrammetry (formerly Vexcel). The main development in this system has been the introduction of a new improved model called the UltraCam-X (UC-X) at the ASPRS 2006 Annual Conference (Gruber 2007). This retains the same basic design as the previous UltraCam-D

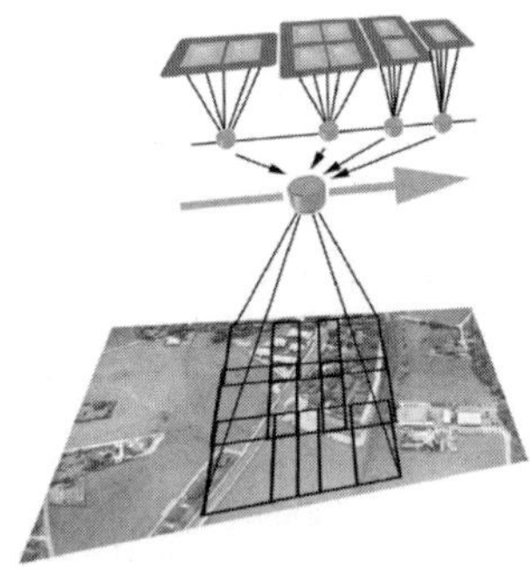

Figure 4.22. Concept of the Microsoft UltraCam airborne digital cameras—showing the arrangement of the four lenses mounted in a line along the flight direction and the multiple CCD area arrays that are used to generate the final panchromatic image (Source: Microsoft Photogrammetry).

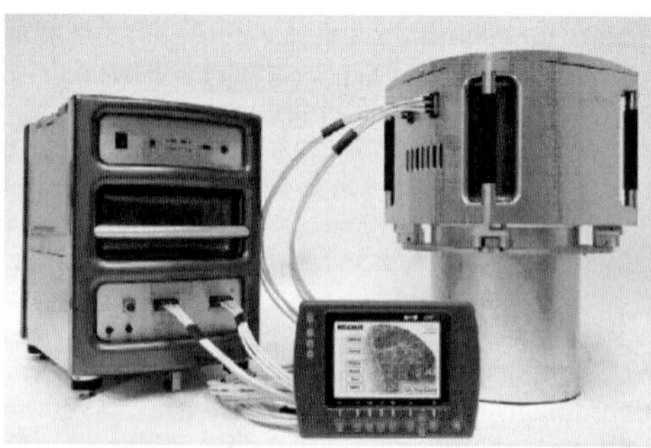

Figure 4.23. The current model of the Microsoft UltraCam-X camera is shown, comprising (from left to right) the control and data storage unit, the user's display screen and the camera unit (Source: Microsoft Photogrammetry).

(UC-D) model with its four lenses and accompanying shutters mounted in-line for the generation of the vertical panchromatic images from the nine component CCD arrays. A further four lenses and shutters, each with a corresponding array, are fitted to generate the small-format multi-band images that are used to colourize the final panchromatic image. However the major change to the system has been the use of higher density CCD arrays, which result in the size of the final merged panchromatic image produced by the new UC-X model being 14,430 × 9420 pixels (= 133 Megapixels) instead of the 11,500 × 7500 pixels (= 84 Megapixels) of the previous UC-D model. Similarly, the size of the final multi-spectral image generated from the four small-format cameras in the UC-X model has been increased to 4992 × 3328 pixels (= 16.6 Megapixels) as compared with the 4008 × 2572 pixels (= 10.3 Megapixels) of the UC-D model. Over 80 UltraCam systems have been sold at the time of writing.

In summary, one can say that the widespread adoption of multiple camera systems based on the use of both small-format and medium-format frame cameras has been one of the most significant developments in airborne digital imaging during the inter-Congress period. This has been due largely, on the one hand, to the lack of availability of large-format area arrays and to the needs of both the operators of digital cameras and the users of the resulting digital imagery to have greater ground coverage within a single frame. The use of readily available small- and medium-format cameras, lenses and arrays has resulted in the creation of numerous loosely-coupled systems that are significantly less expensive than the highly integrated DMC and UltraCam systems that have been purpose-built for photogrammetric applications and continue to sell well.

4.3 AIRBORNE DIGITAL LINE SCANNERS

All those imagers that use linear arrays of detectors as the method of imaging the terrain in digital form require some method of scanning the ground from the airborne platform. In the vast majority of cases, this is implemented through the forward motion of the platform itself—the so-called pushbroom scanning principle—by means of which the scanning takes place in the along-track direction resulting in a continuous strip image of the ground (Petrie 2005). However, an alternative that has been adopted in certain cases recently is to scan the ground in the cross-track direction using a linear array in much the same manner as a panoramic film camera. This panoramic scanning concept been implemented in a few (mainly military) airborne scanners to ensure a wide angular coverage of the ground. Thus the coverage that results from panoramic scanning comprises a sequential series of strip images that are roughly parallel to one another in the cross-track direction to ensure complete coverage of the ground. With all these line scanners, whether of the pushbroom or the panoramic type, each line that appears in the final image has had a different position and attitude in the air at the moment of its exposure. These values need to be measured continuously in-flight using a suitable system, which, in current practice, usually takes the form of an integrated GPS/IMU or GNSS/IMU system. Besides which, large changes in the heading or the tilt of the airborne platform may result either in gaps in the image or in double imaging, so a fast-acting gyro-stabilized mount is, in practice, needed to ameliorate this effect. Such devices (the GPS/IMU and gyro-stabilized mount) are of course used very widely with airborne digital frame cameras though they are not obligatory components of the overall system.

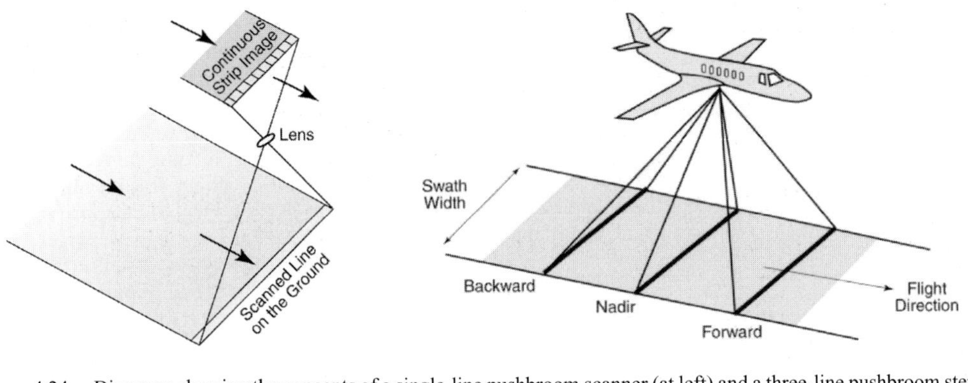

Figure 4.24. Diagrams showing the concepts of a single-line pushbroom scanner (at left) and a three-line pushbroom stereo-scanner (at right) (Drawn by M. Shand).

In contrast to the airborne digital frame cameras, where CMOS area arrays have largely taken over as the imaging sensors fitted within small-format cameras and where the developments in CCD area arrays have resulted in a doubling of the pixel numbers generated in medium-format cameras, there has been little change in the technology of the linear arrays used in line scanners during the inter-Congress timeframe. The longest commercially available CCD linear arrays contain 12,000 to 14,000 detectors, which is much the same as in 2004. As with area arrays, the production of colour images causes complications both in the technology and in the subsequent processing of the image data acquired by the linear arrays. The simplest solution is to utilize tri-linear arrays, which comprise three CCD linear arrays that are mounted adjacent and parallel to one another in the imager's focal plane. Each array is equipped with a suitable filter to generate the red, green and blue (RGB) components of the final colour image. A fourth array and filter can be added to generate a near infra-red (NIR) image. However, this three (or four) line solution to the problem of generating colour images results in offsets being created such that different lines on the ground are being imaged by each of the three (or four) linear arrays at the same instant of time. This offset needs to be compensated for during subsequent image processing. The more sophisticated methods that have been introduced successfully to overcome these difficulties, such as the use of the trichroid or tetrachroid beam-splitting prisms that have been adopted in the Leica Geosystems ADS40, are more complex and more expensive than the simpler tri-linear array solution. However, they do eliminate the need to deal with the offsets that occur with tri-linear arrays.

Following on from these more general remarks, developments in the area of airborne digital line scanners since the 2004 Congress will be reviewed in some more detail under two main headings:

(i) single-line scanners; and
(ii) three-line stereo scanners.

4.3.1 *Single-line digital scanners*

The developments that have taken place within the category of single-line scanners will be reviewed under four main headings or groups: those producing monochrome, colour, hyperspectral and panoramic linescan imagery.

4.3.1.1 *Monochrome pushbroom line scanners*

These pushbroom scanners are equipped with a single CCD linear array that produces a continuous monochromatic strip image of the terrain. They are mainly used for military reconnaissance purposes. There is little new development to report. Typical of these devices is the Thales Optronics Model 8010 scanner from the UK, which is equipped with a 4k CCD linear array and a range of interchangeable lenses of different focal lengths. A very similar device, called Global Scan, is available from COSE in France. This features a 6k linear array of CCD detectors and a comparable range of lenses of differing focal lengths. The main interest in these monochrome scanners comes from military and homeland security agencies for use in monitoring and surveillance activities.

4.3.1.2 *Colour pushbroom line scanners*

These are of a similar design to those in the previous group of monochrome line scanners, but feature a single tri-linear array to produce a continuous strip of colour linescan imagery instead of a single array producing a single strip of monochrome imagery. Again there is only a limited development to report. The well-established Carl Zeiss VOS-60 and VOS-80 scanners both feature a single tri-linear array with 3×6000 CCD detectors having red, green and blue (RGB) filters and are equipped with lenses of $f = 60$ and 80 mm, respectively. However, a new system that does fall into this category is the new colour line scanner from the VeriMap company based in Calgary, Canada. This uses a much longer tri-linear RGB array with 12,000 detectors in conjunction with a single additional CCD linear array that images the ground simultaneously in the near infra-red (NIR) part of the spectrum.

4.3.1.3 *Hyperspectral pushbroom line scanners*

Almost all the scanners that produce hyperspectral line scanner imagery and are in series production make use of a CCD area array as the imaging device. However, this is operated in a pushbroom mode, like those using a linear array. The remaining detectors in the area array collect the hyperspectral (colour) data for each pixel in that single line in hundreds of quite narrow spectral bands or channels. This spectral data is generated using a dispersive prism or grating that acts as an imaging spectrometer, which is placed above the CCD area array. The final result is an "image cube". The older whiskbroom scanning technology that utilized a rotating mirror to scan the ground in the cross-track direction with a linear array that is used to resolve the spectral data for each individual scanned pixel appears to have largely gone out of favour, though it is being used in the ARES hyperspectral scanner that is being constructed for the German Space Agency (DLR) by Integrated Spectronics. This Australian company had previously built the HyMap series of hyperspectral scanners based on the whiskbroom scanning principle. The Argon company (formerly Daedalus, then SenSysTech), which formerly had also produced whiskbroom-type hyperspectral scanners, appears to have largely withdrawn from supplying the commercial hyperspectral imaging market.

While there are quite a substantial number of other hyperspectral scanners that have been constructed individually and are being operated mainly for research purposes, currently there are two main commercial suppliers of airborne hyperspectral pushbroom line scanners: ITRES Research from Calgary, Canada, and the Specim company from Oulu, Finland. In the case of ITRES Research, its CASI series of hyperspectral scanners that cover the visible and near infra-red (VNIR) parts of the spectrum had already been developed successively from the original CASI instrument to the CASI-2 and CASI-3 models. These last two models have been further developed recently into the CASI-550 and CASI-1500 models that have cross-track coverages of 550 and 1500 pixels, respectively, and up to 288 individual spectral channels. The company's new short wave infra-red (SWIR) scanner is the SASI-600, which has a cross-track coverage of 600 pixels and covers the wavelength range 950 to 2450 nm in up to 100 spectral channels. ITRES has also produced its TASI 600 scanner for use over the thermal or long-wave infra-red (LWIR) part of the spectrum. This instrument also has a swath width of 600 pixels and either 32 or 64 spectral channels.

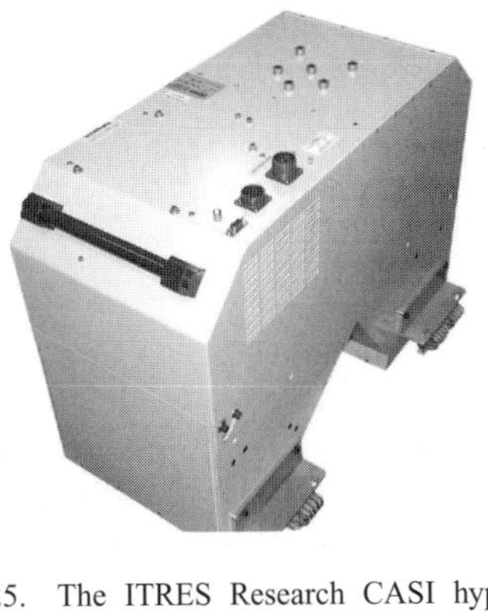

Figure 4.25. The ITRES Research CASI hyperspectral pushbroom line scanner (Source: ITRES Research).

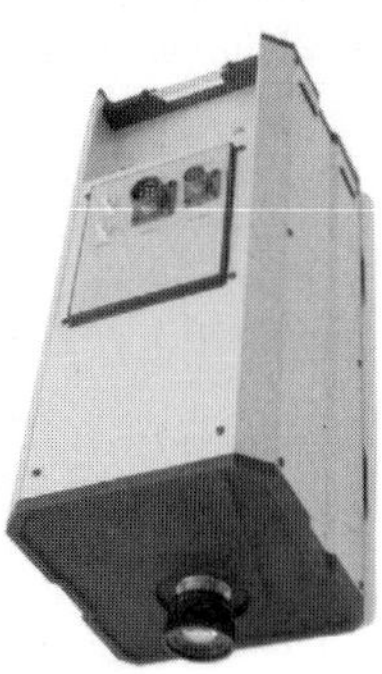

Figure 4.26. The AISA+ hyperspectral pushbroom line scanner produced by Specim (Source: Specim).

The Specim company produces a fairly similar range of hyperspectral scanners. The AISA+ and AISA Eagle scanners produce imagery with respective swath widths of 512 and 1024 pixels in up to 244 spectral channels in the VNIR region. A new development, which is designed specifically for use in UAVs, is the AISA Eaglet, which will provide a swath width of 1900 pixels over the VNIR range with up to 200 spectral channels. The company's SWIR scanner is the AISA Hawk, which features a cross-track coverage of 320 pixels and a spectral coverage of up to 254 channels. Finally, Specim has recently announced its AISA Thermal hyperspectral scanner that will provide a swath width of 320 pixels and will record ground radiation in up to 57 separate spectral channels across the LWIR (thermal) part of the spectrum. This last instrument is scheduled to appear sometime during 2008, perhaps in time for the 2008 Beijing Congress.

Besides these two main system suppliers, there is much interest within Europe in the APEX hyperspectral pushbroom scanner that is being constructed by a Swiss/Belgian team with the support of the European Space Agency (ESA). The optical elements of this scanner are being supplied by the ElOp company from Israel. The APEX scanner will be used as a prototype and development instrument for a future spaceborne hyperspectral imager. It will have a swath width of 1000 pixels and 300 spectral channels covering the VNIR and SWIR parts of the spectrum. APEX has been somewhat delayed in its development, but currently it is scheduled for delivery in 2008, in which case, it will be used for research campaigns starting in 2009.

In summary, one can say that the area of hyperspectral scanners has been one of steady development throughout the inter-Congress period with both the major suppliers of airborne hyperspectral systems introducing pushbroom-type imagers with enhanced capabilities in terms of using larger arrays to ensure wider ground coverage and a greater number of spectral channels. At the same time, the spectral coverage has been extended into the SWIR and LWIR parts of the spectrum. The previous dominant whiskbroom scanning technology in this field has largely been replaced by these scanners based on the pushbroom principle.

4.3.1.4 *Panoramic line scanners*

In this type of scanner, a linear array is used to scan the ground in the cross-track direction instead of in the along-track direction that is used in the pushbroom scanner. This produces a series of panoramic linescan strip images of the ground at right angles to the flight direction. Essentially the methodology that is used is very similar to that of a panoramic film camera, in which a scanning slit is moved rapidly across a film wrapped around a cylindrical surface to expose an image of the ground in the cross-track direction. This allows the acquisition of imagery over a very wide field of view, which can even stretch from horizon-to-horizon. In the digital version of the panoramic scanning principle, the exposure slit is replaced by a linear array producing a monochrome image. Furthermore, just as most panoramic film cameras are used in military reconnaissance aircraft, the same applies to the digital panoramic line scanners. A prominent supplier of these devices is the Goodrich company from the USA. It has built many of these digital panoramic line scanners for the US Air Force in the shape of the SYERS systems mounted on Lockheed U-2 aircraft, the Raptor systems used on Tornado aircraft of the British RAF and the DB-110 systems mounted on F-16 aircraft of the Polish and Greek air forces (Iyengar & Lange 2006). In all these cases, the image data can be transmitted continuously from the aircraft to the ground in near real-time over a radio link.

Another panoramic scanner with a design similar to that of the Goodrich examples is the Thales Optronics Type 8042 scanner from the UK, which is also in use by various air forces. However, there are also some civilian users of these panoramic line scanners. Prominent among these have been the DDP-2 (Direct Digital Panoramic) scanners that have been developed and operated by the ImageAmerica company, which has been acquired by the Google company in July 2007. The resulting linescan imagery, together with the available archive, will be utilized in Google Earth. At the same time, Google announced that it was acquiring the Panoramio Web site that was started in 2005 and allows users to geo-locate and display their own images within Google Earth and Google Maps using spaceborne and airborne imagery as well as maps.

In summary, there is currently a great deal of interest in these digital panoramic line scanners, which are seen as a direct replacement for the panoramic film cameras that are in widespread use on military reconnaissance aircraft.

4.3.2 *Three-line stereo scanners*

As the name suggests, three-line pushbroom scanners are equipped with three quite separate linear arrays that point in the forward, nadir and backward directions, respectively, along the line of flight (Fricker 1999 & 2004). They produce three separate overlapping strip images from which 3-D stereo-models of the ground can be formed. Initially the concept and the technology were developed by the German Aerospace Agency (DLR) and various German industrial aerospace companies. However, the concept could only come into full practical use with the advent of airborne positioning and orientation systems as a vital supporting technology in the shape of the integrated GPS/IMU systems that were developed during the 1990s. DLR then demonstrated that the technology had finally matured with its WAAC and HRSC-A three-line pushbroom scanners that were developed in the mid- to late-1990s. Leica Geosystems then licensed the basic WAAC technology from DLR (Fricker et al. 1999) and then proceeded to design and construct its own ADS40 three-line pushbroom scanner, which was first shown at the ISPRS Congress held in Amsterdam in 2000. By the time of the 2004 Congress, the ADS40 had already become

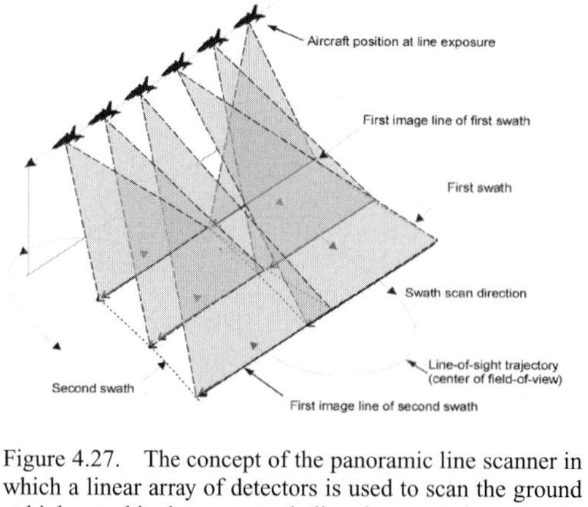

Figure 4.27. The concept of the panoramic line scanner in which a linear array of detectors is used to scan the ground at high speed in the cross-track direction to produce a panoramic linescan image (Source: Goodrich).

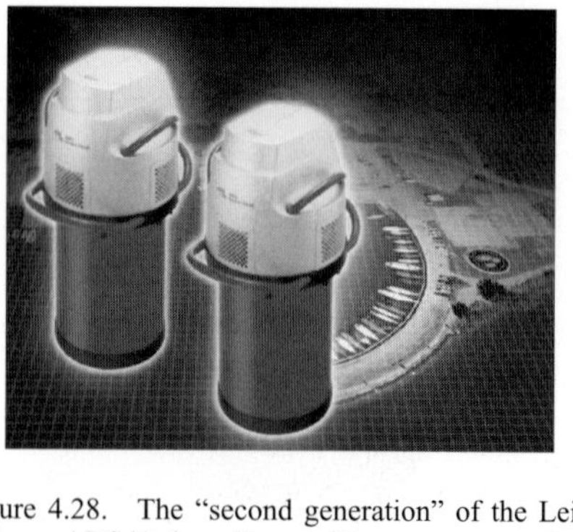

Figure 4.28. The “second generation” of the Leica Geosystems ADS40 three-line pushbroom scanner showing its alternative SH51 and SH52 “sensor heads” (Source: Leica Geosystems).

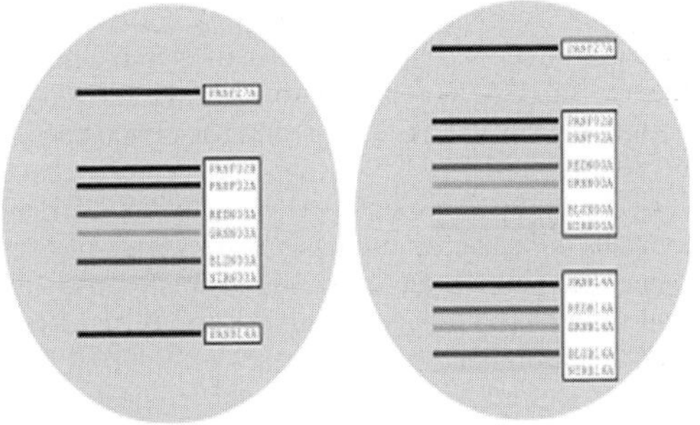

Figure 4.29. Diagrams that show the layouts of the SH51 and SH52 focal plates. The SH51 plate (on left) features a single tetrachroid (four-way beam-splitter) prism placed in the nadir-pointing position, while the SH52 (on right) utilizes two of these prisms; the first in the nadir-pointing position, the second in the backward-pointing position (Source: Leica Geosystems).

well established in the market (Fricker 2004, Flint 2005).

Recent major developments of the ADS40 scanner have been the introduction in 2006 of two alternative "sensor heads"—called SH51 and SH52—in the 2nd generation version of the instrument (Fricker 2007). Each of these heads employs a tetrachroid prism, which acts as a four-way beam-splitter that causes the radiation from a single line on the ground to be split into its respective spectral bands and imaged by the appropriate RGB and NIR linear arrays mounted in the focal plane of the scanner. This also ensures their correct co-registration. The SH51 version utilizes a single example of the tetrachroid prism that is located in the nadir-pointing position, resulting in a total of four spectral images (RGB + NIR) being generated in the vertical direction. The forward- and backward-pointing arrays generate purely panchromatic (monochrome) images. By contrast, the SH52 version utilizes two tetrachroid prisms, the first being placed in the nadir position (as in the SH51), while the second is placed in the backward-pointing position. Each of the two prisms generates four co-registered spectral images and these are supplemented by a single pair of coupled linear arrays generating a pan image in the forward-pointing direction, making a total of twelve arrays in the SH52 version. It is possible to upgrade the SH40 "sensor head" of the original (first generation) version of the ADS40 to either the SH51 or the SH52 head. Besides which, the new electronics introduced with the ADS40 2nd generation imager provide a much improved signal-to-noise ratio ensuring a better quality image through increased sensor sensitivity. Yet another major change to the ADS40 has been the introduction of a new integrated GPS/IMU or GNSS/IMU sub-system, called the IPAS10, to provide direct geo-referencing of the ADS40 linescan image data. This sub-system was developed originally by Terramatics, a small Canadian company that was later acquired by Leica in 2005. This has replaced the original Applanix POS/AV unit that had previously been used in the ADS40.

Another recent addition to the market (in 2007) is the JAS 150 three-line pushbroom scanner built by Jena-Optronik (Georgi et al. 2005). The design of this imager has also been derived from a DLR development, in this case, the HRSC-A series of scanners (Neukum 1999) with which it shares a similar specification and a similar number (nine), size (12,000 pixels) and geometric arrangement of its CCD linear arrays (Petrie 2007). Five of the nine linear arrays provide stereo panchromatic channels with two pairs of forward- and backward-pointing arrays that are set at angles of ±20° and ±12° from the vertical, respectively, while the fifth panchromatic array is nadir-pointing. The remaining four linear arrays are grouped around the nadir-pointing position and provide multi-spectral coverage of the ground in RGB and NIR channels. The offsets of the resulting imagery need to be compensated for during the post-flight processing. The obligatory GPS/IMU sub-system can be either the IGI AEROcontrol or the Applanix POS/AV unit, while the equally obligatory gyro-controlled mount can be any one of the standard units—Somag GSM3000, Intergraph T-AS or Leica Geosystems PAV30—that are widely available on the market.

Another somewhat different approach to the design of a three-line pushbroom scanner is the use of three tri-linear arrays that are placed in the forward, nadir- and backward-pointing directions, respectively,

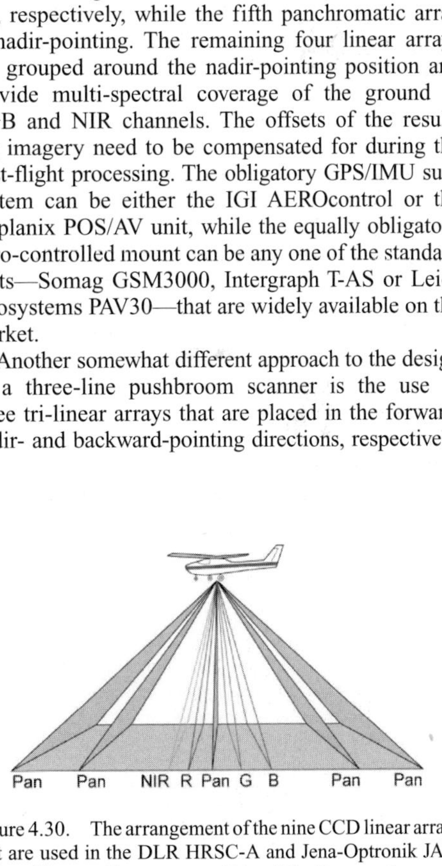

Figure 4.30. The arrangement of the nine CCD linear arrays that are used in the DLR HRSC-A and Jena-Optronik JAS-150 pushbroom line scanners. Four of the five panchromatic lines are placed at ±20° and ±12° to the vertical to point in the forward- and backward-pointing positions respectively, while the fifth line is placed in the nadir-pointing position. The remaining four linear arrays that generate RGB+NIR line-scan images are grouped around the nadir position (Drawn by M. Shand).

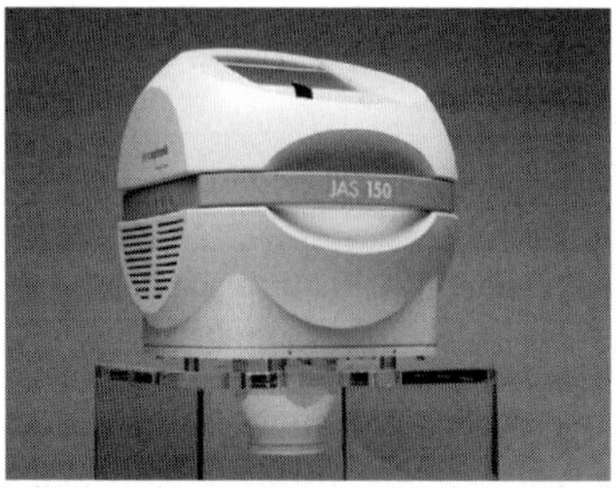

Figure 4.31. The Jena-Optronik JAS-150 pushbroom line scanner (Source: Jena-Optronik).

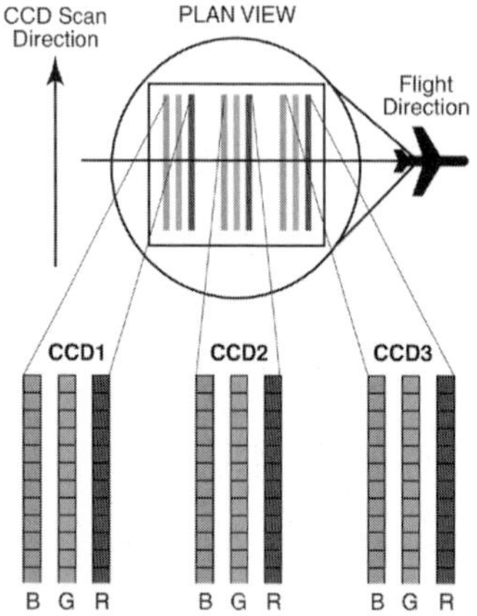

Figure 4.32. The arrangement of the three CCD tri-linear arrays that are utilized in the 3-DAS-1 pushbroom line scanner, constructed by Wehrli Associates and Geosystems (Drawn by M. Shand).

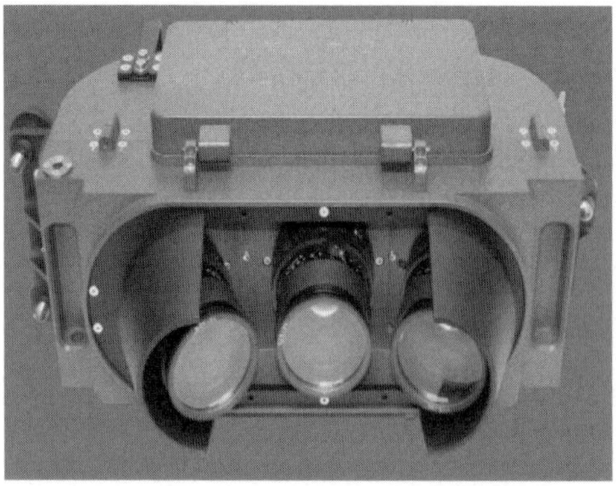

Figure 4.33. The Wehrli Associates 3-DAS-1 three-line pushbroom stereo-scanner features three lenses to produce the backward-, nadir- and forward-pointing images, respectively (Source: Wehrli Associates).

within the focal plane. These tri-linear arrays generate all three overlapping strips of stereo linescan imagery directly in colour, instead of the three overlapping panchromatic images that have been used in the ADS40 and JAS 150 scanners discussed above. This alternative approach was used by the Starlabo company in Japan as the basis of its series of Starimagers or Three Line Scanners (TLS). The different models in the series used tri-linear arrays of varying lengths in combination with lenses of differing focal lengths. Unfortunately, the Starlabo company went out of business early in 2006. The other company to pursue this alternative approach is Wehrli Associates from the USA, which has built its 3-DAS-1 three-line scanner in cooperation with the Geosystems company from Ukraine. The original prototype of this scanner was exhibited at the ISPRS Congress in 2004; a second model was shown at the ASPRS Annual Conference in 2005. However, so far, the 3-DAS-1 scanner has not reached full commercial production. Another model from the Wehrli/Geosystems partnership is the 3-OC model that was introduced in 2006 (Petrie 2006a). This also uses three tri-linear arrays, but with the forward- and backward-pointing arrays both pointing in the along-track direction at an angle of 45° to the vertical instead of the 26° forward- and 16° backward-pointing angles used in the 3-DAS-1 scanner. The 3-OC model is designed specifically to produce continuous oblique linescan imagery of the ground with a view to satisfying the growing market for systematic coverage of the ground using oblique imagery that has already been discussed above in the context of multiple airborne frame cameras.

In summary, airborne three-line pushbroom scanners offer a very viable alternative to digital frame cameras in terms of generating stereo-imagery from which ortho-images and elevation data can be derived. However, so far, this area has remained a near-monopoly of the Leica Geosystems ADS40 product for a variety of commercial rather than technical reasons. In the meantime, the ADS40 scanner continues to be developed in a systematic manner to retain its commanding position in this particular field.

4.4 AIRBORNE RADAR IMAGING SYSTEMS

Apart from a few systems that are operated by government research establishments, the main interest in airborne radar imaging systems is centred on those that are operated by two commercial companies, Intermap Technologies and Fugro EarthData, that are based in North America but operate on a world-wide scale. In the case of Intermap, the inter-Congress period has seen a very considerable increase in its capacity. Prior to this, the company had operated a single Learjet aircraft that was fitted with the company's STAR-3i interferometric SAR (InSAR) system operating in the X-band (λ = 3 cm). The company also operated a TopoSAR system that combines an X-band SAR with a longer wavelength (λ = 45 cm) fully polarized P-band SAR, which it had acquired when it bought the German Aerosensing Radarsysteme company in 2003. Since then, it has developed a new InSAR architecture that combines elements of both the STAR-3i and TopoSAR systems (Mercer 2007). This has resulted in three new systems, STAR-4 (from 2004), STAR-5 and STAR-6 (both from 2007), being added to its operational inventory. The final processed products that result from these STAR systems include an ortho-rectified digital radar image, a digital surface model (DSM) and a digital terrain model (DTM). The models have height accuracies in the range ±0.5 to 3 m for the different products that are derived from SAR data acquired at different flying heights. Since the previous very successful NEXTMap Britain project, which produced a country-wide DEM for Great Britain (Coleman & Mercer 2002), over the last three years, Intermap has embarked on several still larger projects. These include NEXTMap Europe, NEXTMap USA and NEXTMap Indonesia, all of which involve the acquisition of SAR interferometric data over enormous areas of terrain for the production of digital image maps and elevation data.

Besides Intermap, Fugro EarthData has also continued to use its GeoSAR system, which consists of dual X-band and P-band InSAR systems mounted on a Gulfstream jet aircraft. The overall system has now been developed further to incorporate a Leica ALS40 laser scanner that is being operating in a profiling mode to supplement the measured InSAR data. Like the STAR systems, the GeoSAR system has been used to undertake surveys of huge areas of terrain. These include the mapping of large areas of tropical forest in South America carried out on behalf of the National Geospatial-Intelligence Agency (NGA)

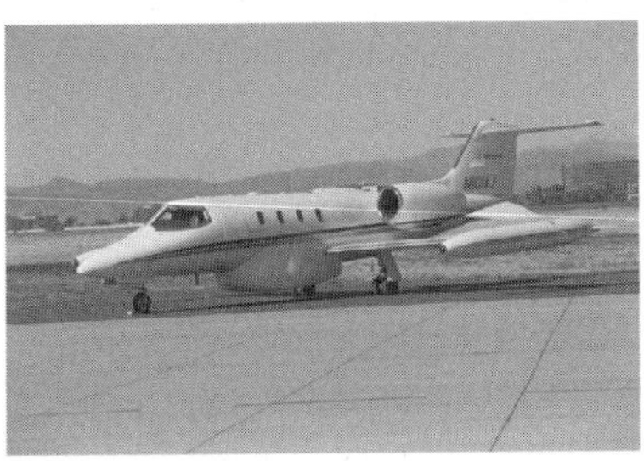

Figure 4.34. This Learjet aircraft is fitted with the STAR-3i interferometric SAR imager, which is housed in the prominent bulge on the underside of the aircraft (Source: Intermap).

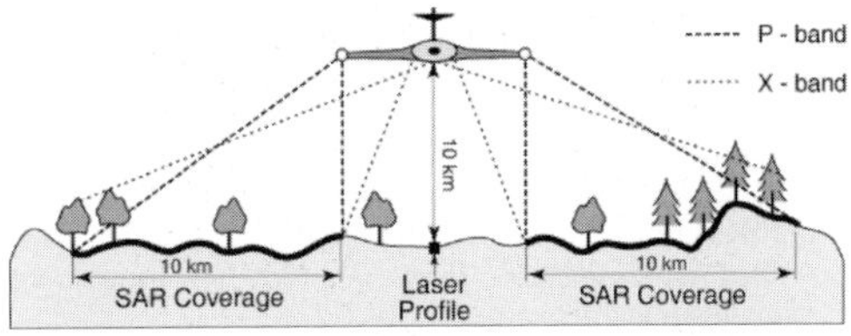

Figure 4.35. A cross-sectional diagram showing the ground coverage achieved by the Fugro GeoSAR airborne radar imaging system using its twin X-band and twin P-band interferometric SAR imagers and its ADS40 laser profiler (Drawn by M. Shand).

and similar surveys in Papua New Guinea funded by the Australian government. In these heavily forested areas, the P-band InSAR system is especially suited to data acquisition with its ability to penetrate the dense vegetation canopy as well as the prevailing cloud cover. Similar remarks apply to the OrbiSAR-1 system with dual X- and P-band SARs, which is operated by the Brazilian Orbisat company over the country's Amazonian forests.

In summary, over the period being covered by this review, there is much more activity in the field of airborne radar imaging systems for mapping than ever before, almost all of it involving the use of interferometric SAR imagery and the production of both image maps and elevation data.

REFERENCES & FURTHER READING

Artes, F. 2004. Medium-Format Digital Cameras Come of Age. *Earth Observation Magazine* 13(6): 8–12.

Artes, F. 2006. Geo-referenced Oblique Aerial Imagery: Pictometry—Intelligent Visual Technology. *GeoInformatics* 9(2): 20–23.

Coleman, M. & Mercer, B. 2002. NEXTMap Britain: Completing Phase 1 of Intermap's Global Mapping Strategy. *GeoInformatics* 5(8): 16–19.

Cramer, M. 2005. Digital Airborne Cameras: Status and Future. *ISPRS Workshop on High-Resolution Earth Imaging for Geospatial Information, Hanover, Germany.*

Flint, D. 2005. Innovations to Increase Productivity of Airborne Sensors. In D. Fritsch (ed.), *Photogrammetric Week '05*: 69–77. Heidelberg: Wichmann.

Fricker, P., Sandau, R. & Walker, S. 1999. Digital Photogrammetric Cameras: Possibilities and Problems. In D. Fritsch & R. Spiller (eds.), *Photogrammetric Week '99*: 71–82. Heidelberg: Wichmann.

Fricker, P. 2004. Three-Line Sensors: Rapid, Efficient Digital Imagery. *Photogrammetric Engineering & Remote Sensing* 70 (12): 1325–1326.

Fricker, P. 2007. Raising the Bar for Multi-Band High-Resolution Airborne Imagery. In D. Fritsch (ed.), *Photogrammetric Week '07*: 71–79. Heidelberg: Wichmann.

Georgi, C., Stognienko, R., Knuth, S. & Albe, G. 2005. JAS: The Next Generation Digital Aerial Scanner. In D. Fritsch (ed.), *Photogrammetric Week '05*: 147–154. Heidelberg: Wichmann.
Gruber, M. 2007. UltraCamX, The New Digital Aerial Camera System by Microsoft Photogrammetry. In D. Fritsch (ed.), *Photogrammetric Week '07*: 137–145. Heidelberg: Wichmann.
Iyengar, M. & Lange, D. 2006. The Goodrich 3rd Generation DB-110 System: Operational on Tactical and Unmanned Aircraft. Airborne Intelligence, Surveillance Reconnaissance (ISR) Systems and Applications - III. *Proceedings of SPIE* Vol. 6209.
Mayr, W. & Ohlhof, T. 2004. Geoinformation Via Parallel Image Engineering and Image Information Mining. *Proceedings of ESA-EUSC Conference: Theory and Applications of Knowledge Driven Image Information Mining with Focus on Earth Observation (ESA SP-553), Madrid, Spain.*
Mercer, B. 2007. National and Regional Scale DEMs Created From Airborne INSAR. *International Archives of Photogrammetry, Remote Sensing and Spatial Information Sciences* 36-3/W49B: 113–117.
Neukum, G. 1999. The Airborne HRSC-A—Performance and Application Potential. In D. Fritsch & R. Spiller (eds.), *Photogrammetric Week '99*: 83–88. Heidelberg: Wichmann.
Petrie, G. 2003. Airborne Digital Frame Cameras: The Technology is Really Improving. *GeoInformatics* 6(7): 18–27.
Petrie, G. 2004. The ISPRS 2004 Technical Exhibition: Airborne Imaging Dominates the Show. *GeoInformatics* 7(7): 30–37.
Petrie, G. 2005. Airborne Pushbroom Line Scanners: An Alternative to Digital Frame Cameras. *GeoInformatics* 8(1): 50–57.
Petrie, G. 2006a. Further Advances in Airborne Digital Imaging: Several New Imagers Introduced at ASPRS. *GeoInformatics* 9(5): 16–23.
Petrie, G. 2006b. IGI's Airborne Systems: An Expanded Product Range! *GeoInformatics* 9(7): 36–41.
Petrie, G. 2007. Jena-Optronik's Imaging Scanners: Earth Observation from Space & the Air. *GeoInformatics* 10(6): 42–46.
Petrie, G. & Walker, A.S. 2007. Airborne Digital Imaging Technology: A New Overview. *The Photogrammetric Record* 22(119): 203–225.
Smith, W. & Merrill, P. 2004. From Obscurity to Mainstream: Oblique Imagery and Orthophotography Combine to Spark New Applications. *Earth Imaging Journal* Nov/Dec. 2004: 26–28.
Toelg, T. 2007. Newest Technologies for Compact Digital Aerial Survey Cameras. In D. Fritsch (ed.), *Photogrammetric Week '07*: 131–136. Heidelberg: Wichmann.
van Achteren, T., Delaure, B. & Everaerts, J. 2006. Instrument Design for the Pegasus HALE UAV Payload. *International Archives of Photogrammetry, Remote Sensing and Spatial Information Sciences* 36-1/W44.

Advances in Photogrammetry, Remote Sensing and Spatial Information Sciences: 2008 ISPRS Congress Book – Li, Chen & Baltsavias (eds)

CHAPTER 5

Close range photogrammetry sensors

Hans-Gerd Maas

ABSTRACT: This chapter gives an overview of the different types of sensors used in close range photogrammetry. It starts with area solid state sensor cameras, which are certainly the most prominent cameras in close range photogrammetry. The focus is on high resolution cameras, with some attention being given to high-speed cameras and omni-directional vision systems. For 3-D data acquisition, multi-ocular camera systems, triangulation systems and novel range cameras will be discussed.

Keywords: Close range photogrammetry, instrument development, area sensor cameras, smart cameras

5.1 INTRODUCTION

Driven by progress in sensor technology, by algorithms and data processing capabilities, the focus of close range photogrammetry has propagated into a wide range of new application fields over the past two decades. A central issue in many developments is the integration of sensor technology with reliable data processing schemes to generate highly automated online or real-time photogrammetric measurement systems. Here we often find the term "image engineering", standing for custom-made solutions based on active and passive sensing and illumination devices combined with task-specific image analysis techniques. For this reason, sensors form an essential component of close range photogrammetry in promising new application fields such as industrial design, production and quality control processes as well as robotics, virtual reality or quantitative biomedical imaging. Here, close range photogrammetry has also seen a large number of successful research projects passing into practical application, providing efficient solutions for new measurement tasks and thus opening new fields of activity and creating new markets for photogrammetrists.

The following sections will give an overview of different types of sensors used in close range photogrammetry, characterize them and discuss their potential. It will start with area solid state sensor cameras, which are certainly the most prominent cameras in close range photogrammetry. The focus will be on high resolution cameras, with side glances at high-speed cameras. Special attention will also be given to omni-directional vision systems. For 3-D data acquisition, multi-camera systems, range cameras and triangulation systems will be discussed. In a short excursion, smart cameras will be touched upon. Purely amateur cameras, webcams and similar devices will not be discussed in detail.

5.2 AREA SENSOR CAMERAS

The most prominent sensor in close range photogrammetry is obviously an area sensor. While the transition from analogue to digital has taken place in aerial photogrammetry during the past few years, film was replaced by solid state sensors in close range photogrammetry much earlier. Area sensor cameras can be divided into several categories: stand-alone cameras versus machine vision type cameras requiring permanent computer connection, high resolution cameras versus high speed cameras or 2-D cameras versus 3-D imaging systems. In the following, the most important types of camera will be presented and discussed.

5.2.1 *SLR-type cameras*

Digital SLR (single lens reflex) cameras (Fig.5.1) can be considered the "workhorse" of close range photogrammetry. They can be used as stand-alone digital image acquisition systems, comprising sensor, optics, data storage, power supply and camera control in

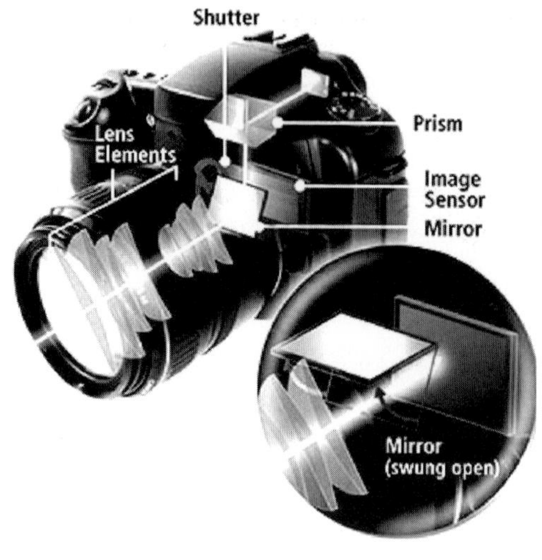

Figure 5.1. Digital SLR-camera principle (Long, 2005). (see colour plate page 491)

one unit. At the time of writing (2007), digital SLR cameras offer sensor formats up to 21 Megapixel, while digital camera backs, which can be adapted to a medium format camera body, offer up to 39 Megapixels. An overview over recent camera models as well as tests, mainly concerning resolution and radiometric quality, can be found on www.dpreview.com.

The obvious advantages of digital SLR-cameras can be seen in their portability and the low system costs due to their character as a amateur or semi-professional mass market devices. They are especially suited for data acquisition in VR model generation, architectural photogrammetry or similar applications. 3-D information is generated by taking images sequentially from two or more viewpoints and performing consecutive photogrammetric multi-image processing. The disadvantages of digital SLR-cameras can be seen in their limited suitability for real-time data processing as a consequence of indirect interfacing. Digital SLR cameras are better suited for offline processing tasks, as they do—unlike machine vision type cameras—usually not write directly into host computer main memory. Moreover, their frame rate is limited and does not exhaust the potential of solid state sensors or system buses.

The accuracy potential of digital SLR cameras is characterized by large format CCD or CMOS sensors, which allow an image coordinate measurement precision down to 1/50 pixel for well defined targets. This accuracy can only be translated into object space after thorough calibration, determining parameters describing the interior orientation, lens distortion and electronic effects. A setback of digital SLR-cameras can often be found in the mechanical instability of the camera body geometry, leading to unstable interior orientation parameters. Maas & Niederöst (1997) reported principal point variations of more than 12 pixels for digital SLR cameras in normal camera handling. These variations form an incompliance with the geometric model of central perspective and may cause severe object space deformations. To (at least partly) compensate for this effect, a bundle adjustment procedure with constrained image variant interior orientation parameters has been presented by Tecklenburg et al. (2001).

5.2.2 *High resolution cameras*

While digital aerial cameras reach their resolution by linear array sensors or by combining many area sensors, close range photogrammetry applications are usually based on single area sensors. Beyond digital SLR cameras, which reach up to 21 Megapixels, digital camera backs with up to 39 Megapixels (7216 × 5412 pixels) are offered on the market. In 2006, Dalsa presented the first sensor to break the 100 Megapixel barrier: their 4″×4″ CCD sensor has 10,560 × 10,560 pixels with a 9 μm pixel size, yielding a total of 111 Megapixels. This greyscale sensor, which is primarily meant for astronomy applications, is still extremely expensive. Generally, as a consequence of the complexity of the solid state sensor production process, the costs of sensors grow exponentially with their spatial resolution.

In the 1980s and 1990s, when the resolution potential of solid state sensors was restricted to a few Megapixels, macro- and micro-scanning schemes were presented: Macro-scanning is achieved by mechanically moving a small sensor over a larger image format, leading to a line- or patch-wise image acquisition. A 94 Megapixel RGB macro-scanning camera based on a 8000 pixel RGB linear array sensor is described in Schneider et al. (2005). Besides the non-simultaneous image acquisition process, the principle suffers from mechanical limitations to the achievable accuracy potential. As an extension of the technique, improving the accuracy potential and enabling patch-wise focusing, Luhmann & Wester-Ebbinghaus (1986) presented the Réseau-scanning camera principle. Micro-scanning (Lenz, 1989) achieves resolution enhancement by shifting a sensor equipped with a small mask by fractions of a pixel. Small shifts can be realized accurately in the x and y image coordinate direction by Piezo elements, leading to a fine-scan within each pixel and a quadrupling of the resolution. Again, the image acquisition process is non-simultaneous, and illumination changes during image acquisition will severely reduce the radiometric quality of images. A combination of macro- and micro-scanning, resulting in an image format of more than 20,000 × 20,000 full RGB pixels, was shown by Lenz (1993).

Many close range photogrammetry applications with primarily geometric task definition prefer greyscale sensors rather than colour sensors. In most cases, greyscale sensors will offer a slightly higher accuracy potential. If colour is needed, e.g. for visualization purposes, it is mostly generated by a Bayer pattern (Bayer 1975).

5.2.3 *High-speed cameras*

While high resolution solid state sensor cameras will usually offer a frame rate of a few frames per second, many dynamic phenomena require substantially higher frame rates. Cameras based on consumer grade standards offer an image rate of 25/50 Hz or 30/60 Hz, respectively. Cameras with an image rate of 100 Hz and more are classified as high speed cameras. This sector has seen a rapid development over the past 25 years, with current cameras typically offering 1000 × 1000 pixel sensors at a frame rate of up to 5000 frames per second (fps) (Fig.5.2). Application fields of high speed cameras are for example in car crash testing, assembly line mechanics, bio-medical motion analysis, ballistics, combustion, the examination of flow phenomena or material testing (Maas & Hampel 2006). One major limitation of high-speed cameras is the data rate. The 5 GHz pixel rate of modern off-the-shelf high speed cameras can only be achieved by multiple readout taps. Most cameras allow for a shift between temporal and spatial resolution at a constant data rate, enabling much higher frame rates at a reduced spatial resolution. Another

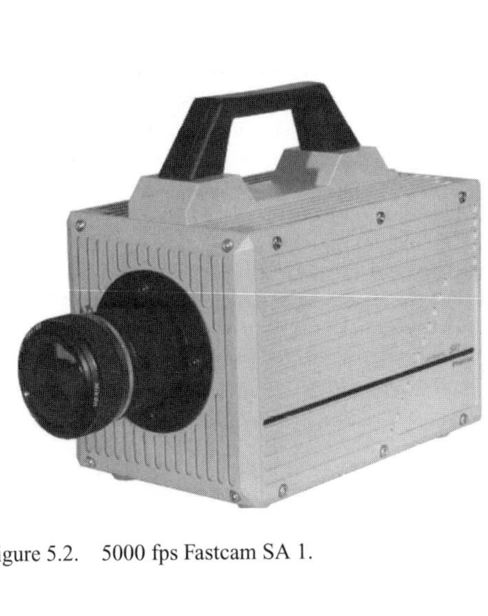

Figure 5.2. 5000 fps Fastcam SA 1.

limitation is obviously light intensity: At imaging rates beyond 1000 fps, cameras are operating in the sub-millisecond exposure time range, requiring an enormous amount of light to warrant sufficient image intensity.

The top temporal resolution of custom-made systems is currently at more than one million frames per second. A 312 × 260 pixel sensor with 1 Mfps, limited to capturing a maximum of 100 successive frames, is shown by Etoh et al. (2003).

5.3 MULTI-OCULAR SYSTEMS

Many applications of close range photogrammetry aim at the determination of 3-D object coordinates of discrete targets. While 3-D coordinates can be obtained by processing multiple images acquired sequentially in static applications, dynamic applications require vision systems consisting of multiple synchronized cameras. Automatic systems will often require using three or more cameras to allow for the reliable solution of matching ambiguities by trinocular or multi-ocular vision techniques (Ayache & Lustman 1987, Maas 1991).

The technology of solid state sensor cameras allows for synchronization down to the nanosecond level. A major issue to be solved in multi-ocular camera systems is data transfer. In many cases, a parallel data transfer from multiple cameras to a host computer will not be possible. In these cases, image sequences may be stored onboard the camera or in specially designed interface devices. In some cases, image data can be processed on the camera, thus completely avoiding image data transfer bandwidth problems.

Figure 5.3 shows an early trinocular 25 Hz camera system used in a 3-D flow measurement system presented by Maas (1991). Dynamic applications with very fast motion may require using multiple high-speed cameras. As an alternative, a multi-ocular system based on one high-speed camera and a multi-mirror system (Fig. 5.4) is shown by Putze (2005). The mirrors image four different views onto the quadrants of the sensor, which can be considered as four virtual cameras.

A high-end system for human motion analysis (by Motion Analysis Corp.) consists of up to 32 synchronized cameras running at 480 fps. To enable real-time processing of this amount of data, the system is limited to tracking retro-reflective markers. Using LED ring-flashes in combination with very short exposure times, the system produces quasi-binary images, which can be processed on the camera, drastically reducing the data rate by transferring image coordinates of a limited number of targets rather than images. When going to even higher imaging rates, system integration is often constricted by the

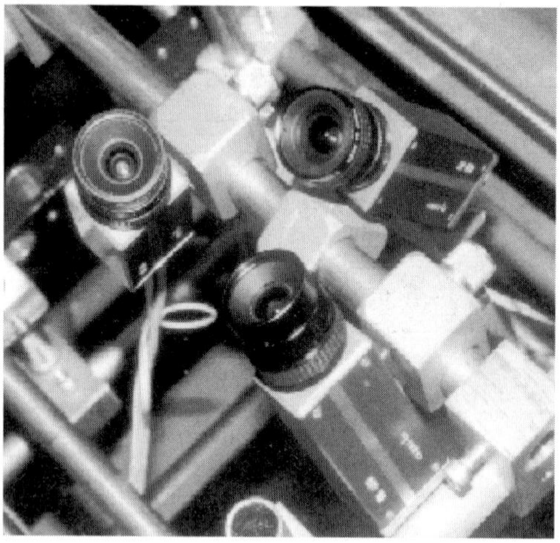

Figure 5.3. Three-camera system (Maas 1991).

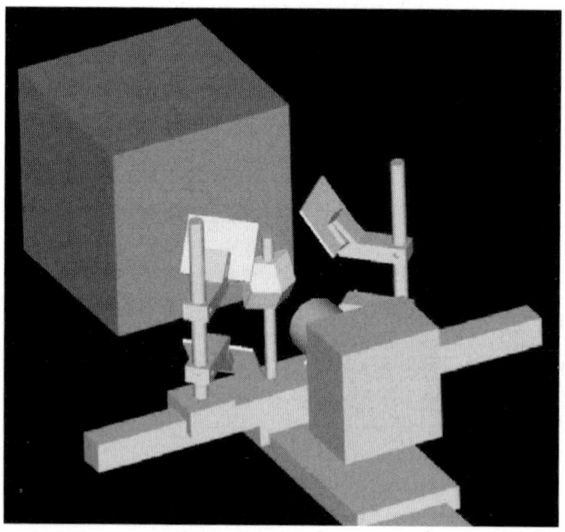

Figure 5.4. Multi-ocular high-speed camera system. (see colour plate page 491)

lack of synchronizability of many high-speed cameras. To solve this problem, Raguse & Heipke (2006) presented a technique for handling asynchronous stereoscopic image sequences by introducing timing parameters into the bundle adjustment.

5.4 ACTIVE TRIANGULATION SYSTEMS

Triangulation is the most widely used optical 3-D measurement technique. As a very rough classification, it can be subdivided into passive and active techniques. While passive systems are realized by multiple cameras or images taken from multiple viewpoints (stereo photogrammetry), active techniques are usually based on the combination of a camera and a projector generating spatio-temporal structured light patterns. A rather simple active triangulation system can be realized by a laser diode with a cylindrical lens, generating a laser light-sheet illuminating a line on the object surface. Imaged under a parallax, the deformations of the line are a measure for depth variations along the illuminated profile (Fig. 5.5).

Depending on the sensor resolution, the object surface quality and the system configuration, a profile height precision potential of up to 1:50,000 of the profile length can be achieved. The step from profiles to surface models can be performed either by translating or rotating the projection system or by moving the object. The latter is achieved inherently in applications on industrial assembly lines, where these techniques are used frequently. Implemented in suitable cameras with on-chip processing capabilities, 3-D surface measurement can be performed at data rates beyond ten million points per second.

A straight-forward extension towards a full-field measurement technique is the projection of multiple lines rather than one. This raises the question of solving ambiguities in the matching between projected and detected lines. A number of solutions has been developed in the past, with the most prominent ones being the coded light approach and hierarchical phase shift methods:

The coded light approach (Altschuler et al. 1979) achieves a very efficient method of coding projected strips by a sequential projection of n binary strip patterns with $2^0 \ldots 2^{n-1}$ vertical black and white line pairs (Fig. 5.6). The pattern can be generated

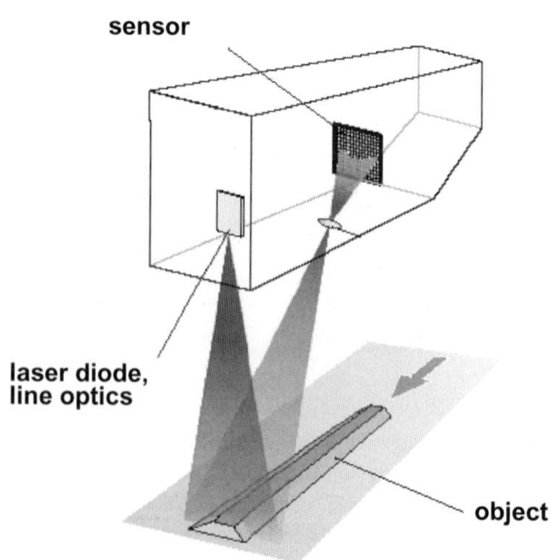

Figure 5.5. Laser light-sheet triangulation principle. (see colour plate page 491)

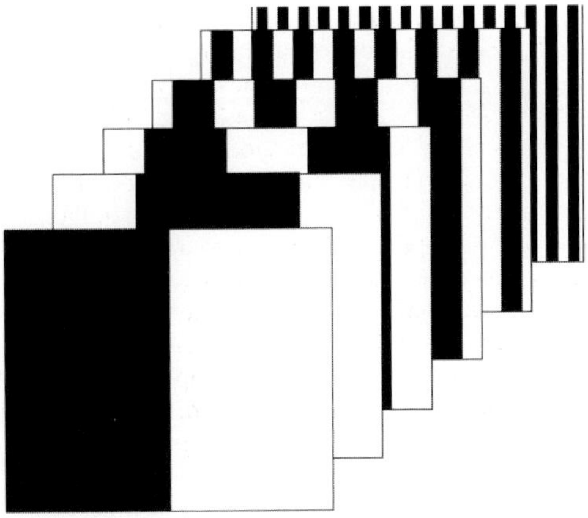

Figure 5.6. Coded light approach binary pattern.

by a off-the-shelf beamer technology. Acquiring one image per projected pattern, a binary code can be read from the image sequence for each pixel, uniquely identifying the projected line belonging to the pixel as a pre-requisite for 3-D coordinate determination. Due to the binary characteristics of the images and the simple code-reading, the method is very fast and hardware implementation friendly. A drawback of the method in its original implementation is its limited accuracy potential due to the lack of subpixel accuracy in line detection.

A better accuracy can be obtained by the projection of a strip pattern with a sinus-modulated intensity (Fig. 5.7). Assuming a Lambertian reflecting object surface of uniform brightness, the intensity of a pixel *g(x,y)* is directly related to the column *x'* of the projected pattern illuminating this pixel. Using interpolation techniques, a subpixel accuracy location in the projected pattern becomes possible, leading to an improved accuracy.

In practice, the assumption of a uniformly bright surface will hardly be fulfilled. This can be solved by a sequential projection of patterns shifted by 0°, 120° and 240°. From the three grey values g_1, g_2, g_3 of a pixel at position (x, y), a phase angle α (as a linear measure for the projector column coordinate x') can be determined together with the surface brightness I_H and the projection contrast I_K:

$$g_1 = I_H + I_k \cdot \cos\alpha$$
$$g_2 = I_H + I_k \cdot \cos(\alpha + 120^\circ)$$
$$g_3 = I_H + I_k \cdot \cos(\alpha + 240^\circ)$$

$$\alpha = \frac{\sqrt{3} \cdot (g_1 - g_2)}{2 \cdot g_1 - g_2 - g_3} \rightarrow x' = \left(n + \frac{\alpha}{360^\circ}\right) \cdot \lambda$$

In this equation system, the multiple of wavelengths n can be determined by a hierarchical approach projecting 2-4 patterns with increasing modulation frequencies sequentially. Hybrid codes, combining the efficient ambiguity solution of the coded light approach with the accuracy of phase-shift methods, have been shown by Malz (1989).

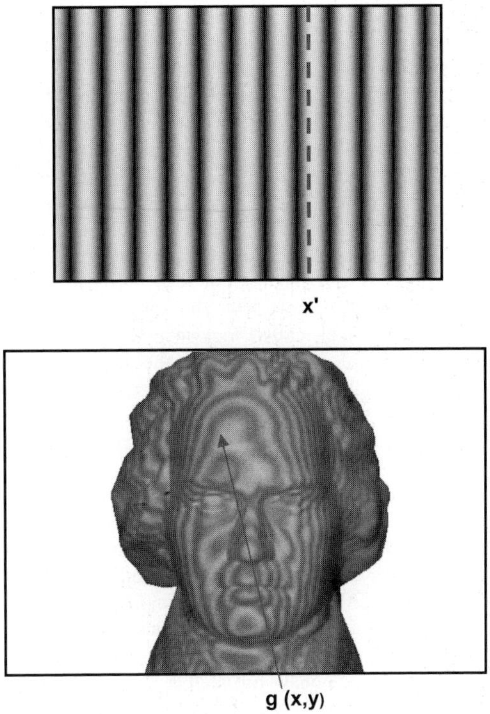

Figure 5.7. Sinus pattern projection.

5.5 RANGE CAMERAS

Several research labs and companies are currently developing range cameras (also called range imagers or 3-D cameras), which are capable of obtaining a grey value image and a simultaneously recorded depth image from a single sensor. These cameras are based on the time-of-flight (tof) measurement principle, deriving distance information from the travel time of emitted light from the instrument to a scene point and back. Like geodetic electro-optical distance meters, range imagers based on the tof measurement can be distinguished into modulation and pulse based methods. Modulation based tof rangefinders measure the phase difference between a modulated light source and the reflected wave. Pulse based rangefinders measure the round-trip time of a single burst of light. Unlike stereoscopic multi-camera systems, range cameras deliver depth information in real-time without the necessity of stereo-matching. Unlike laser scanners, depth is determined for the whole scene simultaneously.

Two European companies, PMD-Technologies and MESA Imaging, offer compact all-solid-state modulation based range cameras based on solid state sensors with CMOS technology and LED arrays (Lange 2000, Oggier et al. 2003). The LEDs (Fig. 5.8) emit light in the near infrared spectrum with about 1 W illumination power, sinus-modulated at a carrier frequency of typically 20 MHz. This light is diffusely reflected by the scene and imaged by the sensor (Fig. 5.9). Every sensor pixel, called lock-in pixel (Büttgen et al. 2005), is able to demodulate the signal and to determine its phase by sampling the returning light amplitude at a very high temporal resolution using photon-mixing devices (Fig. 5.9). The reconstructed phase is proportional to the time of flight and thus to the distance of the reflecting object. The signal-to-noise-ratio (and thus the distance measurement precision) can be improved by a summation of the sampling over a larger number of modulation periods (Büttgen et al. 2005). In addition, the pixel grey value is obtained from the offset of the reconstructed signal. This way, a greyscale image and a depth image are obtained simultaneously for each pixel. The cameras can deliver Cartesian coordinates (X,Y,Z) directly via a USB interface. Current prototypes are limited to a sensor format of up to 176 × 144 pixels with frame rates up to 50 fps.

The phase measurement principle leads to ambiguities at longer ranges, if only one modulation frequency is used. With a modulation frequency of 20 MHz one obtains a non-ambiguity range of 7.5 m. Using lower modulation frequencies and higher illumination

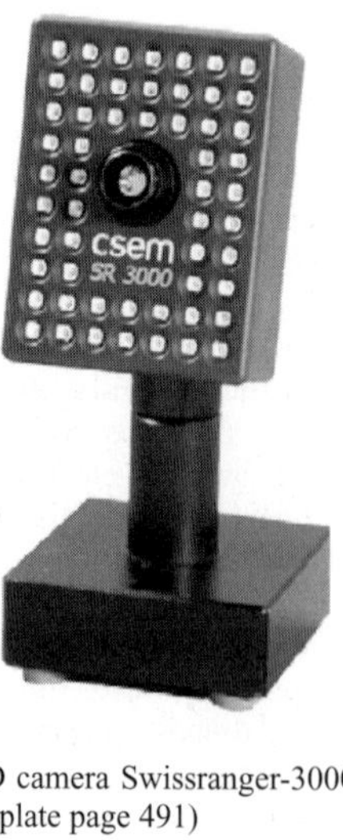

Figure 5.8. 3-D camera Swissranger-3000 (MESA Imaging). (see colour plate page 491)

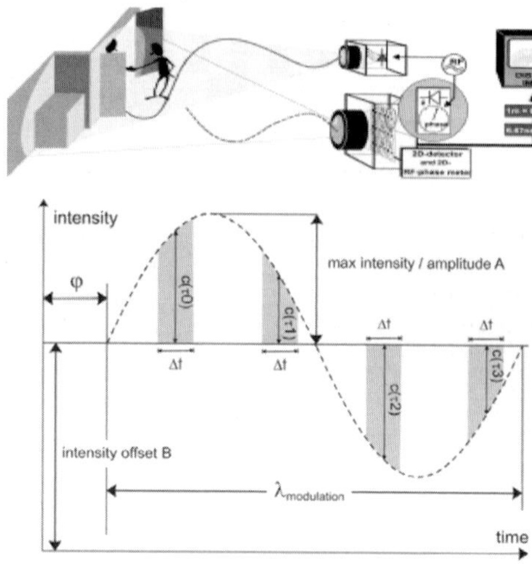

Figure 5.9. Range imaging configuration (Lange, 2000) and pixel phase demodulation (Kahlmann and Ingensand 2005). (see colour plate page 491)

power, longer ranges can be achieved. As a rule-of-thumb, the depth coordinate measurement accuracy is specified as 0.1% of the range, translating into a few millimetres to a few centimetres for typical operation ranges. Further factors influencing the distance measurement precision are surface reflectance properties, the frame rate and the pixel position. Detailed studies on the accuracy potential of a 3-D camera have been shown by Kahlmann & Ingensand (2005). Besides the typical errors of imaging sensors (interior orientation, lens distortion), these studies unveiled a number of additional error sources connected with the distance measurement part. These errors include periodic non-linearities in the phase-distance relation, external and internal temperature-dependencies, fixed pattern noise, object-dependent multi-path effects as well as some unexplained behaviour. Generally, these errors could only partly be eliminated by a calibration; some time- and/or scene-dependent effects have not be modelled satisfactorily so far.

A pulse based tof range imaging technique is shown by Niclass et al. (2005). While the PMD and MESA cameras determine tof indirectly by phase measurement, their principle is based on direct time-of-flight measurement. The scene is illuminated by a cone of pulsed laser light. The reflected light is captured by a 32 × 32 pixel CMOS imager consisting of single photon avalanche diode detectors operating in Geiger mode. Time-of-flight is measured by a CMOS time discriminator based on a time-to-digital converter. The optical detectors enable picosecond time discrimination, corresponding to a depth resolution in the order of a millimetre. The sensitivity of single photon avalanche diode detectors allows the use of inexpensive laser sources with milliwatt power to achieve a maximum range of several metres under strict eye safety operation. The experimental implementation shown by Niclass et al. (2005) uses a 100 mW laser diode and time discriminators with 120 ps resolution. Millimetre resolution is achieved by averaging over up to 10,000 shots. The implementation is also still limited by the fact that depth measurements are performed serially due to the implemented readout circuitry. Skews caused by the serial readout are compensated by referring tof measurements to a reference plane in the background of the scene.

The main advantage of pulse based tof measurement methods over modulated type tof sensors is the lack of ambiguities caused by multiples of the carrier wavelength. Basically, this allows for range of operation distances between a few metres and several kilometres, which can only be achieved by using complex sequential modulation schemes in modulation based sensors.

Despite the error sources of current prototypes and their low sensor format, tof based range cameras form a very interesting option for future use in photogrammetry due to their characteristics as a mono-sensorial 3-D measurement system providing range image sequences directly without any computer processing. Potential application fields could be robotics, automotive navigation and security (vehicle seat occupancy—Goktürk & Rafii 2005), surveillance or human motion analysis (Westfeld 2007).

5.6 COMPARISON OF 3-D TECHNIQUES

Multi-ocular stereoscopic systems, 3-D cameras and triangulation sensors compete in 3-D scene information determination. However, each of the techniques has its strengths and weaknesses, allowing for a clear definition of application field:

- Single camera sequential image acquisition: Determination of discrete 3-D object coordinates in static applications.
- Multi-ocular stereoscopic systems: Determination of discrete 3-D object coordinates in dynamic applications.
- Triangulation sensors: Determination of surface models in static applications or quasi-static deformation measurement.
- 3-D cameras: Determination of surface models in dynamic applications.

5.7 OMNI-DIRECTIONAL VISION SYSTEMS

Besides conventional central perspective digital cameras and laser scanners, unconventional sensor types are becoming popular in close range photogrammetry. Many of these can be summarized in the category of omni-directional vision systems. Omni-directional vision can be obtained by camera bundles, fisheye lenses, hyperbolic mirrors or rotating linear array sensors.

5.7.1 *Panoramic cameras*

Panoramic images can basically be obtained from any amateur digital camera by applying shareware image stitching tools. Beyond this, high resolution digital terrestrial panoramic cameras are offered by several manufacturers. Digital panoramic cameras based on a rotating linear array sensor (Fig. 5.10) capture a 360° horizontal field of view. This can be rather useful for recording landscapes, indoor scenes or city squares in architectural or industrial tasks, such as cultural heritage documentation, as-built documentation or virtual reality data capture.

The vertical opening angle is determined by the length of the sensor and the focal length of the lens. The vertical resolution of the cameras is given by the number of pixels of the sensor, which may exceed 10,000 pixels. The horizontal resolution is also determined by the focal length of the lens and may reach up to 100,000 pixels. This way, a true RGB image format of 1 Gigapixel can be achieved, going clearly beyond the potential of any area sensor.

Using a camera in photogrammetric applications requires a geometric model of the camera. In the case of rotating linear array panoramic cameras, this is the model of cylinder projection. Advanced cylinder projection models, including camera self-calibration

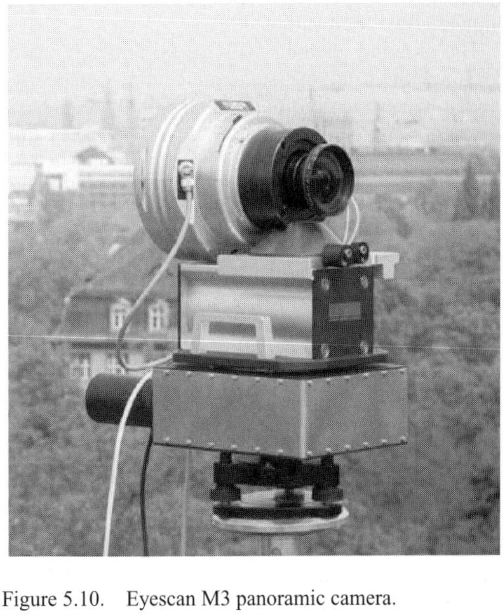

Figure 5.10. Eyescan M3 panoramic camera.

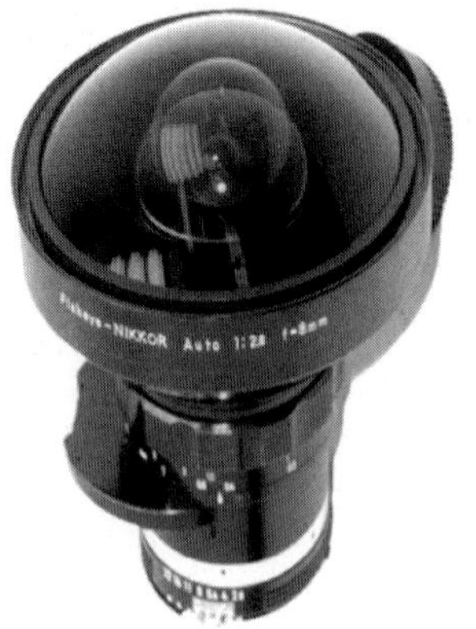

Figure 5.11. Fisheye lens (Nikkor 8 mm).

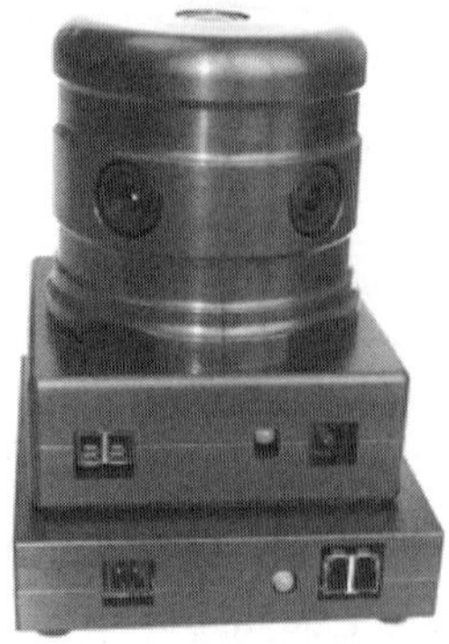

Figure 5.12. Six-camera omni-directional system (pointgrey Ladybug).

parameters allowing for a calibration of a panoramic camera in the 0.1 pixel range, have been presented by Schneider & Maas (2003) and by Parian & Grün (2003).

5.7.2 *Fisheye lenses*

Fisheye lenses (Fig. 5.11) with an opening angle of 180° or more are often used for visualization tasks such as surveillance, the documentation of ceiling frescos in historical buildings or internet presentations of building interiors. Beyond these pure visualization tasks, fisheye lenses may be an interesting tool for photogrammetric measurement systems. Fisheye lenses are for instance being used in mobile mapping systems (van den Heuvel et al. 2006) or for solar radiation analysis (Schwalbe 2005).

The imaging geometry of fisheye lenses deviates considerably from the standard central perspective model. Fisheye lenses are often modelled on the basis of an equi-angular camera model (Ray 1994). A fundamental geometric model for photogrammetric handling of hemispherical fisheye lens images based on an equi-angular camera model is shown in Schwalbe (2005). With a set of additional parameters, the model allows for fisheye image measurements at 0.1 pixel accuracy level.

Full-spherical $360° \times 180°$ imaging can be achieved by the combination of a rotating linear array panoramic camera and a 180° fisheye lens. A mathematical model for handling full-spherical images in photogrammetric applications is shown in Schwalbe & Schneider (2005).

5.7.3 *Poly-dioptric solutions*

Poly-dioptric spherical imaging systems consist of multiple cameras arranged in a way that they cover a full-spherical field of view. Full-spherical imaging can for instance be achieved by a back-to-back configuration of two fisheye cameras. Using standard cameras and lenses, a full-spherical vision system will consist of a larger number of cameras (Fig. 5.12).

5.7.4 *Catadioptric systems*

Catadioptric hemispherical imaging systems consist of a camera looking at a convex mirror. Using a hyperbolic mirror, a hemispherical image can be obtained, which is only confined by a blind spot caused by the camera occlusion. Catadioptric systems can be realized at relatively low cost, but will usually also show a relatively low resolution. They are, for instance, popular in robotics applications (Fig. 5.13). A mathematical model for handling catadioptric systems in measurement applications can be found in Baker & Nayar (1998).

5.8 SMART CAMERAS

Smart cameras offer an interesting option mainly for machine vision tasks such as quality assurance on an assembly line, web inspection, active triangulation or surveillance. A Smart camera consists of a sensor, which is closely integrated with a processor in the camera (e.g. Teschler 2007). CMOS technology even allows simple image processing routines to be performed on the imaging sensor, allowing for extremely fast and parallel image processing. Generally, smart cameras can be considered stand-alone machine vision systems without the necessity of a computer connection. They deliver simple real-time image measurements, which can be used in a process control environment.

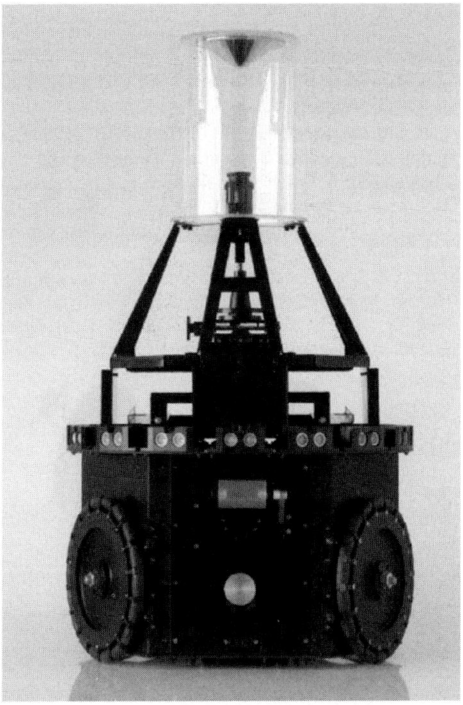

Figure 5.13. RoboCup robot with catadioptric camera (www.idmind.pt).

REFERENCES

Altschuler, M., Altschuler, B. & Taboada, J. 1979. Measuring surfaces spacecoded by a laserprojected dot matrix. *Imaging Applications for Automated Industrial Inspection and Assembly, SPIE Proceedings Series* Vol. 182.

Amiri Parian, J. & Grün, A. 2003. A sensor model for panoramic cameras. In Grün & Kahmen (eds.), *Optical 3D Measurement Techniques VI Volume II*: 130–141.

Ayache, N. & Lustman, F. 1987. Fast and reliable passive trinocular stereovision. *First International Conference on Computer Vision*: 422–427. London: Computer Society Press.

Baker, S. & Nayar, S. 1998. A Theory of Single-Viewpoint Catadioptric Image Formation. *International Journal of Computer Vision* 35, Issue 2: 175–196.

Bayer, B. 1975. Color imaging array. *United States Patent 3,971,065.*

Büttgen, B., Oggier, T., Lehmann, M., Kaufmann, R. & Lustenberger, F. 2005. CCD/CMOS Lock-In Pixel for Range Imaging: Challenges, Limitations and State-of-the-Art. *Proceedings of the First ETH Zürich Range Imaging Research Day.*

Etoh, T., Poggemann, D., Kreider, G., Mutoh, H., Theuwissen, A., Ruckelshausen, A., Kondo, Y., Maruno, H., Takubo, K., Soya, H., Takehara, K., Okinaka, T. & Takano, Y. 2003. An Image Sensor Which Captures 100 Consecutive Frames at 1000000 Frames/s. *IEEE Transactions on Electronic Devices* 50/1: 144–151.

Gokturk, B. & Rafii, A. 2005. An occupant classification system—Eigen shapes or knowledge-based features. *IEEE Computer Society Conference on Computer Vision and Pattern Recognition (CVPR'05)—Workshops*: 57–64.

van den Heuvel, F.A., Verwaal, R. & Beers, B. 2006. Calibration of fisheye camera systems and the reduction of chromatic aberration. *ISPRS Commission V Symposium: Image Engineering and Vision Metrology.*

Kahlmann, T. & Ingensand, H. 2005. Investigation of the 3D-Range-Imaging-Camera SwissRangerTM. In A. Grün & H. Kahme, (eds.), *Optical 3-D Measurement Techniques VII.*

Lange, R. 2000. 3D *Time-of-Flight Distance Measurement with Custom Solid-State Image Sensors in CMOS/ CCD Technology. Dissertation, University of Siegen, Germany.*

Lenz, R. 1989. Digitale Kamera mit CCD-Flächensensor und programmierbarer Auflösung bis zu 2994 × 2320 Bildpunkten pro Farbkanal. In H. Burkhardt, K. Höhne & B. Neumann (eds.), *Mustererkennung 1989*: 411–415. Springer-Verlag.

Lenz, R. & Lenz, U. 1993. New developments in high resolution image acquisition with CCD area sensors. In A. Grün & H. Kahmen (eds), *Optical 3-D Measurement Techniques II* :53–62. Karlsruhe: Wichmann Verlag.

Long, B. 2005. Digital SLR glossary. *Macworld* July 2005.

Luhmann, T. & Wester-Ebbinghaus, W. 1986. Rolleimetric RS—a new digital image processing system. *IAPRS* 26, Part 2.

Maas, H.-G. 1991. Digital Photogrammetry for determination of tracer particle coordinates in turbulent flow research. *Photogrammetric Engineering & Remote Sensing* 57, No. 12: 1593–1597.

Maas, H.-G. & Niederöst, M. 1997. The accuracy potential of large format stillvideo cameras. In S. El Hakim (ed.), *Videometrics V SPIE Proceedings Series Vol. 3174.*

Maas, H.-G. & Hampel, U. 2006. Photogrammetric Techniques in Civil Engineering Material Testing and Structure Monitoring. *Photogrammetric Engineering and Remote Sensing* 72, 1: 39–45.

Malz, R. 1989. Adaptive light encoding for 3-D-sensing with maximum measurement efficiency. In Burkhardt, H., Höhne, K. & Neumann, B. (eds.), *Mustererkennung 1989.* Springer-Verlag.

Niclass, C., Besse, P. & Charbon, E. 2005. Arrays of Single Photon Avalanche Diodes in CMOS Technology: Picosecond Timing Resolution for Range Imaging. *Proceedings of the First ETH Zürich Range Imaging Research Day.*

Oggier, T., Lehmann, M., Kaufmann, R., Schweizer, M., Richter, M., Metzler, P., Lang, G., Lustenberger, F. & Blanc, N. 2003. An all-solid-state optical range camera for 3D real-time imaging with subcentimeter depth resolution (SwissRanger™). *SPIE Proceedings Series* 5249 (65): 534–545.

Putze, T. 2005. Geometric modelling and calibration of a virtual four-headed high speed camera-mirror system for 3-D motion analysis. In A. Grün & H. Kahmen (eds), *Optical 3-D Measurement Techniques VII Vol. II*: 167–174.

Ray, S.F. 1994. *Applied Photographic Optics: Lenses and Optical Systems for Photography, Film, Video and Electronic Imaging, second ed.: 559–563. Oxford: Focal Press.*

Raguse, K. & Heipke, C. 2006. Photogrammetric synchronization of image sequences. *IntArchPhRS* XXXVI, Part. 5: 254–259.
Schneider, D. & Maas, H.-G. 2003. Geometric modelling and calibration of a high resolution panoramic camera. In A. Grün & H. Kahmen (eds), *Optical 3-D Measurement Techniques VII Vol. II*: 122–129.
Schneider, D., Pötzsch, M. & Maas, H.-G. 2005. Accuracy and application potential of the 94 megapixel RGB macroscanning camera Pentacon Scan 5000. *CIPA Archives for Documentation of Cultural Heritage* XX-2005 & *ISPRS International Archives* XXXVI-5/C34: 534–537.
Schwalbe, E. 2005. Geometric Modelling and Calibration of Fisheye Lens Camera Systems. Proceedings 2nd Panoramic Photogrammetry Workshop (Eds. Reulke/Knauer), *International Archives of Photogrammetry, Remote Sensing and Spatial Information Sciences* XXXVI, Part 5/W8.
Schwalbe, E. & Schneider, D. 2005. Design and testing of mathematical models for a full-spherical camera on the basis of a rotating linear array sensor and a fisheye lens. In A. Grün & H. Kahmen (eds), *Optical 3D Measurement Techniques VII I*: 245–254.
Tecklenburg, W., Luhmann, Th. & Hastedt, H. 2001. Camera modelling with image-variant parameters and Finite Elements. In A. Grün & H. Kahmen (eds), *Optical 3-D Measurement Techniques V*.
Teschler, L. 2007. Intelligent design with smart cameras. *Machine Design* 4: 74.
Westfeld, P. 2007. Development of Approaches for 3-D Human Motion Behaviour Analysis Based on Range Imaging Data. In A. Grün & H. Kahmen (eds), *Optical 3-D Measurement Techniques*.

Advances in Photogrammetry, Remote Sensing and Spatial Information Sciences: 2008 ISPRS Congress Book – Li, Chen & Baltsavias (eds)

CHAPTER 6

LIDAR: Airborne and terrestrial sensors

Aloysius Wehr

ABSTRACT: Based on the definition of a LIDAR system, the components used are introduced. Then the ranging units, their ranging principles and the optical-mechanical scanning devices that are applied in today's commercially available LIDARS are presented and their performances are discussed. The ranging principles which are optimal are worked out with regard to airborne and terrestrial laser scanning. A market overview of today's laser scanners is accompanied by an analysis concerning the hardware developments carried out during the last four years. Finally, possible next generation ranging principles are discussed.

Keywords: LIDAR, radar scanning, types of scanning

6.1 INTRODUCTION

Four years ago, radar scanning was already a well established surveying tool and the last four years have been used to develop scanners with advanced range capabilities and new measurement features, e.g. full waveform detection. At that time, however, terrestrial laser scanning was primarily applied for very sophisticated surveying tasks, which could not be realized by conventional surveying tools, because system costs were very high and the laser scanning data processing required specially trained personnel. Due to this fact laser scanning was not widely used and could only be afforded for projects having a large budget for surveying tasks. In the past four years, terrestrial laser scanning has matured to a well established measuring tool due to hardware and software improvements and reduction in prime costs. Therefore, more and more terrestrial laser scanning is applied, e.g. in surveying industrial objects in design and production, in architectural surveying, in cultural heritage surveying, in archaeology and in the forensic field. Due to the wide application possibilities, several companies have come to the market offering terrestrial laser scanning hardware and software. However, looking more closely at the hardware, it becomes very clear that the key system components of the laser scanners are developed and built by only a few manufacturers.

Observing the developments in laser scanning during the past four years, it has become obvious that additional sensors have been integrated, advanced calibration procedures and algorithms have been developed and more user friendly post processing software has been made available. Therefore, in the following paragraphs, first LIDARs (Light Detection and Ranging) are discussed as a system. Then functioning and performances of key units are explained, tables of commercial airborne and terrestrial LIDARs are presented and the application possibilities are discussed.

6.2 SYSTEM ASPECTS

In the beginning of laser scanning, the laser surveying device was regarded as an autonomous item. However, laser scanning has become widely accepted as a surveying device by the advent of user friendly online and post processing software. Furthermore, today LIDARs are equipped with additional sensors that determine the position and the orientation of the laser scanner and with additional imaging units such as multispectral scanners and RGB still or video cameras. Regarding these facts, the author prefers the expression LIDAR system because to obtain valid laser scanning surveying results recording and post processing software is required.

In Table 6.1, the system components of airborne and terrestrial LIDAR systems are compiled. The required units and software for controlling, synchronizing and recording are assumed to be integral parts of the laser scanner. Table 6.1 shows that for terrestrial LIDAR

Table 6.1. System components of airborne and terrestrial LIDAR systems.

	Terrestrial LIDAR			Airborne LIDAR
	static		kinematic	
	Indoor	outdoor		
Components	LIDAR with 2D-scan camera	LIDAR with 2D-scan camera DGPS DGPS	LIDAR with 2D-scan camera video POS POS and optional DMI	LIDARwith 1D-Scan camera video POS POS
Software	matching, filtering, modelling	position and orientation determination, geo-coding, matching, filtering, modelling	position and orientation determination, geo-coding, matching, filtering, modelling	position and orientation determination, geo-coding, matching, filtering, modelling, analysing

systems one must distinguish between static and kinematic surveys. The table makes clear that indoor LIDAR systems comprise only the laser scanner and the processing software in the simplest configuration. The most complex systems are the airborne and kinematic terrestrial LIDAR systems, because here one has to deal with a moving measurement platform. Its instantaneous position and orientation has to be known exactly for each laser measurement point to obtain valid geo-coded 3-D coordinates. The position and orientation is determined by a Position and Orientation System (POS) consisting of a DPGS receiver, an Inertial Measurement Unit (IMU) and a control unit, which controls the DGPS receiver and the IMU and stores the DGPS and IMU data. Normally the POS works independently from the LIDAR. Therefore, LIDAR and POS must be synchronized. This is realized on the basis of the PPS-signal (Pulse Per Second) of the GPS receiver. In order to achieve the required position accuracy of better than 10 cm in the airborne case, reference GPS receivers must be set up on the ground. These stations are used to build up DGPS with carrier phase measurements. The processing of the surveyed data is carried out in the following order. First DGPS positions are computed. This result is used in a separate program that computes an integrated position and orientation solution using the IMU data and the processed DGPS data. Now, the orientation angles in roll, pitch and yaw and the position in three dimensions are available in a rate of, e.g. 100 Hz. These POS data are input for another separate software module, which geo-references the laser scanner data. As the laser scanners have much higher measurement rates—up to 80 kHz—than the POS, interpolation of the POS data is required. Furthermore, the accuracy of the geo-referenced LIDAR data is very dependent on the exact calibration, which is mainly determined by the external orientation differences between IMU and laser scanner. The previously described processing steps are also valid for kinematic terrestrial LIDARs. However, here commercial software is available that permits a so-called tightly coupled solution for the POS data. This means DGPS data and inertial measurement data are processed in one program and, dependent on data quality, either DGPS data support the IMU data or the IMU data support the DGPS carrier phase solution.

Although, this short introduction makes clear that the performance of a LIDAR system is not anymore determined by the performance of the LIDAR sensor itself, the key components of the LIDAR sensor will be explained and their performance will be discussed.

6.3 LIDAR COMPONENTS

According to Figure 6.1, LIDARs are built up of an opto-electronic ranging unit and a scanning device that deflects both the transmitted and the received beam in either one or two dimensions. For a moving sensor platform, one dimensional scanning is sufficient, because the second dimension is realized by the movement of the platform. Stationary laser scanners need scanning in two dimensions to perform 3-D digitization of objects, interiors and outdoor scenes. The ranging unit measures the slant range between the origin of the ranging device and the point of impact of the optical beam on the target's surface. If the instantaneous direction of the beam is known—meaning that the scanning angles are known—3-D coordinates can be computed. Therefore, in the simplest scanning setup (Table 6.1), using only the laser scanner, the slant range and the scanning angles accuracy determine the total accuracy of the system. Ranging and scanning devices will be explained in the following paragraphs.

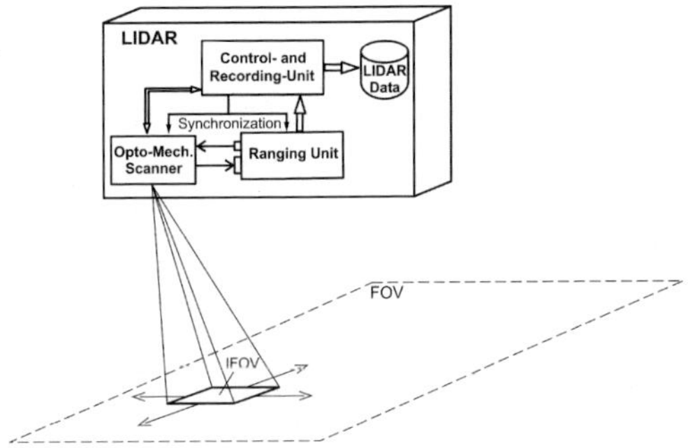

Figure 6.1. LIDAR components.

6.3.1 *Ranging unit*

Figure 6.1 shows that the ranging unit setup corresponds with a monopulse radar setup. This means transmitter and receiver are located at the same position. A laser is typically used as a signal source, because its beam can be highly collimated and the transmitted energy is high enough for precise measurements over longer distances with high measurement rates. At the receiver, the conversion of the optical radiation into an electrical current is realized by sensitive avalanche photodiodes. Most commercial laser scanners operate with ranging units that work by the time of flight principle (Fig. 6.2). The laser emits a very short light pulse of e.g. 5 to 10 ns. After travelling two times the distance R between the

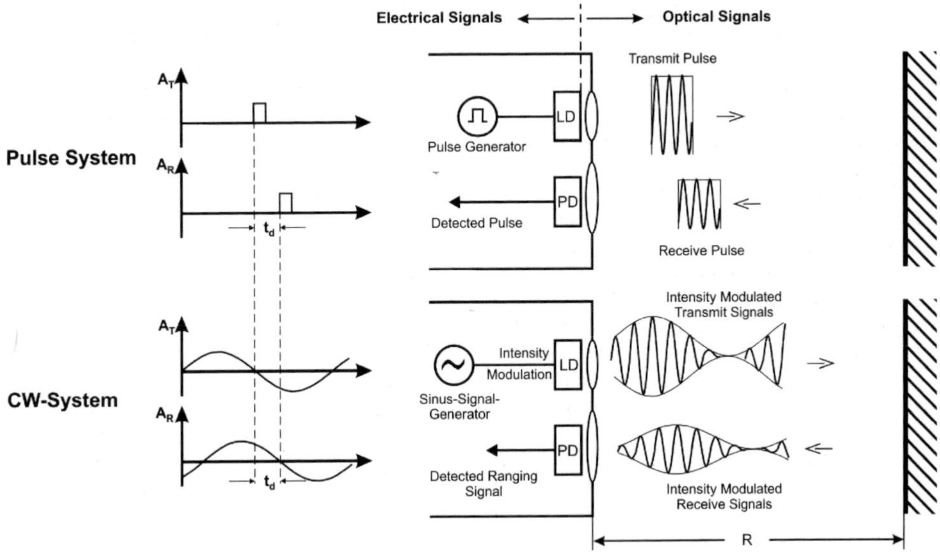

Figure 6.2. Time-of-flight ranging.

transmitting and receiving aperture, the laser pulse is detected, delayed by the travelling time t_d:

$$t_d = \frac{2R}{c} \Leftrightarrow R = \frac{c}{2} \cdot t_d \tag{1}$$

where c is the speed of light. Using this measurement technique, ranging accuracy down to less than 1 cm is possible. For higher accuracy, the phase difference between the transmitted and received signal is measured (Fig. 6.2). Here the intensity of the continuously emitted laser light (Continuous Wave cw) is modulated for example by a sinusoidal signal with frequency ω. In this case the signal is also received, dependent on the measured range and delayed by time t_d. As the phase difference θ is directly proportional to the travelling time t_d with

$$t_d = \frac{\theta}{\omega}, \tag{2}$$

the range R can be calculated by using Equations (1) and (2):

$$R = \frac{c}{2} \cdot \frac{1}{\omega} \cdot \theta \tag{3}$$

As the phase difference can only be measured between 0° to 360°, the maximum unambiguous range R_{unamb} is

$$R_{unamb} = c \cdot \frac{\pi}{\omega} \tag{4}$$

Assuming stable oscillators, so that the signal frequency ω does not cause additional errors and assuming that the speed of light is error free, ranging accuracy is given by:

$$\sigma_{Rpulse} = \frac{c}{2} \cdot \sigma_{t_d} \quad \text{pulse system} \tag{5}$$

$$\sigma_{Rphase} = \frac{c}{2} \cdot \frac{1}{\omega}\sigma_\theta \quad \text{phase difference} \tag{6}$$

Comparing Equation (5) and Equation (6) makes clear that the pulse ranging accuracy σ_{Rpulse} is directly proportional to the accuracy of the travelling time interval measurement σ_{td}. In this case, the ranging accuracy can only be improved by improving the travelling time interval measurement. For the phase difference measuring case, however, an additional physical parameter, the frequency ω of the ranging signal, can be used to improve the ranging accuracy without changing the accuracy of the phase difference measurement. Better accuracy can be achieved easily by increasing the frequency ω of the ranging signal. As the maximum measurable phase difference is 2π, the maximum unambiguous range R_{max} is limited to

$$R_{max} = \frac{c}{2} \cdot \frac{2\pi}{\omega} = \frac{\lambda}{2}, \tag{7}$$

if λ is the wavelength of the ranging signal. This means ranging over distances longer than R_{max} requires the use of additional ranging signals with longer wavelengths. The application of this ranging technique is also known as side tone ranging. LIDARs using the phase difference measurement typically work with two to three ranging tones. The tone with the shortest wavelength determines the accuracy of the ranging unit. Assuming an accuracy of the phase difference measurement of e.g. 0.5 deg and a wavelength of 1 m, a slant range accuracy of 0.7 mm is possible. This example shows that such systems offer very high accuracy but are limited in their maximum range. Therefore, phase difference measurements are only applied in terrestrial surveying systems working over ranges up to 200 m, whereas pulse ranging can be found in LIDARs for close range and airborne applications as well.

The so-called target resolution ΔR_{tar} is another physical parameter that has to be regarded in pulse ranging systems. If the laser beam illuminates more than one target within its spot and the targets are located in different ranges, they must have a range difference greater than ΔR_{tar} to be resolved as separate targets. Figure 6.3 shows the case for two targets. It makes clear that ΔR_{tar} is directly proportional to the pulse width T_p of the laser:

$$\Delta R_{tar} = \frac{c}{2} \cdot T_p \tag{8}$$

This means distinguishing targets with a range difference of 0.5 m requires a pulse width T_p shorter than 3 ns. By using the phase difference measurement principle, it is impossible to detect targets in different ranging intervals, because the resulting ranging signal phase out of the optical detector is the mean phase of all signal phases of scatterers within the illuminated

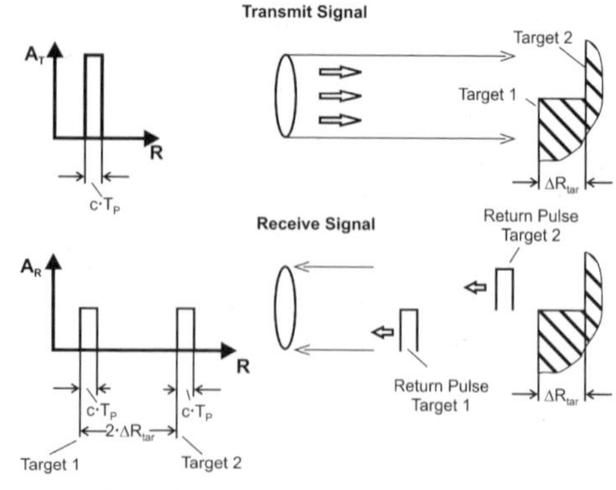

Figure 6.3. Resolving multiple targets.

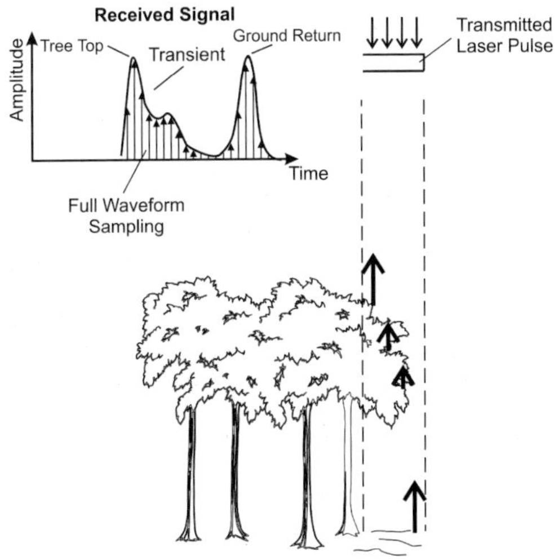

Figure 6.4. Full waveform ranging.

laser spot. The different signal phases of the backscatterers are weighted by their intensity.

As pulse ranging LIDARs have the potential to separate targets with regard to their range and the limiting factor for resolving the targets is the pulse width, systems were developed which transmit very short laser pulses, typically shorter than 10 ns. To resolve as many targets as possible, first the transient of the total return signal of each laser pulse is sampled by very fast analogue to digital conversion circuits and stored on hard disk. This detection process is known as full waveform detection. In a second processing step, the actual ranging is carried out and targets in the different range intervals are distinguished. Figure 6.4 shows the principle of a full waveform return signal of a laser beam hitting a tree and the ground.

6.3.2 *Scanner*

The ranging unit of a LIDAR performs only ranging measurements to the instantaneously illuminated spot. Therefore, the laser beam has to be scanned to achieve surface covering surveys. For terrestrial laser scanning, a two-dimensional scanning pattern is required (Fig. 6.1) whereas a one-dimensional scan is sufficient if the LIDAR is mounted on a moving platform, for example a car, helicopter or airplane. Most commercial scanning devices are based on the physical principle of reflection.

6.3.2.1 *Scanner mechanisms of non-moving terrestrial laser scanners*

Non-moving terrestrial laser scanners need a two-dimensional scanning pattern to achieve a range image. A two-dimensional scan can be realized by, for example, two galvanometer scanners with rotation axes mounted orthogonally to each other (Fig. 6.5).

The advantage of this setup is that the drives can be programmed independently and certain points can be randomly addressed within the field of view of the scanners. The very limited speed of the scanners—especially if larger apertures are required—is the main disadvantage of this concept. Therefore, most manufacturers of terrestrial laser scanners prefer the use of scanning mechanisms that perform a continuous non-accelerated movement of the optical deflecting devices. In this case, in one direction the scanning pattern is typically a line scan realized either by a multi-facetted mirror (Fig. 6.6) or a mirror mounted under 45 degrees at the front face of a shaft (Fig. 6.7). The deflection of the transmitted and the received beam in the second dimension is accomplished by an azimuth rotation of the line scanner.

As the rotation in azimuth requires a much slower angular movement than those of the line scan, the mass which has to be accelerated can be much higher. Therefore, many LIDAR manufacturers prefer to move the complete line scanner including the opto-electronic ranging unit. Combining the shaft scanner with an azimuth scanner, an almost spherical coverage is possible and in the nadir the point density becomes the highest (Fig. 6.8). Combining the multi-facet scanner with an azimuth scanning device, a hemispherical

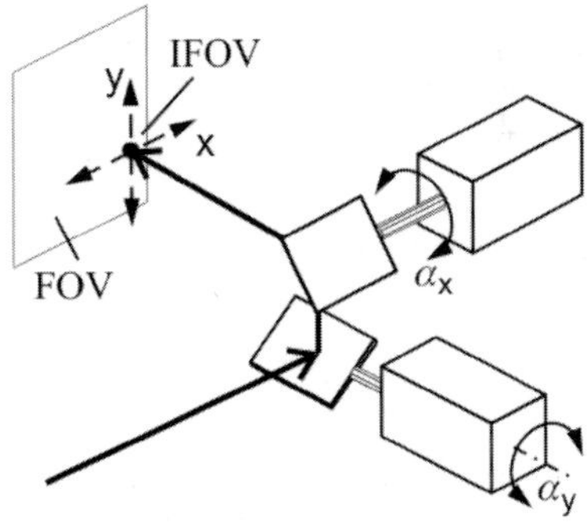

Figure 6.5. Galvanometer scanner.

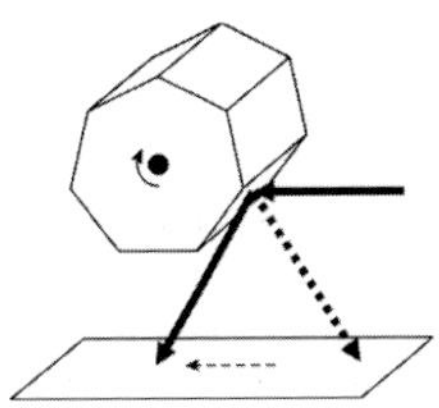

Figure 6.6. Polygon scanner.

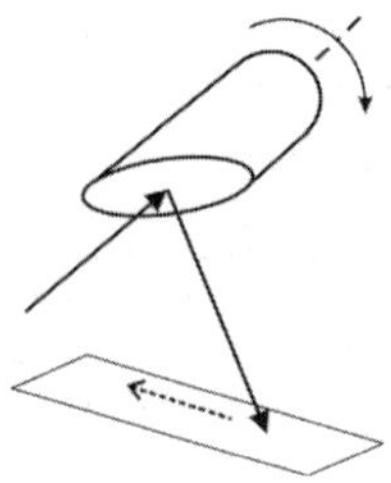

Figure 6.7. Shaft scanner.

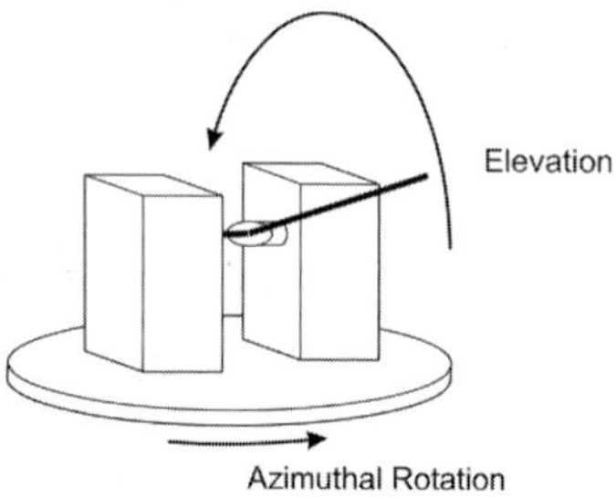

Figure 6.8. Shaft scanner with azimuth rotation.

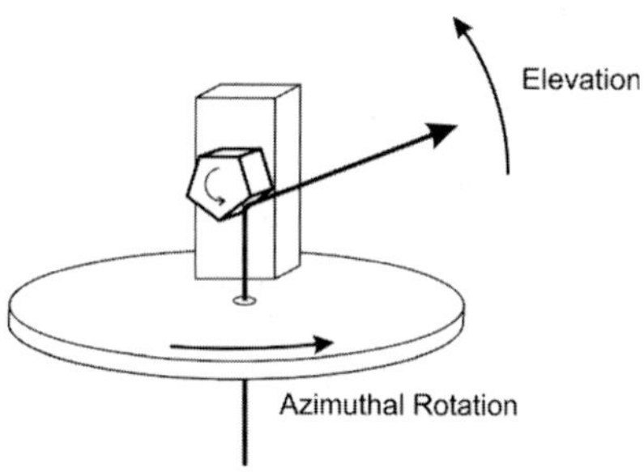

Figure 6.9. Polygon scanner with azimuth rotation.

coverage can hardly be achieved (Fig. 6.9). This kind of scanner setup is used in LIDARs of the company Riegl whereas the shaft scanner with additional azimuth rotation is used typically in Zoller & Froehlich and Faro scanners.

6.3.2.2 *Airborne scanning mechanisms*
In airborne applications, only a one-dimensional scan is required. The second dimension is accomplished by the forward motion of the sensor carrying platform, which normally is an airplane. Here mirror scanners are applied in most cases. Their principles are shown

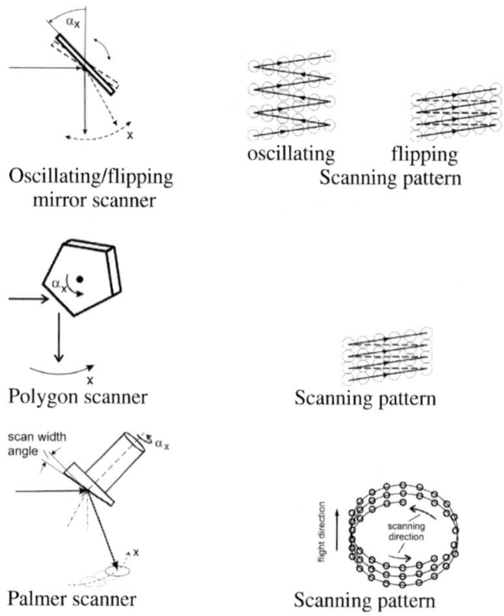

Figure 6.10. Airborne mirror scanners.

in Figure 6.10. They are built up by a multi-facetted mirror scanner (polygon scanner), by an oscillating/flipping mirror or a nutating mirror (Palmer scanner). The oscillating mirror causes a zig-zag scanning pattern if the ranging is carried out during forward and backward motion of the mirror, or a z-shape pattern if ranging is only performed in one moving direction (Fig. 6.10). The nutating mirror at the front side of the shaft, which is tilted 45 degrees against the laser beam, causes a moving elliptical scanning pattern on the ground during flight.

There is one airborne LIDAR system built by TopoSys that uses a so-called fibre scanner. The principle of the setup is shown in Figure 6.11. Fibres are mounted in the focal plane as line arrays in the transmitter and the receiver. As one detector and one laser are used only, the transmitted and received light has to be switched sequentially through the fibres. This is also realized by using a mirror scanner. Here the mirror is mounted on the shaft in such a way that when the shaft rotates the light is imaged on a circle and the instantaneous field of views on this circle is couple either in or out of the optical light. As the apertures of the fibres are very small, the scanning mirror can be designed small too, and therefore very high scanning speeds of up to 415 scans per second are possible. Due to the very high scanning rate, the sampling in the flight direction is much higher than in the across flight direction, because the fibres have finite spacings. Therefore

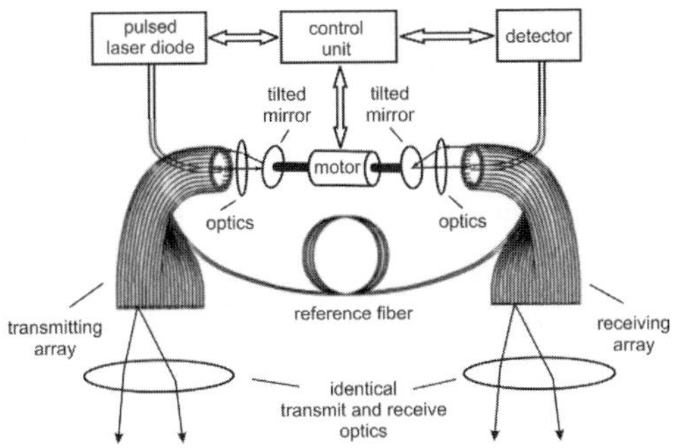

Figure 6.11. Fibre scanner of Harrier (TopoSys).

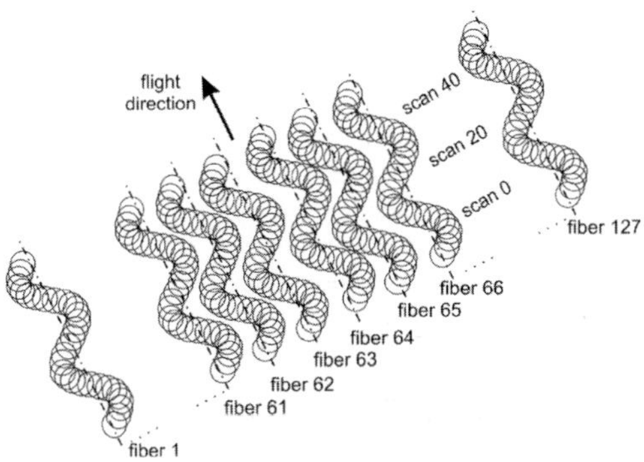

Figure 6.12. Scanning pattern of fibre scanner (TopoSys).

the fibre arrays in the transmitting and receiving path are moved in a swinging mode. This causes a swinging scanning pattern on the ground depicted in Figure 6.12, which achieves a much better coverage. Another advantage besides the high scanning rate is that optical transmit and receive paths are identical and there are no moving parts in front of the optics. All other scanners need an external window, which means additional signal losses. The main drawback is the manufacturing of the fibre scanner, which today still requires much work.

6.3.2.3 *Scanning mechanisms of kinematic terrestrial LIDARs*

For moving terrestrial laser scanners, very similar requirements as for airborne laser scanners can be identified. That is to say, high sampling and scanning rates are demanded in order to achieve high point densities even if the sensor is carried by a moving car or a fast moving train. In most cases, a one-dimensional scan is sufficient, because as in the airborne case the second dimension is given by the forward motion of the sensor carrying platform. Typically shaft or multi-facetted mirror scanners are applied. The shaft scanner is especially useful for surveying the clearance between the engine and the surrounding objects beside and above a railroad track. Here very fast scanning is required because the train should carry out the measurement with a typical speed of up to 100 km/h (Blug 2005). A very advanced system was developed by Fraunhofer Institute for Physical Measurement Techniques (IPM) in Freiburg. (Fig. 6.13).

By using a double sided mirror mounted in a rotating hollow shaft and two opto-electronic transmitting and receiving units, the scanning speed is doubled. The shaft can be rotated up to 300 turns per seconds. This means 600 profiles per second with 3333 measurements for a projected measurement rate of 1 MHz for each ranging unit are possible.

The multi-facetted mirror scanner is applied in the Riegl LIDAR. The company IGI mbH, Germany, together with 3-D Laser Mapping Ltd., United Kingdom, built up the Street Mapper by integrating several Riegl scanners and the IGI's TERRAcontrol POS on a carrier that can be mounted on top of a car. As the scanners look in different directions (Fig. 6.14), the surroundings of the car are surveyed and can be modelled off-line in three dimensions. Several scanners are necessary, because these scanners perform only a line scan with a certain field of view, e.g. 80 degrees.

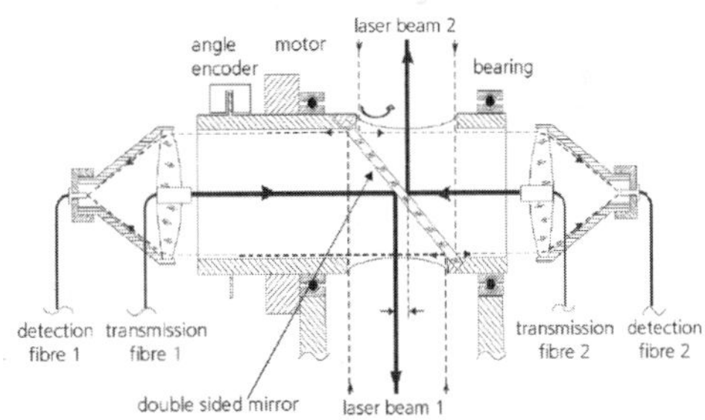

Figure 6.13. Twin head scanner (Fraunhofer Institute; IPM).

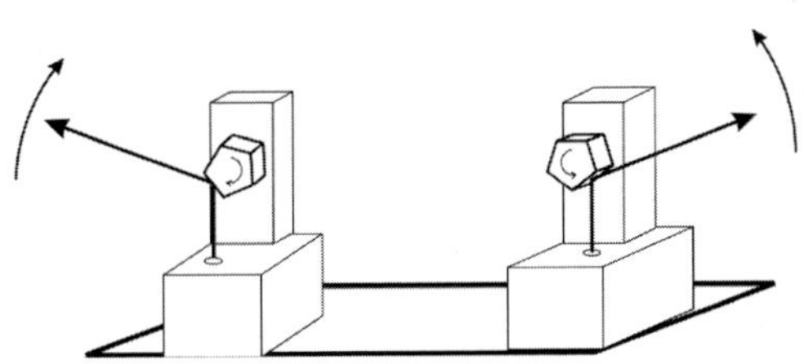

Figure 6.14. Carrier with multi-facetted mirror scanner for kinematic surveying.

6.3.3 *Synchronization and co-registration*

Synchronization is an issue if at least two systems measure independently time dependent physical parameters. This is, for example, the case if the LIDAR surveying platform is moving. Here each surveyed laser measurement point is sampled at a certain time point and at that instant the platform's orientation and position has to be known in order to calculate in post processing geo-referenced laser measurement points. As the position and orientation is measured by an independent measurement system (POS) that has its own time base, LIDAR and POS have to be synchronized. As the position normally is determined by GPS signals, the one pulse per second of GPS and the associated GPS time is used for synchronization of POS and LIDAR (Wehr & Lohr 1999). The synchronization must be extended to all sensors of the platform if, for example, multi-spectral line scanners and cameras are used.

In the case of non-moving platforms and using a still camera as an additional sensor, the associated instantaneous pixel of the still camera has to be known for each laser measurement. The process to carry out these tasks is known as co-registration. It requires mainly the determination of both the external orientations of the LIDAR as well as of the camera.

6.4 LIDAR SYSTEMS

In Section 3 key components of LIDARs were discussed. The different items are selected and integrated depending on the application so that an optimum LIDAR system can be built up. In this section, compilations of typical commercial scanners are presented. In the following it is distinguished between terrestrial laser scanners and airborne laser scanners.

6.4.1 *Terrestrial laser scanners*

A good compilation of terrestrial laser scanners can be found on the web page of GIM International (Lemmens 2007b). An excerpt of the most advanced systems is compiled in Table 6.2 and Table 6.3. Table 6.2 presents laser scanners working with the ranging principle of measuring the phase difference. Table 6.3 shows the parameters of pulse laser scanners. Two tables were prepared to work out better the similarities and differences between the different types of laser scanner systems.

Table 6.2 and Table 6.3 show that all systems carrying out phase difference ranging use shaft scanners, whereas in pulse systems either a rotating mirror, oscillating mirror or rotating polygon mirror scanners are applied. Very remarkable is the fact that the ranges of laser scanners of Table 6.2 are limited to 79 m, whereas most of the pulse systems exhibit range performances of more than 300 m. However, their measurement rates are much lower with a maximum of 50 kHz, while phase difference systems make possible measurement rates up to 500 kHz.

In Table 6.2, the similarities between the HDS6000 and the Imager 5006 are remarkable. Both systems show identical performance. The reason is the key components of the HDS6000 of Leica Geosystems are built by Zoller & Fröhlich. Half of the laser scanners of Table 6.2 can be equipped with an external camera to obtain 3-D RGB-images. Looking at the pulse system, most of them are equipped with an internal camera or even video. The reason for this is that intensity measurements of early pulse systems do not have the performance of the phase difference systems.

Table 6.2. Terrestrial laser scanners with phase difference ranging.

Company	3rd tech	Basis software	Faro europe	Leica geosystems	Zoller & Fröhlich
System	DeltaSphere-3000IR	Surphaser 25 HSX	Photon 80	HDS6000	Imager 5006
Wavelength [nm]	780	690	785	650, 690	650, 690
Measurement rate	17 kHz	190 kHz	120 kHz	500 kHz	500 kHz
Max. range	15 m	38 m	76 m	79 m	79 m
Slant range accuracy	7 mm @ 50 m	5 mm @ 50 m, 90% reflectivity	2.9 mm @ 25 m	6 mm @ 50 m, 18% albedo	4 mm @ 50 m, 20% reflectivity
Spot size	11.8 mm @ 15 m	3 mm @ 10 m	4 mm @ 25 m	14 mm @ 50 m	14 mm @ 50 m
Scanner	shaft scanner	shaft scanner	shaft scanner	shaft scanner	shaft scanner
Max. FOV [deg]	288 × 360	360 × 270	320 × 360	310 × 360	310 × 360
Intensity recording	8 bit	8 bit	9 bit	12 bit	16 bit
Ext. camera	y	n	y/opt. internal colour	n	y

Table 6.3. Terrestrial laser scanners with pulse ranging.

Company	Callidus precision systems	I-Site	Leica geosystems	Optech	Riegl	Trimble
System	CP 3200	4400-LR	ScanStation 2	ILRIS-3DER	LMS-Z390i	GX
Wavelength [nm]	906	905	532	1535	near infrared	532
Measurement rate	1.75 kHz	4.4 kHz	50 kHz	3.5 kHz	11 kHz	5 kHz
Max. range	32 m	700 m	300 m @ 90%	1700 m @ 80%	400 m	200 m @ 35%
Slant range accuracy	5 m	20 mm @ 50 m	4 mm @ 50 m	7 mm @ 50 m	6 mm @ 50 m	1.6 mm @ 50 m
Spot size	–	70 mm @ 50m	6 mm @ 50 m	8 mm @ 50 m	16 mm @ 50 m	3 mm? @ 50 m
Scanner	shaft scanner	scanning mirror	scanning mirror	scanning mirror	Rotating polygon	Oscill. Mirror
Max. FOV [deg]	140 × 360	80 × 360	270 × 360	40 × 360	80 × 360	60 × 360
Intensity recording	Y	10 bit	12 bit	24 bit RGB	12 bit	8 bit
Ext. camera	integr camera	integr camera	internal camera	internal camera	y	internal video

Table 6.4. Combined pulse and cw of the Callidus company.

System	CPW 8000
Wavelength	658 nm
Measurement rate	41 kHz
Max. range	80 m
Slant range accuracy	2 mm @ 30 m
scanner	shaft scanner
Max. FOV [deg]	300 × 360

Regarding both tables, it can be concluded that phase difference laser scanners are preferable in near range. Here high accuracy down to a tenth of a millimetre is possible at high scan rates. Surveys over longer ranges are best carried out with pulsed laser scanners.

Recently the company Callidus developed a laser scanner that uses in parallel the pulse ranging and the phase difference principles. This scanner offers high accuracy only possible by measuring the phase difference and the long unambiguous range of a pulse system. Its technical parameters are compiled in Table 6.4.

During the past four years, more and more terrestrial LIDARs have been applied on moving platforms such as cars and trains. Especially with the rising demand for precise town models, advanced surveying systems are required that can be realized well by LIDAR systems composed of a fast scanning LIDAR, a precise POS and a still or video camera. The POS consists of a DGPS receiver and an inertial measurement unit. To assure exact positions even during short-term GPS reception losses caused by large buildings, a Distance Measuring Indicator (DMI) is used as an additional sensor. Such surveying systems are also known as kinematic LIDAR systems.

The companies 3D Laser Mapping Ltd. and IGI mbH put a kinematic LIDAR system on the market, which is comprised of at least three laser scanners, a 12 channel GPS receiver, an inertial measurement unit offering a data rate of 256 Hz and two video cameras with 0.5 mega pixels each. The laser scanners used are standard laser scanners of the company Riegl with a range of 150 m, carrying out 10,000 measurements per second and performing 100 scans per second with a scanning angle of 80 degrees.

Other companies integrate LIDARs with shaft scanners, which carry out ranging by applying the phase difference measurement principle. Today the fastest laser scanning mechanism for terrestrial kinematic application has already been presented in Section 3.2.3. It is implemented in the clearance profiler scanner (CPS) of IPM. CPS is designed to survey the clearance profile along railroad tracks (Wölfelschneider et al. 2005, Blug et al. 2005). It achieves a measurement rate of up to 2 MHz. Therefore, it can be used on trains travelling at up to 100 km/h. Its range is limited to 20 m. Figure 6.13 shows that the optical signal is relayed via fibres to system electronics. Therefore, the twin-head scanner can be adjusted and assembled independently from the electronics. The use of fibres in the front-end relaxes the electro-magnetic interference problem that is a most important issue when applying laser scanners on trains and construction machines. Ranging is carried out by the phase difference measurement principle.

Although such a scanner has been developed for railroad applications, it is also applicable on cars for surveying towns.

6.4.2 *Airborne LIDARS*

Tables 6.5 and 6.6 present a compilation of eight commercial airborne LIDAR systems with their key technical parameters. The table is based on the market overview presented on the webside of GIM International (Lemmens 2007a). All airborne LIDARs carry out pulse ranging, because they have to perform ranging over much longer slant ranges than terrestrial laser scanners. The FLI-MAP 400 can be used only up to flying heights of 400 m. Due to the limited range, this system is primarily used on helicopters. For surveying with airplanes, LIDARS should be designed for a minimum flying altitude of 600 m. Tables 6.5 and 6.6 make clear that the trend goes to altitudes well above 1000 m.

Comparing today's airborne LIDAR systems with those of four years ago, it becomes obvious that the range performance has been improved and full waveform detection has been implemented. Also the intensity recording has been improved to higher resolutions of up to 16 bits. All systems in Tables 6.5 and 6.6 have comparable ranging accuracy between 1 cm and 3 cm. Only the ALS50-II of Leica has 10 cm. However, one has to remember that it flies up to a height of 6000 m. The author thinks that the 1 cm to 3 cm accuracy for LIDAR-systems presents only the performance of the LIDAR ranging unit, because the POS has lower accuracy.

Six of the eight manufacturers used commercial available POSs, either of the company Applanix or of the company IGI. Only Leica and TopEye have proprietary POSs.

Table 6.5. Airborne laser scanners, part 1.

Manufacturer	Fugro	IGI	Leica geosystems	Optech
Type	FLI-MAP 400	LiteMapper 5600	ALS50-II	ALTM Gemini
Wavelength [nm]	1500	1550	1064	1060
Pulse length [ns]	4	3.5	9 ns	7 ns
Beam divergence [mrad]	0.45	0.5	0.22	0.15
Scanner	rotating mirror	polygon	oscillating. mirror	oscillating. mirror
PRF [kHz]	250	200	150	167 (70 full wavef.)
Max. scan angle [deg]	60	60	75	50
Max. recorded echoes/pulse	4	unlimited	4	4
Pulse sampling frequency	0.025 ps	1 GHz	N/A	1 GHz
Pulse detection	threshold	full waveform	leading edge detection	leading edge detection (full wavef.)
Intensity	11 bit	16 bit	8 bit	12 bit (8 bit full wavef.)
POS	Applanix POS AV410	AERO control	Leica IPAS10	Applanix POS AV610
Range accuracy	3 cm	2 cm	10 cm	3 cm
Point spacing along-track	0.27 m @ 150 km/h	0.6 m @ 150 km/h	0.23 m @ 150 km/h	2 m @ 150 km/h
Point spacing across-track	0.46 @ 400 m AGL	0.6 m	0.24 m @ 1000 m AGL	2.6 m @ 4000 m
Max. flying height	400 m	1800 m	6000 m	4000 m

Table 6.6. Airborne laser scanners, part 2.

Manufacturer	Riegl	BLOM TopEye	TopoSys	TopoSys
Type	LMS-S560-A	TopEye Mk II	Falcon III	Harrier 56
Wavelength [nm]	1500	1064	1550	1550
Pulse length [ns]	4 ns	4 ns	5 ns	4 ns
Beam divergence [mrad]	0.3	1	0.7	0.3
Scanner	Polygon	Palmer	Fibre	polygon
PRF [kHz]	200	50	125	200
Max. scan angle [deg]	60	40	28	60
Max. recorded echoes/pulse	unlimited	unlimited	unlimited or 8	unlimited
Pulse sampling frequency	1 GHz	2 GHz	1 GHz or analogue	1 GHz
Pulse detection	Full waveform	Full waveform	Full waveform Or rising edge	Full waveform
Intensity	16 bit	10 bit	8 bit	16 bit
POS	IGI AERO control/ Applanix POS AV510	Trimble 5700, Honeywell 764 POS GPS TopEye PP	Applanix POS AV510	Applanix POS AV410
Range accuracy	2 cm	1 cm	1 cm	3 cm
Point spacing along-track	0.26 m @ 150 km/h	N/A	0.1 m @ 150 km/h	0.26 m @ 150 km/h
Point spacing across-track	2.9 m @ 1000 m	N/A	0.3 m @ 1000 m	2.9 m @ 1000 m
Max. flying height	1000 m	750 m	2000 m	1000 m

6.5 CONCLUSIONS

This short introduction into LIDAR systems and the overview of commercial ones shows that only two radar ranging principles—pulse and measurement of the phase difference—are applied. During the last four years, the performance has been improved with regard to longer ranges while retaining unchanged the ranging accuracy, sampling and scanning speed. Other ranging methods well established in microwave radar such as FM-CW ranging (Frequency Modulation-Continuous Wave-ranging) or the use of pseudo-noise-sequences (PN-Sequences) have not been considered by commercial developments. Looking at airborne systems, the full waveform sampling technology was implemented recently, mostly as an optional extension of existing systems. This technique was introduced by the Scanning Laser Imager of Canopies by Echo Recovery (SLICER) (Blair et al. 1994) and optimized in the final remote sensing system LVIS (Laser Vegetation Imaging Sensor) (Blair et al. 1999). LVIS was a large footprint (10–30 m) scanning system operating at altitudes up to 10 km. The studies of Blair et al. (1999) work out the great advantage of full waveform sampling if large footprint sensors are applied. Even in dense tropical moist forest, returns within a footprint can be detected that penetrate to the sub-canopy ground level. Today commercial LIDARs work with small laser footprints. Here, ongoing research will make clear if the large effort in full waveform sampling will be in proportion to the gain in information or if detecting a number of discrete returns will meet the requirements of most surveys.

LIDARs with photo counting detection still have not yet found their way into the commercial market, although studies and developments have been carried out since the mid-1990s (Priedhorsky et al. 1996) and this technology was first implemented in ground based systems for carrying out ranging to Earth-orbiting satellites (Degnan 2001). Photon counting LIDARs are denoted as digital systems, because discrete events are counted. Furthermore, much lower pulse energy is required compared to conventional systems, as a small number of photons per pulse are sufficient to obtain valid ranging measurements (Degnan 2002, Degnan et al. 2001, Slatton 2005).

REFERENCES

Blug, A., Baulig, C., Wölfelschneider, H. & Höfler, H. 2005. Novel Geometry for High Speed Outdoor Laser Scanning System. In A. Grün & H. Kahmen (eds.), *Optical 3-D Measurement Techniques VII*, Vol. I: 154–159.

Degnan, J.J. 2001. Unified approach to photon-counting microlaser rangers, transponders, and altimeters. *Surveys in Geophysics* 22: 431–447.

Degnan, J.J. 2002. Photon-counting multikilohertz micro-laser altimeters for airborne and spaceborne topograpic measurements. *Journal of Geodynamics* 34: 503–549.

Degnan, J., McGarry, J., Zagwodzki, T., Dabney, P., Geiger, J., Chabot, R., Steggerda, C., Marzouk, J. & Chu, A. 2002. Design and performance of an airborne multikilohertz photon- counting, microlaser altimeter. *Proceedings of Land Surface Mapping and Characterization using Laser Altimetry, 22–24 Oct. 2001. International Archives of Photogrammetry and Remote Sensing* XXXIV3-W4: 9–16.

Lemmens, M. 2007a. Airborne LIDAR Sensors—Product Survey. *GIM international* 21, 2: 24–27.

Lemmens, M. 2007b. Terrestrial Laser Scanners—Product Survey. *GIM international* 21, 8: 41–45.

Priedhorsky, W.C., Smith, R.C. & Ro, C. 1996. Laser ranging and mapping with a photon-counting detector. *Applied Optics* 35: 441–452.

Slatton, K.C., Carter, W.E. & Shrestha, R. 2005. A simulator for airborne laser swath mapping via photon counting. In R.S. Harmon, J.T. Broach & J.H. Holloway, Jr. (eds.), *Detection and Remediation Technologies for Mines and Mine-like Targets, Proceedings of SPIE* 5794: 12–20.

Wehr, A. & Lohr, U. 1999. Airborne laser scanning—an introduction and overview. *ISPRS Journal of Photogrammetry & Remote Sensing* 54(2–3): 68–82.

Wölfelschneider, H., Blug, A., Baulig, C. and Höfleret, H. 2005. Fast Distance Measurement for Laser Scanning. *Technisches Messen* 72, 7–8: 455–467.

Advances in Photogrammetry, Remote Sensing and Spatial Information Sciences: 2008 ISPRS Congress Book – Li, Chen & Baltsavias (eds)
© 2008 Taylor & Francis Group, London, ISBN 978-0-415-47805-2

CHAPTER 7

Land mobile mapping systems

Naser El-Sheimy

ABSTRACT: Mobile Mapping Systems (MMS), the methodology that integrates digital imaging with direct geo-referencing, has developed rapidly over the past 15 years. This chapter provides a brief history of land-based MMS, including a summary of some of the systems developed up to now. It then details the development of the geo-referencing principle used by MMS and shows how such formulae can be used to determine the accuracies of points measured by the system. The chapter concludes with a closer examination of two examples of land-based MMS.

Keywords: Mobile mapping systems, spatial data collection, multi-sensor integration

7.1 INTRODUCTION

The past 20 years have seen an explosive growth in the demand for geo-spatial data. This demand has numerous sources and takes many forms; however, the net effect is an ever-increasing thirst for data that is more accurate, has higher density, is produced more rapidly, and is acquired less expensively. For mapping and GIS projects, this data has traditionally been collected using terrestrial surveying techniques or by aerial photogrammetric surveys.

Unfortunately, the terrestrial surveying techniques are intrusive and not well suited for rapid or dense data collection. More recently, point-wise GPS data collection systems have become popular. However, these systems—like traditional terrestrial surveys—still require each point of interest to be occupied, and consequently they have not significantly reduced either the cost or time requirements of data collection. On the other hand, until recently aerial photogrammetry in its classical form of film-based optical sensors (analogue) has been widely used for high accuracy mapping applications at all scales. Ground control points (GCPs) were the only required source of information for providing the geo-referencing parameters and suppressing undesirable error propagation. In general, the necessity for GCPs was so evident that all methods of operation relied on them. Even with the major changes in photogrammetry from analogue to analytical and then to the digital mode of operation, it was taken for granted that GCPs were the only source for providing reliable geo-referencing information. External geo-referencing information was therefore labelled as auxiliary data, indicating that they were only useful in minimizing the number of GCPs. The drawback of indirect geo-referencing is the cost associated with the establishment of the GCPs. They usually represent a significant portion of the overall budget. In some cases, this cost can be prohibitive, especially when imagery is to be acquired and geo-referenced in remote areas such as areas found in many developing countries (for more details see Schwarz et al. (1993)). The use of the GCPs also often puts operational constraints on a specific flight mission. This situation changed fundamentally when GPS data were included in the block adjustment of aerial triangulation. In principle, the use of GPS data made block triangulation entirely independent of GCPs. For the first time in the history of photogrammetry, the geo-referencing process became autonomous, as GCPs were not necessarily required any more. However, this is only true for a block triangulation scheme with overlapping images. Other sensors cannot be fully geo-referenced by GPS alone. Well-known examples are pushbroom digital scanners, laser scanners and imaging radar systems, which are important in kinematic mapping applications.

Geo-referencing of pushbroom imagery, for example, requires the instantaneous position and attitude information of each scan line. GCPs alone are not sufficient to resolve all the three positions and three orientation parameters associated with each scan line, of which there may be thousands in a single image. Also, the data rate of most available GPS receivers is not high enough to support the data rate required for airborne pushbroom scanners (about 20–40 lines/sec). Other restrictions of GPS come from the operational environment of kinematic close range terrestrial mapping systems, especially in urban centres where stand-alone GPS, due to frequent signal blockage, cannot reach the accuracy requirements of such applications. Also, for non-imaging sensors such as imaging radar systems, the use of GPS in stand-alone mode will not be feasible, since no support by block formation is easily realized.

During the last decade, direct geo-referencing became possible by integrating GPS and INS, such that all the exterior orientation information has become available with sufficient accuracy at any instant of time (Wright & El-Sheimy 2003, Toth 1997, Schwarz et al. 1993). The integration of GPS/INS puts the geo-referencing of photogrammetric data on a new level and frees it from operational restrictions. Together with digital data recording and data processing, it initiates the era of Mobile Mapping Systems (MMS). These MMS, which can be used as an alternative to both point-wise GPS and traditional techniques of data collection, integrate various navigation and remote sensing technologies on a common aerial or land-based platform. They capitalize on the strengths of the individual technologies in order to increase the efficiency of data collection. They have in common that they integrate a set of sensors mounted on a common platform and synchronized to a common time base. They are typically used in kinematic mode. In principle, they are capable of operating with the data measured on the platform. No other information, such as known ground control, is needed, although it may be included as redundant information. Systems of this type:

- Can be immediately deployed everywhere on the globe without the need for identifying existing ground control.
- Employ a task-oriented system design through integration at the measurement level. Data flow optimization can therefore be a built-in feature.
- Can be equipped with real-time quality control features by including data redundancies in the system design and by using a combination of real-time data processing and expert knowledge to obtain homogeneous results.
- Generally use software geo-referencing to transform the time-dependent measurement process into a sequence of geo-referenced images that can be considered as independent geometric units in post-mission processing.

The most important benefits of MMS are a reduction in both the time and cost of data collection; however, they also have a number of additional advantages. For example, both spatial and attribute information can be determined from the remotely sensed data. Furthermore, data can be archived and revisited, permitting additional data collection without additional field campaigns. Systems with these characteristics are likely to play an increasing role in geomatics engineering.

Table 7.1 shows a comparison of land-based MMS with alternative methods of spatial data collection. Evident in this table is how multi-sensor systems combine the advantages of the individual alternative techniques.

The conceptual layout and data flow of a multi-sensor system for mapping applications is shown in Figure 7.1. The selection of sensors for such a system obviously depends on system requirements, such as accuracy, reliability, operational flexibility and range of applications. The data acquisition module has, therefore, to be designed keeping both the carrier vehicle and the intended applications in mind. The data acquisition module contains navigation sensors and imaging sensors. Navigation sensors are used to solve the geo-referencing problem. Although a number of different systems are used in general navigation, the rather stringent requirements in terms of accuracy and environment make the integration of an INS with GPS receivers the core of any sensor combination for an accurate mobile mapping system for short range applications. This combination also offers considerable redundancy and makes the use of additional sensors for reliability purposes usually unnecessary. However, the addition of an odometer type device, such as the ABS, for close range applications may be useful for operational reasons, as for instance keeping a fixed distance between camera exposures.

Imaging sensors may be subdivided based on the way they contribute to the information about the object space. They may provide descriptive information, as for instance grey scales, or geometric information, as for instance direction or range from the camera to the object. Table 7.2 summarizes the contribution of sensors typically used in multi-sensor systems for mapping applications.

The selected sensor configuration requires a certain data processing sequence. Part of the processing will have to be performed in real-time, such as data compression for the imaging data and initial quality control processing for the navigation data. Most of the data, however, will immediately be stored for post-mission use. In post-mission, the data processing hierarchy is determined by the fact that all mapping data have to be geo-referenced first before they can be used in the integration process. The first step is, therefore, the geo-referencing of all mapping data and their storage in a multimedia database.

Although MMS can be air-based or land-based, the author will, however, mainly use the land-based systems to present the concept for MMS and will refer to

Table 7.1. Comparison of spatial data collection techniques.

Technique	Disadvantages	Advantages
Close range photogrammetry	• Establishing sufficient control is difficult and expensive	• Rapid data collection from images • Data can be archived and revisited • Less intrusive data collection
Total Station terrestrial surveying	• Labour intensive and slow • Not suitable for dense collection over a wide area • Intrusive data collection	High relative accuracy
Point-wise GPS	• Not suitable in urban centres or forested areas • Not suitable for dense data collection over a wide area • Intrusive data collection	• Little skill required • High absolute accuracy
Aerial photogrammetric surveys	• Incomplete coverage—only points visible from the air can be collected • Expensive data collection campaigns	• Rapid data collection from images • Data collection can be partially automated • Data can be archived and revisited • Less intrusive data collection
Land based mobile mapping systems	High initial system cost	• Little or no control required • Rapid and dense data collection over a wide area • More cost-effective data acquisition • Data can be archived and revisited • Less intrusive data collection

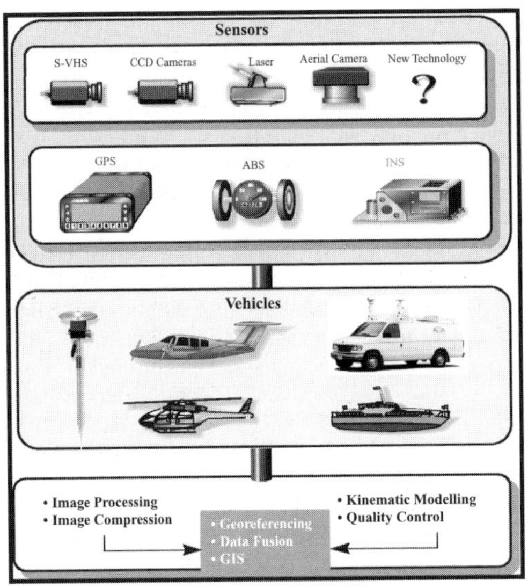

Figure 7.1. Multi-sensor integration for mapping application. (see colour plate page 492)

the airborne systems where there are differences. This chapter will start with a brief review of land-based mobile mapping systems, including their history and their principle of operation. It then provides a closer examination of a number of implementations for both land and aerial systems. Additional reviews of mobile mapping technology, including aerial systems, can be found in Schwarz & El-Sheimy (2005), Grejner-Brzezinska (2004), Mostafa & Schwarz (2001), Li (1997) and Novak (1995).

7.2 HISTORY OF LAND-BASED MULTI-SENSOR SYSTEMS

The first operational land-based MMS was developed by the Centre for Mapping at the Ohio State University. Their system—called GPSVan™—integrated a code-only GPS receiver, two digital CCD cameras, two colour video cameras and several dead-reckoning sensors (Goad 1991, Novak 1991). All components were mounted on a van; the GPS provided the position of the van and the images from the CCD cameras were used

Table 7.2. Summary of mapping related sensors.

Type of sensor	Type of information	Characteristics
Analogue aerial cameras	Descriptive/Geometric	Photos; high geometric accuracy (function of the photo scale).
Multi-spectral linear scanners (Pushbroom)	Descriptive/Geometric	Multi-spectral images, geometric accuracy depends on sensor resolution.
CCD cameras	Descriptive/Geometric	Images; geometric accuracy depends on sensor resolution.
Imaging laser	Descriptive/Geometric	Image + Distance between the object and the sensor.
Laser profiles, Laser scanners	Descriptive/Geometric	Distance between the object and the sensor, scanning angle.
Impulse radar	Descriptive/Geometric	Thickness of objects (mainly used for pavement structure voids).
Ultra sonic sensors	Geometric	Distance between the object and the sensor. Road rutting measurement, and cross-section profile measurement.
GPS positioning and attitude determination	Geometric	High accuracy position and medium accuracy attitude, global reference.
Inertial sensors	Geometric	Low-to-high accuracy relative position and attitude.
Odometers	Geometric	Distances.

to determine the positions of points relative to the van. The dead-reckoning sensors, which consisted of two gyroscopes and an odometer (wheel counter) on each of the front wheels, were primarily used to bridge GPS signal outages. These sensors were also used to provide orientation information for the exposure stations; however, there is little—if any—published information that examines the orientation accuracy of the GPSVan's dead-reckoning sensors and the poor accuracy of similar sensors suggests that the orientation information they provided would have been of marginal quality at best. The two video cameras were used solely for archival purposes and to aid attribute identification; no relative positioning was performed from the video imagery. Using bundle adjustment techniques with relative orientation constraints, the GPSVan was able to achieve relative object space accuracies of approximately 10 cm. Unfortunately, because only carrier phase smoothed code-differential GPS was used, absolute object space accuracies were limited to 1–3 m. Receivers capable of measuring the carrier phase were being manufactured at the time and there is no rational given in the literature for not using them. However, they were likely not used because of their limited availability and high cost.

GPSVan successfully illustrated how land-based multi-sensor systems could improve the efficiency of GIS and mapping data collection; however, the absolute accuracy of the object space points was too poor for many applications, especially when compared with competing technologies. Also, the dead reckoning sensors in the GPSVan were not very suitable for bridging GPS outages. Therefore, further developments of land-based mobile mapping systems focused on improving system reliability and increasing absolute object space accuracies. The obvious technique for improving absolute accuracy was to use carrier-phase differential GPS, while the obvious choice for a more accurate dead-reckoning sensor was a high precision Inertial Measurement Unit (IMU). The use of an IMU has an additional advantage over other types of dead-reckoning sensors in that it also provides high-accuracy orientation information for the exposure stations. Further developments of land-based MMS based on GPS-Van™ or involving the same researchers—including NAVSYS GPS/Inertial Mapping system (GIM) and LambdaTech's GPSVision—all used IMUs as their dead-reckoning sensors (Coetsee et al. 1994). Later independent implementations of land-based MMS added dual frequency carrier phase differential GPS, more accurate IMUs and more sophisticated processing techniques. Examples of some later systems include the VISAT™ system (Schwarz et al. 1993), KiSS™ (Hock et al. 1995) and GI-EYE™ (Brown 1998). The VISAT system, in its final form, is notable because of the large number of imaging sensors it employed. Where previous land-based MMS were simple stereovision systems employing only two forward facing cameras, VISAT had eight digital colour cameras, permitting more flexible data collection and better imaging geometry (El-Sheimy et al. 2007a).

Table 7.3. Implementations of land-based multi-sensor systems (Ellum & El-Sheimy 2002).

Name	Developers	Platform	Sensors		Reference
			Navigation	Mapping	
GPSVan™	Ohio State University	Van, Train	GPS, 2 gyros, 2 odometers (wheel counter)	2 monochrome CCD digital cameras, 2 colour VHS cameras (for archival purposes)	Goad (1991) Novak (1991)
VISAT™	University of Calgary, Absolute Mapping Solutions	Van	Dual frequency GPS, navigation-grade IMU	8 monochrome CCD digital cameras, 1 colour VHS camera (for archival purposes)	Schwarz et al. (1993), El-Sheimt et al. (2007a,b)
GIM™	NAVSYS Corp.	Truck	GPS, low-cost IMU	1 CCD digital camera, 1 VHS camera	Coetsee et al. (1994)
KiSS™	University of the Federal Armed Forces, Munich	Van	GPS, 3 ring-laser gyros, odometer, barometer, inclinometer, compass	2 monochrome CCD digital cameras, colour VHS camera (for archival purposes)	Hock et al. (1995) Sternberg et al. (2001)
TruckMAP™	John E. Chance and Associates, Inc.	Van	Dual-antenna GPS, digital attitude sensor	Reflectorless laser range-finder	Reed et al. (1996)
Gator Communicator™	University of Florida	Person	GPS, digital compass, inclinometer	2 CCD digital cameras	Alexander (1996) Barker-Benfield (2000)
Indoor MMS	National Research Council, Canada	Mobile Robotic Platform	Wheel encoders	8 CCD digital cameras, bi-iris laser scanner (also used for navigation)	El-Hakim et al. (1997)
GI-EYE™	NAVSYS Corp.	Any Land-Based Vehicle	GPS, low-cost IMU	1 CCD digital camera	Sullivan & Brown (2002)
CDSS	Geodetic Institute Aachen	Van	GPS, 2 odometers, barometer	2 monochrome CCD digital cameras	Bäumke & Heimes (2001)
WUMMS	Wuhan Technical University	Truck	GPS, unspecified dead-reckoning sensor	CCD digital camera, laser range finder	Li (1999)
Backpack MMS	University of Calgary	Person	GPS, digital compass, inclinometer	Colour consumer digital camera	Ellum & El-Sheimy (2001)
GPSVision™	Lamda Tech Int. Inc.	Van	GPS, navigation-grade IMU	2 colour CCD digital cameras	Lambda Tech (2007)
MoSES	University of the Federal Armed Forces Munich	Van	GPS, navigation-grade IMU, odometer, barometer, inclinometer	2 CCD digital cameras (possible laser scanner and colour video camera)	Graefe et al. (2001)
Laser scanner MMS	Wuhan Technical University	Truck	GPS	CCD digital camera, laser scanner	Li et al. (2001)

In one important aspect land-based MMS have largely led their airborne counterparts. Namely, from their inception land-based MMS have used digital cameras as their remote sensors. This was possible because of the much smaller camera-to-object distances in land-based MMS when compared to airborne systems. The poor resolution of CCD chips meant that they could not be used in aerial applications without an unacceptable decline in accuracy. Indeed, the resolution of CCD chips has only recently improved to the level that they can be used in airborne mapping systems (and even this is debatable). The use of digital cameras is advantageous because they eliminate the requirement to scan photographs. Consequently, they substantially reduce the period from raw data collection to extracted data dissemination. Digital sensors also simplify automatic point and feature extraction, and allow for more flexible data storage possibilities—for example, the images can be stored in Multi-Media GIS (El-Sheimy et al. 2007b).

A list of some land-based MMS is presented in Table 7.3. Not considered in the table are systems that merely mount single or multiple navigation sensors on a moving platform. Such systems have the same limitation as traditional land-based surveying systems; viz. each point of interest must be occupied. Furthermore, such systems are also significantly less appropriate for GIS data collection because of the requirement to manually record attribute information. Also not included—again, in the table or in this chapter—are systems that merely use GPS with a video camera exclusively for archival purposes. Such systems do not use the imagery for positioning purposes, and are consequently not mobile "mapping" systems.

It is worth emphasizing that GPS was the primary motivator for the development of both land-based and airborne MMS, and the absolute accuracies these systems achieve are still largely dependent upon the performance of the GPS. Indeed, except for some rare indoor MMS, such as that described in El-Hakim et al. (1997), all MMS currently use GPS for their positioning sensor.

7.3 COORDINATE DETERMINATION FROM GEO-REFERENCED PLATFORMS

As mentioned above, MMS integrate navigation sensors and algorithms together with sensors that can be used to determine the positions of points remotely. All the sensors are rigidly mounted together on a platform; the former sensors determine the position and orientation of the platform, and the latter determine the position of points external to the platform. The sensors that are used for the remote position determination are predominantly photographic sensors and thus they are typically referred to as imaging sensors (Schwarz et al. 1993). However, additional sensors such as laser rangefinders (Morin & El-Sheimy 2002) or laser scanners (Li et al. 2001) are also used in MMS and therefore the more general terms of mapping sensors (Li 1997) or relative sensors (Novak 1995) may also be used when referring to the remote sensors. In the following, imaging, mapping, relative and remote sensors are used interchangeably. In the discussion of the history of land-based MMS, it is evident that the platform that holds the sensors is typically a van. However, the use of other platforms such as trains (Blaho & Toth 1995, Sternberg et al. 2001) and even people (Barker-Benfield 2000, Ellum & El-Sheimy 2001) have also been investigated and implemented.

The strength of Mobile Mapping Systems is their ability to directly geo-reference their mapping sensors. A mapping sensor is geo-referenced when its position and orientation relative to a mapping coordinate frame is known. Once geo-referenced, the mapping sensor can be used to determine the positions of points external to the platform in the same mapping coordinate frame. In the direct geo-referencing done by MMS, the navigation sensors on the platform are used to determine its position and orientation. This is fundamentally different from traditional indirect geo-referencing where the position and orientation of the platform are determined using measurements made to control points. These control points are established through a field survey prior to data acquisition, and their establishment is typically expensive and time-consuming. Therefore, eliminating this step results in significant decreases in both the cost and time requirements for data collection. The task of establishing control points is additionally complicated by the fact that both its cost and the time required are frequently difficult to estimate. Also, for many terrestrial surveys, the establishment of sufficient control points is virtually impossible—for example, consider the control requirements to map an entire city using close range photogrammetry. Finally, for some mapping sensors—such as laser scanners or pushbroom CCD arrays—it is not possible to establish control. The use of these sensors is not practical unless direct geo-referencing is performed.

The basis for all geo-referencing formulas—direct or indirect—is a seven parameter conformal transformation where the coordinates of a point in the MMS coordinate frame ($\boldsymbol{r}_p^s$) are related to their coordinates in a mapping coordinate frame ($\boldsymbol{r}_p^m$):

$$\boldsymbol{r}_p^m = \boldsymbol{r}_s^m + \mu_s^m \boldsymbol{R}_s^m \boldsymbol{r}_p^s \quad (1)$$

In the above equation, $\boldsymbol{r}_s^m$ is the position of the mapping sensor in the mapping coordinate frame, and μ_s^m and $\boldsymbol{R}_s^m$ are respectively the scale factor and rotation matrix between the mapping sensor and the mapping coordinate frames. In an MMS, all three parameters are typically measured, albeit indirectly. In actuality, the position of a GPS antenna on the platform and the orientation of an IMU or other attitude-sensing device are measured. The scale factor can be determined directly, for example, from a laser rangefinder or indirectly, for example, using stereo techniques with two images.

Because direct measurements of $\boldsymbol{r}_s^m$, μ_s^m and $\boldsymbol{R}_s^m$ are typically not made, Equation 1 is normally extended to include terms that account for the indirect measurements. Also, MMS are usually moving during or between measurements. Consequently, the position and orientation of the system with respect to the mapping coordinate system are changing with time, and Equation 1 must be modified to reflect this. As an example of both of these changes, the geo-referencing formula for a system integrating a mapping sensor with GPS and an IMU is:

$$\boldsymbol{r}_P^m = \boldsymbol{r}(t)_{GPS}^m + \boldsymbol{R}(t)_{IMU}^m(\boldsymbol{r}_{IMU/s}^{IMU} - \boldsymbol{r}_{IMU/GPS}^{IMU} + \mu_s^m \boldsymbol{R}_s^{IMU} \boldsymbol{r}_p^s) \qquad (2)$$

Figure 7.2 shows the development of this equation, and Table 7.4 describes the meaning of the various parameters and how they are determined. By examining this table, it can be seen that there are many quantities dependent upon either calibration or synchronization. Consequently, total system accuracy critically depends on both of these tasks being correctly performed. To examine the effects of calibration, synchronization, or measurement errors, a first order error analysis can be performed by linearizing the geo-referencing equation and adding a term for the synchronization error. For example, the error analysis for Equation (2) is:

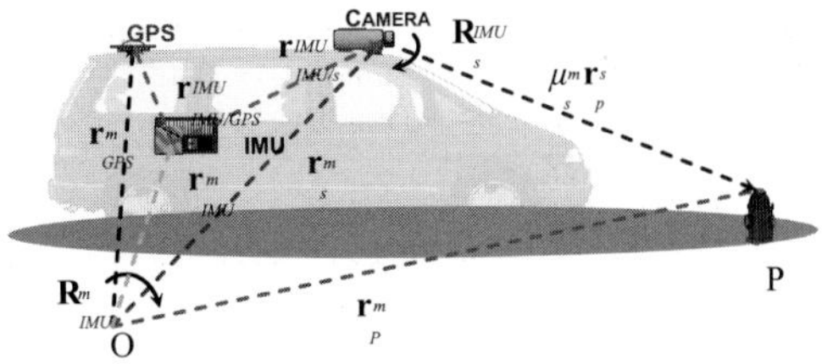

$$\boldsymbol{r}_P^m = \boldsymbol{r}_s^m + \mu_s^m \boldsymbol{R}_s^m \boldsymbol{r}_P^s$$

$$\boldsymbol{r}_P^m = (\boldsymbol{r}_{GPS}^m - \boldsymbol{r}_{IMU/GPS}^m + \boldsymbol{r}_{IMU/s^i}^m) + \mu \boldsymbol{R}_{IMU}^m \boldsymbol{R}_{s^i}^{IMU} \boldsymbol{r}_p^{s^i}$$

$$\boldsymbol{r}_P^m = (\boldsymbol{r}_{GPS}^m - \boldsymbol{R}_{IMU}^m \boldsymbol{r}_{IMU/GPS}^{IMU} + \boldsymbol{R}_{IMU}^m \boldsymbol{r}_{IMU/s^i}^{IMU}) + \mu_s^m \boldsymbol{R}_{IMU}^m \boldsymbol{R}_{s^i}^{IMU} \boldsymbol{r}_p^{s^i}$$

$$\boldsymbol{r}_P^m = \boldsymbol{r}_{GPS}^m + \boldsymbol{R}_{IMU}^m (\boldsymbol{r}_{IMU/s^i}^{IMU} - \boldsymbol{r}_{IMU/GPS}^{IMU} + \mu_s^m \boldsymbol{R}_{s^i}^{IMU} \boldsymbol{r}_p^{s^i})$$

Figure 7.2. Development of geo-referencing formulae.

Performing an error analysis prior to construction of an MMS is an important step in a system's design. Such an analysis is required in order to choose the appropriate components so that the integrated system will be able to meet the desired accuracies.

It should be noted that the position and orientation are typically determined using a GPS and IMU that have been previously integrated. In this case, Equation 2 reduces to

$$\boldsymbol{r}_P^m = \boldsymbol{r}_{IMU}^m + \boldsymbol{R}_{IMU}^m (\boldsymbol{r}_{IMU/s}^{IMU} + \mu_s^m \boldsymbol{R}_s^m \boldsymbol{r}_p^s) \qquad (4)$$

Of course, the missing calibration terms from Equation 2 have not disappeared—they are merely "hidden" in the $\boldsymbol{r}_{IMU}^m$ term, which is the position of the IMU determined from the integrated GPS/IMU navigation

$\delta \boldsymbol{r}_P^m = \delta \boldsymbol{r}(t)_{GPS}^m$	GPS position error
$+\, \delta \boldsymbol{R}(t)_{IMU}^m (\boldsymbol{r}_{IMU/s}^{IMU} - \boldsymbol{r}_{IMU/GPS}^{IMU} + \mu_s^m \boldsymbol{R}_s^{IMU} \boldsymbol{r}_p^s)$	Attitude error
$+\, \boldsymbol{R}(t)_{IMU}^m \delta \boldsymbol{r}_{IMU/s}^{IMU}$	Calibration error in IMU/sensor offset
$-\, \delta \boldsymbol{R}(t)_{IMU}^m (\boldsymbol{r}_{IMU/s}^{IMU} - \boldsymbol{r}_{IMU/GPS}^{IMU} + \mu_s^m \boldsymbol{R}_s^{IMU} \boldsymbol{r}_p^s)$	Calibration error in IMU/GPS offset (3)
$+\, \delta \mu_s^m \boldsymbol{R}(t)_{IMU}^m \boldsymbol{R}_s^{IMU} \boldsymbol{r}_p^s$	Scale error
$+\, \mu_s^m \boldsymbol{R}(t)_{IMU}^m \delta \boldsymbol{R}_s^{IMU} \boldsymbol{r}_p^s$	Calibration error in IMU/sensor differential rotations
$\mu_s^m \boldsymbol{R}(t)_{IMU}^m \boldsymbol{R}_s^{IMU} \delta \boldsymbol{r}_p^s$	Sensor error
$+\, \delta t \left[v(t) + \omega(t)(\boldsymbol{r}_{IMU/s}^{IMU} - \boldsymbol{r}_{IMU/GPS}^{IMU} + \mu_s^m \boldsymbol{R}_s^{IMU} \boldsymbol{r}_p^s) \right]$	Synchronization error.

Table 7.4. Terms in expanded geo-referencing formulae.

Variable	Description	Variable	Description
r_P^m	Position vector of the point of interest in the mapping frame. Unknown.	$r_{IMU/s}^{IMU}$	Vector of position differences between IMU and camera. Determined through calibration and measured in the IMU coordinate frame.
r_p^s	Position vector of the point in relative sensor frame. Measured by the relative sensor.	$r_{IMU/GPS}^{IMU}$	Vector or position differences between IMU and the GPS antenna. Determined through calibration and measured in the IMU coordinate frame.
$r(t)_{GPS}^m$	Position vector of the GPS antenna. Determined using kinematic GPS and measured in the mapping coordinate frame.	μ_s^m	Scale factor from sensor space to object space. Determined directly or using stereo techniques.
$R(t)_{IMU}^m$	Rotation matrix between IMU and map ping coordinate frames. Determined using integrated IMU measurements.	R_s^{IMU}	Rotation matrix between IMU and camera coordinate frames. Determined through calibration.

system. The optimal integration of navigation sensors, particularly GPS and IMU, is a subject of much research interest both inside and outside the mobile mapping community. It is typically done using a 15 state Kalman filter that estimates the errors in the position, velocity, orientation of the integrated system and the errors in the accelerometer and gyroscope biases in the IMU. For more information on GPS/IMU integration, see El-Sheimy et al. (2006), Wang et al. (2002) and Farrell & Barth (1999).

Finally, it is worth noting that Equation 1 is the basis of the collinearity equations used in a photogrammetric bundle adjustment. For an imaging remote sensor, this implies that the estimates of exposure station position and orientation can be input as weighted parameter observations into an adjustment using unified least squares. This technique yields the most accurate object space coordinates.

7.4 SYSTEM CALIBRATION

The accuracy of points measured by direct geo-referencing critically depends on the calibration of the integrated system used for the data collection. Errors in calibration have the same effect as measurement errors, and the importance of an accurate calibration cannot be overestimated. In this section, three elements of integrated system calibration are examined: camera calibration, boresight calibration, and lever-arm calibration. The first element refers to the determination of the camera's interior geometry, the second element refers to the determination of the differential rotations between the camera and the orientation sensor, and the third element refers to the determination of the position offsets between the camera and position sensor.

7.4.1 *Camera calibration*

In MMS, cameras are usually calibrated by self-calibration. There are two reasons for this:

1. The digital cameras prevalent in MMS have relatively unstable interior geometries and, consequently, require more frequent calibrations.
2. The cameras are integrated into the MMS and it is not feasible to remove them for laboratory calibration.

With regard to the latter point, it should be noted that laboratory calibrations can be done on the cameras prior to their inclusion into the MMS. However, this must be done with Point 1 in mind—the camera must have an interior geometry of metric quality. If this is not the case, then the best that can be done with a laboratory calibration is to use such results as weighted parameters in the self-calibration.

In MMS camera self-calibrations—like in virtually all photogrammetric applications—a standard parameter set is used. This set includes the camera's focal length (c) and principal point offsets (xp, yp), two or three terms for symmetric radial lens distortion (k1, k2, k3), and two terms for decentring lens distortion (p1, p2). Additionally, for digital cameras, terms may be added to account for affinity (differential scaling) and shear of the image axes (b1 and b2, respectively). An overview of this standard parameter set can be found in Ellum & El-Sheimy (2001).

7.4.2 *Lever arm calibration*

The simplest and most common method for determining the lever-arm between the GPS antenna and perspective centre of the camera is to measure it using conventional survey methods. Unfortunately, the accuracy of this technique is limited by the inability to directly observe the phase and perspective centres of the antenna and camera, respectively. Without using esoteric measurement procedures, this accuracy is limited to the centimetre level. For many applications, however, this is not sufficient. To achieve this accuracy, it is necessary to relate measurements of the camera to its front nodal point. In film-based cameras, this can be done by making measurements of the camera's fiducial marks. In digital cameras it is not normally possible to make direct measurements in the image plane, and consequently such a procedure is not possible.

An alternative "pseudo" measurement technique is to use the difference in positions determined by GPS observations and positions resulting from a bundle adjustment. However, the accuracy of this technique is dependent upon finding a calibration field that is suitable for both GPS and photogrammetry, i.e. a field that minimizes GPS errors such as multipath, and has dense and well-distributed targets for the photogrammetric measurements. If such a target field can be found, then the offset vector in the camera coordinate frame can be calculated using

$$\boldsymbol{r}^{c}_{GPS} = \boldsymbol{R}^{c}_{M}(\boldsymbol{r}^{M}_{GPS} - \boldsymbol{r}^{M}_{c}) \qquad (5)$$

where $\boldsymbol{R}^{c}_{M}$ is the rotation matrix between the mapping and camera coordinate frames. This matrix, like the exposure position $\boldsymbol{r}^{M}_{c}$, is available from the adjustment. When several exposures are available then the accuracy of the offset vector can be improved by averaging and, if the covariance of one or both position vectors is known, by weighted averaging. Because of the difficulties in obtaining an accurate exposure position in aerial photogrammetric adjustments, this technique is really only practical for land-based systems.

Another method of determining the offsets is to include them in a bundle adjustment as unknown parameters. However, opinion on this approach is mixed. For instance, Mikhail et al. (2001) indicates that the offsets are usually included; while Ackermann, (1992) claims that the offsets cannot be included as they result in singularities in the adjustment. For airborne cases, the truth is somewhere in the middle. The offsets are highly correlated with both the interior and exterior orientation parameters, particularly with the focal length and exposure station position. Because of this correlation, offsets determined in the adjustment are not very accurate, especially the z-offset. For close range photogrammetry, the same conclusion applies, although the use of convergent imagery, to some degree, de-correlates these parameters and makes the recovery of the offsets more reliable. In any case, to include the offsets in the adjustment as unknowns it is necessary to provide parameter observations of the GPS positions. Otherwise the effects of focal length and z-offset cannot be separated (i.e. they are 100% correlated), and the adjustment is rendered singular.

7.4.3 *Boresight calibration*

Boresight calibration essentially refers to the determination of the rotation matrix $\boldsymbol{R}^{c}_{b}$ that relates the axes of the orientation sensor to the axes of the camera. In the most common method of performing this calibration, it is necessary to determine explicitly both the $\boldsymbol{R}^{c}_{M}$ and $\boldsymbol{R}^{b}_{M}$ matrices. This, in turn, requires that a known target field be imaged with the camera and IMU mounted together, the roll, pitch and yaw angles measured by the IMU, and the ω, ϕ and κ angles determined by photogrammetric resection. $\boldsymbol{R}^{c}_{M}$ can then be determined using the latter set of angles, and $\boldsymbol{R}^{b}_{M}$ can be determined using the former set. With both the $\boldsymbol{R}^{c}_{M}$ and $\boldsymbol{R}^{b}_{M}$ rotation matrices available, $\boldsymbol{R}^{c}_{b}$ can be calculated using:

$$\boldsymbol{R}^{c}_{b} = \boldsymbol{R}^{c}_{M}(\boldsymbol{R}^{b}_{M})^{T} \qquad (6)$$

Although this calculation can be done with a single exposure, it is, obviously, advantageous to use multiple exposures and to average the results. Of course, it is not possible to simply average the individual elements of the $\boldsymbol{R}^{c}_{b}$ rotation matrices from each exposure station, as the resulting rotation matrix would almost certainly not be orthogonal. Instead, a set of Euler angles must be extracted from each exposure station's $\boldsymbol{R}^{c}_{b}$ matrix, those angles averaged, and a final $\boldsymbol{R}^{c}_{b}$ reconstructed. In practice, images at the edge of the photogrammetric block are occasionally excluded from the above average calculation because their adjusted attitude parameters are less accurate than those images closer to the middle of the block (Škaloud 1999). An alternative to this procedure is to weight the contribution of each exposure according to its angular standard deviations coming from the adjustment.

Bäumker & Heimes (2001) presented an analogue to the above technique where instead of averaging the angles, an unweighted least squares adjustment was used to estimate small angular misalignments. This was done after the bundle adjustment, and treated both the orientations from the adjustment and the measured orientations as fixed. Unfortunately, this method, while being conceptually more complicated than the one above, will likely give exactly the same results—if not worse, because of the small angle approximation. Also, it is only suitable for determining small misalignments between sensors, and thus requires a reasonably accurate initial estimate for $\boldsymbol{R}_b^c$.

Like the lever-arm calibration, it is also possible to determine the $\boldsymbol{R}_b^c$ matrix by including it in a bundle adjustment as an unknown. In order to do this, the matrix is parameterized using three Euler angles. For example, if α_1, α_2, and α_3 are the three angles, then $\boldsymbol{R}_b^c$ may be described by:

$$\boldsymbol{R}_b^c = \boldsymbol{R}_z(\alpha_3)\boldsymbol{R}_y(\alpha_2)\boldsymbol{R}_x(\alpha_1) \tag{7}$$

Of course, any order of rotation is possible. Regardless of the order, the three angles are included as parameters in the adjustment. A disadvantage of this procedure is that the addition of these angles necessitates rather fundamental changes to the implementation of the adjustment, as the collinearity equations become functions of six angles instead of just three. This, in turn, makes the linearization of the collinearity equations considerably more complex.

Unlike the lever-arm, it is not really possible to determine the misalignment angles between the camera and IMU by making external measurements. The difficulty here is that it is not generally possible to get the orientation of either the IMU's or the camera's axes from external measurements to these sensors. A summary of the techniques for boresight and lever-arm calibration, along with their advantages and disadvantages, is presented in Table 7.5.

7.5 EXAMPLES OF MOBILE MAPPING SYSTEMS

To illustrate the design and operation of MMS, two systems are described below.

7.5.1 *The VISAT system*

VISAT (Fig. 7.3) is a van-based MMS that uses digital cameras, an inertial navigation system and the GPS Satellite system. Its design objectives were as follows:

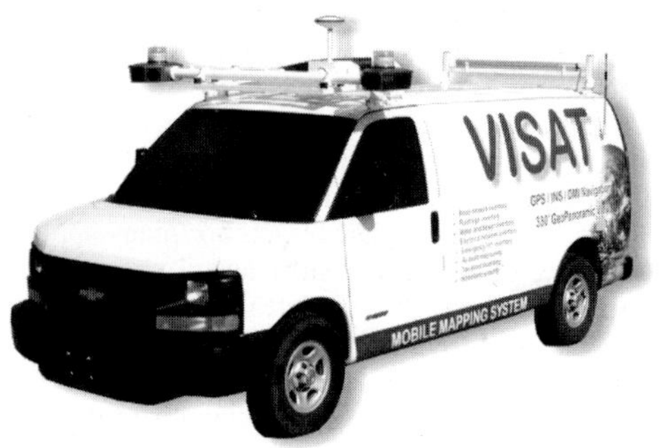

Figure 7.3. VISAT™ Van Mobile Mapping System. (see colour plate page 492)

- Positioning of all visible objects of interest within 30 m of the Van with an RMS accuracy of 0.3 m or better while travelling 110 km/h.
- Automatic data acquisition with real-time quality control.
- Post-mission data processing, with separate modules for geo-referencing, image database management and quality assessment.

The above design goals, all of which were successfully achieved, were selected because a system designed with them in mind would be appropriate for nearly all GIS and mapping data collection activities.

7.5.2 *Components and operation*

The core hardware components of the VISAT™ van are a strap down Inertial Navigation System (SINS), a dual-frequency GPS receiver and a cluster of digital colour cameras. The primary purposes of these components are: the GPS provides the position of the van; the SINS provides the orientation of the van; and the cameras are used for relative positioning from the van. These components, however, also have important secondary functions.

Table 7.5. Summary of techniques for boresight and lever-arm calibration.

Technique	Advantages	Disadvantages
External measurement (lever-arm only)	• Independent from the bundle adjustment	• Difficulty in observing centres of antenna and camera
Post-adjustment averaging	• Does not require any changes to the adjustment model • Conceptually simple	• Requires accurate space resection
Including in adjustment as unknown parameter	• Permits full covariance information to be more easily included	• Correlations with other parameters implies results may not be reliable

For the GPS, these secondary tasks include controlling the long-term error growth of the SINS through the GPS/SINS Kalman filter and providing the precise timing base for all data streams. The secondary tasks of the SINS stem from its ability to be used as a position sensor in addition to an orientation sensor; consequently, these tasks include bridging GPS signal outages, detecting and correcting GPS cycle slips, and precise interpolation between GPS positions. The last task—interpolation between GPS positions—is possible because the SINS provides data at 200 Hz, while the GPS positions and velocities are only available at 1–2 Hz.

In addition to the GPS, SINS and cameras, the VISAT system also integrates a Distance Measuring Instrument (DMI). The pick-up from the DMI is used to trigger the acquisition of the images from the cameras at constant distance intervals defined by the user.

Table 7.6 summarizes the primary and secondary tasks of the sensors in VISAT.

Currently, the VISAT™ imaging component consists of 6 to 12 progressive colour digital cameras (1600 × 1200 pixels or 2048 × 2048 pixels) that provide a 280° to 360° field of view. The images are captured at high sampling rates and can be controlled by either time or travelled distance (usually every 2–7 m). The images are captured while the van is moving at the highway posted speed (up to 110 km/h). Essentially, all navigation and mapping data streams are synchronized to a common time frame using a high frequency synchronized multi-channel clock through data logging software. The data logging software (VISAT™ Log) has an expert module for real-time quality control, which communicates with the system's operator and provides useful information such as length of the survey, distance to master station, directions to specific routes, etc.

Surveying by VISAT consists of three parts, which are essentially the same as for any mobile mapping system:

- Data acquisition.
- Geo-referencing of the images using data from the navigation sensors (GPS and SINS).
- Positioning of objects in a mapping frame using a photogrammetric workstation and two or more geo-referenced images.

All data collected by the VISAT van is post-processed. During the post-processing, the digital images acquired from the VISAT van are geo-referenced using the position and orientation as determined by the GPS and SINS data. The system position and orientation are interpolated at the instants of image exposures and then combined with the system calibration parameters, as described by lever arm and boresight angles, to relate the images to the real world coordinate system. The geo-referenced images are hosted on VISAT™ GeoImage Servers along with the camera calibration that describes the inner orientation of the sensors, and the system calibrations that describe the lever arm and boresight angles. This new generation of servers allows client access via NET Remoting on TCP, HTTP, or Named Pipe channels for desktop, workgroup, or internet deployment and distribution. VISAT™ GeoImage Servers also act as a Universal Description Discovery and Integration Service (UDDI Web Service). The distributed three layers architecture of the server, Login and Security, GeoImage Metadata, and Image File Server architecture can accommodate mega size VISAT Image Libraries by using Network Load Balancing (NLB) and Server Clustering techniques.

The next stage is the extraction of 3-D coordinates from the images. In addition, geometric informa-

Table 7.6. Primary and secondary functions of VISAT sensors.

Sensor	Primary function	Secondary function
GPS	• Determines 3-D space position of the cameras	• Controls the INS error propagation • Provides system synchronization • Gives coordinates in WGS84
INS	• Determines 3-D space orientation of the cameras	• Bridges GPS outages • Corrects GPS cycle slips • Gives precise interpolation between GPS fixes
Cameras	• Two geo-referenced cameras provide the position of objects in 3-D space	• Provides redundancy, i.e. more than two cameras/images of the same object • Can be used to update navigation data through triangulation procedures
DMI	• Triggers the cameras at constant distance intervals	• Updates the INS data if the GPS signal is blocked for periods longer than the INS bridging level required to fix GPS ambiguities (half a cycle)

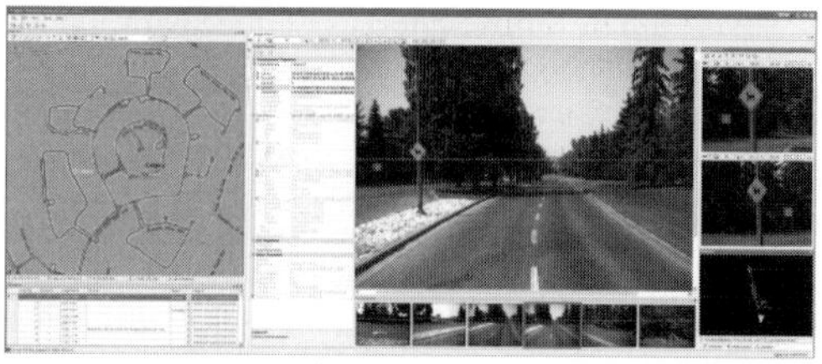

Figure 7.4. VISAT™ station measurement. (see colour plate page 492)

Table 7.7. VISAT system absolute accuracy.

Survey mode	Horizontal	Vertical
GPS/INS—Ideal conditions	0.13 m	0.08 m
INS Only—Worst case conditions	0.29 m	0.19 m

tion and attributes of themed objects such as control points, utility lines and land parcels may be needed to form GIS elements for themed layers. This task is performed using a photogrammetric workstation called the VISAT Station™ that enables the measurement of objects appearing in the images and the generation of GIS elements. VISAT™ Station (Fig. 7.4) is the client application that enables the user to make use of the collected geo-referenced images and perform mapping and GIS editing. The geo-referenced images are accessed from a local file on the user's desktop or by logging on to one of the VISAT™ GeoImage Servers on the enterprise's Intranet or by subscription to public servers on the Internet. Point features are digitized by measuring the point in at least two images. The 3-D point coordinates are obtained using simple photogrammetric intersection. Attribute information can also be collected by assigning the appropriate point symbol for the point feature. VISAT™ data are easily populated into GIS software like ARC-GIS.

7.5.3 *Accuracy*

The objective of any MMS is the determination of 3-D coordinates of all objects within the field of view of the CCD and video cameras. The accuracy of these 3-D coordinates is a function of the complete processing chain, which involves GPS positioning, SINS position/attitude determination, target localization in the images, system calibration, distance between the object and the cameras, and the camera geometry used in the 3-D computation. To check the system accuracy, a test field, established in Calgary, was used.

Thirty well-defined features (e.g. fire hydrants, telephone boxes, etc.) were selected as check points for testing the absolute accuracy of the VISAT system. To obtain independent reference values, each feature was occupied for approximately five minutes using static GPS methods with baselines tied to multiple control points. The survey of the test area was carried out in two of the VISAT survey modes, GPS/INS and INS only. In the GPS/SINS mode, the van collects navigation information from both the GPS receiver and the SINS. This is an ideal operational environment and, in principle, provides the best results due to the continuous GPS signal lock. The INS-only mode provides navigation results independent of the GPS by stopping the system at regular time intervals and updating the velocity channels by zero. This can be a typical situation in urban areas where the GPS signal is blocked by high buildings.

Table 7.7 summarizes the root mean square (RMS) for the two modes of operation.

7.5.4 *Backpack mobile mapping*

The VISAT system described above is perhaps the most rigorous and complete implementation of a van-based MMS and the accuracies it obtains are suitable for all but the most demanding cadastral and engineering applications. However, this accuracy does not come cheaply. The VISAT system and others like it cost well over $200,000 USD. This cost primarily stems from the navigation-grade IMUs used—such sensors cost $90,000 and higher. Additionally, the platforms and other sensors used in the van are also expensive. Because of their high cost, the market for such van-based MMS is rather small, and such systems are "one-off" systems that are operated by the companies or institutions that build them. In effect, this means that while several companies are making a profit using MMS, nobody is making a profit manufacturing them. This also means that the benefits of mobile mapping—in particular the lower costs and greater efficiency of data collection—are not being enjoyed by a wide community. Additionally, the van-based systems are extremely complex and many smaller survey or mapping firms do not have the expertise required to operate them.

To overcome the above-mentioned disadvantages of current van-based MMS, a prototype model of a backpack was developed at the University of Calgary (El-Sheimy & Ellum 2001, Hunter & El-Sheimy 2004). The objective of the backpack MMS was to compete in both accuracy and initial cost with current methods of GIS data collection, while offering the increases in data collection efficiency and flexibility that only an MMS can provide.

7.5.5 *Components and operation*

The positioning sensor used in the Backpack Mobile Mapping System (BMMS) is a dual frequency GPS receiver. GPS is used for the same reasons it is used in nearly all MMS—namely, no other positioning technology offers anywhere near the same accuracy and flexibility at the same cost and size. For a portable MMS, the latter feature is particularly important. The selection of a dual-frequency receiver would appear to run contrary to the design goal of a less expensive system. However, the relatively poor accuracy of the orientation sensors in the Backpack MMS means that the GPS positions are important for controlling the orientation of the photogrammetric networks measured from the images taken by the system. Thus, the increase in cost of a dual frequency receiver is warranted by its significantly better accuracy and reliability. For the orientation sensor, the BMMS uses a Leica Digital Magnetic Compass (DMC). The DMC is lightweight (less than 28 grams), small in size (31 mm × 33 mm × 14 mm), and has low power consumption (runs several hours using a 9V battery). The trade-off for these features is a relatively low accuracy; roll and pitch are only accurate to approximately 0.5°, and azimuth is only accurate to about 1°. However, the effects of this lower accuracy, particularly in azimuth, are largely eliminated by the use of GPS positions in the bundle adjustment used to determine object space coordinates.

To measure orientation, the DMC uses three micro-electromechanical (MEMS) based accelerometers and three magnetic field sensors. The accelerometers sense the direction of the gravity vector from which the roll and pitch angles can be easily calculated. Similarly, the magnetic field sensors sense the components of the Earth's magnetic field from which azimuth can be determined. Because the Earth's magnetic field is the reference, the azimuth angles from the DMC must be corrected for magnetic declination if they are to refer to true north. Fortunately, there are freely available global models of the Earth's magnetic field that are accurate to better than 1°. More accurate regional and local models are also available. The accuracy of the azimuth angles reported by the DMC depends heavily on the degree to which the local magnetic field is being disturbed. Disturbances in the magnetic field can be divided into two categories: softmagnetic, which are caused by nearby magnetic materials, and hardmagnetic, which result from nearby electric fields and permanent magnets. If the sources of these disturbances are fixed relative to the magnetic sensors—such as the camera and GPS antenna on the backpack MMS – then their effect can be removed through calibration. The DMC implements several internal calibration routines that perform both softmagnetic and hardmagnetic calibrations for fixed disturbances. Obviously, disturbances that are not fixed cannot be compensated for and must therefore be avoided. The digital camera used in the prototype BMMS is the Kodak DC260. Consumer digital cameras are ideal for inclusion in portable mobile mapping systems because of their high resolution, low cost and large internal memory. The internal memory of the camera is particularly valuable because it negates the requirement to transfer the images to a logging computer.

The operation of the BMMS is essentially the same as for VISAT or any other MMS. However, because of the low accuracy of the orientation sensor, positioning by direct geo-referencing will give poor results. Therefore, the best mode of operation is to use the measured position and orientation in a bundle adjustment. The bundle adjustment optimally combines the image point measurements with the observed positions and orientations. The adjustment also enables other information, such as a known distance between object space points, or constraints enforcing verticality or zero-height differences, to be included.

Like VISAT, the BMMS also has an associated software package, and, also like VISAT, the software used to manipulate the data collected by the backpack MMS is as important as the MMS itself. Indeed, it is probably more accurate to claim that the software is part of the MMS, as nearly every MMS has had dedicated software developed for it. For the backpack MMS, user-friendly software is vital for it to achieve its goal of bringing MMS to less-specialized users.

The BMMS was tested a number of times through a target field that simulated a "typical" urban environment in which the BMMS would be expected to operate. The target field had nearby vertical structures, pavement, and foliage—in short, somewhat of a worse case environment for GPS. It also had nearby metal buildings and light standards that could influence the azimuth reported by the DMC.

The absolute accuracies of the BMMS in a variety of configurations are shown in Tables 7.8 and 7.9. From these tables, it can be seen that with as few as

Table 7.8. Backpack MMS system absolute accuracy (20 m camera-to-object distance).

Number of image point measurements	Horizontal		Vertical	
	Max	RMS	Max	RMS
3 Images				
5 Image Points	0.13 m	0.10 m	0.10 m	0.06 m
10 Image Points	0.13 m	0.10 m	0.12 m	0.07 m
6 Images				
5 Image Points	0.06 m	0.05 m	0.14 m	0.11 m
10 Image Points	0.07 m	0.05 m	0.17 m	0.11 m

Table 7.9. Backpack MMS system absolute accuracy (40 m camera-to-object distance).

Number of image point measurements	Horizontal		Vertical	
	Max	RMS	Max	RMS
3 Images				
5 Image Points	0.82 m	0.69 m	0.91 m	0.78 m
10 Image Points	0.47 m	0.35 m	0.91 m	0.78 m
6 Images				
5 Image Points	0.81 m	0.69 m	0.76 m	0.59 m
10 Image Points	0.52 m	0.38 m	0.74 m	0.59 m

five image point measurements at a 20 m camera-to-object distance it is possible to achieve mapping accuracies comparable with L1 RTK GPS systems. This accuracy level is something of a benchmark for any MMS as such single frequency GPS systems are currently used in many spatial data collection projects. It is also worth noting that this accuracy exceeds the accuracy of the VISAT system described in the previous section. Unfortunately, the good performance at 20 m is not reflected at 40 m, especially in elevation. The poor performance at 40 m has two causes:

- *Poor imaging geometry*—The exposure stations for this test were nearly collinear. In this arrangement, the GPS positions do not significantly help to define the orientation of the photogrammetric network measured from the MMS's images. The entire network is essentially free to swing about the axis formed by the exposure stations with only the roll and pitch angles from the DMC constraining the rotation.
- *Poor image point measurements*—The images from the camera are stored in JPEG format. Because of the compression used in JPEG images there is a loss of image fidelity. This problem is obviously compounded as the camera-to-object distance increases, and for some points—particularly those in shadows—it is believed that the measurement error could have been as high as two pixels.

The above problems are two important considerations that a user must keep in mind when operating the backpack MMS. However, the first cause—poor imaging geometry—can also be used to illustrate an important advantage of the backpack MMS—namely, the mobility of the backpack MMS means that a user could be guided to the best possible geometry. Additionally, it should be noted that the problems specified above are the same as experienced by any MMS, but that the poorer performance of the orientation sensor in the Backpack MMS means that the errors due to poor imaging geometry or poor image point measurements are amplified. For a more complete description of the Backpack MMS, and for more results, consult Ellum (2001).

7.6 CONCLUSIONS

Looking at the development of land-based MMS over the past 15 years, the following conclusions can be drawn:

- MMS has developed from a topic of academic interest to a commercially viable industry with applications beyond just mapping applications.
- The GPS/INS direct geo-referencing technology has matured, due to the ongoing refinement of the integration techniques and advances in the systems calibration techniques. Miniaturization of inertial sensors will open the door for new systems such as the portable MMS presented in this chapter. It appears that a major step forward is imminent as MEMS sensor technology will result in considerable cost reductions for tactical-grade gyros and in new multi-sensor system concepts.
- The ongoing development of mathematical modelling and advanced post-mission estimation techniques will further increase the accuracy and robustness of the solutions.

REFERENCES

Ackermann, F. 1992. Kinematic GPS Control for Photogrammetry. *The Photogrammetric Record* 14(80): 261–276.

Alexander, J.F. 1996. Gator communicator: design of a hand held digital mapper. *Proceedings of Third Congress on Computing in Civil Engineering*: 1052–1057.

Barker-Benfield, S. 2000. Extra dimension: professor-patented mapping device combines old, new. *The Florida Times-Union*. http://www.jacksonville.com/tuonline/stories/071200/bus_3519070.html. Accessed 15 Jan 2001.

Bäumker, M. & Heimes, F.-J. 2001. New Calibration and Computing Method for Direct Geo-referencing of Image and Scanner Data Using the Position and Angular Data of an Hybrid Navigation System. *Proceedings of OEEPE-Workshop Integrated Sensor Orientation, Hannover, Germany*.

Blaho, G. & Toth, C. 1995. Field experiences with a fully digital mobile stereo image acquisition system. *Proceedings of 1995 Mobile Mapping Symposium, Columbus, OH*: 97–104.

Coetsee, J., Brown, A. & Bossler, J. 1994. GIS data collection using the GPSVan supported by a GPS/Inertial mapping system. *Proceedings of GPS-94, Institute of Navigation (ION), September 20-23, 1994, Salt Lake City, UT.*

Grejner-Brzezinska, D.A., Li, R., Haala, N. & Toth, C.K. 2004. From Mobile Mapping to Telegeoinformatics: Paradigm Shift in Geospatial Data Acquisition, Processing, and Management. *Photogrammetric Engineering & Remote Sensing* (70)2: 197–210.

El-Hakim, S.F., Boulanger, P., Blais, F. & Beraldin, J.A. 1997. A system for indoor 3-D mapping and virtual environments. *Proceedings of SPIE Vol. 3174–Videometrics V*: 21–35. San Diego, CA: The International Society for Optical Engineering (SPIE).

Ellum, C.M. 2001. *The Development of a Backpack Mobile Mapping System*, M.Sc. thesis, December 2001, UCGE Report No. 20159, Department of Geomatics Engineering, The University of Calgary (http://www.geomatics .ucalgary.ca/Papers/Thesis/NES/01.20159.CEllum.pdf).

Ellum, C.M. & El-Sheimy, N. 2001. A mobile mapping system for the survey community. Proceedings of the 3rd International Symposium on Mobile Mapping Technology (MMS 2001). Cario, Egypt. On CD-ROM.

Ellum, C. & El-Sheimy, N. (2002). Land-Based Mobile Mapping Systems. *Photogrammetric Engineering & Remote Sensing* 68, 1: 13–17.

El-Sheimy, N. & Wright, B. 2005. The Development and Testing of a Prototype Mobile Mapping System for Real-time Forest Fire Hot Spot Detection. *Photogrammetric Engineering & Remote Sensing* 71, 4: 461–470. [http://www.asprs.org/publications/pers/2005journal/april/].

El-Sheimy, N., Nassar, S., Shin, E.-H. & Niu, X. 2006. Analysis of Various Kalman Filter Algorithms with Different Inertial Systems for INS/GPS Integrated Systems. *Canadian Aeronautics and Space Journal* 52, 2: 59–67.

El-Sheimy N., Hassan, T. & Lavigne, M. 2007a. Mobile Mapping for Automatic Extraction of Highway 3-D Linear Features. *The 3rd Annual Map Middle East Conference, Dubai, UAE, April 9–11.*

El-Sheimy, N., Hassan, T. & Lavigne, M. 2007b. VISAT: Mapping what you see. *Coordinates* 3, V: 6–11.

Farrell, J.A. & Barth, M. 1999. *The Global Positioning System & Inertial Navigation*. New York: McGraw-Hill, Inc.

Goad, C.C. 1991. The Ohio state university mapping system: the positioning component. *Proceedings of the 47th Annual Meeting of Institute of Navigation (ION), June 10–12, Williamsburg, VA*: 121–124.

Graefe, G., Caspary, W., Heister, H., Klemm, J. & Sever, M. 2001. The road data acquisition system MoSES—determination and accuracy of trajectory data gained with the Applanix POS/LV. *Proceedings of 3rd International Symposium on Mobile Mapping Technology (MMS 2001),. Cario, Egypt. January 3–5, 2001.* On CD-ROM.

He, G., Orvets, G. & Hammersley, R. 1996. Capturing urban infrastructure data using mobile mapping system. *Proceeding of the 52nd Annual Meeting, The Institute of Navigation (ION), June 19–21, 1996, Cambridge, MA*: 667–674.

Hock, C., Caspary, W., Heister, H., Klemm J. & Sternberg, H. 1995. Architecture and design of the kinematic survey system KiSS. *Proceedings of the 3rd International Workshop on High Precision Navigation, Stuttgart, Germany, April, 1995*: 569–576.

Hunter, A. & El-Sheimy, N. 2004. Mobile GIS and Speech Recognition. *Photogrammetric Engineering & Remote Sensing* 70, 7: 851–860.

Lambda Tech. 2007. *Welcome to lambda tech international.* Website:http://www.lambdatech.com. Accessed 7 December 2007.

Lapucha, D., Schwarz, K.-P., Cannon, M.E. & Martell, H. 1990. The use of INS/GPS in a highway survey system. *Proceedings of Position, Location and Navigation Symposium (PLANS 1990), Institute of Electrical and Electronics Engineers (IEEE), Las Vegas, Nevada, March 20–23, 1990*: 413–420.

Li, D., Zhong, S.-D., He, S.-X. & Zheng, H. 1999. A mobile mapping system based on GPS, GIS and multi-sensor. *Proceedings International Workshop on Mobile Mapping Technology, Bangkok, Thailand, April 21–23, 1999*: 1-3-1–1-3-5.

Li, Q., Li, B., Chen, J., Hu, Q. & Li, Y. 2001. 3-D mobile mapping system for road modeling. *Proceedings of 3rd International Symposium on Mobile Mapping Technology (MMS 2001), Cario, Egypt, January 3–5, 2001.* On CD-ROM.

Li, R. 1997. Mobile mapping: an emerging technology for spatial data acquisition. *Photogrammetric Engineering and Remote Sensing* 63, 9: 1085–1092.

Mikhail, E.M., Bethel J.S. & McGlone, J.C. 2001. *Introduction to Modern Photogrammetry.* New York: John Wiley and Sons, Inc.

Mostafa, M.M.R. & Schwarz, K.P. 2001. Digital Image Georeferencing from a Multiple Camera System by GPS/INS. *ISPRS Journal of Photogrammetry and Remote Sensing* 56: 1–12.

Morin, K. & El-Sheimy, N. 2002. Post-mission Adjustment Methods of Airborne Laser Scanning Data. *Proceedings of the FIG/ASPRS Annual Conference, Washington D.C., Apr. 2002.* On CD-ROM.

Novak, K. 1991. The Ohio state university mapping system: the stereo vision system component. *Proceedings of the 47th Annual Meeting of the Institute of Navigation (ION), June 10–12, Williamsburg, VA*: 121–124.

Novak, K. 1993. Data collection for multi-media GIS using mobile mapping systems. *Geomatics Info Magazine (GIM)* 7, 3: 30–32.

Novak, K. 1995. Mobile mapping technology for GIS data collection. *Photogrammetric Engineering and Remote Sensing* 61, 5: 493–501.

Reed, M.D., Landry, C.E. & Werther, K.C. 1996. The application of air and ground based laser mapping systems to transmission line corridor surveys. *Proceedings of Position, Location and Navigation Symposium (PLANS 1996), Institute of Electrical and Electronics Engineers (IEEE), Atlanta, GA, April 22–26*: 444–451.

Rothermel, R.C. 1993. Mann gulch fire: a race that couldn't be won. *United States Department of Agriculture, Forest Service, General Technical Report, INT-299.* Ogden, Utah: Intermountain Research Station.

Schwarz, K.-P., Martell, H.E., El-Sheimy, N., Li, R., Chapman, M.A. & Cosandier, D. 1993. VISAT—a mobile highway survey system of high accuracy. *Proceeding of the Vehicle Navigation and Information Systems Conference, Institute of Electrical and Electronics Engineers (IEEE), Ottawa, Canada, October 12–15*: 467–481.

Schwarz, K-P. & El-Sheimy, N. 2005. Digital Mobile Mapping Systems—state of the Art and Future Trends, *ISPRS Book Series—Advances in Mobile Mapping Technology*: 3–18. London: Taylor & Francis Group. ISBN 978-0-415-42723-4.

Škaloud, J. 1999. Problems in Direct-Georeferencing by INS/DGPS in the Airborne Environment. *ISPRS Workshop on Direct versus indirect methods of sensor orientation. Commission III, WG III/1, Barcelona, Spain, November*: 25–26.

Sternberg, H., Caspary, W., Heister, H. & Klemm, J. 2001. Mobile data capturing on roads and railways utilizing the kinematic survey system KiSS. *Proceedings of 3rd International Symposium on Mobile Mapping Technology (MMS 2001), Cario, Egypt, January 3–5, 2001*. On CD-ROM.

Sullivan, D. & Brown, A. 2002. High Accuracy Autonomous Image Georeferencing Using a GPS/Inertial-Aided Digital Imaging System. *Proceedings of ION National Technical Meeting 2002, San Diego, CA, Jan. 2002*.

Toth, C.K. 1997. Direct Sensor Platform Orientation: Airborne Integrated Mapping System (AIMS). *International Archives of Photogrammetry and Remote Sensing, ISPRS Comm. III* XXXII, part 3-2W3: 148–155.

Wang J., Lee, H.K. & Rizos, C. 2002. GPS/INS Integration: A Performance and Sensitivity Analysis. *4th Symposium on GPS/GNSS, Wuhan, China, 6-8 November 2002*.

Wright, B. & El-Sheimy, N. 2003. Real-Time Direct Georeferencing of Thermal Images For Identification and Location of Forest Fire Hotspots. *6th Conference on Optical 3D Measurement Techniques, Zurich, Switzerland, September 22–25, 2003*.

Advances in Photogrammetry, Remote Sensing and Spatial Information Sciences: 2008 ISPRS Congress Book – Li, Chen & Baltsavias (eds)

CHAPTER 8

Small satellite missions

Rainer Sandau

ABSTRACT: This chapter gives some insight into what small satellites are. It indicates the facts and trends driving the rapid evolution of small satellite missions. The actual status is summarized, taking into account the ground segment, the launch segment and the space segment. The latter is elaborated in more detail. So small satellites of Surrey Satellite Technology Ltd. (SSTL), the small satellite initiatives of the USA and in the rest of the world are addressed, as well as small satellites developed by universities, all with an emphasis on the time period beginning in 2004. Constellations and formations are another aspect considered. The last section deals with some visions about the future of small satellite missions for remote sensing, the new capabilities and the challenges. Furthermore, the matter of success and failure of small satellite missions is discussed. The conclusions show that actually small satellite missions are changing the economics of space.

Keywords: Small satellites, new launchers, space tourism

8.1 INTRODUCTION

At the beginning of the space age all satellites were small – by necessity and not by choice. Fifty years ago, the first artificial satellite, Sputnik-1 (launched 4 October 1957) was a 58 cm diameter aluminium sphere with a mass of 83 kg. The first successful satellite of the USA, Vanguard-1 (launched 17 March 1958) was a 16 cm diameter sphere with a mass of 1.6 kg (Meurer 2006).

The incredible increase of knowledge coming from these small space missions induced a huge amount of new questions and the space projects grew bigger and bigger. This was due to the growing complexity of the missions leading to increasing costs and development times. This trend became true not only in the US and the Soviet Union but also in the other regions entering the space field, such as Western Europe and Asia. The space programmes ran into a kind of cost spiral: higher costs led to fewer missions that led to the demand for higher reliability, which again led to longer schedules and higher costs. The return to smaller missions was initiated by:

- The restriction to dedicated missions with only single instruments or sensor systems optimized to observe specific physical phenomena.
- The reduction in space research budgets.

Small satellite missions can be achieved by using different approaches and methods. One approach is to focus on a single task and use available off-the-shelf technology to build a small satellite system (bus and payload) for the intended remote sensing purpose. Another possible approach is to take full advantage of the ongoing technology developments leading to further miniaturization of engineering components, development of micro-technologies for sensors and instruments, which allow designing dedicated, well-focused Earth observation missions. At the extreme end of miniaturization, the integration of micro-electromechanical systems (MEMS) with microelectronics for data processing, signal conditioning, power conditioning and communications leads to the concept of application specific integrated micro-instruments (ASIM). These micro- and nano-technologies have led to the concepts of nano- and pico-satellites, constructed by stacking wafer-scale ASIMs together with solar cells and antennas on the exterior surface, enabling the concept of space sensor webs.

Further milestones in the cost-effective Earth observation mission developments are the availability and improvement of small launchers, the development of small ground station networks connected with rapid and cost-effective data distribution methods, and cost-effective management and quality assurance procedures.

Since the advent of modern technologies, small satellites using off-the-shelf technologies for missions focused on specific physical phenomena have also been perceived to offer an opportunity for countries with a modest research budget and little or no experience in space technology to enter the field of spaceborne Earth observation and its applications. Small satellite technology is a major mean to bring within the reach of every country the opportunity to operate small satellite Earth observation missions and utilize the data effectively at low costs, as well as to develop and build application-driven missions. It provides the opportunity to conduct or participate in Earth observation missions using small, economical satellites, and associated launches, ground stations, data distributions structures and space system management approaches.

The situation in the field of small satellite missions for Earth observation has matured in the last ten years. This may be, for instance, observed from the topics and the quality of contributions to the series of conferences taking place in Berlin, Logan, at the annual International Astronautical Congress (IAC) or conducted by space agencies like ESA or CNES. Also ISPRS WG I/6 is involved in small satellite related conferences like, for instance, EOSS 2007 in Kuala Lumpur (EOSS 2007).

But what exactly is a small satellite? A small satellite is mainly determined by its mass, volume and power consumption. But often mass is used only. Different entities define or name small satellites in dependency on their products or programmes, for instance:

- ESA: Small 350 kg–700 kg
  Mini 80 kg–350 kg
  Micro 50 kg–80 kg
- EADS Astrium: miniXL 1000 kg–1300 kg
  Mini 400 kg–700 kg
  Micro 100 kg–200 kg
- CNES: Mini 500 kg + P/L (Proteus)
  Micro 120 kg + P/L (Myriade)

To end this confusion, IAA proposed a simplified definition (Sandau 2006b), which is based on the first satellite classification of Martin Sweeting (Sweeting 1991) and may of course be extended to smaller spacecraft (S/C) to come (Kramer 2007, EO 2007).

- mini satellites <1000 kg
- micro satellites <100 kg
- nano satellites <10 kg
- pico satellites <1 kg.

Figure 8.1 refers also to additional features that are essential when discussing small satellite characteristics, such as cost and response time (measures of cost-effectiveness).

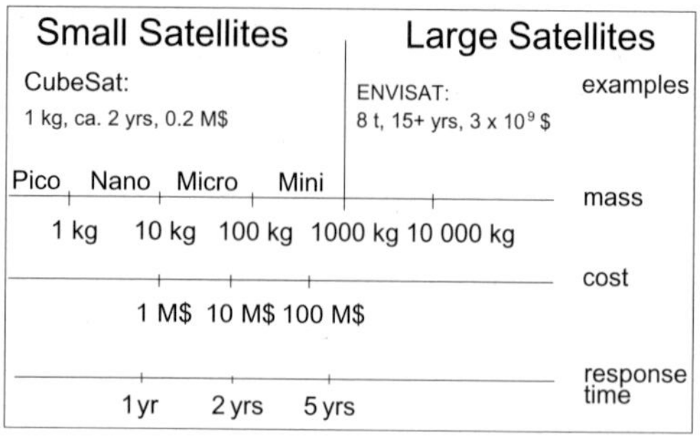

Figure 8.1. Some features of small satellites.

At UNISPACE III (Background Paper 1998), the costs of developing and manufacturing a typical mini-satellite was indicated to be US$ 5–20 million, while the cost of a micro-satellite was correspondingly US$ 2–5 million. The cost of a nano-satellite could be below US$ 1 million (prices as of 1999).

Whereas the development and production time for large satellites is observed to be 15+ years, the corresponding time for minis should be 3–5 years, for micros 1.5 years, for nanos about 1 year, and for picos less than 1 year. Of course, cost and duration figures are to be considered ball-park figures. They are bases on the use of state-of-the-art technology by professional teams. They may deviate considerably if key technology is to be developed and/or the implementation teams are at the beginning of their learning curve (Sandau et al. 2006). Figure 8.1 is complemented by two examples showing the edges of the feature ranges.

8.2 SMALL SATELLITE MISSIONS: FACTS AND TRENDS

There is an increasing need for cost-effective Earth observation missions to meet the information requirements of an almost ever growing range of applications. This is perhaps most clearly seen in the many current moves for international cooperation in the field of environment where measurements from Earth observation satellites are an essential element. This is especially so where we need to acquire, analyse and use data documenting the conditions of the resources and the environment of our Earth on a long term permanent basis. Regional and Continental initiatives such as the ESA-EU GMES (Global Monitoring of Environment and Security) initiative aim to provide timely and quality data, information and knowledge through EO systems. The global dimension of GMES will provide Europe's policy makers with information they need and form an explicit

European contribution to international environmental monitoring endeavours. This kind of approach is pursued in many regions. The question is: how can we implement appropriate space systems in a timely and cost-effective manner?

From Figure 8.1 we can learn that, roughly, the smaller the satellite the less the cost and the response time. This is a strong motivation for the development of small satellite missions. The IAA study (Sandau 2006a) presented the state of the art of small satellite missions and examined several factors that enable one to produce a cost-effective small satellite mission for Earth observation. One of the key enablers in designing a cost-effective mission is availability of key expertise. As the number of successful space-faring nations grows, the pool of expertise available to meet the challenges of small missions grows.

8.2.1 *General facts*

The main advantages of small satellite missions are:

- more frequent mission opportunities and therefore faster return of science and application data.
- larger variety of missions and therefore also greater diversification of potential users.
- more rapid expansion of the technical and/or scientific knowledge base.
- unique possibility to build cost-effective satellite constellations to improve the time resolution for the EO task according to the task dependent necessities or requirements.
- gradual decrease of the performance of a constellation in the case of malfunction of one satellite.
- greater involvement of local and small industry.
- cost-effective replacement of defective satellites.

After some years of global experience in developing low cost or cost-effective Earth observation missions, one may break down the missions into categories like:

- Commercial—Requiring a profit to be made from satellite data or services.
- Scientific/Military—Requiring new scientific/military data to be obtained.
- New technology—Developing or demonstrating a new level of technology.
- Competency demonstration—Developing and demonstrating a space systems competency.
- Space technology transfer/training—Space conversion of already competent engineering teams.
- Engineering competency growth—Developing engineering competence using space as a motivation.
- Education—Personal growth of students via course projects or team project participation.

Large satellite missions and small satellite missions are considered to be complementary rather than competitive. The large satellite missions are sometimes even a precondition for cost-effective approaches.

8.2.2 *Trends*

Small satellite missions are supported by several contemporary trends:

- Advances in electronic miniaturization and associated performance capability.
- The recent appearance on the market of new small launchers (e.g. through the use of modified military missiles to launch small satellites).
- The possibility of 'independence' in space (small satellites can provide an affordable way for many countries to achieve Earth Observation and/or defence capability, without relying on inputs from the major space-faring nations).
- Ongoing reduction in mission complexity as well as in those costs associated with management; with meeting safety regulations, etc.
- The development of small ground station networks connected with rapid and cost-effective data distribution methods.

These trends are complemented by:

- the development of small ground station networks connected with rapid and cost-effective data distribution methods allowing low per satellite operation costs, even for large satellite constellations
- cost-effective management and quality assurance procedures.

If we consider optical EO systems, we have to deal with three types of resolutions: spatial, spectral and time resolution. Due to the progress in many fields of technology, the spatial and spectral resolutions of small satellites are improving, allowing improving performance features, which add to the unique features of small satellite missions: high time resolution depending only on the number of satellites included in the constellation.

8.3 ACTUAL STATUS OF SMALL SATELLITE MISSIONS

In a space mission, all components need to contribute to the cost-effectiveness of the mission; this includes:

- The Ground segment,
- The Launch segment, and
- The Space segment.

8.3.1 *Ground segment*

In the "traditional" approach, all specific tasks like S/C monitoring & control, payload (P/L) data reception and archives, P/L data products and distribution are often assigned to specific facilities. These functions are often supplied separately, with considerable effort required to integrate them, with a "bottom line" of several million Euros. To operate for instance constellations of micro- or nano-satellites, the operation costs have to be low on a per satellite basis. A simple downscaling of such systems for small satellites is not usually cost effective and so more bespoke or *ad hoc* developments that litter the current developer community are not sufficient.

It is envisaged, however, that new missions will return to first principles of ground segment design, tempered with unique insights gained from the development of cost-conscious operational (meteorological) ground systems as well as the current generation of smallsat infrastructure. Using a modular, open architecture based on a data distribution "backbone" and "handlers" to isolate the system from changes and variations in mission protocols and service output requirements, a ground system that would meet the cost, functionality and flexibility requirements of a small satellite mission can be characterized. This thinking is now permeating mission designers and will embrace both Mission Control and Data Processing within a common infrastructure (Sandau 2006a). Key words connected to those qualitative changes are for instance: open systems, automation, "internet" technology, increasing on-board autonomy (this has influence on the space segment!), multi-session operations, ground station networks.

8.3.2 *Launch segment*

Commercial launch services are now available on most launch systems, many of which are new vehicles designed or modified specifically for the international commercial market. The most dramatically shift has been the entry of the Russian and Ukrainian launch systems commercially operated by joint ventures with US or European companies. New launch systems around the world are even beginning to use major components built in other countries, further blurring national divisions.

Access to space of small satellites for Earth observation is still a very limited portion access to space in general. The first limitation is related to the orbit characteristics, as most of the Earth observation is performed at low altitude. Access to space for communication satellites is mostly using the geo-stationary orbit; consequently since the market is dominated by communication satellites, the offer for Earth observation is limited for any secondary payload or piggy-back launch options (Sandau 2006a).

There are a couple of micro launchers on the market, which are used to launch small satellites in dedicated orbits. Table 8.1 gives some features of those micro launchers.

As already mentioned, to launch a small satellite as a secondary payload or piggy back is another option. For instance, the auxiliary payload adapter ASAP of Ariane 5 allows t six small satellites to be launched together with the primary payload.

Generally, some remarkable launch opportunities for small satellites have taken place (Kramer 2007):

- The first dual-payload launch in history took place on 22.06.1960 with the spacecraft Transit-2A and Solrad-1 of NRL, Washington, DC.
- The launch of Oscar-1 (launch 12 December 1961, 4.5 kg) is regarded as the first satellite secondary ride, along with the Discover-36 spacecraft as primary payload. Since then, there have been (infrequent) piggyback launches of small payloads on US launchers.
- In 1989, Arianespace provided a critical catalyst to the micro-satellite community with the introduction of the ASAP platform (first standard interface for small S/C).
- In the 1990s, Russia and Ukraine started to provide affordable multiple small satellite missions into LEO on converted missiles (Dnepr, Zenit-2, Cosmos-3M, Shtil, etc).
- In 2001, Indian Space Research Organisation (ISRO) started with piggyback launches for BIRD (Bi-spectral InfraRed Detection) and PROBA (Project for On-Board Autonomy).

Table 8.1. Micro-launchers (After: Sandau 2006a).

Launcher	Country	Lift-off M(t)	PL Mass LEO	Per/apo km	Inclin. degrees	Price M$
Pegasus XL	USA	22	440	200/200	28.5	10/15.
Shavit	Israel	23	225	370/370	143.4	10/15
Volna	Russia	36	120	200/200	63	nc
Minotaur	USA	36	640	185/185	28.5	12.5
SHTIL	Russia	40	430	200/200	63	0.5
START 1	Russia	47	450	300/300	90	5/10.
VLS 1	Brazil	51	280	400/400	25	6/7.

- Japan (Japan Aerospace Exploration Agency JAXA) and China have also provided occasional secondary launches for small payloads.
- The ESPA ring of AFRL (USA) is a fairly recent development for small S/C. A first launch with multiple small secondary payloads took place on 7 March 2007.

8.3.3 *Space segment*

8.3.3.1 *Small Satellites of SSTL*

The centre of Satellite Engineering (CSER) at the University of Surrey launched its first micro-satellite UoSat-1 on 6 October 1981 as a secondary P/L to NASA's SME (Solar Mesosphere Explorer) mission. This was followed by UoSat-2, secondary P/L to Landsat-5, launched on 1 March 1984 (EE Surrey 1984). Both micro-satellites carried experimental communications and technology demonstrations within the amateur satellite service and operated in orbit successfully for more than eight and five years, respectively. As a consequence, the UoS created a company, SSTL (Surrey Satellite Technology LTD) in 1985, dedicated to small satellite research and development and looking for new concepts in functionality, services and in cost reduction (Kramer 2007, EO 2007). Today, SSTL is regarded as the pioneer of micro-satellite design including a modular and flexible platform, and the development of suitable instruments. Until 2004, SSTL had launched 19 micro-satellites, partly in collaboration with different countries (France, Korea, Portugal, Chile, Thailand, China, Malaysia, USA). Table 8.2 shows the small satellites launched since 2004 and gives an insight to future extensions.

In 1990, SSTL started the first micro-satellite technology transfer programme, providing partnerships and on-the-job training of engineers and scientists of foreign national organizations in cooperative programmes. Many of the mentioned international collaborations included such programmes.

Table 8.2a. SSTL-developed small satellites since 2004, without DMC (Disaster Monitoring Constellation) (EO 2007).

Spacecraft	Launch (end mission)	Payload, objectives, comments	Mass power
TopSat QinetiQ, UK	27 Oct 2005 (ops in 2007)	MicroSat—150 bus (turnkey); technology demonstration for high-resolution imaging; RALCam1 with 2.8 m GSD (Pan), 5.6 m GSD (MS)	115 kg
GIOVE-A (ESA)	28 Dec 2005 (ops in 2007)	Use of newly developed mini-satellite bus, GMP, representing the first element in space for Europe's Galileo Program in MEO (low-cost rapid response approach).	600 kg
CFESat LANL (DOE)	09 Mar 2007	US research technology mission on Microsat—100 bus (turn-key): Radio for VHF/UHF spectrum monitoring for ionospheric and lightning studies, reprogrammable FPGA payloads,	160 kg
RapidEye MDA, (5 S/C)	2007	Microsat—150 bus (turnkey); 5 S/C constellation, commercial MS imaging system in 5 bands at 6.5 m GSD, swath of 78 km;	175 kg 100 W

Table 8.2b. SSTL-developed small satellites since 2004, DMC—initially proposed in 1996 (EO 2007).

Spacecraft	Launch (end mission)	Payload, objectives, comments	Mass power
AlSat-1 CNTS, Algeria	28 Nov 2002 (ops in 2007)	Microsat—100 bus; ESIS (Extended Swath Imaging System) with 32 m resolution MS. 600 km swath:	98 kg
BILSAT-1 TUBITAK, Turkey	27Sep 2003 (Aug 2006)	MicroSat—100 bus; Pan camera: resolution of 12 m, swath of 25 km, FOR: 300 km; MS system: resolution = 26 m, snapshots of 55 km, FOR: 300 km; Coban: 4-band camera with a resolution of 120 m	110 kg
NigeriaSat-1 NASRDA	27 Sep 2003 (ops in 2007)	MicroSat—100 bus; ESIS with 32 m resolution MS, 600 km swath;	98 kg
UK-DMC-1 BNSC, UK	27 Sep 2003 (ops in 2007)	MicroSat—100 bus; ESIS with 32 m resolution MS, 600 km swath; technology demonstrations: commercial internet router,	110 kg
Beijing-1 MoST, China	27 Oct 2005 (ops in 2007)	MicroSat—150: ESIS with 32 m resolution MS, 600 km swath; CMT (China Mapping Telescope): 4 m resolution, 24 km swath, X-band downlink of 40 Mbit/s	168 kg

Table 8.2c. SSTL-developed small satellites since 2004, future extension of DMC with advanced payloads (Baker et al. 2008).

Spacecraft	Launch
Spain-DMC Deimos, Spain	2008
UK-DMC-2 BNSC, UK	2008
KANOPUS 1–3 Russia	2008
NigeriaSat-2 NASRDA	2009
NigeriaSat-X NASRDA	2009

In 1996, SSTL proposed initially a Disaster Monitoring Constellation. The idea was to provide a daily global imaging capability at medium resolution (32 m) in 3-4 spectral bands, for rapid response disaster monitoring and mitigation. By 1999, five subscribers to the DMC idea had signed agreements with SSTL to share observational data of their S/C with every other member in the consortium in the case of a disaster event. The DMC consortium comprises a partnership between organizations in Algeria, China, Nigeria, Turkey and the UK. DMC is the first operational Earth observation constellation providing pushbroom imagery in a wide swath.

The practice of customer team training, as well as the form of the cooperative agreement, has been adopted by other space entities since then.

8.3.3.2 *Small satellite initiatives in the USA*

NASA launched its first mini-satellite IMP-8 (Explorer 50) on 26 October 1973, dedicated to studies of the magnetotail, cosmic rays and solar wind. Until 2004, from different originators like NASA, USAF, DoE, US Navy, Space Imaging, NRO/NRL and Digital Globe, more than 40 mini- and micro-satellites have been launched. Table 8.3 gives a list of small satellites launched since 2004 and planned future missions.

8.3.3.3 *Small satellite developments in the rest of the world*

In this category, the first small satellites EXOS-A (Exospheric Observation Satellite) was launched by the University of Tokyo, Japan, on 2 February 1978 (mass 126 kg). It was together with EXOS-B (launch 16 September 1978, mass 92 kg), the Japanese contribution of the International Magnetospheric Study. Until 2004, nearly 60 small satellites have been launched from different nations including Japan, India, Germany, UK, USA, Sweden, ESA, NASA, Brazil, Korea, Israel, Taiwan, South Africa, Denmark, China, Italy, Cyprus, Russia, Australia, Morocco, Czech Republic and Canada. Table 8.4 lists the small satellites launched since 2004 and planned future launches.

8.3.3.4 *Small satellites developed by universities*

During the 1990s, satellite and payload developments became the programme of choice for challenging, multi-year training courses in quite a few engineering departments at universities throughout the world. The intent was and is to enrich the student training programme, to stimulate interest in a problem-solving multi-disciplinary technical environment, to be imaginative and resourceful, and to take some risk (with ample and essential help from mentors and partners (Kramer 2007). A good amount of enthusiasm and lots of volunteer work by all parties involved are needed to bring such low-cost programme activities to maturity. An invaluable amount of professionalism is gained for all students in such programmes. The main advantage of such programmes is: Students have the "luxury" to fail, something that most in professional space industry must avoid. In fact, failing and learning from failures are some of the most important parts of a student's education. One of the most difficult problems of such small satellite building is to get a launch opportunity as a secondary P/L with some primary P/L. Many things have to be considered for such a piggyback flight including the orbit, the price, and sometimes the long waiting periods. Since the launch of the first university micro-satellite UoSat-1 on 6 October 1981, about 25 more satellites have been launched from different universities in UK, USA, Germany, France, Russia, Israel, South Africa, Italy, Morocco until 2004. Table 8.5 provides some university satellites launched or planned in the time frame 2004-2008.

8.3.3.5 *CubeSats*

The CubeSat idea, concept and programme began in 1999 at SSDL (Space Systems Development Laboratory) of Stanford University under the leadership of Robert J. Twiggs. The CubeSat concept is a first attempt to standardize the bus structure and deployment of pico-satellites for low-cost experiments and applications, in particular for students-built satellites at universities. CubeSat is a name given a cube-shaped pico-satellite design of 10 cm side length and a mass of ≤1 kg. All CubeSats feature a standard form factor and share launches by using standard launch tubes referred to as P-PODS (Poly Picosatellite Orbital Deployers) (Pranajaya et al 2003).

The first CubeSat launches were (Kramer 2007, EO 2007):

- The first multiple CubeSat mission, involving half a dozen picosatellites as secondary payloads, took place on 30 June 2003 on a Rockot KS vehicle of Eurockot from Plesetsk, Russia. The CubeSat S/C are: XI (University of Tokyo), CUTE-I (Tokyo Institute of Technology), CanX-1 of UTIAS/SFL (University of Toronto Institute for Aerospace Studies/Space Flight Laboratory), AAUSat (Aalborg

Table 8.3. Small satellites of the USA since 2004 (EO 2007).

Mission	Sponsor	S/C contractor (kg)	Launch mass (end mission)	Launch date	Comments
XSS-11	AFRL	Lockheed Martin	145	11.04.2005	Rendezvous with inactive S/C (on-orbit servicing)
ST5 (Space Technology 5)	NASA	NASA/GSFC	25 / S/C	22.03.2006 (30.06.2006)	Technology verification of 3 S/C constellation (3 months)
CloudSat	NASA/CSA	BATC	700	28.04.2006 (ops in 2007)	EO: CPR (Cloud Profiling Radar)
CALIPSO	NASA/CNES	Alcatel Alenia Space	635	28.04.2006 (ops in 2007)	EO: (CALIOP, IIR, WFC)
STEREO	NASA	JHU/APL	620/S/C	26.10.2006 (ops in 2007)	2 S/C mission in Sun—Earth connection program
TacSat-2	DoD/AFRL	MicroSat Systems	370	16.12.2006 (ops in 2007)	Technology Demonstration + EO payload
GeneSat-1 (triple-Cube Sat)	NASA/ARC	SCU, Stanford	4.6	16.12.2006 (Feb. 2007 mission completed)	Study the effects of the microgravity environment on biological culture
THEMIS	NASA	Swales (UCB)	128/S/C	17.02.2007	Constellation of 5 Probes in HEO to study magnetospheric substorms
MidSTAR-1	USNA	USNA	120	09.03.2007	Tech demo (ICSat, CFTP)
FalconSat-3	USAFA	SpaceQuest	54	09.03.2007	Tech demo boom deployment
STPSat-1	STP (USAF) AFRL	AeroAstro	170	09.03.2007	EO: SHIMMER, CITRIS, MEPSI
AIM (SMEX mission)	NASA, Hampton University	OSC	210	25.04.2007	Study the causes of Earth's highest-altitude clouds
TacSat-1	OFT, NRL	OSC	132	2007	Tech demo: Copperfield—2 communications package
TacSat-3	NRL, AFRL	Swales Aerospace	<400	2007/8	Hyperspectral imaging mission (VNIR, SWIR)
OCO	NASA/JPL	OSC	407	2008	Global measurements of atmospheric carbon dioxide
ST8	NASA	OSC	194	2009	Tech demos: NGU, Sail-MAST, MLHP, DM)
Glory	NASA	OSC	550	2009/10	Climate research mission: APS, CC, TIM

Table 8.4. Small satellites of the rest of the world since 2004 (EO 2007).

Mission	Sponsor	S/C contractor	Launch mass (kg)	Launch date (end mission)	Comments
FormoSat-2, (ROCSat-2)	NSPO, Taiwan	EADS Astrium	746	20 May 2004 (ops in 2007)	EO (RSI, ISUAL)
DEMETER	CNES France	CNES	129	29 Jun 2004 (ops in 2007)	geophysical parameters (IMSC, IPD, ISL, IAP)
UniSat-3	University of Rome	La Sapienza, GAUSS	12	29 Jun 2004 (ops in 2007)	Tech demos with solar cells
DSP-2	CNSA (China)/ESA	DFHSat, Beijing	330	25 Jul 2004 (ops in 2007)	Study of magnetosphere in geospace

(Continued)

Table 8.4. *(Continued)*

Mission	Sponsor	S/C contractor	Launch mass (kg)	Launch date (end mission)	Comments
PARASOL	CNES, France	CNES	120	18 Dec 2004 (ops in 2007)	EO missin in A-Train of NASA
NanoSat-1	INTA, Spain	INTA	19	18 Dec 2004 (ops in 2007)	Tech demo: OWLS (Optical Wireless Links for intra-Satellite) communications
INDEX, (Reimei)	JAXA	JAXA	72	23 Aug 2005 (ops in 2007)	Tech demo, auroral emissions (EISA, MAC, CRM)
OICETS, (Kirari)	JAXA	NEC Tokyo	570	23 Aug 2005 (ops in 2007)	Inter-orbit laser communication experiments
Monitor-E	KhSC, Russia	KhSC	750	26 Aug 2005 (ops in 2007)	Imaging mission with Pan + MS imagery
CryoSat-1	ESA	EADS Astrium	670	08 Oct 2005	EO (SIRAL, DORIS, LRR) ***launch failure***
TopSat	QinetiQ, RAL, UK	SSTL	115	27 Oct 2005 (ops in 2007)	High resolution Pan + MS imagery
SSETI Express	ESA	European Universities	62	27 Oct 2005	Deployment of 3 CubeSats: UWE-1, NCube-2, XI-V
GIOVE-A	ESA	SSTL	602	28 Dec 2005 (ops in 2007)	Fist S/C of Galileo In-Orbit Validation, secured frequency
CUTE-1.7	Tokyo Inst. Technology	LSS	6	21 Feb 2006 (16 Mar 2006)	Technology demonstrations
FormoSat-3 (ROCSat-3, COSMIC)	NSPO, UCAR	OSC	69/S/C	15 Apr 2006 (ops in 2007)	6 S/C constellation Meteo-rology (IGOR, TIP, CERTO/TBB)
EROS-B	ImageSat, Cyprus	IAI, Israel	350	25 Apr 2006 (ops in 2007)	High resolution Pan imagery (commercial)
COM-PASS-2	Roskosmos, Russia	IZMIRAN, Troitsk	85	26 May 2006 (ops in 2007)	Demonstration of topside sounding techniques
UniSat-4	University of Rome	La Sapienza, GAUSS	10	26 Jul 2006	Technology experiments ***launch failure of Dnepr***
KompSat-2, (Arirang-2)	KARI Korea	EADS Astrium	800	28 Jul 2006 (ops in 2007)	EO (MSC) high resolution Pan + MS imagery
Solar-B, (Hinode)	JAXA NASA, SAO	MELCO	900	23 Sep 2006 (ops in 2007)	Solar physics mission of Japan, USA, & UK (SOT, XRT, EIS)
HIT-Sat	Hokkaido Institute of Technology	HIT, Japan	3	23 Sep 2006	Technology demonstration
SAR-Lupe-1	BMVg, Germany	OHB-System	770/S/C	19 Dec 2006	X-band SAR constellation (5 S/C) of the German Defense Ministry
CartoSat-2	ISRO, India	ISRO	680	10 Jan 2007	High resolution Pan imagery
LAPAN-TUBSAT	LAPAN, Indonesia	TU Berlin	56	10 Jan 2007	Camera with 5 m resolution Camera with 200 m resolution
HY-1B	SOA, Beijing	DFHSat (CAST)	360	11 Apr 2007	Survey of ocean resources
EgyotSat-1	NARSS, Egypt	Yuzhnoye, Ukraine	100	17 Apr 2007	High-resolution MS imager, + S & F communications
SAR-Lupe-2	BMVg, Germany	OHB-System	770 / S/C	02 Jul 2007	X-band SAR constellation (5 S/C) of the German Defense Ministry

(Continued)

Table 8.4. *(Continued)*

Mission	Sponsor	S/C contractor	Launch mass (kg)	Launch date (end mission)	Comments
ZASat-002 Sumbandila-Sat	DST, South Africa	SunSpace, SAC/CSIR	81	2007	Multispectral pushbroom imager (6 bands, 6.25 m resolution, 45 km swath)
CanX-2	UTIAS/SFL, Toronto	UTIAS/SFL	3.5	2007	Tech demo: NanoPS, CMOS/APS imager, Arms. GOE
μLabSat-2	JAXA	JAXA	60 kg	2007	Technology demonstrations, nanosatellite as passenger
Delfi-C3	TU Delft	TU Delft	3	2007	Triple CubeSat demo mission
SMOS	ESA	Alcatel Alenia Space	670	2007	Tech demo in soil moisture & ocean salinity (MIRAS)
RapidEye	RapidEye, Germany	MDA, SSTL	175 / S/C	2007	5 S/C constellation imaging mission (REIS)
PROBA-2	ESA	QinetiQ/ Verhaert, Belgium	130	2007	Science payload for sun monitoring (SWAP, LYRA, TPMU, DSLP), tech. demos
TecSAR	MoD, Israel	IAI/MBT, Israel	300	2007	SAR payload (X-band) on a minisatellite
THEOS	GISTDA, Thailand	EADS Astrium	750	2007	High resolution Pan + MS imagery
STSat-2	MOST, Korea	KAIST, SaTReC	<100	2007/8	Science mission: first launch with Korean-built launcher
RazakSat (MACSat)	ATSB, Malaysia	SaTReCi/, ATSB	190	2007/8	High resolution Pan + MS imagery
X-Sat	NTU, Singapore	DSO, Singapore	120	2008	EO: Iris, PPU
AlSat-2	ASAL, Algeria	EADS Astrium	130	2008	High resolution Pan + MS imagery
OceanSat-2	ISRO, India	ISRO	950	2008	Ocean surface physics (colour, winds, etc.) & atmosphere
SDS-1	JAXA, Japan	JAXA	100	2008	Technology demonstration mission
OceanSat-2	ISRO	ISRO	950	2008	EO (OCM, SCAT, ROSA)
CASSIOPE	CSA, Canada	MDA, Bristol Aerospace	400	2008	Ionospheric research (ePOP), Mail delivery service (CX)
Flying Laptop	University of Stuttgart	IRS	<100	2008	Technology demonstration, EO (MICS, TICS)
PRISM	University of Tokyo	ISSL (UT)	5	2008	Technology pathfinder: use of extensible boom optics system
Jason-2 (OSTM)	CNES/NASA	Alcatel Alenia Space	553	2008	Altimetry (Poseidon-2, AMR, DORIS, TRSR-2, LRA)
ALMASat-1	U. of Bologna, Italy	U. of Bologna	25	2008	Technology demonstration
HAUSat-2	Hankuk Aviation University	Korea	25	2008/9	Technology demonstration of new bus system
Chandrayaan	ISRO, India	ISRO	650	2008	Moon mission of ISRO
PRISMA (2 S/C)	SNSB, Sweden	SSC	150 50	2009	GNC demonstrations: autonomous formation, rendezvous, proximity operations
CryoSat-2	ESA	EADS Astrium	670	2009	EO (SIRAL, DORIS, LRR)

(Continued)

Table 8.4. (*Continued*)

Mission	Sponsor	S/C contractor	Launch mass (kg)	Launch date (end mission)	Comments
SERVIS-2	NEDO, METI	USEF, Japan	900	2009	Tech demo/validation of COTS components
Picard	CNES	CNES	155	2009	Monitoring of solar diameter
Microscope	CNES/ESA	CNES/ONERA	200	2010	Test of equivalence principle, SAGE, DFACS
EnMAP	DLR, GFZ, Germany	OHB System	850	2010	Hyperspectral imaging mission (VNIR, SWIR)
Swarm	ESA	EADS Astrium	500 / S/C	2010	Constellation of 3 S/C to survey the geomagnetic field
LEAP	NCKU, Taiwan	National Cheng Kung University	24	2010/11	Detection of low frequency (ELF/ULF) signals as earthquake precursors
TARANIS	CNES, France	CNES	152	2011	Study of transient event energetic mechanisms (transient luminous emissions)

Table 8.5. Small satellites developed or planned by universities in the time frame 2004–2008 (EO 2007).

Spacecraft	Launch	Comments
UniSat-3 (University of Rome, Italy)	29 Jun 2004	SIC mass of 12 kg, (multiple SIC launch on Dnepr-l vehicle), Uni-Sat-3 is operational as of 2007 (3 years after launch). Payload: technology experiments (solar cell, magnetometer, etc.)
FalconSat-2 (USAFA Colorado)	24 Mar 2006	SIC mass of 19.5 kg, a launch failure on Falcon-l (SpaceX) was experienced, the SIC was 3-axis stabilized combination of solar-pressure spin tapes and hysteresis rods, communications in S-band and VHF
UniSat-4 (University of Rome, Italy)	26 Jul 2006	SIC mass of 10 kg (multiple SIC launch on Dnepr-l vehicle). A launch failure occurred after 2 minutes of flight representing a great setback and disappointment to all institutions involved (22 small satellites -UniSat-4 payload: MPPT (Maximum Peak Power Tracking), COTS cameras, GPS receiver, magnetometers, SIRDARIA (Spacecraft Integrated Reentry Device Aero Resistant Increasing Area), TPS,
LAPAN-TUBSAT TUB +LAPAN (Indonesia)	10 Jan 2007	SIC mass of 56 kg, Payload: commercial high-resolution video camera with a detector of 752 × 582 pixels. The spatial ground resolution is 5 m on a swath width of 3.5 km. A second wide-angle CCD video camera has a resolution of 200 m on a swath of 81 km. Interactive S/C pointing for event monitoring.
FalconSat-3 (USAFA Colorado)	09 Mar 2007	S/C mass of 54 kg, multiple launch on STP-1 flight; demonstration of boom deployment technology; payload: MPACS (Micro Propulsion Attitude Control System), FLAPS (Flat Plasma Spectrometer), and PLANE (Plasma Local Anomalous Noise Environment)
RRISM (ISSL, University of Tokyo)	2008	PRISM (Picosatellite for Remote-sensing and Innovative Space Missions) S/C mass of 5 kg (3-axis stabilized), an extensible boom concept is being used in the optics system to provide sufficient focal length for the imaging task.

(*Continued*)

Table 8.5. (*Continued*)

Spacecraft	Launch	Comments
Flying Laptop (University of Stuttgart, IRS, Germany)	2008	S/C mass of ~100 kg with the objective to demonstrate and qualify new small-satellite technologies for future projects. 3-axis stabilization (micro star tracker) for high-accuracy pointing. Use of an FPGA-based OBC. Payload: MICS (Multispectral Imaging Camera System, TICS (Thermal Infrared Camera System), BRDF (Bi-directional Reflectance Distribution Function) measurements; Demonstration of precipitation measurements in Ka-band, etc. Panoramic camera to provide context colour video imagery of Earth.

University of Denmark), DTUSat (Technical University of Denmark), QuakeSat (Stanford University, Stanford, CA).
- The second multiple spacecraft launch, involving 3 CubeSats (UWE-1, XI-V, and nCube-2), was released/deployed from SSETI-Express, a microsatellite of European students, and itself a secondary payload on a multiple S/C mission. The launch of this S/C mission took place on 27 October 2005 (Cosmos-3M launch vehicle of AKO Polyot from the Plesetsk Cosmodrome, Russia) involving the following S/C: TopSat of QinetiQ (UK) and China-DMC+4 (Beijing-1) of SSTL (UK) as primary payloads. The other secondary payloads on this multi-satellite flight were: SSETI Express (European Students), Mozhayets 5 (Russia), Sinah-1 (Iran) and Rubin-5 (OHB, Bremen, Germany).
- On 26 July 2006, a launch of multiple smallsats on a Dnepr-1 launch vehicle from Baikonur, Kazakhstan (launch provider: ISC Kosmotras), ended in a total launch failure after about two minutes of flight (David 2006).

8.3.3.6 *Constellations and formations*

Small satellites represent the only affordable and meaningful solution for the development of future cooperative distributed space systems (constellations and formations) to provide a new observational dimension in the field of remote sensing (Saylor et al. 2007). Table 8.6 gives some overview of such constellations in operation as well as in planning or development.

8.3.3.7 *Further aspects of the small satellite services*

Today, small satellites are changing the economics of space (Kramer 2007, EO 2007, Sweeting 2007). These S/C embrace cutting edge COTS technology, permitting novel and less-expensive ways to perform meaningful observation missions. In operation and in planning are SAR and hyperspectral imaging missions on small satellites (Bermyn & Dorn 2007, Baker et al. 2007). A major challenge for high-resolution imaging missions is hyperspectral missions on small S/C, as well as the thermal stability and high precision position and attitude control (Sandau 2006b). Some of the ongoing or planned developments are (Kramer 2007, EO 2007):

- The SAR-Lupe constellation (5 SIC, German Defence Ministry) developed by OHB System, Bremen has already two minisatellites in orbit (770 kg satellite). The constellation is planned to be completed in orbit by mid-2008.
- The TecSAR mission of the Israel Defence Ministry (300 kg) is planned for launch in the fall of 2007.
- An example of planned civil minisatellite SAR missions is the AstroSAR-Lite initiative of Astrium Ltd. and SSTL. The goal is to provide low-cost mission solutions with innovative designs to customers in the field of SAR (Synthetic Aperture Radar) observations.
- The PROBA mission of ESA (launch 22 October 2001) is flying CHRIS (Compact High Resolution Imaging Spectrometer), a UK hyperspectral imager developed by SSTL.
- TacSat-3 is a DoD (USA) technology demonstration minisatellite with a planned launch in 2008. It carries ARTEMIS (Advanced Responsive Tactically Effective Military Imaging Spectrometer), a hyperspectral imager of AFRL and the US Army. An equally important aspect of TacSat-3 represents a first implementation of a demonstration of emerging avionics and interface standards, as well as of modular spacecraft bus standards.
- The EnMAP mission of DLR is a small satellite hyperspectral mission (850 kg) under development with a planned launch in 2010.
- ZASat-003 (South African Satellite-003) of SunSpace Ltd. employs a multi-sensor imager for hyperspectral imagery in the region of 400–2350 nm. A launch of the minisatellite is expected in 2010.
- HERO (Hyperspectral Environment and Resource Observer) is a minisatellite mission of CSA (Canadian Space Agency) with a planned launch in 2010 (210 bands in VNIR + SWIR).

Table 8.6. Overview of some satellite constellation missions in remote sensing (EO 2007).

Mission	Launch	Description
GRACE (USA, Germany)	17 Mar 2002	Constellation of 2 minisatellites (432 kg each) in co-planar orbit to measure the Earth's gravity field.
DMC (SSTL)	2002–2005	Constellation of 5 microsatellites to provide daily global medium-resolution imagery (co-planar orbit with cumulative swath)
ST5 (NASA)	22 Mar 2006	Technology verification of a 3 microsatellites constellation for the duration of 3 months.
FormoSat-3/ COSMIC (NSPO, UCAR)	14 Apr 2006	Constellation of 6 microsatellites in 3 orbital planes. Collection of radio occultation data of Earth's atmosphere for operational weather prediction, climate, ionospheric and gravity research
STEREO (NASA)	26 Oct 2006	A 2 satellite constellation (620 kg each) in NASA3 Sun-Earth connection program (heliocentric orbit in the ecliptic plane) to observe the sun in stereo
THEMIS (NASA)	17 Feb 2007	Constellation of 5 Probes in HEO to study magnetospheric substorms, each S/C has a mass of 128 kg
COSMO-SkyMED (ASI)	2007–2009	Constellation of 4 SAR satellites (but 1700 kg each!) in a co-planar orbit to provide SAR imagery in X-band.
SAR-Lupe (Germany)	2006–2008	Constellation of 5 minisatellites in 3 orbital planes providing high- resolution SAR data in X-band. Each S/C has a mass of 770 kg.
RapidEye (Rapid-Eye)	2007 (4th Q)	Constellation of 5 minisatellites (175 kg) in a co-planar orbit to provide high-resolution multispectral imagery (6.5 m) on a daily basis
DMC-2nd generation (SSTL)	2008–2009	Constellation of 3 minisatellites in one orbital plane (Deimos-l, UK-DMC-2, NigeriaSat-2) providing wide-swath medium and high-resolution imagery
Swarm (ESA)	2010	Minisatellite constellation of 3 S/C to provide the best ever survey of the geomagnetic field (multi-point measurements in 3 orbital planes)
GPM (NASA, JAXA)	2010–12	The Global Precipitation Mission is planned with 2 core satellites plus 8 microsatellites. Plus EGPM of ESA.
MMS or MagCon (NASA)	2013	The Magnetospheric MultiScale constellation consists of 4 S/C (in planning)
CICERO (USA, Taiwan)	2013	CICERO (Community Initiative for Continuing Earth Radio Occultation) is in planning phase, a successor to FormoSat-3, - a constellation of a 100 microsatellites

Also SSTL's 3rd DMC generation DMC-3 will include SAR and hyperspectral features and combine it with the constellation advantages.

8.4 THE FUTURE OF SMALL SATELLITE MISSIONS FOR REMOTE SENSING

The IAA study (Sandau 2006a) considered the past experience of the global small satellite community and reviewed and incorporated the work of other studies and bodies that deal with disseminating information about small satellite missions and in promoting the appropriate use of such technology and we have surveyed the state of our current knowledge. But it also brings to light new capabilities as well as challenges that must be addressed in order to produce successful, cost-effective small satellite missions.

8.4.1 *New capabilities*

There are three new developments that may prove to greatly enhance the capabilities of small satellite missions. These are:

1. the convergence of data acquisition and data visualization technologies,
2. the ready availability of new small launchers and the rise of "space tourism", and
3. the development of smaller, lighter, lower power satellites that can act as a constellation or independently.

While there are many other developing technologies that hold promise, these factors may well transform the small satellite enterprise in the next ten years. A mission can be cost-effective and achieve all its measurement requirements without having to actually make all the measurements itself. To put this in concrete terms, NASA has a series of research satellites (Aqua, CloudSat, CALIPSO, PARASOL, Aura and OCO) called the "A Train" that fly in formation. These satellites make individual measurements that support cross platform science. Many of the instruments that image the surface also use ancillary information, such as digital elevation maps, to add context to their products. One could readily envision a small satellite mission that was intended to provide some niche product, such as crop yield forecasting in a particular region. Such a small satellite could produce a very specific measurement, say normalized difference vegetative index (NDVI), which would be corrected for aerosols and clouds using data from the A Train. With such a tightly defined measurement requirement, the spacecraft resource requirements could be quite small and the data system could be designed for "store and forward" operations with the data pushed to the analysis site over the internet. Another approach, as evidenced in SSTL's DMC, is to decrease the ground repeat delay by forming a cooperative that shares data that are produced among the elements of the constellation. Membership is acquired by contracting for the production of an element of the constellation. Each member of the cooperative then gets the benefit of a much shorter revisit time. In short, the economies of scale begin to operate as more members join the cooperative.

Getting into space is still a challenge. During the last ten years there have been more small launchers available and at prices that are quite reasonable compared to the cost of a small satellite. One of the newest and, potentially, most vigorous areas of development of small launchers has come about under the impetus of "space tourism". On 4 October 2004, Burt Rutan and Paul Allen, built and flew the world's first private spacecraft to the edge of space to win the $10 million Ansari X PRIZE. Perhaps the early history of the development of commercial aviation presages the next twenty years of space access. At the turn of the century, air travel was relatively risky and quite expensive. As the commercial market for air transport grew costs dropped as did risk. Now, air transport is so cost-effective that it is used to ship bulky agricultural goods, such as apples, half way around the world at prices that are competitive with local transport and production. To make space tourism viable, the cost of putting a person in space will have to be reduced to of the order of $1M. At those kinds of costs for mass to orbit, small satellite missions will no longer be strongly constrained by launch costs. If we step back from the purely speculative, commercial launch services are now available on most launch systems, many of which are new vehicles designed or modified specifically for the international commercial market. As already mentioned, the most dramatic shift has been the entry of the Russian and Ukrainian launch systems operated as joint ventures with US or European companies. New launch systems around the world are even beginning to use major components built in other countries, further blurring national divisions. This international trend is important because some nations still insist on the use of a "national" launch capability. The increasing availability of these low-cost launchers and the development of dispensers have opened up possibilities for single launches of a constellation as well as individual payloads. The launch of the NASA/DLR GRACE satellites used Eurockot Launch Services, the joint venture owned by Astrium and the Russian company Khrunichev, to place two satellites in a closely controlled formation via a dispenser. This launch was the first commercial use of the Russian SS-19 ICBM, which provides the two booster stages for the ROCKOT launch vehicle with a heritage of 150 flights. At the other end of the cost and mass spectrum, Ariane 5 has been used to launch 6 auxiliary payloads along with the primary Helios satellite. This included Nanosat, Spain's first small satellite, built by the country's INTA national space agency (Instituto Nacional de Técnia Aeroespacial), with a mass of less than 20 kg. In another example, the Cluster mission formed a constellation of four satellites, flying in formation, using two separate launches.

Once the spacecraft is in orbit, the remaining costs are largely associated with operating the spacecraft (including monitoring its health and safety) and collecting the data. As the number of spacecraft increases in a constellation, there would be, without a change in the operations paradigm, a concomitant increase in the costs to operate the constellation. In order to have a cost effective constellation of micro- or nano-satellites, the operations costs have to be low on a per satellite basis especially since some of these constellations are envisaged as consisting of tens or even hundreds of micro- or nano-satellites. Powerful, cheap, microprocessors provide the means for increased autonomy at the individual satellite level and across the constellation. At issue, though, is developing the software to perform these operations and subsequently testing the software so that its operation can be verified before flight. Qualifying these systems for spaceflight will be a challenge that must be addressed.

8.4.2 *Challenges*

The biggest long-term challenge for the small satellite community is that of developing a robust commercial market that supports the infrastructure that has been developed to produce small satellites. Small satellites have appealed to some nations as an instrument of national pride and as a means to focus and enhance the industrial base as well as providing a means of attracting students to a high tech industry. This is, of course, a finite market. After the first few satellites, there has to be reason other than becoming a space-faring nation to invest in, develop, and fly the next space mission and continue the development and training of students. To develop a robust market, small satellite manufacturers must remain relevant and cost-effective. It appears that in many markets, space technology has entered the era of diminishing returns – for example, if you can achieve imagery from space with a spatial resolution of about one metre, do you really gain anything marketable by imaging at one centimetre? This plateau effect means that more vendors can aspire to provide the same product. How many suppliers can the market support? It may be that the market can support more suppliers of imagery if revisit time is a key driver. The user then must draw products from several sources and understand enough about each independent data source so that the desired product can be produced. Raw data products, though, are not likely to capture many more users: tailored products that address specific needs can be supplied by small satellites. The vertical integration of the industry, to provide instruments, data and integrated data products, is likely to spur significant growth.

Until that robust commercial market has been developed, government support will continue to be the financial mainstay of the small satellite community. This situation will remain in force until some economies of scale can be achieved. At this time, SSTL (UK) and RapidEye (Germany) are two notable examples of commercial ventures that have achieved some stability. Also SatrecI (Korea) and SunSpace (South Africa) may serve as examples. They did this by identifying and cultivating a niche market that they are able to address. Much of the small satellite community is still tied to education and research activities – activities that rely on government support. Inter-government cooperative agreements provide the means of broadening the opportunities available to the community. Bureaucracies are averse to risk, however, and small organizations and cooperative agreements are often viewed as risky.

Managing risk is a key problem, then. Since no complex system can be designed and tested against all failure modes, experience is often the best and only guide to making trades. Large organizations tend to have more restrictions on what can fly and may have stringent risk assessment processes. In NASA terms, the confidence in a subsystem or system is called the Technology Readiness Level or TRL of the item. Higher TRLs mean the element has significant flight experience. The highest TRL is assigned to elements with direct flight heritage. Small satellites can be quite effective as platforms to raise the TRL of an element to be used in a latter design. The challenge faced by the small satellite community is to gain a broader acceptance of the notion that TRLs can be raised as an integral part of a mission rather than by implementing a dedicated mission such as the JPL-led Deep Space missions.

Making small satellites more cost-effective calls for new technologies but who then pays to certify these new technologies for spaceflight? There is certainly a higher risk associated with unproven technology. For example, the ready availability of large format detectors at relatively low cost shifts the design choices from being driven by the detector resolution to being driven by other factors such as the interplay between spacecraft stability and off-nadir pointing capability or downlink bandwidth and onboard storage, etc. Can a system be designed that can use these new detectors? How do they behave in space? A small mission is arguably the best way to perform a flight verification because even a failure to operate on orbit, or even to achieve orbit, can still be a successful demonstration from an educational or developmental viewpoint. Cost-sharing between a larger, richer, risk-averse partner and a smaller, poorer more risk-tolerant partner may prove beneficial to both parties.

8.4.3 *Success and failure of small satellite missions*

What makes a mission cost effective? The simplest answer is that the desired end is achieved for a price that is acceptable to all parties. While some mission objectives may only be achieved by the large, complex instruments and spacecraft, there are many uses for small missions. For many potential customers, the best price point is established by sharing risk. If the risk is borne broadly, even a failure to achieve launch can still yield a cost-effective mission because the partners view the educational and infrastructure return as sufficiently high and the other shared aspects of the partnership yield some of the required information. To remain cost-effective in the commercial arena, small missions must be able to incorporate new technologies that reduce costs and improve performance. Small satellite missions face growing competition in regional markets from GPS-based solutions, UAVs, balloons and sensor webs, for example. The chief advantage of satellites is their global access. Exploiting that, and successfully marketing that advantage, will hold the long-term key to keeping small satellites cost-effective.

Assessing whether a mission is successful or not involves many different measures. Assuring that a small-satellite mission is considered successful means that these differing measures must be addressed and considered in the design of the mission. Some of these measures of success are, in fact, much more likely to be fulfilled by a small satellite mission than a large one. For example, students are much more likely to be involved in a small satellite mission. The experience gained in the design, construction, test, flight, operation and data analysis phase of the mission will guarantee "success" in terms of the educational experience of the students. Small satellites can demonstrate new technologies or measurement techniques. If they achieve these goals, they are "successful" even if the scope of the goal is small (for example, a small satellite mission need not inventory the global carbon budget but it could provide a measure of the amount of carbon produced in boreal forest fires). In terms of impact at the national level, a small satellite that is produced by a country may well evoke more pride of ownership than an instrument or participation in a large-scale investigation. In the end, success is subjective: the true measure is whether the programme continues and flourishes.

8.5 CLOSING REMARKS

Small satellites are now changing the economics of space. These S/Cs embrace cutting-edge COTS technology, permitting novel and less-expensive ways to perform meaningful observation missions. In the early 21st century, the spectrum of small satellite services is becoming as impressive as that of their bigger brother satellites, but at considerably reduced costs. Microsatellites experienced an impressive evolution from flying gadgets to real and advanced service providers. In fact, small satellites make it possible to open new fields of services previously considered too expensive, in particular constellation missions providing a good time resolution of the observations. One consequence of this ongoing development may be that small satellites, especially micro-satellites, will change society in a comparable manner to the PC when they entered the mainframe computer market two decades ago.

REFERENCES

Background Paper (No. 9). 1998. Small Satellite Missions—for the Third United Nations Conference on the Exploration and Peaceful Uses of Outer Space. *United Nations Document A/CONF. 184/BP/9*, 26 May 1998.

Baker, A.M., Eves, S., Boland, L., Hodges-Jackson, S. & Sweeting, M.N. 2007. SmallSAT SAR: An Affordable, All-Weather Imaging Solution for Developing Countries. *Proceedings of the International Workshop on Earth Observation Small Satellites for Remote Sensing Applications (eoss2007), Kuala Lumpur, Malaysia, 20–23 Nov. 2007.*

Baker, A.M., Davies, P. & Boland, L. 2008. Moving Towards Commercial Earth Observation Services with Small Satellie Constellations. In R. Sandau, H.-P. Röser & A. Valenuela (eds.), *Small Satellites for Earth Observation–Selected Contributions.* Springer.

Bermyn, J. & Dorn, C. 2007. PROBA Spacecraft Family Small Mission Solutions for Emerging Applications. *Proceedings of the International Workshop on Earth Observation Small Satellites for Remote Sensing Applications (eoss2007), Kuala Lumpur, Malaysia, 20–23 Nov. 2007.*

David, L. 2006. Recent CubeSat Losses Spur Renewed Development. *Space News* 4 Sept. 2006.

Kramer, H.J. 2007. An Overview of Small Satellites in Remote Sensing. *Proceedings of the International Workshop on Earth Observation Small Satellites for Remote Sensing Applications, eoss2007, 20–23 November 2007, Kuala Lumpur, Malaysia.*

Meurer, R.H. 2006. Key Challenges and Opportunities after a 20 Year History of Promoting Small Satellites. *Proceedings of the 4S Symposium: Small Satellite Systems and Services , Chia Laguna Sardinia, Italy, 25–29 Sept. 2006.* ESA SP-618.

EE Surrey. 1984. http://personal.ee.surrey.ac.uk/Personal/L.Wood/UoSAT-2/. Last accessed January 2008.

EO. 2007. http://www.eoportal.org/. Last accessed January 2008.

EOSS. 2007. http://seminar.spaceutm.edu.my/eoss2007/. Last accessed January 2008.

Pranajaya, F.M., Zee, R.E., Lundahl Thomsen, P., Blanke, M., Wisnewski, R., Franklin, L. & Puig-Suari, J. 2003. An Affordable Low-Risk Approach in Launching Research Spacecraft as Tertiary Payloads. *Proceedings of AIAA/USU Conference on Small Satellites, Logan, UT, USA, 13–14 Aug. 2003*: SSC03-VI-1.

Sandau, R. (ed.). 2006a. *International Study on Cost-Effective Earth Observation Missions.* Leiden: Balkema. http://iaaweb.org/iaa/Studies/erthobservation.pdf

Sandau, R. 2006b. Potential and shortcoming of small satellite for topographic mapping. *Proceedings of ISPRS Workshop, Ankara, Turkey, 14–16 Feb. 2006.* http://www.isprs.org/commission1/ankara06/makaleler/Sandau_Small_Satellite.pdf

Sandau, R., Paxton, L. & Esper, J. 2006. Outcomes and Visons of the International Study on Cost-effective Earth Observation Missions. *Proceedings of the 57th IAC (International Astronautical Congress), Valencia, Spain, 2–6 Oct. 2006.* IAC-06-B5.4.04.

Saylor, W.W., Smaagard, K., Nordby, N. & Barnhart, D.J. 2007. New Scientific Capabilities Enabled by Autonomous Constellations of Smallsats. *Proceedings of the 21st Annual AIAA/USU Conference on Small Satellites, Logan, UT, USA, 13–16 Aug. 2007*: SSC07-II-7.

Sweeting, M.N. 1991. Why Satellites are Scaling Down. *Space Technology International*:55–59. (The journal was generally referred to as "Space Technology" which has since been discontinued.)

Sweeting, M.N. 2007. Individual Small Spacecraft and Multiple Satellite based on Constellations for Earth Observation. *Proceedings of the International Workshop on Erth Observation Small Satellites for Remote Sensing Applications (eoss2007), Kuala Lumpur, Malaysia, 20–23 Nov. 2007.*

Advances in Photogrammetry, Remote Sensing and Spatial Information Sciences: 2008 ISPRS Congress Book – Li, Chen & Baltsavias (eds)

CHAPTER 9

Unmanned aerial vehicles for photogrammetry and remote sensing

Jurgen Everaerts

ABSTRACT: Unmanned Aerial Vehicle (UAV) platforms have become a new tool for remote sensing in the past four years. These platforms are complementary to traditional airborne or spaceborne platforms, and they can be operated in circumstances where the traditional platforms cannot, due to cost, lack of flexibility or danger. UAVs need extensive ground support for mission planning, aircraft control and data processing. An overview of the systems in both the airborne and ground segments is given. Using UAVs is not without challenges, since many applicable regulations have not yet been modified to take them into account; this is less of a problem for small, low altitude systems than for high altitude, long endurance platforms. It is expected that a complete integration of UAVs in the airspace will take another five to ten years. Until that time, *ad hoc* solutions will need to be found. A review of the different types of civil UAVs is made; low altitude systems are promising for research in and development of new remote sensing systems and applications. High altitude systems will provide recurrent coverage of regional size. Finally, the use of new systems always leads to new research opportunities.

Keywords: Unmanned platforms, unmanned aerial vehicle (UAV), UAV systems

9.1 INTRODUCTION

In the past four years, the use of unmanned platforms for civil and scientific purposes has received much attention and it has become the subject of research projects around the world from aerospace, remote sensing and telecommunication communities. Some striking examples are the Mars Rovers Spirit and Opportunity that have exceeded all expectations in terms of endurance and science results returned. Closer to Earth, the United States Defense Advanced Research Projects Agency (DARPA) sponsored challenges for autonomous road vehicles (Unmanned Ground Vehicles or UGV) have brought impressive technical advances in a very short time: in the first challenge, in 2004, not a single vehicle was able to run for more than 8 miles; in 2005, five teams completed the 131.6 mile course, four of them in less than 10 hours; and in 2007, the DARPA Urban Challenge saw several cars drive autonomously in simulated urban traffic, with six completing the 60 mile course (DARPA 2007).

Unmanned aerial vehicles (UAVs) have a long tradition in the military world as targets for shooting practice and as platforms for reconnaissance and even for offensive actions. UAVs have come to be used in the civil and scientific world in the past few years as well. However, these systems are totally different in concept from the military ones. They face different challenges and have other opportunities as well. In what follows, military systems are generally not considered. A comprehensive list of close to 800 UAV systems produced in 48 countries is maintained by UVS International (Van Blyenburgh 2007). In parallel to the development of UAV systems and the experiments with UGVs, Unmanned Undersea Vehicles (UUVs) have been used for military, scientific and indusdrial applications.

Although a UAV does not need a pilot in the aircraft, it is not necessarily autonomous. All UAVs are remotely controlled (sometimes called RPV or Remotely Piloted Vehicles) from the ground via real-time command-and-control radio links. Usually at least part of the data (e.g. forward looking video for the pilot) acquired during flight is downlinked to the ground control station (GCS). Since the UAV and GCS cannot function independently, the whole system is called a UAS (Unmanned Aerial System).

9.2 USE OF UAV PLATFORMS

UAV platforms are complementary to the traditional remote sensing platforms, i.e. satellites and manned aircraft. Manned aircraft are well established as platforms for all kinds of remote sensing activities, and they have been used for that purpose for almost 80 years (e.g. Simmons Aerofilms of the UK was established in 1919). Aircraft operate from existing airfields, sharing the airspace with all other air traffic. This leads to fixed costs that are too high for most science projects to bear more than once, and requires the survey size to be large enough to be affordable for research purposes.

Satellites dedicated to Earth observation have been active since the early 1970s (ERTS-1 a.k.a. Landsat 1 was launched in 1972), but not many programmes have offered long term operational services, which makes it difficult to rely on them for commercial projects. Satellite system costs have been borne by government or multinational organizations in all but a few cases, e.g. the RapidEye constellation, expected to be launched in 2008 (RapidEye 2008).

UAV systems draw significant advantages from the fact that they do not carry a pilot: this reduces their weight and complexity significantly, which in turns extends the endurance. Apart from this, UAV based remote sensing systems are perfectly suited for "dull, dirty and dangerous" (UAS Roadmap 2005) missions, opening new research opportunities in areas where temporal resolution is key (e.g. traffic monitoring, crop growth modelling) or where human pilots cannot be sent (e.g. determining the extent of chemical accidents or monitoring volcanic activity).

Low altitude (i.e. below a few 100 metres in most places) UAV systems are small, low to moderate cost, very flexible (can be brought to the survey site by car, do not need highly trained pilots, require almost no ground infrastructure), and aimed at small scale projects (typically the size of a square km), or at experimentation purposes. A new paradigm in remote sensing is coming to life: using systems that are "just-good-enough" instead of using the best.

High altitude (lower stratosphere, above 15 km) UAVs are local geostationary systems in many senses, although they do not image the whole region allocated to them completely instantaneously. They are not affected themselves by weather (except in exceptional circumstances such as hurricanes) or hindered by other air traffic (which is rare at most). This allows them to be flexible (avoiding underlying cloud covered regions; hovering above sites that are affected by crisis situations). This also makes them a perfect choice for telecommunication (USE-HAAS 2006, HAPCOS 2006) as well as for Homeland Security purposes (Lockheed 2008)

Table 9.1 compares these platforms.

9.3 COMPONENTS OF AN OPERATIONAL UAV REMOTE SENSING SYSTEM

As stated before, the Unmanned Aerial System is composed of an airborne UAV and a ground Control Station (GCS) and Data Processing Centre (DPC) segment. Within these segments, several systems are integrated, as shown in Figure 9.1. Although the airborne segment is the most visible part, the heart of the system is in the ground segment that plans the missions and operates the system.

The Airborne segment's systems are:

- The airframe: the mechanical structure that enables flight (wings, fuselage, etc.) and provides room and support for all other systems.
- The propulsion and power generation system. If conventional propulsion is used, a fuel tank needs to be carried; (solar-) electric propulsion requires batteries (or fuel cells).
- The avionics that control the propulsion and control surfaces to execute the flight as directed by the remote pilot. This usually also includes an autopilot system, GPS, altimeter and inertial motion unit. Usually, this is also the central command unit that interfaces with the instrument. On most UAVs, the avionics are equipped with forward- and/or nadir-looking (low resolution video) cameras. This imagery is transmitted to the aircraft control system in the ground segment as part of the telemetry.
- The telecommunication system: this is used to command and control the UAV (via the avionics), to downlink telemetry data (on the status of the airborne systems) and potentially also instrument data.
- One or more instruments, for remote or *in situ* sensing. For short duration flights, on-board data storage is used; for long duration missions, a direct downlink of the data is preferable.

The Ground segment's systems are:

- Mission planning: using telemetry, instrument data, weather data and mission goals, the mission planner determines the best way to achieve the goals. This is then given to the pilot. Mission planning interfaces with Air Traffic Control (ATC) for flight permission.
- Aircraft control: the pilot will monitor the UAV telemetry, and control the aircraft on that basis. Even when the UAV is in autonomous mode, a pilot will usually be required to be present. The pilot communicates with ATC during the mission. In most cases, ATC prefers to receive that communication via the aircraft.
- Telecommunication: this is used to command and control the UAV (via the avionics), to receive

Table 9.1. Properties of traditional and UAV platforms.

	Airborne platforms	Satellite platforms	Low altitude UAV	High altitude UAV
Coverage	**Local to Regional** 10–10,000 km^2	**Global reach** Only low resolution satellites provide a true global coverage; resolution satellites take snap-shots.	**Local** Up to ~1 km^2 Within sight of the controller.	**Regional** Up to 200,000 km^2
Update rate	**Low or non-existing** Multi-year update programs at National Mapping and Cadastral Agencies (NMCAs)	**High** Within days or weeks	**Ad-hoc** No programmes in place	**Weeks-months**
Flexibility	**Flight plans** Needs to be integrated in other air traffic; is dependent on weather	**Orbit constraints** The orbit determines the time of overpass	**Total** Below air traffic, flies whenever required (and if weather allows)	**Flight plans + ad-hoc** Above air traffic, can avoid underlying weather or be sent to crisis events
Quality	**?** Many private companies have high standards, but these may not be interchangeable	**Constant** Satellite observations are usually processed by a very limited number (usually one only) of processing centres.	**?** These systems are mostly under development, their quality will improve.	**Constant** Follows the satellite data processing philosophy
Spatial resolution	**High to Very High** cm to dm	**Low to Medium** m to km	**Very high** Cm	**High** dm
Positional accuracy	**High** Well established methods for sub-pixel accuracy	**Low to medium** Stand-alone precision of several pixels	**High** Well established methods for sub-pixel accuracy	**High** Well established methods for sub-pixel accuracy
Spectral resolution and accuracy	**Not always well controlled** Film is qualitative only; commercial digital systems have wide bands, in photogrammetry this is not an issue.	**Calibrated** Only digital sensors are used on satellites nowadays, all well calibrated.	**?** Depends on the instrument that is carried	**Calibrated** Purpose designed instruments.
Target application(s)	**Mapping, image acquisition**	**Environmental, image acquisition**	**Experimental, crisis management, traffic monitoring**	**Mapping, environmental, crisis management**
System cost (order of magnitude, €)	**1–10 M€**	**10–1000 M€**	**10–1000 k€**	**10–100 M€**

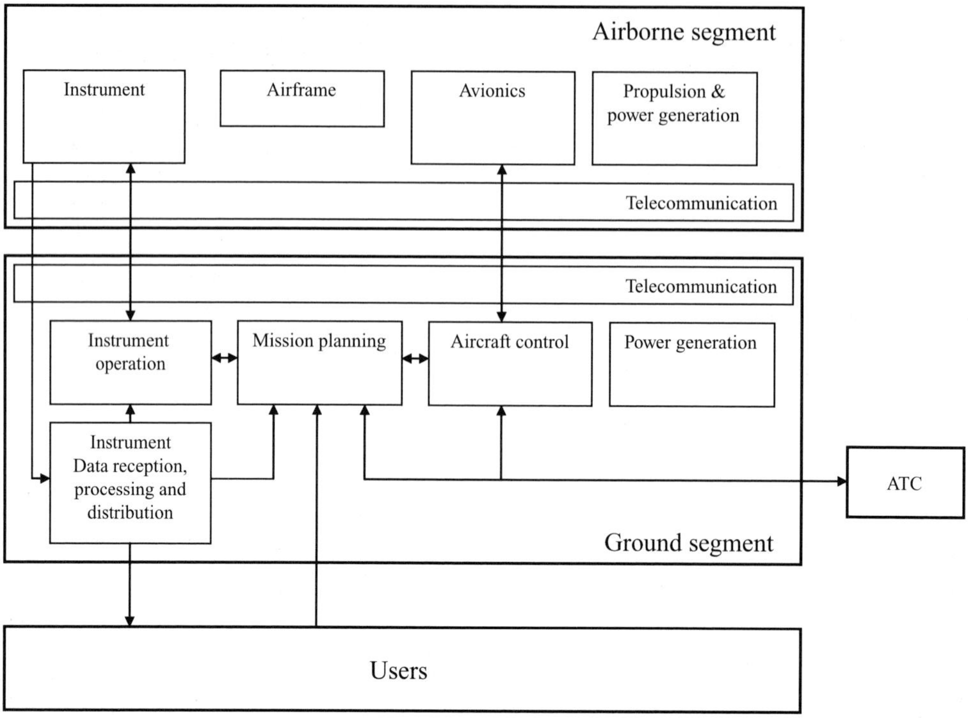

Figure 9.1. The components of an UAV system and their functional interfaces.

telemetry data (on the status of the airborne systems) and potentially also to control the instrument.

- Instrument operation: based on the mission planning, the instrument is operated to collect the data that are required. This may involve setting frame rates, aperture and integration time setting, zooming and pointing (e.g. in a gimballed system).
- Instrument data reception, processing and distribution: the raw data that are received are used for mission planning and instrument operation (processed in a basic way so that the integrity and quality of the data can be assessed); they are then transferred to the processing system. After archiving and applying all necessary processing steps, the results can then be distributed. For high altitude, long endurance (HALE) systems, this processing facility is very much comparable to satellite processing centres.

9.4 CHALLENGES RELATED TO THE USE OF UAV SYSTEMS

A major obstacle so far to the introduction of UAVs for any purpose today is that they have not been integrated into the civil aviation authorities' regulations. For low altitude UAVs, this can be circumvented in many places by qualifying the system as model aircraft; high-altitude systems currently rely on the granting of waivers. In the future, as these systems mature, and due to pressure from military users (who want to integrate their UAVs with other air traffic as well), it is expected that UAVs will be accepted. Unfortunately, accidents with (military!) aircraft have shown that there is much work to be done yet to improve system robustness. For instance, a Belgian B-HUNTER crashed in Kinshasa (Congo) killing a woman in 2006; a US Predator crashed in Nevada in 2007. Both were due to a human error. In Kinshasa, the remote pilot thought the aircraft was not yet in the air when he shut down the engines; in Nevada, the pilot was using a second control station that did not have the exact same functionality. When he tried to lower the landing gear, he actually shut down the engine. Nevertheless, work is ongoing at US FAA, Eurocontrol, EASA, ICAO and NATO.

Another obstacle to easy deployment of UAVs is the need for communication links, both for command and control and for data transmission. Currently, there is no frequency band reserved for UAV operations by the International Telecommunication

Union (ITU), so a UAV operator needs to be able to modify his transmission equipment from country to country. Low altitude systems sometimes use standard WiFi (IEEE 802.11) communication protocols, but this severely limits the range and relative speed of receiver and transmitter, and these protocols have been developed for stationary systems.

9.5 OVERVIEW OF UAV SYSTEMS

Despite the obstacles, many UAV systems have been developed and flown: Table 9.2 shows how the numbers have increased over the past four years. It is evident that this is still very much a military market. The developmental systems are those that so many engineering schools are working on.

9.4.1 *Low altitude platforms*

To guarantee public safety, most low altitude systems are small or designed to fly slowly, and stay within (line-of-) sight of the aircraft controller. Communication links are usually for command-and-control only; storing remote sensing data on board is easy, and it is less complex to land and exchange data storage (memory cards or hard disks) than to transmit all data to a ground receiver.

There is hardly any easier way to realize an unmanned flying platform than to use a tethered balloon or blimp. Although it is definitely low-technology, it has a critical advantage over other platforms: its altitude can be controlled quite easily. It is a platform that is perfectly suited to survey a small area such as an archaeological site (Çabuk 2007) with any type of instrument, and obviously offers a low-risk platform to test instrument prototypes; but it is not inherently stable. The payload capability of lighter-than-air systems is directly related to the balloon's size: at ground level, 1 m^3 of He gas is able to lift 1 kg (including the balloon envelope weight). Another lighter-than-air solution is the untethered blimp, which has also been used in archaeological surveys, or for crisis management (Sanswire 2008).

Table 9.2. Evolution of numbers of UAV projects (Blyenburgh 2007).

	2004	2005	2006	2007
Civil/Commercial	33	55	47	61
Military	362	397	413	491
Dual purpose	39	44	77	117
Research	43	35	31	46
Developmental		219	217	269

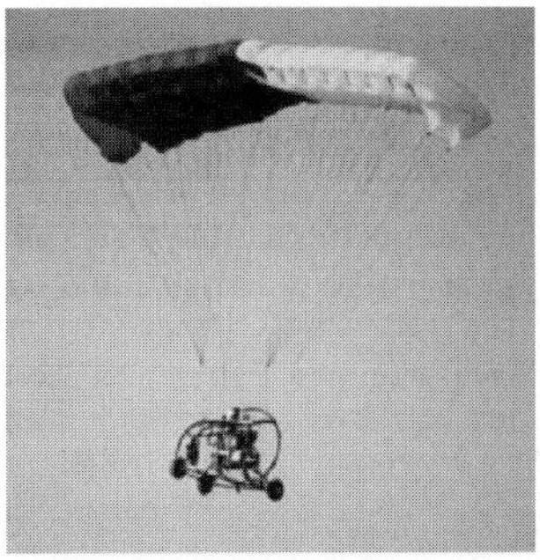

Figure 9.2. A powered paraglider UAV. (see colour plate page 493)

One step up in technology is the powered paraglider. It is powered by an off-the-shelf motor (e.g. chainsaw motor) and has a large payload capacity. Take-off and landing require no more than a small stretch of lawn, but the system is quite sensitive to wind, which may hinder deployment. These systems have nevertheless been deployed in many places around the world. At Bonn University, Hans-Peter Thamm has extensive experience with these (Thamm 2008).

Fixed wing, low altitude systems are usually very small. To generate sufficient lift, they need to fly at relatively high speed, even when carrying a very small payload. Most of these systems are developed in aerospace design groups, as a practical way to demonstrate the aircraft design theory. In this class, the Aerosonde UAV stands out; it has been in service for over 10 years already, was the first UAV to cross the Atlantic Ocean autonomously in 1998, and was recently used to acquire *in situ* data within hurricane Noel in November 2007 (Aerosonde 2008).

Rotary wing (helicopter) UAVs for low altitude are probably the largest segment in the civil UAV market. In Japan, over 2000 of these systems are used in agriculture (spraying pesticides, applying fertilizers, seeding, even ploughing next to remote sensing tasks such as detection of water stress or diseases, harvest decision support and crop loss mapping for insurance). Japanese regulations are noteworthy: these UAV systems are catalogued as "flying ploughs", as part of farming equipment, which has made it possible for the market to develop. Many mature systems are available on the market, with payload capacity from a few kg to several tens of kg.

9.4.2 *Medium altitude platforms*

Hardly any civil UAV systems are designed to operate at medium altitude, in between other air traffic. The integration of these systems in "normal" air traffic is being debated at many levels, but it is not expected to be regulated in the next few years. Most probably,

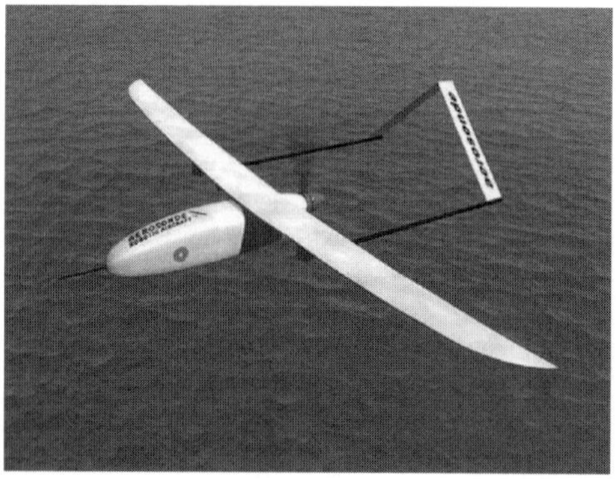

Figure 9.3. The Aerosonde UAV (© Jon Becker, Aerosonde Pty Ltd). (see colour plate page 493)

if allowed to mix with other air traffic, these systems will need to be equipped with "sense-and-avoid" systems that can handle any mid-air crisis situations autonomously. In most countries, medium altitude UAV systems are a liability rather than an opportunity for civil or scientific users.

9.4.3 *High altitude platforms*

Going higher still, above air traffic (12–14 km), there is another space with high potential for unmanned systems. Between 15 and 20 km altitude, wind speeds are minimal and so is turbulence. Remote sensing platforms at that altitude come close to being "stratellites", i.e. local geo-stationary systems, comparable to geo-stationary satellites, but much closer to earth. As for conventional geo-stationary satellites, communication companies are interested in this possibility: it requires less power to receive and transmit digital data over a few 100s of km than over more than 35,000 km, but to be useful, significant payload mass needs to be carried and sufficient power (kilowatts) needs to be available.

High altitude systems need to have a long endurance to make any economic sense: just flying up to 15 km altitude may already require many hours. Truly long endurance (many days to months or longer still) also means that the only realistic source of energy shall be solar (technically, nuclear power is also a possibility). Technical challenges regarding power storage are still important: one can use batteries, but these have a limited capacity (in Wh/kg). Hydrogen fuel cells have a larger capacity, but for long endurance they will have to be reversible (at altitudes above 15 km, relative humidity is almost zero, so there is no water to be collected from the air), and lightweight leak-proof storage tanks for hydrogen and oxygen, both highly reactive, are far from standard items. Both batteries and fuel cells are affected by the low temperatures (the product of the hydrogen fuel cell, water, will freeze), so a part of the capacity will have to be used for self-heating. Finally, transmitting power from the ground to the aircraft (microwave or laser) has been investigated, but it is not sustainable: as stated in Goebel (2006): "In 1987, the Canadian Communications Research Center flew the 'Stationary High Altitude Relay Platform (SHARP)' UAV in a circle at 150 meters (490 feet) above a transmitting antenna. The UAV required 150 watts, and was able to obtain this level of power from the 6 to 12 kilowatt microwave beam."

For telecommunications and security purposes, which require station keeping capability, having a large payload capability and long endurance is imperative. Therefore, these users are looking at stratospheric blimps of truly enormous size. Volumes up to 250,000 m^3 are proposed, and the associated costs are many tens of millions of US dollars. Technical challenges are huge—producing a light-weight He-containing envelope of this size; attaching a large array of flexible solar cells (a large blimp needs a lot of power for station keeping) while the blimp is inflated; and launching the blimp in absolutely windless conditions.

When looking at remote sensing applications, versatility of the platform is important: with an average cloud cover of about 66% worldwide (ISCCP 2005), a successful platform will need to be able to fly to unclouded areas. This can be done with fixed wing aircraft, as was proven by Aerovironment's aircraft Pathfinder and Helios (Goebel 2006); in the past few years, the Zephyr program of QinetiQ has achieved a remarkable track record, culminating in a 54 hour (i.e. flying through two nights) long flight on solar energy only, attaining almost 20 km altitude (QinetiQ 2007). The Zephyr aircraft are extremely light-weight, and are able to carry a small payload only. The Pegasus program of VITO, the Flemish Institute for Technological Research in Belgium, builds on the Zephyr heritage.

9.6 UAV RESEARCH OPPORTUNITIES

UAV platforms can be tools for remote sensing research in many ways.

Low altitude systems give researchers a tool to acquire remote sensing data in a flexible way. So, prototypes of instruments (remote sensing and/or *in situ*) and aiding sensors can be flown and tested, in very short development cycles; and without the need to contract an aerial survey service. This has led many remote sensing research groups to acquire this type of UAV system around the world (Colomina et al. 2007). As a consequence, more thematic research can be started, based on new or existing remote sensing sensors.

New users, such as archaeologists, have already seen the advantage of UAV systems: their areas of interest are usually too small to use traditional systems (due to their high fixed costs), and they have different requirements from traditional remote sensing researchers. Also, in developing countries, low cost platforms are the only way to make remote sensing sustainable (Thamm 2008, Lambers 2007).

In the past few years, the interest in "security" applications and "disaster monitoring" has grown considerably. In a preparatory phase, these applications use traditional remote sensing data. During a crisis (e.g. industrial accident like in Bhopal, 1984 or Chernobyl, 1986; natural disasters like forest fires, flooding, earthquakes), there is an urgent need for real-time information on the actual situation. This can be addressed by UAV systems. Many research projects of this type are ongoing (OSIRIS 2008, SEVESEO 2008).

High altitude platforms (HAPs) can be used as an experimental platform for space instruments. The constraints are similar—mass and power need to be minimized; and since the high altitude platform is able to land, any malfunction can be addressed and design changes can be implemented. The high resolution wide swath digital camera that VITO is building for Pegasus, called MEDUSA (Van Achteren 2006), is funded by the European Space Agency.

HAPs are also interesting for planetary exploration—the conditions at 20 km altitude on earth are comparable to those at low altitude on Mars, for instance.

UAV systems, low and high altitude alike, offer the possibility to explore the temporal domain in remote sensing. This may lead to new applications related to change detection, traffic monitoring, vegetation growth modelling, etc. The boundary between remote sensing, surveillance (border patrol, crowd control, etc.) and crisis management (monitoring forest fires, volcanic activity, industrial accidents, including the integration of remote sensing platforms with sensorwebs) is an area of intense research already, and this is likely to grow. In this case, the remote sensing instruments are considered to be a part of a sensor web, which also includes *in situ* sensors (water level gauges, air quality sensors, etc.) and models (e.g. dispersion models to predict the evolution of a pollution plume in the atmosphere).

UAV systems will need to have sense-and-avoid ability, before they will be able to participate in the civil air traffic. This requires detection of other aircraft and obstacles, and reliably estimating the chances of collision. Already, this is being addressed with image analysis (Carnie 2006).

REFERENCES

Aerosonde. 2008. http://www.aerosonde.com.

Çabuk, A., Deveci A. & Ergincan F. 2007. Improving Heritage Documentation. *GIM International* 21, 9, September 2007.

Carnie, R., Walker, R. & Corke, P. 2006. Image Processing Algorithms for UAV "Sense and Avoid". *Proceedings 2006 IEEE Conference on Robotics and Automation, Orlando, Florida*: 2848–2853.

Colomina, I. et al. 2007. The uVISION project for helicopter-UAV photogrammetry and remote-sensing. *Proceedings of the 7th Geomatic Week, Barcelona, Spain.*

DARPA. 2007. http://www.darpa.mil/grandchallenge (accessed 14 February 2008).

Goebel, G. 2006. *Unmanned Aerial Vehicles.* http://www.vectorsite.net/twuav.html (accessed 14 February 2008).

HAPCOS. 2006. *High Altitude Platforms for Communications and Other Services. EU COST-297 Action.* http://www.hapcos.org (accessed 14 February 2008).

ISCCP. 2005. *International Satellite Cloud Climatology Project.* http://isccp.giss.nasa.gov (accessed 14 February 2008).

Lambers, K., Eisenbeiss, H., Sauerbier, M., Kupferschmidt, D., Gaisecker, Th., Sotoodeh, S. & Hanusch, Th. 2007. Combining photogrammetry and laser scanning for the recording and modelling of the late intermediate period site of Pinchango Alto, Palpa, Peru. *Journal of Archaeological Science* 34(10), 1702–1712.

Lockheed. 2008. *Lockheed Martin High Altitude Airship.* http://www.lockheedmartin.com/products/HighAltitudeAirship/index.html (accessed 14 February 2008).

OSIRIS. 2008. *Open architecture for Smart and Interoperable networks in Risk management based on In-situ Sensors, EU FP6 Integrated Project.* http://www.osiris-fp6.eu (accessed 14 February 2008).

QinetiQ. 2007. http://www.qinetiq.com/home/newsroom/news releases homepage/2006/3rd quarter/QinetiQ s Zephyr UAV achieves flight record.html (accessed 14 February 2008).

RapidEye. 2008. http://www.rapideye.de (accessed 14 February 2008).

Sanswire. 2008. http://www.sanswire.com/sas51/ (accessed 14 February 2008)

SEVESEO. 2008. *EO-Based Services in Support to Industrial and Technological Risk. Management, ESA project.* http://www.seveseo.eu/ (accessed 14 February 2008).

Thamm. 2008. *THE LOW COST DRONE, a new tool for the very high resolution remote sensing.* http://www.rsrg.uni-bonn.de/Projekte/hape_dronen_Webpage/HP_Drone_00_start.html (accessed 14 February 2008).

UAS. Roadmap 2005. *Unmanned Aircraft Systems Roadmap 2005–2030, The Office of the Secretary of Defence, Washington DC, USA.* See http://www.acq.osd.mil/usd/Roadmap%20Final2.pdf (accessed 14 February 2008) for an online copy.

USE-HAAS. 2006. *Study on High Altitude Aircraft (HAAS) and Airships Deployed for Specific Aeronautical and Space Applications. EU FP6-2002-AERO-2 SSA Project Contract No 516081*. http://www.usehaas.org (accessed 14 February 2008).

Van Achteren, T., Delauré, B. & Everaerts, J. 2006. Instrument Design for the Pegasus HALE UAV payload. *The International Archives of the Photogrammetry, Remote Sensing and Spatial Information Science* XXXVI–1/.

Van Blyenburgh, P. (ed.). 2007. *UAS Unmanned Aircraft Systems, The Global Perspective 2007/2008*. Paris: UVS International. See http://www.uvs-info.com (accessed 14 February 2008) for an online copy.

Thamm. 2008. *THE LOW COST DRONE, a new tool for the very high resolution remote sensing*. http://www.rsrg.uni-bonn.de/Projekte/hape_dronen_Webpage/HP_Drone_00_start.html (accessed 14 February 2008).

QinetiQ. 2007. http://www.qinetiq.com/home/newsroom/news_releases_homepage/2006/3rd_quarter/QinetiQ_s_Zephyr_UAV_achieves_flight_record.html (accessed 14 February 2008.

Part III
Data processing and analysis

Advances in Photogrammetry, Remote Sensing and Spatial Information Sciences: 2008 ISPRS Congress Book – Li, Chen & Baltsavias (eds)

CHAPTER 10

Remote sensing signatures: Measurements, modelling and applications

Shunlin Liang, Michael Schaepman & Mathias Kneubühler

ABSTRACT: Signatures from five remote sensing domains—spectral, spatial, angular, temporal and polarization—provide the basis for the description and discrimination of Earth surfaces and their variability. These signatures have been used for a wide range of terrestrial applications. In this chapter, we review the measurements, modelling and applications of these signatures with emphasis on recent advances, and a focus mainly on optical remote sensing.

Keywords: Remote sensing, signature, spectral, spatial, angular, temporal, polarization

10.1 INRODUCTION

The radiation signals (L) received by a remote sensor can be generally represented mathematically as follows:

$$L = f(\lambda, A, t, \Theta, p, \Psi_a, \Psi_s) \tag{1}$$

where λ indicates spectral dependence; A refers to the spatial context and t the temporal variations; Θ represents the angles specifying the illumination viewing geometry and p the degree to which the radiation is polarized; Ψ_a and Ψ_s represent the parameter set describing the atmosphere and the land surface, respectively.

One of the most important objectives of terrestrial remote sensing is to estimate surface properties (Ψ_s) from radiation signals (L) at specific spectral bands, spatial and temporal resolutions, viewing geometry and polarization states. These five domains (spectral, spatial, angular, temporal and polarization) are denoted as remote sensing signatures (Gerstl 1990).

In the following, we review recent advances in acquisition and use of these five signatures in the aspects of measurements, modelling, inversion and applications. Although these signatures are discussed separately, they are ultimately measured, modelled and utilized in combination for the most part. Worldwide conferences, such as the "International Symposium on Physical Measurements and Signatures in Remote Sensing" (2005, China; 2007, Switzerland), organized by us on behalf of the ISPRS Commission VII Working Group 1, reflect the research involving multiple signatures.

10.2 SPECTRAL SIGNATURES

For any given object on the land surface, the amount of solar radiation that is reflected or emitted varies with wavelength. The spectral signatures are the radiation signals collected at different spectral bands that form the basis to classify land surfaces and/or evaluate their geophysical and biophysical properties.

10.2.1 *Measurements*

The spectral properties of land surfaces are measured by either multispectral or hyperspectral sensor systems depending on the number and spectral width of bands. Table 10.1 shows the multispectral bands of some typical moderate resolution imaging radiometers.

Hyperspectral imagers typically acquire hundreds of contiguous spectral bands throughout the visible to thermal infrared spectrum. Multispectral signatures are discrete compared to the contiguous signatures obtained from hyperspectral images. Widely used hyperspectral imagers include the airborne AVIRIS, CASI, AISA and Hymap, as well as the satellite-based Hyperion and CHRIS employed mainly for terrestrial applications.

Table 10.1. Some of the principal sensors useful for terrestrial moderate resolution remote sensing, revised from Townshend & Justice (2002).

Sensor	Platform	Spatial resolution (m)	Swath (km)	Spectral bands (nm)
AATSR	ERS-2	1000	500	480–500
				540–560
				630–690
				795–835
				1550–1700
AVHRR	NOAA–POES	1100	2700	580–680
				725–1000
				1580–1640
				3550–3930
				10,300–11,300
				11,500–12,500
MERIS	Envisat	300	575	660–670
		1200	1150	855–875
MODIS	Terra and Aqua	500		459–479
		500		545–565
		250		620–670
		250		841–876
		500		1230–1250
		500		1628–1652
		500		2105–2155
		1000		3929–3989
		1000		10,780–11,280
		1000		11,770–12,270
POLDER	ADEOS	6000	2400	433–453
				555–575
				660–680
				845–885
SeaWiFs	OrbView2/Sea Star	1100	2800	443–453
		4500		545–565
				660–680
				845–885

Microwave remote sensing uses several wavelengths, which were given code letters during World War II and which remain unchanged to this day (Table 10.2).

Lidar remote sensing relies on active laser systems operating at several wavelengths. Topographic mapping and vegetation monitoring Lidars generally use 1064 nm lasers, while bathymetric systems typically use 532 nm frequency lasers due to the ability of this frequency to penetrate water with less attenuation.

10.2.2 *Modelling*

Spectral signatures can be modelled using three methods: radiative transfer (RT), geometric-optical (GO), and computer simulations. The distinction between RT and GO models is becoming increasingly blurred because hybrid models that integrate RT and GO models have been developed. Computer simulation models require extensive computer resources and processing time and are appropriate for surface radiation simulations. The basic principles and representative models have been discussed by Liang (2004) and different authors of the special issue of Remote Sensing Reviews (Liang & Strahler 2000), and recent advances are reviewed below.

The development of new radiative transfer models has slowed significantly in recent years. An exhaustive review of existing literature resulted in only a few publications describing new RT models in a wide variety of fields. For example, Pitman et al. (2005) applied a numerical RT algorithm to calculate quartz emissivity. Kokhanovsky et al. (2005) developed an approximate snow reflectance model based on the asymptotic solution to the RT equation. Li & Zhou (2004) simulated the snow-surface bidirectional reflectance factor

Table 10.2. Microwave sensor band codes.

Band	Frequency (GHz)	Wavelength (cm)	System applications
Ka	40–26	0.8–1.1	Early airborne radar systems
K	26.5–18.5	1.1–1.7	Early airborne radar systems
X	12.5–8	2.4–3.8	Extensively on airborne systems for military reconnaissance and terrain mapping
C	8–4	3.8–7.5	Common on many airborne research systems (NASA AirSAR) and spaceborne systems (including ERS-1 and 2 and RADARSAT).
L	2–1	15.0–30.0	Used onboard American SEASAT and Japanese JERS-1 satellites and NASA airborne system
P	1–0.3	30.0–100.0	Longest radar wavelengths, used on NASA experimental airborne research system

(BRF) and hemispherical directional reflectance factor (HDRF) of snow-covered sea ice multilayered azimuth- and zenith-dependent plane-parallel RT model.

In the field of vegetation canopy studies, recent efforts are mainly focused on determining the three-dimensional (3-D) structure of the canopy using one-dimensional (1-D) models (Pinty et al. 2004a, Rautiainen & Stenberg 2005, Smolander & Stenberg 2003, Widlowski et al. 2005) or stochastic radiative transfer models (Kotchenova et al. 2003, Shabanov et al. 2005). Liangrocapart & Petrou (2002) developed a two-layer model of the bidirectional reflectance of homogeneous vegetation canopies, taking into account the anisotropic scattering of both the vegetation canopy and the background, such as bare soil or leaf litter. Community efforts to compare some vegetation radiative transfer models are ongoing (Pinty et al. 2004b, Widlowski et al. 2007). Nilson et al. (2003) demonstrated the possible applications of a multipurpose forest reflectance model. Combining leaf radiative transfer models with canopy scale models in various permutations has become more popular and is increasingly used in scaling based approaches (e.g. PROSPECT/GeoSAIL, PROSPECT/FLIGHT in Koetz et al. (2004); PROSPECT/DART (Malenovsky et al. 2008); and other coupled approaches in Bacour et al. (2002)).

The classic GO models essentially characterize the interaction of direct solar radiation with land surfaces. Including the diffuse radiation field in the GO model leads to a hybrid RT/GO model (Peddle et al. 2004). GO models have been used recently for classifying forest types and estimating biophysical parameters (Peddle et al. 2004) and detecting forest structural change (Peddle et al. 2003, Zeng et al. 2008b) from Thematic Mapper (TM) imagery, modelling soil reflectance (Cierniewski et al. 2004), determining the gap fraction of forest canopy (Liu et al. 2004), and estimating woody plant coverage of the grasslands (Chopping et al. 2006), and background reflectance (Canisius & Chen 2007) from multiangular observations. The same principle has also been used for topographic correction of remote sensing imagery in forested terrain (Soenen et al. 2005).

Little progress has been made in developing computer simulation models (e.g. radiosity, Monte Carlo ray tracing), but several studies recently use this approach. For example, Casa & Jones (2005) estimated potato crop biophysical parameters using a look-up table (LUT) created from a ray tracer. Börner et al. (2001) developed an end-to-end multispectral and hyperspectral simulation tool based on the ray tracing principle. Ray-tracing methods have been used to simulate both optical and microwave signatures (Disney et al. 2006) and to estimate forest structural parameters (Kobayashi et al. 2007).

10.2.3 *Signature generation and applications*

The subset of the spectral signatures generated from a set of measurements at different wavebands is usually more valuable for specific applications. Colour compositing using two or three bands for visual interpretation is the classic example. Many quantitative models or image classifications are normally based on a few bands or their combinations. The linear transformation techniques include principal component transformation and Tasselled Cap transformation.

More popular methods are based on vegetation indices (VI). Earlier developed indices have been extensively summarized by Liang (2004). Numerous new indices have been put forth. A comprehensive analysis of broadband and narrowband vegetation indices and their angular sensitivity is described in Verrelst et al. (2008). A brief description of some of these algorithms follows. Gitelson et al. (2003) compared a series of indices and found the following three perform very well:

$$(R_{800} - R_{700})/(R_{800} + R_{700}), \quad R_{860}/(R_{708} * R_{550}) \text{ and } R_{750\text{-}800}/R_{695\text{-}740} - 1 \tag{1}$$

The last index is linearly related to chlorophyll concentrations. Ustin et al. (2008) compared the retrieval capacity of indices and models for the plant pigment system. In a recent study estimating leaf area index (LAI) and crown volume (VOL), Schlerf et al. (2005) demonstrated that linear regression models quantify LAI and VOL accurately using hyperspectral image data. Harris et al. (2005) used the floating-position water band index to estimate leaf water moisture. To assess the water content of vegetation, they also compare leaf water moisture to the normalized difference water index (NDWI) and the moisture stress index (MSI). The normalized difference snow index (NDSI) is an indicator of snow cover (Salomonson & Appel 2004). Chen et al. (2005b) developed a biological soil crust index (BSCI) that exaggerates the difference between biological soil crusts and bare sand, dry plant material or green plant backgrounds. Chikhaoui et al. (2005) proposed a "land degradation" index to characterize land degradation in a small Mediterranean watershed using Advanced Spaceborne Thermal Emission and Reflection Radiometer (ASTER) data and ground-based spectro-radiometric measurements.

Although the conventional multivariate regression analyses using spectral signatures are widely used, different machine learning methods and other advanced statistical analysis techniques show great potential, such as artificial neural network (ANN) methods for estimating various biophysical variables (Fang & Liang 2003, 2005); genetic algorithms for estimating LAI (Fang et al. 2003), regression tree methods for estimating fractional vegetation coverage (Hansen et al. 2002), Bayesian networks for estimating LAI (Kalacska et al. 2005), and support vector machines for estimating LAI from multiangular observations (Durbha et al. 2007).

Use of spectral signatures via physical reflectance or emittance models (see Section 2.2) and optimization methods has been popular recently in terrestrial remote sensing. For example, Gascon et al. (2004) estimated LAI, crown coverage and leaf chlorophyll concentration from SPOT and IKONOS imagery using a 3-D canopy radiative transfer model. To reduce computational requirements, some parametric functions are determined using LUTs created by the 3-D reflectance model. Meroni et al. (2004) applied this algorithm to invert LAI from hyperspectral data. Schaepman et al. (2005) inverted biophysical and biochemical variables from multiangular and hyperspectral remote sensing data using a coupled leaf-canopy-atmosphere radiative transfer model. The multiangle imaging spectroradiometer (MISR) science team also used this method to produce land surface products (Diner et al. 2005). Qin et al. (2008) incorporated the adjoint algorithm of the canopy RT model in the optimization process to speed up computation.

The high computational demands of the optimization approach have led to the use of simpler surface reflectance models rather than forcing optimization algorithm efficiencies. One of the general trends in terrestrial remote sensing is to use simpler empirical or semi-empirical models. The optimization algorithms are used to estimate the parameters in these simple models. The parameters are then related to surface properties. For example, Widlowski et al. (2004) fitted a simple BRDF model to multiangular observations and then linked the surface structural properties to one of the parameters. Chen et al. (2005c) used this approach to map the global clumping index from multiangular observations.

An alternative solution to overcome the high computational demand problem is to use the LUT approach, which has been used for a variety of remote sensing inversion issues, such as atmospheric correction (Liang et al. 2006b), estimating LAI (Koetza et al. 2005) and incident solar radiation (Liang et al. 2006a). In an ordinary LUT approach, the dimensions of the table must be large enough to achieve high accuracy, which leads to much slower on-line searching. Moreover, many parameters must be fixed in the LUT method. To reduce the dimensions of the LUTs for rapid table searching, Gastellu-Etchegorry et al. (2003) developed empirical functions to fit the LUT values so that a table searching procedure becomes a simple calculation of the local functions. Alternatively, Liang et al. (2005) developed a simple linear regression instead of table searching for each angular bin in the solar illumination and sensor viewing geometry. Dorigo et al. (2007) gave an overview of models and LUT optimization techniques used in agro-ecosystem modelling.

Spectral signatures have been used to estimate atmospheric aerosol loadings and water vapour contents that are especially important for atmospheric correction of optical remote sensing imagery. The differential absorption technique is widely used to estimate the total water vapour content of the atmosphere directly from multispectral or hyperspectral imaging systems. The general idea is to utilize one spectral band in the water absorption region (e.g. 0.94 μm) and one or more bands outside of the absorption region. In a recent study, Liang & Fang (2004) applied an ANN to estimate water vapour from hyperspectral data. Miesch et al. (2005) developed a water vapour correction algorithm for hyperspectral data using Monte Carlo simulations. A relatively long history exists for estimating aerosol loadings from remotely sensed imagery using the spectral signatures such as the "dark-object" methods (e.g. Hsu et al. 2004, Liang et al. 1997) and the hyperspectral method (Liang & Fang 2004);

10.3 SPATIAL SIGNATURES

Land surface information is acquired by sensors at different spatial resolutions, which are usually characterized by the pixel size on the ground. Spatial resolution is sometimes expressed in terms of Instantaneous Field of View (IFOV). The spatial signature depends on the spatial resolution of the sensor and its spatial response in addition to the surface variations. Generally speaking, the coarser the spatial resolution, the less variation there is between pixels in the image (Curran 2001, Jupp et al. 1988, 1989, Woodcock & Strahler 1987).

10.3.1 *Measurements*

Current remote sensing data are of different spatial resolutions, such as ultra resolution (e.g. QuickBird, Ikonos), fine resolution (e.g. ETM+, ASTER), medium resolution (e.g. MODIS, MERIS) and coarse resolution. The microwave sensors are also providing finer spatial resolutions from kilometres to metres (Table 10.3).

10.3.2 *Modelling*

Spatial signatures characterize the spatial dependencies of pixel values, which are generally spatially autocorrelated, non-stationary, non-normal, irregularly spaced and discontinuous. Different methods can be used to model and characterize these spatial signatures and we discuss a few typical methods. Although there are also many other methods such as the Moran's I spatial autocorrelation metric for coral reef study (Purkis et al. 2006), lacunarity methods and fractal triangular prisms for classifying urban images (Myint & Lam 2005), they are not addressed here. Additional details are available elsewhere (Lam 2008).

10.3.2.1 *Geo-statistics*

Geo-statistics, as part of spatial statistics, have been widely employed in remote sensing (e.g. Bailey & Gatrell 1995, Cressie 1993, Curran 1988, Goovaerts 1997, Woodcock et al. 1988a, 1988b). The typical geo-statistical measures may include: (i) variogram, (ii) covariance function, (iii) correlogram, (iv) general relative variogram, (v) pairwise relative variogram, (vi) rodogram, and (vii) madogram. Combining GO models with geo-statisitcal methods such as regression kriging or annotated co-kriging substantially improves the elimination of model imperfections and allows the production of spatially continuous information (Zeng et al. 2008a).

10.3.2.2 *Grey level co-occurrence matrix*

Co-occurrence measures use a grey tone spatial dependence matrix to calculate texture values of the image. This is a matrix of relative frequencies in which pixel values occur in two neighbouring processing windows separated by a specified distance and direction. It shows the number of occurrences of the relationship between a pixel and its specified neighbour. This type of spatial structure information is particularly useful for high resolution images.

Table 10.3. Characteristics of some typical microwave sensors, adapted from Shi (2008).

Sensor	Frequency in GHz	Polarization	Incidence In degree	Pixel Resolution in m	Available Time Frame
ERS-1/2	5.3	VV	Fixed at 23°	3.8 to 12.5	Since 1991/1995
ASAR	5.33	Dual-polarization	varying with mode	3.8 to 150	Since 2002
Radarsat-1	5.3	HH	varying with mode	10 to 100	Since 1997
SIR-C/X-SAR	1.25 and 5.3/9.6	Fully polarimetric/VV	varying with data takes	6 to 27	April and October, 1994
JERS-1	1.27	HH	Fixed at 35°	18	1994–1997
PALSAR	1.27	Fully polarimetric, Dual-polarization, and HH	varying with mode	10 to 100	Since Dec. 2005
Radarsat-2	5.3	Fully polarimetric, Dual-polarization, and HH	Varying with mode	3 to 100	2006
Terral-SAR	9.6	VV or Dual-polarization	Varying with mode	1 to 16	2006

Many investigations consider the use of the grey level co-occurrence probability texture features for classification purposes only (Jobanputra & Clausi 2006).

10.3.2.3 *Fourier transformation*

Power spectrum analysis is useful for those images that have regular wave patterns with a constant interval, such as wave patterns of sand dunes. Fourier transformation is applied to determine the power spectrum, which gives the frequency and direction of the pattern. Fourier transformations are typically used for the removal of noise such as striping, spots, or vibration in imagery by identifying periodicities (areas of high spatial frequency).

10.3.2.4 *Wavelet transform*

The wavelet transform can be effectively used as a spatial analysis tool for modelling spatial relationships of remote sensing data. Wavelet analysis is a technique to transform an array of N numbers from their actual numerical values to an array of N wavelet coefficients.

Since the wavelet functions are compact, the wavelet coefficients measure only the variations around a small region of the image data. This property makes wavelet analysis very useful for image processing; the "localized" nature of the wavelet transform allows easy identification of features in the image data such as spikes (e.g. noise or discontinuities), discrete objects, edges of objects, etc. The localization also implies that a wavelet coefficient at one location is not affected by the coefficients at another location in the data. This makes it possible to remove "noise" of all different scales from a signal, simply by discarding the lowest wavelet coefficients.

10.3.2.5 *Markov random field models*

A Markov random field (MRF) is recognized as a powerful stochastic tool to model the joint probability distribution of image pixels in terms of local spatial interactions. A wide range of MRF models have been proposed over the last several decades. Essentially, an MRF model considers an image as a realization of a Markov random field, and is often parameterized by a function, consisting of two basic components: observable variables and model parameters. MRF models can be used not only to extract texture features from image textures but also to model the image segmentation problem. Multiresolution spatial models or approximate kriging methods have been adapted recently to handle massive data sets for imputation and smoothing (Magnussen et al. 2007).

10.3.2.6 *Spatial scaling*

Merging images of different spatial and spectral resolutions to create a high spatial resolution multispectral combination is considered a spatial scaling process. Various approaches have been proposed to integrate image data with different spatial resolutions (Pohl & van Genderen 1998, Ranchin et al. 2003, Vesteinsson et al. 2008).

Spectral (un)mixing is an algorithm that estimates the percentage of each land cover (called endmembers) within each low resolution multi- and hyperspectral pixel (Asner & Lobell 2000, Gross & Schott 1998), which is considered as a spatial downscaling problem. Here we do not attempt to distinguish subpixel mapping, unmixing, or soft classification. Recent developments include unmixing with variable endmembers (Garcia-Haro et al. 2005), the directional mixing method (Garcia-Haro et al. 2006), stochastic mixing model for hyperspectral imagery (Eismann & Hardie 2004, 2005), unmixing using neural networks and wavelets (Mertens et al. 2004) and genetic algorithms (Mertens et al. 2003).

10.3.3 *Applications*

Geo-statistics has been widely applied in quantitative remote sensing (Curran & Atkinson 1998), such as in studying the structure and understanding the nature of spatial variation in remote sensing images (Woodcock et al. 1988b, Ramstein & Raffy 1989); in forestry to analyse forest stand structure (Cohen et al. 1990, St-Onge & Cavayas 1995); in the estimation of structural damage in balsam fir stands (Franklin & Turner 1992); in the sampling of ground data that is to be correlated with remotely sensed data (Atkinson & Emery 1999); and in estimating biomass (Phinn et al. 1996).

Wavelet transformation has been used for image classification (Kandaswamy et al. 2005, Meher et al. 2007, Schmidt et al. 2007), segmentation (De Grandi et al. 2007), compression (Li et al. 2007), noise removal (Chen et al. 2006b), and fusion (Amolins et al. 2007).

Random field models have been used for identifying urban areas from optical imagery (Zhong & Wang 2007), image classification (Luo et al. 2007) and segmentation (Xia et al. 2006), and spatial-temporal urban landscape change analysis using a Markov chain model and the genetic algorithm (Tang et al. 2007).

The spatial co-occurrence matrix has been used for evaluating land development (Unsalan 2007), estimating forest structure variables (Kayitakire et al. 2006), predicting population density (Liu et al. 2006), and discriminating urban features (Myint et al. 2004, Zhang et al. 2003).

Boucher et al. (2006) presented a method that exploits both the temporal and spatial domains of time series satellite data to map land cover changes.

10.4 ANGULAR SIGNATURES

Many spaceborne sensors can observe earth surfaces off-nadir in order to increase swath width and temporal resolution, and to exploit angular variations in the reflectance of various land surfaces. Angular signatures are the observations of surfaces illuminated and measured at different directions. They may be considered as noise in calculating the vegetation indices, for example, and we need to remove the angular dependence. On the other hand, angular signatures can be extremely valuable for estimating land surface variables.

10.4.1 *Measurements*

The spaceborne sensors for near-simultaneous multi-angular acquisition include the NASA MISR flown on the NASA Earth Observing System Terra; the French Space Agency's (CNES) POLarization and Directionality of the Earth's Reflectance (POLDER), flown initially on the Japanese ADEOS satellite series and currently on the CNES Parasol platform; and Sira (UK) Ltd.'s Compact High Resolution Imaging Spectrometer (CHRIS), flown on the European Space Agency (ESA) Proba-1 satellite. Terra was launched in December 1999, Proba in 2001 and Parasol in 2004.

To perform nearly simultaneous observations from space at multiple view angles, two technical solutions may be used: a) multiple optics or b) accurate along track pointing. The Along Track Scanning Radiometer 1 (ATSR-1) flown on ERS-1 was based on solution (a). ATSR-1 was followed by ATSR-2 in 1995 and the Advanced ATSR (AATSR) on ENVISAT in 2001. All ATSR sensors acquire dual-view angle data (approximately 0° and 53° at surface) in four spectral channels for ATSR-1 and seven spectral channels for ATSR-2 and AATSR. The use of the along track scanning technique makes it possible to observe the same point on the Earth's surface at two view angles through two different atmospheric paths within a short period of time.

Multiangular data sets can be acquired sequentially rather than near-simultaneously, and work with these accumulating, or "sequential", systems is also making important contributions to multiangle remote sensing.

10.4.2 *Modelling, inversion and applications*

The basic modelling methods are very similar to those described in Section 2.2, and the inversion methods focus mainly on physical approaches. The recent paper by Diner et al. (2005) in terms of MISR and three review papers on the optical and thermal multi-angular remote sensing (Schaepman 2007, Chopping 2008, Menenti et al. 2008) provide comprehensive discussions. Angular reflectance terminology has seen further standardization efforts, resulting in a comprehensive overview of multiangular data products (Schaepman-Strub et al. 2006). In the following, only a few application examples are introduced.

Angular signatures have been used to estimate aerosol properties (Grey et al. 2006, Martonchik et al. 2002), fractional vegetation coverage (North 2002, Roujean & Lacaze 2002), LAI and FPAR (Casa & Jones 2005, North 2002, Qi et al. 2000, Roujean & Breon 1995, Zhang et al. 2000), canopy clumping index (Chen et al. 2005c), biochemical parameters (Kimes et al. 2002, Kuusk 1998), surface albedo (Schaaf et al. 2008), and land surface skin temperature and emissivity (Coll et al. 2006, Jimenez-Munoz & Sobrino 2007, Menenti et al. 2008, Sobrino et al. 2004).

There are additional studies combining angular signatures with other signature types (Baret & Buis 2008). For example, Kimes et al. (2006) estimated forest structural data from the airborne Laser Vegetation Imaging Sensor and Airborne MISR (AirMISR). Kneubühler et al. (2006) assessed the spectral behaviour of agricultural crop stands using multitemporal angular CHRIS/PROBA data. In fact, angular signatures are almost always combined with spectral signatures (Diner et al. 2005).

10.5 TEMPORAL SIGNATURES

Temporal signatures are derived from time sequences of observations. They are particularly significant for monitoring the Earth's environmental changes.

10.5.1 *Measurements*

The measurement of temporal signatures is synonymous with temporal resolution and other related terms, such as the "repeat" cycle or "revisit time". Temporal resolution depends on various other factors, including the satellite's orbital altitude and the sensor's view angle (which together influence the image's area of coverage, or swath width), the instrument's tilting capabilities and the latitude of the area of interest (Aplin 2006).

The current and planned Landsat-type satellites cover the complete equatorial surface during each orbital repeat cycle by having ground swaths between 120 and 200 kilometres. Their orbital periods, and thus global coverage times, vary from 16 days for Landsat to 22, 24 and 26 days for the Indian, French and Chinese/Brazilian satellites. Taken singly, even these repeat cycles are too long for many applications. The moderate resolution satellites (Table 10.1) have much higher temporal resolutions, typically daily. For

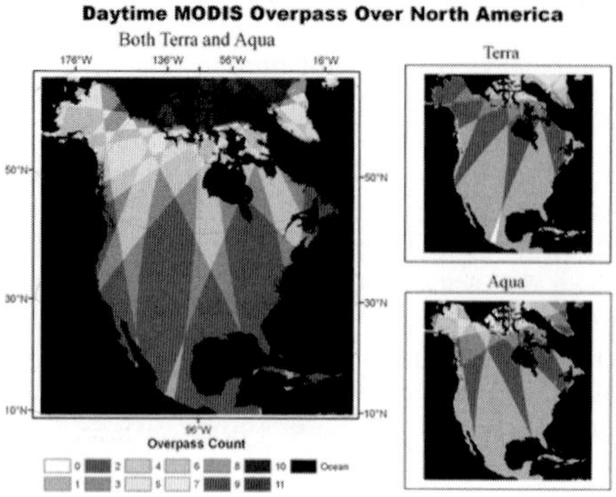

Figure 10.1. Number of MODIS overpass counts over North America based on the Terra and Aqua orbital simulation from GMT 12:00 to 24:00, 16 June 2006 (Wang et al. 2008). (see colour plate page 493)

MODIS, the temporal resolutions, which depend on the geographic location, are shown in Figure 10.1.

To monitor the temporal variations of the dynamic atmospheric system for accurate weather and climate prediction, geo-stationary satellites are operational in geo-stationary orbits and are also useful for land environmental monitoring. The imaging sensors acquire data typically every 15–30 minutes.

10.5.2 *Modelling*

Before reviewing the modelling techniques, it is helpful to discuss three pre-processing techniques, namely, temporal compositing, temporal smoothing, and gap filling.

10.5.2.1 *Temporal compositing*

One of the most common techniques used to obtain cloud-free and spatially coherent images is temporal compositing (Cihlar et al. 1994, Holben 1986). Compositing is a technique that for any given pixel selects the date that best fits a given criterion, for a pre-specified period of time.

The first compositing algorithm developed and most commonly used is the normalized difference vegetation index (NDVI) maximum value composite (Holben 1986). Several other compositing algorithms have been studied, including the Soil Adjusted Vegetation Index (SAVI) maximum value composite (Qi & Kerr 1997), minimum value composite of the red band; NDVI maximum value composite followed by the minimum value of the viewing zenith angle (Cihlar et al. 1994). Cabral et al. (2003) compared five different algorithms using vegetation data, including maximization of NDVI, minimization of the red channel, maximization of NDVI followed by minimization of short-wave infrared, the third lowest value of near-infrared, and the third lowest darkness value (defined as the arithmetic mean of the red and NIR bands). Selection of the third lowest values is based on the assumption that there is a low likelihood of a cloud shadow falling on a given pixel more than twice, over a period of one month.

10.5.2.2 *Temporal smoothing*

Temporal compositing is a simple approach, but if the composite period is long, the land surface does not remain static; and if it is too short, the atmospheric disturbance cannot be removed effectively, especially in cloudy regions. For example, there exist many low quality pixels in 8- or 16-day composite MODIS products (Moody et al. 2005). Several methods, based on interpolation of time series data, have been proposed to remove such noise and to reconstruct high quality NDVI time-series data. These methods can be generally categorized into two types. The first type includes removing noise in the time domain, such as the best index slope extraction (BISE) algorithm (Viovy et al. 1992), the asymmetric Gaussian function fitting approach (Jonsson & Eklundh 2002), the weighted least squares linear regression approach (Sellers et al. 1994), the Savitzky-Golay filter approach (Chen et al. 2005a) and the ecosystem-dependent temporal interpolation technique (Moody et al. 2005). The second type includes noise removal methods in the frequency domain, such as Fourier-based fitting methods (Roerink et al. 2000).

Lu et al. (2007) applied a wavelet-based method to remove the contaminated data from MODIS NDVI, LAI and albedo time-series, with a comparison with the BISE algorithm, Fourier-based fitting method, and the Savitzky-Golay filter method.

10.5.2.3 *Gap filling*

Many space agencies (e.g. NASA) are routinely producing high-level land products, which are usually spatially and temporally discontinuous due to cloud cover, seasonal snow and instrument problems. To enable these products to be used with their various kinds of gaps, it is intuitively appealing to use either temporal or spatial filters. Several mathematical filters have been used to fill gaps in remotely sensed data, such as simple linear interpolation, BISE (Viovy et al. 1992), Fourier transformation (Hermance 2007, Julien et al. 2006, Roerink et al. 2000), Asymmetric Gaussian filter (Jonsson & Eklundh 2002), Savitzky-Golay filter (Chen et al. 2005a), or locally adjusted cubic-spline capping method (Chen et al. 2006a). These methods have been used mainly to restore the NDVI profile (Cihlar 1996, Sellers et al. 1994), but they can also be used for LAI with some adjustments.

Moody et al. (2005) tried to combine both temporal and spatial methods and developed an ecosystem-dependent temporal interpolation technique to fill missing or snow-covered pixels in the MODIS albedo data product. Fang et al. (2007, 2008) have developed a data filtering algorithm, called the temporal spatial filter (TSF), which integrates both spatial and temporal characteristics for different plant functional types.

10.5.2.4 *Functional representation*
Fisher (1994a,b) described the NDVI temporal profile using an empirical statistical model for agriculture crops, each crop corresponding to a set of coefficients. The formula is a double logistic function with five coefficients (k, c, p, d, q) and two constant values (*vb* and *ve*):

$$NDVI(t) = vb + \frac{k}{1+\exp[-c(t-p)]} - \frac{k+vb-ve}{1+\exp[-d(t-q)]} \tag{2}$$

where t is the time variable representing the day of the year, January 1 is set zero, k is related to the asymptotic value of NDVI, c and d (day^{-1}) denote the slopes at the first and second inflection points, p and q (day) are the dates of these two points, *vb* and *ve* are the NDVI values at the beginning and the end of the growing season. This function was used to decompose the temporal NDVI profile of a mixed pixel into the temporal NDVI profiles of several specific covers within the mixed pixel.

The temporal profiles of NDVI from AVHRR have also been presented by the classic Fourier function:

$$NDVI(t) = a_0 + \sum_{i=1}^{I}\left[a_i\cos\left(\frac{2\pi i}{T}t\right) + b_i\sin\left(\frac{2\pi i}{T}t\right)\right] \tag{3}$$

where T is the period of record. The low-order finite Fourier series has been used for classifying land surface cover types (Liang 2001).

A simplified version has been used for smoothing NOAA NDVI data (Hermance et al. 2007, Jonsson & Eklundh 2004):

$$\begin{aligned} f(t) &= c_1 + c_{2+}t + c_3t^2 + c_4\sin(\omega t) + c_5\cos(\omega t) \\ &\quad + c_6\sin(2\omega t) + c_7\cos(2\omega t) \\ &\quad + c_8\sin(3\omega t) + c_9\cos(3\omega t) \end{aligned} \tag{4}$$

where $\omega = 6\pi/N$.

Another representation is the modified double-Gaussian curve

$$f(t) = A_1e^{-(t-t_{01})/\Delta t_1} + A_2e^{-(t-t_{02})/\Delta t_2} + a\cdot t + b \tag{5}$$

where the $(a\cdot t + b)$ is the "soil line" with parameters *a* and *b*.

10.5.3 *Applications*

Land surface change detection has to rely largely on the temporal signatures and some of the references can be found in the recent review papers (Coppin et al. 2004, Lam 2008).

These data smoothing approaches have also been utilized in a variety of other applications. The BISE algorithm has been used to classify vegetation and forest types (Xiao et al. 2003). The Fourier-based fitting approach has been employed to derive terrestrial biophysical parameters (Moody & Johnson 2001). Vegetation seasonality information has been extracted by use of Asymmetric Gaussian function fitting methods (Jonsson & Eklundh 2002). The ecosystem-dependent, temporal-interpolation technique can provide researchers with a snow-free, land-surface albedo product (Moody et al. 2005). Koetz et al. (2007) used multi-temporal CHRIS/PROBA data for LAI estimation based on an RTM inversion approach. The temporal signatures from the polar orbiting satellite data have been used recently for estimating aerosol optical depth (Liang et al. 2006b, Zhong et al. 2007) and incident solar radiation (Liang et al. 2006a, Liang et al. 2007).

Before finishing this section, we have to mention the terrestrial applications of temporal signatures from geo-stationary satellite data, including mapping surface albedos (Govaerts & Lattanzio 2007, Govaerts et al. 2004, Wang et al. 2007), land surface skin temperature (Oku & Ishikawa 2004, Sun et al. 2004), and solar radiation (Pinker et al. 2007, Wang et al. 2007, Zheng et al. 2007).

10.6 POLARIZATION SIGNATURES

Polarization is the property of electromagnetic waves that describes the direction of the electric field and is used to distinguish between the different directions of oscillation of electromagnetic waves propagating in the same direction. Polarization by scattering is observed as light passes through the atmosphere. Solar light reflected by natural surfaces is partly polarized. The surface polarized reflectance is highly anisotropic, and varies between zero close to the backscattering direction and a few per cent in the forward scattering direction. For a given observation geometry, the polarization is about twice as large over the pixels classified as "desert" as those over vegetated surfaces.

Polarization signatures represent the vector nature of the optical field across a scene. While the spectral signature is more about surface inherent properties, polarization signature tells us about surface features, shape, shading and roughness. Polarization tends to provide information that is largely uncorrelated with spectral and intensity images, and thus has the potential to enhance many fields of optical metrology.

An earlier review of the understanding and prospects of using polarized light was given by Herman & Vanderbilt (1997).

10.6.1 *Measurements*

In outdoor measurements, the most rapid variations of polarization with wavelength result from atmospheric spectral features. In the visible to shortwave-IR spectrum, there is strong variation with atmospheric aerosol content.

Polarized reflectance measurements have been made in the laboratory (Talmage & Curran 1986), at the surface (Breon et al. 1995a, Shibayama & Watanabe 2007), and from aircraft (Deuze et al. 1993) and space from POLDER. Despite the short life of the POLDER instrument, a unique set of observations on the global distribution of polarized reflectance was obtained. In addition to its multispectral and multidirectional capabilities, the POLDER radiometer is able to measure the polarization status of the reflected light at three wavelengths (443, 670 and 865 nm).

Before the launch of the ADEOS platform, the only spaceborne measurements of the polarized light reflected by the Earth were acquired using a photographic device from the space shuttle (Egan et al. 1991, Roger et al. 1994).

Polarization signatures are extremely valuable in microwave remote sensing. Horizontal polarization is much more sensitive to changes in the soil dielectric constant than vertical polarization measurements and information on the polarization difference has been used to characterize the vegetation in several studies and algorithms (Jackson 2008). The RADAR systems have evolved from single polarization to multi-polarization (Shi 2008), as shown in Table 10.3.

Polarization is also useful in Lidar remote sensing. For example, the presence of significant cross-polarized light relative to a linearly polarized transmitter can indicate the presence of ice in clouds or non-spherically shaped dust particles in the atmosphere. Polarized Lidars have been developed to measure Stokes parameters of backscattered light in studies of forest and Earth surface properties, and to enhance contrast in the Lidar detection of fish (Tyo et al. 2006).

10.6.2 *Modelling*

Modelling of polarization signatures is mainly based on radiative transfer theory in the vector form. Many publically available RT codes are in the vector format. A widely used radiative transfer code (6S) in optical remote sensing has a vector version now (Kotchenova & Vermote 2007, Kotchenova et al. 2006).

It is believed that polarized light is generated at the surface by specular reflection on the leaf surfaces or the soil facets. This hypothesis has been used to elaborate analytical models for the polarized reflectance of vegetation (Rondeaux & Herman 1991, Vanderbilt & Grant 1985) and bare soils (Breon et al. 1995b). For example, Nadal & Breon (1999) developed a simple polarized reflectance function to fit the POLDER data:

$$R_p(\theta_s,\theta_v,\varphi) = \rho\left[1-\exp\left(-\beta\frac{F_p(\alpha)}{\mu_s+\mu_v}\right)\right] \tag{6}$$

where $F_p(\alpha)$ is the specular reflectance of the Fresnel equation with the incident angle α, ρ and β are two parameters to be adjusted on the measurements for each cover type. ρ is the "saturation" value, whereas ρβ is the slope of the linear relationship toward the large scattering angles. θ_s ($\mu_s = \cos(\theta_s)$) and θ_v ($\mu_v = \cos(\theta_v)$) are the solar and viewing zenith angles, φ is the relative azimuth angle between the solar illumination and viewing directions.

10.6.3 *Applications*

Polarization signatures can be treated as noise or signal. For most optical sensors, polarization effects are considered noise that requires correction.

Due to mirrors, gratings and prisms, the radiometric response function depends on the polarization of the incoming light. A correct radiometric calibration therefore requires knowledge of both the polarization properties of the instrument (measured before launch) and the actual polarization of the incoming light (measured for each observed pixel). Schutgens & Stammes (2002, 2003) developed the radiometric calibration methods for the polarization-sensitive spaceborne instruments for the correct retrieval of data products that require absolute radiances. Sun & Xiong (2007) analysed the MODIS visible bands sensitive to polarization of incident light.

Levy et al. (2004) demonstrated that disregarding polarization introduces little error into global and long-term averages aerosol retrieval, yet can produce very large errors on smaller scales and individual aerosol retrievals. This correction has been incorporated into the latest MODIS aerosol retrieval algorithm (Levy et al. 2007).

The polarization signature, characterized by the degree of polarization and the polarization direction, may contain some information about the surface such as its roughness (Curran 1981), water content (Curran 1978), crop types (Shibayama & Akita 2002), or the leaf inclination distribution (Rondeaux & Herman 1991, Shibayama & Watanabe 2007). Shibayama et al. (2007) linked the polarization of reflected light from crop canopies measured in the field with the leaf inclination angle.

Kokhanovsky et al. (2007) inter-compared the aerosol optical thickness (AOT) at 0.55 μm retrieved using different satellite instruments and algorithms based on the analysis of backscattered solar light for a single scene over central Europe. These algorithms are based on spectral, angular or polarization signatures. Studies demonstrate the great potential of polarization signatures in estimating AOT (Hasekamp & Landgraf 2007). Kacenelenbogen et al. (2006) analysed the relationship between daily fine particle mass concentration (PM2.5) and columnar aerosol optical depth derived from POLDER.

10.7 CONCLUSIONS

Five remote sensing signatures have been evaluated in terms of their measurement, modelling and application potential. These signatures were discussed separately in this chapter, but as has been demonstrated they are used jointly in many cases. Integration of data sets from multiple sources with different signatures will continue to play a critical role in future remote sensing research. In particular, advanced data assimilation schemes (Liang & Qin 2008), evidential reasoning (e.g. Sun & Liang 2008) and bridging scaling gaps from local to biome level will be of increasing interest.

ACKNOWLDEGMENTS

We acknowledge the support of all WG VII/1 members and participants of the spectral signatures conference series contributing to the advancement of this research field. We gratefully acknowledge support from NASA, ESA and CAS, substantially contributing to these efforts.

REFERENCES

Amolins, K., Zhang, Y. & Dare, P. 2007. Wavelet based image fusion techniques—An introduction, review and comparison. *ISPRS Journal of Photogrammetry and Remote Sensing* 62: 249–263.

Aplin, P. 2006. On scales and dynamics in observing the environment. *International Journal of Remote Sensing* 27: 2123–2140.

Asner, G.P., & Lobell, D.B. 2000. A biogeophysical approach for automated SWIR unmixing of soils and vegetation. *Remote Sensing of Environment* 74: 99–112.

Atkinson, P. & Emery, D. 1999. Exploring the relation between spatial structure and wavelength: Implications for sampling reflectance in the field. *International Journal of Remote Sensing* 20: 2663–2678.

Bacour, C., Jacquemoud, S., Leroy, M., Hautceur, O., Weiss, M., Prevot, L., Bruguier, N. & Chauki, H. 2002. Reliability of the estimation of vegetation characteristics by inversion of three canopy reflectance models on airborne POLDER data. *Agronomie,* 22: 555–565.

Bailey, T.C. & Gatrell, A.C. 1995. *Interactive Spatial Data Analysis*. Harlow: Longman.

Baret, F. & Buis, S. 2008. Estimating Canopy Characteristics from Remote Sensing Observations: Review of Methods and Associated Problems. In S. Liang (ed.), *Advances in Land Remote Sensing: System, Modelling, Inversion and Application, Chapter 7*: 173–201. New York: Springer.

Börner, A., Wiest, L., Keller, P., Reulke, R., Richter, R., Schaepman, M. & Schläpfer, D. 2001. SENSOR: A tool for the simulation of hyperspectral remote sensing systems. *ISPRS Journal of Photogrammetry and Remote Sensing* 55: 299–312.

Boucher, A., Seto, K.C. & Journel, A.G. 2006. A novel method for mapping land cover changes: Incorporating time and space with geostatistics. *IEEE Transactions on Geoscience and Remote Sensing* 44: 3427–3435.

Breon, F.M., Tanre, D., Lecomte, P. & Herman, M. 1995a. Polarized Reflectance of Bare Soils and Vegetation—Measurements and Models. *IEEE Transactions on Geoscience and Remote Sensing* 33: 487–499.

Breon, F.M., Tanre, D., Lecomte, P. & Herman, M. 1995b. Polarized reflectance of bare soils and vegetation: Measurements and models. *IEEE Transactions on Geoscience and Remote Sensing* 33: 487–499.

Cabral, A., De Vasconcelos, M.J.P., Pereira, J.M.C., Bartholome, E. & Mayaux, P. 2003. Multi-temporal compositing approaches for SPOT-4 VEGETATION. *International Journal of Remote Sensing* 24: 3343–3350.

Canisius, F. & Chen, J.M. 2007. Retrieving forest background reflectance in a boreal region from Multi-anglo Imaging SpectroRadiometer (MISR) data. *Remote Sensing of Environment* 107: 312–321.

Casa, R. & Jones, H.G. 2005. LAI retrieval from multiangular image classification and inversion of a ray tracing model. *Remote Sensing of Environment* 98: 414–428.

Chen, J., Jonsson, P., Tamura, M., Gu, Z.H., Matsushita, B. & Eklundh, L. 2005a. A simple method for reconstructing a highquality NDVI time-series data set based on the Savitzky-Golay filter. *Remote Sensing of Environment* 91: 332–344.

Chen, J., Zhang, M.Y., Wang, L., Shimazaki, H. & Tamura, M. 2005b. A new index for mapping lichen-dominated biological soil crusts in desert areas. *Remote Sensing of Environment* 96: 165–175.

Chen, J.M., Deng, F. & Chen, M.Z. 2006a. Locally adjusted cubic-spline capping for reconstructing seasonal trajectories of a satellite-derived surface parameter. *IEEE Transactions on Geoscience and Remote Sensing* 44: 2230–2238.

Chen, J.M., Menges, C.H. & Leblanc, S.G. 2005c. Global mapping of foliage clumping index using multi-angular satellite data. *Remote Sensing of Environment* 97: 447–457.

Chen, J.S., Lin, H., Shao, Y. & Yang, L.M. 2006b. Oblique striping removal in remote sensing imagery based on wavelet transform. *International Journal of Remote Sensing* 27: 1717–1723.

Chikhaoui, M., Bonn, F., Bokoye, A.I. & Merzouk, A. 2005. A spectral index for land degradation mapping using ASTER data: Application to a semi-arid Mediterranean catchment. *International Journal of Applied Earth Observation and Geoinformation* 7: 140–153.

Chopping, M. 2008. Terrestrial Applications of Multiangle Remote Sensing. In S. Liang (ed.), *Advances in Land Remote Sensing: System, Modelling, Inversion and Application, Chapter 5*. New York: Springer.

Chopping, M.J., Su, L.H., Laliberte, A., Rango, A., Peters, D.P.C. & Martonchik, J.V. 2006. Mapping woody plant cover in desert grasslands using canopy reflectance modelling and MISR data. *Geophysical Research Letters* 33, Art. No. L17402.

Cierniewski, J., Gdala, T. & Karnieli, A. 2004. A hemispherical-directional reflectance model as a tool for understanding image distinctions between cultivated and uncultivated bare surfaces. *Remote Sensing of Environment* 90: 505–523.

Cihlar, J. 1996. Identification of contaminated pixels in AVHRR composite images for studies of land biosphere. *Remote Sensing of Environment* 56: 149–163.

Cihlar, J., Manak, D. & Diorio, M. 1994. Evaluation of Compositing Algorithms for Avhrr Data over Land. *IEEE Transactions on Geoscience and Remote Sensing* 32: 427–437.

Cohen, W., Spies, T. & Bradshaw, G. 1990. Semivariograms of digital imagery for analysis of conifer canopy structure. *Remote Sensing of Environment* 34: 167–178.

Coll, C., Caselles, V., Galve, J.M., Valor, E., Niclos, R. & Sanchez, J.M. 2006. Evaluation of split-window and dual-angle correction methods for land surface temperature retrieval from Envisat/Advanced Along Track Scanning Radiometer (AATSR) data. *Journal of Geophysical Research-Atmospheres* 111: Art. No. D12105.

Coppin, P., Jonckheere, I., Nackaerts, K., Muys, B. & Lambin, E. 2004. Digital change detection methods in ecosystem monitoring: a review. *International Journal of Remote Sensing* 25: 1565–1596.

Cressie, N. 1993. *Statistics for Spatial Data*. New York: John Wiley and Sons, Inc.

Curran, P. 1981. The Relationship between Polarized Visible-Light and Vegetation Amount. *Remote Sensing of Environment* 11: 87–92.

Curran, P. & Atkinson, P. 1998. Geostatistics and remote sensing. *Progress in Physical Geography* 22: 61–78.

Curran, P.J. 1978. Photographic Method for Recording of Polarized Visible-Light for Soil Surface Moisture Indications. *Remote Sensing of Environment* 7: 305–322.

Curran, P.J. 1988. The semi-variogram in remote sensing: an introduction. *Remote Sensing of Environment* 3: 493–507.

Curran, P.J. 2001. Remote sensing: Using the spatial domain. *Environmental and Ecological Statistics* 8: 331–344.

De Grandi, G.D., Lee, J.S. & Schuler, D.L. 2007. Target detection and texture segmentation in polarimetric SAR images using a wavelet frame: Theoretical aspects. *IEEE Transactions on Geoscience and Remote Sensing* 45: 3437–3453.

Deuze, J.L., Breon, F.M., Deschamps, P.Y., Devaux, C., Herman, M., Podaire, A. & Roujean, J.L. 1993. Analysis of the Polder (Polarization and Directionality of Earths Reflectances) Airborne Instrument Observations over Land Surfaces. *Remote Sensing of Environment* 45: 137–154.

Diner, D.J., Martonchik, J.V., Kahn, R.A., Pinty, B., Gobron, N., Nelson, D.L. & Holben, B.N. 2005. Using angular and spectral shape similarity constraints to improve MISR aerosol and surface retrievals over land. *Remote Sensing of Environment* 94: 155–171.

Disney, M., Lewis, P. & Saich, P. 2006. 3D modelling of forest canopy structure for remote sensing simulations in the optical and microwave domains. *Remote Sensing of Environment* 100: 114–132.

Dorigo, W.A., Zurita-Milla, R., de Wit, A.J.W., Brazile, J., Singh, R. & Schaepman, M.E. 2007. A review on reflective remote sensing and data assimilation techniques for enhanced agroecosystem modelling. *International Journal of Applied Earth Observation and Geoinformation* 9: 165–193.

Durbha, S.S., King, R.L. & Younan, N.H. 2007. Support vector machines regression for retrieval of leaf area index from multiangle imaging spectroradiometer. *Remote Sensing of Environment* 107: 348–361.

Egan, W.G., Johnson, W.R. & Whitehead, V.S. 1991. Terrestrial Polarization Imagery Obtained from the Space—Shuttle—Characterization and Interpretation. *Applied Optics* 30: 435–442.

Eismann, M.T. & Hardie, R.C. 2004. Application of the stochastic mixing model to hyperspectral resolution, enhancement. *IEEE Transactions on Geoscience and Remote Sensing* 42: 1924–1933.

Eismann, M.T. & Hardie, R.C. 2005. Hyperspectral resolution enhancement using high-resolution multispectral imagery with arbitrary response functions. *IEEE Transactions on Geoscience and Remote Sensing* 43: 455–465.

Fang, H., Kim, H., Liang, S., Schaaf, C., Strahler, A., Townshend, G.R.G. & Dickinson, R. 2007. Spatially and temporally continuous land surface albedo fields and validation. *Journal of Geophysical Research-Atmosphere, 112, Article Number* D20206, doi:20210.21029/22006JD008377.

Fang, H. & Liang, S. 2003. Retrieve LAI from Landsat 7 ETM+ Data with a Neural Network Method: Simulation and Validation Study. *IEEE Transactions on Geoscience and Remote Sensing* 41: 2052–2062.

Fang, H. & Liang, S. 2005. A hybrid inversion method for mapping leaf area index from MODIS data: experiments and application to broadleaf and needleleaf canopies. *Remote Sensing of Environment* 94: 405–424.

Fang, H., Liang, S. & Kuusk, A. 2003. Retrieving Leaf Area Index (LAI) Using a Genetic Algorithm with a Canopy Radiative Transfer Model. *Remote Sensing of Environment* 85: 257–270.

Fang, H., Liang, S., Townshend, J. & Dickinson, R. 2008. Spatially and temporally continuous LAI data sets based on a new filtering method: Examples from North America. *Remote Sensing of Environment* 112: 75–93.

Fischer, A. 1994b. A simple model for the temporal variations of NDVI at regional scale over agricultural countries.

Validation with ground radiometric measurements. *International Journal of Remote Sensing* 15: 1421–1446.

Fisher, A. 1994a. A model for the seasonal variations of vegetation indices in coarse resolution data and its inversion to extract crop parameters. *Remote Sensing of Environment* 48: 220–230.

Franklin, J. & Turner, D.L. 1992. The application of a geometric optical canopy reflectance model to semiarid shrub vegetation. *IEEE Transactions on Geoscience and Remote Sensing* 30: 293–301.

Garcia-Haro, F.J., Camacho-de Coca, F. & Melia, J. 2006. A directional spectral mixture analysis method: Application to multiangular airborne measurements. *IEEE Transactions on Geoscience and Remote Sensing* 44: 365–377.

Garcia-Haro, F.J., Sommer, S. & Kemper, T. 2005. A new tool for variable multiple endmember spectral mixture analysis (VMESMA. *International Journal of Remote Sensing* 26: 2135–2162.

Gascon, F., Gastellu-Etchegorry, J.P., Lefevre-Fonollosa, M.J. & Dufrene, E. 2004. Retrieval of forest biophysical variables by inverting a 3-D radiative transfer model and using high and very high resolution imagery. *International Journal of Remote Sensing* 25: 5601–5616.

Gastellu-Etchegorry, J.P., Gascon, F. & Esteve, P. 2003. An interpolation procedure for generalizing a look-up table inversion method. *Remote Sensing of Environment* 87: 55–71.

Gerstl, S.A.W. 1990. Physics concepts of optical and radar reflectance signatures A summary review. *International Journal of Remote Sensing* 11: 1109–1117.

Gitelson, A.A., Gritz, Y. & Merzlyak, M.N. 2003. Relationships between leaf chlorophyll content and spectral reflectance and algorithms for non-destructive chlorophyll assessment in higher plant leaves. *Journal of Plant Physiology* 160: 271–282.

Goovaerts, P. 1997. *Geostatistics for Natural Resources Evaluation*. New York: Oxford University Press.

Govaerts, Y.M. & Lattanzio, A. 2007. Retrieval error estimation of surface albedo derived from geostationary large band satellite observations: Application to Meteosat-2 and Meteosat-7 data. *Journal of Geophysical Research-Atmospheres* 112, Art. No. D05102.

Govaerts, Y.M., Lattanzio, A., Pinty, B. & Schmetz, J. 2004. Consistent surface albedo retrieval from two adjacent geostationary satellites. *Geophysical Research Letters* 31, Art. No. L15201.

Grey, W.M.F., North, P.R.J., Los, S.O. & Mitchell, R.M. 2006. Aerosol Optical Depth and Land Surface Reflectance from Multi-angle AATSR Measurements: Global Validation and Inter-sensor Comparisons. *IEEE Transactions on Geoscience and Remote Sensing* 44: 2184–2197.

Gross, H.N. & Schott, J.R. 1998. Application of spectral mixture analysis and image fusion techniques for image sharpening. *Remote Sensing of Environment* 63: 85–94.

Hansen, M., DeFries, R.S., Townshend, J.R.G., Sohlberg, R., Dimiceli, C. & Carroll, M. 2002. Towards an operational MODIS continuous field of percent tree cover algorithm: Examples using AVHRR and MODIS data. *Remote Sensing of Environment* 83: 303–319.

Harris, A., Bryant, R.G. & Baird, A.J. 2005. Detecting near-surface moisture stress in Sphagnum spp. *Remote Sensing of Environment* 97: 371–381.

Hasekamp, O.P. & Landgraf, J. 2007. Retrieval of aerosol properties over land surfaces: capabilities of multiple-viewing-angle intensity and polarization measurements. *Applied Optics* 46: 3332–3344.

Herman, M. & Vanderbilt, V. 1997. Polarimetric observations in the solar spectrum for remote sensing purposes. *Remote Sensing Reviews* 15: 35–57.

Hermance, J.F. 2007. Stabilizing high-order, non-classical harmonic analysis of NDVI data for average annual models by damping model roughness. *International Journal of Remote Sensing* 28: 2801–2819.

Hermance, J.F., Jacob, R.W., Bradley, B.A. & Mustard, J.F. 2007. Extracting phenological signals from multi-year AVHRR NDVI time series: Framework for applying high-order annual splines with roughness damping. *IEEE Transactions on Geoscience and Remote Sensing* 45: 3264–3276.

Holben, B.N. 1986. Characteristics of maximum-value composite images for temporal AVHRR data. *International Journal of Remote Sensing* 7: 1435–1445.

Hsu, N.C., Tsay, S.C., King, M.D. & Herman, J.R. 2004. Aerosol properties over bright-reflecting source regions. *IEEE Transactions on Geoscience and Remote Sensing* 42: 557–569.

Jackson, T. 2008. Passive Microwave Remote Sensing for Land Applications. In S. Liang (ed.), *Advances in Land Remote Sensing: System, Modelling, Inversion and Application, Chapter 2*: 9–18. New York: Springer.

Jimenez-Munoz, J.C. & Sobrino, J.A. 2007. Feasibility of retrieving land-surface temperature from ASTER TIR bands using two-channel algorithms: A case study of agricultural areas. *IEEE Geoscience and Remote Sensing Letters* 4: 60–64.

Jobanputra, R. & Clausi, D.A. 2006. Preserving boundaries for image texture segmentation using grey level co-occurring probabilities. *Pattern Recognition* 39: 234–245.

Jonsson, P. & Eklundh, L. 2002. Seasonality extraction by function fitting to time—series of satellite sensor data. *IEEE Transactions on Geoscience and Remote Sensing* 40: 1824–1832.

Jonsson, P. & Eklundh, L. 2004. TIMESAT—a program for analyzing time-series of satellite sensor data. *Computers and Geosciences* 30: 833–845.

Julien, Y., Sobrino, J.A. & Verhoef, W. 2006. Changes in land surface temperatures and NDVI values over Europe between 1982 and 1999. *Remote Sensing of Environment* 103: 43–55.

Jupp, D.L.B., Strahler, A.H. & Woodcock, C.E. 1988. Autocorrelation and regularization in digital images I. Basic theory. *IEEE Transactions on Geoscience and Remote Sensing* 26: 463–473.

Jupp, D.L.B., Strahler, A.H. & Woodcock, C.E. 1989. Autocorrelation and regularization in digital images II. Simple image models. *IEEE Transactions on Geoscience and Remote Sensing* 27: 247–258.

Kacenelenbogen, M., Leon, J.F., Chiapello, I. & Tanre, D. 2006. Characterization of aerosol pollution events in France using ground-based and POLDER-2 satellite data. *Atmospheric Chemistry and Physics* 6: 4843–4849.

Kalacska, M., Sanchez-Azofeifa, A., Caelli, T., Rivard, B. & Boerlage, B. 2005. Estimating leaf area index from satellite imagery using Bayesian networks. *IEEE Transactions on Geoscience and Remote Sensing* 43: 1866–1873.

Kandaswamy, U., Adjeroh, D.A. & Lee, A.C. 2005. Efficient texture analysis of SAR imagery. *IEEE Transactions on Geoscience and Remote Sensing* 43: 2075–2083.

Kayitakire, F., Hamel, C. & Defourny, P. 2006. Retrieving forest structure variables based on image texture analysis and IKONOS-2 imagery. *Remote Sensing of Environment* 102: 390–401.

Kimes, D., Gastellu-Etchegorry, J. & Esteve, P. 2002. Recovery of forest canopy characteristics through inversion of a complex 3D model. *Remote Sensing of Environment* 79: 320–328.

Kimes, D.S., Ranson, K.J., Sun, G. & Blair, J.B. 2006. Predicting lidar measured forest vertical structure from multi-angle spectral data. *Remote Sensing of Environment* 100: 503–511.

Kneubühler, M., Koetz, B., Huber, S., Schopfer, J., Richter, R. and Itten, K.I. 2006. Monitoring vegetation growth using multitemporal CHRIS/PROBA data. *IEEE Geoscience and Remote Sensing Symposium (IGARSS 2006)*: 2677–2680. IEEE Denver (USA).

Kobayashi, H., Suzuki, R. & Kobayashi, S. 2007. Reflectance seasonality and its relation to the canopy leaf area index in an eastern Siberian larch forest: Multi-satellite data and radiative transfer analyses. *Remote Sensing of Environment* 106: 238–252.

Koetz, B., Baret, F., Poilve, H. & Hill, J. 2005. Use of coupled canopy structure dynamic and radiative transfer models to estimate biophysical canopy characteristics. *Remote Sensing of Environment* 95: 115–124.

Koetz, B., Kneubühler, M., Huber, S., Schopfer, J. and Baret, F. 2007. Radiative transfer model inversion based on multi-temporal CHRIS/PROBA data for LAI estimation. In M.E. Schaepman, S. Liang, N.E. Groot & M. Kneubühler (eds.), *10th Intl. Symposium on Physical Measurements and Spectral Signatures in Remote Sensing, Intl. Archives of the Photogrammetry, Remote Sensing and Spatial Information Sciences* XXXVI, Part 7/C50, 344–349. ISSN 1682-1777.

Kokhanovsky, A.A., Aoki, T., Hachikubo, A., Hori, M. & Zege, E.P. 2005. Reflective properties of natural snow: Approximate asymptotic theory versus in situ measurements. *IEEE Transactions on Geoscience and Remote Sensing* 43: 1529–1535.

Kokhanovsky, A.A., Breon, F.M., Cacciari, A., Carboni, E., Diner, D., Di Nicolantonio, W., Grainger, R.G., Grey, W.M.F., Holler, R., Lee, K.H., Li, Z., North, P.R.J., Sayer, A.M., Thomas, G.E. & von Hoyningen-Huene, W. 2007. Aerosol remote sensing over land: A comparison of satellite retrievals using different algorithms and instruments. *Atmospheric Research* 85: 372–394.

Kotchenova, S.Y., Shabanov, N.V., Knyazikhin, Y., Davis, A.B., Dubayah, R. & Myneni, R.B. 2003. Modelling lidar waveforms with time-dependent stochastic radiative transfer theory for remote estimations of forest structure. *Journal of Geophysical Research-Atmospheres* 108, Art. No. 4484.

Kotchenova, S.Y. & Vermote, E.F. 2007. Validation of a vector version of the 6S radiative transfer code for atmospheric correction of satellite data. Part II. Homogeneous Lambertian and anisotropic surfaces. *Applied Optics* 46: 4455–4464.

Kotchenova, S.Y., Vermote, E.F., Matarrese, R. & Klemm, F.J. 2006. Validation of a vector version of the 6S radiative transfer code for atmospheric correction of satellite data. Part I: Path radiance. *Applied Optics* 45: 6762–6774.

Kuusk, A. 1998. Monitoring of vegetation parameters on large areas by the inversion of a canopy reflectance model. *International Journal of Remote Sensing* 19: 2893–2905.

Lam, N. 2008. Methodologies for Mapping Land Cover/Land Use and its Change. In S. Liang (ed.), *Advances in Land Remote Sensing: System, Modelling, Inversion and Application, Chapter 13*: 341–367. New York: Springer.

Levy, R.C., Remer, L.A. & Kaufman, Y.J. 2004. Effects of neglecting polarization on the MODIS aerosol retrieval over land. *IEEE Transactions on Geoscience and Remote Sensing* 42: 2576–2583.

Levy, R.C., Remer, L.A., Mattoo, S., Vermote, E.F. & Kaufman, Y.J. 2007. Second-generation operational algorithm: Retrieval of aerosol properties over land from inversion of Moderate Resolution Imaging Spectroradiometer spectral reflectance. *Journal of Geophysical Research-Atmospheres* 112.

Li, B., Jiao, R.H. & Li, Y.C. 2007. Fast adaptive wavelet for remote sensing image compression. *Journal of Computer Science and Technology* 22: 770–778.

Li, S.S. & Zhou, X.B. 2004. Modelling and measuring the spectral bidirectional reflectance factor of snow-covered sea ice: An intercomparison study. *Hydrological Processes* 18: 3559–3581.

Liang, S. 2001. Land cover classification methods for multi-year AVHRR data. *International Journal of Remote Sensing* 22: 1479–1493.

Liang, S. 2004. *Quantitative Remote Sensing of Land Surfaces*. New York: John Wiley and Sons, Inc.

Liang, S., Fallah-Adl, H., Kalluri, S., JaJa, J., Kaufman, Y.J. & Townshend, J.R.G. 1997. An operational atmospheric correction algorithm for Landsat Thematic Mapper imagery over the land. *Journal of Geophysical Research-Atmospheres* 102: 17173–17186.

Liang, S. & Fang, H. 2004. An improved atmospheric correction algorithm for hyperspectral remotely sensed imagery. *IEEE Geoscience and Remote Sensing Letters* 1: 112–117.

Liang, S. & Strahler, A. 2000. Land surface Bidirectional Reflectance Distribution Function (BRDF): Recent advances and future prospects. *Remote Sensing Reviews* 18: 83–551.

Liang, S. & Qin, J. 2008. Data assimilation methods for land surface variable estimation. In S. Liang (ed.), *Advances in Land Remote Sensing: System, Modelling, Inversion and Application*: 313–339. New York: Springer.

Liang, S., Zheng, T., Liu, R., Fang, H., Tsay, S.C. & Running, S. 2006a. Mapping incident Photosynthetically Active Radiation (PAR) from MODIS Data. *Journal of Geophysical Research-Atmospheres* 111, D15208, doi:15210.11029/12005JD006730.

Liang, S., Zhong, B. & Fang, H. 2006b. Improved estimation of aerosol optical depth from MODIS imagery over land surfaces. *Remote Sensing of Environment* 104: 416–425.

Liang, S.L., Stroeve, J. & Box, J.E. 2005. Mapping daily snow/ice shortwave broadband albedo from Moderate Resolution Imaging Spectroradiometer (MODIS): The improved direct retrieval algorithm and validation with

Greenland *in situ* measurement. *Journal of Geophysical Research-Atmospheres* 110, Art. No. D10109.
Liang, S.L., Zheng, T., Wang, D.D., Wang, K.C., Liu, R.G., Tsay, S.C., Running, S. & Townshend, J. 2007. Mapping high-resolution incident photosynthetically active radiation over land from polar–orbiting and geostationary satellite data. *Photogrammetric Engineering and Remote Sensing* 73: 1085–1089.
Liangrocapart, S. & Petrou, M. 2002. A two-layer model of the bidirectional reflectance of homogeneous vegetation canopies. *Remote Sensing of Environment* 80: 17–35.
Liu, J.C., Melloh, R.A., Woodcock, C.E., Davis, R.E. & Ochs, E.S. 2004. The effect of viewing geometry and topography on viewable gap fractions through forest canopies. *Hydrological Processes* 18: 3595–3607.
Liu, X.H., Clarke, K. & Herold, M. 2006. Population density and image texture: A comparison study. *Photogrammetric Engineering and Remote Sensing* 72: 187–196.
Lu, X., Liu, R., Liu, J. & Liang, S. 2007. Removal of Noise by Wavelet Method to Generate High Quality Temporal Data of Terrestrial MODIS Products. *Photogrammetric Engineering and Remote Sensing* 73: 1129–1139.
Luo, J., Ming, D., Shen, Z., Wang, M. & Sheng, H. 2007. Multi-scale information extraction from high resolution remote sensing imagery and region partition methods based on GMRF-SVM. *International Journal of Remote Sensing* 28: 3395–3412.
Magnussen, S., Naesset, E. & Wulder, M.A. 2007. Efficient multiresolution spatial predictions for large data arrays. *Remote Sensing of Environment* 109: 451–463.
Malenovsky, Z., Martin, E., Homolova, L., Gastellu-Etchegory, J.P., Zurita-Milla, R., Schaepman, M.E., Pokorny, R., Clevers, J.G.P.W. & Cudlin, P. 2008. Influence of woody elements of a Norway spruce canopy on nadir reflectance simulated by the DART model at very high spatial resolution. *Remote Sensing of Environment* 112: 1–18.
Martonchik, J.V., Diner, D.J., Crean, K.A. & Bull, M.A. 2002. Regional Aerosol Retrieval Results From MISR. *IEEE Transactions on Geoscience and Remote Sensing* 40: 1520–1531.
Meher, S.K., Shankar, B.U. & Ghosh, A. 2007. Wavelet-feature-based classifiers for multispectral remote sensing images. *IEEE Transactions on Geoscience and Remote Sensing* 45: 1881–1886.
Menenti, M., Jia, L. & Li, Z.L. 2008. Multi–angular Thermal Infrared Observations of Terrestrial Vegetation. In S. Liang (ed.), *Advances in Land Remote Sensing: System, Modelling, Inversion and Application, Chapter 4*: 51–93. New York: Springer.
Meroni, M., Colombo, R. & Panigada, C. 2004. Inversion of a radiative transfer model with hyperspectral observations for LAI mapping in poplar plantations. *Remote Sensing of Environment* 92: 195–206.
Mertens, K.C., Verbeke, L.P.C., Ducheyne, E.I. & De Wulf, R.R. 2003. Using genetic algorithms in sub-pixel mapping. *International Journal of Remote Sensing* 24: 4241–4247.
Mertens, K.C., Verbeke, L.P.C., Westra, T. & De Wulf, R.R. 2004. Sub-pixel mapping and sub-pixel sharpening using neural network predicted wavelet coefficients. *Remote Sensing of Environment* 91: 225–236.
Miesch, C., Poutier, L., Achard, W., Briottet, X., Lenot, X. & Boucher, Y. 2005. Direct and inverse radiative transfer solutions for visible and near-infrared hyperspectral imagery. *IEEE Transactions on Geoscience and Remote Sensing* 43: 1552–1562.
Moody, A. & Johnson, D.M. 2001. Land-surface phenologies from AVHRR using the discrete fourier transform. *Remote Sensing of Environment* 75: 305–323.
Moody, E.G., King, M.D., Platnick, S., Schaaf, C.B. & Gao, F. 2005. Spatially complete global spectral surface albedos: Value-added datasets derived from terra MODIS land products. *IEEE Transactions on Geoscience and Remote Sensing* 43: 144–158.
Myint, S.W. & Lam, N. 2005. Examining lacunarity approaches in comparison with fractal and spatial autocorrelation techniques for urban mapping. *Photogrammetric Engineering and Remote Sensing* 71: 927–937.
Myint, S.W., Lam, N.S.N. & Tyler, J.M. 2004. Wavelets for urban spatial feature discrimination: Comparisons with fractal, spatial autocorrelation & spatial co-occurrence approaches. *Photogrammetric Engineering and Remote Sensing* 70: 803–812.
Nadal, F. & Breon, F.M. 1999. Parameterization of surface polarized reflectance derived from POLDER spaceborne measurements. *IEEE Transactions on Geoscience and Remote Sensing* 37: 1709–1718.
Nilson, T., Kuusk, A., Lang, M. & Lukk, T. 2003. Forest reflectance modelling: Theoretical aspects and applications. *Ambio* 32: 535–541.
North, P.R.J. 2002. Estimation of f(APAR), LAI & vegetation fractional cover from ATSR-2 imagery. *Remote Sensing of Environment* 80: 114–121.
Oku, Y. & Ishikawa, H. 2004. Estimation of land surface temperature over the Tibetan Plateau using GMS data. *Journal of Applied Meteorology* 43: 548–561.
Peddle, D.R., Franklin, S.E., Johnson, R.L., Lavigne, M.B. & Wulder, M.A. 2003. Structural change detection in a disturbed conifer forest using a geometric optical reflectance model in multiple-forward mode. *IEEE Transactions on Geoscience and Remote Sensing* 41: 163–166.
Peddle, D.R., Johnson, R.L., Cihlar, J. & Latifovic, R. 2004. Large area forest classification and biophysical parameter estimation using the 5-Scale canopy reflectance model in multiple forward-mode. *Remote Sensing of Environment* 89: 252–263.
Phinn, S., Franklin, J., Hope, A., Stow, D. & Huenneke, L. 1996. Biomass distribution mapping using airborne digital video imagery and spatial statistics in a semi-arid environment. *Journal of Environmental Management* 47: 139–164.
Pinker, R.T., Li, X., Meng, W. & Yegorova, E.A. 2007. Toward improved satellite estimates of short-wave radiative fluxes—Focus on cloud detection over snow: 2. Results. *Journal of Geophysical Research-Atmospheres* 112.
Pinty, B., Gobron, N., Widlowski, J.L., Lavergne, T. & Verstraete, M.M. 2004a. Synergy between 1-D and 3-D radiation transfer models to retrieve vegetation canopy properties from remote sensing data. *Journal of Geophysical Research-Atmospheres* 109, Art. No. D21205 .
Pinty, B., Widlowski, J.L., Taberner, M., Gobron, N., Verstraete, M.M., Disney, M., Gascon, F., Gastellu, J.P., Jiang, L., Kuusk, A., Lewis, P., Li, X., Ni-Meister, W., Nilson, T., North, P., Qin, W., Su, L., Tang, S., Thompson, R., Verhoef, W., Wang, H., Wang, J., Yan, G. & Zang, H. 2004b. Radiation Transfer Model Intercom-

parison (RAMI) exercise: Results from the second phase. *Journal of Geophysical Research-Atmospheres* 109, Art. No. D06210, ISI:000220622400004.
Pitman, K.M., Wolff, M.J. & Clayton, G.C. 2005. Application of modern radiative transfer tools to model laboratory quartz emissivity. *Journal of Geophysical Research-Planets* 110, Art. No. E08003.
Pohl, C. & van Genderen, J.L. 1998. Multisensor image fusion in remote sensing: concepts, methods and applications. *International Journal of Remote Sensing* 19: 823–854.
Purkis, S.J., Myint, S.W. & Riegl, B.M. 2006. Enhanced detection of the coral Acropora cervicornis from satellite imagery using a textural operator. *Remote Sensing of Environment* 101: 82–94.
Qi, J. & Kerr, Y. 1997. On current compositing algorithms. *Remote Sensing Reviews* 15: 235–256.
Qi, J., Kerr, Y.H., Moran, M.S., Weltz, M., Huete, A.R., Sorooshian, S. & Bryant, R. 2000. Leaf area index estimates using remotely sensed data and BRDF models in a semiarid region. *Remote Sensing of Environment* 73: 18–30.
Qin, J., Liang, S., Li, X. & Wang, J. 2007. Development of the adjoint model of a canopy radiative transfer model for sensivity study and inversion of leaf area index. *IEEE Transactions on Geoscience and Remote Sensing*, revised.
Ramstein, G. & Raffy, M. 1989. Analysis of the structure of radiometric remotely sensed images. *International Journal of Remote Sensing* 10: 1049–1073.
Ranchin, T., Aiazzi, B., Alparone, L., Baronti, S. & Wald, L. 2003. Image fusion—the ARSIS concept and some successful implementation schemes. *ISPRS Journal of Photogrammetry and Remote Sensing* 58: 4–18.
Rautiainen, M. & Stenberg, P. 2005. Application of photon recollision probability in coniferous canopy reflectance simulations. *Remote Sensing of Environment* 96: 98–107.
Roerink, G.J., Menenti, M. & Verhoef, W. 2000. Reconstructing cloudfree NDVI composites using Fourier analysis of time series. *International Jour. of Remote Sensing* 21: 1911–1917.
Roger, J.C., Santer, R., Herman, M. & Deuze, J.L. 1994. Polarization of the Solar Light Scattered by the Earth Atmosphere System as Observed from the United States. Shuttle. *Remote Sensing of Environment* 48: 275–290
Rondeaux, G. & Herman, M. 1991. Polarization of light reflected by crop canopies. *Remote Sensing of Environment* 38: 63–75.
Roujean, J.-L. & Breon, F.-M. 1995. Estimating PAR absorbed by vegetation from bidirectional reflectance measurements. *Remote Sensing of Environment* 51: 375–384.
Roujean, J.L. & Lacaze, R. 2002. Global mapping of vegetation parameters from POLDER multiangular measurements for studies of surface-atmosphere interactions: A pragmatic method and its validation. *Journal of Geophysical Research-Atmospheres,* 107.
Salomonson, V.V. & Appel, I. 2004. Estimating fractional snow cover from MODIS using the normalized difference snow index. *Remote Sensing of Environment* 89: 351–360.
Schaaf, C., Martonchik, J., Pinty, B., Govaerts, Y., Gao, F., Lattanzio, A., Liu, J., Strahler, A. & Taberner, M. 2008. Retrieval of Surface Albedo from Satellite Sensors. In S. Liang (ed.), *Advances in Land Remote Sensing: System, Modelling, Inversion and Application, Chapter 9*: 219–243. New York: Springer.
Schaepman-Strub, G., Schaepman, M.E., Painter, T.H., Dangel, S. & Martonchik, J.V. 2006. Reflectance quantities in optical remote sensing—definitions and case studies. *Remote Sensing of Environment* 103: 27–42.
Schaepman, M.E. 2007. Spectrodirectional Remote Sensing: From Pixels to Processes. *International Journal of Applied Earth Observation and Geoinformation* 9: 204–223.
Schaepman, M.E., Koetz, B., Schaepman-Strub, G. & Itten, K.I. 2005. Spectrodirectional remote sensing for the improved estimation of biophysical and chemical variables: Two case studies. *International Journal of Applied Earth Observation and Geoinformation* 6: 271–282.
Schlerf, M., Atzberger, C. & Hill, J. 2005. Remote sensing of forest biophysical variables using HyMap imaging spectrometer data. *Remote Sensing of Environment* 95: 177–194.
Schmidt, F., Doute, S. & Schmitt, B. 2007. WAVANGLET: An efficient supervised classifier for hyperspectral images. *IEEE Transactions on Geoscience and Remote Sensing* 45: 1374–1385
Schutgens, N.A.J. & Stammes, P. 2002. Parametrisation of Earth's polarisation spectrum in the ultra-violet. *Jour. of Quantitative Spectroscopy and Radiative Transfer* 75: 239–255.
Schutgens, N.A.J. & Stammes, P. 2003. A novel approach to the polarization correction of spaceborne spectrometers. *Journal of Geophysical Research-Atmospheres,* 108, NO. D7, 4229, doi:10.1029/2002JD002736.
Sellers, P., Tucker, C., Collatz, G., Los, S., Justice, C., Dazlich, D. & Randall, D. 1994. A global 1 by 1 NDVI data set for climate studies. Part 2: The generation of global fields of terrestrial biophysical parameters from the NDVI. *International Journal of Remote Sensing* 15: 3519–3546.
Shabanov, N.V., Huang, D., Yang, W., Tian, B., Knyazikhin, Y., Myneni, R.B., Ahl, D.E., Gower, S.T., Huete, A.R., Arogao, L.E.O.C. & Shimabukuro, Y.E. 2005. Analysis and optimization of the MODIS leaf area index algorithm retrievals over broadleaf forests. *IEEE Transactions on Geoscience and Remote Sensing* 43: 1855–1865.
Shi, J. 2008. Active Microwave Remote Sensing Systems and Applications to Snow Monitoring. In S. Liang (ed.), *Advances in Land Remote Sensing: System, Modelling, Inversion and Application, Chapter 3*: 19–49. New York: Springer.
Shibayama, M. & Akita, S. 2002. A Portable Spectropolarimeter for field crop canopies—Distinguishing species and cultivars of fully developed canopies by polarized light. *Plant Production Science* 5: 311–319.
Shibayama, M. & Watanabe, Y. 2007. Estimating the mean leaf inclination angle of wheat canopies using reflected polarized light. *Plant Production Science* 10: 329–342.
Smolander, S. & Stenberg, P. 2003. A method to account for shoot scale clumping in coniferous canopy reflectance models. *Remote Sensing of Environment* 88: 363–373.
Sobrino, J.A., Soria, G. & Prata, A.J. 2004. Surface temperature retrieval from Along Track Scanning Radiometer 2 data: Algorithms and validation. *Journal of Geophysical Research-Atmospheres* 109, Art. No. D11101.

Soenen, S.A., Peddle, D.R. & Coburn, C.A. 2005. SCS + C: A modified sun-canopy-sensor topographic correction in forested terrain. *IEEE Transactions on Geoscience and Remote Sensing* 43: 2148–2159.

St-Onge, B.A. & Cavayas, F. 1995. Estimating forest stand structure from high resolution imagery using the directional variogram. *International Journal of Remote Sensing* 16: 1999–2021.

Sun, D.L., Pinker, R.T. & Basara, J.B. 2004. Land surface temperature estimation from the next generation of Geostationary Operational Environmental Satellites: GOES M-Q. *Journal of Applied Meteorology* 43: 363–372.

Sun, J.Q. & Xiong, X.X. 2007. MODIS polarization—sensitivity analysis. *IEEE Transactions on Geoscience and Remote Sensing* 45: 2875–2885.

Sun, W. & Liang, S. 2008. Methodologies for mapping plant functional types. In S. Liang (ed.), *Advances in Land Remote Sensing: System, Modelling, Inversion and Application*: 369–393. New York: Springer.

Talmage, D.A. & Curran, P.J. 1986. Remote Sensing Using Partially Polarized Light. *International Journal of Remote Sensing* 7: 47–64.

Tang, J., Wang, L. & Yao, Z. 2007. Spatio-temporal urban landscape change analysis using the Markov chain model and a modified genetic algorithm. *International Journal of Remote Sensing* 28: 3255–3271.

Tyo, J.S., Goldstein, D.L., Chenault, D.B. & Shaw, J.A. 2006. Review of passive imaging polarimetry for remote sensing applications. *Applied Optics* 45: 5453–5469.

Unsalan, C. 2007. Measuring land development in urban regions using graph theoretical and conditional statistical features. *IEEE Transactions on Geoscience and Remote Sensing* 45: 3989–3999.

Ustin, S., Asner, G., Gamon, J., Huemmerich, K., Jacquemoud, S., Zarco-Tejada, P. & Schaepman, M.E. 2008. Retrieval of Quantitative and Qualitative Information about Plant Pigment Systems from High Resolution Spectroscopy. *Remote Sensing of Environment.*

Vanderbilt, V.C. & Grant, L. 1985. Plant canopy specular reflectance model. *IEEE Transactions on Geoscience and Remote Sensing* 23: 722–730.

Verrelst, J., Schaepman, M.E., Kötz, B. & Kneubühler, M. 2008. Angular sensitivity of vegetation indices derived from CHRIS/PROBA data in two Alpine eco-systems. *Remote Sensing of Environment*, in press.

Vesteinsson, A., Aanaes, H., Sveinsson, J.R. & Benediktsson, J.A. 2008. Spectrally Consistent Pansharpening. In S. Liang (ed.), *Advances in Land Remote Sensing: System, Modelling, Inversion and Application, Chapter 11*: 293–311. New York: Springer.

Viovy, N., Arino, O. & Belward, A.S. 1992. The Best Index Slope Extraction (Bise)—a Method for Reducing Noise in Ndvi Time-Series. *International Journal of Remote Sensing* 13: 1585–1590.

Wang, K., Liang, S., Zheng, T. & Wang, D. 2007. Simultaneous estimation of surface photosynthetically active radiation and albedo from GOES. *Remote Sensing of Environment*, revised.

Widlowski, J.L., Pinty, B., Gobron, N., Verstraete, M.M., Diner, D.J. & Davis, A.B. 2004. Canopy structure parameters derived from multi-angular remote sensing data for terrestrial carbon studies. *Climatic Change* 67: 403–415.

Widlowski, J.L., Pinty, B., Lavergne, T., Verstraete, M.M. & Gobron, N. 2005. Using 1-D models to interpret the reflectance anisotropy of 3-D canopy targets: Issues and caveats. *IEEE Transactions on Geoscience and Remote Sensing* 43: 2008–2017.

Widlowski, J.L., Taberner, M., Pinty, B., Bruniquel-Pinel, V., Disney, M., Fernandes, R., Gastellu-Etchegorry, J.P., Gobron, N., Kuusk, A., Lavergne, T., Leblanc, S., Lewis, P.E., Martin, E., Mottus, M., North, P.R.J., Qin, W., Robustelli, M., Rochdi, N., Ruiloba, R., Soler, C., Thompson, R., Verhoef, W., Verstraete, M.M. & Xie, D. 2007. Third Radiation Transfer Model Intercomparison (RAMI) exercise: Documenting progress in canopy reflectance models. *Journal of Geophysical Research-Atmospheres* 112, Art. No. D09111.

Woodcock, C.E. & Strahler, A.H. 1987. The factor of scale in remote sensing. *Remote Sensing of Environment* 21: 311–332.

Woodcock, C.E., Strahler, A.H. & Jupp, D.L.B. 1988a. The use of variograms in remote sensing: Part I. Scene models and simulated images. *Remote Sensing of Environment* 25: 323–348.

Woodcock, C.E., Strahler, A.H. & Jupp, D.L.B. 1988b. The use of variograms in remote sensing: Part II. Real digital images. *Remote Sensing of Environment* 25: 349–379.

Xia, Y., Feng, D. & Zhao, R.C. 2006. Adaptive segmentation of textured images by using the coupled Markov random field model. *IEEE Transactions on Image Processing* 15: 3559–3566.

Xiao, X.M., Braswell, B., Zhang, Q.Y., Boles, S., Frolking, S. & Moore, B. 2003. Sensitivity of vegetation indices to atmospheric aerosols: continental—scale observations in Northern Asia. *Remote Sensing of Environment* 84: 385–392.

Zeng, Y., Huang, J., Wu, B., Schaepman, M.E., de Bruin, S. & Clevers, J. 2008a. Comparison of two canopy reflectance models inversion for mapping forest crown closure using imaging spectroscopy. *Canadian J. of Rem Sens* (in press).

Zeng, Y., Schaepman, M.E., Wu, B., Clevers, J.G.P.W. & Bregt, A. 2008b. Quantitative forest canopy structure assessment using an inverted geometric-optical model and up-scaling. *International Journal of Remote Sensing*, in press.

Zhang, Q., Wang, J., Gong, P. & Shi, P. 2003. Study of urban spatial patterns from SPOT panchromatic imagery using textural analysis. *International Journal of Remote Sensing* 24: 4137–4160.

Zhang, Y., Tian, Y., Knyazikhin, Y., Martonchik, J.V., Diner, D.J., Leroy, M. & Myneni, R.B. 2000. Prototyping of MISR LAI and FPAR algorithm with POLDER data over Africa. *IEEE Transactions on Geoscience and Remote Sensing* 38: 2402–2418.

Zheng, T., Liang, S. & Wang, K.C. 2007. Estimation of incident PAR from GOES imagery. *Journal of Applied Meteorology and Climatology*, in press.

Zhong, B., Liang, S. & Holben, B. 2007. Validating a new algorithm for estimating aerosol optical depths from MODIS imagery. *International Journal of Remote Sensing* 28: 4207–4214.

Zhong, P. & Wang, R.S. 2007. A multiple conditional random fields ensemble model for urban area detection in remote sensing optical Images. *IEEE Transactions on Geoscience and Remote Sensing* 45: 3978–3988.

*Advances in Photogrammetry, Remote Sensing and Spatial Information Sciences:
2008 ISPRS Congress Book – Li, Chen & Baltsavias (eds)*
 ISBN 978-0-415-47805-2

CHAPTER 11

Geometric modelling of linear CCDs and panoramic imagers

Karsten Jacobsen

ABSTRACT: With only a few exceptions, optical spaceborne cameras are CCD-line cameras. The different orientation methods such as the mathematical correct geometric reconstruction, the sensor oriented, bias corrected rational polynomial coefficients (RPCs) and the approximate solutions such as 3-D affine transformation, direct linear transformation and terrain dependent RPCs are described with their potential, problems and required number and distribution of control points. Aerial CCD-line scanner systems today are in use as three-line scanners, usually supported by GPS and inertial measurement units (IMU). The aerial triangulation of three line scanners, required for reliable and more precise results, is discussed. Panoramic imagers have been used in space by the CORONA and the KVR1000 systems. Their high resolution images are still important today. In addition, aerial panoramic imagers are in use mainly for military applications. The geometric handling of panoramic images by transformation to perspective geometry and the determination of the platform motion influence by self calibration with special additional parameters is explained.

Keywords: Image geometry, orientation, self calibration, optical satellite images, aerial CCD-line scanner images, panoramic cameras

11.1 SPACEBORNE SYSTEMS

11.1.1 *Introduction*

Very high resolution optical space images like those from IKONOS, QuickBird, OrbView-3, Cartosat-2, Kompsat-2, Resource DK1, EROS B and World-View-1 today compete with classical aerial images. The information content of the space images can be compared with aerial photos having a scale up to 1:25,000, but the geometric conditions should also be on a similar level, requiring satisfactory mathematical handling. Images close to the original, initially named as level 1A, and images projected to a specified object surface, originally named as level 1B, and even epipolar images are available. The geometric handling has to respect the given image product.

11.1.2 *Geometry of combined CCD-lines*

Spaceborne CCD-line sensors are usually based on a combination of individual CCD-lines merged together, based on a combination of sub-images, corrected by sensor calibration (Fig. 11.1) (Jacobsen 1997). These merged images are still named original images, because the real original images are not available to the user.

The merged images in theory are only correct for the reference height H0 (Fig. 11.2); for another height such as H1 or H2, a mismatch of neighbouring sub-scenes occurs. For instance, for IRS-1C and

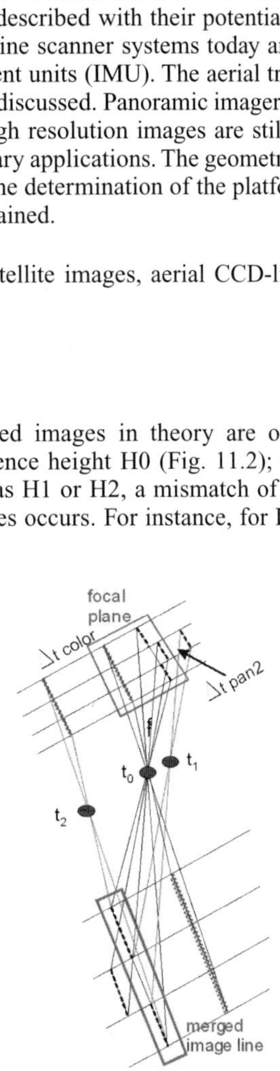

Figure 11.1. Combination of CCD-sensors to a homogenous virtual CCD-line.

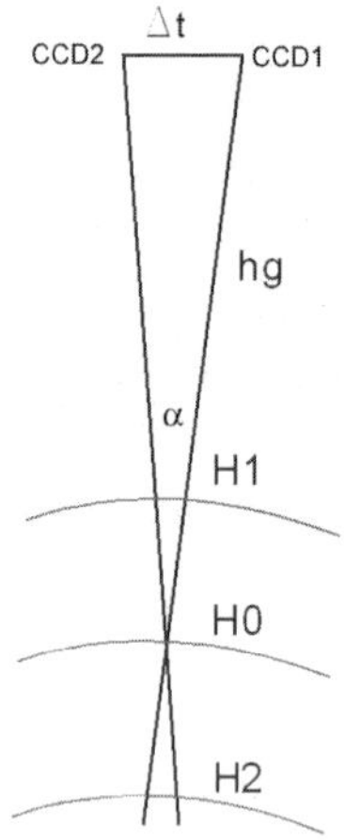

Figure 11.2. Mismatch of CCD-lines as a function of height and reference height.

IRS-1D, a mismatch of 1 pixel appears for height differences of 450 m against the reference height H0 and for QuickBird a 1 pixel mismatch will appear at a height difference of 2.8 km; that means it is never important.

11.1.3 *Image geometry and image products*

The classical optical satellites have a fixed orientation in relation to the satellite orbit during imaging. Today the imaging satellites are equipped with reaction wheels or control moment gyros, allowing a permanent, fast and precise change of the view direction. So usually the images are not taken in the orbit direction just based on the movement of the satellite: they are scanned directly in relation to the specified ground window. This may be in a north-south direction, but for stereo imaging it is often also in an east-west direction. IKONOS is equipped with a second CCD-line, allowing also the scan with the TDI-sensor against the movement in the orbit; that means from south to north. Not all satellites are able to generate a sufficient image quality because of the speed of the satellite in the orbit; they are extending the imaging time by asynchronous or slow down mode (Fig. 11.3). So the sampling rate of QuickBird is limited to 6900 lines/sec, with the 0.61 m ground sampling distance (GSD). This would correspond to 4.2 km/sec orbit speed, but the satellite has a foot print speed of 7.1 km/sec, requiring a slow down factor of 1.7. OrbView-3 and EROS have to use the asynchronous mode also to obtain sufficient image quality. This asynchronous mode has to be respected by the orientation procedure.

For IKONOS, only images projected to a surface with constant height are distributed as Geo-images (Fig. 11.4). There is still confusion with the product names – the expression level 1B is used for QuickBird Basic Imagery (original images) while the expression level 1B traditionally is used for projected images. For QuickBird, in addition to the original images (Basic Imagery) and the images projected to a surface with constant height (OrthoReady Standard), images are also projected to the rough DEM GTOPO30 (Standard Imagery) for distribution. For OrbView-3, at first only the original images (OrbView Basic) could be ordered but now projected images are also available as OrbView Geo.

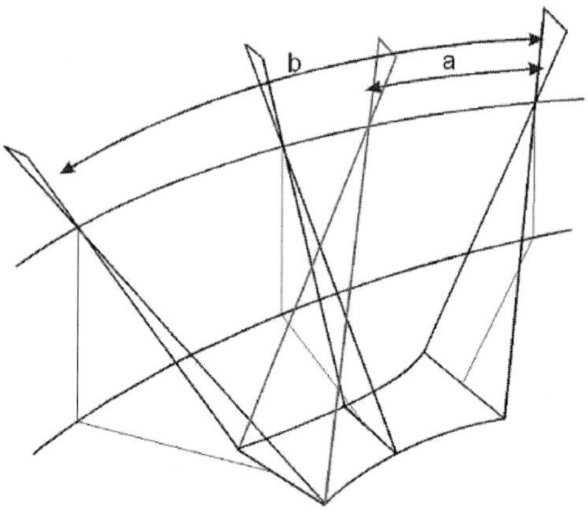

Figure 11.3. Asynchronous image mode (slow down mode) by permanent change of view direction during imaging—extension of imaging time. Slow down factor = b/a.

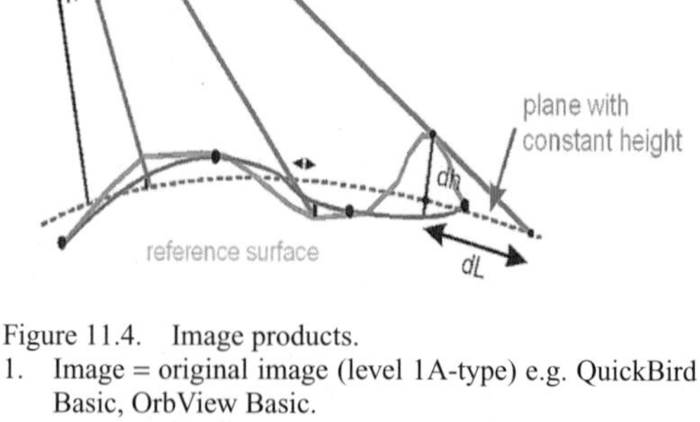

Figure 11.4. Image products.

1. Image = original image (level 1A-type) e.g. QuickBird Basic, OrbView Basic.
2. Projection to plane with constant height (level 1B-type) e.g. IKONOS Geo, QuickBird OR Standard.
3. Swinging line = QuickBird Standard—related to the rough DEM GTOPO30.

11.1.4 *Orientation of CCD-line images.*

11.1.4.1 *Sensor oriented RPCs*

Sensor oriented Rational Polynomial Coefficients (RPCs) from the satellite image vendors are describing the location of image positions as a function of the object coordinates (longitude, latitude, height) by the ratio of polynomials (Grodecki 2001). The sensor related RPCs are based on the direct sensor orientation of the satellite together with information about the inner orientation and have an accuracy depending upon the quality of the direct sensor information. Third order polynomials with 20 coefficients are used, so with 80 coefficients the relation of the image coordinates to the object coordinates can be described (Formula 1).

Based on the sensor oriented RPCs, a terrain relief correction can be made; that means the value dL in Figure 11.4 together with the direction of this vector are computed. This can be made for level 1A- and level 1B-type images. The resulting object positions can be related to ground control points by two-dimensional transformation leading to the bias corrected RPC solution. For IKONOS, a simple shift of the terrain relief corrected scene to control points is usually sufficient; for other sensors, or old IKONOS images without the information of the reference height, a two-dimensional affinity transformation of the computed object coordinates to the control points is required.

$$xij = \frac{Pi1(X,Y,Z)j}{Pi2(X,Y,Z)j} \qquad yij = \frac{Pi3(X,Y,Z)j}{Pi4(X,Y,Z)j}$$

$$\begin{aligned} Pn(X, Y, Z)j = {} & a_1 + a_2 * Y + a_3 * X + a_4 * Z + a_5 * Y * X \\ & + a_6 * Y * Z + a_7 * X * Z + a_8 * Y^2 \\ & + a_9 * X^2 + a_{10} * Z^2 + a_{11} * Y * X * Z \\ & + a_{12} * Y^3 + a_{13} * Y * X^2 + a_{14} * Y * Z^2 \\ & + a_{15} * Y^2 * X + a_{16} * X^3 * + a_{17} * X * Z^2 \\ & + a_{18} * Y^2 * Z + a_{19} * X^2 * Z + a_{20} * Z^3 \end{aligned}$$

Formula 1. Rational polynomial coefficients.
xij, yij = normalized scene coordinates
X,Y = normalized geographic object coordinates
Z = height

11.1.4.2 *Terrain dependent RPCs*

The relation of the scene to object coordinates can be approximated by RPCs also just based on control points. Of course, it is not possible to determine 80 orientation unknowns—this would require at least 40 ground control points per scene, so only the most important RPCs are computed. The number of chosen unknowns is quite dependent upon the number and three-dimensional distribution of the control points. This method cannot be controlled just by the residuals at the control points. Some commercial programs offering this method do not use any statistical checks for high correlation of the unknowns, making the correct handling very dangerous. A test with a commercial program using this method was leading with IKONOS data to control point residuals below 1 m and did not mark any problems. Even in an area within the range of the control points, discrepancies in the range of 50 m appeared at check points and outside the range of the control points, discrepancies exceeding 500 m. This method cannot be controlled; it has to be avoided and it is absolutely not a serious solution. In general, available orientation information should be used to reduce the required number and distribution of control points.

11.1.4.3 *Three-dimensional affine transformation*

Like the terrain dependent RPCs, the three-dimensional affine transformation is not using any available sensor orientation information. The 8 unknowns for the transformation of the object point coordinates to the image coordinates have to be computed based on control points located not in the same plane (Formula 2) (Hanley et al. 2002). At least 4 well distributed control points are required. The 3-D affinity transformation is based on a parallel projection, which is approximately given in the orbit direction but not in the direction of the CCD-line.

$$\begin{aligned} x_{ij} &= a1 + a2 * X + a3 * Y + a4 * Z \\ y_{ij} &= a5 + a6 * X + a7 * Y + a8 * Z \end{aligned}$$

Formula 2. 3-D affine transformation.

The not precise mathematical model of parallel projection is not a problem for a narrow field of view if in addition the height differences in the object space are not very large. For large height differences and unknown slow down mode, extended formulae are introduced by Jacobsen (2007b). Formula 3 respects the perspective geometry in the CCD-line direction and also a possible asynchronous imaging mode.

$$\begin{aligned} x_{ij} &= a1 + a2 * X + a3 * Y + a4 * Z + a9 * X * Z \\ &\quad + a10 * Y * Z \\ y_{ij} &= a5 + a6 * X + a7 * Y + a8 * Z + a11 * X * Z \\ &\quad + a12 * Y * Z \end{aligned}$$

Formula 3. Extended 3-D affine transformation.

For the handling of original images, a further extension has been made (Formula 4), which respects that the scene limits projected to the ground may not be parallel to each other.

$$\begin{aligned} x_{ij} &= a1 + a2 * X + a3 * Y + a4 * Z + a9 * X * Z \\ &\quad + a10 * Y * Z + a13 * X * X \\ y_{ij} &= a5 + a6 * X + a7 * Y + a8 * Z + a11 * X * Z \\ &\quad + a12 * Y * Z + a14 * X * Y \end{aligned}$$

Formula 4. Extended 3-D affine transformation for original images.

11.1.4.4 *Direct Linear Transformation (DLT)*
Like the 3-D affine transformation, the DLT is not using any pre-information about image orientation. The 11 unknowns for the transformation of the object coordinates to the image coordinates (Formula 5) have to be determined with at least 6 control points. The small field of view for high resolution satellite images, together with the limited object height distribution in relation to the satellite flying height, is causing more problems with correlation of unknowns as in the 3-D affine transformation. The DLT is based on a perspective image geometry that is available only in the direction of the CCD-line. There is no justification for the use of this method for the orientation of satellite images having more unknowns than required by other solutions.

$$xij = \frac{L1 * X + L2 * Y + L3 * Z + L4}{L9 * X + L10 * Y + L11 * Z + 1}$$

$$yij = \frac{L5 * X + L6 * Y + L7 * Z + L8}{L9 * X + L10 * Y + L11 * Z + 1}$$

Formula 5. DLT transformation.

11.1.4.5 *Reconstruction of imaging geometry*
A mathematically correct method of scene orientation is the reconstruction of the imaging geometry. For level 1B-type images, for the geo-referenced scene centre or the first line, the direction to the satellite is available in the image header data. This direction can be intersected with the orbit of the satellite published with its Keppler elements. Depending upon the location of an image point, the location of the corresponding projection centre in the satellite orbit and the view direction can be computed, respecting the Earth rotation and the slow down factor of asynchronous imaging. So the view direction from any ground point to the corresponding projection centre can be reconstructed. This method requires the same number of control points as the sensor oriented RPC-solution; that means it can be used also without control points if the direct sensor orientation is accepted as accurate enough. Otherwise it requires the same two-dimensional transformation of the terrain relief corrected object points to the control points as the sensor oriented RPCs. A slightly different geometric model has been developed by Toutin (Toutin 2003).

In general, similar orientation methods as used for the level 1-B type images can be used for the orientation of original images (level 1A-type), but the handling of original images is more difficult—they are not corrected for effects of high frequency satellite rotations. The level 1B-type images are geo-coded and only need a terrain relief correction and a 2-D improvement of the location, which is close to a datum problem. For this reason, the basic conditions for the approximate solutions are more difficult for original images.

The scene orientation of original space images by geometric reconstruction is not new. At first it had been developed for SPOT images. For instance in Jacobsen (1997), the image geometry is reconstructed based on the given view direction, the general satellite orbit and a few control points. Based on control points, the attitudes and the satellite height are improved. The X- and Y-locations of the projection line are fixed because they are nearly numerically dependent upon the view direction. In addition, two additional parameters are required for image affinity and angular affinity. The affinity may be caused by incorrect information about flying height, while the angular affinity may be caused by an imprecise perpendicular orientation of the CCD-line to the orbit. Three control points are necessary for these six unknowns. Additional parameters can be introduced if geometric problems exist. Only for scenes with totally unknown orientation will the full sensor orientation with 6 orientation elements be adjusted together with the necessary additional parameters. This requires a good vertical distribution of control points; for flat areas the full orientation cannot be computed. Other solutions use the given sensor orientation together with some required correction parameters. In general, more control points with a good three-dimensional distribution are required if the existing sensor orientation information is not to be used and the view direction is computed by means of control points. The orientation of the original images also can be made with sensor oriented RPCs in an iterative manner. It has to be improved by means of control points, corresponding to the bias corrected RPC solution (Grodecki 2001).

11.1.4.6 *Comparison of results*
The different orientation methods have been compared with different types of satellite images. As typical examples the results achieved in the Zonguldak test area are shown (Jacobsen 2007c). Zonguldak is located in Turkey at the Black Sea. The control points of the mountainous area range from sea level up to 440 m, so optimal conditions for the approximate orientation solutions exist.

The bias corrected RPC solution leads even with just one control point to a root mean square error of the independent check points of 0.75 m, corresponding to 0.75 GSD. More control points do not improve the result. With only one control point, the bias correction is limited to a simple shift. This result has been confirmed by several IKONOS orientations. (Figure 11.5). The geometric reconstruction leads to 0.9 m up to 1.0 m root mean square errors of the check

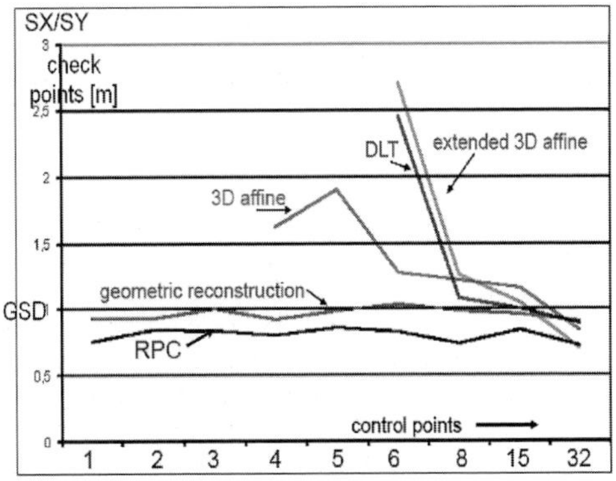

Figure 11.5. Orientation of IKONOS images, Zonguldak (mountainous), 1.0 m GSD root mean square discrepancies at independent check points as a function of number of control points used. (see colour plate page 494)

point coordinate components and is also nearly independent of the control point number. An orientation with terrain dependent RPCs could not be accepted because of poor accuracy in areas not directly neighboured to the control points. The 3-D affine transformation requires at least four control points and is not reaching the same accuracy level. Here a higher number of control points improves the result. For the DLT and the extended 3D affine solution, at least six control points are required and only the result with 15 control points is close to the level of the geometric reconstruction and the bias corrected, sensor oriented RPC solution.

The orientation of a QuickBird scene in the same area leads to similar results, with the exception that a two-dimensional affine transformation is required after the terrain relief correction, so at least 3 control points are necessary for the orientation of a QuickBird scene. The geometric reconstruction and the bias corrected RPC-solution is on the same level of approximately 0.9 GSD root mean square errors of independent check points. Even with a high number of control points, the 3-D affine transformation does not reach the same accuracy level (Figure 11.6). This can be explained by the larger field of view of QuickBird and the asynchronous imaging mode. Only with the extended 3-D affine transformation (Formula 3) with at least eight control points is the orientation on the same accuracy level as the mathematical correct solutions.

With OrbView-3 images no accuracy on the GSD-level can be reached as in the shown example (Fig. 11.7). OrbView-3 has a projected pixel size of 2 m, but with 50% over-sampling 1 m GSD. This may be the reason for just reaching approximately 1.6 GSD accuracy at independent check points. The results achieved with the approximate orientation methods are much larger than the bias corrected, sensor oriented RPC-solution. Only with the 3-D affine transformation for original images (Formula 4) can root mean square differences of 3 up to 2 GSD be reached with at least eight control points.

Several high and very high resolution images have been oriented by Jacobsen (2007c). All orientations listed in Table 11.1 are based on geometric reconstruction or bias corrected RPC solutions. Under usual conditions, sub-pixel accuracy has been reached. In the case of KOMPSAT-1 and IRS-1C,

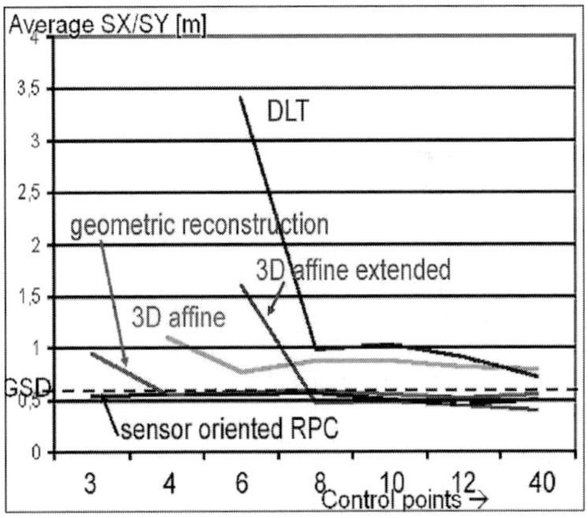

Figure 11.6. Orientation of QuickBird, Zonguldak, 0.62 m GSD, root mean square discrepancies at independent check points as a function of control point number. Exception: 40 GCPs = discrepancies at control points. (see colour plate page 494)

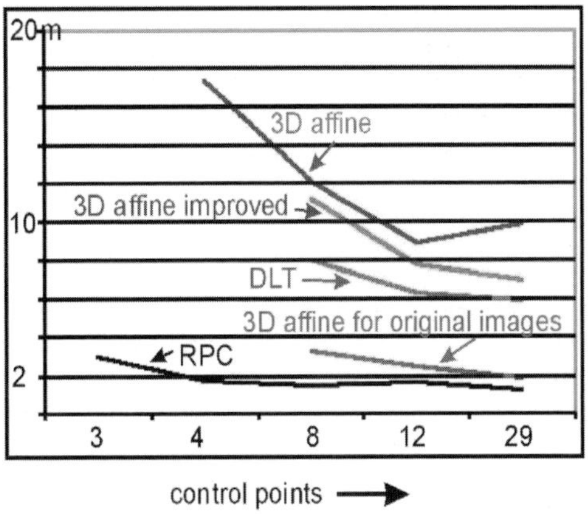

Figure 11.7. Orientation of OrbView-3 Basic, Zonguldak—root mean square differences at independent check points (vertical direction), depending upon number of control points used. (see colour plate page 494)

Table 11.1. Root mean square differences of independent check points (RMSX / RMSY) based on scene orientation with geometric reconstruction or bias corrected RPC solution.

	Level type	GSD [m]	RMSX RMSY [GSD]
ASTER, Zonguldak	A	15	0.7
KOMPSAT-1, Zonguldak	A	6.6	1.3
SPOT, Hannover	A	10	0.5
SPOT 5, Zonguldak	A	5	1.0
SPOT 5, Zonguldak	B	5	1.0
SPOT HRS, Bavaria	A	5×10	0.7/1.1
IRS-1C, Hannover	A	5.7	0.9
IRS-1C, Zonguldak	B	5.7	1.6
Cartosat-1, Warsaw	B	2.5	0.6
OrbView-3, Zonguldak	A	1	1.3
IKONOS, Zonguldak	B	1.0	0.7
QuickBird, Zonguldak	B	0.61	0.8

Zonguldak, the achieved accuracy was limited by the control point quality. In the case of OrbView-3, the accuracy exceeds the GSD, but it is below the projected pixel size. No significant difference between the orientation of level 1A-type and level 1B-type scenes can be seen. Under operational conditions with well defined control points, pixel- or even sub-pixel-accuracy can be reached.

11.2 AIRBORNE SYSTEMS

11.2.1 *Introduction*

Airborne CCD-line scanners at first were used for classification purposes. Their images have a different exterior orientation for every CCD-line, but within each CCD-line there is a stable geometry (Fig. 11.8). This is still better than scanning systems based on diodes, having a different orientation for every pixel. Without additional orientation information from GPS and inertial measurement units (IMU), the georeference can only be estimated using a high number of control points. For this reason, polynomial models have mostly been used. Polynomial solutions cannot be called image orientation; they are only fitting the homogenous, but geometric distorted, image to reference points. Usually they are not respecting the influencing digital elevation model. Only the geometrically more precise methods are described.

For photogrammetric application, only three-line-scanners are in use such as HRSC from DLR, Leica ADS40, the former Starlabo Starimager, Wehrli 3-DAS-2 and Jenoptronic JAS 150. The basic principle was developed as DPA by Otto Hofmann at Messerschmidt, Bölkow, Blohm, Munich around 1970 (Hofmann & Navé 1982). The first practical application came with the MOMS satellite and it was

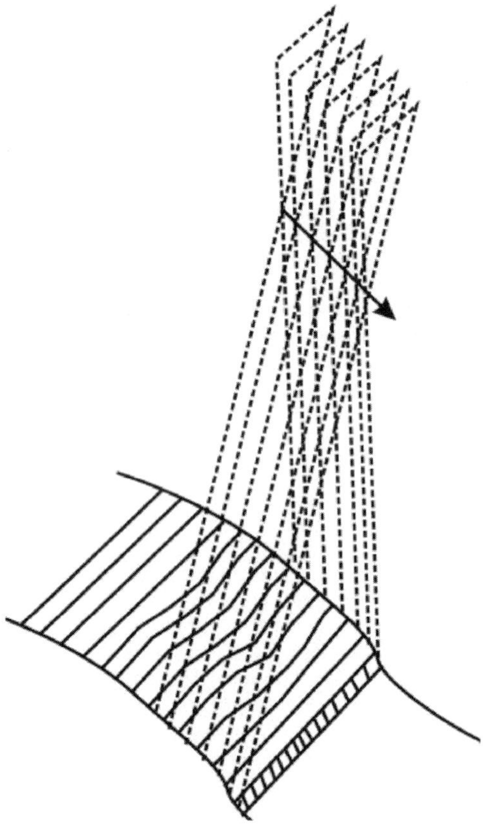

Figure 11.8. CCD-line image.

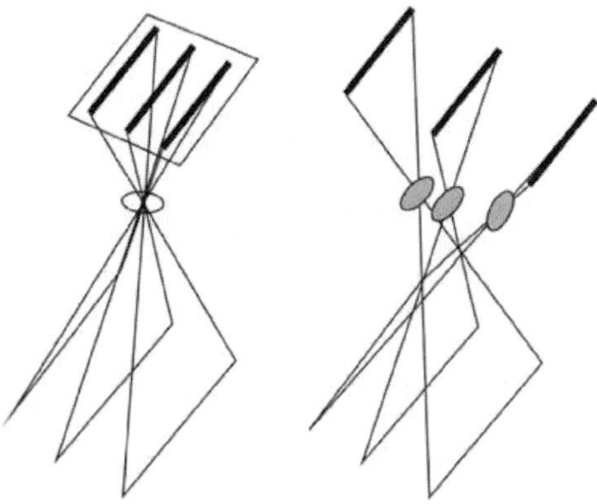

Figure 11.9. 3-line-scanner with 1 focal plane (left) and 3-line scanner with 3 optics (right).

also included as HRSC in the failed first German Mars-mission. After this, the German Aerospace Centre DLR modified the HRSC for aerial application. Supported by DLR, Leica developed the ADS40.

These systems are operated together with kinematic GPS and IMU. Three-line-scanners may have separate CCD-lines on one focal plane or they may have a combination of sub-cameras, each with just a line (Fig. 11.9). The mathematical handling is more or less the same, because a camera system has a known geometric relation to the sub-cameras. The recording interval of the IMU is short enough to guarantee by linear interpolation no loss of accuracy for the orientation determination of any individual CCD-line. But the orientation may not be accurate or reliable enough or the orientation devices may fail.

Leica ADS40 images allow a mapping based on stereoscopic view by human operator. The direct sensor orientation is not precise enough for a stereoscopic view without disturbing y-parallaxes, and so an adjustment using tie points is required (Fricker 2001).

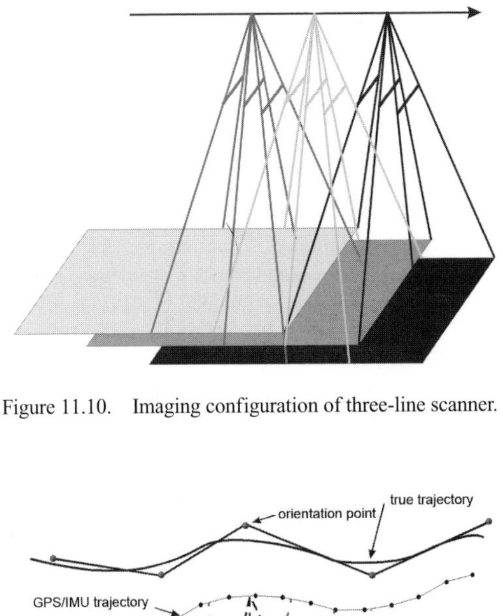

Figure 11.10. Imaging configuration of three-line scanner.

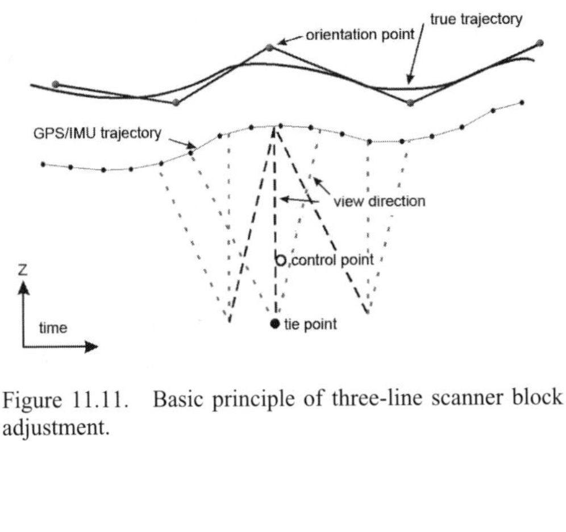

Figure 11.11. Basic principle of three-line scanner block adjustment.

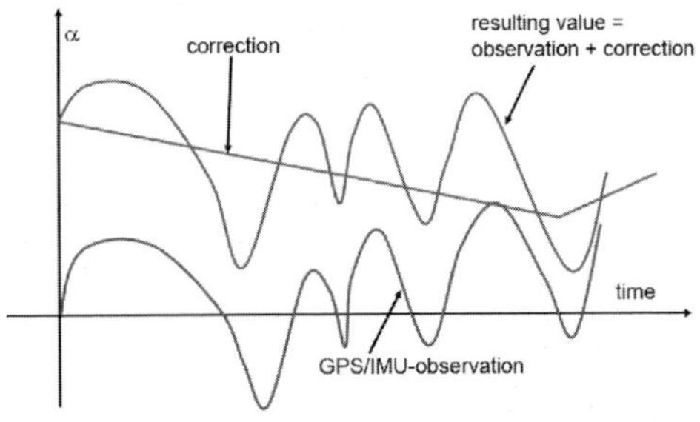

Figure 11.12. Correction of trajectory—high frequency observation with low frequency correction. (see colour plate page 494)

11.2.2 *Aerotriangulation*

With line scanner images, no classical bundle block adjustment is possible because of missing fixed connection of neighbouring lines, but the orientation is only changing continuously, so groups of CCD-lines can be joined together to "orientation images" with an orientation presented by "orientation fixes" (Hofmann & Nave 1982, Ohlhof 1995). Joining neighboured CCD-lines to orientation images, stabilized with tie points to the images of the other view direction and supported by control points (Fig. 11.11), allows also approximate orientation determination of the whole system without the support of GPS and IMU. Of course, if no support by GPS and IMU is available, quite a few more control points are required. By adjustment of GPS / IMU supported three-line scanner data (Fig. 11.10), only the low frequency data are improved. The high frequency data are still unchanged (Fig. 11.12).

The aerotriangulation of the three-line sensors is possible without ground control if it is supported by GPS and IMU, but for reliability reasons and for the determination of the datum, control points are required for practical applications. Supported by control points, a standard deviation of all coordinate components is possible on the sub-pixel level (Cramer 2006). Without control points, the offsets cannot be guaranteed.

11.3 PANORAMIC IMAGERS

11.3.1 *Introduction*

Panoramic cameras are scanning the object from one side to the other with perspective geometry only in the line perpendicular to the scan direction (Fig. 11.13). They have been used in space by the US CORONA system (McDonald 1997) from 1959 up to 1972 and the Soviet/Russian KVR1000, from the air (Slama 1980) and also on the Earth (Reulke et al. 2004).

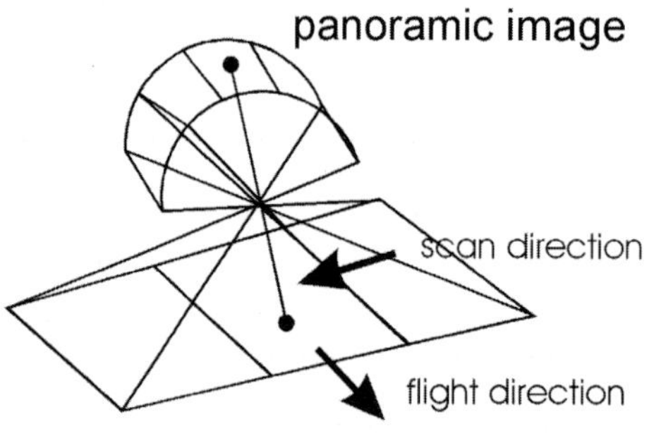

Figure 11.13. Principle of panoramic camera.

11.3.2 *Panoramic geometry*

Even if the object area is flat, the image scale is changing depending upon the relation of the focal length and the oblique distance. Across the view direction, the scale change is 1/cosν with ν as the nadir angle and in the view direction it is $1/\cos^2\nu$ During scanning, the aircraft or satellite is moving, causing a deformation of the image geometry (Fig. 11.14).

Like the CCD-line images, every image line perpendicular to the scan direction has a different exterior orientation, so in aerial application the support of GPS/IMU has advantages. For the mathematical handling, the collinearity equation has to be modified for the changing scale and the changing exterior orientation. Another possibility is a correction of the panoramic image to perspective geometry without influence of the platform motion (Fig. 11.15).

$x' = x - (y/f - x/r^2) * P22$

$y' = y - (y/f - y/(f^2 + y^2)) \cdot P22$

$x' = x - \arctan y/x \cdot P23 \qquad y' = y$

$x' = x - \sin(y/300.) \cdot P24 \qquad y' = y$

$x' = x \qquad y' = y - \sin(y/300) \cdot P25$

$x' = x - \sin(y/150.) \cdot P26 \qquad y' = y$

Formula 6. Special additional parameters for handling panoramic images.

The aircraft and the scanning speed is often unknown, so the influence of the platform motion has to be

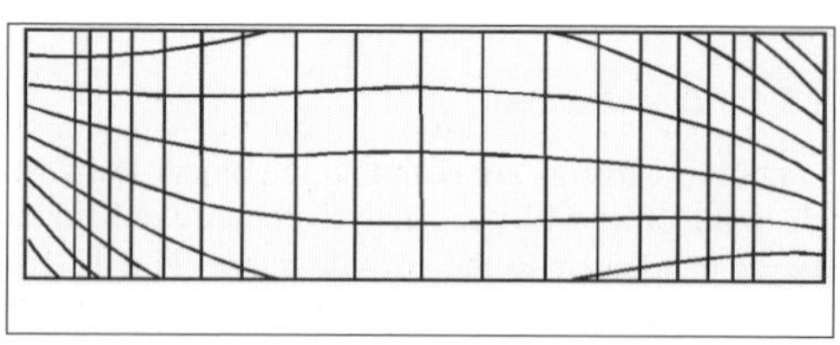

Figure 11.14. Panoramic image of a square grid in the object space.

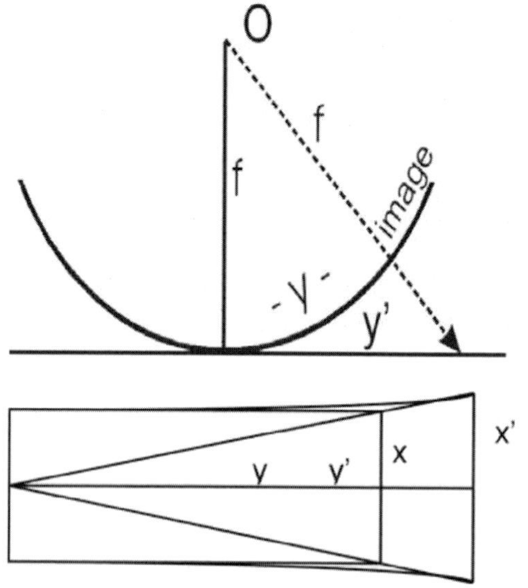

$$y' = f \cdot \tan\frac{y}{f}$$

$$x' = \frac{x}{f} \cdot \sqrt{f^2 + y^2}$$

Figure 11.15. Transformation of panoramic image to perspective geometry.

determined by self calibration with additional parameters in the orientation process or by bundle adjustment. In the Hannover program system BLUH, special additional parameters are available for the handling of panoramic images (Jacobsen 1988), which can be used after pre-correction to perspective geometry (Fig. 11.16).

11.3.3 *Aerial panoramic cameras*

In the example of the block adjustment of aerial panoramic images, shown in Figure 11.17, the scale in the image centre is 1:10,000, at the border with a nadir angle of 49° it is 1:15,300 across the view direction and 1:23,600 in the view direction. In the centre, the base to height relation is 1:12 while it is 1:18.4 at the border. Object points determined in a perspective aerial model have a homogenous accuracy in X and Y and the vertical accuracy is approximately the horizontal accuracy multiplied by the height to base relation; this is not the case for panoramic models. In panoramic image models with a nadir angle of 45°, the horizontal accuracy in the scan direction is the same as the vertical accuracy. With larger nadir angles, the vertical accuracy is better than the accuracy in the scan direction (Fig.11.18).

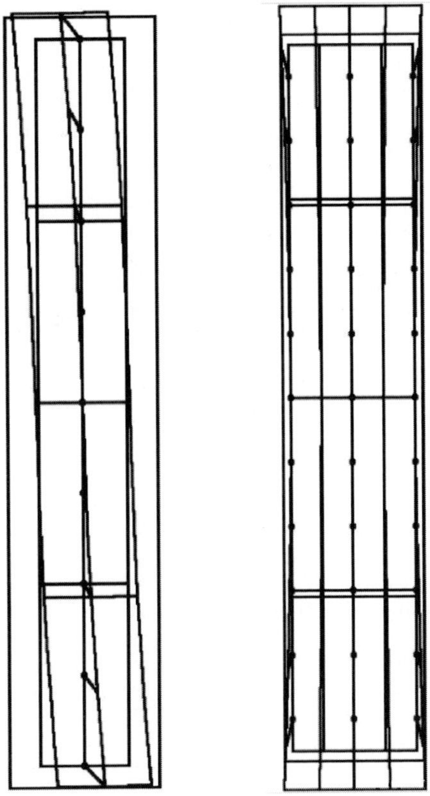

Figure 11.16. Left: dynamic effect of panoramic aerial image determined by self calibration with additional parameters.
Right: effect of transformation panoramic image to perspective geometry.

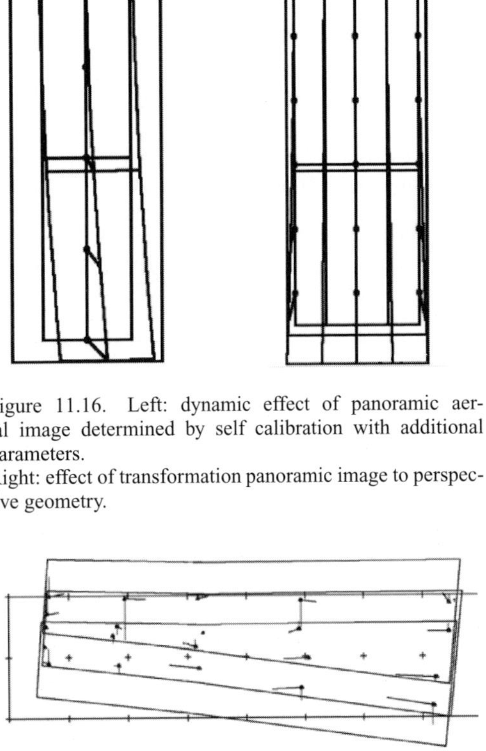

Figure 11.17. Block of 3 panoramic images (from centre of image—left, up to end—right).

$$SX = \frac{hg}{f \cdot \cos \upsilon} \cdot \sigma o$$

$$SY = \sqrt{\frac{hg}{f \cdot \cos \upsilon + SZ \cdot \tan \upsilon}} \cdot \sigma 0$$

$$SZ = \frac{hg^2}{f \cdot b \cdot \cos \upsilon} \cdot \sigma o$$

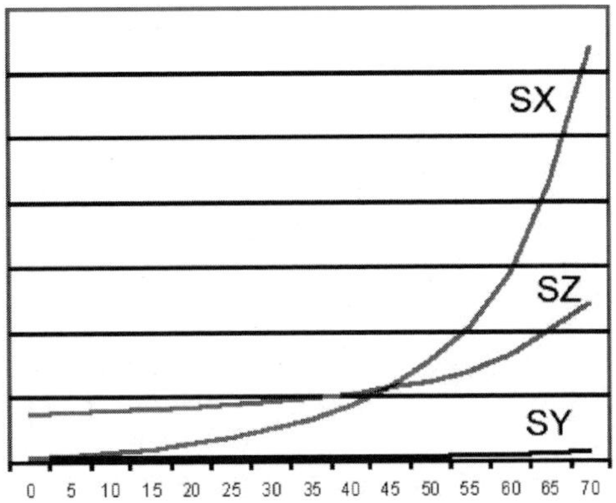

Figure 11.18. Relative accuracy of object coordinates determined in panoramic model with base to height relation 1:12 as function of nadir angle.
horizontal = nadir angle [°]

Formula 7. Standard deviation of object coordinates determined by panoramic model.
Hg = flying height f = focal length ν = nadir angle
σo = standard deviation of image coordinates
b = base length

11.3.4 *Spaceborne panoramic cameras*

The same handling as shown above with aerial panoramic images is possible with satellite panoramic images, but the conditions are better than in aerial images. With +/–34.5°, the largest nadir angle for a CORONA KH4B image is smaller than for aerial systems and the sensor movement is much better known than for aerial systems.

The most often used CORONA camera is the KH4B having a focal length of 627 mm and a film size of 55.37 mm × 756.92 mm. For this format, the difference from perspective geometry is not exceeding dx = 3.9 mm and dy = 53.4 mm, for the large image format in the y-direction. The systematic image errors (difference of mathematical model against reality), based on the additional parameters shown in Formula 6 plus affinity, for a CORONA stereo pair is shown in Figure 11.19. The typical S-shape caused by the sensor movement during imaging is obvious; in addition a stronger affinity can be seen. The scale difference between x and y image coordinates can be explained by the inaccurate focal length caused by the imprecise film motion during imaging. With the above mentioned handling, digital elevation models based on the convergent arranged KH4B stereo camera combination can be determined with at least a relative vertical accuracy of up to 5 m (Schneider et al. 2001).

The orientation of the Soviet / Russian KFA1000 is similar to the handling of CORONA images, with

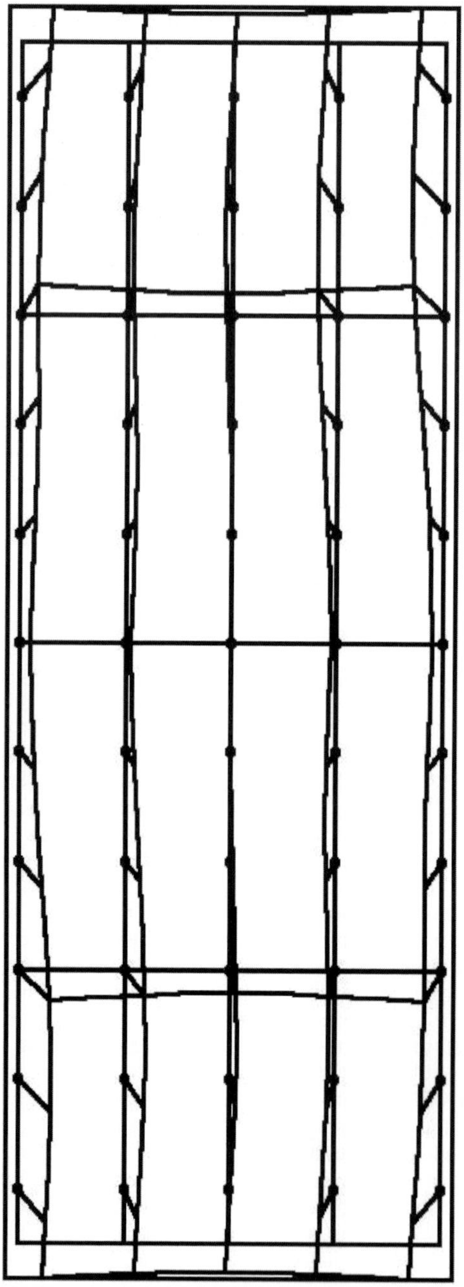

Figure 11.19. Systematic image errors of a CORONA stereo pair. Largest vector = 440 μm.

the difference that the KVR1000 was not used in a stereo combination. Instead, the Soviet Union used the combination of the panoramic KVR1000 together with perspective stereo models of the TK350.

11.3.5 *Terrestrial panoramic cameras*

The mathematical handling of terrestrial panoramic cameras is much simpler. The camera axis is not moving during imaging. In the vertical direction, perspective geometry exists, while in the horizontal scan direction the viewing angle is a linear function of the time, corresponding to the image coordinate component (Reulke et al. 2004). For precise point determination such terrestrial panoramic cameras need a satisfactory calibration, which is also possible by self calibration with additional parameters.

11.4 CONCLUSIONS

Today existing high resolution space imagery is competing with aerial photographs requiring similar orientation accuracy. Only with geometric correct orientation methods like geometric reconstruction and sensor oriented, bias corrected rational polynomial coefficients is this guaranteed. The approximate image orientation models like the 3-D affine transformation, direct linear transformation and terrain dependent RPCs require more and three-dimensional, well distributed control points; in addition they cannot guarantee the accuracy in any case.

Today aerial CCD-line scanner systems are in use as three-line scanners, usually supported by GPS and inertial measurement units (IMU). Based on GPS and IMU in theory no adjustment and no control points are required, but aerial triangulation solves the datum problem, guarantees the reliability and reduces disturbing y-parallaxes in the stereo models.

Panoramic images can be transformed into perspective geometry and the effect of the sensor motion during exposure can be handled by self calibration with special additional parameters. Also terrestrial panoramic imagers are in use; because of the fixed positioning during orientation, the geometric model is simple, and the system calibration may be improved by self calibration.

REFERENCES

Cramer, M. 2006. The ADS40 Vaihingen/Enz geometric performance test. *IJPRS* 60, 6: 363–374.

Dial, G. & Grodecki, J. 2002. IKONOS Accuracy without Ground Control, Pecora 15/Land Satellite Information IV/ISPRS Com. I, Denver 2002. *Int. Arch. PhRS* XXXIV, part 1: 31–35.

Gervaix, F. 2002. Aerotriangulation: auch für den ADS40 Luftbildsensor? *Photogrammetrie, Fernerkundung, Geoinformation* 2002: 85–91.

Grodecki, J. 2001. IKONOS Stereo Feature Extraction—RPC Approach. *ASPRS annual conference St. Louis*.

Fricker, P. 2001. ADS40—Progress in digital aerial data collection. *Photogrammetric Week 01*: 105–116. Heidelberg: Wichmann Verlag.

Hanley, H.B., Yamakawa, T. & Fraser, C.S. 2002. Sensor Orientation for High Resolution Imagery, Pecora 15/ Land Satellite Information IV / ISPRS Com. I, Denver. *Int. Arch. PhRS* XXXIV, part 1: 69–75.

Hofmann, O. & Navé, P. 1982. DPS—a digital photogrammetric system for producing digital elevation models and orthophotos by means of linear array scanner imagery. ISPRS Com III, Helsinki. *Int. Arch. PhRS* 24–III: 216–227.

Jacobsen, K. 1988. Handling of Panoramic and Extreme High Oblique Photographs in Analytical Plotters, ISPRS congress Kyoto 1988. *Int. Arch. PhRS* XXVII, B2: 232–237.

Jacobsen, K. 1997. Calibration of IRS–1C PAN–camera. *Joint Workshop "Sensors and Mapping from Space", Hannover 1997*. http://www.ipi.uni–hannover.de/ (Feb. 18th, 2008).

Jacobsen, K. 2007a. Digital Height Models by Cartosat–1, ISPRS Hannover Workshop 2007. *Int. Arch. PhRS* XXXVI Part I/W51.

Jacobsen, K. 2007b. Geometry of Digital Frame Cameras, *ASPRS Annual Conference, Tampa 2007*.

Jacobsen, K. 2007c. Orientation of high resolution optical space images. *ASPRS Annual Conference, Tampa 2007*.

McDonald, R. (ed.) 1997. *CORONA between the sun & the earth*. ASPRS. ISBN 1-57083-041-X.

Ohlhof, T. 1995. Block triangulation using three-line images. *Photogrammetric Week 1995*: 197–200. Heidelberg: Wichmann Verlag.

Passini, R. & Jacobsen, K. 2004. Accuracy Analysis of Digital Orthophotos from Very High Resolution Imagery. ISPRS Congress, Istanbul 2004. *Int. Arch. PhRS* XXXV, B4: 695–700.

Reulke, R., Wehr, A. & Griesbach, D. 2004. High resolution mapping using CCD–line camera and laser scanner with integrated position and orientation system, ISPRS congress Istanbul. *Int. Arch. PhRS* XXXV, B3: 506–511.

Schneider, T., Seitz, R., Förster, B. & Jacobsen, K. 2001. Remote Sensing Based Parameter Extraction for Erosion Control Purposes in the Loess Plateau of China. *ISPRS Workshop High Resolution Mapping from Space 2001*.

Slama, C. (ed.). 1980. *Manual of Photogrammetry*, Fourth edition, chapter 4.2.2 Panoramic Cameras: 196–207. ASPRS. ISBN 0-937294-01-2.

Toutin, Th. 2003. Error tracking in IKONOS geometric processing using a 3D parametric modeling. *Photogramm. Eng. Remote Sens.* 69, 1: 43–51.

Advances in Photogrammetry, Remote Sensing and Spatial Information Sciences: 2008 ISPRS Congress Book – Li, Chen & Baltsavias (eds)

CHAPTER 12

DSM generation and deformation measurement from SAR data

Michele Crosetto & Paolo Pasquali

ABSTRACT: This chapter addresses two interferometric applications of Synthetic Aperture Radar (SAR) data, and in particular the phase of complex SAR images. The first is deformation measurement and monitoring, which is based on Differential Interferometric SAR (DInSAR) techniques. The second is Digital Surface Model (DSM) generation, which is based on interferometric SAR (InSAR) techniques. The chapter describes the most important SAR data sources, which include spaceborne, airborne and terrestrial sensors. Then the most innovative algorithmic developments in these two fields are discussed, and the most relevant research fields. The wide range of applications of both InSAR and DInSAR are illustrated by providing a set of references to relevant published works. Finally, the chapter provides an overview of the new SAR sensors and missions, which because of their spatial resolution and revisiting capabilities are expected to foster new research activities in the fields of SAR interferometry, and to drive the development of new SAR application fields.

Keywords: SAR, interferometry, InSAR, DInSAR, DSM, deformation

12.1 INTRODUCTION

This chapter addresses two main applications of SAR data. The first one, deformation measurement, is based on the DInSAR techniques, while the second one, DSM generation, is based on the InSAR techniques. A comprehensive review of SAR interferometry is given by Rosen et al. (2000), while Crosetto et al. (2005b) provide a review of the state-of-the-art of DInSAR. The primary objective of this chapter is to review the recent advances in InSAR and DInSAR techniques. Considering approximately the last four years, this work highlights the most innovative developments, both from the algorithmic and application viewpoints.

This chapter includes two main parts: one devoted to deformation measurements, and the other to DSM generation. The deformation monitoring part includes the following sections. Section 12.2 briefly reviews the three main sources of DInSAR data, i.e. spaceborne SAR sensors, airborne SAR sensors, and ground-based or terrestrial SAR sensors. Section 12.2.1 provides an overview of the important developments made in the last few years for DInSAR processing and analysis. Section 12.2.2 addresses the DInSAR quality issues, and the important topic of DInSAR validation. Section 12.2.3 mentions a set of relevant DInSAR applications, and finally Section 12.2.4. briefly describes the key characteristics of the new SAR sensors and missions. The DSM generation includes the following sections. Section 12.3 briefly reviews the most recent advances in DSM generation from InSAR data; Section 12.3.1 provides an overview of the achievements of the SRTM mission; Section 12.3.2 summarizes the other important development made in the last few years for DSM generation from spaceborne data; and Section 12.3.3 addresses the status of DSM and DTM generation from airborne SAR sensors. Finally, the Conclusions summarize the key outcomes of this review.

12.2 DINSAR FOR DEFORMATION MEASUREMENT

The DInSAR techniques for deformation measurement are mainly based on SAR data acquired by the following spaceborne sensors: ERS-1/2, ASAR-ENVISAT, RADARSAT-1, JERS and ALOS-PALSAR. ERS-1 and 2 (C-band) have an important historical archive, which ranges from 1991 to date. ERS-2 is expected to be operated until 2011. The ASAR sensor (C-band) onboard ENVISAT represents an important interferometric data source, which is expected to be operated until 2013. Several research projects based on ERS and Envisat data are described in http://eopi.esa.int/esa/esa. RADARSAT-1 (C-band) is another important data source for DInSAR studies, e.g. see Lu et al. (2003) and Ferretti et al. (2007). Finally, the Japanese JERS and ALOS PALSAR, both of them working in L-band, represent valuable data sources for the specific characteristics of L-band, like the interesting performances in vegetated areas (in L-band the temporal decorrelation is lower than that of C-band), and the fact that L-band is less prone to aliasing in the presence of big deformations.

Airborne SAR platforms potentially represent a complementary source of data, which could take advantage of the flexibility of airborne sensors with respect to the satellite ones (track direction, revisiting time, spatial resolution). However, the application of DInSAR to airborne data encounters some important technical limitations, and it is mainly limited to experimental research, e.g. see Reigber & Schreiber (2003) and Perna et al. (2008).

Ground-Based SAR systems (GB-SAR) have gained more and more importance in recent years. Different groups have developed terrestrial SAR systems for research purposes, e.g. see Leva et al. (2003), Tarchi et al. (2003), Luzi et al. (2004) and Pipia et al. (2007). In addition, it is worth mentioning the development of other types of sensors, e.g. the bistatic SAR Interferometer proposed by Lopez-Dekker et al. (2007). The GB-SAR systems offer operational monitoring capability in emergency scenarios, e.g. see the example of the Stromboli volcano for tsunami early warning (Antonello et al. 2004). Very recently an Italian company, IDS (www.ids-spa.it), started selling the first commercial GB-SAR systems.

12.2.1 *New algorithms for DInSAR processing*

A large part of the early DInSAR results were achieved by using the standard DInSAR configuration, i.e. by analysing a single differential interferogram derived from a pair of SAR images. This is the simplest DInSAR configuration, which often is the only one that can be implemented, due to the limited SAR data availability. A remarkable improvement in the quality of the DInSAR results is given by the DInSAR methods that make use of large sets of SAR images acquired over the same deformation phenomenon. These techniques, hereafter called Advanced DInSAR (A-DInSAR) techniques, represent an outstanding advance with respect to the standard ones, both in terms of deformation modelling capabilities and quality of the deformation estimations. The output of an A-DInSAR analysis usually includes the estimation of the average deformation velocity over the observed period, and the estimation of the temporal evolution of deformation (deformation time series). The so-called "topographic correction" is an additional important parameter, which plays a key role in the advanced geo-coding of the A-DInSAR outputs, e.g. Lanari et al. (2004).

Several A-DInSAR approaches have been proposed in recent years after the publication of the Permanent Scatterers technique by the Politecnico of Milan (Ferretti et al. 2000). These A-DInSAR approaches are usually referred to as Permanent or Persistent Scatterers Interferometry (PSI). It is worth noting that the term Permanent Scatterers is directly associated with a technique patented by the Politecnico di Milano (Italy), and which now is exclusively licensed to TRE (www.treuropa.com), a spin-off company of this university. Therefore, the second term, Persistent Scatterers Interferometry, is usually employed to indicate all the techniques of this type. Most of the A-DInSAR or PSI approaches, e.g. Ferretti et al. (2001), Mora et al. (2003), Werner et al. (2003) and Duro et al. (2005), model the temporal evolution of the deformation with linear functions. Crosetto et al. (2005a) model the deformation by stepwise linear functions, whose parameters are computed by least squares adjustment. Colesanti et al. (2003) address the detection of both time-uniform and seasonal deformation phenomena. The so-called Small Baseline Subset (SBAS) technique (Berardino et al. 2002, Lanari et al. 2004, Pepe et al. 2005) represents a deformation model-free A-DInSAR approach. Recently different new A-DInSAR approaches have been proposed. Kampes & Hanssen (2004) adapted the LAMBDA method, part of an integer least squares estimator developed for the ambiguity resolution in GPS, to the problem of PSI. Van Leijen & Hanssen (2007) describe the use of adaptive deformation models in PSI. An improvement of the PSI performances can be achieved by using 3-D phase unwrapping procedures. Pepe et al. (2007) describe a phase unwrapping procedure implemented in two steps, to exploit both the spatial characteristics and the temporal relationships among the interferograms. Crosetto et al. (2007) describe another approach, named 2+1-D unwrapping, which couples the classical 2-D phase unwrapping with 1-D phase unwrapping in time. Another example of 3-D phase unwrapping is described in Hooper & Zebker (2007).

Another important recent development in the field of DInSAR (and InSAR in general) is the extension of the Interferometric techniques to data acquired using the ScanSAR mode. After the first developments presented by Monti Guarnieri and Prati (1996) and the large-scale application of the technique to the short-duration SRTM mission, the first results obtained from long-duration missions have been presented by Holzner & Bamler (2002) and Monti Guarnieri et al. (2003) for Radarsat-1 and ENVISAT

ASAR data, respectively. The ScanSAR acquisition mode allows the extension of the imaged swath in the range direction with a corresponding loss of resolution in the azimuth direction. An example of a differential interferogram obtained from ENVISAT ASAR Wide-Swath (ScanSAR) data over an area of 200 by 200 km is showed in Figure 12.1.

The interest in ScanSAR DInSAR techniques is strengthened by the introduction of specific strategies to minimize the burst-desynchronization issues during the planning of the acquisitions, as presented in Rosich (2007) and Rosich et al. (2007).

In the last part of this section, we mention some other relevant research work made in the field of DInSAR for deformation measurement. Perissin & Ferretti (2007) make a comprehensive analysis of the characteristics of urban targets that behave as persistent scatterers, identifying six main typologies of urban SAR targets: ground-level and elevated backscatterers, simple and resonating dihedrals, poles and trihedrals. Taking into account the opportunistic measurement made by A-DInSAR (the persistent scatterers represent only a small fraction of all measured pixels), this type of study is important to understand the behaviour of the targets exploited by these techniques. Finally, it is worth mentioning new techniques, named multi-dimensional SAR imaging techniques, to monitor deformations at high resolution on dominant targets (Lombardini 2005, Fornaro & Serafino 2006, Serafino et al. 2007). These techniques can separate different scatterers interfering in the same resolution cell, and estimate their relative velocities.

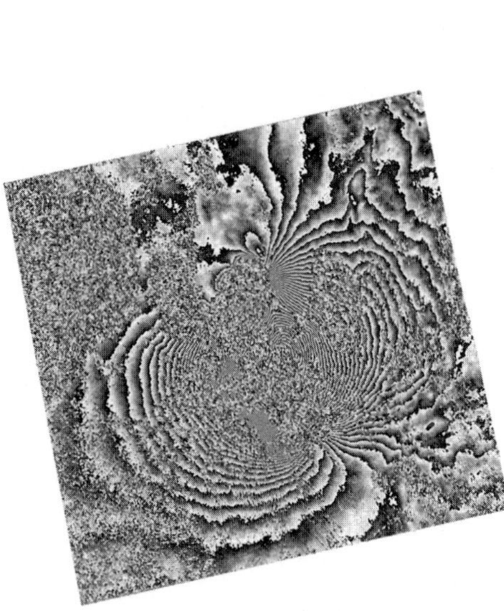

Figure 12.1. ASAR co-seismic deformation pattern (Ethiopia), after Wright et al. (2007). Each black-white cycle corresponds to 2.8 cm of Line-of-Sight deformation. Size of the image: about 200 by 200 km.

12.2.2 *Quality issues and DInSAR validation*

In the last few years, there has been an increasing interest in the quality of the parameters estimated by A-DInSAR techniques, and in the validation of the A-DInSAR results. Ketelaar et al. (2007) describe a reliability assessment of PSI using independent overlapping SAR tracks. Through datum connection of multiple tracks, they evaluate precision and reliability of PSI estimates. Ferretti et al (2007) present the results of a blind experiment using two pairs of artificial dihedral reflectors. They demonstrate that DInSAR measurements can allow a displacement time series estimation with sub-millimetre accuracy, both in the horizontal and vertical directions. Another interesting validation experiment based on artificial corner reflectors is described in Marinkovic et al. (2007). Several A-DInSAR validation results have been published in the last few years, e.g. Colesanti et al. (2003b), Casu et al. (2006), Teatini et al. (2007), Raetzo et al. (2007) and Crosetto et al. (2008). It is important to underline that the validation results usually concern the average deformation velocity, while in general there is a lack of validation results for the deformation time series. Concerning the quality of the parameters estimated by the A-DInSAR techniques, it is important to keep in mind that the A-DInSAR performances may vary significantly, depending on the type of deformation phenomenon at hand. Under the best conditions (high PS density, long-term motion, availability of large SAR data stacks, almost linear deformation, slow deformation rates, etc.) the accuracy of the A-DInSAR estimates can be in the region of 1 mm/yr. However, these results cannot in general be extrapolated to other conditions, e.g. in the presence of low PS density, high deformation rates, etc.

Concerning PSI validation, it is worth mentioning some of the most important projects funded by the European Space Agency in the last few years. The first one is the PSIC4 (Persistent Scatterers Interferometry Codes Cross Comparison and Certification for Long term differential interferometry) project, where eight teams specialized in PSI processed a large stack of SAR images over the same test site. The site was located around the mining area of Gardanne, in Provence (France). The PSIC4 project represents a unique experiment for the number of PSI teams involved and for the quality of the available ground truth data, which involves more than 1000 levelling benchmarks measured once or twice per year. However, it is worth emphasizing that the PSIC4 mining test site includes abrupt non-linear movements with magnitudes that range from centimetres up to some decimetres, i.e. these are non-ideal characteristics from the viewpoint of C-band PSI with the temporal acquisition capabilities of ERS and Envisat. The PSIC4 results are available in the project website, http://earth.esa.int/psic4. Within the GMES Terrafirma project (www.terrafirma.eu.com), the PSIC4 dataset was further analysed in the "Provence Intercomparison" study, which involved the inter-comparison of the products of the four Terrafirma PSI teams involved in the PSIC4 project, see www.terrafirma.eu.com/Provence_intercomparison.htm. Finally, the "Terrafirma Validation Project" represents a PSI validation exercise run within GMES Terrafirma, which is focused on the analysis of the four PSI service providers of this project. The results of this project, which involves both inter-comparison of processed slant-range outputs, and the comparison of the geo-coded PSI products against ground truths, are available at http://terrafirma.eu.com/Terrafirma_validation.htm.

12.2.3 *Relevant DInSAR applications*

Since the first description of the technique, which was based on L-band SEASAT SAR data (Gabriel et al. 1989), the great potential of DInSAR for land deformation applications has been recognized. Since then, many DInSAR applications have been developed, and the capability of these techniques has been extensively documented. A great contribution to this success certainly comes from the spectacular results achieved in different fields of the geo-sciences, with tens of papers published in *Nature* and *Science*. Another important contribution comes from the activities of some small commercial companies that provide A-DInSAR deformation monitoring services, e.g. see the results of the GMES Terrafirma project (www.terrafirma.eu.com), and Ferretti et al. (2007). Finally it is important to underline the capability of A-DInSAR and PSI to monitor deformations over very large areas, e.g. Casu et al. (2007).

Some examples of the most relevant A-DInSAR application fields are described below. Seismology represents one of the most active fields, including coseismic studies (e.g. Fialko et al. 2005), postseismic deformation studies (Jónsson et al. 2003, Gourmelen & Amelung 2005), and the monitoring of a seismic and interseismic tectonic events (Colesanti et al. 2003a, Fialko 2006). Vulcanology is another relevant application field, e.g. Salvi et al. (2004), Wicks et al. (2006). Different DInSAR based studies have been conducted in the field of glaciology (Joughin et al. 2004, Rignot & Kanagaratnam 2006). Landslides are another important application field; see Colesanti et al. (2003b), Hilley et al. (2004), Dixon et al. (2006) and Raetzo et al. (2007).

Applications of ScanSAR DInSAR technologies are presented in Wright et al. (2007) for earthquakes and Anderssohn & Kaufmann (2007) tectonics.

12.2.4 *New sensors and missions*

During 2007 four new generation satellite-based SAR sensors were successfully lunched: TerraSAR-X, COSMO-SkyMed-1,COSMO-SkyMed-2,andRADARSAT-2. These new sensors are expected to consolidate the already developed DInSAR applications, and open new application fields. In particular, TerraSAR-X and RADARSAT-2 are expected to contribute to the establishment of a commercial Earth observation market based on SAR imagery. The first sensor is the German TerraSAR-X, which was lunched on 15 June 2007 and is the first satellite providing continuously global SAR data in the X-Band. Its Commissioning Phase is planned to end in December 2007. TerraSAR-X operates in three different basic modes: the Spotlight mode with 10 × 10 km scenes at a resolution of 1–2 m; the Stripmap mode with 30 km wide strips at a resolution between 3 and 6 m; and the ScanSAR mode with 100 km wide strips at a resolution of 16 m. TerraSAR-X provides single or dual polarized data, and a revisiting time of 11 days; for details see www.dlr.de/tsx/start_en.htm.

The Italian COSMO-SkyMed sensors represent another important potential source of high-resolution SAR data (www.asi.it/SiteEN/News.aspx? Area=Osservare+la+Terra). COSMO-SkyMed is an innovative Earth observation programme, which involves the launch of a constellation of four SAR satellites, operating in the X-band. The first satellite was launched on 7 June 2007, while the second on 12 December 2007. The entire constellation will be completed by 2009. As for the TerraSAR-X sensor, the spatial and temporal resolutions of COSMO-SkyMed sensors are expected to considerably extend the deformation monitoring capability in urban areas, where the X-band experiences the least temporal decorrelation. However, it is important to underline that COSMO-SkyMed is conceived as a dual use programme, to meet both civil and defence objectives. The actual proportion of acquisitions devoted to civilian uses will represent a key issue for an affective exploitation of its data for deformation monitoring DInSAR applications.

RADARSAT-2 (C-Band) was successfully launched on 14 December 2007; for details see www.radarsat2.info. It offers a spatial resolution that ranges from 3 to 100 m, and left- and right-looking operation to get an increased re-visit time The different types of polarization (HH, HV, VV and VH) are expected to support a wide range of thematic applications (especially in agriculture and forestry).

12.3 DSM GENERATION BASED ON INSAR TECHNIQUES

The generation of Digital Elevation Models (DEMs) and Digital Surface Models (DSMs) has been among one of the first applications of techniques based on the exploitation of the phase information of SAR data that has come to maturity. One of the major drawbacks of the technology, related to temporal decorrelation of the InSAR data, has been addressed either by minimizing/suppressing the temporal separation of the two or more acquisitions exploited for the DMS generation phase or by selecting different frequencies/polarizations.

The different approaches available for spaceborne and airborne platforms are discussed in Sections 12.3.1, to 12.3.2 and 12.3.3, respectively.

12.3.1 *Spaceborne InSAR DSM generation: SRTM*

After the success of the ERS-1/2 Tandem mission, the issue of temporal decorrelation with respect to DSM generation from spaceborne SAR data has been directly tackled by the Shuttle Radar Topography Mission (http://www2.jpl.nasa. gov/srtm/), e.g. see Farr (2001) and Rabus et al. (2003), whose design allowed the unique acquisition of space-based single-pass interferometric data, suitably at best for DSM generation.

The mission, flown on board of the Space Shuttle in February 2000, was designed to obtain elevation radar data on a near-global scale, to generate the most complete high-resolution digital topographic database of Earth. SRTM's C-Band and X-Band radars covered most of Earth's land surface that lies between 60 degrees north and 56 degrees south latitude, or about 80 per cent of Earth's land mass. NASA has recently released version 2 of the SRTM-C digital topographic data (also known as the finished version). Version 2 is the result of a substantial editing effort by the National Geospatial Intelligence Agency and exhibits well-defined water bodies and coastlines and the absence of spikes and wells (single pixel errors), although some areas of missing data (voids) are still present. The Version 2 directory also contains the vector coastline mask derived by NGA during the editing, called the SRTM Water Body Data. The 3 arc-seconds (about 90 m) data can be obtained at reproduction cost, or freely downloaded.

After the availability of the first SRTM data, much effort has been invested to evaluate the accuracy of these data. Starting from the definition of the accuracy assessment methodologies and tests on limited areas, as for example in Breit et al. (2002), Heipke et al. (2002), Audenino et al. (2003), Sun et al. (2003), Bourgine et al. (2005) and Carabajal et al. (2006), the validation has finally been carried out to assess the accuracy of the whole dataset, as in Salamonowicz (2005), Gorokhovich et al. (2006), Rodriguez et al. (2006) and Berry et al. (2007). The complementarity between C and X has been studied also in Marschalk et al. (2004) and Hoffmann et al. (2006)

A peculiar but important issue to be tackled for the generation of the SRTM-C final DEM has been the filling of the voids showing up in the processed SAR data, mainly in areas with severe topography or very low backscatter. This aspect has been covered in a number of publications, e.g. Dowding et al. (2004), Ham et al. (2005), Kuuskivi et al. (2005), Hofton et al. (2006), Grohman et al. (2006), Ling et al. (2007), Luedeling et al. (2007), proposing different methods, based on pure interpolation or incorporating also height information derived from other datasets. One completely filled SRTM dataset (SRTM DEM Version 3, generated by CGIAR) is for example available at http://srtm.csi.cgiar.org/.

This currently unique dataset is suitable as input layer for a large number of potential applications, and many examples have been shown in different disciplines. The big contribution of the SRTM to these application domains may be seen in the availability of a dataset with very good (considering its spatial resolution) height accuracy and consistent characteristics over a very large area, allowing studies focussing not only on limited regions (where DEMs at higher resolution / height accuracy might be obtained). Examples of the use of SRTM for geomorphology have been shown by Guth (2006) and Yastikli et al. (2006) and for volcanology by Wright et al. (2006). Hydrology studies may take great advantage from the use of SRTM data, as shown by Ling et al. (2005), Kiel et al. (2006), Ludwig et al. (2006) and Valeriano et al. (2006). Forestry research is covered in Kellndorfer et al. (2004), Simard et al. (2006) and Walker et al. (2007). The use of SRTM in archaeology for the detection of ancient settlement mounds has been shown by Menz et al. (2006), while Surazakov et al. (2006) showed the use of SRTM data for estimating volume change of glaciers.

12.3.2 *Spaceborne InSAR DSM generation: new and future missions*

Looking at improving the results obtained from the SRTM mission, and exploiting existing high- or very-high resolution SAR missions, research is currently focusing on the definition of satellite SAR missions that would complement existing missions with the (major) purpose of generating very high resolution DSMs. The idea, originating among others from

Moccia et al. (2000) and Massonnet (2001), is to build satellite constellations that, using either passive on active SAR satellites that orbit around a master sensor, provide single-pass capabilities that eliminate temporal decorrelation issues.

This possibility has been studied in Krieger et al. (2003), Moccia et al. (2005), Krieger et al. (2005), Eineder et al. (2006) and Krieger et al. (2007) and will result in the TanDEM-X (Zink et al. 2007) mission, to be launched in 2009, that will exploit the TerraSAR-X mission with a second satellite, orbiting around the first one, that would provide a continuous combination of along- and across-track baseline, suitable for currents monitoring and DSM generation, respectively. The availability of a DTED-Level 3 DSM is expected about 3 years after the launch of the Tandem-X mission. A similar concept is under development for the COSMO-SkyMed mission, as documented by Coletta et al. (2006).

The missions considered in these studies correspond to high frequency (X-Band) sensors. Some initial examples of such very-high resolution DEMs, although in repeat-pass configuration over very stable areas, have been shown by Eineder et al. (2007) and Meta et al. (2007). One example of such products is shown in Figure 12.2, where it is compared with SRTRM data

An alternative to such solutions, looking at stable scattering mechanisms at lower frequencies, may be seen in the exploitation of ALOS PALSAR L-Band data that, thanks to the lower frequency (and hence higher penetration in the vegetated areas) of the PALSAR system show higher temporal correlation (and hence lower phase noise) compared to the C-Band and X-Band mentioned systems, as for example in Shimada (2007). One example of such a product is shown in Figure 12.3, compared with SRTRM data.

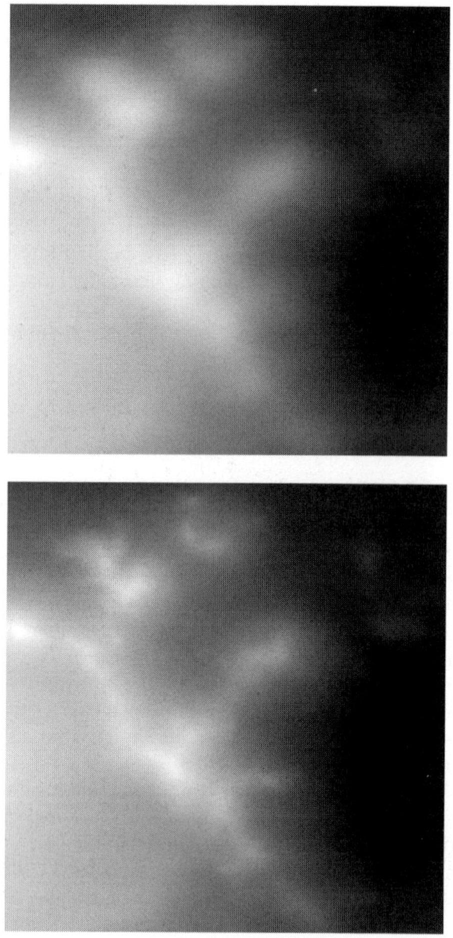

Figure 12.2. Comparison between an SRTM (top) and a TerraSAR-X DEM (bottom); Bolivia area. Data © Infoterra, processing sarmap / CREASO.

12.3.3 *Airborne InSAR DSM and DTM generation*

Airborne SAR interferometry has solved the issue of temporal decorrelation thanks to the employment of two of more SAR antennas looking at the same illumination area simultaneously. At least three commercial companies may be quoted as offering operational services and products for high- and very-high resolution DSM generation based on airborne interferometric SAR systems, as described in Wheeler et al. (2000), Rombach et al. (2003) and Li et al. (2004). Additional information can be obtained from http://www.earthdata.com/servicessubcat.php?subcat=ifsar, http://www.intermap.com/ and http://www.orbisat.com.br/, respectively. These systems can be employed in mapping of very large areas, as for example in Mercer (2007).

One of the peculiarities of these systems is the capability of providing not only DSM products but also Digital Terrain (or bald earth) Models. This feature is obtained either by exploiting dual-frequency acquisitions (e.g. in Earthdata's geoSAR and Orbisat's OrbiSAR), or through appropriate post-processing techniques, as in Wang et al (2001). One example of a DTM and a DSM product of the same area is shown in Figure 12.4.

This possibility is of particular interest in forestry and environmental applications, where for example it is very important to be able to estimate tree heights, as shown in Balzter et al. (2003), Andersen et al (2004), Izzawati et al. (2006) and Balzter et al. (2007); the tree height may in this case be obtained as direct differences between the DSM and DTM height values.

Beside the cited operational systems, several research and experimental airborne SAR systems are

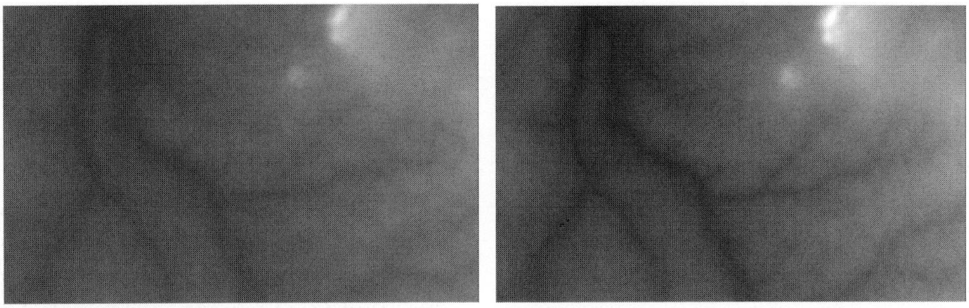

Figure 12.3. Comparison between an SRTM (left) and a PALSAR DEM (right); Malawi area. After Pasquali et al. (2007).

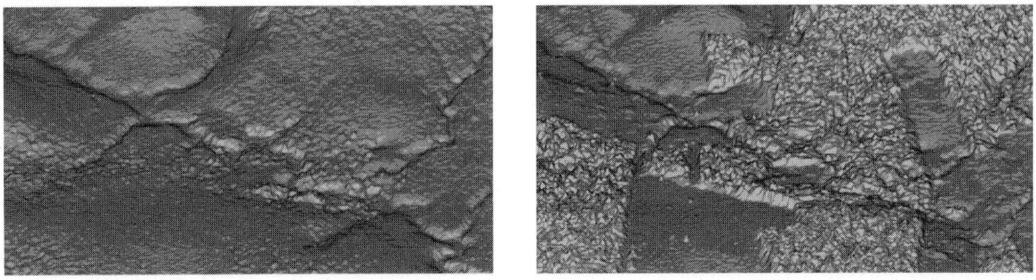

Figure 12.4. P-Band Digital Terrain Model 3-D representation (on the left) respect to a X-Band Digital Surface Model 3-D representation (on the right). Image courtesy of OrbiSat da Amazonia S.A.

currently operated. With the increase of the resolution of these systems, the motion compensation step to be performed during the focusing remains a key issue, and issues like the dependency of the motion compensation on the topography of the imaged area, addressed for example in Prats et al. (2007), and the motion compensation of repeat-pass airborne data (Angelliaume et al. 2007) are current important open research topics.

12.4 CONCLUSIONS

In this chapter the most innovative algorithmic and application developments of InSAR and DInSAR techniques achieved over the last four years have been highlighted. Considering the DInSAR techniques for deformation measurement, the main SAR data sources have been briefly reviewed. Satellite SAR sensors are by far the most important data source for DInSAR deformation studies, and in the forthcoming years this will be reinforced by the advent of different new spaceborne SAR sensors. Airborne SAR platforms and GB-SAR systems represent interesting complementary data sources, especially for the flexibility offered by these sensors. In particular, the latter have experienced an important development in the last years, and there will be a further expansion in the near future. From the algorithmic viewpoint, in the last few years there has been a remarkable improvement, resulting in increasing performances of the DInSAR techniques, especially those based on redundant SAR data stacks (A-DInSAR and PSI techniques). Several of the proposed procedures have reached a high degree of maturity, and are used in an operational way by specialized added-value companies. In addition, it is worth mentioning the new application opportunities that have been opened by the advent of ASAR wide-swath interferometry. In this context of advanced algorithmic development, the quality and validation issues play a relevant role in the field of DInSAR. The chapter provides a set of references to works and projects related to these issues. In addition, the above development is clearly testified by the wide range of DInSAR applications published in high level research journals. Finally, the advent of new and advanced SAR sensors, four of them lunched since June 2007, is expected to drive a further increase in research activities and the development of new applications. It is important to underline that some of the new SAR sensors, like TerraSAR-X and COSMO-SkyMed-1 and 2, offer resolutions of 1 m, which are comparable to those of the high resolution optical satellites.

Considering the InSAR techniques for DSM generation, great achievements have been obtained from the processing of the data from the SRTM mission that provided a unique dataset for its accuracy and consistency. The current focus is now in the development of spaceborne missions that would provide results and characteristics competing with those offered by airborne systems. Developments of such solutions would probably enlarge in any case the potentialities of exploiting SAR data for routine operations in environment monitoring of different geophysical parameters.

REFERENCES

Andersen, H.-E., McGaughey, R., Reutebuch, S., Schreuder, G., Agee, J. & Mercer, B. 2004. Estimating Canopy Fuel Parameters in a Pacific Northwest Conifer Forest Using Multifrequency Polarimetric IFSAR. *Proceedings of XXth ISPRS Congress, July 12–23, 2004, Istanbul.*

Anderssohn, J. & Kaufmann, H. 2007. Monitoring surface deformation in active margin settings with the ScanSAR technique using Wide Swath ENVISAT data. *Proceedings of Envisat Symposium 2007, 23–27 April 2007, Montreux (Switzerland).*

Angelliaume, S., Oriot, H. & Xavier Dupuis, X. 2007. The X-band RAMSES multi-pass interferometry campaign. *Proceedings of Fringe 2007, 26–30 November 2007, Frascati (Italy).*

Antonello, G., Tarchi, D., Casagli, N., Farina, P., Guerri, L. & Leva, D. 2004. SAR interferometry from satellite and ground-based system for monitoring deformations on the Stromboli volcano. *Proceedings of IGARSS 2004, Vol. 1, September 2004*: 633–636.

Audenino, P., Rognant, L. & Chassery, J. 2003. Qualification of SRTM DEM. A first approach toward an application dependant qualification framework. Geoscience and Remote Sensing Symposium, 2003. *IGARSS '03 Proceedings* 5: 3082–3084. IEEE International.

Balzter, H., Rowland, C., Milne, R., Stebler, O., Patenaude, G., Dawson, T. & Saich, P. 2003. Potential of polarimetric SAR interferometry for forest carbon accounting. Geoscience and Remote Sensing Symposium, 21–25 July 2003. *IGARSS '03 Proceedings* 3: 1945–1947. IEEE International.

Balzter, H., Rowland, C.S. & Saich, P. 2007. Forest canopy height and carbon estimation at Monks Wood National Nature Reserve, UK, using dual-wavelength SAR interferometry. *Remote Sensing of Environment* 108, 3: 224–239.

Berardino, P., Fornaro, G., Lanari, R. & Sansosti, E. 2002. A new algorithm for surface deformation monitoring based on small baseline differential SAR interferograms. *IEEE Transactions on Geoscience and Remote Sensing* 40, 11: 2375–2383.

Berry, P.A.M., Garlick, J.D. & Smith, R.G. 2007. Near-global validation of the SRTM DEM using satellite radar altimetry. *Remote Sensing of Environment* 106, 1: 17–27.

Bourgine, B. & Baghdadi, N. 2005. Assessment of C-band SRTM DEM in a dense equatorial forest zone. *C. R. Geosciences* 337, 14: 1225–1234.

Breit, H., Knopfle, W., Adam, N., Eineder, M., Suchandt, S. & Rabus, B. 2002. SRTM X-SAR DEM of Europe-Results and algorithmic improvements. Geoscience and Remote Sensing Symposium, 2002. *IGARSS '02. 2002, 24–28 June 2002* vol.1: 155–157. IEEE International.

Carabajal, C.C. & Harding, D.J. 2006. SRTM C-band and ICESat Laser Altimetry Elevation Comparisons as a Function of Tree Cover and Relief. *Photogrammetric Engineering & Remote Sensing* 72, 3: 287–298.

Casu, F., Manzo, M. & Lanari, R. 2006. A quantitative assessment of the SBAS algorithm performance for surface deformation retrieval from DInSAR data. *Remote Sensing of Environment* 102: 195–210.

Casu, F., Manzo, M., Pepe, A., Mazzarella,G. & Lanari, R. 2007. On the capability of the SBAS-DInSAR technique to investigate deformation phenomena of large areas with low resolution data. *Proceedings of Envisat Symposium 2007, 23–27 April 2007, Montreux (Switzerland).*

Colesanti, C., Ferretti, A., Novali, F., Prati, C. & Rocca, F. 2003a. SAR monitoring of progressive and seasonal ground deformation using the Permanent Scatterers Technique. *IEEE Transactions on Geoscience and Remote Sensing* 41, 7: 1685–1701.

Colesanti, C., Ferretti, A., Prati, C. & Rocca, F. 2003b. Monitoring landslides and tectonic motions with the Permanent Scatterers Technique. *Eng. Geol.* 68: 3–14.

Coletta, A., Galeazzi, C., Caltagirone, F., Scorzafava, E., Amorosi, L., Caliò, E., Notarantonio, A. & Moccia, A. 2006. Interferometric Missions: Applications within ASI Space Programs. *Proceedings of SpaceOps 2006, June 2006, Rome (Italy).*

Crosetto, M., Crippa, B. & Biescas, E. 2005a. Early detection and in-depth analysis of deformation phenomena by radar interferometry. *Eng Geol* 79(1–2): 81–91.

Crosetto, M., Crippa, B., Biescas, E., Monserrat, O., Agudo, M. & Fernández, P. 2005b. Land deformation monitoring using SAR interferometry: state-of-the-art *Photogrammetrie Fernerkundung und Geoinformation* 6: 497–510.

Crosetto, M., Monserrat, O., Muniganti, P., Sawicka, M., Rossi, G., Calcagni, L. & Crippa, B. 2007. A 2+1D phase unwrapping for advanced DInSar data processing. *Proceedings of Fringe, 26–30 November 2007, Frascati (Italy).*

Crosetto, M., Bisecas, E., Duro J., Closa, J. & Arnaud, A. 2008. Quality assessment of advanced interferometric products based on time series of ERS and Envisat SAR data. *Photogrammetric Engineering and Remote Sensing* (in press).

Dixon, T., Amelung, F., Ferretti, A., Novali, F., Rocca, F., Dokka, R., Sella, G., Kim, S., Wdowinski, S. & Whitman, D. 2006. Subsidence and flooding in New Orleáns. *Nature—Brief communications* 441: 587–588.

Dowding S., Kuuskivi, T. & Li, X. 2004. Void Fill of SRTM Elevation Data - Principles, Processes and Performance. Proceedings of ASPRS Conference on Images to Decision: Remote Sensing Foundation for GIS Applications, September 12–16, 2004, Kansas City, Missouri.

Crosetto, M, Gili, J.A., Biescas, E., Agudo, M., Monserrat, O. & Crippa, B. 2005. High Resolution Differential Interferometry using time series of ERS and ENVISAT SAR data. *Proceedings of the 6th. Geomatic Week, February 2005, Barcelona, Spain.* (CD-ROM).

Eineder, M., Krieger, G. & Roth, A. 2006. First Data Acquisition and Processing Concepts for the TanDEM-X Mission. *Proceedings of From Sensors to Imagery, ISPRS Commission I Symposium, Paris, July 3–6, 2006.*

Eineder, M., Adam, N. & Yague-Martinez, N. 2007. First TerraSAR-X Interferometry Evaluation. *Proceedings of Fringe 2007 Workshop, Frascati (Italy), 26–30 November 2007.*

Farr, T. 2001. The shuttle radar topography mission. *Proceedings of IGARSS 2001.*

Ferretti, A., Prati, C. & Rocca, F. 2000. Nonlinear subsidence rate estimation using permanent scatterers in differential SAR interferometry. *IEEE Transactions on Geoscience and Remote Sensing* 38(5): 2202–2212.

Ferretti, A., Prati, C. & Rocca, F. 2001. Permanent scatterers in SAR interferometry. *IEEE Transactions on Geoscience and Remote Sensing* 39(1): 8–20.

Ferretti, A., Prati, C., Rocca, F. & Young, B. 2007a. 1,500 RADARS A scenes for PSInSAR™: the Italian case study. *Proceedings of Advanced SAR Workshop 2007, 11–13 September, Vancouver (BC).*

Ferretti, A., Savio, G., Barzaghi, R., Borghi, A., Musazzi, S., Novali, F., Prati, C. & Rocca, F. 2007b. Submillimeter Accuracy of InSAR Time Series: Experimental Validation. *IEEE Transactions on Geoscience and Remote Sensing* 45, 5: 1142–1153.

Fialko, Y., Sandwell, D., Simona, M. & Rosen, P. 2005. Three-dimensional deformation caused by the Bam, Iran, earthquake and the origin of shallow slip deficits. *Letters to Nature* 435: 295–299.

Fialko, Y. 2006. Interseismic strain accumulation and the earthquake potential on the southern San Andreas fault system. *Letters to Nature* 441: 968–971.

Fornaro, G. & Serafino, F. 2006. Imaging Single and Double Scatterers in Urban Areas via SAR Tomography. *IEEE Transaction on Geoscience and Remote Sensing* 44(12): 3497–3505.

Gabriel, A.K., Goldstein, R.M. & Zebker, H.A. 1989. Mapping small elevation changes over large areas: differential radar interferometry. *J Geophys Res* 94 (B7): 9183–9191.

Gorokhovich, Y. & Voustianiouk, A. 2006. Accuracy assessment of the processed SRTM-based elevation data by CGIAR using field data from USA and Thailand and its relation to the terrain characteristics. *Remote Sensing of Environment* 104, 4: 409–415.

Gourmelen, N. & Amelung, F. 2005. Postseismic Mantle Relaxation in the Central Nevada Seismic Belt. *Science* 310: 1473–1476.

Grohman, G., Kroenung., G. & Strebeck, J. 2006. Filling SRTM voids: The delta surface fill model. *Photogrammetric Engineering and Remote Sensing* 72, 3: 213–216.

Guth, P. 2006. Geomorphometry from SRTM: Comparison to NED. *Photogrammetric Engineering & Remote Sensing* 72, 3: 269–278.

Ham, A., Rupe, C., Kuuskivi, T. & Dowding, S. 2005. A Standardized Approach to Phase Unwrap Detection/ Removal and Void Fill of the Shuttle Radar Topography Mission (SRTM) Data. *SRTM Workshop, 14–16 June 2005, Reston, Virginia.*

Heipke, C., Koch, A. & Lohmann, P. 2002. Analysis of SRTM DTM—Methodology and practical results. *ISPRS Commission IV Symposium in Ottawa, July 2002.*

Hilley, G., Bürgmann, R., Ferretti, F., Novali, F. & Rocca, F. 2004. Dynamics of Slow-Moving Landslides from Permanent Scatterer Analysis. *Science* 304: 1952–1955.

Hoffmann, J. & Walter, D. 2006. How Complementary are SRTM-X and - C Band Digital Elevation Models? *Photogrammetric Engineering & Remote Sensing* 72, 3: 261–268.

Hofton, M., Dubayah, R., Bryan Blair, J. & Rabine, D. 2006. 279 Validation of SRTM Elevations Over Vegetated and Non-vegetated Terrain Using Medium-Footprint Lidar *Photogrammetric Engineering & Remote Sensing* 72, 3: 279–286.

Holzner, J. & Bamler, R. 2002. Burst-mode and ScanSAR *interferometry. IEEE Transaction on Geoscience and Remote Sensing* 40 (9): 1917–1934.

Hooper, A. & Zebker, H.A. 2007. Phase unwrapping in three dimensions with application to InSAR time series. *J. Opt. Soc.* 24: 2737–2747.

http://www2.jpl.nasa.gov/srtm/, last visited on March, 12th 2008.

http://srtm.csi.cgiar.org/, last visited on March, 12th 2008.

http://eros.usgs.gov/conferences/SRTM/index.html, *SRTM Workshop, June 14 – 16, 2005, Reston, Virginia,* last visited on March, 12th 2008.

http://www.earthdata.com/servicessubcat.php?subcat=ifsar, last visited on March, 12th 2008.

http://www.intermap.com/, last visited on March, 12th 2008.

http://www.orbisat.com.br/, last visited on March, 12th 2008.

Izzawati, Wallington, E.D. & Woodhouse, I.H. 2006. Forest height retrieval from commercial X-band SAR products. *IEEE Transactions on Geoscience and Remote Sensing* 44, 4: 863–870.

Jónsson, S., Segall, P., Pedersen, R. & Björnsson, G. 2003. Post-earthquake ground movements correlated to pore-pressure transients. *Letters to Nature* 424: 179–183.

Joughin, I., Abdalati, W. & Fahnestock, M. 2004. Large fluctuations in speed on Greenland's Jakobshavn Isbræ glaciar. *Letters to Nature* 432: 608–610.

Kampes, B.M. & Hanssen, R.F. 2004. Ambiguity Resolution for Permanent Scatterer Interferometry. *IEEE Transaction on Geoscience and Remote Sensing* 42, 11: 2446–2453.

Ketelaar, G., van Leijen, F., Marinkovic, P. & Hanssen, R.F. 2007. Quality: Multi-TrackPS-InSAR: Datum Connection and Reliability Assessment. *Proceedings of Envisat Symposium 2007, 23–27 April 2007, Montreux (Switzerland).*

Kellndorfer, J., Walker, W., Pierce, L., Dobson, C., Fites, J.A, Hunsaker, C., Vona, J. & Clutter, M. 2004. Vegetation height estimation from Shuttle Radar Topography Mission and National Elevation Datasets. *Remote Sensing of Environment* 93, 3: 339–358.

Kiel, B., Alsdorf, D. & LeFavour, G. 2006. Capability of SRTM C and X Band DEM Data to Measure Water Elevations in Ohio and the Amazon. *Photogrammetric Engineering & Remote Sensing* 72, 3: 313–320.

Krieger, G., Fiedler, H., Mittermayer, J., Papathanassiou, K. & Moreira, A. 2003. Analysis of multistatic configurations for spaceborne SAR interferometry. *IEE Proceedings—Radar, Sonar and Navigation* 150, 3: 87–96.

Krieger, G., Moreira, A., Hajnsek, I., Werner, M., Fiedler, H. & Settelmeyer, E. 2005. The TanDEM-X Mission Proposal. Proceedings of the ISPRS Hannover Workshop 2005: High-Resolution Earth Imaging for Geospatial Information, Hannover, Germany, May 17 20, 2005.

Krieger, G., Moreira, A., Fiedler, H., Hajnsek, I., Werner, M., Younis, M. & Zink, M. 2007. TanDEM-X: A Satellite Formation for High-Resolution SAR Interferometry. *IEEE Transactions on Geoscience and Remote Sensing* 45(11, Part 1): 3317–3341.

Kuuskivi T., Lock, J., Li X., Dowding, S. & Mercer, B. 2005. Void Fill of SRTM Elevation Data: Performance Evaluations. *Proceedings of ASPRS 2005 Annual Conference, 7–11 March 2005, Baltimore, Maryland.*

Lanari, R., Mora, O., Manunta, M., Mallorqui, J.J., Berardino, P. Sansosti, E. 2004. A small-baseline approach for investigating deformations on full-resolution differential SAR interferograms. *IEEE Transactions on Geoscience and Remote Sensing* 42(7): 1377–1386.

Leva, D., Nico, G., Tarchi, D., Fortuny, J. & Sieber, A.J. 2003. Temporal analysis of a landslide by means of a ground-based SAR interferometer. *IEEE Transactions on Geoscience and Remote Sensing* 41: 745–752.

Li, X., Tennant, K. & Lawrence, G. 2004. Three-Dimensional Mapping with Airborne Ifsar Based Star Technology—Intermap's Experiences. *Proceedings of XXth ISPRS Congress, 12–23 July 2004, Istanbul.*

Ling, F., Zhang, Q.-W. & Wang, C. 2005. Comparison of SRTM Data with other DEM sources in Hydrological Researches. Proceedings of the 31st ISPRS International Symposium on Remote Sensing of Environment, 20–24 May 2005, Saint Petersburg.

Ling, F., Zhang, Q.W. & Wang, C. 2007. Filling voids of SRTM with Landsat sensor imagery in rugged terrain. International Journal of Remote Sensing 28, 2: 465–471.

Lombardini, F. 2005. Differential tomography: a new framework for SAR interferometry. *IEEE Transactions on Geoscience and Remote Sensing* 43: 37–44.

Lopez-Dekker, P., Merlano, J.C., Duque, S., Sanz-Marcos, J., Aguasca, A. & Mallorqui, J.J. 2007. Bistatic SAR Interferometry Using Envisat and a Ground Based Receiver: First Results. *Proceedings of Envisat Symposium 2007, 23–27 April 2007, Montreux (Switzerland).*

Lu, Z., Wright, T. & Wicks, C. 2003. Deformation of the 2002 Denali Fault Earthquakes, Mapped by Radarsat-1 Interferometry. *EOS* 84(41): 430–431.

Ludwig, R. & Schneider, P. 2006. Validation of digital elevation models from SRTM X-SAR for applications in hydrologic modelling. *ISPRS Journal of Photogrammetry and Remote Sensing* 60, 5: 339–358.

Luedeling, E., Siebert, S. & Buerkert, A. 2007. Filling the voids in the SRTM elevation model—A TIN-based delta surface approach. *ISPRS Journal of Photogrammetry and Remote Sensing* 62, 4: 283–294.

Luzi, G., Pieraccini, M., Mecatti, D., Noferini, L., Guidi, G., Moia, F. & Atzeni, C. 2004. Ground-based radar interferometry for landslides monitoring: atmospheric and instrumental decorrelation sources on experimental data. *IEEE Transactions on Geoscience and Remote Sensing* 42: 2454–2466.

Marinkovic, P., Ketelaar, G., van Leijen, F. & Hanssen, R. 2007. InSAR quality control: analysis of five years of corner reflector time series. *Fringe 2007 Workshop, Frascati (Italy), 26–30 November 2007.*

Marschalk, U., Roth, A., Eineder, M. & Suchandt, S. 2004. Comparison of DEMs derived from SRTM/X- and C-band. Geoscience and Remote Sensing Symposium, 2004. *Proceedings of IGARSS '04* vol.7: 4531–4534. IEEE International.

Massonnet, D. 2001. The interferometric cartwheel: a constellation of passive satellites to produce radar images to be coherently combined. *International Journal of Remote Sensing* 22, 12: 2413–2430.

Menze, B.H., Ur, J.A. & Sherratt, A.G. 2006. Detection of Ancient Settlement Mounds—Archaeological Survey Based on the SRTM Terrain Model. *Photogrammetric Engineering & Remote Sensing* 72, 3: 321–330.

Mercer, B. 2007. The NextMap Program—National Scale DEMs Created from Airborne InSAR. Proceedings of the ISPRS Hannover Workshop 2007: High-Resolution Earth Imaging for Geospatial Information, Hannover, Germany, 29 May–1 June 2007.

Meta, A., Prats, P., Steinbrecher, U., Scheiber, R. & Mittermayer, J. 2007. First TOPSAR image and interferometry results with TerraSAR-X. *Proceedings of Fringe 2007 Workshop, Frascati (Italy), 26–30 November 2007.*

Moccia, A., Chiacchio, N. & Capone, A. 2000. Spaceborne bistatic synthetic aperture radar for remote sensing applications. *International Journal of Remote Sen*sing 21, 18: 3395–3414.

Moccia, A. & Fasano, G. 2005. Analysis of Spaceborne Tandem Configurations for Complementing COSMO with SAR Interferometry, EURASIP. *Journal on Applied Signal Processing* 20: 3304–3315.

Monti Guarnieri, A & Prati, C. 1996. ScanSAR focusing and interferometry. *IEEE Transactions on Geoscience and Remote Sensing* 34: 1029–1038.

Monti Guarnieri, A., Guccione, P., Pasquali, P. & Desnos, Y.L. 2003. Multi-mode ENVISAT ASAR interferometry: techniques and preliminary results. *IEE Proceedings on Radar, Sonar and Navigation* 150: 193–200.

Mora, O., Mallorquí, J.J. & Broquetas, A. 2003. Linear and nonlinear terrain deformation maps from a reduced set of interferometric SAR images. *IEEE Transactions on Geosciences and Remote Sensing*: 2243–2253.

Pasquali, P., Monti Guarnieri, A., D'Aria, D., Costa, L., Small, D., Jehle, M. & Rosich, B. 2007. ALOS PALSAR Verification Processor. *Proceedings of Envisat Symposium 2007, 23–27 April 2007, Montreux (Switzerland).*

Pepe, A., Sansosti, E., Berardino, P. & Lanari, R. 2005. On the Generation of ERS/ENVISAT DInSAR Time-Series via the SBAS technique. *IEEE Transactions on Geoscience and Remote Sensing Letters* 2: 265–269.

Pepe, A., Manunta, M., Mazzarella, G. & Lanari, R. 2007. A Space-Time Minimum Cost Flow Phase Unwrapping Algorithm for the Generation of Persistent Scatterers Deformation Time-Series. *Proceedings of IGARSS 2007, 23–27 July 2007, Barcelona, (Spain).*

Perissin, D. & Ferretti, A. 2007. Urban-Target Recognition by Means of Repeated Spaceborne SAR Images. *IEEE Transcations on Geoscience and Remote Sensing* 45, 12: 4043–4058.

Perna, S., Wimmer, C., Moreira, J. & Fornaro, G. 2008. X-Band Airborne Differential Interferometry: Results of the OrbiSAR Campaign over the Perugia Area. *IEEE Transcations on Geoscience and Remote Sensing* (in press).

Pipia, L., Aguasca, A., Fabregas, X., Mallorqui, J.J., Lopez-Martinez, C. & Marturia, J. 2007. Mining Induced Subsidence Monitoring in Urban Areas with a Ground-Based SAR. *Urban Remote Sensing Joint Event, 11–13 April 2007*: 1–5.

Prats, P., Camara de Macedo, K.A., Reigber, A., Scheiber, R. & Mallorqui, J.J. 2007. Comparison ofTopography- and Aperture-Dependent Motion Compensation Algorithms for Airborne SAR. *IEEE Geoscience and Remote Sensing Letters* 4(3): 349–353.

Raetzo, U., Wegmüller, U., Strozzi, T., Marks, F. & Farina, P. 2007. Monitoring of Lumnez landslide with ERS and Envisat data. *Proceedings of Envisat Symposium 2007, 23–27 April 2007, Montreux (Switzerland).*

Rabus, B., Eineder, M., Roth, A. & Bamler, R. 2003. The shuttle radar topography mission-a new class of digital elevation models acquired by spaceborne radar. *ISPRS Journal of Photogrammetry and Remote Sensing* 57, 4: 241–262

Reigber, A. & Schreiber, R. 2003. Airborne Differential SAR Interferometry: First Results at L-Band. *IEEE Transcations on Geoscience and Remote Sensing* 41(6): 1516–1520.

Rignot, E. & Kanagaratnam, P. 2007. Changes in the Velocity Structure of the Greenland Ice Sheet. *Science* 311: 986–990.

Rodríguez, E., Morris, C.S. & Belz, J.E. 2006. A Global Assessment of the SRTM Performance. *Photogrammetric Engineering & Remote Sensing* 72, 3: 249–260.

Rombach, M. & Moreira, J. 2003. Description and applications of the multipolarized dual band OrbiSAR-1 InSAR sensor. *Proceedings of the 2003 International Radar Conference* 5: 245–250.

Rosen, P.A., Hensley, S., Joughin, I.R., Li, F.K., Madsen, S.N., Rodriguez, E. & Goldstein, R.M. 2000. Synthetic Aperture Radar Interferometry. *Proceedings of the IEEE* 88(3): 333–382.

Rosich, B. 2007. Optimising ENVISAT ASAR interferometry opportunities by orbit maintenance. *Proceedings of Envisat Symposium 2007, 23–27 April 2007, Montreux (Switzerland).*

Rosich, B., Monti Guarnieri, A., D'Aria, D., Duesmann, B., Navas, I. &. Vazzana, S. 2007. ASAR Wide Swath Mode Interferometry: Optimisation of the scan pattern synchronisation. *Proceedings of Envisat Symposium 2007, 23–27 April 2007, Montreux (Switzerland).*

Salamonowicz, P. 2005. Comprehensive Assessment of the Shuttle Radar Topography Mission Elevation Data Accuracy. *SRTM Workshop, June 14–16 2005, Reston, Virginia.*

Salvi, S., Atzori, S., Tolomei, C., et al. 2004. Inflation rate of the Colli Albani volcanic complex retrieved by the permanent scatterers SAR interferometry technique. *Geophys. Res. Letters* 31(12): L12606. doi:10.1029/2004GL020253

Serafino, F., Fornaro,G., Pauciull, A., et al. 2007. MultiD SAR Focusing for Accurate Target location and Tracking of Slow Movements: Results and Validation. *Proceedings of Envisat Symposium 2007, 23–27 April 2007, Montreux (Switzerland).*

Shimada, M. 2007. Coherence dependency of the PALSAR POLinSAR on forest in Japan and Amazon. *Proceedings of Geoscience and Remote Sensing Symposium, IGARSS, Barcelona, 23–28 July 2007*: 2636–2639. IEEE International.

Simard, M., Zhang, K., Rivera-Monroy, V.H., Ross, M.S., Ruiz, P.L., Castañeda-Moya, E., Twilley, R.R. & Rodriguez, E. 2006. Mapping Height and Biomass of Mangrove Forests in the Everglades National Park with SRTM Elevation Data. *Photogrammetric Engineering & Remote Sensing* 72, 3: 299–312.

Sun, G.; Ranson, K.J.; Kharuk, V.I. & Kovacs, K. 2003. Validation of surface height from shuttle radar topography mission using shuttle laser altimeter. *Remote Sensing of Environment* 88, 4: 401–411.

Surazakov, A. & Aizen, V. 2006. Estimating Volume Change of Mountain Glaciers Using SRTM and Map-Based Topographic Data. *IEEE Transactions on Geoscience and Remote* Sensing 44, 10, Part 2: 2991–2995.

Tarchi, D., Casagli, N., Fanti, R., et al. 2003. Landslide monitoring by using ground-based SAR interferometry: An example of application to the Tessina landslide in Italy. *Eng Geol* 68(1–2): 15–30.

Teatini, P., Strozzi, T., Tosi, L., Wegmüller, U., Werner, C. & Carbognin, L. 2007. Assessing short- and long-time displacements in the Venice coastland by synthetic aperture radar interferometric point target analysis. *J. Geophys. Res.* 112, F01012, doi: 10.1029/2006JF000656.

Valeriano, M.M., Kuplich, T.M., Storino, M., Amaral, B.D., Mendes, J.N. & Lima, D.J. 2006. Modeling small watersheds in Brazilian Amazonia with shuttle radar topographic mission-90 m data. *Computers & Geosciences* 32, 8: 1169–1181.

Van Freek, L.J. & Hanssen, R.F. 2007. Persistent Scatterer Interferometry Using Adaptive Deformation Models. *Proceedings of Envisat Symposium 2007, 23–27 April 2007, Montreux (Switzerland).*

Walker, W.S., Kellndorfer, J.M. & Pierce, L.E. 2007. Quality assessment of SRTM C- and X-band interferometric data: Implications for the retrieval of vegetation canopy height. *Remote Sensing of Environment*106, 4: 428–448.

Wang, Y., Mercer, B., Tao, V.C, Sharma, J. & Crawford, S. 2001. Automatic generation of bald earth digital elevation models from digital surface models created using airborne IFSAR. *Proceedings of 2001 ASPRS Annual Conference, St. Louis, Missouri, USA., 23–27April 2001.*

Wheeler, K. & Hensley, S. 2000. The GeoSAR airborne mapping system. Alexandria, VA, Radar Conference, 2000. *IEEE 2000 International, 7–12 May 2000*: 831–835.

Werner, C., Wegmüller, U., Strozzi, T. & Wiesmann, A. 2003. Interferometric point target analysis for deformation mapping. *Proceedings of IGARSS 2003, Vol. 7, July 2003*: 4362–4364.

Wicks, C., Thatcher, W., Dzurisin, D. & Svarc, J. 2006. Uplift, thermal unrest and magma intrusion at Yellowstone caldera. *Letters to Nature* 440: 72–75.

Wright, R., Garbeil, H., Baloga, S.M. & Mouginis-Mark, P.J. 2006. An assessment of shuttle radar topography mission digital elevation data for studies of volcano morphology. *Remote Sensing of Environment* 105, 1: 41–53.

Wright, T., Pasquali, P., D'Aria, D., Monti Guarnieri, A. & Hamling, I. 2007. Monitoring deformation in Afar, Ethiopia, using Envisat Wide-Swath Interferometry. *Proceedings of Envisat Symposium 2007, 23–27 April 2007, Montreux (Switzerland).*

Yastikli, N., Kocak, G. & Buyuksalih, G. 2006. Accuracy And Morphological Analyses of GTOPO30 and SRTM X -C Band DEMs in the Test Area Istanbul. *Proceedings of the ISPRS Ankara Workshop on Topographic Mapping from Space, 14–16 February 2006, Ankara.*

Zink, M., Fiedler, H., Krieger, G. & Moreira, A. 2007. The TanDEM-X Mission. *Proceedings of Fringe 2007, 26–30 November 2007, Frascati (Italy).*

*Advances in Photogrammetry, Remote Sensing and Spatial Information Sciences:
2008 ISPRS Congress Book – Li, Chen & Baltsavias (eds)*

CHAPTER 13

Early stages of LiDAR data processing

Norbert Pfeifer & Jan Böhm

ABSTRACT: The first processing step of Light Detection and Ranging (LiDAR) data is the orientation of the acquired point cloud. This applies to the terrestrial as well as to the airborne case. In the terrestrial application, registration, that is alignment to a common reference frame, must be distinguished from geo-referencing. A reference to a geodetic coordinate system is not always needed in terrestrial applications, as sometimes a local coordinate system is sufficient. Special attention is paid to registration approaches that require neither markers nor additional orientation sensors. In the airborne case, a simultaneous orientation and determination of calibration parameters on the job is becoming standard. In the airborne case, the classification into ground points (or pixels for elevation images) and off-terrain is also standard procedure and necessary for almost all applications.

Keywords: LiDAR, ALS, TLS, orientation, registration, geo-referencing, point cloud, filtering, DTM

13.1 INTRODUCTION

Laser scanning provides—in the first instance—point clouds. From each scanning position a point (cloud) is acquired in the local coordinate system of the sensor. The first processing step is therefore the orientation of the individual scans and transformation into a superior coordinate system.

In the terrestrial case, typically more than one standpoint is required to fully capture the geometry of a scene. This corresponds to multi-station configurations also encountered in terrestrial photogrammetry and classical surveying. We do not treat the case of kinematic scanning here, i.e. scanning from a mobile platform, which is dealt with in Chapter II-4. Therefore, we are not required to reconstruct a trajectory of the scanner movement but rather a single position and orientation with respect to a common or global reference frame. Thus, orientation of terrestrial laser scan (TLS) data is similar to photogrammetric relative or absolute orientation.

We note that calibration of terrestrial laser scanners became a research topic during the last four years, and initial results have been achieved. As the number of contributions is still comparatively small, giving an overview of the relevant literature seems currently premature. We refer to the work of Lichti (2006) in which an overview of existing approaches is given. On-the-job calibration may very well be integrated tightly with orientation in the up-coming four years.

Before obtaining the first point cloud in airborne laser scanning, the raw measurements of angles and ranges are fused with the flight path observations with GPS and IMU. This enables the computing of the raw point cloud, i.e. irregularly distributed 3-D points without a classification. By applying methods of strip adjustment, the instrument calibration—in photogrammetric terms its interior orientation—can be improved. Also some deficiencies in the determination of the flight path can be overcome with the methods provided by strip adjustment. Both can lead to point clouds with higher precision. Finally, strip adjustment provides reliable assessments of accuracy, especially if over-determination is taken into account, which is absent in direct geo-referencing (Skaloud 2006).

"Filtering", the next processing step in ALS, is a first action of classification of the point cloud. Often it is delivered directly to the customer by the service provider. It is a standard procedure and will be described here, too. For filtering, the points are grouped into ground points and off-terrain points, originating from reflections on, e.g. the vegetation canopy, house roofs, urban furniture, etc. It enables the computing of a digital terrain model (DTM). The first processing steps described here enable therefore the computing of a high quality digital terrain model (DTM). It should be noted that the methods of filtering can be applied to airborne data (standard case) but also to terrestrial laser scanning point clouds.

The state of the art of and the latest contributions since the ISPRS congress in Istanbul 2004 to orientation of multiple scans in TLS will be discussed in Section 13.2. In Section 13.3, strip adjustment will be presented. First the problem will be defined more clearly, followed by an indication of the state of the art and latest contributions, and concluded by a discussion. The same structure is applied for filtering in the subsequent section. Finally, a conclusion will be drawn.

13.2 REGISTRATION OF TERRESTRIAL LIDAR

13.2.1 *Problem statement*

Commonly several separate stations are required to entirely capture the geometry of a scene with a terrestrial laser scanner. This might be necessary because of a limited field of view of the sensor used, the extent of the object, self occlusion of the object or occlusions caused by other objects. Before the data can be passed down the processing pipeline to succeeding steps, such as meshing and modelling (see Chapter IV-6), the alignment of the stations into a common reference frame has to be performed, usually referred to as registration. Mathematically we seek the rigid body transformation defining the six parameters of translation and rotation, which transforms from the sensor coordinate system to the common or global coordinate system. If the common coordinate system is a geodetic coordinate system, this also solves the problem of geo-referencing. The problem of registration is thus very similar to the absolute orientation of a stereo model encountered in photogrammetry. However, for range sensors we do not need to determine seven transformation parameters, since the scale is fixed.

As frequently as registration occurs across different fields of application, as varied are the approaches for its solution. The proposed approaches can usually be traced back to either classical surveying, photogrammetry or computer vision. The approaches differ in their prerequisites on the scene, prior knowledge, extra sensor measurements and the level of automation and robustness. The approaches can be categorized into (1) marker-based approaches, which require the placement of markers in the scene, either as control or tie points; (2) sensor-based approaches, which require additional sensors to be attached to the scanner in order to directly determine the position and orientation of the sensor; and (3) data-driven approaches, which use the geometry and other properties of the acquired data to determine the transformations in between scans.

13.2.2 *Marker-based registration*

In classical surveying the idea of known control points is predominant. Control points are typically represented by artificial targets and are linked to some geodetic coordinate system. For historical reasons, 3-D control points in a geodetic global system usually do not exist. The control information is separated into a horizontal and a vertical control network.

As more and more terrestrial laser scanners are equipped with compensators and optical plummets, they can be precisely levelled and set up over known points. Reflector-like targets can be set up over additional control or tie points and their coordinates measured with the scanner. The registration is then no different from the adjustment of a network of total stations.

If the scanner cannot be levelled precisely, the separate vertical and horizontal control information has to be transferred onto three-dimensional control points, typically using a total station. When enough control points are placed in the area covered by the scans, the sought after transformation parameters can be computed in a network adjustment procedure. See Figure 13.1 for a typical example of a marker used in terrestrial laser scanning.

If the geo-referencing of the data is not required, the control information is dispensable. The markers placed in the overlapping areas of the scans can simply be used as tie points. In this case, the registration is computed as a free network adjustment. Marker-based registration of laser scan data can also be combined with photogrammetry, as shown by Al-manasir & Fraser (2006).

The marker-based registration is expected to achieve the highest accuracy. Consistently, accuracies in the range of one millimetre to a few millimetres can be achieved. A prerequisite to such high accuracies is the proper design of the network, which should follow classical geodetic rules for network design.

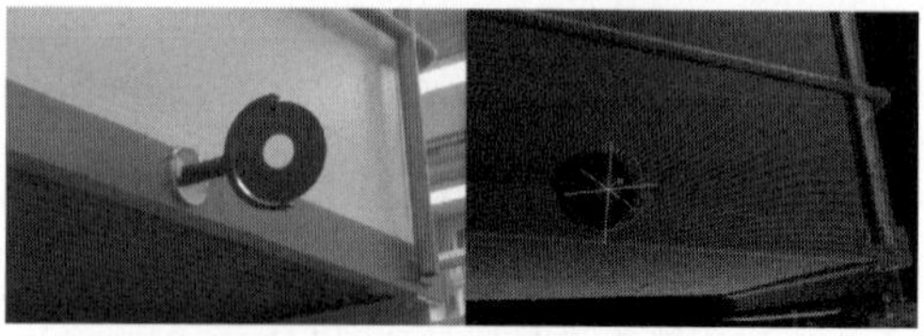

Figure 13.1. An example of a marker used for terrestrial laser scanning. A photograph of the marker is shown on the left, the corresponding point cloud to the right.

Practical considerations and actual accuracy values are given by Bornaz et al. (2003).

The development of appropriate software tools has been done mainly in the commercial domain and today these tools are included in the software packages of most scanner manufactures, especially when the manufacturer offers a solution integrated with a line of classical surveying equipment.

However accurate, it must be noted that marker-based approaches require extra effort for the placement and measurement of the targets. The placement of such targets may be prohibited in certain sites. The availability of a reference measurement system, such as a total station, might be a further criterion that prohibits the use of the classical solution. For these reasons, marker-less approaches are of high interest both from a practical point of view and from an algorithmic point of view. Marker-less approaches for the registration of terrestrial LiDAR have been a major topic of research within the photogrammetric community during recent past years.

13.2.3 *Sensor-based registration*

Well-known in aerial LiDAR, sensor-based orientation involves the integration of additional sensor measurements with the LiDAR measurements. Typically, a GPS sensor and an IMU are integrated for estimation of position and orientation parameters. This technology has also successfully been transferred to mobile mapping systems (Chapter II-4). The technology can also be used in static terrestrial laser scanning. However, since the sensor is not moving, no trajectory is to be determined, but rather a single position and orientation for each station. The quality of the registration depends heavily on the quality of the orientation sensors used.

Direct orientation of terrestrial laser scans has been proposed by Schuhmacher & Boehm (2005). They include GPS, compass and pan-tilt sensors to directly measure position and orientation of the laser scanner and thus determine the absolute pose of the data. Due to the low cost sensors used, the system is mainly suited to providing initial approximations for later refinement.

13.2.4 *Data-driven registration*

Data-driven registration attempts to find the transformation parameters in between the stations from the sensed LiDAR data itself. Several methods are known for point cloud registration, most developed in the computer vision and computer graphics communities. A recent review and accuracy comparison of selected methods was performed by Salvi et al. (2007). A previous comparison was given by Dalley & Flynn (2001). In the following, we will concentrate on the *de facto* standard method and recent innovations presented within the photogrammetric community.

13.2.4.1 *Iterative closest point*

The most important group of algorithms is the Iterative Closest Point (ICP) algorithms. The ICP was originally proposed by Besl & McKay (1992) for the registration of digitized data to an idealized geometric model, a typical task in quality inspection. Other than the name might suggest, ICP can be used for the registration of geometric data in different representations. The authors explicitly list point sets, line segment sets, implicit curves, parametric curves, triangle sets, implicit surfaces and parametric surfaces as possible representations. The algorithm starts from an approximate initial transformation of the two datasets. Then, for each point of the first geometric dataset, the closest point of the second dataset is determined. From these correspondences, a refined transformation is computed. This simple scheme of finding correspondences and computing a transformation is iterated until convergence is reached. See Figure 13.2 for an example of the application of the ICP to the registration of terrestrial LiDAR data.

In its most basic form, the ICP is restricted to pair-wise registration of fully overlapping datasets. Several extensions and improvements to the original algorithm have been proposed since its original publication. An overview is given by Eggert et al. (1998) (we also recommend the extended version of the article, which is available online) and Rusinkiewicz & Levoy (2001). It shall be noted that the ICP has progressed

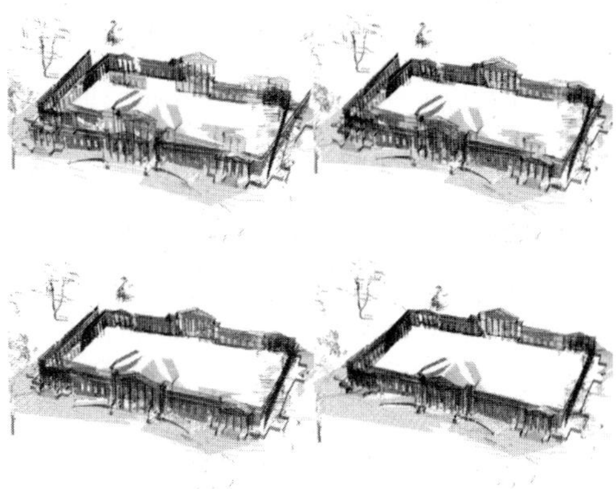

Figure 13.2. Example for the application of the ICP algorithm. Starting in the top left corner, the approximate alignment of five scan stations is shown. In the next images, the results of the ICP after one, two and five iterations are shown.

to be the dominant registration algorithm for multi-station point cloud alignment.

While a review of all extensions to ICP is beyond the scope of this chapter, two extensions suggested within the photogrammetric community shall be mentioned. Gühring (2001) extends the ICP to a multi-view registration algorithm and uses an iterative estimation scheme to compute the transformation parameters, taking into account accuracy information of the measured data. 3-D least squares matching developed by Gruen & Akca (2005) can also be regarded as a variant of ICP expressed in the framework of least squares matching.

13.2.4.2 *Further iterative methods*

Chen & Medioni (1991) have presented an independent variant of the ICP algorithm. Their algorithm was specifically designed to handle range images and minimizes point distances along their normal vector. This approach has proven superior to classical ICP in some cases (Salvi et al. 2007).

Another iterative approach is the Normal Distribution Transform (NDT) originally developed by Biber & Straßer (2003) for two-dimensional data in the context of autonomous robot navigation. The space is subdivided into a regular grid of cells. For each cell, the probability that there is a point at a certain position in the cell is modelled by a normal distribution. Using a numerical optimization scheme, the two-dimensional transformation in between two scans is found. The approach was later extended to three-dimensional data by Ripperda & Brenner (2005) by making horizontal sections across scans and computing the NDT across all sections. A more general three-dimensional variant of the NDT is given by Magnusson et al. (2007).

The iterative methods attempt to converge to the most accurate solution. In practice, algorithms based on the ICP concept have proven to provide excellent results, rivalling marker-based methods. However, results can be corrupted by lack of overlap, lack of geometric features, variations in point sampling density, outliers and local minima. It is therefore desirable to initialize the iterative methods close to the final solution. This can be achieved through manual interaction, e.g. the selection of a minimum set of corresponding points. The desire for automation motivates the need for coarse registration algorithms, which do not require approximation.

13.2.4.3 *Coarse registration*

There is a vast number of approaches for coarse registration. We refer the reader to existing reviews on the topic (Campbell & Flynn 2001, Huber & Hebert 2003). Here we concentrate on approaches published recently in the photogrammetric community specific to terrestrial LiDAR.

The Extended Gaussian Image (EGI) for registration of laser scans was proposed by Dold (2005) and Makadia et al. (2006) based on the concept introduced by Horn (1984). While the EGI originally is only useful for finding the orientation, this information can be used to reduce the search for the full pose to a search for the translation in a three-dimensional parameter space.

Feature-based registration approaches extract features from the measured data localized at a point or a region of the scan. They attempt to match features in between separate scans and compute the transformation from these correspondences. A number of different suggestions for useful features are known, especially from the area of small volume scanners, such as stripe projection scanners and laser triangulation scanners.

A popular approach is to compute curvature of the surface and extract extremal points as suggested by Thirion (1996). Instead of explicitly computing curvature, one can also use the variation of normal vectors in a local neighbourhood as suggested by Pingi (2005). Curvature and normal vectors were also used for the registration of terrestrial laser data by Bae & Lichti (2004). Alshawabkeh et al. (2006) use curvature to extract edges, which they suggest as features for registration. Barnea & Filin (2008) have developed an interest operator to detect key points based on the first-order derivatives of the range data. Their operator must also be seen in context of approaches that extract features from intensity data described below.

The above mentioned curvature-based methods have proven successful on objects such as statues, reliefs and vases. In the case of terrestrial laser scans, especially of urban scenes where planar surfaces and sharp edges are dominating, they might fail. Therefore methods based on planar regions have been developed. Planar segmentation for feature extraction is used by Dold & Brenner (2006) and also von Hansen (2006). The extracted planar patches are matched across the scans to retrieve the pose parameters. While the former use a region growing approach to extract planar patches, the latter uses a hierarchical partitioning into cells of the point cloud and an RANSAC approach to determine dominant planes within the cell. Both approaches have been demonstrated on urban scenes. Rabbani et al. (2007) have suggested integrating registration with object reconstruction. Thereby the approach is no longer limited to planar patches, but can use any reconstructed geometric shape as tie elements. The approach has been demonstrated using simple surfaces such as planes, cylinders, spheres and tori extracted from an industrial scene.

When the geometry of the scanned data does not provide enough information for the registration or the shapes are too complex to be sufficiently described by shape features, additional information acquired along with the range information can be employed. A source of information that has become quite popular recently is intensity or texture information acquired simultaneously with the range data. This idea has a long history and was probably introduced by Roth (1999). In the context of terrestrial laser scanning, intensity features extracted from a co-registered camera were proposed by Wendt (2004). Feature points are extracted using the SUSAN operator only in planar regions of the scan to ensure maximum visibility across scans. Since the image is co-registered with the range data, three-dimensional coordinates can be interpolated for each feature point. In order to match feature points across scans, an orthographic patch is extracted in the neighbourhood of the point. The author suggests the Metropolis algorithm to perform the matching. The approach was later refined in Wendt (2007). Rectified image patches were also used in the aforementioned approach by Dold & Brenner (2006) to support planar patch matching by image correlation, when geometric information was insufficient to determine all transformation parameters.

As improved interest operators became available in computer vision, these were immediately used to extract better key points for registration. The SIFT developed by Lowe (2004) was used for range data registration by Bendels et al. (2004). They extract key points from the intensity images of a small volume triangulation scanner. They take special care to interpolate correct three-dimensional coordinates corresponding to the two-dimensional key points. For time-of-flight laser scanners, SIFT was used by Böhm & Becker (2007) directly on the intensity of the backscattered laser light (see Figure 13.3 for an example) and by Barnea & Filin (2007) on co-registered colour images.

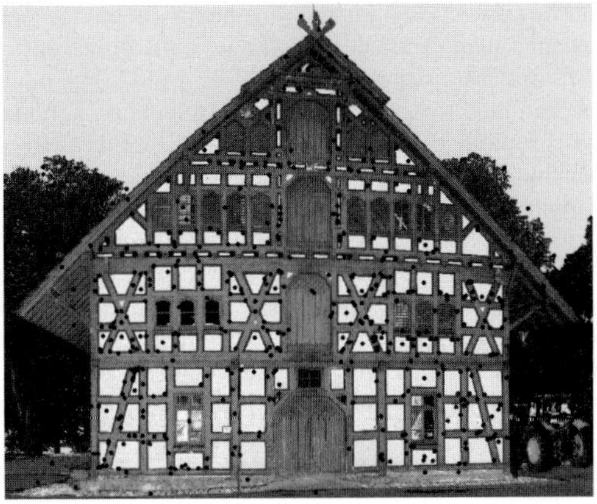

Figure 13.3. Example of feature points extracted from the intensity data of a terrestrial laser scan using the SIFT.

13.2.5 *Discussion*

Looking at practical reports of projects, we can clearly conclude that ICP has demonstrated its capability for reliable and accurate registration of point clouds. The wealth of publications on the topic of coarse registration since the ISPRS congress in Istanbul 2004, however, clearly shows the practical demand for these approaches and also the scientific challenges. The broad range of suggested solutions reflects the broad range of applications and sensors, each posing individual obstacles. The development is driven in general by the desire for automation to overcome cumbersome manual interaction, a major bottleneck in the execution of large projects.

The demand for automation is also reflected by the fact that the major portion of publications is on marker-less registration, a necessity to reduce manual labour in the field. An interesting trend is the fact that many authors suggest including additional data such as intensity or even additional sensors and thereby work towards data fusion—eventually towards a fusion of terrestrial laser scanning and terrestrial photogrammetry.

Unfortunately many of the approaches were developed unrelated to each other. The community is currently missing an attitude towards benchmarking new algorithms. From the side of ISPRS working groups, we should start (and continue in part) to provide reference datasets on which new developments can be tested. We acknowledge that it is difficult to find suitable datasets and thereby common ground in between the broad range of approaches, but consider the attempt worthwhile.

13.3 STRIP ADJUSTMENT IN ALS

13.3.1 *Problem outline*

Measurement of laser scanning point clouds from aerial platforms requires that the position and angular attitude of the platform are known quasi continuously. This is opposed to photo flights, where the position has to be known at the exposure positions only. With a pulse repetition rate (PRR) of 200kHz, 200,000 range and scanner angle measurements are made per second, and for each one the sensor coordinate system has its own exterior orientation. A lack of calibration or errors in the flight path determination lead to errors in the ground point coordinates, which often become visible in the overlapping area of neighbouring strips (see Figure 13.4).

The equipment for direct geo-referencing, i.e. the "direct" observation of the dynamic exterior orientation, is often a combination of the GNSS (global navigation satellite system) and an inertial measurement unit (IMU). A GPS antenna is mounted on top

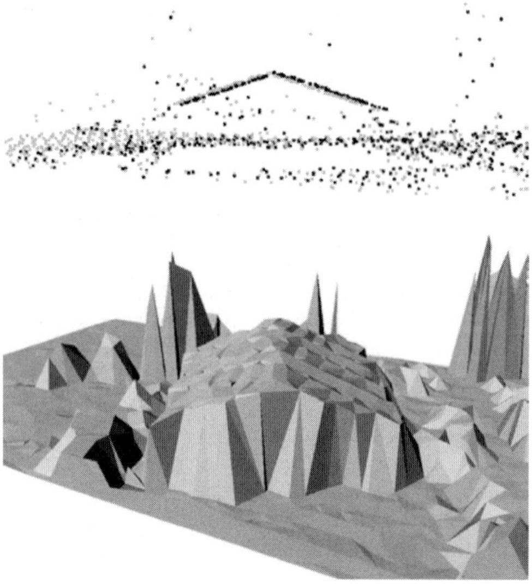

Figure 13.4 Points from ALS from two overlapping strip. The offset between the strips is visible at the roof by the jagged lines in the perspective view showing a triangulation of the data (top). The lower panel shows a side view, with points of one strip in grey, and from the other strip in black. Data courtesy of Section of Optical and Laser Remote Sensing, Delft University of Technology.

of the aircraft, whereas the IMU is rigidly mounted to the sensor platform. The vector from the GPS antenna phase centre to the laser firing point, or more precisely the point of reflection at the mirror, is called the GPS offset vector or lever arm. It is measured after installing a laser scanner in the aircraft, which is routinely performed with total stations.

In combination with a terrestrial GPS reference station (up to 30 to 50 km distant from the aircraft), the flight path is reconstructed in post-processing. The measurement rate of GPS receivers is in the range of 1Hz to 10Hz, which is lower than the PRR by a factor of 10,000 or more. Aircraft for ALS are typically flown at a speed of approx. 60 ms^{-1}, but especially helicopters may fly slower. The intermediate positions of the sensor along the flight path and its angular attitude are determined by the IMU with its accelerometers and gyros. Measurement frequency of the IMU is in the range of 100Hz up to 2kHz. The observations of exterior orientation are joined in a Kalman filter and the flight path can be determined with an accuracy of ±5 to 15 cm under good conditions (Skaloud 2006). The constant rotation angles between the IMU coordinate system and the laser scanner coordinate system is determined by performing a calibration flight, following suggested flight patterns by laser scanner system producers. Such a calibration flight may be used to improve the GPS offset vector, too. Unknown stability of the calibration is one reason for applying those methods also "on the job", i.e. during the execution of a commercial project. To increase accuracy, some form of ground control is necessary, leading to the methods of strip adjustment. Also errors in the flight path determination (e.g. temporally wrong ambiguity fixes in the GPS processing) may be detected and corrected by those methods.

The mathematical model relating the primary measurements of the laser scanner, i.e. range r and scan angle α, with the time-dependent exterior orientation of the sensor system, expressed by the position of the antenna phase centre (x_0, y_0, z_0) and the sensor attitude angles ω, φ, and κ, to the ground point (x, y, z), is (e.g. Filin 2003):

$$\begin{pmatrix} x \\ y \\ z \end{pmatrix} = \begin{pmatrix} x_0 \\ y_0 \\ z_0 \end{pmatrix} + \mathbf{R}_{\omega\varphi\kappa} \left(\mathbf{t} + \mathbf{R}_{\mathrm{m}} \begin{pmatrix} 1 & 0 & 0 \\ 0 & \cos\alpha & \sin\alpha \\ 0 & -\sin\alpha & \cos\alpha \end{pmatrix} \begin{pmatrix} 0 \\ 0 \\ -r \end{pmatrix} \right) \quad (1)$$

In the above equation, $\mathbf{t}$ is the GPS antenna offset and $\mathbf{R}_{\mathrm{m}}$ is the rotation between the IMU coordinate system and the laser scanner coordinate system. The rotation angles are summarized in the matrix $\mathbf{R}_{\omega\varphi\kappa}$ accounting for the rotation of the body frame to the global frame. All measurement acquired by the multi sensor system have to be synchronized, which is possible with the GPS time signal. Further details, e.g. the transformation between geometrically (GPS) and physically (IMU) defined coordinate systems or from geocentric to national coordinate systems, are not considered here.

13.3.2 *State of the art in 2004*

The state of the art of strip adjustment will be presented in two subsections[1]. First, the mathematical models will be presented. They describe how points observed by airborne laser scanning can be "improved". The models can be distinguished into those relying on a model of the multi sensor system, and those concentrating on the visible deficiencies (i.e. discrepancies in overlapping strips) directly and removing them as well as possible without considering a mathematical model of the measurement process. In either case, methods for observing the differencing between the point clouds of overlapping strips are necessary. This is—to some extent—independent of the mathematical models and will therefore be treated in a separate subsection.

[1]The state of the art review is based on Pfeifer et al. (2005).

13.3.2.1 *Mathematical models of strip adjustment*

The approaches to laser strip adjustment can be categorized into two groups. The methods from the first group use only the observed discrepancies in the laser scanner data points from two overlapping strips. They remove the effects of calibration of flight path errors and shall therefore be called "effect-correcting", as opposed to the "sensor system"-driven approaches explained later. Correction functions are determined for each strip, and the parameters of these functions are chosen in order to minimize the discrepancies.

$$\mathbf{p}'_{i,j} = \mathbf{p}_{i,j} + \mathbf{c}_j(\mathbf{p}_{i,j}) \quad (2)$$

Here $\mathbf{p}_{i,j} = (x_{i,j}, y_{i,j}, z_{i,j})$ is the i-th laser point measured in strip j, and $\mathbf{c}_j$ is the correction function for strip j. The point corrected after strip adjustment is $\mathbf{p}'_{i,j}$.

The functions $\mathbf{c}_j$ may only be shift vectors, $\mathbf{c}_j = (\Delta x_j, \Delta y_j, \Delta z_j)$ and independent of the location within the strip. Crombaghs et al. (2000) and Kraus & Pfeifer (2001) make another simplification and apply the correction function to the height component alone $c_j = (0, 0, \Delta z_j(x,y))$, using a linear function (vertical offset and tilts in and across flight direction), and polynomials, respectively. The approach of Kraus & Pfeifer (2001) allows the correction of shorter wavelength deformations, too. A method that is not restricted to vertical correction, but also removes discrepancies in planimetry, was developed by Kilian et al. (1996), where the function **c** has parameters for constant offset and time dependent drifts for shift in and rotation around the three coordinate axes, requiring that the time of the measurement is known. Vosselman & Maas (2001) describe a similar method, mentioning that this model does not allow the correction of short time effects caused by the limited GPS accuracy. Knowledge of the measurement time is not required but replaced by parameterization along the strip axes.

The second group of methods is based on a model of the sensor system (Eq. 1), relating each point to its original observations:

$$\mathbf{p}_{i,j} = \mathbf{f}(\mathbf{O}(t_i), \mathbf{R}(t_i), r_i, \alpha_i, \mathbf{s}) \quad (3)$$

where t_i is the measurement time, and $\mathbf{O}(t_i)$ and $\mathbf{R}(t_i)$ are the origin and the attitude of the platform, determined from GPS and IMU measurements. The laser scanner observations are the range r_i, and the angle measurement α_i. The vector **s** describes the system parameters (e.g. offset between the GPS antenna and the origin of the platform). In the adjustment, the corrected laser points become:

$$\mathbf{p}'_{i,j} = \mathbf{f}(\mathbf{O}(t_i) + \Delta\mathbf{O}, \mathbf{R}(t_i) + \Delta\mathbf{R}, r_i + \Delta r, \alpha_i + \Delta\alpha, \mathbf{s} + \Delta\mathbf{s}) \quad (4)$$

The Δ–terms can be constants, functions dependent on time, scale factors, or take other forms. In Burman (2002), the unknowns (the "Δ-terms") are a constant offset and a time dependent drift for ΔO, and ΔR (IMU–sensor misalignment and IMU drift). In Filin (2003), additionally an IMU offset, a range offset and a scan angle error are considered. In Filin (2003) the capability of least squares adjustment of the mathematical model is also exploited to study the ability to recover different system parameters. It is shown that it is necessary to have surfaces with different expositions (i.e. not only horizontal surfaces), and surfaces with different aspects.

Thiel & Wehr (1997) have published calibration procedures for palmer scanners. Such a scanner is not performing beam deflection across the flight direction only, but scans along a circle. Together with the forward motion of the aerial platform, the terrain is covered twice, once in the forward looking half of the circle, and later in the backward looking half. Discrepancies between the two observations over the same area can be used to compute calibration parameters. It is thus possible to compute pitch, heading and roll offset angles.

These models can be used in two scenarios. Firstly they can be used to calibrate the entire system. In this case, the unknowns, e.g. the offset between GPS antenna phase centre and the reference point of ranging, i.e. a constant ΔO, are usually determined for the entire block. Secondly, they can be used on-the-job to determine small offsets between calibration values and actual values or to handle drift effects, e.g. the IMU drift for each strip, i.e. a linear, time-dependent $\Delta R_j(t)$.

13.3.2.2 *Measurement of discrepancies*

The mathematical models require also observations as input for strip adjustment, but so far only methods for correcting the points have been presented. For that, the discrepancies between overlapping strips, or between a strip and a ground control area, are exploited. However, there are considerably different approaches for obtaining these.

Kager & Kraus (2001) use schema points (like photogrammetric "von Gruber" points, but many more in the strip direction) for finding tie features. Suitable tie features are searched in one strip in a spiral pattern starting from each schema point. The requirements are that the inclination of an adjusting plane to the point and its neighbours is small and also that the standard deviation of the plane fit is small. The representative elevation in the strip is the height of the plane. The height of the adjusting plane in the overlapping strip then provides a measure of the height discrepancy.

In Vosselman (2002a) and Vosselman (2002b) methods for automatic measurement of discrepancies with linear features are described. In this approach, special tie features are extracted in elevation data and—independently—in intensity data. For the first case, it is suggested to model the linear feature with an analytical model in profile sections. For finding lines in the reflectance data, the response of a grey value edge in laser scanning data is modelled. A line in planimetry can be used to measure an offset in one direction, a line in 3-D space, e.g. the ridge of a house, can be used to measure an offset in the vertical and one horizontal direction.

In Maas (2001), a method for matching in a TIN structure is explained. Also the pulse reflectance data may be used, not only the 3-D location of the point. Burman (2002) applies this method to every *n*-th point (e.g. $n = 1000$). The TIN is not always a truthful representation of the measured objects (e.g. houses and terrain), but is also influenced by shadowing effects and above ground objects may be represented in the TIN surface wider than they are in reality. This has to be considered, especially when applying TIN matching (Maas 2001). In this method no reduction of noise is performed for the discrepancy observations, but on the other hand each point can potentially serve to observe a discrepancy.

It should be mentioned that manual methods for measurements of discrepancies were also suggested in strip adjustment, e.g. by the early work of Kilian et al. (1996).

13.3.3 *Contributions to strip adjustment in ALS in the period 2004–2008*

In Filin & Vosselman (2004), segmentation is applied to the overlapping part of the laser scanner strips in order to find homologous patches. The 3-D points from either strip are forced to lie in the segmented surface patches. For each strip, an offset vector is determined. It should be noted that there is an equivalence in determining a GPS offset vector per strip (sensor system point of view, **t** in Eq. 1) and an effect correcting approach with $\mathbf{c}_j = (\Delta x_j, \Delta y_j, \Delta z_j)$ (Eq. 2). The parameters of the patches are updated between the iterations of the strip adjustment, i.e. they are computed new from the corrected points.

In Kager (2004), the mathematical model has time dependent polynomials for ΔO, ΔR and a constant IMU-sensor misalignment. Corrections are also determined for the observed beam deflection angles across and in the flight direction and the range. The angle and range measurement are used as observations to four (corner) points of a tie patch. These points are considered as unknowns, but observed by the range and angle measurements. The parameters of a tie patch are determined from the corner points, where each strip including the patch contributes four points. Of course, the corner points from the different strips do not have to coincide, but the (planar) patch forces the points to lie in the same surface. The parameters of the patches are determined simultaneously with the system parameters. The patches are found automatically by first sorting the points in a matrix like structure, with the columns parallel to the flight path and the rows across it. Then the points are analysed strip wise in a moving window of rows, looking for planar patches. In Kager (2004), strip height difference models are also suggested as a rigorous method for detection of systematic errors (Figure 13.5). The method has been used in a number of commercial projects and precision of a few centimetres on well defined (i.e. hard) surfaces can be reached.

Wehr et al. (2006) have further developed their methodology for calibrating palmer scanners. They use control areas with known ground coordinates and tie patches, identified by the radiometric properties. The Institute of Navigation from Stuttgart University is also the designer of this laser scanner (ScaLARS). The parameters suggested for calibration are all physically explainable: IMU-misalignment, GPS offset vector, range finder constant offset and scale, and beam direction. The latter describes a mismatch between the mirror centre of rotation and the laser pointing direction before it hits the beam. They conclude that the calibration works well, but not all discrepancies observed on the ground can be explained by the calibration procedure. Therefore, a more detailed physical sensor model should be developed.

Skaloud & Lichti (2006) describe a rigorous approach for estimating a subset of typical calibration parameters, namely the IMU misalignment and a range finder constant offset. In overlapping areas, planes with different inclination are extracted manually. The plane parameters and the calibration parameters are estimated simultaneously.

Fries (2006) describes a rigorous mathematical model also estimating the calibration parameters and the tie patch parameter simultaneously, like Skaloud & Lichti (2006) and Kager (2004). Next to the relative orientation between the coordinate frames of the components (GPS, IMU, laser scanner), the "internal"

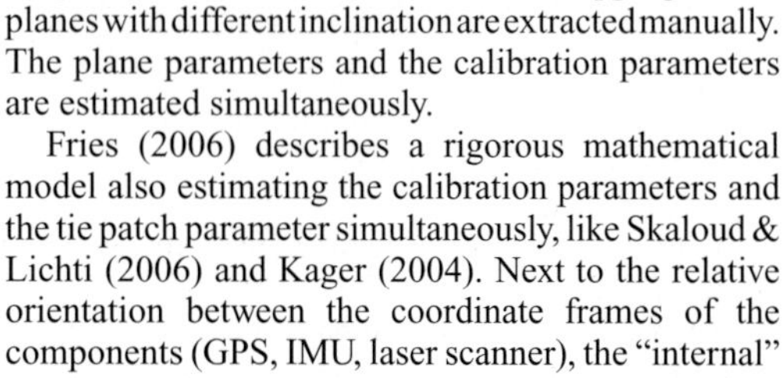

Figure 13.5. Left: The background shows airborne laser data acquired over a glacier. In the overlaps of adjacent strips, the height differences of the strips are shown. Dark tunes refer to differences larger than 30 cm. Right: Principle of airborne laser scanning strip adjustment based on tie patches (Kager 2004). (see colour plate page 494)

laser scanner parameters estimated are a range finder offset, scan angle offset (index) and the scan angle scale. Tie patches are determined by a method similar to the method of Kager (2004).

Csanyi & Toth (2007) investigated the accuracy potential of airborne laser scanning and note that engineering grade accuracy (a few centimetres) can only be achieved by adding ground control. By simulation they conclude that optimal targets are rotationally invariant with a diameter of about 1 m for a point density of 5 points/m^2. The targets have a bright interior and a black ring around them. Mathematical models for transforming the extracted target centres to the measured ground point coordinates included a vertical shift (producing acceptable results) and the spatial similarity transformation providing the best results.

13.3.4 *Discussion*

Application of strip adjustment is becoming a standard procedure. This hypothesis is also supported by looking at the papers of authors closer to hardware producers (Fries 2006, Wehr et al. 2006).

The positioning errors are currently the dominating source of errors (Csanyi & Toth 2007), whereas the laser ranging accuracy is in the order of 2 cm to 3 cm in optimal cases. Airborne laser scanning strip adjustment is a possibility to shift the accuracy and precision of the final point cloud towards the ranging precision of the laser range finder (Csanyi & Toth 2007, Kager 2006). Strip adjustment can also be seen as a method for checking the internal quality of a point cloud. The height differences, especially between overlapping strips, excluding forested areas, are a rigorous method of analysing systematic errors.

There is no unique mathematical model in strip adjustment yet. Some authors prefer to use only parameters that can be explained physically, whereas others also allow polynomials as correction functions. These empirically identified parameters can be compared to distortion parameters in on-the-job calibration of digital images, which are only partly due to the optical system, and simultaneously model systematic behaviour of unknown origin.

13.4 FILTERING OF LASER SCANNING DATA

13.4.1 *Problem statement*

Given are the points $p_i = (x_i, y_i, z_i)$ measured by ALS. These points are considered to be free of systematic errors and therefore obtained by a system with proper calibration or after the procedure of strip adjustment. There can be some additional information like the echo number (e.g. 1st, 2nd, last), an intensity measure, information obtained from a digital colour image and associated with the point, or full waveform information. Those points p_i, which are under the consideration of the measurement precision on the terrain surface, shall be found.

This formulation looks at the task from a classification point of view. It can also be applied to rasterized representations of the data. Such a raster can be computed, e.g. by averaging all elevations within a grid cell. Alternatively, the problem can be formulated as (directly) obtaining a DTM from the given set of points, without interest in a classification. For either point of view, "filtering" is not trivial. Ground features and small objects may have the same geometrical appearance and only additional attribute information can provide the correct result in such cases. On the other hand, attribute information only does not provide a correct classification either. It is especially not the case that the last of multiple echoes is always a ground point. It can also come from an intermediate storey in a forest, between canopy and ground. Using only last pulse points for ground extraction is, however, an approach used frequently as a starting point.

The derivation of a DTM from laser scanning point clouds involves a number of steps. First, the points have to be classified into echoes from the terrain and off-terrain points. Terrain interpolation methods can be applied to these points. To obtain high quality terrain models, structural feature lines need to be extracted. Finally, in order to keep the detailed information provided by dense airborne laser scanning point clouds, but reduce the volume of data to allow efficient handling, a qualified thinning strategy becomes necessary. This step will not be considered in this section, as it does not consider obtaining a DTM, but only improving its representation in terms of storage space. Thinning is also an application dependent step and guiding parameters are provided by the application (e.g. maximum error tolerance or r.m.s. error). Suitable thinning for applications in hydraulics is presented by Mandlburder et al. (2007).

13.4.2 *State of the art in 2004*

Filtering algorithms can be grouped according to different methodologies (e.g. Sithole & Vosselman 2004). Here, the approaches are grouped into morphological filters, densification approaches, surface based filters and segmentation based filters. The first group looks "only" in a limited neighbourhood for determining the fate of a point. The densification algorithms have a concept of the terrain surface and start from a coarse representation to a finer and finer one. The surface based filters work in a different direction. First, many (or all) points are considered to belong to the surface, and iteratively points not belonging to the ground are excluded. Finally, the segmentation based approaches do not operate on the points directly, but first form larger compounds, i.e. segments. A segment can only entirely belong to the ground or not. Then, a rule system is applied to classify the segments on the basis of some segment features.

The morphological filters take their name from mathematical morphology for grey value images. The grey values refer to the elevation (Kilian et al. 1996), but also formulations on point clouds are possible (Vosselman 2000). The structure element (i.e. a small "image") describes admissible height differences for a certain horizontal distance. In Vosselman (2000), the admissible height difference depends on the horizontal distance. If one point is a ground point, then the following reasoning is applied to find further ground points. For smaller distances between a ground point and its neighbouring points, smaller height difference can be accepted. The admissible height difference grows with increasing horizontal distance. The height differences are stored in the structure element as a function of the horizontal distance. This structure element is placed at each point and off-terrain points can be identified as having larger height differences than admissible. The structure element itself can be obtained from assumptions on terrain slope or from point clouds with a known classification (training data). In the first case, the structure element is an inclined line. A variant of the morphological filtering is described by Sithole (2001). There the structure element is linear, but its slope is a function of terrain slope itself. In steeper areas, higher admissible height differences are allowed. The information on terrain shape is reconstructed by an averaging process over large areas where the influence of tree and building height is decreased. Kilian et al. (1996) use multiple structure elements with different horizontal sizes. The admissible height difference is constant for each structure element. They apply the structure elements for the morphological opening operation. With different structure element sizes, different objects (small cars, big buildings) are tackled. Zhang et al. (2003) describe a comparable method using different window sizes and allowed height differences. They propose a method for choosing these parameters. Lohmann et al. (2000) apply the morphological operations Erosion and Dilation to replace raster terrain elevations with filtered elevations.

The filters of the second group are the progressive filters. More and more points are classified as ground points. All the approaches in this group require an initialization with a low number of sure ground points. Axelsson (2000) suggests using the lowest points of large cells in a regular, quadratic layout. These points are triangulated forming the first, coarse approximation of the terrain, which is the reference surface for the subsequent step. In each triangle, a further ground point is found by analysing the height differences of the unclassified points to the triangle surface. Also the angles between the triangle and the edges between vertices and the new points are analysed. If a point has offsets below the threshold values, it is considered a ground point, and the next triangle face is investigated. The triangulation is thus getting denser and denser, representing the ground surface more truthfully after each insertion of a ground point. von Hansen & Voegtle (1999) use different offset measures and take as starting points the lower part of the convex hull of the point set. Sohn & Dowman (2002) apply the progressive densification with two different steps. First a "downward step" is taken and points below the current triangulation are added. This is followed by the "upward step", where one or more points above each triangle are added.

As mentioned above, the surface based filters do not add points (as the densification approaches) but rather remove points to obtain the classification. Therefore, they start with a surface model approximating all points. This model allows the computation of residuals, i.e. the distances of the points to the surface model itself. The choice of the surface model, i.e. its flexibility, etc., is the crucial part of these algorithms. Points lying above the surface are unlikely to be ground points and are—depending on the specific filter—downweighted or removed for the next iteration and surface calculation step. Kraus & Pfeifer (1998) developed a method known as robust filtering. They use linear prediction (kriging) with individual point accuracies as the surface model. A weight function is used to compute weights from the residuals. It is asymmetric with respect to zero, and gives high weights to points on or below the surface and small weights to points above it. These weights are transformed into accuracy (low weight = poor accuracy), which can be considered in the kriging approach. Points with good (low value) accuracy have more influence on the run of the surface, while points with small weights have less influence. After the surface computation, the residuals are computed anew and iteration continues. Pfeifer et al. (2001) embed this method in a hierarchical setup. This improves performance on large buildings and reduces computation time. Elmqvist (2001) uses a surface-snake-approach, where the inner forces determine the surface stiffness and the external forces are a negative gravity. Iteration starts with a horizontal surface below all points. Following the external forces, it first moves upwards to reach the points, but the inner forces (stiffness) prohibits the surface (membrane) to bend up to the points on the vegetation canopy or house roofs. Brovelli et al. (2002) have a similar approach analysing the residuals obtained from an averaging spline surface. They combine it also with edge extraction, looking for objects surrounded by edges. If these areas are elevated they are additionally removed. With this step, the idea of segmentation is already incorporated.

The fourth group of filters are the segmentation filters. Obviously, they have their advantages in areas where distinct segments do exist. These are areas strongly influenced by building activity. Nardinocchi et al. (2003) apply a region growing technique for obtaining the segments by analysing height differences. The geometric and topologic situation of the segments (height differences between neighbouring segments, etc.) are input for a set of rules where the output is a classification of the segments into terrain, buildings and vegetation. Jacobsen & Lohmann (2003) also use region growing (in eCognition) for obtaining the segments. They apply measures like "compactness" and height differences to neighbouring segments, in order to classify the segments including a terrain class. In an earlier work, Schiewe (2001) uses maximum and average gradients for classifying the segments. The segmentation based methods work not directly on the points but on larger compounds and are therefore less susceptible to noise.

13.4.3 *Contributions to ALS filtering in the period 2004–2008*

13.4.3.1 *Filter test*

An experimental comparison of the performance of some filters is described by Sithole & Vosselman (2004). They investigate filtering quality by interactive filtering of a point cloud using human knowledge and aerial images. The data was provided by EuroSDR [www.eurosdr.net]. Eight test sites were picked featuring urban and rural scenes, "steep slopes, dense vegetation, densely packed buildings with vegetation in between, large buildings (a railway station), multi-level buildings with courtyards, ramps, underpasses, tunnel entrances, a quarry (with breaklines), and data gaps" (Sithole & Vosselman 2004).

Researchers were asked to participate in the test, and eight groups responded. The test participants filtered the data sets with their own algorithms and the classified ground points were compared to the manual reference by G. Sithole. If the interactive ground point selection is correct, then this comparison measures the quality of filtering. The errors (discrepancies between different classifications) are free of flight path determination or DTM interpolation errors.

They conclude that most filters produce satisfctory results in moderately sloped terrain with little vegetation and smaller buildings. The most difficult scenes were "complex cityscapcs with multi-tier buildings, plazas", etc. and "discontinuities in the bare Earth" (Sithole & Vosselman 2004). Furthermore, surface-based filters tended to provide better results, but it should be noted that classification based filters did not take part in the comparison.

The data sets, i.e. the unfiltered point clouds and the manual classification, became to some extent a standard for benchmarking new algorithms. It is, for example, used by Akel et al. (2007) to assess the filtering quality of their development.

13.4.3.2 *New filter algorithms*

Sithole (2005) constructs segments that allow natural persons to walk from one point in a segment to each other point in the segment. For the streets and open level areas in a city, this is indeed the case, omitting effects like data bounds. This segmentation is first performed along lines in different directions. "Lines" have a certain width and the lines of each direction entirely cover the total area. Height differences, jumps and horizontal distance between points along the lines are used for classifying the segments as "raised", "lowered", and "terraced" with respect to each other. As each point is in lines of different directions, planimetrically extended segments can be formed. The relation between neighbouring segments (raised, etc.) is used for classifying the segments into "ground" and other classes.

Tovari et al. (2005) use a region growing approach to build segments. The homogeneity criterion for growing is the normal vector similarity. The normal vectors are estimated in a first step, i.e. before filtering. Then, a surface is interpolated, like in the surface based approaches and residuals are computed for each point. A representative residual is computed for each segment (the mean residual), and depending on that residual a weight is computed as in the robust filtering. This weight is assigned to all points in the segment and the surface is computed anew. This iterative procedure can be seen as a combination of a surface based and segmentation based approach.

The method of Kobler et al. (2007) is intended to generate a high quality DTM in steep mountainous areas covered by "heterogeneous forests". This heterogeneity may lead to a high variation in the penetration rate and therefore also to a high variation in the number of ground echoes. A two-stage method is suggested. In the first step, the initial filtering stage, negative outliers and most, but not necessarily all, off-ground points are removed. Existing techniques, e.g. morphological filters, are used with parameters preventing deletion of ground points. In the second stage, the DTM is produced from the initially filtered point cloud. For this task, multiple ground elevation estimates at DTM grid points are interpolated from surrounding ground returns. These elevation estimates are generated from multiple independent samples taken from the initially filtered point cloud. TINs are used for computing the elevation of the DTM grid points from the selected point sample. At each grid point, the lowest elevation plus a constant is the final DTM elevation.

Akel et al. (2007) present a surface based method enriched by segmentation. As surface model orthogonal polynomials are used, replacing the functional model of surface description of the robust interpolation with these orthogonal polynomials. They argue that a global surface based approach cannot capture fine details, and therefore use the result obtained so far as a starting point in a region growing approach, which is controlled by normal vector similarity. This enables the maintenance of surface edges, which are rounded off when using only a surface based approach. As mentioned above, the reference data of Sithole & Vosselman (2004) is used for evaluating the result.

13.4.3.3 *Exploitation of full waveform information*

In conventional as opposed to full waveform laser scanning systems discrete return times are given for the received echoes. This leads to discrete ranges and therefore points in 3-D. In full waveform ALS systems, the echoes are digitized as a function of time (or equivalently: range). This is typically performed with a sampling interval of 1ns, corresponding to a range interval of 15 cm. Discrete points are obtained by analysing the echo function, e.g. looking for the maxima (Wagner et al. 2006). The emitted pulse has a certain length (e.g. 5ns or 1.5 m), and the received echoes have the same length or are widened in comparison. The echo width therefore holds information on the backscattering surface. If it is vertically spread, the echo becomes wider. This can be the case in inclined terrain or on house roofs, but for small footprint systems the widening due to target inclination is very small. Low vegetation and tree canopies, but also other non-ground objects, exhibit notable height variation within the footprint (e.g. 0.5 m diameter).

Doneus & Briese (2006) have shown that the echo width can thus be used for a first classification of the measurements. Points originating from widened echoes cannot be ground points and can be removed beforehand. Especially in forest scenes with the presence of low vegetation, the filtering results are superior to filtering based on only the points (Figure 13.6).

In Mandlburger et al. (2007), the method is embedded in the robust filtering. The echo widening is used to compute an *a priori* weight, and points with larger echo width obtain lower weights. Because of this, the necessary number of iterations in the robust filtering is decreased and improvement on edges is reported.

13.4.3.4 *Break lines*

A DTM can be computed after filtering of the point cloud. Various methods exist for this task, which will not be detailed here. The surface can either be represented as a regular grid, which is an efficient storage structure because only the heights and not the planimetric position of the points in the regular matrix structure have to be stored explicitly. An alternative is the TIN (triangular irregular network) data structure, where the raw points are triangulated. As a consequence, no reduction of measurement noise is performed. In any case, no specific terrain features are explicitly represented and stored, such as peaks, sinks, break lines or mountain ridges. As laser scanning samples in a regular pattern (in the angular domain of the scanner itself), no explicit objects can be measured. High quality terrain representations, however, rely on those features. Additionally, representing features allows terrain analysis and also efficient thinning of the data preserving the relevant details.

One of the first approaches was presented by Brügelmann (2000), who introduced a full processing chain starting from an ALS elevation image leading to smooth vector break lines. After the DTM grid computation, curvature analysis and thresholding, together with non-maxima suppression is applied for finding break lines. The height is interpolated from the DTM.

Brzank et al. (2005) present an approach for the extraction and modelling of pair-wise structure lines in coastal areas described in a cross section by a hyperbolic tangent function. Initial values for the location of the break lines are obtained from the analysis of the elevation image with image processing tools (edge detection). The profile function is then fitted to the original points in a patch, which is shifted along the line in an overlapping manner (see Figure 13.7).

A similar approach was presented by Briese (2004) requiring one manually picked start edge per line. This approach does not require the computation of a regular raster or the filtering of the data as a first step. The model of the break line consists of two planes which have an intersection in the break line itself. An initial approximation of the break line is necessary to separate the points into "left" and "right" points, each contributing to one plane. The plane parameters are estimated for one patch using robust adjustment with a weight function similar to that of robust interpolation in surface based filtering. After successful fitting, the procedure continues in either direction of the found break line with overlapping patches. Figure 13.8 shows an example.

13.4.4 *Discussion*

Filtering of airborne laser scanning point clouds is a matured technology. New algorithms are published, and measured against existing references. Another observation is the diversification of filtering. New algorithms use segmentation and are therefore suited for areas under strong human influence. Other algorithms

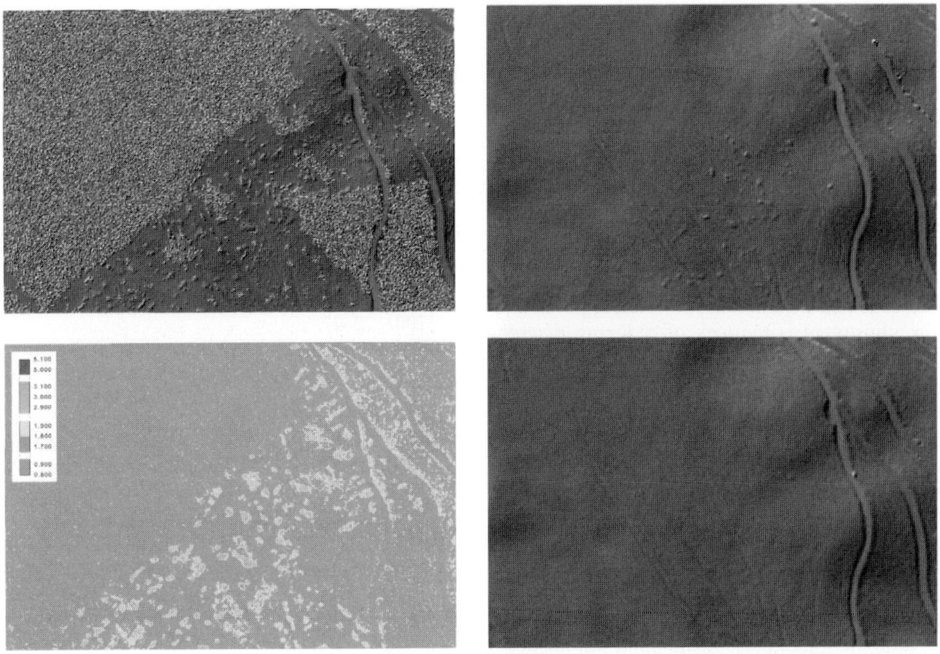

Figure 13.6. From top left to lower right: The digital surface model based on the ALS data; automatically computed terrain model with hierarchic robust interpolation (remaining vegetation artifacts); echo width shown, with darker tones referring to wider echoes; terrain model computed with hierarchic robust interpolation after first removing point with large echo width. From Doneus and Briese (2006). (see colour plate page 495)

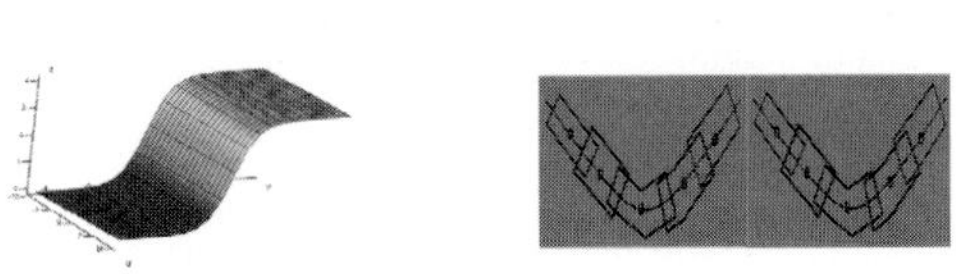

Figure 13.7. Left: Hyperbolic tangent function as model of an edge pair. Right: Overlapping break line patches. Both from Brzank et al. (2005). (see colour plate page 495)

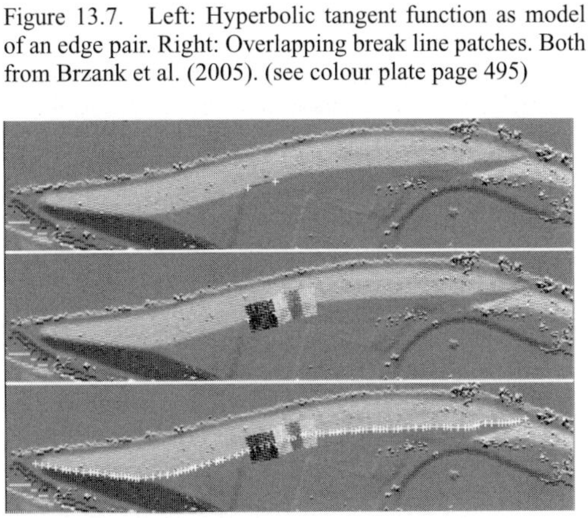

Figure 13.8. Break line growing. Top: Shading of a surface model and start-segment; Middle: ALS points within the start patch (black) and within the three subsequently extrapolated patch areas in forward direction; Bottom: Result of the break line growing procedure (white crosses). From Briese (2004)

are targeting other special environments, e.g. mountain forests. The classification accuracy is high, although obvious errors remain in non-trivial cases (Sithole & Vosselman 2004). This is solved by manually improving the automatic filter results in commercial projects.

One approach for increasing the automation in filtering is a tighter integration with classification of the entire point cloud, possibly with aerial images. Exploiting additional attribute information, especially the full waveform as suggested by Doneus & Briese (2006), is a valuable alternative. The advantage is that the data is coming from one sensor only, alleviating problems of different epochs, viewing angles, etc. Future systems with multiple wavelength lasers may provide better means for object classification, thus also supporting the task of filtering.

13.5 CONCLUSIONS

In this contribution the early steps of data processing in terrestrial and airborne laser scanning have been discussed: orientation, along with calibration in the airborne case (strip adjustment) and filtering. There have been a number of contributions from many different authors from different institutions, showing the interest and need for research in these fields.

For the registration of terrestrial LiDAR data the wealth of publications raises the expectation for fully automated procedures to come into practice soon. However, we strongly encourage benchmarking new developments on common datasets.

Filtering is the task of classifying the acquired point cloud into ground points and off-terrain points (reflections from vegetation canopies, cars, buildings, etc.). It has been investigated since approximately 1996 and the methods are mature. There is still a number of competing methods and new sensor developments are integrated into the filtering approaches.

Strip adjustment is the task of reducing discrepancies in overlapping strips acquired by airborne laser scanning and discrepancies at control surfaces. These discrepancies originate from poor calibration or deficiencies in the flight path computation. Different sensors with different scanning mechanisms require different mathematical models, but generally the process of identifying the best suited model is still going on. It has been shown that accuracy of a few centimetres ($\leq \pm$ 5cm) can be achieved under favourable circumstances (hard smooth surfaces, high point density, low flying height, etc.).

It is only possible to speculate about the developments in the coming four years. We expect a tighter integration of imaging and laser scanning, for calibration, orientation and (terrain) model reconstruction.

REFERENCES

Al-Manasir, K. & Fraser, C.S. 2006. Registration of terrestrial laser scanner data using imagery. *Photogrammetric Record* 21(115): 225–268.

Alshawabkeh, Y., Haala, N. & Fritsch, D. 2006. 2D-3D feature extraction and registration of real world scenes. *ISPRS Image Engineering and Vision Metrology*: 32–37.

Akel, N.A., Filin, S. & Doytsher, Y., 2007. Orthogonal polynomials supported by regional growing segmentation fort he extraction of terrain from Lidar data. *Photogrammetric Engineering and Remote Sensing* 73(11): 1253–1266.

Axelsson, P. 2000. DEM generation from laser scanner data using adaptive TIN models. *International Archives of Photogrammetry and Remote Sensing* 33(B4/1): 110–117.

Bae, K.-H. & Lichti, D.D. 2004. Automated registration of unorganised point clouds from terrestrial laser scanners. *International Archives of Photogrammetry and Remote Sensing* XXXV, Part B5: 222–227.

Barnea, S. & Filin, S. 2007. Registration of terrestrial laser scans via image based features. *ISPRS Workshop on Laser Scanning and SilviLaser*: 32–37.

Barnea, S. & Filin, S. 2008. Keypoint based autonomous registration of terrestrial laser point-clouds. *ISPRS Journal of Photogrammetry & Remote Sensing* (In press).

Bendels, G.H., Degener, P., Wahl, R., Körtgen, M. & Klein, R. 2004. Image-Based Registration of 3D-Range Data Using Feature Surface Elements. *The 5th International Symposium on Virtual Reality, Archaeology and CulturalHeritage (VAST 2004)*: 115–124. The Eurographics Association.

Besl, P.J. & McKay, N.D. 1992. A Method for Registration of 3-D Shapes. *IEEE Transactions on Pattern Analysis and Machine Intelligence* 14(2): 239–256.

Biber, P. & Straßer, W. 2003. The Normal Distributions Transform: A New Approach to Laser Scan Matching. *IEEE International Conference on Intelligent Robots and Systems,* 3: 2743–2748.

Bornaz, L., Lingua, A. & Rinaudo, F. 2003. Multiple scan registration in LiDAR close range applications. *International Archives of the Photogrammetry, Remote Sensing and Spatial Information Sciences* XXXIV, Part 5/W12, 2003: 72–77.

Böhm, J. & Becker, S. 2007. Automatic Marker-free Registration of Terrestrial Laser Scans using Reflectance Features. *Proceedings Optical 3–D Measurement Techniques VIII* Volume I: 338–343.

Briese, C. 2004. Three-dimensional modelling of breaklines from airborne laser scanner data. *International Society for Photogrammetry and Remote Sensing (ISPRS), XXth ISPRS Congress, Istanbul, Turkey.*

Brzank, A., Lohmann, P. & Heipke, C. 2005. Automated extraction of pair wise structure lines using airborne laserscanner data in coastal areas. *International Archives of Photogrammetry and Remote Sensing* XXXVI, 3/W19.

Burman, H. 2002. Laser strip adjustment for data calibration and verification. *International Archives of Photogrammetry and Remote Sensing* XXXIV, 3: 67–72.

Campbell, R. & Flynn, P. 2001. A Survey of Free-Form Object Representation and Recognition Techniques. *Computer Vision and Image Understanding* 81(2): 166–210.

Chen, Y. & Medioni, G. 1991. Object modeling by registration of multiple range images. *IEEE International Conference on Robotics and Automation*: 2724–2729.

Crombaghs, M., Brügelmann, R. & de Min, E. 2000. On the adjustment of overlapping strips of laseraltimeter height data. *International Archives of Photogrammetry and Remote Sensing* XXXIII, 3A: 230–237.

Crombaghs, M., Oude Elerink, S., Brügelmann, R. & de Min, E. 2002. Assessing height precision of laser altimetry DEMs. *International Archives of Photogrammetry and Remote Sensing* XXXIV: 85–90.

Dalley, G. & Flynn, P.J. 2001. Range Image Registration: A Software Platform and Empirical Evaluation. *3rd International Conference on 3-D Digital Imaging and Modeling (3DIM 2001)*: 246–253.

Dold, C. 2005. Extended Gaussian images for the registration of terrestrial scan data. *Proceedings of the ISPRS Workshop Laser scanning 2005*: 180–185.

Dold, C. & Brenner, C. 2006. Registration of terrestrial laserscanning data using planar patches and image data. *IAPRS* XXXVI, Part 5: 78–83.

Doneus, M. & Briese, C. 2006. Digital terrain modelling for archaeological interpretation within forested areas using full-waveform laserscanning. *7th internation symposium on virtual reality, archaeology and cultural heritage, VAST.*

Eggert, D.W., Fitzgibbon, A.W. & Fisher, R.B. 1998. Simultaneous Registration of Multiple Range Views for Use in Reverse Engineering of CAD Models. *Computer Vision and Image Understanding* 69(3): 253–272.

Elmqvist, M. 2001. Ground estimation of lasar radar data using active shape models. *OEEPE workshop on airborne laser scanning and interferometric SAR for detailed digital elevation models, Stockholm, Sweden.*

Filin, S. 2002. Surface clustering from airborne laser scanning data. *International Archives of Photogrammetry and Remote Sensing* XXXII, 3A: 119–124.

Filin, S. 2003. Recovery of systematic biases in laser altimetry data using natural surfaces. *Photogrammetric Engineering & Remote Sensing* 69(11): 1235–1242.

Filin, S. & Pfeifer, N. 2005. Neighborhood systems for airborne laser data. *Photogrammetric Engineering & Remote Sensing* 71(6).

Filin, S. & Vosselman, G. 2004. Adjustment of airborne laser altimetry strips. *International Archives of Photogrammetry and Remote Sensing* XXXV, B3: 5.

Jacobsen, K. & Lohmann, P. 2003. Segmented filtering of laser scanner DSMs. *International Archives of Photogrammetry and Remote Sensing* 34 (3/W13).

Gruen, A. & Akca, D. 2005. Least squares 3-D surface and curve matching. *ISPRS Journal of Photogrammetry and Remote Sensing* 59(3): 151–174.

Gühring, J. 2001. Reliable 3-D surface acquisition, registration and validation using statistical error models. *3DIM 2001, Quebec City, Canada*: 224–231.

Horn, B.K.P. 1984. Extended Gaussian images. *Proceedings of IEEE* 72(12): 1671–1686.

Huber, D. & Hebert, M. 2003. Fully automatic registration of multiple 3-D data sets. *Image and Vision Computing* 21(7): 637–650.

Kager, H. 2004. Discrepancies between overlapping laser scanning strips—simultaneous fitting of aerial laser scanner strips. *International Archives of Photogrammetry and Remote Sensing* XXXV, B/1: 555–560.

Kager, H. & Kraus, K. 2001. Height discrepancies between overlapping laser scanner strips—simultaneous fitting of aerial laser scanner strips. In A. Grün and Kahmen, (eds.), *Fifth Conference on Optical 3–D Measurement Techniques, Vienna, Austria:* 103–110.

Kager, H. 2006. The importance of exact geo-referencing of airborne Lidar data. *Natural Resource Management: Going Critical* 10(4): 1–10.

Kilian, J., Haala, N. & Englich, M. 1996. Capture and evaluation of airborne laser scanner data. *International Archives of Photogrammetry and Remote Sensing* XXXI, B3: 383–388.

Kraus, K. & Pfeifer, N. 1998. Determination of terrain models in wooded areas with airborne laser scanner data. *ISPRS Journal of Photogrammetry and Remote Sensing* 53: 193–203.

Kraus, K. & Pfeifer, N. 2001. Advanced DTM generation from LIDAR data. *International Archives of Photogrammetry and Remote Sensing* XXXIV, 3/W4: 23–30.

Lichti, D. 2007. Error modelling, calibration and analysis of an AM–CW terrestrial laser scanner system. *ISPRS Journal of Photogrammetry and Remote Sensing* 61(5): 307–324.

Lohmann, P., Koch, A. & Schaeffer, M. 2000. Approaches to the filtering of laser scanner data. *International Archives of Photogrammetry and Remote Sensing* 33(B3/1): 534–541.

Lowe, D.G. 2004. Distinctive image features from scale-invariant keypoints. *International Journal of Computer Vision* 60(2): 91–110.

Maas, H.-G. 2001. On the use of pulse reflectance data for laserscanner strip adjustment. *International Archives of Photogrammetry and Remote Sensing* XXXIV, 3/W4.

Mandlburger, G. & Briese, C. 2007. Using Airborne Laser Scanning for Improved Hydraulic Models. *Proceedings of Congress on Modeling and Simulation—MODSIM07, Christchurch, New Zealand.*

Mandlburger, G., Briese, C. & Pfeifer, N. 2007. Progress in Lidar sensor technology—chance and challenge for DTM generation and data adminstration. *Photogrammetric Week 2007*. Wichmann.

Magnusson, M., Lilienthal, A. & Duckett, T. 2007. Scan registration for autonomous mining vehicles using 3D-NDT: Research Articles. *Journal of Field Robotics Archive* 24(10): 803–827.

Makadia, A., Patterson, A. & Daniilidis, K. 2006. Fully Automatic Registration of 3-D Point Clouds. *Computer Vision and Pattern Recognition* 1: 1297– 1304.

Nardinocchi, C., Forlani, G. & Zingaretti, P. 2003. Classification and filtering of laser data. *International Archives of Photogrammetry and Remote Sensing* 34(3W13).

Pfeifer, N., Oude Elberink, S. & Filin, S. 2005. Automatic tie element detection for laser scanner strip adjustment. *International Archives of Photogrammetry and Remote Sensing* XXXVI, 3/W19.

Pfeifer, N., Stadler, P. & Briese, C. 2001. Derivation of digital terrain models in the SCOP++ environment. *OEEPE workshop on airborne laserscanning and interferometric SAR for detailed digital elevation models, Stockholm, Sweden.*

Pingi, P., Fasano, A., Cignoni, P., Montani, C. & Scopigno, R. 2005. Exploiting the scanning sequence for automatic registration of large sets of range maps. *Computer Graphics Forum* 24(3): 517–526.

Rabbani, T., van den Heuvel, F. & Vosselman, G. 2007. An integrated approach for modelling and global registration of point clouds. *ISPRS Journal of Photogrammetry and Remote Sensing* 61(6): 355–370.

Ripperda, N. & Brenner, C. 2005. Marker-Free Registration of Terrestrial Laser Scans Using the Normal Distribution Transform, 3D-ARCH 2005. *International Archives of Photogrammetry and Remote Sensing* XXXVI, 5/W17.

Roth, G. 1999. Registering two overlapping range images. *Proceedings of the Second International Conference on 3-D Digital Imaging and Modeling*: 191–200.

Rusinkiewicz, S. & Levoy, M. 2001. Efficient variants of the ICP algorithm. *Proceedings of the Third Intl. Conf. on 3-D Digital Imaging and Modeling*: 142–152.

Salvi, J., Matabosch, C., Fofi, D. & Forest, J. 2007. A review of recent range image registration methods with accuracy evaluation. *Image and Vision Computing* 25(5): 578–596.

Schiewe, J. 2001. Ein regionen-basiertes Verfahren zur Extraktion der Geländeoberfläche aus Digitalen Oberflächen—Modellen. *Photogrammetrie, Fernerkundung, Geoinformation* 2: 81–90.

Schuhmacher, S. & Boehm, J. 2005. Geo-referencing of terrestrial laserscanner data for applications in architectural modeling 3D–ARCH 2005. *International Archives of Photogrammetry and Remote Sensing* XXXVI, 5/W17.

Sithole, G. 2001. Filtering of laser altimetry data using a slope adaptive filter. *International Archives of Photogrammetry and Remote Sensing* 34(3/W4): 203–210.

Sithole, G. 2005. Segmentation and Classification of Airborne Laser Scanner Data, Dissertation, TU Delft, ISBN 90 6132 292 8. *Publications on Geodesy of the Netherlands Commission of Geodesy* 59.

Sithole, G. & Vosselman, G. 2004. Experimental comparison of filter algorithms for bare-Earth extraction from airborne laser scanning point clouds. *ISPRS Journal of Photogrammetry and Remote Sensing* 59: 85–101.

Skaloud, J. 2006. Reliability of Direct Geo-referencing Phase 1: An overview of the current approaches and possibilities. *EuroSDR Official Publication* 51.

Skaloud, J. & Lichti, D. 2006. Rigorous approach to bore-sight self-calibration in airborne laeser scanning. *ISPRS Journal of Photogrammetry and Remote Sensing* 61(1): 47–59.

Sohn, G. & Dowman, I. 2002. Terrain surface reconstruction by the use of tetrahedron model with the MDL criterion. *International Archives of Photogrammetry and Remote Sensing* 34(3A): 336–344.

Thiel, K.-H. & Wehr, A. 1999. Calibration procedures of the imaging laser altimeter and data processing. *ISPRS joint workshop "Sensors and Mapping from Space". Hannover, Germany.*

Thirion, J. 1996. New feature points based on geometric invariants for 3-D image registration. *Int. J. Computer Vision* 18(2): 121–137.

von Hansen, W. & Voegtle, T. 1999. Extraktion der Geländeoberfläche aus flugzeuggetragenen Laserscanner-Aufnahmen. *Photogrammetrie, Fernerkundung, Geoinformation* 4: 229–236.

von Hansen, W. 2006. Robust Automatic Marker-free Registration of Terrestrial Scan Data. *International Archives of Photogrammetry and Remote Sensing* XXXVI, Part 3 (PCV06): 105–110.

Vosselman, G. 2000. Slope based filtering of laser altimetry data. *International Archives of Photogrammetry and Remote Sensing* 33, B3/2: 935–942.

Vosselman, G. 2002a. On the estimation of planimetric offsets in laser altimetry data. *International Archives of Photogrammetry and Remote Sensing* XXXIV, 3A: 375–380.

Vosselman, G. 2002b. Strip offset estimation using linear features. *3rd International Workshop on Mapping Geo–Surfical Processes using Laser Altimetry, Columbus, Ohio, USA: 9.*

Vosselman, G. & Maas, H.-G. 2001. Adjustment and filtering of raw laser altimetry data. *Proceedings of OEEPE Workshop on Airborne Laserscanning and Interferometric SAR for Detailed Digital Terrain Models, Stockholm, Sweden.*

Wagner, W., Ulrich, A., Ducic, V., Melzer, T. & Studnicka, N. 2006. Gaussian Decomposition and Calibration of a Novel Small-Footprint Full-Waveform Digitising Airborne Laser Scanner. *ISPRS Journal of Photogrammetry and Remote Sensing* 60(2): 100–112.

Wehr, A., Schiele, O. & Kleusberg, A. 2006. CW-Laser-Scanning: state of the art at INS and future multi spectral sensors. *ISPRS workshop "Lidar workshop: laser scanner for aerial and terrestrial applications", Istanbul, Turkey.*

Wendt, A. 2004. On the Automation of the Registration of Point Clouds Using the Metropolis Algorithm. *IAPRS* XXXV, Part 3: 106–111.

Wendt, A. 2007. A concept for feature based data registration by simultaneous consideration of laser scanner data and photogrammetric images. *ISPRS Journal of Photogrammetry and Remote Sensing* 62(2): 122–134.

Zhang, K., Chen, S., Whitman, D., Shyu, M., Yan, J. & Zhang, C. 2003. A progressive morphological filter for removing nonground measurements from airborne LIDAR data. *IEEE Transactions on Geoscience and Remote Sensing* 41(4): 872–882.

Advances in Photogrammetry, Remote Sensing and Spatial Information Sciences: 2008 ISPRS Congress Book – Li, Chen & Baltsavias (eds)

CHAPTER 14

Pan-sharpening for improved information extraction

Yun Zhang

ABSTRACT: This chapter provides a review of recent research publications on the incorporation of pan-sharpening techniques for improved information extraction published since 2004. A brief overview is also given on the development process of pan-sharpening techniques and early applications of pan-sharpening for improved information extraction to provide some context information. Research publications since 2004 indicated that the research focus has shifted from traditional fusion of SPOT Pan and Landsat TM or SPOT XS to the fusion of Ikonos, QuickBird and Landsat ETM+ Pan and MS images for improved information extraction. The studies demonstrated that a well pan-sharpened MS image could provide more information for improved information extraction, at lease for improved visual interpretation. From the evaluation results of individual pan-sharpening techniques, however, divided comments could be seen for the majority of fusion techniques depending on the data sets used and the applications involved, except for the PCI Pansharp fusion technique which received overall favourable observations in different studies. In terms of classification methods used, object-oriented classification was most frequently investigated and achieved more accurate classification results than Maximum Likelihood classification in majority of the studies. However, it was also observed that the current object-oriented classification is still not capable of effectively utilizing the information existing in high resolution MS or pan-sharpened MS images. It is found that much more room still exists for the improvement of information extraction techniques rather than the image fusion techniques. For the improvement of information extraction techniques, object-oriented approaches exhibit significantly more development potential than pixel-based approaches.

Keywords: Image fusion, pan-sharpening, information extraction, pixel-based classification, object-based classification, segmentation, evaluation

14.1 INTRODUCTION

For Earth observation, more than 320 satellite missions have either been launched or been planned (Earth Observation Resources 2008). For civilian applications, around 80 medium or high resolution satellites, with ground resolutions of better than 39 m, have been developed or are in the development process. Among the medium or high resolution satellites, more than 30 optical satellites are already in orbit and more than 25 other satellites are planned for launch before 2011 (Stoney 2006, Earth Observation Resources 2008). Table 14.1 gives an overview of the spatial and spectral characteristics of most optical satellites with a spatial resolution of equal to or better than 32 m.

From Table 14.1 it can be seen that most optical satellites (more than 70%) simultaneously collect both panchromatic (Pan) images with a higher spatial resolution and multispectral (MS) images with a lower spatial resolution. The other remaining optical satellites collect either high spatial resolution Pan images or high spectral resolution MS images but mostly with medium spatial resolution. The highest spatial resolution reaches 0.41 m by the Pan images of the potential GeoEys-1 satellite to be launched in 2008, and the highest spectral resolution of widely used satellite images reaches 14 spectral bands provided by the ASTER sensor on board the TERRA satellite.

For remote sensing applications, both high spatial resolution and high spectral resolution are often desired to achieve more detailed and more accurate information acquisition. Therefore, the need for effective fusion of Pan and MS images (also called pan-sharpening) is tremendous to improve information extraction. To meet this need, research on using pan-sharpening for improved information extraction has

Table 14.1. Spatial resolutions and spectral bands of current and future, medium and high resolution, optical satellite systems, edited according to Zhang & Kerle (2007) and Earth Observation Resources (2008).

	Spatial resolution (metres) and (# bands)						
		MS					
Optical satellite	PAN	VNIR	SWIR	MWIR	TIR	Swath (km)	Year launch
Landsat 5		30 (4)	30 (2)		120 (1)	185	1984
SPOT 2	10	20 (3)				60	1990
IRS-P2		36.4 (4)				74	1994
IRS-1C	5.8	23.5 (3)	70.5 (1)			70, 142	1995
IRS-1D	5.8	23.5 (3)	70.5 (1)			70, 142	1997
SPOT 4	10	20 (3)	20 (1)			60	1998
Landsat 7	15	30 (4)	30 (2)		60 (1)	185	1999
CBERS 1 and 2	20	20 (4)				113	1999, 2003
Ikonos 2	1	4 (4)				11	1999
Terra/ASTER		15 (3)	30 (6)		90 (5)	60	1999
KOMPSAT-1	6.6					17	1999
EROS A1	1.9					14	2000
MTI		5(4),20 (3)	20 (3)	20 (2)	20 (3)	12	2000
Quickbird 2	0.6	2.5 (4)				16	2001
SPOT 5	2.5–5	10 (3)	20 (1)			60	2002
IRS-P6 / ResourceSat-1	6	6 (3), 23.5 (3)				24, 70, 140	2003
DMC-AlSat1		32 (3)				600	2002
DMC-BILSAT-1	12	28 (4)				25, 55	2003
DMC-NigeriaSat 1		32 (3)				600	2003
UK-DMC		32 (3)				600	2003
OrbView-3	1	4 (4)				8	2003
ROCSat2 / FormoSat-2	2	8 (4)				24	2004
KOMPSAT-2	1	4 (4)				15	2004
DMC-Beijing-1	4	32 (3)				24, 600	2005
IRS-P5/CartoSat-1	2.5					30	2005
ALOS	2.5	10 (4)				35, 70	2005
RazakSat	2.5	5 (4)				20	2006
Resurs DK-1	1	3 (3)				28.3	2005
TopSat	2.5	5 (3)				25	2005
RapidEye A–E	6.5	6.5 (5)				78	2007
WorldView-1	0.5					17.5	2007
EROS B – C	0.7	2.8				16	2008
WorldView-2	0.5	1.8 (8)				16	2008
GeoEye-1	0.41	1.64 (4)				15	2008
Plèiades-1 and 2	0.7	2.8 (4)				20	2008, 2009
CBERS 3 and 4	5	20 (4), 40	40 (2)		80	60, 120	2008, 2011

been active for more than a decade (e.g. Munechika et al. 1993, Printz et al. 1997, Zhang 2001, Shaban & Dikshit 2002, Kiema 2002, Shaban & Dikshit 2002, Nichol & Wong 2005, Forsythe 2005, Colditz et al. 2006, and Lewinski 2006).

This chapter provides, through literature review, an overview of the image fusion techniques developed for Pan and MS fusion (pan-sharpening), research activities on incorporation of pan-sharpening for improved information extraction, and findings on the benefits and limitations of pan-sharpening for information extraction. The focus of the review is given to those research papers published since 2004.

14.2 PAN-SHARPENING TECHNIQUES

Since the SPOT satellite first provided 10 m Pan and 20 m MS images in 1986, much research has been done to develop effective image fusion techniques to sharpen MS images using Pan images. A comprehensive review of different early stage fusion methods can be found in Pohl & Van Genderen (1998), a general description of frequently used image fusion methods can be found in Zhang (2004), and a review of recently frequently discussed wavelet based image fusion techniques can be found in Amolins et al. (2007).

Pohl & Van Genderen (1998) divided the image fusion techniques into three processing levels, i.e. pixel, feature and decision based fusion levels.

- Pixel based fusion directly fuses individual pixels of input images to produce a fused product.
- Feature based fusion needs to extract features or segments of interest first, and then performs a fusion according to the characteristics of the features extracted.
- Decision based fusion requires the identification or categorization of individual features in individual input images through a classification method, and the features are then combined according to certain decision rules.

Comparing the three levels of fusion techniques, pixel based fusion needs the least intervention from the operator and has the best potential to keep the original information of the input images in the fused product. Therefore, this paper will just focus on pixel based fusion techniques for improved information extraction.

14.2.1 *Early stage pan-sharpening techniques*

In general, the pixel based image fusion techniques reviewed by Pohl & Van Genderen (1998) can be seen as early stage or traditional image fusion methods. Pohl & Van Genderen (1998) grouped the fusion methods into two groups: (1) colour related techniques, and (2) statistical/numerical methods. The first group comprised IHS (Intensity, Hue, Saturation) technique, HSV (Hue, Saturation, Value) technique, and variations. The second group included techniques like PCA, regression, arithmetic operations such as image differencing and ratios, and wavelet based techniques. During that period, IHS fusion was a predominant method in image fusion, whereas other methods were still in their early development stage, in which certain characteristics of mathematical functions were directly utilized to integrate the information from one image into another in a straightforward fashion. Most of the fusion methods could produce reasonable fusion results for images available at that time (mostly SPOT Pan, SPOT XS (MS) and Landsat TM (MS)). But certain colour distortions remained in the fused results and some even had visible artifacts. Empirical work was usually required in the fusion process to find out the best fusion solution by empirically selecting optimal data sets, adjusting individual bands, and/or choosing and testing different fusion methods.

14.2.2 *Frequently used pan-sharpening techniques*

According to the application popularity, Zhang (2004) grouped the traditional fusion methods into four groups: (1) IHS fusion and the variations, (2) PCA (Principal Components Analysis), (3) arithmetic combinations, and (4) wavelet based fusion.

- IHS fusion uses the IHS colour transform to convert a colour image from the RGB (Red, Green, Blue) colour space into the IHS colour space. The intensity (I) band is replaced by a high resolution Pan image in the IHS space. The fused image is obtained through a reverse IHS transform converting the Pan image together with the hue (H) and saturation (S) bands back to RGB space (Shettigara 1992).
- PCA fusion uses the same band substitution concept, but the first principal component of the PCA transform of the MS image is usually replaced by the high resolution Pan. Through a reverse PCA transform, the Pan image can be fused into the low resolution MS bands (Shettigara 1992).
- Arithmetic combination based fusion methods include Brovey Transform, SVR (Synthetic Variable Ratio), and RE (Ratio Enhancement) techniques. The basic procedure of the Brovey Transform is, first, multiplying each of the three MS bands by the high resolution Pan, and then dividing each of the three Pan-MS products by the sum of the three MS bands. The SVR and RE techniques are similar, but involving more sophisticated calculations for the MS sum for better fusion quality (Zhang 1999).
- Wavelet based fusion uses a wavelet transform to decompose a high resolution Pan image into a low resolution Pan image with corresponding high resolution features. The low resolution Pan is then replaced by each band of the low resolution MS image, respectively. The wavelet fused MS image is formed through a reverse wavelet transform converting each of the low resolution MS bands together with the high resolution features from the Pan back to high resolution space, i.e. injecting the high resolution features into each of the MS bands respectively (Yocky 1996, Garguet-Duport et al. 1996, Ranchin & Wald 2000).

14.2.3 *New pan-sharpening techniques*

Because the traditional or early stage image fusion methods experienced significant colour distortions in the fusion of the Pan and MS images of the new satellites launched in and after 1999 (such as Ikonos, Landsat 7, and QuickBird), Zhang (2004) introduced a new, fully automatic image fusion technique developed at the University of New Brunswick (UNB), Canada—the UNB Pansharp (also called PCI Pansharp as it was incorporated into PCI Geomatica).

- The UNB Pansharp (or PCI Pansharp) uses the least squares technique to best adjust the grey values of the fused images according to those of the original Pan and MS images. It also utilizes a set of statistic approaches to estimate the grey value relationship between all of the input Pan and MS bands to automate the fusion process. The new fusion technique solved two major problems in image fusion—colour distortion and operator dependency (Zhang 2004).

To overcome the significant colour distortions in the fusion of the new satellite images, many other image fusion methods have also been developed in recent years, among which most methods involve a wavelet transform in the fusion process. Amolins et al. (2007) provided a review and comparison of the wavelet based fusion approaches. They categorized the wavelet based fusion methods into three types of schemes according to the processing characteristics, i.e. (1) basic schemes (containing early stage or traditional techniques), (2) hybrid schemes, and (3) advanced schemes.

- Basic schemes apply the wavelet transform directly to the Pan or both Pan and MS images using a decimated algorithm and use a simple method, such as substitution or addition, to inject spatial detail. Representative fusion techniques are, for example, those introduced by Garguet-Duport et al. (1996), Yocky (1996), Ranchin & Wald (2000) and Li et al. (2002). They belong to early stage pan-sharpening techniques mainly dealing with SPOT Pan and Landsat TM (or SPOT XS) fusions.
- Hybrid schemes apply either an IHS or PCA transform to the MS image, and then apply the wavelet transform to the I or PC1 component and the Pan image. Spatial detail is injected using a simple method. Representative techniques are those reported by Nunez et al. (1999), Ming & Shunjun (2003), Gonzalez-Audicana et al. (2004), Zhang & Hong (2005), and Gonzalez-Audicana et al. (2005). These schemes were mainly developed to deal with colour distortion problems. The majority of them were tested with the new satellite images.
- Advanced schemes apply the wavelet transform directly to both Pan and MS images but use a more sophisticated method to inject spatial detail. These methods may involve thresholding based on different characteristics, such as average gradient or local standard deviation, or may require additional information. Often, advanced schemes use a non-separated algorithm. Wavelet fusion techniques introduced by Aiazzi et al. (2002), Garzelli & Nencini (2005), Wu et al. (2005) and Otazu et al. (2005) belong to this category. They were also developed to deal with colour distortion problems, mainly tested with the new satellite images.

14.3 PAN-SHARPENING FOR INFORMATION EXTRACTION

14.3.1 *Early stage applications*

Early applications of image fusion techniques for improved information extraction mainly involved IHS, PCA and arithmetic combination techniques for fusing optical Pan and MS images, mostly dealing with SPOT Pan and Landsat TM or SPOT SX images. For example, Munechika et al. (1993) employed a RE (Ratio Enhancement) fusion technique to fuse SPOT Pan and Landsat TM for improving land cover classification; Zhang (2001) used SVR (Synthetic Variable Ratio) fusion to fuse SPOT Pan and Landsat TM images for building extraction; Kiema (2002) applied an IHS technique to SPOT Pan and Landsat TM fusion for topographic object extraction; and Shaban & Dikshit (2002) employed HPF (High Pass Filter) method to fuse SPOT Pan and MS images for urban objects classification.

Lau et al. (2000) investigated the influence of different traditional image fusion techniques on the accuracy of information extraction. IHS, PCA and HPF fusion techniques were employed to fuse SPOT Pan and MS images. Unsupervised classification based on ISODATA clustering and supervised Maximum Likelihood Classification (MLC) were performed to classify the original SPOT MS images and pan-sharpened SPOT MS images. Accuracy analysis demonstrated that image fusion could bring a certain degree of improvement in classification accuracy.

14.3.2 *Recent applications*

In recent years (after 2004), the use of traditional image fusion techniques, such as IHS, PCA and Brovey, can still be seen in some publications for classification accuracy improvements. However, for information extraction from the new satellite images, such as Landsat 7, Ikonos and QuickBird, the PCI Pansharp fusion technique has been increasingly used. The use of wavelet based fusion techniques is also increasing, because traditional image fusion techniques cannot effective preserve the spectral information in the fusion of Pan and MS images of the new satellites.

An overview of different applications found in the literature review is given below, which is sorted according to the fusion techniques used. The application area of each study is given in the corresponding subtitle.

14.3.2.1 *Use of traditional fusion methods*

In recent research publications, traditional image fusion techniques can still been seen, but the application direction has been shifted to the attempts to fuse the Pan and MS images of Landsat ETM+, Ikonos, and QuickBird for improved information extraction. The applications or investigations are for example:

14.3.2.1.1 Influences on classification accuracies
Colditz et al. (2006) investigated the influences of several fusion techniques on classification accuracy of Landsat ETM+ images. The image fusion techniques PCA, Brovey, HSV (Hue, Saturation, Value) (Mohanty & Majumdar 2003), MMT (Multisensor Multiresolution Technique) (Zhukov et al. 1999), AIF (Adaptive Image Fusion) (Smith 1996) and Wavelet (ERDAS Imagine), (King & Wang 2001) were used to fuse the ETM+ Pan and MS images (Figure 14.1). Maximum Likelihood classifier, object-based classification and support vector machines were employed to classify the pan-sharpened ETM+ images (Figure 14.2). It was found that every fusion approach distorted spectral signatures at least slightly, limiting the application potential of those classification techniques, which rely on the absolute spectral values. However, among all of the classification results, the AIF fusion approach showed the best with a major improvement compared to the reference, especially along object edges. Acceptable results were achieved by Wavelet, MMT, and PCA, whereas Brovey and HSV performed poorly and could not be recommended for classification of fused imagery. In terms of classifications techniques, Maximum Likelihood classification achieved the best classification accuracy with an average overall accuracy of 80% for most fusion methods, whereas the average overall accuracies of object-based classification and support vector machines were 75%.

14.3.2.1.2 Change detection
Shah et al. (2007) evaluated several traditional image fusion techniques for the fusion of Ikonos Pan and MS images for change detection. The fusion techniques tested were PCA, HPF, three different ratio based fusions from Pradines (1986), Price (1987) and Shen et al. (1994), Wavelet (ARSIS) (Ranchin & Wald, 2000), SVR (Synthetic Variable Ratio) (Zhang 1999), and PCA-Wavelet (Winter & Winter 2002). Maximum Likelihood classifier was applied to the classification of the pan-sharpened Ikonos images into four classes (roofs, roads, low vegetation and trees). According to classification accuracy assessments, the authors found that only PCA, SVR and HPF achieved a higher overall accuracy than that of the reference image (original Ikonos MS). The maximum overall accuracy obtained from PCA fused Ikonos image reached 94%.

14.3.2.2 *Use of PCI pansharp*
For improving the interpretation capacity and classification accuracy of the new satellite images from Landsat 7, Ikonos and QuickBird, PCI Pansharp

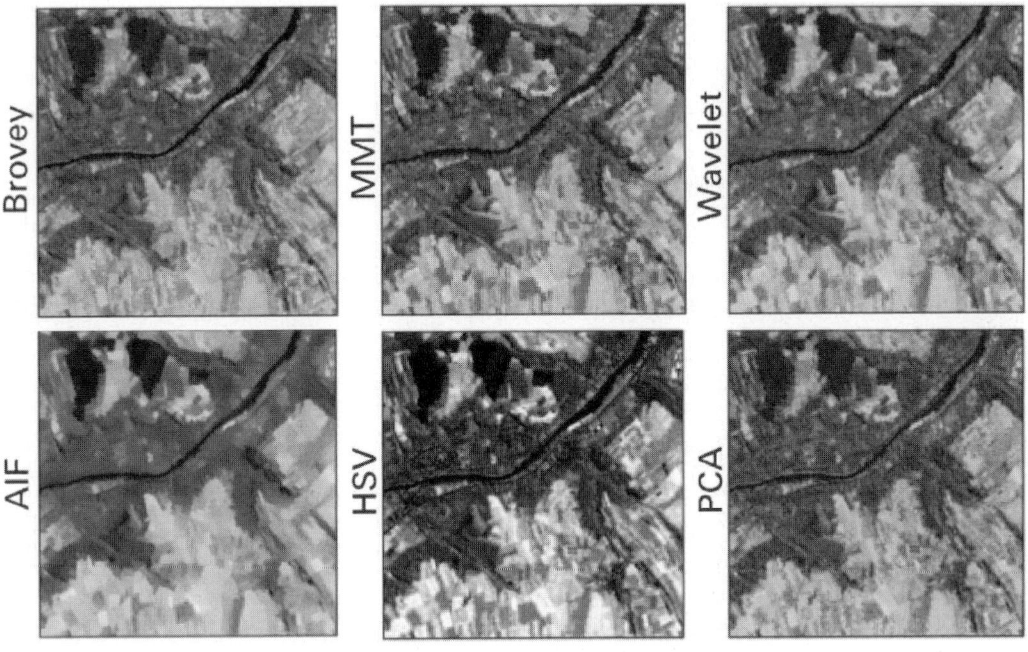

Figure 14.1. Landsat ETM+ Pan-MS fusion results of different fusion approaches. (Adopted from Colditz et al. 2006, rotated). (see colour plate page 496)

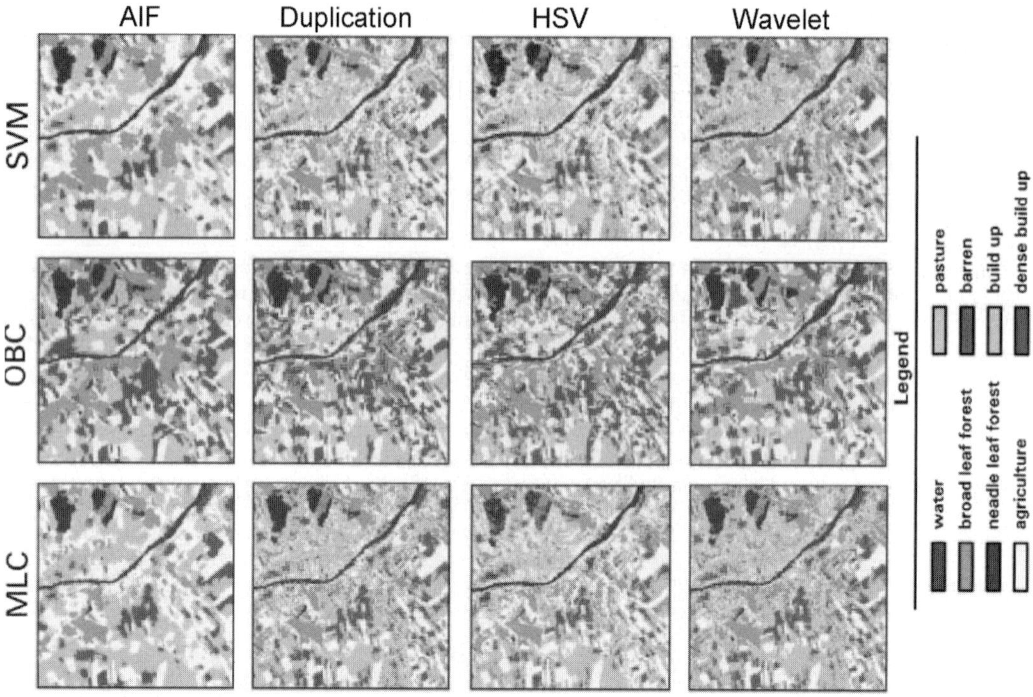

Figure 14.2. Selected classification results from Landsat ETM+ Pan-MS fusions for maximum likelihood classifier (MLC), object based approach (OBC), and support vector machine (SVM) classifications. Brovey, MMT and PCA were not shown due to similarity to other approaches (adopted from Colditz et al. 2006, rotated). (Note: "Duplication" was the original MS image but with duplicated pixels resampled from 30 m to 15 m using a nearest neighbour method). (see colour plate page 496)

(or UNB Pansharp) (Zhang 2002, Zhang 2004) has been increasingly used, because the IHS, PCA and arithmetic combination fusion techniques have encountered significant colour distortions (Zhang 2002, Nichol & Wong 2005, Forsythe 2005). Some application examples involving PCI Pansharp are given below.

14.3.2.2.1 Landslide monitoring

To effectively monitor landslide changes, Nichol & Wong (2005) introduced a multi-resolution approach to identify multi-scale landslide changes. SPOT MS images were classified using Maximum Likelihood classification to identify large scale landslides. Pan-sharpened Ikonos MS images were used to identify small landslides. To find out the best Ikonos fusion results for the monitoring, IHS, Brovey, SFIM (Smoothing filter-based modulation) (Liu 2000) and PCI Pansharp (Zhang 2002) were used to fuse the Ikonos Pan and MS images. According to the fusion results comparison, the pan-sharpened Ikonos image using PCI Geomatica software was judged to be the most satisfactory, reaching the quality comparable to an aerial ortho-photograph of 1:10,000 scale (Figure 14.3). It was found that PCI Pansharp enabled detailed interpretation of landslides and associated environmental features.

14.3.2.2.2 Road extraction

To overcome the limitations of using stand-alone MS images or Panchromatic images for road extraction, Zhang et al. (2004) introduced a new approach, which integrates image fusion, multispectraland feature extraction into the road extraction. QuickBird Pan and MS images (Figure 14.4.a and 14.4.b) were fused using PCI Pansharp. Unsupervised fuzzy K mean clustering method was used to classify the pan-sharpened QuickBird image (Figure 14.4.c), obtaining a binary road image (Figure 14.5.a). Edge detection was performed to extract road edges (Figure 14.5.b) from the original Pan image (Figure 14.4.a). After an edge-aided segmentation and a shape-based classification, the final road network was extracted (Figure 14.5c). For accuracy assessment of the road extraction, road completeness and correctness (Heipke et al. 1997) were used. The completeness value reached 0.94, and the correctness value was 0.98. The results demonstrated that the proposed method achieved significantly higher accuracy than those of Pan based feature extraction (e.g. Hinz et al. 2001), multispectral classification (e.g. Shackelford & Divis 2003), and MS and Pan integrated classification (e.g. Granzow 2001).

14.3.2.2.3 Urban sprawl monitoring

For urban expansion monitoring, Forsythe (2005) used PCI Pansharp to fuse Landsat ETM+ Pan and MS images collected in different years. It was proved that the Landsat-7 Pan and MS data can be enhanced through data fusion to provide 15 m spatial resolution multispectral data. A combined unsupervised classification and image differencing change detection process was conducted, which allowed for the monitoring of urban development. The results

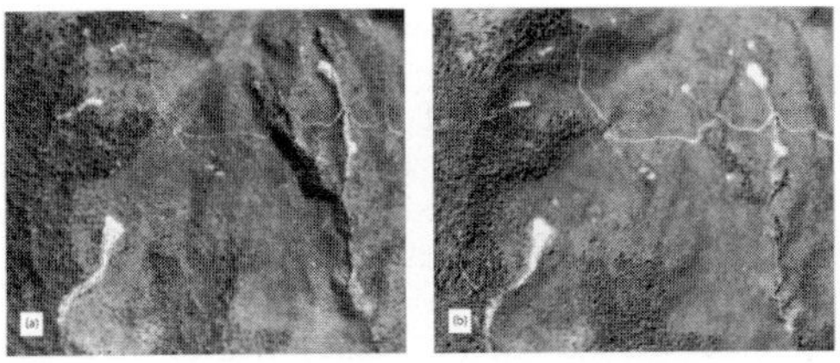

Figure 14.3. Comparison between aerial photograph and pan-sharpened Ikonos MS image showing landslide scars and patches of disturbed ground. (a): Aerial ortho-photograph (1:10 000); (b): Pan-sharpened Ikonos image (PCI Pansharp). (Adopted from Nichol & Wong 2005, subset). (see colour plate page 497)

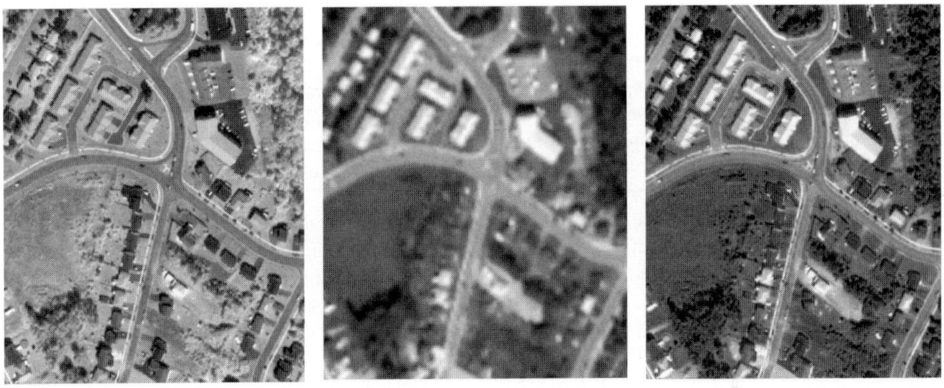

(a) (b) (c)

Figure 14.4. Fusion of QuickBird Pan and MS images for road extraction. (a) Original Pan image; (b) Original MS image; (c) Pan-sharpened MS image using UNB Pansharp (Adopted from Zhang et al. 2004, subsets). (see colour plate page 497)

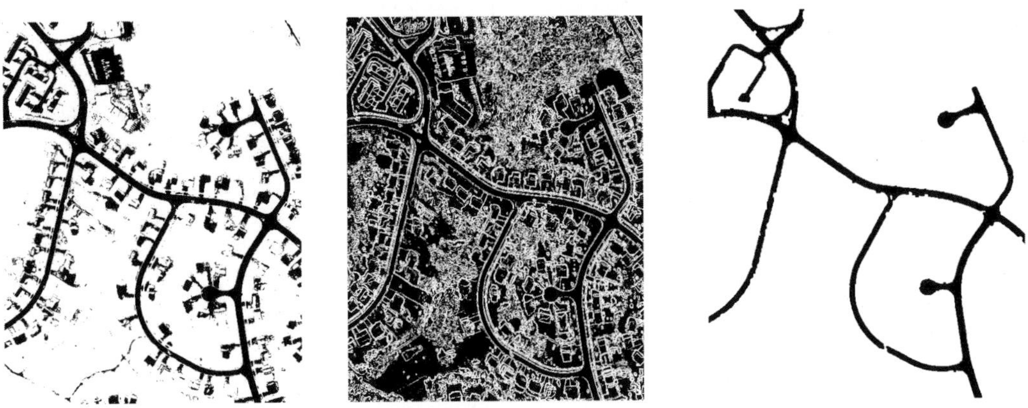

(a) (b) (c)

Figure 14.5. Intermediate and final road extraction results of the proposed approach. (a) Road networks classified from pan-sharpened QuickBird MS image using fuzzy k mean clustering method; (b) Road edges detected from original Pan image using Robert edge detector; (c) Final road network extracted after edge-aided classification. (Adopted from Zhang et al. 2004).

demonstrated that the pan-sharpened ETM+ images enabled the discrimination of finer change detail than the original ETM+ MS images did.

14.3.2.2.4 Object-oriented classification

Lewinski (2006) applied PCI Pansharp to Landsat ETM+ Pan and MS fusion for object-oriented classification. eCognition software was used for object oriented classification. The segmentation rules were specified based on the original and fused ETM+ data sets. A total of 18 land-use classes were defined and classified. The results demonstrated that the pan-sharpened ETM+ MS and original ETM+ Pan data did not have significant influence on the improvement of land-use class separation in the classification process; however, they were useful for image segmentation process.

14.3.2.2.5 Supervised segmentation

To improve the object segmentation accuracy and reduce the tedious work in iterative selection of segmentation parameters (Scale, Shape and Smoothness) as suggested by eCognition, Maxwell & Zhang (2005) developed a Fuzzy-based Supervised Segmentation (FSS) approach to automatic selection of optimum segmentation parameters for eCognition. PCI Pansharp was used to fuse the Pan and MS images to improve the segmentation detail.

The FSS approach was tested using pan-sharpened QuickBird MS images, which contained the steps as illustrated in Figure 14.6: (a) selection of objects of interest in pan-sharpened imagery; (b) initial segmentation using eCognition to obtain sub-segments; (c) selection of sub-segments to train the FSS system; (d) formation of the desired object segment model using the outer boundaries of the selected sub-segments; (e) convergence test using segmentation result from the first iteration of the FSS process; and (f) convergence test using the segmentation result of the third FSS iteration.

As the estimated parameters from the third iteration (Figure 14.6f) satisfied the desired segmentation, they were accepted as optimum segmentation parameters and input into eCognition to segment the entire image, achieving ideal segmentation results over the entire image in a very fast fashion. eCognition's trial-and-error segmentation approach was also conducted (Figure 14.7b and 14.7c) to compare with those of the FSS approach (Figure 14.6f and 14.7a). A noticeable accuracy improvement was observed and the segmentation efficiency was significantly increased (from hours or days to minutes or tens of minutes).

14.3.2.3 *Use of wavelet fusion*

14.3.2.3.1 Snow mapping

Sirguey et al. (2007) implemented the ARSIS wavelet fusion concept (Ranchin & Wald 2000) to fuse the two MODIS 250 m spectral bands into its five 500 m bands to test the influence of the band fusion on the accuracy improvement of snow mapping in mountainous environments. Snow maps derived from 15 m ASTER imagery, collected from the same platform (TERRA) as the MODIS imagery, were used as reference to compare with the snow maps derived from the fused MODIS bands. Various metrics were used to assess the benefit of the proposed method for snow mapping. It was found that the enhanced snow map exhibited great benefit for environmental and hydrological applications in steep terrain.

14.3.2.3.2 Urban land cover mapping

Boloorani et al. (2006) examined the influence of wavelet fusion of Ikonos Pan and MS images on urban land cover mapping using pixel-based and object-based classification approaches. The DWT (Discrete Wavelet Transform) method provided by ERDAS 8.7 was used to fuse Ikonos Pan and MS. Due to spectral artefacts in the fused data, a low-pass filtering was performed to reduce the artefacts, but the sharpness of the fused data was also reduced (Figure 14.8). Object-oriented classification of eCogni-tion and Maximum Likelihood classification were carried out to classify both the original Ikonos MS image and the wavelet fused Ikonos MS image. Accuracy assessment showed a general classification accuracy increase with the fused data, compared to that of the original data. And, the object-oriented classification achieved a more accurate classification result than the traditional Maximum Likelihood classification.

14.3.2.3.3 Land cover class interpretation

Balcik & Sertel (2007) examined the potential of wavelet fusion of Landsat ETM+ Pan and MS images for spatial and spectral enhancement of different land cover classes. The Wavelet Resolution Merge tool of ERDAS Imagine 8.7 was used to fuse the ETM+ Pan and MS images containing different land cover classes. Visual interpretation, spectral profiles and correlation coefficients were used to evaluate the fused ETM+ results. Analysis results demonstrated that the wavelet based image fusion had the ability to improve the spatial resolution with minimal spectral distortion compared to the original MS image. The wavelet fused EMT+ image presented better spatial details than the original ETM+ MS image. However, the spatial detail of the wavelet fused images was not as good as that of the original Pan images (Figure 14.9).

14.3.2.4 *Use of other fusion methods*

14.3.2.4.1 Object-oriented classification

Ehlers et al. (2006) investigated fusion techniques for segment based image analysis. The Ehlers fusion

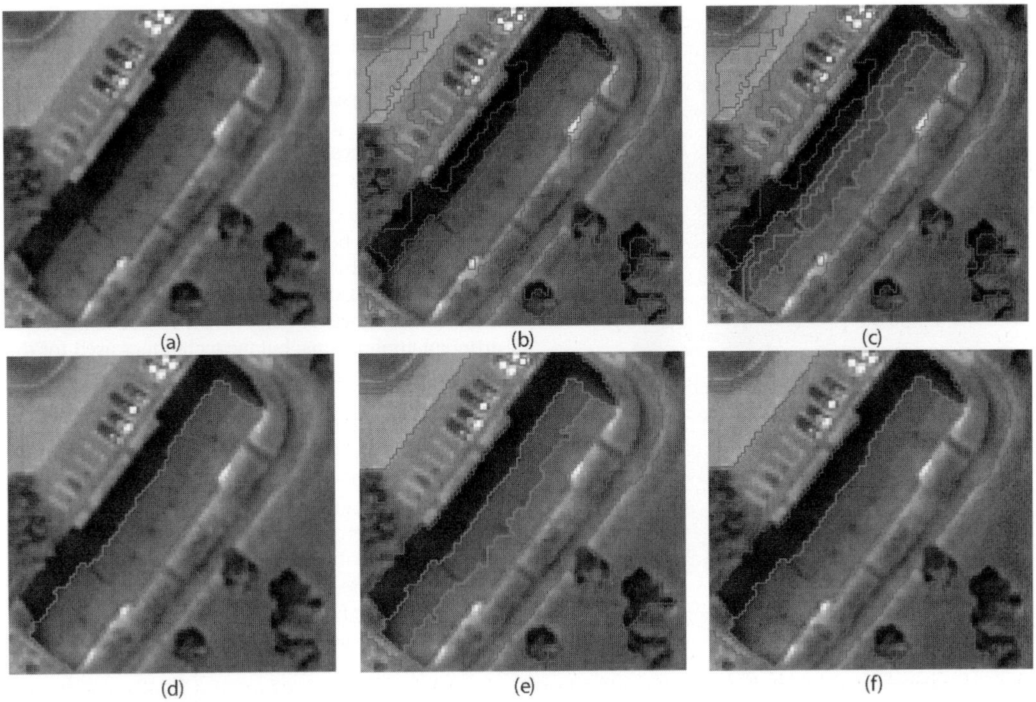

Figure 14.6. Process of the proposed FSS approach and the corresponding segmentation results in pan-sharpened QuickBird image (PCI Pansharp). (a) Object of interest in a training area; (b) Initial segmentation with sub-segments of an object; (c) Sub-segments selected for training the Fuzzy System; (d) Desired segment model formed by the selected sub-segments; (e) Segmentation result using the estimated segmentation parameters from the first iteration; (f) Segmentation result using the estimated, segmentation parameters from the final iteration (the 3rd iteration). (Adopted from Maxwell & Zhang 2005). (see colour plate page 498)

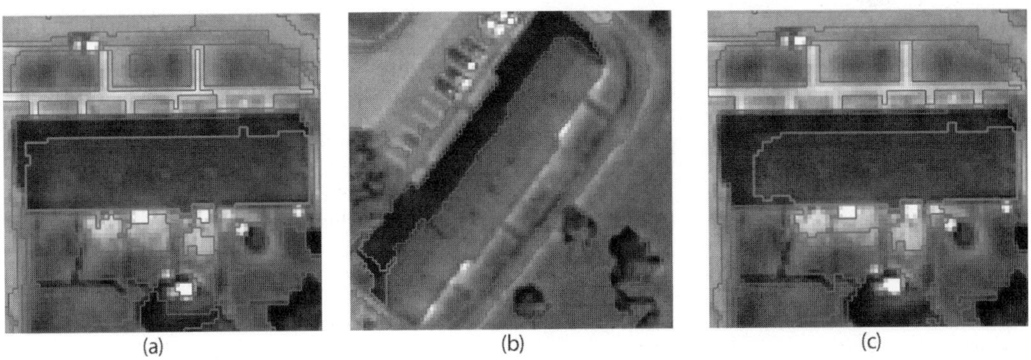

Figure 14.7. Comparison of segmentation results from the proposed FSS approach (a) and eCognition's trial-and-error approach (b) and (c) in pan-sharpened QuickBird image (the segmentation result of the FSS approach for the building in (b) can be found in Figure 14.6f. (Adopted from Maxwell & Zhang 2005). (see colour plate page 498)

technique, which involved an IHS transform and a Fast Fourier Transform (FFT), was used to fuse SPOT 5 Pan (5 m) (Figure 14.10a) and Landsat ETM+ MS (Figure 14.10b) resulting in SPOT-ETM fusion result (Figure 14.10c). The SPOT-ETM image was then classified using eCognition's object-oriented approach (Figure 14.10d). The authors concluded that the Ehlers fusion showed much better results than all other standard and newly proposed fusion techniques However, the benefits of the Ehlers

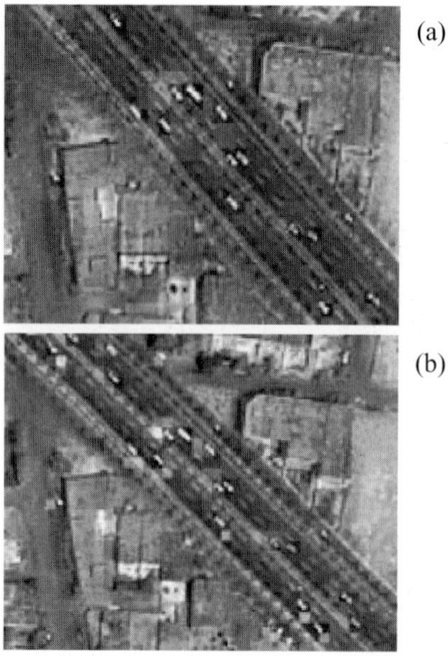

Figure 14.8. The wavelet fused Ikonos image using ERDAS Imagine 8.7 software. (a) Low-pass filtered fused data to remove artifacts; (b) Fused data without low-pass filtering (adopted from Boloorani et al. 2006). (see colour plate page 498)

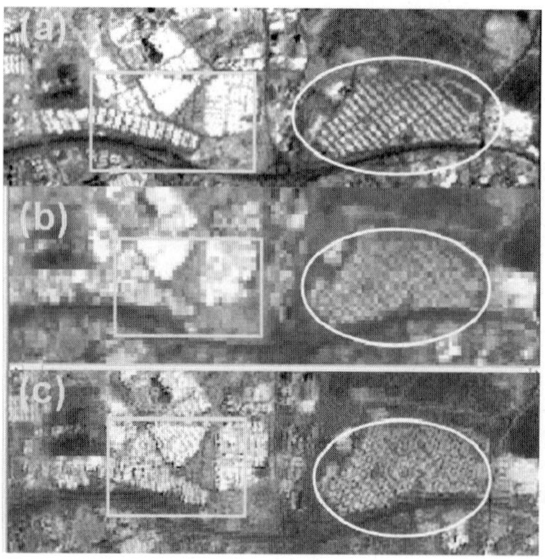

Figure 14.9. Landsat ETM+ Pan and MS fusion using the Wavelet Resolution Merge tool of ERDAS Imagine. (a) Original Pan image; (b) Original MS image; (c) Pan and MS fused image. (Adopted from Balcik & Sertel 2007). (see colour plate page 499)

fusion did not automatically lead to enhanced classification results using segment based classification approach. Therefore, the authors introduced a 'cookie cutter' approach for the integration of high resolution RGB and low resolution hyperspectral image data, and a decision based fusion technique to combine high resolution Pan data with MS images for improved classification. Ehlers (2007) presented similar results, but with the fusion of Ikonos Pan and SPOT MS images.

14.3.2.5 *Fusion methods unspecified*

Several further publications were found in the literature review which aimed at using image fusion to improve information extraction. The authors evaluated the influence of pan-sharpened MS images for different applications, but the techniques used for the fusion were not specified. They are for example:

14.3.2.5.1 Fire fuels mapping

To support fire fuels mapping, Stefanacci (2006) introduced the development of advanced digital mapping methodologies. The project attempted to identify whether automated stand delineation and fuels mapping could be performed consistently using a readily available data source at reasonable price. Pan-sharpened Landsat 7 and QuickBird images (fusion method unspecified) were used for the mapping. Image classification using eCognition proved that the pan-sharpened images were a cost-effective source for accurate inventories of fire fuels and associated risks.

14.3.2.5.2 Land cover classification in mountainous areas

Bishop & Congalton (2006) used pan-sharpened Landsat EMT+ images (fusion method unspecified) to evaluate the effect of terrain normalization on classification accuracy. ISODATA based unsupervised classification, maximum likelihood based supervised classification, and hybrid classifications were performed to classify the pan-sharpened ETM+ images with and without terrain flattening. Seven land cover classes were classified. Moderate classification accuracies were achieved for both images with and without terrain flattening. It was found that the topographic flattening did not significantly improve map accuracy.

14.3.2.5.3 Tree species identification

Hájek (2006) conducted an in-depth analysis of Ikonos MS and pan-sharpened Ikonos MS for knowledge-based classification (eCognition). After certain pre-processing, the object-oriented classification was firstly tested on the 4 m Ikonos MS and then applied to the 1-meter pan-sharpened Ikonos imagery (fusion method unspecified). The classification accuracy was assessed according to GIS field reference data. The accuracy statistics were calculated from the assembled error matrix. The results of the 4 m and 1 m Ikonos data were

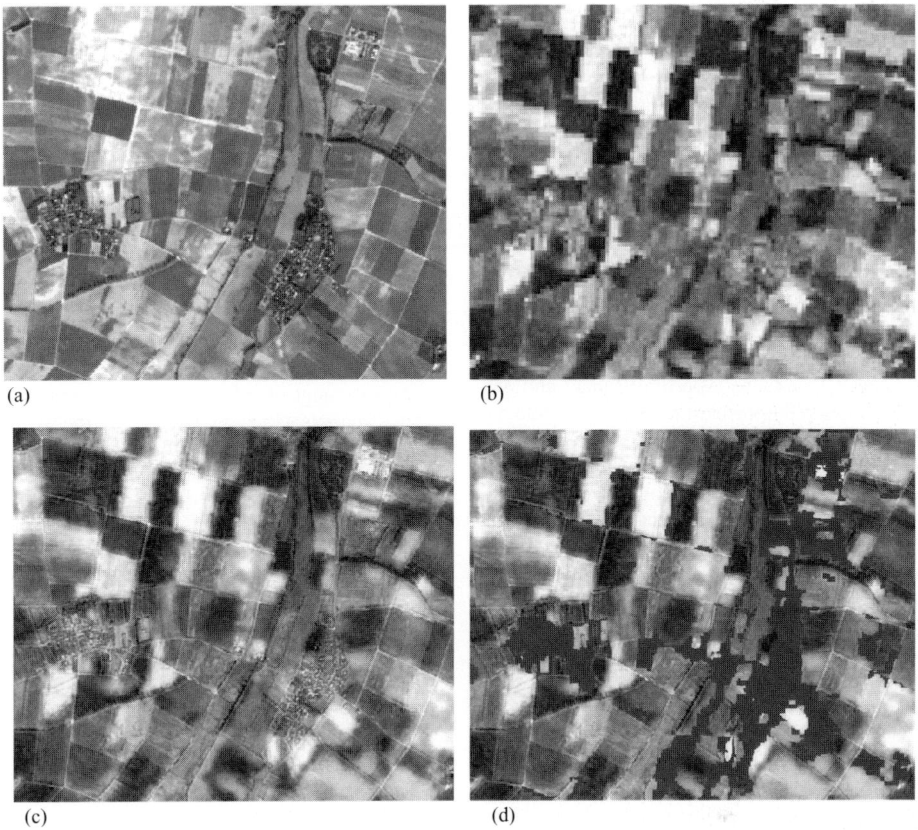

Figure 14.10. SPOT-5 Pan and Landsat ETM+ MS image fusion and object-oriented classification. (a) Original SPOT-5 Pan image, 5 m pixel size; (b) Landsat ETM+ image (bands 1, 2, and 3) co-registered to the SPOT-5 image and resampled to 5 m pixel size; (c) SPOT-ETM fused image using IHS-FFT integrated (Ehlers) fusion approach; (d) Result after a subsequent segmentation and classification using eCognition (settlement in red color) (adopted from Ehlers et al. 2006, subsets). (see colour plate page 499)

compared by means of Overall Accuracies and Kappa Index of Agreement. The developed rules based on fuzzy membership functions were found highly convertible and suitable for semi-automated application. Both data types (Ikonos MS and pan-sharpened Ikonos MS) had their benefits and should be utilized for different forest management tasks. It was found that for the purpose of tree species identification very good results could be achieved by the combination of object-oriented approach and the topo-corrected Ikonos imagery.

14.3.2.5.4 Crop classification

To assess the classification accuracies of different spatial resolutions for crop classification, Ozdarici & Turker (2006) classified a series of multi-resolution images, consisting of SPOT4 XS (20 m), SPOT5 XS (20 m), Ikonos MS (4 m), QuickBird MS (2.44 m), and pan-sharpaned QuickBird MS (0.61 m, fusion method unspecified) covering an agricultural area, using a field-based classification technique. For conducting the field-based (pre-field) classification, field boundaries were first digitized. The statistical measures (mean, median, and mode values) for each field were calculated from the selected common spectral bands (G, R, and NIR) of each image, forming a set of new mean, median and mode images in which each field contains an unique mean, median or mode value. A Maximum Likelihood classification (MLC) was then carried out to classify the new mean, median and mode images for six different crop types. After calculating the classification accuracies, it was observed that similar results were obtained for the images with different spatial resolutions, when the mean images were classified. However, the pan-sharpened QuickBird 0.61 m image provided the highest overall accuracy, 85.2%, when the median images were classified. On the other hand, the lowest overall accuracy was found to be 42.9%, for the classified median image derived from the SPOT4 XS.

14.4 DISCUSSION AND CONCLUSION

With the focus on recent research in image fusion for improved information extraction, papers published in remote sensing journals and conferences since 2004 were reviewed. Due to page limit, applications on the fusion of SAR (Synthetic Aperture Radar) and MS images for improved information extraction could not be included. Many research papers on the development of new image fusion techniques (mostly wavelet based fusions) have also been published since 2004. But they are out of the scope of this chapter, so that they were not included either. Therefore, this chapter manly focused on the use of pan-sharpening (fusion of Pan and MS) techniques for improved information extraction. To give some context information, image fusion techniques and their early stage applications were also briefly introduced.

The intention of this chapter was to include as many related papers as possible. Due to access limitation to some conference proceedings, some related papers may not have been found and included; however, the papers reviewed in this chapter can still provide general information on the development trend, advantages and limitations of pan-sharpening techniques for information extraction, and the capacity of currently frequently used information extraction techniques.

Readers can also gain useful information on advantages and limitations of individual pan-sharpening techniques for information extraction through the evaluation comments of individual papers and/or direct comparison of the fusion results presented by individual papers.

Because different information extraction techniques were used to extract information from different images for different applications, readers can also gain information on the suitability of individual classification techniques for information extraction from images of different spatial resolutions. To facilitate the review and comparison, the evaluation comments on fusion techniques and classification techniques by individual authors are summarized in Table 14.2, together with the corresponding application purpose and data used.

From Table 14.2 we can see that different pan-sharpening techniques were applied to different images for different applications. Therefore, it is not surprising that different evaluation results were observed. However, some general development trend and common agreements between the authors can still be found when we take a closer look at individual applications and compare between the papers.

14.4.1 *Influence of pan-sharpening techniques*

In terms of the influence of pan-sharpening techniques on information extraction, the findings can be summarized as follows:

- Most of the authors found that pan-sharpening techniques did bring noticeable benefit to the improvement of information extraction, including those papers which did not specify the fusion techniques used.
- In the evaluation of individual fusion techniques, some evaluation discrepancies exist, but PCI Pansharp received overall favourable observations.
  - For traditional fusion techniques, it was observed by one paper that AIF, Wavelet, MMT and PCA techniques could improve classification accuracy for ETM+ Pan and MS fusion (Figure 14.1); whereas another paper found that PCA, SVR and HPF performed better for improving classification accuracy of Ikonos Pan and MS fusion. Wavelet fusion was found to be good for ETM+ fusion, but not acceptable for Ikonos fusion. Brovey and HSV were evaluated by one study and could not be recommended for classification.
  - For PCI Pansharp, all of five papers observed favourable fusion results for information extraction, which involved the fusion of Ikonos, QuickBird and ETM+ Pan and MS images (Figs. 14.3, 14.4 14.6 and 14.7). One paper compared the fusion of Ikonos Pan and MS images using PCI Pansharp, SFIM, IHS and Brovey, and found that PCI Pansharp produced the most satisfactory fusion results comparable to 1:10,000 aerial photography.
  - For wavelet fusion techniques, it was found that the different wavelet fusion methods evaluated had the ability to improve classification accuracy. But, one paper reported obvious artefacts in the pan-sharpened Ikonos results (Figure 14.8), and another paper observed the visible loss of spatial information in the pan-sharpened ETM+ results compared to original ETM+ Pan (Figure 14.9).
  - For other fusion methods, the Ehlers fusion was introduced and used to fuse SPOT-5 Pan with ETM+ MS and other images. The authors stated that their fusion results (Figure 14.10) showed much better than all other fusion methods.

Table 14.2. Image fusion techniques and classification techniques used for different information exactions and the evaluation comments by individual authors.

Traditional fusion methods

Paper	Application	Image used	Fusion method	Classification method	Evaluation comments	
					Fusion	Classification
Colditz et al. (2006)	Influences on classification accuracies	Landsat ETM+	AIF, Wavelet (ERDAS), MMT, PCA, Brovey, HSV	Maximum Likelihood, Object-based, Support vector	– Every fusion approach distorted spectral signatures at least slightly, limiting the application of those classification techniques which rely on absolute spectral values – AIF showed the best with a major classification improvement compared to that from original MS – Wavelet, MMT and PCA achieved acceptable results – Brovey and HSV cannot be recommended for classification	– Maximum Likelihood classification achieved the best overall accuracy, on average 80% – Object-based classification and support vector machines reached the average overall accuracies of 75%
Shah et al. (2007)	Change detection	Ikonos	PCA, SVR, HPF, Pradines-Ratio, Price-Ratio, Shen-Ratio, ARSIS-Wavelet	Maximum Likelihood	– PCA, SVR and HPF achieved a higher overall accuracy than that of the original Ikonos MS – PCA achieved the maximum overall accuracy, 94% – Other fusions achieved lower accuracy that the original MS	

PCI Pansharp

Paper	Application	Image used	Fusion method	Classification method	Evaluation comments	
					Fusion	Classification
Nichol &ww Wong (2005)	Landslide monitoring	Ikonos	PCI Pansharp, SFIM, IHS, Brovey	Maximum Likelihood, + Interpretation	– PCI Pansharp was judged to be the most satisfactory, reaching the quality of an aerial photograph of 1:10000 scale – PCI Pansharp enabled detailed interpretation of landslides and associated environmental features	

(Continued)

Table 14.2. (*Continued*)

					Evaluation comments	
Paper	Application	Image used	Fusion method	Classification method	Fusion	Classification
Zhang et al. (2004)	Road extraction	QuickBird	PCI Pansharp	Fuzzy K mean + Post classification	– The proposed method with PCI Pansharp achieved significantly higher accuracy than those of Pan based feature extraction, MS classification, and MS and Pan integrated classification	
Forsythe (2005)	Urban sprawl monitoring	Landsat ETM+	PCI Pansharp	Unsupervised classification	– The fused image enabled the discrimination of finer change detail than the original ETM+ MS images	
Lewinski (2006)	Object-oriented classification	Landsat ETM+	PCI Pansharp	Object-oriented (eCognition)	– The fusion was useful for image segmentation process – The fusion did not have significant influence on the improvement of land-use class separation in the classification process.	
Maxwell & Zhang (2005)	Supervised segmentation	QuickBird	PCI Pansharp	Supervised Segmentation + eCognition	– The fused image allowed to extract detailed object outlines	– With the supervised segmentation, a noticeable accuracy improvement was observed and the segmentation efficiency was significantly increased

Wavelet fusion

					Evaluation comments	
Paper	Application	Image used	Fusion method	Classification method	Fusion	Classification
Sirguey et al. (2007)	Snow mapping	MODIS	ARSIS wavelet		– The enhanced snow map exhibited great benefit for environmental and hydrological applications in steep terrain.	

Boloorani et al. (2006)	Urban land cover mapping	Ikonos	Wavelet (ERDAS)	Object-oriented (eCognition), Maximum Likelihood	– The fused data showed a general classification accuracy increase	– Object-oriented classification achieved more accurate result than Maximum Likelihood Classification
Balcik & Sertel (2007)	Land cover class interpretation	Landsat ETM+	Wavelet (ERDAS)	Interpretation + spectral profiles + correlation coefficients	– he fusion had the ability to improve the spatial resolution with minimal spectral distortion – However, the spatial detail of the fused images was not as good as that of the original Pan images	

Other fusion methods

					Evaluation comments	
Paper	Application	Image used	Fusion method	Classification method	Fusion	Classification
Ehlers et al. (2006)	Object-oriented classification	SPOT5 Pan Landsat ETM+ MS	Ehlers fusion	Object-oriented (eCognition)	– The fusion showed much better results than all other standard and newly proposed fusion techniques	– The benefits of the fusion did not automatically lead to enhanced classification results

Unknown fusion methods

					Evaluation comments	
Paper	Application	Image used	Fusion method	Classification method	Fusion	Classification
Stefanacci (2006)	Fire fuels mapping	Landsat ETM+, QuickBird	Unspecified	eCognition	– Pan-sharpened images were a cost-effective source for accurate inventories of fire fuels and associated risks	
Bishop & Congalton (2006)	Land cover classification	Landsat EMT+	Unspecified	ISODATA, Maximum Likelihood, Hybrid	– Moderate classification accuracies were achieved for both images with and without terrain flattening	
Hájek (2006)	Tree species identification	Ikonos	Unspecified	Object-based (eCognition)	– Both Ikonos MS and pan-sharpened Ikonos MS had their benefits and should be utilized for different forest management tasks	– Very good results could be achieved for tree species identification by combining object-oriented approach and pan-sharpened Ikonos imagery

(*Continued*)

Table 14.2. (*Continued*)

Paper	Application	Image used	Fusion method	Classification method	Evaluation comments: Fusion	Evaluation comments: Classification
Ozdarici & Turker (2006)	Crop classification	QuickBird	Unspecified	Pre-field classification (based on Maximum Likelihood)	– The pan-sharpened QuickBird image provided the highest overall accuracy, 85.2%, when its median image was classified – The SPOT4 XS had the lowest overall accuracy, 42.9%, when its median image classified – Similar classification results were obtained for all the images, when their mean images were classified	

- When visually comparing the pan-sharpened images published in individual papers, it can also be found that the fusion results of PCI Pansharp are outstanding. Readers can compare the images published in the original papers or the images cited in this paper (Figs. 14.1, 14.3, 14.4, 14.6, 14.7, 14.8, 14.9 and 14.10).

14.4.2 *Influence of information extraction techniques*

In terms of information extraction methods, Maximum Likelihood classification, object-oriented classification, and visual interpretation were most frequently used and discussed in the papers. It can be seen in Table 14.2 that

- close to 50% of the studies used or evaluated object-oriented classification, more than 30% used or evaluated Maximum Likelihood classification, and around 20% directly or indirectly used visual interpretation;
- one study found that the Maximum Likelihood classification achieved better accuracy than object-oriented classification for pan-sharpened ETM+ images;
- another study observed a better accuracy with object-oriented classification for pan-sharpened Ikonos images compared to Maximum Likelihood classification;
- an additional one indicated that object-oriented classification was a great tool for pan-sharpened Ikonos images; and
- two others found that the currently available commercial tool for object-oriented classification was not sufficient for utilizing the information existing in the pan-sharpened images, and suggested further development of new techniques for effective information extraction from pan-sharpened imagery.

14.4.3 *Overall findings*

Overall, it can be concluded that a well pan-sharpened image can significantly increase the image information for information extraction, at least for visual interpretation. However, obvious limitations still exist in currently available information extraction techniques, including Maximum Likelihood classification, object-oriented classification, and other information extraction techniques. The achievable information extraction accuracy is still low, normally between 50% and 80%. Only some simple classifications can achieve an accuracy of 90% or above. Comparing pixel-based classification, such as Maximum Likelihood and other techniques, object-oriented classification exhibits much better potential for classification of high resolution MS images, including pan-sharpened MS images.

According to the author's observation through literature review and first hand research in the field, pan-sharpening techniques may still have some room for further improvement to increase the accuracy of information extraction and interpretation. However, much more room still remain in the information extraction techniques for further improvement. The capability of current classification techniques is still far behind the ability of the human interpretation, i.e. much information, especially spatial information, existing in a MS image can be extracted by visual interpretation, but can still not be properly classified by a computer algorithm. To better utilize the information exists in a high-resolution MS image or pan-sharpened MS image, more effort needs to be given to the development of more effective information extraction techniques, either through pixel-based approaches, object-based (or oriented) approaches, or other approaches. However, the development of effective information extraction techniques is not an easy task. It has been a very challenging research topic for decades. The recently emerging object-based approaches, though still having many problems, have shown a promising direction for effective information extraction. More research may need to be focused on the development in this area.

REFERENCES

Aiazzi, B., Alparone, L., Baronti, S. & Garzelli, A. 2002. Contextdriven fusion of high spatial and spectral resolution images based on oversampled multiresolution analysis. *IEEE Transactions on Geoscience and Remote Sensing* 40(10), 2300–2312.

Amolins, K., Zhang, Y. & Dare, P. 2007. Wavelet based image fusion techniques—an introduction, review and comparison. *ISPRS Journal of Photogrammetry and Remote Sensing* 62: 249–263.

Balcik, F.B. & Sertel, E. 2007. Wavelet-based image fusion of Landsat ETM images: a case study for different landscape categories of Istanbul. *ISPRS Commission VII, WG2 & WG7, Conference on Information Extraction from SAR and Optical Data, with Emphasis on Developing Countries, 16–18 May 2007 Istanbul.*

Bishop, J.B. & Congalton, R.G. 2006. An Evaluation Of The Effect Of Terrain Normalization On Classification Accuracy Of Landsat Etm+ Imagery. *ASPRS 2006 Annual Conference, Reno, Nevada, May 1–5, 2006.*

Boloorani, A.D., Erasmi, S. & Kappas, M. 2006. Urban land cover mapping using object/pixel-based data fusion and Ikonos images. *ISPRS VII Midterm Symposium, 2006, Enschede, the Netherlands, 8–11 May 2006.*

Colditz, R.R., Wehrmann, T., Bachmann, M., Steinnocher, K., Schmidt, M., Strunz, G. & Dech, S. 2006. Influence of Image Fusion Approaches on Classification Accuracy – A Case Study. *International Journal of Remote Sensing* 27: 3311–3335.

Earth Observation Resources. 2008. http://directory.eoportal.org/res_p1_Earthobservation.html (last access 25 February 2008)

Ehlers, M., Greiwe, A. & Tomowski, D. 2006. On segment based image fusion. *1st International Conference on Object–based Image Analysis (OBIA 2006), Salzburg University, Austria, July 4–5, 2006.*

Ehlers, M. 2007. Segment based image analysis and image fusion. *ASPRS 2007 Annual Conference, Tampa, Florida, May 7–11, 2007.*

Forsythe, K.W. 2005. Monitoring Rapid Urban Expansion: A Case Study of Calgary, Alberta, Canada. *3rd International Symposium Remote Sensing and Data Fusion Over Urban Areas (URBAN 2005), Tempe, AZ, USA, March 14–16 2005.*

Garguet-Duport, B., Girel, J., Chassery, J. & Pautou, G. 1996. The use of multiresolution analysis and wavelets transform for merging SPOT panchromatic and multispectral image data. *Photogrammetric Engineering & Remote Sensing* 62(9):1057–1066.

Garzelli, A. & Nencini, F. 2005. Interband structure modeling for pansharpening of very high-resolution multispectral images. *Information Fusion* 6(3): 213–224.

Gonzalez-Audicana, M., Saleta, J.L., Catalan, R.G. & Garcia, R. 2004. Fusion of multispectral and panchromatic images using improved IHS and PCA mergers based on wavelet decomposition. *IEEE Transactions on Geoscience and Remote Sensing*, 42(6): 1291–1299.

Gonzalez-Audicana,M., Otazu, X., Fors, O. & Seco, A. 2005. Comparison between Mallat's and the 'à trous' discrete wavelet transform based algorithms for the fusion of multispectral and panchromatic images. *International Journal of Remote Sensing* 26(3): 595–614.

Granzow, E. 2001. Automatic Feature Recognition and Extraction from Remote Sensing Imagery. *National Consortia on Remote Sensing in Transportation.* U.S Department of Transportation.

Hájek, F. 2006. Object analysis of Ikonos XS and pan-sharpened imagery in comparison for purpose of tree species estimation. *First International Conference on Object-based Image Analysis (OBIA 2006), Salzburg University, Austria, July 4–5, 2006.*

Heipke, C., Mayer, H., Wiedemann, C. & Jamet, O. 1997. Evaluation of Automatic Road Extraction. *International Archives of Photogrammetry and Remote Sensing* 32: 47–56.

Hinz, S., Baumgartner, A., Mayer, H., Wiedemann, C. & Ebner, H. 2001. Road Extraction Focussing on Urban Areas. In E.P. Baltsavias, A. Gruen & L.V. Gool (eds.), *Automatic extraction of Man-Made Objects from Aerial and Space Images (III)*: 255–266. Lisse: Balkema Publishers.

Kiema, J.B.K. 2002. Texture analysis and data fusion in the extraction of topographic objects from satellite imagery. *International Journal of Remote Sensing* 23: 767–776.

King, R.L. & Wang, J. 2001. A wavelet based algorithm for pan sharpening Landsat 7 Imagery. *IEEE Geoscience and Remote Sensing Symposium, IGARSS'01*, 2: 849–851.

Lau, W., King, B. & Li, Z. 2000. The Influences of Image Classification by Fusion of Spatial–oriented Image. *International Archives of Photogrammetry and Remote Sensing* XXXIII, Part B7: 752–759.

Lewinski, S. 2006. Applying fused multispectral and panchromatic data of Landsat ETM+ to object oriented classification. *Proceedings of the 26th EARSeL Symposium, New Developments and Challenges in Remote Sensing, May 29–June 2, 2006, Warsaw, Poland.*

Li, S., Kwok, J.T. & Wang, Y. 2002. Using the discrete wavelet frame transform to merge Landsat TM and SPOT panchromatic images. *Information Fusion* 3(1): 17–23.

Liu, J.G. 2000. Smoothing filter based intensity modulation: a spectral preserve image fusion technique for improving spatial details. *International Journal of Remote Sensing* 21: 3461–3472.

Maxwell, T. & Zhang, Y. 2005. A fuzzy logic approach to optimization of segmentation of object-oriented classification. *Proceedings of SPIE 50th Annual Meeting—Optics & Photonics 2005, San Diego, California, USA, 31 July—4 August 2005.*

Ming, L. & Shunjun, W. 2003. A new image fusion algorithm based on wavelet transform. *Proc. 5th International Conference on Computational Intelligence and Multimedia Applications, Xi'an, China, 27–30 September 2003: 154–159.*

Mohanty, K.K. & Majumdar, T.J. 2003. Comparison of adaptive filter, IHS and Fourier transform based techniques for IRS-1C PAN/LISS-III data merging. *International Journal of Remote Sensing* 24: 1969–1976.

Munechika, C.K., Warnick, J.S., Salvaggio, C. & Schott, J.R. 1993. Resolution enhancement of multispectral image data to improve classification accuracy. *Photogrammetric Engineering and Remote Sensing* 59(1): 67–72.

Nichol, J. & Wong, M.S. 2005. Satellite remote sensing for detailed landslide inventories using change detection and image fusion. *International Journal of Remote Sensing* 26: 1913–1926.

Nunez, J., Otazu, X., Fors, O., Prades, A., Pala, V. & Arbiol, R. 1999. Multiresolution-based image fusion with additive wavelet decomposition. *IEEE Transactions on Geoscience and Remote Sensing* 37(3): 1204–1211.

Otazu, X., Gonzalez-Audicana, M., Fors, O., Nunez, J. 2005. Introduction of sensor spectral response into image fusion methods. Application to wavelet-based methods. *IEEE Transactions on Geoscience and Remote Sensing* 43(10): 2376–2385.

Ozdarici, A. & Turker, M. 2006. Field-based classification of agricultural crops using multi-scale images. *First International Conference on Object-based Image Analysis (OBIA 2006), Salzburg University, Austria, July 4–5, 2006.*

Pohl, C. & Van Genderen, J.L. 1998. Multisensor image fusion in remote sensing: concepts, methods and applications. *International Journal of Remote Sensing* 19(5): 823–854.

Pradines, D. 1986. Improving SPOT image size and multispectral resolution. *Proceedings SPIE Conference Earth Remote Sensing using the Landsat Thematic Mapper and SPOT Systems 660: 78–102.*

Price, J.C. 1987. Combining panchromatic and multispectral imagery from dual resolution satellite instruments. *Remote Sensing of Environment* 21(2): 119–128.

Prinz, B., Wiemker, R. & Spitzer, H. 1997. Simulation of high resolution satellite imagery form multispectral airborne scanner imagery for accuracy assessment of fusion algorithms. *Proceedings of the ISPRS Joint Workshop 'Sensors and Mapping from space', Hannover, Germany, October 1997.*

Ranchin, T. & Wald, L. 2000. Fusion of high spatial and spectral resolution images: The arsis concept and its implementation. *Photogrammetric Engineering and Remote Sensing* 66(1): 49–61.

Shaban, M.A. & Dikshit, O. 2002. Evaluation of the merging of SPOT multispectral and panchromatic data for classification of an urban environment. *International Journal of Remote Sensing* 23: 249–262.

Shackelford, A.K. & Davis, C.H. 2003. A combined fuzzy pixel-based and object-based approach for classification of high-resolution multispectral data over urban areas. *IEEE Transactions on Geoscience and Remote Sensing* 41: 2354–2363.

Shah, C.A. & Quackenbush, L.J. 2007. Analyzing multi-sensor data fusion techniques: a multi-temporal change detection approach. *ASPRS 2007 Annual Conference, Tampa, Florida, May 7–11, 2007.*

Shen, S.S., Lindgren, J.E. & Payton, P.M. 1994. Panchromatic band sharpening of multispectral image data to improve machine exploitation accuracy. *Proceedings of the SPIE—The International Society for Optical Engineering* 2304: 124–131.

Shettigara, V.K. 1992. A generalized component substitution technique for spatial enhancement of MS images using a higher resolution data set. *Photogrammetric Engineering and Remote Sensing* 58: 561–567.

Sirguey, P., Mathieu, R., Arnaud, Y., Khan, M.M. & Chanussot, J. 2007. Improved Resolution for the Detection of Snow With MODIS Using Wavelet Fusion. *IGARSS 2007, July 23–27, Barcelona, Spain.*

Smith, D.M. 1996. Speckle reduction and segmentation of Synthetic Aperture Radar images. *International Journal of Remote Sensing* 17: 2043–2057.

Stefanacci, J. 2006. Automated stand delineation and fire fuels mapping. *ASPRS 2006 Annual Conference, Reno, Nevada, May 1–5, 2006.*

Stoney, W.E. 2006. ASPRS guide to land imaging satellites. Updated for the NOAA Commercial Remote Sensing Symposium, Key Trends and Challenges in the Global Marketplace (http://www.asprs.org/news/satellites/ASPRS_DATABASE_090706.pdf, last access Feb. 23, 2008)

Winter, M.E. & Winter, E.M. 2002. Resolution enhancement of hyperspectral data. *IEEE Aerospace Conference Proceedings* 3: 1523–1529.

Wu, J., Huang, H., Qiu, Y., Wu, H., Tian, J. & Liu, J. 2005. Remote sensing image fusion based on average gradient of wavelet transform. *IEEE International Conference on Mechatronics and Automation, 29 July–1 August 2005*: 1817–1821.

Yocky, D.A. 1996. Multiresolution wavelet decomposition image merger of Landsat Thematic Mapper and SPOT panchromatic data. *Photogrammetric Engineering & Remote Sensing* 62(9): 1067–1074.

Zhang, Y, Wang, R. & Munroe, K. 2004. Multi-resolution and multi-spectral image fusion for urban object extraction. *Proceedings of XXth ISPRS Congress, Commission III, 12–24 July 2004, Istanbul, Turkey*: 960–966.

Zhang, Y. & Kerle, N. 2007. Satellite remote sensing for near-real time data collection. In Zlatanova & Li (eds.), *Geospatial Information Technology for Emergency Response*: 75–102. London, New York, Singapore: Taylor & Francis.

Zhang, Y. 1999. A new merging method and its spectral and spatial effects. *International Journal of Remote Sensing* 20(10): 2003–2014.

Zhang, Y. 2001. Detection of Urban Housing Development by Fusing Multisensor Satellite Data and Performing Spatial Feature Post-Classification. *International Journal of Remote Sensing* 22: 3339–3355.

Zhang, Y. 2002. Problems in the fusion of commercial high-resolution satellite as well as Landsat 7 images and initial solutions. *ISPRS Proceedings of Symposium on Geospatial Theory, Processing and Applications, Ottawa, Canada, 8–12 July 2002.*

Zhang, Y. 2004. Understanding Image Fusion. *Photogrammetric Engineering and Remote Sensing* 70: 657–661.

Zhang, Y. & Hong, G. 2005. An IHS and wavelet integrated approach to improve pan-sharpening visual quality of natural colour IKONOS and QuickBird images. *Information Fusion* 6(3): 225–234.

Zhukov, B., Oertel, D., Lanzl, F. & Reinhaeckel, G. 1999. Unimixing-based multisensor multiresolution image fusion. *IEEE Transactions on Geoscience and Remote Sensing* 37: 1212–1225.

Advances in Photogrammetry, Remote Sensing and Spatial Information Sciences: 2008 ISPRS Congress Book – Li, Chen & Baltsavias (eds)

CHAPTER 15

Object extraction and attribution from hyperspectral images

Freek van der Meer, Harald van der Werff, Mark van der Meijde, Frank van Ruitenbeek, Chris Hecker & Steven de Jong

ABSTRACT: Spectral reflectance in the visible and near-infrared offers a rapid and inexpensive technique for determining the mineralogy of samples and obtaining information on chemical composition. Absorption-band parameters such as the position, depth, width and asymmetry of the feature have been used to quantitatively estimate composition of samples from hyperspectral field and laboratory reflectance data. The parameters have also been used to develop mapping methods for the analysis of hyperspectral image data. This has resulted in techniques providing surface mineralogical information (e.g. classification) using absorption-band depth and position; however, no attempt has been made to prepare images of the absorption-band parameters. In this chapter, a simple linear interpolation technique is proposed in order to derive absorption-band position, depth and asymmetry from hyperspectral image data. It is demonstrated that these absorption band maps provide basic information for surface compositional mapping. Next, we look at stratified image analysis approaches and data integration techniques. Lastly we use contextual information for image analysis. A full contextual image analysis approach is presented in the form of a template matching technique. Future challenges in the field of hyperspectral remote sensing are to fully explore spatial as well as spectral image characteristics for image classification, to couple image products to subsurface information, and to include temporal spectral variability as a means of characterizing geo-objects.

Keywords: Hyperspectral images, spectral matching, reflectance spectroscopy, contextual image analysis

15.1 INTRODUCTION: HYPERSPECTRAL REMOTE SENSING

Imaging spectrometers acquire imagery in many, narrow and contiguous spectral bands with the aim of collecting "image radiance or reflectance spectra" that can be compared with field or laboratory spectra of known materials. Imaging spectrometry has been widely used in geological mapping, specifically in so-called hydrothermal alteration systems. These are areas where the composition of host rocks is altered through the circulation of hot fluids giving rise to the formation of new mineral assemblages in a predefined order in (3-D) space. Surface mineralogical information can be derived from imaging spectrometer data by comparing imaged reflectance spectra of unknown composition to data from spectral libraries. This comparison is mostly done on a pixel-by-pixel basis. In general, a matching is done to express the similarity between the unknown pixel spectrum and known spectra from spectral libraries. As a result, in geology, information on surface mineralogy can be derived from imaging spectrometry data, which in turn can be incorporated into geologic models. Field and laboratory spectra have been used to relate absorption features to chemical composition of samples in both the areas of soil science and mineralogy, as well as in the area of vegetation science. For the analysis of hyperspectral image data, there are several techniques available to derive surface composition (e.g. surface mineralogy) from a combination of absorption-band position and depth. However, no such technique provides spatial information on the variation of absorption-band depth, position and shape despite the fact that these parameters are of vital use in quantitative surface compositional mapping.

Quantitative estimates of mineralogical composition and chemical analysis on the basis of spectroscopic data have been demonstrated by many authors. In this chapter, we evaluate pixel based methods of hyperspectral data analysis to (1) derive surface compositional information, and (2) to derive absorption feature characteristics. We will highlight the intrinsic problems and solutions to these problems for these methods. Thereafter we will discuss contextual analysis techniques as an alternative approach. This chapter is written largely from a geological perspective on the present and future use of hyperspectral data. Although some remarks may be considered universal, it should be noted that not all presented results are equally applicable to other thematic fields. We also do not aim at providing a thorough and detailed review of the entire field, but merely try to sketch the development of hyperspectral remote sensing in geology providing a framework for future studies.

15.2 PIXEL-BASED METHODS FOR SPECTRAL MATCHING

15.2.1 *Introduction*

Most analytical techniques to handle hyperspectral data work on a pixel-by-pixel basis, either to (1) match known spectra with unknown pixel spectra to derive surface composition, or (2) to derive absorption band parameters that can yield input to empirical models.

15.2.2 *Spectral matching techniques*

Spectral matching techniques used for compositional mapping are various. Techniques to process hyperspectral imagery in order to obtain surface compositional information on a pixel-by-pixel basis for the entire image are reviewed by (Van der Meer et al. 2001). Techniques that specifically use absorption band position and depth include (1) the Relative Absorption Band-Depth (RBD) approach of Crowley et al. (1989), (2) the Spectral Feature Fitting (SFF) technique of Clark et al. (1990) and (3) the TRICORDER (Crowley & Swayze 1995) and TETRACORDER algorithms developed at the USGS spectral laboratory. These techniques work on so-called continuum removed reflectance spectra, thus acknowledging that the absorption in a spectrum has two components: a continuum and individual features. Crowley et al. (1989) developed a method of mineral mapping from imaging spectrometer using Relative Absorption Band-Depth Images (RBD) generated directly from radiance data. In essence, RBD images provide a local continuum correction removing any small channel to channel radiometric offsets, as well as variable atmospheric absorption and solar irradiance drop off for each pixel in the data set. To produce a RBD image, several data channels from both absorption band shoulders are summed and then divided by the sum of several channels from the absorption band minimum. The resulting absorption band-depth image gives the depth of an absorption feature relative to the local continuum, which can be used to identify pixels having stronger absorption bands indicating that these may represent a certain mineral.

Spectral feature fitting (embedded in the ENVI software) uses continuum removed pixel spectra, which are compared to continuum reference spectra of known mineralogy. A least squares fit is calculated band by band between each reference end-member and the unknown (continuum removed) pixel spectra. A "Scale" image is produced for each end-member selected for analysis by first subtracting the continuum-removed spectra from one, thus inverting them and making the continuum zero. A large scale-factor is equivalent to a deep spectral feature, while a small scaling factor indicates a weak spectral feature. A least squares fit is then calculated band by band between each reference end-member and the unknown spectrum. The total root mean square (RMS) error is used to form an RMS error image for each end-member. The ratio of the scale image and the RMS image provides a "Fit" image that is a measure of how well the unknown spectrum matches the reference spectrum on a pixel-by-pixel basis. The Tricorder and its successor Tetracorder use spectral matching algorithms carried out in a two step process. First, the local spectral slope (the "continuum") is estimated and removed both from reference and observed spectra. Next, the identification of materials from their spectra is constrained by (1) the goodness of fit of a spectral feature to a reference, (2) reflectance level, (3) continuum slope, and (4) presence or absence of key ancillary spectral features. The Tetracorder uses these reference continuum-removed-spectral features to compute a weighted fit between unknown spectra and known library spectra. By means of an expert system approach, surface compositional information is derived and results are validated. These mapping methods described all produce validated surface compositional information (mostly mineralogical maps), however do not provide information on the absorption band position, depth and asymmetry on a pixel by pixel basis although these parameters are used in the matching performed.

Since absorption band parameters are of importance to quantitative reflectance spectroscopy, attempts have been made to develop linear estimation methods to derive absorption band parameters from hyperspectral image data. Another spectral matching technique is the Cross correlogram spectral matching (CCSM; Van der Meer & Bakker 1997); an approach toward mineral mapping from imaging spectrometer data using the cross correlogram of pixel and reference spectra. A cross correlogram is constructed by calculating the cross correlation at different match positions between a test spectrum (i.e. a pixel spectrum) and a reference spectrum (i.e. a laboratory mineral spectrum or a pixel spectrum known to represent a mineral of interest) by shifting the reference spectrum over subsequent channel positions. The statistical significance of the cross correlation coefficient can be assessed by a student's t-test and the skewness can be calculated as an estimator of the goodness-of-fit. The cross correlogram for a perfectly matching reference and test spectrum is a parabola around the central matching number ($m = 0$) with a peak correlation of 1. Deviations from this shape indicate a different surface mineralogy. Mineral mapping on a pixel-by-pixel basis is achieved by extracting three parameters from the cross correlograms and combining these into a statistical estimate of the goodness of fit of the two spectra compared: the correlation coefficient at match position zero, the moment of skewness (based on the correlation differences between match numbers of equal but reversed signs, e.g. $m = 4$ and $m = -4$), and the significance (based on a student t-test testing the validity of the correlation coefficient at $m = 0$). In order to evaluate the surface mineralogy maps, a root mean square error assessment procedure is proposed in Van der Meer & Bakker (1997), in which the error is calculated from the difference between the calculated pixel cross correlogram and the ideal cross correlogram calculated for the reference.

The most used mapping method in hyperspectral remote sensing is the spectral angle mapper. The spectral angle mapper calculates the spectral similarity between a test reflectance spectrum and a reference reflectance spectrum assuming that the data is correctly calibrated to apparent reflectance with dark current and path radiance removed. In this approach, the spectra are treated as vectors in a space with dimensionality equal to the number of bands, n.

The outcome of the spectra angle mapping for each pixel is an angular difference measured in radians. The main problem using the spectral angle mapper is the appropriate selection of the threshold to classify the derived rule images. Since the angle is neither a physical nor a statistical measure, there is no statistically/physically sound approach to do so. Some problems related to spectral matching techniques are illustrated in Figure 15.1. Here the correlation is calculated between a calcite laboratory spectrum (from the JPL 160 mineral library available in ENVI) and a dolomite spectrum and an iron rich dolomite spectrum. We calculate the correlation coefficient for parts of the spectrum starting with the first 800 (of a total of 826 channels of the Beckman spectrometer) channels and subsequently diminishing this number by steps of 50 channels such that at the end only the correlation between the 50 channels in the long

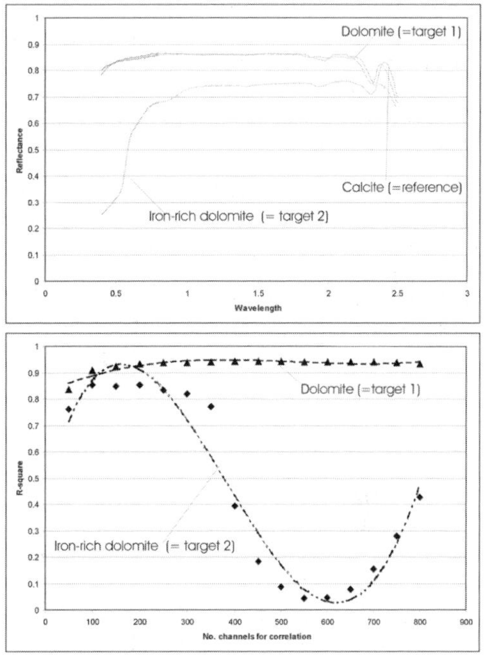

Figure 15.1. Correlation between calcite (reference) and dolomite (target 1) and iron-rich dolomite (target 2). See text for discussion.

wavelength portion of the SWIR are calculated. For the calcite versus dolomite spectrum, we notice that the correlation is high (0.93) and diminishing to 0.84 when using only the last 50 channels of the spectrometer. This can be explained by the fact that calcite and dolomite have an absorption band around 2.3 micron, which changes only from 2.35 to 2.3 micron for calcite versus dolomite.

Hence the significant difference in the two spectra is in this part of the spectrum. In fact, the graph shows that including other wavelengths only adversely influences the matching result. The same can be observed when looking at the correlation graph of the calcite versus the iron-rich dolomite spectrum. However, this graph shows various correlation maxima and minima because the shape of the graphs are similar in parts of the spectrum but differ, particularly in the visible part of the spectrum. Hence including irrelevant information severely damages the matching result. Another point arises in thresholding the correlation to perform a classification. How to objectively determine a threshold? This is a particular problem when using the SAM technique because the angles will vary non-linearly with the per cent spectrum mixed into a mixture.

15.2.3 *Absorption-band parameter estimation*

Alternative approaches fully exploit the spectral properties of absorption bands in hyperspectral data sets. The following absorption-band parameters calculated from continuum-removed spectra are often used: (1) the absorption-band position, (2) the absorption-band depth, and (3) the absorption-band asymmetry. The absorption band parameter definition as discussed above assumes nearly continuous (contiguous) spectral data, whereas imaging spectrometers acquire data in a large number of discrete spectral bands. The band centre wavelength position is usually employed for further calculations, thus providing spectral measurements on a discrete number of wavelength positions characterizing an absorption feature. To accommodate this, a simple linear method is proposed to calculate the absorption feature parameters from image data. Figure 15.2 graphically explains the procedure followed.

First, for the absorption feature of interest, the image bands are determined that would serve as the shoulders of the absorption feature. By definition, there is a short wavelength shoulder (shoulder 2, denoted by S_2 in Fig. 15.2) and a long wavelength shoulder (shoulder 1, denoted S_1 in Fig. 15.2). Next, the data are continuum removed using the mentioned shoulders as starting and ending points. Subsequently, two bands are selected as the absorption points which are used in the interpolation (points A_1 and A_2 in Fig. 15.2). From the interpolated wavelength, position (e.g. the wavelength of maximum absorption) can be found by interpolating between the shoulders and absorption points in the spectrum as

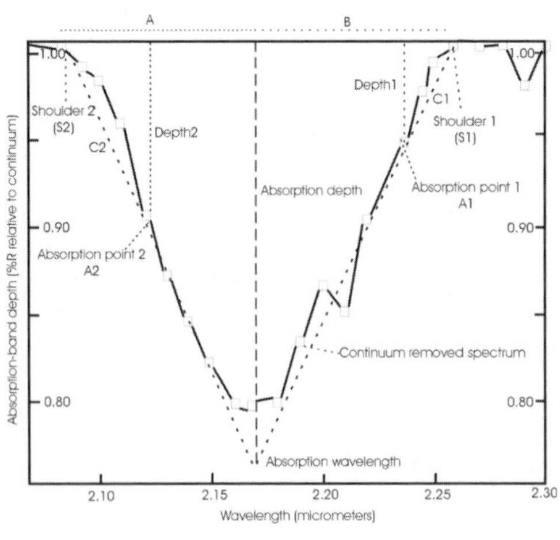

Figure 15.2. Definition of the parameters used in the linear interpolation of absorption feature characteristics.

$$Absorption\_wavelength = -\left[\frac{C_1}{C_1+C_2} * (A_1 - A_2)\right] + A_1 \quad (1)$$

The associated absorption-band depth is derived as

$$Absorption\_depth = \left[\frac{S_1 - absorbtion\_wavelength}{S_1 - A_1}\right] * Depth_1 \quad (2)$$

An example of absorption bands calculated for a HyMAP data set acquired over a sedimentary sequence is given in Figure 15.3.

15.3 STRATIFIED ANALYSIS OF HYPER-SPECTRAL DATA

15.3.1 *Data stratification*

An alternative approach to thematic analysis of hyperspectral data is stratified analysis. The aim would be to stratify the data on known thematic data layers prior to analytical approaches being applied to the data. In many cases, field or thematic map data is available. This data can thus be used for data processing. One can imagine clustering a data set on mapping units

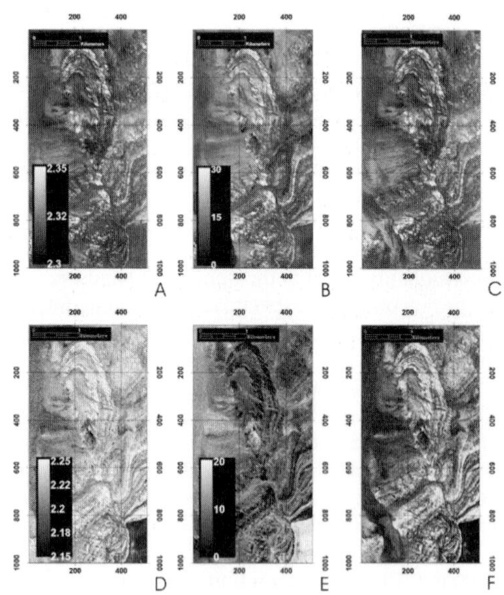

Figure 15.3. Absorption band parameters calculated from HyMAP data over a sedimentary terrain (A = carbonate absorption band wavelength position, B = depth of carbonate absorption band, C = asymmetry of the carbonate absorption band, D = absorption band wavelength for OH features in SWIR, E = depth of OH feature, F = asymmetry of the OH feature).

and starting the analysis within these natural boundaries. Furthermore, various data products from hyperspectral data sets can be integrated using statistical techniques. As an example we look at Probe-1 (also known as HyMAP) from an area in California (Santa Barbara) known for the presence of oil/gas seeps. These seeps can be detected in hyperspectral imagery either through (1) anomalous spectral behaviour of the vegetation, or (2) through mineral alteration in the soil. When evaluating the imaged area in terms of vegetation properties in relation to the lithology, we can clearly see the necessity for performing a stratified analysis. We calculated (1) the NDVI, (2) the carter stress ratio of band 695 nm over band 420 nm, (3) the carter stress ratio of band 695 nm over band 760 nm, and (4) the red edge inflection point. The so-called Carter ratio's (Carter 1994) are band ratios that are sensitive to vegetation stress. We computed these two ratios, 695/420 and 695/760, on a vegetation masked image (that is on pixels that have a NDVI value of greater than 0.5). Similarly we calculate the red edge inflexion point. In Table 15.1 we summarize the results of these calculations for number of key lithologies in the imaged area. The Monterey formation is the oil-producing unit.

15.3.2 *Example from Santa Barbara (Ca.)*

Hence, we first separated the data into a vegetation and a soil component by simply masking the data using the NDVI. Next, we stratified the data on the known geological units since these not only affect the soil type and hence the vegetation type but there is also a strong link between seepage and lithology. The main oil reservoir rock at some areas outcrops thus allowing seepage of oil to the surface. Seeps are also more likely to appear in gas form when sitting close to the host reservoir rock. The main derivative of interest for gas seepage in the vegetated areas is the red edge position, which can be regarded as an indicator of health status of the vegetation. In the soil component, we singled out various mineral maps (using spectral matching techniques described earlier) in particular

Table 15.1. Typical values of vegetation parameters for selected lithologies in the Santa Barbara imaged area.

	Vegetation		Stress ratio	
Lithology	NDVI	Red edge	695/760	695/420
Repetto	0.2–0.3	725	<0.2	0.65–1.25
Monterey	0.6–0.8	712	0.3–0.5	0.5–0.75
Sisquoc	0.5–0.8	715	0.2–0.3	0.99–1.2
Sespe	0.3–0.5	721	0.1–0.3	1.25–1.5

iron oxides and enrichment in kaolinite and carbonate, which are indicative for seepage. These data layers for each unit were integrated using statistical approaches. The resulting spatial patterns were integrated with known oil and gas seeps using a spatial data integration approach (SDI). The basic proposition, the mathematical hypotheses that we want to test in this study—the presence or absence of an undiscovered oil/gas reservoir—is formulated. The probabilistic theory embedded in the favourability function allows this proposition to be addressed.

The favourability function method requires the two assumptions that: (i) past occurrences of a given type (i.e. a clearly identified type of process) can be characterized by sets of layers of supporting spatial data, and that (ii) new discoveries of the same type will occur in the future under similar circumstances. That is, the present should be able to predict the future. Consider a probability at any A, prob{Fp}, for the proposition Fp. This prior probability, a pixel p will contain a future reservoir, is obtained by

$$\mathrm{Prob}\{Fp\} = \text{size of F/size of A} \tag{3}$$

where F denotes the unknown areas that will be reservoirs within A and "size of B" represents the size of the surface area covered by any sub-area B in A. Prob$\{F_p\}$ has the same value for all p. The purpose of the modelling is to see how the probability at p will be changed as we observe the m pixel values at p. At pixel p, the pixel value $v_l(p)$ of the l_{th} layer is c_l, which is one of the n_l classes (map units), $\{1, 2, \dots, n_l\}$. Consider a set of all pixels whose value in the 1st layer is c_l. The set is the thematic class in the 1st layer whose pixel value is c_l. The set is denoted by $A_l c_l$ and it is one of the non overlapping nl sub-areas $\{A_{11}, A_{12}, \dots, A_l n_l\}$ in the 1st layer. Assume that the occurrences at each pixel p can be expressed as the joint conditional probability

$$Prob\{F_b / C_c, C_2 \dots C_m\} \tag{4}$$

that p will be a future undiscovered oil/gas reservoir assuming that the p contains the m values, $(c_1, c_2 \dots cm)$. When eq.4 is approximately similar to Prob$\{F_p\}$ we can state that the pixel values at the p, $(c_1, c_2 \dots c_m)$ do not add any useful information as to whether the pixel is a undiscovered reservoir.

However, if

$$Prob\{F_b / C_c, C_2 \dots C_m\} >> Prob\{F_b\} \tag{5}$$

or

$$Prob\{F_b / C_c, C_2 \dots C_m\} << Prob\{F_b\} \tag{6}$$

then the pixel values, (c_1,c_2...c_m) provide very significant information and they are highly correlated either positively or negatively with occurrences. We assume that c_1, will contain an undiscovered oil/gas reservoir and we assume that the c_1,...c_m are conditionally independent given the condition F_p, (p will contain an undiscovered oil/gas reservoir). Hence, under the above conditional independence assumption, the joint conditional probability becomes

$$\frac{Prob\{C_1\}\ldots Prob\{C_m\}}{\Pr ob\{C_1,\ldots C_m\}} Prob\{F_b\} \frac{\Pr ob\{F_b/C_1\}}{\Pr ob\{F_b\}}\cdots\frac{\Pr ob\{F_b/C_m\}}{\Pr ob\{F_b\}} \quad (7)$$

Under the conditional independence assumption, this joint conditional probability can be expressed in terms of three components. The first component, the ratio of prob{c_1}... prob{c_m} and prob{c_1,...,c_m} consists of the probabilities related to the input spatial data. The second component, the prior probability Prob{F_p}, is the probability that a pixel p will be an undiscovered reservoir prior to having any evidence.

In Figure 15.4, we provide an interpretation of the patterns observed. Particularly the predicted probability map for the mineral alteration provides spectacular results. The image clearly shows an elliptical pattern of high probability values around the actual seep location forming what appears to be a halo. In the centre of the halo, low values give the appearance of a clogged area similar to the resulting patterns described in edge-leakage. The vegetation stress analysis in this area did reveal less information, which is due to the fact that few pure vegetation pixels are found and that large parts over the anomalous area are relatively devoid of vegetation.

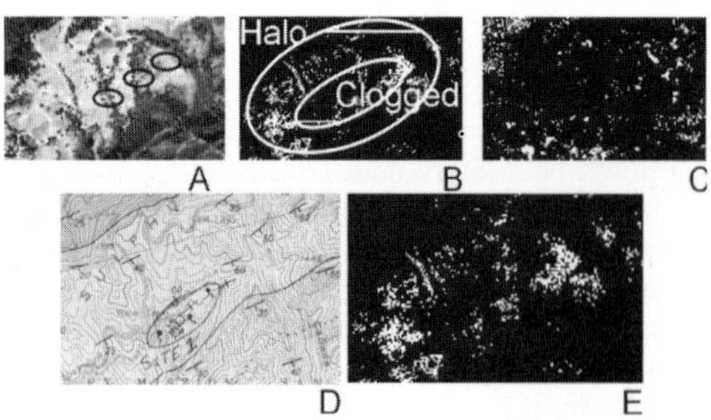

Figure 15.4. Results and interpretation. False colour image of probe bands 16, 9, 4 in RGB with seeps (A), SDI probability minerals with interpretation of the anomaly (B) SDI probability vegetation (C), geological map (D) and SDI probability minerals with annotation. All probability maps display P > 0.8. (see colour plate page 500)

15.4 CONTEXTUAL ANALYSIS OF HYPER-SPECTRAL IMAGERY

15.1.6 *Stratified approaches*

Stratified approaches are one step in the direction of using prior and thematic information in the analysis of hyperspectral data sets. A next step would be to include spatial and contextual information. Pixel-based methods described earlier have in common that they yield surface compositional information (e.g. in geological applications this is often surface mineralogy) that has to be further translated into a geological model (which involves understanding the spatial context of surface mineralogy). Furthermore, these models bypass the common notion in geology that spatial (contextual) information provides valuable information to infer the transitional nature of geological context. In other words, field geologists working in, for example, hydrothermal alteration systems use mineral paragenesis and the mineral assemblage in their mapping. When identifying a certain suite of minerals at a location, not only do they know in which part of the alteration system they are spatially, but also from this information they can infer which mineralogic transition they are about to discover when progressing through the terrain. Hence, not only do they use information on the surface mineralogy, but also they use neighbourhood information to unravel the geologic history and alteration system of an area.

15.1.7 *True contextual image analysis*

To combine both spectral as well as spatial information in the analysis of hyperspectral imagery, we have developed a spatial spectral algorithm; the template matching algorithm (Van der Werff et al. 2007). A template consists of a one- or two-dimensional array that is filled on both sides of a central pixel with spectra (or spectral derivatives such as ratios, etc.) that characterize certain cover types of interest. The template is matched with an image by moving the kernel or a filter over the image and calculating parameters at each pixel. At each pixel, the template is also rotated and the following parameters are calculated: the best template fit, the worst template fit, the mean template fit, the variance in template fit, the mean variance in fit of individual pixels, the variance in variance in fit of individual pixels, and the optimal scan angle, which is the angle belonging to the best template fit.

The Template-fit is the total value for N number of pixels; a mean fit is derived as:

$$\overline{F}_p = \frac{\Sigma_{p=0}^{N} F_{(p)}}{N} \quad (8)$$

The variance in pixel fit, the Template fit Variance layer, is calculated for the template as:

$$V_p = \frac{\Sigma_{p=0}^{N}\left(F_{(p)} - \overline{F}_p\right)^2}{N} \tag{9}$$

When the template is rotated, for every orientation (every scan angle) of the template the extreme values are stored as "Best Template Fit" and "Worst Template Fit". The optimal scan angle is set to the scan angle (best orientation) that gave "Best Fit Template"; otherwise variance in template fit cannot be determined. The "Mean Template Fit" is calculated using:

$$\overline{F}_T = \frac{\Sigma_{a=0}^{A}\overline{F}_{p(a)}}{A} \tag{10}$$

where A is the number of scan angles. The variance in template fit is calculated by:

$$V_T = \frac{\Sigma_{a=0}^{A}\left(\overline{F}_{p(a)} - \overline{FT}\right)^2}{A} \tag{11}$$

Figure 15.5 shows the principle of template matching whereas Figure 15.6 shows some results of template matching on synthetic image lines showing the improved capabilities for detecting crisp and fuzzy boundaries.

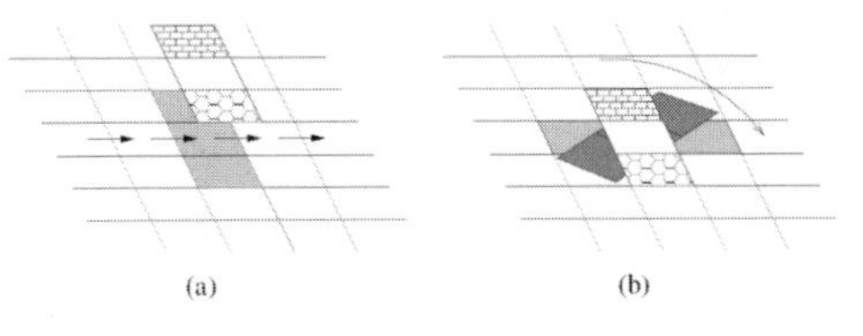

Figure 15.5. Principle of template matching. A micro image is defined on the basis of spectral characteristics of two classes with crisp or fuzzy boundaries (A) and swept over the hyperspectral image to calculate a spatial correlation (B).

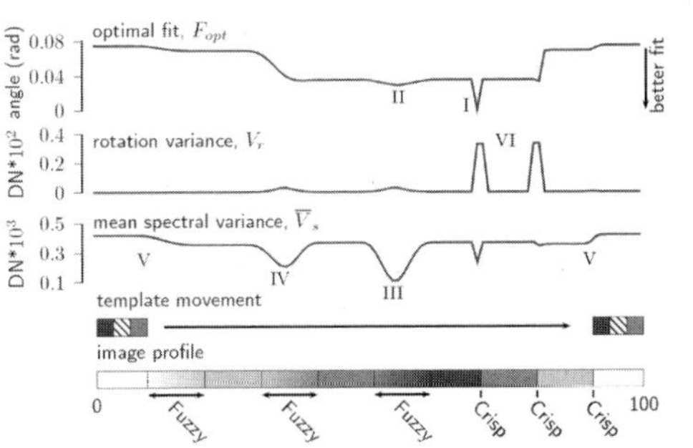

Figure 15.6. Results of template matching on synthetic image line showing the improved capabilities for detecting crisp and fuzzy boundaries.

15.5 CHALLENGES

In this chapter we have sketched a development in the field of hyperspectral remote sensing for geological applications from pixel based to context based approaches. Early work relied on statistical matching of pixel spectra to library spectra to come to image products useful for geology. In the early 2000s, the work started to include stratified approaches, data integration approaches and geo-inversion. More recently researchers looked at segmentation and contextual approaches to couple spectral as well as spatial information in hyperspectral data. Most image processing techniques that are available to analyse hyperspectral data by-pass three important aspects of remote sensing: (1) neighbourhood information is not incorporated and hence the common knowledge formulated in geo-statistics that nearby control points are more likely to contain information more similar than far apart control points, (2) surface information is not coupled to shallow subsurface information, although many problems are in need of a 3-dimensional approach of study, and (3) the change over time of the spectral signature of certain Earth surface materials that contains information on the type (and use) of the material at hand that may allow a better classification and or discrimination (from other) of materials is not used in sub-pixel classification. The challenge ahead lies in a full incorporation of the 3-D space and the 4-D temporal space to characterize image elements.

15.6 CONCLUSIONS

Reflectance spectroscopy has been used to derive estimates of soil/rock geochemistry and foliar biochemistry using, to date, mostly field and laboratory spectroscopic techniques. These techniques most often make use of absorption feature characteristics (e.g. absorption band wavelength position, depth and asymmetry), which are combined with geochemical analysis in a multi-linear regression to find empirical relationships with the chemistry of a sample. Absorption features have also been used as input to spectral feature fitting techniques that allow mapping surface composition (mainly mineralogy) from hyperspectral image data. However, by this process, the absorption feature analysis is confined to input data in a feature fitting. A simple linear interpolation method is introduced to estimate absorption-band parameters from hyperspectral image data. By applying this hyperspectral data it has been demonstrated that absorption feature maps correspond favourably with the main alteration phases characterizing the systems studied. Thus, the derived feature maps allow enhancement of the analysis of airborne hyperspectral image data for surface compositional mapping. The next step is to expand in the direction of stratified analysis of hyperspectral data. This allows the incorporation of prior information into the analysis of hyperspectral data sets. The last example we present is using spatial information in full contextual analysis of hyperspectral data sets. In this case study, we deploy Markov chain analysis to include the repetitive patterns in a sedimentary sequence for the study of hyperspectral data sets. Overall with this chapter, we wish to advocate including both prior knowledge as well as spatial contextual information as an essential step forward in the science of hyperspectral remote sensing. An overview of contextual image analysis approaches is found in De Jong & Van der Meer (2004).

REFERENCES

Carter, C.A. 1994. Ratios of leaf reflectances in narrow wavebands as indicators of plant stress. *International Journal of Remote Sensing* 15: 697–703.

Clark, R.N., Gallagher, A.J. & Swayze, G.A. 1990. Material absorption band depth mapping of imaging spectrometer data using the complete band shape least-squares algorithm simultaneously fit to multiple spectral features from multiple materials. *Proceedings of the Third Airborne Visible/Infrared Imaging Spectrometer (AVIRIS) Workshop*: 176–186. Los Angeles: JPL Publication 90–54.

Crowley, J.K., Brickey, D.W. & Rowan, L.C. 1989. Airborne imaging spectrometer data of the Ruby mountains, Montana: mineral discrimination using relative absorption band-depth images. *Remote Sensing of Environment* 29: 121–134.

Crowley, J.K. & Swayze, G.A. 1995. Mapping minerals, amorphous materials, environmental materials, vegetation, water, ice, and other materials: The USGS Tricorder Algorithm. *Summaries of the Fifth Annual JPL Airborne Earth Science Workshop*: 39–40. Los Angeles: JPL Publication 95–1.

De Jong, S. & Van der Meer, F. 2004. *Remote sensing image analysis: including the spatial domain*. Dordrecht: Springer-Kluwer Academic Publishers.

Van der Meer, F. & Bakker, W. 1997. Cross Correlogram Spectral Matching (CCSM): application to surface mineralogical mapping using AVIRIS data from Cuprite, Nevada. *Remote Sensing of Environment* 61: 371–382.

Van der Meer, F., De Jong, S. & Bakker, W. 2001. Imaging spectrometry: Basic analytical techniques. In F. van der Meer & S. de Jong (eds), *Imaging Spectrometry: Basic Principles and Prospective Applications*: 17–61. Dordrecht: Kluwer Academic Publishers.

Van der Werff, H., Van Ruitenbeek, F., Van der Meijde, M., Van der Meer, F., De Jong, S. & Kalubandara, S. 2007. Rotation-variant template matching for supervised hyperspectral boundary detection. *IEEE Geoscience and Remote Sensing Letters* 4(1): 70–74.

Yang, H., van der Meer, F., Bakker, W. & Tan, Z.J. 1999. A back-propagation neural network for mineralogical mapping from AVIRIS data. *International Journal of Remote Sensing* 20: 97–110.

Advances in Photogrammetry, Remote Sensing and Spatial Information Sciences: 2008 ISPRS Congress Book – Li, Chen & Baltsavias (eds)

CHAPTER 16

Automated extraction of roads, buildings and vegetation from multi-source data

Helmut Mayer, Stefan Hinz & Uwe Stilla

ABSTRACT: This chapter gives an overview of novel approaches and developments in the area of automated object extraction from multi-source data, i.e. mostly images, but also laser and Synthetic Aperture Radar (SAR) data, from the ground, from the air and from space, since the ISPRS congress in Istanbul in 2004. It is organized firstly according to the major object types, namely roads and traffic, buildings and finally vegetation. Additionally, we have devoted one section to two new promising modelling techniques, namely appearance based and generative modelling. For both, an intuitive introduction and relevant literature are given. We also emphasize the trend towards stochastic modelling and automated learning. For all three, it is shown how they are employed for buildings, roads and trees. Before ending the overview with conclusions, we present the state of the art of the evaluation of object extraction approaches as this is an important prerequisite for practical implementation.

Keywords: Object extraction, roads, buildings, vegetation, multi-source data, generative modelling, appearance-based modelling, stochastical modelling learning, evaluation

16.1 INTRODUCTION

The interval since the ISPRS congress in Istanbul in 2004 has seen a couple of noteworthy developments in the field of automated extraction of topographic objects, most notably roads and buildings, but also vegetation. The goal of this contribution is not a comprehensive overview, but it focuses on a couple of issues we consider important and for which we think we can provide useful insights.

As the term "multi-source data" is not inherently clear, we first define our scope to be aerial images, satellite images, close range images, airborne laser, terrestrial laser and finally satellite and aerial Synthetic Aperture Radar (SAR). Our focus is on roads in satellite and aerial images (Section 16.2) as well as on buildings in aerial and terrestrial images, and laser scanner data (Section 16.3). Also vegetation is seen to be more and more important in urban settings, as shown by a number of recent publications (Section 16.4).

Because we consider the advances in appearance based and generative stochastical modelling the most interesting developments in the field, we have dedicated Section 16.5 to them, describing the basic developments as well as the application to the extraction of topographic objects. Some of these issues were from our point of view nicely addressed in the proceedings of the conference Photogrammetric Image Analysis (PIA) we recently organized (Stilla et al. 2007). One focus there was facades of buildings, which is also reflected in Section 16.5.

In spite of all the advances in the field, there is still a large gap between the theoretical work on fully automated object extraction and practical applications (Mayer 2008). Due to lack of space, but also due to its scope inside the whole book, this contribution addresses more theoretical work. Yet, we advocate comparative evaluation to show what fully automated approaches are capable of, but also to make clear which of the approaches have the largest potential for practice (Section 16.6). We finally end up with conclusions.

16.2 ROADS IN SATELLITE AND AERIAL IMAGES

For roads there has been basic work carried out on a more detailed analysis of the scale-space behaviour and also on the level of road markings (Section 16.2.1). Due to the widespread availability of high quality digital multi-spectral data, they have been analysed and found useful for road extraction (Section 16.2.2). There is also more and more interest in roads in urban areas as presented in Section 16.2.3 and it was demonstrated how to map roads or road like features by tracking vehicles (Section 16.2.4). With the recent technological advances in airborne and spaceborne SAR, road extraction from SAR data has also been given considerable attention. Because of the different imaging geometry of SAR and the different reflection characteristics in the microwave domain, adapted models and extraction strategies for road extraction are necessary. These are addressed in Section 16.2.5.

Please note that our emphasis is on full automation of the extraction throughout this chapter. Semi-automatic approaches or approaches incorporating prior information from GIS road databases are thus only covered in special cases in the following.

16.2.1 *Scale space behaviour of roads*

The importance of incorporating the scale space behaviour of objects into automatic extraction approaches has been recognized since the 1990s. It has been shown that different levels of image resolution can be linked to certain levels of abstraction of knowledge. For road extraction, usually two (Baumgartner et al. 1999, Hinz & Baumgartner 2003) but in some special cases also three (Mayer 1998, Hinz 2004a) different scales have been used. While the coarse scale, i.e. 2 to 4 m, allows the extraction of roads based on their fundamental characteristics and functional properties—such as curvilinear shape, network characteristics and optimum routes between certain places—the fine scale, i.e. less than 50 cm, enables detailed analysis and precise geometric delineation of the road layout. The extraction benefits from using both scales, since initial hypotheses can be efficiently formed in the coarse scale, which are validated and tested in the fine scale. Vice versa, intermediate results obtained in the fine scale can be evaluated with respect to their importance for the global road network involving functional properties of roads described in the coarse scale.

Recent developments deal with the automatic adaptation of specific parts of the road model for multiple scales. In contrast to the earlier approaches mentioned above, where the particular scales were chosen more or less empirically, Heuwold (2006), Heuwold & Pakzad (2006) and Heuwold et al. (2007) aim at generalizing this issue for a quasi-continuous scale space. Their focus is on parallel linear objects such as road markings and road sides similar to the work by Mayer & Steger (1998) as well as Hinz & Baumgartner (2003), yet they extend the analysis to 2-D linear scale space. The results shown in Heuwold (2006) are promising for the case of parallel linear structures as they often appear in the context of road extraction, but they also show that this kind of fundamental research is far from being solved in general.

16.2.2 *Multi-spectral classification for road extraction*

The acquisition of multi-spectral data in the visible domain, not only from airborne sensors but since 1999 also from spaceborne sensors, has reached a resolution regime in which multi-spectral analysis can substantially support road extraction. For instance, Mena & Malpica (2005) use colour information in the course of accumulating road evidence with different statistical parameters: i) supervised classification, employing the Mahalonobis distance; ii) comparison of local colour distributions with trained distributions using the Bhattacharyya distance; and iii) Haralick features derived from the co-occurrence matrix. Classifying an image based on these features yields potential road segments, which are cleaned and skeletonized into road centre axes. In the same spirit, Mohammadzadeh et al. (2006) employ fuzzy clustering and classification before morphological filtering. Zhang & Couloigner (2006) apply a segmentation approach using unsupervised K-means clustering. The regions obtained are evaluated and filtered with the help of generic shape descriptors. As above, the final road centrelines are derived by grouping and skeletonization. In contrast to these approaches, Ziems et al. (2007) use colour information to better identify false alarms when determining potential errors in existing road databases, e.g. if GIS road axes run through fields or bushes. To this end, a statistical analysis of the colour distribution derived from potential road areas in comparison with trained distributions is carried out.

Common classification approaches like those mentioned above yield good hypotheses for roads if enough undisturbed "pavement pixels" are present. Clearly, the most dominating factors are therefore the image resolution and the complexity of the imaged scene. It is thus not surprising that the above approaches show limitations when applied to spaceborne multispectral images—also if pan-sharpened data is involved.

Dealing with mixed pixels is thus an important issue if roads are to be mapped using satellite images. To cope with this, Bacher & Mayer (2004, 2005) developed a two-step strategy. First, training information for a supervised classification is obtained from an initial step of road extraction with very strict parameter settings. To this end, the "line-based" approach of Wiedemann & Hinz (1999), as well as the more "edge-based" approach (Baumgartner et al. 1999), are applied. The results are fed into a multispectral classification to generate a so-called roadclass-image.

This image is subsequently used together with the red (R)-, green (G)-, blue (B)-, and near infra-red (IR)-channel. The final road extraction is done with the line-based approach of Wiedemann & Hinz (1999), but now incorporating all channels simultaneously. By means of this strategy, the linear properties of roads in each channel are exploited and—if available—supplemented with area-based colour information.

While most of the previous approaches show a distinct separation between colour processing (like pixel-based classification) in order to obtain initial hypotheses and subsequent road specific processing to filter and refine the results, future research will more and more integrate colour processing directly into road extraction modules. Subpixel precise extraction of edges and lines, which is one of the basic tools of road extraction, has been adapted to colour images for some years already (Steger 2000). This provides an ideal basis to more deeply integrate colour features into road extraction.

16.2.3 *Road extraction in urban areas*

The extraction of roads in complex environments is one of the challenging issues in photogrammetry and computer vision, since many issues related to automatic scene interpretation are involved. Factors greatly influencing the scene complexity are, for instance, the number of different objects, the degree of their interrelations, and the variability of both. Moreover, each factor—and thus the scene complexity—is related to a particular scale. To account for these factors, detailed modelling, contextual reasoning, multi-view image exploitation and internal evaluation have proven to be of great importance over the past years.

Hinz (2004a, 2004b) employs a detailed scale-dependent road and context model to deal with the high complexity of this type of scene. The corresponding extraction strategy is subdivided into three levels: level 1 comprises the analysis of context, i.e. the segmentation of the scene into the urban, rural and forest areas as well as the analysis of context relations, e.g. the determination of shadow areas and the detection of vehicles; level 2 includes the detection of homogeneous ribbons as preliminary road segments in coarse scale, collinear grouping thin bright road markings in fine scale and the construction of lanes and carriageways from groups of road markings and road sides; level 3, finally, completes the road network by fusing road segments detected in overlapping images, iteratively closing gaps in the extraction and exploiting the network characteristics to generate a topologically complete road network. A key feature of the approach is the incorporation of a scheme for internal evaluation. Hypotheses generated during extraction are internally evaluated so that their relevance for further processing can be assessed. This facilitates decisions inherently appearing during the extraction. The results achieved with the approach show that it is possible to extract roads in complex environments, i.e. the extraction is also possible when the appearance of the roads is heavily affected by other objects.

Recent developments are directed towards improved segmentation techniques to obtain better initial hypotheses. Grote et al. (2007) use normalized graph-cuts to integrate various radiometric and topological characteristics into the segmentation process. Further improvements seem possible if more precise height information is available. An approach that heavily relies on height data is described in Clode et al. (2007). Primary road hypotheses are generated by classification employing colour intensities and height gradients. The result is then vectorized by convolving a complex-valued disk (so-called Phase Coded Disk) with the image. The Phase Coded Disk basically represents the local features of the road model. Centreline and width of the road are obtained from the magnitude image while the direction is determined from the corresponding phase image.

Despite encouraging results, there are still many fundamental questions to be solved for road extraction in urban areas. Which type of modelling is appropriate to capture the variability of the object classes, especially in the light of the success of appearance-based approaches (Section 5 below)? Which relations between objects can support road extraction, and which are more or less clutter? Is it possible to design a strategy that adapts itself to the given extraction without losing control over the computational load? Is it better to start with a monolithic strategy and incorporate dynamic elements or use generic search algorithms and apply heuristics to control the search space?

16.2.4 *Road mapping through car tracking and activity analysis*

Especially with the increasing availability of airborne videos and highly overlapping photogrammetric image sequences, models and strategies comprising temporal features gain more and more interest. These show great potential to add very valuable additional information to the geometric and radiometric properties of roads. In addition to this, first approaches for road mapping by activity analysis and car tracking were recently developed. We also note that, the other way around, airborne monitoring of vehicle activity in urban areas may be decisively improved by exploiting road information (Stilla et al. 2004).

The work by Pless (2006), for instance, aims at detecting temporal changes in stabilized airborne videos.

To this end, no specific object model for roads or traffic is employed. Instead, Pless (2006) developed a generic scheme to discern static background from active foreground on the basis of eigenvalue analysis. Especially in the case of dense traffic, active image regions correspond to the main roads. While this approach is mainly designed for the analysis of inner city areas, the Bayesian car tracking system by Koch et al. (2006) is able to fuse multiple car tracks from different flight paths. This system is particularly interesting since it is based on a comprehensive Bayesian model for the characteristics of the SAR sensor used as well as the detection and fusion scheme. By this means, the inherent uncertainties in the physical and mathematical sensor modelling and track hypothesis generation can be consistently handled. As a final step, the—potentially interrupted—car tracks are transformed into object space and fused to eventually delineate a precise and topologically intact road network.

The recent work in this field of research shows the importance of developing models and strategies that combine evidence from various sources in a sound statistical framework. The evidence may stem, for example, from multiple views, different sensors or external data. The need for and the potential of further developments in this area were recognized some years ago, but it will be even more important in the future as a consequence of the need to better exploit the potential of new airborne and—in particular—spaceborne sensor systems such as RapidEye, Worldview-1, TerraSAR-X, TanDEM-X or Radarsat-2.

16.2.5 *Road extraction from Synthetic Aperture Radar data*

Due to its all-weather capability and its day and night imaging potential, SAR has become an alternative for mapping applications. However, road extraction from SAR images has to cope with the inevitable consequences of the side-looking geometry of SAR, most prominently occlusions caused by shadow and layover. These effects make road extraction complicated, especially in urban areas, but also in open and rural scenes.

To deal with these effects, Negri et al. (2006) utilize a very stringent road model, describing roads as dark, straight lines that meet at intersections. Also additional information can be considered in order to compensate for possible gaps: Wessel & Hinz (2004) and Amberg et al. (2005) introduce context information to this end. Bright features and their contextual relationships are incorporated in the road extraction procedure. Detected vehicles and rows of building layover, as well as metallic scattering caused by road signs, are further indicators of roads.

It has also been shown that the use of multi-aspect SAR images, i.e. illumination from different directions, improves road extraction. Multi-aspect SAR images contain both redundant and complementary information. A sophisticated fusion step provides the ability to combine information from different sensors, the result of which is more accuracy and better information than acquired from one sensor alone. Tupin et al. (2002), dell'Acqua et al. (2003) and Hedman et al. (2007) used this for both real and simulated SAR scenes. To better cope with inherent uncertainties and conflicting hypotheses, Hedman et al. (2007) embed the whole process into a Bayesian framework for linear feature extraction, evaluation and fusion. In addition, background information in the form of local and global context is also incorporated on a statistical basis.

In general, the results achieved with SAR data are promising, yet they cannot keep up with those obtained from optical data of similar spatial resolution. However, due to its all-day and all-weather capabilities, SAR offers unique potential for applications in the context of natural hazards or civil crises. Hence, future efforts will also be directed towards near-real-time processing and rapid mapping with SAR.

16.3 BUILDINGS

Much recent work on building extraction makes use of the distinct 3-D characteristics of buildings clearly visible in digital surface models (DSM), mostly from aerial laser scanner data, but partially also generated by matching aerial images (Section 16.3.1). Section 16.3.2 presents new work on image data, e.g. analysing the scale-space behaviour of building features. Finally, in cities there is a tendency towards much more detailed 3-D models for the facades also, which can only be generated from terrestrial data (Section 16.3.3).

16.3.1 *Detection and delineation of buildings in digital surface models*

Arefi & Hahn (2005) show how to distinguish between terrain and off-terrain points as a first step towards building extraction by means of geodesic morphological filtering. Dorninger & Nothegger (2007) present an approach for 3-D segmentation, which can make full use of high resolution 3-D points (more than twenty per square metre) from laser scanning or image matching showing also at least partially the vertical facades. The basic idea is a full 3-D representation of planes and clustering in parameter space. Results for laser and points from image data show that very high quality models can be generated. Finally, Rottensteiner (2006) gives an approach for the consistent estimation of building parameters by considering geometrical regularities using soft constraints. False hypotheses can be eliminated by means of robust estimation.

Tarsha-Kurdi et al. (2007) compare model- and data-driven approaches for light detection and ranging (LIDAR) data. Basically, the former is seen to consist of models from a library while data-driven is mostly regarded as synonymous with polyhedra. Thus, data-driven approaches are more flexible, yet there is no way to avoid modelling structures that are not buildings.

Durupt & Taillandier (2006) show how by combining surface and cadastral data it is possible to operationally generate building models automatically. Results for hundreds of buildings support their claim. A specific application is given in a paper by Rehor & Bähr (2006): the goal is to decide by means of classification if buildings in laser-scanner data are damaged, e.g. by an earthquake, and what kind of damage has occurred.

The gap of missing structures of facades in classical nadir airborne laser data and missing structures from roofs in terrestrial laser data can be closed by capturing the scene with an oblique looking laser scanner (Hebel & Stilla 2007). For capturing the roof and all facades of buildings, a multi aspect acquisition and registration of point clouds is necessary. Full waveform LIDAR provides an intensity value for each laser point and allows a direct texturing of the scene.

16.3.2 *Building extraction from aerial images*

Drauschke et al. (2006) analyse the scale-space behaviour of features describing buildings. It was found that there is no optimal scale, but the features are mostly stable over a larger scale-interval so that a small number of scales is sufficient for building extraction.

Zebedin et al. (2006, 2007) show how facade planes derived from dense 3-D points can be refined by sweeping the planes and projecting them into all views where they are visible. This does not only lead to very good estimates for the plane, but also for the texture as demonstrated by a number of convincing examples.

Boudet et al. (2006) present a way to conduct quality control for given 3-D building models based on digital aerial images. Measures such as multi-image correlation are employed to decide by means of classification if a roof facet complies with quality specifications.

16.3.3 *Detailed 3-D models for buildings in cities from terrestrial data*

Detailed 3-D models of buildings in cities from very high resolution terrestrial data have become a focus only recently. The work can basically be split into two groups: work focusing on geometry, i.e. planes for facades or modelling by dense depth maps; and work trying to understand the semantics of (parts of) the scene. This section is mainly focusing on the former with only one approach (Becker & Haala 2007) from the second group. The rest of the work focusing on semantics is given in Section 4 on appearance based and generative modelling.

Ikeuchi et al. (2004) employ an omni-directional camera and focus on image based rendering (IBR). 2-D information from a map is employed together with epipolar matching between the images to generate 3-D textured models of buildings. IBR restricts the practical applicability as new 3-D objects need to be integrated in the images used for interpolation. A more flexible approach, as yet restricted to planes, is given in a paper by Wilczkowiak et al. (2005). Parallelepipeds are employed to model constraints such as orthogonality and parallelism. The duality between shape parameters of parallelepipeds and calibration parameters of cameras is utilized for the 3-D reconstruction from single images after manually defining projected points and constraints. Results also for a larger number of images show that this can be a good basis for a semi-automatic extraction of planar models (polyhedra).

A system particularly suitable for the visual 3-D reconstruction of urban areas is presented in papers by Akbarzadeh et al. (2006) and Mordohai et al. (2007). It builds on the work by Nistér (2004) and Pollefeys et al. (2004). Images are taken from a car. While in Akbarzadeh et al. (2006) the orientation is based on the images as well as on GPS and INS (inertial navigation system), Mordohai et al. (2007) only employ GPS and INS as this is faster and they found the high quality Applanix system used to be sufficient for this task. Yet, they note that for light-weight UAVs (unmanned aerial vehicles), orientation based only on images and GPS is an important option. The main part of the work is on the generation of high quality surface models. The images used are below one megapixel, but a given part of the scene can usually be seen in four up to 14 images. The algorithm uses plane sweeping with a number of differently oriented planes making use of all given images. To avoid wrong matches due to occlusions, the images are organized in two sets representing the two sides of the central image and the solution for the better side is finally used. The implementation makes heavy use of the graphical processing unit (GPU) and thus results can be computed in real-time, i.e. as fast as the images are acquired. The results are visually impressive. Mordohai et al. (2007) also give first results for a quantitative evaluation. For one building captured with 3000 images, a median error of 2.6 cm and an average error of 6.6 cm were achieved with 83% of the points within 5 cm of the ground truth. The completeness was 73%.

A general problem encountered when trying to generate more semantically meaningful entities consists of the determination of geometrical primitives such as planes, spheres, cylinders and cones from 3-D point clouds. The latter might stem from laser scanning or by image matching. A particularly sophisticated and extremely fast approach to determine geometrical primitives based on Random Sample Consensus (RANSAC) (Fischler and Bolles 1982) is given by Schnabel et al. (2007).

Finally, Becker & Haala (2007) is the only approach aiming at a semantic interpretation that is explained in this section as it does not use the techniques presented in Section 5. It combines data from a laser scanner with image data. The facade plane is assumed to be given together with the vertical direction. The plane is decomposed into rectangular tiles, which are analysed to obtain rectangular configurations combining the laser and the image data. Finally, detailed 3-D structures for the windows (mullions and transoms) are generated via image matching.

16.4 VEGETATION

16.4.1 *Extraction of trees from aerial and terrestrial laser scanner data*

Conventional airborne LIDAR, which records the first and the last pulse, was successfully applied to retrieve forest parameters such as tree height, crown diameter, number of stems, stem diameter and basal area on the tree level (Hyyppä et al. 2004). Also, tree species classification became feasible with first/last pulse scanning systems providing a high point density (Brandtberg 2007). Recently, the successful application of novel small footprint full waveform systems to tree species classification was reported (Reitberger et al. 2006) using advantageously the intensity and the pulse width. Approaches to tree species classification are usually based on single tree segmentation that separates the tree crown from the outer geometry of the forest surface. The above methods have in common that they reconstruct—at least locally—the Canopy Height Model (CHM) to find the local maximum as the best guess for the stem position and delineate a segment polygon as the tree crown.

Reitberger et al. (2007) segment single trees by a watershed algorithm and an approach to stem detection that clusters hierarchically potential stem reflections and reconstructs the stem with a RANSAC-based adjustment. The analysis of waveform data in the tree structure shows that intensity and pulse width significantly help to discriminate stem points, crown points and ground points. Moreover, the mean intensity of stem points turned out to be the most salient feature for the discrimination of coniferous and deciduous trees.

Work on tree extraction from terrestrial laser scanner data is scarce. A particular exception is the paper by Barnea et al. (2007). They segment images with the Mean-Shift algorithm and then classify trees using, e.g. a cornerness measure, via k-nearest neighbours and fine tune via Graph Cuts. First results for one data set show the basic feasibility of the approach.

16.4.2 *Image-based tree modelling*

An approach combining airborne laser scanner and image data to position treetops in 3-D is presented by Korpela (2006, 2007). The basic idea is to restrict multiple image matching to a 3-D search volume given by a LIDAR surface model.

Quan et al. (2006) present an approach for the semi-automatic construction of realistic looking tree models from close-range images. It is based on a high quality structure from motion and dense depth estimation approach, as well as image segmentation, and allows automatic recovery of the geometry of all leaves by fitting a deformable leaf model. The quality of the derived models is good, but the manual effort is high and it only works well for relatively large leaves and thick branches compared to the resolution/pixel-size. Some of these shortcomings are overcome by the work of Tan et al. (2007) employing shape patterns of visible branches of parts of a tree to reconstruct missing parts of the tree.

Another recent approach for the extraction of trees from image sequences employs a stochastical generative approach and is thus presented in Section 16.5 below.

16.5 APPEARANCE BASED AND STOCHASTICAL GENERATIVE MODELLING

While the focus for modelling and object extraction has been on bottom-up feature-based methods and (perceptual) grouping for at least two decades, recently there has been a tendency towards top-down generative as well as appearance based modelling. Particularly for the former, modelling of the uncertainty of data and especially models has become a focus of research. Appearance-based and generative modelling are employed for roads as well as traffic and even more intensively for buildings and particularly their facades. There is also recent work on the 3-D reconstruction of unfoliaged deciduous trees, avoiding the difficult matching of branches between images.

16.5.1 *Appearance based modelling*

An important reason for the popularity of appearance-based modelling is arguably the Scale-Invariant Feature Transform (SIFT) by David Lowe (2004). The basic idea is to extract point-structures from the image, possibly over different scales, and then use the image function around the points in the form of patches to represent parts of objects. By additionally describing the spatial arrangement of these points for an object, a flexible description for textured, compact, more or less rigid objects is obtained, which can be generated more or less automatically from images and possibly a segmentation of the object, i.e. no 3-D model of the object is needed.

Key questions for the patches are how to normalize them for rotation and scale and particularly how to compare patches from a new image and from the model. In the most simple case, one can use (normalized) cross correlation, but the scheme used in the SIFT operator, which is based on (normalized) orientation histograms, has proven to be rather successful (Mikolajczyk & Schmid 2005). Another issue is the arrangement of the local patches. It is often detected robustly by the Hough-transform.

In addition to the work by Lowe, Agarwal et al. (2004) in particular have raised the awareness of the potential of appearance-based modelling. While they show that it is possible to recognize cars in side views at different scales, their work is lacking an efficient way to deal with occlusions. For this, it is helpful if the appearance-based approach does not just localize (the centre of) an object, but segment the object. Thus, a more fine-grained decision about the acceptance or rejection of hypotheses is possible. An important contribution is by Leibe & Schiele (2004a, b) on Implicit Shape Models (ISM), which employs probabilistic reasoning for the decisions. Finally, posing the problem of appearance-based recognition in a probabilistic framework, Fei-Fei et al. (2006) have shown that it is possible to learn the representation for a large number of classes (101) from only a few data when using a weak general prior for initialization.

As the description of the objects can be generated more or less automatically from images, because the texture is used as an additional information source, and due to a robust modelling, e.g. using the Hough transform, possibly together with a principled probabilistic formulation, appearance-based approaches overcome many of the limitations of approaches based on aspect graphs from the 1990s. For the latter, 3-D models are a prerequisite and the handling of occlusions is difficult. Yet, though appearance-based models can be rather powerful, one has to note that their application relies on a number of assumptions, e.g. for a road, there are for larger portions neither characteristic image points nor can the spatial arrangement be described in terms of points.

16.5.2 *Generative stochastical modelling*

In 2000, David Mumford announced the "dawning of the age of stochasticity" (Mumford 2000). Since then, generative stochastical models have made clear their potential for object extraction.

The basic idea of generative modelling is to take a 3-D geometrical model including its physical characteristics and to project it into the image space using (at least approximately) known imaging parameters. Usually, the generated predicted and the real images will be different due to different parameters in imaging, but even more of the model. The goal is to modify the parameters in a way to make the predicted images more and more similar to the real images. As the parameter space is usually vast and with many local minima, stochastical sampling is used to find a solution. In many cases variations of Markov Chain Monte Carlo (MCMC) (Neal 1993) is employed to find a meaningful solution.

Yet, a big problem with MCMC is that it is limited to one set of parameters. It is not possible to compare solutions with different numbers of parameters, e.g. the fit of a plane with three parameters to the fit of a sphere with four parameters. Fortunately, Green (1995) introduced Reversible Jump (RJ) MCMC, allowing a change in the number, but also the type, of parameters.

Dick et al. (2004) have shown the potential of generative stochastical modelling for topographic object extraction. In particular, they have demonstrated that one can interpret buildings and especially their facades by stochastically varying and projecting models into images.

An important advantage of generative stochastical models is their ability to show what is actually modelled by means of simulation. The 3-D physical model consists partly of probability distributions, i.e. a window is about 1.2 to 1.5 m high with a probability of 20%. By sampling the distributions, one obtains instances of the given objects. One can "verify" a model by generating a larger number of instances by random sampling and showing that they resemble the real objects. If they do not, the model is wrong. This is in contrast to many other types of modelling where one can only judge if the final recognition results are good, though one does not know if one has actually correctly modelled the object. Dick et al. (2004) employ RJMCMC, as only by the reversible jumps is it possible to switch between different numbers of objects, e.g. change the numbers of windows for a facade.

To generate 3-D models of buildings similar to those given in the images, the 3-D models are projected into the geometry of the given images based on the usually known parameters of the cameras. This results in simulated images, which are compared to the given images giving likelihoods and together with suitably chosen priors posterior probabilities. By stochastic variation, different models are generated using RJMCMC and the best is chosen according to the posterior probability.

16.5.3 *Roads and traffic*

The arguably scientifically most interesting work on road extraction is by Stoica et al. (2004) using generative stochastical modelling. They employ a Gibbs point process framework. A data term describing the roads in the image in the form of statistical hypothesis tests is linked to a prior stating that roads form a connected network of lines, favouring aligned segments and penalizing superposition. An estimate for the road network is found by minimizing an energy function employing RJMCMC. As for Dick et al. (2004) above, an interesting feature of their approach is that one can sample the prior distribution. If this results in road-like structures, one knows that the structures that one describes in the model are actually road-like. Results for SPOT and aerial optical images, as well as ERS radar data, show the feasibility of the approach.

Another approach, which uses statistics, but for learning road structures and particularly changes in their characteristics in semi-automated road tracking, is by Zhou et al. (2007). Road trackers are learned as weighted combinations of training profiles and changes in the road structures are detected in the form of novelties by means of one class support vector machines (SVM). Results on aerial images with 1 m resolution and also by comparison to ground truth data show that the learning helps, but also makes clear that the approach is still far from modelling all details existing in the real data.

Finally, Nguyen et al. (2007) show how cars can be detected in aerial images with a ground resolution of about 10 cm based on an appearance-based model and a learning approach, namely boosting. The latter converts (boosts) a weak learning algorithm into a stronger one. An on-line version of boosting is employed together with "active" learning: The user only labels informative examples (positive or negative), which have not been labelled correctly by the current system. Results on ground truth data show a correctness of 80% for a completeness of about 80%. Kluckner et al. (2007) have extended the above approach by means of a stereo matcher to check hypothesis. This enables the automatic learning of a high quality model very close to the hand-generated model from only one car given by the user.

16.5.4 *Buildings with focus on facades*

Facade interpretation from terrestrial images, wide-baseline image sequences, or laser scanner data based on stochastical generative modelling has been a focus of research since the seminal paper by Dick et al. (2004) as described in Section 5.2 above. The results shown are impressive though restricted to a limited number of objects (facades, windows, columns, etc.), as no automatic learning is used to generate the models.

Alegre & Dallaert (2004), Brenner & Ripperda (2006), as well as Ripperda & Brenner (2007) all focus on the hierarchical nature of the arrangement of objects on facades. They recursively split facades employing a formal grammar deriving a structural description in the form of a derivation tree. The splits are controlled stochastically based on RJMCMC. Ripperda & Brenner (2007) particularly employ image as well as depth (terrestrial laser) data. The distributions for splits rely on colour or depth changes for larger parts of the facade as well as self-similarity, as measured by autocorrelation for individual windows.

Müller et al. (2007) and Van Gool et al. (2007) present two approaches for single images of facades with weak and strong perspective effects that make use of the hierarchical nature of the arrangement of objects on facades as in the paragraph above, but not in a stochastical way. In both cases, first the vertical and a horizontal vanishing point are determined. Müller et al. (2007) employ images of facades with no or weak perspective effects, which are rectified based on the information about the vanishing points. On the rectified images, symmetries are detected based on mutual information. They are used to split the facades into floors and tiles. The individual tiles are first split separately into smaller rectangles and then translational symmetries are used to obtain a robust subdivision. In a next step, 3-D objects from a library are matched via a generative approach to the rectangles and finally the depth of groups of rectangles is determined manually. Van Gool et al. (2007) are focusing on images with strong perspective. Chain wise similar feature points are employed to determine groups and from them symmetries as well as—assuming horizontal and vertical repetitions—also the calibration of the camera. Given the calibration, the repetition in the groups as well as the depth values are optimized via graph cuts, regularized by a smoothness constraint for the depth. The resulting depth map is finally refined by means of a shape prior preferring rectangles. A large number of results for both approaches looks very promising. The authors note that the use of multiple images might be helpful.

Mayer & Reznik (2007) employ multiple images to automatically generate facade planes oriented vertically by determining the vertical vanishing point. They use information in patches around corners in the rectified facade images to delineate windows in an appearance-based way using ISM (Section 5.1) on dark facades only. A generative approach simulating windows by black rectangles is employed for bright facades. Reznik & Mayer (2007) utilize ISM coherently for the appearance-based detection as well as for the delineation of windows on any kind of facade. The windows are grouped into rows and columns and model selection based on Akaike's Information Criterion (AIC) (Akaike 1973) is employed to compare different groupings. Individual windows always lead to the best likelihood as they can adapt to the individual shapes. Only by taking into account the lower number of parameters for rows, columns, etc., do they prevail. The 3-D extent of the windows is determined via plane sweeping.

Finally, the paper by Ortner et al. (2006) is one of the few examples where generative stochastical modelling in the form of a point process is employed for aerial data, particularly height data generated by laser scanning and by matching of images. The model consists of rectangles for which height profiles parallel to two of their sides are used to define the likelihood. Additionally, interactions are defined as priors, such that buildings exclude each other and are aligned. The model is sampled by means of RJMCMC and results show potential, though computation times in the range of hours for a megapixel image are a prohibitive factor for practical use in the near future.

16.5.5 *Trees*

Huang & Mayer (2007a, b) present a generative stochastical approach for the 3-D extraction of the branching structure of unfoliaged deciduous trees from urban image sequences. The trees are modelled in 3-D by means of L-systems (Prusinkiewicz & Lindenmayer 1990). By this means, realistically looking images of trees can be generated and visualized by means of computer graphics. The idea of extracting given trees in images is to vary the parameters stochastically with MCMC to generate 3-D trees whose simulated images look similar to the given images. To restrict the search space, the trunk is extracted as a vertical straight line and matched employing trifocal tensors (Hartley & Zisserman 2003). The comparison between given and simulated images is accomplished via cross correlation. By means of this generative stochastical procedure, the approach overcomes the complexity and uncertainty of extracting and matching branches as lines in several images due to weak contrast, background clutter and particularly the varying order of branches when projected into different images. First results show that the approach is basically feasible.

16.6 EVALUATION

While evaluation has been intensively discussed in the 1980s and 1990s (Price 1986, Förstner 1996), it has meanwhile reached an outstanding status. In areas such as stereo matching (Scharstein & Szeliski 2002), it is practically impossible to get a paper accepted by a high-level conference without evaluating the presented approach. In the following, we distinguish two types of evaluation: i) external evaluation, and ii) internal evaluation, the latter sometimes also being referred to as self-diagnosis. Both internal and external evaluation yield quantitative results, which should be independent of a human observer as much as possible.

The aim of internal evaluation is to estimate the geometric and semantic accuracy of the extracted objects during extraction. This information can only be derived from redundancies within the underlying data or the incorporated object knowledge. In this context, "object knowledge" means knowledge that is purely described by the object model and not by other external data. Therefore, internal evaluation is strongly linked to the extraction process. The results of internal evaluation are particularly important if the extraction results are combined with other data, e.g. if they are fused with results from other extraction approaches or if they are used for the update of GIS data. On the other hand, they are also very useful for guiding a human operator during post-editing the results of an automatic extraction. If the system provides the operator with confidence values characterizing its performance, the time needed for inspecting the results can be reduced significantly (Fischler & Heller 1998).

Concerning internal evaluation, only a little research has been conducted so far. Many approaches derive confidence values from low level features such as lines or edges. In the following steps, the values are propagated and aggregated, eventually providing a basis for the final decision about the presence of the desired object. For instance, Tupin et al. (1999) apply this to the interpretation of SAR satellite scenes. There are also attempts to validate the decisions by interpretation of the local context, which enhances the reliability of the results (Ruskoné 1996). However, these evaluations are typically based on local criteria like quality measures for adjacency, parallelism or collinearity, and thus they are hardly able to assess the extraction quality of the whole object. An attempt to achieve a highly independent evaluation is made by Hinz & Wiedemann (2004). They propose to utilize "knowledge redundancies", e.g. in the form of object properties that have not been used during the extraction. This concept is applied to two different road extraction systems and, based on a test series on aerial images, it is made clear how results with attached confidence values can increase system efficiency for practical applications. It is shown that the developed internal evaluation is a very helpful tool to support and guide an operator through the road network by pointing to such parts where problems or uncertainties occurred during the extraction. It also shows great potential to improve an automatic extraction, mainly due to it being less sensitive to improper parameter settings.

In contrast to internal evaluation, external evaluation determines quality measures referring to externally given objects. This is usually carried out by a comparison of the extraction results with reference data of superior quality. External evaluation is therefore independent of the extraction approach. The results of the external evaluation can be used for a comparison of different extraction approaches and to learn of the strengths and weaknesses of such approaches in practical applications. The recent EuroSDR test on road extraction (Mayer et al. 2006) is a typical example. Also for building extraction from aerial images and laser scanner data, a EuroSDR test (Kaartinen et al. 2005) has been conducted and Pfeifer et al. (2007) report on a comparison of two approaches for building footprint extraction from airborne laser scanner data.

Typically, normalized measures like completeness, correctness and geometric RMS (root mean square) have been introduced to evaluate an extraction approach on a very general basis (Heipke et al. 1998). Other generic measures that can be inferred from them (and vice versa) are detection rate, false alarm rate, or branching factor (Shufelt 1999). In general, evaluation schemes are not free of parameters. For the evaluation of road axes, the width of the reference's tolerance buffer is a typical parameter. The conclusions drawn from an external evaluation must thus be seen in the light of the particular goals and thus parameter settings of the evaluation procedure. Unfortunately, this fact is sometimes heavily underestimated.

Another important point of external evaluation concerns the increasing detail of the object models used, which naturally also calls for more detailed validation schemes. The evaluation methodology for road networks presented by Wiedemann (2003) may serve as an example of this. Besides the above-mentioned measures, which typically are calculated for the validation of road centre axes, functional properties of the road network are also evaluated. More specifically, Wiedemann (2003) developed a methodology to validate topological features of a road network by comparing network junctions, length-weighted connected components, as well as optimum paths calculated for the reference and the extracted network. The topological measures add very valuable information to the overall assessment of an automatic extraction since they indicate to what extent a particular extraction approach is able to produce results that comply with the original purpose of roads, i.e. connecting various places that are important in daily life.

A similar situation appears for the evaluation of other complex objects, especially buildings. Often, the data sets are compared pixel by pixel, the shift of corner points being taken as the basis for the comparison (Kaartinen et al. 2005), or only the area of the reference building footprint is compared to the extraction yielding a percentage of coverage. All this, however, does not allow a statement as to whether the outline has been delineated correctly or whether certain important parts of a building have been extracted or missed, let alone that the shape has been preserved.

16.7 CONCLUSIONS AND OUTLOOK

The four years since the ISPRS Istanbul congress in 2004 have seen a large number of novel approaches evolve for the automated extraction of buildings, roads and vegetation. We have shown that there are particularly trends towards appearance-based and generative approaches modelling the uncertainty of the models and the data by stochastic means, including learning.

As the models used are partly of very different complexity, the problem of model selection arises. For this, e.g. Akaike's Information Criterion (AIC) (Akaike 1973), or Minimum Description Length (MDL) (Rissanen 1978, 1987) can be employed. A comparison between both for geometrical problems is given by Schindler & Suter (2006). Yet, some proponents argue that a full Bayesian approach with appropriate priors would avoid the model selection problem. Unfortunately, the correct priors need to be known for this.

While generative modelling is conceptually attractive, it can be rather inefficient. One way to mitigate this is by integrating generative—top-down and discriminative—bottom-up modelling. Tu et al. (2005) have presented a way of doing this in a coherent way. Finally, Geman et al. (2002) have shown how probabilistic modelling can be made sound and efficient by means of a hierarchical composition of objects.

REFERENCES

Agarwal, S., Awan, A. &. Roth, D. 2004. Learning to detect objects in images via a sparse, part-based representation. *IEEE Transactions on Pattern Analysis and Machine Intelligence* 26(11): 1475–1490.

Akaike, H. 1973. Information theory and an extension of the maximum likelihood principle. *Second International Symposium on Information Theory*: 267–281.

Akbarzadeh, A., Frahm, J.-M., Mordohai, P., Clipp, B., Engels, C., Gallup, D., Merrell, P., Phelps, M., Sinha, S., Talton, B., Wang, L., Yang, Q., Stewenius, H., Yang, R., Welch, G., Towles, H., Nistér, D. &. Pollefeys, M. 2006. Towards urban 3-D reconstruction from video. *International Workshop on Urban Reconstruction*: 1–6.

Alegre, F. &. Dallaert, F. 2004. A probabilistic approach to the semantic interpretation of building facades. *International Workshop on Vision Techniques Applied to the Rehabilitation of City Centres*: 1–12.

Amberg, V., Coulon M., Marthon P. & Spigai, M. 2005. Improvement of road extraction in hihg.resolution SAR databyacontext-basedapproach. *InternationalGeoscience and Remote Sensing Symposium, 2005* 1: 490–493.

Arefi, H. &. Hahn, M. 2005. A morphological reconstruction algorithm for separating off-terrain points from terrain points in laser scanning data. *International Archives of the Photogrammetry, Remote Sensing and Spatial Information Sciences* 35 (3/W19): 120–125.

Bacher, U. &. Mayer, H. 2004. Automatic road extraction from IRS satellite images in agricultural and desert areas. *International Archives of the Photogrammetry, Remote Sensing and Spatial Information Sciences* 35(B3): 1055–1060.

Bacher, U. &. Mayer, H. 2005. Automatic road extraction from multispectral high resolution satellite images. *International Archives of the Photogrammetry, Remote Sensing and Spatial Information Sciences* 36(3/W24): 29–34.

Barnea, S., Filin, S. &. Alchanatis, V. 2007. A supervised approach for object extraction from terrestrial laser point clouds demonstrated on trees. *International Archives of the Photogrammetry, Remote Sensing and Spatial Information Sciences* 35(3/W49A): 135–140.

Baumgartner, A., Steger, C., Mayer, H., Eckstein, W. &. Ebner, H. 1999. Automatic road extraction based on multi-scale, grouping, and context. *Photogrammetric Engineering & Remote Sensing* 65(7): 777–785.

Becker, S. &. Haala, N. 2007. Refinement of building facades by integrated processing of LIDAR and image data. *International Archives of the Photogrammetry, Remote Sensing and Spatial Information Sciences* 36(3/W49A): 7–12.

Boudet, L., Paparoditis, N., Jung, F., Martinoty, G. &. Pierrot-Deseilligny, M. 2006. A supervised classification approach towards quality self-diagnosis of 3-D building models using digital aerial images. *International Archives of the Photogrammetry, Remote Sensing and Spatial Information Sciences* 35(3): 136–141.

Brandtberg, T. 2007. Classifying individual tree species under leaf-off and leaf-on conditions using airborne LIDAR. *ISPRS Journal of Photogrammetry and Remote Sensing* 61: 325–340.

Brenner, C. &. Ripperda, N. 2006. Extraction of facades using RJMCMC and constraint equations. *International Archives of the Photogrammetry, Remote Sensing and Spatial Information Sciences* 36(3): 155–160.

Clode, S., Rottensteiner, F., Kootsookos, P. &. Zelniker E. 2007. Detection and vectorization of roads from LIDAR data. *Photogrammetric Engineering & Remote Sensing* 73: 517–536.

dell'Acqua, F., Gamba, P. &. Lisini, G. 2003. Improvements to urban area characterization using multitemporal and multiangle SAR images. *IEEE Transactions on Geoscience and Remote Sensing* 4(9): 1996–2004.

Dick, A., Torr, P. &. Cipolla, R. 2004. Modelling and interpretation of architecture from several images. *International Journal of Computer Vision* 60(2): 111–134.

Dorninger, P. &. Nothegger, C. 2007. 3-D segmentation of unstructured point clouds for building modeling. *International Archives of the Photogrammetry, Remote Sensing and Spatial Information Sciences* 35(3/W49A): 191–196.

Drauschke, M., Schuster, H.-F. & Förstner, W. 2006. Detectability of buildings in aerial images over scale-space. *International Archives of the Photogrammetry, Remote Sensing and Spatial Information Sciences* 35(3): 7–12.

Durupt, M. & Taillandier, F. 2006. Automatic building reconstruction from a digital elevation model and cadastral data: An operational approach. *International Archives of the Photogrammetry, Remote Sensing and Spatial Information Sciences* 35(3): 142–147.

Fei-Fei, L., Fergus, R. & Perona, P. 2006. One-shot learning of object categories. *IEEE Transactions on Pattern Analysis and Machine Intelligence* 28(4): 594–611.

Fischler, M. & Bolles, R. 1981. Random Sample Consensus: A paradigm for model fitting with applications to image analysis and automated cartography. *Communications of the ACM* 24(6): 381–395.

Fischler, M. & Heller, A. 1998. Automated techniques for road network modeling. *Proceedings of Image Understanding Workshop '98*: 501–516.

Förstner, W. 1996. 10 pros and cons against performance characterization of vision algorithms. *Workshop on Performance Characteristics of Vision Algorithms*: 13–29.

Geman, S., Potter, D. & Chi, Z. 2002. Composition systems. *Quarterly of Applied Mathematics* LX: 707–736.

Green, P. 1995. Reversible Jump Markov Chain Monte Carlo computationandBayesianmodeldetermination. *Biometrika* 82(4): 711–732.

Grote, A., Butenuth, M. & Heipke, C. 2007. Road extraction in suburban areas based on normalized cuts. *International Archives of Photogrammetry, Remote Sensing and Spatial Information Sciences* 36(3/W49A): 51–56.

Hartley, R. & Zisserman, A. 2003. *Multiple view geometry in computer vision—second edition.* Cambridge, UK: Cambridge University Press.

Hebel, M. &.Stilla, U. 2007. Automatic registration of laser point clouds of urban areas. *International Archives of the Photogrammetry, Remote Sensing and Spatial Information Sciences* 36(3/W49A): 13–18.

Hedman, K., Hinz, S. & Stilla, U. 2007. Road extraction from SAR multi-aspect data supported by a statistical context-based fusion. *Proceedings of IEEE-ISPRS Workshop URBAN 2007*, (on CD).

Hedman, K., Wessel, B. & Stilla, U. 2005. A fusion strategy for extracted road networks from Multi-Aspect SAR Images. *International Archives of the Photogrammetry, Remote Sensing and Spatial Information Sciences* 36 (3/W24): 185–190.

Heipke, C., Mayer, H., Wiedemann, C. & Jamet, O. 1998. External evaluation of automatically extracted road axes, *Photogrammetrie—Fernerkundung—Geoinformation (PFG)* 2: 81–94.

Heuwold, J. 2006. Verification of a methodology for the automatic scale-dependent adaptation of object models. *International Archives of the Photogrammetry, Remote Sensing and Spatial Information Sciences* 36(3): 173–17.

Heuwold, J. & Pakzad, K. 2006. Automatic scale-dependent adaptation of variable object models. *International Archives of the Photogrammetry, Remote Sensing and Spatial Information Sciences* 36 (W40): 65–70.

Heuwold, J., Pakzad, K. & Heipke, C. 2007. Scale behaviour prediction of image analysis models for 2-D landscape objects. *International Archives of Photogrammetry, Remote Sensing and Spatial Information Sciences* 36 (3/W49A): 43–50.

Hinz, S. 2004a. *Automatische Extraktion urbaner Straßennetze aus Luftbildern.* PhD thesis, Deutsche Geodätische Kommission **C** (580). Munich, Germany.

Hinz, S. 2004b, Automatic road extraction in urban scenes—and beyond. *International Archives of Photogrammetry, Remote Sensing and Spatial Information Sciences* 35 (B3): 349–354.

Hinz, S. & Wiedemann, C. 2004. Increasing efficiency of road extraction by self-diagnosis. *Photogrammetric Engineering and Remote Sensing* 70(12): 1457–1466.

Huang, H. & Mayer, H. 2007. Extraction of 3-D unfoliaged trees from image sequences via a generative statistical approach. *Pattern Recognition—29th DAGM Symposium*: 385–394. Berlin, Germany: Springer.

Huang, H. & Mayer, H. 2007. Extraction of the 3D branching structure of unfoliaged deciduous trees from image sequences. *Photogrammetrie—Fernerkundung—Geoinformation* 6/07: 429–436.

Hyyppä, J., Hyyppä, H., Litkey, P., Yu, X., Haggren, H., Rönnholm, P., Pyysalo, U., Pikänen, J. & Maltamo, M. 2004. Algorithms and methods of airborne laser scanning for forest measurements. *International Archives of Photogrammetry, Remote Sensing and Spatial Information Sciences* 36(8/W2): 82–89.

Ikeuchi, K., Sakauchi, M., Kawasaki, H. & Sato, I. 2004. Constructing virtual cities by using panoramic images. *International Journal of Computer Vision* 58(3): 237–247.

Kaartinen, H., Hyyppä, J., Gülch, E., Vosselman, G., Hyyppä, H., Matikainen, L., Hofmann, A.D., Mäder, U., Persson, Å., Söderman, U., Elmqvist, M., Ruiz, A., Dragoja, M., Flamanc, D., Maillet, G., Kersten, T., Carl, J., Hau, R., Wild, E., Frederiksen, L., Holmgaard, J. & Vester, K. 2005. Accuracy of 3D city models: EuroSDR comparison. *International Archives of Photogrammetry, Remote Sensing and Spatial Information Sciences* 36(3/W19): 227–232.

Kirchhof, M., Jutzi, B. & Stilla, U. 2007. Iterative processing of laser scanning data by full waveform analysis. *ISPRS Journal of Photogrammetry & Remote Sensing* 63(1): 99–114.

Kluckner, S., Pacher, G., Grabner, H., Bischof, H. & Bauer, J. 2007. A 3-D teacher for car detection in aerial images. *Eleventh IEEE International Conference on Computer Vision, Workshop on 3D Representation for Recognition (3dRR-07)*: 1–8.

Koch, W., Koller, J. & Ulmke, M. 2006. Detection of moving objects in airborne thermal videos. *ISPRS Journal of Photogrammetry and Remote Sensing* 61(3–4): 197–208.

Korpela, I. 2006. 3-D tree top positioning by multiple image matching of aerial images in a 3-D search volume bounded by LIDAR surface models. *International Archives of the Photogrammetry, Remote Sensing and Spatial Information Sciences* 35(3): 191–196.

Korpela, I. 2007. 3-D tree top positioning by multiple image matching of aerial images in a 3-D search volume bounded by LIDAR surface models. *Photogrammetrie—Fernerkundung—Geoinformation*, 1/07: 35–44.

Leibe, B. & Schiele, B. 2004a. Combined object categorization and segmentation with an Implicit Shape Model. *ECCV'04 Workshop on Statistical Learning in Computer Vision*: 1–15.

Leibe, B. & Schiele, B. 2004b. Scale-Invariant object categorization using a scale-adaptive mean-shift search. *Pattern Recognition—DAGM 2004*: 145–153. Berlin, Germany: Springer.

Lowe, D. 2004. Distinctive image features from scale-invariant keypoints. *International Journal of Computer Vision* 60 (2): 91–110.

Mayer, H. 1998. Automatische Objektextraktion aus digitalen Luftbildern. Habilitation. *Deutsche Geodätische Kommission* C (494). Munich, Germany.

Mayer, H. 2008. Object extraction in photogrammetric computer vision. *ISPRS Journal of Photogrammetry and Remote Sensing* (in press).

Mayer, H., Hinz, S., Bacher, U. & Baltsavias, E. 2006. A test of automatic road extraction approaches. *International Archives of the Photogrammetry, Remote Sensing and Spatial Information Sciences* 36(3): 209–214.

Mayer, H. & Reznik, S. 2007. Building facade interpretation from uncalibrated wide-baseline image sequences. *ISPRS Journal of Photogrammetry and Remote Sensing* 61(6): 371–380.

Mayer, H. & Steger, C. 1998. Scale-Space Events and Their Link to Abstraction for Road Extraction. *ISPRS Journal of Photogrammetry and Remote Sensing* 53(2): 62–75.

Mena, J. & Malpica, J. 2005. An automatic method for road extraction in rural and semi-urban areas starting from high resolution satellite imagery. *Pattern Recognition Letters* 26: 1201–1220.

Mohammadzadeh, A., Tavakoli, A. &.Valadan Zoej, M.J. 2006. Road extraction based on fuzzy logic and mathematical morphology from pan-sharpened IKONOS images. *The Photogrammetric Record* 21(113): 44–60.

Mordohai, P., Frahm, J.-M., Akbarzadeh, A., Clipp, B., Engels, C., Gallup, D., Merrell, P., Salmi, C., Sinha, S., Talton, B., Wang, L., Yang, Q., Stewenius, H., Towles, H., Welch, G., Yang, R., Pollefeys, M. & Nistér, D. 2007. Real-time video-based reconstruction of urban environments. *International Archives of the Photogrammetry, Remote Sensing and Spatial Information Sciences* 36(5/W47): 1–8.

Müller, P., Zeng, G., Wonka, P. & Van Gool, L. 2007. Image-based procedural modeling of facades. *SIGGRAPH—ACM Transactions on Graphic* 26(3), Article 85: 1–9.

Mumford, D. 2000. The dawning of the age of stochasticity. *Mathematics: Frontiers and Perspectives—American Mathematical Society: 1–23.*

Neal, R. 1993. *Probabilistic inference using Markov Chain Monte Carlo methods. Technical Report CRG-TR-93-1.* Department of Computer Science, University of Toronto.

Negri, M., Gamba, P., Lisini, G. & Tupin, F. 2006. Junction-aware extraction and regularization of urban road networks in high-resolution SAR images. *IEEE Transactions on Geoscience and Remote Sensing* 44(10), 2962–2971.

Nguyen, T., Grabner, H., Gruber, B. & Bischof, H. 2007. On-line boosting for car detection from aerial images. *IEEE International Conference on Research, Innovation and Vision for the Future (RIVF'07):* 87–95.
Nistér, D. 2004. An efficient solution to the five-point relative pose problem. *IEEE Transactions on Pattern Analysis and Machine Intelligence* 26(6): 756–770.
Pfeifer, N., Rutzinger, M., Rottensteiner, F., Muecke, W. & Hollaus, M. 2007. Extraction of building footprints from airborne laser scanning: Comparison and validation techniques. *Urban Remote Sensing Joint Event, IEEE*: 1–9.
Pless, R. 2006. Detecting roads in stabilized video with the spatio-temporal structure tensor. *Geoinformatica* 10(1): 39–56.
Pollefeys, M., van Gool, L., Vergauwen, M., Verbiest, F., Cornelis, K. & Tops, J. 2004. Visual modeling with a hand-held camera. *International Journal of Computer Vision* 59(3): 207–232.
Price, K. 1986. Anything you can do, I can do better (No you can't). ... *Computer Vision, Graphics, and Image Processing* 36(2–3): 387–391.
Prusinkiewicz, P. & Lindenmayer, A. 1990. T*he algorithmic beauty of plants*. Berlin, Germany: Springer.
Quan, L., Tan, P., Zeng, G., Yuan, L., Wang, J. & Kang, S.B. 2006. Image-based plant modeling. *SIGGRAPH—ACM Transactions on Graphics* 25(3): 772–778.
Rehor, M. & Bähr, H.-P. 2006. Segmentation of damaged buildings from laser-scanning data. *International Archives of the Photogrammetry, Remote Sensing and Spatial Information Sciences* 35(3): 67–72.
Reitberger, J., Krzystek, P. &.Stilla, U. 2006. Analysis of full waveform LIDAR data for tree species classification. *International Archives of the Photogrammetry, Remote Sensing and Spatial Information Sciences* 36(3): 228–233.
Reitberger, J., Krzystek, P. & Stilla, U. 2007. Combined tree segmentation and stem detection using full waveform LIDAR data. *International Archives of Photogrammetry, Remote Sensing, and Spatial Information Sciences* 36 (3/W52): 332–337.
Reznik, S. & Mayer, H. (2007). Implicit Shape Models, model selection, and plane sweeping for 3D facade interpretation. *International Archives of the Photogrammetry, Remote Sensing and Spatial Information Sciences* 36(3/W49A): 173–178.
Ripperda, N. & Brenner, C. 2007. Data driven rule proposal for grammar based facade reconstruction. *International Archives of the Photogrammetry, Remote Sensing and Spatial Information Sciences* 36(3/W49A): 1–6.
Rissanen, J. 1978. Modeling by shortest data description. *Automatica* 14: 465–471.
Rissanen, J. 1987. Minimum-Description-Length principle. *Encyclopedia of Statistical Sciences* 5: 523–527. New York, USA: John Wiley & Sons, Inc.
Rottensteiner, F. 2006. Consistent estimation of building parameters considering geometric regularities by soft constraints. *International Archives of the Photogrammetry, Remote Sensing and Spatial Information Sciences* 35(3): 13–18.
Ruskoné, R. 1996. *Road network automatic extraction by local context interpretation: application to the production of cartographic data. PhD. Thesis. Université Marne-La-Valleé.*
Scharstein, D. & Szeliski, R. 2002. A taxonomy and evaluation of dense two-frame stereo correspondence algorithms. *International Journal of Computer Vision* 47(1): 7–42.
Schindler, K. & Suter, D. 2006. Two-view multibody structure-and-motion with outliers through selection. *IEEE Transactions on Pattern Analysis and Machine Intelligence* 28(6): 983–995.
Schnabel, R., Wahl, R. & Klein, R. 2007. Efficient RANSAC for point-cloud shape detection. *Computer Graphics Forum* 26(2): 214–226.
Shufelt, J. 1999. Geometric constraints for object detection and delineation. Boston, USA: Kluwer Academic Publishers.
Steger, C. 2000. Subpixel-precise extraction of lines and edges. *International Archives of Photogrammetry and Remote Sensing* 33(B3): 141–156.
Stilla, U., Hinz, S., Hedman, K. & Wessel, B. 2007. Road extraction from SAR imagery. In *Remote Sensing of Impervious Surfaces*. Boca Raton, FL, USA: Taylor & Francis.
Stilla, U., Mayer, H., Rottensteiner, F., Heipke, C. & Hinz, S. (eds.). 2007. Photogrammetric Image Analysis. *International Archives of the Photogrammetry, Remote Sensing and Spatial Information Sciences* 36(3/W49A).
Stilla, U., Michaelsen, E., Soergel, U., Hinz, S. &. Ender, H.-J. 2004. Airborne monitoring of vehicle activity in urban areas. *International Archives of the Photogrammetry, Remote Sensing and Spatial Information Sciences* 35(B3): 973–979.
Stoica, R., Descombes, X. & Zerubia, J. 2004. A Gibbs point process for road extraction from remotely sensed images. *International Journal of Computer Vision* 57(2): 121–136.
Tan, P., Zeng, G., Wang, J., Kang, S. & Quan, L. 2007. Image-based tree modeling. SIGGRAPH—*ACM Transactions on Graphics* 26(3), Article 87: 1–7.
Tarsha-Kurdi, F., Landes, T., Grussenmeyer, P. & Koehl, M. 2007. Model-driven and data-driven approaches using LIDAR data: Analysis and Comparison. *International Archives of the Photogrammetry, Remote Sensing and Spatial Information Sciences* 35(3/W49A): 87–92.
Tu, Z., Chen, X., Yuille, A. & Zhu, S.-C. 2005. Image parsing: Unifying segmentation detection and recognition. *International Journal of Computer Vision* 63(2): 113–140.
Tupin, F., Bloch, I. & Maitre, H. 1999. A first step toward automatic interpretaion of SAR images using evidential fusion of several structure detectors. *IEEE Transactions on Geoscience and Remote Sensing* 37(3): 1327–1343.
Tupin, F., Houshmand, B. & Datcu, M. 2002. Road detection in dense urban areas using SAR imagery and the usefulness of multiple views. *IEEE Transactions on Geoscience and Remote Sensing* 40(11): 2405–2414.
Van Gool, L., Zeng, G., Van den Borre, F. & Müller, P. 2007. Towards mass-produced building models. *International Archives of the Photogrammetry, Remote Sensing and Spatial Information Sciences* 36(3/W49A): 209–220.
Wessel, B. & Hinz, S. 2004. Context-supported road extraction from SAR imagery: Transition from rural to built-up areas. *Proceedings of EUSAR 2004, Ulm, Germany*: 399–402.
Wiedemann, C. & Hinz, S. 1999. Automatic extraction and evaluation of road networks from satellite imagery. *International Archives of Photogrammetry and Remote Sensing* 32(3/2W5): 95–100.

Wilczkowiak, M., Sturm, P. & Boyer, E. 2005. Using geometric constraints through parallelepipeds for calibration and 3D modeling. *IEEE Transactions on Pattern Analysis and Machine Intelligence* 27(2): 194–207.

Zebedin, L., Klaus, A., Gruber, B. & Karner, K. 2006. Facade reconstruction from aerial images by multi-view plane sweeping. *International Archives of the Photogrammetry, Remote Sensing and Spatial Information Sciences* 35(3): 31–36.

Zebedin, L., Klaus, A., Gruber, B. & Karner, K. 2007. Facade reconstruction from aerial images by multi-view plane sweeping. *Photogrammetrie—Fernerkundung—Geoinformation* 1/07: 17–24.

Zhang, Q. & Couloigner, I. 2006. Automated road network extraction from high resolution multi-spectral imagery. *ASPRS 2006 Annual Conference, Reno, Nevada* (on CD).

Zhou, J., Cheng, L. & Bischof, W.F. 2007. Online learning with novelty detection in human-guided road tracking. *IEEE Transactions on Geoscience & Remote Sensing* 45 (12), 3967–3977.

Ziems, M., Gerke, M. & Heipke C. 2007. Automatic road extraction from remote sensing imagery incorporating prior information and colour segmentation. *International Archives of Photogrammetry, Remote Sensing and Spatial Information Sciences* 36(3/W49A): 141–148.

Advances in Photogrammetry, Remote Sensing and Spatial Information Sciences: 2008 ISPRS Congress Book – Li, Chen & Baltsavias (eds)

CHAPTER 17

Processing of multitemporal data and change detection

Haigang Sui, Qiming Zhou, Jianya Gong & Guorui Ma

ABSTRACT: Processing of multitemporal images and change detection has been an active research field in remote sensing for decades. Although plenty successful applications have been reported on the monitoring and detecting of environmental change, there are enormous challenges in applying multitemporal imagery to derive timely information on the Earth's environment and human activities. In recent years, great progress has been observed to overcome technological obstacles by the development of new platforms and sensors. The wider availability of large archives of historical images also makes long-term change detection and modelling possible. Such a development stimulates further investigations to develop more advanced image processing methods and new approaches to handling image data in the time dimension. This chapter reviews the current progress in the processing of multitemporal data and change detection. The nature of the environmental change on the Earth's surface is analysed first, followed by a review of the development of remote sensing change detection technology in the following four aspects: image preprocessing for change detection, the classification of change detection methods and approaches, methods of remote sensing change detection, and accuracy assessment in multitemporal image processing. This chapter tries to find the comprehensive solution from multi-source data, integrated processes and intelligent methods with the assistance of prior knowledge. It also points out the challenges that change detection is facing and possible countermeasures, in the hope of deepening the research into change detection technology for remote sensing images.

Keywords: Remote sensing, change detection, classification framework, accuracy assessment

17.1 INTRODUCTION

We are living in a changing world. In the Earth environment, natural and human induced environmental changes occur in time and space. Effective detection and modelling of such change in the context of geospatial information technology are typically termed change detection, which often includes the detection of changes of the objects on the ground or more general change of the environmental background (Richard et al. 2005, Bovolo & Bruzzone 2007). Timely and correct detection of changes of surface features is extremely important for better understanding of relationships and interactions between human and natural phenomena, and lays the foundation for better decision making (Lu et al. 2004). Remotely sensed data have been the most important data source for studying the human living environment in the past 40 years and the large collection of past and present remote sensing imagery makes it possible to analyse spatio-temporal patterns of environmental elements and impact of human activities in past decades. Research has been widely reported on the methodology of remote sensing change detection and monitoring (e.g. Singh 1989, Lu et al. 2004).

Change detection is the process of identifying differences in the state of an object or phenomenon by observing it at different times (Singh 1989). The basic premise in using remotely sensed data for change detection is that changes in the objects of interest will result in changes in reflectance values or local textures that are separable from changes caused by other factors such as differences in atmospheric conditions, illumination and viewing angles, and soil moisture (Deer 1999). The objective of change detection is to compare spatial representation of two points in time by controlling all variances caused by differences in variables that are not of interest and to measure changes caused by differences in the variables of interest (Green et al. 1994). That often means the goal of a change detection method is to detect "significant" change while rejecting "unimportant" ones (Richard et al. 2005).

Aimed at different applications, large numbers of change detection algorithms and theoretical models have been researched from various aspects by scholars worldwide, such as algebraic methods (Prakash & Gupta 1998, Sohl 1999), classification methods (Jensen et al. 1995, Munyati 2000), object-oriented methods (Coppin & Bauer 1996, Heikkonen & Varjo 2004, Walter 2004, Baudouin et al. 2006), time series analysis (Eastman & Fulk 1993), change trajectory analysis (Zhou et al. 2008) and visual analysis (Sunar 1998, Asner et al. 2002), and so on. Due to the importance of monitoring change of the Earth's surface features, processing of multitemporal imagery and change detection have been an active research field in remote sensing. New methods and techniques are constantly invented and developed (Lu et al. 2004), and they have been widely used in fields such as detecting, monitoring and modelling changes of landuse and land cover (LULC), forest, ecosystems, natural hazards and urban and built-up areas, GIS updating and target monitoring and tracking (e.g. Pilon et al. 1988, Vogelmann 1988, Roy et al. 1991, Sader et al. 1991, Alwashed & Bokhari 1993, Coppin & Bauer 1994, Jha & Unni 1994, Muchoney & Haack 1994, Ridd & Liu 1998, Foody & Boyd 1999, Liu & Zhou 2005).

The development of change detection technology has been reviewed by numerous previous works such as Singh (1989), Mouat et al. (1993), Coppin & Bauer (1996), Jensen (1996), Carlotto (1997), Deer (1999), Yuan et al. (2000), Lu et al. (2004) and Richard et al. (2005). Most of these works, however, focus their discussion only on the fields of remote sensing. It has been generally agreed that change detection is a complicated and integrated process. No existing approach is optimal and applicable to all cases. Furthermore, at present automatic change detection is difficult in real-time applications. Although in recent years stimulated by application new approaches are continuously emerging, it is still doubtful whether they can completely solve the change detection problem. The present literature has reviewed change detection approaches and presented many useful conclusions, but there is a lack of systematic analysis on the change detection theories and methods and comprehensive solutions are ignored.

This chapter does not intend to discuss specific methods or algorithms. Instead, it attempts to find a comprehensive solution based on the review of the current state-of-the-art change detection techniques. Firstly, the nature of the changes on the Earth's surface is analysed, followed by discussions on change detection techniques from four aspects, namely, image preprocessing for change detection, the contents of change detection, change detection methods and algorithms, and accuracy assessment for change detection. The challenges that change detection is currently facing and possible countermeasures are also discussed.

17.2 THE NATURE OF THE CHANGES ON THE EARTH'S SURFACE

The concept of change detection covers a broad range of applications. In general, questions related to change detection can be considered in three groups, namely, "if it changed" (*I*), "what is changed" (*W*) and "how it changed" (*H*). To answer the "*I*" question, change/no change detection is required. For the "*W*" question, "from-to" change detection needs to be used. When considering the "*H*" question, the process of the change is addressed by time series and change trajectory analysis.

Obviously the answers to the above become increasingly demanding on the data and processing techniques from those to the "*I*" questions to those to the "*H*" questions. Methods and techniques have also been developed accordingly. For example, some techniques such as image differencing can only provide answers to the "*I*" questions, while others such as post-classification comparison may also help to answer the "*W*" question. The "*H*" questions demand the most from data and processing techniques, requiring techniques such as change trajectory analysis based on time series established from multitemporal images.

It is important to note that the better change detection is based on the better understanding about the process and driving force of the change. The change on the Earth's surface is generally presented in four forms including species change of objects, expansion, shrinkage or shape variation, position change, and fragmentation or combination (Lu et al. 2004). Use of the *a priori* knowledge about the causes, processes and consequences of the change can improve the automation of change detection techniques for more practical implementation and application (Lunetta & Elvidge 2000, Zhang 2004).

All changes on Earth have their own laws. A good understanding about these processes makes it possible to forecast the time, location and progress of the change. To simplify the discussion, we classify the changes on the Earth's surface into three general categories according to their causes, that is, the changes directly caused by natural forces, the changes caused by the activities of animals and plants, and human-induced changes. Different change has different change type and representation type. Table 17.1 gives

Table 17.1. Causes and impacts of the changes on earth's surface.

Cause	Category	Typical changes	Information or knowledge for change detection
by natural forces	violent natural disasters	earthquake, storm, flood, fire, landslide, volcano, rainstorm, snow, hail and frost	geographical distribution of the risk areas, repeating cycle, and intensity; environmental factors, bearing bodies of the disaster and mechanism of their interaction and so on
	slow natural change on earth's surface	tectonic movement, drainage evolution	main reasons and factors leading to the changes, evolution model and so on
by animals and plants	growth and senescence of vegetation	size, colour, and shape in different growth stages; size of area and range by expansion and shrinkage of vegetated areas	geographical distribution of vegetation, growth phenology and spectral characteristics at different phases and expansion model of vegetation and so on
	activities of animals	damage done to vegetation and man-made objects by animals	geographical distribution of animals and their habitats, life growth and reproduction pattern, and so on
by human activities	direct impact	built-up areas, mass movement, deforestation, wars	environmental planning, sustainable development strategy and so on
	indirect impact	global warming, ground subsidence, extreme weather conditions, desertification	environmental feedback on human-induced changes and so on

the auses and impacts of the changes on the Earth's surface and useful information and knowledge for change detection.

However, not all changes on the Earth's surface can be detected by remote sensing. Limited by the spatial, temporal, spectral and radiometric resolution of remotely sensed images, much information, such as changes that are small in size or location (spatially undetectable), short in time (temporally undetectable) and similar in spectral signature (spectrally undetectable) cannot be extracted. It is clear that changes to be detected by remote sensing must meet the requirements on spatial, temporal and spectral resolution.

Change detection is not an isolated process of image processing. From the technical point of view, better understanding of the causes and processes of the change on the Earth's surface can provide useful guidance for the selection of data, technical approaches and methods. To minimize the uncertainty in change detection results, one should attempt to maximize the use of available information and knowledge such as features and objects extracted from images, existing basic geographic information and geographical theorems. From the application point of view, change detection methods provide better technical assistance for the study of processes of the earth system, which in turn can help planning and management for sustainable development. Obviously, these two aspects can supplement each other.

17.3 IMAGE PREPROCESSING FOR CHANGE DETECTION

The process of change detection with remote sensing imagery generally consists of such steps as data selection, image preprocessing, feature and object identification and extraction, comparison and identification of the change, and accuracy assessment (Richard et al. 2005, Coppin et al. 2004). A successful implementation of a remote sensing change detection analysis requires careful considerations about the platform and sensor system, environmental characteristics and image processing methods (Lu et al. 2004). The temporal, spatial, spectral and radiometric resolutions of remotely sensed data and important environmental factors such as atmospheric and soil moisture conditions and phonological characteristics have significant impacts on the success of a remote sensing change detection project (Jensen 1996, Weber 2001, Lu et al. 2004). Although requirements on image preprocessing may vary among different change detection methods, data selection, multitemporal radiometric corrections and image registration are important and indispensable steps.

17.3.1 *Data selection*

The first step in investigating changes of objects on the Earth's surface is to choose data sources that match the target features (Weber 2000). Selection of appropriate image data sources is one of the important prerequisites for remote sensing change detection. Considerations in data selection consist of three aspects, namely, selection of images for detection, selection of reference data, and selection of multi-source data. In order to obtain suitable data sets, it is necessary to investigate the study area, to analyse spatial distribution, spectral characteristics and temporal scale of the changing objects, and to identify available remotely sensed data. Jensen & Cowen (1999) have reported an investigation on the features of an urban area and its suburbs, and specified the minimum requirement on the spatial resolution of remote sensing images for adequate information extraction. Pu et al. (1988) and Cheng et al. (2000) also investigated the optimal image spatial resolution for change detection on numerous objects.

To compare multitemporal images, it is often suggested, whenever possible, to select images acquired from the same type of sensors, with the same spectral and spatial resolutions, and at the same seasonal timeframe in order to minimize unwanted variances due to the changing factors such as sun angle, and seasonal and phenological differences (Coppin & Bauer 1996, Biging et al. 2000). Another problem worth considering is the quality of the acquired images. For example, if 20% or more of the study area is covered by cloud, the image data is typically regarded as not useful. Some other factors may also be considered, such as soil moisture that may alter the spectral signature of the features under investigation (Zhao et al. 2003). In reality, however, data selection is often restricted by many practical limits, such as availability of image data, costs of acquisition and atmospheric conditions. Therefore, for real-world applications, one often has to face the trade-off between the constraints of available resources versus ideal data selection with consistent spatial and spectral resolution and radiometric properties.

In practice, the reference data should not have a lower spatial resolution than that of the detection image. The spectral resolution of the reference data should also match the spectral bands of the detection image (Solberg et al. 1996). The temporal resolution is relative to the change cycle. Lunetta et al. (2004) studied the temporal resolution impact on land change detection and found that the accurate detection of land cover change needs an observation at least every three or four years. They further stated that results can be significantly improved if the time interval is reduced to one or two years.

Although change detection based on bi-temporal images is still the most common approach, use of integrated multitemporal, multi-source data is more preferred with the rapid increase of data availability. In general, change detection using multi-source data is based on four approaches, namely, integrating data with multiple spatial resolutions (Mayaux & Lambin 1997, Zhou et al. 2004), integrating multitemporal (or time series) data (Lambin & Linderman 2006, Alexandre et al. 2006), fusing data with multiple spectral resolutions (Gong 1994, Solberg et al. 1996, Pohl & van Genderen 1998, Chen et al. 2004, Zhang 2005), and integrating remotely sensed data with other geographical data sets (such as map layers and digital terrain models). Studies are also reported on approaches based on the combination of these approaches (Zhou et al. 2008).

Using data with multiple spatial resolutions, changes can be detected in multiple scales to allow rapid responses in combating some rapidly spreading disasters (Lambin & Linderman 2006). Data with multiple temporal resolutions are usually helpful in exploring the causes and consequences of the change for better understanding and forecasting of the processes of moving objects (Martin 1989, Charbonneau et al. 1993, Michener & Houhoulis 1997, Petit & Lambin 2001). Data fusion techniques help to integrate images from different sensors so as to further explore the spectral characteristics of identified ground objects that can be identified by image features. In addition, the image features can complement and verify each other, which in turn make for better results in change detection. The integration of remotely sensed data and other geographical data sets makes full use of the existing data sources for locationing and acquiring attributes of objects, which is an indispensable process in change detection in urban areas.

Generally speaking, making comprehensive use of multiple spatial, spectral and temporal remote sensing images and GIS data can improve the results of change detection. However, when considering the costs of obtaining such data and the extra workload in processing images from different sensors, the optimization of data selection is a crucial process, which attempts to maximize the variance shown by the multi-source data but at the same time to minimize data redundancy. In this regard, one needs to answer two basic questions, i.e. which combination of multi-source data is the best for a specific application, and to what degree can the change detection issue be solved for a given application and multi-source data sets.

17.3.2 *Radiometric correction*

Radiometric conditions during image acquisition may be affected by factors such as solar altitudes, look angles, atmospheric and meteorological conditions, and coverage of cloud, rain or snow. In order to filter out such variations, radiometric correction of remotely sensed imagery is needed if the digital numbers of the images are going to be compared (Leonardo et al. 2006).

Radiometric correction is to remove or reduce the inconsistency between the values recorded by sensors and the spectral reflectance and radiance of the objects. Radiometric correction encompasses absolute and relative radiometric corrections (Elvidge et al. 1995). The absolute radiometric correction mainly rectifies the radiation distortion that is irrelevant to the radiation features of the object surface and is caused by the state of the sensors, solar illumination, and dispersion and absorption of the atmosphere. The typical methods consist of adjusting the radiation value to the standard value with the transmission code of atmospheric radiation (Kaufman 1988, Tanre et al. 1990), with spectral curves in the lab (Smith et al. 1990), and with dark object and transmission code of radiation (Teillet & Fedosejevs 1995), and rectifying the scene by removing the dark objects (Chavez 1988). Since it is expensive and often impractical to survey the atmospheric parameters and ground objects at the present time, and almost impossible to gain those data from the past, the absolute radiometric correction is difficult to implement for most situations in reality. For the relative radiometric correction, on the other hand, an image is regarded as reference. The radiation features of other images can then be adjusted to match this reference. Methods taking this approach consist of correction by histogram regularization (Chavez & Mackinnon 1994) and correction with fixed object (Casselles & Garcia 1989, Hall et al. 1991, Schott et al. 1988). The relative radiometric correction can remove or reduce the variations caused by atmosphere, sensors, and other noises between images acquired in different conditions. Moreover, its algorithms are simple and easy to implement, so that relative radiometric correction has been widely employed (Andrefouet et al. 2001, Song et al. 2001).

At present, the radiometric correction is typically undertaken in the image preprocessing for change detection. Methods used include image regression (Yang & Lo 2000), pseudo-invariant features (Casselles & Garcia 1989), dark set and bright set normalization (Schott et al. 1988), no-change set radiometric normalization (Hall et al. 1991), histogram matching (Chavez & Mackinnon 1994), second simulation of the satellite signal in the solar spectrum (Vermote et al. 1997), and so on.

It should be pointed out that radiometric correction is not necessary for all change detection methods. Generally speaking, the radiometric correction is necessary in multi-sensor, multitemporal land cover change analysis (Leonardo et al. 2006), if the image digital numbers (DNs), or their transformations such as vegetation indices, are to be compared. Studies have shown, however, that if features or objects to be compared are extracted from classification processes, i.e. with the post-classification comparison methods, radiometric correction is not necessary (Kawata et al. 1988, Zhou et al. 2008).

17.3.3 *Image registration*

Precise image-to-image registration is essential for change detection. The importance of accurate geometric registration of multitemporal imagery is obvious because largely spurious results of change detection will be produced if there is mis-registration (John et al. 1992, Dai & Khorram 1998, Stow 1999, Verbyla & Boles 2000, Carvalho 2001, Stow & Chen 2002). If high registration accuracy is not achieved, a great deal of false change area in the scene will be shown by image displacement. It is generally agreed that geometric registration accuracy at the sub-pixel level is acceptable.

John et al. (1992) studied the influence that the impact of the geometric registration error has on the change detection results by simulating the MODIS data with Landsat MSS image. Dai & Khorram (1998) have analysed the relationship between the registration accuracy of TM images and change detection. They found that a registration error of about 0.19 pixels may cause the accuracy of change detection to fall by 10%. Based on deviation theory, Zhang (2005) studied the influence of geometric registration error on change/no-change detection using Landsat 7 multispectral and SPOT 4 panchromatic images and reported that the error of a pixel in geometric registration may cause the detection accuracy to fall by 45%. He further estimated that if a change detection accuracy of 90% is targeted, the accuracy of geometric registration needs to be higher than 0.22 pixels. Bovolo & Bruzzone (2007) analysed the polar coordinates distribution of the change vector with the CVA (change vector analysis)-based polar coordinate change detection framework, and found that the distribution of no-change cases is greatly influenced by image content and registration error.

Although the geometric registration accuracy at the sub-pixel level is generally agreed to be necessary for change detection, it is doubtful whether this suits all cases with various data sources and detection targets. Further studies on this are certainly required. In reality, it is often difficult to achieve high registration accuracy with multitemporal images, especially those from different sensors, due to factors such as imaging models, acquisition conditions, curvature and rotation of the Earth. In mountainous regions and urban areas, in particular, general image registration methods are ineffective and ortho-rectification is needed (Meyer et al. 1993, Gooch & Chandler 1998, Chandler 1999).

Although high accuracy geometrical registration is required for pixel-based methods (e.g. algebra and post-classification comparison methods), it may be unnecessary for some other methods. For feature-based change detection methods such as the object-oriented method, the so-called "buffer detection" algorithm can be employed to compare the extracted features or objects. In this way, the harsh requirement of accurate registration can be avoided (Li et al. 2002). However, at present little research has been done on the extent of tolerable registration error for different kinds of objects. In order to avoid image registration impact on change detection, an alternative approach is to develop a new algorithm that integrates image registration and change detection simultaneously. The core idea of this approach is to make full use of no-change objects as foundations for image registration, iteratively registering images and detecting change until given conditions convergence. Therefore, accumulation of registration errors in traditional methods can be overcome so as to improve the detection accuracy. The main disadvantage of this approach is that the algorithm is complex and more difficult to implement than traditional ones. Zhang (2005) studied the polygon-based method for solving image registration and change detection integrally, which regards change detection as a matching process between GIS and area object features of remote sensing images. Experiments showed that this method is capable of overcoming the accumulation of registration errors to obtain better results than traditional methods.

17.4 THE CONTENTS OF CHANGE DETECTION

Over the past years, researchers have put forward large numbers of change detection techniques and summarized or classified them from different viewpoints (Li et al. 2002, Lu et al. 2004, Richard et al. 2005). However, because change detection techniques involve comprehensive and complicated analysis of images of multi-spatial, multitemporal, multispectral resolutions and geographic data, currently no classification is sufficient for comprehending the whole research content. Being different from previous works, in this study we propose a classification scheme focusing on a higher level framework for change detection techniques, based on five aspects, namely, detection purpose, detection data, detection dimension, detection time scale and detection contents (Fig. 17.1).

According to detection purpose, change detection in remote sensing can be classified into three types: identifying change/no change (*I*—refer to Section 2; also apply to below) (Angelici et al. 1977, Pilon et al. 1988, Chavez & Mackinnon 1994, Sunar 1998, Coppin & Bauer 1994); identifying from-to change (*W*) (Li & Yeh 1998, Dai & Khorram 1999, Petit et al. 2001, Woodcock et al. 2001, Liu & Lathrop 2002); and identifying change process and trajectories (*H*) (Lambin & Linderman 2006, Zhou et al. 2008). The first type (the *I*-type) mainly detects location and areal extent of change, with no concern about the attribute of the change. Typical *I*-type change detection methods include image differencing, image ratioing, vegetation index comparison, principal components analysis (PCA), and so on. The second type (the *W*-type) detects not only the location and areal extent of change but also the origin and result of the change. It is typically used in application domains such as landuse or land cover change and in GIS data updating. The *W*-type change detection algorithms include post-classification comparison, CVA, and so on. In addition to the above contents, the third type (the *H*-type) also detects the process and trajectory of the change. At the highest level of change detection, the *H*-type change detection is most difficult, as it requires more effective change detection algorithms and more comprehensive data collection.

For data sources, change detection can be categorized into three types, namely, images versus images (Prakash & Gupta 1998, Bruzzone et al. 1999, Sohl 1999, Bovolo & Bruzzone 2007), images versus graphics (Johnsson 1994, Theseira et al. 2002) and multi-source data integration (Drury & Walker 1987, Bonham-Carter et al. 1988, Venkataraman et al. 2000). Of these three types, the first is the most widely used and more intensively studied. The second type typically detects changes by comparing existing GIS data with new images. This kind of change detection is mainly handled by feature-based methods (Gautama et al. 2007). Multi-source data integration has proven to be an effective means to improve change detection accuracy and to make full use of the existing data resources. It is shown as an important trend in the development of change detection techniques.

Regarding dimension, change detection can be divided into 2-D change detection (Gautam & Chennaiah 1985, Singh et al. 1990, Manavalan et al. 1995, Rigina et al. 1999, Read & Lam 2002, Weng 2002, Lu et al. 2004) and 3-D change detection (Li et al. 2002, John et al. 2005). At present, 2-D change detection is more emphasized and widely conducted. However, in some applications such as urban GIS updating and monitoring of identified objects, the altitude change of objects needs to be detected. In situations of this kind, it is necessary to employ 3-D change detection methods. In recent years, the rapid development of 3-D reconstruction techniques such as digital photogrammetry, LIDAR and RADAR interferometry has provided more efficient ways for 3-D change detection. Xia (2006) reported the research on change detection algorithms of stereo-images and strategies for Digital Elevation Model (DEM) updating, which can provide change detection results, updated DEM and exterior orientation elements of the new images at the same time.

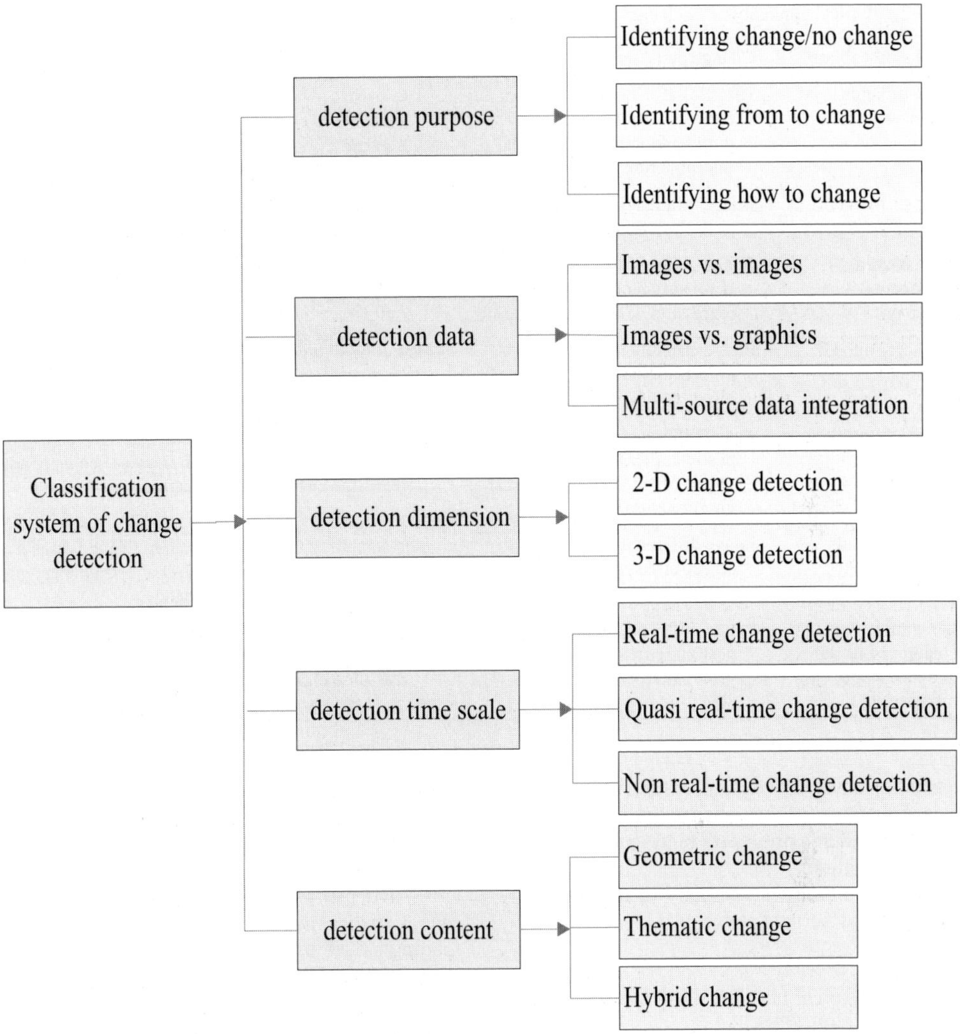

Figure 17.1. Classification framework of change detection.

Referring to the time scale, change detection can be specified as real-time change detection (Ivanov et al. 2000), quasi real-time change detection (Javier et al. 2003, David 2005) and non real-time change detection (Coppin & Bauer 1996, Heikkonen & Varjo 2004, Walter 2004, Lu et al. 2005). The first is mainly used in detecting changes of moving objects. With increasing use of video techniques, this kind of change detection is obtaining more attention, especially in the military domain. Quasi real-time change detection is commonly used for emergency services, such as detecting and evaluating damage caused by fires, landslides, floods, tropical cyclones, volcanic activities and earthquakes. Non real-time change detection is used to detect change with a comparatively slow pace, such as changes of urban buildings, forests, rivers and landuse/land cover. At present, most change detection applications belong to this category.

Considering contents, change detection can be grouped into geometric change (AI-Khudhairy 2005), thematic change (Coppin & Bauer 1996, Chalifoux et al. 1998, Lyon et al. 1998, MacLeod & Conglton 1998, Sunar 1998, Arbogast & Brown 1999), and hybrid change (Munyati 2000). Geometric change detection focuses on changes of the geometric position and shape of detected objects, such as newly built roads or buildings. Thematic change detection applies to specific subjects such as landuse change, change

Table 17.2. Typical applications of change detection methods and classification attributes.

Typical applications	Change detection methods	Classification attributes
Change of landuse or land cover	Post-classification comparison, indirect classification of multitemporal images, visual interpretation	The *W*- or *H*-type change detection, image versus image, 2-D change, non real-time change, hybrid change
Change of forest or vegetation	Image differencing, ratioing, post-classification comparison	The *I*-type change detection, image versus image, 2-D change, non real-time change, thematic change
Natural hazards monitoring	Image differencing, ratioing, model (fire occurrence index), regression, visual interpretation	The *I*-type change detection, image versus image, 2-D change, quasi real-time change, geometric change
City change and GIS updating	Image differencing, ratioing, object-oriented method, visual interpretation	The *I*- or *W*-type change detection, image versus image or image versus graphics, 2-D or 3-D change, non real-time change, hybrid change
Object monitoring and tracking	Image differencing, ratioing, post-classification comparison, object-oriented method, visual interpretation	The *I*- or *H*-type change detection, image versus image or image versus graphics, 2-D or 3-D change, non real-time change, hybrid change
Investigation and monitoring for military applications	Image differencing, ratioing, object-oriented method, visual interpretation, time-series analysis	The *I*- or *W*-type change detection, image versus image or image versus graphics, 2-D or 3-D change, quasi real-time change, hybrid change
Other applications (e.g. changes of wetland and landscape, crop monitoring)	Image differencing, ratioing, post-classification comparison, visual interpretation, object-oriented method	The *I*- or *W*-type change detection, image versus image or image versus graphics, 2-D or 3-D change, quasi real-time change, hybrid change

of vegetation cover, city expansion, and so on. Hybrid change detection analyses changes in both location/shape and attributes of objects to get more complete change information.

Table 17.2 presents a summary of the classification system as described above.

17.5 CHANGE DETECTION METHODS AND ALGORITHMS

In past years, a variety of change detection techniques have been developed and efforts have been made to produce comprehensive summaries and reviews of these methods. As one of the pioneer works, Singh (1989) classified change detection methods into two types, namely, classification comparison and direct comparison. Deer (1999) proposed a classification of three categories, including pixel-based, feature-based and object-based change detection. According to image registration and data sources, Li et al. (2002) classified change detection methods into two types among seven methods. More recent reviews were made by Lu et al. (2004) and Richard et al. (2005). Lu et al. (2004) generalized the change detection methods into seven types, namely, arithmetic operation, transformation, classification comparison, advanced models, GIS integration, visual analysis and some other methods. They also discussed preprocessing of change detection, choice of the threshold value and accuracy assessment in detail. Richard et al. (2005) summarized change detection methods in general terms and propose a classification of direct difference, statistical hypothesis testing, predictive models, shading model, and background modeling, and so on. He emphasized the classification of change detection methods but did not solely focus on those in remote sensing.

Obviously, different classification benchmarks result in different classification methods. Based on the former classification methods, this chapter classifies change detection from its essence. Change detection approaches can be characterized into two broad groups, namely, bi-temporal change detection and temporal trajectory analysis (Coppin et al. 2004). The former measures changes based on a two-epoch timescale, i.e. the comparison between two dates. The latter analyses the changes based on a continuous timescale, i.e. the focus of the analysis is not only on what has changed between dates, but also on the progress of the change over the period. At present, most change detection methods belong to the bi-temporal change detection approach. Almost all classifications for change detection algorithms are based on bi-temporal change detection with little attention on temporal trajectory analysis.

For bi-temporal change detection, detection algorithms can be attributed to one of the three approaches, namely, directly comparing different data sources (direct comparison method), comparing extracted information (post-analysis comparison method) and integrating all data sources into a uniform model (uniform modelling method). The detection elements of the direct comparison method include pixel, basic image features and transformed features. The texture features (Chen 2002) and edge features (Rowe & Grewe 2001) are always taken as basic image features. For multispectral remotely sensed images, transformation is often an important procedure. In order to make the multispectral images comparable with each other, transformation is often needed to present image features. Sometimes these transformations can be quite complex, e.g. Kauth-Thomas transformation (Liu & Ma 1999), vegetation index (Lyon et al. 1998), PCA (Byrne et al. 1980), independent component analysis (ICA) (Zhong & Wang 2006), canonical correlation analysis (CCA) (Zhang 2004), and so on. The detection elements of the post-analysis comparison method mainly include objects extracted from images. Based on two of the most widely-used methods for object extraction, namely, image classification and feature extraction, comparison between objects after classification and feature extraction are typical for the post-analysis comparison method.

According to modelling strategies, the uniform modelling method can be categorized as modelling for detection methods and modelling for detection process. Compared with the bi-temporal change detection, the temporal trajectory analysis lays more emphasis on discovering the trend of change by constructing the 'curves' or 'profiles' of multitemporal data. From the viewpoint of processing methods, temporal trajectory analysis can be decomposed into bi-temporal change detection and then relative post-processing is implemented after the bi-temporal change detection. On the other hand, the so-called long time-series analysis method can be employed for temporal trajectory analysis. Another important application of temporal trajectory analysis is real-time change detection such as video image sequences analysis.

Considering the above and for ease of discussion, we hereby present the methods of change detection in seven categories (Fig. 17.2), namely, direct comparison (Coppin & Bauer 1996, Lobo 1997, Atkinson & Lewis 2000, Townshend et al. 2000), classification (Jensen et al. 1995, Munyati 2000), object-oriented methods (Coppin & Bauer 1996, Heikkonen & Varjo 2004, Walter 2004, Baudouin et al. 2006), model method, time-series analysis (Zhou et al. 2008), visual analysis (Sunar 1998, Asner et al. 2002) and the hybrid method.

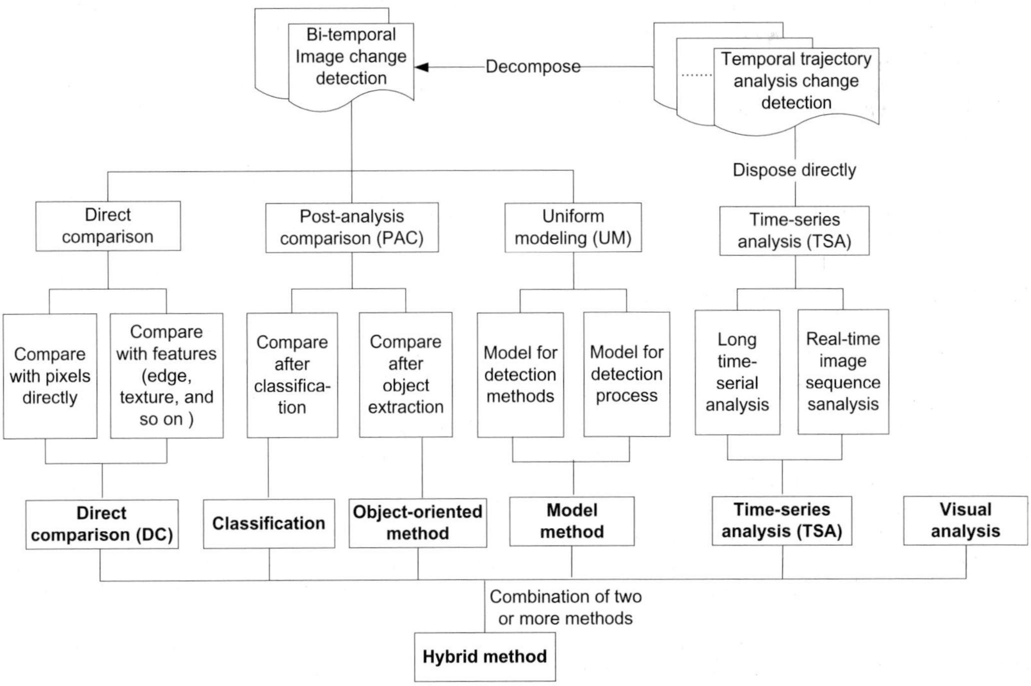

Figure 17.2. Classification concept of change detection algorithms.

17.5.1 *Direct comparison*

The direct comparison method compares the grey values (or digital numbers—DNs) directly by algebraic operations. Image differencing and image ratioing are the simplest and most widely used methods. Other methods include CVA, correlation coefficient method, inner product analysis, and cross correlation, and so on. Lu et al. (2004) also took transformation as a category of change detection methods. However, strictly speaking, the main goal of transformation is to reduce data redundancy between spectral bands, so that it alone does not qualify as a change detection method. Similarly, models such as the spectral mixture models, reflectance model and biophysical models are developed for converting the image reflectance values to physically based parameters. They form preconditions for change detection in specific applications where derived physical parameters are compared. In order to improve change detection performance and efficiency, the direct comparison method can choose the integrations of multiple-features (such as the multiple combinations of the texture, matrix and edge features) to replace single pixel value (Hegarat-Mascle & Seltz 2004). It can also select other features (such as vegetation indices and soil indices) and those features after transformation (such as the output from principal component analysis, independent component analysis, canonical correlation transformation, and wavelet transformation, and so on) to conduct algebraic comparison.

The selection of threshold is the key and has been considered difficult in the direct comparison method. Two methods are often used for the selection of thresholds (Singh 1989, Deer 1999, Yool et al. 1997, Lu et al. 2004). The first is the interactive or manual trial-and-error procedure in which an analyst interactively adjusts the thresholds and evaluates the resulting image until satisfied. The second is selection according to some statistical measures such as a standard deviation from class mean. In recent years, some advanced approaches have been proposed such as Bayes theory (Bruzzone & Prieto 2000), neural network (Gonzalo 2006), Markov model (Kasetkasem & Varshney 2002) and mathematical morphology (Wang et al. 2005). These methods, however, are rarely adopted in practical applications because of their complexity. An approach based on the integration of multiple methods for the selection of thresholds using the Markov random field (MRF) fusion method has also been reported (Farid & Yakoub 2006).

17.5.2 *Classification*

The classification method consists of post classification comparison and direct multitemporal image classification. The core idea of this method is to classify remotely sensed images into different categories and provide a matrix of change information. This method is typically applied to such applications as landuse or land cover change, forest or vegetation change, change of urban or built-up areas and changes of wetlands (Jensen et al. 1995, Munyati 2000). The advantage of this method is that it is not only capable of providing the solution to the *I*-type change detection, but also suitable for the *W*-type change detection. Since the method does not make direct comparison between image DNs, it does not need extensive image preprocessing (e.g. radiometric calibration) as required by the above direct comparison method. The disadvantages of the post classification comparison method include the greater demand on computing and labelling, and the difficulty in estimating propagated errors in change detection since it is often difficult to adequately assess classification accuracies due to the lack of simultaneous historical reference data.

The accuracy of the change detection using the classification method is dependent upon the accuracy of the image classification. A high quality classification often requires a large amount of high quality training data. Since the collection of appropriate multitemporal ground reference data is usually a difficult and expensive task (Bruzzone & Prieto 2000), when such high quality training data are not available, the classification method often gives some unsatisfactory change detection results. In many applications, when it is not possible to obtain sufficient ground reference data, the use of unsupervised classification is required. In recent years, research into artificial neural network (ANN), fuzzy logic, Bayesian network and other new mathematical theories for change detection has been reported to provide better solutions for this (Abuelgasim et al. 1999, Dai & Khorram 1999, Gopal & Woodcock 1999, Woodcock et al. 2001, Liu & Lathrop 2002, Florence et al. 2006, Ouyang et al. 2006, Bovolo & Bruzzone 2007).

17.5.3 *Object-oriented method*

The object-oriented method is also called the object-based comparison method. The basic strategy is to extract objects from multitemporal images using image segmentation and other feature extraction algorithms and distinguish the changes between corresponding objects. Typically this method is applied to applications such as change detection of man-made objects on high resolution images, urban data updating and military reconnaissance. This method compares the extracted objects directly, so it is not sensitive to data noise and geometric distortion. Two critical issues need to be addressed, namely, the methods to compare objects and to identify the changes between objects. Sui (2002) proposed the buffer detection algorithm for object comparison. Ancillary information, e.g. digital maps from GIS, can be used in the object-oriented method. When detecting geographical objects, data from a GIS database can provide great help to detect changes. Work with the focus on the integration of GIS for change detection include, for example, Lo & Shipman (1990), Weng (2002), Yang & Lo (2002), Sui (2002) and Zhang (2005). The powerful GIS functions provide efficient tools for multi-source data processing and change detection analysis, so that one can expect more work taking this approach as a generic trend in change detection (Lu et al. 2004).

The critical challenge of the object-oriented method is the object detection and segmentation. Because feature and object extraction from images is often difficult and prone to error, in practice this method is mostly applied to image-to-map and is rarely used for image-to-image change detection. To overcome the difficulty of object extraction, a knowledge-based system is often employed (Zhang 2005). Another approach is to improve the method of image segmentation, so that more efficient and accurate object extraction can be achieved (Flanders et al. 2003).

17.5.4 *Model method*

The model method is a more application-oriented change detection method. It is different from general change detection methods such as direct comparison and classification and can perceive the essential rules of application issues and take changes as the core elements to construct mathematical models. The basic strategy is to adopt uniform models that integrate all procedures and methods. Generally this method can be grouped into two kinds, namely, models based on processing approaches (approach-based models) and models based on the course of processing (process-based models). The advantage of the model method is that it considers all factors for change detection and can obtain best results. It is simple to apply and can be used for *H*-type change detection problems. However, the greatest challenge in this approach is how to construct suitable models. Even in existing suitable models, the application problem is so complex that many of the parameters are difficult to obtain.

In general, the approach-based models are to model the change contents according to certain postulate conditions, which often closely simulate the actual situations for given applications. Hussin et al. (1991) used an area production model to detect forest cover changes. Morisette et al. (1999) used generalized linear models to detect land cover change. Yue et al. (2002) used a curve-theorem-based approach for change detection in the Yellow River Delta. Read & Lam (2002) reported that the spatial statistics models such as fractal dimension can potentially be used to detect land cover changes. Because images can be regarded as random fields, probability statistics theory and models such as MRF-Gibs (Kasetkasem & Varshney 2002), Bayesian network (Ouyang et al. 2006), support-vector machine and kernal-based methods (Hassiba 2006) are often used. Other models such as predictive models (Skifstad & Jain 1989), shading models (Liu et al. 1998, Li et al. 2002) and background models (Ivanov et al. 2000) have also been developed.

The process-based models are to model the process of change detection. All procedures of the change detection are integrated into a uniform framework and are considered as a whole. Li et al. (2002) argued that it is necessary to integrate image registration and change detection to minimize the impact of geometric distortion. Two kinds of process-based model are reported. One detects change between new non-rectified images and old digital orthomaps (DOM) and digital elevation models (DEM). The other is three-dimensional change detection between new multi-overlapped photos and old digital line graphics (DLG), DEM, DOM and digital raster graphics (DRG) products. The core idea of these models is to take image registration, parameters calculation, change detection, 3-D reconstruction (for 3-D change detection) and utilization of ancillary information into a whole process in which different processing results verify each other until the ultimate balance and/or given conditions can be satisfied. Theories on error, uncertainty, fuzzy set, optimization and so on are often adopted for these models. For on-orbit change detection applications, the process-based model is a vital method. It should be pointed out that it is possible and necessary to combine the approach-based with process-based model methods. However, at present few works has been reported on this approach.

17.5.5 *Time series analysis*

Time series analysis is mainly used for temporal trajectory analysis. In contrast to bi-temporal change detection, the temporal trajectory analysis is usually based on low spatial resolution images such as AVHRR and MODIS, which have a high temporal resolution. The trade-off of using these images, however, is the loss of spatial details that makes auto-classification very difficult, so that the temporal trajectory analysis is commonly restricted to, for example, vegetation dynamics in large areas (Lambin & Ehrlich 1997), or change trajectories of individual land cover classes (Mertens & Lambin 2000). Quantitative parameters such as normalized difference vegetation index (NDVI) or area of given land cover class are often used as the dependent variables for the establishment of change trajectories.

Two approaches are common for time series analysis, namely, long-time serial analysis and real-time image sequences analysis. Liu (2004) has categorized time series change into secular trend change, seasonal fluctuation, periodical fluctuation and irregular fluctuation. At present, standard PCA (Eastman & Fulk 1993) and change vector analysis (CVA) (Malila 1980) are widely adopted methods that are based on time-series curves of remotely sensed data for analysing land cover change. For non-real-time, long-time serial analysis change detection, a dynamic Bayesian network seems a good tool. The method uses the time-series dynamic data to produce reliable probabilistic reasoning, which can make both static and dynamic analysis on the change detection of remote sensing images. Ouyang et al. (2006) has studied the theory and methods of applying dynamic Bayesian networks to detect multitemporal land cover changes in Beijing using three temporal TM images.

Although traditional video change detection is based on general video monitoring that is widely applied in banks, traffic control and manufacturing (Kirchhof & Stilla 2006, Celik et al. 2007), techniques of videography have now been applied to aero-survey and spaceborne remote sensing. The real-time image sequences analysis can therefore be used for temporal trajectory analysis. Video monitoring sequence images have high sampling rates in time with large data volumes. Moreover, the difference between two sequence images can be quite large, and sometimes the image quality can be quite poor. To efficiently apply video monitoring sequence images to change detection, new challenges are faced to adopt techniques applied in video monitoring fields such as optic flow techniques (Woelk & Koch 2004), Kalman filtering and tracking and background modelling (Celik et al. 2007).

17.5.6 *Visual analysis*

Although many automatic or semi-automatic change detection methods have been developed and implemented, visual analysis is still irreplaceable. In general, a human interpreter has a great capability to distinguish chromatic colours and comprehend image properties such as texture, size, shape, image pattern and spatial associations. To further help the interpretation of change, computer visual and analytical enhancement can make it easier for image comparison and change analysis. Methods that are commonly employed in visual analysis include false colour composites, band replacement, mixed displays and alternate display. Although in general visual analysis is time-consuming and inefficient, especially for large area change detection applications, it is simple and easy to learn and operate. Under many circumstances, there may be no better methods than visual analysis. It is also used as supplementary to other methods (Sunar 1998, Asner et al. 2002), especially in some applications where results from semi-automatic algorithms need to be compared, identified and verified by visual analysis (Lu et al. 2004).

17.5.7 *Hybrid methods*

A hybrid method refers to a combination of two or more methods mentioned above for change detection. There are two typical approaches for this, namely, to use different detection methods in different detection phases or procedures (procedure-based hybrid analysis) and to use different detection methods separately and then analyse their results comprehensively (result-based hybrid analysis). The advantage of this method is that it makes full use of the virtues of many algorithms and obtains better change detection results than a single method.

For the procedure-based hybrid analysis, the combination of classification and algebraic methods (Michener & Houhoulis 1997, Warner 2005) is often employed. For instance, the image differencing techniques may be used to identify the area with significant spectral changes and then the post-classification comparison method is applied for the *W*-type change detection. The combination of classification and object-oriented method is also becoming more widely adopted. Jiang & Ram (2003) presented a procedure for efficiently retrieving and representing shapes of lakes on remotely sensed imagery using supervised classification, object recognition and parametric contour tracing. Zhou et al. (2008) reported a method that combines post-classification and time series analysis to model the categorical land cover change trajectories. Zhang et al. (2002) used road density and TM spectral information to detect urban land change by combining post-classification comparison and image differencing methods. Research has also shown that a combination of two change detection techniques, such as image differencing and PCA, NDVI and PCA, or PCA and CVA, could improve the change detection results (Lu et al. 2004). For the result-based hybrid analysis, decision-making fusion strategies such as voting (Zhan et al. 2000), spatial analysis (Liu & Zhou 2005) and fuzzy logic (Yeary et al. 2007) are widely applied.

Many case studies have shown that the hybrid approach satisfactorily enhanced the results obtained by individual methods (Pilon et al. 1988, Fan et al. 2001). However, the optimum selection of method combination for given applications remains in question, so that the typical trial-and-error procedure has to be employed. This leads to subjective and sometimes quite arbitrary selection of methods, resulting in highly application-restricted combinations. Moreover, the fusion of the procedure-based and the result-based hybrid analysis is also worth further investigation especially for the change detection that relies on multi-data sources.

To summarize, Table 17.3 shows the types, typical methods, advantages, disadvantages and development prospects of change detection methods.

Aiming at different applications, a large number of change detection methods and techniques have been implemented and tested. It is commonly agreed, however, that there is no single method suitable for all application cases. Even for some specific applications, the choice of the best method is often difficult, as it is largely dependent upon data sources, image quality, the geography of the study area and experience of the analyst. Because of the difficulty in identifying a suitable method, in practice different change detection techniques are often tested and compared to provide a best result based on the accuracy assessment or qualitative assessment (Lu et al. 2004).

It should be pointed out that present change detection algorithms are lacking a rigorous theoretical basis. Most of methods are only data-driven at the pixel level, while many applications require knowledge-driven methods at the feature level. The correlative information between new and historical images has not been fully utilized and spatial data mining techniques are rarely adopted to extract useful information and knowledge from old spatial databases to support change detection. Most existing methods regard change detection as one or multiple steps in data processing without considering the uniform solution, which requires incorporating other steps such as image registration and classification. In addition, 3-D change detection methods are seldom considered, which may have great impact on the detection accuracy, especially in the urban area. What is more, many steps in the processing procedure still rely on the interference from an experienced human operator. Full automation in change detection still has a long way to go.

17.6 ACCURACY ASSESSMENT FOR CHANGE DETECTION

The accuracy of change detection depends on many factors, including precise geometric registration and calibration or normalization, availability and quality of ground reference data, the complexity of landscape and environment, methods or algorithms used, the analyst's skills and experience, and time and cost restrictions (Lu et al. 2004, Richard et al. 2005). Shao (2006) summarized the main errors in change detection including errors in data (e.g. image resolution, accuracy of location and image quality), errors caused by preprocessing (the accuracy of geometric correction and radiometric correction), errors caused by change detection methods and processes (e.g. classification and data extraction error), errors in field survey (e.g. accuracy of ground reference) and errors caused by post processing (e.g. change identification, removal of the small map patches, and vectorization).

Accuracy assessment techniques in change detection originate from those of remote sensing image classification. It is natural to extend the accuracy assessment techniques for processing a single time image to that of bi-temporal or multitemporal images. Among various assessment techniques, the most efficient and widely used is the error matrix (i.e. confusion table) of classification. Based on this matrix, Biging et al. (2000) proposed the error matrix for the accuracy assessment of change detection, and similar approaches have been reported by Pilon et al. (1988), Green et al. (1994), Kwarteng & Chavez (1998), Mas (1999) and Gong & Mu (2000). In recent years, some alternative methods are also used in analysing and evaluating change detection of land cover. Nemmour & Chibani (2006) proposed the fuzzy error matrix (FEM) and receiver operating characteristics (ROC). Morisette & Khorram (2000) used accuracy assessment curves to analyse satellite-based change detection. Lowell (2001) developed an area-based accuracy assessment method for analysis of change maps. Owing to the fact that many steps and processes are involved in change detection, an error propagation model needs to be adopted (Martin 1989, Charbonneau et al. 1993, Michener & Houhoulis 1997, Mertens & Lambin 2000, Carmel et al. 2001, Petit et al. 2001).

Since it is often difficult, if not impossible, to collect reliable multitemporal ground reference datasets, some alternative approaches have to be employed for the assessment of change detection accuracy. There are three general approaches to obtain the ground references for bi-temporal change detection, namely, field survey with the assistance of historical GIS data, simultaneously, or within the time proximity, high-resolution images, and visual interpretation. Each of the three methods has its own advantages and disadvantages, and they are dependent upon the application. Biging et al. (2000) made recommendations for accuracy assessment of change detection. They identified the factors that affect accuracy assessment, presented a sampling design, analysed the applicability of various assessment methods and illustrated possible applications. For long-time temporal trajectory change detection, it is more difficult to obtain the ground reference data. When the ground reference is unavailable, the consistency check rules are often employed to assess accuracy. Liu & Zhou (2004) proposed an accuracy analysis of time series of remote sensing change detection by rule-based rationality evaluation with post-classification comparison. At the present time, accuracy assessment of

Table 17.3. Summary of change detection methods.

Type	Typical Methods	Advantages	Disadvantages	Development prospects
Direct comparison	Image differencing, image ratioing, vegetation index differencing, CVA, PCA, ICA, wavelet transformation, tasselled cap, biophysical parameter method.	Simple and easy to implement and interpret results. Some methods can reduce dimension and data redundancy efficiently. Different features can be extracted and this reduces detection difficulty for some specific application.	It is sensitive to preprocessing including radiometric correction and image registration and threshold selection is difficult. It only provides *I*-type change detection. It is often puzzled to select a method because it is easily influenced by image noise and application cases.	Considering utilization of multi-features including image-level features and application-level features. Developing more efficient transformation methods to reduce data redundancy and hybrid threshold selection algorithm.
Classification	Post-classification comparison, direct classification of multitemporal images, spectrum-fitting method, spectrum matching filtering method.	Many classification algorithms can be directly used. It can provide detailed matrix of change information and accuracy assessment is easy. It minimizes impacts of atmospheric, sensor and environmental differences.	Classification accuracy directly influences the accuracy of change detection. It is time-consuming to create classification results and a professional operator is necessary. It is difficult and expensive to obtain appropriate multitemporal ground reference.	Developing efficient and robust classification methods with more attention to object-oriented classification and techniques such as SVM, kernel, etc.
Object-oriented comparison	Object comparison and detection, integrated GIS and remote sensing method.	It is insensitive to noise and preprocessing. It is a robust method when objects can be appropriately extracted from images. It is often used as a key step for urban data updating.	It is very difficult to automatically extract objects. Semi-automatic extraction is time-consuming and the results depend upon the analyst's experience in image segmentation.	It is important to apply new image segmentation techniques and knowledge-based systems to improve extraction level. It is a good idea to combine GIS and other geographical information with the extraction.
Model methods	Approach-based models (predictive models, shading model, MRF, etc.) and process-based models.	They can minimize the impacts of noise and preprocessing. It is simple to apply and can be used for the *H*-type change detection.	It is difficult to create models. Whether the model is correct or not and the accuracy of the model directly determine the results of change detection.	For approach-based models it is important to integrate image processing and application models. For process-based models, it is necessary to create a whole framework to integrate part or all processing procedures.

Time series analysis	Dynamic Bayesian network, real-time image sequences change detection.	It can detect non-real-time temporal trajectory change and real time changes. It can be used for the *H*-type change detection and for discovering the change processes.	It is complex and time-consuming to process. To verify the accuracy, it is difficult and even impossible to obtain all historical ground data.	Efficient long-time analysis for specific application is necessary. Video monitoring and processing techniques are helpful for the real-time image sequences change detection.
Visual analysis	False-colour composite, band displacement method, alternate display method, mixed display method	Human experience and knowledge can be used and it provides intuitive understanding for the results. Multi-data sources can be integrated and analysed at one time.	The results depend on the analyst's skill in image interpretation. Main work is manual or semi-automatic and it is time-consuming.	It is important to use hardware equipment to improve speed and also necessary to provide computer-human interaction tools to assist the algorithm of automatic extraction.
Hybrid methods	Procedure-based hybrid analysis and result-based hybrid analysis.	Making full use of virtues of many algorithms can improve accuracy of detection results to a certain extent.	It is difficult to select best combination of methods and processes. It is complex with low efficiency.	Comprehensive analysis from the integration of processes and methods.

change detection is mainly based on pixel and little work has been done to assess accuracy for feature-level or object-level change detection. Furthermore, more effort is needed on the accuracy assessment for change detection of long-time sequence images, especially when the ground reference data is insufficient or impossible to obtain. For these cases, the accuracy assessment methodology needs to be considered from a new viewpoint.

17.7 DISCUSSION AND RECOMMENDATIONS

For change detection, the starting point is to understand the driving factors and processes of the Earth's surface changes. Changes of the objects on the Earth's surface have their time-space evolvement processes. Better understanding of these change processes can assist in the retrieval of objects and changes and improve change detection accuracy. However, it is often difficult to fully understand change processes. In recent years, data mining, knowledge discovery, knowledge engineering and other advanced techniques have been employed to acquire such understanding.

To study changes of objects on the Earth's surface, data sources need to be selected to match the target features of the objects. Multi-source data especially with different characteristics are often needed for change detection. However, in practice it is often not easy to select multi-source data because of unknown relationships between change detection and fusion of multi-source data. Precise radiometric correction and image registration are prerequisites for many change detection methods, but they may be less critical for some other methods. For example, the post-classification comparison method is insensitive to radiometric correction and feature-level change detection is less demanding on sub-pixel geometric registration.

All change detection techniques are characterized by five attributes, namely, detection purpose, detection data, detection dimension, detection time scale and detection contents. From the point of view of methodologies, change detection algorithms are classified into seven categories: comparison, classification, object-oriented method, model method, time-series analysis, visual analysis, and hybrid methods. In practice, selection and comparison of different algorithms are needed for an application project. The hybrid methods have a promising potential but they face the challenge of algorithm design. The ideal solution should provide a flexible framework to integrate multi-source data, processing procedures, intelligent method and existing geographical knowledge.

To achieve real time, automated and high accuracy change detection, the further development of methods and algorithms can be summarized as below:

1. From the point of view of data sources, the method should be flexible and effective to integrate multi-source data such as multitemporal, multi-resolution and multispectral remotely sensed images and GIS data sets. This can make up the deficiencies in geometric relationships and spectral identification between the image change and change on the surface of the Earth so as to improve detection accuracy.
2. From the point of view of data processing, it should provide a uniform framework to integrate image preprocessing, image registration, parameter calculation, 3-D reconstruction, change detection and other related procedures to minimize the impact of error propagation caused by the separation of each processing step.
3. From the point of view of detection algorithms, it is necessary to adopt pattern recognition, data mining, evolution computation and other new theories and techniques to improve intelligent change detection methods.
4. From the point of view of assisting knowledge, it is important to make full use of all existing knowledge to reduce the uncertainties and improve the accuracy of change detection.

With the continuous development of computer networks, space technology and remote sensing, demands for automated, real-time and in-orbit processing become more urgent than ever before for change detection. To achieve this, the challenges to the technology include the full automation of image registration, image matching, feature extraction, image interpretation, image fusion, data-cleaning, image classification, and data mining and knowledge discovery from a GIS database.

ACKNOWLEDGMENTS

This research is supported by Natural Science Foundation of China (60602013), National Key Basic Research and Development Program of China (2006CB701304), Hong Kong Research Grants Council Competitive Earmarked Research Grant (HKBU2029/07P) and Hong Kong Baptist University Faculty Research Grant (FRG/06-07/II-76).

REFERENCES

Abuelgasim, A.A., Ross, W.D., Gopal, S. & Woodcock, C.E. 1999. Change detection using adaptive fuzzy neural networks: environmental damage assessment after the Gulf War. *Remote Sensing of Environment* 70: 208–223.

Ai-Khudhairy, D.H.A., Caravaggi, L. & Giada, S. 2005. Structural damage assessments from IKONOS data using change detection, object-oriented segmentation, and classification techniques. *Photogrammetry and Remote Sensing* 7: 825–837.

Alexandre, B., Karen, C.S. & Andre G.J. 2006. A novel method for mapping land cover changes: incorporating time and space with geostatistics. *IEEE Transactions on Geoscience and Remote Sensing* 44(11): 3427–3435.

Alwashed, M.A. & Bokhari, A.Y. 1993. Monitoring vegetation changes in Al Madinah Saudi Arabia, using Thematic Mapper data. *International Journal of Remote Sensing* 14(2): 191–197.

Andrefouet, S., Muller-Karger, F.E., Hochberg, E.J., HU, C. & Carder, K.L. 2001. Change detection in shallow coral reef environments using Landsat 7 ETM+ data. *Remote Sensing of Environment* 78: 150–162.

Angelici, G., Brynt, N., & Friendman, S. 1977. Techniques for land use change detection using Landsat imagery. *Proceedings of the 43rd Annual Meeting of the American Society of Photogrammetry and Joint Symposium on Land Data Systems, Falls Church, VA, USA*: 217–228. Bethesda, MD: American Society of Photogrammetry.

Arbogast, A.F. & Brown, D.G. 1999. Digital photogrammetric change analysis as applied to active coastal dunes in Michigan. *Photogrammetric Engineering and Remote Sensing* 65(4): 467–474.

Asner, G.P., Keller, M., Pereira, R., & Zweede, J.C. 2002. Remote sensing of selective logging in Amazonia—assessing limitations based on detailed field observations. Landsat ETMz, and textural analysis. *Remote Sensing of Environment* 80: 483–496.

Atkinson, P.M. & Lewis, P. 2000. Geostatistical classification for remote sensing: An introduction. *Computers and Geosciences* 26: 361–371.

Baudouin, D., Patrick, B. & Pierre, D. 2006. Forest change detection by statistical object-based method. *Remote Sensing of Environment* 102: 1–11.

Biging, G.S., Colby, D.R. & Congalton, R.G. 2000. Sampling systems for change detection accuracy assessment. In R.S. Lunetta & C.D. Elvidge (eds.), *Remote Sensing Change Detection: Environmental Monitoring Methods and Applications*: 281–308. London: Taylor & Francis.

Bonham-Carter, G.F., Agterberg, F.P. & Wringh, D.F. 1988. Integration of geological data sets for gold exploration in Nova Scotia. *Photogrammetric Engineering and Remote Sensing* 54: 1585–1592.

Bovolo, F. & Bruzzone, L. 2007. A theoretical framework for unsupervised change detection based on change vector analysis in the polar domain. *IEEE Transactions on Geoscience and Remote Sensing* 45(1): 218–235.

Bruzzone, L. & Prieto, D. 2000. Automatic analysis of the difference image for unsupervised change detection. *IEEE Transactions on Geoscience and Remote Sensing* 38(3): 1171–1182.

Bruzzone, L., Prieto, D.F. & Serpico, S.B. 1999. A neural-statistical approach to multitemporal and multisource remote-sensing image classification. *IEEE Transactions on Geoscience and Remote Sensing* 37: 1350–1359.

Byrne, G.F., Crapper, P.F. & Mayo, K.K. 1980. Monitoring land cover change by principal component analysis of multitemporal Landsat data. *Remote Sensing of Environment* 10: 175–184.

Carlotto, M.J. 1997. Detection and analysis of change in remotely imagery with application to wide area

surveillance. *IEEE Transactions on Image Processing* 6: 189–202.
Carmel, Y., Dean, D.J. & Flather, C.H. 2001. Combining location and classification error sources for estimating multitemporal database accuracy. *Photogrammetric Engineering and Remote Sensing* 67: 865–872.
Carvalho, D.E. 2001. *Mapping and monitoring forest remnants: A multiscale analysis of spatio-temporal data.* PhD thesis. Wageningen University, Netherlands.
Casselles, V. & Garcia, M.L. 1989. An alternative simple approach to estimate atmospheric correction in multitemporal studies. *International Journal of Remote Sensing* 10: 1127–1134.
Celik, T., Demirel, H., Ozkaramanli, H. & Uyguroglu, M. 2007. Fire detection using statistical color model in video sequences. *Journal of Visual Communication and Image Representation* 18: 176–185.
Chadwick, J., Dorsch, S., Glenn, N., Thackray, G. & Shilling, K.S. 2005. Application of multitemporal high-resolution imagery and GPS in a study of the motion of a canyon rim landslide. *ISPRS Journal of Photogrammetry and Remote Sensing* 59: 212–221.
Chalifoux, S., Cavayas, F. & Gray, J.T. 1998. Mapping-guided approach for the automatic detection on Landsat TM images of forest stands damaged by the spruce budworm. *Photogrammetric Engineering and Remote Sensing* 64: 629–635.
Chandler, J. 1999. Effective application of automated digital photogrammetry for geomorphological research. *Earth Surface Processes and Landforms* 24: 51– 63.
Charbonneau, L., Morin, D. & Royer, A. 1993. Analysis of different methods for monitoring the urbanization process. *Geocarto International* 1: 17–25.
Chavez, P.S. 1988. An improved dark-object subtraction technique for atmospheric scattering correction of multispectral data. *Remote Sensing of Environment* 24: 459–479.
Chavez, P.S. & Mackinnon, D.J. 1994. Automatic detection of vegetation changes in the southwestern United States using remotely sensed images. *Photogrammetric Engineering and Remote Sensing* 60: 571–583.
Chen, D., Stow, D.A. & Gong, P. 2004. Examining the effect of spatial resolution and texture window size on classification accuracy: an urban environment case. *International Journal of Remote Sensing* 25: 2177–2192.
Chen, X. 2002. Using remote sensing and GIS to analyze land cover change and its impacts on regional sustainable development. *International Journal of Remote Sensing* 23: 107–124.
Cheng, K.S., Yeh, H.C. & Tsai, C.H. 2000. An anisotropic spatial modeling approach for remote sensing image rectification. *Remote Sensing of Environment* 73(1): 46–54.
Coppin, P. & Bauer, M. 1994. Processing of multitemporal Landsat TM imagery to optimize extraction of forest cover change features. *IEEE Transactions on Geoscience and Remote Sensing* 32: 918–927.
Coppin, P. & Bauer, M. 1996. Digital change detection in forest ecosystems with remote sensing imagery. *Remote Sensing of Environment* 13: 207–304.
Coppin, P., Jonckheere, I., Nackaerts, K. & Muys, B. 2004. Digital change detection methods in ecosystem monitoring: A review. *International Journal of Remote Sensing* 25 (9): 1565–1596.
Dai X.L. & Khorram, S. 1998. The effects of image misregistration on the accuracy of remotely sensed change detection. *IEEE Transactions on Geoscience and Remote Sensing* 36(5): 1566–1577.
Dai, X.L. & Khorram, S. 1999. Remotely sensed change detection based on artificial neural networks. *Photogrammetric Engineering and Remote Sensing* 65: 1187–1194.
David, A.L. 2005. Multispectral Land Sensing: Where From, Where to? *IEEE Transactions on Geoscience and Remote Sensing* 43(3): 414–421.
Deer, P. 1999. *Digital Change Detection Techniques: Civilian and Military Application.* London: Taylor & Francis.
Drury, S.A. & Walker, A.S.D. 1987. Display and enhancement of gridded aeromagnetic data of the Solway Basin. *International Journal of Remote Sensing* 8: 1433–1444.
Eastman, J.R. & Fulk, M. 1993. Long sequence time series evaluation using standardized principal components. *Photogrammetric Engineering and Remote* Sensing 59: 991–996.
Elvidge, C.D., Yuan, D., Weerackoon, R.D. & Lunetta, R.S. 1995. Relative radiometric normalization of Landsat Multispectral Scanner (MSS) data using an automatic scattergram-controlled regression. *Photogrammetric Engineering and Remote Sensing* 61: 1255–1260.
Fan, H., Ma, A. & Li, J. 2001. Case study on image differencing method for land use change detection using thematic data in Renhe District of Panzhihua. *International Journal of Remote Sensing* 5(1): 75–80.
Farid, M. & Yakoub, B. 2006. Markovian fusion approach to robust unsupervised change detection in remotely sensed imagery. *IEEE Geoscience and Remote Sensing Letters* 3(4): 1–5.
Flanders, D., Hall-Beyer, M. & Pereverzoff, J. 2003. Preliminary evaluation of eCognition object-based software for cut block delineation and feature extraction. *Canadian Journal of Remote Sensing* 29(4): 441–452.
Florence L., Sabine V. & Eleonore, W. 2006. Fuzzy multitemporal landuse analysis and mine clearance application. *Photogrammetry and Remote Sensing* 11: 1245–1253.
Foody, G.M. & Boyd, D.S. 1999. Detection of partial land cover change associated with the migration of the inter-class transitional zone. *International Journal of Remote Sensing* 20(14): 2723– 2740.
Gautam, N.C. & Chennaiah, G.C. 1985. Landuse and land cover mapping and change detection in Tripura using satellite Landsat data. *International Journal of Remote Sensing* 6: 517–528.
Gautama, S., Bellens, R., De Tre, G. & Philips, W. 2007. Relevance criteria for spatial information retrieval using error-tolerant graph matching. *IEEE Transactions on Geoscience and Remote Sensing* 45(4): 810–817.
Gong, P. 1994. Integrated analysis of spatial data from multiple sources: an overview. *Canadian Journal of Remote Sensing* 20: 349–359.
Gong. P. & Mu, L. 2000. Error detection in map databases: a consistency checking approach. *Geographic Information Sciences* 6: 188–193.
Gonzalo, P. 2006. A Hopfield neural network for image change detection. *IEEE Transactions on Neural Networks* 17(5): 1250–1264.
Gooch, M.J. & Chandler, J.H. 1998. Optimization of strategy parameters used in automated digital elevation model generation. In D.N.M. Donoghue (ed.), *International*

Archives of Photogrammetry and Remote Sensing. ISPRS, Data Integration: Systems and Techniques, Cambridge 32(2): 88–95.

Gopal, S. & Woodcock, C.E. 1999. Artificial neural networks for detecting forest change. In C.H. Chen (ed.), *Information Processing for Remote Sensing*: 225–236. Singapore: World Scientific Publishing Co.

Green, K., Kempka, D. & Lackey, L. 1994. Using remote sensing to detect and monitor land cover and landuse change. *Photogrammetric Engineering and Remote Sensing* 60: 331–337.

Hall, F.G., Strebel, D.E., Nickeson, J.E. & Goetz, S.J. 1991. Radiometric rectification: toward a common radiometric response among multidate, multisensor images. *Remote Sensing of Environment* 35: 11–27.

Hassiba, N. & Youcef, C. 2006. Multiple support vector machines for land cover change detection: An application for mapping urban extensions. *ISPRS Journal of Photogrammetry and Remote Sensing* 61: 125–133.

Hegarat-Mascle, S.L. & Seltz, R. 2004. Automatic change detection by evidential fusion of change indices. *Remote Sensing of Environment* 91: 390–404.

Heikkonen, J. & Varjo, J. 2004. Forest change detection applying Landsat Thematic Mapper difference features: a comparison of different classifiers in boreal forest conditions. *Forest Science* 50: 579–588.

Hussin, Y.A., Reich, R.M. & Hoffer, R.M. 1991. Estimating slash pine biomass using radar backscatter. *IEEE Transactions on Geoscience and Remote Sensing* 29: 427–431.

Ivanov, Y., Bobick, A. & Liu, J. 2000. Fast lighting independent background subtraction. *International Journal of Computer Vision* 37(2): 199–207.

Javier, H., Jose, I.B., Paul, L.R., Alessandro, P., Franco, M. & Sandro, S. 2003. Monitoring landslides from optical remotely sensed imagery: the case history of Tessina landslide, Italy. *Geomorphology* 54: 63–75.

Jensen, J.R. & Cowen, D.C. 1999. Remote sensing of urban/suburban infrastructure and socioeconomic attributes. *Photogrammetric Engineering and Remote Sensing* 65: 611–622.

Jensen, J.R. 1996. *Introductory Digital Image Processing: a Remote Sensing Perspective*. Englewood Cliffs, NJ: Prentice-Hall.

Jensen, J.R., Rutchey, K., Koch, M. & Narumalani, S. 1995. Inland wetland change detection in the Everglades Water Conservation Area 2A using a time series of normalized remotely sensed data. *Photogrammetric Engineering and Remote Sensing* 61(2): 199–209.

Jha, C.S. & Unni, N.V.M. 1994. Digital change detection of forest conversion of a dry tropical Indian forest region. *International Journal of Remote Sensing* 15(13): 2543– 2552.

Jiang, L. & Ram, M.N. 2003. A shape-based approach to change detection of lakes using time series remote sensing images. *IEEE Transactions on Geoscience and Remote Sensing* 41(11): 1035–1037.

Johnsson, K. 1994. Segment-based landuse classification from SPOT satellite data. *Photogrammetric Engineering and Remote Sensing* 60: 47–53.

Kasetkasem, T. & Varshney, P.K. 2002. An image change detection algorithm based on Markov random field models. *IEEE Transactions Geoscience and Remote Sensing* 40(8): 1815–1823.

Kaufman, Y.J. 1988. Atmospheric effect on spectral signature. *IEEE Transactions on Geoscience and Remote Sensing* 26: 441–451.

Kawata, Y., Veno, S. & Kusaka, T. 1988. Radiometric correction for atmospheric and topographic effects on Landsat MSS images. *International Journal of Remote Sensing* 9: 729–748.

Kirchhof, M. & Stilla, U. 2006. Detection of moving objects in airborne thermal videos. *ISPRS Journal of Photogrammetry and Remote Sensing* 61: 187–196.

Kwarteng, A.Y. & Chavez, P.S. 1998. Change detection study of Kuwait City and environs using multitemporal Landsat Thematic Mapper data. *International Journal of Remote Sensing* 19: 1651–1662.

Lambin, E.F. & Ehrlich, D. 1997. Land cover changes in sub-Saharan Africa (1982–1991): application of a change index based on remotely sensed surface temperature and vegetation indices at a continental scale. *Remote Sensing of Environment* 61, 181–200.

Lambin, E.F. & Linderman, M. 2006. Time Series of Remote Sensing Data for Land Change Science. *IEEE Transactions on Geoscience and Remote Sensing* 44(7): 1926–1928.

Leonardo, P., Franceisco, G., Jose, A., Sobrino, Juan, C., Jimenez, M. & Haydee, K. 2006. Radiometric correction effects in Landsat multi-date/mult-sensor change detection studies. *International Journal of Remote Sensing* 2: 685–704.

Li, D., Sui, H. & Xiao, P. 2002. Automatic Change Detection of Geo-spatial Data from Imagery. *International Archives of Photogrammetry and Remote Sensing, Xi'an, China* 34(2): 245–251.

Li, L., Maylor, K. & Leung, H. 2002. Integrating intensity and texture differences for robust change detection. *IEEE Transaction on Image Processing* 11(2): 105–112.

Li, X. & Yeh, A.G.O. 1998. Principal component analysis of stacked multitemporal images for the monitoring of rapid urban expansion in the Pearl River Delta. *International Journal of Remote Sensing* 19: 1501–1518.

Liu, H. & Zhou, Q. 2004. Accuracy analysis of remote sensing change detection by rule-based rationality evaluation with post-classification comparison. *International Journal of Remote Sensing* 25(5): 1037–1050.

Liu, H. & Zhou, Q. 2005. Establishing a multivariate spatial model for urban growth prediction using multitemporal images. *Computers, Environment and Urban Systems* 29(5): 580–594.

Liu, H. 2004. *Studies of Arid land Classification and Lake Monitoring Based on MODIS Remote Data.* PhD Dissertation, Fujian Agriculture and Forestry University, China (in Chinese).

Liu, S. & Ma, J. 1999. Masking and multi-factor step by step orthogonal transformation method to extract lithology information. *Journal of Remote Sensing* 4: 295–299 (in Chinese).

Liu, S., Fu, C. & Chang, S. 1998. Statistical change detection with moments under time–varying illumination. *IEEE Transaction on Image Processing* 7: 1258–1268.

Liu, X. & Lathrop, R.G. 2002. Urban change detection based on an artificial neural network. *International Journal of Remote Sensing* 23(12): 2513–2518.

Lo, C.P. & Shipman, R.L. 1990. A GIS approach to landuse change dynamics detection. *Photogrammetric Engineering and Remote Sensing* 56: 1483–1491.

Lobo, A. 1997. Image segmentation and discriminant analysis for the identification of land cover units in ecology. *IEEE Transactions on Geoscience and Remote Sensing* 35: 1136–1145.

Lowell, K. 2001. An area-based accuracy assessment methodology for digital change maps. *International Journal of Remote Sensing* 22: 3571–3596.

Lu D., Mausel, P., Brondizio, E. & Moran, E. 2004. Change detection techniques. *International Journal of Remote Sensing* 25(12): 2365–2407.

Lu, D., Mausel, P., Batistella, M. & Moran, E. 2005. Land cover binary change detection methods for use in the moist tropical region of the Amazon: A comparative study. *International Journal of Remote Sensing* 26(1): 101–114.

Lunetta, R. S., Johnson, D. M., Lyon, J. G., & Crotwell, J. 2004. Impacts of imagery temporal frequency on land cover change detection monitoring. *Remote Sensing of Environment* 89(4): 444–454.

Lunetta, R.S. & Elvidge, C.D. (eds.). 2000. *Remote Sensing Change Detection: Environmental Monitoring Methods and Applications*. London: Taylor & Francis.

Lyon, G.J., Yuan, D., Lunetta, R.S. & Elvidge, C.D. 1998. A change detection experiment using vegetation indices. *Photogrammetric Engineering and Remote Sensing* 64: 143–150.

Macleod, R.D. & Congalton, R.G. 1998. A quantitative comparison of change detection algorithms for monitoring eelgrass from remotely sensed data. *Photogrammetric Engineering and Remote Sensing* 64(3): 207–216.

Malila, W. 1980. Change vector analysis: an approach for detecting forest changes with Landsat. *Proceedings of the 6th Annual Symposium on Machine Processing of Remotely Sensed Data, West Lafayette, IN, 3–6 June 1980*: 326–335. West Lafayette, IN: Purdue University Press.

Manavalan, P., Kesavasamy, K. & Adiga, S. 1995. Irrigated crops monitoring through seasons using digital change detection analysis of IRD–LISS 2 data. *International Journal of Remote Sensing* 16: 633–640.

Martin, R.G. 1989. Accuracy assessment of Landsat-based visual change detection methods applied to the rural-urban fringe. *Photogrammetric Engineering and Remote Sensing* 55: 209–215.

Mas, J.F. 1999. Monitoring land cover changes: a comparison of change detection techniques. *International Journal of Remote Sensing* 20: 139–152.

Mayaux, P. & Lambin, E.F. 1997. Tropical forest area measured from global land cover classifications: Inverse calibration models based on spatial textures. *Remote Sensing of Environment* 57: 29–43.

Mertens, B. & Lambin, E.F. 2000. Land cover-change trajectories in southern Cameroon. *Annals of the Association of American Geographers* 90: 467–494.

Meyer, P., Itten, K.I, Kellenberger, T., Sandmeier, S. & Sandmeier, R. 1993. Radiometric corrections of topographically induced effects on Landsat TM data in an alpine environment. *ISPRS Journal of Photogrammetry and Remote Sensing* 48(4): 17–28.

Michener, W.K. & Houhoulis, P.F. 1997. Detection of vegetation changes associated with extensive flooding in a forested ecosystem. *Photogrammetric Engineering and Remote Sensing* 63: 1363–1374.

Morisette, J. T., Khorram, S. & Mace, T. 1999. Land cover change detection enhanced with generalized linear models. *International Journal of Remote Sensing* 20: 2703–2721.

Morisette, J.T. & Khorram, S. 2000. Accuracy assessment curves for satellite-based change detection. *Photogrammetric Engineering and Remote Sensing* 66: 875–880.

Mouat, D.A., Mahin, G.C. & Lancaster, J. 1993. Remote sensing techniques in the analysis of change detection. *Geocarto International* 2: 39–50.

Muchoney, D. M. & Haack, B.N. 1994. Change detection for monitoring forest defoliation. *Photogrammetric Engineering and Remote Sensing* 60(10): 1243– 1251.

Munyati, C. 2000. Wetland change detection on the Kafue Flats, Zambia, by classification of multitemporal remote sensing image dataset. *International Journal of Remote Sensing* 21: 1787–1806.

Nemmour, H. & Chibani, Y. 2006. Fuzzy neural network architecture for change detection in remotely sensed imagery. *International Journal of Remote Sensing* 27(3): 705–717.

Ouyang, Y., Ma, J. & Dai, Q. 2006. Study on Dynamic Bayesian Networks for Multitemporal Remote Sensing Change Detection. *Journal of Remote Sensing* 7: 440–448.

Petit, C., Scudder, T. & Lambin, E. 2001. Quantifying processes of land cover change by remote sensing: resettlement and rapid land cover changes in south-eastern Zambia. *International Journal of Remote Sensing* 22: 3435–3456.

Petit, C.C. & Lambin, E.F. 2001. Integration of multi-source remote sensing data for land cover change detection. *International Journal of Geographical Information Science* 15: 785–803.

Pilon, P.G., Howarth, P.J. & Bullock, R.A. 1988. An enhanced classification approach to change detection in semi-arid environments. *Photogrammetric Engineering and Remote Sensing* 54(12): 1709–1716.

Pohl, C. & Genderen, J.L. 1998. Multisensor image fusion in remote sensing: concepts, methods, and applications. *International Journal of Remote Sensing* 19: 823–854.

Prakash, A. & Gupta, R.P. 1998. Landuse mapping and change detection in a coal mining area - a case study in the Jharia coalfield, India. *International Journal of Remote Sensing* 19: 391–410.

Pu, J., Yang, S., Ding, J. & Zhou, L. 1988. The primary application study of Spot image in Tangshan urban area. *Remote Sensing of Environment China* 3 (3): 163–172.

Read, J.M., & Lam, N.S.N. 2002. Spatial methods for characterizing land cover and detecting land cover changes for the tropics. *International Journal of Remote Sensing* 23: 2457–2474.

Radke, R.J., Andra, S., Al-Kofahi, O. & Roysam, B. 2005. Image Change Detection Algorithms: A Systematic Survey. *IEEE Transactions on Image Processing* 14(3): 294–307.

Ridd, M.K. & Liu, J.J. 1998. A comparison of four algorithms for change detection in an urban environment. *Remote Sensing of Environment* 65(2): 95–100.

Rigina, O., Baklanov, A., Hagner, O. & Olsson, H. 1999. Monitoring of forest damage in the Kola Peninsula, Northern Russia due to smelting industry. *Science of the Total Environment* 229: 147–163.

Rowe, N.C. & Grewe, L.L. 2001. Change detection for linear features in aerial photographs using edge-finding. *IEEE Transaction on Geoscience and Remote Sensing* 39(7): 1608–1612.

Roy, P.S., Ranganath, B.K., Diwakar, P.G., Vohra, T.P.S., Bhan, S.K., Singh, I.J. & Pandian, V.C. 1991. Tropical forest mapping and monitoring using remote sensing. *International Journal of Remote Sensing* 12(11): 2205–2225.

Sader, S.A., Powell, G.V.N. & Rappole, J.H. 1991. Migratory bird habitat monitoring through remote sensing. *International Journal of Remote Sensing* 12(3): 363–372.

Schott, J.R, Salvaggio, C. & Volchok, W.J. 1988. Radiometric scene normalization using pseudo invariant features. *Remote Sensing of the Environment* 26: 1–16.

Shao F. 2006. *Research on Change Information Extraction of Remote Sensing Image Based on ANN.* Master Dissertation, Shandong University of Science and Technology, China (in Chinese).

Singh, A. 1989. Digital change detection techniques using remotely-sensed data. *International Journal of Remote Sensing* 10(6): 989–1003.

Singh, S., Sharma, K.D., Singh, N. & Bohra, D.N. 1990. Temporal change detection in uplands and gullied areas through satellite remote-sensing. *Annals of Arid Zone* 29: 171–177.

Skifstad, K. & Jain, R. 1989. Illumination independent change detection for real world image sequences. *Computer Visualization, Graphics and Image Processing* 46(3): 387–399.

Smith, M.O., Ustin, S.L., Adams, J.B. & Gillespie, A.R. 1990. Vegetation in deserts: A regional measure of abundance from multispectral images. *Remote Sensing of the Environment* 31: 1–26.

Sohl, T.L. 1999. Change analysis in the United Arab Emirates: an investigation of techniques. *Photogrammetric Engineering and Remote Sensing* 65(4): 475–484.

Solberg, A.H.S., Taxt, T. & Jain, A.K. 1996. A Markov random field model for classification of multisource satellite imagery. *IEEE Transactions on Geoscience and Remote Sensing* 34: 100–112.

Song, C., Woodcock, C.E., Seto, K.C., Lenney, M.P. & Macomber, S.A. 2001. Classification and change detection using Landsat TM data: when and how to correct atmospheric effects? *Remote Sensing of Environment* 75: 230–244.

Stow, D.A. & Chen, D.M. 2002. Sensitivity of multitemporal NOAA AVHRR data of an urbanizing region to landuse/land cover changes and misregistration. *Remote Sensing of Environment* 80: 297–307.

Stow, D.A. 1999. Reducing the effects of misregistration on pixel-level change detection. *International Journal of Remote Sensing* 20: 2477–2483.

Sui, H. 2002. *Automatic Change Detection for Road-Networks Base on Features.* Ph.D dissertation, Wuhan University, China (in Chinese).

Sunar, F. 1998. An analysis of changes in a multi-date data set: a case study in the Ikitelli area. Istanbul, Turkey. *International Journal of Remote Sensing* 19(2): 225–235.

Tanre, D., Deroo, C., Duhaut, P., Herman, M. & Morcrette, J.J. 1990. Description of computer code to simulate the satellite signal in the solar spectrum: the 5S code. *International Journal of Remote Sensing* 2: 659–668.

Teillet, P.M. & Fedosejevs, G. 1995. On the dark target approach to atmospheric correction of remotely sensed data. *Canadian Journal of Remote Sensing* 21(4): 374–387.

Theseira, M.A., Thomas, G. & Sannier, C.A.D. 2002. An evaluation of spectral mixture modeling applied to a semi-arid environment. *International Journal of Remote Sensing* 23: 687–700.

Townshend, J.R.G., Justice, C.O., Gurney, C. & McManus, J. 1992. The impact of misregistration on change detection, *IEEE Transactions on Geo-science and Remote Sensing* 30(5): 1054–1060.

Townshend, J.R.G., Huang, C., Kalluri, S.N.V., Defries, R.S., Liang, S. & Yang, K. 2000. Beware of per-pixel characterization of land cover. *International Journal of Remote Sensing* 21: 839–843.

Venkataraman, G., Madhavan, B.B., Ratha, D.S., Antony, J.P., Goyal, R.S., Banglani, S. & Roy, S.S. 2000. Spatial modeling for base metal mineral exploration through integration of geological data sets. *Natural Resources Research* 9: 27–42.

Verbyla, D.L. & Boles, S.H. 2000. Bias in land cover change estimates due to misregistration. *International Journal of Remote Sensing* 21: 3553–3560.

Vermote, E., Tanre, D., Deuze, J. L., Herman, M. & Morcrette, J.J. 1997. Second simulation of the satellite signal in the solar spectrum, 6S: an overview. *IEEE Transactions on Geoscience and Remote Sensing* 35: 675–686.

Vogelmann, J.E. 1988. Detection of forest change in the green mountains of Vermont using multispectral scanner data. *International Journal of Remote Sensing* 9(7): 1187– 1200.

Walter, V. 2004. Object-based classification of remote sensing data for change detection. *ISPRS Journal of Photogrammetry and Remote Sensing* 58: 225–238.

Wang, Z.Y. & Li, W.Q. 2005. Detecting fast motion objects based on mathematical morphology. *Computer Engineering and Science* 1(27): 27–29.

Warner, T. 2005. Hyperspherical direction cosine change vector analysis. *International Journal of Remote Sensing* 26(6): 1201–1215.

Weber, K.T. 2001. A method to incorporate phenology into land cover change analysis. *Journal of Range Management* 54: A1–A7.

Weber, P. 2000. Initial performance validation for the multispectral thermal imager. In *Proceedings of SPIE 4030*: 2–9.

Weng, Q. 2002. Land use change analysis in the Zhujiang Delta of China using satellite remote sensing, GIS and stochastic modelling. *Journal of Environmental Management* 64: 273–284.

Woelk, F. & Koch, R. 2004. Fast monocular Bayesian detection of independently moving objects by a moving observer. In C.E. Buelthoff, H.H. Giese, M.A. Schoelkopf & B. Rasmussen (eds.), *Proceedings of DAGM Symposium, Lecture Notes in Computer Science* 3175: 27–35.

Woodcock, C.E., Macomber, S.A., Pax-Lenney, M. & Cohen, W.B. 2001. Monitoring large areas for forest change using Landsat: generalization across space-time and Landsat sensors. *Remote sensing of Environment* 78: 194–203.

Xia, S. 2006. *Terrain Change Detection and Data Updating with Arial Imagery.* PhD dissertation, Wuhan University, China (in Chinese).

Yang, C.J. & Lo, C.P. 2000. Relative radiometric normalization performance for change detection from multi date satellite images. *Photogrammetric Engineering and Remote Sensing* 66(8): 967–980.

Yang, X. & Lo, C.P. 2002. Using a time series of satellite imagery to detect land use and land cover changes in the Atlanta, Georgia metropolitan area. *International Journal of Remote Sensing* 23(9): 1775–1798.

Yeary, M., Nemati, S., Yu, T.Y. & Wang, Y. 2007. Tornadic time-series detection using Eigen analysis and a machine intelligence-based approach. *IEEE Geoscience and Remote Sensing Letters* 4(3): 335–339.

Yool, S.R., Makaio, M.J. & Watts, J.M. 1997. Techniques for computer assisted mapping of rangeland change. *Journal of Range Management* 50: 307–314.

Yuan, D., Elvidge, C.D. & Lunetta, R.S. 2000. Survey of multispectral methods for land cover change analysis. In R.S. Lunetta & C.D. Elvidge (eds.), *Remote Sensing Change Detection: Environmental Monitoring Methods and Applications*: 21–39. London: Taylor & Francis.

Yue, T., Chen, S., Xu, B., Liu, Q., Li, H., Liu, G. & Ye, Q. 2002. A curve-theorem based approach for change detection and its application to Yellow River Delta. *International Journal of Remote Sensing* 23(11): 2283–2292.

Zhan, X., Defries, R., Townshend, J.R.G., Dimiceli, C., Hansen, M., Huang, C. & Sohlberg, R. 2000. The 250 m global land cover change product from the Moderate Resolution Imaging Spectroradiometer of NASA's Earth Observing System. *International Journal of Remote Sensing* 21: 1433–1460.

Zhang, L. 2004. *Change Detection in Remotely Sensed Imagery Using Multivariate Statistical Analysis*. PhD Dissertation, Wuhan University, China (in Chinese).

Zhang, Q., Wang, J., Peng, X., Gong, P. & Shi, P. 2002. Urban built-up land change detection with road density and spectral information from multitemporal Landsat TM data, *International Journal of Remote Sensing* 23: 3057–3078.

Zhang, X. 2005. *The Theory and Methods of Change Detection Based on Remotely Sensed Imagery and GIS*. PhD Dissertation, Wuhan University, China (in Chinese).

Zhao, Y., Liao, Y., Zhang, B. & Lai, S. 2003. Monitoring technology of salinity in water with optical fibre sensor. *Journal of Lightwave Technology* 21(5): 1334–1338.

Zhong, J. & Wang, R. 2006. Multitemporal remote sensing change detection based on independent component analysis. *International Journal of Remote Sensing* 27(9): 2055–2061.

Zhou, Q., Li, B. & Zhou, C. 2004. Studying spatio-temporal patterns of landuse change in arid environment of China. In Z. Li, Q. Zhou & W. Kainz (eds.), *Advances in Spatial Analysis and Decision Making*:189–200.

Zhou, Q., Li, B. & Kurban, A. 2008. Trajectory analysis of land cover change in arid environment of China. *International Journal of Remote Sensing* 29(4): 1093–1107.

Part IV
Data modelling, management and visualization

Advances in Photogrammetry, Remote Sensing and Spatial Information Sciences: 2008 ISPRS Congress Book – Li, Chen & Baltsavias (eds)

CHAPTER 18

Spatio-temporal modelling

Wolfgang Kainz & Xinming Tang

ABSTRACT: When we represent phenomena of the real world in a GIS, we need to design a model that describes these phenomena in the best way for our particular purposes. This process is called spatial modelling. We distinguish between two fundamental views on spatial phenomena, field and object. Various models exist to deal with time in spatial phenomena. Spatial models are implemented in spatial databases. In the process of database design, we go through several modelling steps to produce database schemas from high level software and hardware independent conceptual and logical schemas to the implementation with particular database software. To ensure a logically consistent representation of spatial features in a database, we need topology. This chapter introduces the concepts of real world phenomena, their abstraction and representations in spatial databases. Concepts of space and time relevant for spatial modelling are discussed and the fundamental principles of database design are presented. Topology and spatial relationships relevant to spatio-temporal modelling are explained.

Keywords: spatial modelling, concepts of space and time, spatio-temporal models, topology, database design, spatial relations

18.1 INTRODUCTION

We live in a constantly changing world. What we perceive with our senses are processes, states and events that happen or exist in the real world. They may be natural or man-made, and we call them real world phenomena. In our brains, we process the input from our senses, which leads to mental models, learning, cognition and knowledge. Everything that happens in the real world and that we process through our senses leads to models of reality. Usually, we agree on common principles and interpretations derived from experience, cultural background and research or learning that enable us to reach a consensus on the perception of real world phenomena.

It is evident that for many phenomena such an agreement is obvious because of the same educational, cultural or societal context, or simply by common sense. For instance, there is hardly any dispute about the perception of a tree, water body or mountain, although we give them different names and meanings according to our languages and cultural backgrounds. The essence, however, is well understood.

We will not discuss here the fundamental philosophical questions of “What can we know?” or “How real is reality?” They have been the subjects of philosophical contemplation since the dawn of mankind, and the answers largely depend on the philosophical or religious orientation of the persons involved. What we state here, however, is that in our minds we create models of the real world. The assumption (based on experience and common sense) is that within a given context the individual mental models coincide to such an extent that we perceive the same reality. To represent this reality in information systems databases, we apply processes of ordering, classifying, structuring and mapping, commonly known as spatial modelling.

18.2 REAL WORLD PHENOMENA AND THEIR ABSTRACTIONS

In the real world, we distinguish between natural and man-made phenomena. Natural phenomena exist independently from human actions and are subject mainly to the laws of nature. Examples are the landscape (topography), weather, or natural processes that shape and influence the environment. Man-made phenomena are objects that have come into existence by human activities through construction or building activities.

Based on these phenomena we develop high-level abstract models for particular purposes and applications. Features are abstractions of phenomena. They populate the models that are usually organized in thematic layers. Examples of such models are a cadastre, topography, soil, hydrology, land cover or landuse.

The fact that these models are abstractions of the real world can be illustrated by the example of a cadastre. Let us assume that a cadastre is a legal and organizational framework for the handling of land. It is a very important, clearly described and understood concept. Yet, we do not see a cadastre when we look around us. What we see are real world phenomena such as buildings, roads, fences around pieces of land and people. A cadastre abstracts from certain phenomena and their relationships to create something new that is real in a given context.

Layers are an ordering principle for real world phenomena. Again, when we look around us, we do not see the world in layers. Yet, we are used to organizing phenomena of the real world in such a way that we classify them according to a perceived purpose or characteristic into subsets, i.e. layers, that allow us to deal with them efficiently. An abstraction of phenomena to fields and objects as well as their implementation are treated later in this chapter.

In order to conceptualize mental models of the real world, we need to categorize the phenomena we observe. These phenomena exist in space and time and have therefore a spatial (geometric) and a temporal extent. They possess certain thematic characteristics that allow us not only to refer to them in terms of spatio-temporal but also according to thematic information. The thematic information of real world phenomena is the basis for the definition of layers.

Real world phenomena are related to each other and they are subject to various processes that change, modify or influence them in many different ways. Features are representations of real world phenomena in a spatial model. Features have relationships among each other and related processes. Spatial data are representations of features in a database. They are represented as objects and fields with suitable data structures. Topology exposes certain spatial relationships and algorithms used to process the data. Through spatial analysis and query, we extract spatial information that is used for reasoning and acquiring knowledge. Figure 18.1 illustrates this process.

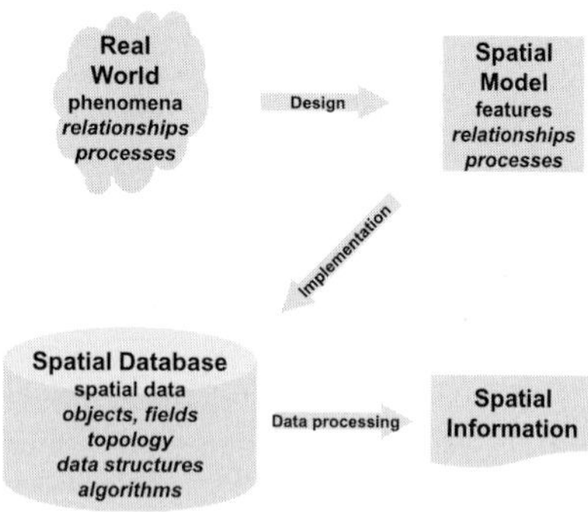

Figure 18.1. The process from real world phenomena to spatial models, databases and spatial information.

18.3 CONCEPTS OF SPACE AND TIME

Space and time are two closely related concepts that have been the subject of philosophical and scientific consideration since the dawn of mankind. The physical space in which humans live is the three-dimensional (Euclidean) space as a frame of reference for our senses. Of all possible spaces, this is the space that is illustrative and that we perceive as being real. Time is a measure for change in our immediate experience. Usually we assume time to be of a continuous linear nature extending from the past, through the present into the future.

Space and time (at least as we perceive them) are so well known and appear to be given beyond any doubt that we hardly ever contemplate their structure and characteristics in our everyday lives. When we deal with information systems that process and manipulate spatio-temporal data, we need a clear and well-understood model of space and time. The following sections describe how concepts of space and time developed in Western philosophy and science. We discuss them according to three epochs: (i) pre-Newtonian concepts, (ii) Newtonian and classical concepts, and (iii) contemporary concepts of space and time. A detailed discussion of space and time in spatial information systems can be found in Langran (1992), Stock (1997), Couclelis (1999) and Peuquet (2002).

18.3.1 *Pre-Newtonian concepts of space and time*

The concepts of space and time of this epoch are mainly dominated by the ideas of Greek philosophers about the logical conditions for things to change and the structure of the world in which change occurs.

Heraclitus (around 500 BC) of Ephesus (western Turkey) studied the problem of change, i.e. how can the identity of things be preserved when they change. He stated that everything flows, nothing remains, and the only thing that really exists is change (processes). "Everything flows" and "We cannot step into the same river twice" are attributed to him.

At about the same time, Parmenides of Elea (southern Italy) developed a completely opposite philosophy of the non-existence of the void. He postulates (through deductive reasoning) that change does not exist and that the real world (the real being) is plenum (a solid complete compact being), immutable, without change and eternal. A void (or empty space, i.e. something non-existent) does not exist. What we perceive as change is a delusion of our senses. His ideas were further developed and "proven" by his student ZENO.

One of ZENO's famous "proofs" that change and movement cannot exist is known as the race between Achilles and the tortoise. The tortoise gets a head-start and begins the race at point B, whereas Achilles starts from point A. When Achilles reaches point B, the tortoise already has moved on to point C. When Achilles reaches point C, the tortoise has moved on to another point, and so on. The lead of the tortoise gets smaller and smaller, *ad infinitum*. We get an infinite number of (smaller and smaller) leads.

The argument is as follows: In order to reach the tortoise, Achilles must catch up an infinite number of (finite in length) leads. It is impossible to run this infinite number of short distances, because Achilles would have to run infinitely far (or forever). Therefore, it is impossible for Achilles to catch the tortoise. Since we can easily catch a tortoise when we would run against it, we end up with a paradox. This "proves" that the assumption that movement and change are real leads to contradictions. Therefore, movement and change are impossible.

The fallacy of the argument lies in the fact that ZENO does not know that an infinite geometric series may have a finite sum under certain conditions.

DEMOCRITUS (460–370 BC) did not accept the non-existence of change as postulated by PARMENIDES. Space is an absolute and empty entity existing independently from the atoms that fill the space. Atoms are indivisible real things; they are immutable and eternal, and have different size and weight. There is no empty space within atoms. An atom is a plenum. Objects are formed as a collection of atoms. The importance of the Atomist theory is evident in today's modern particle physics.

Greek mathematics was strongly dominated by the Pythagorean number theoretic approach. It was essentially arithmetic based on (philosophical) properties of numbers, counting and the ratios between numbers. The discovery of irrational numbers such as the square root of 2 (the length of the diagonal in the unit square) shook the foundation of Greek mathematics that was based on counting in natural units. The need for a truly geometrical description of the world became apparent.

The great philosopher PLATO (427–347 BC) and one member of his school, EUCLID (at 300 BC), laid the foundation for a new geometric modelling of the real world. This geometric model of matter is based on right triangles as atoms and solids built from these triangles. Matter consists of four elements: earth, air, fire and water. Each element is made of particles, i.e. solids (the so-called Platonic solids) that in turn are made from triangles.

In his book, The Elements, EUCLID developed a mathematical theory of geometry that remained valid until the late 19th century. Euclidean geometry was considered a true description of our physical world until it was discovered that many consistent geometric systems are possible, some of them non-Euclidean, and that geometry is not a description of the world but yet another formal mathematical system with no necessary reference to real world phenomena.

18.3.2 *Classical concepts of space and time*

The time between the classical Greek period and the rise of modern science was dominated by the philosophy and teaching of ARISTOTLE (384–322 BC). According to his ideas, empty space is impossible, and time is the measure of motion with regard to what is earlier and later. Space is defined as the limit of the surrounding body towards what is surrounded. Space is a place of all things that occupy a place in space. Following from this approach, space can be conceptualized in two possible ways:

- *Absolute space*. Space as a set of places. It is an absolute real entity, the container of all things. Its structure is fixed and invariable. Generally, this is considered the space as described by the Euclidean geometry.
- *Relative space*. Space is a system of relations. It is the set of all material things, and relations are abstracted from them. Space is a property of things or things have spatial properties.

The rise of modern science took shape in the works of COPERNICUS (heliocentric system: the sun seen as the centre of our planetary system), KEPLER (mathematical foundation of the heliocentric system), and GALILEI (foundations of mechanics) in the 16th and 17th centuries.

NEWTON (1643–1727 AD) was a brilliant scientist (dynamic theory) and philosopher. In his philosophy, he was an outspoken proponent of the concept of absolute space, although it strictly contradicts his dynamics theory. The concept of absolute space remained dominant until the late 19th century.

LEIBNIZ (1646–1716) on the other hand proposed the concept of relative space. For him, space is a system of relationships between things. It is interesting to note that both NEWTON and LEIBNIZ are the founders of mathematical calculus.

One of the greatest philosophers, KANT (1724–1804), claimed that space and time are not empirical physical objects or events. They are merely *a priori* true intuitions, not developed by experience, but used by us to relate and order observations of the real world. Space and time have empirical reality (they are absolute and *a priori* given) and transcendental idealism (they belong to our conceptions of things but are not part of the things). We cannot know anything about the things as such. In this regard, KANT can be regarded as a proponent of absolute space, yet in a more sophisticated way than the previous philosophical approaches.

18.3.3 *Contemporary concepts of space and time*

The development of modern physics (such as field theories, theory of relativity, and quantum theory) and mathematics (non-Euclidean geometries) lead to the conclusion that traditional Euclidean geometry (describing the three-dimensional space of our perception) is only an approximation of the real nature of the world. The field theories (FARADAY and MAXWELL) lead to the assumption that space is not empty, but is filled with energy. Therefore, a material existence of space is strongly supported by these theories.

As a consequence of the special and general theory of relativity by EINSTEIN (1879–1955), space and time cannot anymore be considered as two separate entities. We speak of space-time, which is considered a four-dimensional curved space that can only be described by non-Euclidean geometry. Quantum mechanics states the principle of uncertainty and the discrete character of matter and energy. It has become evident that the space of our perception is not necessarily identical to the microscopic (sub-atomic) space or the space of cosmic dimensions.

Other classifications of space are physical vs. commonsense space or global vs. local space. Physical space leans more towards geometric space, e.g. for robotics, whereas commonsense space usually addresses spatial relations in human thinking and languages. Global space can be considered to be a kind of background (or container) for objects in that space. These objects might have relationships and properties of a (relative) local nature (Vieu 1997).

18.3.4 *Concepts of space and time in spatial information systems*

Spatial information is always related to geographic space. This is the space beyond the human body, space that represents the surrounding geographic world. Within such space, we move, we navigate in it, and we conceptualize it in different ways. Geographic space is the space of natural and man-made features of the geographic world. Geographic information system technology is used to manipulate objects in geographic space and to acquire knowledge from spatial facts.

Geographic space is distinct from tabletop space. In other words, objects that are smaller than us, objects that can be moved around on a tabletop, are not subject of our interest. The human understanding of space, influenced by language and cultural background, plays an important role in how we design and use tools for the processing of spatial data.

In the same way as spatial information is always related to geographic space, it relates to geographic time, the time whose effects we observe in the changing geographic world around us. We are less interested in pure philosophical or physical considerations about time or space-time, but more in the observable spatio-temporal effects that can be described, measured and stored in information systems.

18.4 SPATIAL MODELLING

As mentioned in the previous sections, we always create models of the real world in our minds. When we want to acquire, store, analyse, visualize and exchange information about the real world, we use other media and means than just interpersonal communication. We need representations of our mind models, i.e. models of the real world that can be used to acquire, store, analyse and transfer information about real world data (spatial models).

In general, modelling can be described as creating a structure preserving mapping (morphism) from a domain to a co-domain. In our case, the domain is the real world, and the co-domain is a spatial model. Such a mapping normally creates a "smaller" (i.e. abstracted, generalized) image of the original. Structure preserving means that the elements of the co-domain (spatial features) behave in the same (however simplified or abstracted) way as the elements of the domain (spatial phenomena). Figure 18.2 illustrates the principle of spatial modelling.

As mentioned above, we consider space to be the three-dimensional Euclidean space of our common sense. All phenomena exist in this space and undergo changes, which we perceive as the passing of time. In this sense, time is modelled indirectly as changes in (spatial or thematic) attributes of the features.

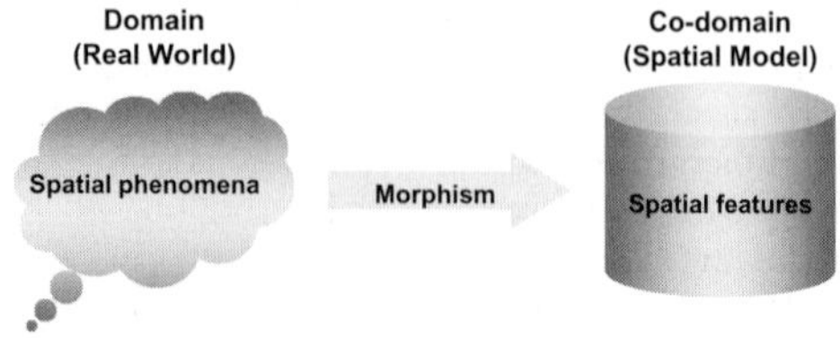

Figure 18.2. Spatial modelling is a structure preserving mapping from the real world to a spatial model.

We call a spatial model that also considers time a spatio-temporal model. Sometimes such a model is addressed as four-dimensional giving the impression that time is the fourth dimension in addition to the three spatial dimensions. However, it is better to call it spatio-temporal instead of four-dimensional. First, time is not of the same type as the spatial dimensions; it has a distinctly different quality. Secondly, the term "four-dimensional" only makes sense when we always would consider three spatial dimensions. In most of the cases, however, the spatial model only considers two dimensions.

As discussed above, we usually apply the order principle of layers to structure our views of the world. Depending on the nature of the phenomena, we use either field-based or object-based spatial data models.

The most common realizations of spatial data models are—in historic sequence—maps and databases. Both have distinct characteristics, advantages and disadvantages. Whereas maps usually have been used to picture real world phenomena, databases can be used to represent real and virtual worlds.

Real worlds are subsets of the reality that we perceive. Virtual worlds are computer generated "realities" that exist only as possibilities with no counterpart in the real world. Yet, we can visualize them, navigate through them and perceive them as "real" (therefore the term virtual reality). There is no difference between real and virtual world models with regard to their representation. The only difference is that the former are models of something that exists in the real world and the latter are models of something that exists only in a virtual (physically non-existent) world.

18.4.1 *Spatial models*

Among spatial data models, we can distinguish two major types, field- and object-based models. Field-based models consider spatial phenomena to be of a continuous nature where in every point in space a value of the field can be determined. Examples of such phenomena are temperature, barometric pressure or elevation. Object-based models consider space to be populated by well distinguishable, discrete, bounded objects with the space between objects potentially being empty. Examples are a cadastre with clearly identifiable objects like parcels and buildings.

Field versus object can be viewed—in general terms—as a manifestation of the philosophical conception of *plenum* versus atomic space, or as in modern physics, wave versus particle. In GIS, fields are usually implemented with a tessellation approach, objects with a (topological) vector approach. The following sections briefly illustrate the two different model types. A more detailed technical treatment of GIS can be found in Worboys & Duckham (2004).

18.4.1.1 *Field-based models*

The underlying space for a field-based model is usually taken as the two- or three-dimensional Euclidean space. A field is a computable function from a geometrically bounded set of positions (in 2-D or 3-D) to some attribute domain. Computable means that for every position within the geometric bounds a value can be determined by either measurement or by computation. A field-based model consists of a finite collection of such fields (Fig. 18.3).

Fields can be discrete, continuous and differentiable. Discrete fields represent features with boundaries; continuous fields are used for features where the underlying function is continuous, such as for temperature, barometric pressure or elevation. If the field function is differentiable, we can even compute the slope at every position.

It is important to note that our definition of field is completely independent of implementation considerations. This means that we have not considered a data structure to represent them. It is, however, obvious that some data structures seem to be the logical choice for certain types of field.

Though geometrically bounded, the domain of a field is still an infinite set of positions. Computers cannot easily represent field values for all these positions, so we must accept an approximation. The standard way to obtain this is to finitely represent the geometrically bounded space through a subdivision into a regular or irregular tessellation, consisting either of square (cubic) or triangular (tetrahedral) parts. These individual parts are called locations and are often approximated by points.

Our finite approximation of positions into locations leads to some forms of interpolation that must be dealt with. The field value at a location can be interpreted as one for the complete tessellation cell, in which case the field is discrete, not continuous or differentiable. Some convention is needed to state which value prevails on cell boundaries; with square cells, this convention often says that lower and left boundaries belong to the cell.

Another option is to interpret the field value at a location as representative only for some position

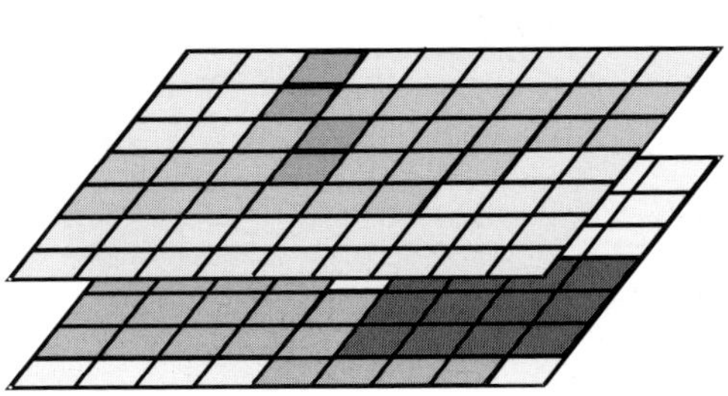

Figure 18.3. Two data layers in a field-based model.

within the cell. Again that position is fixed by convention, and may be the cell centroid or, for instance, its lower left corner. Field values for positions other than these must be computed through some form of interpolation function, which will use one or more nearby field values to compute the value at the requested position. This allows representing continuous and differentiable functions.

To represent spatial features using a field-based approach, we have to perform the following steps:

- Define or use a suitable model for the underlying space (tessellation).
- Find suitable domains for the attributes.
- Sample the phenomena at the locations of the tessellation to construct the spatial field functions.
- Perform analysis, i.e. compute with the spatial field functions.

18.4.1.2 *Object-based models*

Object-based models decompose the underlying space into identifiable, describable objects that have spatial, thematic and temporal attributes, as well as relationships among each other (Fig. 18.4). The space outside the objects is empty. Examples of objects are buildings, cities, towns, districts or countries; attributes are, for instance, the number of floors, population, boundary or area.

In a GIS database implementation, objects are represented as a structured collection of geometric primitives (points, lines, polygons and volumes) under geometric, thematic and topological constraints.

Object-based models are discrete models. Operations in the model always refer to the manipulation of individual objects or sets of objects. Manipulations concern the spatial, thematic, topological or temporal domain. Accordingly, they are realized through geometric, attribute manipulation or topological operations. Topology plays a major role in object-based models. It is the "language" that allows us to specify and enforce consistency constraints for spatial databases. We will deal with it later in this chapter.

The majority of object-based models are two-dimensional. Recently, three-dimensional data models have been proposed. Their topology is more difficult to handle than for the standard two-dimensional cases.

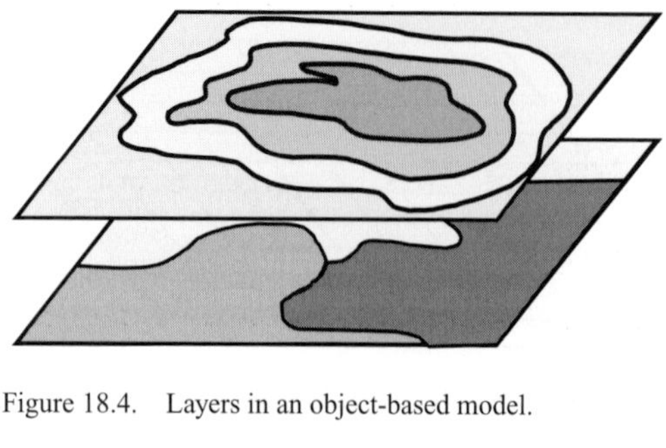

Figure 18.4. Layers in an object-based model.

18.4.2 *Spatio-temporal models*

Beside geometric, thematic and topological properties, spatial data also possess temporal characteristics. It is, for instance, interesting to know who were the owners of a land parcel in 1980 or how land cover of a given piece of land changed over the last 20 years.

Spatio-temporal data models are data models that can also handle temporal information in spatial data. Several models have been proposed.

Before we describe the major characteristics of the spatio-temporal data models, we need a framework to describe the nature of time itself. Time can be characterized according to the following properties:

Time density. Time can be discrete or continuous. Discrete time is composed of discrete elements (seconds, minutes, hours, days, months or years). In continuous time, for two points in time, there is always another point in between. We can also structure time by events (points in time) or states (time intervals). When we represent states by intervals bounded by nodes (events), we can derive temporal relationships between events and states such as "before", "overlap", "after", etc.

Dimensionality of time. Valid time (or world time) is the time when an event really happens. Transaction time (or database time) is the time when the event is recorded in the database.

Time order. Time can be linear, extending from the past to the present and into the future. We know also branching time (possible scenarios from a certain point in time onwards) and cyclic time (repeating cycles such as seasons or days of a week).

Measures of time. A chronon is the shortest non-decomposable unit of time that is supported by a database (e.g. a millisecond). The life span of an object is measured by a (finite) number of chronons. Granularity is the precision of a time value in a database (e.g. year, month, day, second, etc.). Different applications require different granularity. In cadastral applications, time granularity can be a day; in geological mapping, time granularity is more likely to be in the order of thousands of years.

Time reference. Time can be represented as absolute (fixed time) or relative (implied time). Absolute time marks points on the timeline where events happen (e.g. 6 July 1999 at 11:15 p.m.). Relative time is indicated relative to other points in time (e.g. "yesterday", "last year", "tomorrow", which is all relative to "now", or "two weeks later", which may be relative to an arbitrary point in time).

In spatio-temporal data models, we consider changes of spatial and thematic attributes over time. In data analysis, we can keep the spatial domain fixed and look only at the attribute changes over time for a given location in space. We would, for instance, be interested how land cover changed for a given location or how the landuse changed for a given land parcel over time, provided its boundary did not change.

On the other hand, we can keep the attribute domain fixed and consider the spatial changes over time for a given thematic attribute. In this case, we could be interested to see which locations were covered by forest over a given period.

Finally, we can assume both the spatial and attribute domain being variable and consider how an object changed over time. Here, the problem of object identity becomes apparent. When does a change or movement cause an object to disappear and become a new one? In the following, we describe the major characteristics of some popular spatio-temporal models. Useful references concerning these topics can be found in Langran (2002), Peuquet & Duan (1995) and Worboys (1994).

18.4.2.1 *Space-time cube model*

This model is based on a two-dimensional space (spanned by the *x*- and *y*-axis) whose features are traced through time (along the *z*-axis) thereby creating a space-time cube (Fig. 18.5). The traces of objects through time create a worm-like trajectory in the space-time cube. This model potentially allows absolute, continuous, linear, branching and cyclic time. It supports only valid time. The attribute domain is kept fixed and the spatial domain variable.

18.4.2.2 *Snapshot model*

In the snapshot model, layers of the same theme are time-stamped. For every point in time t_i that we would like to consider, we have to store a layer and assign the time to it as an attribute (Fig. 18.6). We do not have any information about the events that caused different states between layers. This model is based on a linear, absolute, discrete time. It supports only valid time and multiple granularities. The spatial domain is fixed (field-based) and the attribute domain is variable.

18.4.2.3 *Space-time composite model*

The space-time composite model starts with a two-dimensional situation (plane or layer) at a given start time. Every change of features that happens later is projected onto the initial plane and intersected with the existing features, thereby creating an incrementally built polygon mesh (Fig. 18.7). Every polygon in this mesh has its attribute history stored with it. The space-time composite model is based on linear, discrete, relative time. It supports both valid and transaction time, and multiple granularities. It keeps the attribute domain fixed and the spatial domain variable.

18.4.2.4 *Event-based model*

In an event-based model, we start with an initial state and record events (changes) along the time line (Fig. 18.8). Whenever a change occurs, an entry is recorded. This is a time-based model. The spatial and thematic attribute domains are secondary. The model

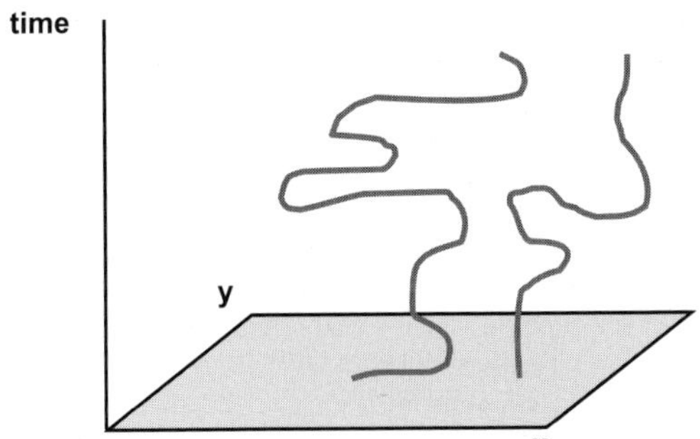

Figure 18.5. Space-time cube.

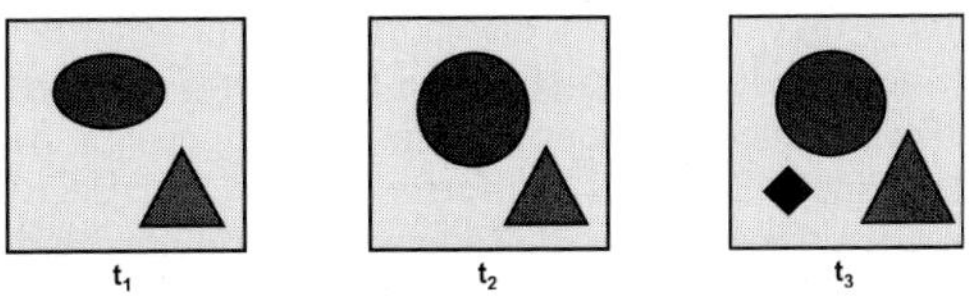

Figure 18.6. Snapshot model.

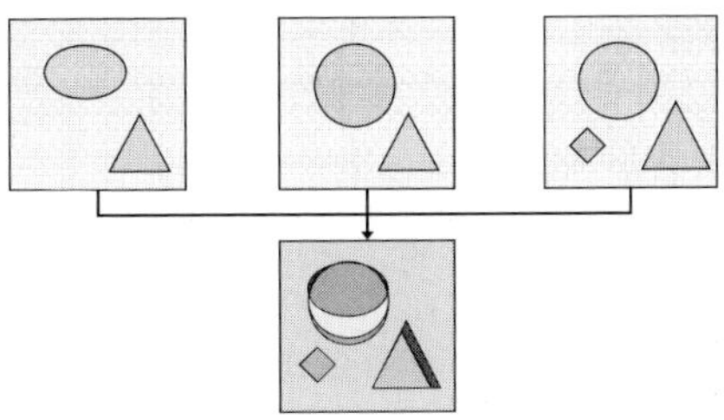

Figure 18.7. Space-time composite model.

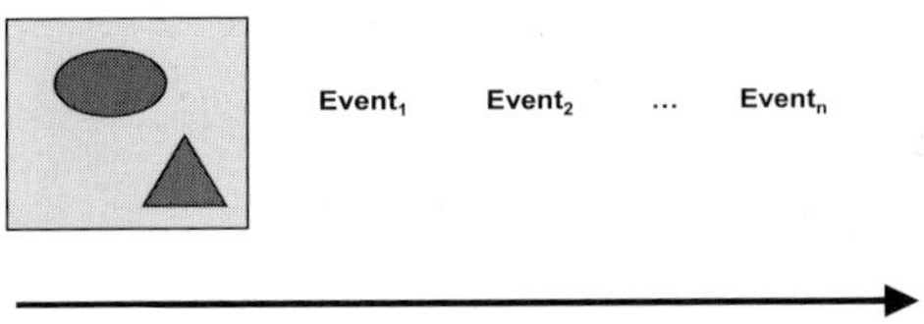

Figure 18.8. Event-based model.

is based on discrete, linear, relative time, and supports only valid time and multiple granularity.

18.4.2.5 *Spatio-temporal object model*
This model is based on spatio-temporal objects (ST-objects) that are a complex of ST-atoms (STA). Both objects and atoms have a spatial and a temporal extent. Figure 18.9 shows three objects O_1, O_2, and O_3 at times t_1, t_2, t_3. The model is based on discrete, absolute, linear time, and supports valid and transaction time as well as multiple granularities.

18.4.3 *Implementations of spatio-temporal models*

The result of spatial modelling is usually implemented in a spatial database as a model of the real world (Fig. 18.10). Historically, maps were the first analogue models of the real world, followed by spatial databases that have taken over the role as data storage and repository.

18.4.3.1 *Maps*
The best known (conventional) models of the real world are maps. Maps have been used for thousands of years to represent information about the real world. Today, they are usually drawn on paper or other permanent media and function as data storage and visualization media. Their conception and design has developed into an art and science with a high degree of scientific sophistication and artistic craftsmanship.

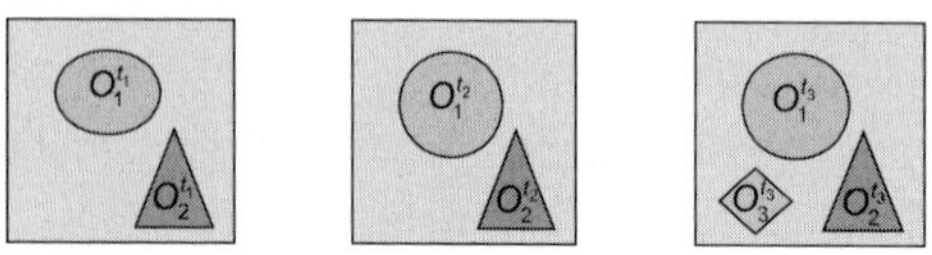

Figure 18.9. Spatio-temporal object model.

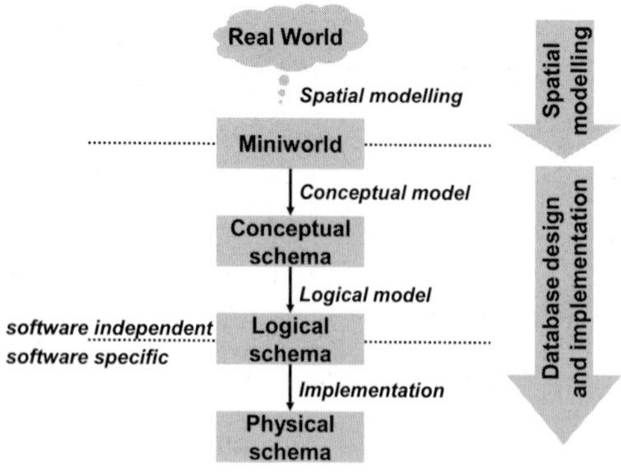

Figure 18.10. Spatial modelling and database design.

Maps have proven to be extremely useful models of reality for many applications in various domains.

Yet, maps are two-dimensional (flat) and static. It is not easy to visualize three-dimensional dynamic features without considerable abstractions in the spatial and temporal domains. We distinguish between topographic and thematic maps. Other cartographic products that are not maps are often used to represent three-dimensional and dynamic phenomena. These are, for instance, block diagrams, animations and panoramic views.

A disadvantage of maps is that they are restricted to two-dimensional static representations, and that they are always displayed at a given scale. The map scale determines the spatial resolution of the graphic feature representation. The smaller the scale, the less detail a map can show. The accuracy of the primary data, on the other hand, puts limits to the scale at which a map can be drawn. The selection of a proper map scale is one of the first and most important steps in a map conception.

A map is always a graphic representation at a certain level of detail, which is determined by the scale. The process to derive less detailed representations from a detailed one is called map generalization (or cartographic generalization).

Cartography as the science and art of map making functions as an interpreter translating real world phenomena into correct, clear and understandable representations for our use.

With the advent of computer systems, analogue cartography became digital cartography. It is important to note that whenever we speak about cartography today, we implicitly assume digital cartography. The use of computers in map making is an integral part of modern cartography. The role of the map changed accordingly. Increasingly, maps are losing their role as data storage. This role is taken over by databases. What remains is the visualization function of maps.

18.4.3.2 *Databases*
Spatial databases store representations of spatial phenomena in the real world to be used in a geographic information system (Rigeaux et al. 2002). In the design of a database, we distinguish between different levels of definition. A language that we use to define the database is called a data model; each level typically has its own data model. They will be discussed in the next section.

The assumption for the design of a spatial database is that spatial phenomena exist in a two- or three-dimensional Euclidean space. All phenomena have various relationships among each other and possess spatial (geometric), thematic and temporal attributes. Phenomena are classified into thematic layers depending on the purpose of the database. This is usually described by a qualification of the database as, for example, cadastral, topographic, landuse or soil database.

The representations of spatial phenomena (spatial features) are stored in a scale-less and seamless manner. Scale-less means that all coordinates are world coordinates given in units that are normally used to reference features in the real world (geographic coordinates as latitude and longitude, or some other spatial reference system). A seamless database does not show map sheet boundaries or other partitions of the geographic space other than imposed by the spatial features themselves.

It must be noted, however, that scale plays a role when data are captured from maps as data source. Here, the scale of the source map determines the accuracy of the feature coordinates in the database. Likewise, the accuracy of measurements in field surveys determines the quality of the data. If the coordinates are given in units other than geographic coordinates, information concerning the spatial reference system should also be present.

The advantages of databases as real world models are that they are scale-less, potentially three-dimensional, dynamic and seamless. It is easy to query a database, and to combine data from different layers (spatial joins or overlays). Spatio-temporal databases consider not only the spatial and thematic but also the temporal extent of the features they represent.

It is important to note here that we use the term "model" with different connotations. A "spatial model" is a representation of a number of phenomena in the real world, usually to enable some type of administration, computation and/or simulation. It is the result of the activity of "spatial modelling".

A "data model", on the other hand, is a database language used as a tool in database design, in an activity called "data modelling". The result of that activity is not a data model, but a database schema, an abstract definition of the contents of the (future) database. A database schema can be viewed as a special kind of real world model, but it is abstract because it identifies only types of things in that real world, and not the things themselves.

18.5 DATABASE DESIGN, TOPOLOGY AND SPATIO-TEMPORAL RELATIONSHIPS

Spatial data are computer representations of spatial features. A modelling language for a GIS database is a spatial data model. It is used in the design of spatial databases. A spatial database holds a digital representation of the real world, sometimes called a digital landscape model (DLM).

A DLM is the basis for spatial analysis and manipulation of spatial data. The feature space for a spatial database is a geometric space in which features at various levels of detail are modelled. A good data model should allow for multiple representations of spatial features. The transformation of a representation from a detailed to a less detailed version is called model generalization. This is different from cartographic generalization where a graphic representation (digital or analogue) at a smaller scale is derived from a large scale dataset or map under certain merely graphic constraints.

We are particularly interested in properties of the geometric space that remain invariant under certain transformations. This is important to guarantee a consistent representation of the features in a database. These properties are related to topology, which will be discussed later in this section.

Some features possess a degree of uncertainty in either their thematic or spatial extent. For example, soil types do not have crisp boundaries. Linguistic expressions of vagueness or uncertainty such as "moderate slope" or "close to a lake" are often a better analysis approach than clear cut class boundaries. We have to devise proper means to take care of these uncertainties in our spatial models. Fuzzy logic, a branch of modern mathematics, provides the tools.

18.5.1 *Database design*

In designing a database, we have to pass through several design phases. The result of every phase is a schema of the respective level. A schema is a representation of the database as described with a particular data model. This approach is called the ANSI/SPARC architecture; it comprises four levels (Elmasri & Navathe 2004).

A database is generally thought to serve multiple users or user groups. They may have different perceptions of the data stored. At this level, each user (group) is supported with its own external view of the database. We may define an external view as a personalized conceptual database schema. There will be as many external views as there are users and user groups. Domain experts mainly do modelling at this level.

All external views are merged into a single conceptual schema of the database. This is usually done with high-level semantic data models such as the entity-relationship (E-R) model, the extended E-R model, or object-oriented data models. The basic constructs of the E-R model are entity types, attribute types and relationship types. Instances of the types populate the database. A conceptual schema is implementation-independent and not related to any particular database management software. It provides an answer to the question what features are represented in the database.

The conceptual schema is transformed into a logical schema using one of the logical data models. Currently, the most popular one is the relational data model, which is based on relational algebra. Most commercial database implementations provide support for this model. It is easy to understand because it is based on relations—sets of records—that have a straightforward implementation as tables. The logical schema is meant to provide the definition of a redundancy-free dataset. Each fact should be stored only once in the database.

A physical schema is the result of the implementation of the logical schema with particular database management software. Table 18.1 summarizes this architecture.

18.5.2 *Topology*

Topology is the branch of mathematics that deals with properties of spaces that remain invariant (i.e. do not change) under certain transformations. A simple example illustrates this.

Assume you have some geometric figures that are drawn on a sheet of rubber (Fig. 18.11 left). Now, take

Table 18.1. Data models and schemas in database design.

Schema	Models used to derive the schema
External views	Depending on different user perspectives, a subset of the real world is defined and described (spatial modelling).
Conceptual schema	A synthesis of external views to create a conceptual schema making use of semantic data modelling techniques such as the entity-relationship approach.
Logical schema	Transformation of the conceptual schema into a logical schema using the relational model. Emphasis is on redundancy removal.
Physical schema	This is the mapping of the logical schema into data structures and algorithms. It is normally not acces sible or visible to the user. Its emphasis is on processing speed.

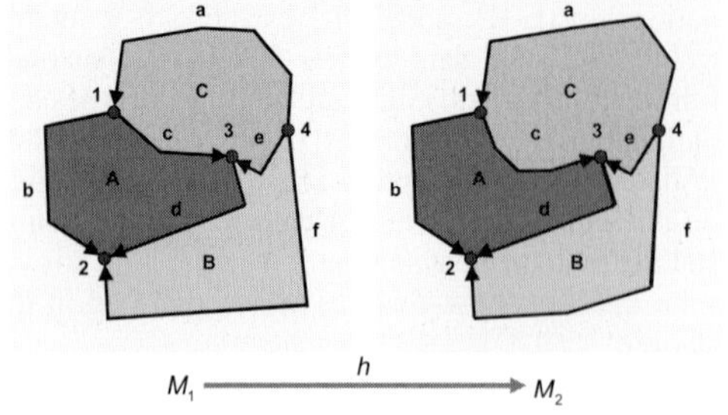

Figure 18.11. Rubber sheet transformation.

the sheet and pull on its edges, but do not tear or break it (Fig. 18.11 right). The figures will change in shape and size. Some properties, however, do not change.

The neighbourhood relationships between A, B and C remain, and the boundary lines have the same start and end points. The areas are still bounded by the same boundary lines, only their shapes and the length of the perimeters have changed.

These relationships are invariant under a continuous transformation. The properties are called topological properties, and the transformation is called a topological mapping (or homeomorphism). The result is topologically equivalent (or homeomorphic) to the original dataset.

The mathematical properties of the geometric space used for spatial data can be described as follows:

- The space is a three-dimensional Euclidean space where for every point we can determine its three-dimensional coordinates as a triple (x, y, z) of real numbers. In this space, we can define points, lines, polygons and volumes as geometric primitives of the respective dimension. A point is a zero-dimensional, a line a one-dimensional, a polygon a two-dimensional and a volume is a three-dimensional primitive.
- The space is a metric space, i.e. we can always compute the distance between two points according to a given distance function (metric).
- The space is a topological space, i.e. for every point there exists a neighbourhood that is topologically equivalent to an interval (one-dimensional), a disk (two-dimensional) or a ball (three-dimensional), depending on the dimension of the space in use. These neighbourhoods allow us to define open sets that under certain conditions define a topology on the space.

We can structure space by simple topological subspaces that are easy to handle and can be used as building blocks for spatial data. They are called simplexes and are the simplest geometric shapes of a respective dimension: point (0-simplex), line segment (1-simplex), triangle (2-simplex) and tetrahedron (3-simplex). A collection of simplexes that satisfy certain conditions is called a simplicial complex (Fig. 18.12).

It is common to use homeomorphic images of simplexes (cells) in spatial data models. Cells are nodes (0-cells), arcs (1-cells), polygons (2-cells) and volumes (3-cells). A collection of cells (or cell complex) is the topological equivalent of a collection of spatial features (Fig. 18.13).

In many applications, we will use a two-dimensional space only, i.e. the usual Euclidean plane. In fact, models for three-dimensional spatial data are much more difficult than two-dimensional models (Mesgari 2000).

The consistency of a spatial dataset is important for any processing and analysis that might be applied. The model must provide means to define and enforce topological consistency in a database. Topology provides us with a set of rules that can easily be applied to any two-dimensional spatial dataset.

If we can show that the following rules are satisfied for every element in the dataset, we know that it is a topologically consistent two-dimensional configuration.

1. Every arc must be bounded by two nodes (start node and end node).
2. For every arc, there exist two polygons (left polygon and right polygon).
3. Every polygon has a closed boundary consisting of an alternating sequence of nodes and arcs.
4. Around every node, there exists an alternating closed sequence of arcs and polygons (umbrella).
5. Arcs do not intersect except at nodes.

In mathematical terms, these rules guarantee that the configuration is a two-dimensional manifold where every point has a neighbourhood equivalent to an open disk. These rules are valid only for a two-dimensional cell complex. They cannot be applied without additions and modifications to higher dimensions. Figure 18.14 illustrates the closed boundary condition (rule 3) for polygon A as closed sequence of node 1—arc c—node 3—arc d—node 2—arc b—node 1.

Figure 18.15 shows rule 4 ("umbrella" condition) for node 3 as closed sequence of arc c—polygon C—arc e—polygon B—arc d—polygon A—arc c.

Interior and boundary are properties of spatial features that remain invariant under topological mappings. This means that under any topological mapping, the interior and the boundary of a feature remain unbroken and intact.

18.5.3 *Spatial relationships*

Whereas relationships between simplexes or cells define consistency constraints for spatial data, we can use the topological properties of interior and boundary to define relationships between spatial features. Since the properties of interior and boundary do not change under topological mappings, we can investigate their possible relations between spatial features.

Let us assume two spatial regions A and B. Both have their respective boundary and interior. When we consider all possible combinations of intersections between the boundaries and the interiors of A and B, we know that these will not change under any topological transformation. From these intersection patterns, we can derive eight mutual spatial relationships between two regions (Egenhofer & Franzosa

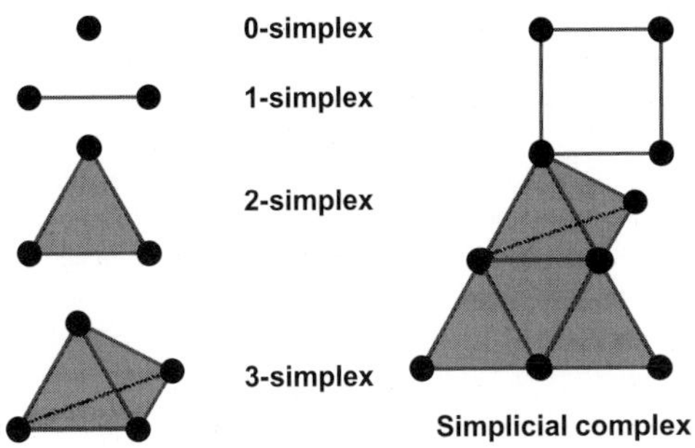

Figure 18.12. Simplexes and simplicial complex.

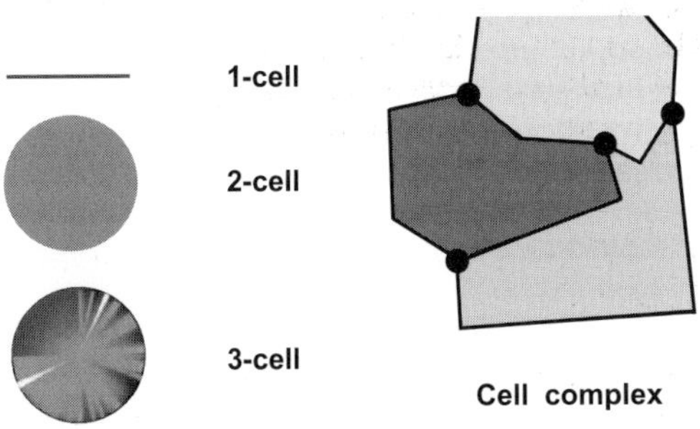

Figure 18.13. Cells and a cell complex.

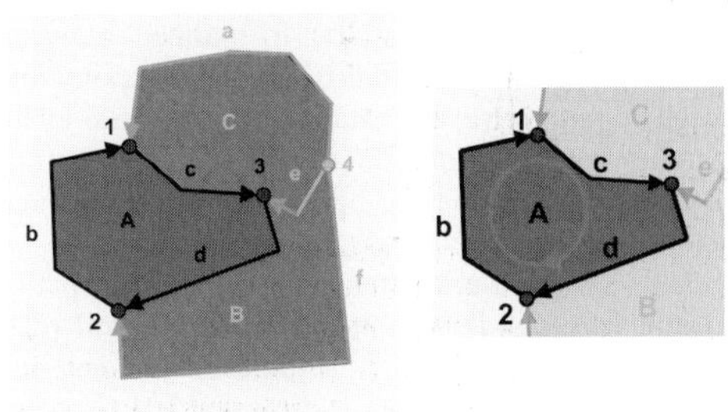

Figure 18.14. Closed boundary condition.

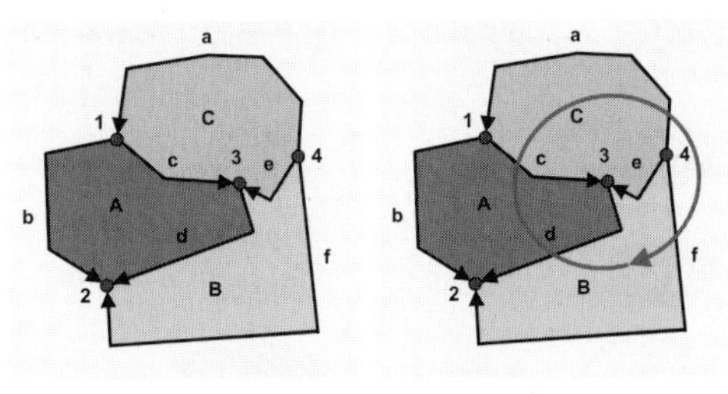

Figure 18.15. Topological consistency rule 3 ("umbrella" condition).

1991). It is also possible to consider other types of relationships based on order structures (Kainz et al. 1993) or relationships based on fuzzy topology (Tang 2004, Tang et al. 2006). If, for instance, the boundary of A intersects the boundary of B, and the interiors of A and B do not intersect, we say that A and B meet. Figure 18.16 shows all possible eight spatial relationships: disjoint, meet, equal, inside, covered by, contains, covers and overlap. These relationships can be used, for instance, in queries against a spatial database.

Networks, like road or river networks, are best modelled as a one-dimensional subset (or skeleton) of a cell complex. The topological relationships are then reduced to incidence relationships between edges (arcs) and nodes. Such a structure is also called a graph. Graph theory, although closely related to topology, has developed as an independent mathematical discipline. A special type of graph frequently used in GIS is a planar graph. It is completely embedded in the plane such that no two edges intersect except at nodes.

18.5.4 *Temporal and spatio-temporal relationships*

Event and state are two basic elements in temporal databases. Event reflects the time of a feature change and we often treat it as a time point. On the other hand, state represents the relatively invariable period of a feature and is recorded as time interval. Similar to topological relationships among spatial features, temporal relationships can be formalized between events and states. Three kinds of temporal topological relationships event-event, state-state, and state-event play a key role in temporal databases. Thirteen kinds of temporal relations have been formalized between two states (Allen 1984, Roshannejad 1995).

Spatio-temporal relationships exist among spatio-temporal features. If both spatial and temporal relationships are considered, it is possible to describe change processes based on spatio-temporal relationships. For example, an "expand" process can be the combination of spatially "contains" and temporally "meets with" of a spatial feature (Figure 18.17).

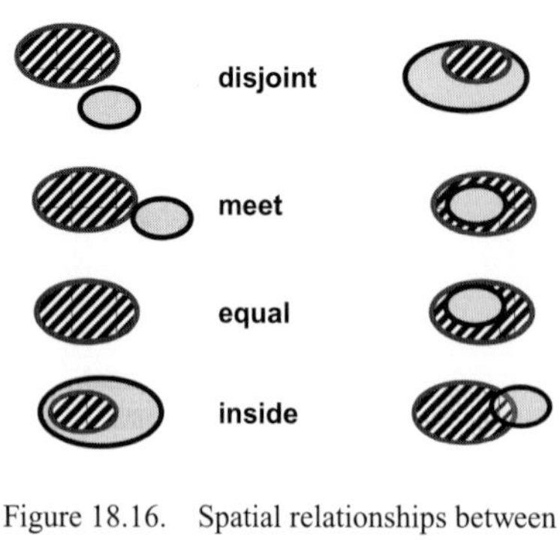

Figure 18.16. Spatial relationships between two regions.

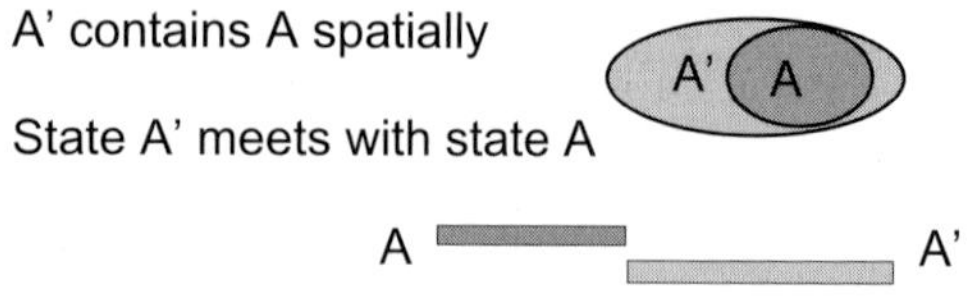

Figure 18.17. An example of spatio-temporal relation.

18.6 SUMMARY

Geographical information systems process spatial information. The information is derived from spatial data in a database. To work sensibly with these systems, we need models of spatial information as a framework for database design. These models address the spatial, thematic and temporal dimensions of real world phenomena. An understanding of the principal concepts of space and time rooted in philosophy, physics and mathematics is a necessary prerequisite to develop and use spatial data models.

We know of two major approaches to spatial data modelling: the analogue map approach and spatial databases. Today, the function of maps as data storage (maps as databases) is increasingly being taken over by spatial databases. In databases, we store representations of relevant phenomena of the real world. These representations are abstractions according to selected spatial data models. We know two fundamental approaches to spatial data modelling: field-based and object-based models. Both have their merits, advantages and disadvantages for particular applications.

Consistency is an important requirement for every model. Topology provides us with the mathematical tools to define and enforce consistency constraints for spatial databases, and to derive a formal framework for spatial relationships among spatial objects.

Spatial data not only possess spatial and thematic attributes, but extend also into the temporal domain. A model of time for spatial information is an important ingredient for any spatial data model, thus leading to what we call spatio-temporal data models.

REFERENCES

Allen, J.F. 1984. Towards a general theory of actions and time. *Artificial Intelligence* 23: 123–154.

Couclelis, H. 1999. Space, time, geography. In P.A. Longley, M.F. Goodchild, D.J. Maguire, & D.W. Rhind (eds.), *Geographical Information Systems, Vol. 1. Second edition*: 29–38. John Wiley & Sons.

Egenhofer, M.J. & Franzosa, R. 1991. Point-set topological spatial relations. *International Journal of Geographic Information Systems* 5(2): 161–174.

Elmasri, R. & Navathe, S.B. 2004. *Fundamentals of database systems*. Addison Wesley.

Kainz, W., Egenhofer, M.J. & Greasley, I. 1993. Modeling spatial relations and operations with partially ordered sets. *International Journal of Geographic Information Systems* 7(3): 215–229.

Langran, G. 1992. *Time in geographic information systems*. London: Taylor & Francis.

Mesgari, M.S. 2000. *Topological cell-tuple structures for three-dimensional spatial data*. PhD Thesis University of Twente and ITC, ITC Dissertation Number 74.

Peuquet, D.J. 2002. *Representations of space and time*. New York: The Guildford Press.

Peuquet, D.J. & Duan, N. 1995. An event-based spatiotemporal data model (ESTDM) for temporal analysis of geographical data. *International Journal of Geographic Information Systems* 9(1): 7–24.

Rigeaux, P., Scholl, M. & Voisard, A. 2002. *Spatial databases with applications to GIS*. Academic Press.

Roshannejad, A.A. & Kainz, W. 1995. Handling Identities in Spatio-Temporal Databases. *Technical papers of the 1995 ACSM/ASPRS Annual Convention & Exposition, February 27–March 2, Charlotte, North Carolina*: 119–126. Bethesda, MD: American Society for Photogrammetry and Remote Sensing and American Congress on Surveying and Mapping.

Stock, O. (ed). 1997. *Spatial and Temporal Reasoning*. Dordrecht, The Netherlands: Kluwer Academic Publishers.

Tang, X.M. 2004. Spatial object modeling in fuzzy topological spaces with applications to land cover change. PhD Thesis University of Twente and ITC, ITC Dissertation Number 108.

Tang, X.M., Fang, Y. & Kainz, W. 2006. Fuzzy topological relations between fuzzy spatial objects. In L. Wang, L. Jiao, G. Shi, X. Lu & J. Liu (eds.), *Proc. Third International Conference on Fuzzy Systems and Knowledge Discovery, Xi'an, China, September 24–28, 2006, (Lecture Notes in Artificial Intelligence, 4223)*: 324–333. Berlin, Heidelberg: Springer-Verlag.

Worboys, M. 1994. A unified model for spatial and temporal information. *The Computer Journal* 37(1): 26–34.

Worboys, M. & Duckham, M. 2004. *GIS a computing perspective, Second edition*. CRC Press.

Vieu, L. 1997. Spatial Representation and Reasoning in Artificial Intelligence. In O. Stock (ed), *Spatial and Temporal Reasoning*: 5–41. Dordrecht, The Netherlands: Kluwer Academic Publishers.

Advances in Photogrammetry, Remote Sensing and Spatial Information Sciences: 2008 ISPRS Congress Book – Li, Chen & Baltsavias (eds)

CHAPTER 19

Multi-scale modelling and representation of geospatial data

Zhilin Li

ABSTRACT: Scale is one of the most important but unsolved issues in various scientific disciplines. This chapter tackles the multi-scale issues in modelling and representation of geo-spatial data. First concepts of scale, multi-scale and transformation in scale are clarified. Then discussions are conducted on the multi-scale modelling and representation of digital map data, digital terrain models (DTM) and digital images. For digital maps, the development of low-level operations/algorithms and high-level rules are reviewed and a set of 40 operations are identified. For digital image and DTM data, the techniques for achieving multi-scale representation are described, multi-scale effects are illustrated with cases and the determination of optimum scale is also addressed.

Keywords: Multi-scale modelling, representation, resolution, spatial data

19.1 INTRODUCTION

Scale is an old issue in geo-sciences. In mapping, maps are produced at certain scales, e.g. 1:10,000 and 1:100,000. At different scales, the level of detail represented on a map is different. The derivation of smaller-scale maps from larger-scale maps through some operations is a multi-scale issue and is called "*map generalisation*". In geography, there is a similar issue. Normally, geographical data are sampled in small enumeration units. And in some applications, such data need to be aggregated to a larger enumeration unit. However, the statistical results will be different when the aggregation is carried out in different ways. This issue is called the "*modifiable areal unit problem*" (MAUP) (Openshaw 1984) although some researchers have reservation on this issue (Tobler 1989). In remote sensing image processing, there is an issue of aggregating images from finer resolution to coarser resolution. Indeed, there is also a similar issue in all geo-sciences, such as digital terrain modelling, geomorphology (e.g. de Boer 1992), oceanography (e.g. Stommel 1963), soil science (e.g. Hillel & Elrick 1990), biology (e.g. Haury et al. 1977), biophysics (e.g. Friedl 1997), social sciences (e.g. Dovers 1994), hydrology (e.g. Blöschl & Sivapalan 1995), environmental sciences (e.g. Bian 1997), etc.

Indeed scale issue has recently attracted great attention from geo-science communities. In as early as the 1980s, multi-scale had been recognized as a fundamental issue in spatial data handling. In 1983, a small group of leading scientists in the US was convened by NASA to define critical research areas in spatial data handling and multi-scale representation was identified as one of them (Marble 1984). Since then, this topic has become part of the international research agenda in spatial information sciences. Many researchers have advocated research on this topic (Abler 1987, Rhind 1988). As a result, a number of books devoted to scale issues have recently been published, including *Scaling Up in Hydrology Using Remote Sensing* edited by Stewart et al. (1996), *Scale in Remote Sensing and GIS* edited by Quattrochi & Goodchild (1997), *Scale Dependence and Scale Invariance in Hydrology* edited by Sposito (1998), *Modelling Scale in Geographical Information Science* edited by Tate & Atkinson (2001), *Scale and Geographic Inquiry: Nature, Society, and Method* edited by Sheppard & McMaster (2004), *Algorithmic Foundation of Multi-Scale Spatial Representation* authored by Li (2007) and *Generalisation of geographic information: Cartographic modelling and applications* by Mackaness et al. (2007).

Also within both the ISPRS and ICA (International Cartographic Association), there has been a working group (or commission) on multi-scale representation for many years. In spite of such international efforts, scale is a still a major unsolved issue in geo-spatial data handling. As digital map data, digital image data and digital terrain models (DTM) are the three types of geo-spatial data which are of major concern within the ISPRS, this chapter will provide an overview of multi-scale modelling and representation of these types of data.

19.2 BASIC CONCEPTS OF SCALE AND MULTI-SCALE

In order to facilitate the discussion on multi-scale modelling and representation, some basic concepts need to be clarified.

19.2.1 *The concepts of scale*

There is a *spatial scale* and a *temporal scale*. In both domains, the scale may range from micro to macro (see Fig. 19.1) and may be in different levels of measurement (i.e. nominal, ordinal, interval and ratio). In this chapter, only spatial scale is discussed.

In the communities of photogrammetry, remote sensing and geo-spatial information science, people are interested in reality (phenomena) neither at micro nor at micro scales. Instead, they are interested in those in the middle part of the scale spectrum, i.e. *geo-scale*.

There is an argument that "digital data sets don't have any scale but have only resolution". This is true in the sense that digital data sets can be plotted (and manipulated) at any scale one wishes. However, scale is also implied by the accuracy level of the digital data sets. For example, the height accuracy of the 1:50,000 DEM of China is expected to range from 4 m (flat areas), 7 m (hilly areas), 11 m (mountainous areas) to 19 m (highly mountainous areas). On the other hand, accuracy is affected by resolution. Therefore, the scale of a set of geo-spatial data should be defined by a set of parameters, possibly:

- cartographic ratio (or magnitude of study area)
- accuracy, and
- resolution.

19.2.2 *The concepts of multi-scale*

By *multi-scale* one may mean different things such as multi-resolution, multi-ratio (cartographic ratio), variable ratio and variable resolution. In addition, *level of details* (LoD) is another term used for multi-scale.

For a single representation, it is possible to have the following four combinations (Fig. 19.2):

a. Uniform cartographic ratio with uniform resolution,
b. Uniform cartographic ratio but with variable resolution,
c. Variable cartographic ratio but with uniform resolution, and
d. Variable cartographic ratio with variable resolution.

For multi-scale modelling and representation, cartographic ratio and resolution are still the two parameters to be considered. As in the case of single representation, both cartographic ratio and resolution could be varied or uniformly changed. As a result, there are nine possible types of change for any given spatial representation, as shown in Figure 19.3:

a. Same cartographic ratio and same resolution (i.e. as original one),
b. Same cartographic ratio but with different (but uniform) resolution,

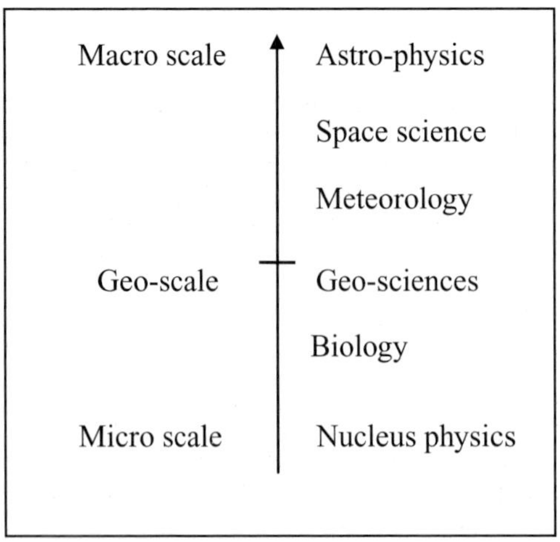

Figure 19.1. Spectrum of spatial scale, from micro to macro.

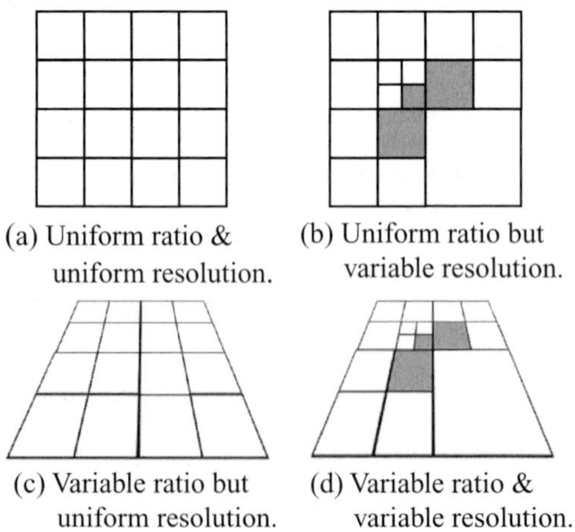

Figure 19.2. Combinations of cartographic ratio and resolution.

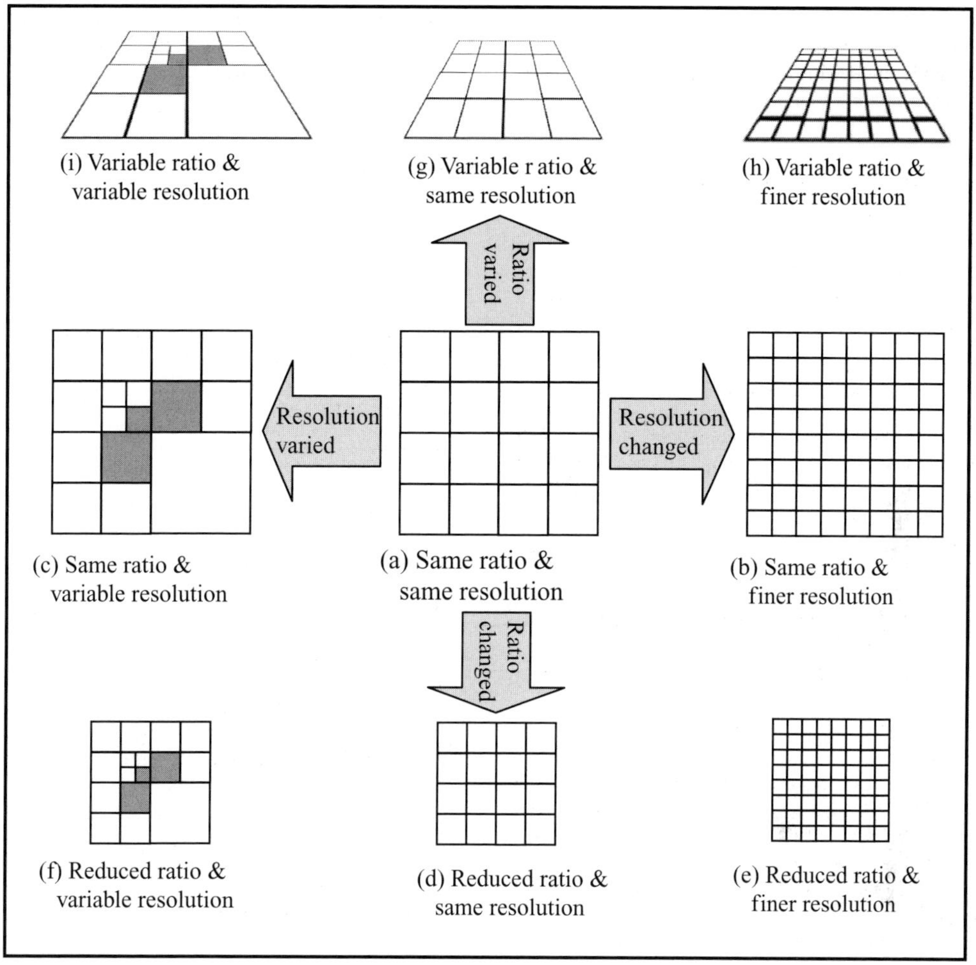

Figure 19.3. Nine types of possible scale changes for a spatial representation.

c. Same cartographic ratio but with variable resolution,
d. Different (but uniform) cartographic ratio but with same resolution,
e. Different (but uniform) cartographic ratio with different (but uniform) resolution,
f. Different (but uniform) cartographic ratio with variable resolution,
g. Variable cartographic ratio with same resolution,
h. Variable cartographic ratio with different (but uniform) resolution, and
i. Variable cartographic ratio with variable resolution.

A representation with variable cartographic ratio is normally generated for visualization with a perspective view and not for numerical analysis. Variable resolution is usually employed to represent different level of detail, if there is such a need. However, the most commonly used multi-scale representation is either "different (but uniform) cartographic ratio with different (but uniform) resolution", or "different (but uniform) cartographic ratio with variable resolution".

19.2.3 *Transformation in scale*

Transformation in scale means scaling here, which is a process to change the cartographic ratio or resolution of data. Scaling-up means to make the resolution coarser (or cartographic ratio smaller) while scaling-down means to make the resolution finer (or cartographic ratio larger). Scaling-up is also termed upsampling or upsizing while scaling-down is also downsampling or downsizing.

In Euclidean space, an increase (or decrease) in cartographic ratio will cause an increase (or decrease) in length, in area and in volume in a 3-D space. However, the shape and complexity of a feature remains unchanged. Figure 19.4 is an example of scale reduction in a 2-D Euclidean space. The graphic representation of the feature at Scale 2 is a 2 times reduction of that at Scale 1 and the graphic representation at Scale 3 is a 4 times reduction of that at Scale 1. In such a transformation process, the (area) size of the graphic representation is reduced by 2^2 and 4^2 times, respectively. When the graphic representation at scale 3 is increased by 4 times, the enlarged graphics is identical to the original one shown at scale 1. That is, the transformations are reversible.

However, in the *fractal geographical space*, it has long been discovered that different length values will be obtained for a coastal line represented on maps at different scales. The length measured from smaller scale maps will be shorter than if the same unit size (at map scale) is used for measurement. This is because different levels of reality (e.g. the Earth's surface with different degrees of abstraction) have been measured. Indeed, at a smaller scale, the size of the graphic representation is reduced. At the reduced graphics, the complexity of the graphics is still retained at the same level as before. As a result, if the graphic is enlarged back to the size at the larger scale, the level of complexity of the representation will appear to be much reduced. In other words, in geographical space, the level of complexity cannot be recovered by an increase in scale. Figure 19.5 illustrates the effect of an increase and decrease in scale on spatial representation in geographical space. It shows that the transformations in scale in such geographical space are not reversible.

Transformation in scale could be carried at discrete scales (e.g. 1:10,000 to 1:50,000 and then to 1:500,000) or at a continuous scales. The results could be either with isotropic (metric multi-scale) representation or view-dependent (i.e. variable scale). A classification is given in Figure 19.6. Metric multi-scale representation for map production, termed *generalization,* is carried out at discrete scale while zooming requires a continuous transformation in scale, theoretically speaking. However, in practice, such a theoretical zooming is not easy to implement at the current stage and even giant Internet companies can only offer a few discrete scales (or resolution) in their systems (e.g. Google Earth and Virtual Earth).

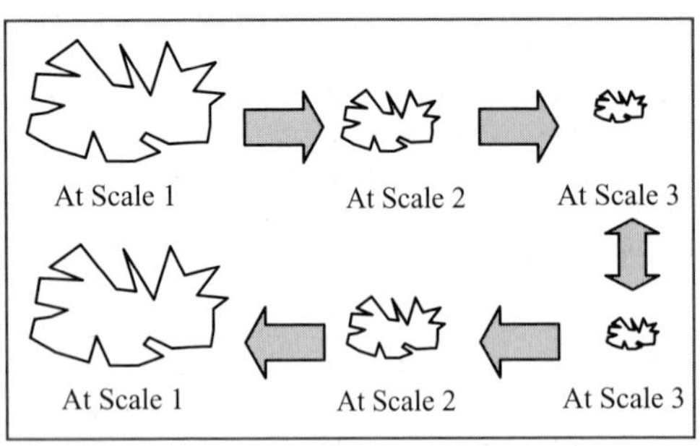

Figure 19.4. Scale change in Euclidean space: a reversible process.

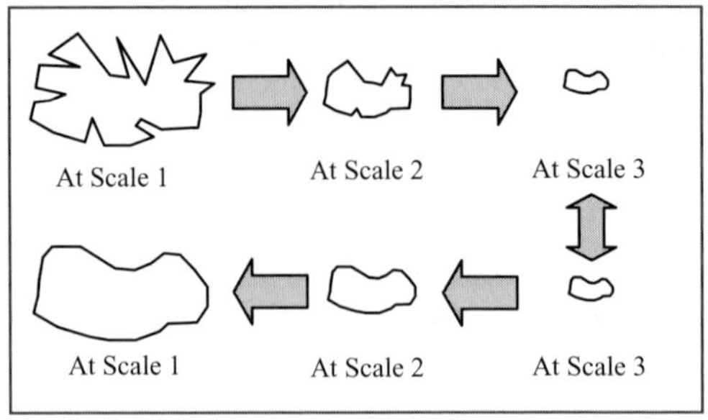

Figure 19.5. Scale change in 2-D geographical space: lost complexity not recovered.

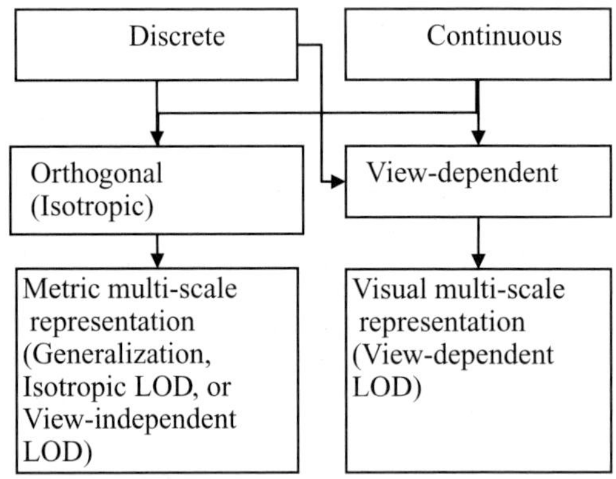

Figure 19.6. Alternative approaches for multi-scale representation of DTM (Li et al. 2005).

19.3 MULTI-SCALE MODELLING AND REPRESENTATION OF DIGITAL MAPS

The multi-scale issue related to digital map is the derivation of maps at a smaller scale from those at a larger scale, which is a scaling-up process, traditionally termed as map generalization. This is required for map production, map revision and real-time zooming in a geographic information system. A review of historical development in digital map generalization has been given by Li (2007b) and thus section emphasizes the more recent developments.

In generalization for map production, one applies geometric transformations to make the large-scale maps simplified to suit the representation at a smaller scale. In map updating, one needs to first establish logical links between the spatial objects in the databases at different scales (called multiple representation databases—MRDB) so that any change at larger-scale databases can be propagated into smaller-scale databases, and then apply geometric transformations to those objects to be updated in smaller-scale databases. The topic of MRDB will be addressed in another chapter in this volume (Sester 2008), and thus is omitted in this chapter.

As a map contains (geo)metric information (related to position, size and shape), thematic information (related to the types and importance of features) and relational information (about spatial relationship between neighbouring features), there must be three types of transformation involved in digital map generalization: geometric, thematic and relational.

The *geometric transformations* are achieved by some *operations,* called *generalization operations* here. The thematic information can be used to form rules, which are in turn used to control generalization operations. Relational information can be used as constraints for the geometric transformation.

19.3.1 *Low-level geometric transformations*

In classic textbooks, only very few operations were available, e.g. "classification, induction, simplification and symbolization" by Robinson et al (1984) and "aggregation, combination, simplification, displacement and selective omission" by Keates (1989). However, these operations are too general to be computerized.

From the late 1980s, researchers tried to identify more concrete operations. Beard & Mackaness (1991) identified 8 operations. McMaster & Shaea (1992) have identified a list of 12 operations. A summary of such work in the 1980s and 1990s is given in Table 19.1 (modified from Su 1997). However, these operations are still too general to be computerized. Indeed, the same term may mean different things in different context. For example, "*typification*" is often referred to the re-arrangement of buildings (e.g. from 4 rows × 4 columns to 3 × 3). However, this term has also been used to mean the reduction of "S" bends, e.g. from 5 bends to 3, along a road (Plazanet et al. 1995).

Recently, Li (2007a) has carried out a systematic study of generalization operations and differentiated 6 sets with a total of 40 operations, as shown in Table 19.2. He also makes available at least one algorithm for each of these 40 generalization operations.

During the last four years, researchers continued to produce new algorithms (Basaraner & Selcuk 2004, Haunert & Sester 2007, Li et al. 2004a, Li et al. 2004b, Steiniger & Meier 2004, Bader et al. 2005, Burghardt 2005, Gokgoz 2005, Sester 2005, Matuk et al. 2006, Ai et al. 2006, Haunert & Sester 2007, Qian et al. 2007, Touya 2007). Efforts have also been spent on the evaluation of generalization operations (e.g. Cheng & Li 2006) and improvements of some well-known algorithms such as Perkal (1966), Douglas & Peucker (1973), Li & Openshaw (1992), Christensen (2000), Zhou & Bertolotto (2006) and Zhu et al. (2007).

19.3.2 *High-level rules and integrated optimization*

Algorithms are used to achieve desirable geometric transformations. More elegant algorithms could be developed if sound theories are available. The natural principle by Li & Openshaw (1993) serves for such a purpose, which is based on the natural generalization process observed in our daily life.

Algorithms can be employed whenever there is a need. The question of "when needed" should be answered by rules that are formalized using thematic information and other knowledge. The work by Töpfer & Pillewizer (1966) might be considered as the first attempt at knowledge acquisition through analysis of topographic maps. They formulated the so-called Principle of Selection or Radical Law to express the relationship between map scale and the number of features represented on maps.

Research on rule-based generalization systems started in the later 1980s (Nickerson & Freeman 1986) and was highlighted by a symposium called "Towards a Rule Base for Map Generalization" held in Syracuse in April 1990 (Buttenfield & McMaster 1991). Researchers continue to make efforts on formalization of rules (Gao et al. 2004, Trailandier 2007). On the other hand, an alternative term, generalization constraint, has become more popular in recent literature (Steiniger & Weibel 2005, Burghardt et al. 2007, Harrie & Weibel 2007). Such constraints have similar functions as generalization knowledge does. The latter describes the restrictions and specifications that should be supported in the generalization process from a macro and schematic view, whereas the former defines various concrete conditions and situations to control specific aspects of an object, a group of objects or a whole map (Gao et al. 2004). In fact, some constraints may be expressed in terms of spatial relations, i.e. topological relations, metric relations (distance, direction) and order relation. Continued efforts are also spent on spatial relations (Yan et al. 2006, Mascret et al. 2006, Deng et al. 2007).

Table 19.1. Operations for geometric transformation developed in 1980s and 1990s.

Researchers / Operations	Robinson et al. (1984)	Keates (1989)	DeLucia & Black (1987)	McMaster & Monmonior (1989)	Beard & Mackaness (1991)	McMaster & Shea (1992)
Agglomeration			√			
Aggregation			√	√		√
Amalgamation				√		√
Classification	√				√	√
Coarsen					√	
Collapse			√	√	√	√
Combination		√			√	
Displacement		√		√	√	√
Enhancement				√		√
Exaggeration		√			√	√
Induction	√					
Merge				√		√
Omission		√		√	√	
Refinement			√	√		√
Selection		√			√	
Simplification	√	√	√	√		√
Smoothing				√		√
Symbolisation	√					
Typification						√

Table 19.2. A set of 40 operations for multi-scale representation of digital map data (Li 2007b).

Features		Main Operations	Sub-Operations	Number
Point	Individual points	Displacement		3
		Elimination		
		Magnification		
	A class of points	Aggregation		5
		Regionalization		
		Selective Omission		
		(Structural) simplification		
		Typification		
Line	Individual lines	Displacement		8
		Elimination		
		(Scale-driven) generalization		
		Partial modification		
		Point reduction		

(*Continued*)

Table 19.2 (*Continued*)

Features		Main Operations	Sub-Operations	Number
		Smoothing	Curve-fitting	
			Filtering	
		Typification		
	A class of line	Collapse	Double-to-single line	6
			Ring-to-point	
		Displacement		
		Enhancement		
		Merging		
		Selective omission		
Area	Individual areas	Collapse	Area-to-point	10
			Area-to-line	
			Partial	
		Displacement		
		Exaggeration	Directional thickening	
			Enlargement	
			Widening	
		Elimination		
		(Shape) simplification		
		Split		
	A class of areas	Aggregation		8
		Agglomeration		
		Amalgamation		
		Dissolving		
		Merging		
		Relocation		
		(Structural) simplification		
		Typification		

At this stage, the key issue is the application of these relation models to digital map generalization (Monnot et al. 2007).

Recently, in Europe, attention is shifted to the enrichment of databases (e.g. Neun et al. 2004, Luescher et al. 2007), application of agent (e.g. Lamy et al. 2002, Galanda & Weibel 2002, Gaffuri 2006) and optimization methods (Harrie 1999, Hojholt 2000, Sester 2000, Ware et al. 2003, Burghardt 2005) and web-based generalization service (e.g. Neun & Burghardt 2005, Foerster & Stoter 2006). Another interesting development is the generalization of 3-D structures, especially buildings and 3D-city models (Li 2007a, Meng & Forberg 2007).

19.4 MULTI-SCALE MODELLING AND REPRESENTATION OF DIGITAL IMAGES

There are a number of multi-scale issues in remote sensing image processing and applications (Raffy 1994, Cao & Lam 1997, Marceau & Hay 2000). However, the most fundamental issues are scaling techniques, scaling effect and optimum scale determination.

19.4.1 *Scaling of digital images*

Scaling is a process for various image processes such as image matching, feature extraction, image segmentation, image compression, image fusion, image rectification and image classification.

Scaling-up is an aggregation process. The commonly used solution is the pyramid structure. That is to aggregate $n \times n$ (e.g. 2×2) pixels into a new one (Fig. 19.7). The brightness of the new image could be the simple average of the $n \times n$ pixels or a function of these $n \times n$ pixels. Wavelets have also been employed to decompose an image into a pyramid structure (Mallat 1989). Another commonly used method is to smooth the image by a low-pass filter while the pixel size is kept unchanged.

However, when the resolution of the new image is not an integral times the original image (e.g. from 2 m to 3 m), then resampling using some interpolation technique is employed. Many interpolation techniques are available and the commonly used are nearest neighbour, bilinear and bicubic. The same techniques can also be applied to the scaling-down (called de-aggregation) process.

Scaling is a matured process used in digital image processing and there is not much new development. However, to maintain the sharpness of the new image (e.g. Ramanjaneyulu & Rao 2002), to make multi-resolution image processes (e.g. registration between two images) more reliable (Chen & Varshney 2004) and to minimize aggregation effects in multi-scale image processing (e.g. rectification) (Gluckman & Nayar 2001) are still not easy tasks.

19.4.2 *Scaling effect and optimum scale determination*

Another topic is to investigate the effect of scaling on image quality and various applications, and then to determine an optimum scale for a particular type of application. For example, Markham & Townshend (1981) investigated the land-cover classification accuracy as a function of image resolution; Townshend & Justice (1988) addressed the selection of the spatial resolution of satellite sensors required for global monitoring of land transformations; Benson & MacKenzie (1995) researched the effects of sensor spatial resolution on landscape structure parameters; O'Neill et al. (1996) discussed the effect of scale on landscape pattern; Usery et al. (2004) carried out a study on the effect of resampling and resolution for remote sensing and other geo-spatial data on watershed modelling; Ahla et al. (2005) studied the effects of aggregated land cover data on estimating the net primary production (NPP); Bo et al. (2005) explored the scale effect on land cover mapping.

To determine an optimum scale for a particular analysis, one could resample image data into various resolutions, and then compare results obtained from images with all these resolutions with a benchmark. The best result will correspond to the optimum resolution. This process could also be made self-adaptive. It means that the resampling and analysis will automatically stop when certain conditions are met. These conditions could be the maximization and/or minimization of a parameter or some parameters, or the difference between a parameter or some parameters obtained in this round and in the previous round being within a given range.

Of course, it would be more desirable to have theoretical models to determine the optimum scale (resolution) for a given application. A lot of effort has been spent on this topic (e.g. Atkinson & Curran 1995, 1997, Ju et al. 2005). Many approaches have been suggested, such as fractal dimension (Xia & Clarke 1997) and variograms (Woodcock et al. 1988, Oliver 2001, Atkinson & Aplin 2004). However, perhaps

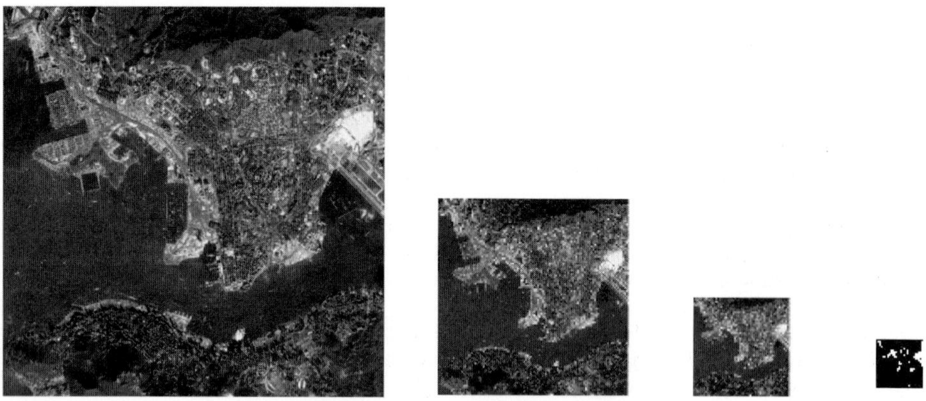

Figure 19.7. Scaling-up: Aggregation of digital images. (see colour plate page 500)

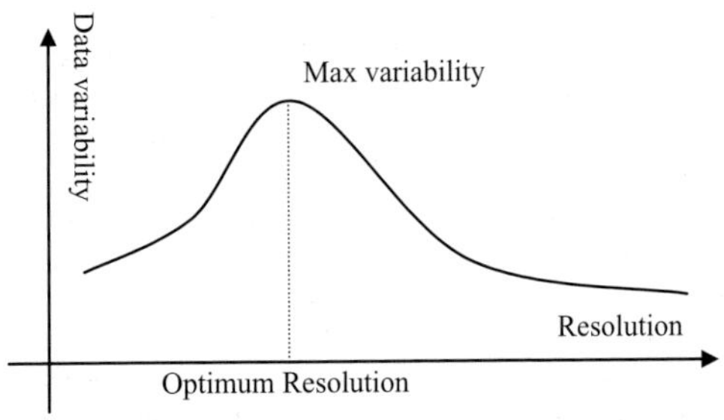

Figure 19.8. Data variability for the determination of optimum resolution.

the most interesting one is the so-called *local variance method*, which was proposed by Woodcock & Strahler (1987). The working principle is illustrated in Figure 19.8. The procedure is as follows:

- To plot local variance as a function of resolution;
- To find the resolution with maximum variance and regard it as the optimum resolution for this set of data.

The local variance is defined as the average of the variances within a moving window (resolution) passing through the entire area. The reason behind this method is that if the spatial resolution is considerably finer than the objects in the area, most of the measurements will be highly correlated with their neighbours and thus the local variance will be low. On the other hand, if the objects studied approximate the size of the resolution, the values tend to be different from each other and therefore the local variance increases (Cao & Lam 1997).

19.5 MULTI-SCALE MODELLING AND REPRESENTATION OF DTM DATA

Similar to digital image data, the most fundamental issues for DTM data are techniques for multi-scale representation, modelling scaling effect and optimum scale determination.

19.5.1 *Multi-scale representation of DTM*

As discussed in Section 2.3, multi-scale representation of DTM data could be produced at discrete scales or at continuous scales and the results could be either be with an isotropic representation or view-dependent (i.e. variable scale). The view-dependent representation, often termed as level of details (LoD), is normally for visualization and will be omitted in this chapter. Those who are interested in this topic could refer to the recent book by Luebke et al. (2003). The discrete metric multi-scale representation is similar to the pyramid structure for digital images and no further discussion on this topic will be given.

For continuous metric multi-scale representation, based on different kinds of philosophy, three approaches can be identified, i.e. critical-points-based, smoothing-based and scale-driven.

The critical-points-based approach is built upon the assumption that some points on a terrain surface are more important than others and the fidelity of the terrain surface is largely preserved even if only those important points are kept. The resultant data set may be not in a regular grid anymore. Another reason behind this approach is that some height values (such as the peaks of hills) are not allowed to change in some kinds of geographical analysis. Critical-points-based approach will produce a data set with variable resolution. Those algorithms for reduction of line points mentioned in Section 3, e.g. Douglas-Peucker algorithm (Douglas & Peucker 1973), may all be applicable in this case. Some specially designed techniques for selection of VIP (very important points) are also available (Chen & Guevara 1987, Li et al. 1998).

The smoothing-based approach is built upon the assumption that small variations are the high-frequency parts of the terrain surface. Therefore, filtering techniques can be used to filter out the high-frequency parts. Alternatively, a smooth surface using a certain order of polynomial function may be fitted to the original data points to represent the terrain surface. These two techniques will not change the resolution of the original DTM. If there is a need to change the resolution, aggregation and/or resampling may be used. Skeleton- and wavelet-based techniques are also in use (Ai 2004, Zakšek & Podobnikar 2005). In the case of smoothing, there is a need to determine the window size. In other words, the window size is not directly related to the scales. To solve this problem, Li & Openshaw (1993) formulated a *Natural Principle* for objective generalization, which objectively links the window size to the source (input) data scale and result (output) data scale. Figure 19.9 illustrate the application of Li-Openshaw's natural principle for use in multi-scale representation of DTM data. A more detailed description of such a process is described Li & Li (1999).

Another new development is to make the DTM simplification on a 2-D plane (Matuk 2006). The steps are as follows: (a) to slice a 3-D DTM into a series of contours, (b) to produce the skeleton of each contour, (c) to simplify the skeleton by retraction, and (d) to recover the (smoothed) DTM surface from smoothed contour lines. Another interesting development is the DTM generalization that takes the importance of objects into account (Kremeike 2004). With this approach, important objects such as roads can be enhanced artificially (e.g. widened) (Filin et al. 2007).

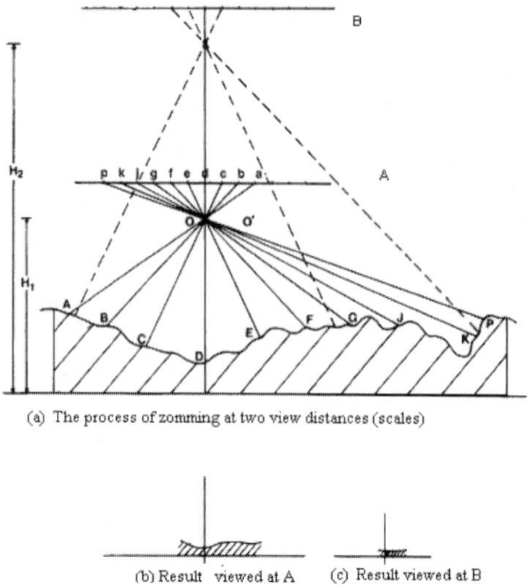

Figure 19.9. The *natural principle* applied to a terrain surface (Li & Openshaw 1993).

19.5.2 *Scaling effect and optimum scale determination*

The modelling of scale (resolution) effect has been a major topic in digital terrain analysis (Gallant 2006). The normal strategy is

- to resample original data into a series of new sets with different resolutions,
- to perform terrain analysis with these sets of data, and
- to find some relationship between resolution and analysis results.

The analysis could be carried out for the accuracy of DTM, the parameters computed from DTM, the features extracted from DTM and the DTM-based application models. A classic example is the investigation of the effect of sampling interval (resolution of measured data) on DTM accuracy by Li (1990, 1992), who found that such a relationship is linear for composite data (i.e. grid plus feature data, including peaks, pits, break lines, ridge lines and ravines) but becomes second order for regular grid data only. In a similar way, the effect of DEM resolution on the various terrain parameters and DTM-based applications have been investigated, e.g. on slope (Chen et al. 2006), on a number of terrain parameters and attributes (Gao 1998, Albani et al. 2004, Zhu 2006, Deng et al. 2007), on topographic index (Brasington & Richards 1998), on watershed model (Wolock & Price 1994), on flood modelling (Haile & Rientjes 2005), on digital soil mapping (Zhu 2006), on the accuracy of distributed rockfall models (Dorren & Heuvelink 2004), and on terrain classification (Cheng et al. 2004).

As one can understand, it is not enough to model the scale effects. Indeed, it would be more desirable to find an optimum scale (resolution) for a particular analysis or application, if it exists. This could be carried out through experimental testing or through theoretical analysis. Within ISPRS, from the early 1980s, researchers spent effort on the determination of optimum sampling intervals for a DTM project (Li 1990). In DTM interpretation, Florinsky & Kuryakova (2000) investigated the variation of correlation coefficients between terrain attributes (slope, aspect, curvature, etc.) and targeted biophysical properties (e.g. soil moisture). Then they found the resolution range where such a variation is relatively stable as the optimum resolution. But their findings cannot be simply generalized to other applications because optimum scale is application-dependent. Therefore, determination of optimum scale continues to attract attention from various areas, e.g. terrain stability mapping (Tarolli & Tarboton 2006) and digital soil survey (Smith et al. 2006).

It might also be interesting to conduct an investigation into the applicability of these techniques for optimum scale determination in digital image processing in DTM-based analysis.

19.6 CONCLUDING REMARKS

In this chapter, a number of issues on multi-scale modelling and analysis have been tackled. First, a discussion clarified the concepts of scale and multi-scale. Then scale issues in modelling and representation of digital maps, digital images and digital terrain models (DTM) are discussed. In modelling and representation of digital maps, the major issues include operations and algorithms for geometric transformation, multi-representation databases and high-level rules to control geometric transformation. In modelling and representation of digital images and DTM, the major issues include techniques for aggregation and de-aggregation, modelling of scale effect and the determination of optimum scale.

Scale is one of the most fundamental issues in the modelling and representation of geo-spatial data. Indeed, scale is everywhere in science, like God in religion. Therefore, it should be treated as a basic dimension. The proper incorporation of scale in spatial modelling and representation and the determination of optimum scale for different applications are issues worthy of great attention.

ACKNOWLEDGEMENTS

This paper is supported by a State 973 project of China (2006CB701304) and the Research Council of the Hong Kong (PolyU 5221/07E). The constructive comments by Monika Sester are very much appreciated.

REFERENCES

Abler, R. 1987. The National Science Foundation National Center for Geographic Information and Analysis. *International Journal of Geographical Information Systems* 1(4): 303–326.

Ahl, D., Gower, S., Mackay, D., Burrows, S., Norman, J. & Diak, G. 2005. The effects of aggregated land cover data on estimating NPP in northern Wisconsin. *Remote Sensing of Environment* 97(1): 1–14.

Ai, T.H. 2004. A generalization of contour line based on the extraction and analysis of drainage system. *International Archive of Photogrammetry and Remote Sensing* XXXV(IV/3) (CR–Rom).

Ai, T., Liu, Y. & Chen, J. 2006. The hierarchical watershed partitioning and data simplification of river network. In R. Andreas, W. Kainz & G. Elmes (eds.), *Progress in Spatial Data Handling*: 617–632. Berlin: Springer.

Albani, M., Klinkenberg, B., Ansison, D.W. & Kimmins, J.P.K. 2004. The choice of window size in approximating topographic surface from digital elevation models. *International Journal of Geographical Information Science* 18(6): 577–593.

Atkinson, P.M. & Aplin, P. 2004. Spatial variation in land cover and choice of spatial resolution for remote sensing. *International Journal of Remote Sensing* 25(18): 3687–3702.

Atkinson, P.M. & Curran, P.J. 1995. Defining an optimal size of support for remote sensing investigations. *IEEE Transactions on Geoscience and Remote Sensing* 33,3: 768–776.

Atkinson, P.M. & Curran, P.J. 1997. Choosing an appropriate spatial resolution for remote sensing investigations. *Photogrammetric Engineering and Remote Sensing* 63(12): 1345–1351.

Atkinson, P.M. & Tate, N.J. 2000. Spatial scale problems and geo-statistical solutions: a review. *Professional Geographer* 52(4): 607–623.

Bader, M., Barrault, M. & Weibel, R. 2005. Building displacement over a ductile truss. *International Journal of Geographical Information Science* 19(8–9): 915–936.

Basaraner, M. & Selcuk, M. 2004. An attempt to automate generalization of buildings and settlement areas in topographic maps. *International Archives of Photogrammetry and Remote Sensing* XXXV(B4). (CD-ROM).

Beard, M.K. & Mackaness, W. 1991. Generalization operators and supporting structures. *Proceedings Auto Carto* 10: 29–45.

Benson, B.J. & MacKenzie, M.D. 1995. Effects of sensor spatial resolution on landscape structure parameters. *Landscape Ecology* 10: 113–120.

Bian, L. 1997. Multiscale nature of spatial data in scaling up environmental models. In D. Quattrochi & M. Goodchild (eds.), *Scale in Remote Sensing and GIS*: 13–26. CRC Press.

Blöschl, G. & Sivapalan, M. 1995. Scale issues in hydrological modelling: A review. *Hydrological Processes* 9: 251–290.

Bo, Y.C., Wang, J.F. & Li, X. 2005. Exploring the scale effect in land cover mapping from remotely sensed data: the statistical separability-based method. *Proceedings of IGARSS, Seoul, Korea, 2005*: 3884–3887.

Brasington, J. & Richards, K. 1998. Interactions between model predictions, parameters and DTM scales for topmodel. *Computers & Geosciences* 24(4): 299–314.

Burghardt, D. 2005. Controlled line smoothing by Snakes. *GeoInformatica* 9(3): 237–252.

Burghardt, D., Schmid, S. & Stoter, J. 2007. Investigations on cartographic constraint formalization. *ICA Workshop on Progress in Automated Map Generalization, 2–3 August 2007, Moscow*. http://ica.ign.fr/Moscow/program.php (last accessed on 25 Feb 2008).

Buttenfield, B.P. & McMaster, R.B. (eds.). 1991. *Map generalization: Making Rules for Knowledge Representation*. London, UK: Longman Scientific and Technical.

Cao, C. & Lam, N. 1997. Understanding the scale and resolution effects in remote sensing and GIS. In D.A. Quattrochi & M.F. Goodchild (eds.), *Scale in Remote Sensing and GIS*: 57–72.

Chen, H. & Varshney, P.K. 2004. Size-dependent image resampling for mutual information based remote sensing image registration. *Proceedings of 2004 IEEE International Geoscience and Remote Sensing Symposium, 20–24 Sept. 2004, Anchorage, Alaska*. (CD-ROM).

Chen, N., Tang, G.A. & Wang, Y.M. 2006. The influence of resolutions of DEM on the accuracy of the slope information. *International Symposium on Terrain Analysis and Digital Terrain Modelling, 23–25 November 2006, Nanjing, China*. (CD-ROM).

Chen, Z. & Guevara, J., 1987. Systematic selection of very important points (VIP) from digital terrain model for constructing triangular irregular network. *Auto Carto* 8: 50–56.

Cheng, T. & Li, Z.L. 2006. Effect of generalization on area features: A comparative study of two strategies. *The Cartographical Journal* 43(2): 157–170.

Cheng, T., Fisher, P. & Li, Z.L. 2004. Double vagueness: Uncertainty in multi-scale fuzzy assignment of duneness. *Geo-spatial Information Science* 7(1): 58–66.

Christensen, A. H. 2000. Line generalization by waterline and medial-axis transformation: Success and Issues in an implementation of Perkel's proposal. *The Cartographic Journal* 26(1): 19–32.

de Boer, D.H. 1992. Hierarchies and spatial scale in process geomorphology: a review. *Geomorphology* 4: 303–318.

DeLucia, A.A. & Black, R.B. 1987. A comprehensive approach to automatic feature generalisation. *Proceedings of 13th International Cartographic Conference, October 1987, Morelia, Mexico, 4*: 169–192.

Deng, M., Li, Z.L. & Chen, X.Y. 2007. Extended Hausdorff distance for spatial objects in GIS. *International Journal of Geographical Information Science* 21(3–4): 459–475.

Deng, Y., Wilson, J.P. & Bauer, B.O. 2007. DEM resolution dependencies of terrain attributes across a landscape. *International Journal of Geographical Information Science* 21(1–2): 187–213.

Dorren, L.K. & Heuvelink, G.B.M. 2004. Effect of support size on the accuracy of distributed rockfall model.

International Journal of Geographical Information Science 18(6): 595–609.
Douglas, D.H. & Peucker, T.K. 1973. Algorithms for the reduction of the number of points required to represent a digitized line or its caricature. *The Canadian Cartographer* 10(2): 112–122.
Dovers, S.R. 1994. A framework for scaling and framing policy problem in sustainability. *Ecological Economy* 12: 93–106.
Filin, S., Abo Akel, N., Kremeike, K., Sester, M. & Doytsher, Y. 2007. Interpretation and Generalization of 3D Landscapes From Lidar Data. *Cartography and Geoinformation Science* 34(3): 231–243.
Florinsky, I.V. & Kuryakova, G.A. 2000. Determination of grid size for digital terrain modeling in landscape investigations—exemplified by soil moisture distribution at a micro-scale. *International Journal of Geographical Information Science* 14: 795–814.
Friedl, M.A. 1997. Examining the effects of sensor resolution and sub-pixel heterogeneity of spectral vegetation indics: Implications for biophysical modelling. In D. Quattrochi & M. Goodchild (eds.), *Scale in Remote Sensing and GIS*: 113–140. CRC Press.
Foerster, T. & Stoter, J. 2006. Establishing an OGC Web Processing Service for generalization processes. *9th ICA Workshop on Map Generalization and Multiple Representation, 25 June 2006, Portant/Vancouver.* http://ica.ign.fr/.
Gaffuri, J. 2006. How to merge optimization and agent-based generalization techniques in a single generalization model. *9th ICA Workshop on Map Generalization and Multiple Representation, 25th June 2006, Portant/Vancouver*. http://ica.ign.fr/.
Galanda, M. & Weibel, R. 2002. An Agent-based Framework for Polygonal Subdivision Generalization. In D. Richardson & P. van Oosterom (eds.), *Advances in Spatial Data Handling*: 121–136. Berlin: Springer–Verlag.
Gluckman, J., Shree, K. & Nayar, S. 2001. Rectifying Transformations That Minimize Resampling Effects. *IEEE Computer Society Conference on Computer Vision and Pattern Recognition (CVPR'01)* 1: 211–217.
Gao, J. 1998. Impact of sampling intervals on the reliability of topographic variables mapped from grid DEMs at a micro-scale. *International Journal of Geographical Information Science* 12: 875–890.
Gao, W.X., Gong, J.Y. & Li, Z.L. 2004. Thematic knowledge for the generalization of land-use map. *The Cartographic Journal* 41(3): 245–252.
Gallant, J. 2006. Multiscale methods in terrain analysis. Keynotes Speech. *International Symposium on Terrain Analysis and Digital Terrain Modelling, 23–25 November 2006, Nanjing, China.* (CD-ROM).
Gokgoz, T. 2005. Generalization of contours using deviation angles and error bands. *The Cartographic Journal* 42(2): 145–156.
Haile, A.T. & Rientjes, T.H.M. 2005. Effects of Lidar DEM resolution in flood modelling: a model sensitivity study for the city of Tegucigalpa, Honduras. *ISPRS WG III/3, III/4, V/3 Workshop—Laser scanning 2005, September 12–14, 2005, Enschede, the Netherlands.* (CD-ROM).
Harrie, L. 1999. The Constraint Method for Solving Spatial Conflicts in Cartographic Generalization. *Cartography and Geographic Information Science* 26(1), 55–69.
Haunert, J.-H. & Sester, M. 2007. Area collapse and road centerlines based on straight skeletons. *GeoInformatica* doi:10.1007/s10707-007-0028- (On-line).
Haury, L., McGowan, J. & Wiebe, O. 1977. Patterns and processes in time-space scales of plankton distribution. In J. Steele (ed.), *Spatial Pattern in Planton Communities*: 277–327. NY and London: Plenum Press.
He, H.S., Ventura, S.J. & Mladenoff, D.J. 2002. Effects of spatial aggregation approaches on classified satellite imagery. *International Journal of Geographical Information Science* 16: 93–109.
Hillel, D. & Elrick, D. 1990. Scaling in Soil Physics: Principles and Applications. *Soil Science Society of America* Vol.25.
Hojholt, P. 2000. Solving space conflicts in a map generalization: Using a finite element method. *Cartography and Geographic Information Science* 27(1): 65–73.
Ju, J.C., Gopal, S. & Kolaczyk, E.D. 2005. On the choice of spatial and categorical scale in remote sensing land cover classification. *Remote Sensing of Environment* 96: 62–77.
Keates, J. 1989. *Cartographic Design and Production. Second Edition.* UK: Longman Scientific.
Kremeike, K. 2004. Generalization of Dense Digital Terrain Models while Enhancing Important Objects. *International Archives of Photogrammetry & Remote Sensing* 35(4). (CD-ROM).
Lamy, S., Ruas, A., Demazeau, Y., Jackson, M., Mackaness, W. & Weibel, R. 2002. The Application of Agents in Automated Map Generalization. *Joint ISPRS/ICA workshop on Multi-Scale Representations of Spatial Data, 7–8 July 2002, Ottawa, Canada.* http://www.geo.unizh.ch/ICA/
Li, J. & Ai, T. 2007. The DEM generalization based on the simplification of valley system. *Geoinformatics 2007, 25–27 May 2007, Nanjing, China, Proceedings of the SPIE,* Volume 6753.
Li, Z.L. 1990. *Sampling Strategy and Accuracy Assessment for Digital Terrain Modelling.* Ph.D. Thesis, University of Glasgow, 298pp.
Li, Z.L. 1992. Variation of the the accuracy of digital terrain models with sampling interval. *The Photogrammetric Record* 14(79): 113–128.
Li, Z.L. 2007a. *Algorithmic Foundation of Multi-scale Spatial Representation.* Bacon Raton: CRC Press (Taylor & Francis Group).
Li, Z.L. 2007b. Digital map generalization at the age of enlightenment: A review of the first forty years. *The Cartographic Journal* 44(1): 80–93.
Li, Z. L. & Li, C. 1999. Objective representation of DTM surface in scale dimension. *Proceedings of Joint ISPRS Commission Workshop on Dynamic and Multi Dimensional GIS, Beijing, China:* 17–22.
Li, Z.L. & Openshaw, S. 1992. Algorithms for automated line generalisation based on a natural principle of objective generalisation. *International Journal of Geographic Information Systems* 6(5): 373–389.
Li, Z.L. & Openshaw, S. 1993. A natural principle for objective generalisation of digital map data. *Cartography and Geographic Information System* 20(1): 19–29.
Li, Z.L., Zhu, Q. & Gold, C. 2005. *Digital Terrain Modelling: Principles and Methodology*. Boca Raton: CRC Press (Taylor & Francis Group).

Li, Z.L., Khoshelham, K., Ding, X. & Zheng, D. 2004a. Empirical mode decomposition (EMD) transform for spatial analysis. In Z. Li, Q. Zhou & W. Kainz (eds.), *Advances in Spatial Analysis and Decision Making*: 19–30. Lisse: A.A. Balkema.

Li, Z.L., Lam, K. & Li, C.M. 1998. Effect of compression on the accuracy of DTM. *Geographic Information Science* 4(1–2): 37–43.

Li, Z.L., Yan, H., Ai, T. & Chen, J. 2004b. Automated Building Generalization Based on Urban Morphology and Gestalt Theory. *Journal of Geographical Information Science* 18(5): 513–534.

Luebke D., Reddy, M., Cohen, J., Varshney, A., Watson, B. & Huebner, R. 2003. *Level of Detail for 3D Graphics*. San Francisco: Morgan Kaufmann Publishers.

Luescher, P. Burghardt, D. & Weibel, R. 2007. Ontology-driven enrichment of spatial databases. *ICA Workshop on Progress in Automated Map Generalization, 2–3 August 2007, Moscow.* http://ica.ign.fr/Moscow/program.php (last accessed on 25 Feb 2008).

Mackaness, W., Ruas, A. & Sarjakoski, T. (eds.). 2007. *Generalisation of geographic information: Cartographic modelling and applications.* Amsterdam: Elsevier.

Mallat, S.G. 1989. A theory for multiresolution signal decomposition: the wavelet representation. *Pattern Analysis and Machine Intelligence* 11: 674–693.

Marble, D. 1984. Geographic information systems: An overview. *Proceedings of Pecora 9, Sioux Falls, S.D.*:18–24. Reprinted In D. Peuquet, & D.F. Marble (eds.), *Introductory Readings in Geographic Information Systems*: 8–17. London: Taylor and Francis.

Markham, B.L. & Townshend, J.R.G. 1981. Land Cover Classification Accuracy as a Function of Sensor Spatial Resolution. *Proceedings of 15th International symposium on Remote Sensing of Environment*: 1075–1090.

Mascret, A., Devogele, T., Le Berre, I. & Henaff, A. 2006. Coastal matching process based on the discrete Frechet distance. In R. Andreas, W. Kainz & G. Elmes (eds.), *Progress in Spatial Data Handling*: 383–400. Berlin: Springer.

Matuk, K. 2006. *Feature-based terrain model simplification.* PhD Thesis, Dept. of Land Surveying and Geo-Informatics, Hong Kong Polytechnic University. 138pp.

Matuk, K., Gold, C. & Li, Z.L. 2006. Skeleton based contour line generalization. In R. Andreas, W. Kainz & G. Elmes (eds.), *Progress in Spatial Data Handling*: 643–658. Berlin: Springer.

McMaster, R. & Monmonior, M. 1989. A conceptual framework for quantitative and qualitative raster-mode generalisation. *Proceedings of GIS/LIS'89, Orlando, Florida*: 390–403.

McMaster, R.B. & Shea, K.S. 1992. *Generalization in Digital Cartography*. Washington DC: Association of American Geographers.

Marceau, D.J. & Hay, G.J. 2000. Remote sensing contribution to the scale issue. *Canadian Journal of Remote Sensing* 25(4): 349–366.

Meng, L. & Forberg, A. 2007. 3D Building Generalization. In W. Mackaness, A. Ruas & T. Sarjakoski (eds.), *Generalisation of Geographic Information: Cartographic Modelling and Applications*: 211–231. Amsterdam: Elsevier.

Monnot, J-L., Lee, D. & Hardy, P. 2007. Topological constraints, actions and reflexes for generalization by optimization. *ICA Workshop on Progress in Automated Map Generalization, 2–3 August 2007, Moscow.* http://ica.ign.fr/Moscow/program.php (last accessed on 25 Feb 2008).

Nickerson, B. & Freeman, H. 1986. Development of a rule-based system for automatic map generalization. *Proceedings of the Second International Symposium on Spatial Data Handling, 5–10 July 1986, Seattle, USA*: 537–556.

Neun, M. & Burghardt, B. 2005. Webservices for an Open Generalization Research Platform. *8th ICA Workshop on Map Generalization and Multiple Representation, 7–8 July 2006,* A Coruna. http://ica.ign.fr/.

Neun, M., Burghardt, D. & Weibel, R. 2004. Data Enrichment for Adaptive Generalisation. 6th *Workshop on Progress in Automated Map Generalization, Leicester (UK), 20–21 August 2004.*

Oliver, M., A. 2001. Determining the spatial scale of variation in environmental properties using the variogram. In N.J Tate & P.M. Atkinson (eds.), *Modelling Scale in Geographical Information Science*: 193–220. Chichester: Wiley.

O'Neill, R.V., Hunsaker, C.T. & Timmins, S.P., Jackson, B.L., Jones, K.B., Riitters, K.H. & Wickham, J.D. 1996. Scale problems in reporting landscape pattern at the regional scale. *Landscape Ecology* 11(3): 169–180.

Openshaw, S. 1984. *The Modificable Areal Unit Problem.* CATMOG # 38. Norwick, England: Geo Books.

Perkal, J.D. 1966. *An attempt at objective generalisation.* (Translated by W. Jackowshi from Proba obiektywnej generalizacji, Geodezia I Kartografia, Tom VII, Zeszyt 2, 1958. 130–142). *Michigan Inter-University Community of Mathematical Geographers Discussion Paper No. 10.* Dept. of Geography, University of Michigan, USA.

Plazanet, C., Affholder, J.G. & Fritsch, E. 1995. The importance of geometric modeling in linear feature generalization. *Cartography and Geographic Information Systems* 22(4): 291–305.

Qian, H., Meng, L. & Zhang, M. 2007. Network simplification based on the algorithm of polarization transformation. *Proceedings of ICC2007, 4–9 August 2007, Moscow.* (CD-ROM).

Quattrochi, D. & Goodchild, M. (eds.). 1997. *Scale in remote sensing and GIS.* Boca Raton: CRC Press.

Raffy, M. 1994. Change of scale theory: a capital challenge for space observation of Earth. *International Journal of Remote Sensing* 15(12): 2353–2357.

Ramanjaneyulu, M. & Rao, K. 2002. A novel technique to resample high resolution remote sensing satellite data. *Proceedings of 2004 IEEE International Geoscience and Remote Sensing Symposium, 24–28 June, Canada.* (CD-ROM).

Rhind, D. 1988. A GIS research agenda. *International Journal of Geographical Information Systems* 2(1): 23–28.

Robinson, A.H., Sale, R., Morrison, J.L. & Muehrcke, P.C. 1984. *Elements of Cartography (5th Edition).* New York: Wiley.

Sester, M. 2000. Generalization based on Least Squares Adjustment. *International Archives of Photogrammetry and Remote Sensing* XXXIII(B4): 931–938.

Sester, M. 2005. Optimizing Approaches for Generalization and Data Abstraction. *International Journal of Geographic Information Science* 19(8–9): 871–897.

Sester, M. 2008. Multiple Representation Databases. (In this volume).

Sheppard, E. & McMaster, R. (eds.). 2004. *Scale and Geographic Inquiry: Nature, Society, and Method*. Malden: Blackwell Publishing.

Smith, M.P, Zhu, A.X., Burt, J. E. & Stiles, C. 2006. The effects of DEM resolution and neighborhood size on digital soil survey. *Geoderma* 137(1–2), 58–69.

Sposito, G. (ed.). 1998. *Scale Dependence and Scale Invariance in Hydrology*. Cambridge: Cambridge University Press.

Steiniger, S. & Meier, S. 2004. Snakes: A technique for line smoothing and displacement in map generalization. *ICA Workshop on Generalization and Multiple Representation, 20–21 August 2004, Leicester, UK.*

Stewart, J.B., Engman, E.T., Feddes, R.A. & Kerr, Y. (eds.). 1996. *Scaling up in hydrology using remote sensing*. Chichester: John Wiley & Sons.

Stommel, H. 1963. Varieties of oceanographic experience. *Science* 139: 572–576.

Su, B. 1997. *Morphological Transformations for Generalization of Spatial Data in Raster Format.* PhD Thesis, Curtin University of Technology, Australia. 236pp.

Tarolli, P. & Tarboton, D.G. 2006. A new method for determination of Most Likely Initiation Points and the evaluation of Digital Terrain Model scale in terrain stability mapping. *Hydrology and Earth System Sciences* 3: 395–425.

Tate, N. & Atkinson, P. (eds.). 2001. *Modelling Scale in Geographical Information Science.* Chichester: John Wiley & Sons.

Töpfer, F. & Pillewizer, W. 1966. The principles of selection. *The Cartographic Journal* 3(1):10–16.

Touya, G. 2007. River Network Selection based on Structure and Pattern Recognition. *Proceedings of ICC2007, 4–9 August 2007, Moscow.* (CD-ROM).

Townshend, J. & Justice, C. 1988. Selecting the spatial resolution of satellite sensors required for global monitoring of land transformations. *International Journal of Remote Sensing* 9: 187–236.

Trailandier, P. 2007. Automatic knowledge revision of a generalization system. *ICA Workshop on Progress in Automated Map Generalization, 2–3 August 2007, Moscow.* http://ica.ign.fr/Moscow/program.php (last accessed on 25 Feb 2008).

Steiniger, S. & Weibel, R. 2005. Relations and Structures in Categorical Maps. *8th ICA Workshop on Map Generalization and Multiple Representation, 7–8 July 2006.* http://ica.ign.fr/.

Usery, E., Finn, M.P., Scheidt1, D.J., Ruhl1, S., Beard1, T. & Bearden, M. 2004. Geospatial data resampling and resolution effects on watershed modeling: A case study using the agricultural non-point source pollution model. *Journal of Geographical Systems* 6(3): 289–306.

Ware, J.M., Jones, C.B & Thomas, N. 2003. Automated map generalization with multiple operators: a simulated annealing approach. *International Journal of Geographical Information Science* 17(8): 743–769.

Wolock, D.M. & Price, C.V. 1994. Effects of digital elevation model and map scale and data resolution on a topography-based watershed model. *Water Resources Research* 30: 3041–3052.

Woodcock, C.E. & Strahler, A.H. 1987. The factor of scale in remote sensing. *Remote Sensing of Environment* 21: 311–332.

Woodcock, C.E., Strahler, A.H. & Jupp, D.L.B. 1988. The use of variograms in remote sensing II: Real digital images. *Remote Sensing of Environment* 25: 349–379.

Xia, Z.G. & Clarke, K. 1997. Approaches to scaling of geospatial data. In D.A. Quattrochi & M.F. Goodchild (eds.), *Scale in Remote Sensing and GIS*: 309–360. CRC Press.

Yan, H., Chu, Y., Li, Z.L. & Guo, R. 2006. A quantitative description model for direction relations based on direction groups. *GeoInformatica* 10(2): 177–195.

Zaksek, K. & Podobnikar, T. 2005. An effective DEM generalization with basic GIS operations. *8th ICA WORKSHOP on Generalisation and Multiple Representation, July 7–8th, 2005.* http://aci.ign.fr/Acoruna/Papers/Zaksek_Podobnikar.pdf. (last accessed on 28 Feb 2008).

Zhou, M. & Bertolotto, M. 2005. Efficiently Generating Multiple Representations for Web Mapping. *Web and Wireless Geographical Information Systems, Lecture Notes in Computer Science* 3833: 54–65. Berlin: Springer.

Zhu, A.X. 2006. Neighborhood size of terrain derivatives and its impact on digital soil mapping. *International Symposium on Terrain Analysis and Digital Terrain Modelling, 23–25 November 2006, Nanjing, China.* (CD-ROM).

Zhu, K., Wu, F., Wang, H. & Zhu, Q. 2007. Improvement and assessment of Li-Openshaw algorithm. *Proceedings of ICC2007, 4–9 August 2007, Moscow.* (CD-ROM).

*Advances in Photogrammetry, Remote Sensing and Spatial Information Sciences:
2008 ISPRS Congress Book – Li, Chen & Baltsavias (eds)*

CHAPTER 20

Multiple representation databases

Monika Sester

ABSTRACT: Geo-spatial data are acquired for different purposes and thus based on different data models. This leads to different data sets. In order to allow for an integrated use of data from different sources, methods for an integrated representation of those data are proposed. Such a representation is called Multiple Representation Data Base (MRDB) and consists of the different data sets on the one hand, and links between the individual objects in the different data sets. Such an explicit linking allows for a direct access of corresponding objects; possibly, also the differences (transitions) between the data sets are known and can be exploited. Links can be established using semantic and geometric matching techniques or known rules to derive one data set from another. There are several areas that benefit from such a multiple representation scheme. The first is the integrated analysis of different data sets, leading to a mutual exchange of information. The second is the possibility to propagate updates from one data set to the linked ones. Also, in the context of multi-scale-representation, it can help to efficiently visualize data on small display devices.

Keywords: MRDB, multi-scale databases, generalization, matching, update, information fusion

20.1 INTRODUCTION

Many geographic data sets existing worldwide reflect the diversity in how we perceive and model the world around us (Mustière & van Smaalen 2007). These data sets have different levels of detail (e.g. individual buildings vs. building blocks) or thematic contents (e.g. roads modelled as topographic objects with name and road class vs. roads modelled with rich traffic information in a navigation data set). A Multi- representation-database (MRDB) is a spatial database used to integrate and store data of different type, scale, thematic granularity, time, or accuracy, which relate to the same physical object or phenomenon. This includes data stemming from different applications and origin, as well as data presenting and symbolizing the spatial phenomena differently. Also, data of different types can be integrated, e.g. raster and vector data (Butenuth et al. 2007), as well as 3-D data, surface data and multimedia information like video or clips. In the MRDB, the connections between individual objects in the different representations are explicitly stored.

Depending on the type of representations linked, one can specialize MRDB into multi-scale (or multi-resolution) databases. In this case, objects of different scale are linked with each other. MRDB is mostly being addressed in the context of vector data and multi-scale representations. Thus, in this chapter there will be a concentration on this issue, although also the general case will be addressed. Figure 20.1 shows an example of an MRDB with three levels of detail of settlement information and explicit links between individual entities.

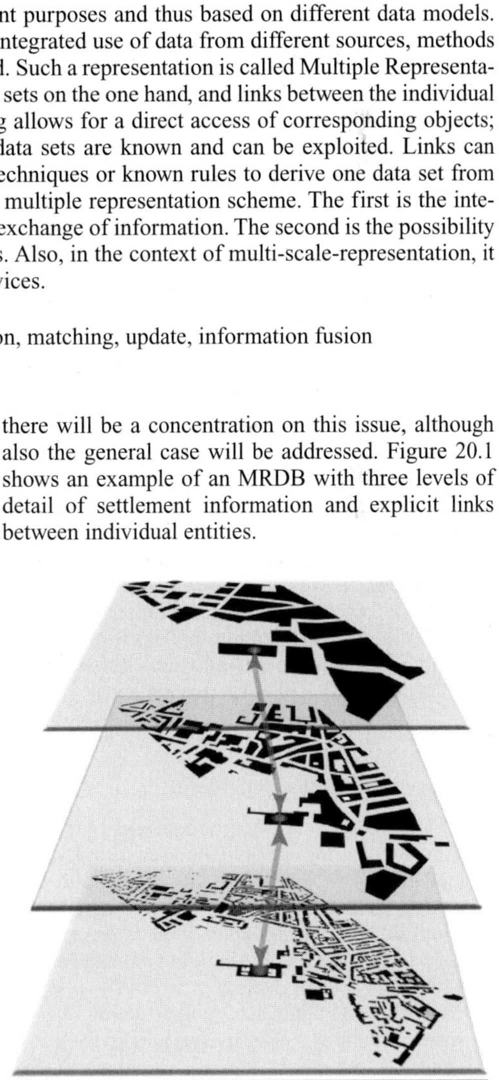

Figure 20.1. Example of an MRDB in three levels of detail; explicit links are shown for one object.

National mapping agencies usually provide multiple representations of their topographic data sets: on the one hand, digital landscape models (DLM) are available in different scales—corresponding to a multi-scale data representation; on the other hand, digital cartographic models (DCM) are derived from these DLMs in order to provide a readable map. Similarly, the DLMs can be linked with other data sets like geological maps, leading to a multiple-representation data base.

The remainder of this chapter is structured as follows: first, there will be a presentation of possible data models for MRDB. Then the benefits of such a data structure will be listed. Among them, three applications will be described in more detail. A summary and an outlook conclude the chapter.

20.2 DATA MODELS FOR MRDB

In order to model and represent different representations of the same object, data models and corresponding data structures have to be designed. A popular way for modelling is described by Vagenot et al. (2002) and Balley et al. (2004). They propose different ways to represent two mono-represented data sets in an integrated multiple-representation scheme:

1. The first solution is to create only one entity type with the attributes of both original representations.
2. The second solution is to create a new superclass with the attributes that are common to both representations. The two individual entity types represent sub-types created with an is-a-relationship, and thus inherit the attributes of the superclass.
3. The third way is to explicitly link the two original entities with a relationship "corresponds-to", which expresses semantically that the two entities relate to the same physical phenomenon.

In the last case, the relationships may have different cardinalities, which depend on the type of object classes and objects that are linked:

1. 1:1-relationship indicates the equivalence relationship and can, for example, be used to link a road modelled in a topographic data set with the same road modelled in the navigation data set. Also, it could link a building represented as an area in a cadastral data set with a building represented as a point in an address database.
2. 1:n-relationship indicates an aggregation relation, where one entity in one database is linked to several entities in the other database, or vice versa. This is the general case in multi-scale representation, e.g. when some buildings are linked to one built-up area.
3. n:m-relationships are needed, when several entities of one data set are linked to a different number of entities in the other data set. This is the case when generalizing buildings in a cartographic database with the typification operation. Also, n:m-relationships occur in the case of matching different road databases, when the segmentation of the roads into different individual parts differs in the two data sets (Walter & Fritsch 1999, Mantel & Lipeck 2004).

Another option for storing multiple representations of data that is, however, only possible for multi-*scale* representations is an implicit storage, which covers all the different representations in a special data structure. One example is the BLG (Binary Line Generalization)-tree (van Oosterom & Schenkelaars 1995), which is based on the Douglas-Peucker-Algorithm (Douglas & Peucker 1973): the different levels in the tree correspond to an incremental refinement of the details of linear objects. Similarly, the GAP-tree proposed by van Oosterom (1995) allows the representation of areal objects in different levels of detail. Sester & Brenner (2004) propose a elementary vocabulary to code geometric and topologic changes of objects when going from one scale to the next. This vocabulary allows the description of the hierarchical decomposition in terms of scale. This is especially beneficial for the progressive transmission of incrementally more and more detail of a geometry via a possibly limited bandwidth channel.

In the remainder of the chapter, the data model that is based on explicit correspondence-relations between homologous objects in the two data sets is used as the data structure.

20.3 BENEFITS OF AN MRDB

There are several benefits of an MRDB, which relate both data producers and data users.

1. For data producers like national mapping agencies one benefit is the ease of maintenance of several data sets in a consistent manner. This allows on the one hand to reduce redundancy in data storage, and also to ensure and preserve consistency, especially in update processes: if one object is updated in one data set, the effects can be propagated to the related data sets using the links. This also allows an incremental update, i.e. a restriction of the effects of local changes locally. This application will be described in more detail below. Also, it is

possible to check the quality of the integrated data and identify possible errors. When integrating different measurements relating to the same object, its quality can be improved.

2. Users benefit from an adequate representation of the data for their applications. This is especially important when the data is accessed via a limited bandwidth channel, when the user only wants to pay for as much information as he/she needs for the application at hand. This presumes that there are mechanisms available that are able to determine the appropriate scale for the application and the user device. Such functionality is especially important for the visualization of spatial data on small-display devices.
3. Multi-scale data inspection: the availability of linked representation of objects in different scales allows for an "information drilling" (Bernier & Bédard 2007), which means that individual objects can be inspected at higher level of detail than their surrounding objects.
4. Mutual exchange of information: data sets that are integrated in an MRDB can access information of linked entities; thus one data set can be enriched with the information from the corresponding data set.
5. Finally, when data are provided in an MRDB, an integrated analysis of a spatial problem is possible, taking all the relevant data into account. This has been demonstrated for routing on different data sets by Volz (2005). Also, a hierarchical analysis can be conducted as shown in Gabay & Sester (2002). Information of one data set can be accessed by querying another data set, e.g. the number of inhabitants of a city area can be queried on the city level, if the city objects have links downwards to urban blocks and those, in turn, to the individual buildings, which finally contain the number of people living in it as attributes (Hampe 2007).

20.4 ESTABLISHMENT OF LINKS IN AN MRDB

The determination of links of an MRDB depends on the underlying data sets that have to be linked. Already existing data sets can be linked using semantic and geometric matching techniques. If data sets of different thematic and geometric resolution have to be connected, the links can be generated by generalization. In this case, the links also imply functional dependencies between the two data sets that can later be used in the update process.

20.4.1 *Generation of links by matching*

Links can be established using semantic and geometric matching processes. These processes have to be available for all geometric primitives, namely points, lines and polygons. There are different approaches for each case. For all matching situations, the knowledge about the semantically comparable object classes is presumed to be known. Matching point objects can be achieved taking geometric distances into account. This approach presumes that the position and orientation of the data sets is the same. The allowed distances can vary depending on the object types. Other approaches try to use the local context of point objects as additional criterion (e.g. Samal et al. 2004). In this case, the requirement of known orientation can be relaxed. Matching linear objects has been investigated by many researchers. There are approaches that extend the point-based approach by using appropriate buffers. Furthermore, geometric characteristics of the lines can also be taken into account (e.g. length, orientation, number of points). In addition, relations of the features can be used (e.g. Walter & Fritsch 1999, Mantel & Lipeck 2004, Shi & Meng 2006). Matching polygon objects can use the degree of overlap (e.g. measured by overlay ratios or the symmetric difference), as well as form parameters (Hild 2003). If the matched data sets are of different scale, the matching procedures have to take these possible differences into account (see e.g. Mustière 2006). Due to typical differences in object modelling, not only 1:1-relationsships between corresponding objects occur, but more often there will be 1:n or n:m-relations. The semantic of the 1:1-link is a general "corresponds-to"-relationship, which is mutual. Depending on the type of matching process involved, more detailed relationships can be derived, e.g. containment (1:n-relation) or n:m-relation.

20.4.2 *Generation of links by generalization*

Generating links by generalization processes has the advantage that the functional dependencies between the corresponding data sets are explicitly known and reproducible; thus the semantics of the links is known: through the process it is clear, how the objects in the target data set are represented. Generalization operations used are algorithms for model generalization (aggregation, area collapse), but also for cartographic generalization (e.g. simplification, displacement, typification). Usually, the type of generalization operation also determines the type of relationship between corresponding objects. Also in this linking process, all cardinalities of relations can occur, i.e. 1:1 (e.g. in case of simplification or displacement), 1:n (e.g. in case of aggregation), n:m (e.g. in case of typification).

20.5 APPLICATIONS OF AN MRDB

In the following, some applications will be described that make use of the MRDB structure.

20.5.1 *Incremental update*

If the connections between the different representations are known in terms of a functional relationships—as it is the case with data of different scales, where generalization functions link the objects—then there is the possibility to reduce the time and cost for updating the connected data sets in the future: in the ideal case, only the most detailed landscape model has to be updated manually, the derived data sets (both DLMs and DCMs) can be updated automatically by exploiting the links between the corresponding objects. The advantage of this incremental generalization approach is that an automatic update process can be implemented, which keeps the different models in a consistent state and only has to treat the affected objects. This is a strong motivation for national mapping agencies (NMA's) to implement an MRDB.

The concept of incremental generalization was introduced by Kilpeläinen & Sarjakoski (1995). The basic idea of incremental update is that the influences of the update action can be restricted locally and thus can operate on a limited area and thus on a limited number of objects, and is therefore less time consuming than the full processing of the whole data set. The goal is that it produces the same results as a complete generalization. This is possible if the generalization process can be partitioned into "modules", which can be executed in a defined order. Modules can either be different thematic categories, e.g. buildings, but also spatially limited areas, e.g. mesh generated by road network, see also Galanda (2003). Harrie & Hellström (1999) present a system for the propagation of updates that is based on object specific rules. After introducing an object, a re-generalization is triggered. This approach cannot, however, guarantee that the results of an incremental update are the same as a complete generalization of the dataset. Ellsiepen (2006) proposes a rich data structure to store the links between a generalized object and other objects together with additional information: the generalization function used the original object(s) it was derived from, as well as the objects that have influence on it. A similar method is presented by Skogan & Skagestein (2005). Their approach is based on so-called productions that are stored in the datasets. A production is composed of an identifier of the generalization rule that is used to create a target object, the identifiers of objects in the source dataset that are examined by this rule and the identifier of the object in the target dataset that is created. By this the history of the complete generalization is documented. Influences of updates can be found by analysing the productions in which changed objects are involved. A disadvantage of this method is that a high amount of additional data has to be maintained. Haunert (2005) therefore proposed to exploit the information that is implicitly given with the generalization rules: often the rule that is used for the generalization of a feature evaluates its direct neighbourhood. Then, the information about the dependencies can be derived from the topology and does not need to be stored explicitly.

In the following, two cases for determining the "modules" or the local context of a change are presented. The local context determines the objects that are affected by the change of an object. There are different approaches to determine this local context, which can be distinguished for the two MRDB generation principles described above.

20.5.1.1 *Case 1: Links generated by aggregation function*

If the links were generated by a generalization process, the local context of an object correspondence can be determined by inspecting all objects that are involved or affected by a generalization operation. The generalization process has to be repeated locally for the affected objects.

Haunert (2005) investigated the problem of incremental update in the case of area aggregation. He could show that the MRDB structure can be exploited to a high degree to restrict the influence area of a local change: the possible affected object can be incrementally expanded if objects on a higher aggregation level are included in the processing. Figure 20.2 shows some screenshots of the incremental expansion of the boundary area.

The example shows how the boundary area (in blue) is incrementally expanded, when the changed object (in red) is aggregated to an object of the boundary area. The expansion process stops when the changed object is adequately generalized, i.e. in this case, when the aggregation process has finished. In the example, of all 5461 objects only 348 had to be analysed, i.e. 6% of all objects. This corresponds to a saving of 94%.

20.5.1.2 *Case 2: Links generated by matching or other generalization functions*

In the first case, the aggregation function implicitly defined the maximum possible range of influence of an object on its neighbours when being aggregated. Thus, the simple principle of incrementally expanding the boundary objects could be applied. This is, however, not possible for the case when other generalization operations have been applied, e.g. typification or displacement; then objects not present in the final representation could also have an influence on the generalization result. In such cases, therefore,

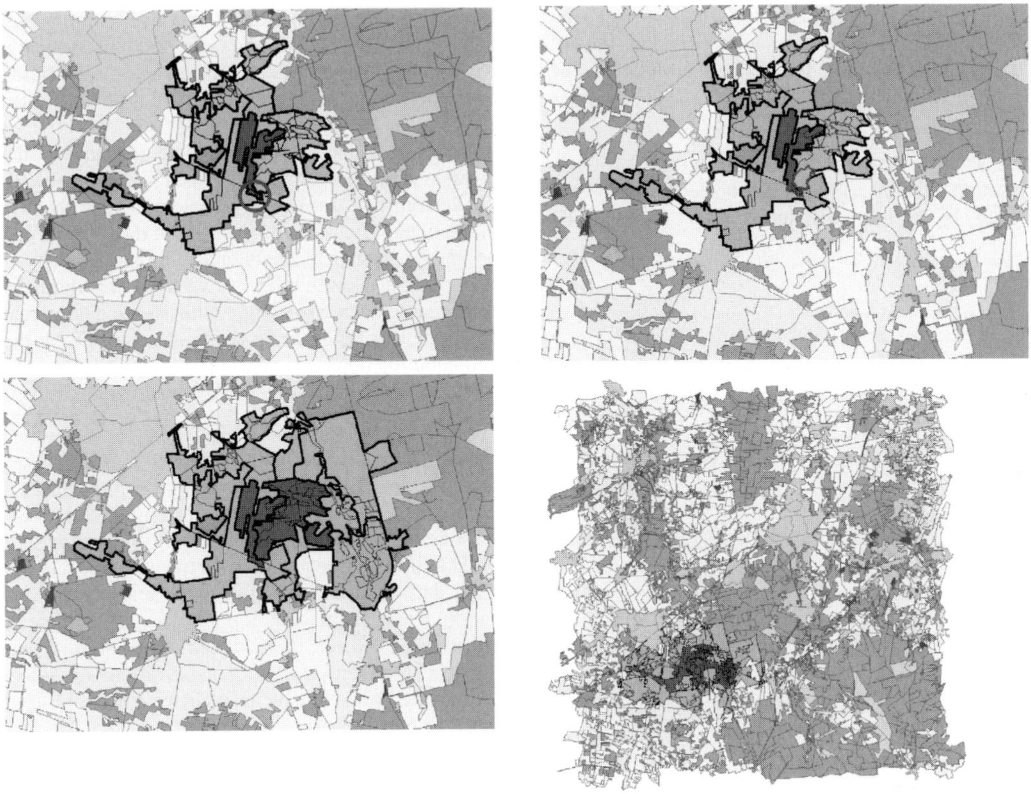

Figure 20.2. Incremental expansion of boundary of maximally affected objects; whole area with possibly affected objects (in blue). (see colour plate page 500)

all affected and affecting objects have to be stored as proposed by Skogan or Ellsiepen. In the case of matching, however, the situation is even worse, as these influences are unknown—the only fact that is available is the correspondence relationship between linked objects. Thus, for such cases other approaches for the incremental update have to be provided.

In the typical case of data sets linked by matching, not all objects have a matching partner, as the data set only coincides in some of the object types (e.g. in the case of matching a geological with a topographic data set, only some features will be in common in both data sets). Figure 20.3 shows the corresponding objects of two data sets. Some of the objects have direct partners in the other data set (indicated by solid lines), there are, however, also objects which do not, and thus are linked with the superior mesh of the partner object (dotted lines).

20.5.1.3 *Local context*

The local context determines the objects that are affected by the change of an object. In the case of links generated by matching, the local context of an object correspondence can be determined indirectly—from the matching process itself. Matching objects create a displacement vector field, which describes the geometric deformations an object undergoes when being transformed to the other data set. Objects that do not have a corresponding partner in the other data set can be transformed using their local context, which can be determined taking their local relationships to the matched objects into account. This can be topologic relations like containment (e.g. lake lies within forest) or connectivity (e.g. minor road joins major road). Using a topologic transformation, a rubber sheeting operation, these relationships are preserved (see Fig. 20.3).

20.5.1.4 *Update propagation operations*

In general, there are three possible update types in a database: MODIFY, INSERT or DELETE. In general there is no single INSERT, DELETE or MODIFY operation in a topological consistent database. If one object is changing then all adjacent neighbouring objects are changing too! e.g. the insertion of a new road brings about a modification of the superior road to which the new object is linked to; thus there is a MODIFY-operation together with the INSERT-operation.

20.5.1.4.1 Operation MODIFY
In case an object is modified (see Figure 20.4: Update situations: modification of red object (left), insertion of red object (right), deletion of object (bottom.), first of all it has to be checked if the modification only relates to semantics or also to geometry. If the semantic changes have influences on the geometric appearance of the objects (e.g. a change of a road type could have a change of the signature with as a consequence), or if there was a change in geometry (e.g. change of the position and flow of a river), then the geometry in the derived data set has to be determined. If there is a functional relationship between the data sets, it can be applied again (e.g. smoothing with Douglas-Peuker-Algorithm and given tolerance value). When the links were generated by matching, rubber sheeting is applied using the displacement vector field generated by the neighbouring objects in order to determine the geometric shape in the corresponding data set.

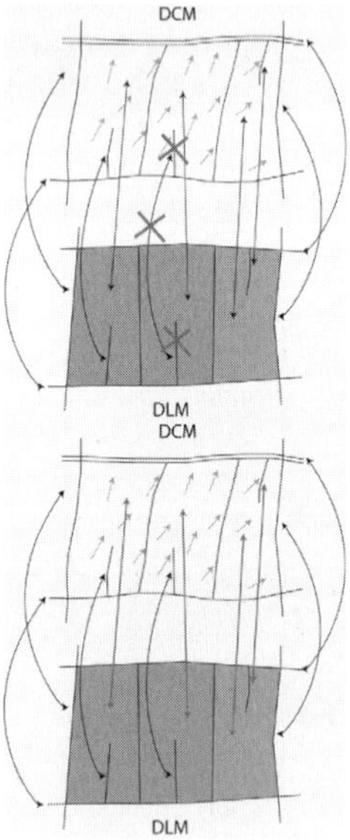

Figure 20.3. Left top: matched Objects (solid lines), objects without direct partner in other data set are linked to mesh they are lying in (dashed lines); top right: displacement vector field generated by directly matched objects; lower left: propagated geometries of objects without direct partner using rubber sheeting.

20.5.1.4.2 Operation INSERT
If a new object is inserted (see Fig. 20.4, right), it has to be assigned to the superior mesh it is lying in. This ensures that the global context in terms of the corresponding displacement vector field is preserved. In order to transfer the geometry of the new object to the other data set, the existing displacement vector field can be used. In this way the topology of the data is preserved; however, it could lead to geometrical distortions (see Section 0). If generalization is applied, the affected objects have to be determined and used in the re-generalization process with the same parameters, taking all the objects, including the new one, into account.

20.5.1.4.3 Operation DELETE
If an object is deleted (see Fig. 20.4, bottom), then this object will be deleted together with its corresponding object in the second data set. Also, consequences of these actions have to be taken. This is the case when the object had formed a constraint for a related object earlier. As an example, a road was responsible for separating two areas of similar object type. If this road

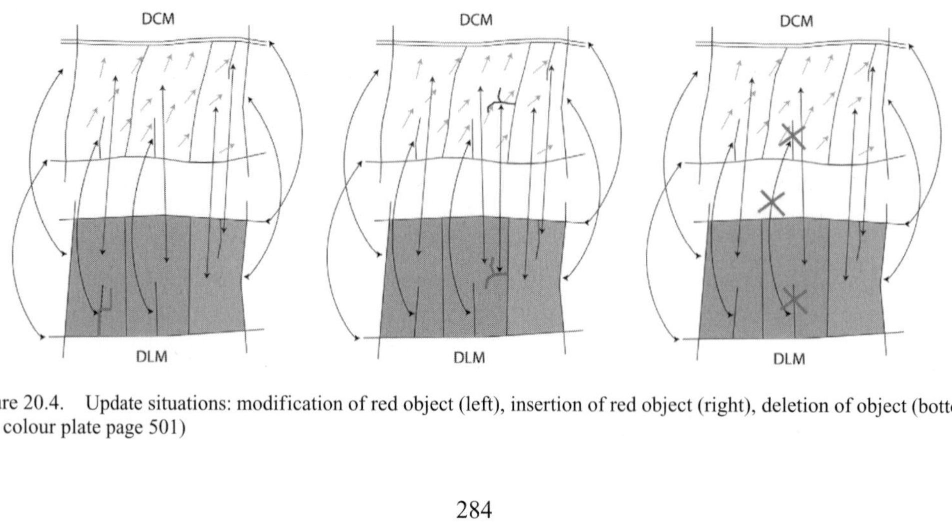

Figure 20.4. Update situations: modification of red object (left), insertion of red object (right), deletion of object (bottom). (see colour plate page 501)

is deleted, then, in turn, the areas can be merged. It also has to be checked if the object was used to determine the vector field earlier. If yes, then a new vector field has to be determined and the objects in the local neighbourhood have to be transformed anew.

20.5.1.5 *Examples and discussion of rubber sheeting approach*

The benefit of this approach is the fact that the connections between corresponding objects are explicitly known. Based on these links, the update is relatively simple. Furthermore, the relation function (i.e. transformation between the two data sets) is mutual, and thus can be applied in both directions. This means that objects of data set 1 can be transferred to data set 2 and vice versa.

Figure 20.5 shows an example of a situation before and after application of the displacement vector field. The lake is nicely adjusted in its local environment.

The drawback of using the displacement vector field is twofold: first of all, it could lead to deformations of objects which is especially severe for object with distinct shape (e.g. circular or rectangular). (see Figure 20.6). Furthermore, the update can lead to the effect that spatial conflicts occur between too close objects.

20.5.1.6 *Solving spatial conflicts with local generalization operation*

In cases of local conflicts or deformation of objects, we propose to include a local generalization process instead of rubber sheeting, namely a displacement that can take object characteristics into account. All affected areas and also their neighbouring objects have to be taken into account; the neighbourhood has to be expanded until a boundary of superior objects is found that has the effect of being able to limit the influence range of the generalization process. Superior objects are usually linear objects of topographic importance that partition the space, such as roads and rivers. All possibly affected objects are then fed into the displacement process, in our case the program PUSH (Sester 2005). The program is able to model the behaviour of objects in terms of stiffness

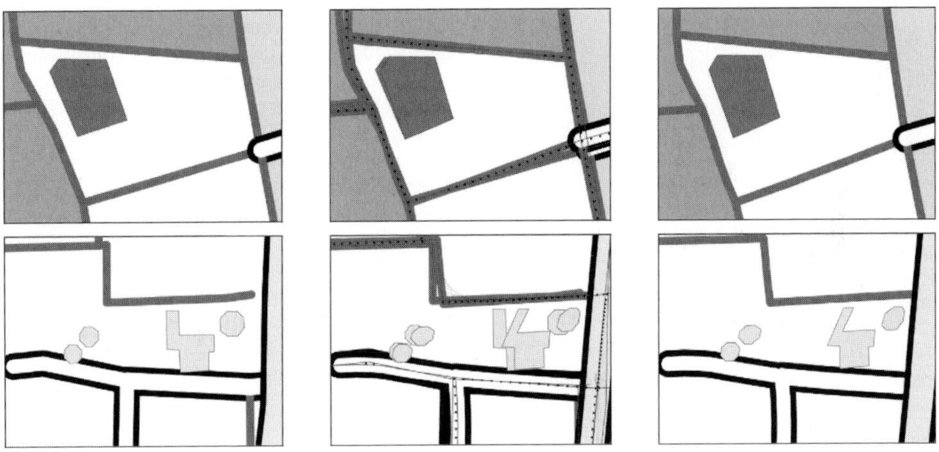

Figure 20.5. Inserting lake: situation before (left) and after applying rubber-sheeting (right); middle: both situations and displacement vectors. (see colour plate page 501)

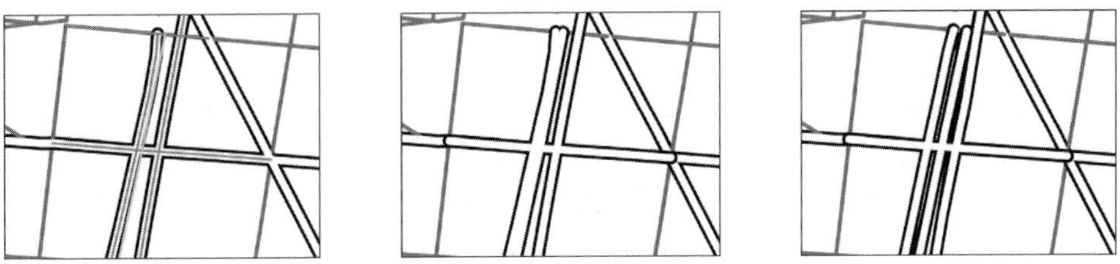

Figure 20.6. Insertion of new road (blue) (left); conflicts after applying signature (right); situation after conflict resolution with displacement (bottom). (see colour plate page 501)

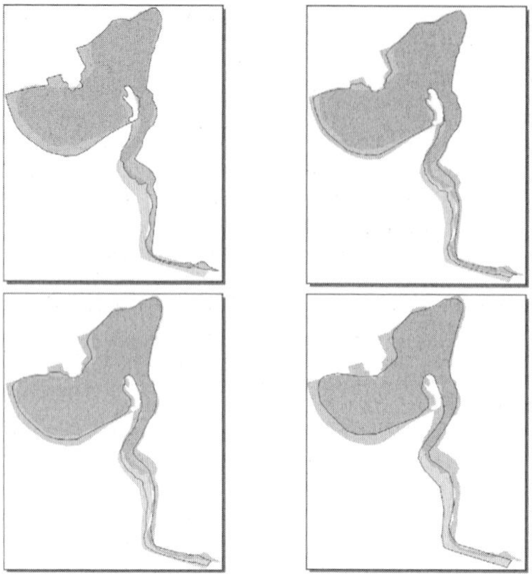

Figure 20.7. Incremental morphing of green and red geometry. The weighting scheme between red and green changes from top left to lower right: 1:0, 0.6:0.4, 0.4:0.6, 0:1. (see colour plate page 502)

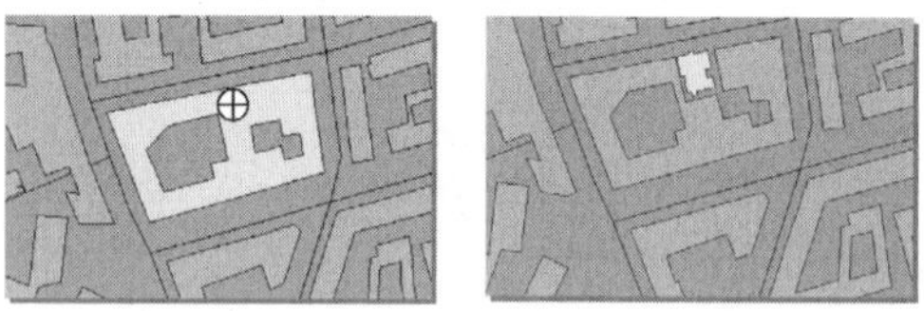

Figure 20.8. Visualization of a landmark in a small scale view: left: marking whole building block; right: delineating precise, detailed object geometry of landmark. (see colour plate page 502)

and positional variability. Figure 20.7 visualizes that the insertion of a new road (in blue) leads to spatial conflicts with its neighbouring roads, as the space is too small to adequately visualize all three roads. A displacement with the given vector field in this case cannot solve this conflict. Therefore, in that local context area, the objects undergo a displacement operation, which leads to the desired result of representing all objects adequately.

20.5.2 *Geometric information fusion*

For integrated analysis processes, it is important that the underlying information is unambiguous. If, however, the information stems from different sources, typically there will be geometric discrepancies among the corresponding object entities. Thus, there is a need for geometrically fusing or homogenizing geometric shapes. This can be accomplished using morphing methods. An approach by von Gösseln & Sester (2006) takes the quality of the individual objects into account and creates a target geometry as a weighted mean. This flexibly allows to either fully adapt one data set to the other, or to generate an intermediate geometry by averaging the geometries of the two input objects. In Figure 20.8 examples for the adjustment with different weighting schemes are given. In this way, the linked objects can be fused and thus generate a unique object representation for the given application.

20.5.3 *Web-applications: adaptive visualization*

Using an MRDB in a Web-context has been described by several researchers (e.g. Sarjakoski & Sarjakoski 2007), mainly with the issue of providing data with good graphic quality and adequate resolution especially for small mobile displays. Due to the lack of online generalization operations, data of different scales are pre-processed and stored in an MRDB. The applications, in turn, access that data resolution which is most appropriate for the problem at hand. An MRDB allows for an object-wise approach and an incremental refinement of individual objects instead of re-loading whole data layers when zooming in. It also enables the display of multiple levels of detail in one presentation, as known from fish-eye projections. One possibility is to use a continuous change of the scale as shown e.g. by Harrie et al. (2002); the other option is to use discrete changes of scales as shown by Bernier & Bédard (2007). The following figure is taken from Hampe & Elias (2004): an important landmark object has to be visualized in a larger context. At this scale the level of detail is rather coarse, so that individual buildings are no longer shown, but aggregated building blocks, instead. Using an MRDB, a visual information drilling is possible by showing the landmark in higher resolution in its coarser environment (Figure 20.8).

20.6 SUMMARY AND CONCLUSION

An MRDB is a powerful structure to manage and use multiple representations of data. The structure allows the representation and integrated handling of multiplicities of data that can be geometric, thematic or temporal, but also relating to data type. The different representations are connected by explicit links that allow for an easy query and access of the related objects. This explicit link structure, however, also presents an overhead in data storage. Also, it has to be standardized in order to be exploited by follow-up processes, especially in the context of web-services. Hampe & Intas (2005) presented a first approach to extend existing OGC services to also include MRDB structures.

There are first experiences that show the benefits of such a data structure, especially when it comes to the consistent management and update of NMA data. However, there is an ongoing discussion whether to fully store the link-structure in an MRDB or to create and exploit the links on-the-fly when needed. This has to be investigated in the near future.

The general advantage of MRDB—whether the links are stored explicitly or created on-the-fly—is its ability to exploit the benefits of existing data sets in a consistent manner. This implies the seamless integration of different data sets and also includes the mutual transfer of updates between connected data sets. This is an important property that is becoming more vital with the growing availability of data sets in spatial data infrastructures.

ACKNOWLEDGEMENTS

Parts of the research presented in this chapter have been funded by the Federal Agency of Cartography and Geodesy in Germany (BKG), the German Science Foundation (DFG), and the EU (GiMoDig-project IST-200-30090). The support is gratefully acknowledged.

REFERENCES

Badard, T. 1999. On the automatic retrieval of updates in geographic databases based on geographic data matching tools. *Proc. of the 19th Int. Cartographic Conference, Ottawa*: 47–56.

Balley S., Parent, C. & Spaccapietra S. 2004. Modelling geographic data with multiple representation. *International Journal of Geographical Information Science* 18 (4): 327–352.

Bernier, E. & Bédard, Y. 2007. A Data Warehouse Strategy for on-Demand Multiscale Mapping. In W. Mackaness, A. Ruas & L.T. Sarjakoski (eds.), *Generalization of Geographic Information: Cartographic Modelling and Applications*: 177–198. Elsevier.

Butenuth, M., Gösseln, G. v., Heipke, C., Lipeck, U., Sester, M. & Tiedge, M. 2007. Integration of heterogeneous geospatial data in a federated database. *ISPRS Journal of Photogrammetry and Remote Sensing* 62, 5: 328–346.

Douglas, D. & Peucker, T. 1973 Algorithms for the reduction of the number of points required to represent a digitized line or its caricature. *The Canadian Cartographer* 10(2): 112–122.

Ellsiepen, M. 2006. Strategies for automatic and interactive generalization of topographic vector data over several scales (in German). *DGfK-Workshop "Aktuelle automationsgestützte Verfahren zur Generalisierung topographischer Daten" (in German), Bonn, 22–23.11.2006.*

Gabay, Y. & Sester, M. 2002. *Forming And Utilizing Communication Between Two Spatial Representations At Different Scales—A Demonstration.* Institute of Cartography and Geoinformatics.

Galanda, M. 2003. *Automated Polygon Generalization in a Multi Agent System.* PhD Thesis, Department of Geography, University of Zurich, Switzerland.

Hampe, M. & Intas, S. 2006. Extension of the OGC Web Feature Service Standard for Multiple Representation Data. In Kainz & Pucher (eds.), *Proceedings of the ISPRS TCII Symposium, IAPRS Vienna Austria*: 49–54.

Hampe, M. 2007. *Integration einer multiskaligen Datenbank in eine Webservice-Architektur.* Dissertation. Wissenschaftliche Arbeiten der Fachrichtung Geodäsie und Geoinformatik der Leibniz Universität Hannover, 2007.

Hampe, M. & Elias, B. 2004. Integrating Topographic Information and Landmarks for Mobile Applications. In G. Gartner (ed.), *Location Based Services & Telecartography, Proceedings of the Symposium 2004, Geowissenschaftliche Mitteilungen* 66: 147–157. TU Wien.

Harrie, L. & Hellström, A.-K. 1999. A Prototype System for Propagating Updates between Cartographic Data Sets. *The Cartographic Journal* 36(2): 133–140.

Harrie, L., Sarjakoski, L.T. & Lehto, L. 2002. A Mapping Function for Variable-Scale Maps in Small-Display Cartography. *Journal of Geospatial Engineering* 4(2): 111–123.

Haunert, J.-H. 2005. Link based Conflation of Geographic Datasets. *ICA Workshop on Generalisation and Multiple Representation, 7–8 July 2005, La Coruña, Spain.*

Hild, H. 2003. *Automatische Georeferenzierung von Fernerkundungsdaten.* Deutsche Geodätische Kommission, Reihe C, Nr. 562, München.

Kilpeläinen, T. & Sarjakoski, T. 1995. Incremental generalization for multiple representations of geographic objects. In J.C. Muller, J.P. Lagrange & R. Weibel (eds.), *GIS and Generalization: Methodology and Practice*: 209–218. London: Taylor & Francis.

Mantel, D. & Lipeck, U.W. 2004. Matching Cartographic Objects in Spatial Databases. *IAPRS* 35, Part B4: 172–176.

Mustière, S. 2006. Results of experiments on automated matching of networks at different scales. *IAPRS* 36 (Part 2/W40): 92–100.

Mustière, S. & v. Smaalen, J. 2007. Database Requirements for Generalization and Multiple Representation. In W. Mackaness, A. Ruas, L.T. Sarjakoski (eds.), *Generalization of Geographic Information: Cartographic Modelling and Applications*: 113–136. Elsevier.

Samal, A., Seth, S. & Cueto, K. 2004. A feature-based approach to conflation of geospatial sources. *International Journal of Geographic Information Science* 18 (5): 459–490.

Sarjakoski, L.T. 2007. Conceptual Models of Generalization and Multiple Representation. In W. Mackaness, A. Ruas, L.T. Sarjakoski (eds.), *Generalization of Geographic Information: Cartographic Modelling and Applications*: 11–35. Elsevier.

Sarjakoski, T., & Sarjakoski, L.T. 2007. A Real-Time Generalization and Map Adaptation Approach for Location-Based Services. In W. Mackaness, A. Ruas, L.T. Sarjakoski (eds.), *Generalization of Geographic Information: Cartographic Modelling and Applications*: 137–159. Elsevier.

Sester, M. 2005. Optimizing Approaches for Generalization and Data Abstraction. *International Journal of Geographic Information Science*19, Nr. 8–9: 871–897.

Sester, M. & Brenner, C. 2004. Continuous Generalization for Visualization on Small Mobile Devices In P. Fisher (ed.), *Proceedings of the International Symposium on Spatial Data handling (SDH), Leicester*: 469–480.
Shi, W. & Meng, L. 2006. Some ideas for integrating multidisciplinary spatial data. *IAPRS* 36 (Part 2/W40): 30–35.
van Oosterom, P. 1995. The GAP-tree, an approach to 'on-the-fly' map generalization of an area partitioning, in: In J.C. Muller, J.P. Lagrange & R. Weibel (eds.), *GIS and Generalization: Methodology and Practice*: 120–147. London: Taylor & Francis.
van Oosterom, P. & Schenkelaars, V. 1995. The Development of an Interactive Multi-Scale GIS. *International Journal of Geographical Information Systems* 9(5): 489–507.
Volz, S. 2005. *Modellierung und Nutzung von Relationen zwischen Mehrfachrepräsentationen in Geo-Informationssystemen*. PhD-thesis, Universität Stuttgart, Fakultät für Luft- und Raumfahrtechnik und Geodäsie.
Walter, V. & Fritsch, D. 1999. Matching spatial data sets: a statistical approach. *International Journal of Geographical Information Science* 13(5): 445–473.

Advances in Photogrammetry, Remote Sensing and Spatial Information Sciences: 2008 ISPRS Congress Book – Li, Chen & Baltsavias (eds)

CHAPTER 21

Dynamic GIS

Christopher M. Gold, Darka Mioc & François Anton

ABSTRACT: Traditional GIS has focused on static 2-D or polygon data structures for cartographic data, and grids or TINs for terrain. More recent versions have extended this somewhat, but, as far as we know with commercial products, the underlying structures remain at most 2.5-D and static. This chapter attempts to demonstrate, and illustrate with dynamic GIS prototype systems and applications, how the lowest level of spatial data storage and algorithms may be extended towards a fully dynamic or even kinetic GIS. If dynamic algorithms are added to GIS then further possibilities for various spatial data applications open up. Adding kinetic algorithms to GIS allows the manipulation of moving points within the spatial data structure, opening up opportunities for managing continuous change over time, collision detection, fluid flow and continuous line-drawing within a topologically valid space. The data structure that allows the implementation of dynamic and kinetic algorithms in GIS while maintaining the topology of the spatial objects is based on the Voronoi diagram and its dual Delaunay Triangulation. Many traditional operations in GIS are greatly simplified by this proximal spatial model, and several are demonstrated in this paper. The Voronoi data structure and its dual, the Delaunay triangulation, may also be modified incrementally or kinetically, and we demonstrate with the dynamic 3-D Voronoi diagram the possibilities of applications in oceanography, geology and similar disciplines normally out of reach of the static 2-D GIS.

Keywords: Dynamic GIS, topology maintenance, updating, Voronoi diagram

21.1 INTRODUCTION

Previous research on dynamic GIS has been summarized in the "Dynamic GIS" chapter of the recently published Handbook of GIScience (Albrecht 2007). GIScience research on dynamic GIS had mostly concentrated on attempts to add the temporal dimension to existing GIS models (Albrecht 2007). The snapshot data model (Langran 1992) was the earliest representation of time in GIS and the snapshots can be used as a collection over time to represent a 4-D space-time cube (Pequet 2001). Object-orientation gave a theoretical basis for a new approach in handling time by time-stamping objects and their attributes (Stefanakis 2003, Yuan 2001). Another approach was to look at cellular automata and agent-based modelling (ABM) systems in an attempt to capture the dynamics of spatial behaviour (Albrecht 2007). However, ABM systems could not overcome the fundamental shortcomings of GIS (Raper 2000). Albrecht (2007) concluded that although the discipline of Geography had adapted to the notion of process almost 100 years ago, it failed completely to scrutinize the fundamental role of time.

Another drawback is that the theoretical models coming from the GIS scientific community have not been implemented in commercial GI systems, as recent developments have shown (Goodchild 2006). Furthermore, examining recent GIS development by comparison with some of the early theoretical advances (in particular *naïve geography*), Goldchild (2006) stated that very little has come of these ideas. The bulk of existing research on dynamic GIS has largely ignored the parallel research developments in geomatics, computer science and computational geometry (Gold 2006). Interestingly enough, the major GIS vendors have implemented many of the research results originating from computer science and geomatics into their recent software releases (ESRI 2003, 2005).

Nadi & Delavar (1995) summarized spatio-temporal data as static or dynamic, with dynamic data being real-time, near-real-time or time-stamped. For a temporal GIS, they considered temporal databases, temporal visualization and representation, and temporal analysis. For temporal modelling, they considered time-stamping (snapshot model, space-time composite, spatio-temporal objects) and event based or domain-oriented spatio-temporal modelling. Under implementation they considered road traffic volume simulation, vehicle movement simulation, graphical change modelling, path analysis and animation.

Chen et al. (2001) in an early review stated that: "Dynamic and multi-dimensional GIS in fact means non-traditional GIS." This is still the case. They pointed out that "dynamic" could apply to data acquisition, data models, data retrieval, process modelling, visualization, change detection and others.

Thus "dynamic" is still not a consistent term when used about GIS. It may refer to database management of time-varying objects, temporal change of viewpoint as in "Google Earth", simulation of temporal processes such as rainwater runoff, the visualization of objects moving over a static map, the actual map updating process itself, or the true simulation within the computer of the spatial relationships of real-world moving objects.

Various papers have discussed the raster or sphere-based QTM model for the representation of irregular or moving objects and their spatial adjacency. These may use Voronoi diagrams, or perhaps cellular automata. Examples include Chen et al. (2003), Pang & Shi (2002) and Shi & Pang (2000). Yi et al. (2001) used 9-intersection modelling to track topology changes.

Pan et al. (2001) treated "dynamic" as change of viewpoint for visualization. Claramunt et al. (2000) looked at the cartographic visualization of urban traffic. Hardy (2000) used dynamic map objects to preserve the topological connectivity of cartographic road segments. de Vasconcelos et al. (2002) and Pebesma et al. (2000) considered a "dynamic GIS" for Eulerian runoff simulation, whereas others have used "free-Lagrange" techniques for simulation where particle adjacency needs to be updated as they move. As an example of topological concerns outside GIS, Ganovelli et al. (2000) maintained the topological relations of soft objects in surgery simulation. Even with these various application domains, advanced research in the domain of dynamic and even kinetic GIS coming from the field of geomatics (Gold 2006, van Osteroom 2006 and Zlatanova et al. 2004) does not easily fit in a mainstream view of GIscience.

A convenient summary of current interest in dynamic GIS (from a geomatics perspective) may be obtained from the last two ISPRS Workshops in dynamic and multidimensional GIS. In the 2005 workshop (Gold 2005), Mostafavi (2005), Grum (2005) and Karimipour et al. (2005) discussed problems of the relative spatial position of objects moving within the two-dimensional map—usually a single point at a time, but in the case of Mostafavi (2005) for multiple moving points. Ledoux & Gold (2005), Dakowicz & Gold (2005) and Tse et al. (2005) were concerned with the incremental construction or updating of 3-D models. In the 2007 workshop (Jiang & Zhao 2007), Gold (2007) discussed 2-D dynamic map topology maintenance, Goralski & Gold (2007) covered 3-D visualization for maritime navigation, Boguslawski & Gold (2007) described 3-D topology updating, and Dakowicz & Gold (2007) showed finite-difference flow modelling. Wu & Wang (2007) described updating algorithms for constrained triangulations, Chen et al. (2007) discussed incremental topological updating of cadastral databases, Zheng et al. (2007) described the updating of the topology of a road network graph, and Zeng et al. (2007) described change-only updating of road network databases. Thus, the main topics of interest among all these researchers involved in ISPRS workshops in dynamic and multidimensional GIS have been: moving point location within the map (or flow between cells); topological map updating; and 3-D map object construction, editing and visualization—including topology. We can thus see an increasing interest in the hard problems of 2-D and 3-D incremental topology change—especially for map updating, but also for 3-D modelling and for simulation.

The rest of this review paper will focus on attempts to resolve the "hard" issues of topology maintenance within a GIS as it attempts to track real-world moving objects, or else as these relationships change during manual or automated updating. This has been the authors' particular interest for several years, and all the issues are not yet resolved, so inevitably there will be frequent reference to their own work, as well as to those few other GIS workers actively involved in developing algorithms for dynamic and kinetic GIS topology maintenance.

21.1.1 *Data structures and algorithms*

The recent developments of the major commercial GISs have shown that the use of data structures and algorithms seems to be crucial for the implementation of advanced functionalities in commercial GIS. For example, the most recent "on the fly" computations implemented in the recent releases of major commercial GIS are heavily relying on the underlying data structures (ESRI 2003). Other examples include the topology maintenance on the fly, map object history, LOD, terrain feature models, etc (ESRI 2005). The underlying data structures and algorithms used in current commercial GIS were originally developed by researchers from the fields of geomatics, computer science and computational geometry (Gold 1992, De Floriani et al. 2004, van Oosterom et al. 2002) and their early prototypes demonstrated the usefulness and the efficiency of dynamic data structures and algorithms for GIS.

21.2 DYNAMIC GIS

The word dynamic can have many meanings in GIS. The simplest use of dynamic might be the case of dynamically generated maps, for example on a web site or a GPS-based navigation system. Here the issue is to receive the request and compile the map from standard objects and layers as quickly as possible. A relatively easy extension of this issue is to add the locations of moving vehicles, for example, as points on this map, e.g. for fleet monitoring and delivery services. (Note that there is often no interaction of these vehicles with the map itself—their locations are merely plotted. However, for GPS-based navigation there is often some form of road-following algorithm involved.)

A somewhat more advanced approach assigns ranges of temporal existence to these entities and permits queries about snapshots in time (Marchand et al. 2005, Langran 1992). Essentially, this is a database approach[1], where the objects retrieved by the temporal query are displayed on the map.

Where the map objects themselves change over time, traditional GIS is frequently used to produce snapshots of attribute values at some particular date, usually for polygon maps and raster images (Albrecht 2007). Various map-matching techniques are then used to examine the differences. Note that this is usually done for field-type data, where attribute values exist throughout the map area. In the case of polygon maps, the boundaries usually remain unchanged between snapshots—if this is not the case, then the problem becomes more difficult. Dynamic fields, such as temperature, are usually managed as sequences of snapshots, using raster images. These images can be compared using traditional image analysis techniques. Typical questions are concerned with the values, changes and history of attributes at specific locations.

When we consider discrete map objects as opposed to fields, the appropriate techniques are related to database management methods, tracking the existence of individual objects over time, or perhaps the co-existence of two or more. If an object's status changes, then an event has taken place, which may also be treated as a database entity.

When objects are elastic, their boundaries may change, as for clouds or forest fires, and the usual approach is to process them as fields, e.g. using image analysis. However, in this case the object itself is not the primary entity, but becomes an attribute attached to particular pixels. The same situation holds if the object as a whole moves—as in the case of clouds. Definitions here can become difficult if clouds split or merge: how are individual objects defined? What happens if there are internal variations, such as fire temperature? Treating the whole map as a field or image is often the most practical, but the object definition is lost.

In this chapter we consider that the primary problems occur with questions of spatial object proximity, both static and dynamic, where objects exist as defined entities with their spatial location as a fundamental part of their definition—they lose their meaning if they have no position. However, as we will show, spatial proximity or adjacency is not easy to define, or to maintain under object movement.

21.2.1 *Objects and fields*

GIS has traditionally separated the mapping of objects, and of fields, and has used separate data structures for each. In addition, as shown in the previous discussion, it is not always clear when we are talking about objects, and when about fields. Perhaps the simplest distinction is based on what may be the most fundamental GIS query: "What is here?" when pointing at some map location. For fields there is always an answer (perhaps temperature or population density) within the boundary of the map. For objects the answer may be "nothing"; e.g. for a map of houses the answer will be "House" or "Not House". Thus for fields, with complete map coverage, usually by pixels, polygons or triangles, there is some implicit or explicit definition of adjacency where two cells meet at a common boundary.

This is not present for discrete unconnected objects, creating difficulties when mapping of moving objects is desired. While no interaction between a moving object and its neighbours may be needed for simple applications, any attempt at real-world simulation needs to maintain proximity relationships between them in order to prevent collision. It should be noted that this applies for dynamic (insertion and deletion) situations as well as kinetic ones involving point movement. Another fundamental GIS operation—interpolation—is of this type: a query point is considered to have been inserted at the desired location; its set of neighbours obtained; and an attribute (e.g. elevation) estimate calculated from this set. In all these dynamic or kinetic cases, a set of neighbouring objects is required, based on spatial proximity.

Many techniques have been developed to obtain this set, often based on metric distance (Gold 1992). However, many inconsistencies have been noted where pure distance has been used (Jones et al. 1995). A more consistent approach has been developed

[1] See the paper: C.M. Gold. An event-driven approach to spatio-temporal mapping. *Geomatica* 50: 415–424, 1996 for a discussion of change of what—time, space, attribute?

more recently (Jones & Ware 1998) based on a field spatial model—an object set is converted to a field model by calculating a proximal map: each object has an associated polygon representing the portion of the map closer to that object than to any other. This partitions the map into polygons and thus adjacency relationships can be defined based on the common boundaries. Thus the "What is here?" query is transformed to "What is closest to here?" As a query of a map-covering field, an answer is always obtainable—and the adjacency data structure is identical to that for other field models. (Rasters are a special case where the N-E-S-W relationships between adjacent cells need not be preserved explicitly.) These proximal maps—called Thiessen polygons, Dirichlet cells or Voronoi Diagrams—give a spatial model providing a unifying concept between objects and fields as well as adjacency information—or connectivity information, or topology as it is frequently called in GIS. The closest topological notion from adjacency is connectedness. We will call this adjacency information topology in the remainder of this chapter.

21.2.2 *Geometry and topology*

It is traditional in GIS to talk about geometry (coordinate information) and topology (connectivity information) separately when talking about vector data consisting of points, lines and polygons. More complex objects (e.g. building outlines) are composed of these elements, with attributes attached. The same approach may be used for space-covering polygon tessellations, where boundaries are shared by adjacent polygons, etc.

This requires the description of the relationships between several classes of entities, which may be cartographic objects (in that attributes may be associated with them) or else merely entities linking these objects (perhaps a triangle edge in a TIN). A variety of tessellation data structures have been proposed over the years, and perhaps the most effective are edge data structures where the edges provide the linking between points, as well as the relationships with the polygons on each side (Guibas & Stolfi 1985, Weiler 1985).

Where we are referring to discrete objects (e.g. houses along roads), these structures are inadequate, as they apply only to connected graphs—hence the traditional difficulties with islands within polygons. In order to take advantage of the connectivity properties of a graph, our map needs to be always connected—usually, but not necessarily, as a tessellation. There are at least three ways to achieve this: with a triangulation, with a Voronoi diagram, or with the dual graph of a polygon set.

The most common example of primal and dual graphs is the relationship between a Voronoi diagram (see Aurenhammer 1991) and its dual Delaunay triangulation, where each triangle edge is perpendicular to a Voronoi edge. However, this dual graph (usually a triangulation) is equally appropriate for expressing the relationships between adjacent arbitrary polygon sets, e.g. for administrative districts (see Gold 1994 and Yang & Gold 1993).

21.2.3 *Map objects*

A simple triangulation (TIN) used for terrain modelling (Okabe et al. 1992) is often used to specify the spatial relationships between discrete data points: here the attribute information (the coordinates) is associated primarily with the points, although an attribute such as colour may be associated with the triangular polygons. Only occasionally is attribute information associated with the edges. Thus if we wish to separate our attributes (including coordinates) from our topology, an edge data structure is an advantage.

If we are attempting to tessellate more complex discrete objects (Shewchuk 1996), such as houses, we run into difficulties as the building boundaries are clearly part of the cartographic object while exterior triangle edges serve merely to give the relationships to adjacent buildings. This becomes particularly acute when building edges are sufficiently long that a standard Delaunay triangulation of the corner points fails to preserve the building edge, and a manually-inserted constrained triangulation edge is required (Jones et al. 1995). This is clearly different from the rest of the edges, and must have a label (attribute) attached. This breaks down our separation between objects (having attributes, including coordinates) and our pure topology (adjacency information).

A second approach involves the construction of a Voronoi diagram of all the cartographic objects (which are themselves composed of cartographic fragments such as line segments and points). Each cartographic object is composed of several cartographic fragments. The internal relationships between these, as well as the external relationships between separate objects, are expressed by the dual triangulation. Thus, the Delaunay triangulation contains the topology, completely separated from the cartographic objects and their descriptions.

The difficulties with this approach have been that the algorithms (Roos 1990) for the construction of the line-segment Voronoi diagram are extremely complex (Sugihara et al. 2000) and sensitive to the limitations of computer floating-point arithmetic (Shewchuk 1997). Held (2001) stated that it took him ten years to achieve this for a static algorithm. Gold et al. (1995) constructed a line-segment Voronoi diagram of unconnected line segments. Imai (1996) constructed an incremental algorithm for line segments and closed polygons, where a maximum of two line segments meet at a vertex. Karavelas (2004) described a robust incremental algorithm (without deletion). Yang & Gold (1995) developed an incremental line-segment Voronoi diagram algorithm, but it suffered from a lack of arithmetic robustness. Gold & Dakowicz (2006) built a dynamic line-segment Voronoi diagram algorithm, where points and line segments may be inserted, intersected and deleted—which is particularly useful for a dynamic GIS. The method was based on maintaining the adjacency relationships of the Voronoi cells as the pen object was moved within the map—a kinetic algorithm. The difficulties associated with this approach are primarily concerned with preserving the correct order of tangent points around the Delaunay circumcircles at locations where line segments meet. A new iterative algorithm was developed (Anton & Gold 1997, Anton et al. 1998) that converged on the correct solution from an initial condition while preserving the necessary order of the generator locations around the circumcircle. However, no proof of robustness was obtained.

In order to obtain the desired separation of cartography and topology, these line segments needed to be distinguished from the triangle edges. This was achieved by creating line-segment Voronoi generators, in addition to the point generators, again by moving the pen (Gold et al. 1995). In this case, however, the result was not an arbitrary triangle edge but a line-segment object, and both point and line objects were connected by the triangulation topology.

21.2.4 *Dynamic algorithms and data structures for GIS*

The number of dynamic algorithms for maintaining GIS data structures is limited, because they are usually based on a line-intersection spatial model, rather than a tessellation. The primary tessellation model that can be used in GIS is the Constrained Delaunay triangulation (Jones et al. 1995), which is usually static, in that all vertices are added first to give a simple Delaunay triangulation (Chew 1989), and then constrained edges are added to give enforced boundaries (Ware et al. 1998). Deletion of these edges is not addressed, and even point deletion in the simple Delaunay triangulation is only recently described (Devillers 1999, Mostafavi et al. 2003). As previously mentioned, there is no satisfactory separation of cartography and topology.

Our own approach to resolving these conflicting requirements was first of all to insist on a tessellation model, as this preserved the topological requirement of the ordering of edges around nodes. The Voronoi diagram/Delaunay triangulation approach is fairly obvious as it gives a stable tessellation (and thus topology) from arbitrary vertex locations. Incremental insertion and deletion algorithms were used to give a dynamic solution (Gold 1992).

Since line segments are also required to define cartographic features, and not just points, a kinetic model was developed, which allowed the Voronoi diagram/Delaunay triangulation to be maintained as the pen point moved through the tessellation: a line is the locus of a moving point. While the dynamic model could be developed with only the CCW and InCircle geometric predicates used in Computational Geometry (Guibas & Stolfi 1985, and Shewchuk 1997), giving a guaranteed robustness, the kinetic model requires additional tests to handle near-degenerate cases, especially point collision.

An alternative model, the kinetic Constrained Delaunay triangulation, was also constructed in this fashion. As with the line-segment Voronoi diagram, any constrained edge/line segment could be deleted by reversing the movement of the pen. Thus, building boundaries could be inserted and deleted as specific, labelled triangle edges, although complete separation of objects and topology is not achieved—individual triangle edges may be either.

In either the constrained Delaunay triangulation or the line-segment Voronoi diagram, all the update operations used have their inverses, as point movement may expand or contract the trailing line (Mioc et al. 1999). Preserving the topological relationships during construction means that potential collisions may be detected in advance, and the appropriate join operations implemented. This is simplified as the lines and their proximal regions are embedded in the two-dimensional space, guaranteeing that, for example, one Voronoi diagram line segment may detect an imminent collision and form the appropriate junction that preserves the correct node and region ordering around the junction point.

21.3 QUAD-EDGE BASED VORONOI DATA STRUCTURE

The algorithm used to construct the Voronoi vertices has been described in Guibas & Stolfi (1985). The boundaries between the regions of this tessellation form a net (the Voronoi diagram), whose dual graph (the Delaunay triangulation) stores the spatial adjacency (topology) relationships among objects. Within such a dynamic Voronoi spatial data structure, such as developed by Gold (1990), the map objects (points and/or line segments) are stored as nodes of the dual spatial adjacency (topology) graph: the Delaunay triangulation. The underlying data structure used is the Quad-Edge data structure (Guibas & Stolfi 1985).

The Quad-Edge data structure is used for computing the line Voronoi diagram (Gold & Dakowicz 2006), which is the basis of the dynamic Voronoi data structure for points and line segments. The Quad-Edge data structure was introduced by Guibas & Stolfi (1985) as a primitive topological structure for the representation of any subdivision on a two-dimensional

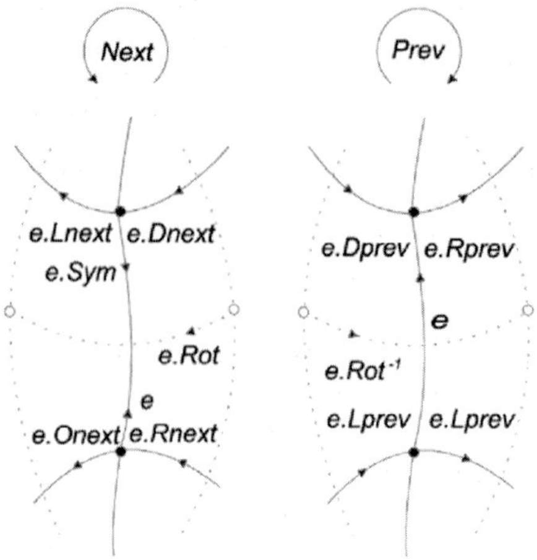

Figure 21.1. The edge functions (adapted from Guibas & Stolfi 1985).

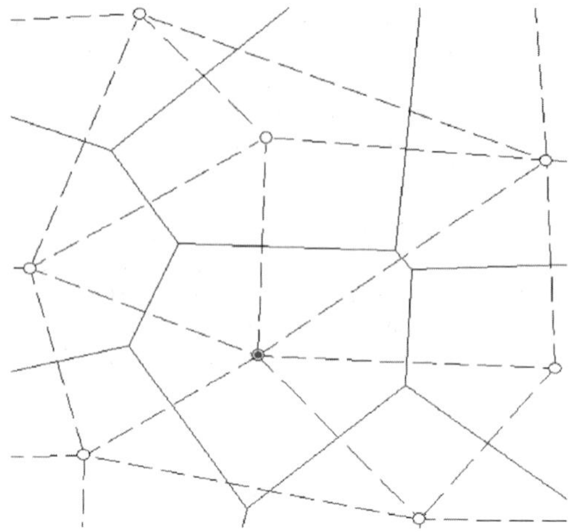

Figure 21.2a. A simple Voronoi diagram.

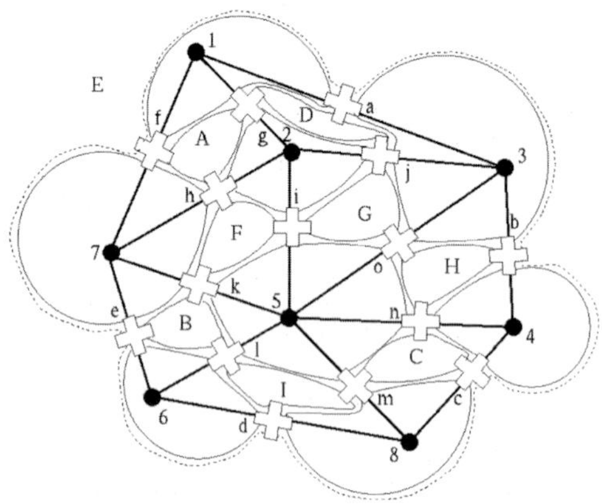

Figure 21.2b. The quad-edge data structure corresponding to the Voronoi diagram.

manifold. The Quad-Edge data structure is the implementation of an edge algebra (Guibas & Stolfi 1985), which is the mathematical structure that defines the topology of any pair of dual subdivisions on a two-dimensional manifold. In the context of the application of the Quad-Edge data structure to the computation of Voronoi diagrams, both a primal planar graph (the Voronoi diagram) and its dual graph (the Delaunay triangulation) are stored in the Quad-Edge data structure—see Guibas & Stolfi (1985).

Guibas & Stolfi (1985) developed a convenient mathematical structure for representing the topological relationships among edges of a pair of dual subdivisions on a two-dimensional manifold[2]. A subdivision of a manifold M is a partition S of M into three finite collections of disjoint parts, the vertices (denoted by *VS*), the edges (denoted by *ES*) and the faces (denoted by *FS*) with the following properties:

- every vertex is a point of M,
- every edge is a line of M,
- every face is a disk of M,
- the boundary of every face is a closed path of edges and vertices.

A directed edge of a subdivision *P* is an edge of *P* together with a direction along it (see page 80 in Guibas & Stolfi 1985). Since directions and orientations can be chosen independently, for every edge of a subdivision there are four directed, oriented edges (Guibas & Stolfi 1985). For any oriented directed edge *e* we can define unambiguously its vertex of origin *e Org*, its destination, *e Dest*, its left face, *e Left*, and its right face, *e Right*. The flipped version *e Flip* of an edge *e* is the same un-oriented edge taken with opposite orientation and same direction. The symmetric of *e*, *e Sym* corresponds to the same undirected edge with the opposite direction but the same orientation as *e*.

Edge functions (see Figure 21.1) allow the traversal of the pair of dual subdivisions. The next edge with the same origin, *e Onext* is defined as the one immediately following *e* (counterclockwise) in the ring of edges out of the origin of *e* (see Fig. 21.1). The next counterclockwise edge with the same left face, denoted by *e Lnext*, is defined as the first edge we

[2] A two-dimensional manifold is a topological space with the property that every point has an open neighbourhood which is a disk.

encounter after *e* when moving along the boundary of the face *F* = *e Left* in the counterclockwise sense as determined by the orientation of *F*.

Figure 21.2a shows a Voronoi diagram (plain lines) and its dual Delaunay triangulation (dashed lines). Figure 21.2b shows the corresponding Quad-Edge data structure, where (a,b,c,...) are Quad-Edges, (1,2,3,...) are Delaunay vertices, and (A,B,C,...) are Voronoi regions. As shown in Figures 21.2a and 21.2b, each branch of the Quad-Edge is part of a loop around a Delaunay vertex/Voronoi face, or around a Delaunay triangle/Voronoi vertex (see Fig. 21.2b).

21.3.1 *Delaunay/Voronoi quad edge equivalence*

Two subdivisions *S* and *S*' are said to be the *dual* (Guibas & Stolfi 1985) of each other if for every directed and oriented edge *e* of either subdivision there is another edge *e Dual* (that is defined as the dual of *e* and isolated by parenthesis in the following expressions) of the other subdivision such that:

- the dual of *e Dual* is *e*: (*e Dual*) *Dual* = *e*,
- the dual of the symmetric of *e* is the symmetric of *e Dual*: *e Sym Dual* = (*e Dual*) *Sym*,
- the dual of the flipped version of *e* is the symmetric of the flipped version of *e Dual*: *e Flip Dual* = (*e Dual*) *Flip Sym*,
- moving counter-clockwise around the left face of *e* in one subdivision is the same as moving clockwise around the origin of *e Dual* in the other subdivision:
- *eLnext Dual* = *(e Dual) Onext*$^{-1}$.

The dual of an edge *e* is the edge of the dual subdivision that goes from the (vertex corresponding to the) left face of *e* to the (vertex corresponding to the) right face of *e* but taken with orientation opposite to that of *e*. The definition of the dual of an edge allows us to define the operation *Rot*: the rotated version of an edge *e* is the dual of *e* directed from *e Right* to *e Left* and oriented so that moving counterclockwise around the right face of *e* corresponds to moving counterclockwise around the origin of *e Rot*. More concisely, *eRot* = *eDualFlipSym* = *eFlipDual*.

21.3.2 *Edge algebra*

Edge algebra (Guibas & Stolfi 1985) is the mathematical structure used for representing simultaneously a pair of dual subdivisions, (in our use of the Quad-Edge data structure, the Delaunay triangulation and the Voronoi Diagram). It captures all the topological properties of a subdivision (Guibas & Stolfi 1985). The topology of the subdivision is completely determined by its edge algebra, and vice versa. All the edge functions can be expressed using three basic primitives, *Flip*, *Rot*, and *Onext* described above (Guibas & Stolfi 1985). An edge algebra is (Guibas & Stolfi 1985), an abstract algebra (*E*, *E**, *Onext*, *Flip*, *Rot*) where *E* and *E** are arbitrary finite sets (of edges), and *Onext*, *Rot*, and *Flip* are mappings on *E* and *E**.

The Quad-Edge traversal operations are based on the edge algebra (*E*, *E**, *Onext*, *Flip*, *Rot*), and their expression as composition of the basic primitives (Guibas & Stolfi 1985).

21.3.3 *Basic topological operations in the quad-edge data structure*

The main advantage of the Quad-Edge data structure is that all the construction and modification of planar graphs can be done using two basic topological operators (see Table 21.1 and Fig. 21.3), and the

Table 21.1. Basic quad-edge topological operators.

Operation	Description
e: = MakeEdge[]	Creates an edge e to a newly created data structure representing an empty manifold
Splice[a,b]	Joins or separates the two edge rings a Org and b Org, and independently, the two dual edge rings a Left and b Left

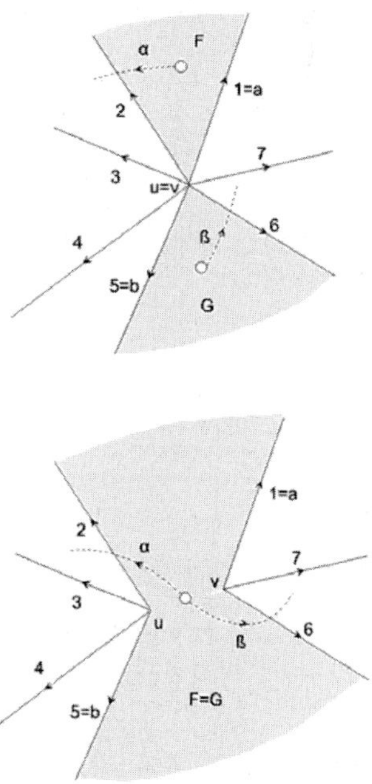

Figure 21.3. The Splice topological operator.

complex topological operations built from these two basic topological operators. These complex topological operations are presented in the Table 21.2.

The Quad-Edge topological operators modify the graph locally (Mioc et al. 2007).

21.3.4 *The atomic actions on the dynamic voronoi data structure*

The map state changes are produced by map commands (Gold 1992) composed of atomic actions (Mioc et al. 1998). Each atomic action in the map command executes the geometric algorithm for addition, deletion or change of map objects and corresponding Voronoi cells.

The atomic actions are:

- The Split action (see Fig. 21.4) inserts a new point (see Fig. 21.5a and 21.5b) into the structure by splitting the nearest point from the pointed location into two points.
- The Merge action (see Fig. 21.4) deletes the selected point by merging it with its nearest neighbour.
- The Switch action is performed when a point moves (Guibas et al. 1991) and a topological event occurs (i.e. the moving point enters or exits a circle circumscribed to a Delaunay triangle, switching[3] the common boundary of two adjacent triangles (see Fig. 21.6).
- The Link action (see Fig. 21.9) adds a line segment[4] between the points obtained after a Split action. The Link action must occur after a Split action, and adds a line segment (see Fig. 21.10) between the point selected for splitting and the newly created point.

Table 21.2. Complex quad-edge topological operators.

Operation	Description
e: = Connect[a,b]	Adds a new edge e connecting the destination of a to the origin of b, in such a way that a Left = e Left = b Left
DeleteEdge[e]	Disconnects the edge e from the rest of the data structure
Swap[e]	Rectifies e in order to respect the empty circumcircle criterion

[3] The *Switch* action will be used in the construction of the *Move* action. The *Move* (topological event) action (see Fig. 21.7) moves the selected point from its current position to a new position or until the next topological event (see Fig. 21.8).

[4] A line segment is composed of two half-line segments, whose Voronoi regions are on each side of the line segment, having the line segment as a common boundary (see Fig. 21.10).

- The Unlink action (see Fig. 21.9) removes the selected line segment. The Unlink action must occur before a Merge action, and removes the line segment between the selected point and its nearest object.

These actions compose the set of atomic actions of the dynamic spatial Voronoi data structure (Mioc et al. 2006). We can see on Figure 21.10 that with the addition of the new line the Voronoi diagram is modified only locally.

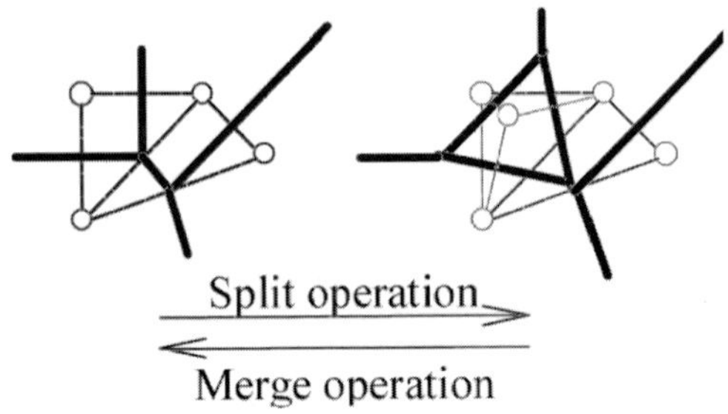

Figure 21.4. The topological change due to the Split and Merge operations.

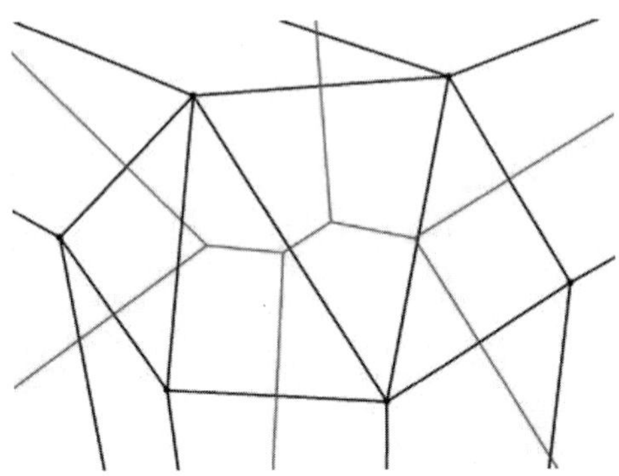

Figure 21.5a. Voronoi diagram before a new point is added.

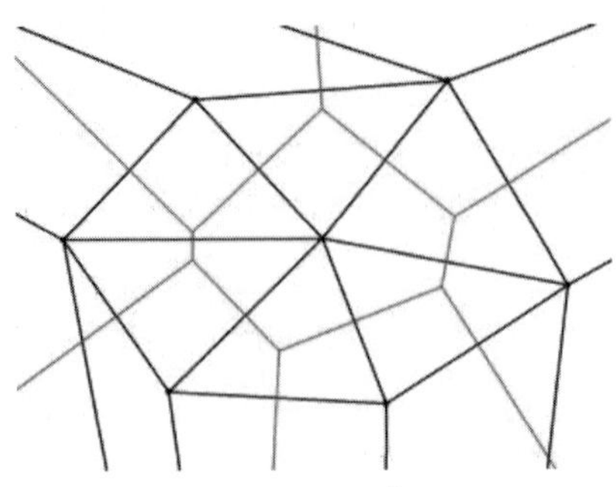

Figure 21.5b. Point insertion in the Voronoi diagram.

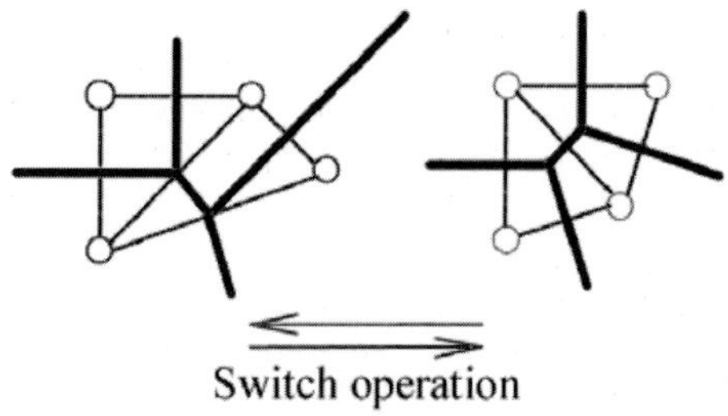

Figure 21.6. The topological change due to the switch operation.

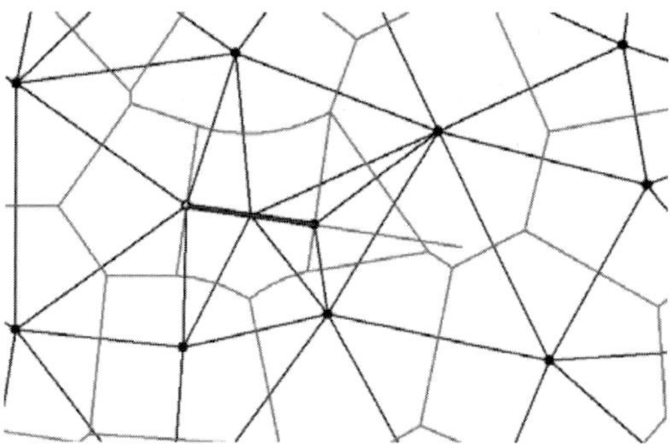

Figure 21.7. Moving Point in the dynamic Voronoi diagram.

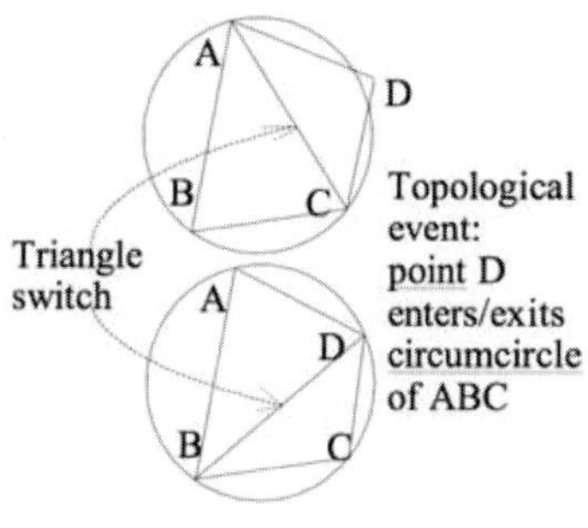

Figure 21.8. Triangle switch.

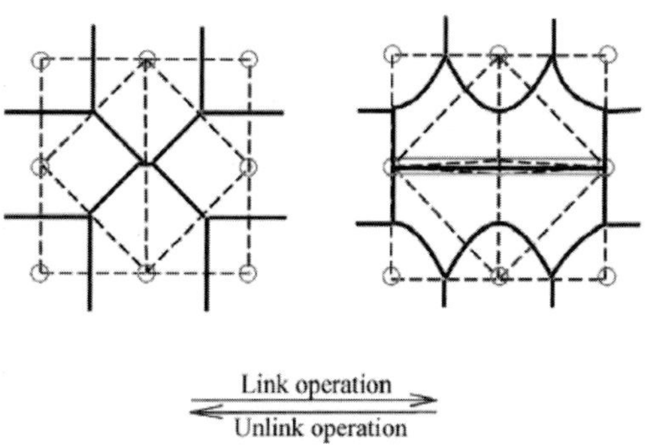

Figure 21.9. The topological change due to the link and unlink operations.

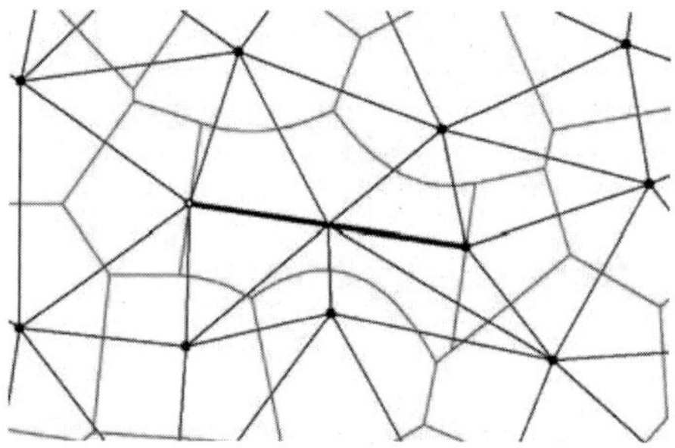

Figure 21.10. Line insertion in the Voronoi diagram.

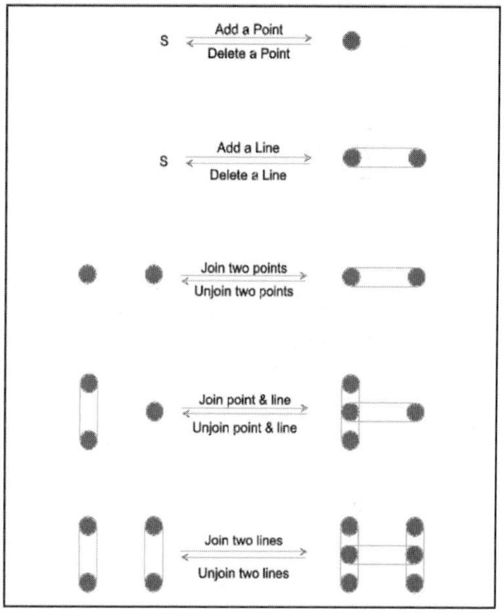

Figure 21.11. Map updates in the Voronoi diagram.

21.3.5 *The dynamic map updates*

The atomic actions are the basis upon which dynamic map updates (see Fig. 21.11) have been built.

All the map update commands (Gold 1992) of this dynamic Voronoi data structure are complex operations composed of atomic actions (Mioc et al. 2006). The composition of atomic actions into map commands is provided by syntactic rules. In the next sections we will show the dynamic GIS applications developed using Quad-edge based Voronoi data structure.

21.4 HIERARCHICAL STRUCTURES

The techniques described so far are graph-theoretical methods, which are simple to store in computer main memory, but which may be difficult to preserve when their size exceeds the available space. In this case, graph-partitioning methods are needed to save spatially-adjacent portions of the network on particular disc pages. Gold & Angel (2006) described a hierarchical method for Delaunay triangulations that appears to handle that problem. With the hierarchical method, a simple walk through a triangulation provides a reasonable point location algorithm, e.g. Gold et al. (1977). If we are dealing with very large maps the algorithm can be speeded up by having one or more levels of a Delaunay hierarchy, where higher levels may consist of subsets of the points in the lower levels. Having located the nearest point in a higher level, a downward link leads to the equivalent point in the lower level, and the process continues (Devillers 2002). The Voronoi equivalent view of this method gives higher level cells that contain the point generators of lower level cells, generating a spatial geographic hierarchy following Christaller (1966). We can also decide that some level may be designated as a disc page, and duplicate elements of an edge data structure (e.g. the Quad-Edge of Guibas & Stolfi 1985) may be used to permit navigation across page faults (Gold & Angel 2006). The graph structure can be maintained dynamically. This approach may reduce the need for special spatial database structures to manage large graphs.

21.5 APPLICATIONS

Applications of dynamically-maintained spatial data structures include interactive terrain modelling, interactive map construction (both for polygon maps and discrete objects), integrated structures for modelling fields as well as objects (see Fig. 21.12, 21.13a and 21.13b), and simulation of object movement within the map space (Gold & Dakowicz 2006).

The original concept used here is that, as a curve (line) is the locus of a moving point, then segments are drawn by maintaining the topology of a single moving point as it moves through the topological network (visualized as either the Voronoi diagram or Delaunay triangulation).

With the dynamic Voronoi diagrams a new type of GIS system for maritime navigation safety is developed. The system takes advantage of the properties of the kinetic Voronoi diagram, and uses the Quad-Edge data structure for the real-time maintenance of the spatial relationships of ships and other navigational objects (see Fig. 21.14). The locations of ships are updated using a standard onboard AIS transponder, and moving-points Voronoi algorithms. The spatial relationships are used for collision detection and avoidance (Goralski & Gold 2007). The system is

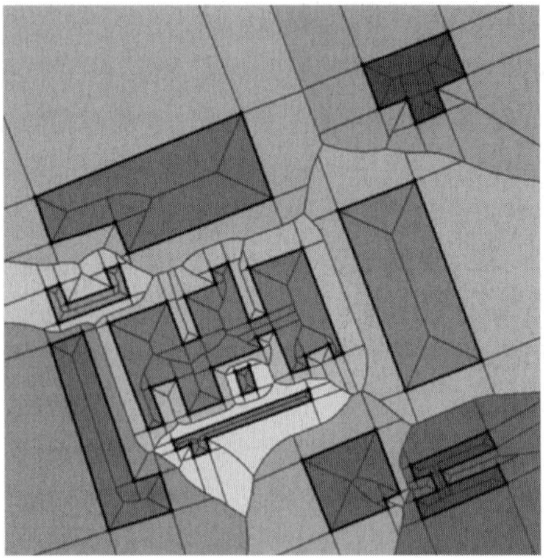

Figure 21.12. Line segment Voronoi diagram. (see colour plate page 502)

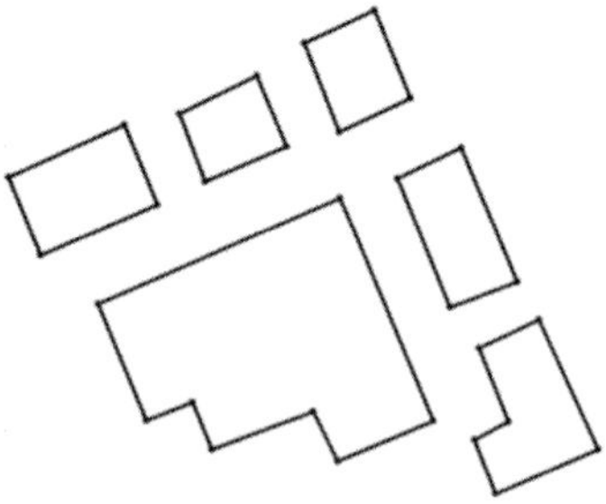

Figure 21.13a. The map of buildings to digitize.

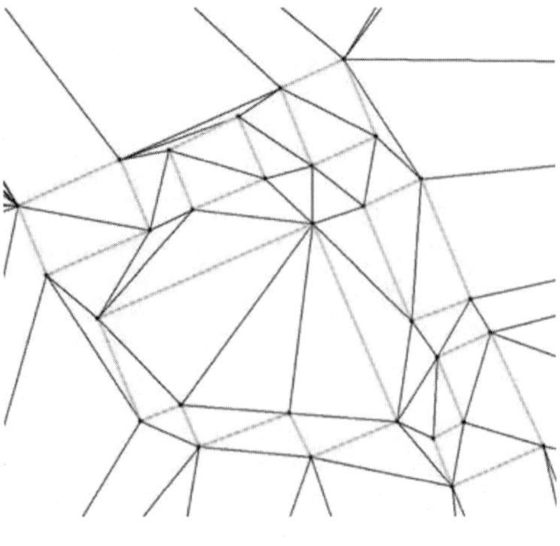

Figure 21.13b. Constrained Voronoi diagram.

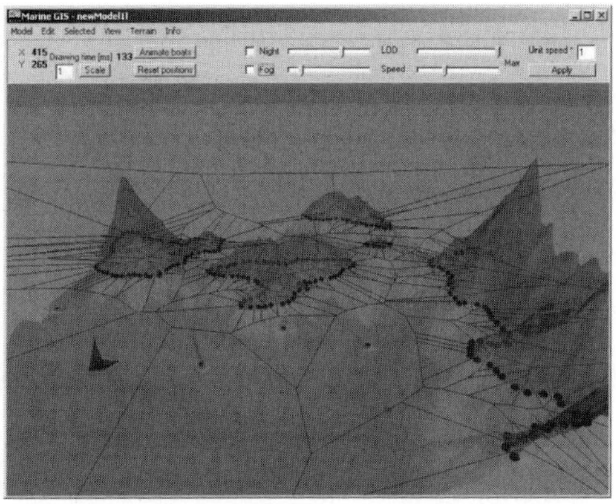

Figure 21.14. Dynamic Voronoi diagrams for marine GIS. (see colour plate page 502)

aimed at tackling the main cause of marine accidents (human errors) by providing navigational aid and decision support to navigators.

21.6 CONCLUSIONS

In this paper we started with a summary of recent algorithmic work in the development of a dynamic GIS. The main topics of interest have been moving point location within the map, 2-D topological map updating, and 3-D map object construction, including topology. We have then focused on one approach to local 2-D map topology maintenance, and attempted to show the importance of the dynamic and even kinetic algorithms and data structures needed for the implementation of a dynamic GIS.

Our approach is based on the dynamic maintenance of a tessellated spatial model—the Voronoi diagram for line segments and its corresponding Delaunay triangulation. The line Voronoi diagram has definite advantages for dynamic GIS applications, and facilitates a kinetic structure for map updating and simulation. One of the advantages is that the moving-point Delaunay triangulation/Voronoi diagram model approximates human thinking (Mioc et al. 2006), and manages collision detection, snapping and intersection at the data input stage by maintaining a topology based on a complete tessellation. The use of the Constrained Delaunay triangulation in GIS allows the simulation of edges, and not just points, with only minor changes to the moving-point model, but at the cost of confusing map objects and topological entities. We have shown that the line-segment Voronoi diagram provides a better approach for modelling the spatial relationships of compound map objects built from points and line segments than the Constrained Delaunay triangulation.

Until recently it has been difficult to develop a robust implementation for the line-segment Voronoi diagram. We believe that this method is now viable for dynamic GIS, and in many cases it could replace the Constrained Delaunay triangulation. However, whichever method is used, the concept of using the moving point as a pen, permitting interactive navigation within the map under construction, together with the ability to delete and add line segments as desired in the construction and updating process, appears to be a very useful approach.

Resolution of these issues opens up a variety of elegant solutions to dynamic GIS: incremental and interactive updating of the cartographic product with only local changes to the topology and hence to the properties just described. In addition, as described in Gold & Yang (1993) and Mioc et al. (1999), all updates may be time-stamped, and hence the map may be updated or rolled back to any desired date.

ACKNOWLEDGEMENTS

We would like to acknowledge the financial support of the EU Marie-Curie Chair in GIS at the University of Glamorgan.

REFERENCES

Albrecht, J. 2007. Dynamic GIS. In J. Wilson & S. Fotheringham (eds.), *The Handbook of Geographic Information Science*: 436–446. London: Blackwell.

Anton, F. & Gold, C.M. 1997. An iterative algorithm for the determination of Voronoi vertices in polygonal and non–polygonal domains. *Proceedings of the 9th Canadian Conference on Computational Geometry (CCCG'97), Kingston, ON, Canada*: 257–262.

Anton, F., Snoeyink, J. & Gold, C.M. 1998. An iterative algorithm for the determination of Voronoi vertices in polygonal and non–polygonal domains on the plane and the sphere. *Proceedings, 14th European Workshop on Computational Geometry (CG'98), Spain*: 33–35.

Aurenhammer, F. 1991. Voronoi Diagrams—A Survey of a Fundamental Geometric Data Structure. *ACM Computing Surveys* 23(3): 345–405.

Boguslawski, P. & Gold, C.M. 2007. Augmented Quad-edge—3D Data Structure for Modelling of Building Interiors. *Proceedings, 5th ISPRS Workshop on Dynamic and Multi-dimensional GIS, Urumchi, China, 28–29 August 2007*: 51–54.

Chen, J., Li, Z.L. & Jiang, J. 2001. Dynamic and Multi-dimensional GIS: an Overview. *Third ISPRS Workshop on Dynamic and Multi-dimensional GIS, Bangkok*: 52–58.

Chen, J., Zhao, X.S. & Li, Z.L. 2003. An algorithm for the generation of Voronoi diagrams on the sphere based on QTM. *Photogramm. Eng. Remote Sens.* 69(1): 79–89.

Chen, J., Zhou, X. & Li, Z. 2007. Event-driven Incremental Updating of Cadastral Spatial Database Considering Topological Integrity. *Proceedings, 5th ISPRS Workshop on Dynamic and Multi-dimensional GIS, Urumchi, China, 28–29 August 2007*: 185–195.

Chew, P. 1989. Constrained Delaunay Triangulations. *Algorithmica* 4: 97–108.

Christaller, W. 1966. (translation). *Central Places in Southern Germany*. Prentice-Hall, NJ.

Claramunt C., Jiang, B. & Bargiela, A. 2000. A new framework for the integration, analysis and visualisation of urban traffic data within geographic information systems. *Transportation Research Part C: Emerging Technologies* 8, 1–6: 167–184.

Dakowicz, M. & Gold, C.M. 2005. Interactive TIN modification with a cutting tool. *Proceedings, 4th ISPRS Workshop on Dynamic and Multi-dimensional GIS, Pontypridd, Wales, UK*: 5–9.

Dakowicz, M. & Gold, C.M. 2007. Finite Difference Method Runoff Modelling Using Voronoi Cells. *Proceedings, 5th ISPRS Workshop on Dynamic and Multi-dimensional GIS, Urumchi, China, 28–29 August 2007*: 55–60.

De Floriani, L., Magillo, P., Puppo, E. & Sobrero, D. 2004. A Multi-Resolution Topological Representation for Non-manifold Meshes. *Computer-Aided Design Journal* 36(2): 141–159.

Devillers, O. 2002. The Delaunay Hierarchy. *Int. J. Foundations of Computer Science* 13, 2: 163–180.

Devillers, O. 1999. On deletion in Delaunay triangulations. *The 15th Annual ACM Symposium on Computational Geometry*: 181–188.

ESRI. 2003. ArcGIS: Working with Geodatabase Topology. *ESRI White Paper*: 3–4.

ESRI. 2005. GIS Topology. *ESRI White Paper*: 3–10.

Ganovelli, F., Cignoni, P., Montani, C. & Scopigno, R. 2000. Multiresolution model for soft objects supporting interactive cuts and lacerations. *Computer Graphics Forum* 19, 3: C-271–C-281.

Gold, C.M. 1990. Spatial Data Structures—The Extension from One to Two Dimensions. In L.F. Pau (ed.), *Mapping and Spatial Modelling for Navigation, NATO ASI Series F, No. 65*: 11–39. Berlin: Springer-Verlag.

Gold, C.M. 1992. An object-based dynamic spatial data model, and its applications in the development of a user-friendly digitizing system. *Proceedings of the Fifth International Symposium on Spatial Data Handling, Charleston*: 495–504.

Gold, C.M. 1994 Three approaches to automated topology, and how computational geometry helps. *Proceedings of the Sixth International Seminar on Spatial Data Handling, Edinburgh, Scotland*: 145–158.

Gold, C.M. (Ed.), 2005. *Proceedings 4th ISPRS Workshop on Dynamic and Multi-dimensional GIS, Pontypridd, Wales, UK*.

Gold, C.M. 2006. What is GIS and What is Not? *Transactions in GIS* 10(4): 505–519.

Gold, C.M. 2007. Dynamic Cartography Using Voronoi/Delaunay Methods. *Proceedings 5th ISPRS Workshop on Dynamic and Multi-dimensional GIS, Urumchi, China, 28–29 August 2007*: 41–47.

Gold, C.M. & Angel, P. 2006. Voronoi Hierarchies. *Proceedings, GIScience, Munster, Germany, LNCS 4197*: 99–111.

Gold, C.M., Charters, T.D. & Ramsden, J. 1977. Automated contour mapping using triangular element data structures and an interpolant over each triangular domain. *Proceedings: Sigraph '7. Computer Graphics* Vol. 11: 170–175.

Gold, C.M. & Dakowicz, M. 2006. Kinetic Voronoi—Delaunay drawing tools. *Proceedings, 3rd. International Symposium on Voronoi Diagrams in Science and Engineering, Banff, Canada, July 2006*: 76–84.

Gold, C.M., Remmele, P.R. & Roos, T. 1995. Voronoi Diagrams of line segments made easy. *Proceedings, 7th Canadian Conference on Computational Geometry, Quebec, QC, Canada*: 223–228.

Gold, C.M. & Yang, W. 1993. The design of an urban GIS to manage frequent spatial updates. In D. Du, J. Chen, B. Forester & X. Shi (eds), *Proceedings AUSIA'93 Advances in Urban Spatial Information and Analysis, Wuhan, China*: 1–7.

Goodchild, M.F. 2006. GIScience Ten Years After Ground Truth. *Transactions in GIS* 10(5): 687–692.

Goralski, I.R. & Gold, C.M. 2007. Maintaining the Spatial Relationships of Marine Vessels Using the Kinetic Voronoi Diagram. *Proceedings ISVD 2007, Glamorgan, UK, July 9–11 2007*: 84–90.

Goralski, I.R. & Gold, C.M. 2007. The development of a Dynamic GIS for Maritime Navigation Safety. *Proceedings 5th ISPRS Workshop on Dynamic and Multi-dimensional GIS, Urumchi, China, 28–29 August 2007*: 47–50.

Grum, E. 2005. The least risk path in 3D: the 'Use case' of indoor navigation. *Proceedings 4th ISPRS Workshop on Dynamic and Multi-dimensional GIS, Pontypridd, Wales, UK*: 47–51.

Guibas, L., Mitchell, J.S.B. & Roos, T. 1991. Voronoi Diagrams of moving points in the plane. Proceedings, 17th International Workshop on Graph Theoretic Concepts in Computer Science, Fischbachau, Germany. *Lecture Notes in Computer Science* 70: 113–125. Berlin: Springer-Verlag.

Guibas, L. & Stolfi, J. 1985. Primitives for the manipulation of general subdivisions and the computation of Voronoi Diagrams. *Transactions on Graphics* 4: 74–123.

Hardy P. 2000. Laser-Scan Active Objects and Dynamic Topology for Spatial Data Re-Engineering and Rich Data Modelling. *Dagstuhl Seminar 011910, May 2000*.

Held, M. 2001. VRONI: an engineering approach to the reliable and efficient computation of Voronoi Diagrams of points and line segments. *Computational Geometry, Theory and Application* 18(2): 95–123.

Imai, T. 1996. A topology oriented algorithm for the Voronoi Diagram of polygons. *Proceedings, 8th Canadian Conference on Computational Geometry, Carleton University Press, Ottawa, Canada*: 107–112.

Jiang, J. & Zhao, R. (eds.). 2007. *Proceedings, 5th ISPRS Workshop on Dynamic and Multi-dimensional GIS, Urumchi, China, August 2007*.

Jones, C.B., Bundy, G.L. & Ware, J.M. 1995. Map generalization with a triangulated data structure. *Cartography and Geographic Information Systems* 22(4): 317–331.

Jones, C.B. & Ware, J.M. 1998. Proximity Search with a Triangulated Spatial Model. *Computer Journal* 41(2): 71–83.

Karavelas, M.I. 2004. A robust and efficient implementation for the segment Voronoi Diagram. *International Symposium on Voronoi Diagrams in Science and Engineering 2004*: 51–62.

Karimipour, F., Delavar, M.R., Frank, A.U. & Rezayan, H. 2005. Point in polygon analysis for moving objects. *Proceedings, 4th ISPRS Workshop on Dynamic and Multi-dimensional GIS, Pontypridd, Wales, UK*: 68–72.

Langran, G. 1992. *Time in Geographic Information Systems*. Taylor and Francis.

Ledoux, H. & Gold, C.M. 2005. Interpolation as a tool for the modelling of three-dimensional geoscientific datasets. *Proceedings, 4th ISPRS Workshop on Dynamic and Multi-dimensional GIS, Pontypridd, Wales, UK:* 79–84.

Li, J., Jun Chen, J., Zhao R., Ma, L. & Chen, H. 2007. Calculating Topological Spatial Relations Based on Region Algebra: The Thinking in Adjacency Relationship. *Proceedings, 5th ISPRS Workshop on Dynamic and Multi-dimensional GIS, Urumchi, China, 28—29 August 2007*: 67–79.

Marchand, P., Bédard, Y. & Brisebois, A. 2005. Considerations for the Optimization of Analysis and Implementation of Spatio-Temporal Exploration and Analysis. *International Society for Photogrammetry and Remote Sensing (ISPRS) Workshop WG II/5, II/6, IV/1 and IV/2 Joint Workshop on Spatial, Temporal and Multi-Dimensional Data Modelling and Analysis, 5–8 September, Cardiff, United Kingdom.*

Mioc, D., Anton, F., Gold, C.M. & Moulin, B. 1998. Spatio-temporal change representation and map updates in a dynamic Voronoi data structure. *Proceedings of the Eight International Symposium on Spatial Data Handling, Vancouver, Canada*: 441–452.

Mioc, D., Anton, F., Gold, C.M. & Moulin, B. 1999. "Time travel" Visualization in a Dynamic Voronoi Data Structure. *Cartography and Geographic Information Science* 26: 99–108.

Mioc, D., Anton F., Gold, C.M. & Moulin, B. 2006. Map updates in a dynamic Voronoi data structure. *Proceedings, ISVD 2006, Banff, Alberta, CANADA, July 2–5 2006*: 264–269.

Mioc, D., Anton, F., Gold, C.M. & Moulin, B. 2007. Reversibility of the Quad-Edge operations in the Voronoi data structure. *Proceedings, ISVD 2007, Glamorgan, UK, July 9–11 2007.*

Mostafavi, M.A. 2005. Nearest Neighbour queries for mobile objects in a global dynamic environment. *Proceedings 4th ISPRS Workshop on Dynamic and Multi-dimensional GIS, Pontypridd, Wales, UK*: 128–133.

Mostafavi, M., Gold, C.M. & Dakowicz, M. 2003. Dynamic Voronoi/ Delaunay Methods and Applications. *Computers and Geosciences* 29(4): 523–530.

Nadi, S. & Delavar, M.R. 1995. Spatio-Temporal Modeling of Dynamic Phenomena in GIS. *Computers, Environment and Urban Systems* 19, 2: 107–115.

Okabe, A., Boots, B. & Sugihara, K. 1992. *Spatial Tessellations—Concepts and Applications of Voronoi Diagrams.* Chichester: John Wiley and Sons.

Oosterom, P. van. 2006. Constraints in spatial data models in a dynamic context. In J. Drummond, R. Billen, E. Joao & D. Forrest (eds.), *Dynamic and mobile GIS: investigating changes in space and time, (Innovations in GIS, 10):* 104–137. London: Taylor & Francis, CRC Press.

Oosterom, P. van, Stoter, J., Quak, W. & Zlatanova, S. 2002. The balance between geometry and topology. In D. Richardson & P. van Oosterom (eds.), *Advances in Spatial Data Handling*: 209–224. Berlin: Springer-Verlag.

Pan, M.Y., Li, Y.F. & Zhao, D.P. 2001. Application of VRML in a Dynamic and Multi-dimensional Digital Harbour. *Third ISPRS Workshop on Dynamic and Multi-dimensional GIS, Bangkok*: 224–226.

Pang, M.Y.C. & Shi, W. 2002. Development of a Process-Based Model for Dynamic Interaction in Spatio-Temporal GIS. *GeoInformatica* 6, 4: 323–344.

Pebesma, E.J., Karssenberg, D. & de Jong, K. 2000. The stochastic dimension in a dynamic GIS. In: J.G. Bethlehem & P.G.M. van der Heijden (eds.), *Compstat 2000, Proceedings in Computational Statistics*: 379–384. Heidelberg: Physica-Verlag.

Peuquet, D.J. 2001. Making Space for Time: Issues in Space-Time Representation. *Geoinformatica* 5(1): 11–32.

Qian, J., Zhou, Q. & Hou, Q. 2007. Comparison of Pixel-based and Object-oriented Classification Methods for Extracting Built-up Areas in Aridzone. *Proceedings, 5th ISPRS Workshop on Dynamic and Multi-dimensional GIS, Urumchi, China, 28–29 August 2007*: 163–171.

Raper, J. 2000. *Multidimensional geographic information science*. London: Taylor and Francis.

Roos, T. 1990. Voronoi Diagrams over Dynamic Scenes. *Proceedings, Second Canadian Conference on Computational Geometry, Ottawa*: 209–213.

Shewchuk, J.R. 1996. Triangle: Engineering a 2D Quality Mesh Generator and Delaunay Triangulator. *First Workshop on Applied Computational Geometry (Philadelphia, Pennsylvania)*: 124–133. Association for Computing Machinery.

Shewchuk, J.R. 1997. Adaptive Precision Floating-Point Arithmetic and Fast Robust Geometric Predicates. *Discrete and Computational Geometry* 18(3): 305–363.

Shi, W. & Pang, M.Y.C. 2000. Development of Voronoi-based cellular automata—an integrated dynamic model for Geographical Information Systems. *International Journal of Geographical Information Science* 14, 5: 455–474.

Stefanakis, E. 2003. Modelling the History of Semi-structured Geographical Entities. *International Journal of Geographic Information Science* 17(6): 517–546.

Sugihara, K., Iri, M., Inagaki, H. & Imai, T. 2000. Topology-oriented implementation—an approach to robust geometric algorithms. *Algorithmica* 27(1): 5–20.

Tse, R.O.C., Dakowicz, M., Gold C.M. & Kidner, D. 2005. Building reconstruction using LIDAR data. *Proceedings, 4th ISPRS Workshop on Dynamic and Multi-dimensional GIS, Pontypridd, Wales, UK*: 156–161.

Vasconcelos de, Maria J.P., Gonçalves, A., Catry, F.X., Paúl, J.U. & Barros, F. 2002. A working prototype of a dynamic geographical information system. *International Journal of Geographical Information Science* 16(1): 69–91.

Ware, J.M. & Jones, C.B. 1998. Conflict Reduction in Map Generalization Using Iterative Improvement. *GeoInformatica* 2(4): 383–407.

Weiler, K. 1985. Edge-based data structures for solid modeling in a curved surface environment. *IEEE Comput. Graph. Appl.* 5(1): 21–40.

Wu, L. & Wang, Y. 2007. Updating Algorithms for Constraint Delaunay TIN. *Proceedings, 5th ISPRS Workshop on Dynamic and Multi-dimensional GIS, Urumchi, China, 28–29 August 2007*: 135–139.

Yang, W. & Gold, C.M. 1993. An outline of a dynamic solution to spatio-temporal windowing in an urban GIS. In D. Du, J. Chen, B. Forester & X. Shi (eds), *Proceedings, AUSIA'93, Advances in Urban Spatial Information and Analysis, Wuhan, China*: 36–44.

Yang, W. & Gold, C.M. 1995. Dynamic spatial object condensation based on the Voronoi Diagram. In J. Chen, X. Shi & W. Gao (eds), *Proceedings, Fourth International Symposium of LIESMARS'95—Towards three-dimensional, temporal and dynamic spatial data modelling and analysis, Wuhan, China*: 134–145.

Yi, S.Z., Zhong, Y., Zhou, L.Z., Chen, J. & Liu, Q.L. 2001. A Spatio-temporal Data Model for Moving Area Phenomena. *Third ISPRS Workshop on Dynamic and Multi-dimensional GIS, Bangkok*: 333–338.

Yuan, M. 2001. Representing Complex Geographic Phenomena in GIS. *Cartography and Geographic Information Science* 28(2): 83–96.

Zeng, H.Y., Tian, H.Z., Huang, M.J., Chen, S.S. & Zhong, W.K. 2007. An Updating Method of Geospatial Data Based on Traffic Flow Analysis. *Proceedings, 5th ISPRS Workshop on Dynamic and Multi-dimensional GIS, Urumchi, China, 28–29 August 2007*: 203–206.

Zlatanova, S., Rahman, A.A. & Shi, W. 2004. Topological models and frameworks for 3D spatial objects. *Journal of Computers & Geosciences* 30(4): 419–428.

Advances in Photogrammetry, Remote Sensing and Spatial Information Sciences: 2008 ISPRS Congress Book – Li, Chen & Baltsavias (eds)

CHAPTER 22

Semantic integration of heterogeneous geospatial information

Marinos Kavouras & Margarita Kokla

ABSTRACT: Semantic interoperability among heterogeneous geo-spatial sources is a key research issue for Geographic Information Science, as in various other domains. It requires the semantic integration of information created according to different conceptualizations of geographic reality and therefore entails the mapping between the underlying ontologies, thus leading to the notion of ontology integration. Despite the recent ontological advances in information science, there is great confusion in fundamental notions, such as ontologies and semantics, as well as the way these notions are involved in the integration process. This chapter presents a framework for the disambiguation of these issues and the thorough analysis of the semantic integration process. The emphasis lies in the semantic integration of ontologies. Semantic integration is analysed in three sub-processes: (a) semantic information extraction, (b) ontology comparison and reconciliation, and (c) integration. Based on the framework, an evaluation of existing integration methodologies and tools is also presented.

Keywords: Semantic interoperability, ontology comparison, integration

22.1 INTRODUCTION

Semantic interoperability among heterogeneous geospatial sources is a key research issue for Geographic Information Science (Vckovski 1998, Goodchild et al. 1999), as in various other domains. The semantic aspects of interoperability are also considered highly relevant and important to GIScience (Shepherd 1991, Bishr 1998, Frank & Raubal 1998, Gahegan 1999, Harvey 1999, Sheth 1999, Uitermark et al. 1999, Kuhn 2002, Rodríguez & Egenhofer 2003, Kokla & Kavouras 2005). However, despite the criticality of the research issue, not many solid approaches exist.

In general, interoperability may be defined as "the ability of 'systems' or 'products' to operate effectively and efficiently in conjunction, on the exchange and reuse of available resources, services, procedures and information, according to the intended use of their providers, in order to fulfil the requirements of a specific task" (Kavouras & Kokla 2008). The above definition adopts a wide conception of "systems", which may refer to software systems, organizations, infrastructures, administration services, legal frameworks, etc. The exchange and reuse of available resources "according to the intended use of their providers" presupposes the understanding and analysis of information placing emphasis on its meaning, thus leading to the notion of semantic interoperability.

Semantic interoperability is the most complicated and difficult kind of interoperability to achieve. Bishr (1998) defined six interoperability levels ranging from the lowest level of network protocol communication to the highest, most complicated semantic level. Semantic interoperability (and interoperability in general) deals with the exchange and reuse of existing information sources and therefore presupposes the integration of these sources while preserving their semantics. Semantic integration entails the resolution of heterogeneities in the conceptualization of similar or related information.

Different conceptualizations of geographic reality exist, even for fundamental geographic concepts such as rivers, forests, transportation networks, etc. For example, the concept "forest" is conceptualized differently in related scientific fields such as forestry, environmental science, cadastre, land use management, etc. These differences are shifted to the corresponding representations of the conceptualizations (e.g. nomenclatures, taxonomies, ontologies, etc.). Conceptualization differences are caused by various factors, which raise different views and partitioning of reality (Kavouras & Kokla 2008):

- Perspective/interest.
- Disciplinary training.
- Methodology.

* Granularity.
* Ethnic, cultural or social views.
* Human cognitive diversity.

For example, a discipline poses a specialized, scientific view of reality, which differs from common sense conceptualizations. In addition, the methodology employed may also set specific constraints in the way reality is conceptualized and partitioned, e.g. land cover classifications based on photo interpretation capabilities. These differences in the conceptualization of geographic reality further complicate the semantic integration process.

The remainder of the chapter is organized as follows. Section 22.2 analyses the reasons for the confusion around the semantic integration process and associated notions. An analysis of existing integration frameworks is presented in Section 22.3. Section 22.4 presents a framework for the disambiguation of the issues regarding the semantic integration of ontologies followed by an evaluation of existing integration methodologies and tools (Section 22.5). These issues are thoroughly analysed in Kavouras & Kokla (2008). Finally, Section 22.6 presents the conclusions of this work.

22.2 REASONS FOR THE CONFUSION IN SEMANTIC INTEGRATION ISSUES

Much confusion is raised by the term "semantic integration" owing to the unconformity in both its constituents: "semantics" and "integration".

Semantics is a critical research issue, which is the focal point of study of many diverse scientific fields, such as philosophy, linguistics, cognition and information science. However, despite the attention that these terms have received in literature, there is no consensus on their meaning, their constituents, their usefulness and implementation. Notions such as "concepts", "properties", "relations", "semantics", etc. acquire different meanings in different scientific fields.

This may be attributed to the different perspectives adopted by the diverse scientific fields. For example, concerning semantics and to some extent ontologies as well, there is a difference between "rich" or "deep" semantics, which aim at the univocal definition of concepts and "shallow" or "specification" semantics, which are related to database issues (e.g. the representation of the entities and their attributes in a database schema). Generally two main perspectives may be distinguished (Table 22.1).

Table 22.1. Two perspectives of semantic integration.

	"Higher" ontological perspective	"Lower" implementation perspective
Focuses on	Definition of the semantics of concepts	Information formalization and processing
Deals with	Meaning, con cepts, properties, relations	Structures, formats, encodings, representation languages
Resolves	Semantic heterogeneities	Explication heterogeneities

* A "higher" ontological perspective deals with the proper and accurate definition and semantic representation of reality. This perspective focuses on the formation of concepts, meaning, properties and relations as well as the resolution of semantic heterogeneities among different conceptualizations.
* A "lower" implementation perspective deals with information formalization and processing, focusing on design and database issues (e.g. the specification of classes and attributes that describe an application) and the resolution of heterogeneities among different specifications, i.e. structures, formats, encoding, representation languages, etc.

The above distinction seems unambiguous in the sense that the two perspectives have distinct but supplementary competence. However, notions dealt with by the first perspective are systematically used to describe notions from the second. For example, the term "properties" is often adopted to describe database attributes and the term "semantics" to describe the specification of a class based on database attributes.

Another reason for the confusion around the semantic integration process is that the term "integration" is confused with several similar processes, such as alignment, mapping, merging, unification, fusion, coordination, etc. These processes are not clearly and univocally explicated, and the terms are loosely used, thus impeding any attempt to compare and evaluate corresponding approaches. Furthermore, there is also a plethora of different perspectives to integration *per se*. "Database integration" (Parent & Spaccapietra 1998), "schema integration" (Batini et al. 1986, Rahm & Bernstein 2001), "data integration" (Lenzerini 2002), "ontology-based integration" (Wache et al. 2001), "ontology integration" (Calvanese et al. 2001), "semantic integration" (Gruninger & Kopena 2003, Doan et al. 2004).

Database integration refers to the integration of information sources, and usually also incorporates a process for associating their underlying schemata, i.e. the schematic description of the database in the form of a conceptual schema. These information sources usually conform to a certain conceptualization of reality specified by an ontology. An ontology explicates the knowledge of a domain by determining the domain concepts, their meaning, their specification and their mutual relations. Therefore, attempts to compare, combine or associate knowledge presupposes the integration of the underlying ontologies. In certain cases, the same ontology-conceptualization of reality is adopted by the underlying databases, usually called the reference ontology. In that case, the reference ontology is used to consolidate the integration process. However, the majority of information sources do not comply with a single ontology, necessitating the integration of the underlying ontologies, i.e. ontology integration. Furthermore, since ontologies describe the semantics of the domain concepts, semantic integration is also ontology-dependent. Usually, ontology integration deals with existing ontologies, in order to support information exchange and reuse. However, sometimes it is necessary to take into account ontology evolution for the case of ontologies that change over time.

22.3 ANALYSIS OF EXISTING INTEGRATION FRAMEWORKS

In order to reduce the confusion and vagueness around the integration process, several integration frameworks have been developed. They usually aim at disambiguating a specific integration process, e.g. integration on the web (Uschold & Gruninger 2002) and ontology-based integration (Wache et al. 2001) or a similar process, e.g. alignment (Bouquet et al. 2004) and combination (Klein 2001). Furthermore, existing frameworks focus on specific dimensions of the integration process. Some of them are: integration architecture (Pinto et al. 1999, Calvanese et al. 2001, Uschold & Gruninger 2002), representation languages (Wache et al. 2001), types of heterogeneities resolved (Klein 2001, Wache et al. 2001, Bouquet et al. 2004), the role of ontologies in the process, for example committing to a common reference ontology or providing mappings among multiple ontologies (Tamma & Visser 1998, Wache et al. 2001), and query mechanisms (Calvanese et al. 2001).

Although the scope of developing an integration framework is to disambiguate the process and to facilitate the evaluation of integration methodologies, existing frameworks do not agree on the terms and parameters pervading the integration process, even for fundamental issues, such as the types of heterogeneities between ontologies. Furthermore, they adopt a specific perspective and partially deal with some dimensions of the integration process. This is a reasonable approach because it is not possible to analyse every aspect of such a complicated process; instead, some specific issues may be more thoroughly analysed. Nevertheless, it seems that the "semantic" dimension of the integration process is neglected, since most of the integration frameworks focus on implementation issues and the resolution of explication heterogeneities (data structures, formats, encodings, etc.); semantic heterogeneities are usually equated with the simpler case of naming heterogeneities. However, the identification and resolution of semantic heterogeneities is a critical issue for the integration of geo-spatial ontologies, since geo-spatial concepts are not properly and fully defined by simple specification elements such as database attributes and therefore their comparison and integration is a more complicated process.

For this reason, in what follows, a framework for the semantic integration of multiple ontologies is presented, laying emphasis on issues relative to the geo-spatial domain such as the definition of semantics of geo-spatial concepts, the extraction of semantics from existing information sources, the identification and resolution of semantic heterogeneities, and the parameters that delineate the integration process. The framework is then applied to analyse and evaluate different integration methodologies and tools, both general and geo-spatial.

22.4 DEVELOPMENT OF A SEMANTIC INTEGRATION FRAMEWORK

The framework for the semantic integration of geo-spatial ontologies is based on certain assumptions regarding the nature of geo-spatial ontologies, the sources of semantics, the types of heterogeneities, the user involvement and the role of ontologies in the integration. Three sub-processes are distinguished (Fig. 22.1): (a) semantic information extraction, (b) ontology comparison and reconciliation, and (c) integration. The two processes that precede integration are usually studied independently in literature, but their collaboration is important in order to globally describe the semantic integration process. They are also necessary in order to retain the "semantic" aspect of the integration process. Semantic integration cannot be accomplished without the prerequisite steps of semantic information extraction and heterogeneity identification and resolution.

Kavouras & Kokla (2008) identify six principal characteristics for the description of the three sub-processes of semantic integration (Fig. 22.2):

- assumptions: especially regarding the source of semantics and the objective of the process
- semantic level addressed: refers to the degree to which the process deals with the semantics of the information sources and the richness of the semantic elements used

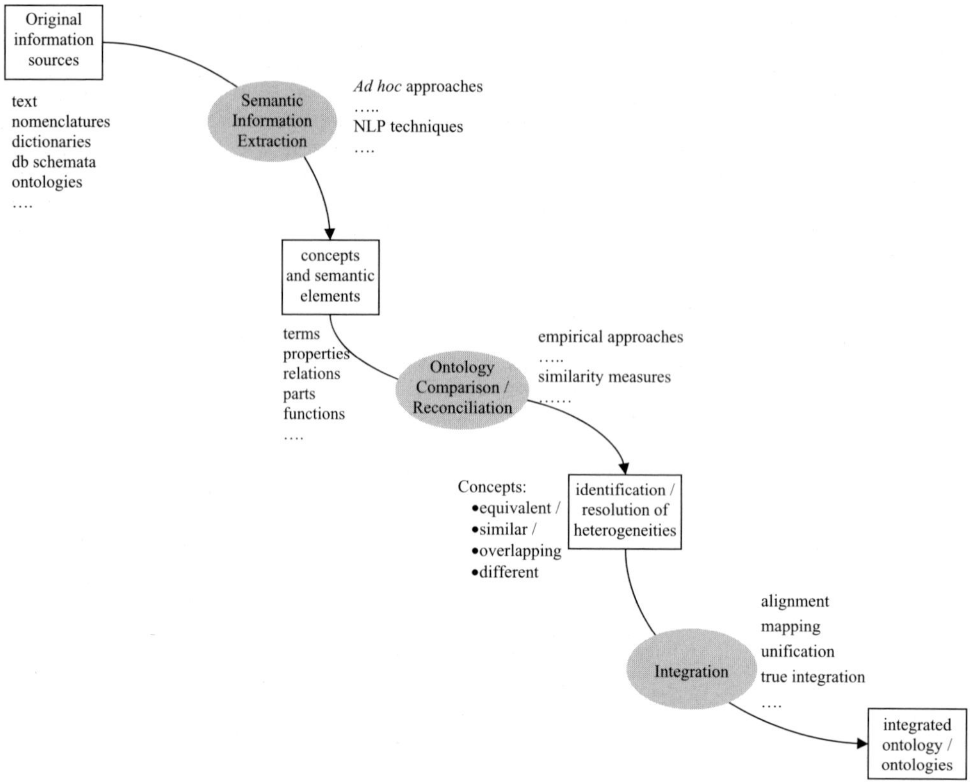

Figure 22.1. The three sub-processes of semantic integration.

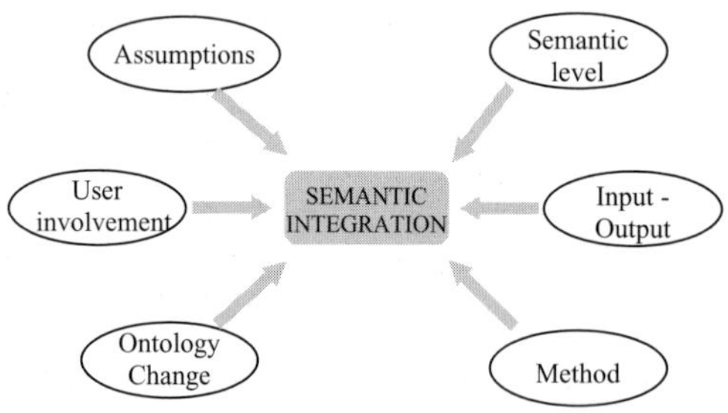

Figure 22.2. Principal characteristics for the description of semantic integration.

- input-output components: are related to the requirements and the objective of the process and the method used
- method: is a determinant of the semantic weight of the process; it is affected by the available semantic elements and other restrictions of the original information sources
- ontology change: the degree to which the process alters the original ontologies
- user involvement: the degree to which the process is determined by subjective user choices or is objectively performed.

These principal characteristics will be used in order to analyse the three sub-processes of semantic integration and subsequently evaluate different integration methodologies and tools.

The first sub-process refers to the extraction and formalization of the semantics of information from the original information sources. Generally different types of information sources may be used, such as text, nomenclatures, dictionaries, database schemata, ontologies, etc. In the geo-spatial domain, existing information sources are usually terminological ontologies; they describe (a part of) the geo-spatial domain based on concept terms and usually definitions and their mutual relations. Sometimes concept definitions are accompanied by characteristics of varying semantic importance; they can be defined either superficially (e.g. database attributes) or in-depth (e.g. semantic properties). The methods used to extract semantic information from available information sources vary in their sophistication ranging from empirical *ad hoc* approaches to more advanced Natural Language Processing (NLP) techniques, which apply specific methods to perform the process in an explicit way. Their output is usually a set of semantic elements (properties and relations) that describe the concepts of the information source.

The second sub-process concerns the identification and reconciliation of heterogeneities among the original information sources. It is based either on the output of the semantic extraction sub-process (concept terms and semantic elements), or on other available elements (e.g. concept terms, attributes, parts and subsumption relations). The output of this sub-process is the identification of similarities and heterogeneities among original concepts and preferably a suggestion for the reconciliation of heterogeneities. Methods of varying sophistication and automation may be applied for the second sub-process, as for the first one. The comparison is usually performed according to similarity measures, whereas reconciliation is a more complicated process, which can be performed either empirically or systematically. Most ontology integration tools compare concepts on the basis of their terms and subsequently ask the user to decide on the relation between two similar terms, taking into account the comparison result.

Finally, the third sub-process constitutes the core of the semantic integration process. Integration is a general term broadly used in conjunction with other similar but not explicitly defined terms, such as alignment, fusion, coordination, mapping, merging, unification, compatibility, etc. All these terms determine the way resource ontologies are interconnected to reach the final result. They vary in the degree of alteration they cause to the resource ontologies, the final result (one or many ontologies) and the use of a reference ontology that acts as a guide for the integration process. For example, mapping refers to the discovery and matching of equivalent concepts among the resource ontologies (Klein 2001). It does not induce any change to the resource ontologies, nor does it impose any reference ontology. Alignment is a mapping process among concepts of the original ontologies, which makes them compatible with each other. Unification (Sowa 2000) on the other hand enforces the original ontologies to become compatible with each other, usually through a reference ontology. The output is a single ontology whereas the original ones are altered and cannot be used any longer. True integration (Sowa 2000) is a process that results in a single ontology without, however, altering the original ones. It is preferred in the geo-spatial domain when the original ontologies will continue to be used after the integration takes place. This is a common case in the cooperation of large organizations, which integrate their ontologies for certain projects yet they do not wish to alter them definitely. In order to clarify the confusion around the different integration terms and approaches, Kavouras & Kokla (2008) identify and analyse ten principal directions pursued in existing integration approaches, such as manual *ad hoc* mappings, intuitive mappings based on "light" lexical information, extensional mappings, grounding to a single shared ontology, etc.

22.5 ANALYSIS AND EVALUATION OF EXISTING APPROACHES

The proposed semantic integration framework may be used to guide the analysis and evaluation of integration approaches in order to facilitate the selection of the appropriate approach or a combination of approaches for a specific integration task according to the needs, the available elements and the desirable result. In the following, some indicative integration approaches, both general and geo-spatial, are analysed based on the proposed framework.

PROMPT (Noy & Musen 2002, 2003) is a tool for ontology alignment and merging. The merging process generates a single ontology whereas alignment results in accordant, but separately maintained ontologies. The two processes are performed semi-automatically, i.e. the resulting heterogeneities are revealed to the user, reparative actions are recommended and user feedback is required to achieve the final result. PROMPT comprises the second and third semantic integration sub-processes, i.e. ontology comparison and reconciliation, and integration. The ontology comparison and the consecutive suggestions for the resolution of heterogeneities are based on concept terms. When concept terms are identical, it is suggested that the concepts are merged into a single one. For similar but not exactly the same concept terms, a weaker merging is suggested. The operations that are provided by PROMPT are: merging of concepts and attributes, copy of a concept with or without its super-concepts and attributes, addition or removal of a super-concept, etc.

KRAFT (Visser et al. 1998, 1999) is a project that performs all three integration sub-processes. Information extraction exploits different information sources, such as text, databases and ontologies. KRAFT uses text-based analysis to extract information from the sources, agents to translate the information to KRAFT language and ontology creation processes to generate a single ontology while preserving the mappings to the original information sources. Information extraction is based on text analysis techniques, which result in a set of concepts that describe the domain. The extracted concepts are used to build a reference ontology, which is connected to a top-level ontology. The reference ontology is defined as a concept hierarchy; concepts are described by terms, attributes and axioms. The process is performed semi-automatically, since the user is asked to enrich the ontology with extra concepts, attributes and axioms. A mapping process associates the concepts of the original ontologies with those of the reference ontology; thus the original ontologies are not altered and may continue to be used independently.

MOMIS (Bergamaschi et al. 2001, Beneventano et al. 2003, Beneventano & Bergamaschi 2004) also uses a semi-automatic method to merge original ontologies into a single ontology. MOMIS also performs all three integration sub-processes. In the first step, the system checks the original information sources, which are in the form of schemata, for similar objects. Then, clustering techniques are applied to create groups of similar objects from different information sources, which will subsequently form the merged ontology. This ontology is further enriched with semantic relations drawn from WordNet. These relations are validated by the user who may also add additional relations for concepts and their attributes.

GLUE (Doan et al. 2001, 2002) is a tool that implements the second and third sub-processes in a semi-automatic way. Concept comparison is performed using multi-criteria similarity measures in order to identify similar concepts among the original ontologies; the comparison is based on instances, concept terms, attributes, attribute values and subsumption relations. GLUE follows an ontology mapping approach to integration, since it establishes mappings among similar concepts based on a set of domain-dependent and domain-independent constraints.

Geo-spatial approaches to integration primarily deal with architectural issues based on the specification of mappings among the local schemata of the underlying Geographical Information Systems (GIS) and a common ontology. ISIS (Interoperable Spatial Information System) (Leclercq et al. 1999, Jouanot et al. 2003) implements a multi-agent architecture to provide interoperability among different GIS. An object-oriented spatial data model is used to capture the thematic and spatial information and resolve heterogeneities through the definition of a common context, called reference context. A similar architectural approach is followed by Cruz et al. (2002) who establish a shared ontology to act as a guide for the mapping of local schemata. The approach is based on the knowledge of a domain expert who undertakes the mapping of concepts between the shared ontology and the local schemata. Ram et al. (2002) proposed an architecture and a web-based prototype, called GeoCosm, for the integration of local schemata into a global schema, which is also guided by a shared ontology, the management of metadata and the generation of queries. The above architectural approaches may be applied when a "lighter" integration of different geographic information schemata is desirable. In that case, the local schemata conform to a common ontology and the mapping is performed not directly among the local schemata, but indirectly through the local ontology. These approaches do not comprise an extraction process, whereas the identification and reconciliation of heterogeneities is based solely on expert knowledge.

MIGI (Metadata Integration and Geodata Integrity) (Hakimpour & Geppert 2001, 2002, Hakimpour & Timpf 2001) is a semi-automatic approach to schema integration. The semantics of concepts are described by formal ontologies, i.e. they are defined using logical axioms. These intentional definitions are used to identify similarity relations (equality, specialization, overlapping and disjoint) among schema elements. Then, the approach uses the similarity relations to build the global schema and generates mappings between the global schema and the local schemata.

ONTOGEO (Kokla & Kavouras 2001, 2002, 2005, Kavouras & Kokla 2002) is a project that focuses on the semantic aspect of the ontology integration process. It performs semantic integration of geo-spatial ontologies, including all three sub-processes. The first sub-process extracts semantic properties and relations and their values from definitions of geo-spatial concepts using NLP techniques. The formalization of these semantic elements is used by the second sub-process to identify similarities and heterogeneities among concepts. A conceptual analysis process called Semantic Factoring is used to reconcile heterogeneities among concepts. The third sub-process performs true integration, since it generates a single ontology without altering the original ones. The integrated ontology is generated by an algorithm based on the mathematical theory Formal Concept Analysis (FCA).

Besides approaches with a focus on the integration process, there are other approaches, which deal with some peripheral yet important aspects of the integration process. An example is the similarity measure for the comparison of similar concepts among different ontologies proposed by Rodríguez & Egenhofer (2003). The model takes into account different constituents of concept representation such as terms, semantic relations (is-a and part-whole) among concepts, and distinguishing features further classified as functions, parts and attributes. This approach may be used to guide the comparison and reconciliation process when the representation of concepts comprises the above elements. However, in order for the similarity result to be useful in an integration endeavour, it should be supplemented by an evaluation process to determine the type of mapping among similar concepts.

22.6 CONCLUSIONS

The primary objective of this review is to clarify the vague scene of the integration process, placing emphasis on the semantic aspects. A framework for the semantic integration was presented in order to disambiguate notions such as semantics, ontologies and integration. The framework aims at exhibiting the complexity of the semantic integration process, its defining parameters, and the sub-processes it comprises in order to facilitate the analysis and comparison of existing as well as future approaches based on several criteria such as: (a) the available semantic elements, (b) the type of integration pursued, and (c) the desired degree of user involvement-interaction.

Furthermore, the framework assists in the selection of the appropriate approach, or a combination of approaches, for a specific integration endeavour. Since it is not possible to devise an all-inclusive approach to semantic integration for every need and application, the point is to digest the multitude of existing approaches in order to achieve methodological interoperability. For that reason, the literature review included some representative approaches, both general and geo-spatial, in order to demonstrate the usefulness of the proposed framework for the analysis and evaluation of integration approaches.

For example, if the aim is to assist an expert in mapping ontologies represented as concept hierarchies whose concepts are loosely defined by their terms, then a user-assisting tool such as PROMPT or KRAFT may be applied. For performing schema integration when the semantics of schema elements are described by formal ontologies, a tool such as MIGI may be used to merge the local schemata and derive a global schema. If the ontology concepts are represented by their definitions and a stronger type of integration with minimal user intervention is sought, then an approach such as ONTOGEO is more suitable. Other integration tasks may be dealt with by the combination of approaches; for example, the use an NLP technique for semantic information extraction, the application of a similarity measure for ontology comparison and the implementation of a tool for integration.

REFERENCES

Batini, C., Lenzerini, M. & Navathe, S.B. 1986. A comparative analysis of methodologies for database schema integration. *ACM Computing Surveys* 18(4): 323–364.

Beneventano, D., Bergamaschi, S., Guerra, F. & Vincini, M. 2003. Synthesizing an integrated ontology. *IEEE Internet Computing Magazine* 7(5): 42–51.

Beneventano, D. & Bergamaschi, S. 2004. The MOMIS methodology for integrating heterogeneous data sources. *Presented at IFIP World Computer Congress, Toulouse France, 22–27 August*, http://dbgroup.unimo.it/prototipo/paper/ifip2004.pdf (accessed 3 May 2007).

Bergamaschi, S., Castano, S., Beneventano, D. & Vincini, M. 2001. Semantic integration of heterogeneous information sources. Special Issue on Intelligent Information Integration, Data and Knowledge Engineering. *Elsevier Science* 36(1): 215–249.

Bishr, Y. 1998. Overcoming the semantic and other barriers to GIS interoperability. *International Journal of Geographical Information Science* 12(4): 299–314.

Bouquet, P., Euzenat, J., Franconi, E., Serafini, L., Stamou, G. & Tessaris, S. 2004. Specification of a common framework for characterizing alignment. *Deliverable D2.2.1 in KWEB EU-IST-2004-507482.*

Calvanese, D., De Giacomo, G. & Lenzerini, M. 2001. A framework for ontology integration. *Proc. 1st Semantic Web Working Symposium (SWWS 2001), Stanford, USA, 30–31 July 2001*: 303–316.

Cruz, I.F., Rajendran, A., Sunna, W. & Wiegand, N. 2002. Handling semantic heterogeneities using declarative agreements. *Proc. 10th ACM international symposium on Advances in geographic information systems, McLean, VA, 8–9 November 2002*: 168–174. New York: ACM Press.

Doan, A., Domingos, P. & Halevy, A. 2001. Reconciling schemas of disparate data sources: a maching-learning approach. *Proc. ACM SIGMOD Conference, Santa Barbara, California, USA, 21–24 May 2001*: 509–520.

Doan, A., Madhavan, J., Domingos, P. & Halevy, A. 2002. Learning to map between ontologies on the semantic web. *Proc. 11th International WWW Conference (WWW2002), Honolulu, Hawaii, USA, 7–11 May 2002*: 662–673.

Doan, A., Noy, N. & Halevy, A. 2004. Introduction to the special issue on semantic integration. *ACM SIGMOD Record* 33(4): 11–33.

Frank, A. & Raubal, M. 1998. Specifications for interoperability: formalizing image schemata for geographic space. *Proc. 8th International Symposium on Spatial Data Handling, SDH'98, Vancouver, Canada, 11–15 July 1998:* 331–348.

Gahegan, M. 1999. Characterizing the semantic content of geographic data, models, and systems. In M. Goodchild, M. Egenhofer, R. Fegeas & C. Kottman (eds.), *Interoperating Geographic Information Systems*: 71–84. Norwell, MA: Kluwer Academic Publishers.

Goodchild, M.F., Egenhofer, M., Fegeas, R. & Kottman, C. (eds.) 1999. *Interoperating Geographic Information Systems*. Boston, MA: Kluwer Academic Publishers.

Gruninger, M. & Kopena, J. 2003. Semantic integration through invariants. *Workshop on Semantic Integration at ISWC-2003, Sanibel Island, FL, 20 October 2003.*

Hakimpour, F. & Geppert, A. 2001. Resolving semantic heterogeneity in schema integration: an ontology based approach. *Proc. International Conference on Formal Ontology in Information Systems FOIS-2001, 17–19 October 2001*: 297–308.

Hakimpour, F. & Geppert, A. 2002. Global schema generation using formal ontologies. *Proc. 21st International Conference on Conceptual Modeling (ER2002), Tampere, Finland, 7–11 October 2002*: 307–321. Springer-Verlag.

Hakimpour, F. & Timpf, S. 2002. A step towards geodata integration using formal ontologies. *Proc. 5th AGILE Conference on Geographic Information Science, Palma de Mallorca, Spain, 25–27 April 2002: 55–60.*

Harvey, F. 1999. Designing for interoperability: overcoming semantic differences. In M.F. Goodchild, M.J. Egenhofer, R. Fegeas & C.A. Kottman (eds.), *Interoperating Geographic Information Systems*: 85–98. Boston: Kluwer Academic Publishing.

Jouanot, F., Cullot, N. & Yétongnon, K. 2003. A Solution for Contextual Integration Based on the Calculation of a Semantic Distance. *Proc. 5th International Conference on Enterprise Information Systems (ICEIS 2003), France, April 22–26*: 483–486.

Kavouras, M. & Kokla, M. 2002. A method for the formalization and integration of geographical categorizations. *International Journal of Geographical Information Science* 16(5): 439–453.

Kavouras, M. & Kokla, M. 2008. *Theories of Geographic Concepts: Ontological Approaches to Semantic Integration*. Boca Raton, FL: CRC Press.

Klein, M. 2001. Combining and relating ontologies: an analysis of problems and solutions. In A. Gomez-Perez, M. Gruninger, H. Stuckenschmidt & M. Uschold (eds.), *Proc. IJCAI'01- Workshop on Ontologies and Information Sharing, Seattle, USA, 4–5 August 2001.*

Kokla, M. & Kavouras, M. 2001. Fusion of top-level and geographic domain ontologies based on context formation and complementarity. *International Journal of Geographical Information Science* 15(7): 679–687.

Kokla, M. & Kavouras, M. 2002. Extracting latent semantic relations from definitions to disambiguate geographic ontologies. *Second International Conference on Geographic Information Science (GIScience 2002), Boulder, CO, USA, 25–28 September 2002.*

Kokla, M. & Kavouras, M. 2005. Semantic information in geo-ontologies: extraction, comparison, and reconciliation. *Journal on Data Semantics* 3534: 125–142.

Kuhn, W. 2002. Modeling the semantics of geographic categories through conceptual integration. In M. Egenhofer & D.M. Mark (eds.), *Proc. Second International Conference on Geographic Information Science (GIScience 2002), Boulder, CO, USA, 25–28 September 2002*: 108–118.

Leclercq, E., Benslimane, D. & Yétognon, K. 1999. ISIS: A semantic mediation model and an agent based architecture for GIS interoperability. *Proc. International Symposium Database Engineering and Applications (IDEAS), 2–4 August 1999*: 87–91. IEEE.

Lenzerini, M. 2002. Data integration: a theoretical perspective. *Proc. 21st ACM SIGACT SIGMOD SIGART Symposium on Principles of Database Systems (PODS 2002), Madison, Wisconsin, USA, 3–5 June 2002*: 233–246.

Noy, N.F. & Musen, M.A. 2002. PromptDiff: a fixed-point algorithm for comparing ontology versions. *Proc. Eighteenth National Conference on Artificial Intelligence (AAAI-2002), Edmonton, Alberta*: 744–750.

Noy, N.F. & Musen, M.A. 2003. The PROMPT suite: interactive tools for ontology merging and mapping. *International Journal of Human-Computer Studies* 59(6): 983–1024.

Parent, C. & Spaccapietra, S. 1998. Issues and approaches of database integration. *Communications ACM* 41(5): 166–178.

Pinto, H.S., Gomez-Perez, A. & Martins, J.P. 1999. Some issues on ontology integration. In V.R. Benjamins, B. Chandrasekaran, A. Gomez-Perez, N. Guarino & M. Uschold (eds.), *Proc. IJCAI99 Workshop on Ontologies and Problem Solving Methods: Lessons Learned and Future Trends, Stockholm, Sweden, 2 August 1999*: 7.1–7.12. Amsterdam: CEUR Publications.

Rahm, E. & Bernstein, P.A. 2001. *On matching schemas automatically, Technical Report MSR-TR-2001-17*. Redmond: Microsoft Research, Microsoft Corporation.

Ram, S., Khatri, V., Zhang, L. & Zeng, D.D. 2002. GeoCosm: A semantics-based approach for information integration of geo-spatial data. In H. Arisawa & Y. Kambayashi (eds.), *Proc. of ER 2001 Workshops, Jokohama, Japan, 27–30 November 2001*, LNCS 2465: 152–165. Berlin Heidelberg: Springer-Verlag.

Rodríguez, A. & Egenhofer, M. 2003. Determining semantic similarity among entity classes from different ontologies. *IEEE Transactions on Knowledge and Data Engineering* 15(2): 442–456.

Shepherd, I.D.H. 1991. Information integration and GIS. In D. Maguire, M. Goodchild & D. Rhind (eds.), *Geographical Information Systems: Principles and Applications*: 337–360. London: Longman.

Sheth, A. 1999. Interoperability and spatial information theory. In M. Goodchild, M. Egenhofer, R. Fegeas & C. Kottman (eds.), *Interoperating Geographic Information Systems*: 5–29. Dordrecht, the Netherlands: Kluwer.

Sowa, J.F. 2000. *Knowledge Representation: Logical, Philosophical, and Computational Foundations*. Pacific Grove: Brooks/Cole.

Tamma, V.A.M. & Visser, P.R.S. 1998. Integration of heterogeneous sources: towards a framework for comparing techniques. *Proc. the "6 Convegno dell'Associazione Italiana per l'Intelligenza Artificiale (AI*IA), Workshop su Strumenti di organizzazione ed accesso intelligente per informazioni eterogenee", Padova, Italy, 23 September* 1998: 89–93.

Uitermark, H., Oosterom, P.V., Mars, N. & Molenaar, M. 1999. Ontology-based geographic dataset integration. *Proc. Workshop on Spatio-Temporal Database Management, Edinburgh, Scotland, 10–11 September 1999*: 60–79.

Uschold, M. & Gruninger, M. 2002. Creating semantically integrated communities on the World Wide Web. *Workshop on the Semantic Web, Co-located with WWW 2002, Honolulu, HI, 7 May 2002.*

Vckovski, A. 1998. *Interoperable and Distributed Processing in GIS*. London: Taylor & Francis.

Visser, P.R.S, Jones, D., Bench-Capon, T.J.M. & Shave, M.J.R. 1998. Assessing heterogeneity by classifying ontology mismatches. *In N. Guarino (ed.), Formal Ontology in Information Systems: 148–162.* Amsterdam: IOS Press.

Visser, P.R.S, Jones, D., Bench-Capon, T.J.M., Shave, M.J.R. & Diaz, B.M. 1999. Resolving ontological heterogeneity in the KRAFT Project. *Proc. 10th International Conference on Database and Expert Systems Applications, Florence, Italy, 30 August—3 September 1999*: 668–677. London: Springer-Verlag.

Wache, H., Vogele, T., Visser, U., Stuckenschmidt, H., Schuster, G., Neumann, H. & Hubner, S. 2001. Ontology-based integration of information—A survey of existing approaches. In H. Stuckenschmidt (ed.), *Proc. IJCAI-01 Workshop: Ontologies and Information Sharing, Seattle, WA, 4–5 August 2001*: 108–117.

Advances in Photogrammetry, Remote Sensing and Spatial Information Sciences: 2008 ISPRS Congress Book – Li, Chen & Baltsavias (eds)

CHAPTER 23

3-D Data modelling and visualization

Sabry El-Hakim

ABSTRACT: This chapter intends to provide general knowledge of the processes and techniques required to create detailed photo-realistic shape representation from point clouds and to visually interact with it in a non-immersive or immersive environment. Image sequences or range sensors may be used to capture the 3-D data. The presented work comes from computer graphics, computer vision, computational geometry and photogrammetry, all of which tackle many of the same issues in search of that illusive optimum solution. Analysing and classifying methods from the huge number of multi-discipline publications is a challenging task. Thus, only a selected few are reviewed to focus on the established and practically tested approaches and to identify the persistent problems. Reading and comprehending the listed references is the key to fully grasping this subject. Only methods to create detailed geometric and appearance models are considered here; thus image-based rendering, which skips the geometric model, surveying and computer-aided design techniques that create merely sparse models and other forms of representation, such as movies, panoramas and orthophotos, will not be covered.

Keywords: 3-D data modelling, visualization, surface reconstruction

23.1 INTRODUCTION

Richly detailed models, in geometry, colour and lighting, of real world objects are essential for many interactive visualization applications ranging from manufacturing design to virtual museums and augmented environments. For these models to be useful, they must attempt to satisfy the following requirements:

1. The resolution of the rendered model should ideally match what is perceptible by the human eyes on a real visit or physical interaction.
2. For the acquired data, accuracy and noise level must be sufficiently high, otherwise all subsequent processing on this data will be negatively affected. It also has to guarantee that the accuracy of the final model matches the application specifications.
3. Data collection and modelling tools should be highly automated in all operations in order to complete a given project in a reasonable time.
4. Generate efficient sized models, without loss of desired details, so they can be viewed interactively by the application software and hardware.
5. Overall low cost.

Currently no single system or software can claim to satisfy all the requirements for all applications. The current state of the art imposes a compromise since it is usually impractical, for example, to acquire data with resolution matching the smallest detail on a surface in a realistic time, or that we can achieve high accuracy and resolution with full automation at low cost all at once. Even if we can acquire data at a very high resolution, the reconstruction process will almost always compel us to simplify the surface representation in order to handle the data and allow interactive visualization. Also, except for the simplest of objects, the collected data will undoubtedly be incomplete due to occlusions and problematic areas such as sharp edges or, for some sensors, surface material properties. Therefore, the goal of many research activities is to find solutions to these continuing issues.

There are many ways of creating a virtual representation, such as panoramas or Image-Based Rendering (IBR), but a textured 3-D geometric model that is a truthful computer representation of the real object is the most desirable as it allows unrestricted interactive visualization and manipulation. This offers the user the freedom to choose any viewpoint at a variety of lighting conditions, but the model must have dense 3-D data (high resolution meshes) on all surfaces to guarantee a realistic experience close up. Models without fine geometric details and surface irregularities will exhibit smooth flat-looking surfaces and silhouettes that are easily detected by the human eye, even with rich texture maps. Only accurate representation of all the fine geometric and colour details and the correct reflectance properties of surface material can photo-realistic rendering be achieved.

The remainder of this chapter is organized as follows. The next section introduces the main processing steps of modelling and visualization 3-D data. Since data capture is an important part of the modelling pipeline, a short analysis will be given in the third section, but without sensor characteristics or calibration, which are covered in Part II of this book. A review of geometric surface reconstruction is presented in the fourth section, followed by issues in appearance modelling for enhancing the photo-realism of the 3-D model. The sixth section focuses on the problem of how to interact in real-time with large 3-D models, while the seventh section gives a short introduction to virtual environments and augmented and mixed reality.

23.2 PROCESS OVERVIEW

The data used to generate 3-D models comes from sensors that closely capture the geometry and appearance of objects and scenes. But regardless of the sensor and the technique used for data acquisition, there are some primary steps that should be followed to create a realistic 3-D model suitable for interactive visualization (Fig. 23.1). The steps are closely interconnected and may overlap or iterate. However, there is no single system that can handle the whole pipeline; thus we have to deal with multiple tools that may or may not interact perfectly. Some operations are still in development and have no established commercial solutions, particularly for large sets of complex data. Tools that create models from scratch or modify existing ones, such as Maya® or 3dsmax®, are general purpose and highly interactive. Others that create models from captured data, such as PolyWorks® or Geomagic®, are again general purpose and assume the data has already been approximately aligned. They remain complex to use and cannot deal with huge data sets or address many visualization issues. So performing all the operations in Figure 23.1 is typically restricted to specially trained and experienced users, and can be costly and time-consuming undertakings. Note that most operations are of course dependent on the form and properties of the point data. Sparse points, for example, from feature-based matching or low resolution scanning may be difficult to model without extra information about object topology. On the other hand, densely sampled data lends itself to automated techniques without additional information besides the point coordinates. In the subsequent sections, we give more details on each of the processing steps, focusing on the latter case—reconstruction from sufficiently dense data such as that from dense image-matching or high-resolution scanning.

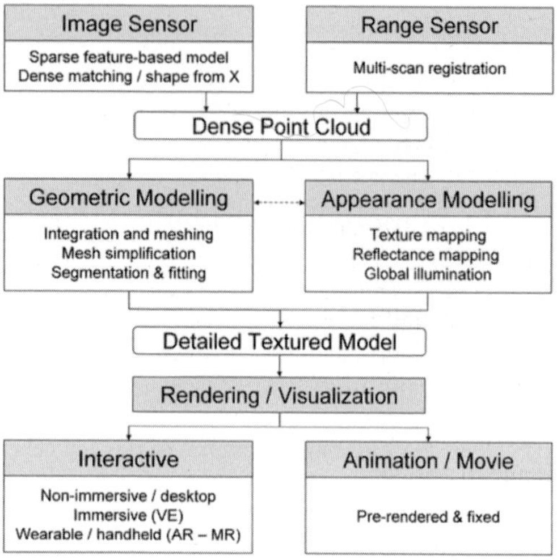

Figure 23.1. From 3-D capture to effective realistic models.

23.3 3-D SHAPE CAPTURE TECHNIQUES

An overview of image-based and range-based 3-D shape capture techniques is given. Image-based methods are highly desirable as they entail low cost acquisition devices and short on-site time; however, they may not capture all the fine geometric details and remain highly interactive. Increasing the level of detail in image-based models is a very active area of research in computer vision. Several techniques are now available for automatic camera position and orientation and 3-D point extraction (e.g. Pollefeys et al. 2004). Interactive modelling of basic shapes (Section 23.3.1) is often needed as a first step to fully automated shape reconstruction (Section 23.3.2) and adding fine details from stereo matching (Section 23.3.3) or shape from shading (Section 23.3.4). Range based, or active 3-D, can capture large numbers of 3-D points automatically but have other drawbacks such as cost, portability, scalability and flexibility (Section 23.3.5).

23.3.1 *Image-based capture of basic shapes*

The Façade system (Debevec et al. 1996) has been developed to create 3-D models of buildings from a small number of photographs. The basic geometric shape is first recovered interactively using models of polyhedral elements. In this step, the actual size of the elements and camera pose are computed assuming that camera intrinsic parameters are known. Since the assumed shapes determine camera poses and all 3-D points, the results are accurate, as the real structure elements match those assumed shapes. El-Hakim (2002) designed an approach to model regular shapes such as blocks, arches, columns and windows using a small number of interactively measured seed points; then the remaining points completing those shapes are automatically added. Some software tools are commercially available, but they are not flexible enough or automated, thus they can usually be used for only simple shapes.

23.3.2 *Fully automated shape modelling from images*

3-D reconstruction from images without any user intervention remains experimental and the robustness in various applications is not yet convincing. Available techniques rely on constraints from *a priori* assumptions about surface shapes and relationships between surfaces (e.g. Dick et al. 2004, Grossmann & Santos-Victor 2005). Without any knowledge about the object, the modelling procedure may still be automated using constraints derived recursively from the data, such as visibility constraints (Hilton 2005). Only lab tests and selected sections of some structures have been fully automatically modelled and there is yet no complete evaluation of the different approaches.

23.3.3 *Fine geometric details from stereo matching*

Brown et al. (2003) give an overview of advances in most stereo matching techniques and Scharstein & Szeliski (2002) focused on the performance evaluation of dense area-based stereo techniques. Seitz et al. (2006) extended the evaluation to multi-image matching. Feature-based methods work best with scale-invariant features extracted with specially designed image operators, such as the Scale Invariant Feature Transform (SIFT) operator (Lowe 2004). Area-based methods work directly on the pixels using one image window as a template and searching a second image for the best matching window. Larger windows give better support for the matching; on the other hand the larger the window, the larger the error from foreshortening (image neighbourhood of corresponding pixels look different in shape and size in the two images, particularly for wide baselines and large angular motions) and from assuming that all points in the window have the same disparity. Most stereo matching techniques assume Lambertian surfaces, where appearance does not change with the viewpoint, which is not the case with most non-matt surfaces. Lack of textures and occlusions are also major problems. Using a short baseline may solve many problems but it is impractical for large objects and it is highly sensitive to noise (small errors in image space results in large errors in 3-D space). Using constraints, such as the epipolar line, is a proven way of increasing the success of image matching. Constraints may also come from interactively measured seed points, assumptions about depth smoothness, surface shape from an approximate model, and using more than two images. Combining feature-based and area-based matching is another strong constraint (Gruen et al. 2004). Using sub-pixel precision is important in creating more accurate smooth surfaces compared to single-pixel accuracy that creates discrete or stepped depth maps. Research is still very active and convincing results from practical data, rather than typical lab data, are becoming available (Remondino & Zhang 2006).

23.3.4 *Shape from shading, texture and photometric stereo*

Shape From Shading or SFS (Horn & Brooks 1989, Zhang et al. 1999) can be an option to add fine details to 3-D image-based models. Although the method received a great deal of attention in computer vision research, it lacks success in determining the absolute 3-D shape of objects in actual applications due to its ill-posed formulation and its unrealistic assumptions, such as the camera looking orthogonally at a Lambertian surface and there being only one single light source located at infinity. Formulation based on viscosity solutions can reduce the ill-posed problem (Prados & Faugeras 2005). More reliable and precise results were obtained when both disparity and shading cues were used, by integrating stereo matching with SFS (Cryer et al. 1995, Fassold et al. 2004). Some techniques use SFS to provide constraints for stereo matching (Worthington 2002). A related technique, photometric stereo (Woodham 1980), uses a fixed viewing direction while the light path is varied between successive images. This solves the under-determination of SFS. Another cue about shape or surface gradient can be driven from texture (Kanatani & Chou 1989). The way the texture of known geometry, e.g. a grid of parallel lines or dots, is distorted on the surface can be used to extract its shape. Since the technique requires dense known textures on smoothly shaped surfaces, its application is limited.

23.3.5 *Shape from range sensors*

Range sensors can provide highly detailed and accurate representation of most shapes (Blais 2004).

Combined with colour information, either from the sensor itself or from a digital camera, a realistic looking model can be created. However, at least for now they are costly, usually bulky, hard to use, and influenced by surface properties. Also, as a range sensor is intended for a specific range and volume, one designed for close range is not suitable for medium or long range. Range sensors, mainly laser scanners, have been used for many years to acquire geometry for industrial parts for inspection and reverse engineering (Varady et al. 1997), city modelling (Fruh & Zakhor 2003) and heritage objects and sites (Beraldin et al. 2002). Bernardini & Rushmeier (2002) give a review of creating 3-D models based on laser scanning data. Here we focus on the steps required before surface reconstruction (Section 5).

The raw data first needs to be processed for errors and outlier removal, noise reduction and some hole filling (Weyrich et al. 2004). The next step is the aligning or the registration of multiple scans. Due to object size, shape and occlusions, it is usually necessary to use multiple scans from different locations and directions to cover every surface at the desired resolution or level of detail. The resolution here may be defined as the smallest distance between the captured points. Aligning all the scans requires significant effort and affects the final accuracy of the 3-D model. It is usually performed in two steps: (1) device-based or data-based (manual or automatic) initial global alignment followed by (2) a more precise Iterative Closest Point (ICP) technique that takes into consideration the different point accuracy (see Salvi et al. (2007) for review). Once the scans are aligned, they must be integrated to remove redundant points, which is part of the reconstruction phase. Texture acquired by a separate digital camera is registered with the geometric model using points common between the 2-D image and the 3-D model. In effect, this is finding the camera pose using the model points as control points. This must be done for every image unless the camera is fixed to the scanner; then it need only be done once.

23.4 GEOMETRIC SURFACE RECONSTRUCTION

Raw data, or the registered point clouds, need additional processing, usually called surface reconstruction, to describe the geometry of an object before it can be useful for applications. Surface reconstruction is closely related to the method of surface representation. Details of geometric modelling of free form objects and object representation methods can be found in Campbell & Flynn (2001) and Farin (1990). We can either model the visible surfaces or create solids. We will focus here on the visible surface-based techniques, which are the most widely used for applications requiring visualization since surfaces are the only visible parts.

23.4.1 *Surface representation*

A surface consists of elements (points) and the way they are connected. Non-geometrical information such as colour and physical properties may also be attached to the elements. A surface can be reconstructed by joining together a collection of small surface patches, or manifolds. The manifold, which may also be called a compact surface, is a Euclidean surface in which each edge has exactly two adjacent faces and the faces connected to a vertex form a fan of triangles. To create such manifolds, we need efficient data structures. There are many ways to design a data structure or a representation suitable for object reconstruction (Bolle & Vemuri 1991). The representation can be explicit/parametric (splines or triangular meshes) or implicit/volumetric (voxels, sweeps or Binary Spatial Partition—BSP-trees). Both representations have advantages and disadvantages; thus, for each specific geometric problem the more efficient representation should be chosen. In efficiency, volumetric methods are generally simpler and can handle noise, redundancy and spurious data better than parametric methods. But in accuracy relative to input data, parametric representations are closely or perfectly matched to the 3-D data while the volumetric ones, such as voxels, depend on the voxel size and tend to smooth small details. Since both representations may be needed at different stages of the geometric modelling, a transformation routine between the two representations is required.

Except for the simplest of objects, for the surfaces to be accurately described parametrically, for example by splines (Guo 1997), they are segmented into patches that require multi-functions joined in a smooth way. However, this creates a problem at the boundaries of the patches; thus much of the published research focuses on how to achieve continuous transitions. Triangle meshes are a valuable alternative to spline surfaces, since their simplicity allows for more flexible and highly efficient processing. Moreover, the consequent use of triangle meshes as surface representation avoids error-prone conversions, e.g. from splines to mesh, needed for visualization. Implicit or volumetric methods (Lorensen & Cline 1987, Curless & Levoy 1996) are effective either as an intermediate step in data processing or as a representation method for some applications like medical imaging. Other applications, such as object design, which reverse engineer manufactured parts, require parametric piecewise functional representation to be useful for Computer Aided Design (CAD).

23.4.2 *Surface reconstruction techniques*

The registered multiple views obtained by the techniques described in Section 3 will have overlapping regions. Therefore, the first step in surface reconstruction is to integrate the data in those regions to create a non-redundant surface mesh (Dorai et al. 1998) and then construct a manifold that accurately approximates the shape of the object. Integration is more than a simple averaging problem since the points from the different views will have different qualities depending on the view direction and the distance between the sensor and the surface. This means applying different weights to different points. Points from a scan direction close to the surface normal should have higher weight than those away from it. Points close to the boundary should have less weight than those away from the boundary. Also patches with higher sampling rates should override those with lower rates.

There are several reconstruction techniques available (for reviews, see Bolle & Vemuri (1991) and Varady et al. (1997) and for tutorials see ACM-SIGGRAPH and EUROGRAPHICS courses), but ideally the technique should be able to use all the data, handle noise, provide uncertainty measures, remove redundancy, fill holes, create efficient size models, and have no restriction on object shapes. Delaunay triangulation (Boissonnat 1984) is the basis for many techniques. However, they have limitations on the surface type and are inefficient if applied without optimization. Errors and noise in the data may result in the wrong adjacency and thus wrong shape and even a degenerate solution. Several robust techniques have been developed to deal with noisy data (Edelsbrunner & Mücke 1990, Yoon et al. 2007). The more general sculpting techniques such as alpha shapes (Edelsbrunner & Mucke 1994) are often used for their high accuracy and ability to work on non-convex surfaces.

The models may be constructed from a range image, i.e. points organized in a grid (Soucy & Laurendeau 1992, Turk & Levoy 1994) or irregular/unorganized points (Boissonnat 1984, Hoppe et al. 1992, Edelsbrunner & Mucke 1994, Mencl & Muller 1997, Ohtake et al. 2005). Construction from organized points takes advantage of the connectivity information in the range image, while meshing of unorganized points requires determining the adjacency of triangle faces and vertices. Sometimes tessellation is applied to the range data to identify areas that should not be connected by polygons using a distance threshold.

After the mesh has been reconstructed, some repairing is often needed to fill cracks and holes and fix incorrect triangles and non-manifold and degenerate surface parts (Borodin et al. 2002, Liepa 2003, Ju 2004, Bischoff et al. 2005). These errors result in visible faults, and cause lighting blemishes due to the incorrect normals. The model will also be unsuitable for reverse engineering or replicas. Another problem is due to the fact that an object is rarely sampled optimally. Some areas, such as edges and high curvature surfaces, are usually under-sampled and end up joined by a transitional surface rather than a sharp edge, while flat areas are often over-sampled. The nature of manufactured objects dictates that edges and sharp corners be accurately modelled. Dey et al. (2001) proposed a technique for automatic detection and correction of such sampling problems, while Luebke et al. (2002) survey simplification techniques needed to deal with over-sampling. Subdivision techniques (Catmull & Clark 1978, Zorin et al. 1996) are useful for under-sampled areas. The triangles in these areas are subdivided into smaller triangles with points shifted according to pre-set rules. Other methods to sharpen edges in meshes are proposed by Wang (2006) and Lai et al. (2007).

As noted above, for many reconstruction operations, segmentation is necessary to divide the surface into sets appropriate for surface fitting and modelling complex objects. This is particularly important for applications such as reverse engineering for manufacturing and CAD. Several segmentation and fitting techniques are available, including those for free form and deformable objects (Attene et al. 2006).

23.5 PHOTO-REALISM AND APPEARANCE MODELLING

The issue of high quality appearance has received less attention than geometric modelling, at least in photogrammetry. Photo-realistic rendering, defined as having no difference between the rendered model and a photograph taken from the same viewpoint, goes much further than simply draping static imagery over geometry. Proper lighting and measurement of surface reflection properties must be included. Lensch et al. (2003) give an overview of the problem of accurate measuring of such properties while Gösele et al. (2000) show how to generate high quality imaging by controlled illumination. Including non-photo realistic or artistic rendering in parts of the presentation to enhance the experience or stress a point has also been advocated by Strothotte & Schlechtweg (2002).

23.5.1 *Basic texture quality issues*

Aliasing: When textured models are viewed from a long distance, aliasing, moiré effect, jagged lines, and other visible abnormalities may result. This is because when the surface is far away, a displayed pixel on that surface may be associated with several texels from the texture map. MIP-maps (from the Latin: Multum In Parvo), a sequence of textures each of which is a progressively lower resolution image of the original (by a factor of 2), are commonly used, especially in computer games, to solve this problem (Williams 1983). Texture memory caching and bilinear or tri-linear filtering to remove aliasing from the created images are also needed. Since real-time re-sampling and filtering can be computationally expensive, MIP-maps may be pre-computed and stored. Most recent hardware supports MIP-mapping operations, but actual managing and set up of the caching scheme is up to the application.

Blurring or stretching: Surfaces appearing far away or at oblique angles in the image will be covered by a small image part. Since this part has to be stretched to cover the 3-D surface, blur and mark elongation will take place. View-dependent texture mapping, by taking sufficient images and selecting the most appropriate image depending on the location of the surface, can solve this problem (Debevec et al. 1998). In this method, the displayed view of a surface part in the scene selects the image that looks at this part from the direction closest to the desired view. In order to prevent sudden transition during manipulation, the different images must be blended or morphed together.

Low dynamic range: The range of brightness in the scene cannot be captured in a single exposure by current digital cameras. This causes loss of details in the dark areas and saturation in the bright areas if both exist in the same scene. Techniques are available to recover a great deal of the scene dynamic range (Debevec & Malik 1997). Most techniques require taking several images of the same view at various exposure times and combining them to create a single high-dynamic-range image.

23.5.2 *Texture distortions from geometric sources*

The textures on the models can be visibly affected by even small geometric errors. Weinhaus & Devich (1999) cover geometric distortions, mostly those resulting from the various transformations between image space and object space. The sources of geometric errors affecting texture quality may be grouped in three categories.

Mapping between polygon plane and image plane: The correct mapping is given by a projective transform. Since most 3-D viewers do not use this transform, distortions arise, mainly on large polygons (Fig. 23.2b). The distortion will be less perceptible if the surface is parallel to the image plane and/or the polygon size is very small. To correct this distortion, parameters of a projective transform are determined, using the 3-D coordinates of the polygon, and applied to re-sample or warp the texture within each polygon (Fig. 23.2c). The easiest texture warping method is point sampling in which the texture value nearest to the desired sample point is used. This usually results in aliasing or noise; thus we need to apply sophisticated filters such as space variant filters whose shape varies as they move across the image.

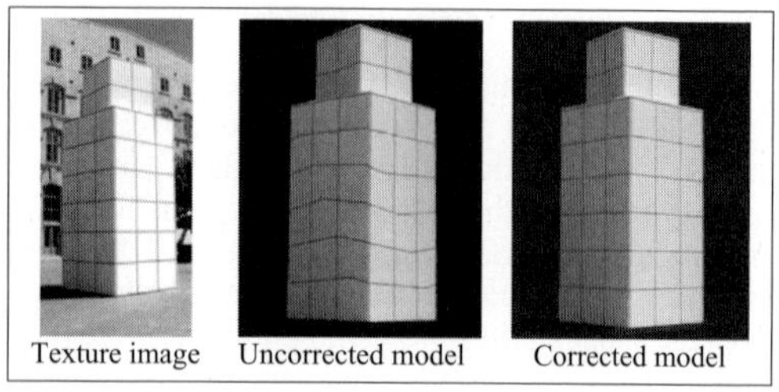

Figure 23.2. Geometric texture error and its correction.

Camera calibration and orientation: Proper calibration with lens distortion must be applied otherwise visible errors, like line discontinuity, will be visible at common edges of adjacent polygons mapped from different images. It is also important that the internal calibration parameters remain valid for all images by either keeping camera settings constant or applying self-calibration. External camera pose, or registration, with respect to the geometric model must also be accurately determined.

Surface reconstruction assumptions: Deviations of the mesh of polygons from the underlying true object surface give rise to texture errors. Since a single polygon represents a plane, applying large polygons on surfaces with even slight curvature will result in features being projected on the wrong surface or disappearing altogether, particularly near corners and surface intersections.

23.5.3 *Seamless combination of textures*

Distortions from radiometric effects are usually visible along common edges of adjacent polygons mapped from different images. Due to variations in lighting, surface specularity and camera gain settings, sensed colour and intensity for a segment shown in images taken from separate positions will not match. When these images are stitched together on the model, seams and discernible patches will become visible. The digitized brightness value for a pixel is not a measure of the radiance (the amount of light energy reflected from the surface). There is an unknown non-linear mapping between the pixel value and the radiance. This non-linearity is largest near the saturated part of the response curve. Knowing the response function allows the merging of images taken at different exposure settings, different angles, or by different cameras. The inverse of the response function is used to recover the scene radiance, up to a scale (Debevec & Malik 1997). The response function for a camera can be determined once and for all in a controlled environment, just like a lab calibration, and then applied to any image taken by that camera. When the response function is not available, seamless texture maps can be achieved by two different approaches.

Texture blending: This technique uses the weighted average of pixel values of polygons in overlapping sections between adjacent textures. Pulli et al. (1998) applied a weighting scheme with three different weights: directional weight, sampling-quality weight, and feathering weight. Directional weight favours rays originating from views whose viewing direction is close to that of the virtual camera. The sampling-quality weight reflects how well a ray samples the surface. Each pixel is assigned a weight that is defined as the cosine of the angle between local surface normal and the direction from the surface point to the camera. Feathering weight hides artefacts due to differences in lighting between images. Most blending techniques require a model with a uniform polygon or triangle size and shape and use this to select the size of the overlapped region. If the size of that region is too small, the seams between regions will not be completely removed. On the other hand, if the region is too large important details may be blurred. To alleviate this, Baumberg (2002) developed an automatic multi-band approach to preserve the high-frequency detail in the overlapping region that does not require uniform triangle size. Other varieties of texture blending can be found in Bernardini et al. (2001), Rocchini et al. (2002) and Wang et al. (2001).

Colour adaptation: El-Hakim et al. (1998) used a per-image correction (called global colour adaptation) and a per-polygon correction (termed local colour adaptation). Global colour adaptation estimates colour offsets between images. Colour differences along the border of adjacent regions of polygons are minimized, e.g. by least squares adjustment. Local colour adaptation modifies the colours of each polygon to ensure smooth transitions to all adjacent polygons. This is not straightforward since each edge of a polygon will have a different offset as it borders a different polygon. This requires an iterative solution to impose a gradual change to the colours within a polygon. A plane is fitted to all colour offsets observed at the borders to minimize these differences. The process is repeated until there are no noticeable changes. The user has to decide which is better for the case at hand: texture blending, which better reduces artefacts at adjacent polygons but tends to blur the textures or colour adaptation, which produces sharper texture maps but does not completely correct the texture differences, or both.

23.5.4 *Lighting and reflectance*

The texture image contains whatever illumination existed at imaging time. Ideally this illumination should be removed and replaced by dynamic illumination consistent with the rendering point of view (Triccoli & Allen 2008). Lighting a computer generated model entails specifying the illumination model and light position, intensity, direction and colour (Lee et al. 2004). The lighting simulation should also go beyond a traditional single-point light source and should take into account surface reflectance properties as well as secondary reflectance from other surfaces. Global illumination techniques, where all objects in the scene affect light simulation, can greatly increase the level of realism compared to a single light source (Shirley & Morley 2003, Damez et al. 2003). Also, the interaction between the incident illumination and the reflective properties of a surface should be taken into consideration since colour varies depending on how the light is reflected from the surface. The surface Bi-directional Reflectance Distribution Function (BRDF) describes how much light is reflected from a certain surface. The interaction between illumination and BRDF is usually obtained by convolution (Ramamoorthi & Hanrahan 2004). Since the incident illumination on a surface is scattered in several directions, the BRDF is a 4-D function that describes which fraction of that light is reflected in a given direction, where the direction is set with respect to the surface normal. This remains an active area of research in computer graphics and several techniques are available (e.g. Rusinkiewicz 1998, Debevec et al. 2000, Goesel et al. 2003, Kautz et al. 2003).

23.5.5 *Bump and displacement mappings*

Even when proper texturing and lighting are applied, the rendered surface may look unrealistically smooth if it has no bumps or scratches, particularly on large polygons. Several techniques have been developed to improve realism at the visualization phase at the end of the rendering pipeline rather than pre-modelling every wrinkle or small-scale geometric detail. This is efficient in memory use, takes advantage of graphics hardware, and takes into account the viewing direction. These methods increase the perceived geometrical details at runtime by adjusting pixel coordinates or normal directions, or both, to account for variation in surface bumps without increasing the complexity (polygon count) of the existing geometric model.

Bump mapping (Blinn 1978, Rushmeier 1997) introduces per-pixel small variations in surface normal directions to add roughness and a more realistic look to the lighted surface. However, since this is only a shading effect, flatness can still be seen on silhouettes. Another technique, displacement mapping (Cook 1984), uses recursive subdivision of the initial model triangles then displaces the pixel coordinates of the created vertices. Displacement is carried out along the normal of the original surface according to a displacement map that is a function of the textures. A variation of this approach, named view-dependent displacement mapping (Wang et al. 2003), considers the displacement as a ray-tracing problem and computes it as a function of the viewing angle, which allows for efficient rendering of self shadows and occlusions. However, this requires significant computation increase at runtime. Different implementations of bump and displacement mappings are included in most commercial rendering tools, and current graphic cards have some support for both.

23.6 INTERACTIVE VISUALIZATION

This section gives some details on rendering issues affecting smooth navigation through the model. The ability to interact with 3-D models is a continuing problem due to the fact that model sizes are increasing at a faster rate than computer hardware advances. The rendering algorithm should be capable of delivering images at real-time frame rates, at least 20 frames-per-second, even for very large number of polygons and with large high resolution textures. Garland (1999), Luebke et al. (2002), DeFloriani & Magillo (2002), Akenine-Moeller & Haines (2002) and Dietrich et al. (2007) cover many aspects of this subject. Here we discuss the main factors affecting the performance of a rendering system, mainly hardware limitations and model size.

23.6.1 *Hardware limitations*

Processing speed on a given hardware is affected by: the number of transformations that can be performed per second; the number of triangles that can be rendered with texture per second; the available main and texture memory; the disk and network access speed (bandwidth); and data transfer rate from main memory to graphic card memory. Some of these factors will affect only start up or loading time. For example, significant computations are needed initially for the rendering software to build an internal representation of the scene graph. Also high bandwidth is crucial to rapidly load the model and texture in memory at start up. Humphreys et al. (2001) and Strengert et al. (2004) used parallel processors or clusters to cope with large models. Lario et al. (2005) used onboard video memory to store geometry information to reduce data transfer from main memory to graphic card, which is the main bottleneck.

23.6.2 *Size of the geometric model*

Besides advanced hardware, visibility-based polygon decimation and simplification, and memory management techniques are necessary for real-time interaction with large models. Hierarchical multi-resolution Level Of Detail (LOD), where objects far from the viewpoint are rendered with pre-sampled lower resolution representations, is a standard technique that is supported by most scene graph libraries. Cohen et al. (2003) developed a low-level lightweight API system for geometric LOD that is integrated in the OpenGL library. Progressive meshes (Hoppe 1996) minimize the size of the output mesh given an allowed maximum error. Hoek & Damon (2004) review automatic model simplification and run-time LOD management techniques. They discuss LOD framework, LOD management, LOD simplification models and metrics for error evaluation. In addition to applying an LOD algorithm, we may also need to apply visibility culling. This approach skips objects that are: outside the viewing frustum; face away from the viewer or; occluded by other objects or surfaces (Cohen-Or et al. 2003). Once visibility is determined, Hidden Surface Removal (HSR) is performed with a variant of the classic Z-buffer algorithm.

Huge models (Levoy et al. 2000, Borgeat et al. 2007) containing hundreds of millions or even billions of polygons need more creative techniques to handle them. Aliaga et al. (1999) presented a system for rendering very complex 3-D models at interactive rates. It selects a subset of the model as preferred viewpoints and partitions the space into virtual cells. Each cell contains near geometry rendered using LOD and visibility culling, and far geometry rendered as a textured depth mesh. GigaWalk (Baxter et al. 2002) is a system for interactive walkthrough of huge complex environments. It combines occlusion culling and LOD and uses two graphics pipelines with one or more processors. Cignoni et al. (2004) developed the Adaptive Tetra-Puzzles system where for each frame the best-fit variable resolution LOD is selected according to the view frustum and the requested visualization accuracy. The selection is taken from a pre-constructed multi-resolution hierarchy of geometric model patches. Geo-morphing of LOD (GoLD) is a view-dependent real-time technique for multi-resolution models (Borgeat et al. 2005). It uses geo-morphing to smoothly interpolate between both geometric and texture patches composing a hierarchical LOD structure to maintain seamless continuity between adjacent patches.

23.6.3 *Size of the texture*

If the active texture does not fit into graphic card memory then swapping will occur, resulting in noticeable performance degradation. In this case, the viewing application must concoct a texture caching routine that decides on which texture to keep or move into memory and which to move to RAM or to system virtual memory (Dumont et al. 2001). It is rare that all textures are needed for any one frame; thus, to take advantage of this, two techniques are suitable: view frustum culling, which skips rendering and loading into memory parts of the model that are outside the view frustum; and MIP-mapping, which was originally developed to reduce aliasing (Section 5.1). To control the file size and the rendering speed more precisely, it is better to divide the scene into smaller uniform groups. Wei (2004) proposed a tile-based texture-mapping algorithm by which it only physically stores a small set of texture tiles instead of one large texture. Another approach is adaptive texture maps (Kraus & Ertl 2002), which are texture maps with locally adaptive resolution and an adaptive boundary of the map's domain instead of a rectilinear domain. The locally adaptive resolution permits continuous simplification, and allows efficient representation of fine details rather than increasing the resolution of the whole texture map. Adaptive boundaries considerably reduce memory costs if the texture data contains large colour-uniform regions.

23.7 VIRTUAL ENVIRONMENTS, AUGMENTED AND MIXED REALITY

Models are often interactively visualized in a non-immersive desktop environment. However, there are applications that require an immersive Virtual Environment (VE) where a special user interface is needed for interaction (Hand 1997, Dachselt & Hubner 2007). Special glasses or head-mounted gear are usually needed and the relationship between the viewer's hand and head and the model must be continuously measured. This is a different experience from desktop environments and should give a realistic feeling of being present in the environment (Draper et al. 1998). Although many advances have taken place in VE, it is still in its infancy and there is no comprehensive analysis or understanding of all the factors that contribute to a fully realistic experience of being present in the environment. Another problem is that there are very few extremely realistic and navigable 3-D models available. However, VE has currently established itself in some high-end applications such flight simulations and training. In those applications, true feeling of presence is critical; therefore the cost of such systems remains very high to provide the necessary computational power and realistic displays. In addition to our vision, other senses such as hearing, touch or haptic, and smell may also be computer controlled to enhance the realism.

Compared to a VE, which is completely non-real, Augmented Reality (AR) enhances the user's perception of and interaction with the real world through supplementing parts of the real world with 3-D virtual objects that appear to coexist in the same space as the real world (Bimber & Raskar 2005, Azuma et al. 2001). AR systems must: (1) blend virtual in a real environment, (2) be real-time interactive, and (3) flawlessly register the worlds in 3-D. Traditionally, AR displays are hand-held or eye-worn, although other forms of optical and projection technologies are emerging. In general, AR is more challenging than VE since it has to deal with the real world as well as the virtual. The accurate calibration of the optical components and the continued real-time registration between the two are the most critical issues for AR. This also means the constant tracking of both the user's and real objects' positions using some precise electromagnetic, mechanical or optical sensors or devices. Research in AR is usually focused on improving the accuracy of registration and tracking techniques.

Mixed Reality (MR) is a more general technology where the surrounding environment can be virtual or real (Freeman et al. 2005, Egges et al. 2007). The idea is to have a world where the virtual and the real are merged in any combination to best suite the application. For example, virtual characters can be integrated into a real world, which is augmented with virtual parts and where the real user is interacting with all. While VE has diverse applications, from tourism to training, AR has specialized applications such as industrial, medical and museum and heritage applications. MR is used mostly in games with some limited applications in heritage and education.

23.8 CONCLUDING REMARKS

This chapter summarized the different ways to capture, process and visualize 3-D shape information about real objects and sites. It is obvious that there is no single approach or sensor that can work effectively in all situations. Therefore, one must select the optimum sensor and methodology, or a combination of those, for a given application. This is a difficult task due to the lack of comprehensive performance analysis and comparisons traceable to specification standards such as those available for surveying and Coordinates Measuring Machines (CMM). Various criteria must be analysed, mainly geometric accuracy, percentage of erroneous measurements, level of detail captured, missed parts, visual quality of the 3-D model and the total effort necessary to create such models. We also need to evaluate the fully completed

3-D reconstruction, not just individual views. An issue to take into account is the different behaviour that 3-D shape capture systems exhibit for different application parameters, such as object size, shape and surface material, lighting conditions, angle of view and visibility constraints. Designing and performing controlled experiments with all 3-D systems under various parameters and configurations and on different objects and sites should be the subjects of further research. For now, designing a 3-D modelling and visualization project is an experimental process that often involves trade-offs between conflicting performance measures and requires experience and extensive training to select and apply the best suited 3-D system(s) to a practical application.

REFERENCES

Akenine-Moeller, T. & Haines, E. 2002. *Real-time rendering (2nd edition)*. A. K. Peters Ltd, ISBN:1568811829.

Aliaga, D., Cohen, J., Wilson, A., Baker, E., Zhang, H., Erikson, C., Hoff, K., Hudson, T., Stürzlinger, W., Bastos, R., Whitton, M., Brooks, F. & Manocha, D. 1999. MMR: An interactive massive model rendering system using geometric and image-based acceleration. *Symposium on Interactive 3D Graphics, Atlanta, USA, 26 April—2 May*: 199–206.

Attene, M., Katz, S., Mortara, M., Patane, G., Spagnuolo, M. & Tal, A. 2006. Mesh segmentation—a comparative study. *IEEE Int. Conf. Shape Modelling and Applications, Matsushima, Japan, 14–16 June*: 14–25.

Azuma, R., Baillot, Y., Behringer, R., Feiner, S. & MacIntyre, B. 2001. Recent advances in augmented reality. *Computers and Graphics* 21(6): 34–47.

Baumberg, A. 2002. Blending images for texturing 3D models. *British Machine Vision Conference (BMVC), Cardiff, UK, 2–5 Sept.*: 404–413.

Baxter III, W.V., Sud, A., Govindaraju, N.K. & Manocha, D. 2002. GigaWalk: Interactive walkthrough of complex environments. *University of North Carolina at Chapel Hill Technical Report TR02-013.*

Beraldin, J.-A., Picard, M., El-Hakim, S.F., Godin, G., Valzano, V., Bandiera & Latouche, C. 2002. Virtualizing a Byzantine crypt by combining high-resolution textures with laser scanner 3D Data. *Virtual Systems and Multimedia (VSMM'02), Gyeogju, South Korea, 25–27 Sept.*: 3–14.

Bernardini, F., Martin, I. M. & Rushmeier, H. 2001. High-quality texture reconstruction from multiple scans. *IEEE Trans. Visualization and Computer Graphics* 7(4): 318–332.

Bernardini, F. & Rushmeier, H. 2002. The 3D model acquisition pipeline. *Computer Graphics Forum* 21(2): 149–172.

Bimber, O. & Raskar, R. 2005. *Spatial Augmented Reality.* Wellesley, MA.

Bischoff, S., Pavic, D. & Kobbelt, L. 2005. Automatic restoration of polygon models. *ACM Transactions on Graphics* 24(4): 1332–1352.

Blais, F. 2004. Review of 20 years of range sensor development. *J. Electronic Imaging* 13(1): 232–240.

Blinn, J.F. 1978. Simulation of wrinkled surfaces. *ACM SIGGRAPH Computer Graphics* 12(3): 286–292.

Boissonnat, J.-D. 1984. Geometric structures for three-dimensional shape representation. *ACM Transactions on Graphics* 3(4): 266–286.

Bolle, R. & Vemuri, B.C. 1991. On three-dimensional surface reconstruction methods. *IEEE Trans. Pattern Analysis and Machine Intelligence* 13(1): 1–13.

Borgeat, L., Godin, G., Blais, F., Massicotte, P. & Lahanier, C. 2005. GoLD: Interactive display of huge coloured and textured models. *ACM Trans. On Graphics* 24(3): 869–877.

Borgeat, L., Godin, G., Massicotte, P., Poirier, G., Blais, B. & Beraldin, J.-A. 2007. Visualizing and analyzing the Mona Lisa. *IEEE Computer Graphics and Applications* 27(6): 60–68.

Borodin, P., Novotni, M. & Klein, R. 2002. Progressive gap closing for mesh repairing. *Advances in Modelling, Animation and Rendering, Bradford, UK, 1–5 July*: 201–213.

Brown, M. Z., Burschka, D. & Hager, G. D. 2003. Advances in computational stereo. *IEEE Transactions on Pattern Analysis and Machine Intelligence* 25(8): 993–1008.

Campbell, R.J. & Flynn, P.J. 2001. A survey of free-form object representation and recognition techniques. *Computer Vision and Image Understanding* 81(2): 166–210.

Catmull, E. & Clark, J. 1978. Recursively generated B-Spline surfaces on arbitrary topological meshes. *Computer Aided Design* 10(6): 350–355.

Cignoni, P., Ganovelli, F., Gobbetti, E., Marton, F., Ponchio, F. & Scopigno, R. 2004. Adaptive Tetra-Puzzles: Efficient out-of-core construction and visualisation of gigantic multi–resolution polygonal models. *ACM Trans. on Graphics* 23(3): 796–803.

Cohen-Or, D., Chrysanthou, Y., Silva, C.T. & Durand, F. 2003. A survey of visibility for walkthrough applications. *IEEE Trans. Visualization and Computer Graphics* 9(3): 412–431.

Cohen, J., Luebke, D., Duca, N. & Schubert, B. 2003. GLOD: Level of Detail for the Masses. *John Hopkins U. Tech. Report: JHU-CS–GL03-4.*

Cook, R.L. 1984. Shade trees. *ACM Computer Graphics* 18(3): 223–231.

Curless, B. & Levoy, M. 1996. A volumetric method for building complex models from range images. *SIGGRAPH '96, New Orleans, USA, 4–9 August*: 303–312.

Cryer, J.E., Tsai, P.-S. & Shah, M. 1995. Integration of shape from shading and stereo. *Pattern Recognition* 28(7): 1033–1043.

Dachselt, R. & Hubner, A. 2007. Virtual environments: three-dimensional menus: a survey and taxonomy. *Computers and Graphics* 31(1): 53–65.

Damez, C., Dmitriev, K. & Myszkowski, K. 2003. State of the art in global illumination for interactive applications and high-quality animation. *Computer Graphics Forum* 21(4): 55–77.

Debevec, P.E., Taylor, C.J. & Malik, J. 1996. Modelling and rendering architecture from photographs: A hybrid geometry and image-based approach. *SIGGRAPH'96, New Orleans, USA, 4–9 August*: 11–20.

Debevec, P.E. & Malik, J. 1997. Recovering high dynamic range radiance maps from Photographs. *SIGGRAPH'97, LA, CA, USA, 3–8 August*: 369–378.

Debevec, P.E., Yu, Y. & Borshukov, G. 1998. Efficient View-Dependent Image-Based Rendering with Projective Texture-Mapping. *9th Eurographics Rendering Workshop, Vienna, Austria, 29 June –1 July*: 105–116.

Debevec, P.E., Hawkins, T., Tchou, C., Duiker, H.-P., Sarokin, W. & Sagar, M. 2000. Acquiring the reflectance field of a human face. *SIGGRAPH'2000, New Orleans, USA, 4–9 August, 23–28 July*: 145–156.

DeFloriani, L. & Magillo, P. 2002. Multiresolution mesh representation: Models and data structures. In M. Floater, *Tutorials on Multiresolution in Geometric Modelling*: 363–418. Springer-Verlag.

Dey, T.K., Giesen, J., Goswami, S., Hudson, J., Wenger, R. & Zhao, W. 2001. Undersampling and oversampling in sample based shape modelling. *IEEE Visualization, San Diego, CA, 21–26 October*: 83–90.

Dick, A., Torr, P. & Cipolla, R. 2004. Modelling and interpretation of architecture from several images. *Int. Journal of Computer Vision* 60(2): 111–134.

Dietrich, A., Gobbetti, E. & Yoon, S.-E. 2007. Massive-model rendering techniques: a tutorial. *IEEE Computer Graphics and Applications* 27(6): 20–34.

Dorai, C., Wang, G., Jain, A.K. & Mercer, C. 1998. Registration and integration of multiple object views for 3D model construction. *IEEE Trans. Pattern Analysis and Machine Intelligence* 20(1): 83–89.

Draper, J.V., Kaber, D.B. & Usher, J.M. 1998. Telepresence. *Human Factors*, 40(3): 354–375.

Dumont, R., Pellacini, F. & Ferwerda, J.A. 2001. A perceptually-based texture caching algorithm for hardware-based rendering. *Eurographics Workshop on Rendering, London, UK, 25–27 June*: 246–256.

Edelsbrunner, H. & Mücke, E.P. 1990. Simulation of simplicity: a technique to cope with degenerate cases in geometric algorithms. *ACM Trans. on Graphics* 9(1): 66–104.

Edelsbrunner, H. & Mucke, E.P. 1994. Three-dimensional alpha shapes. *ACM Transactions on Graphics* 13(1): 43–72.

Egges, A., Papagiannakis, G. & Magnenat-Thalmann, N. 2007. Presence and interaction in mixed reality environments. *The Visual Computer* 23(5): 317–333.

El-Hakim, S.F., Brenner, C. & Roth, G. 1998. A multi-sensor approach to creating accurate virtual environments. *ISPRS Journal for Photogrammetry & Remote Sensing* 53(6): 379–391.

El-Hakim, S.F. 2002. Semi-automatic 3d reconstruction of occluded and unmarked surfaces from widely separated views. *ISPRS Commission V Symposium, Corfu, Greece, 2–6 September*: 143–148.

Farin, G. 1990. *Curves and surfaces for computer aided geometric design*. San Diego, CA: Academic Press Professional, Inc.

Fassold, H., Danzl, R., Schindler, K. & Bischof, H. 2004. Reconstruction of archaeological finds using shape from stereo and shape from shading. *9th Computer Vision Winter Workshop (CVWW), 2–6 February, Piran, Slovenia*: 21–30.

Freeman, R., Steed, A. & Zhou, B. 2005. Rapid scene modelling, registration and specification for mixed reality systems. *ACM Virtual Reality Software and Technology, Monterey, California, 7–9 November:* 147–150.

Früh, C. & Zakhor, A. 2003. Constructing 3D city models by merging aerial and ground views. *IEEE Computer Graphics and Applications* 23(6): 52– 61.

Garland, M. 1999. Multiresolution modelling: survey & future opportunities. *Eurographics '99—State of the Art Reports, Milan, Italy, 7–11 Sept.:* 111–131.

Gösele, M., Heidrich, W., Lensch, H. & Seidel, H.-P. 2000. Building a photo studio for measurement purposes. *Vision, Modeling, and Visualization, Saarbrucken, Germany, 22–24 November*: 231–238.

Grossmann, E. & Santos-Victor, J. 2005. Least-squares 3D reconstruction from one or more views and geometric clues. *Computer Vision and Image Understanding* 99: 151–174.

Gruen, A. Remondino, F. & Zhang, L. 2004. Photogrammetric reconstruction of the Great Buddha of Bamiyan. *The Photogrammetric Record* 19, 107: 177–199.

Guo, B. 1997. Surface reconstruction: from points to splines. *Computer Aided Design*, 29(4): 269–277.

Hand, C. 1997. A survey of 3-D interaction techniques. *Computer Graphics Forum* 16(5): 269–281.

Hilton, A. 2005, Scene modelling from sparse 3D data. *Image and Vision Computing* 23: 900–920.

Hoppe, H., DeRose, T., Duchamp, T., McDonald, J. & Stuetzle, W. 1992. Surface reconstruction from unorganized points. *SIGGRAPH'92, Chicago, USA, 27–31 July*: 71–78.

Hoppe, H. 1996. Progressive meshes. *SIGGRAPH'96, New Orleans, USA, 4–9 August*: 99–108.

Horn, B.K.P. & Brooks, M.J. (eds.) 1989. *Shape from Shading*. Cambridge, Mass: McGraw Hill.

Humphreys, M., Eldridge, M., Buck, I., Everett, M., Hanrahan, P. & Stoll, G. 2001. WireGL: a scalable graphics system for clusters. *SIGGRAPH'01, Los Angeles, USA, 12–17 August*: 129–140.

Ju, T. 2004. Robust repair of polygonal models. *SIGGRAPH'04, Los Angeles, USA, 8–12 August*: 888–895.

Kanatani, K., Chou, T.C. 1989. Shape from texture: general principle. *Artificial Intelligence* 38(1): 1–48.

Kautz, J., Lensch, H.P.A., Goesele, M., Lang, J. & Seidel, H.-P. 2003. Modelling the world: the virtualization pipeline. *12th Int. Conf. on Image Analysis and Processing (ICIAP' 03), Mantova, Italy, 17–19 September*: 166–175.

Kraus, M. & Ertl, T. 2002. Adaptive texture maps. *Eurographics Conf. on Graphics Hardware, Aire-la-Ville, Switzerland, July*: 7–15.

Lai, Y.-K., Zhou, Q.-Y., Hu, S.M., Wallner, J. & Pottmann, H. 2007. Robust feature classification and editing. *IEEE Trans. Visualisation and Computer Graphics* 13(1): 34–45.

Lario, R., Pajarola, R. & Tirado, F., 2005. Cached geometry manager for view-dependent LOD rendering. *13th Int. Conference in Central Europe Computer Graphics, Visualization and Computer Vision, Plzen, Czech Republic, February*: 9–16.

Lee, C.H., Hao, X. & Varshney, A. 2004. Light collages: lighting design for effective visualization. *IEEE Visualisation'2004, 10–15 October, Austin, USA*: 281–288.

Lensch, H.P.A., Kautz, J., Gösele, M., Heidrich, W. & Seidel, H.-P. 2003. Image-based reconstruction of spatial appearance and geometric detail. *ACM Transactions on Graphics* 22(2): 234–257.

Levoy, M., et al. 2000. The digital Michelangelo project: 3D scanning of large statues. *SIGGRAPH 2000, New Orleans, USA, 4–9 August*: 131–144.

Liepa, P. 2003. Filling holes in meshes. *Eurographics/ACM SIGGRAPH symposium on Geometry processing (SGP), Aachen, Germany, 23–25 June, Volume 43*: 200–207.

Lorensen, W.E. & Cline, H.E. 1987. Marching cubes: a high resolution 3D surface reconstruction algorithm. *ACM Computer Graphics* 21(4): 163–169.
Lowe, D. 2004. Distinctive image features from scale invariant keypoints. *Int. Journal of Computer Vision* 60(2): 91–110.
Luebke, D., Reddy, M., Cohne, J., Varshney, A., Watson, B. & Huebner, R. 2002. *Level of detail for 3D graphics*. Morgan-Kaufmann Publishers.
Mencl, R. & Muller, H. 1997. Interpolation and approximation of surfaces from three-dimensional scattered data points. *Scientific Visualization, Dagstuhl, Germany, 9–13 June*: 223–232.
Ohtake, Y., Belyaev, A. & Seidel, H.-P. 2005. An integrating approach to meshing scattered point data. *Eurographics/ ACM SIGGRAPH symposium on Geometry processing (SGP), Vienna, Austria, 2–4 July*: 61–69.
Pollefeys, M., Van Gool, L., Vergauwen, M., Verbiest, F., Cornelis, K., Tops, J. & Koch, R. 2004. Visual modelling with a hand-held camera. *Int. Journal of Computer Vision* 59(3), 207–232.
Prados, E. & Faugeras, O. 2005. A generic and provably convergent shape-from-shading method for orthographic and pinhole cameras. *Int. Journal of Computer Vision* 65(1–2): 97–125.
Pulli, K., Abi-Rached, H., Duchamp, T., Shapiro, L.G. & Stuetzle, W. 1998. Acquisition and visualization of coloured 3-D objects. *Int. Conference on Pattern Recognition, Brisbane, Australia, 16–20 August*: 99–108.
Ramamoorthi, R. & Hanrahan, P. 2004. A signal–processing framework for reflection. *ACM Transactions on Graphics* 23(4): 1004–1042.
Remondino, F. & Zhang, L. 2006. Surface reconstruction algorithms for detailed close-range object modelling. *ISPRS Commission III Symposium, Bonn, Germany, 20–22 September*: 117–123.
Rocchini, C., Cignoni, P., Montani, C. & Scopigno, R. 2002. Acquiring, stitching and blending diffuse appearance attributes on 3D models. *The Visual Computer* 18: 186–204.
Rushmeier, H., Taubin, G. & Gueziec, A. 1997. Applying shape from lighting variation to bump map capture. *Eurographics Rendering Workshop, Saint-Etienne, France, 16–18 June*: 35–44.
Rusinkiewicz, S. 1998. A new change of variables for efficient BRDF representation. *Eurographics Rendering Workshop, Vienna, Austria, 29 June—1 July*: 11–22.
Salvi, J, Matabosch, C., Fofi, D, Forest, J. 2007. A review of recent range image registration methods with accuracy evaluation. *Image and Vision Computing* 25(5): 578–596
Scharstein, D. & Szeliski, R. 2002. A Taxonomy and evaluation of dense two-Frame stereo correspondence Algorithms. *Int. J. Computer Vision* 47(1–3): 7–42.
Seitz, S.M., Curless, B., Diebel, J., Scharstein, D. & Szeliski, R. 2006. A comparison and evaluation of multi-view stereo reconstruction algorithm. *CVPR'06, New York, USA, 17–22 June*: 519–528.
Shirley, P. & Morley, R.K. 2003. *Realistic Ray Tracing. Second edition*. A K Peters Ltd. ISBN 1-56881-198-5.
Soucy, M. & Laurendeau, D. 1992. Multi-resolution surface modelling from multiple range views. *IEEE Computer Society Conference on Computer Vision and Pattern Recognition, Champaign, Illinois, 15–18 June*: 348–353.
Strengert, M., Magallon, M., Weiskopf, D., Gothe, S. & Ertl, T. 2004. Hierarchical visualization and compression of large volume datasets using GPU clusters. *Eurographics Symposium on Parallel Graphics and Visualization, Grenoble, Francs, 10–11 June*: 1–7.
Strothotte, T. & Schlechtweg, S. 2002. *Non-Photorealistic computer graphics: modelling, rendering and animation*. San Francisco: Morgan Kaufmann.
Troccoli, A. & Allen, P. 2008. Building illumination coherent 3D models of large-scale outdoor scenes. *Int. J. Computer Vision* 78(2–3): 261–280.
Turk, G. & Levoy, M. 1994. Zippered polygon meshes from range images. *SIGGRAPH '94, Orlando, USA, 24–29 July*: 311–318.
Varady, T., Martin, R.R. & Cox, J. 1997. Reverse engineering of geometric models—an introduction. *Computer Aided Design* 29(4): 255–268.
Wang, L., Kang, S.B., Szeliski, R. & Shum, H.-Y. 2001. Optimal texture map reconstruction from multiple views. *Computer Vision and Pattern Recognition (CVPR01), Hawaii, 11–13 December*: 347–354.
Wang, L., Wang X., Tong X., Lin S., Hu S., Guo B. & Shum, H.-Y. 2003. View-dependent displacement mapping. *SIGGRAPH'03, San Diego, USA, 27–31 July*: 334–339.
Wang, C.L. 2006, Incremental reconstruction of sharp edges on mesh surfaces. *Computer-Aided Design* 38(6):689–702.
Woodham, R.J. 1980. Photometric method for determining surface orientation from multiple images. *Optical Engineering* 19(1): 139–144.
Worthington, P.L. 2002. Novel view synthesis using needle-map correspondence. *British Machine Vision Conference (BMVC2002), Cardiff, UK, 2–5 September*: 718–727.
Wei, Li-Yi. 2004. Tile-based texture mapping on graphics hardware. *Eurographics Conf. On Graphics Hardware, Grenoble, France, 29–30 August*: 55–64.
Weinhaus, F.M. & Devich, R.N. 1999. Photogrammetric texture mapping onto planar polygons. *Graphical Models and Image Processing* 61(1): 61–83.
Weyrich, T., Pauly, M., Keiser, R., Heinzle, S., Scandella, S. and Gross, M. 2004. Post-processing of scanned 3D surface data. *Eurographics Symposium on Point-Based Graphics, Zurich, Switzerland, 2–4 June*: 85–94.
Williams, L. 1983. Pyramidal parametrics. *Computer Graphics* 17: 1–11.
Yoon, M. Lee, Y., Lee, S., Ivrissimtzis, I.P., Seidel, H.-P. 2007. Surface and normal ensembles for surface reconstruction. *Computer-Aided Design* 39(5): 408–420
Zhang, R., Tsai, P.-S., Cryer, J.E. & Shah, M. 1999. Shape from shading: A survey. *IEEE Trans. PAMI* 21(8): 690–706.
Zorin, D., P. Schröder & Sweldens, W. 1996. Interpolating subdivision for meshes with arbitrary topology. *SIGGRAPH '96, New Orleans, USA, 4–9 August*: 189–192.

Part V
Applications

Advances in Photogrammetry, Remote Sensing and Spatial Information Sciences: 2008 ISPRS Congress Book – Li, Chen & Baltsavias (eds)

CHAPTER 24

Spatial data infrastructures and clearinghouses

Costas Armenakis

ABSTRACT: The concepts and development of Spatial Data Infrastructures (SDIs) have been in continuous evolution since the middle 1990s. SDIs have been evolved from centralized repositories of geo-spatial data to Internet based virtual networks of distributed harmonized spatial databases providing web-based easy discovery, access and visualization to enormous amounts of geo-spatial data. The concepts and principles for establishing SDIs and their functionality are presented. They have been built on interoperable data, software and hardware systems, open international standards and specifications, policies and technologies and enable seamless views of heterogeneous data sources through interoperability and integration. The main components of SDIs are the geo-spatial data, the data catalogues and metadata and the clearinghouse portal offering services for accessing the data. Forms of spatial databases, such as spatial warehouses and data marts and their architectural configurations, are discussed. As SDIs evolve from spatial repositories to user-driven enterprises providing also services, applications and data processing, the role of the data mining process is presented as well as the relation between SDIs and the location based services. As such, SDIs are influenced by the technological information technology (IT) and communications developments, where geo-spatial data and information are widely used. Location-based services can rely on SDIs to provide access to on-line location based databases and information for mobile applications leading to ubiquitous SDIs, anytime and for everyone. The development of SDIs varies around the world and is not balanced. Development of global SDI will facilitate the addressing of global geo-related issues.

Keywords: Spatial data infrastructures (SDIs), clearinghouses, interoperability, metadata, spatial data warehouses, data marts, data mining, location based services (LBS)

24.1 INTRODUCTION

The concepts and development of the Spatial Data Infrastructures (SDIs) have been in continuous evolution since the middle 1990s. The establishment of SDIs was driven by: a) the spatial data itself, characterized by high generation costs, duplication, digital availability, inconsistency and increasing volume; b) the expanding Internet capabilities and capacity; and c) the need to easily access and use the spatial data and information for knowledge-based decision making at various levels of government. While the same reasons are still valid today, there are additional factors that motivate the further development and implementation of SDIs. These include the "democratization" of geo-spatial information by both users and traditionally non-geo-spatial technologies, the development of new geo-spatial-enabled technologies and applications, the need of geo-spatial related services, the evolution of geomatics science, data acquisition technologies and systems, the wide use of the Internet and the World Wide Web (WWW), the computer, information and communications technologies, the regionalization and globalization of geo-related issues, and the budgetary pressures on the traditional geo-spatial data providers (e.g. national mapping organizations) for cost-benefit review of their operations. These recent developments are changing the traditional role of SDIs, from an Internet-based virtual warehouse (repository) for access, visualization and delivery of geo-spatial data to a virtual geo-spatial service-oriented user driven enterprise (Fig. 24.1), delivering not only data but services, applications and solutions in support of data providers, service providers and application developers using interoperable and reusable components (CGDI 2004).

In the last 15 years, significant developments have been made in establishing Spatial Data Infrastructures. The development of SDIs was initially initiated at national level and then expanded to regional,

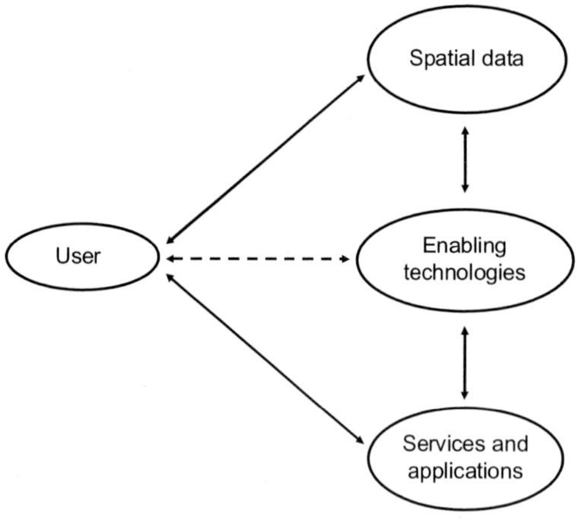

Figure 24.1. Relations between user and SDI.

international and global levels. The development and implementation of SDIs is not well balanced across the globe. Examples of national SDIs are the Australian Spatial Data Infrastructure (Trinder 2005), the Canadian Geo-spatial Data Infrastructure (CGDI 2005a), the SDI of China (Li & Wu 2004), the US National Spatial Data Infrastructure (Nebert 2006, US NSDI 2007). Examples of regional SDI are the Infrastructure for Spatial Information in the European Community (INSPIRE 2007), the Antarctica SDI (AntSDI 2007), and the Geographic Information Technology (GIT) Barents project (GIT Barents 2007). The Global Mapping project is an example of a global SDI initiative (Global Map 2007). Collaboration among national and regional SDIs programmes continues. The Global Spatial Data Infrastructure Association (GSDI) is an organization of organizations, agencies, industry and individuals, promoting international cooperation for the development of local, national and international spatial data infrastructures to better address socio-economic and environmental issues (GSDI 2008).

24.2 SPATIAL DATA INFRASTRUCTURES (SDIs)—CONCEPTS AND PRINCIPLES

Spatial Data Infrastructures (SDIs) provide a comprehensive framework on which geo-spatial science, geographic data and information and the Internet are brought together. The SDI is the geo-spatial state of a region and makes this spatial information and related geo-spatial services and applications available, meaningful and useful to all users over the Internet. It provides the basis to support discovery, access, visualization, dissemination, coordination, use/re-use and sharing of geo-spatial information and related services for socio-economic development and knowledge-based decision-making. The world infrastructure implies a reliable and supportive environment that enables access to the geo-spatial data based on standard communication protocols and specifications.

24.2.1 *Building elements of a SDI*

An SDI comprises the following building elements (CGDI 2005a,b), (Fig. 24.2):

Fundamental and non-fundamental collections of geo-spatial data: This is mainly the core content and the heart of the SDI. It consists of set of geo-referenced data—fundamental data upon which other geo-referenced thematic layers can tie upon. Up-to-date and accurate data are top priority in a SDI. Geo-spatial data are used directly as data sources or for data analysis. The value of these data was one of the main factors for establishing a SDI. The fundamental datasets, also known as framework/core/base/reference data, are continuous and fully integrated geo-spatial data that provide the reference and context information for the SDI. They consist of data layers, such as geodetic control, road network, orthoimagery, hydrographic network, geographical names, administrative boundaries, elevation data and cadastral information upon which all the other geo-spatial data are horizontally and vertically aligned. Therefore, their role is vital and these are authoritative, quality, accurate and up-to-date data. SDIs also include non-fundamental geo-referenced datasets such as thematic dataset layers that describe geo-related characteristics, such as vegetation cover or geology. An example is the Atlas of Canada that provides a collection of maps and related information about Canada. It offers thematic data such as freshwater and population distribution via the Canadian Geo-spatial Data Infrastructure (CGDI) web services (CGDI 2005a).

Technologies: The operability of an SDI relies on the enabling tools and technologies that provide the supportive infrastructure environment. This technological environment is the Internet, the development of open standards, the emergence of Extensible Markup Language (XML) and Geography Markup Language (GML) that allow the sharing of structured information across the Internet and the encoding for the transport and storage of geo-spatial information, including both the geometry and properties of geographic features. It also includes the computer and database technology, communications with emphasis on wireless technology, the data generation systems and the GIS systems.

Users: The users of a SDI refer to the groups that provide the data, the tools and technologies and the end-users. So the user group of a SDI consists of the

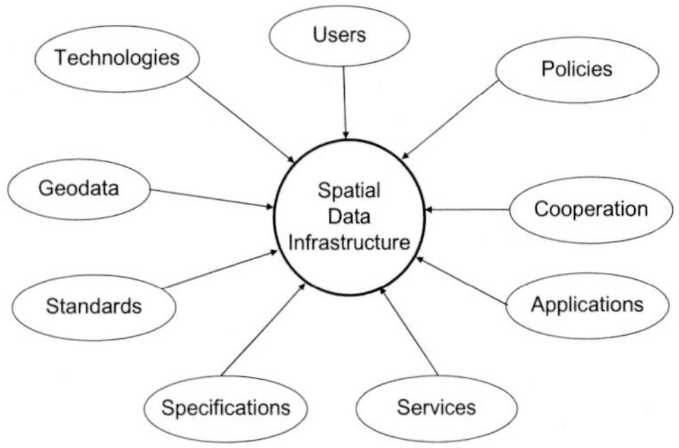

Figure 24.2. Building elements of a SDI.

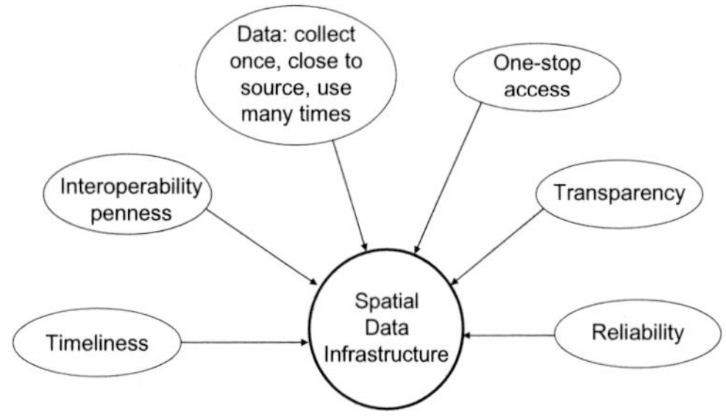

Figure 24.3. Implementation principles of a SDI.

data consumers, the data providers and the application developers.

Standards and specifications: Common procedures and standards serve the interoperability and the sharing of data, service and applications across the SDI. Examples are the use of Web Map Service for data visualization, Web Feature Service for data manipulation, Metadata for data description and Styled Layer Descriptor for cartographic data representation. Open and interoperable standards are being developed by organizations such as the Technical Committee 211 on Geographic Information/Geomatics of the International Organization for Standardization (ISO/TC211), the Open Geo-spatial Consortium (OGC), and the World Wide Web Consortium (W3C).

Services and applications: Services allow the user to access the data contents of the SDI via the web and the applications provide the capability to process and analyse the geo-spatial data.

Cooperation: SDI is founded on a distributed network of collaborating partners. SDI is built on collaborating relationships among data providers, technology developers, academia, private and public sectors.

Policies: Implementation of SDI relies on the development of policies related to the data (e.g. access, sharing and copyrights), implementation principles and approaches, partnerships and responsibilities of the SDI's participants.

24.2.2 *Principles for implementing a SDI*

The implementation of a SDI is governed by the following main principles (CGDI 2005a, McKee 2003) (Fig. 24.3):

Collection of data once and closest to the source, and use them many times: To avoid duplication and overlap and facilitate database maintenance, the geo-spatial data should be collected once by the corresponding data provider that is responsible for the collection of the specific data set, its quality and its updating. This results in a distributed network of data sources. The data can then be used and re-used.

Single point of access (one-stop access) for geo-spatial data: The geo-spatial data are accessed by a web portal that serves as a single gateway to a virtual distributed network of data and services accessible at many locations and maintained by the responsible data and services providers.

Interoperability and openness: Heterogeneous resources are accessed, used/re-used, and exchanged in a mode that facilitates the sharing of data and services independently of systems and data forms. SDI therefore is based on interoperable and open standards and specifications that allow the transparent communication of heterogeneous systems by a common application interface.

User-driven: The role of SDI is changing from being a data provider operation to a service provider business-model offering geo-spatial data-based services and applications driven by the users' needs. Thus, the SDI is a dynamic entity that responds to changing and new requirements and business applications for information and service delivery to the customer base.

Reliability: SDI requires to provide reliable and quality data and services and to ensure uninterrupted access to critical components by offering robustness and redundancy.

Timeliness: Access to data, information, services and applications offered by the SDI has to be timely to support on-time and real-time requests by the user community.

Transparency: To serve the wide user community, a SDI should operate in a simple and intuitive manner without requiring the user to understand the complexities of the underlying structures and technologies of the SDI system.

24.2.3 *Characteristics of SDI*

SDIs offer discovery, access, visualization and sharing of numerous distributed sources of geo-spatial information using the Internet. They minimize redundancy and the cost of geo-spatial data and enable seamless views of heterogeneous data sources through interoperability and integration. SDIs promote the use of common and open standards and specifications for geo-spatial data, services and applications and offer common definitions for geo-spatial data and information (semantics interoperability). They promote collaborative activities and facilitate the link of various SDIs at different spatial scales and government levels. They are transparent to the users by offering uncomplicated and easily used interfaces and access to data, services and applications.

24.2.4 *SDI system functionality and components*

SDIs provide access to on-line location based databases and information through an Internet-based portal and contribute to the development of location-based services (LBS) and mobile applications, thus enabling a change from a static SDI to dynamic spatio-temporal information environments and services. A typical operational scenario for a SDI is based on the user request. The user's main purpose is to access a specific data set or to get an answer to a geo-related query by accessing a website geo-spatial portal. Through the web services the user will need to search, discover, access and process the data. This is served through descriptive information about the data and services—metadata—and data catalogues and is a clearinghouse function leading to the data. The data are usually located in distributed databases. They can be organized as centralized geo-spatial data repositories—spatial data warehouses and as smaller scale theme-oriented data warehouses called data marts. The geo-spatial data is the core content of the SDI consisting of the framework fundamental data and the additional thematic geo-referenced data sets. All these components operate together in an environment of open and interoperable standards and specifications. Lately, the functionality of SDIs has expanded to extract valid and usable knowledge from the geo-spatial data through the data mining process. Figure 24.4 shows the main components of a SDI.

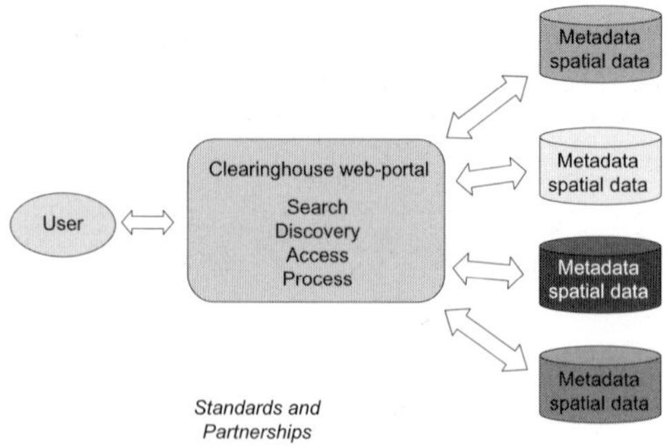

Figure 24.4. Main components of a SDI. (see colour plate page 503)

24.2.5 *Interoperability*

Interoperability is an important element in the development of SDIs. It enables the portability and sharing of data and systems by providing the ability to access various heterogeneous resources using a single interface. Interoperability is based on using common communication and standardization systems and allows users to (CGDI 2005b):

- find data, services and applications, across the virtual Internet-based SDIs
- use tools and information independently of the platforms that supports them
- be independent of single vendor processing environment.

Interoperability is applied at different levels, such as:

- data interoperability, where data and services are shared independently of data formats
- semantic interoperability, for matching relevant data from different applications by sharing a common understanding of the data descriptions, conceptual models and operations
- network interoperability, for allowing the communication between the various components over the network
- interface interoperability, for allowing client applications to execute procedures on remote systems.

Non-interoperability not only increases the technology risks but most importantly the real-world risks, where similar geo-spatial information is used by various groups to respond in critical situations, such as emergency response and disaster management (Reichardt 2004). Interoperability is ensured by using common and open standards and specifications.

24.3 CLEARINGHOUSES AND DATA CATALOGUES

Today we collect and generate large volumes of geo-spatial data. Much of this becomes part of the data content of the SDI. There are two main requirements to be met: the need by the user to discover and access this information, and the need to effectively distribute the information by the data provider. The data provider organizes the information and describes it in a manner that facilitates the Internet search and navigation for geo-spatial data performed by the user. Discovery and access of the data is facilitated by the spatial

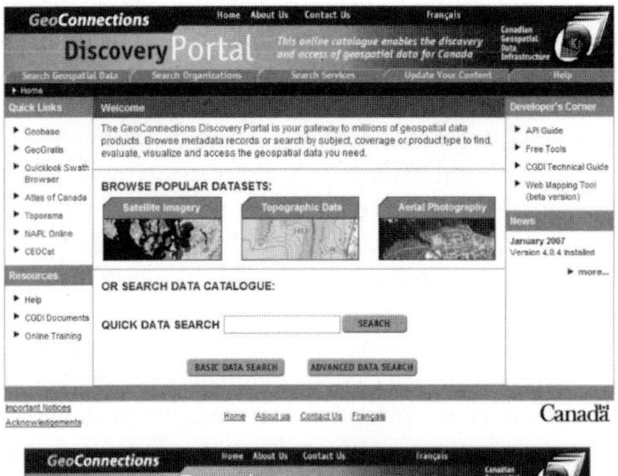

Figure 24.5. Examples of data discovery portals—Canadian GeoConnections. (see colour plate page 503) (from: http://geodis cover.cgdi.ca/, accessed 21 Feb. 2008)

data clearinghouses, which are discovery web-portals serving as the gateway to the geo-spatial data and information. The discovery portal assists the user to perform searches for data, either by listing the various data sets accessible using hyperlinks or by offering searches for geo-spatial data based on location, data type, theme, keywords, subject, time period and database listings. Figure 24.5 shows the interfaces of the Canadian GeoConnections for the general and the advanced data discovery portals.

The discovery and access is performed with the geo-spatial data catalogues, which use the description of the data—their metadata—for the spatial search. Essentially, spatial clearinghouses support the search for geo-spatial data over the Internet and allow the user to search and find what data are available and then access and visualize these data to meet their needs based on links to the actual data or through order forms. Thus, the data catalogue systems are not only file locators but offer also additional functionalities such as access, viewing, downloading and data ordering. The clearinghouse could be a single centralized server but usually it is a distributed system of servers available across multiple participating sites of the specific SDI (Steven 2005). The OpenGIS Catalogue Services Specification defines common interfaces to discover, browse and query metadata about data, services and other potential resources and is based on the communication protocol ISO 23950/ ANSI Z39.50 (Nebert 2006).

24.4 METADATA

"Meta" in Greek means change. Metadata therefore is about changes in the data and the term describes the various changes of the data since their creation. Metadata is the descriptive information about the data. This information is of a general nature, like the map lineage kept about the map series of an organization, but also includes spatial and attribute properties of the data sets. The metadata information serves both the data documentation (data inventory) and the description of the data (data usefulness). Metadata represent the what, where, when, who, why, and how questions of the data or service being documented, thus providing the user with a complete description and history of the data or service. Metadata is where a search query from the clearinghouse application is directed in order to initially discover and then access and retrieve the required data. Their functionality in a SDI is to determine the data availability, suitability, quality, accessibility, transfer and processing.

Metadata can be generated at the various stages of the data generation, such as at the data acquisition, product generation, theme layers and individual features. There are three different levels of metadata (GSDI 2004):

- Discovery metadata that provide general information about the data documentation and its geographical coverage, and which covers the queries on the nature, content and scope of the data sources.
- Exploration metadata that provide sufficient information and details on the applicability of the data to address application oriented queries and ensure the suitability of the data for solving a given problem.
- Exploitation metadata that provide specific information and details on the data properties and characteristics for proper use of the data in an application.

The required minimum information to be provided by a SDI is that of the discovery level metadata. This information allows the user to just discover the nature and content of the data but not any details on the data. This is part of the required standards for metadata. The standards for metadata ensure consistency in the metadata types and content. This makes the metadata information more reliable and universal and ensures quick and proper comparisons of the data. Conformance with the metadata definition standards (e.g. ISO 19115 Metadata as XML) ensures the interoperability across the SDI network and ensures that the users can discover, understand and share data by finding and comparing common details regarding the data.

24.5 DATA MINING, SPATIAL DATA WAREHOUSES AND DATA MARTS

The volume of geo-spatial data in SDI is continuously increasing, thus making the interpretation of the data more difficult. The changing role of SDI is also to provide services for data exploration and analysis by interpreting and evaluating the patterns in the geo-spatial data. Spatial data mining is the process of discovering and extracting spatial patterns from large spatial databases. Spatial data mining contributes to the spatial knowledge discovery process. Spatial data are characterized by their spatial complexity, their spatial relationships and the neighbouring correlations (Shekhar et al. 2004). Therefore, data mining for geo-spatial data is a more complex process than dealing with aspatial data sets.

Data mining involves the application of data analysis and discovery algorithms that detect various spatial patterns of the data (Fayyad et al. 1996). They could describe the characteristics of the data or they can be of a predictive type. The data patterns could reveal spatial clustering, spatial neighbouring associations, detection of spatial trends, spatial characterization and classification (Ester et al. 2001). Geo-spatial data mining tools are built on the concepts of aspatial data mining, spatial databases and spatial statistics and serve both the data exploratory stage and the decision-making process. The data organization and structures in the data warehouses enhance the implementation of data mining operations.

The role of SDI is changing from a data delivery network to also providing spatial decision support through the development of direct data access services (Radwan et al. 2004). This obviously has an impact on the management of geo-spatial data. According to its founding principles, SDIs are based on a network of federated distributed databases and focus on its purpose, scope and policies. The architecture of the data storage and management at the various locations is usually different, based on the location of the data sources, the amount of data and the data properties (accuracy, currency, criticality, etc.) (Fisher 2005). The geo-spatial data in a SDI are to be integrated, interoperable and consistent and are stored in the data stores. Data stores may be on-line or off-line data repositories used to manage the spatial data. A spatial data warehouse is a data store that stores, manages the data and provides seamless access to the data.

Data warehouses have evolved from the transactional (record-based) legacy database systems. A data warehouse is an enterprise-oriented, integrated, time variant and non-volatile collection of data in support of management's decisions (Bedard et al. 2001). They are designed to perform analysis of the data. In general, the data have been imported from heterogeneous data sources and stored at different granularity levels to support data analysis for decision-making. Data warehouses are enterprise systems and are centralized as they are the single and homogeneous sources of the data. As such a data warehouse is considered an authoritative data repository. The data warehouse supports on-line analytical processing (OLAP), which is different from the on-line transaction processing (OLTP) supported by the traditional transactional databases (Chaudhuri & Dayal 1997). OLAP can be extended to a spatial OLAP (SOLAP), which manipulates the spatial, thematic and temporal dimensions of the data using operations such as drill-up/roll-up (increase level of aggregation), drill-down (decrease level of aggregation), swap/pivot (re-orientation of the multidimensional view of the data) and slice and dice (selection and rearranging of the data) (Chaudhuri & Dayal 1997, Bédard et al. 2007). OLAP and SOLAP tools serve data exploration and analysis and for data knowledge discovery (data mining).

Building and maintaining a geo-spatial data warehouse in an organization is not an easy task and it takes time until its final architectural schema is defined. To overcome the various delays, organizations are putting into place geo-spatial data marts, which are specialized data stores. A data mart can be defined as a specialized, subject-oriented, highly aggregated data repository. These "mini" data warehouses can usually be implemented in shorter time, are more restricted in scope, serve a special purpose, and are built from subsets of the data warehouse. Hybrid approaches are also possible.

Access to the data warehouses and data marts is similar to accessing centralized or decentralized databases (Bédard et al. 2001). The users can access a data warehouse directly or can access data marts created by the data warehouse. It is also possible to have a solution without a data warehouse, where there are only data marts to be accessed (Fig. 24.6).

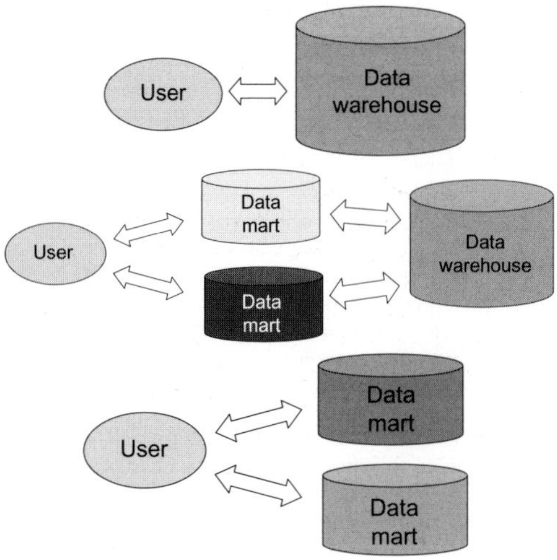

Figure 24.6. Typical architectures for accessing data warehouses and data marts. (see colour plate page 503)

24.6 SPATIAL DATA INFRASTRUCTURES AND LOCATION BASED SERVICES (LBS)

Because of their mobile nature, mobile computing and mobile communications are becoming geo-spatial information dependent. The developments are on the flow of information, wireless operations, nano-satellites and nano-sensors; and in spatial decision support systems by the users. The proliferation of data collection tools, from satellite sensors and GPS receivers to camera-equipped mobile phones, and the resulting availability of huge amounts of geo-spatial data, together with heightened consumer expectations (e.g. reliable car navigation systems) have created a rapidly increasing demand and need for timely and accurate geo-spatial information. SDIs are developed largely due to their Internet-based accessibility. The current Internet mobility affects also the SDIs as location-based services (LBS) are continuously growing. LBS are wireless application services accessible by mobile users using their spatial mobile location. OGC defines LBS as wireless-IP service that uses geographic information to serve a mobile user (OGC 2004).

Location, directory, proximity and navigation are typical location-based services. The basic components of LBS are: the mobile user, the position of the user, the mobile computing device, the content of geo-spatial information, the communication network, and the service provider (Reichenbacher 2007). The decision making is based on the position of the user and the timely availability of the geo-spatial data and data analysis. Obviously, there is a strong relation between SDI and LBS as SDI's role is to provide geo-spatial data and services and can have a significant impact on the LBS functionality and operations (Sarjakoski et al. 2002).

To define the SDI's contribution to LBS, the specific characteristics of the LBS to be considered are (Smith et al. 2004, OGC 2004):

- most of the LBS users are not familiar or trained in geo-spatial sciences and technology
- data, services and decision making are time and space critical
- mobile computing devices have small displays and limited processing power
- current communication networks have limited bandwidths.

Thus, SDIs should offer a certain level of simplicity in the technology and avoid "GIS complexity". Data and services should be scalable and quality data are required. The information has to be at various levels of detail and resolution. Real-time response is required for "instant" decision making. Lightweight approaches are to be implemented and the exchange of information should be in compact form. The interoperability should be ensured among all the technologies involved in the LBS. The GeoMobility Server (GMS) is an OGC open platform for location-based application services (Zlatanova & Verbree 2003).

24.7 CLOSING REMARKS

SDIs have been evolving in the last decade or so from centralized repositories of geo-spatial data to an Internet based virtual network of distributed harmonized spatial databases providing web-based easy discovery, access and visualization to enormous amounts of geo-spatial data. They have been built on interoperable data, software and hardware systems, open international standards and specifications, policies and technologies. SDIs are accessed via an Internet-based portal, a web-based gateway to the geo-spatial data. The data are accessed via spatial data clearinghouses based on their metadata and data catalogues, which provide a coordinated and efficient data search, navigation, discovery, access and data processing. An overview of the various SDI elements and components has been presented along with their respective functionality.

SDIs are formed at national and international levels. However, their development varies around the world depending on national policies and availability of digital geo-spatial data. Further progress in establishing world-wide geo-spatial information frameworks is expected based on the United Nations Geographic Information Working Group (UNGIWG) initiative on building the UN Spatial Data Infrastructure (UNSDI 2008). UNSDI aims to support decision making at various levels by enabling access, retrieval and dissemination of geo-spatial data and services in a rapid and secure way. To move forward in the implementation of SDIs, the following four key issues have been identified (Masser 2005): a) creation of appropriate SDI governance structures, b) access to data has to be further facilitated, c) capacity and capabilities building, and d) continue the emphasis on data interoperability. Development of global SDI will facilitate the addressing of global geo-related issues (Steven et al. 2005).

Spatial data warehouses and data marts were also discussed, as well as the data mining process leading the data exploration and analysis for knowledge discovery in the spatial databases. The latter contributes to the changing role of SDIs from geo-spatial data providing systems to enterprise-like approaches for providing value to the users (Morales & Radwan 2004, Hjelmager et al. 2005). This value extends from data to services to applications and data processing focused on the user and the decision-making process. The interoperability will extend beyond data and semantics to include policies as well. In addition, SDIs will contribute to addressing pressing issues such as demographics, biodiversity, environment, land management, natural resources, energy management, and risk and emergency management. Further, the Global Earth Observation System of Systems (GEOSS), which is based on standards and protocols similar to SDIs, will enhance the SDIs by including more remotely sensed data and real-time environmental measurements based on terrestrial sensors (Nebert 2008).

The wider use of geo-spatial data and information by the technological IT and communications developments are affecting SDIs. SDIs will influence and be influenced by mass market on-line mapping services, such as Microsoft Virtual Earth and Google Earth and the distributed web-based technologies. Location-based services can rely on SDIs to provide access to on-line location based databases and information for mobile applications leading to ubiquitous SDIs, anytime and for everyone.

REFERENCES

AntSDI. 2007. *The Antarctic spatial data infrastructure.* http://www.antsdi.scar.org/ (accessed Feb. 21, 2008).

Bédard, Y., Merrett, T. & Han, J. 2001. Fundamentals of spatial data warehousing for geographic knowledge discovery. *Geographic Data Mining and Knowledge Discovery, Research Monographs in GIS Ch.* 3: 53–73. Taylor & Francis.

Bédard, Y., Rivest, S. & Proulx, M. 2007. Merging GIS and Business Intelligence for Spatial Data. In H.A. Karimi (ed.), *Encyclopaedia of Geoinformatics.*

Chaudhuri, S. & Dayal, U. 1997. An overview of data warehousing and OLAP technology. *ACM SIGMOD Record* 26: 65–74.

CGDI. 2004. *A developer's guide to the CGDI: Developing and publishing geographic information, data and associated services.* GeoConnections, Canada. http://www.geoconnections.org/en/resourcelibrary/keyStudiesReports (accessed 21 Feb. 2008).

CGDI. 2005a. *Canadian Geospatial Data Infrastructure—Achieving the Vision of the CGDI.* GeoConnections, Canada. http://www.geoconnections.org/en/resourcelibrary/keyStudiesReports (accessed Feb. 21, 2008).

CGDI. 2005b. *Canadian Geospatial Data Infrastructure—Architecture Description, v2.0.* GeoConnections, Canada. http://www.geoconnections.org/en/resourcelibrary/keyStudiesReports. (accessed Feb. 21, 2008).

Ester, M., Kriegel, H.-P. & Sander, J. 2001. Algorithms and applications for spatial data mining. *Geographic Data Mining and Knowledge Discovery, Research Monographs in GIS*: 53–73. Taylor & Francis.

Fayyad, U.M., Piatetsky-Shapiro, G. & Smyth, P. 1996. Knowledge discovery and data mining: Towards a unifying framework *Proc. 2nd Int. Conf. on Knowledge Discovery and Data Mining, Portland, Oregon*: 82–88.

Fisher, B. 2005. *Server architecture models for the national spatial data infrastructures (NSDI), OGC White Paper.* http://www.opengeospatial.org/pressroom/papers. (accessed 21 Feb. 2008).

GIT Barents. 2007. *Geographic Information Technology within the Barents Region*, http://www.gitbarents.fi/ (accessed 21 Feb. 2008).

Global Map. 2007. *International Steering Committee for Global Mapping.* http://www.iscgm.org/cgi–bin/fswiki/wiki.cgi. (Accessed 21 Feb. 2008).

GSDI. 2004. *Developing spatial data infrastructures: The SDI Cookbook*, Nebert, D.D. (ed.), v2.0. http://www.gsdi.org/docs2004/Cookbook/cookbookV2.0.pdf. (accessed 21 Feb. 2008).

GSDI. 2008. *Global Spatial Data Infrastructure Association.* http://www.gsdi.org. (accessed 21 Feb. 2008).

Hjelmager, J., Delgado, T., Moellering, H., Cooper, A., Danko, D., Huet, M., Aalders, H. & Martynenko, A. 2005. Developing a modeling for the spatial data infrastructure. *Proceedings International Cartographic Conference ICA 2005, 9–16 July 2005, A Coruna, Spain.* (CD ROM).

INSPIRE. 2007. Directive 2007/2/EC of the European Parliament and of the Council for establishing an Infrastructure for Spatial Information in the European Community. (INSPIRE). *Official Journal of the European Union* L 108/1. http://www.inspire–geoportal.eu/index.htm. (accessed 21 Feb. 2008).

Li, D. & Wu, H. 2004. On developing spatial data infrastructure of China. *IAPRS* XXXV, Part B4.

Masser, I. 2005. The future of spatial data infrastructures. *ISPRS Workshop on Service and Application of Spatial Data Infrastructure*, ISPRS XXXVI, 4/W6.

McKee, L. 2003. The spatial web, An Open GIS Consortium (OGC). *OGC White Paper.* http://www.opengeospatial.org/pressroom/papers. (accessed 21 Feb. 2008).

Morales, J. & Radwan, M. 2004. Architecting distributed geo–information services: beyond data infrastructures. *IAPRS* XXXV, Part B4.

Nebert, D. 2006. Development and status of the U.S. National Spatial Data Infrastructure: Concepts and components. *OGC TC/PC Meeting, June 26–30, 2006, Edinburgh, UK.* http://edina.ac.uk/ogcconference/interop_pres/NSDI06–DouglasNebert.ppt (accessed 10 Dec. 2007).

Nebert, D. 2008. Global Earth Observing System–of–Systems. GEOSS): A Global SDI Framework. *10th International Conference for Spatial Data Infrastructure, February 25–29, 2008, St. Augustine, Trinidad.* http://www.gsdi.org/gsdi10/papers/TS31.3paper.pdf. (accessed 21 Feb. 2008).

OGC. 2004. *OpenGIS Location Services. (OpenLS): Core Services, OGC 03–006r3.* http://www.opengeospatial.org/standards/olscore. (accessed 21 Feb. 2008).

Radwan, M., Alvarez, L., Onchaga, R. & Morales, J. 2004. The changing role of the geodata infrastructure: From a data delivery network to a virtual enterprise supporting complex services. *IAPRS* XXXV, Part B2.

Reichardt, M. 2004. *The havoc of non–interoperability, An Open GIS Consortium.* (OGC) White Paper, http://www.opengeospatial.org/pressroom/papers (accessed Feb. 21, 2008).

Reichenbacher, T. 2007. Adaptation in mobile and ubiquitous cartography. In W. Cartwright, M.P. Peterson & G. Gartner (eds.), *Multimedia Cartography, 2*nd *edition*: 383–397. Springer.

Sarjakoski, T., Tiina Sarjakiski, L., Lehto, L., Sester, M., Illert, A., Nissen, F., Rystedt, B. & Ruatsalainen, R. 2002. Geospatial info–mobility services—A challenge for national mapping agencies. *IAPRS & SIS* 34(4): 356–360.

Shekhar, S., Zhang, P., Huang, Y. & Vatsavai, R.R. 2004. Trends in Spatial Data Mining, In. H. Kargupta, A. Joshi, K. Sivakumar & Y. Yesha (eds.), *Data Mining: Next Generation Challenges and Future Directions.* AAAI/MIT Press.

Smith, J., Mackaness, W., Kealy, A. & Williamson, I. 2004. Spatial data infrastructure requirements for mobile location journey planning. *Transactions in GIS* 8(1): 23–44.

Steven, A.R. 2005. The US national Spatial Data Infrastructure: What is new? *ISPRS Workshop on Service and Application of Spatial Data Infrastructure, ISPRS* XXXVI, 4/W6.

Steven, A.R., Onsrud, H.J. & Rao, M. 2005. Global spatial data infrastructure. (GSDI): Encouraging SDI development internationally. *ISPRS Workshop on Service and Application of Spatial Data Infrastructure, ISPRS* XXXVI, 4/W6.

Trinder, J.C. 2005. The evolution of ISPRS activities in the spatial information sciences: Developments in SDI in Australia. *ISPRS Workshop on Service and Application of Spatial Data Infrastructures, ISPRS* XXXVI, 4/W6.

UNSDI. 2008. U*nited Nations Spatial Data Infrastructure.* http://www.ungiwg.org/unsdi.htm (accessed 21 Feb. 2008).

US NSDI. 2007. *National Spatial Data Infrastructure.* http://www.fgdc.gov/nsdi/nsdi.html (accessed 21 Feb. 2008).

Zlatanova, S. & Verbree, E. 2003. Technological developments within 3D location–based services. *International Symposium and Exhibition on Geoinformation 2003, October 13–14, 2003, Shah Alam, Malaysia*: 153–160.

Advances in Photogrammetry, Remote Sensing and Spatial Information Sciences:
2008 ISPRS Congress Book – Li, Chen & Baltsavias (eds)

CHAPTER 25

Web mapping/GIS services and applications

Songnian Li

ABSTRACT: The field of web mapping and GIS has gone through some tremendous changes over the last few years. Driven by the constant updating of web and other related technologies, as well as the significant revolution of web mapping caused by mainstream IT vendors such as Google, Yahoo and Microsoft, we will see more great changes ahead of us. This chapter presents a technical review of web mapping development over the last decade or so and examines the current status of the development. Special emphasis is put on how web mapping or GIS has evolved with the technology revolution and how the newly emerged technologies might reshape the development in the field. The chapter ends with discussions of some important issues and challenges as well as potential directions for future development.

Keywords: GIS, mapping, internet, web based, distributed, services, application

25.1 INTRODUCTION

The Internet and the World Wide Web (the *Web* or *WWW*) continuously hold as many as people can imagine possibilities for new forms of communications, innovations and new businesses. According to the survey from Netcraft Inc. and the estimate from *WorldWideWebSize.com*, there are now about 150,000,000 web sites with an estimated 25 billions web pages as of November 2007. While there is no survey being found on counting the number of web mapping sites, services or applications, the *ProgrammableWeb.com* provides statistics based on the mashup tags, which shows that as of November 2007 about 40% of all mashups counted are mapping related.

Web technologies and the continuous development of web mapping and geographical information system (GIS) technologies, services and applications have substantially increased and enhanced the open use of GIS in a number of directions including: (1) spatial data access and dissemination, (2) spatial data exploration and visualization, (3) spatial data processing, analysis and modelling, (4) collaborative spatial decision support using public participatory GIS, and (5) integration of web-based geo-spatial services in mainstream and enterprise computing processes and environments. Especially with the recent emerging of Google Maps and Yahoo! Maps, to name but two, the public awareness and enthusiasm in web mapping and, to a lesser extent, GIS, have never been so great. These new, emerging web map services have drawn mass public attention, which demonstrates the success of bringing the use of maps and mapping/GIS services into people's daily lives.

The rich collection of application programming interfaces (API) provided by these technologies, such as Google Map API and Yahoo Map API, allow for easy customized web map services and inclusion of maps into hundreds and thousands of websites by developers and end users. Specially, the availability of these APIs enables easy "mashup" of web applications and services that combines contents from one or more online sources, and allows users to add their own content, which is overlaid on the base maps provided by API vendors.

The new phenomena of web mapping stand in contrast to more proprietary GIS systems such as ArcIMS and GeoMedia WebMap. Those proprietary systems perform powerful and valuable functions as well, but their impact on societal and public awareness of GIS and mapping is different and less pervasive. For traditional GIS vendors, the challenge is "how to take advantages of these new developments while continuing to leverage existing technology investments and the specialized functionality they provide" (Batty 2007a). However, current web mapping and GIS mostly, if not all, do not provide functionalities towards better analysis and modelling of geographically-referenced data, either vendor provided or user added.

This chapter presents a technical review of web mapping and GIS. Efforts are made to search most of the research journal papers, trade magazines, technical reports, professional publications and web papers relevant to the subject. Through a critical review, the chapter summarizes the state-of-the-art technological development in web mapping and GIS, discusses related social and organizational issues, and speculates on some future directions. This chapter also tries to answer several important questions: How has web mapping/GIS evolved with the web technology development? Have we realized or gone beyond those expectations laid out a decade ago when web mapping became an increasingly hot agenda? What are the next steps?

The rest of the chapter starts with the concept of web mapping to clarify terminology, understand the role of maps in web mapping, and the types of services web mapping offers. This is followed by a review section briefly examining the evolution of web mapping development. Section 25.4 explores web mapping technologies, focusing on the new emerging ones, followed by an examination of the current status of web mapping in Section 25.5. The final section presents issues, challenges and some future directions.

25.2 CONCEPT OF WEB MAPPING

Web mapping refers to the delivery of maps and the related mapping functions on the Web. Strictly speaking, web mapping is not web GIS in the sense that web mapping does not provide sufficient spatial analytical and modelling functions, if at all; nor does it provide direct handling (e.g. editing) of map data. However, in reality it is not surprising to read and hear the terms such as "web mapping", "web-based mapping", "web GIS", "Web-based GIS", "Internet mapping", and "Internet GIS" from literature and during trade shows and conferences. A search from the Internet, as shown in Table 25.1, can verify this diversity of terms used.

Both the Internet and the Web are considered as a platform (O'Ralliy 2005). However, the Web is the platform actually running on top of the Internet. Web or web-based applications need to follow web standards and protocols such as hyper text markup language (HTML) and use a web browser as the application interface or interface container. Internet or Internet-based applications, on the other hand, do not need to comply with web standards and protocols, nor do they necessarily use a web browser. Since the

Table 25.1. Search results of web mapping related terms.

Key words	Number of returned items
"web mapping"	59,900
"web GIS"	60,400
"web-based mapping"	64,000
"web-based GIS"	38,600
"web-enabled mapping"	661
"web-enabled GIS"	2,780
"Internet mapping"	159,000
"Internet GIS"	128,000
"Internet-based mapping"	19,600
"Internet-based GIS"	16,100

*Key words are entered into Google search as a whole with quotations. The results were valid on 8 December 2007.

Web is part of the Internet, it is quite normal for many people to refer their web-based mapping/GIS as "Internet-based".

Further to the above, recent developments in web mapping have resulted in some new terms, under the umbrella term—Web 2.0. Web 2.0 presents a second generation of web-based communities, applications and hosted services, which does not refer to an update to any technical specifications, but to changes of the ways software developers and end users use the Web (Wikipedia "Web 2.0"). It promotes: 1) creativity, collaboration, and sharing between users; 2) the move to the Internet and the Web as platforms (O'Reilly 2005); 3) mass social networking channels; and 4) user generated and distributed web contents. The characteristics of Web 2.0 technologies lie in their ability to foster innovations and "assembling" systems, services, and websites by integrating features, services and data from distributed and, very often, independent developers.

One such term is neogeography. Believed to be coined by Di-Ann Eisnor, co-founder of Platial, Inc., it is a new branch of geography that may consist of practices, theories, techniques and software tools that are different from the way the traditional GIS or professional cartographers handle maps and geo-spatial data. The way has simply been transformed into mashing up various web resources to create maps (Van Wyngaarden & Waters 2007). The term has emerged partly as a result of the vast amount of public interest in online mapping and using geo-spatial technologies.

Another new term that is becoming popular is the so-called map mashup. A map mashup is similar to other types of mashups that are defined as a website or web application that uses contents or services from more than one source to create a completely new service (Wikipedia, "mashup"). As a special type of mashup, map mashups combine at least one map data source with added information to create a new web map or mapping application (Pietroniro & Fichter 2007). It is argued that map mashup is a user-controlled web mapping.

While it may be necessary to maintain the distinction between "Web-based" and "Internet-based", the boundary between web mapping and web GIS may become more blurred and may disappear, as more and more analytical and modelling functions are built into current web mapping. Unless otherwise specified, this chapter will use web mapping throughout its discussions.

25.2.1 *Types of web maps*

Maps have played an indispensable role in most web mapping services and applications. In addition to their traditional function of representing an abstraction of a selected part of reality, maps are now taking extra roles in web mapping (Kraak 2004). For spatial data distribution, maps can be used as a graphical index for data search and a media for preview (snapshot). For spatial retrieval and analysis, maps are used as interfaces for either performing the required operations or presenting the results. Maps also are widely used as a metaphor for linking other types of location-based information such as HTML documents, audio, video and photos.

Kraak (2001) classifies web maps into different categories based on the way the map is displayed and the use of the map in web mapping, i.e. static vs. dynamic and interactive vs. view only maps. Arguing about "additional possibilities regarding distributed data sources, collaborative maps, personalized maps, etc.", Wikipedia under "web mapping" presents a more detailed list of potential types of web maps. However, the boundaries between those identified types are somehow not clear, with overlaps. In short, the current classifications of web maps are all based on a mix of format and use.

Whatever classification is used, the maps used in web mapping can be displayed in one of the three basic formats: static raster image, raster image with animation, and vector map. A static map is a raster image either pre-prepared and stored on the server or created on the fly according to the user's request. A dynamic map is a raster image, but with built-in animation features. A vector map follows a feature model, storing and displaying discrete objects. As further discussed in the next section, the selection of a type of map suitable for web mapping depends on the level of interaction, individual or collaborative use, dynamics and timeliness.

25.2.2 *Types of web mapping services*

Web mapping can provide many different types of services to a wide variety of applications. Plewe (1997) provides a first attempt at summarizing early web mapping services into raw data download, static map display, metadata search, dynamic map browser, data preprocessing, web-based GIS query and analysis, and net-savvy GIS. Dragicevic (2004) states a more abstraction level of categorization towards web-based GIS, including spatial data access and dissemination, spatial data exploration and geo-visualization, and spatial data processing, analysis and modelling. A wealth abundance of other publications also offer discussions on various types of web mapping services. Examples include Dawe (1996), Hardie (1998) and Peng & Tsou (2003).

While early web mapping was primarily data driven, focusing on spatial data dissemination, map view and geographic visualization, simple spatial query and analysis, recent developments have shown a trend, service-centric, towards providing web-based mapping and GIS services. Another important change is the shift from traditional organization and domain based services to mass, public oriented services. MapQuest web mapping service is considered as the first of this kind targeting the general public to provide location and routing services, which are more than just distributing, viewing, visualizing and simply querying spatial data.

Both early and recently-developed web mapping services are still short of providing analytical and modelling functions necessary to qualify as web GIS. Partly because of the technological limitations such as Internet bandwidth, software performance, interface design, security and stability, the major impediments are perhaps related to the lack of motivation from vendors and developers, suitable and practical business models, the need from user communities, and solutions to data issues. The design and development of web mapping services have long been based on the assumption that its audience is different from those of the traditional desktop mapping/GIS software. As such, web mapping services should be developed in consideration of users who do not have much knowledge about mapping and GIS, level of functionality, easiness of interface, network speed, and data limitations. Apparently, if web mapping is to be developed with more GIS functions for the same group of users, all these aspects need to be carefully studied.

Table 25.2 summarizes the existing services provided by web mapping and the expected

Table 25.2. Web mapping services and their characteristics.

Service type	Characteristics
Data discovery and dissemination catalogue;	Simply raw data download; preview Search data based on metadata Location-based search; Interface for data clearinghouses
Access, exploration and visualization	Static map viewing; Dynamic map browsing; Geographic visualization
Data processing, query and analysis	Attribute information query; Spatial search and query; Routing and trip planning; Overlay, buffering and pattern analysis; Easy process-driven modelling
Social mapping	Geo-referencing user added content; Managing social networks and contents

services as more and more analytical and modelling functions are included. While social mapping may not be qualified as a separate type of web mapping service, it does require special attention due to recent advances in user controlled online mapping, as will be further discussed in Sections 25.5 and 25.6.

25.3 EVOLUTION OF WEB MAPPING

Through the 1980s and early 1990s, the Internet became the world's largest computer network—a network of computer networks available to a wider user community than just academic and research communities. Supported by standard network protocols and software, the Internet overcomes the incompatibility problems inherent in wide area networks and allows "mass" access to information and services available across this "wide-area" network. Within the same time period, the Web was invented in 1989. According to Wikipedia under "*World Wide Web*", the Web is a system of interlinked, hypertext documents supplied by a collection of numerous web servers, application servers, and database servers through computers connected to the Internet. The core of the Web is the "hyperlink", embedded in web pages to link documents containing text, graphics, multimedia (e.g. audios, videos, animations), and more. Web browsers, as the software for viewing such hyperlinked documents, provide the largest multimedia applications over the Web.

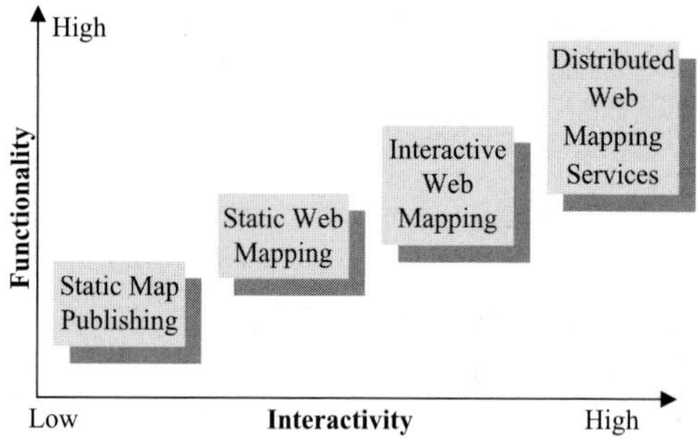

Figure 25.1. Web mapping development path, modified after Peng & Tsou (2005).

Web mapping development, to a large extent, has always been tied up with innovations and developments of web technology, architecture models, specifications, and standards. Among them, web browsers, web servers, and hypertransfer text protocol (HTTP) and HTML specifications played a very important role in early web mapping. It is widely accepted that web mapping development has gone through a few stages from static map publishing through static web mapping and interactive web mapping to distributed web/GIS services (Peng & Tsou 2005). Figure 25.1 illustrates how the interactivity and functionality have been improved along this development path over years.

25.3.1 *Overview of history*

When the first graphical web browser and the HTML 1.0 specifications were developed in early 1990s, there was no support for embedding in-line images in web pages. Accordingly, even the maps in static raster image format could not be posted on the Web. This was changed when the National Center of Supercomputing Applications (NCSA) released its Mosaic browser in early 1993, which extended HTML 1.0 to allow inline images in HTML documents. As such, early web maps were primarily static raster images inserted into the web pages. Figure 25.2 illustrates a typical example of such maps. It is interesting to see how direction information was put together with the map as one raster image due to the lack of support of user interaction and routing.

The availability of early interactive maps and mapping on the Web started after the release of HTML 2.0, which introduced a few new tag elements including the image map and form. The image map feature allowed clickable maps to be embedded in the web page. With a list of predefined image coordinates, different parts of a map could be hyperlinked to different things including another map at a larger scale.

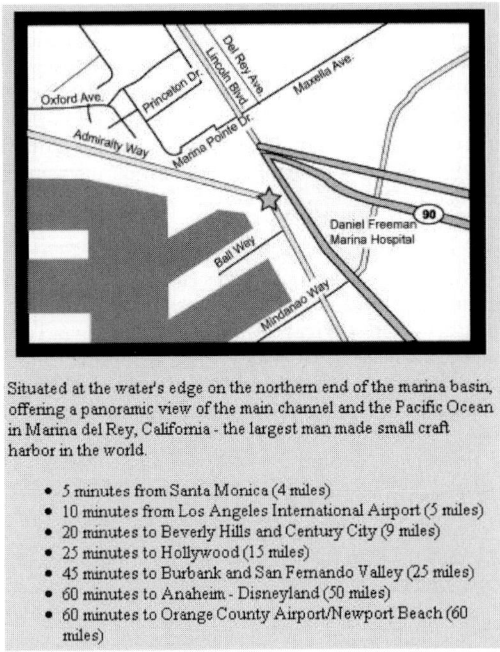

Figure 25.2. Map embedded in web page as a raster image (unknown source).

For example, a number of clicks on a "clickable" world map, depending on the level of details desired, could lead to the display of country maps through regional maps to city maps. While the clickable map (image map) did help to add very basic interaction into early web maps, all map images had to be pre-prepared and stored on the server.

The form feature of HTML 2.0 and its latest versions allow users to send information to the server through interacting with web pages. Each form may have input (text and check box, radio and submit button), text area (multi-line text input), and selection (a drop-down list) elements. Using the form, early interactive web mapping sites enabled users to pick up map layers, decide scale level, input attribute for queries, or disable or enable the display of certain map layers.

With the help of databases and server-side programs, mostly common gateway interface (CGI) and Perl scripts, the maps could be created on the fly based on user's input (Hardie 1998). The first web-based interactive map, developed by Xerox PARC, used the image map feature (Putz 1994). Implemented as CGI/Perl scripts on the server, the scripts accepted requests for map rendering and returned an HTML document including an in-lined graphics interchange format (GIF) image of the requested map. While the rendered maps were still static images, the map viewer allowed user's interaction to zoom the maps to the selected point.

Between 1995 and 2005, the major effort had been put on using different web technologies to improve the interactivity and functionality of the web mapping. JavaScript, Applets (e.g. Java applets and ActiveX controls), and different types of plug-ins (e.g. scalable vector graphics and WebCGM) were among the major technologies used to improve interaction between end users and web mapping interfaces (client). On the server-side, while CGI and Perl languages continued to be used to develop server-side scripts for connecting web server, map server and databases and for creating mapping logics, more effort was put on enhancing web mapping functionality by: 1) developing server applications and middleware servers using Java, C, C++ and C# programming languages; 2) integrating data from multiple, distributed databases; and 3) adding support of spatial data in commercial RDBMS (e.g. Oracle Spatial and IBM DB2 Spatial Extender).

At the same time, the initiatives of open geo-spatial data standards and software specifications and to a lesser extent open source GIS developments had had significant impact on web mapping. Among them, the most influential ones perhaps were the Web Map Services (WMS) and Web Feature Service (WFS) specifications from the Open Geospatial Consortium (OGC) and the open source MapServer environment originally developed by the University of Minnesota. During this period, web mapping developers were mainly concerned about a number of issues (Hardie 1998, Limp 1999, Lowe 1999), such as:

1. How to design an effective yet intuitive web mapping interface and site?
2. The use of image maps, ActiveX controls, Java applets, or software plug-ins to maximize the user interaction and, at the same time, balance the performance.
3. Incompatibility between different versions of the same technology, and between different technologies, e.g. web browser compatibility.

Early web mapping aimed at serving some specific application areas, or at most providing published maps for the general public. This has been changed since MapQuest released its online mapping with location maps and routing services in 1996. Following MapQuest's success, Google, Yahoo and Microsoft all released their online mapping services to the public, which further increased the use of web mapping. These newly emerged developments feature free map data with increasing geographic coverage, overlapped satellite image view, and most importantly the APIs that the developers and even end users can use to create their own web mapping applications accessing the data provided by the data sources associated with the API.

Table 25.3. List of major historic events in web mapping.

Year	Event/Development
1993	the 1st interactive web map was developed (Putz 1994)
1994	the 1st online atlas was released by Canada
1996	the release of MapQuest online mapping services
1996	the 1st version of Intergraph GeoMedia WebMap was released
1996	the 1st edition of ESRI MapObjects was released
1997	the 1st version of ESRI commercial ArcView Internet Map Server (IMS) was released
2000	the 1st version of Microsoft MapPoint was released
2005	the beta version of Yahoo! Maps was released, followed by its official release 9 months later
2005	the beta version of Google Maps was released, followed by its official release 6 months later
2007	new MapQuest beta was released

There are a number of projects, publications and websites providing information about GIS history, or GIS timelines. Examples include University College London, CASA GiSTiMELine project, *GISdevelopment.net* Time Line (with an Indian Flavour), and the book on The History of Geographic Information Systems (Foresman 1997). Software vendors also maintain their own timelines, documenting their in-house development such as ESRI on its *gis.com* (http://www.gis.com/timeline/). However, there are no sources dedicated to web mapping and GIS. Wikipedia, under the term "web mapping", provides an incomplete history of web mapping with no mentioning of web GIS. Nevertheless, Table 25.3 shows a list of major events related to web mapping and GIS according to the above sources and literature search.

25.3.2 *Development from traditional GIS industry*

The GIS industry played an active role in the growth of web mapping and GIS. By 1998, almost all of the major GIS players in the market had some kind of Internet GIS server software product (Limp 1997, Culpepper 1998, Hardie 1998). These vendors and their products, at the time, included Autodesk (MapGuide), Intergraph (GeoMedia WebMap), Mapinfo (ProServer), ESRI (MapObjects IMS and ArcView Internet Map Server), among others. All these packages provided web access interfaces, either HTML-based or using some client-side web technologies such as JavaScript, ActiveX Control, or Java. However, they were different in map rendering (graphics vs. vector), map or data access, client-side computing requirements, level of spatial analysis supported, and data/platform restrictions (Culpepper 1997). In addition, all of these products required a certain level of customization effort through using their natively supported programming languages (Limp 1998), i.e. the web mapping application developers needed to program some server-side components to work with both HTTP server and web map server to support a particular applications.

While there is no room in this chapter to include the examination of more recent development in the GIS industry, interested readers may check out Peng & Tsou (2005, Chapter 8), Bonnici (2005), Wikipedia (under "Comparison of GIS software"), and some trade magazines such as GeoWorld, Geospatial Solutions, etc.

25.4 WEB MAPPING TECHNOLOGY

As discussed in the previous section, web mapping development has always followed the progress of web technologies and other related GIS technologies, which is arguably a technology driven process. Among these technologies, client-server technology has long dominated the development of web mapping since it started. Over the last few years, web service technologies have emerged, and greatly changed the landscape of web mapping. While it is difficult to go over all of these technologies because the number of them is almost infinite (Wikipedia, "*web mapping*"), this section tries to capture some of them that are either fundamental to, or have had an important influence on, web mapping.

25.4.1 *Client-server technologies*

The core of client-server technologies is the client-server model, which has been the basic architecture for most distributed computing. Depending on the number of middle tiers in between the client and the *server*, the architecture based on this model can be 3-tier or n-tier, generally termed as multi-tier (see Fig. 25.3[a]). While the basic 2-tier model may simply split the systems into presentation and logic component tiers, it is the multi-tier architectures that have been commonly used for making more interactive, reliable and scalable geo-spatial data processing and services available over the Web (Worboys & Duckham 2004).

For years, web mapping has relied heavily on the client-server model, a centralized paradigm. Numerous web mapping systems with a variety of architecture designs have been developed (Seggern 1994, Plewe 1997, Coleman 1999, Peng & Tsou 2003). These designs differ in their adoption of thin/thick

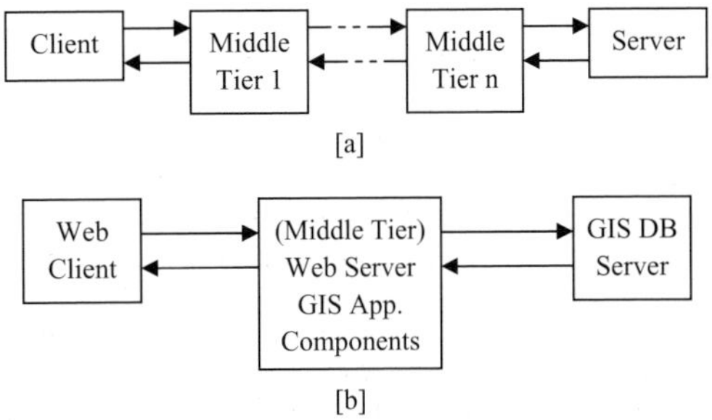

Figure 25.3. Conceptual view of the client-server model.

client strategy and the distribution of data access and processing tasks over the different tiers involved. For example, a 3-tier system may have the presentation logic on the client, a set of web mapping/GIS logic components on one server, and the data on a database server, as shown in Figure 25.3[b]. A common feature shared by all these designs is the concept of "centralization", meaning that all services and data are located and controlled by one or a few dedicated central servers.

Under the client-server architecture, technologies used for web mapping have been grouped into client-side technologies and server-side technologies. On the client side, technologies such as dynamic HTML (DHTML), document object model (DOM), JavaScript, ActiveX controls, Java Applets, and plug-ins have been widely used to develop different types of viewers embedded in web browsers to improve the client interfaces facilitating more interaction between end users and the interfaces or web maps themselves (Peng & Tsou 2005). On the server side, there are mainly three types of technologies: 1) web server extensions, called web map server (WMS) or Internet map server (IMS), including CGI scripts, active server page (ASP) scripts, VBScript, and Java Servlets; 2) web map application servers, middleware; and 3) spatial databases and database servers (Peng & Tsou 2005, Wikipedia "web mapping"). For a comprehensive discussion about the use of these technologies in web mapping, please see Plewe (1997) and Peng & Tsou (2005).

25.4.2 *Web service technologies*

In the later 1990s and early 2000s, a major paradigm shift occurred in distributed computing principles towards service orientation, open standards integration, collaboration, and virtualization (Joseph et al. 2004), which has significantly influenced the development of web mapping and GIS. The concepts and/or technologies of web services (WS), grid computing, peer-to-peer (P2P) architecture, and service-oriented architecture (SOA) are among those contributing to this shift. Although the boundaries between these technologies are sometimes blurring, all together they have moved web mapping and GIS further towards a service-oriented and decentralized paradigm.

25.4.2.1 *Web services*

A web service is a self-contained, self-describing, modular software component or application accessible over a network (e.g. Internet) based on its service description expressed in web service description language (WSDL). Web services must be loosely coupled, location transparent, and protocol independent (Endrei et al. 2004). The service consumer has no need to know the implementation details of the service.

Web service technology is based on a number of open technologies and/or specifications, including extensible markup language (XML), simple object access protocol (SOAP), universal description, discovery and integration (UDDI), and web services description language (WSDL). This technology has shifted the Web from an information sharing and distribution platform to a service platform.

A web service system consists of three collaboration entities: service provider, service requester, and service register, and three basic operations: publish, find and bind, as illustrated in Figure 25.4 (Gottschalk et al. 2002). The service provider creates services and publishes service descriptions to the server register for the service requester to search and find the required service. The service requester then communicates with the service provider and invokes the required service.

Under the web service architecture, the web is used for delivering not just geo-spatial data, but geo-processing functionality that can be created or wrapped in interoperable software components (Anderson & Moreno-Sanchez 2003).

25.4.2.2 *Service Oriented Architecture (SOA)*

Service-oriented architecture is an approach for building distributed systems that utilizes common services to deliver application functions as services to either end-user applications or other services. The key benefits of adopting SOA are incremental deployment and maintenance, reuse of services, and exposure of

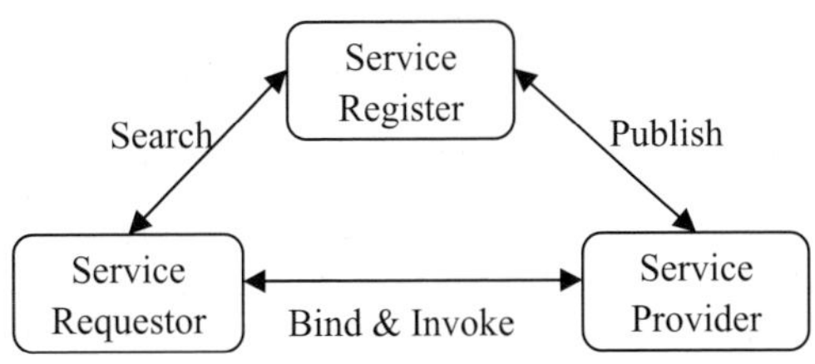

Figure 25.4. Basic web service architecture.

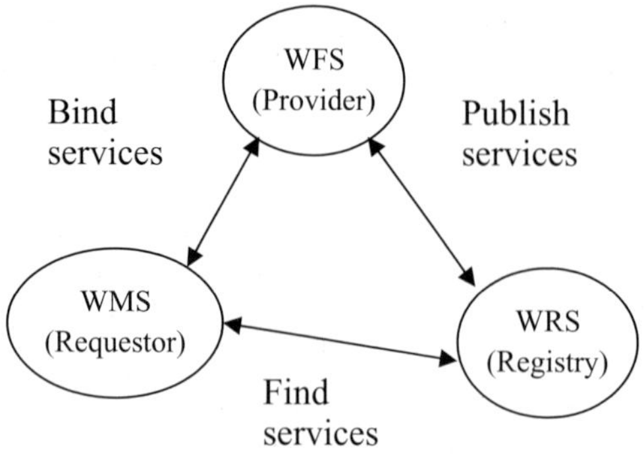

Figure 25.5. Service-oriented architecture using OGC web services.

existing resources. SOA is beginning to shift historically application-centric paradigms to process-centric ones, where processes are essentially "a collection of services invoked in a particular sequence with a particular set of rules, to meet a business requirement" (Endrei et al. 2004). Web services described above have now been widely accepted as one of the technologies that can be used to implement service-oriented architecture. The combination of these two technologies will certainly bring new opportunities to web mapping development.

Open Geo-spatial Consortium (OGC) Web Services and its open web map initiatives represent a most recent and excellent example of how web services technology and SOA may be combined to develop better web mapping applications (Harrison & Reichardt 2001). Figure 25.5 illustrates an example of service-oriented architecture using OGC Web Services.

25.4.2.3 *Grid computing*

Grid computing came from two important inventions: distributed computing and the Internet, where multiple computers connected to each other using Internet protocols to simulate computing power that used to be offered by supercomputers (Yao 2006). The key concept is to split the whole computing process into many sub processes that can be run in parallel on multiple computers distributed over the Internet. To distinguish it from others, grid computing must utilize distributed computing resources not administrated centrally, follow open standards, and achieve a non-trivial quality of services (Foster 2002).

Grid computing has the potential of being an enabling technology that will provide the backbone for the next generation GIS and applications. For discussions on how grid computing can benefit distributed GIS, or so-called "Grid-enabled or Grid GIS", and issues in designing and developing such systems, see Shi et al. (2002), Guan (2004), Shen et al. (2003), and Sun et al. (2005). These early studies indicate the application of grid computing in GIS can help move the traditional "data sharing" focused distributed GIS towards more on geo-processing sharing and distribution.

Little effort has been made on how grid computing can help improve web mapping and GIS, especially in providing more analytical and modelling power. Grid computing, following SOA, provides grid services that are essentially service interfaces associated with grid resources and is implemented based on web technologies. A grid service differs from a web service in that the former is transient—dynamically created and destroyed—and the latter is persistent as long as the service description is still accessible and the service itself is still hosted (Jacob et al. 2005, p118). To make a grid service a potentially transient web service based on grid protocols using WSDL, an Open Grid Services Architecture (OGSA) model may be used (see Foster et al. (2002) for further information on OGSA).

25.4.2.4 *Peer-to-Peer (P2P)*

P2P represents a special kind of grid computing that targets on desktop PCs sitting on "the edge of networks", instead of high-end computers. As opposed to the client-server model, a peer-to-peer network relies on all participating computers (peers) acting as both a server and a client (see Fig. 25.6). Attention has been paid to this architecture due to a number of advantages, such as better utilization of client side computing resources, less bandwidth requirements of the server, easier for sharing between heterogeneous systems, improved scalability, resource aggregation, and potentially easier to design (Hyde 2002, Milojicic et al. 2003, Judd 2005). The key of a P2P system is to facilitate sharing of peers' computer resources (e.g. storage and computer process unit) and data content, collaboration and user control.

According to Hyde (2002), P2P architecture can be classified into four categories:

- Pure P2P: without any central server, completely decentralized.
- Registry P2P: with a metadata server acting as a registrar for new participants.

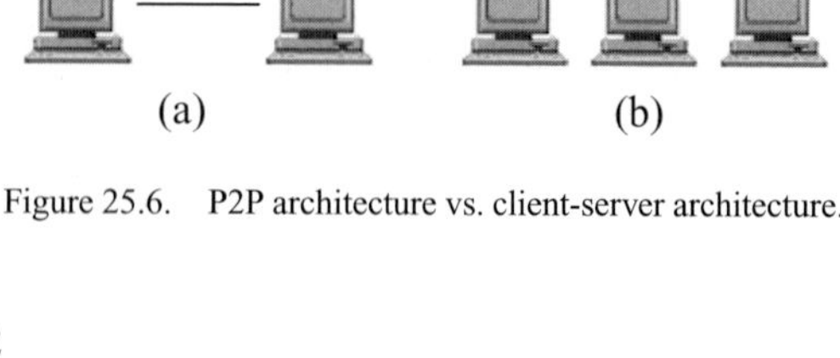

Figure 25.6. P2P architecture vs. client-server architecture.

- Coordination P2P: with a metadata server not only for registering participants but also coordinating activities between participants.
- Hierarchical P2P: with a hierarchy of metadata servers to handling overwhelming number of simultaneous collaborating participants.

P2P is very appealing to GIS data sharing similar to sharing big movies and audio files over the Internet using programs such as BitComet. When combined with web services, it also has the great potential in building novel web mapping/GIS services in P2P environments because both web services and P2P technologies alleviate the complexities of each other (Samtani & Sadhwani 2002). Research efforts have investigated querying and discovering geo-spatial web services using the MBR information to cluster and index the services (Ma et al. 2006), building P2P based GIS web services (Guan et al. 2004), dynamic execution of GIS web services (Chen et al. 2005), and building web services on top of a kind of P2P architecture (Panatkool & Laoveerakul 2002).

The adoption of P2P in GIS in general and web mapping/GIS in particular is slow, attributed to a number of factors. First, P2P technology itself is still facing many challenges; for example, security concerns (e.g. firewalls make the P2P design harder), "losing control" of systems and data (Judd 2005), IP problems in handling of NAT (network address translation) boxes. Second, P2P is not suitable for the situations where one or more large central databases are needed and require accountability and quality assurance of certain authorities, which is the case for most web mapping and GIS services.

25.4.3 *Emerging technologies for web mapping*

Over the last few years, a number of new web technologies have been developed, and subsequently been embraced by web mapping communities. These technologies include services oriented access protocol (SOAP), Asynchronous JavaScript and XML (Ajax), really simple syndication (RSS), web map service API, etc. While these technologies may be categorized under client-server technology, web service technology, or both, they are worth discussing in this separate section due to their significant influence on recent web mapping developments, as will be further elaborated in Section 25.5.

25.4.3.1 *Ajax—Asynchronous JavaScript and XML*

As a web development technique for creating interactive web applications (Wikipedia "*Ajax (programming)*"), Ajax is more a web application model (Merrill 2007), which comprises several technologies mainly including extensible hypertext markup language (XHTML), cascading style sheets (CCS), document object model (DOM), extensible markup language (XML), and server or browser side scripting languages or technologies such as XMLHttpRequest, IFrame and JavaScript. Ajax technology enables the change of the part of web page content without reloading the whole display and behaviours of the existing web page. This is realized through asynchronously requesting extra data from the server and loading it into the web page display in the background.

Web applications based on the Ajax model eliminate the traditional start-stop-start-stop nature of interaction by introducing an intermediary, i.e. an Ajax engine, between the user and the server (see Fig. 25.7). Residing on the client side, the Ajax engine is responsible for both rendering the interface and communicating with the server (Garrett 2005). It allows the user's interaction to happen asynchronously, meaning that the user can continue interacting with the application interfaces while the engine is communicating with the server for extra data. The users of the web application do not need to wait and do nothing while the server is processing user's requests. The bandwidth consumption for web applications using Ajax technology tends to be less. Ajax also allows the adjustment of the map extent accordingly to the screen resolution and the browser window dimension during run-time.

Ajax has great potential in proving more responsive web mapping services and performance enhancement (Tsou 2005, Cha et al. 2007), especially towards providing more analytical and modelling capability. Web mapping communities have already started using this

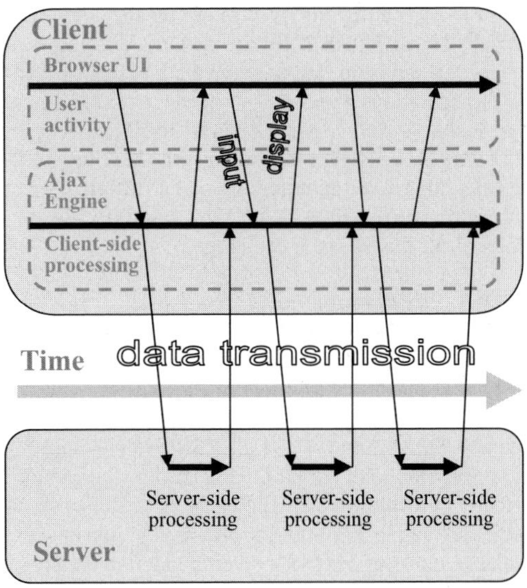

Figure 25.7. The asynchronous pattern of an Ajax application, after Garrett (2005).

technology in developing, for example, Google Maps, a generic integration framework of integrating the Ajax model into the browser-based GIS web services systems (Sayar et al. 2006), and improved interfaces of commercial web mapping software (e.g. ESRI ArcIMS 9.2).

25.4.3.2 *SOAP—services oriented access protocol*

Originally standing for "simple object access protocol", SOAP is a protocol rather than a technology for exchanging XML-based messages over computer networks (Wikipedia "*SOAP*"). Over the Internet, a SOAP message is transmitted through HTTP protocol, and also can be transmitted by other ways (such as JMS or E-mail). SOAP has a number of advantageous features including platform and language independency, XML-based, simple and extensible, W3C standard, and the ability to get around firewalls. However, some of these features may also pose potential security concerns such as firewall passing.

SOAP provides a format of messages sent between Internet/Web applications. The messages are normally encoded as an ordinary XML document that contains the following elements: Envelope, Header, Body, and Fault. While the Envelope and Body elements are required, the other two are optional. The Envelope element is the key to identify the XML documents as a SOAP message; the Fault element provides the necessary information about errors that occurred while processing the message. All the elements above are declared in the default namespace for the SOAP envelope.

Based on SOAP specifications, APIs can be developed for interacting with web mapping services. Web services based SOAP APIs are described using web services description language (WSDL) files, including the operations provided by a service and the message format (using XML Schema) they can receive and address. As such, distributed web mapping services can be invoked remotely by using Simple Object Access Protocol (SOAP)—XML based messaging protocol for the message exchange.

25.4.3.3 *RSS—really simple syndication*

RSS is a lightweight XML format designed for linking and sharing headlines and other web content. RSS facilitates a syndication process for the automated updates of web content such as web page and blog changes (Reed 2006). Reading RSS content requires an aggregator, also called a RSS reader, feed reader or feed aggregator, which allows users to subscribe to a RSS feed by either entering the feed's link into the reader or clicking an RSS icon embedded in web content that initiates the subscription process (Wikipedia "*RSS*"). The aggregator can be web-based, desktop-based, or mobile-based. An easy way to understand how RSS works is to try out some aggregators such as web-based Google aggregator (a Google account is required).

At the first look, RSS does not seem to have anything to do with web mapping. It is GeoRSS that builds the link between RSS and web mapping. GeoRSS is an emerging standard for RSS feeds to also be described by location or geo-tagged in a standardized way in which the location is encoded with enough simplicity and descriptive power to satisfy most needs to describe the location of web content (Reed 2006). GeoRSS has two encodings: GeoRSS simple and GeoRSS GML (geography markup language). The major differences between these two lie in their supported features, especially on coordinate reference systems, and extensibility. Currently, GeoRSS has not been endorsed by OGC as a standard, but it has become a *de facto* standard accepted by major online mapping service companies such as Microsoft Virtual Earth, Google Maps and Yahoo Maps.

GeoRSS is primarily for overlapping RSS feeds with dynamic contents on the maps. It also has the potential for geographic search, e.g. traffic news within 50 km of my home. An interesting example of GeoRSS application is GeoFeeder, software developed by *BRIGHTisolutions*. The software can convert vector data (points, lines and polygons) from ESRI, Google, MapInfo, Autodesk, and GML file formats to GeoRSS format. For example, an airport map in shape file format may be converted into GeoRSS xml with selected attributes, and subsequently fed onto Google maps so the airports and their attributes are marked on the Google map. A trial version of the software may be downloaded from the company's site at: http://www.brightisolutions.com/. In short, GeoRSS holds promise to put the rich geo-spatial content directly in the hands of the consumer community (Reed 2006).

25.5 STATUS OF CURRENT DEVELOPMENTS

Current developments in web mapping are exciting, yet somehow controversial to the traditional concepts, theories and understandings. Reviewing all details of these developments in a chapter like this is impossible. Instead, a review of major developments and their impact are presented in this section.

25.5.1 *Content and knowledge mapping*

Different from traditional web mapping, the development in this direction focuses on providing free online mapping services to the public mass and using web maps as the base to geo-reference and overlay user-added content and knowledge to the map. Enabling technologies, such as web map service API, have also been developed to allow developers and even amateurs to quickly create their own web mapping applications, or mashing up map data sources and services with others. While the development may be influenced by the concept of Web 2.0, it has definitely been led by mainstream information technology (IT) vendors, including MapQuest, Google, Yahoo, Microsoft, and others. Some important characteristics of this development are:

1. The focus has been shifted to connecting the islands of knowledge through locations (Fischer 2007).
2. The role and definition of the map has been changed (Kraak 2004, Van Wyngaarden & Waters 2007).
3. Use of geographic information is increasingly becoming part of everyone's daily life.

The above mentioned four major market players all provide free online map services, with street maps, satellite/aerial image or hybrid view. Google and MapQuest maps do not have a hybrid view, but they allow users to turn on/off street name labels overlapped on image view. Only Yahoo! Maps and Microsoft Live Search Map offer traffic information. Google has recently added terrain view into Google Maps. Both Google and Yahoo provide an overview for ease navigation. On the editing side, Microsoft Live Search Map allows adding of pushpin, while Google Maps provide three types of annotation tools: placemark (similar to pushpin), line and polygon. For collaboration, all map services enable users to email maps, but Google has recently added a collaboration function that allows users to invite other users to collaboratively edit a map using annotation tools. Other functions are common to all these services. It is worth mentioning that Microsoft Live Search Map includes a bird eye view, which is intended to be available for 80% of U.S. and European population, and will expand to Asian (Batty 2007a).

Overall speaking, it is hard to say which of these map services is definitely better than others. While they all share some common functions, each of them has some unique features for capturing the market (see Table 25.4 for a summary). MapQuest, though the first of this kind, seems to be slow in this race of functionality and capacity. For example, the one form entry box for address is only available in its recent release of new beta version in October 2007. Based on the count of monthly unique visits to map services by Compete, Inc. (more recent data can be found at www.compete.com), Yahoo Map and Google Map seem to take the lead in the market (see Fig. 25.8). In addition, all of these map services are struggling with data coverage (street maps and images). As a mass of map and image data are required by these players, many private data acquisition and processing companies have been acquired (Fischer 2007). It is likely that more private companies will be involved.

Building on these map services and data sources, web map APIs have been developed and released by all of these vendors, through which access to their map data sources and embedding maps into web pages became possible. A map API normally provides utilities and services for manipulating maps and adding data content to maps. Provided for free, each map API is tied up with its specific data sources. For example, Google Map API works only with Google Maps data. Map service APIs are intended for web mapping application developers, although they may be used by amateurs if they only want to embed maps into a website or develop a simple web mapping application. And these free APIs are not open source because developers cannot obtain the source codes.

Using map APIs to create web mapping applications has a number of advantages and disadvantages. Base map data is available and tied up with the API. It is easy to mash up maps with other data sources

Table 25.4. Functionality comparison of four map services.

	Google	MapQuest	Microsoft	Yahoo
View				
Street map	X	X	X	X
Satellite image	X	X	X	X
Hybrid view			X	X
Label on/off image	X	X		
Bird eye view			X	
Live traffic			X	X
Navigation				
Zoom In/Out	X	X	X	X
Pan by dragging	X	X	X	X
Overview	X			X
Mapping				
Driving direction	multiple	multiple	A–B	multiple
Reverse route	X	X	X	X
Shortest time		X	X	
Shortest distance		X	X	
Avoid tolls		X		
Avoid highways	X	X		
Action				
Permlink	X	X	X	X
Email map	X	X	X	X
Send to cell	X	X	X	X
Print	X	X	X	X
Save to my map	X	?	X	X
Map API				
Availability	X	X	X	X
Access key/ID	X	X		X
Rate limit	yes	yes	yes	varies

*This table is a modified and updated version of the one from Gruber (2006).

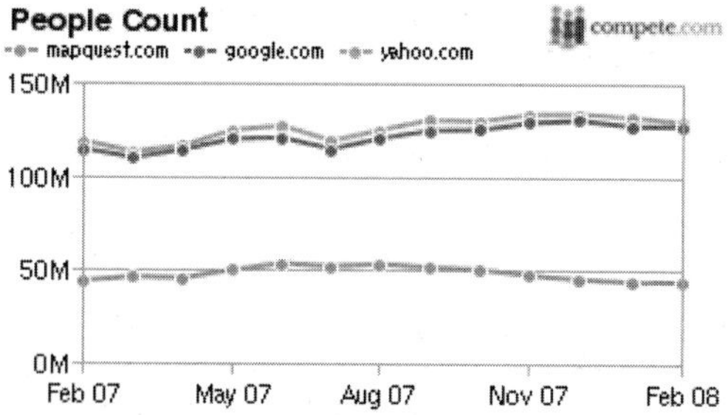

Figure 25.8. Monthly count of accesses to different map services.

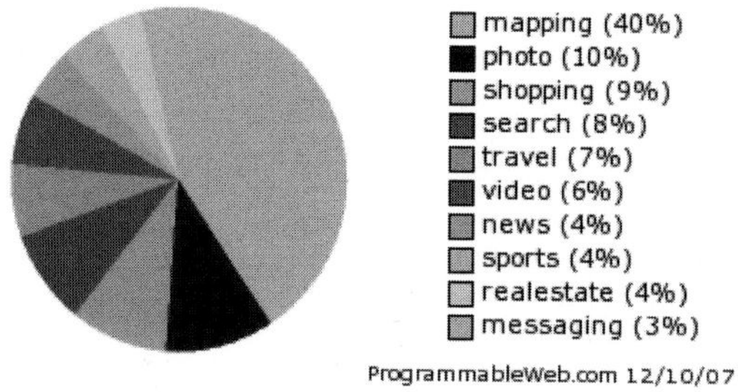

Figure 25.9. Mashup statistics by the Programmable.com. (see colour plate page 504)

available on the Web through their service APIs such as Flickr API and YouTube API. Furthermore, the creation of web mapping application is fast. However, current map APIs do not allow integration of map data from other sources which are not supported by map API, nor do they allow adding of user's local spatial data. The utilities and services provided by map APIs are still limited to geo-coding user added content to point, line and polygon marks, etc. More complex mapping and GIS functions cannot be realized.

The map API and its associated map data services are provided for free under some terms of use, which vary from one provider to another. Some of these terms need to be considered if one plans to use the API to develop web mapping applications. These include:

- *Rate limit*: controlled based on the number of access or queries to map data services through your application created using the API; for example, Yahoo has a varied rate limit depending on which set of APIs you use. Except for Yahoo! Maps Simple API, all other APIs currently have a rate limit of 50,000 queries per IP every 24 hours.
- *Advertising*: all vendors reserve the right to include advertising in the map images provided through the API service, with or without notice prior to the commencement of doing so.
- *Charge of API use*: all vendors currently provide services for free with a condition that they may be used only for services that are generally accessible to consumers without charge. An access key or ID is required for map APIs.

The open maps APIs provided for free by these vendors are the major driving force for the success of map mashup. Figure 25.9 shows the statistics of various mashups created up to 10 December 2007 (completed by *programmableweb.com*), which include the top 10 categories of mashup websites. As seen in the figure, the map-based mashup accounts for a large proportion (40%) of the all counted mashups (the actual number at the time of this writing is 1440). This does not include other types of mashups that are not tagged as "mapping". With the less or little requirements for programming skills and the increased number of mashup editors, creating map mashups will become easier, resulting in more to come.

Map mashups have played a significant role in mapping user-added contents and knowledge, sharing multimedia data like photos and videos, and building social networks. A good example of map mashups for social networking is the Facebook Friend Mapper (needs a Facebook account to access), created using Facebook and Google Maps APIs. In addition, there are numerous map mashups in business and commercial sectors, such as real estate, banking, retailing, etc. Governments also created many map mashups to deliver their information services. While most of government-created map mashups provide a one way communication, the mashup prototyped by Gong (2007) using Yahoo! Maps API and Flickr API allows the public to report the dead birds linking to the West Nile disease.

25.5.2 *Virtual earth*

Virtual Earth, or virtual globe, is a 3-D software model or presentation of the Earth (Wikipedia "*virtual globe*"). It can also be considered as a technology that builds, displays, visualizes and shares 3-D Earth models in a broader sense. Over the last few years, we have seen several virtual Earth technologies become available online, including Google Earth, Microsoft Virtual Earth, NASA World Wind, and Skyline Globe. Except for Microsoft Virtual Earth, all other technologies require the installation of software on the local computer, which needs to be run outside the web browser to access online 3-D models. Microsoft Virtual Earth uses a browser add-on, with one time

Table 25.5. Virtual earth web access information.

Name	URL
Google Earth	http://earth.google.com/
Virtual Earth 3D	http://maps.live.com/
(Tech preview)	http://preview.local.live.com/
NASA World	http://worldwind.arc.nasa.gov/
Skyline Globe	http://www.skylineglobe.com/
Leica Virtual Explorer	http://gi.leica-geosystems.com/LGISub1x250x0.aspx

installation. Based on its Virtual Earth technology, Microsoft also released a very interesting technology preview, allowing people to experience driving in the Virtual Earth environment. While a detailed review of these technologies is beyond the scope of this chapter, Table 25.5 provides the website address of each technology. Further discussions on these technologies may be found in Williamson (2007) and the vendor's website.

One of the important and more advanced developments in this direction is 3-D collaboration. Skyline Globe provides a synchronous collaborative visualization capability. This allows users at different locations to share the same view of the model at the same time, and explore the 3-D view. For example, it permits a user to open a collaboration session on local or server and invite other users by emails to join the session. The collaboration functions include chat, participant list, virtual cursor, free-hand line drawing, and text annotation. Leica Virtual Explorer Collaboration, as an add-on of the Leica Virtual Explorer server, used to allow worldwide users to simultaneously join collaboration sessions to collectively navigate, edit and mark up Leica Virtual Explorer scenes. However, the status of this extension is not clear at the moment of writing. Recently, Unype (http://www.unype.com/) developed an interesting social networking application using Google Earth, which includes chat and private message functions.

Other technologies such as Google Earth and Microsoft Virtual Earth allow users to collaborate in asynchronous mode by "build and share". Users can simple share through emailing their view of the model, possibly with their own annotation or added 3-D models. Since the collaboration is by nature the trends of the Web, as these technologies become increasingly mature we will certainly see more and more collaboration features built onto these Virtual Earth platforms.

The concept of Virtual Earth is very important for the future development of Web GIS. While the development of virtual earth technologies has gone beyond the traditional web mapping domain, it nevertheless provide new horizons to advance the level of geo-spatial technology use.

25.5.3 *Open source and standards, GIS vendors*

Open source is a method to develop software, often called open source software (OSS), which allows anyone to freely view, share, use and modify its source code (Simon 2005). Since GRASS, a flagship open source GIS, was invented, the geo-spatial community has embraced this new method with great enthusiasm. This is evidenced by many open GIS and mapping projects, for examples in Caldeweyher et al. (2006) and Raghavan et al. (2006), and several newly emerged consortia such as OSGeo Foundation (www.osgeo.org).

The *opensourcegis.org* has maintained a long list of open source GIS software developed over the years. Among them, MapServer, GeoTools and most recently GeoServer are notable open source tools for web mapping/GIS. Interestingly, these software tools by now are either certified by OGC or have implemented OGC web services specifications. So far, the open source based web mapping/GIS software is mainly as web map engine or server, WebGIS, web mapping application environment, or a combination of them. In contrast to open web map APIs provided by the IT vendors discussed above, open source web mapping/GIS provides free alternatives, which are technically compatible with the services offered by the vendors. However, putting together an open source web mapping application requires much more effort in programming than the effort needed to create a map mashup using vendor-supplied APIs (Ramsey 2006).

On the other hand, current efforts on open specifications and standards related to web mapping are coordinated by the Open Geo-spatial Consortium (OGC), the World Wide Web Consortium (W3C) and the International Standards Organization (ISO). While W3C is mainly responsible for those Web specifications used in developing web mapping/GIS applications such as HTML, the role of ISO is to endorse open specifications as ISO standards. So far, ISO has 58 standards related to geo-spatial information under TC211; for example, metadata, Web map server interface, Geography Markup Language, and Web Feature Service standards.

As of December 2007, OGC has 41 open specifications and has registered 406 compliant or implementing products (www.opengeospatial.com). An important initiative of OGC related to web mapping is the OGC Web Services (OWS), now in its Phase 5. The initiative works on a number of threads including Sensor Web Enablement (SWE), Geo Processing Workflow (GPW), Agile Geography, Compliance Testing (CITE), and CAD/GIS/BIM. The effort closely relates to some of the new emerging technologies discussed above, such as web services, service-oriented architecture, SOAP, etc., and will complement existing web mapping and web map service specifications. It should be noted that these threads are also in line with the developments discussed in Section 5.1.

OGC standards complement many IT standards used in web mapping. For example, for the web mapping services using SOAP technology to define the process of exchanging messages, OGC standards provide complement in "defining how the content of available services are determined as well as the specific structure of information requests and replies" (Batty 2007b). Furthermore, the combination of open specifications and open source software has the potential to provide geo-processing capabilities not currently seen in commercial web-based GIS or free web map services (Anderson & Moreno-Sanchez 2003). Open standards and open source solutions provide significant benefits to many application sectors, especially governments.

However, one has to understand that neither proprietary nor OSS is a panacea (Simon 2005), meaning that the continuing development of open source solutions will not necessarily diminish the further development of traditional vendor-based solutions, at least not in the near future. Fortunately, OGC standards do not discriminate open source and proprietary.

While continuing the constant upgrading of their traditional desktop GIS products, GIS vendors have also responded to the emerging trend to deliver their GIS across organizations and on more platforms. The shift from PC-based solutions towards more server-based ones provides basic infrastructure for vendors to develop more open web services. The current development by the GIS industry in the area of web mapping/GIS has basically followed two paths: commercial server software and subscription-based services.

The subscription-based mapping/GIS services essentially offer online map services similar to those offered by IT vendors, acting as application service provider (ASP). This kind of service, such as ESRI ArcWeb services, also provides service APIs that developers can use to integrate mapping functions and GIS content into their own browser-based applications. So, what is the difference? Let us use so-called enterprise map mashup as an example. In order to create an enterprise mashup, mostly we will need to use a base map, a geo-coder, and a Web interface, as illustrated in Figure 25.10. If the data sources (database), geo-coding services, and web interface created by APIs are all free, it is free! However, this is not the case of web mapping services provided by traditional GIS vendors such as ArcWeb and MapPoint (Ramsey 2006).

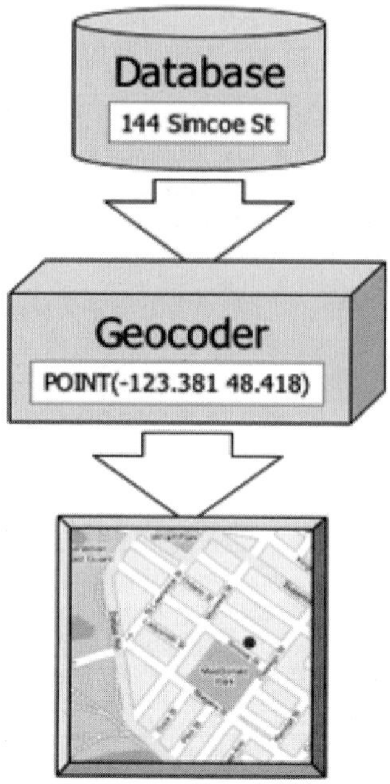

Figure 25.10. Enterprise mashup reading data from the corporate database.
Source: Ramsey (2006).

25.6 ISSUES, CHALLENGES AND FUTURE DIRECTIONS

25.6.1 *Social and organizational issues: data policies, quality, pricing, privacy and ownership*

25.6.1.1 *Data quality and integration*

Data and data integration still remain as big issues for web mapping/GIS developers, although the current developments in distributed databases, open access specifications and so on have greatly eased the associated problems. Before Web 2.0, web services, virtual earth and the like, the problems used to relate to different data formats, models, metadata, and even different standards when dealing with data integration. Now, they have gone beyond this scope and move on to semantics, ontology, integration of CAD data, etc.

Data quality is more problematic to the recent free online map services than the data used to be provided by government for free or by data providers for a fee. Study carried out by Nivala et al. (2007) indicates that all free map services used in his study have various problems with data coverage, accuracy and currency, as well as metadata information. Typical problems found include browsing to a display showing "data not available", coarse resolution after zooming into a certain level, mismatched data, and most importantly, no information about the quality of the data viewed. This has great implications for how these services should be used and what applications may be built based on these services. While free maps may be good for social mapping and providing general location-based information, it may be questioned if used for some mission critical applications such as emergency planning and response.

As some of these vendors offer services for supporting restricted access web applications such as Google Enterprise Map API, a question needs to be asked: can government services rely on the data services provided by non-government organizations? The answer is yes and no. In his recent position paper, Ball (2007) raises the concern of the potential dangers of relying on non-government organizations for the depiction of our shared reality, causing public outcry. Reliance on free commercial mapping services and abdicating government's role of map data creator and data quality arbiter will cost. However, if the intention of using these free map services is to provide non-mission critical services to its citizens, why should the government not use them?

25.6.1.2 *Data policies, privacy and ownership*

The new developments under the Web 2.0 have raised many questions on data. Who controls the data? What about the protection of privacy related to the collected data? Who owns the data? Previously, this did not seem to be a major problem when the mapping services were provided by an organization or company for a specific application purpose. With the free mapping services, everyone can freely add his/her own data provided the terms of use are not violated. The data may be uploaded to a number of servers from where the services are provided. Alternatively, the application, such as map mashup, may divert the user-added content to the server hosting the application. While service API providers normally have clear terms of use on data policies, privacy and ownership, the grey area is in those applications or mashups developed using the APIs. As such, mashup developers need to maintain the balance between the protection of intellectual property rights, consumer's privacy and the publicity as well as free flow of information.

The free map services have put a lot of pressure on traditional GIS vendors who offer web mapping/GIS software. However, "free" does not come without conditions. Most providers have clauses in their terms of use, stating that the services, maps and API, may be used only for personal, non-commercial purpose. In addition, the "no warranty on data" limits these services further in comparison to GIS vendor's solutions.

As stated by Cho (1998): "In providing instant access to vast amounts of data, GIS also provide the opportunity to abuse, to misinform and to invade the privacy of individuals on a greater scale than ever before." Although mapping service providers have clear use policies, for those who want to develop real-time mapping applications over the Web using their map APIs, some terms related to adding real-time tracking data obtained from any tracking systems and devices to the maps are itchy. The level of restrictions varies from one vendor to another. The following two excerpts from Google and Yahoo may serve as examples to show such differences:

- *Google*: "may not use the Service with any products, systems, or applications installed or otherwise connected to or in communication with vehicles for or in connection with: (a) real time route guidance (including without limitation, turn-by-turn route guidance and other routing that is enabled through the use of a sensor); (b) any systems or functions for automatic or autonomous control of vehicle behaviour; or (c) dispatch, fleet management or similar applications."
- *Yahoo*: "shall not use the Yahoo! Maps APIs with location information that is less than 6 hours old and derived from a GPS device or any other location sensing device"; and "shall not use the Yahoo! Maps APIs with location information derived from a GPS device or any other location sensing device where such information was not uploaded to your application or service directly by the end user."

25.6.2 *Technical issues: functionality, interfaces, usability and quality of services*

25.6.2.1 *Web mapping vs. web GIS*

Current web mapping software and services have not been proliferated in providing the analytical capabilities of traditional GIS (Ball 2006), nor modelling functions towards deriving geographic information, leading to new knowledge. GIS software, as a decision making tool, still remains the domain of relatively smaller group of users who are technically savvy and have some kind of basic knowledge of GIS (Van Wyngaarden & Waters 2007).

A question very often asked is to what extent the web mapping system is to be developed to qualify for a web GIS. A few research projects have tried to include spatial analysis and geo-processing functions in web mapping applications or as they called, web GIS (Kinston et al. 2000, Anderson & Moreno-Sanchez 2003). The functions are provided in a step-by-step fashion. Wright (2002) suggested a direction of developing multi-level Web-to-database interfaces for web GIS, to enable the customized access interfaces, so the extent of mapping/GIS functional coverage and design may be separated according to the user group. The concept is similar to the interface customization available in most desktop GIS, where functional components are already part of the package. With the service-oriented architecture, it is possible to customize multi-level interfaces at the web client side by compositing need-only services for a specific level of users.

Open standards and specifications are essential for developing more powerful web mapping programs towards web GIS using SOA-based web services. If web map services follow different standards and interface specifications, putting them together to provide integrated services will be problematic. The problem already existed with current map mashup practices and technologies. For example, using Yahoo! Maps API, Flickr API and flash technology to mash up a web application reporting dead birds potentially carrying West Nile diseases experienced problems because Flickr API does not support direct interaction with flash components (Gong 2007).

An close examination of the existing IT vendors' map APIs reveals some problems not only need to be considered by the developers interested in mashing up web mapping applications using them, but also providing lessons to other types of web map services. Some of the problems are 1) not supporting the integration of geo-spatial data from other data services, i.e. real-time spatial overlay is not possible; 2) not supporting loading of local spatial data; and 3) no spatial analysis functions provided except for measurements.

25.6.2.2 *Interfaces and functionality*

The new interfaces provided by Google Maps, Yahoo! Maps and so on have greatly changed the users' perception of web mapping/GIS, including those professional users and GIS experts. This is both good and bad news for the traditional GIS industry. On the one hand, it may put pressures on them to develop better software interfaces to match the level of ease currently enjoyed by ordinary users, yet to maintain the balance between the interface and the traditionally-rich GIS functions. On the other hand, traditional GIS vendors can learn lessons from this wonderful user experience of interfaces.

However, there are still many problems in interfaces designed for web mapping, especially those targeting the general public. As described in Section 2, maps have a role as interface, map metaphor. The interface of web mapping programs hence has two parts: the interface of the website holding map displays and the map itself. As such, map design becomes part of web mapping/GIS interface design. Currently, meaningful cartographic symbolization is still of concern from both professional cartographer and user's perspectives. In many cases, the symbols, colours, use of space, etc. are not clear, meaningful, and consistent, e.g. inconsistent presence of zooming tools and scale bars (Nivala et al. 2007).

Another subtle but not trivial issue with current web mapping is the lack of accessibility support. According to *WebExhibits.org*, approximately 5% to 8% of the men and 0.5% of the women of the world are born colour blind, not mentioning the population of other groups who have severer assessing disabilities. Accessibility has big implications in web mapping, especially under the current pervasive web mapping development. For example, we all are used to colourful maps now, but think about those colour blind people. Little research has been done in this area so far.

25.6.2.3 *Usability and quality of services*

One of problems in today's web mapping and GIS is the lack of sufficient usability studies. Nivala et al. (2007) recently conducted a study on the usability problems of some popular free map service websites including Google Maps, MSN Maps, MapQuest, and *Multimap.com*. Using "a tourist is planning to visit London" as the user case, the study found a total of 343 unique usability problems distributed among the map sites, with Multimap.com being marginally the biggest contributor.

Many published papers include a section on usability testing or study to examine the usefulness of their specific web mapping/GIS systems or applications. However, in most cases their studies were based on a simple user study approach, i.e. having a small group of users to actually use the systems while keeping a close watch on their use, response and behaviours. More empirical studies using formal usability engineering techniques such as the one used in Nivala et al. (2007) are needed to identify barriers and build a better knowledge base to guide future developments.

Finally, a closely-related issue is the quality of services. As we continue to move into a service-oriented web mapping paradigm, the quality of services provided online will be of concern to both service providers and users. The quality of online services depends on many factors such as network performance (Coleman 2005), throughput and response time (Peng & Tsou 2005), user experience, and, to a less extent, the effectiveness of supporting underlying business processes. With some of the new technologies discussed above, technical quality of services has certainly been greatly improved compared to ten or five years ago. However, more study is needed in qualifying quality of services in a business or use context.

25.6.3 *Future directions*

Ten years ago, people, primarily GIS and mapping professionals, predicted that web mapping and GIS would get into the direction of more data-rich, interactive client/server systems that overcame technical limitations with the increased industry interest and user communities (Plewe 1997, Hardie 1998). Three or four years ago, the expectations went to distributed GIS and services with component-based applications, which are similar to the current situation (Peng & Tsou 2005). So what will be next?

As we have witnessed over the years, technologies, especially web related ones, have played an indispensable role in shaping the development of web mapping and GIS, and technology development is highly unpredictable. In this sense, future directions are always hard to precisely predict. Nevertheless, the future developments in web mapping or GIS may be towards:

25.6.3.1 *Service-oriented, workflow-enabled services*

As the service-driven paradigm continues to dominate the web mapping/GIS development, it is expected workflow principles and technologies will become increasing important. Workflow can help to achieve better understanding of business and service processes and automate the service execution process. It is expected that either workflow principles or engines will be seamlessly integrated into the web mapping service-oriented architecture.

25.6.3.2 *More analytical, functional-rich and collaborative*

We will see more traditional GIS functions being integrated into web mapping/GIS applications in a way transparent to their users, using distributed analytical and geo-processing services coordinated by well-defined workflows. Analysis other than simple access and visualization will be the trend for the whole web community (Burners-Lee 2007). The Web was born with collaboration in mind. As the research and development in the area of collaborative GIS continues to generate solid outcomes in theories, designs, and insights on related social issues (Balram & Dragicevic 2006), we will soon see pervasive collaborative web mapping/GIS tools with rich analytical functions to support various levels of decision support.

25.6.3.3 *More open source based*

Although the debate between open source and proprietary solutions will continue, it is expected to see the increasing adoption of open source based web mapping/GIS solutions. Focused on web services, the next generation web mapping/GIS will be more interoperable. Open standards are the key to ensure interoperability, while open source is likely to provide complete implementation of open standards. According to Simon (2005): "OSS and open standards are intimately connected and that the inherent value of open-source adoption may be attributable in large part to the embodiment of open standards in OSS."

25.6.3.4 *Social mapping and networking*

The development of social mapping and geo-referencing user-added and user-controlled data content will continue to be an important area for web mapping researchers to focus. Accordingly, data policies, privacy and ownership issues as well as technical challenges such as searching by location rather than keyword and location intelligence will be important research areas.

REFERENCES

Anderson, G. & Moreno-Sanchez, R. 2003. Building web-based spatial information solutions around open specifications and open source software. *Transactions in GIS* 7(4): 447–466.

Ball, M. 2006. Position: moving beyond platforms and services. *GeoWorld* 12(2006): 8.

Ball, M. 2007. Position: data want to be free. *GeoWorld* 5(2007): 8.

Balram, S. & Dragicevic, S. 2006. *Collaborative Geographic Information Systems*. London: Idea Group Pulishing.

Batty, P. 2007a. TECHTIME: Here comes Microsoft. *GeoWorld* 1(2007): 42.

Batty, P. 2007b. TECHTIME: Are web services, SOA and OGC important? *GeoWorld* 4(2007): 38.

Berners-Lee, T. 2007. *Future of the World Wide Web. Testimony to the United States House of Representatives, Internet and Telecommunications Subcommittee, March 1, 2007.*

Bonnici, A.M. 2005. *Web GIS software comparison framework (draft)*. Geomatics Dept, Sir Sandford Fleming College, Oct 2005.

Caldeweyher, D., Zhang, J. & Pham, B. 2006. OpenCIS—open source GIS-based web community information system. *International Journal of Geographical Information Science* 20(8): 885–898.

Cha, S.J., Hwang, Y.Y., Chang, Y.S., Kim, K.O. & Lee, K.C. 2007. Integrating Ajax into GIS web services for performance enhancement. *Lecture Notes in Computer Science* 4488(2007): 562–568.

Chen, L., Ma, X., Chen, G., Sun, Y. & Feng, X. 2005. A peer-to-peer architecture for dynamic executing GIS web service composition. *Proc. of 2005 IEEE International* 2: 979–982.

Cho, G. 1998. *Geographic Information Systems and the Law*. New Jersey: John Wiley & Sons.

Coleman, D.J. 1999. Geographical information systems in networked environments. In P.A. Longley, M.F. Goodchild, D.J. Maguire & D.W. Rhind (eds.), *Geographical Information Systems: Principle and Technical Issues*: 317–329. New York: John Wiley & Sons, Inc.

Culpepper, B. 1998. Weave maps across the web: 1998 edition. *GeoWorld* 11(1998): 46–52.

Dawe, P. 1996. *An investigation of the Internet for spatial data distribution*. M. Eng. Report, Department of Geodesy and Geomatics Engineering, University of New Brunswick, Fredericton, N.B., Canada

Dragicevic, S. 2004. The potential of web-based GIS. *Journal of Geographical Systems* (2004)6: 79–81.

Endrei, M., Ang, J., Arsanjani, A., Chua, S., Comte, P., Krogdahl, P., Luo, M. & Newling, T. 2004. *Patterns: service-oriented architecture and web services.* IBM Redbooks. http://www.redbooks.ibm.com/. (Last accessed on December 29, 2007).

Fischer, F. 2007. Digital globes of knowledge and information. *GeoInformatics* 2007(Apr/May): 52–54.

Foresman, T. (ed.). 1997. *The history of geographic information systems.* Prentice Hall.

Foster, I. 2002. *What is the grid? A three point checklist.* Argonne National Laboratory & University of Chicago. http://www-fp.mcs.anl.gov/~foster/Articles/WhatIsTheGrid.pdf (Last accessed on December 29, 2007)

Foster, I., Kesselman, C., Nick, J.M. & Tuecke, S. 2002. *The Physiology of the Grid: an open grid services architecture for distributed systems integration.* A research paper (draft), subject to continuous revision, June 2002.

Garrett, J.J. 2005. *Ajax: a new approach to web applications. adaptive path Blog, February 18, 2005.* http://www.adaptivepath.com/ideas/essays/archives/000385.php. (Last accessed on December 29, 2007)

Gong, J. 2007. *Mashing up web-based participatory GIS: potential for future e-government services.* Unpublished M.A.Sc. Thesis, Ryerson University, October 2007.

Gottschalk, K., Graham, S., Kreger, H. & Snell, J. 2002. Introduction to Web services architecture. *IBM Systems Journal* 41(2): 170–177.

Gruber, F. 2006. *Comparing the mapping services.* http://www.techcrunch.com/2006/04/17/comparing-the-mapping-services.

Guan, Q. 2004. Grid-enabled Urban-CA GIS. *UCGIS Assembly 2004, Adelphi, Maryland, USA, Oct 2004.*

Guan, J.H., Zhou, S.G., Wang, L.C. & Bian F.L. 2004. Peer-to-peer based GIS web services. *Proc. of the 20th ISPRS Annual Congress, Istanbul, Turkey, July 2004.*

Hardie, A. 1998. The development and present state of web-GIS. *Cartography* 27(2): 11–26.

Harrison, J. & Reichardt, M. 2001. *Introduction to OGC Web Services.* OGC White Paper, May 2001.

Hyde, D.C. 2002. How new peer to peer developments may affect collaborative systems. *Proceedings of the 3rd International Symposium on Collaborative Technologies and Systems, San Antonio, Texas, January 28, 2002.*

Jacob, B., Brown, M., Fukui, K. & Trivedi, N. 2005. *Introduction to grid computing.* IBM Redbooks. http://www.redbooks.ibm.com/, Dec 2005. (Last accessed on December 29, 2007)

Joseph, J., Ernest, M. & Fellenstein, C. 2004. Evolution of grid computing architecture and grid adoption models. *IBM Journal* 43(4): 624–645.

Judd, D.D. 2005. Geocollaboration using peer-peer GIS. *Direction Magazine* 05(2005) http://www.directionsmag.com/printer.php?article_id=850. (Last accessed on December 29, 2007)

Kingston, R., Carver, S., Evans, A. & Turton, I. 2000. Web-Based Public Participation Geographical Information Systems: An Aid To Local Environmental Decision-Making. *Computers, Environment and Urban Systems* 24(2): 109–125.

Kraak, M.J. 2001. Settings and needs for web cartography. In M.J. Kraak & A. Brown (eds.), *Web Cartography*: 3–4. New York: Francis and Taylor.

Kraak, M.J. 2004. The role of the map in a Web-GIS environment. *Journal Geographical Systems* 6: 83–93.

Limp, W.F. 1997. Weave maps across the web. GeoWorld 9: 46–55.

Limp, W.F. 1998. Weave maps across the web. *Adams Business Media* 11 :46–52.

Limp, W.F. 1999. Mapping hits warp speed on the World Wide Web. *GeoWorld* 9.

Lowe, J.W. 1999. Designing an effective GIS web site. *Geo Info Systems* 10: 36–44.

Ma, X., Li, G., Xie K. & Shuai, M. 2006. A peer-to-peer approach to Geospatial Web Services discovery. *Proc. of the 1st international conference on Scalable information systems, Hong Kong, May 2006.*

Merrill, D. 2006. *Mashups: the new breed of web app: an introduction to mashups.* IBM DeveloperWorks, http://www.ibm.com/developerworks/xml/library/x-mashups.html. (Last accessed on December 29, 2007)

Milojicic, D.S., Kalogeraki, V., Lukose, R., Nagaraja, K., Pruyne, J., Richard, B., Rollins, S. & Xu, Z. 2003. *Peer-to-peer computing.* HP Laboratories Palo Alto, Hewlett-Packard Company.

Nivala, A.M., Brewster, S. & Sarjakoski, L.T. 2007. Usability problems of web map sites. *Proc. of the International Cartographic Conference, Moscow, August 2007.*

O'Reilly, T. 2005. *What Is Web 2.0: design patterns and business models for the next generation of software.* An article published on O'Reilly Media, Inc. (www.oreilly.com).

Panatkool, A. & Laoveerakul, S. 2002. Decentralized GIS web services on grid. *Proc. of the Open source GIS—GRASS users conference, Trento, Italy, September 2002.*

Peng, Z.R. & Tsou, M.H. 2003. *Internet GIS: distributed geographic information services for the internet and wireless network.* Hoboken: John Wiley & Sons, Inc.

Peng, Z.R. & Zhang, C. 2005. Services-oriented architecture in Internet GIS. *GIS Development* (online) 9(10).

Pietroniro, E. & Fichter, D. 2007. Map mashups and the rise of amateur cartographers and mapmakers. *ACMLA Bulletin* 2007(127): 26–30.

Plewe, B. 1997. *GIS Online: information retrieval, mapping, and the Internet.* Albany: OnWord Press.

Putz, S. 1994. Interactive Information Services Using World-Wide Web Hypertext. *Computer Networks and ISDN Systems* 27 (2): 273–280.

Raghavan, V., Masumoto, S. & Hastings, D. 2006. Free and open source software for geoinformatics—present status and future prospects. *International Symposium on Geoinformatics for Spatial Infrastructure Development in Earth and Allied Sciences, Ho Chi Minh, Vietnam, Nov 2006.*

Ramsey, P. 2006. Mashing up the enterprise. *Geospatial Solutions* 4(2006).

Reed, C. (ed.). 2006. An introduction to GeoRSS: a standards based approach for geo-enabling RSS feeds. *OGC White Paper, Open Geospatial Consortium Inc. July 19, 2006.*

Samtani, G. & Sadhwani, D. 2002. Web Services and Peer-to-Peer Computing: Companion Technologies. *Web Services Architect* (online journal) May 1, 2002.

Sayar, A., Pierce, M. & Fox, G. 2006. Integrating AJAX approach into GIS visualization web services. *Proc. of International Conference on Internet and Web Applications and Services/Advanced International Conference on Telecommunications, Guadeloupe, French Caribbean, Feb 2006.*

Seggern, M. 1994. The enterprise GIS: a client/server approach using distributed relational databases to create a multi location/multi application GIS. *Proc. of the 5th European Conf. and Exhibition on Geographical Information Systems*, Paris, France, March 1994.

Shen, Luo, J., Zhou, C., Cai, S., Zheng, J. Chen, Q., Ming, D. & Sun, Q. 2003. Architecture design of grid GIS and its applications on image processing based on LAN. *Information Sciences* 166(2004): 1–17.

Shi, Y., Shortridge, A. & Bartholic, J. 2002. Grid computing for real time distributed collaborative geoprocessing. *Proc. of the Symposium on Geospatial Theory, Processing and Applications, Ottawa, Canada, July 2002.*

Simon, K. 2005. The value of open standards and open-source software in government environments. *IBM System Journal* 44(2): 227–238.

Sun, Q., Chi, T., Wang, X., Zhong, D. 2005. Design of middleware based grid GIS. *Proc. of Geoscience and Remote Sensing Symposium, July 2005.*

Tsou, M.H. 2005. The big changes in Internet GIS. *GIS Development* (online) 9(10).

Van Wyngaarden, R. & Waters, N. 2007. An unfinished revolution: gaining perspective on the future of GIS. *GeoWorld* 9: 22–25.

Williamson, M. 2007. QUICKATKEREVIEW: Microsoft Virtual Earth. *GeoWorld* 4: 36–37.

Worboys, M. & Duckham, M. 2004. *GIS: a computing perspective.* Boca Raton: CRC Press.

Wright, D. 2002. *USGIS Research priorities: pervasive computing*. http://www.ucgis.org/.

Yao, C.H. 2006. *Grid computation—the fastest supercomputer in the World.* CSA Discovery Guides, November 2006. http://www.csa.com/discoveryguides/grid/review.php.

Zimmermann, H. 1980. OS1 reference model-the ISO model of architecture for open systems interconnection. *IEEE Transactions on Communications* 28(4): 425–432.

Advances in Photogrammetry, Remote Sensing and Spatial Information Sciences: 2008 ISPRS Congress Book – Li, Chen & Baltsavias (eds)

CHAPTER 26

Updating geospatial databases from images

Christian Heipke, Peter A. Woodsford & Markus Gerke

ABSTRACT: This chapter reviews the process of updating geo-spatial databases from images. After a general discussion of updating and its relation to data quality updating, we discuss the generic system architecture for updating geo-spatial databases. We then reflect upon the role of automation in our field, before giving examples of two state-of-the-art developments, one for automatic data verification and change capture from images and a second for automatic data validation. The chapter concludes with some observations for future developments.

Keywords: System architecture, updating, data quality, images

26.1 INTRODUCTION

Geo-spatial information, i.e. information about objects and facts with spatial reference, is an essential part of the national and international infrastructure for the information society. It is estimated that some 80% of our daily decisions rely on geo-spatial information. Geographic information systems (GIS), which allow for the acquisition, storage, manipulation, analysis, visualization and dissemination of geo-spatial information, are therefore of prime interest to society at large. Due to the demands of a number of modern applications—we mention only car navigation and disaster management as two very obvious examples—the updating cycles of the past, amounting to years, are no longer acceptable. Today's applications require much more current information. Thus the need arises for frequent updates of geo-spatial information.

The currency of a database counts among the most important key components of the quality description of the data set. Every application based on spatially referenced data or geo-data requires some knowledge about their quality or at least an idea of the consequence of possible errors and the risk associated with these errors.

Sonnen has highlighted the importance of data quality as spatial data moves into enterprise environments (or mapping moves to spatial information):

"Data quality is a problem we need to address if we in the geo-spatial industry expect to be a part of the enterprise IT picture. Our most pressing need is a simple, reliable way to answer: 'Are these data fit for this purpose?' each time spatial data are merged or shared in an enterprise system" (Sonnen 2007).

Sonnen has also highlighted the fact that data quality issues may be resolved or exacerbated within each data management function. Careful attention to quality issues in the design of data management workflows is required to minimize problems.

Data quality is usually described by a certain set of measures, which express comprehensive and useful criteria. These should enable the user to compare the quality of different data sets. Therefore, quality measures are part of standards or specifications from ISO (International Standards Organisation) and OGC (Open Geo-spatial Consortium). ISO 19113:2002 (ISO 2002) establishes the principles for describing the quality of geographic data and specifies components for reporting quality information. It also provides an approach to organizing information about data quality. Rather than going into detail about these specifications, we start with a subdivision of quality measures into two categories:

- consistency with respect to the data model, also called logical consistency
- consistency with respect to the real world, i.e. consistency of data and reality within the scope of the model.

We refer to the first category as logical consistency since it can be checked by logical reasoning without any comparison of the data to the real world. We can perform a complete and fully automatic data validation for logical consistency using solely the data set and the specifications of the data model. Format specifications, topological constraints, uniqueness of identifiers and domains of attribute values are typical relevant criteria.

For the second category, a comparison of data and reality is required. Basically the comparison can be performed by means of current sensor data or field work. A complete comparison of data and reality requires significant effort and cost, but in return it furnishes all the necessary update information.

Updating a database should obviously not lead to a loss of data quality. Therefore, any change made to the data in the database in order to maintain or restore the consistency with respect to the real world has also to be validated for logical consistency. Thus, both aspects must be considered when updating a geo-spatial database.

The chapter is organized as follows. After a description of the updating task, we discuss the generic system architecture for updating geo-spatial databases. We then reflect upon the role of automation in our field, before giving examples of two state-of-the-art developments, one for automatic data verification and change capture from images and a second for automatic data validation. The chapter concludes with some observations for future developments.

26.2 THE UPDATING TASK

26.1.1 *Verification and data capture*

Updating a database refers to the task of comparing two or more data sets (one representing the current state of a database, the others representing some more recently acquired data) with the aim of detecting and capturing changes, and to import these changes into the database, while keeping the new data logically consistent with the existing data. Note that in this text we use the term "updating" as a synonym for "revision". Updating tasks that need to be supported are the creation, deletion, splitting and merging of objects, and the modification of their geometric, topologic, thematic and temporal descriptions. Basically three tasks are being performed during updating: (a) in a verification step data existing in the database are checked against the new source data for correctness and geometric accuracy; (b) in the following data capture step, new or changed objects are added to the database; and (c) the new state of the database is checked for logical consistency, either stepwise or in a final process. By means of such updating, the database is constantly adapted to the changes of the landscape. Updating is thus closely related to temporal issues in GIS.

In general it will be necessary to use multiple data sources for updating a GIS database, and the second data set can take the form of imagery, results from a field survey, or data acquired from some other source. From a photogrammetric point of view, and for other reasons, updating from images is most attractive (Heipke 2004). The challenge here is to automate all three updating tasks.

26.1.2 *Periodic and continuous updating and the role of images*

While traditionally most topographic information was captured from aerial images, the much shorter updating cycles of today have resulted in major changes in the processing chain. Often field data are incorporated into the database as soon as they become available; thus we move from a periodic to a continuous updating process. Obviously, in many countries man-made changes have gone through a detailed planning process prior to being executed, and information about these changes can be provided by the planning authority, although it is useful only after translation into the schema of the topographic database and field verification. New information sources currently being brought into use range from using large groups of people commonly collecting information via the internet (such as ratings of all sorts) to using taxi driver tracks recorded by GPS as information for the road network, or having the postman deliver information about building changes.

While continuous updating is a very attractive alternative, there is a danger that the data quality slowly degrades over time. Therefore, we believe that continuous updating from data sources such as those mentioned above should be complemented with a periodic verification of the complete database. We also argue that for verification aerial and satellite images are prime sources of information. In addition, they can be used to identify areas that have changed, even though the actual capture of the changed data is then performed on the ground.

26.1.3 *Incremental updating and versioning*

Two important concepts for the update of geo-spatial information are incremental update and versioning. A methodology for updating geographic databases using map versions is given in Peerbocus et al. (2002). Users often link the core geo-spatial information to some application data of their own and thus create value-added information. In order not to lose the key links between the core data and the application data, once a new version of core data becomes available, it is mandatory to provide "change only" information. In this way, users are able to incrementally update their own data sets only in those areas where change has actually occurred. Also, updating is often done in parallel by different operators, possibly using mobile equipment or in distributed environments. In this case, versioning allows exclusive write access to parts of the database to be given to different users and to create various spatially non-overlapping versions. In a second step, these different versions have to be merged to generate a consistent new data set. Incremental updating and versioning can also be used to record time series of events.

26.1.4 *Updating and new generation data models*

Another fundamental factor influencing updating issues is the emergence of a new generation of data models, designed to be more capable and to serve multiple purposes; in short to realize the goal of 'store and update once, use many'. A recent EuroSDR Workshop (EuroSDR 2006) explored this trend, the reasons for it and the current state-of-the-art. It revealed an encouraging degree of convergence and consensus in the emerging pattern (reliance on standards, feature/object based models, use of persistent unique identifiers, use of UML (Unified Modelling Language) as a design tool, use of XML/GML (Extensible Markup Language/Geographic Markup Language) as delivery vehicle) and some research issues, particularly in dealing with formal semantics. These trends are not restricted to Europe and are evident globally.

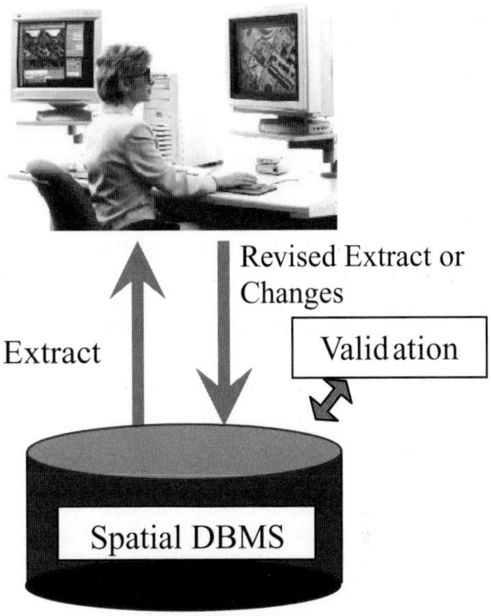

Figure 26.1. File-based data exchange: several cycles are needed, resulting in a very low efficiency. (see colour plate page 504)

26.3 A GENERIC SYSTEM ARCHITECTURE FOR UPDATING GEO-SPATIAL DATABASES

The first and perhaps most important observation with respect to system architectures is a major trend to database-centric architectures. This centralizing tendency, together with the implications of more complex data models, has profound implications for update (Woodsford 2004). Update processes need to be 'data model-aware', and to be tightly coupled with validation services, to avoid costly and lengthy error detection-correction cycles. Whilst it is possible to replicate such validation services in each update client (field, photogrammetric systems, imagery change detection systems) as models become more complex, and business rules come to the fore, such an approach becomes more costly and difficult to sustain. Traditional approaches, with several lengthy iterations to detect and correct errors (see Fig. 26.1) become less and less viable. Note that while Figure 26.1 (and also Fig. 26.2) show a human operator who carries out data verification and change capture, these tasks can at least be partly automated (see also Section 26.4).

Good results have been achieved by close-coupling of a photogrammetric client with an object-oriented database (e.g. BAE Systems SOCETSET with Laser-Scan (now 1Spatial), Gothic (Edwards et al. 2000) or with a geo-spatial database (SOCET SET with ESRI ArcGIS), (BAE Systems 2007). This tightly-coupled approach removes the need for repeated re-validation of changes and lengthy delays.

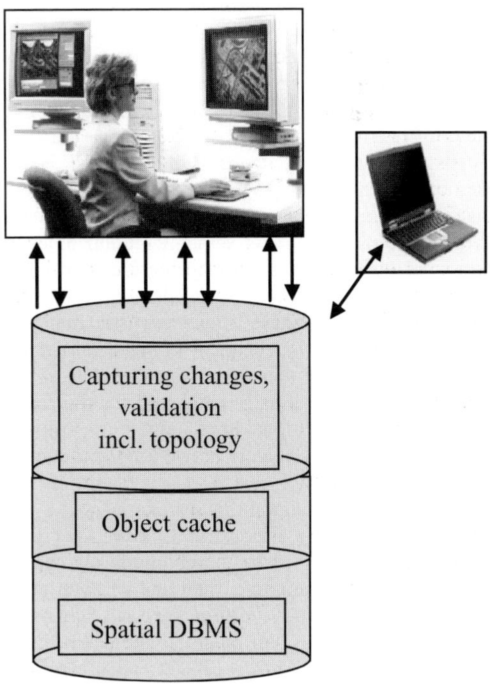

Figure 26.2. Direct link with multi-tier architecture. (see colour plate page 504)

Modern multi-tier architectures extract the changes and the validation procedures including, for example, topology checking, into a middle tier component, for ease of creation, maintenance and scalable access by all client processes (see Fig. 26.2). Service-oriented architectures take this one step further, to a distributed services architecture.

26.4 AUTOMATIC UPDATE OF GEO-SPATIAL DATABASES USING IMAGERY

26.1.5 *Automatic verification and capture of change*

One of the most time-consuming tasks in updating is the verification of the existing data and the detection and capture of change. In the following, we assume that we use up-to-date aerial or satellite imagery for this task. Then, both verification and change detection and capture are prime candidates for automation, and much effort has been spent towards this goal. Indeed, automatic and semi-automatic feature extraction has been a focus of international research in photogrammetry and computer vision for a few decades (e.g. Baltsavias et al. 2001, Heipke et al. 2004, Mayer 2004, Mayer et al. 2006). As a consequence, the results are now starting to enter the commercial market. Obviously, algorithms give particularly good results if applied to well-defined application areas. One of the reasons is that in principle an object extraction task needs additional knowledge in the form of appropriate models or data, which can more easily be formulated for restricted situations.

Historically, knowledge is often buried inside data or hidden in people's heads. This results in serious problems in keeping such knowledge up to date. Progress towards rigorous semantics contributes to removing ambiguities and to storing the knowledge and expertise of the organization where everyone can contribute to it and share it, as enterprise metadata that is portable and independent of specific datasets and systems. Today, techniques developed in artificial intelligence are usually employed for the representation of this additional knowledge in a computer readable form, e.g. semantic networks or production rules.

A critical component in this development is a language to enable logical constraints to be specified. Such a language needs to be unambiguous, logical and portable, compact, intuitive, quantitative, web compatible, declarative and refinable. For a full discussion of these requirements and potential choices of these languages, see e.g. Watson (2007). There are currently several candidates to consider as a knowledge representation language (RDF - Resource Description Framework, OWL—Web Ontology Language, XML Rules/SWRL—Semantic Web Rule Language). None as yet covers all the functionality needed in the geo-spatial domain. This area has received much attention of late through initiatives such as the Semantic Web community (W3C—World Wide Web Consortium 2004a, b) and rapid progress can be anticipated. Also for image interpretation tasks, such knowledge-based systems have proven to be a suitable framework for representing knowledge about objects and exploiting it during the recognition process.

Since any automatic feature extraction algorithm will show a certain error rate, it should be integrated in an interactive workflow leaving final decisions to a human operator. For achieving an efficient workflow, the algorithms have to be equipped with appropriate and reliable self-diagnostics allowing the operator to concentrate on situations where the automatic procedure fails. Walter (2004), for instance, developed a system that supports the operator in quality control of ATKIS (**A**mtlich **t**opographisch-**k**artographisches **I**nformations **s**ystem (Authoritative Topographic Cartographic Information System)) region and line objects by automatically extracting land cover classes from satellite imagery by multi-spectral classification, and comparing them to the corresponding ATKIS objects. He used prior information derived from the existing ATKIS dataset to define training sets for a supervised classification. ATKIS objects showing a high probability of differences to the extracted object classes are indicated as presumed changes. They are visualized to support the human operator's final interactive analysis.

Another example for GIS update from images is the ATOMI (Automated reconstruction of Topographic Objects from aerial images using vectorized Map Information) project of ETH Zürich and the Swiss Federal Office of Topography (Zhang 2004), in which road data were extracted from high resolution colour stereo images and were compared to cartographically generalized vector data in order to update the latter. Champion (2007) and Gladstone et al. (2007) are concerned with updating building information, whereas Ruiz et al. (2007) search for regular patterns as indicators for certain object classes before comparing the results to existing map data, primarily in agricultural areas. A very good overview of existing approaches is given by Steinocher & Kressler (2006).

26.1.6 *Automatic data validation*

Given the database schema and the logical constraints expressed in a rules language, logical consistency can be validated automatically by rules-based processing following the FACT-PATTERN-ACTION dynamic: Given some facts, if they meet any of the patterns/rules, they perform the defined action(s). FACTs are a known data source. PATTERNs are the business rules

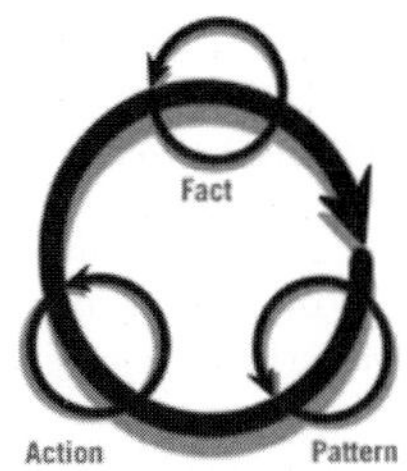

Figure 26.3. The Fact-Pattern-Action dynamic.

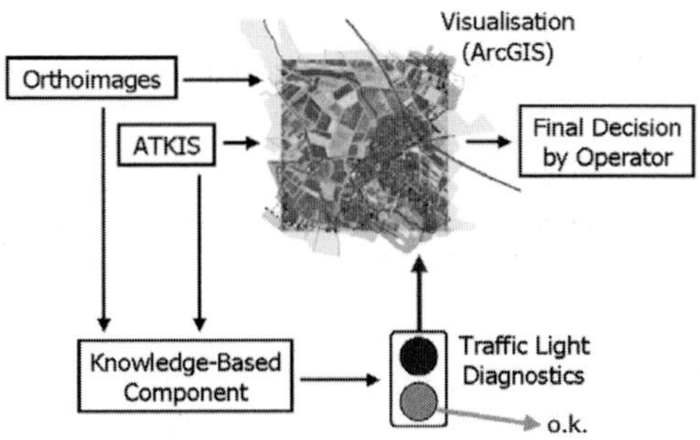

Figure 26.4. Workflow of quality control. (see colour plate page 504)

that the data source obeys or should obey. ACTIONs happen as a result of PATTERNs being applied to FACTs (see Fig. 26.3). Typically, ACTIONS in the validation context are automatic 'fix-ups' or reports on the basis of which targeted user action can be taken to correct errors.

26.5 WIPKA-QS: A SYSTEM FOR AUTOMATIC VERIFICATION AND CHANGE CAPTURE

In this section, we briefly describe our developments towards a specific system for data verification and capture of change. The development is embedded in a broader concept of the knowledge-based workstation WiPKA developed for the Federal Office of Cartography and Geodesy in Germany (BKG), which provides functionality from photogrammetry, GIS and cartography for the acquisition and maintenance of ATKIS. While other such systems exist (see above), the description is intended to show what can be achieved with today's technology (for more details, see Busch et al. 2004, 2006).

The basic concept relies on GeoAIDA, a knowledge-based system for image interpretation based on semantic networks developed at the Institut für Informationsverarbeitung (TNT), Leibniz Universität Hannover (Bückner et al. 2002). In our approach, the comparison of the ATKIS DLMBasis and up-to-date aerial and satellite images comprises two steps, namely verification and the capture of change. Verification is characterized by the following features:

- The image analysis algorithm makes use of the information stored in the GIS to detect the image object.
- If there is a certain degree of consistency of image features and information from the GIS, the object is accepted.
- Otherwise the object is labelled as rejected.

For the subsequent step of change capture, information of new objects not yet stored in the dataset or information about changes of old objects have to be extracted from the image. Thus, it seems reasonable to start with the verification step.

The whole process is integrated in an interactive workflow in which the human operator can focus on those objects where the algorithms detect ambiguous situations and takes the final decisions (see Fig. 26.4). The results of the automatic procedure are passed to the human operator in the form of a so-called traffic light diagnostics, i.e. the results are displayed by means of red and green colour for each GIS object, where green means the object could be verified and red means the object could not be verified. Since the human operator decides on acceptance or rejection in the case of the red objects only, the decision of the automatic procedure has to be reliable, in particular for objects labelled green.

We are currently able to extract and verify road data (Gerke 2006, Ziems et al. 2007), to distinguish between built-up areas, forest and agricultural areas (Müller 2007) and to distinguish between cropland and grassland (Helmholz et al. 2007). Input in all cases are black-and-white orthophotos of approximately 0.4 m ground resolution; we can also process colour and high resolution satellite images. Although the system has been developed with German landscapes in mind, we have demonstrated that it can also handle scenes from other geographic areas.

Extensive testing has shown that we have obtained a gain in productivity of approximately a factor of 3. This means that while we paid special attention to obtain a very small amount of false positives (objects which we classified as correct, but which were indeed wrong), we are able to roughly process two thirds of the objects automatically and correctly.

26.6 THE RADIUS STUDIO EXAMPLE

In this section, a commercial system for checking the logical consistency of a given GIS data set is briefly described in order to demonstrate the state-of-the-art in data validation. We use the example of Radius Studio from 1Spatial as we are most familiar with this system.

Radius Studio is an implementation of a rules-based processing environment. It can be deployed as an instance of a generic multi-tier architecture as shown in Figure. 26.5, or as a Web Service.

The rules-base is a set of conditions that objects from the data store should satisfy. A rule in Radius Studio is a tree of predicates against which objects can be tested. Rules are expressed in a form independent of the schema of any particular data store.

This means they can easily be re-used with different data sources. Before formally defining the rules for use within Radius Studio, they must be articulated and understood. A wide range of circumstances are encountered. The rules may be defined in text form, perhaps in conjunction with a logical data model or feature catalogue. They may be formally expressed in an ontology language such as OWL, in which case they can be used directly (by interfacing with the open source Jena ontology library (Jena 2007)). More often the rules are not explicit or formalized, but exist in the form of knowledge held by domain experts.

Radius Studio provides an intuitive web-based interface for defining rules and building up a rules-base in a tree structured form. For more details, see Woodsford (2007) and 1Spatial (2007).

An action is a procedure (or set of procedures) to be applied to one or more objects, usually when they are found to violate a particular rule. Actions are expressed in a form independent of the schema of any particular data store, so that they can easily be re-used with different sources of data. Actions are defined using a similar graphical user interface as for defining rules, but can also include operations such as assignment, conditionals and sequencing, object creation and deletion, and report generation. Actions can be applied to all the objects from a data store, or in a more targeted manner by use of action maps.

An action map is an ordered list of (rule, action) pairs. When an action map is invoked, objects are checked against the rules. If the object does not conform to that rule, then the associated action is invoked. Action maps are often used to associate fixes to problems with non-conforming objects.

Radius Studio can be used to measure the quality of data in the sense of measuring the degree of conformance of the data to a rules-base. A document or data store can be processed and a report generated with results of the conformance test at both a summary and individual feature level. The final results of the conformance tests are obtained in the form of metadata, which is compliant with the conceptual model of ISO 19115 Metadata (ISO 19115:2003) and encoded in the form recommended in ISO 19139.

Radius Studio can also be deployed using SOAP (Simple Object Access Protocol) Web Service interfaces for the purposes of validating features remotely. In outline, the web services are used to first define a sequence of data processing tasks called a session. The session is then run and rules are asserted against the data. Progress is monitored and finally the results of the conformance test at both a summary and individual feature level are obtained (Watson 2007). Increasingly, Radius Studio is being deployed in rich and potentially dynamic workflow environments such as are enabled by enterprise workflow technologies like BPEL (Business Process Execution Language, see Fig. 26.6) (OASIS 2007). The main drivers for the adoption of such Service Oriented Architectures (SOAs) are that they link computational resources and

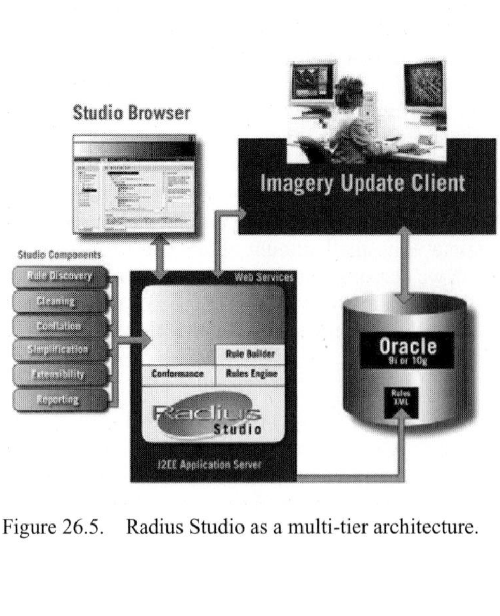

Figure 26.5. Radius Studio as a multi-tier architecture.

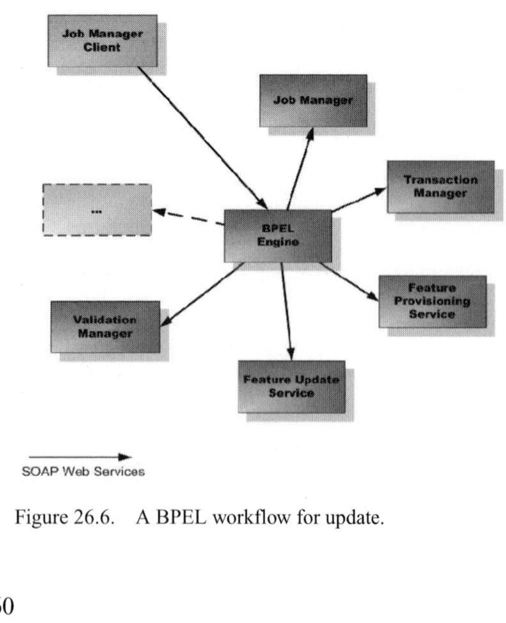

Figure 26.6. A BPEL workflow for update.

promote their reuse so as to help businesses respond more quickly and cost-effectively to changing market conditions.

26.7 SUMMARY AND CONCLUSIONS

In this chapter, we described the updating of geo-spatial databases using aerial and satellite images. The focus was on the updating process as such, rather than individual steps or algorithms. We conclude by emphasizing that images play an essential role in updating geo-spatial databases, because only from images can one obtain an overview of the whole area and is thus able to consistently capture and import into the database all relevant changes. While it can be difficult to obtain imagery at frequent intervals, e.g. due to clouds, terrestrial approaches are much more flexible with respect to suitable conditions for data capture. However, they can only capture individual changes and thus risk violating neighbourhood and/or overall consistency. Thus, the two approaches should be used as complementary methods rather than in competition with each other.

We showed with the help of two examples that while updating is a very time-consuming process, approaches exist and are used in practice that allow significant time and cost savings due to a refined integration of the work carried out by human operators and automatic algorithms based on image analysis and rules-processing. We believe that it is along this way that further progress will be made, if the general trends in computer science, knowledge representation and geo-spatial standards are appropriately taken into account.

ACKNOWLEDGEMENTS

We gratefully acknowledge funding from BKG, and in particular the support by Dietmar Grünreich for the WiPKA development.

REFERENCES

1Spatial. 2007. Radius Studio. www.1spatial.com/products/radius_studio/.

BAE Systems. 2007. *Geo-spatial eXploitation Products*. socetset.com/about_gxp_products.htm.

Baltsavias, E., Grün, A. & van Gool, L. (eds.). 2001. *Automatic Extraction of Man-Made Objects from Aerial and Space Images (III)*. Lisse/Abingdon/Exton(PA)/Tokyo: A.A. Balkema Publishers.

Bückner, J., Pahl, M., Stahlhut, O. & Liedtke, C.-E. 2002. A knowledge-based system for context dependent evaluation of remote sensing data. In L.J.v. Gool (ed.), *DAGM-Symposium, Lecture Notes in Computer Science* 2449: 58–65. Zurich, Switzerland: Springer.

Busch, A., Gerke, M., Grünreich, D., Heipke, C., Liedtke, C.-E. & Müller, S. 2004. Automated verification of a topographic reference dataset: system design and practical results. *International Archives of the Photogrammetry, Remote Sensing and Spatial Information Sciences* XXXV, Part B2: 735–740.

Busch, A., Gerke, M., Grünreich, D., Heipke, C., Helmholz, P., Liedtke, C.-E. & Müller, S. 2006. Automated verification of a topographic dataset using Ikonos imagery. *International Archives of the Photogrammetry, Remote Sensing and Spatial Information Sciences* XXXVI, Part 4: 134–139.

Champion, N. 2007. 2D change detection from high resolution aerial images and correlation digital surface models. *International Archives of the Photogrammetry, Remote Sensing and Spatial Information Sciences* XXXVI, Part 3/W49A: 197–202.

Edwards, D., Simpson, J. & Woodsford, P. 2000. Integration of photogrammetric and spatial information systems. *International Archives of Photogrammetry and Remote Sensing* 33(B2): 603–609.

EuroSDR. 2006. Feature/Object Data Models Workshop. *EuroSDR Official Publication* 49.

Gerke, M. 2006. *Automatic quality assessment of road databases using remotely sensed imagery*. PhD thesis, Deutsche Geodätische Kommission. Reihe C, Dissertationen, Nr. 599.

Gerke, M., Butenuth, M., Heipke, C. & Willrich, F. 2004. Graph supported verification of road databases. *ISPRS Journal of Photogrammetry and Remote Sensing* 58(3–4): 152–165.

Gladstone, C.S., Sanchez Hernandez, C. & Holland, D. 2007. Detecting topographic change using digital photography and digital surface models. *Proceedings RSPSoc 2007, Newcastle.*

Heipke, C. 2004. Some requirements for geographic information systems: a photogrammetric point of view. *PE&RS* (70) 2: 185–195.

Heipke, C., Pakzad, K., Willrich, F. & Peled, A. 2004. Integration of geodata and imagery for automated update of spatial databases. *ISPRS Journal of Photogrammetry and Remote Sensing* 58(3–4): 127–128.

Helmholz, P., Gerke, M. & Heipke C. 2007. Automatic discrimination of farmland types using Ikonos imagery. *International Archives of the Photogrammetry, Remote Sensing and Spatial Information Sciences* XXXVI, Part 3/W49A: 81–86.

ISO 19113. 2002. *Geographic information—Quality principles*. Available at:www.iso.org/iso/en/CatalogueDetailPage.Catalogue Detail? CSNUMBER=26018.

Jena. 2007. Available at: jena.sourceforge.net/.

Mayer, H. 2004. Object extraction for digital photogrammetric workstations. *International Archives of the Photogrammetry, Remote Sensing and Spatial Information Sciences* XXXV, Part B2: 414–422.

Mayer, H., Baltsavias E. & Bacher, U. 2006. Automated extraction, refinement, and update of road databases from imagery and other data. *Report Commission 2 on Image Analysis and Information Extraction, European Spatial Data Research—EuroSDR, Official Publication* 50: 217–280.

Müller, S. 2007. Extraktion baulich geprägter Flächen aus Fernerkundungsdaten zur Qualitätssicherung flächenhafter Geobasisdaten. PhD thesis, Institut für Informationsverarbeitung, Leibniz Universität Hannover.

OASIS. 2007. WS-BPEL 2.0—*Web Services Business Process Execution Language*. Available at: www.oasis-open.org/committees/download.php/22036/wsbpl-specificationdraft%20candidate%20CD%20Jan%2025%2007.pdf.

Peerbocus, M.A., Geneviève, J. & Badard, T. 2002. A methodology for updating geographic databases using map versions. *Proceedings Spatial Data Handling 2002, July 9–12, Ottawa, Ontario, Canada, International Geographic Union,* (on CD-ROM).

Ruiz, L., Recio, J. & Hermonsilla, T. 2007. Methods for automatic extraction of regularity patterns and its application to object-oriented image classification. *International Archives of the Photogrammetry, Remote Sensing and Spatial Information Sciences* XXXVI, Part 3/W 49A: 117–121.

Sonnen, D. 2007. Emerging Issue: Spatial Data Quality. *Directions Magazine* January 2007, available at: http://www.directionsmag.com/article.php?article_id=2372&trv=1.

Steinocher, K. & Kressler, F. 2006. EuroSDR change detec. tion report. *EuroSDR Official Publication Series No. 50*: 111–182.

Walter, V. 2004. Object-based classification of remote sensing data for change detection. *ISPRS Journal of Photogrammetry and Remote Sensing* 58(3–4): 225–165.

W3C. 2004a. *Web Ontology Language (OWL)*. www.w3.org/TR/owl-features/.

W3C. 2004b. *Semantic Web*. www.w3.org/2001/sw/.

Watson, P. 2007. Formal Languages for Expressing Spatial Data Constraints and Implications for Reporting of Quality Metadata. *5th International Symposium on Spatial Data Quality, ISSDQ 2007, Enschede, NL. ISPRS Working Group II/7*, available at: www.itc.nl/ISSDQ2007/proceedings/Session%205%20Dissemintion%20and%20Fitness%20for%20Use/paper%20Paul_Watson[1].pdf.

Woodsford, P.A. 2004. System Architecture for Integrating GIS and Photogrammetric Data Acquistion. *International Archives of the Photogrammetry, Remote Sensing and Spatial Information Sciences* XXXV, Part B2: 758–765.

Woodsford, P.A. 2007. Spatial Database Update and Validation—Current Advances in Theory and Technology. *ISPRS Archives* XXXVI, Part 4/W54:1–7. Available at http://isprs-wg41.nsdi.gov.cn/Xinjiang/Proceedings/papers/1-7%20P.%20A.%20Woodsford%20.pdf.

Zhang, C. (2004). Towards an operational system for automated updating of road databases by integration of imagery and geodata. *ISPRS Journal of Photogrammetry and Remote Sensing* 58, 3–4: 166–186.

Ziems, M., Gerke, M. & Heipke, C. 2007. Automatic road extraction from remote sensing imagery incorporating prior information and colour segmentation. *International Archives of the Photogrammetry, Remote Sensing and Spatial Information Sciences* XXXVI, Part 3/W49A: 141–147.

[all WWW urls validated on Nov. 27 2007].

Advances in Photogrammetry, Remote Sensing and Spatial Information Sciences: 2008 ISPRS Congress Book – Li, Chen & Baltsavias (eds)

CHAPTER 27

Applications in cultural heritage documentation

Petros Patias, Pierre Grussenmeyer & Klaus Hanke

ABSTRACT: The importance of cultural heritage documentation is well recognized. ICOMOS and the International Society of Photogrammetry and Remote Sensing (ISPRS) joined efforts in 1969 and created the International Committee for Heritage Documentation, CIPA. This chapter reviews many of the successful recordings of cultural heritage from recent years.

Keywords: Cultural heritage, CIPA, quality evaluation, recording techniques

27.1 STATEMENT OF APPLICATION PROBLEMS IN CULTURAL HERITAGE DOCUMENTATION

Cultural heritage is a testimony of past human activity and, as such, cultural heritage objects exhibit great variety in their nature, size and complexity, from small artefacts and museum items to cultural landscapes, from historic buildings and ancient monuments to city centres and archaeological sites.

Cultural heritage around the globe suffers from wars, natural disasters and human negligence. The importance of cultural heritage documentation is well recognized and there is an increasing pressure to document our heritage both nationally and internationally. This has alerted international organizations to the need for issuing guidelines describing the standards for documentation. Charters, resolutions and declarations by international organizations underline the importance of documentation of cultural heritage for the purposes of conservation works, management, appraisal, assessment of the structural condition, archiving, publication and research. Important ones include the International Council on Monuments and Sites, ICOMOS (ICOMOS 2007) and UNESCO, including the famous Venice Charter, The International Charter for the Conservation and Restoration of Monuments and Sites, 1964, (UNESCO 2007).

This suite of documentation requirements, as stated by the international agreements, imposes important technical restrictions and dictates specifications, which should be always borne in mind when recording strategies are designed and followed. The common features of the above documents are, briefly:

- recording of a vast amount of four-dimensional (i.e. 3-D plus time) multi-source, multi-format and multi-content information, with stated levels of accuracy and detail;
- digital inventories in 3-D and, as far as available, dated historical images;
- management of the 4-D information in a secure and rational way, making it available for sharing and distribution to other users; and
- visualization and presentation of the information in a user-friendly way, so that different kinds of users can actually retrieve the data and acquire useful information, using Internet and visualization techniques.

Recognizing these documentation needs, ICOMOS and the International Society of Photogrammetry and Remote Sensing (ISPRS) joined efforts in 1969 and created the International Committee for Heritage Documentation, CIPA, (CIPA 2007). CIPA's main objective is to provide an international forum and focal point for efforts in the improvement of all methods for surveying of cultural monuments and sites. The combination of all aspects of photogrammetry with other surveying methods is regarded as an important contribution to recording and monitoring cultural heritage, to the preservation and restoration of any valuable architectural or other cultural monument, object or site, and to provide support to architectural, archaeological and other art-historical research.

Architectural photogrammetry is nearly as old as photography itself. In 1858 the German architect A. Meydenbauer developed photogrammetric techniques for the documentation of buildings and established the first photogrammetric institute in 1885. In the same year, the ancient ruins of Persepolis were the first archaeological objects recorded photogrammetrically.

Since then, photogrammetric techniques have been constantly developing (Patias 2007, Patias 2004, Patias 2001, Patias & Peipe 2000, Atkinson 1996, Karara 1989, Carbonnel 1989, Ogleby & Rivett 1985) on all fronts: hardware, software, procedures. Over the recent past, there has been a movement from optical-mechanical to purely digital hardware and from high-end to low-budget software, (e.g. Tsioukas et al. 2004, Pateraki et al. 2002, Hanke 2001, Georgiadis et al. 2001 Mills et al. 2000, Glykos et al. 1999, Dequal et al. 1999). New procedures are constantly developing: the use of wide angles or even panoramic cameras (e.g., Arias et al. 2004, Pöntinen 2004, Haggrén et al. 2004, Reulke et al. 2005); fusion of photogrammetric data with data from other sensors, especially laser scanners, (e.g. Fryer et al. 2005, Böhler 2005, Balletti et al. 2004, Lingua et al. 2003, Guarnieri et al. 2004, Barber et al. 2002, Beraldin et al. 2002); use of non-metric sensors like off-the-shelf digital cameras, (e.g. Patias et al. 1998, Peipe & Stefiani 2005, Sechidis et al. 1999) or video recorders; the production of specialized products like digital rectification, (e.g. Hemmleb 1999, Wiedemann et al. 2000); monoplotting, (e.g. Karras et al. 1996); unwrapping developable surfaces, (Karras et al. 1996, 1997, 2001); true orthophoto production, (e.g. Boccardo et al. 2001, Grammatikopoulos et al. 2004, Balletti et al. 2005); augmented reality products, (e.g. Agosto et al. 2004, Grün et al. 2004, Gemenetzis et al. 2001, Koussoulakou et al. 1999, 2001a, 2001b) and virtual museums, (e.g. Boulanger et al. 2000).

For large objects (monuments, sites, etc.), architectural mapping, regular topographic surveys and laser scanning are used along with photogrammetric techniques. In many cases a combination of methods is used, especially the coupling of photogrammetry with laser scanning. For the latter, the user can refer to Böhler (2005) and to Balletti et al. (2004) for a valuable comparison of the two techniques.

In addition, alternative systems have been proposed: for example, hybrid geodetic-photogrammetric stations coupled with laser ranging and digital cameras, (e.g. Kakiuchi & Chikatsu 2000, Tauch & Wiedemann 2004).

To the pure photogrammetric processes, one should also add other quasi-photogrammetric or alternative vision techniques, often used for small size objects such as coins, tools and instruments, everyday items, jewellery, clothes furniture, ornaments, weapons, musical instruments, paintings and frescos, and statues. Such techniques are for example :

- shape from silhouettes (e.g. Tosovic et al. 2002);
- shape from structured light (e.g. Tosovic et al. 2002, Salvi et al. 2004, Rocchini et al. 2001, Gühring 2001);
- shape from motion (video cues) (e.g. Enciso et al. 1996, Chiuso et al. 2000, Pollefeys et al. 2000);
- shape from shading (Zhang et al. 1999);
- shape from texture (Forsyth 2002);
- shape from focus/defocus (Schechner 2000, Favaro 2002).

Besides close-range photogrammetric techniques, the documentation of large sites often requires coupling of close-range images with aerial photography or satellite imagery. During the last decade, especially with the rise of very high resolution satellite sensors, myriads of applications have been successfully reported (e.g. Georgoula et al. 2004).

Finally, other specialized techniques for heritage documentation often couple and enhance the value of their deliverables with photogrammetric techniques. To this end, one should mention technologies such as:

- Ground Penetrating Radar.
- Automated monitoring systems.
- Infrared, Ultraviolet, X-ray and microwave sensors.
- Spectroscopy.

27.2 CURRENT STATE-OF-THE-ART RESEARCH

Choosing the appropriate technology (sensor, hardware, software), the appropriate procedures, designing the production workflow and assuring that the final output is in accordance with the set technical specifications is a challenging matter in cultural heritage documentation. For this, the leading parameters are the size and the complexity of the object and the level of accuracy required. These are the major factors, which crucially influence the procedure to be followed and, sometimes, even the viability of photogrammetry itself.

Besides size and complexity, other factors may influence the optimal method to be chosen. These include: the required accuracy and resolution; accessibility of the object and availability of ideally located vibration-free observation stations; availability of instruments and power supply; and the possibility of touching the object and permission to use the selected method.

27.2.1 *Sensors*

27.2.1.1 *General remarks*

For applications in Cultural Heritage Documentation, the palette of sensors was for a long time determined by the use of expensive and specialized equipment, i.e. film or glass plate based metric cameras, stereo cameras, semi-metric cameras (Hanke et al. 2002). Nowadays image acquisition systems for CH applications are normally based on digital sensors. Analogue metric cameras for special purposes (e.g. high accuracy) and semi-metric analogue cameras are nevertheless still in use.

Beside this, archives with thousands of analogue images of CH sites and objects still exist all over the world, which demand the need of precise digitization of film, photo prints or even glass plates (Hanke et al. 2006) for further digital photogrammetric workflow.

In addition to these digital image based sensors (satellite, aerial and close range), different kinds of range based sensors like 3-D scanners (airborne, mid and close range), IR-bolometers, thermal systems (Costantino et al. 2007, Herraez Boquera et al. 1997), geophysical sensors and many more are used to collect a whole spectrum of information on cultural heritage objects and sites (Patias 2003).

The challenge in Cultural Heritage applications is therefore the sometimes very complex data fusion from these heterogeneous sources.

27.2.1.2 *Digital cameras*

The number of available digital cameras suitable for heritage documentation is immense. The sensors can be either CCD or CMOS with reflex or view finder, interchangeable or fix lenses. As long as temporary stability and eventually later calibration is assured, all kinds of the mentioned camera types are widely used for heritage documentation purpose. During the last years the nominal resolution of sensors increased significantly. So 10 to 16 Megapixel cameras can be seen as standard and professional cameras or digital camera backs deliver up to 39 million pixels per image even from full frame sensors (e.g. Fig. 27.1).

Almost every application in cultural heritage documentation is different and very often needs special adaptation of equipment and procedures.

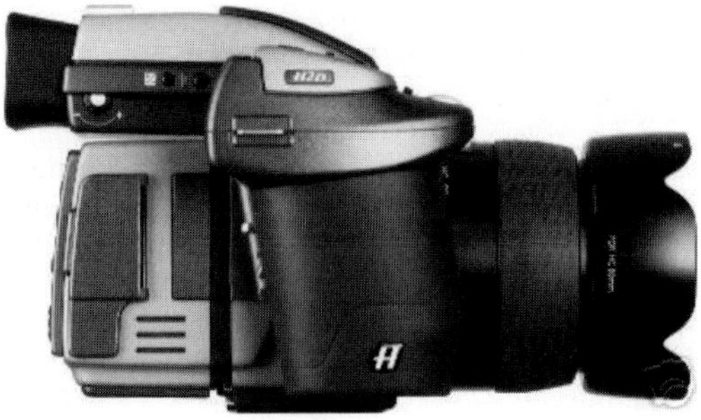

Figure 27.1. Hasselblad digital camera with 39 megapixel full-frame sensor.

Even non-standard imaging principles such as panoramic cameras (Schneider et al. 2004, Maas et al. 2004, Luhmann 2004, Amiri Parian et al. 2006), fisheye systems (Kedzierski et al. 2007. van den Heuvel et al. 2006), catadioptric imaging systems (e.g. Svoboda et al. 2002, Lhuillier 2007), rotating cameras (Remondino et al. 2004) and other omni-directional vision techniques (Geyer et al. 2003) are used.

Suitable software for calibration (Remondino et al. 2004, Karras et al. 2001, Remondino et al. 2006, van den Heuvel et al. 2006, Amiri Parian et al. 2006, Shortis et al. 2006, Ethrog 2006, Al-Ajlouni et al. 2006) and for the further evaluation workflow (Fangi 2006) has been developed.

To ensure better representation of image details under bad lighting conditions (e.g. bright sky versus shadows) "high dynamic range images" (HDRI) come to practical applications. Such digital images, resulting from a set of photographs taken with a range of exposures, do have a much larger dynamic range than film-based images, so there is more detailed radiometric information available (Biber et al. 2005, El-Hakim et al. 2003).

27.2.1.3 *3-D scanners*

In addition to the photographic imaging systems, 3-D scanners play an important role in heritage documentation nowadays. The variety starts from airborne laser scanning systems (LIDAR) for terrain/surface modelling and for site mapping and archaeological prospection, followed by mid-range scanners for data acquisition of facades or entire buildings, ensembles or sites and goes down to close range 3-D scanners for detailed documentation of e.g. archaeological finds, artefacts and sculptures (Boehler 2002).

The spectrum of technology includes range sensors with time-of-flight, phase difference or triangulation using laser points, fringe or other patterns projection (Przybilla et al. 2007, Hanke et al. 2004). In airborne laser scanning applications increasingly full-waveform analysis is used (Doneus et al. 2006, 2007).

Some of these terrestrial sensors are equipped with additional cameras for target selection on the one hand and even texturing of the resulting point cloud. Some scanners allow the exterior mounting of a camera with fixed and/or calibrated relative orientation to the laser sensor (Fig. 27.2). Data fusion is an important topic for that purpose (Drap et al. 2003, Alshawabkeh et al. 2004).

Again the adaptation of the equipment is very often necessary for cultural heritage applications.

27.2.1.4 *Spatial imaging total stations*

Tachometry is traditionally used in projects for the measurement of control networks and control points, and the reflector-less technology is full of interest in Cultural Heritage. The latest total stations (POB 2007) include

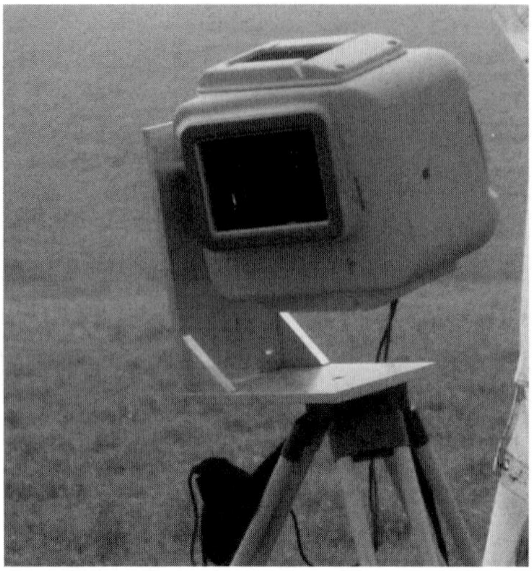

Figure 27.2. Trimble GX scanner with self-assembled adapter for excavations.

now the combining of 3-D scanning, optical positioning and video technologies into a single sensor, and extend the range of applications. Although there is limited resolution in the built-in cameras, the Trimble VX^{TM} spatial station and Topcon Imaging Total Station can be considered as promising sensors for documentation.

27.2.2 *Procedures and software*

27.2.2.1 *Procedures for the photogrammetric restitution*

The procedures for the processing of digital images are well established (Kasser 2001). Three types of procedures (Fig. 27.3) can be considered for Cultural Heritage applications: single-photo, stereo- or multi-photo arrangements. Single photos are usually sufficient for near planar object surfaces and more than two photos for a higher accuracy and reliability of the product (Patias 2004).

For photogrammetric documentation of architecture, simple rules are to be observed to obtain a stable bundle block arrangement (Waldhaeusl & Ogleby 1994). Digital images are referenced from object points or targets placed in the field during the exterior orientation process. New developments towards automation based on line-photogrammetry for efficient reconstruction of detailed building models from one or more digital close-range images are presented in Heuvel (2003). Most of the Digital Photogrammetric Workstations software packages are proposed for the processing of aerial vertical imagery only (GIM International 2007b), and their use may not be effective for terrestrial close range applications. An overview

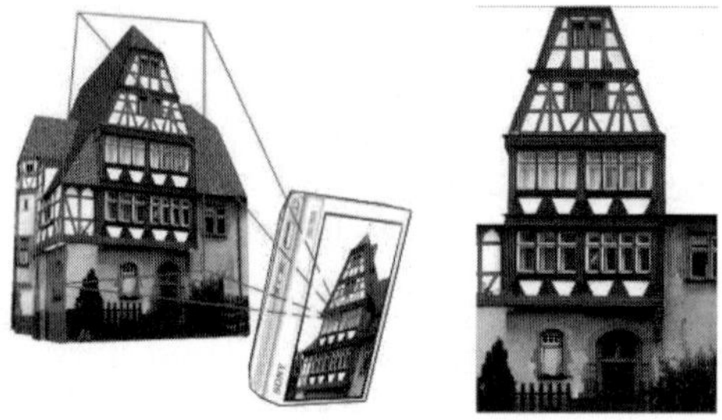

Figure 27.3(a). Single and rectified image.

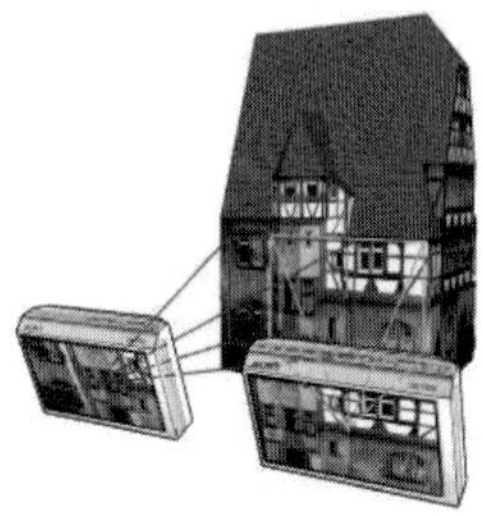

Figure 27.3(b). Stereo-pair (parallel images) allowing stereorestitution.

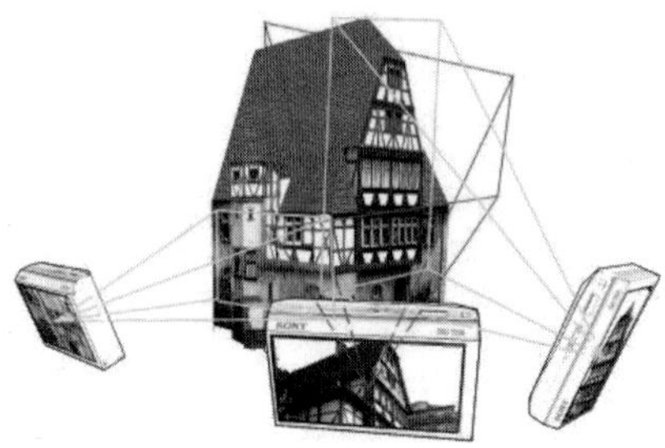

Figure 27.3(c). Multi-photo arrangement for bundle restitution.

of software packages for close range applications is given at www.simple3D.com.

27.2.2.2 *Procedures for terrestrial laser scanning*

A number of separate scans from different locations are usually required to ensure full coverage of the object, structure or site. When collected, scans are based on an arbitrary coordinate system. To use several scans together, their position and orientation must be changed so that each scan uses a common coordinate system. This process is known as cloud point alignment, or registration (English Heritage 2007). This traditional scanning methodology is based on measurements of common targets or spheres to relate

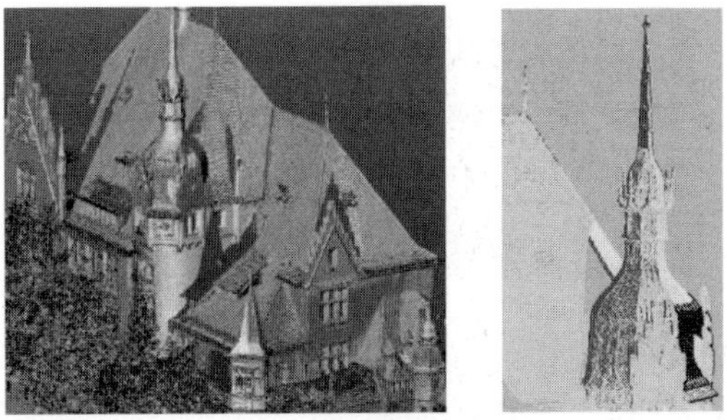

Figure 27.4. Different scans (colors) registered onto the same co-ordinate system. (see colour plate page 505)

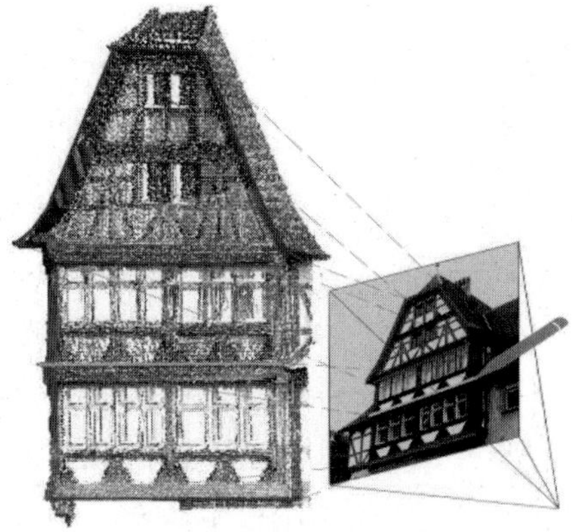

Figure 27.5. Monoplotting 3-D coordinates and lines onto the image in relation to the point cloud mesh. (see colour plate page 505)

multiple scans together or to relate measurements to the existing control network. At least three common targets are required between two point clouds in order to merge them. Some scanners propose a new workflow close to the traditional total station survey workflow (Lemmon & Biddiscombe 2006).

The scanner is thus set up over known points and provides a direct relationship to existing ground points (Hanke et al. 2006). Measurements between stations commonly referred to as traverse measurements, provide an instant relationship between multiple stations and allow the user to view homogenous data in the field without additional post processing (Fig. 27.4). The quality and extent of the survey data can thus be verified in the field, reducing survey time or further re-measurements.

Most of the manufacturers propose both 3-D scanners and software packages with functionalities of registration/orientation methods, automatic detection of tie points, integration of data from different scans, real time visualization during scanning, fly around pan and zoom, geo-referencing, fitting of primitives (GIM International 2007a).

The deliverables are processed from the collected point cloud information into more useful products as unrefined or edited meshes, rendered images, 2-D/3-D drawings or animations. More links to software packages for processing point clouds are listed at www.simple3D.com (section software).

27.2.2.3 *Combination of photogrammetry and laser scanning*

In many cases, a combination of photogrammetry and laser scanning methods is used. The reader can refer to Böhler (2005) and Balletti et al. (2004) for a valuable comparison between the two techniques (Patias 2006). The latest spatial imaging total stations also deliver the same type of data: point clouds and oriented images. Object restitution is thus possible by monoplotting 3-D points and lines onto the image linked to the meshed point cloud (Fig. 27.5).

27.2.3 *Case studies*

27.2.3.1 *Case study 1: The cave of cylcop polyphemus from Odyssey*

27.2.3.1.1 Background and scope of documentation

Based in northern Greece, the cave of the cyclop Polythemus, famous from the Odyssey, is an explored underground cave (Fig. 27.6), not opened to the public up until now.

The purpose of the project is the 3-D documentation of the cave for the preservation of its historical and natural value as well as its touristic exploitation.

27.2.3.1.2 Instrumentation – procedures

The Trimble GS200 Laser scanner (360° × 60°) has been used for the scanning. Seven scans (Fig. 27.7) have been performed in total, providing a mesh of 15 mm to cover the whole interior surface.

The Leica TPS403 total station has been used for measurements of the 13 retro-reflective spherical control points with an accuracy of 3 mm + 2 ppm.

Auxiliary equipment, used during field work, included a portable energy generator, a laptop and two light projectors.

HDS Leica Cyclone was used for the registration of the point clouds with a final accuracy of 1 cm.

27.2.3.1.3 Deliverables

The deliverables are the following sections (Fig. 27.8) and a virtual 3-D walk-in video.

27.2.3.1.4 Credits

The project has been carried out by the Aristotle University, Thessalonki, Greece, Laboratory of Photogrammetry and Remote Sensing, Faculty of Rural and Surveying Engineering, Prof. Petros Patias. patias@auth.gr.

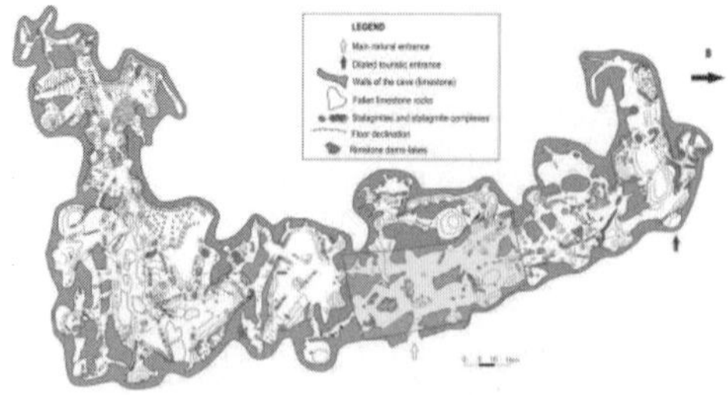

Figure 27.6(a). From the whole cave only the denoted part has been documented. (see colour plate page 505)

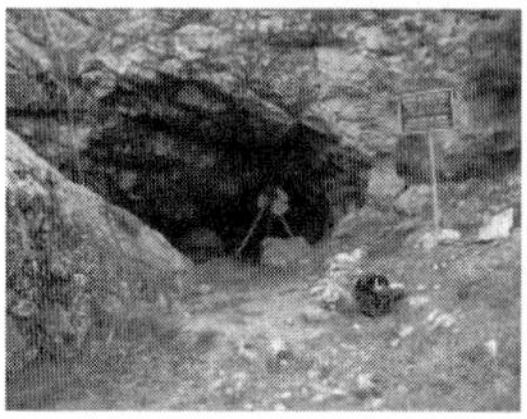

Figure 27.6(b). The cave entrance. (see colour plate page 505)

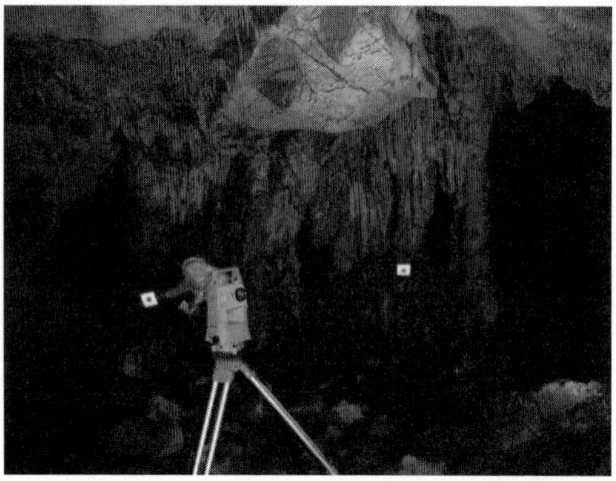

Figure 27.6(c). The cave interior. (see colour plate page 505)

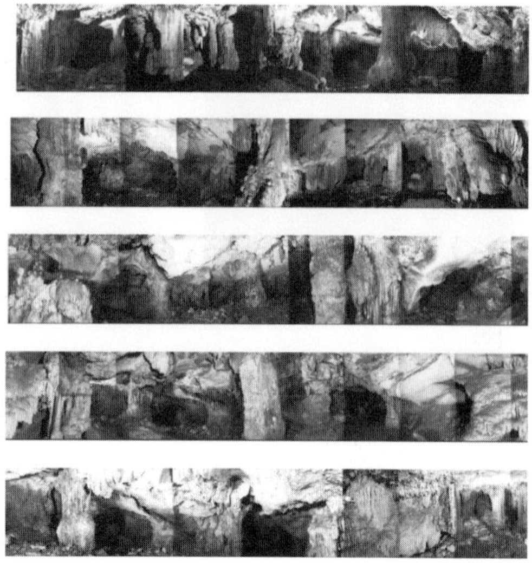

Figure 27.7(a). Panoramic views of the individual scans. (see colour plate page 506)

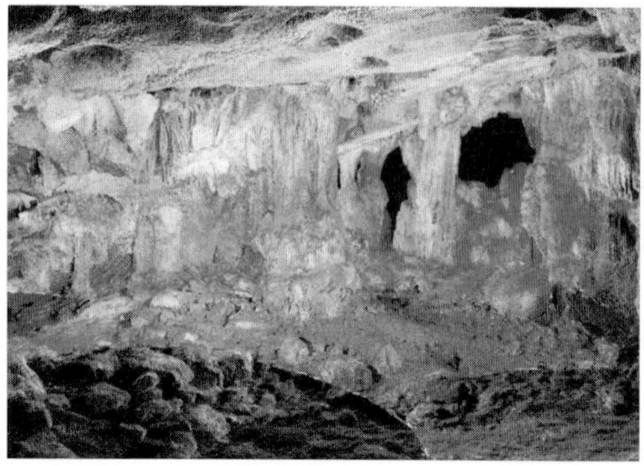

Figure 27.7(b). Part of the final point cloud. (see colour plate page 506)

27.2.3.2 *Case study 2: Archaeological prospection—the iron age hill fort of purbach*

27.2.3.2.1 Background and scope

The Iron Age hill fort consists of linear and non-linear above ground earthwork features with varying preserved heights (Fig. 27.9). It is covered by a forest with varying degrees of understorey, but contains cleared parts, which are partly overgrown with dense bushes and partly covered with clearance piles (Fig. 27.10). It represents a range of environments and therefore it is an ideal case study to test the applications of full-waveform Airborne Laserscanner (ALS) data (Doneus et al. 2007).

27.2.3.2.2 Instrumentation – procedures

A RIEGL Airborne Laser Scanner LMS-Q560 was used to collect the full-waveform ALS data. The scan was performed when the snow was completely melted and the trees were still leafless. Flight altitude was about 600 m above ground, which resulted in a laser footprint size of 30 cm on the ground. A total area of 9 km^2 was covered with a scan angle of 45 degrees by 26 parallel flight tracks, which had a width

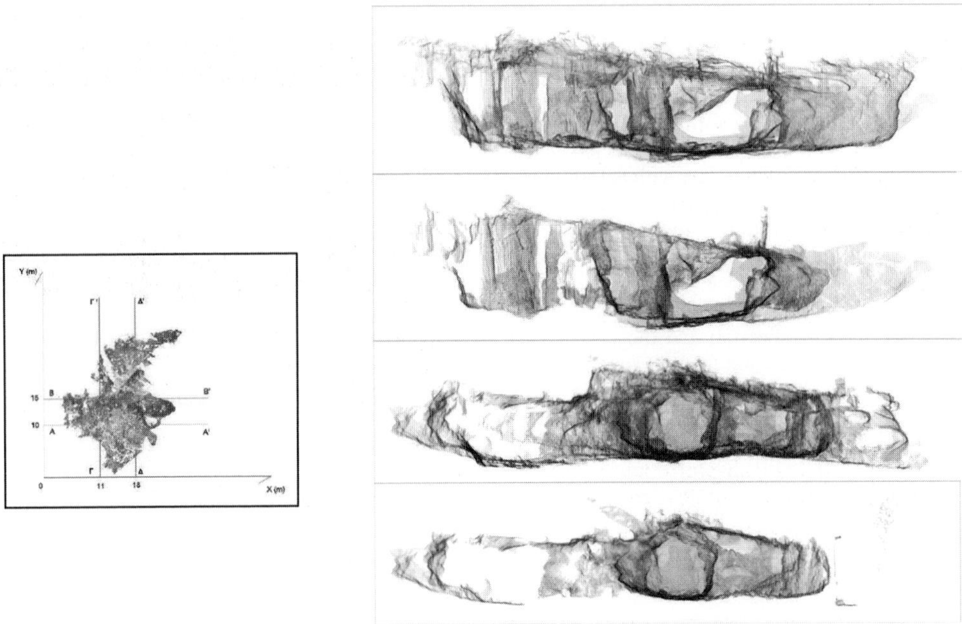

Figure 27.8(a). Four cross sections. (see colour plate page 507)

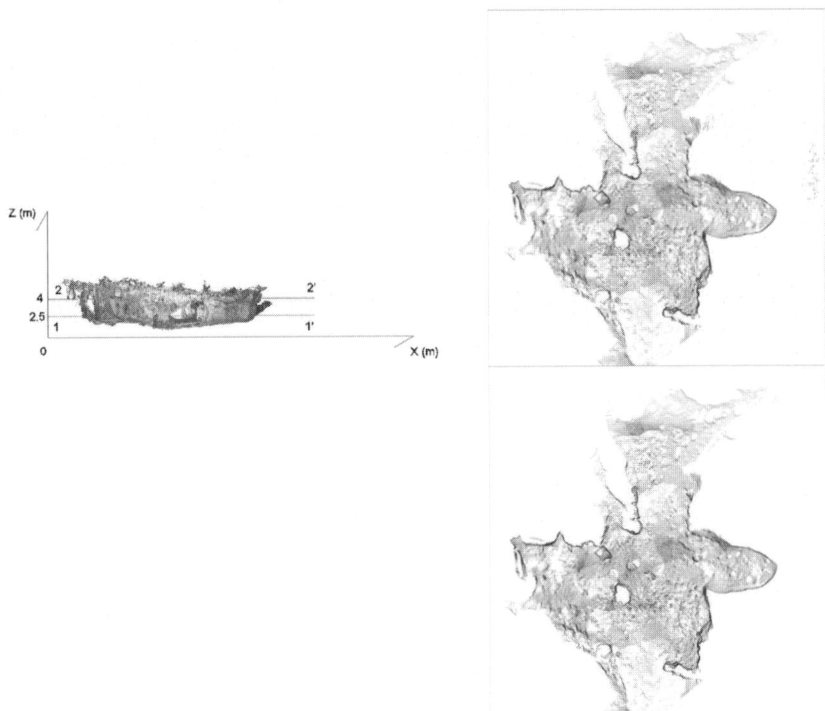

Figure 27.8(b). Two plan sections. (see colour plate page 507)

of approximately 500 m and an overlap of 50%. The real scan rate was 66 kHz resulting in an overall mean point density of eight measurements per m^2.

Using the original full-waveform raw data, a Gaussian decomposition was performed. Based on this procedure the 3-D coordinates of all detected echoes are determined and stored together with the additional information (amplitude, echo width) determined from the fitted Gaussian distribution functions.

Initially, all echoes, except the last echo, were discarded as these will always be off-terrain. The next stage was to discriminate between points that represent actual terrain reflection and those which are derived from other off-terrain features (notably dense vegetation).

Besides a better range discrimination in post-processing, the width of each returning echo, which is derived during the full-waveform processing, can indicate whether a last echo should be classified as terrain or off-terrain. Consequently, the last echo data could be filtered to remove all points with an echo width above a certain threshold.

The remaining point cloud will encompass only those points that were identified as being the last returning echo of a single laser pulse coming from a more or less solid ground surface. It should be noted that there will remain a substantial number of off-terrain points in this data. To solve this problem, the technique of robust interpolation with an eccentric and unsymmetrical weight function was used. This process consists of an iterative procedure based on Kriging within a hierarchic framework, where different strategies can be constructed by consecutively applying four main processing steps: thin-out, interpolate, filter, and sort-out.

Because it is possible to apply the four processing steps iteratively and change various parameters at

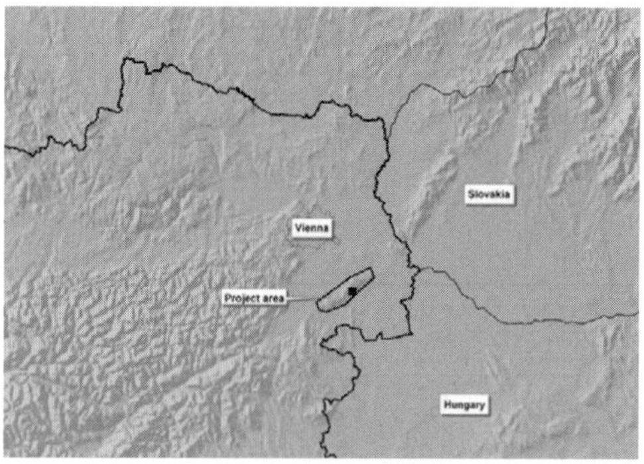

Figure 27.9. Map showing the eastern part of Austria with the "Leithagebirge" depicted as polygon. The black square marks the area of the test scan above the Iron Age hill fort of Purbach.

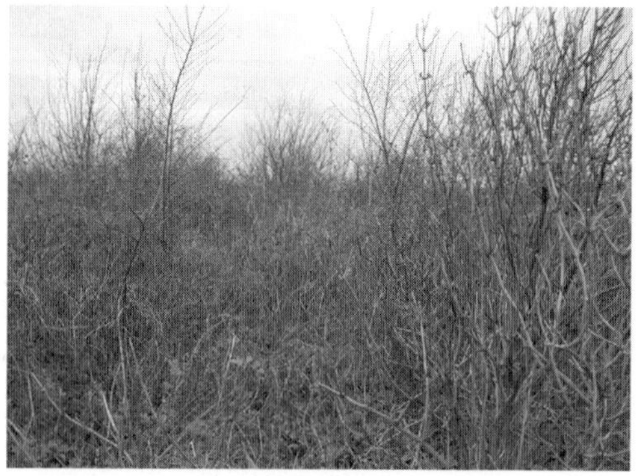

Figure 27.10. Photograph of part of the scanned area taken from the ground: cleared settlement area, which is now under a dense cover of bushes.

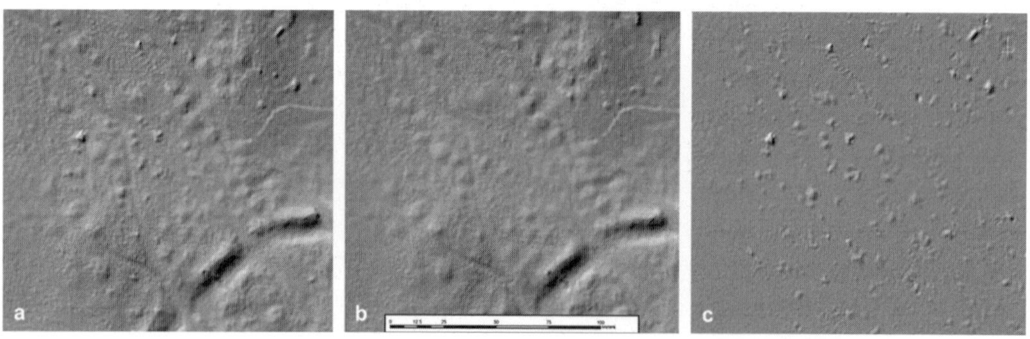

Figure 27.11. (a) DTM derived from filtered last echo data. Low vegetation and clearance piles are still represented in the DTM. (b) DTM derived from last echo data after removing points with large echo widths and filtering. (c) Subtraction of both images. The elevations indicate removed low vegetation and clearance piles.

each step, strategies can be defined for almost any type of situation (e.g. urban, open area, forest).

27.2.3.2.3 Deliverables

The filtering procedure resulted in a DTM, which shows a detailed map of the topography under the forest canopy. Once again it should be noted that while this is still a model of the topography, the post-processing has significantly reduced the gross and systematic errors in the model (Fig. 27.11). The 5.5 hectare large hill fort is surrounded by a massive rampart, which to the northwest is accompanied by a deep moat that is still 2 m in depth. The height difference from the crest of the rampart to the bottom of the ditch varies between 7 and 10 m (Fig. 27.12a). Besides the massive ramparts, more subtle archaeological features are identifiable. Inside the hill fort, three concentric linear structures (Fig. 27.12b) have emerged, which follow the contours. In a cross section, they look like terraces but are, according to Ulbrich, the remains of ramparts cutting off the fore-fort. At the highest point within the hill fort there is a 15 by 20 m area of slightly depressed terrain, which could be the remains of a cistern (Fig. 27.12c—area marked by a hatched line).

The results were compared with the mapping of Ulbrich. Both DTM and topographical maps corresponded well. The positional accuracy differences are approximately 10 cm across the image. The main features, such as banks, ditches, large round barrows, and even the traces of former excavations across the ramparts, are all depicted in Ulbrich's map and can be seen in the ALS-derived DTM (Fig. 27.13).

27.2.3.2.4 Credits

The project has been carried out by Michael Doneus, Martin Fera and Martin Janner from the Department of Prehistoric and Mediaeval Archaeology, University of Vienna, as well as Christian Briese, Christian Doppler Laboratory for Spatial Data from Laser Scanning and Remote Sensing, Institute of Photogrammetry and Remote Sensing, Vienna University of Technology, Vienna, Austria and supported by the Austrian Science Fund (FWF) under project no. P18674-G02.

27.2.3.3 *Case study 3: The "Pontonniers" international high school in Strasbourg*

27.2.3.3.1 Background and scope of documentation

The Pontonniers International high school is a high class eclecticist building from the 18th century located in the very extended German district of Strasbourg, with handsome and ornate buildings, towers, turrets and multiple round and square angles (Fig. 27.14).

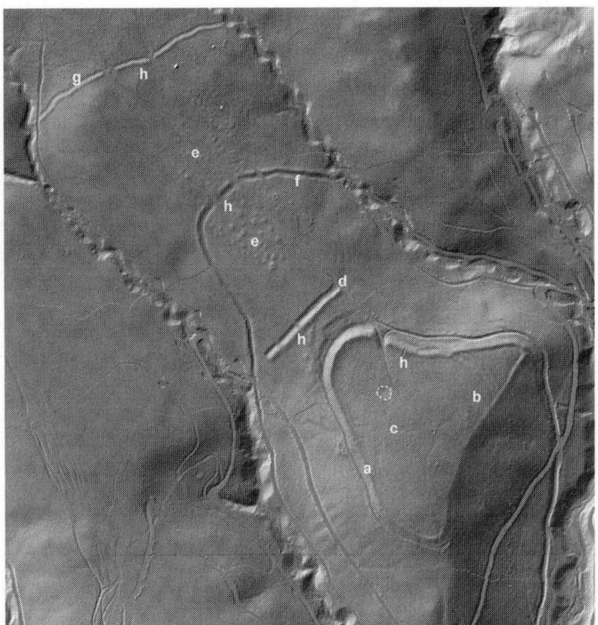

Figure 27.12. Final DTM after filtering using the theory of robust interpolation with an eccentric and unsymmetrical weight function used within a hierarchic framework. Illumination is from northwest. The letters are referred to in the text.

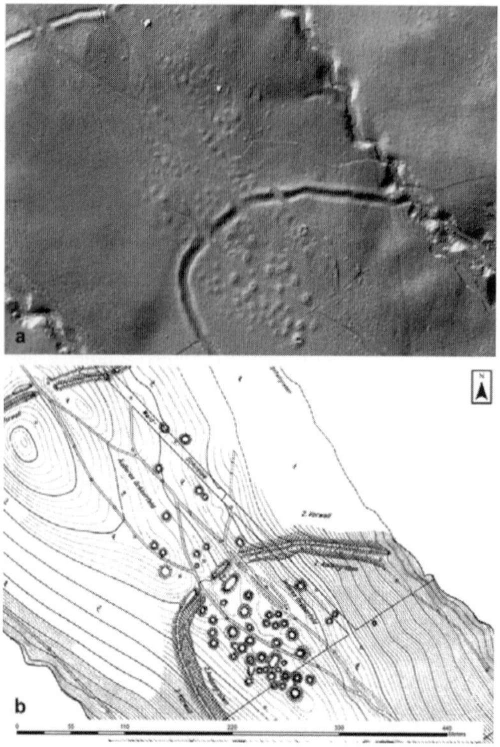

Figure 27.13. (a) Zoomed view to the area of the round barrows in the final DTM. The individual round barrows of the graveyard can now be seen in detail. Most of them show traces of looting in the form of shallow depressions. (b) The same area as mapped in the 1960s (© BEV 2007).

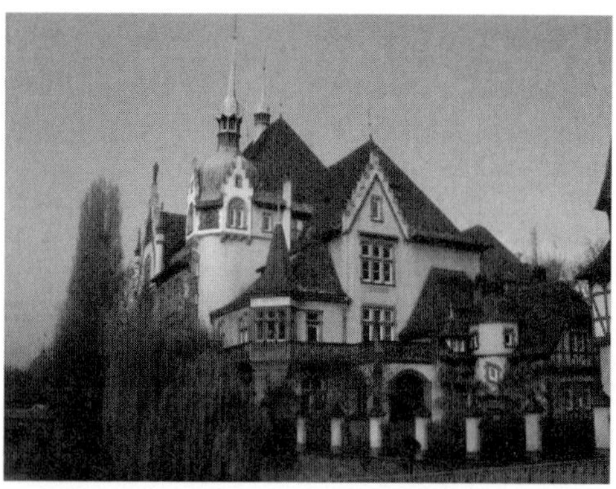

Figure 27.14(a). Overall view of the main building. (see colour plate page 506)

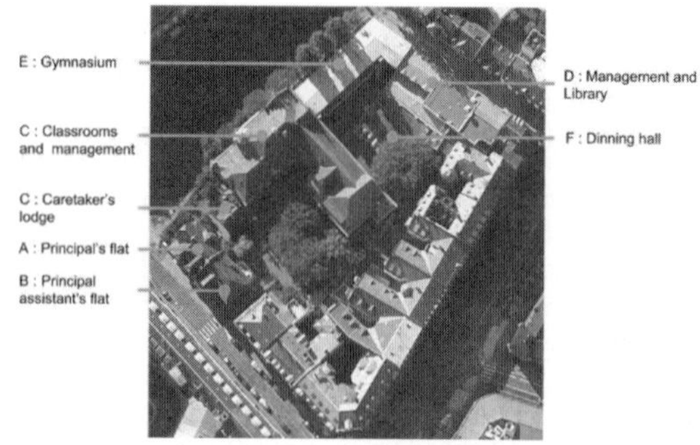

Figure 27.14(b). Aerial view of the site. (see colour plate page 506)

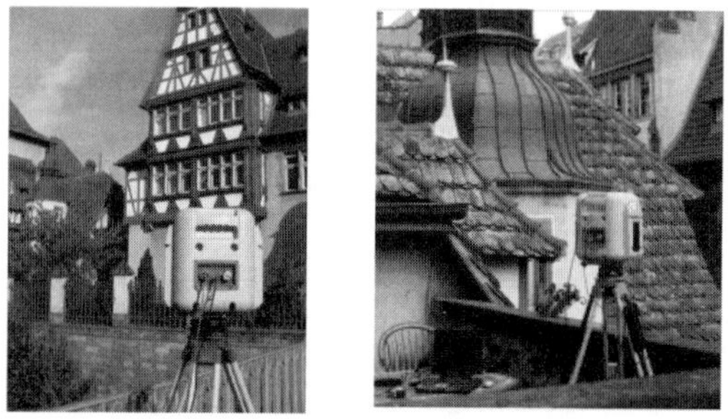

Figure 27.14(c). 3-D scanning system (Trimble GX). (see colour plate page 506)

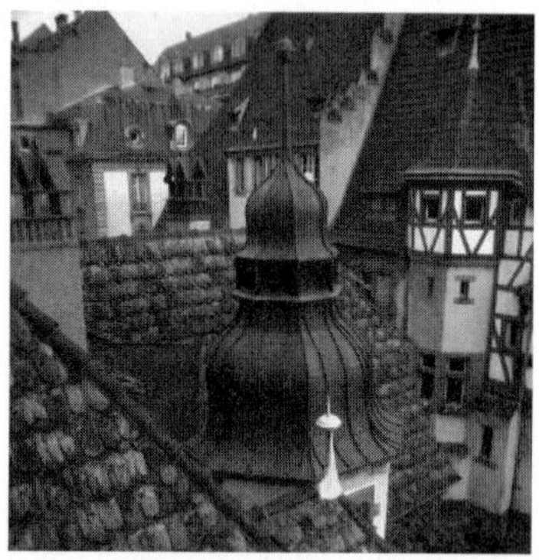

Figure 27.14(d). Roof details. (see colour plate page 506)

In 2006, the Alsace Region Institution asked for building documentation in order to proceed to a complete diagnosis to conform to the current regulation (detection of asbestos and lead, fire safety, heating systems, electricity, accessibility, etc.) as well as to insure the future of the patrimony and among others the general maintenance of roofs and facades. The diagnosis and statements campaigns are usually based on maps and photos, which are used to locate damaged elements. However, because of the possibilities offered by photogrammetric and laser scanning techniques, these methods have been integrated in the diagnosis process.

27.2.3.3.2 Instrumentation—procedures

An accurate control network has been set up in a first stage, by GPS measurements and total station survey workflow.

The Trimble GX terrestrial laser scanner (360° × 60°) has been used for the scanning. The scanner uses an auto focus method for the laser, which proved to be very useful mainly for close range applications. This feature guarantees a constantly small laser spot even at different distances within a scan. The built in video camera is used for both framing of the area of interest as well as colouring the point cloud (Fig. 27.15b).

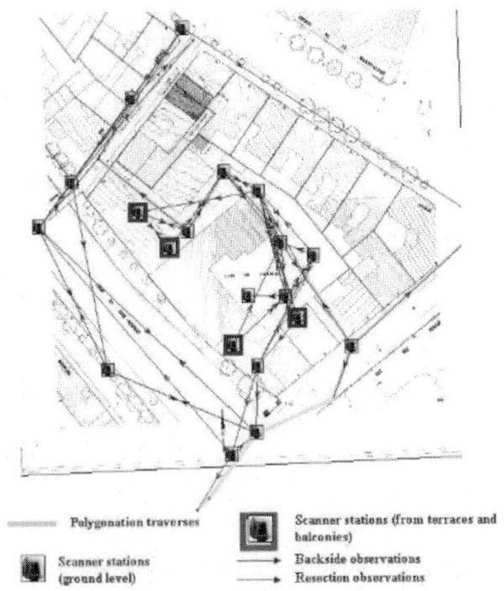

Figure 27.15(a). Ground control network used for the laser scanning survey. (see colour plate page 508)

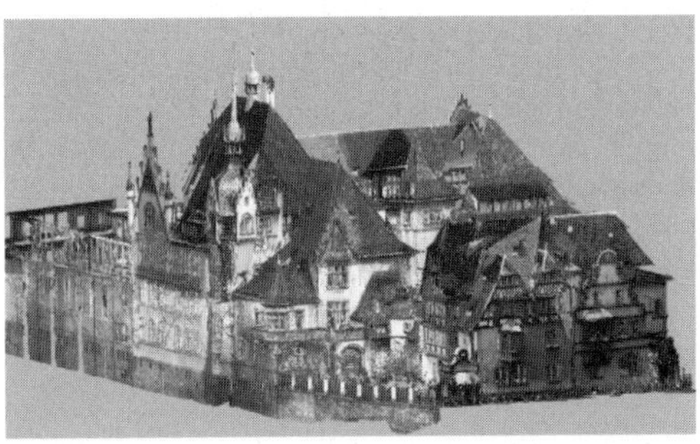

Figure 27.15(b). Final point cloud (with colours from the built-in video camera of the scanner). (see colour plate page 508)

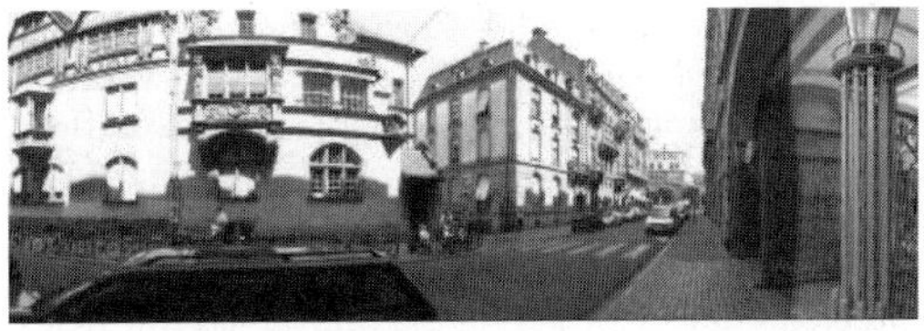

Figure 27.15(c). Panoramic image computed from the video of the scanner. (see colour plate page 508)

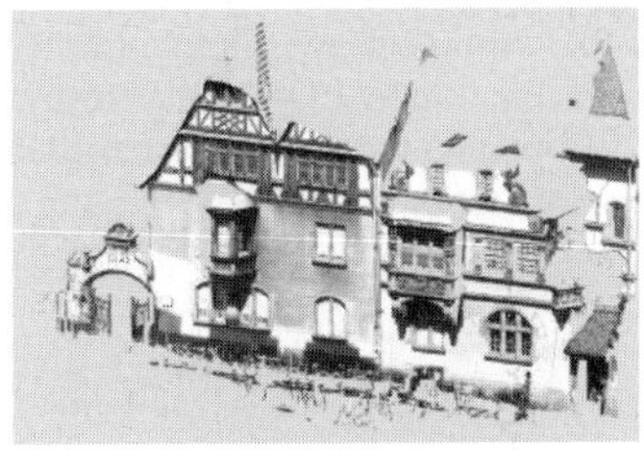

Figure 27.15(d). Point cloud (raw data) displayed with true colour pixels. (see colour plate page 508)

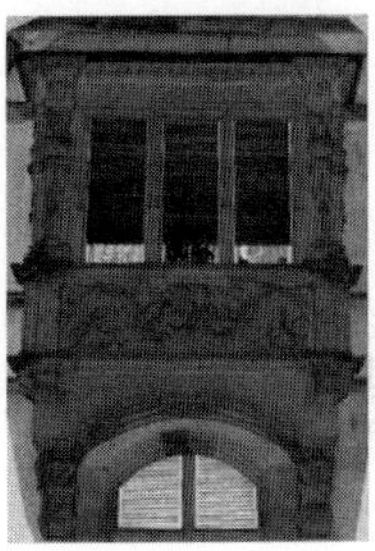

Figure 27.15(e). Detail of the façade (grey scale intensity image), 1 point/1cm. (see colour plate page 508)

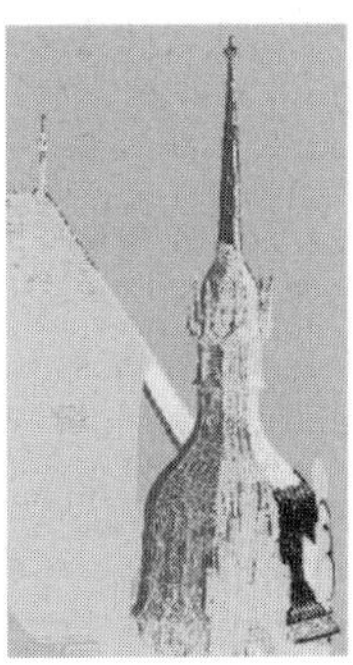

Figure 27.15(f). Control of the model by merging objects from different stations. (see colour plate page 508)

Each part of the site has been scanned twice from at least two different stations. The survey workflow (close to the traditional total station survey workflow) has been used to measure geo-referenced point clouds and avoid any other registration process. The scanner stations have been defined by taking into account problems due to the vegetation, traffic, narrow roads, height of surrounding buildings, hours of sunshine, visibility of reference points. An overview of the different stations network (approx. 20) is shown in Figure 27.15a.

Aerial digital images (4080 × 4076 pixels) from a 2004 campaign were available (overview in Fig. 27.14b). Multi-image photogrammetric techniques based on Canon EOS 5D terrestrial images (4368 × 2912 pixels) with different calibrated objectives have been carried out (Fig. 27.16), and additional photos from an aerial work platform have been taken.

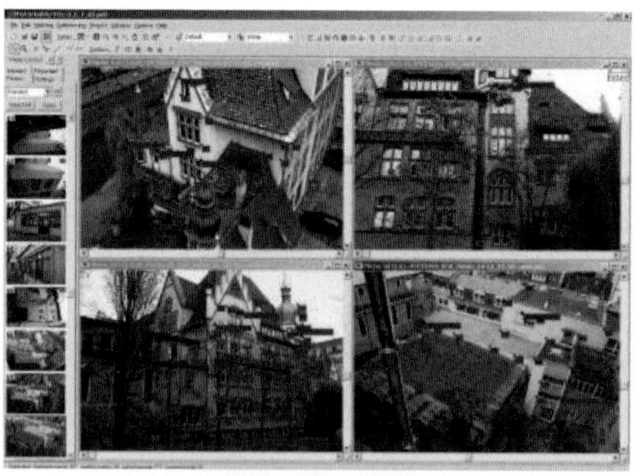

Figure 27.16(a). Set of photos in PhotoModeler's interface. (see colour plate page 508)

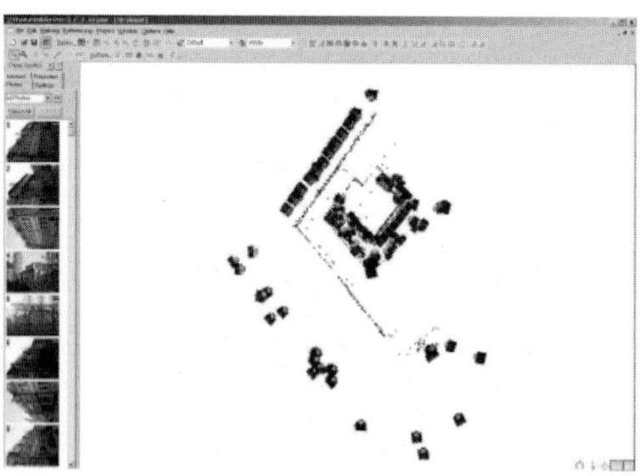

Figure 27.16(b). Overall view of the photo stand points around the site. (see colour plate page 508)

Multi-photo software has been used to orient all the images thanks to control points located around the site and in the inner courtyard. Control points on the façades have been measured by a reflector-less total station. The 3-D geometric points measured in the orientation stage were chosen with their further use for the generation of the orthophotos of the façades in mind (Fig. 27.17).

27.2.3.3.3 Deliverables

No detailed plotting was required because this kind of work is time consuming and expensive. 3-D restitutions are currently replaced by orthophotos, which can be easily imported in AutoCAD© (Fig. 27.17). The work of the different companies involved in the project was specified in separate layers, in order to allow the drawing of the details of their diagnosis.

A 3-D model of the buildings of the site seems essential for further analysis and studies, providing geo-referenced data for the user and visualization in order to know the state of the buildings and to plan the management of future works in the Pontonniers International high school. The link with a database describing the patrimony's referential is used to find existing maps, overall photos as well as zooms on details and other alphanumeric data, by browsing in the 3-D model (Fig. 27.18b).

This 3-D model is used as a browsing interface in the Information System of the Pontonniers high school in Strasbourg for handling all the available data. For the managers of the site, the model is also used to modify and update the database. The information system is based on an EasyPHP server, allowing the creation of databases in MySQL, and interaction between the databases within PHP software packages. The 3-D model is described in VRML, mostly used for the visualization of 3-D scenes on the Web. The realization of the 3-D model precedes any following step. An anchor is defined for each face, which helps for its identification and its interaction. Any face of the model can thus be accessed, by consultation or in order to modify the corresponding tables in the MySQL database.

27.2.3.3.4 Credits

This project as been carried out by the Photogrammetry and Geomatics Group of INSA (Graduate School of Science and Technology), Strasbourg (France), Prof. Pierre Grussenmeyer, Dr. Elise Meyer and Dr. Emmanuel Alby, in close cooperation with M. Rampazzo of the Alsace Region Institution.

27.2.3.4 *Case study 4: The medieval fortress Kufstein*

27.2.3.4.1 Background and Scope of documentation

The fortress of Kufstein (Festung Kufstein, see Fig. 27.19) is a jewel in the lower region of Tyrol,

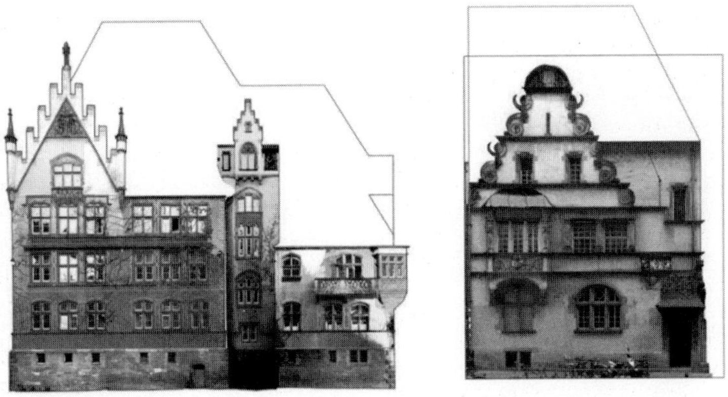

Figure 27.17. Examples of orthophotos imported into the AutoCAD© software to be used as reference documentation for all providers in charge of the diagnosis. (see colour plate page 509)

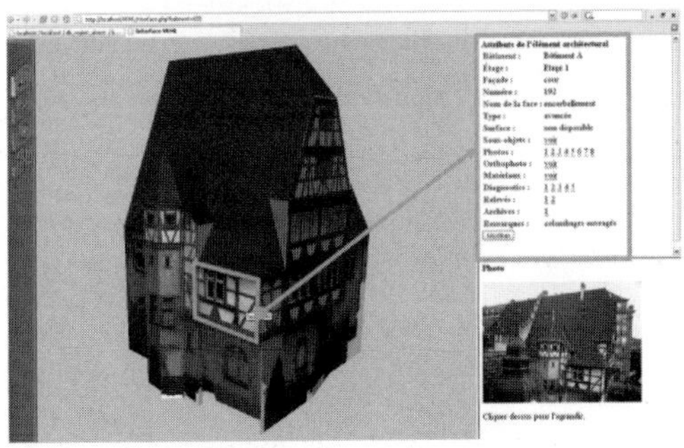

Figure 27.18(a). Interface allowing access to the documents of a building. (see colour plate page 509)

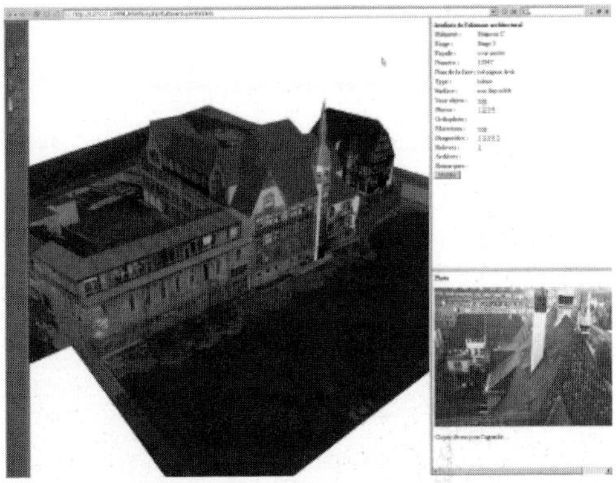

Figure 27.18(b). View of the global 3-D model showing the whole site in the web interface. The access to the information of each building is possible as shown in Figure 27.18(b). (see colour plate page 509)

Austria, and a landmark of the district capital of Kufstein.

Some years ago for further maintenance of the medieval fortress it was necessary to start a serious renovation of the whole complex of the various buildings. To record and document the situation before and maybe compare it to the status after the renovation a photogrammetric recording was initiated.

27.2.3.4.2 Instrumentation—procedures

A private surveying company in Austria—well experienced in the task of documentation of cultural heritage—has taken over the job of façade documentation by means of analytical plotters leading to conventional façade plans (line drawings). The photos had been taken using a helicopter and a metric camera Zeiss UMK 10/13/18 (Fig. 27.20) to record the outer part of the monument and the rocky hill on which it is based. About 180 metric images have been taken from this helicopter platform with an average overlap of 80%.

Because it was not possible to reach the interior parts such as yards and paths within the fortress from the helicopter, a digital camera Olympus E 10 with a resolution of over 4 Million Pixel was calibrated and used to record the interior and hidden parts (Fig. 27.21). In addition to the regular lens, a wide angle adapter was used to achieve an all together focal length of 7 mm (equals 28 mm with a small format camera). Over 350 digital photos have been used to set up the 3-D model of the medieval fortress (Fig. 27.22) and to get good overlap, even in the problematic parts of the medieval fortress.

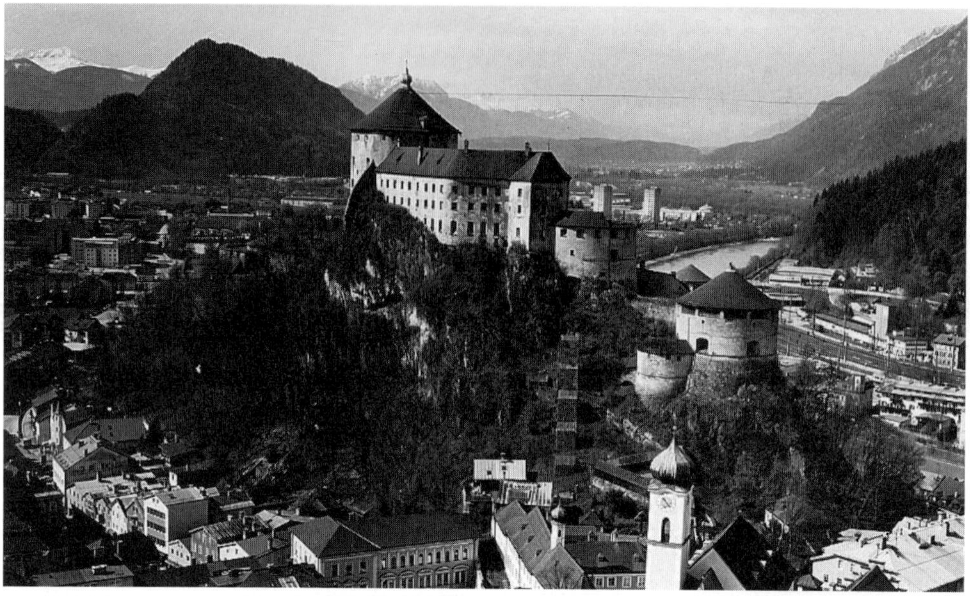

Figure 27.19. Attitude of the fortress of "Kufstein", Tyrol, Austria (overview).

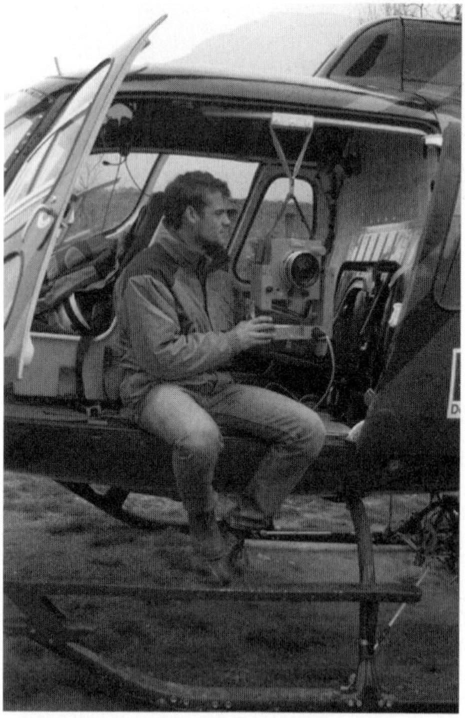

Figure 27.20. Zeiss UMK mounted in helicopter. (see colour plate page 509)

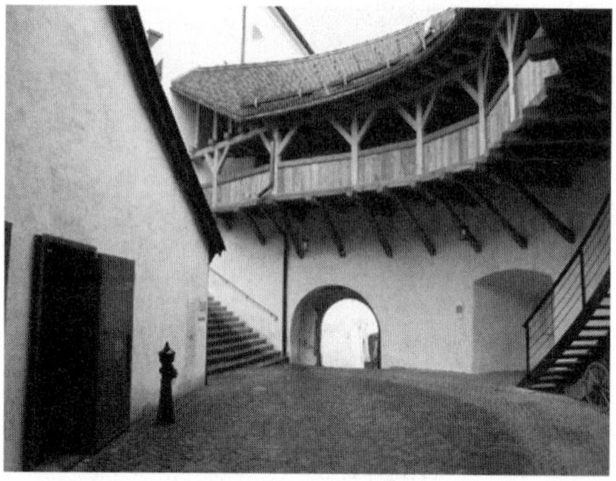

Figure 27.21. Digital photo of interior yard. (see colour plate page 509)

To get a consistent solution of the entire object, all elements and camera orientations were computed during a photogrammetric bundle adjustment where a probable and homogenous result of the whole measurement system could be achieved.

27.2.3.4.3 Deliverables

The deliverables were a digital photo-textured 3-D model of the entire fortress and its surrounding rocks (Fig. 27.23).

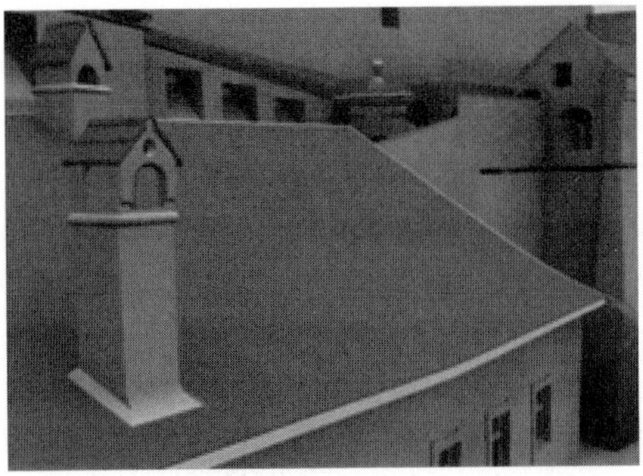

Figure 27.22. Level of detail of the reconstruction (without texture).

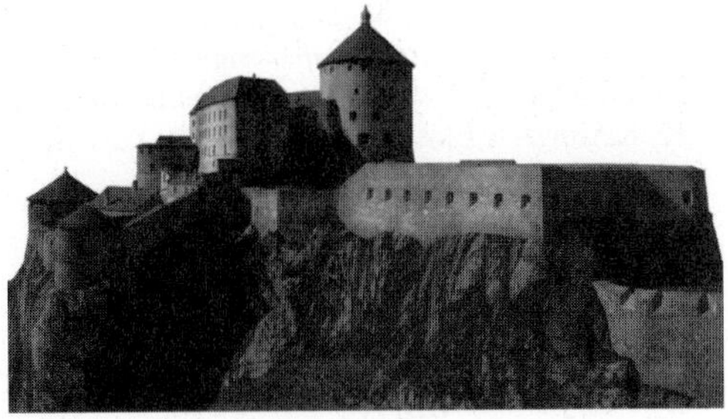

Figure 27.23. Partial view of the textured 3-D model. (see colour plate page 510)

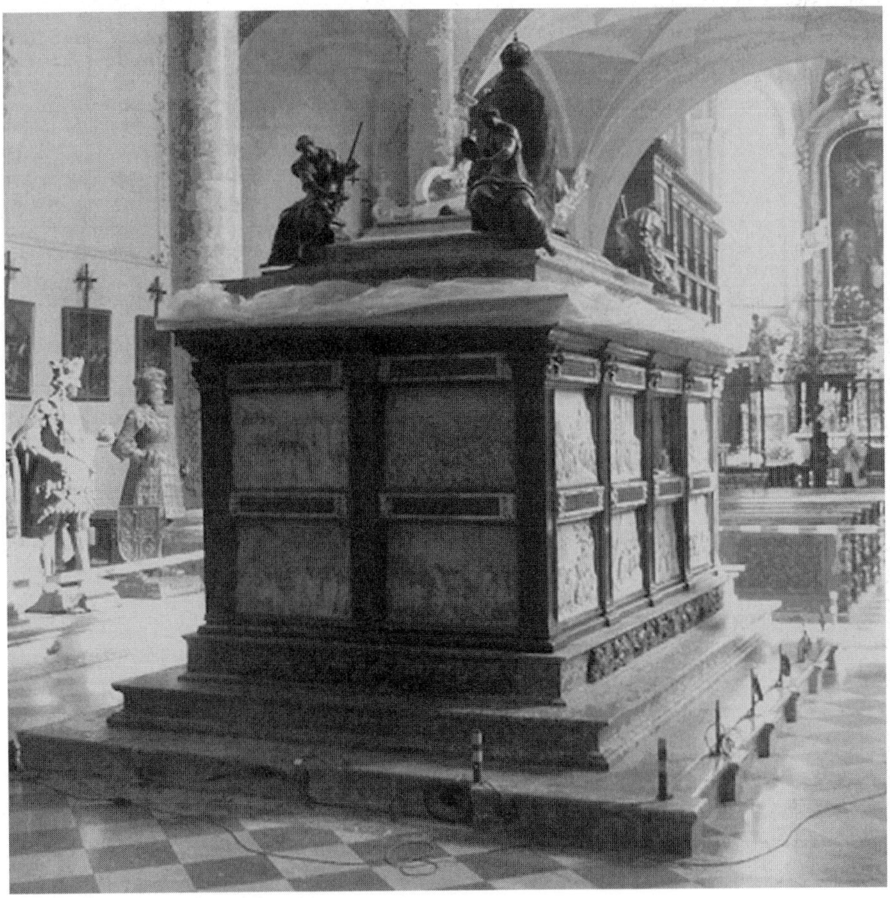

Figure 27.24. Total view of the cenotaph during the measurement work—for the first time ever without lattices and glass plates.

27.2.3.4.4 Credits

The project has been carried out by the Surveying and Geoinformation Unit of the University of Innsbruck, Austria, Prof. Klaus Hanke and Dipl.-Ing. Michael Oberschneider, in close cooperation with Dipl.-Ing. Josef Linsinger (Linsinger & Partner ZT-GmbH, St.Johann/P., Austria).

27.2.3.5 *Case study 5: The cenotaph of emperor Maximilian I*

27.2.3.5.1 Background and Scope of documentation

The Hofkirche at Innsbruck with the empty tomb (cenotaph) of Emperor Maximilian I (Fig. 27.24) is probably the most important art-historical monument in the region of Tyrol, Austria. It was built between 1555 and 1565 under Emperor Ferdinand I who was the grandson of Maximilian and the brother of the famous German Emperor Karl V. The cenotaph with the statue of the kneeling Emperor dominates the centre of the nave of the church. The cenotaph itself has an extent of 6.4 m × 4.5 m × 3.3 m and consists of a frame of black marble in which the 24 reliefs of white marble (each approx. 80 cm x 45 cm) are embedded in two horizontal rows. These reliefs (Fig. 27.25) show scenes from the life of the Emperor Maximilian I. They have a level of detail (Fig. 27.26) within the range of 0.1 mm and had to be documented in particular and with the highest precision available. On the cover of the tomb, the kneeling figure of the Emperor is central, surrounded by representations of the four basic virtues (Fig. 27.27), which are arranged at the four corners (Fig. 27.28). All mentioned figures are made of dark bronze.

27.2.3.5.2 Instrumentation—procedures

A common geodetic control point system was installed for both methods. Photogrammetric work consisted of stereo pairs and separate colour images. 3-D scanning was accomplished with a MENSI S25 for the overall structure and a GOM ATOS II at high resolution for

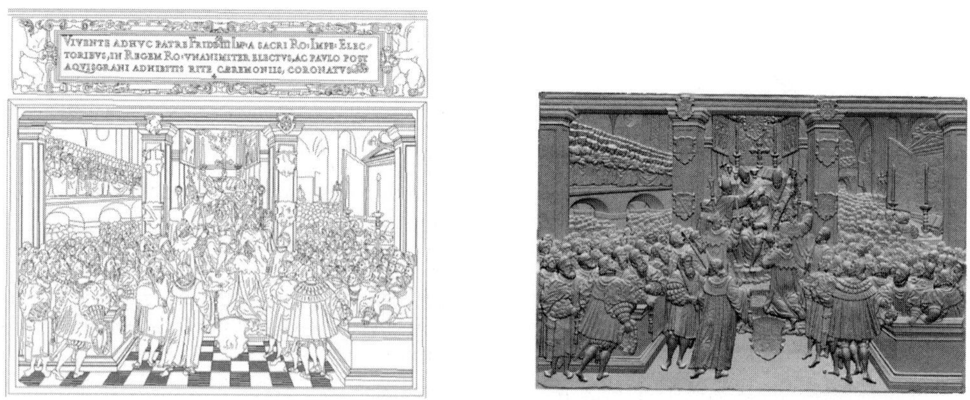

Figure 27.25. One of the relief plates. Left: Result of stereo plotting. Right: Virtual model from scanned data.

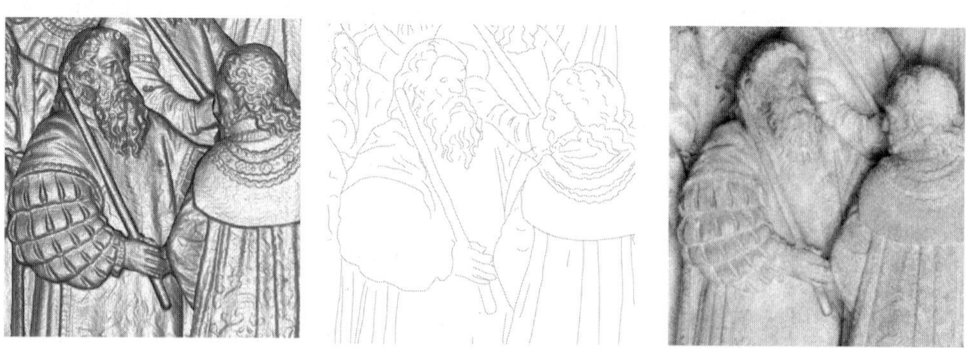

Figure 27.26. Detail from relief plate above (Fig. 27.25), about 10 cm × 10 cm in reality. Left: photograph, centre: result of stereo plotting, right: virtual model from scanned data.

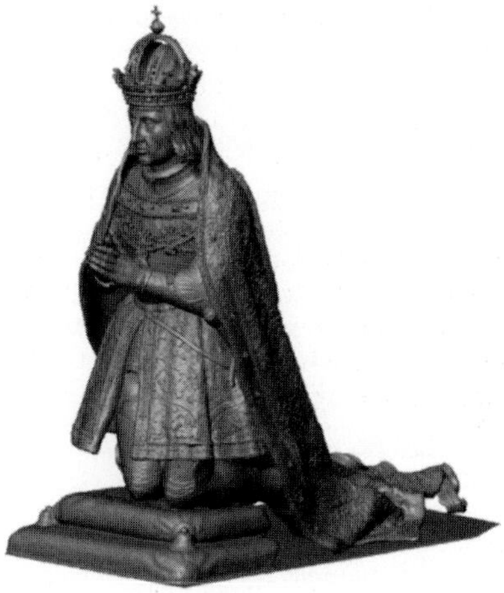

Figure 27.27. Kneeling Maximilian I. Virtual model from scanned data.

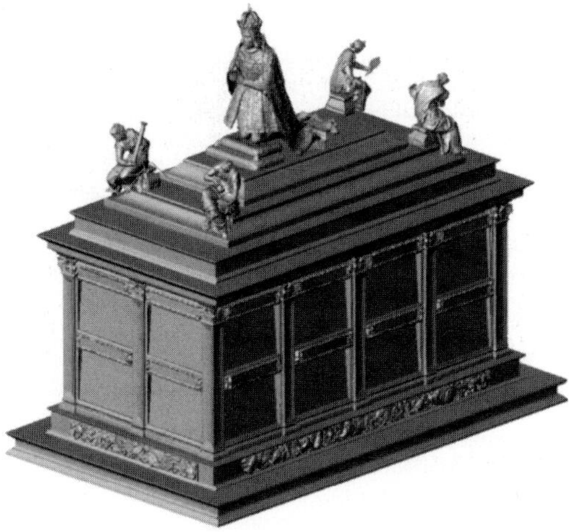

Figure 27.28. Overview 3-D model of the cenotaph.

the relief plates. Line plots from the photogrammetric stereo models did not really give an adequate representation of the object. 3-D visualization using the scanning results can achieve a much better impression of the complicated geometry after the data modelling. In order to model the complex geometry, it is necessary to use huge amounts of data.

27.2.3.5.3 Deliverables
The deliverables were the digital 3-D model of the entire object and more detailed models of the marble plates and statues.

27.2.3.5.4 Credits
The project has been carried out by i3 Mainz, Germany, Prof. Wolfgang Boehler and his team, the Surveying and Geo-information Unit of the University of Innsbruck, Austria, Prof. Klaus Hanke, in close cooperation with Dipl.-Ing. Josef Linsinger (Linsinger & Partner ZT-GmbH, St.Johann/P., Austria), WESTCAM Datentechnik GmbH, Mils and the local authorities (Land Tyrol).

27.3 CONCLUSION AND OUTLOOK

New work habits, new research opportunities, and even new forms of institutions force photogrammetry to offer its services in many aspects. Photogrammetry provides large amounts of highly detailed, very accurate, geo-referenced, 3-D vector and texture data, with stereo-viewing abilities and metadata information. This constitutes its comparative advantage over other techniques and procedures.

Evaluating the current status and envisaging the future evolution of architectural and archaeological applications of photogrammetry, one could note that:

- Classic technology is very mature and the applications based on it are straightforward. Therefore, there are many different applications reported, and this has a favourable impact on the end-users, since it attracts their attention.
- New technology is entering the picture at a growing rate and this drives innovative research. This fact gives rise to rapidly emerging new concepts and, as a spin-off, it attracts participation from related disciplines.
- New issues are entering the research agenda, like standardization issues, systems for quality management, intellectual property issues. It is quite important to note that, although conformance to intrinsic quality measures (i.e. standards) will always be necessary, it is only one part of the story. Quality can only be determined by "fitness for use". Ultimately, quality evaluation needs to include user demands. In building market positions, this means that one should be able to distinguish different groups of users and recognize typologies of quality demands.

REFERENCES—SUGGESTED BIBLIOGRAPHY

Agosto, E., Porporato, C. & Rinaudo, F. 2004. Stereo photomap and 3D realistic model of city centers. *Proc. International Workshop on Vision Techniques applied to Rehabilitation of City Centres, Lisbon, Portugal, 25–27 October 2004. Workshop CD.*

Al-Ajlouni, S.S. & Fraser, C.S. 2006. Zoom-dependent calibration for consumer grade-cameras. *The ISPRS International Archives of Photogrammetry, Remote Sensing and Spatial Information Sciences* 36(5): 20–25.

Alshawabkeh, Y. & Norbert Haala, N. 2004. Integration of digital photogrammetry and laser scanning for heritage documentation. *The ISPRS International Archives of Photogrammetry and Remote Sensing* 35(B5).

Amiri Parian, J. & Gruen, A. 2006. Self-calibrating network analysis for panoramic cameras by heuristic simulation. *The ISPRS International Archives of Photogrammetry, Remote Sensing and Spatial Information Sciences* 36(5): 230–235.

Arias, V., Kuntz, T.R., Richards, H., Watson, R.P. & Van Der Elst, J. 2004. Documentation and virtual modeling of archaeological sites using a non-metric extreme wide-angle lens camera. *International Archives of Photogrammetry and Remote Sensing* 35(B5): 824–827.

Atkinson, K.B. 1996. *Close Range Photogrammetry and Machine Vision*. Scotland: Whittles Publishing.

Balletti, C., Guerra, F., Vernier, P., Studnicka, N., Riegl, J. & Orlandini, S. 2004. Practical comparative evaluation of an integrated hybrid sensor based on Photogrammetry and Laser Scanning for Architectural Representation. *International Archives of Photogrammetry and Remote Sensing* 35(5): 536–541.

Balletti, C., Guerra, F., Lingua, A. & Rinaudo, F., 2005. True digital orthophoto of the San Marco basilica in Venice. International workshop on Vision Techniques for Digital Architectural and Archaeological Archives, (Ancona, Italy, 1–3 July). *International Archives of Photogrammetry, Remote Sensing and Spatial Information Sciences* 34(5/W12): 43–48.

Barber, D., Mills, J., & Bryan, P. 2002. Experiences of laser scanning for close range structural recording. *Proceedings CIPA WG 6: International Workshop on Scanning for Cultural Heritage Recording, Corfu, Greece, 1–2 Sept.*: 121–126.

Beraldin J.-A., Picard, M., El-Hakim, S., Godin, G. Latouche, C., Valzano, V. & Bandiera, A. 2002. Exploring a Byzantine crypt through a high-resolution texture mapped 3D model: Combining range data and photogrammetry. *Proceedings CIPA WG 6: International Workshop on Scanning for Cultural Heritage Recording, Corfu, Greece, 1–2 Sept.*: 65–72.

Biber P., Fleck S., Wand M., Staneker D. & Straßer W. 2005. First experiences with a mobile platform for flexible 3D model acquisition in indoor and outdoor environments—The Waegele. *International Archives of Photogrammetry, Remote Sensing and Spatial Information Sciences* 36(5/W17).

Boccardo, P., Dequal, S., Lingua, A. & Rinaudo, F. 2001. True digital orthophoto for architectural and archaeological applications. International Workshop on Recreating the Past-Visualization and Animation of Cultural Heritage, Ayutthaya, Thailand, 26 Feb.–1 Mar. *International Archives Photogrammetry and Remote Sensing* 34(5C15): 50–55.

Boehler, W. (ed.). 2002. Scanning for Cultural Heritage Recording. Workshop co-organized by CIPA and ISPRS Comm. V. *The International Archives of Photogrammetry and Remote Sensing* XXXIV, Part5, Com. V.

Böhler, W. 2005. Comparison of 3D scanning and other 3D measurement techniques. *Proc. International Workshop on Recording, Modeling and Visualization of Cultural Heritage, Centro Stefano Franscini, Monte Verità, Ascona, Switzerland, 22–27 May 2005*: 89–99.

Boulanger, P. & Taylor, J. 2000. The Virtual Museum—Data generation and Visualization. *International Archives of Photogrammetry and Remote Sensing, 33(Special Sessions, Supplement)*: 4–9.

Carbonnell, M. 1989. *Photogrammétrie appliquée aux relevés des monuments et des centres historiques, ICCROM Rome*. International Centre for the Study of the Preservation and the Restoration of Cultural Property.

Chiuso, A., H. Jin, P. Favaro & S. Soatto, 2000. MFm: 3D Motion and Structure from 2D Motion Causally Integrated Over Time: Implementation. In D. Vernon (ed.), *Computer Vision—ECCV 2000, Lecture Notes in Computer Science* 1843: 734–750.

Costantino D., Angelini M.G. & Caprino, G. 2007. Rapid approach of integrated survey for the conservative analysis of pictures. *International Archives of Photogrammetry, Remote Sensing and Spatial Information Sciences* 35 (5/C53) and *The CIPA International Archives for Documentation of Cultural Heritage* 21(2007): 247–252.

CIPA. 2007. http://cipa.icomos.org (accessed March 2008).

Dequal, S., Lingua, A. & Rinaudo, F. 1999. A new tool for architectural photogrammetry: The 3D Navigator, XVII CIPA International Symposium, Olinda, Brazil October 3–6, 1999. *International archives of Photogrammetry and Remote Sensing* XXXII Part 5C2 http://cipa.icomos.org/fileadmin/papers/olinda/99c503.pdf.

Doneus, M., Briese ,Chr., Fera, M. & Janner, M. 2008. Archaeological prospection of forested areas using full-waveform airborne laser scanning. *Journal of Archaeological Science* 35, Issue 4, April 2008: 882–893.

Doneus, M. & Briese, C. 2006. Digital terrain modelling for archaeological interpretation within forested areas using full-waveform Laserscanning. *The 7th International Symposium on Virtual Reality, Archaeology and Cultural Heritage VAST (2006), October/November 2006.*

Drap, P., Sgrenzaroli, M., Canciani, M., Cannata, G. & Seienturier, J. 2003. Laser scanning and close range photogrammetry: towards a single measuring tool dedicated to architecture and archaeology. *The ISPRS International Archives of the Photogrammetry, Remote Sensing and Spatial Information Sciences* 34(5/C15): 629–634.

El-Hakim, S.F., Gonzo, L., Picard, M., Girardi, St., Simoni, A., Paquet, E., Viktor, H. & Brenner C., 2003. Visualisation of Highly Textured Surfaces. *The 4th International Symposium on Virtual Reality, Archaeology and Intelligent Cultural Heritage VAST (2003).*

English Heritage. 2007. *3D Laser Scanning for Heritage: advice and guidance to users on laser scanning in archaeology and architecture*. Product code 51326. Available at www.heritage3D.org (accessed March 2008).

Enciso, R., Zhang, Z. & Viéville, T. 1996. Dense Reconstruction Using Fixation and Stereo Cues. *Project RobotVis, World Automation Congress 1996.*

Ethrog, U. 2006. CCD camera calibration based on the sun's images *The ISPRS International Archives of Photogrammetry, Remote Sensing and Spatial Information Sciences* 36(5): 109–113.

Fangi, G. 2006. Investigation on the suitability of the spherical panoramas by Realviz Stitcher for metric purposes. *The ISPRS International Archives of Photogrammetry, Remote Sensing and Spatial Information Sciences* 36(5).

Favaro, P. 2002. *Shape from Focus/Defocus.* Washington University Dpt. of Electrical Engineering, Electronic Signals and Systems Research Lab, June 25, 2002. http://homepages.inf.ed.ac.uk/rbf/CVonline/LOCAL_COPIES/FAVARO1/dfdtutorial.html

Forsyth, D.A. 2002. Shape from texture without boundaries. *Proceedings of the 7th European Conference on Computer Vision* Part III: 225–239.

Fryer, J., Chandler, J. & El-Hakim, S. 2005. Recording and modelling an aboriginal cave painting: with or without laser scanning? *International Archives of Photogrammetry, Remote Sensing and Spatial Information Sciences* 36(5W17): 8 p.

Gemenetzis, D., Georgiadis, H. & Patias, P. 2001. Virtuality and Documentation. International Workshop on Recreating the Past-Visualization and Animation of Cultural Heritage, Ayutthaya, Thailand, 26 Feb.-1 Mar. *International Archives Photogrammetry and Remote Sensing* 34(5C15): 159–164.

Georgiadis, Ch., Tsioukas, V., Sechidis, L., Stylianidis, E. & Patias, P. 2000. Fast and accurate documentation of archaeological sites using in-the-field photogrammetric techniques. *International Archives of Photogrammetry and Remote Sensing* 33(Supplement B5): 28–32.

Georgoula, O., Kaimaris, D., Tsakiri, M. & Patias, P. 2004. From the aerial photo to high resolution satellite image: Tools for the archaeological research. *Int. Archives of Photogrammetry and Remote Sensing* 35(B7):1055–1060.

Geyer, C. & Daniilidis, K. 2003. Omnidirectional video. *Visual Computer* 19: 405–416.

GIM International. 2007a. Product survey on Terrestrial Laser Scanners. *The Global Magazine for Geomatics* August 2007, Volume 21, Issue 8.

GIM International. 2007b. Product survey on Digital Photogrammetric Workstations. *The Global Magazine for Geomatics* December 2007, Volume 21, Issue 12.

Glykos, T., Karras & G. Voulgaridis, G. 1999. Close-range photogrammetry within a commercial CAD package. ISPRS WG V/2 & WG5 Workshop: Photogrammetric Measurement, Object Modeling and Documentation in Architecture and Industry, Thessaloniki, Greece, 7–9 July. *International Archives of Photogrammetry and Remote Sensing* 32(5W11): 103–106.

Grammatikopoulos, L., Kalisperakis, I., Karras, G. Kokkinos, T. & Petsa, E. 2004. On automatic orthoprojection and texture-mapping of 3D surface models. *International Archives of Photogrammetry and Remote Sensing* 35(5): 360–365.

Grün, A., Remondino, F. & Zhang, L. 2004. Photogrammetric Reconstruction of the Great Buddha of Bamiyan, Afghanistan. *The Photogrammetric Record* 19(107): 177–199.

Grussenmeyer, P., Alby, E., Meyer, R. & Rampazzo, M. 2006. 3D building model as an interface for a Web Information System. Case study of the Pontonniers high school in Strasbourg. ISPRS Comm. V Symposium, Dresden, Sept. 25–27, Germany. *Int. Arch. of Photogrammetry and Remote Sensing and Spatial Information Sciences* 36(5).

Guarnieri, A., Vettorea, A., El-Hakim, S. & Gonzo, L. 2004. Digital photogrammetry and laser scanning in cultural heritage survey. *International Archives of Photogrammetry and Remote Sensing* 35(5): 154–158.

Gühring, J. 2001. Dense 3D surface acquisition by structured light using off-the-shelf components. *Videometrics and Optical Methods for 3D Shape Measurement, Proceedings SPIE* 4309: 220–231.

Haggrén, H., Junnilainen, H., Järvinen , J., Nuutinen, T., Laventob, M. & Huotarib, M. 2004. The use of panoramic images for 3D archaeological survey. *International Archives of Photogrammetry and Remote Sensing* 35(5): 958–963.

Hanke, K. 2001. Accuracy Study Project of Eos Systems' PhotoModeler, Final Report. http://www.photomodeler.com/pdf/hanke.pdf

Hanke, K. & Grussenmeyer, P. 2002. Architectural Photogrammetry: *Basic theory, Procedures, Tools. September 2002, ISPRS Commission 5 Tutorial, Corfu.*

Hanke, K., Grussenmeyer, P., Grimm-Pitzinger, A. & Weinold, Th. 2006. First Experiences with the Trimble GX Scanner. ISPRS Comm. V Symposium, Dresden, Sept. 25–27, Germany. *Int. Arch. of Photogrammetry and Remote Sensing and Spatial Information Sciences* 36(5).

Hanke, K. & Oberschneider, M. 2002. The Medieval Fortress Kufstein, Austria—an Example for the Restitution and Visualization of Cultural Heritage. *International Archives of Photogrammetry and Remote Sensing* 34(5): 530–533.

Hanke, K. & Linsinger, J. 2003. Festung Kufstein. In T. Luhmann (ed.), *Nahbereichsphotogrammetrie in der Praxis-Beispiele und Problemlösungen*: 39–42. Heidelberg: Wichmann.

Hanke, K. 2004. Restitution and Visualization of the Medieval Fortress Kufstein. *Proc. International Workshop on Vision Techniques applied to Rehabilitation of City Centres, Lisbon, Portugal, 25–27 October 2004*: Workshop CD.

Hanke, K. & Moser, M. 2006. Scanning of Antonio Gaudí's Original Images of his Draft for a Church in Barcelona . *The ISPRS International Archives of Photogrammetry, Remote Sensing and Spatial Information Sciences* 36(5).

Hanke, K. & Boehler, W. 2004. Recording and Visualization of the Cenotaph of German Emperor Maximilian I. *The ISPRS International Archives of Photogrammetry and Remote Sensing* 35(B5).

Hemmleb, M. 1999. Digital Rectification of historical images. *Proc. XVII CIPA Symposium, Mapping and Preservation for the New Millenium, Olinda, Brazil, 3–6 Oct.* http://cipa.icomos.org/fileadmin/papers/olinda/99c806.pdf (accessed March 2008).

Herraez Boquera, J., Navarro Esteve, P. & Lerma Garcia, J.L. 1997. Integration of normal colour and color infrared emulsions for the identification of pathologies in architectural heritage using a digital photogrammetric system. *International Archives of Photogrammetry and Remote Sensing* 32(5C1B): 240–245.

Heuvel, F.A. van den, Verwaal, R. & Beers, B. 2006. Calibration of Fisheye camera systems and the reduction of Chromatic Aberration. *The ISPRS International Archives of Photogrammetry, Remote Sensing and Spatial Information Sciences* 36(5).

Heuvel, F.A. van den. 2003. *Automation in Architectural Photogrammetry; Line-Photogrammetry for the Reconstruction from Single and Multiple Images.* Ph.D. thesis of the Delft University of Technology. Publications on Geodesy 54, ISBN 90 6132 281 2, NCG, Netherlands Geodetic Commission, Delft, The Netherlands.

ICOMOS. 2007. ICOMOS Charters, Resolutions and Declarations. http://www.international.icomos.org/charters.htm (accessed March 2008).

Kakiuchi, T. & Chikatsu, H. 2000. Construction of stereo vision system for 3D objects modelling. *International Archives Of Photogrammetry And Remote Sensing* 33(B5/1): 414–421.

Karara, H.M., (ed.). 1989. *Non-Topographic Photogrammetry, 2nd edition.* American Society for Photogrammetry and Remote Sensing, Science and Engineering Series, Virginia, USA.

Karras, G.E. & Mavrommati, D. 2001. Simple Calibration Techniques for Non-Metric Cameras. *The ISPRS International Archives of Photogrammetry and Remote Sensing* XXXIV-5/C7 (ISSN 1682-1750) and The CIPA International Archives for Documentation of Cultural Heritage XVIII-2001 (ISSN 0256-1840): 39–46.

Karras, G., Petsa, E., Dimarogona, A. & Kouroupis, S. 2001. Photo-Textured Rendering of Developable Surfaces in Architectural Photogrammetry. In B. Fisher, K. Dawson-Howe & C. O'Sullivan (eds.), *Proc of International Symposium on Virtual and Augmented Architecture (VAA01), Dublin, 21–22 June 2001*:147–158. Berlin: Springer Verlag.

Karras, G., Patias, P. & Petsa, E. 1996. Monoplotting and photo-unwrapping of developable surfaces in architectural photogrammetry. *International Archives of Photogrammetry and Remote Sensing* 31(5): 290–294.

Karras, G., Patias, P., Petsa, E. & Ketipis, K. 1997. Raster projection and development of curved surfaces. *International Archives of Photogrammetry and Remote Sensing*, 32(5C1B): 179–185.

Kasser, M. & Egels, Y. (eds.). *Digital Photogrammetry*. Contains chapter on Architectural Photogrammetry, contributed by Grussenmeyer, P., Hanke, K., Streilein, A. New York and London: Taylor & Francis. ISBN 0-7484-0945-9 (paperback) and ISBN 0-7484-0944-0 (hardback).

Kasser, M. & Egels, Y. (eds.). *Photogrammétrie Numérique*. (in French) Contains chapter on Architectural Photogrammetry, contributed by Grussenmeyer, P., Hanke, K., Streilein, A. Paris: Hermes Science Publishing. ISBN 2-7462-0311-1

Kedzierski, M. & Walczykowski, P. 2007. Fisheye Lens Camera System Application to Cultural Heritage Data Acquisition. XXI. International CIPA Symposium, 01–06 October 2007, Athens, Greece. *The ISPRS International Archives of Photogrammetry, Remote Sensing and Spatial Information Sciences 5/C53* (ISSN 1682-1750) and *The CIPA International Archives for Documentation of Cultural Heritage XXI-2007 (ISSN 0256-1840)*: 809–811.

Koussoulakou, A., Sechidis L. & Patias, P. 1999. Virtual inverse photogrammetry. ISPRS WG V/2 & WG5 Workshop: Photogrammetric Measurement, Object Modeling and Documentation in Architecture and Industry, Thessaloniki, Greece, 7–9 July. *International Archives of Photogrammetry and Remote Sensing* 32(5W11): 111–117.

Kousoulakou, A., Patias, P., Sechidis, L. & Stylianidis, E. 2001b. Desktop cartographic augmented reality: 3D mapping and inverse photogrammetry in convergence. *Proc. 20th International Cartographic Association Conference, Beijing, 6–10 Aug.*: 2506–2513.

Lemmon, T. & Biddiscombe, P. 2006. Laser Scanners Improve Survey Workflow. *Professional Surveyor Magazine*, March 2006: 14–18.

Lingua, A., Rinaudo, F., Auditore, G., Balletti, C., Guerra, F. & Pilot, L. 2003. Comparison of surveys of St. Mark's square in Venice. 18th CIPA International Symposium 2003, Antalya, Turkey, 30 Sept.–4 Oct. *International Archives Photogrammetry Remote Sensing* 34(5/C15). http://cipa.icomos.org/fileadmin/papers/antalya/71.pdf.

Lhuillier, M. Toward Flexible 3D Modeling using a Catadioptric Camera. *CVPR 2007–International Conference on Computer Vision and Pattern Recognition, 17–22 June 2007*, ISBN: 1-4244-1180-7:1–8.

Luhmann, Th. 2004. A Historical Review on Panorama Photogrammetry. ISPRS Workshop on Panorama Photogrammetry, Dresden. *International Archives of Photogrammetry and Remote Sensing* XXXIV, Part 5/W16. http://www.tudresden.de/fghgipf/photo/PanoramicPhotogrammetryWorkshop2004/Proceedings.htm (last access March 2008).

Maas, H.-G. & Schneider, D. (eds.). 2004. Panoramic Photogrammetry. ISPRS Workshop on Panorama Photogrammetry, Dresden. *International Archives of Photogrammetry and Remote Sensing* XXXIV, Part 5/W16. http://www.tudresden.de/fghgipf/photo/PanoramicPhotogrammetryWorkshop2004/Proceedings.htm (last access March 2008).

Meyer, E., Grussenmeyer, P., Perrin, J.-P., Durand, A. & Drap, P. 2006. Intra-site Level Cultural Heritage Documentation: Combination of Survey, Modeling and Imagery Data in a Web Information System. In M. Ioannides, D. Arnold, F. Niccolucci & K. Mania (eds.), *Short papers from the joint event CIPA/VAST/EG/EuroMed 2006, 30 October–4 November 2006, Nicosia, Cyprus*: 129–134. EPOCH Publication. ISBN-10: 963 8046 74 0.

Mills, J., Peirson, G., Newton, I. & Bryan, P. 2000. Photogrammetric investigation into the suitability of desktop image measurement software for architectural recording. *International Archives of Photogrammetry and Remote Sensing* 33(B5/2): 525–532.

Ogleby, C. & Rivett, L. 1985. *Handbook of Heritage Photogrammetry.* Australian Heritage Commission, Special Australian Heritage Publication Series No.4, Australian Government Pub. Service, Canberra.

Pateraki, M., Baltsavias, E. & Patias, P. 2002. Image combination into large virtual images for fast 3D modelling of archaeological sites. *International Archives of Photogrammetry and Remote Sensing* 34(5): 440–445.

Patias, P., Sylaiou, S., Sechidis, L., Grussenmeyer, P., Meyer, E., Landes, T. & Alby, E. 2006. *A proposed low-cost system for 3D archaeological documentation.* In M. Ioannides, D. Arnold, F. Niccolucci & K. Mania (eds.), *Proceedings of the 7th International Symposium on Virtual Reality, Archaeology and Cultural Heritage VAST (2006), Nicosia, Cyprus*: 145–149. The Eurographics Association.

Patias, P. 2006. Cultural Heritage Documentation, ISPRS Comm. VI Special Interest Group "Technology Transfer Caravan". *Proceedings of the International Summer School "Digital Recording and 3D Modeling", Aghios Nikolaos, Crete, Greece, 24–29 April 2006*. In workshop CD.

Patias, P. 2001. Caring for the Past, Aiming at the Future: Plans and Policy of ISPRS Commission V. Keynote Address, International Workshop on Recreating the Past - Visualization and Animation of Cultural Heritage, (Ayutthaya, Thailand, 26 Feb.–1 Mar.). *International Archives Photogrammetry and Remote Sensing 34(5C15)*: 1–4.

Patias, P. 2004. *Overview of applications of close-range Photogrammetry and Vision techniques in Architecture and Archaeology, In C. McGlone (ed), Manual of Photogrammetry, 5th edition*: 1044–1050. ISBN 1-57083-071-1. American Society of Photogrammetry Publ., Maryland, USA.

Patias, P. 2003. Information in a Culture-Alert Society. Can Photogrammetry measure the Visual Perception of Works of Art ? *International archives of Photogrammetry, Remote Sensing and Spatial Information Sciences Proceedings of the Workshop* XXXIV-6/W12. On workshop CD.

Patias, P. 2007. Cultural Heritage Documentation, Chapter 9. In J. Fryer, H. Mitchell & J. Chandler (eds.), *Applications of 3D measurement from images*: 225–250, Whittle Publishing Co. ISBN 10: 1-870325-69-9, ISBN 13: 978-1-870325-69-1.

Patias, P. & Peipe, J. 2000. Photogrammetry and CAD/CAM in Culture and Industry—An ever changing paradigm. *International Archives Photogrammetry and Remote Sensing* 33(B5): 599–603.

Patias, P., Stylianidis, E. & Terzitanos, K. 1998. Comparison of simple off-the-self and of-wide-use 3D modelling software to strict photogrammetric procedures for close-range applications. *International Archives of Photogrammetry and Remote Sensing* 32(5): 628–632.

Peipe, J. & Stephani, M. 2003. Performance evaluation of a 5 Megapixel digital metric camera for use in Architectural Photogrammetry. *International Archives of the Photogrammetry, Remote Sensing and Spatial Information Sciences* 34(5/W12): 259–261.

POB. 2007. www.pobonline.com. *Product surveys of Point of Beginning*, accessed: December 2007.

Pollefeys, M., Vergauwen, M. & Van Gool, L. 2000. Automatic 3D Modeling From Image Sequences. *International Archives of Photogrammetry and Remote Sensing* 33(B5): 619–626.

Pöntinen, P. 2004. On the geometrical quality of panoramic images. *International Archives of Photogrammetry and Remote Sensing* 35(5): 82–87.

Przybilla H.-J. & Peipe J. 2007. 3D Modeling of Heritage Objects by Fringe Protection and Laser Scanning Systems XXI. International CIPA Symposium, 01–06 October 2007, Athens, Greece. *The ISPRS International Archives of Photogrammetry, Remote Sensing and Spatial Information Sciences* XXXVI-5/C53 (ISSN 1682-1750) and *The CIPA International Archives for Documentation of Cultural Heritage XXI-2007 (ISSN 0256-1840)*: 615–618.

Remondino, F. & Börlin, N. 2004. Photogrammetric Calibration of Image Sequences Acquired with a Rotating Camera. ISPRS Workshop on Panorama Photogrammetry, Dresden. *International Archives of Photogrammetry and Remote Sensing* XXXIV, Part 5/W16. http://www.tudresden.de/fghgipf/photo/PanoramicPhotogrammetryWorkshop2004/Proceedings.htm (last access March 2008).

Remondino, F. & Fraser,C. 2006. Digital Camera Calibration Methods: Considerations and Comparisons. Commission V Symposium, Dresden, 25–27. Sept. 2006 The *ISPRS International Archives of Photogrammetry, Remote Sensing and Spatial Information Sciences Proceedings* XXXVI, Part 5: 266–272. ISSN 1682–1750.

Reulke, R., Scheibe, K. & Wehr, A. 2005. Integration of digital panoramic camera and laser scanner data. *Proc. International Workshop on Recording, Modeling and Visualization of Cultural Heritage, 22–27 May 2005. Centro Stefano Franscini, Monte Verità, Ascona, Switzerland.* (available on the workshop CD)

Rocchini, C., Cignoni, P., Montani, C., Pingi, P. & Scopigno, R. 2001. A low cost 3D scanner based on structured light. *Computer Graphics Forum* 20(3): 299–308.

Salvi, J., Pages, J. & Batlle, J. 2004. Pattern codification strategies in structured light systems. *Pattern Recognition* 37(4): 827– 849.

Schechner, Y. 2000. Depth from Defocus vs. Stereo: How different really are they? *International Journal of Computer Vision* 89: 141–162.

Schneider, D., Maas, H.-G. 2004. Development and Application of an Extended Geometric Model for High Resolution Panoramic Cameras. The *ISPRS International Archives of Photogrammetry and Remote Sensing* XXXV PART B5. Hardcopy proceedings of the XXth ISPRS Congress in Istanbul, Turkey. ISSN:1682–1750, and softcopy proceedings, ISSN: 1682–1777.

Sechidis, L., Georgiadis, C. & Patias, P. 1999. Transformations for "Calibrated Image" creation using distortion models Measurements and Processing in Archaeology and Architecture. ISPRS WG V/2 & WG5 Workshop, Photogrammetric Measurement, Object Modeling and Documentation in Architecture and Industry (Thessaloniki, Greece, 7–9 July). *International Archives of Photogrammetry and Remote Sensing* 32(5W11): 215–222.

Shortis, M.R., Bellman, C.J., Robson, S., Johnston, G.J. & Johnson, G.W. 2006. Stability of zoom and fixed lenses used with digital SLR cameras Commission V Symposium, Dresden, 25–27 Sept. 2006. The *ISPRS International Archives of Photogrammetry, Remote Sensing and Spatial Information Sciences Proceedings* XXXVI, Part 5: 285–290. ISSN 1682-1750.

Svoboda, T. & Pajdla, 2002. Epipolar Geometry for Central Catadioptric Cameras. *International Journal of Computer Vision* 49(1): 237.

Tauch, R. & Wiedemann, A. 2004. Generation of digital surface models for architectural applications with Archimedes 3D. *International Archives of Photogrammetry and Remote Sensing* 35(B5): 350–353.

Tosovic S., Sablatnig R. & Kampel M., 2002. On combining shape from silhouette and shape from structured light. In H. Wildenauer & W. Kropatsch (eds.), *Proc. 7th Computer Vision Winter Workshop*: 108–118.

Tsioukas, V., Patias, P. & Jacobs, P. 2004. A Novel System For The 3d Reconstruction of Small Archaeological Objects. *International Archives of Photogrammetry and Remote Sensing* 35(B5): 815–818.

UNESCO. 2007. *Venice Charter, 1964.* http://www.international.icomos.org/charters/venice_e.htm

Waldhaeusl, P. & Ogleby, C. 1994. 3x3-Rules for Simple Photogrammetric Documentation of Architecture. *International Archives of Photogrammetry and Remote Sensing* XXX, Part 5, Melbourne, 1994: 426–429.

Wiedemann, A., Hemmleb, M. & Albertz, J. 2000. Reconstruction of historical buildings based on images from the Meydenbauer archives. *International Archives of Photogrammetry and Remote Sensing* 33(B5/2): 887–893.

Zhang, R., Tsai, P-S., Cryer, J.E. & Shah, M. 1999. Shape from Shading: A Survey. *IEEE Transactions on Pattern Analysis and Machine Intelligence, August 1999* 21(8): 690–706.

Advances in Photogrammetry, Remote Sensing and Spatial Information Sciences: 2008 ISPRS Congress Book – Li, Chen & Baltsavias (eds)
 ISBN 978-0-415-47805-2

CHAPTER 28

Natural disaster management: Activities in support of the UN system

Piero Boccardo & Fabio Giulio Tonolo

ABSTRACT: The ready availability and cost-effective management of dependable geo-spatial information is central to raising the operational efficiency of emergency management activities, especially of events hitting developing countries, more vulnerable to calamities and less prepared to face them. This chapter aims to show how an approach based on geomatics techniques may improve the effectiveness of early impact analysis of natural disasters. Since the United Nations system is deeply involved in international emergency response, the first part of the chapter will discuss the main UN activities related to geographic data management. Afterwards, some applications dealing with the use of geomatics techniques in support of early impact activities will be described, making use of, among other examples, the joint project between ITHACA (a non-profit organization formed by university and research centres) and the UN World Food Programme (WFP), formally supported by ISPRS.

Keywords: Disaster management, United Nations, SDI, early impact analysis, UAV

28.1 INTRODUCTION

28.1.1 *Natural disasters in the world*

As widely reported by the scientific and humanitarian environment and the media, the world is facing a dramatic increase in emergencies caused by natural disasters. In fact, the number of events and affected people, not to mention economic losses, has increased by five times in the last 30 years (Fig. 28.1; CRED 2007).

The global population growth pressure forces people to live in areas historically prone to disasters, while global warming seems to exacerbate climatic related hazards. The result is an increase in floods that strike countries with little capacity to cope with new and exceptional destructive events.

The humanitarian and United Nations systems are therefore called upon to intensify their efforts to improve their capacity to provide support to the countries in need and to be better prepared to intervene. Such a request came from, amongst others, the UN Secretary General on various occasions.

28.1.2 *UN facing natural disasters*

In this new globalized world, where changes are a constant, the United Nations must meet challenges and growing demands from its member states. Its Secretariat, agencies, funds and programmes, also in partnership with non-governmental organizations, must deliver services in places to those people most in need. The World Food Programme (WFP) of the United Nations is in the front line of humanitarian operations. It is the biggest UN Agency and responds to more than 120 emergencies per year worldwide [http://www.wfp.org]. According to the UN reform, WFP is also the leader of logistics for UN

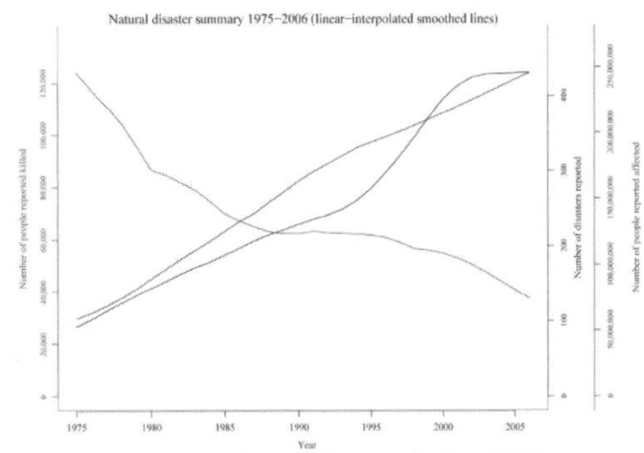

Figure 28.1. Natural disaster summary 1975–2006 (CRED 2007). (see colour plate page 510)

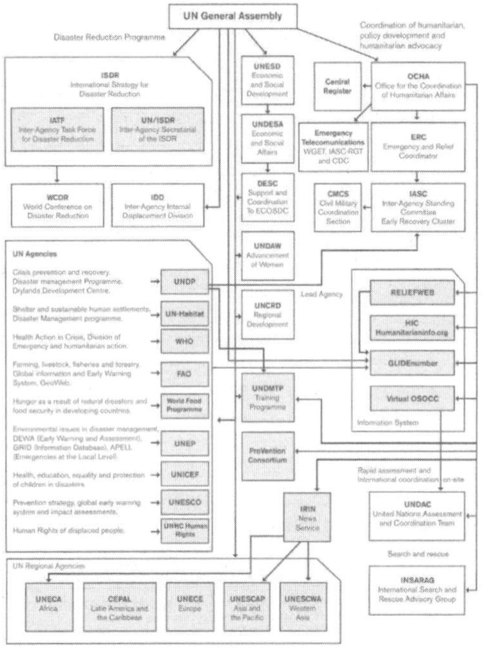

Source: *Max Lock Centre.*

Figure 28.2. Disaster management in the UN System.

and international bodies during emergency response operations (Fig. 28.2). WFP initiated a process to reinforce its capacity to be a leading force in the area of emergency response, improving its Information Management capacity in support of emergency preparedness and response. To do so, an agreement of collaboration with the recently formed Information Technology for Humanitarian Assistance Cooperation and Action (ITHACA) Centre has been signed and joint collaboration started in February 2007.

28.2 GEOGRAPHIC INFORMATION SUPPORTING DISASTER MANAGEMENT

28.2.1 *UN and geographic information management*

From peace-keeping to pandemics, from environment to economic development, the ready availability and cost-effective management of dependable geo-spatial information is central to raising the operational efficiency of the United Nations, in terms of short-term emergency response capacities, long-term risk reduction, development and environmental protection activities—the three pillars of sustainability. The production and use of geo-spatial information within the United Nations has been accomplished historically by its component organizations in accordance with their individual needs and expertise. This has resulted in multiple efforts, reduced opportunities for sharing and reuse of data, and an unnecessary cost burden for the United Nations as a whole.

The need to establish greater system coherence for the applications and exchange of geo-spatial data for UN activities was first highlighted in the UN Geographic Information Strategic Plan (UNGISP) prepared in 2001–2002. The possible future need to develop a Spatial Data Infrastructure (SDI) to assist in this regard was also raised.

Spatial Data Infrastructure, a concept that dates back to the early 1990s, was first introduced by countries to support geographic information exchange, standards adoption and asset sharing across national information networks. The principal objective was to strengthen decision making for sustainable economic development.

Early national initiatives confirmed that the discovery, ready access to, evaluation and dependable utilization of geo-spatial information was greatly facilitated by an underlying infrastructure of policies, technologies, data, common standards, practices, protocols and specifications that collectively make up a 'Spatial Data Infrastructure'. SDIs are all about re-use: re-use of data, re-use of technical capabilities, re-use of skills developed, and re-use of invested intellectual effort and capital. Re-use minimizes the initial system wide investment needed from cooperators to benefit fully from spatial data and information, 'sharing not wearing' the costs and helping to realize more rapid returns on investment. Implementing a SDI also means learning from the experience of others and avoiding pitfalls.

28.2.2 *UNGIWG*

Formed in 2000, UNGIWG (United Nations Geospatial Information Working Group) comprises a network of UN professionals and others working in the fields of cartography and geo-spatial information management science to address common geo-spatial issues: maps, boundaries, data exchange, standards, naming conventions and location. It also works directly with non-governmental organizations, research institutions and industry to develop and maintain common geographic databases and geo-spatial technologies to enhance guidance and operational capabilities. UNGIWG has reported periodically to the UN Chief Executive Board (CEB).

The establishment of UNGIWG was the culmination of several years of efforts to bring together professionals in the fields of cartography and geographic information science within the United Nations system to address common issues affecting the work of the Organization in these fields.

UNGIWG receives the support of many United Nations departments, specialized agencies, programmes and organs, as well as endorsements and pledges of support from various governmental and non-governmental organizations, research institutions and industry. This high level of interest from within and outside the system reflects the importance of this effort and the need for the United Nations to take a leadership role in this area.

The overarching objective of UNGIWG is to promote the use of geographic information within the United Nations system and member states for better decision making. UNGIWG aims at attaining the following:

- improving the efficient use of geographic information for better decision making
- promoting standards and norms for maps and other geo-spatial information
- developing core maps to avoid duplication
- building mechanisms for sharing, maintaining and assuring the quality of geographic information
- providing a forum for discussing common issues and emerging technological changes.

Since its inception, UNGIWG has been laying the foundations for a UNSDI (United Nations Spatial Data Infrastructure). The accomplishments of its Task Groups in particular indicated that refinement and refocusing of current activities and UNGIWG organization could deliver the components for such a system (Henricksen 2007).

During the 7th UNGIWG Plenary meeting held in Santiago in November 2006, it was recognized that the development of a common vision and understanding on UNSDI was a priority. In order to build on common interests, the UNGIWG Secretariat consulted the UN agencies on the activities carried out within the agency that could benefit from the UNSDI. UNGIWG meets annually in Plenary Session and reports periodically to the UN Chief Executive Board (CEB) on progress made and priority issues related to the following Task Group activities:

- International and Administrative Boundaries: Among the different core data layers at the global level, special priority has been given to produce a worldwide consistent coverage of:

  1. International boundaries at scales of 1:1; 1:5; 1:10 and 1:25 million. The dataset is available on request to UN Staff from DPKO/UNCS.
  2. Second administrative level boundaries (SALB), including historic changes: SALB is coordinated by WHO. The objective of the project is to improve the availability of information about administrative boundaries down to the second sub-national level. The SALB data set consists of national mapping agencies (NMA) contact information, coded historic changes and digital maps that can be downloaded at no cost on a country by country basis. The SALB project is a good illustration of the potential that exists when the UN and non-UN communities do collaborate and share their data and technical skills.

- Core Geo-Database: This Task Group inventories and selects global datasets for adoption as UN Spatial Data Infrastructure core datasets. It has prepared a comprehensive report on the following:

  1. Boundaries: national, coastal, administrative.
  2. Human Health: boundaries, facilities.
  3. Human Population: centres, distribution.
  4. Infrastructure: roads, railways, airports, harbours, navigation.
  5. Bathymetry, Topography and Digital Elevation Model.
  6. Hydrology: drainage, water bodies, watersheds.

- Remote Sensing: Satellite imagery and other remotely sensed data are now an integral part of most members' geo-spatial analysis. This Task Group facilitates finding and accessing very high resolution imagery. It promotes sharing and purchasing data, through the UN Systems Contract and other agreements, to improve their use. The UN has been using global Landsat and other satellite imagery since the mid-1970s in order to analyse the changes that have occurred during the past three decades. The goal is to integrate remote sensing within GIS to have a better tool for decision making.
- Interoperable Services: International standards are needed to share and use geo-spatial information within the UN Spatial Data Infrastructure. It allows members to develop and strengthen their respective geo-spatial capacities in a compliant and interoperable manner. As a result, a substantial amount of geo-spatial data can be shared between agencies, thus avoiding duplication.
- GIS Map Production Guidelines: This Task Group is developing guidelines for making maps using GIS software with the intention of standardizing maps scales, smallest mapping units, file interchange formats, map projections, pixel size of raster maps, etc. It has prepared technical guidelines for high resolution and low resolution scale maps.
- Global Navigation Satellite Systems: This new Task Group works to harmonize and facilitate field data collection activities undertaken by the different UN agencies. It also reviews training materials developed by various agencies to produce a harmonized set of training tools, and surveys the use of global positioning devices within each agency at country level to prioritize training activities.

In terms of applications, it has been developing an open source software tool, the FAO-sponsored GeoNetwork (FAO 2004), that facilitates single point of entry access to geo-referenced databases, cartographic products and related metadata located at globally distributed information sources. GeoNetwork users includes so far the cooperative efforts of FAO, WFP, WHO, UNEP and UN-OCHA to develop a more universally applicable ISO and OGC-standards-based information management tool for United Nations users, member states and UN partners.

General outcomes are:

- enhances geo-spatial data access and sharing within and between distributed organizations
- avoids unnecessary duplication of data assets and results in overall information management efficiencies and cost savings for users
- increases cooperation and coordination of data collection and description while preserving data and information ownership
- provides a wide community of spatial information users with easier and timelier access to available digital data including thematic maps
- spatial visualization using derived cartographic outputs of this kind assists communications between technicians and decision makers, and provides a basis for more informed decisions in many situations.

28.2.3 *UN-SPIDER*

Recently, according to resolution 61/110 of 14 December 2006, the United Nations General Assembly agreed to establish the United Nations Platform for Space-based Information for Disaster Management and Emergency Response—UN-SPIDER as a new United Nations programme, with the following mission statement: "Ensure that all countries have access to and develop the capacity to use all types of space-based information to support the full disaster management cycle."

Whereas there have been a number of initiatives in recent years that have contributed to making space technologies available for humanitarian and emergency response, such as the International Charter Space and Major Disasters, UN-SPIDER is the first to focus on the need to ensure access to and use of such solutions during all phases of the disaster, including the risk reduction phase, which will significantly contribute to an increasing reduction in loss of lives and property (UN 2008). The new programme will achieve this by focusing on being a gateway to space information for disaster management support, serving as a bridge to connect the disaster management and space communities and being a facilitator of capacity building and institutional strengthening, in particular for developing countries.

UN-SPIDER is being implemented by the United Nations Office for Outer Space Affairs (UNOOSA) as an open network of providers of space-based solutions to support disaster management activities. Besides Vienna (where UNOOSA is located), the programme will also have offices in Beijing (China) and Bonn (Germany).

28.3 USE OF GEOMATICS FOR NATURAL DISASTER MANAGEMENT: EXAMPLES OF APPLICATIONS

28.3.1 *Worldwide experience*

Several UN agencies (WFP, UNOSAT), Space agencies (ASI, CSA, CNES, DLR/ZKI, ESA, ISRO), research centres (SERTIT, ITHACA), National Civil Protection Services and other organization are currently working on crisis management support through an approach based on geomatics techniques (Henry et al. 2002, Voigt et al. 2007, Bendea et al. 2007a, Unosat 2008).

In the following paragraphs, case study and state-of-the-art applications will be outlined.

28.3.2 *WFP-SDI development (Ajmar 2007)*

Accurate, easily accessible geographic information is crucial to good decision making in humanitarian operations. The main aim of the WFP-SDI project is the development and implementation of a global database and of related rules and procedures in order to store, manage and exchange geo-spatial data for WFP and other actors in the humanitarian sector. The WFP-SDI, currently in the final stage of development and under operational test, aims to be an efficient tool for storing, querying and manipulating large amounts of global geographic information and spatial data for analysis and visualization purposes. The project is coordinated by ITHACA, a non-profit organization with a view, as aforesaid, to collaborate with the UN WFP in support of disaster management. The joint project, mainly aiming to improve the capacity of WFP and the broader international community in early impact analysis, is officially supported by ISPRS.

The database integrates and updates data coming from multiple sources (Fig. 28.3), providing each user with the latest datasets, thus avoiding duplications and mistakes. The data models will be shaped according to specific needs and demands of the user groups within the organization.

The data will be accessed through user-friendly web-based platforms (with the possibility of remote access), and will be integrated in the automated dynamic web-based maps for operational support

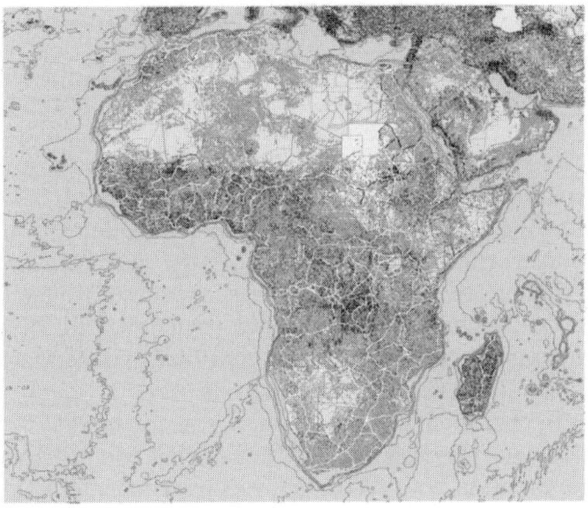

Figure 28.3. Test dataset of the African continent. (see colour plate page 510)

(Logistics Capacity Assessments, Contingency Plans, etc.). These data will be also available to GIS practitioners in-house (Vulnerability Analysis and Mapping, Emergency Preparedness and UN Joint Logistic Center) and, possibly, to partners. Data will also provide a significant source of information for the Early-Warning and Early-Impact Analysis services.

The SDI has three main components:

- Data infrastructure (how data and attributes are defined and stored).
- Data management (the technology to allow multiple users to query, retrieve, manipulate, edit and save data in a common repository).
- Services (what the users can get from the data repository).

At the moment the proposed solution goes in the direction of suggesting an architecture based on a solid back end and a flexible, interoperable and customized front end. A structured and solid back end to grant data reliability and integrity is an assumption deriving from the interpretation of the user needs analysis process, which highlighted that the data managing environment must be a distributed one. The constraint related to low performance internet connection requires also the development of a solution for disconnected data management using database replica and guided solutions for data reconciliation. Re-use and re-organization of currently managed datasets have been priorities in the data modelling phase, together with direct access to open geographic sources (SRTM, archive satellite images, etc.) without any need for data pre-processing. The development of representation rules and map templates allow the creation of a "lowest common denominator" for geographic analysis and mapping, in support of decision making during emergencies.

The defined data model is implemented as an ESRI Workgroup Geo-database based on Oracle 10 g as DBMS. That choice was guided by functionality analysis and by the fact that all three GIS units operating inside WFP (ODAP, VAM and UNJLC) already use the ESRI platform for GIS data management.

28.3.2.1 *Data model definition and consolidation*

Based on the data structure submitted and approved by WFP units (ODAP, VAM and UNJLC), a data model has been consolidated and implemented using an ORACLE 10 g database as the DBMS platform. The database schema is composed of two different entities:

- Base data includes basic geographic and alphanumeric information, in order to produce basic geographic outputs and analysis;
- Transportation is the translation of conceptual schema developed by UNJLC into a logical relational schema (the implemented version is 1.2).

Data representation rules have been associated with data layers inside the geo-database, following OCHA guidelines (http://ochaonline2.un.org/Default.aspx?alias=ochaonline2.un.org/mapguidelines).

28.3.2.2 *Users definition*

Three different roles have been defined, according to the users schema shown in Fig. 28.4.

The Database administrator is in charge of:

1. Adding and removing users to/from the database server;
2. Managing geo-database and user security;
3. Creating and deleting geo-databases;
4. Attaching and detaching geo-databases;
5. Backup and restoring geo-databases;
6. Upgrading geo-databases;
7. Compressing geo-databases; and
8. Starting, stopping, and pausing the database server.

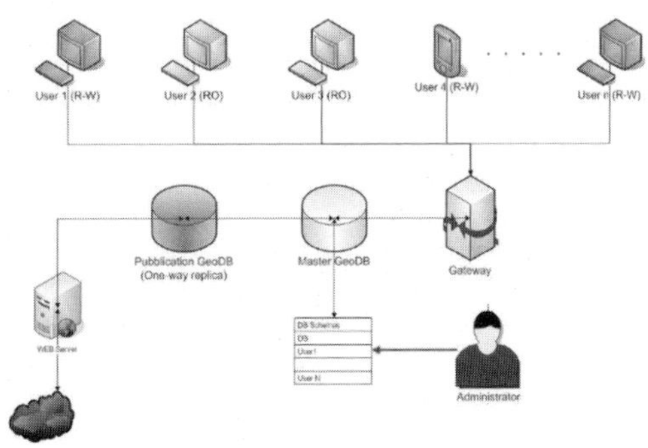

Figure 28.4. Users schema.

The database user with read/write privileges can select and edit all existing data and create new geo-database objects such as feature classes, while the database user with read only privileges can select from every table in the geo-database.

Currently, only one database user with read/write privileges has been created for every structure involved in the data model testing phase. It must be verified that any single structure works on a public version, granting to the master database administrator the right to manage data reconciliation.

28.3.2.3 *Data distribution*

According to the different features of the data users, a proposal for a data distribution schema is presented. Four different levels of users and relative privileges have been identified. That consideration is used as starting point for privileges definition over the whole geo-database, adding only a component of read-only users that access the data through a read-only replica and WebGIS services and applications; the DBMS of that replica may be discussed and can be an Open Source platform (Postgress/PostGIS).

The system architecture is made of two Oracle database instances, one acting as master database and another hosting a full two-way replica for data production; that solution, together with adequate back-up policies, should grant data protection and integrity.

28.3.2.4 *WFP-SDI future development*

Some major tasks must be planned for completing the development of an SDI at WFP level:

- Updating of database schema in order to include more information components:
  1. Image catalogue for acquisitions performed during emergencies, and related data analysis results (in final stage of definition);
  2. Anything may come as a result of the testing phase of the current data model.
- Users privileges refinement:
  1. Fine user privileges definition, acting, if needed, at dataset or feature level (Table 28.1);
  2. Creation and management of data view to filter unnecessary data for some users (i.e. some of the transportation dataset attributes are not necessary for base data analysis and visualization).
- Fine database tuning
- Web services and applications allowing an interoperable level of access and management of the database;
- Distributed computing methods, in order to improve system performance using an available network of computers (ITHACA, UNOSAT, WFP, etc.);
- Full interaction with Geo-network platform.

28.3.3 *Early impact analysis for flood events*

Flood events have a huge impact on human beings, settlements and ways of life (Henry 2002). The support generally requested in the early impact activities after flood events is conceived in provision of analytical products (maps and/or geo-referenced data) showing the impact of the flood on population and on infrastructures, with emphasis on the road network. Furthermore, additional analysis is requested to facilitate the setting up of priorities in undertaking field emergency needs assessment and finally in supporting midterm food security analysis.

In operative terms, map support is required at two different stages: at the first state of alert and just after the event. In both cases, the response time after help-request triggering is short, lasting generally from 24 to 48 hours.

In the first state of alert, before the flood event, small scale maps indicating the areas that will be potentially flooded (the so-called *floodable areas*) are required; in this phase the floodable areas are defined and archive satellite data (radar, multispectral, high resolution) about the zone of interest (acquired before the flood) are collected.

In the second phase, after the flood hits, the following four points must be answered:

- which areas have been physically hit by the flood (the so-called flooded area*s*);
- which areas could be hit directly or indirectly by the flood effect (the so-called affected areas);
- which populations are suffering the effects of the flood (the so-called affected population);
- what damage has occurred to the infrastructure.

In order to derive this information, it is necessary to receive updated satellite data acquired after the event and to apply classifications, photo interpretation techniques and GIS procedures, as shown in the next paragraphs.

According to these requirements, a devoted methodology should encompass the steps high-lighted in the procedure showed in Fig. 28.5.

28.3.3.1 *Definition of floodable areas*

During the early phases of humanitarian crisis management after a flooding event, it is crucial to produce a global assessment at a small scale, identifying areas and infrastructures that may have been damaged. Fast production of this information will allow the organization to perform more precise and time consuming assessment (high resolution satellite or aerial image acquisition and interpret-ation, land surveys, etc.) only in specific areas and after having defined priorities.

A very simple approach, using globally available geographic data and geo-processing tools, is to use

Table 28.1. Possible users definition.

		Visualization	Interaction	Download	Edit	Commit
Public		limited	limited	no	no	no
Humanitarian Partners	Logs	yes	yes	maps + reports	Attributes only	no
	GIS	yes	yes	yes	yes	no
UNJLC/WFP	Logs	yes	yes	maps + reports	Attributes only	no
	GIS	yes	yes	yes	yes	yes

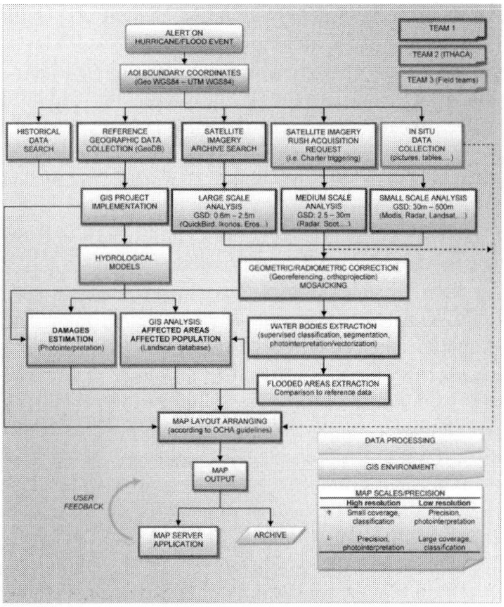

Figure 28.5. A proposal of early impact procedures for flood events.

a CostDistance function over a DEM, employing the water network as a reference. The CostDistance request is basically a flooding request, but it accumulates a penalty for flooding across cells whose elevation exceeds a fixed limit. Thus, by picking the cells of zero cost, it is possible to find all cells reachable by crossing no elevations above the above mentioned limit. Iterating the procedure until a defined maximum elevation value, based on available information, it is possible to simulate the evolution of a flood event. To simulate a flood more realistically, there is the possibility of adding a small positive value to the cost themes; in that case the extent of the flood is limited to n cells by selecting CostDistance values less than n times the small value. A potential problem with this approach is that no flood maintains a constant elevation: it too flows and its elevation descends as the river descends. For that reason, different kinds of water flow models have been developed to significantly increase the reliability of those estimates. The goal is to predict flood effects using hydrologic and hydraulic modelling. Several research topics are involved such as rainfall-runoff modelling, river hydraulic modelling and mapping of hydraulic hazard. The main expected results are:

- delineation of floodable areas for different rain rates;
- delineation of flood extent related to precipitation forecast.

Defining potentially floodable areas can be useful in the Early Impact phase for the identification of the priority areas where to concentrate the analysis. Moreover, the outcomes of the models allow the identification of the available access to the area and the potentially affected population. The development of this kind of flood forecast system can provide enough time for the appropriate actions, being a valid tool for emergency preparedness. The method follows the typical steps of floodplain analysis: data collection and preparation, models development and execution, floodplain mapping. Topography, landuse and soil type information have to be integrated with other kinds of data obtained through GIS analysis and rainfall values. These data are the input for the development of the hydrologic and the hydraulic modelling that allow the extension of the expected flood to be obtained.

Figure 28.6 shows the floodable areas obtained using an implementation of the described approach and compared to the real extent of the flood (Mozambique, February 2007) as detected by satellite active sensors.

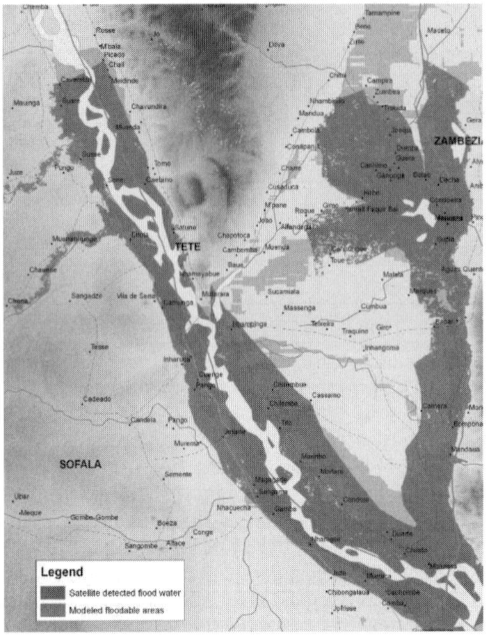

Figure 28.6. Modelled floodable areas overlapped on the real extent of the flood in Mozambique in 2007. (see colour plate page 511)

28.3.3.2 *Definition of flooded areas*

According to the aforesaid, the results of the floodable areas analysis phase can be useful to identify areas and infrastructures that may have been damaged, allowing a priority list to be set of the specific areas that require a more precise assessment. The same areas should be the target for rush satellite imagery acquisition requests, generally performed by the International Charter "Space and Major Disaster". The International Charter is a unified system of space data acquisition and delivery to those populations affected by natural or man-made disasters through authorized users (V.A. 2000).

If the Charter is triggered, normally, in a few days, different kinds of satellite images are available, such as radar images and high resolution ones, suitable for the definition of the flooded areas.

Definition of flooded areas through radar images

As widely reported, radar images enable the easy identification of water bodies; therefore they are the main input data for flood analyses (Aduah et al. 2007, Henry et al. 2003, Schumann et al. 2007). On the other hand, they are affected by geometric distortions (layover, foreshortening and radar shadows), which are hard to model, especially in mountain regions. In detail, the radar images acquired by satellite platforms (Envisat, Radarsat, ERS,...) before and after the event are used for the definition of flooded areas: the request for an archive radar image can be skipped if reliable and updated water bodies data are available, the so-called NWED (Normal Water Extent Database) (Wang et al. 2002). Both images have to be geo-referenced through attitude and position information (generally supplied as metadata) exploiting the sensor, depending on procedures provided by commercial software. In order to improve the mapping accuracy, it is possible to perform a further image-to-image geo-referencing, through a simple polynomial transformation of the radar image preceding the flood with respect to the following one.

From a radiometric point of view, it is generally advisable to apply a preliminary despeckle filter on the images, to reduce the noise that affects radar images. Then, the areas presenting as water can be spotted on both images, exploiting their reflective behaviour towards the electromagnetic radiation emitted by the radar sensor, being roughly that of a specular surface. It turns out that water can be easily identified (Fig. 28.7b), being characterized by low radiometric values (Fig. 28.7a).

By using change detection techniques, it is therefore possible to isolate the flooded areas, distinguishing them from water bodies, as shown in Figure 28.7c. Classified images often suffer from a lack of spatial coherency (speckles or holes in classified areas). In order to improve the topology of the classified flooded areas, it is possible to further process the results by applying filtering algorithms aimed at removing isolated water pixels by using blob grouping and clumping adjacent similar water areas together through the use of morphological operators. It is then advisable to mask the final result using the floodable areas layers that will allow the removal of classification errors, mainly due to radar shadows and speckle noise.

The German TerraSAR-X System and the Italian Cosmo-Skymed, which were launched in 2007, have extended the civilian SAR availability to the 1 m domain, allowing very high resolution all weather image acquisitions (Guzzi 2008, Voigt et al. 2007).

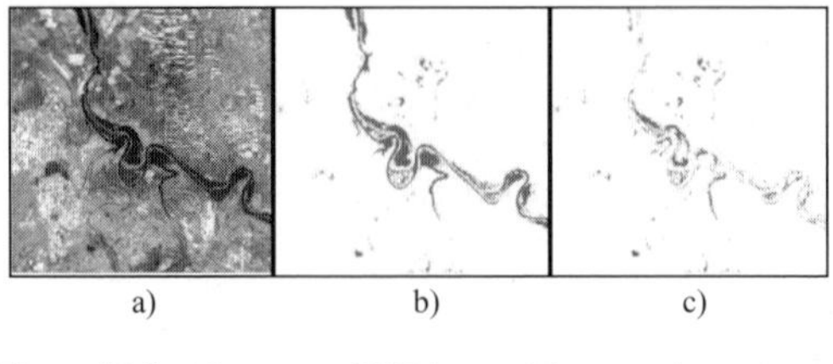

a) b) c)

Figure 28.7. Post-event SAR image (a), extraction or water bodies (b), identification of flooded areas (c).

Definition of flooded areas through multi-spectral data classification
A different approach for the definition of flooded areas is based on multispectral data processing. For several reasons, the MODIS sensor is generally used for large-scale flood monitoring (Brakenridge et al. 2003, Voigt et al. 2007, Aduah 2007). The MODIS (Moderate Resolution Imaging Spectroradiometer) mission grants a worldwide coverage, providing daily images and derived products that are completely free of charge. Furthermore, low geometric resolution (250-500-1000 m) MODIS data allows a regional view of the observed phenomena. Therefore, the use of MODIS data permits a multi-temporal small scale analysis of the flood event evolution in the areas of interest. The detection of water bodies and flooded areas is the result of a classifying procedure of MODIS primary reflectance data (MOD02—MODIS Level 1B Calibrated, Geo-located Radiances).

The MOD02 product is distributed by NASA to users with a low processing level (swath format), which performs only a first geometric correction step and which contains some residual geometric distortions due to the system's peculiar scanning geometry. Therefore, a geometric data correction step is accomplished using a specific software tool, the MRTSwath tool. In order to correctly compare muti-temporal MODIS data, the images are also preventively radiometrically pre-processed and co-registered. The atmospheric correction of the reflectance data is performed through the simplified Dark Subtraction approach. Unlike radar data, the usability of passive optical remote sensing data is restricted by cloud cover. Since MODIS data for the areas of interest are available daily, temporal composite techniques are used in order to reduce cloud cover presence.

For classification purposes (Fig. 28.8d, e, f), specific radiometric indexes, useful for water bodies and flooded area detection and for cloud effects reduction, are defined by using Red (620–670 nm) and Near-Infrared (841–876 nm) MODIS bands (which are the only bands with spatial resolution of 250 m). These indexes are mostly based on particular band ratios. Common indexes, such as the NDVI (Normalized Difference Vegetation Index) and other ones defined by authors according to the specific areas of interest, are considered.

Using reliable water bodies data, it is therefore possible to isolate only flooded areas. Moreover, it is possible to create synthetic products showing temporal evolution of flood events and the different flooded area distributions.

28.3.3.3 *Definition of affected population*
The number and distribution of potentially affected populations is another major type of information

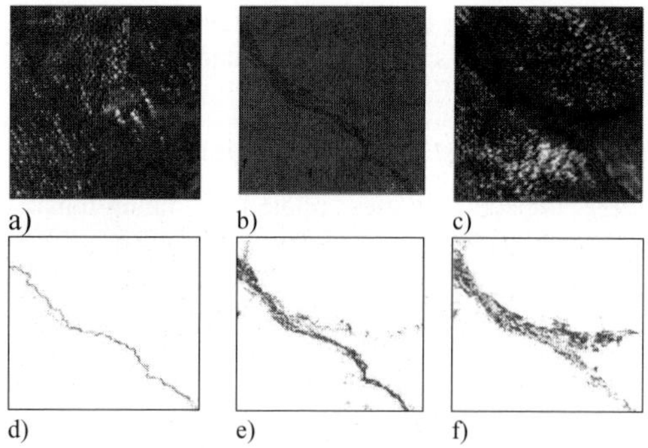

Figure 28.8. Mozambique 2007 flood event evolution. Original MODIS data (false colour): January 16-day composite (a), early February 16-day composite (b), late February 16-day composite (c) and corresponding identified water bodies and flooded areas - (d), (e) and (f). (see colour plate page 511)

requested by managers of humanitarian help distribution. A rapid estimate of this parameter can be obtained crossing potentially flooded and effectively flood affected areas with population distribution data; one globally consistent source of that kind of data is the LandScan Global Population dataset (Dobson et al. 2000).

Use of the GIS Zonal Statistics function is a way of calculating the statistics on values of a raster (as Landscan data is) within the zones of another dataset: the affected areas in that case. That procedure allows an estimate of people living in potentially or currently flooded areas, or in areas isolated by floods. On-field assessment, even on statistical samples, should permit estimates to be corrected.

28.3.3.4 *Definition of the damage to the infrastructure*
The use of very high resolution satellite data (geometric resolution varying from 0.6 m to 2.5 m) permits the creation of specific products showing the flood impact on the infrastructure, with particular emphasis on the road network. This phase aims to provide information on accessibility in the context of the emergency response operations undertaken by WFP. As a matter of fact, in the food supplying process after a flood event, WFP often has to face troubles related to transportation and food aid distribution; it is therefore necessary to know quickly the condition of roads, tracks and trails where food trucks should run.

In order to perform an assessment of damage caused by flood events, common photo interpretation and change detection techniques are used. Moreover, the use of image fusion procedures, such as pan-sharpening techniques, can improve object identification. In this phase, photos acquired *in situ* can also support satellite image interpretation.

28.3.4 *Low-cost UAV for photogrammetric surveys*

Early impact activities related to environmental emergencies are mainly devoted to delivering geo-referenced information and to produce thematic maps that can be useful to the people operating in the field. Satellite imagery is the main input data required to extract information such as flooded areas or damaged infrastructures. If satellite data are not available or not suitable to supply radiometric and geometric information, *in situ* missions must be considered. For this purpose, a suitable approach is to use "low cost" mini-UAV (International UAV Community 2006). For further details on UAV, please refer to Chapter II-6.

28.3.4.1 *Application example*

In the following tests, a photogrammetric survey using a mini-UAV prototype will be shown. The prototype is based on the MH2000 UAV model developed and patented by a research group within the Department of Aerospace Engineering (DIASP) of the Politecnico di Torino, led by Prof. F. Quagliotti and G. Guglieri,

The platform MH2000 (Fig. 28.9) is characterized by a conventional configuration: fixed wing, tail-less integrated wing-fuselage, tractor propeller driven, all-electric power. The aircraft is equipped with the Micropilot MP2128[g] autopilot that allows autonomous flights and provides a real-time attitude of flight.

The current configuration of the UAV allows carrying onboard digital sensors for video and imagery acquisition. The sensors are mounted in a pod (Fig. 28.10) placed on the belly of the fuselage and which can be remotely controlled through a direct link to the autopilot board. A real-time video downlink of the over-passed areas is made possible by the camcorder aimed at the digital camera display.

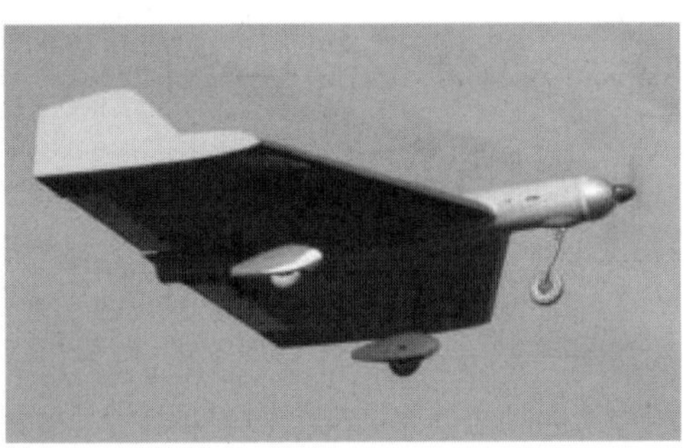

Figure 28.9. The UAV prototype ITHACA 01 during the first test flight. (see colour plate page 511)

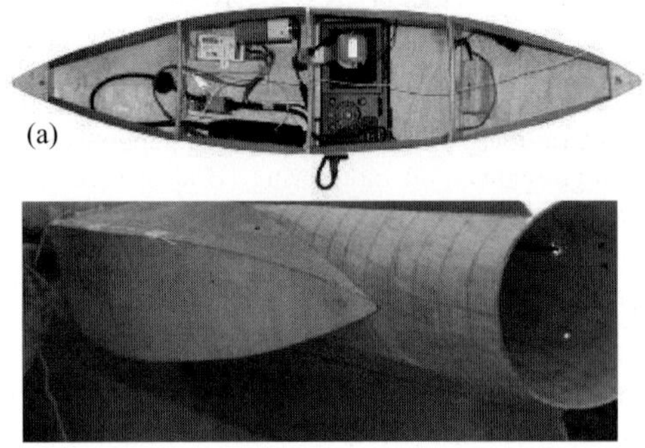

(a)

(b)

Figure 28.10. (a) Pod with Ricoh GR digital, and (b) pod position under the UAV. (see colour plate page 512)

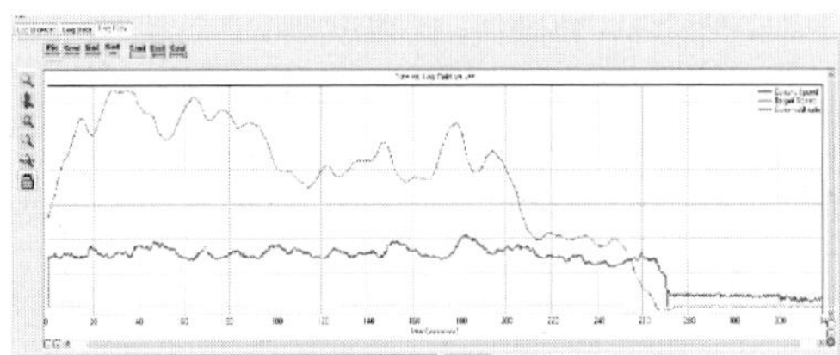

Figure 28.11. UAV speed and elevation plots.

At the moment, a photogrammetric survey over an area affected by a natural disaster has not been carried out. Thus, a case study over an Italian archaeological test site (ruins of a Roman theatre) will be shown (Bendea et al. 2007). The photogrammetric flight plan has been designed setting the UAV speed at 15 m/s, which means a shooting time interval of about 1,5 s (60 m) and 2,5 s (100 m). Topographic surveys have been performed in the two areas in order to produce a network of reference points (RP) with respect to WGS84-IGB00 datum. The RP are represented by 12 natural and 16 man-made target points. The detailed survey focused on inaccessible points such as roof or wall corners in order to obtain a RP network with a suitable elevation range. The average accuracy of points is 0.04 m, suitable for validation purposes according to the expected precision of the survey.

The UAV was remotely driven by a pilot on the ground (for safety reasons) while the presence of the autopilot allowed position and attitude data to be logged for post-processing purposes and for real-time monitoring. Flight parameters such as position, pitch, roll and yaw (Fig. 28.11) can be further processed for a possible direct geo-referencing approach.

The link between the digital camera and the autopilot allowed acquisition of frames automatically.

Shooting time interval was set equal to 2 seconds through the GCS HORIZON^{mp} software. Image acquisition started automatically at a flying altitude of 25 m.

Image processing was carried out using the Leica Photogrammetric Suite software according to the following scheme:

- Ground Control Points measurement (manual);
- Check Points measurement (manual);
- Tie points measurement (automatic);
- Bundle block adjustment.

The test was performed using three frames acquired over the theatre (Fig. 28.12).

The measurement of tie points and control points were carried out manually. The results of the triangulation are in the Tables 28.2 and 28.3. The achieved accuracy is the expected one and suitable for the goal of the project.

The obtained 3-D accuracy is suitable for mapping purposes at very large map scales. At the moment, the photogrammetric test is still in progress: the aim is to perform a larger triangulation using several frames for DSM extraction and orthoimage production tests.

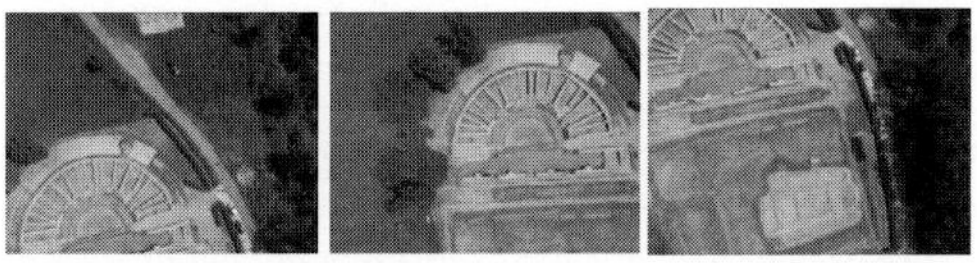

Figure 28.12. Images processed for the first orientation test. (see colour plate page 512)

Table 28.2. GCP residuals.

Point ID	D_X	D_Y	D_Z
200	–0.02	0.00	0.01
10st	0.08	0.04	0.00
11st	0.03	–0.01	–0.07
13rt	–0.04	0.02	0.09
15rt	–0.04	0.06	–0.02
16rt	–0.01	–0.03	–0.01
17rt	0.00	–0.07	0.00
Mean	0.00	0.00	0.00
Mean Abs	0.03	0.03	0.03
Sqm	0.04	0.04	0.05

Table 28.3. CP residuals.

Point ID	D_X	D_Y	D_Z
14rt	–0.04	0.03	–0.13
12rt	–0.09	–0.01	0.03
19rt	–0.03	0.00	–0.08
Mean	–0.05	0.01	–0.06
Mean Abs	0.05	0.02	0.08
Sqm	0.03	0.02	0.08

28.4 CONCLUSION

This first part of the chapter shows that geographic data are a key issue for disaster management, according to the number of mentioned initiatives dealing with geographic data within the UN system and the need for a consolidated UNSDI coming from the operative agencies. It is widely recognized that much effort is going into an approach based on geomatics techniques that may improve the effectiveness of early impact analysis used by the humanitarian officers to guide or support field operations.

Some example of ongoing early impact projects involving the UNWFP—the world's largest humanitarian agency—are presented. The advantages derived from the use of geomatics techniques are clearly visible.

ACKNOWLEDGEMENTS

The authors thank all the people working in the described projects, specifically the ITHACA team, the researchers of Politecnico di Torino (DITAG and DIASP) and the UN WFP staff.

REFERENCES

Aduah, M., Maathuis, B. & Ali Hussin, Y. 2007. Synergistic use of optical and radar remote sensing for mapping and monitoring flooding system in Kafue flats wetland of southern Zambia. *ISPRS Commission VII meeting, Instanbul.*

Ajmar, A. 2007. WFP SDI in support to a UN SDI. Preliminary proposal. http://www.ungiwg.org/docs/unsdi/ WFP-SDI_ Proposal.pdf

Bendea, H., Boccardo, P., Dequal, S., Di Sabato, F., Giulio Tonolo, F., Marenchino, D. & Perez, F. 2007. ITHACA project: a new approach for disaster management. *Joint CIG/ISPRS Conference on Geomatics for Disaster and Risk Management, Toronto.* (On CD-ROM).

Bendea, H., Chiabrando, F., Giulio Tonolo, F. & Marenchino, D. 2007. Mapping of archaeological areas using a low-cost UAV. The augusta bagiennorum test site. *XXI International CIPA Symposium, 01-06 October 2007, Athens.*

Bendea, H., Boccardo, P., Dequal, S., Giulio Tonolo, F. & Marenchino, D. 2007. New technologies for mobile mapping. *The 5th international symposium in Mobile Mapping Technology, Padua.* (On CD-ROM).

Brakenridge, G. & Anderson, E. 2003. Satellite gaging reaches: A strategy for MODIS-based river monitoring. *9th Int. Symp. Remote Sens., Int. Soc. Opt. Eng. (SPIE), Crete*: 479–485.

Dobson, J., Bright, E., Coleman, P., Durfee, R. & Worley, B. 2000. LandScan: A global population database for estimating populations at risk. *Photogramm. Eng. Remote Sens* 66, 7: 849–857.

Eisenbeiss, H. 2004. A mini unmanned aerial vehicle (UAV): system overview and image acquisition. *International Workshop on "Processing and visualization using high-resolution imagery", Pitsanulok.*

FAO. 2004. *FAO Spatial Information Infrastructure. GeoNetwork. Status Report.* SDRN. ftp://ftp.fao.org/unfao/bodies/coaim/pdf/geonetwork.pdf.

Gianinetto, M., Villa, P. & Lechi, G. 2006. Post-flood damage evaluation using Landsat TM and ETM+ data integrated with DEM. *IEEE Transactions on Geoscience and Remote Sensing* 44, 1: 236–243.

Gianinetto, M. & Villa, P. 2006. Monsoon Flooding Response: A Multi-Scale Approach To Water-Extent Change Detection. *International Archive of the Photogrammetry, Remote Sensing and Spatial Information Sciences.* (On CD-ROM).

Guzzi, R. 2007. Cosmo SkyMed constellation: state of the art and new perspectives of remote sensing. *11 Conferenza Nazionale ASITA, Torino* 1, XIX–XXVIII.

Henricksen, B. 2007. *A UNSDI Vision, Implementation Strategy and Reference Architecture.* UNGIWG. www.ungiwg. org/docs/unsdi/UNSDI_Compendium_13_02_2007.pdf.

Henry, J., Fellah, K., Clandillon, S., Allenbach, B. & De Fraipont, P. 2002. *Earth observation and case-based systems for flood risk management*: 1496–1498. 0-7803-7536-X. IEEE.

Henry, J., Chastanet, P., Fellah, K. & Desnos, Y. 2003. *ENVISAT Multi-Polarised ASAR Data for flood Mapping*: 1136–1138. 0-7803-7929-2/03. IEEE.

International UAV Community. 2006. *Referenced UAV systems.* UAV SYSTEMS: Global Perspective 2006. www.uvs-info.com.

Mansourian, A., Rajabifard, A. & Javad, M. 2005. SDI conceptual modeling for disaster management. *ISPRS Workshop on Service and Application of Spatial Data Infrastructure, Hangzhou* XXXVI (4/W6): 125–130.

Schumann, G., Hostache, R., Puech, C. & Hoffmann, L. 2007. High-resolution 3-D flood information from radar imagery for flood hazard management. *IEEE transactions on geoscience and remote sensing* 45, 6: 1715–1725.

UN. 2006a. *Delivering as One. Report of the High-level Panel on UN System-wide Coherence in the areas of Development, Humanitarian Assistance and the Environment.* http://www.un.org/events/panel/resources/pdfs/HLP-SWC-FinalReport.pdf.

UN. 2006b. *Investing in the UN: For A Stronger Organization Worldwide.* New York: UN Department for Public Information. http://www.un.org/reform/investinginun/pdfs/reformbook.pdf.

UN. 2008. *Report on activities carried put in 2007 in the framework of the UN Platform for Space-Based Infromation for Disaster Management and Emergency Response.* http://www.unoosa.org/pdf/reports/ac105/AC105_899E.pdf.

UNOSAT. 2008. Special review 2007. *E-news.* http://unosat.web.cern.ch/unosat/emagazine/eUNOSAT_07 12.pdf.

V.A. 2000. *Charter On Cooperation To Achieve The Coordinated Use Of Space Facilities In The Event Of Natural Or Technological Disasters.* http://www.disasterscharter.org/ charter_e.html.

Voigt, S., Kemper, T., Riedlinger, T., Kiefl, R., Scholte, K. & Mehl, H. 2007. Satellite image analysis for disaster and crisis-management support. *IEEE transactions on geoscience and remote sensing* 45, 6: 1520–1528.

Wang, S., Liu, Y., Yi, Z. & Wei, C. 2002. Study on the method of extablishment of normal water extent database. *Flood monitoring using remote sensing*: 2048-2050. 0-7803-7536-X. IEEE.

Witte, J., Reinicke, W., Heid, E., Furtwängler, C. & Okrob, M. 2005. *Business UNusual: Facilitating United Nations Reform Through Partnerships.* Berlin: Global Public Policy Institute (GPPi).

Advances in Photogrammetry, Remote Sensing and Spatial Information Sciences: 2008 ISPRS Congress Book – Li, Chen & Baltsavias (eds)

CHAPTER 29

Environmental sensing and human health

Stanley A. Morain & Amelia M. Budge

ABSTRACT: Remote sensing (RS), geographic information systems (GIS) and spatial information sciences (SIS) have exploded into health science communities in the past four years. Much of the interest in these geo-spectral and geo-spatial technologies has focused on linking satellite observations of environmental attributes that impact health outcomes, integrating Earth observations with physical Earth system models and with human health statistics, and providing environmental records that facilitate bio-statistical analyses in long term disease surveillance, tracking and decision making. This chapter reviews the origins of environmental sensing applied to human health and technological progress during the 2004–2008 Quadrennium.

Keywords: Environmental sensing, human health, health surveillance, public health

29.1 INTRODUCTION

Remote sensing, geographic information systems and spatial information sciences applied to human health were formally added to TC-VIII, WG2 in 2004. Consequently, the authors have chosen to present a general overview of the subject to introduce readers to its foundations. Special reference has been given to air quality and respiratory health; however, the references are more comprehensive in scope. Major concepts and trends are highlighted in Section 29.2. Sections 29.3, 29.4 and 29.5 address advancements in health surveillance (Section 29.3), an overview of selected programmes and activities related to air quality for health (Section 29.4) and reported data and technology needs and gaps (Section 29.5). Section 29.6 focuses on recommendations from major research sectors.

29.2 ENVIRONMENT AND HEALTH

Public health has been defined as "the science and art of preventing disease, prolonging life and promoting health through organized efforts of society" (Eisenberg et al. 2001, p. 230). It is concerned primarily with health and disease in populations, complementing medical and nursing concerns for the health of individuals. Its chief responsibilities are directed towards monitoring the health of a population, identifying its health needs, fostering policies that promote health and evaluating health services.

29.2.1 *Relating environment to human health*

Everyone's health is impacted by their exposures to microscopic mineral and chemical particles and by organisms bound in air, water and soil systems (Varmus et al. 2003, Kaya et al. 2004, Guptill & Moore 2005, Kaiser 2005, Matthias 2005, WMO 2005, Griffin 2007, Griffin et al. 2007). The health status of individuals and populations is determined by complex interactions between internal (genetic) factors (not treated here) and external (acquired) environmental factors. Many of the external factors can be infectious and contagious throughout whole populations. For example, ozone can lead to chronic asthma in children and elders; viruses are known to generate pandemic influenza; bacteria cause intestinal problems; and pollen promotes hay fever. At the root of all vectorborne diseases (e.g. malaria, plague, dengue fever, cholera) there is a virus or bacterium that can be lethal and can be spread.

For airborne contaminants, an example of how natural and anthropogenic biogeochemical materials are spread throughout the environment is presented in Fig. 29.1. Similar mechanisms exist in nature for spreading waterborne and vectorborne materials.

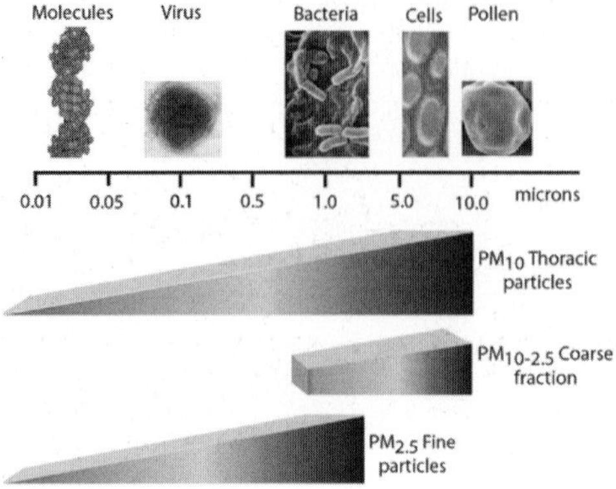

Figure 29.1. Particulate size distribution and related biophysical impacts. Modified from Kaiser (2005). (see colour plate page 512)

Clearly, air quality is a critical parameter for health officials because global and regional atmospheric circulation patterns can deliver chronic and lethal pollutants and contaminants to virtually all places on Earth (Griffin 2004).

Many environmental factors affecting health can be monitored from Earth observing platforms. One of the challenges for integrating these data into human health practice is to demonstrate that such observations actually improve model predictions of environmental conditions that are known to trigger health responses. The body of medical and epidemiological knowledge linking environmental exposures to health outcomes has grown rapidly since the 1980s. Important contributions during the Congress period include Gloster & Alexandersen (2004), Vineis (2004), Wu et al. (2004), Yu et al. (2004), Becker et al. (2005), Cringoli et al. (2005), Grattan et al. (2005), Park et al. (2005), Sulaiman et al. (2005), Selinus et al. (2005), Kuehn (2006) and Schlesinger et al. (2006). Through these known and suspected linkages, it is increasingly clear to scientists and health authorities worldwide that satellite observations will play a significant role in alerting populations to impending health consequences of immediate and short duration, as well as alert earth scientists to prolonged environmental changes that will cycle over many generations.

In the context of the Earth system, environmental health cannot be separated completely from human health. However, they are not the same. Their communities of practice are driven by different educational requirements, skill sets, technologies and models. Environmental health includes not only the health and sustainability of natural ecosystems such as biodiversity and ecosystem services, but also the environments of urban and village landscapes, individual home and building environments, and of proximal biological and chemical contaminants in direct contact with individuals. Studies of environmental health require atmospheric physics and chemistry, water chemistry, geophysics, soils, biology and biochemistry. Public health, on the other hand, is only partly concerned with the effects natural environments have on populations. Officials are concerned with statistical measures over the entire range of a population's risk factors, among them demographics, life style, socioeconomic status and nutrition, access to health care, exposure rates and genetic heritage (Dearry 2005, Brilliant 2007). While these officials are aware of the potential consequences of environmental factors impacting human health, their data gathering structures (hospital admissions, emergency room visits) most often do not require reference to environmental factors that might have triggered a health outcome.

These two broad communities of practice (environmental health and human health) merge their concerns precisely at the medical level where there is too little infrastructure, too few trained personnel, and too little time to fully assess causes and effects. Physicians, school nurses, emergency responders, clinicians and others in healthcare professions have been trained to diagnose a patient's "chief complaints" or manifested physical condition. Integrating geo-spectral and geospatial data and information into systems that allow reliable statistical inferences or that enable issuance of public health alerts based on these data are just beginning to emerge (Lang 2000, Morain & Budge 2006a, Griffin 2007).

29.2.2 *Connecting earth observations to human health*

Human health cannot be monitored directly from satellite altitudes because disease transmission pathways are usually not direct between the observed environment and resulting health effects. However, Earth environments that harbour potential human health threats can be observed from space. Atmospheric dust and aerosol measurements, for example, have been collected for four decades by progressive generations of space observatories. The technologies for these observations continue to improve and the derived products and algorithms from them are gradually finding applications among health surveillance communities (Morain & Budge 2006b). For additional background literature, see Ashbolt (2004), Gauderman et al. (2004), Kellogg et al. (2004), Pope (2004), Stetzenbach et al. (2004), CDC (2005), Centeno et al. (2005), Gyan et al. (2005), Naumova & MacNeill (2005), Noji (2005), Park et al. (2005), Sultan et al. (2005), Tobias et al. (2005), Zanobetti & Schwartz (2005), Hansell et al. (2006), Horwell & Baxter (2006), Matthias et al. (2006) and Pascual et al. (2006).

Table 29.1. Classification of water-related diseases.

Category	Comments
Water-borne diseases	Caused by drinking water contaminated by human or animal faeces or urine containing pathogenic bacteria or viruses. These include cholera, typhoid, amoebic and bacillary dysentery, and other diarrhoeal diseases.
Water-washed diseases	Caused by poor personal hygiene. These include scabies, trachoma and flea-, lice- and tick-borne diseases in addition to the majority of water-borne diseases that are also water-washed.
Water-based diseases	Caused by parasites found in intermediate organisms living in water. These include dracunculiasis, schistosomiasis and some other helminths.
Water-related diseases	Transmitted by insect vectors that breed in water (e.g. mosquitoes). These include dengue, filariasis, malaria, onchocerciasis, trypanosomiasis and yellow fever.

Source: Eisenberg et al. (2001).

Tying environmental attributes to population health (proximal factors) usually begins by studying health outcomes of individuals and by tracing their whereabouts and activities (distal factors) that led them to being exposed. For example, it has long been known that water-related diseases come in several categories and are spread through different pathways. Therefore, monitoring water environments from space provides a crucial beginning to planning health interventions; but they do not necessarily forecast a disease outbreak. Table 29.1 from Eisenberg et al. (2001) highlights these categories and pathways.

29.2.3 *Health issues and costs*

Earth system scientists face a challenge in modelling complex biological, chemical and physical processes at the ecosystem level, and in finding quantitative measures for tracking potential environmental health outcomes (ICSU Scoping Group 2007). An equal challenge is to extract from this knowledge the consequent flow of pathogens and chemicals through airborne, soilborne, waterborne and vectorborne mechanisms, and to translate findings into actionable human health scenarios that point to populations at risk. The knowledge base for tracking health-threatening pathogens varies by nation, region and continent; but the science and technologies for observing, measuring and modelling Earth systems is global. The grand challenge is to add health professionals to efforts that merge environmental surveillance with human health syndromes (a medical term for a variety of a patient's "chief complaints" that together suggest categories of possible diseases).

Until the late 20th century, most experts believed that humid tropical regions were the home for infectious diseases. However, it is apparent now that the northern mid-latitudes are home to growing numbers of emerging and re-emerging infectious diseases (Epstein 1997, Binder et al. 1999, Morens et al. 2004, Fauci et al. 2005, Kuske 2006) and that an integrated global observing strategy is required to monitor these changing patterns (Kennel et al. 1997). The most heavily populated areas of North America, Western Europe and Japan represent potential health threats. This is a triple threat because the area coincides geographically with the highest concentration of air travellers to global destinations; is home to societies adding the highest concentrations of industrial emissions and biological contaminants into the environment; and is the hemisphere over which there is ample evidence for global climate change. There is a persistent ring of hemispheric aerosols around the northern mid-latitudes contributed by industrialized societies. The Earth's wind and ocean circulation systems also play a role in the rising incidence of respiratory diseases like chronic asthma, myocardial infarction (MI), tuberculosis, severe acute respiratory syndrome (SARS) and influenza.

Public health issues are mounting as quickly as the realization that Earth's environments are changing. Dust storms so large that they can be animated using time series satellite imagery can be seen moving across Asia, over the Pacific Ocean, and into North America. Similarly, dust picked up by winds over North Africa is carried to the Caribbean. Fine particulate matter ($PM_{2.5}$) in these storms pose serious health threats for persons with respiratory diseases (Griffin 2007). This is a global concern, one that has captured the attention of several international organizations including the World Health Organization (WHO), the International Council for Science (ICSU) and the Group on Earth Observations (GEO). While the medical community recognizes the adverse effects that $PM_{2.5}$ can have in patients with respiratory conditions, they have lacked timely and reliable information for issuing dust storm warnings.

29.2.4 *Societal awareness of sensing for health*

People and Pixels (Liverman et al. 1998) was among the earliest publications to draw humankind into satellite remote sensing. For almost four decades in the 20th century, scientific literature in remote sensing focused on applications in agriculture, forestry, rangeland, hydrology and mineral exploration. After *People and Pixels*, interest shifted to societal issues like food security, environmental health, public health, disasters and hazards, and most recently to homeland security and counter-terrorism. Because of their immense policy and humanitarian implications, remote sensing and geo-spatial programmes moved quickly into predicting the health consequences of climate change, air and water quality degradation, and natural disasters. Additional stimulants for these concerns included: (1) the World Summits on Sustainable Development; (2) the Committee on Earth Observations (CEOS); (3) formation of the Group on Earth Observations (GEO) in 2003; (4) the series of severe natural regional disasters between 2004 and 2008; and (5) early media attention to the Fourth Assessment Report of the Intergovernmental Panel on Climate Change (IPCC) (In press).

29.2.4.1 *World summits on sustainable development*

The first principle to emerge from the United Nations Conference on Environment and Development in Rio de Janeiro (UN 1992) was that "Human beings are at the centre of concerns for sustainable development. They are entitled to a healthy and productive life in harmony with nature". This is an early global statement linking environmental health to human health. Ten years later at the 2002 World Summit, the Johannesburg Plan of Implementation was adopted (UN 2004). In this Plan, paragraphs 54 and 56 refer specifically to human health issues that are addressable via satellite Earth observations. As stated in the plan (p. 31): "there is an urgent need to address the causes of ill health, including environmental causes, and their impact on development, and to reduce environmental health threats."

29.2.4.2 *Committee on Earth Observation Satellites (CEOS)*

CEOS was formed in 1984 and consists of representatives from over 40 space agencies and participating national and international organizations that build, launch, operate and maintain satellite sensor payloads. Their primary purpose has been to foster coordination among nations in these activities as a cost-effective means for avoiding duplication, optimizing sensing programmes across the range of applications and forming partnerships among programmes having common interests to reduce costs. CEOS recognizes that obtaining long-term commitments for Earth observing programmes for societally-based applications is difficult in the absence of clear statements of need from user communities like human health. Space agency funding is almost always driven by each nation's scientific, industrial and economic needs. In an effort to develop these clear statements and application requirements, CEOS was instrumental in promoting a ministerial-level summit that subsequently became known as The Group on Earth Observations (GEO).

29.2.4.3 *Group on Earth Observations (GEO)*

An Earth observation summit was held in the United States in 2003. Its purpose was "to promote the development of a comprehensive, coordinated, and sustained Earth observation system of systems among governments and the international community". GEO emerged from the first summit with a Declaration signed by ministers or senior representatives from over 30 nations and 20 international organizations. Subsequent summits in 2004, 2005 and 2007 brought more nations and organizations to the table. ISPRS was among these international participants, and has been an active contributor. By November 2007, 73 nations and over 63 international organizations were members. A 10 year implementation plan was developed to identify societal benefit areas (SBAs) and to stimulate communities of practice in each of them to evolve statements of need and target user communities. In addition, GEO stimulated activation of elements forming a Global Earth Observation System of Systems (GEOSS) and some early adopting communities of practice for utilizing it. Human health is among the nine identified SBAs and within these is considered to be a cross-cutting theme. As a result, member nations, supported by their environmental protection agencies and human health infrastructures, began to assess their resources and to take steps toward collaborative enterprises (Schmidt 2005, Dowman & Murai 2008, this volume).

29.2.4.4 *Natural disasters*

If further stimuli were needed for fast-tracking environment/health applications, they were provided abundantly by nature itself. Three, from among many, are mentioned to illustrate that the health of all people is subject to events driven by environmental forces that can be observed and measured, but which may never be controlled or prevented. On 26 December 2004, a severe earthquake with its epicentre near Banda Aceh spawned a series of tsunamis that devastated coastal areas from Indonesia to east Africa, reportedly killing 170,000 people. On 29 August 2005, winds from Hurricane Katrina caused a storm surge that breeched the levees protecting the U.S. city of New Orleans with subsequent flooding that cost almost 2000 lives. For those who remained in the city and survived, health effects included dehydration, food poisoning, hepatitis, cholera, tuberculosis and typhoid fever. Potential longer term exposures were projected due to petrochemicals in the flood waters and uncontrolled populations of mosquitoes transmitting diseases like yellow fever, malaria and west Nile virus. In June and

July 2007, middle and northern England were flooded by persistent rains that surpassed England's worst floods in living memory. Hundreds of thousands were without drinking water when municipal treatment works were inundated. Several hospitals were evacuated, crops were destroyed and thousands were stranded on motorways. In all, it was estimated that these rains would cost £2 billion.

29.2.4.5 *Intergovernmental panel on climate change*

Framing the series of natural disasters was the IPCC's Fourth Assessment Report on climate change (IPCC, in press). Already in the international news, every natural disaster punctuated the report's findings, stimulating media attention and political response. Results from all of the assessments have confirmed that the Earth is gradually warming. There is irrefutable evidence that an important contributor is humankind itself. Ecosystem alterations, reductions of biodiversity and rising levels of greenhouse gases are all exacerbated by human impacts on global environments. The by-products of these impacts are being linked now to human health as responses to degrading environmental quality in the air, on land and in drinking water.

29.3 ADVANCES IN HEALTH SURVEILLANCE

The years leading to ISPRS 2004 Quadrennium are characterized by a steadily growing number of scientific articles reporting project-based research that document the roles environmental monitoring play in human health surveillance. During the 2004 Quadrennium, there was a trend towards aggregating efforts into longer term programmes having higher levels of funding and technology integration. Among the enabling technologies permitting aggregation were: (1) syndromic surveillance systems; (2) interoperable, open source technologies; (3) concepts for virtual satellite constellations and sensor webs; and (4) steps toward implementing a global system of systems to utilize data and information for SBA applications.

29.3.1 *Disease surveillance systems*

There are two principal approaches to disease surveillance: data mining and syndromic surveillance. Data mining is an approach for extracting information from hospital records and from health authority databases. Current algorithms are not very robust and the process is time consuming and often misleading (that is, it may result in many false positives that over-estimate the health threat). It does not appear to have applications for detecting diseases or epidemics when time is critical. Moreover, data mining techniques typically do not include critical sources of information provided by veterinarians, school nurses and first-responders. More than anything, the approach is hindered by an absence of standardized reporting criteria and search terms.

Syndromic surveillance is based on chief complaints recorded by clinicians in real-time, including physicians, hospital nurses, school nurses, emergency ward staff, veterinarians and first responders. These trained professionals record patient attributes like fevers, rashes, lesions, nausea and soreness, which characterize syndromes that alert them to possible diseases. If the same syndrome appears at several localities across a community or within a school over a short period of time, the health care community can react swiftly to understand the geo-spatial domain over which it is reported. Telecommunication, augmented by web based services, then allows the medical community at all scales of reference to share information, marshal resources, direct medical examinations and inaugurate interventions and treatments.

Among the more relevant citations describing health surveillance approaches are: Buehler et al. (2004), Fairchild & Bayer (2004), Heymann & Rodier (2004), Naumova et al. (2005), World Bank (2005), Zelicoff & Bellomo (2005) and Eben (2007).

29.3.2 *Interoperable open source technology*

The Open Geo-spatial Consortium (OGC) made technological advances during the 2004 Quadrennium. Simple Object Access Protocol (SOAP) streamlines data storage, data processing, product generation and product delivery from multiple sources to almost any machine. Figure 29.2 illustrates the basic concept (Budge et al. 2006). SOAP enables flat file datasets (such as tabular health data and hospital records) to be retrieved and integrated with raster and vector data so they can be analysed statistically and visualized for direct output to health alert systems; or shared through a client interface to global health authorities for aggregation and analysis at national and regional levels. Earth observation data are integrated in this system via assimilation of data sets into one or more operational numerical models. Outputs from the models represent inputs in SOAP.

Products can be integrated into multiple existing clients in the larger framework of a Service Oriented Architecture (SOA) (see also Li (2008), this volume). In the Public Services area (left side of Fig. 29.2), new datasets can be integrated into existing client interfaces that include the Mapping Client, WMS Client, Tabular Data Client and Analysis Client. Both the Mapping Client and the WMS Client allow users to visualize those raster images derived from the SOAP system. While the WMS Client acts as an image viewer for routine GIS operations (e.g. zoom, pan and overlay with user-selected vector data), the

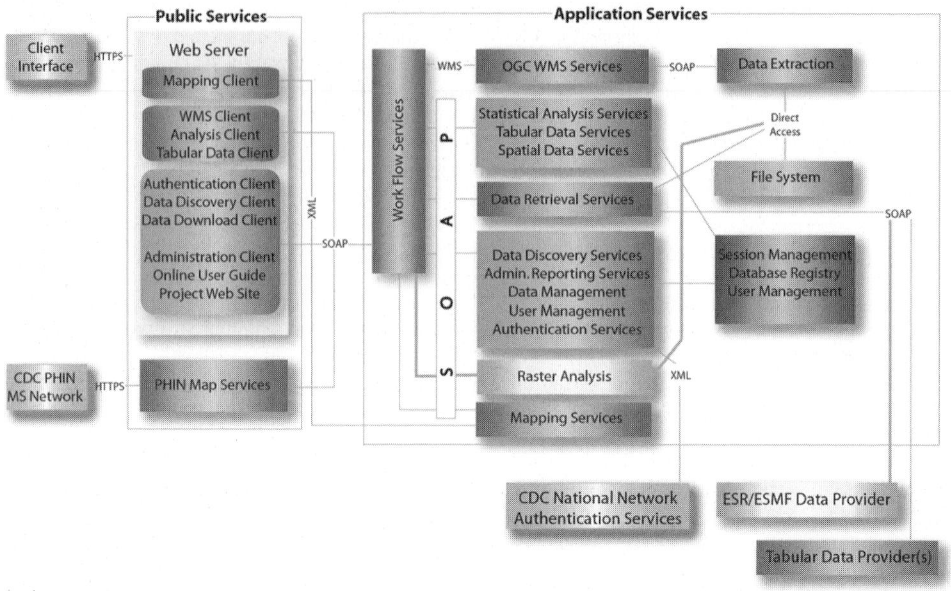

Figure 29.2. Example of a services oriented architecture featuring (on the right) applications services for integrating health data, model outputs and raster/vector data and (on the left) public services through a client interface for a wide variety of data users. This architecture is being developed for possible use in the United States as part of the Environmental Public Health Tracking System.

Mapping Client performs advanced GIS functions. Users have access to ArcGIS functions like buffering, distance measurements, layer attribute extraction and querying. They also can access customized raster processing models that are built on the server to deliver unique products to broader communities.

For further entrée to the literature on GIS and environmental modelling see: Eubank (2004), McKensie (2004), Nuckols et al. (2004), Poulstrup & Hansen (2004), Reid et al. (2004), Barrett et al. (2005), Brooks et al. (2005), Gebre-Michael et al. (2005), Yin et al. (2005), Ayotte et al. (2006), NRC (2006), Bian & Liebner (2007) and Yin et al. (2007).

29.3.3 *Advanced sensor systems*

Space-faring nations continue to launch Earth observatories to test the ability of experimental sensors for monitoring interactions between air, land and water systems. Environmental monitoring is essential for assimilating sensor products into numerical models for forecasting human health threats (Beck et al. 2002, Morain & Sprigg 2007). In 2005, CEOS recognized that the implementation plan for GEOSS required an innovative process whereby the aims and goals of national and regional observing requirements could be aggregated into global capabilities without adding significantly to operational costs. This reasoning led to the concept of a virtual constellation. Its aim is to acquire relevant environmental attributes for all of GEO's societal benefit areas (SBAs). As shown in Fig. 29.3, it consists currently of geo-stationary weather satellites and low Earth orbit satellite systems operated by China, EUMETSAT, India, Japan, Republic of Korea, the Russian Federation, SPOT Image and the USA.

While it is elegant in concept, there are numerous technical challenges in its design and function, and many institutional barriers that are driven by individual national interests and economies (OSTP 2007). There has been some continuity in observations (some even for decades), but for SBAs most observations have been *ad-hoc*[1]. Implementation of the concept requires defining statements of space-based requirements from communities of practice in all of GEO's SBAs. Among the nine SBAs, human health is not as mature as others, and this should be a focus for ISPRS in its 2008 Quadrennium. One of the significant advances during the 2004 Congress period is that experimental data sets and analysis systems have been evaluated by

[1] Source: The CEOS Virtual Constellation Concept. V0.4, 10th March 2006 (draft).

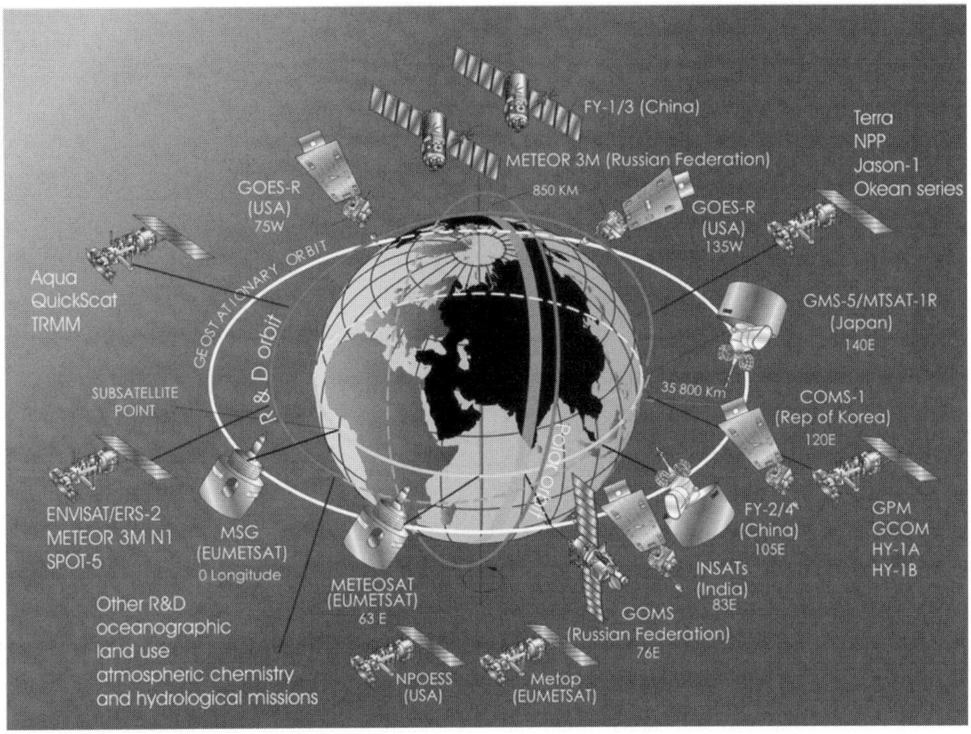

Figure 29.3. Sample configuration for a virtual sensor web. Courtesy, Robert Chen, GEO Secretariat.

both earth scientists and health science professionals in preparation for future operational sensors.

Sensor webs are especially well suited to environmental monitoring and control. They consist of ground-based, aerial and satellite sensors, GPS coordinates of objects and features, and real-time GNSS readings that form wireless communication networks. This architecture allows every sensor to communicate with every other sensor throughout the web forming continuous, temporally synchronized records. These capabilities are evolving new application areas in human health exposures to environmental conditions. They will allow longitudinal studies (that is, long-term studies of individuals or target groups) to assess the evolution of health outcomes with age, track disease transmission pathways as environmental changes occur, and provide coherent time series data for epidemiological research.

29.3.4 *A system of systems*

ISPRS should be an active force for advancing the roles of space science and GIS in public health, and for promoting the creation of a truly integrated community of practice. Not all human health conditions can be assisted by environmental data from satellites or from ground monitoring systems; but almost all health issues are subject to investigation through geographic information systems (Lang 2000). To function at the patient and population levels, there must be seamless, end-to-end integration of human health and environmental data collection, processing and forecasting, storage and retrieval, analysis and application sectors. In broad outline, these sectors include earth scientists who design relevant sensors for environmental attributes useful in public health surveillance, a sensor web that supplies customized products for surveillance, ground networks of environmental monitors that verify and validate space observations, flexible web-based services that enhance decision support systems, accessible telecommunication networks that disseminate reliable information to decision makers, and appropriate health care communities that diagnose and serve their constituents while maintaining individual rights to privacy.

Some of the more obvious health communities of practice served by a global system of systems are shown in Fig. 29.4. The small ovals represent broad sectors of health care services that operate from top-down regional and national levels to local levels.

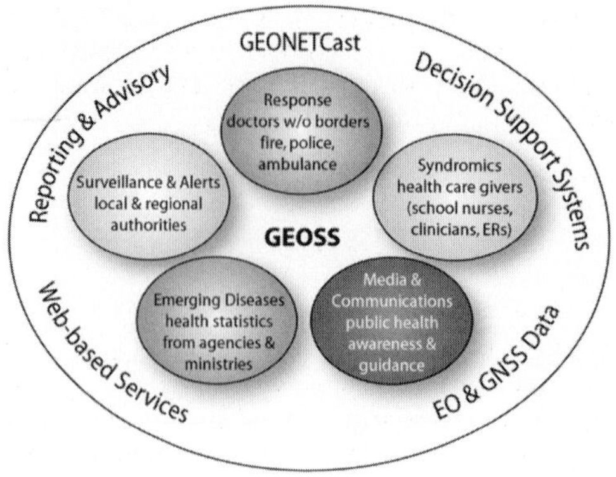

Figure 29.4. Health communities of practice and the information types needed to integrate their functions.

They all have different needs for local and provincial data that must be aggregated medically and geospatially from the bottom-up for interpretation and analysis. The words surrounding the small ovals represent uses for the data and information. In 2008, the communities of practice and the uses for satellite-based data and information remain disparate. The world relies heavily on electronic news media for evolving health threat information like Severe Acute Respiratory Syndrome (SARS) or avian flu outbreaks; but the media cannot yet serve every rural village or isolated population in the developing world.

GEONETCast is a more science-based, integrated system supported by national space agencies and weather services. Potentially it can reach these remote populations to both supply information about their changing local environments and return environmental conditions at those locales for aggregation into regional health advisories. The system is predicated on having a stable network of weather satellites in a sensor web with other Earth observatories, all having different functions, but all having cross-cutting utility for health applications. GEONETCast is a near real-time, global network of satellite-based data dissemination systems designed to distribute space-based, airborne, and *in situ* data, metadata and products to diverse communities. It is a Task in the GEO Work Plan led by EUMETSAT, the United States, China and WMO. Products are being made available to user communities, including data from GOES, MODIS, NOAA-NESDIS and others.

29.4 ACTIVITIES AND PROGRAMMES

A variety of activities and projects coupling air quality and human health have emerged in recent years. Some of these are integrating satellite observations with ground-based and airborne observations of atmospheric conditions. Others are forming consortia to implement demonstration pilot projects that integrate Earth observations with health, epidemiological and socio-economic data. Several workshops have been organized to examine impacts of Earth observations on mitigating health related issues and events. Though many of these projects and activities are initiated by governments and consortia, they are aggregating under the umbrella of GEO as contributions to GEOSS. The scientific unions are also addressing health and the environment via their Science for Health and Well-being committee.

Many air quality initiatives, programmes and projects were begun during the Congress period. Several of these have evolved into high profile programmes and projects that merit elaboration. One is THORPEX, a World Meteorological Organization (WMO) programme that was established in 2003 as a 10-year international global atmospheric research and development effort to reduce and mitigate natural disasters. The aim is to transform weather forecasts into specific information that supports decisions for health advisories and other societal outcomes. THORPEX has become the focus of the GEO Health Task HE-06-03 (Forecasting Health Hazards) with its demonstration project on developing a multi-hazard early warning and response system in West Africa.

Global Monitoring for Environment and Security (GMES) is a joint initiative between ESA and the European Commission. Of particular note is the GMES PROMOTE project that aims to improve air quality forecasting using satellite technology. The goal is to construct and deliver a sustainable and reliable operational service to support decisions on atmospheric policy issues concerning stratospheric ozone depletion, surface ultraviolet exposure, air quality and climate change. The Cambridge Environmental Research Consultants are using information from PROMOTE in their airTEXT product, which is a free service for persons diagnosed with asthma and other respiratory ailments in London. It provides timely information about peak periods of localized air pollution across the London metropolitan area so that residents can take appropriate action.

GEOMON is a project that aims to integrate existing programmes into a single interoperable system for Europe. It is a first step towards building an integrated pan-European Atmospheric Observing System (EAOS) for systematic observations of long-lived greenhouse gases, reactive gases, aerosols and stratospheric ozone. One of its objectives is to integrate satellite observations with ground-based and airborne observations to quantify and understand temporal changes in atmospheric conditions. Thirty-eight European institutions are participating in this project.

In the United States, the Environmental Protection Agency, NASA, the Centers for Disease Control and Prevention (CDC), and the National Oceanic and Atmospheric Administration (NOAA) are contributing resources to a variety of health related projects. The Environmental Public Health Tracking Network (EPHTN) is an emerging national system for assembling and distributing public health data collected by local authorities. These data are augmented by EPA ground-based air and water quality sensors, and by NASA and NOAA model outputs derived from geosynchronous and geo-stationary satellites. These integrated products are designed to be accessible through web-based SOAP services to provide health authorities at all levels with reliable and accurate environmental sensing and health information.

Within the broad federal structure are many agency-specific programmes that support university-based projects relating to human health. Among these are HELIX, GeoMedStat, PHAiRS and 3D-AQS. Collectively, these initiatives, programmes and projects (and many others throughout the world) aim to provide information on atmospheric conditions and events affecting human health. A major challenge remains with distributing data and information to decision and policy makers, especially those in extremely remote locations.

29.4.1 *HELIX*

Health and Environment Linked for Information Exchange is a project led by NASA's Marshall Space Flight Center. It is focused on air quality and asthma in the five county area of metropolitan Atlanta, Georgia.

29.4.2 *GeoMedStat*

This is a project led by the University of Mississippi Medical Center. It is an asthma/air pollution surveillance system that links medical data with geospatial attributes, census data and satellite air quality observations.

29.4.3 *PHAiRS*

Public Health Applications in Remote Sensing is a NASA-sponsored cooperative project with the University of New Mexico and University of Arizona. It nests a dust entrainment model into a numerical atmospheric model to forecast dust episodes that lead to non-communicable chronic respiratory diseases.

29.4.4 *3D-AQS*

The Three Dimensional Air Quality System is a project led by the University of Maryland Baltimore County. It is integrating MODIS and CALIPSO data with *in situ* air quality and GIS data sets to develop temporal, 3-D monitoring capabilities of Eastern U.S. and transatlantic air quality episodes.

29.5 NEEDS AND GAPS

A remarkable development in environmental sensing for health is the realization among world leaders that subtle trends in climate change impact their people and the vitality of their economies. A rich and medically diverse literature accrued throughout the mid-1990s drawing connections between temperature, rainfall, humidity, soil moisture, biodiversity and related attributes to a variety of diseases (see among many others: Cross et al. 1996, Dister et al. 1997, Hassan et al. 1998, Malone et al. 1994, Pope et al. 1993, Roberts et al. 1996, Thompson et al. 1996a,b). Many of these studies demonstrated what the medical community suspected as environmental connections to health conditions and diseases, but the results were not compelling in statistically valid cause and effect relationships. Through clinical studies and the rise of disease surveillance technologies, health research communities are beginning to establish statistical correlations and probabilities of environmental factors that influence health responses.

There is still a major gap in the medical professions for including aerial and satellite data products in their research designs. In the 73 chapters of the 2nd edition of Disease Control Priorities in Developing Countries (World Bank Group, on-going dates), there is no mention of satellite-acquired environmental data as a component of environmental surveillance for disease detection, monitoring or intervention; nor is there any mention of GIS or GNSS technologies. One can imagine several reasons for this, among them being:

1. there is a poor congruence between the timing of general disease outbreaks and the timing of satellite overpasses recording quantitative conditions that might have triggered the outbreak;
2. many chronic conditions and diseases take years to evolve in target groups. In these situations there are often few retrospective, and almost no quantitative, satellite-acquired data and model outputs that can contribute to robust verification or validation; and
3. there are few longitudinal studies of specific environment/health responses, and almost none that employ high spatial resolution environmental data to make statistically convincing correlations between environmental cause and resulting health effect.

In short, the biggest gap in technology is for the earth science community to integrate products from various sensors in the virtual constellation and sensor webs. If these integrated products were available, they would still need to be assimilated into appropriate operational models that deliver meaningful inputs for disease monitoring systems. Indeed, one of the noteworthy advances to appear in human health applications during the Quadrennium has been the assimilation of satellite observations into an operational atmospheric model whose outputs are interoperable with GIS databases (Morain & Sprigg 2005, Morain & Sprigg 2007). Fig. 29.5 shows incremental improvements in modelling a dust storm event in the south-western U.S., before and after assimilation of EO data, and after migration of these data to a higher resolution numerical model. The modelled outputs are 72-hour forecasts of airborne mineral dust that can be superimposed on data from ground observation networks and made available on-line as temporal animations in standard GIS formats. When operational, health professionals across the model domain will be able to access the forecast every day for making health-related decisions. This same system is designed to provide longer term data for developing health correlations and statistical associations.

The following subsections represent a sample of data needs, gaps and recommendations recognized by joint groups of Earth scientists and medical professionals. Most of these ideas are germinated in workshops aimed at introducing communities of practice to new technologies and frames of reference, and to exchange technical information pertinent to the others' understanding of new fields of application. Most often, before effective communication can occur, it is necessary for collaborators to learn the other's vocabularies and lexicons. While many of the suggestions have been developed with specific national or health care sectors in mind, they also are valid in a general sense.

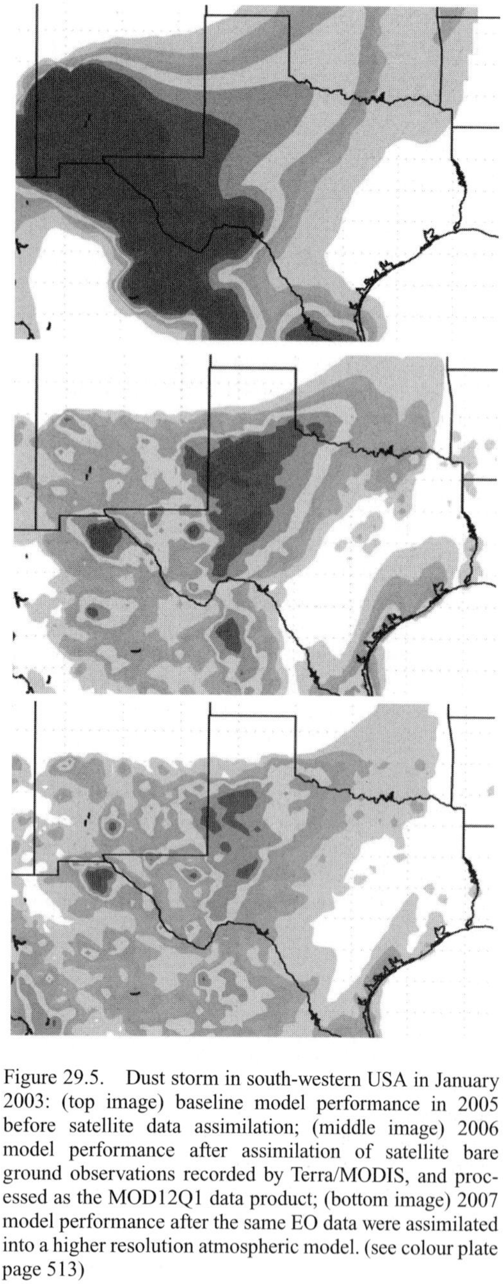

Figure 29.5. Dust storm in south-western USA in January 2003: (top image) baseline model performance in 2005 before satellite data assimilation; (middle image) 2006 model performance after assimilation of satellite bare ground observations recorded by Terra/MODIS, and processed as the MOD12Q1 data product; (bottom image) 2007 model performance after the same EO data were assimilated into a higher resolution atmospheric model. (see colour plate page 513)

29.5.1 *Project design framework*

The design framework for integrated surveillance is crucial, especially in temporal and spatial contexts, as explained by Naumova (personal communication). For employing remote sensing data in public health research, it is essential to: (1) build a conceptual basis for integrating surveillance and monitoring technologies with on-the-ground health care records; (2) formulate a sound hypothesis; (3) develop a conceptual framework; (4) select candidates for measures of exposure; and (5) outline properties of health outcomes clearly.

Compatibility among health outcome measures and exposure measures derived from aerial or satellite observations can be achieved analytically by using normalization schemes, inclusion of proper indicators in modelling procedures, stratification and other techniques. Typically, time-referenced health outcome data are recorded as daily, weekly, monthly, quarterly or annual counts of health-related events. Temporal revisit cycles for satellite sensors range from daily to bi-monthly for sun synchronous platforms, and as fine as twice hourly for geo-stationary platforms.

Complete, uninterrupted temporal overlap between health and exposure data sets is required for at least three reasons: (1) many statistical methods have serious limitations if there are gaps in the time series; (2) temporal completeness expands the array of analytical techniques available for analysis; and (3) because sample sizes available for an analysis are relatively smaller in a joint time series, discontinuities between exposure measures and observed health outcomes lower the ability to detect a statistically valid effect. For proper linking of environmental data sets with health outcomes, it is important to perform a basic time unit alignment. The spatial alignment also has to be carefully addressed. Health outcomes are typically aggregated from specific geographic areas and differ by population coverage. Therefore, these data have to be geo-referenced to ensure proper alignment with environmental data. Temporal discontinuities are therefore an inherent challenge for human health applications.

29.5.2 *Data needs and technology gaps*

The above framework can be achieved by addressing specific data needs (Ver Ploeg & Perrin 2004, Schmidt 2005, Tinkle et al. 2007). Key among these are: (1) using best practices for measuring data quality assurance and control; (2) validating correlations with health outcomes; (3) confirming that aerial or satellite data represent accurately exposures on the ground; (4) using the finest temporal and spatial resolution data sets available for pollution measures; and (5) for respiratory diseases, striving to identify, characterize and measure chemical species in air pollution.

In general, it is accepted by respiratory, cardiovascular and developmental biology health specialists that aerial and satellite data could help fill gaps in existing exposure data. These research specialists suggest that demonstration projects should be designed to correlate long-term hospital admissions and emergency room visits for asthma, MI, arrhythmia, heart failure and stroke with remotely sensed measures of airborne particulates and bio-allergens. Accordingly, longitudinal studies of acute and chronic sufferers should compare evolutionary advances in environmental data collection strategies with on-going data from *in situ* networks to verify and validate results using statistical analyses, and through these means to improve the spatial and temporal resolution of surveillance systems.

29.5.3 *Sample research priorities*

29.5.3.1 *National Research Council*

The United States National Research Council (NRC 2007) published research priorities in three broad areas: (1) Earth material exposure assessments; (2) improved risk-based hazard mitigation; and (3) health risks associated with human alterations of ecosystems as well as those occurring naturally. Often there are weak links in health risk assessments, because biogeophysical processes coupled with anthropogenically generated Earth materials are difficult to quantify. Health risk interventions require better understanding of the source, fate, rate, transport and bioavailability of potentially hazardous Earth materials. Advancements in exposure assessments therefore should include: (a) developing quantitative knowledge of the geospatial context of particulates and aerosol mixtures and associated disease vectors; (b) determining the biogeochemical cycling of trace elements in water and soils with respect to low-dose chronic exposures through drinking water and food; (c) determining the distribution, survival and transfer of pathogens in plants and soils; (d) identifying newly discovered pathogens and pharmaceuticals as emerging contaminants; and (e) quantifying health risks posed by transport through ecosystems (Selinus et al. 2005).

Improving hazard mitigation is important in the context of rapidly evolving local health effects. The first priority should be to integrate earth science, engineering, emergency response and public health disciplines to develop better response scenarios.

Health risks associated with natural and human-modified ecosystems focus on resource extraction, waste disposal, invasive species and similar ecosystem changes. Research priorities include: (a) investigating geomorphic and hydrological changes in the context of disease emergence and transmission; and (b) determining the health effects associated with water quality changes in both surface and groundwater supplies.

29.5.3.2 *World Bank Group*

Under the rubric of tools for public health surveillance, Nsubuga et al. (2007) address three areas for research: (1) surveillance programmes for developing countries; (2) new approaches for standardizing methods; and (3) major technology questions. In the first area, the authors refer to capacity building and infrastructure, listing four major needs: (a) developing and retaining a competent, motivated cadre of health workers; (b) developing and adopting data standards for information system architectures and syndromic surveillance systems (Lober et al. 2004); (c) promoting global policies and practices for international surveillance; and (d) creating useful software (Dean 2000).

For approaches and standardization of methods, they advocate suggestions made by Romaguera et al. (2000): (a) inaugurating an integrated disease surveillance and response (IDSR) for infectious diseases; (b) developing syndromic surveillance for terrorism and emergency response as described by CDC (2004); and (c) advancing clinician-based surveillance methods to enhance diagnostic accuracy and increase timeliness for recognizing outbreaks and interventions, as express by Swaminathan et al. (2001).

Major surveillance technology questions are: (a) how to use data for forecasting temporal and spatial analysis aberration detection; (b) how to conduct surveillance for multiple competing risk factors that lead to a single condition (for example, MI as a result of exposure to high concentrations of $PM_{2.5}$, smoking, hypertension or overweight); (c) how to interpret ecological data relative to data collected on individuals (Greenland 2004); (d) how to measure accuracy and cost-effectiveness for alternative approaches to surveillance (for example active vs. passive vs. integrated; clinician-based vs. data mining); (e) how to link data sources effectively (for example hazard, exposure, outcome data for environmental diseases); and (f) how to strengthen evidence-based decision-making cultures in ministries of health and finance.

29.6 SUMMARY AND RECOMMENDATIONS

Increasingly, environmental impacts on public health are being traced to elevated concentrations of contaminants attributable to economic, social and human pressures on landscapes. What is needed to link environmental health to public health is a community of medically trained professionals who have an appreciation of processes that impact environments and that may in turn influence affected populations. Such a community, working with the community of Earth observing and GIS-trained scientists, would transition use of environmental health information out of space science into routine health care practice. During the 2004 Quadrennium, significant progress has been made to bring the communities of environmental health and public health into joint practice for the benefit of humanity.

To serve in public health, authorities need broad medical training and an appreciation of processes that impact environments that impose secondary impacts on affected populations. Using satellite-acquired data and imagery to study environmental health has many immediate rewards, but these rewards will not be realized until environmental data are integrated into health care provider and health care authority communities of practice.

Someday epidemiologists and health care professionals will be equipped better to understand the consequences of Earth processes affecting health outcomes that are unfolding across all geographic scales. The role these consequences and outcomes play in human health should have the highest priority for science and government. Humans are constantly in the way of pathogens, particles and chemical compounds in both natural and managed environments. To protect and improve their socioeconomic and political infrastructures, all national and international institutions should support environmental knowledge integrated with human health outcomes.

29.6.1 *Institutional recommendations*

- The International Council for Science, working with international and national bodies, GEO and the United Nations should convene an International Year (or Years) of Health and Environment (Blockstein & McManus 2008).
- At national, regional, and local levels, authorities working with their respective science infrastructures should organize summit-level programmes to manage, share, integrate and synthesize data and information among environmental and health disciplines.

29.6.2 *Scientific and technical recommendations*

- Spatial, spectral and radiometric performance for each intended application.
- Minimum instrument performance specifications (S/N, band co-registration, and radiometric accuracy/sensitivity).
- Technical specifications for inter-sensor calibration (satellite-to-satellite and satellite-to-ground).
- Definition of common calibration practices.
- Specification of data formats and inter-operability between acquisition and distribution networks.

ACKNOWLEDGMENTS

This contribution has been sponsored by NASA's Office of Space Science and Applications under Agreement No. NNSO4AA19A, John Haynes project manager. The authors wish to thank also Dr. Elena Naumova, Tufts University; members of the research team for Public Health Applications in Remote Sensing (PHAiRS): Dr. William Sprigg, University of Arizona, and Drs. Karl Benedict and William Hudspeth, University of New Mexico.

REFERENCES

Ashbolt, N.J. 2004. Microbial contamination of drinking water and disease outcomes in developing regions. *Toxicology* 198: 229–238.

Ayotte, B.T., Nolan, J.R., Nuckols, K.P., Cantor, G.R., Robinson, D., Baris, L., Hayes, Karagas, M., Silverman D.T. & Lubin J. 2006. Modeling the probability of arsenic in groundwater in New England as a tool for exposure assessment. *Environ. Sci. and Tech.* 40: 3578–3585.

Barrett, C.L., Eubank, S.G. &. Smith, J.P. 2005. If smallpox strikes Portland. *Scientific American* March: 54–61.

Beck, L.R., Lobitz, B.M. & Wood, B.L. 2000. Remote sensing and human health: New sensors and new opportunities. *Emerg. Infect. Dis.* 6(3).

Becker, S., Mundandhara, S., Devlin R.B. & Madden, M. 2005. Regulation of cytokine production in human alveolar macrophages and airway epithelial cells in response to ambient air pollution particles: Further mechanistic studies. *Toxicol. Appl. Pharmacol.* 207(2 suppl.): 269–275.

Bian, L. & Liebner, D. 2007. A network model for dispersion of communicable diseases. *Trans. in GIS* 11(2): 155–173.

Binder, S., Levitt A.M., Sacks J.J. & Hughes, J.M. 1999. Emerging infectious diseases: Public health issues for the 21st century. *Science* 284: 1311–1313.

Blockstein, D.E. & McManus, K.M. (eds.). 2008. *Integrating Environment and Human Health: A Report of the Seventh National Conference on Science, Policy and the Environment.* Washington, D.C.: National Council for Science and Environment.

Brilliant, L. 2007. Climate, Poverty, and Health. *7th Annual John H. Chaffee Memorial Lecture on Science and the Environment.* Washington D.C.: National Council for Science and the Environment.

Brooks, J.P., Tanner, B.D., Gerba, C.P., Haas, C.N. & Pepper, I.L. 2005. Estimation of bioaerosol risk of infection to residents adjacent to a land applied biosolids site using an empirically derived transport model. *J. Appl. Microbio* 98: 397–405.

Budge, A.M., Benedict, K.L. & Hudspeth, W. 2006. Developing web-based mapping services for public health applications. In S. Nayak, S.K. Pathan & J.K. Garg (eds.), *Proceedings Geospatial Databases for Sustainable Development. Int. Soc. Photogr and Rem. Sens.* 36(IV): 565–569.

Buehler, J.W., Hopkins, R.S., Overhage, J.M., Sosin, D.M. & Tong, V. 2004. Framework for evaluating public health surveillance systems for early detection of outbreaks: Recommendations from CDC Working Group. *Morbidity and Mortality Weekly Report, Recommendations and Reports* 53(RR-5): 1–11.

CDC (Centers for Disease Control and Prevention). 2004. Syndromic surveillance: Reports from a national conference, 2003. *Morbidity and Mortality Weekly Report* 53(suppl.): 18–22.

CDC (Center for Disease Control and Prevention). 2005. *Third National Report on Human Exposure to Environmental Chemicals.* Atlanta, GA: National Center for Environmental Health.

Centeno, J.A., Gray, M.A., Mullick, F.G., Tchounwou, P.B. & Tseng, C.H. 2005. Arsenic in drinking water and health effects. In T.A. Moore, A. Black, J.A. Centeno, J. Harding & D. Trumm (eds.), *Metal Contaminants in New Zealand—Sources, Treatments, and Effects on Ecology and Human Health*: 415–439. Christchurch, NZ: Resolutionz Press.

Cringoli, G., Ippolito, A. & Taddei, R. 2005. Advances in satellite remote sensing in pheno-climatic features for epidemiological applications. *Parassitologia* 47(1): 51–62.

Cross, E.R., Newcomb, W.W. & Tucker, C.J. 1996. Use of weather data and remote sensing to predict geographic and seasonal distribution of *Phlebotomus paptasi* in southwest Asia. *Am. J. Trop. Med. Hyg.* 54: 530–536.

Dean, A.D. 2000. Computerizing public health surveillance systems. In S.M. Teutsch & R.E. Churchill (eds.), *Principles and Practices of Public Health Surveillance*: 229–252. New York: Oxford University Press.

Dearry, A. 2005. *Testimony before the Subcommittee on Oversight and Investigations of the Committee on Energy and Commerce, United States House of Representatives, 109th Congress, Second Session.* Office of Legislative Policy and Analysis.

Dister, S.W, Fish, D., Bros, S., Frank, D.H. & Wood, B.L. 1997. Landscape characterization of peridomestic risk for lyme disease using satellite imagery. *Am. J. Trop. Med. Hyg.* 57: 687–692.

Dowman, I. & Murai, S. 2008. *International cooperation and capacity building.* This volume.

Eben, K. 2007. Biosense or biononsense? *The Scientist* 21(4): 32–42.

Eisenberg, N.S., Bartram, J. & Hunter, P.R. 2001. A Public health perspective for establishing water-related guidelines and standards. In L. Fewtrell & J. Bartram (eds.), *Water Quality: Guidelines, Standards and Health*: 229–256. London: IWA Publishing.

Epstein, P.R. 1997. Climate, ecology, and human health. *Consequences: The Nature and Implications of Environmental Change* 3(2): 3–19.

Eubank, S. 2004. Modeling disease outbreaks in realistic urban social networks. *Nature* 429: 180–184.

Fairchild, A.L. & Bayer, R. 2004. Ethics and the conduct of human health surveillance. *Science* 303: 631–632.

Fauci, A.S., Touchette, N.A. & Folkers, G.K. 2005. Emerging infectious diseases: A 10-year perspective from the National Institute of Allergy and Infectious Diseases. *Emerg. Infect Dis.* 11(4): 519–525.

Gauderman, W.J., Avol, E., Gilliland, F., Vora, H., Thomas, D., Berhave, K., McConnell, R., Kuenzli, N., Lurmann, F., Rappaport, E., Margolis, H., Bates, D. & Peters, J. 2004. The effect of air pollution on lung development from 10–18 years of age. *N. Engl. J. Med.* 351(11): 1057–1067.

Gebre-Michael, T., Malone, J.B. & McNally, K. 2005. Use of geographic information systems in the development of prediction models for *Onchocerciasis* control in Ethiopia. *Parassitologia* 47(1): 135–144.

Gloster, J. & Alexandersen, S. 2004. New directions: Airborne transmission of foot-and-mouth disease virus. *Atmos. Environ.* 38: 503–505.

Grattan, J., Rabartin, R., Self, S. & Thordarson, T. 2005. Volcanic air pollution and mortality in France, 1783–1784. *Comtes Rendus Geoscience* 337: 641–651.

Greenland, S. 2004. Ecologic inference problems in the analysis of surveillance data. In R. Brookmeyer and D.F. Stroup (eds.), *Monitoring the Health of Populations*: 315–340. New York: Oxford University Press.

Griffin, D.W. 2004. Terrestrial microorganisms at an altitude of 20,000 m in Earth's atmosphere. *Aerobiologia* 20: 135–140.

Griffin, D.W. 2007. Atmospheric movement of microorganisms in clouds of desert dust and implications for human health. *Clinical Microbiology Reviews* 20(3): 459–477.

Griffin, D.W., Kubilay, N., Koçak, M., Gray, M.A., Borden, T.C. & Shinn, E.A. 2007. Airborne desert dust and aeromicrobiology over the Turkish Mediterranean coastline. *Atmos. Environ.* 41: 4050–4062.

Guptill, S.C. & Moore, C.G. 2005. Investigating vector-borne and zoonotic diseases with remote-sensing and GIS. In O. Sellinus, B.J. Alloway, J.A. Centeno, R.B. Finkelman, R. Fuge, U. Lindh & P. Smedley (eds.), *Essentials of Medical Geology*: 645–665. London: Elsevier Academic Press.

Gyan, K., Henry, W., Lacaille, S., Laloo, A., Lamesee-Eubanks, C., McKay, S., Antoine, R.M. & Monteil, M.A. 2005. African dust clouds are associated with increased pediatric asthma accident and emergency admissions on the Caribbean island of Trinidad. *Int. J. Biometerol.* 49: 371–376.

Hansell, A.L., Horwell, C.J. & Oppenheimer, C. 2006. The health hazards of volcanoes and geothermal areas. *Occup. Environ. Med.* 63: 149–156.

Hassan, A.N., Beck, L.R. & Dister, S. 1998. Prediction of villages at risk for *filariasis* transmission in the Nile Delta using remote sensing and geographic information technologies. *J. Egypt. Soc. Parasitol.* 28: 75–87, 119–131.

Heymann, D.L. & Rodier, G. 2004. Global surveillance, national surveillance, and SARS. *Emerg. Infect. Dis.* 10(2): 173–175.

Horwell, C.J. & Baxter, P.J. 2006. The respiratory health hazards of volcanic ash: A review for volcanic risk mitigation. *Bull. Volcanol.* 68: 1–24.

ICSU Scoping Group. 2007. *Towards a systems analysis approach to health and wellbeing in the changing urban environment: A report of a CSPR ad hoc scoping group on human health. Paris*: International Council for Science.

IPCC (Intergovernmental Panel on Climate Change). *Fourth Assessment Report (AR4).* In press.

Kaiser, J. 2005. Mounting evidence indicts fine particle pollution. *Science* 307(1717): 1858–1861.

Kaya, S., Sokol, J. & Pultz, T.J. 2004. Monitoring environmental indicators of vector-borne disease from space: A new opportunity for RADARSAT-2. *Can J. Rem. Sens.* 30(3): 560–565.

Kellogg, C.A., Griffin, D.W., Garrison, V.H., Peak, K.K., Royall, N., Smith, R.R. & Shinn, E.A. 2004. Characterization of aerosolized bacteria and fungi from desert dust events in Mali, West Africa. *Aerobiologia* 20: 99–110.

Kennel, C.F., Morel, P. & Williams, G.J. 1997. Consequences: The nature and implications of environmental change 3(2): 21–31.

Kuehn, B.M. 2006. Desertification called global health threat. *JAMA.* 295(21): 2463–2464.

Kuske, C.R. 2006. Current and emerging technologies for the study of bacteria in the outdoor air. *Curr. Opin. Biotechnol* 17: 291–296.

Lang, L. 2000. *GIS for health organizations.* Redlands, CA: ESRI Press.

Li, S. 2008. *Web mapping/GIS services and applications.* This volume.

Liverman, D., Moran, E.F., Rindfuss, R.R. & Stern, P.C. 1998. *People and Pixels: Linking Remote Sensing and Social Science*. Washington, D.C.: National Academy Press.

Lober, W.B., Trigg, L. & Karras, B. 2004. Information system architectures for syndromic surveillance. *Morbidity and Mortality Weekly Report* 53(Suppl.): 203–208.

Malone, J.B. Huh, O.K., Fehler, D.P., Wilson, P.A., Wilensky, D.E. & Holmes, R.A. 1994. Temperature data from satellite images and the distribution of *schistosomiasis* in Egypt. *Am. J. Trop Med. Hyg.* 50: 714–722.

Matthias, A.D. 2005. Monitoring near-surface air quality. In J.F. Artiola, I.L. Pepper & M. Brusseau (eds.), *Environmental Monitoring and Characterization*: 164–181. San Diego, CA: Elsevier Academic Press.

Matthias, A.D., Musil, S.A. & Bohn, H.L. 2006. Physical-chemical characteristics of the atmosphere. In J.F. Artiola, I.L. Pepper & M. Brusseau (eds.), *Environmental Monitoring and Characterization*: 46–57. San Diego, CA: Elsevier Academic Press.

McKensie, F.E. 2004. Smallpox models as policy tools. *Emer. Infect. Dis.* 10: 2044–2047.

Morain, S. & Sprigg, W. 2005. *PHAiRS initial benchmark report for public health. Cooperative Agreement NNS04AA19A.* U.S. National Aeronautics and Space Administration (NASA).

Morain, S.A. & Budge, A.M. 2006a. Science data products for public health decision support. *IGARSS, 2006, Denver Colorado, July 31–August 4.*

Morain, S.A. & Budge, A.M. 2006b. Integrating Earth observation data into geospatial databases that support public health decisions. In S. Nayak, S.K. Pathan & J.K. Garg (eds.), *Int. Arch. of Photogr. Rem. Sens. and Spat. Info. Sci.* 36(Part 4-B): 570–574.

Morain, S.A. & Sprigg, W.A. 2007. *PHAiRS verification and validation report, Cooperative Agreement No. NNSO4AA19A.* U.S. National Aeronautics and Space Administration (NASA).

Morens, D.M., Folkers, G.K. & Fauci, A.S. 2004. The challenge of emerging and re-emerging infectious diseases. *Nature* 430: 242–249.

Naumova, E.N. & MacNeill, I.B. 2005. A system for "signature" forecasting and early outbreak detection. *Environmetrics* 16: 749–766.

Naumova, E.N., Christodouleas, J., Hunter, P.R. & Syed, Q. 2005. Effect of precipitation on seasonal variability in *Cryptosporidiosis* recorded by the North West England Surveillance System in 1990–1999. *J Water Health* 3(2): 185–196.

Noji, E.K. 2005. Disasters: Introduction and state of the art. *Epidemio. Rev.* 27: 3–8.

NRC. 2006. *Beyond Mapping: Meeting National Needs Through Enhanced Geographic Information Science.* Washington, D.C.: The National Academies Press.

NRC. 2007. *Earth Materials and Health: Research Priorities for Earth Science and Public Health.* Washington, D.C.: The National Academies Press.

Nuckols, J.R., Ward, M.H. & Jarup, L. 2004. Using geographic information systems for exposure assessment in environmental epidemiology studies. *Environ. Health Perspec.* 112: 1007–1015.

Nsubuga, P., White, M.E., Thacker, S.B., Anderson, M.A., Blount, S.B., Broome, C.V., Chiller, T.M., Espitia, V., Imtiaz, R., Sosin, D., Stroup, D.F., Tauxe, R.V., Vijayaraghavan, M. & Trostle, M. 2007. Public health surveillance: A tool for targeting and monitoring interventions. In *World Bank Group, Disease Control Priorities in Developing Countries (2nd Edition): Chapter 53 Disease Control Priorities Project.* http://www.dcp2.org/pubs/DCP, last visited October 2007.

OSTP. 2007. *A Plan for A U.S. National Land Imaging Program. Future of Land Imaging Interagency Working Group.* Washington, D.C.: The National Science and Technology Council, Executive Office of the President of the United States.

Park, B.J., Sigel, K., Vaz, V., Komatsu, K., McRill, C., Phelan, M., Colman, T., Comprie, A.C., Comrie, D.W., Warnock, D.W., Galgiani, J.N. &. Hajjeh, R.A. 2005. An epidemic of *coccidioidomycosis* in Arizona associated with climatic changes. *J. Infect. Dis.* 191: 1981–1987.

Park, J.W., Lim, Y.H., Kyung, S.Y., An, C.H., Lee, S.P., Jeong, S.H. & Yu, Y.S. 2005. Effects of ambient particulate matter on peak expiratory flow rates and respiratory symptoms of asthmatics during Asian dust periods in Korea. *Respirology* 10: 470–476.

Pascual, M. Ahumada, J.A., Chaves, L.F., Rodó, X. & Bouma, M. 2006. Malaria resurgence in the East African highlands: Temperature trends revisited. *Proc. of the National Academy of Sciences, USA* 103: 5829–5834.

Pope, K.O., Rejmánková, E., Savage, H.M., Arredondo-Jimenez, J.M., Rodríguez, M.H. & Roberts, D. 1993. Remote sensing of tropical wetlands for malaria control in Chiapas, Mexico. *Ecol. Apps.* 4: 81–90.

Pope, C.A. III. 2004. Air pollution and health. *N. Engl. J. Med.* 351(11): 1132–1133.

Poulstrup, A. & Hansen, H.L. 2004. Use of GIS and exposure modeling as tools in a study of cancer incidence in a population exposed to airborne dioxin. *Environ. Health Perspec.* 112: 1032–1036.

Reid, J.S. Prins, E.M., Westphal, D.L., Schmidt, C.C., Richardson, K.A., Christopher, S.A., Eck, T.F., Reid, E.A., Curtis, C.A. & Hoffman, J.P. 2004. Real-time monitoring of South American smoke particle emissions and transport using a coupled remote sensing/box-model approach. *Geophys. Res. Lett.* doi: 10.1029/2203GL018845.

Roberts, D.R., Paris, J.F., Manguin, S., Harbach, R.E., Woodruff, R. & Rejmánková, R. 1996. Predictions of malaria vector distribution in Belize based on multispectral satellite data. *Am. J. Trop. Med. Hgy.* 54: 304–308.

Romaguera, R., German, R.R., and Klaucke, D.N. 2000. Evaluating public health surveillance. In S.M. Teutsch & R.E. Churchill (eds.), *Principles and Practices of Public Health Surveillance*: 176–193. New York: Oxford University Press.

Schlesinger, P., Mamane, Y. & Grishkan, I. 2006. Transport of microorganisms to Israel during Saharan dust events. *Aerobiologia* 22: 259–273.

Schmidt, C.W. 2005. Global Earth observations for health. *Environmental Health Perspectives* 113(11): 738–9.

Selinus, O., Alloway, B., Centeno, J.A., Finkelman, R.B., Fuge, R., Lindh, U. & Smedley, P. 2005. *Essentials of Medical Geology: Impacts of the Natural Environment on Public Health.* Amsterdam: Elsevier Academic Press.

Stetzenbach, L.D., Buttner, M.P. & Cruz, P. 2004. Detection and enumeration of airborne biocontaminants. *Curr. Opin. Biotechnol.* 15: 170–174.

Sulaiman, I.M., Hira, P.R., Zhou, L., Al-Ali, F.M., Al-Shelahi, F.A, Shweiki, H.M., Iqbal, J., Khalid, N. & Xiao, L. 2005. Unique endemicity of *Cryptosporidiosis* in children in Kuwait. *J. Clin. Microbiol.* 43(6): 2805–2809.

Sultan, B., Labadi, K., Guegan, J.F. & Janicot, S. 2005. Climate drives the meningitis epidemics onset in West Africa. *PLoS Med.* 2: e6.

Swaminathat, B., Barrett, T.J., Hunter, S.B. and Tauxe, R.V. 2001. PulseNet: The molecular subtyping network for foodborne bacterial disease surveillance, United States. *Emerg. Infect. Dis.* 7(3): 382–389.

Thompson, D.F., Malone, J.B., Harb, M., Faris, R., Huh, O.K. & Buck, A.A. 1996a. *Emerg. Infect. Dis.* 2: 234–235.

Thompson, M.C., Connor, S.J., Milligan, P.J.M. & Flasse, S.P. 1996b. The ecology of malaria as seen from Earth-observation satellites. *Ann. Trop. Med. Parasitol.*90: 243–264.

Tinkle, S., Gant, M., Humble, M., Foley, G., Garcia, V. & Bond, A. 2007. *Integrated Earth observations: Application to air quality and human health.* Washington D.C.: United States Environmental Protection Agency, Office of Research and Development. EPA/600/R-007.

Tobias, H.J., Schafer, M.P. Pitesky, M., Fergenson, D.P., Horn, J., Frank, M. & Gard, E.E. 2005. Bioaerosol mass spectrometry for rapid detection of individual airborne Mycobacterium tuberculosis H37Ra particles. *Appl. Environ. Microbiol.* 71: 6086–6095.

UN. 2004. *United Nations Department of Economic and Social Affairs, Division of Sustainable Development. Plan of Implementation of the World Summit on Sustainable Development, Johannesburg.* (see especially pages 31–33). http://www.un.org/esa/sustdev/documents/WSSD_POI_PD/last visited July 25, 2007.

Varmus, H., Klausner, R., Zerhouni, E., Acharya, T., Daar, A.S. & Singer, P.A. 2003. Grand challenges in global health. *Science* 302: 398–399.

Ver Ploeg, M. & Perrin, E. (eds.). 2004. *Eliminating Health Disparities: Measurement and Data Needs.* Washington, D.C.: The National Academies Press.

Vineis, P. 2004. A self-fulfilling prophecy: Are we underestimating the role of the environment in gene-environment interaction Research? *Int. J. Epidemiol.* 33 945–946.

Wu, P.C., Tsai, J.C., Li, F.C., Lung, S.C. & Su, H.J. 2004. Increased levels of ambient fungal spores in Taiwan are associated with dust events from China. *Atmos. Environ.* 38: 4879–4886.

World Bank. 2005. *Public health surveillance toolkit.* http://survtoolkit.org, last visited, July, 2007.

World Bank Group. *On-going dates. Disease Control Priorities in Developing Countries (2nd ed.). Disease Control Priorities Project.* http://www.dcp2.org/pubs/DCP. last visited October 23, 2007.

WMO (World Meteorological Organization). 2005. *THORPEX: A World Weather Research Programme.* WMO-No. 978.

Yin, D., Nickovic, S., Barbaris, B., Chandy, B. & Sprigg, W.A. 2005. Modeling wind-blown desert dust in the southwestern United States for public health warning: A case study. *Atmos. Environ.* 39(33): 6243–6254.

Yin, D., Nickovic, S. & Sprigg, W.A. 2007. The impact of using different land cover data on wind-blown desert dust modeling: Results in the southwestern United States. *Atmos. Environ.* 41(10): 2214–2224.

Yu, I.T.S., Li, Y., Wong, T.W., Tam, W., Phil, M., Chan, A.T., Lee, J.H.W., Leung, D.Y.C. & Ho, T. 2004. Evidence of airborne transmission of the Severe Acute Respiratory Syndrom virus. *N. Engl. J. Med.* 350: 1731–1739.

Zanobetti, A. & Schwartz, J. 2005. The effect of particulate air pollution on emergency admissions for myocardial infarction: A multicity case-crossover analysis. *Environ. Health Perspec.* 113: 978–982.

Zelicoff, A.P. & Bellomo, M. 2005. *Microbe: Are We Ready for the Next Plague?* New York: AMACOM Books.

Advances in Photogrammetry, Remote Sensing and Spatial Information Sciences: 2008 ISPRS Congress Book – Li, Chen & Baltsavias (eds)

CHAPTER 30

Industrial applications of photogrammetry

Thomas Luhmann & Stuart Robson

ABSTRACT: Photogrammetric methods and systems for three-dimensional industrial metrology have gained increasing importance over the past few years. Changes within industry from largely mechanical to digital reference systems have been paralleled by continued developments in digital photogrammetric systems. Key enhancements are in routinely achieved measuring accuracy, the degree of automation and ease-of-use. Practical systems can be categorized as mobile, on-line or off-line. Systems with stationary and process-integrated measuring systems have also become available. In most cases, applications focus on the measurement of discrete 3-D point coordinate data, often supported by appropriate point targets. Surface measurement systems that use fringe projection approaches or area-based pattern matching are increasingly used. Successful applications of industrial photogrammetry are very diverse and include: highly accurate quality control in the aerospace industries; integrated systems in manufacturing; and the recording and documentation of industrial facilities. Standardized recommendations for acceptance and re-verification of optical 3-D measuring systems have recently become available, providing an accepted basis for objective and traceable accuracy figures.

Keywords: Engineering, industrial, metrology, close range, accuracy, precision, reliability.

30.1 INTRODUCTION

30.1.1 *Historical review*

With the exception of a few earlier tests and studies (Fig. 30.1), the industrial use of close range photogrammetry can be regarded as commencing at the beginning of the 1980s. Initial developments focused on the orientation and calibration of arbitrary (convergent) image configurations by bundle adjustment. In contrast to aerotriangulation in aerial photogrammetry, bundle adjustment programs for close range applications require the integration of additional (geodetic) observations, different approaches for self-calibration of a variety of cameras, use of arbitrarily oriented object coordinate systems and, last but not least, automated procedures for the generation of initial values in complex image blocks. Basic methods and the first operational program systems were developed during this period.

Available systems utilized purpose built analogue (photographic) imaging systems and image measuring devices capable of high accuracy image recording and automated point measurement. Systems developed by Geodetic Services (GSI) provide exemplary examples of the peak of the available technologies. The combination of large format metric cameras (CRC-1, CRC-2, Figure 30.2a) with a high-precision digital comparator (AutoSet-1) and powerful software packages provided a complete system for measurement in the aerospace and antennae industries (Fraser & Brown 1986). With statistical precision of up to 1:500,000, and the capability to measure many hundreds of target locations, this technology was able to compete with theodolite measuring systems that were, at that time, the reference for high precision industrial metrology of large objects (Staiger 1992).

Based on the work of Wester-Ebbinghaus, Rollei brought the photogrammetric reseau camera Rolleiflex SLX, followed by Rolleiflex 6006/6008, into the market. This camera showed significant improvements compared to conventional metric cameras, in terms of flexibility, ergonomic handling and price-to-performance ratio. The idea of the reseau, used for the definition and numerical reconstruction of the image plane, allowed the use of standard photo cameras for metrological applications. The instrumental summit of the reseau technology was achieved with Rollei's large format camera LFC and the precision digital comparator RS1 (Reseau-Scanner, Fig. 30.2b) (Luhmann & Wester-Ebbinghaus 1988, Dold 1997).

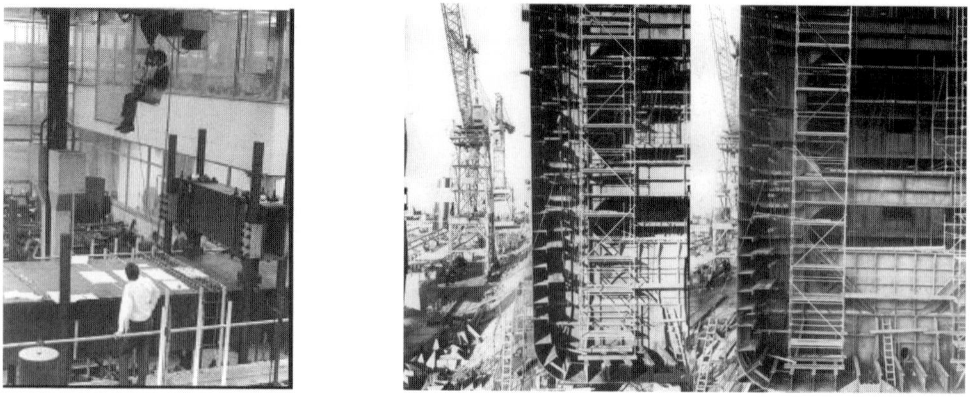

Figure 30.1. Left: Out of plane monitoring of a box girder bridge (Harding 1974). Right: Stereo pair for dimensional quality control of the mid ship section of a super tanker (Newton 1974).

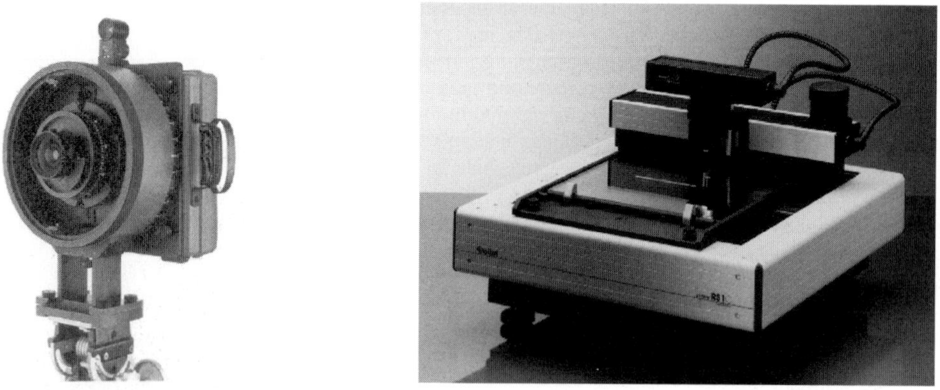

a) GSI CRC-1.

b) Rollei Reseau scanner RS1-C.

Figure 30.2. Analogue large format camera and precision scanner.

The development of digital image processing methods pushed the degree of automation and accuracy of point measurement and matching further ahead. Analysis of retro-target image quality (e.g. Robson & Shortis 1998) and methods for precise detection and measurement of cross-shaped and circular (elliptical) targets have been developed (e.g. Dold 1996, Shortis et al 1995). Multi-image self calibration procedures were also refined and optimized for the increasing volumes of images and target measurements that could be collected and processed (e.g. Fraser 1984, Fraser & Shortis 1990). Approaches for image-based and object-based stereo and multi-image matching founded on least squares solutions were also investigated during that time (e.g. Gruen 1985, Heipke 1991, Schneider 1991). With the Zeiss InduSurf system, the first commercial photogrammetric surface measuring system was established in an industrial application (Fig. 30.3).

Around 1985 the available opto-electronic imaging technology was strongly limited in terms of resolving power and image quality. As a consequence, intermediate imaging solutions utilizing high-resolution scanning cameras were developed. Examples are the micro-scanning cameras (ProgRes 3012, 4608 × 3480 pixels), reseau scanning systems (Rollei RSC, 4500 × 4500 pixels) and macro-scanning (Zeiss UMK HighScan, 15,000 × 11,000 pixels). Due to the required stationary image acquisition, long image acquisition times and poor price-performance-ratio, these instruments were not able to gain a significant market presence.

The key breakthrough was achieved with the invention of so-called still-video cameras. The Kodak

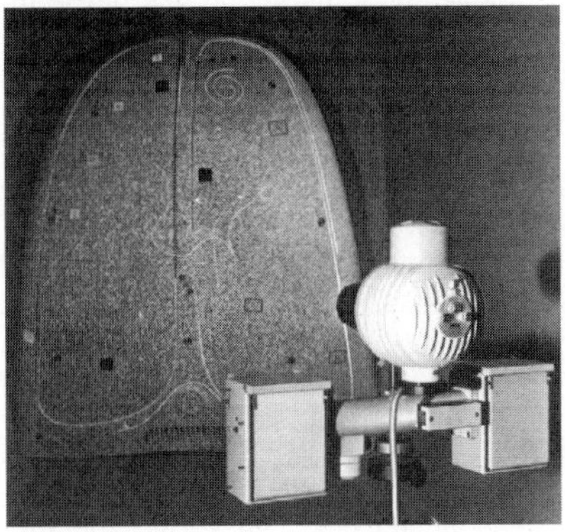

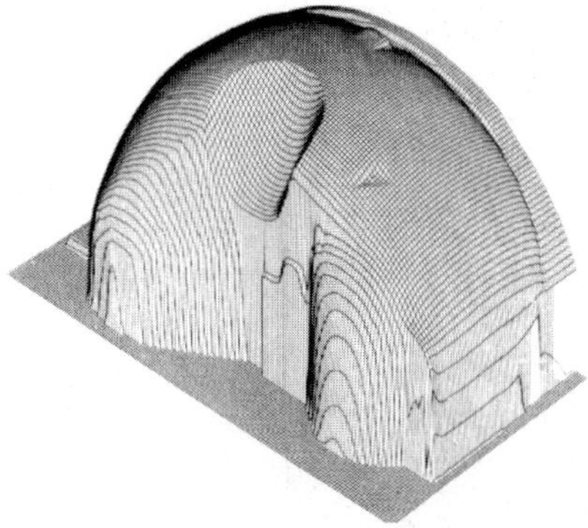

a) Stereo camera and projector b) 3-D model of a wheel housing

Figure 30.3. Surface measuring system Zeiss InduSurf.

DCS 420 SLR camera offered a pixel resolution of 1500 × 1000 pixels and integrated data storage and battery. This camera initiated the increased use of digital photogrammetric off-line systems that allowed for fast hand-held image acquisition and direct data processing on (mobile) PCs. Devices of this type were able to fulfil the demand of fast, automated and accurate photogrammetric point measurement whilst maintaining a user friendly interface to the collection of images.

30.1.2 *State of the art*

In 1995, the Kodak DCS 460/660 digital SLR camera (3000 × 2000 pixel with 9 μm × 9 μm detector size, Fig. 30.4a) became available and was found to be extremely suited to industrial metrology. Using circular retro-reflective targets, image measurement uncertainties of better than 1/20 of a pixel (about 0.5 μm) could be achieved under factory conditions. Given an appropriate image geometry (see Section 30.3), relative accuracies of 1:50,000 to 1:100,000 were routinely reported. It is interesting to note that digital cameras with more than 20 million pixels (Fig. 30.4b) are now routinely available.

For faster or dynamic processes, digital video and high-speed camera systems are available. Controlled through a fast computer interface (e.g. CameraLink, Giga Ethernet), sensors with more than 1500 × 1000 pixels and frame rates of 1000 Hz are commercially available, conferring the opportunity to carry out dynamic photogrammetry of high speed events.

In photogrammetry, the term "metric camera" is used to signify imaging systems that are calibrated by the manufacturer and designed to provide constant

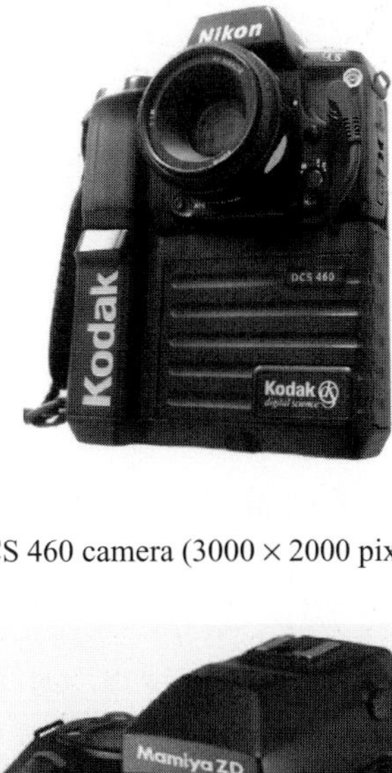

a) Kodak DCS 460 camera (3000 × 2000 pixels).

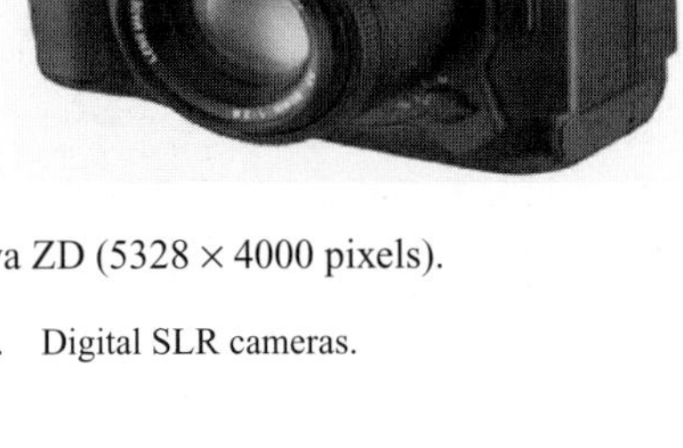

b) Mamiya ZD (5328 × 4000 pixels).

Figure 30.4. Digital SLR cameras.

camera geometry over a long period of time. Today, only a few systems exist in this category, for example the GSI INCA 3 (Fig. 30.5a) and the AXIOS 3-D CamBar stereo system (Fig. 30.5b). Other systems are usually calibrated prior, during or after object measurement by self calibrating bundle adjustment. In such cases, camera parameters can be derived either at the actual moment of object measurement or under very similar circumstances.

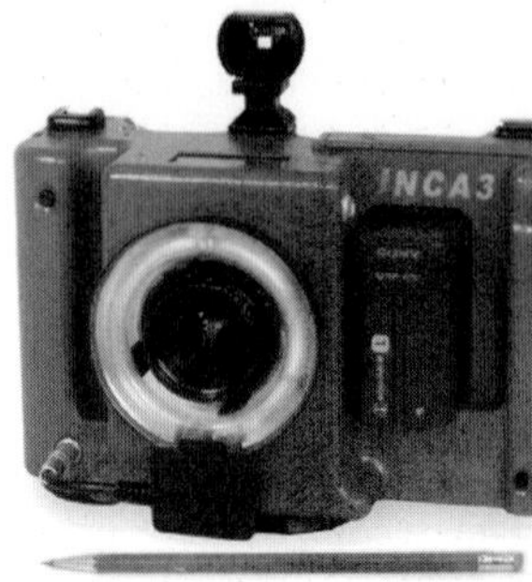

a) GIS INCA 3 photogrammetric metric camera.

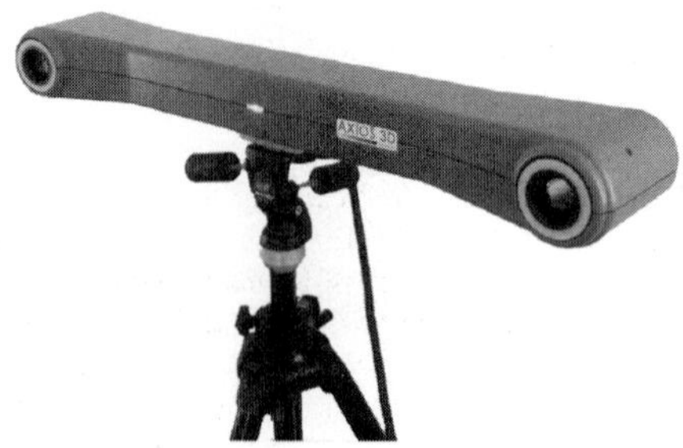

b) AXIOS 3-D CamBar stereo camera system.

Figure 30.5. Metric photogrammetric cameras.

Usually, the established parameters of interior orientation (c, x'_0, y'_0, A_1, A_2, A_3, B_1, B_2, C_1, C_2 or K_1, K_2, K_3) are determined. According to the individual design and characteristics of the camera, sensor and lens in use, these parameters can be estimated with different significance. Changes in camera geometry during an image sequence can be modelled by image-variant calibration. Unflatness of the image sensor can be described by a finite element correction grid (Tecklenburg et al. 2001). The calibration of zoom lenses with changing focal lengths is possible if the accuracy requirements are not high (Cronk et al. 2006). Finally, chromatic imaging errors can be modelled and compensated (Luhmann et al. 2006).

Based on the above mentioned camera devices diverse multi-image processing systems are available allowing for interactive image measurements, often with interfaces to CAD or metrology software systems. Examples include AICON 3-D Studio, Phocad and Phidias. These solutions are oftefn classified as off-line photogrammetric systems since the first stage of the process is a sequential image acquisition, which is followed by a subsequent step of image processing and object reconstruction. However, developments of so-called intelligent cameras with integrated processor and image measurement are creating an intermediate stage on the way to on-line systems. Examples include the GSI INCA and AICON WheelWatch systems.

If at least two synchronized cameras observe the same object, assuming given parameters for interior and exterior orientation, the direct determination of 3-D information is possible. Based on this principle, two-camera and multiple-camera systems are available that use a tactile probe with calibrated targets

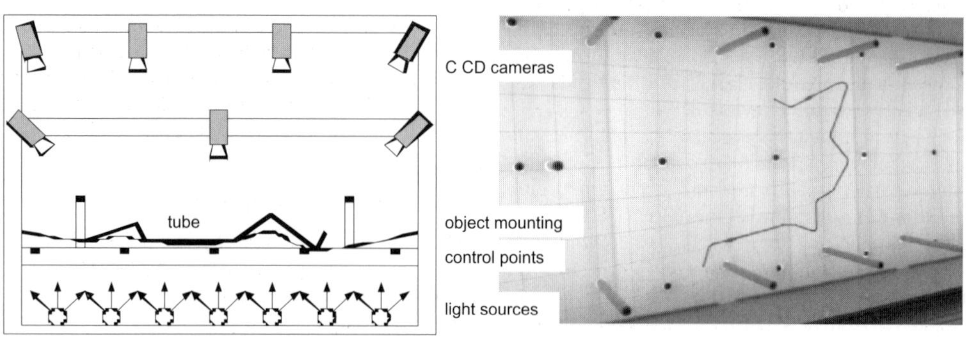

Figure 30.6. Multi-image on-line system for tube measurements (AICON TubeInspect).

that can be imaged and measured in real-time by the cameras (e.g. Metronor, Fig. 30.11a). The principle can even be adapted to the use of a single camera if the orientation of the probe is calculated by space resection (Luhmann 2005). The camera can also be integrated into the probe itself, observing an external field of reference points (Fig. 30.11b).

Multi-camera systems can be configured for special applications, for example the integrated process control or robot control. The measurement of tubes (Bösemann 1996) (Fig. 30.6) provides an interesting example of an automated system that optimizes illumination and camera placement in order to measure a wide range of different engineered tubes without targets.

Additional overviews for mathematical and technical backgrounds and references to the development of close range photogrammetry are given. The reader is referred to Luhmann et al. (2006), Luhmann (2003), Dold (1997) and Atkinson (1996). Examples of recent applications are given by way of example throughout this chapter, but the reader is directed to Luhmann et al. (2006) and Fryer et al. (2007) for a more comprehensive set of examples.

30.2 PHOTOGRAMMETRIC SYSTEMS FOR POINT MEASUREMENT

30.2.1 *Off-line photogrammetry*

In most industrial applications, the 3-D measurement of object points is performed by artificial targets or by tactile probing. For targeting, white circular markers are most successful, utilized either as retro-reflective material, self-illuminating devices or simply printed onto paper or folio. If these targets are combined with a computer readable coded pattern, not only can the target centre be measured to sub-pixel accuracy but the point identification (point number) of the target can also be decoded automatically. An example of such coded targets can be seen in Figs. 30.8 and 11b. In order to relate target locations to key features on the object of interest, targeted points or target equipped mechanical adapters have to be designed and configured. Fig. 30.7 demonstrates an example of a set of adapters designed for use in the aerospace industry for recording the shape of a wing root.

A typical off-line system will consist of a high resolution digital camera with integrated data storage, usually provided in the form of memory cards or mini hard disks capable of recording several hundred images, which may be required for more complicated networks. Data processing is carried out on a (mobile) computer with a suitable data interface (e.g. FireWire, USB). The photogrammetric image processing and 3-D point coordination process is typically carried out according to the following sequence:

Image point measurement
All images are searched for dot-shaped patterns that serve as candidates for real targets. If coded targets are detected, measured image points are stored with their corresponding point number. Otherwise image coordinate measurements for each identified target image point are stored with sequential point numbers.

Calculation of approximate values
Target image points that are identified by their target code provide known correspondences over all images. If at least five homologous points are detected for an image pair, a local 3-D model coordinate system can be created by relative orientation. Using epipolar geometry, all other (non-coded) points can be matched. Further images or models can be added to this initial model until all images are provisionally oriented. If points are identified through the target coding process as control points, the local model system can be transformed into a global object coordinate system. At the end of this procedure all images will be approximately oriented, and all target image points appearing on two or more images will have approximate 3-D coordinates. Alternative strategies for approximate values are explained in Luhmann et al. (2006).

Bundle adjustment with self-calibration
A subsequent bundle adjustment provides all exterior orientations, adjusted 3-D object coordinates and camera parameters (Granshaw 1980). Hence, the camera model (interior orientation) can be calculated exactly for the time of image acquisition. Given a strong network configuration, the detection and elimination of gross errors (outliers) will be undertaken within the bundle adjustment process. Such outliers might be caused by poor quality target images or faulty identification and point matching.

Output of results
The calculated object points can be processed in many different ways. Typically, these are used for machining (Fig. 30.8), integrated into CAD environments (Fig. 30.9) or programs for deformation analysis (Fig. 30.10) or comparison with nominal values.

30.2.2 *On-line photogrammetry*

On-line systems are based on simultaneous synchronized multi-image acquisition of the object space. Photogrammetric coordinate measuring systems using tactile probing are widely used (Fig 30.11a). Usually a hand-held probing device is used to measure the object. It consists of a probing tip and a minimum

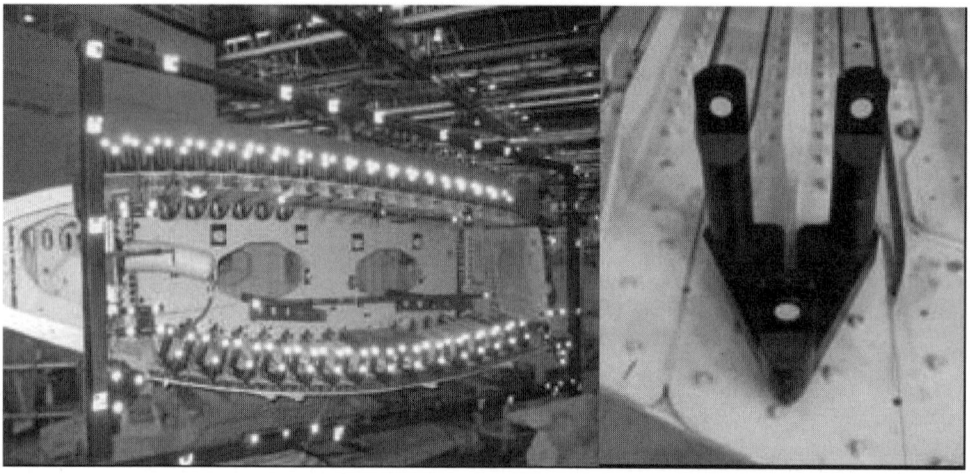

Figure 30.7. Left: Airbus wing root fitted with targets for photogrammetric measurement. Right: detailed view showing target design. (Courtesy of Airbus UK). (see colour plate page 513)

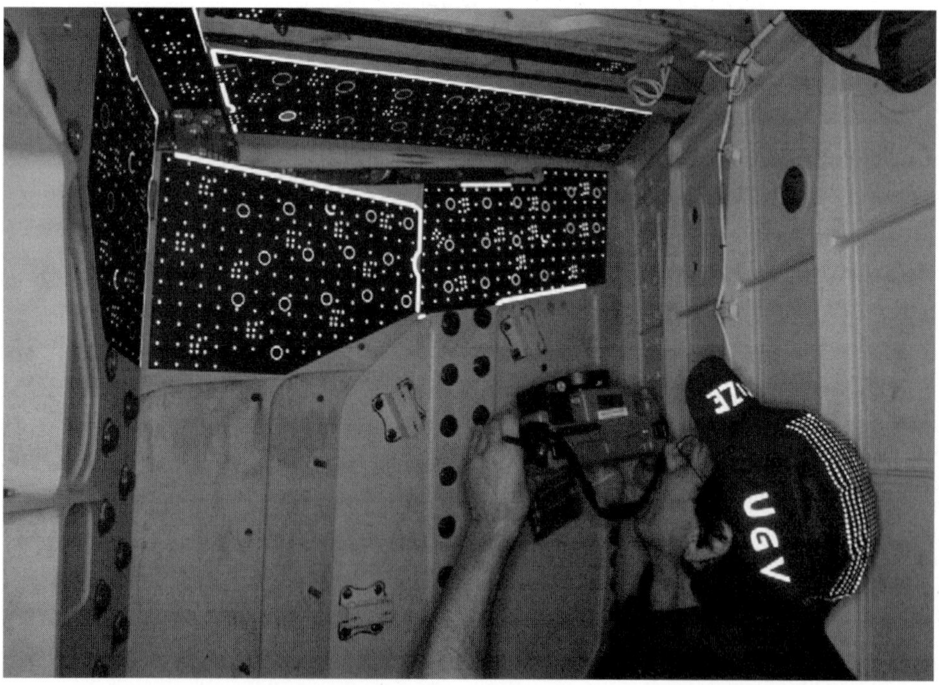

Figure 30.8. Photogrammetric measurement of the junction of the wing and fuselage in an Airbus A380 aircraft using a GSI V-STARS system. The photogrammetric system triggers an image processing function to automatically generate a 3-D surface and a 5 axis machining tool path. This process produces a titanium part that will fit in the corners. (Copyright: Airbus). (see colour plate page 513)

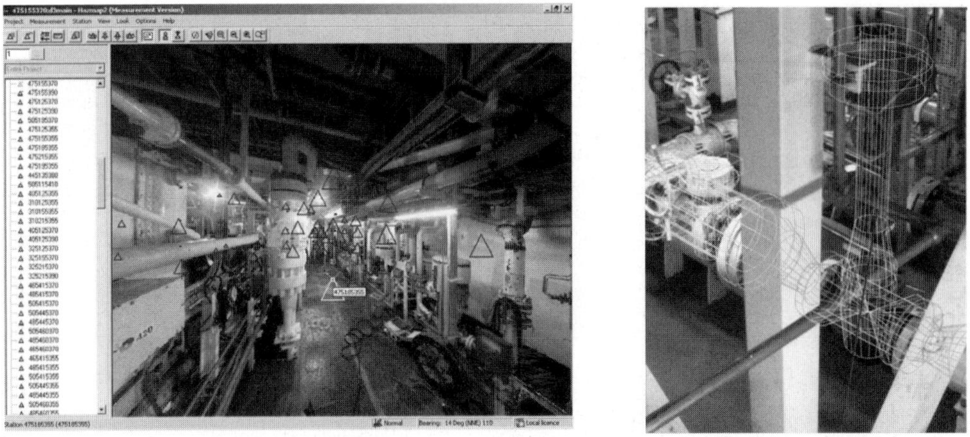

Figure 30.9. Left: Example sub-image from the panoramic "Hazmap" site measurement system showing further panoramic image locations as triangles. Right: Example integration with CAD where a new pipe design has been integrated into the Hazmap environment for clash detection. (Images Courtesy of Hazmap). (see colour plate page 514)

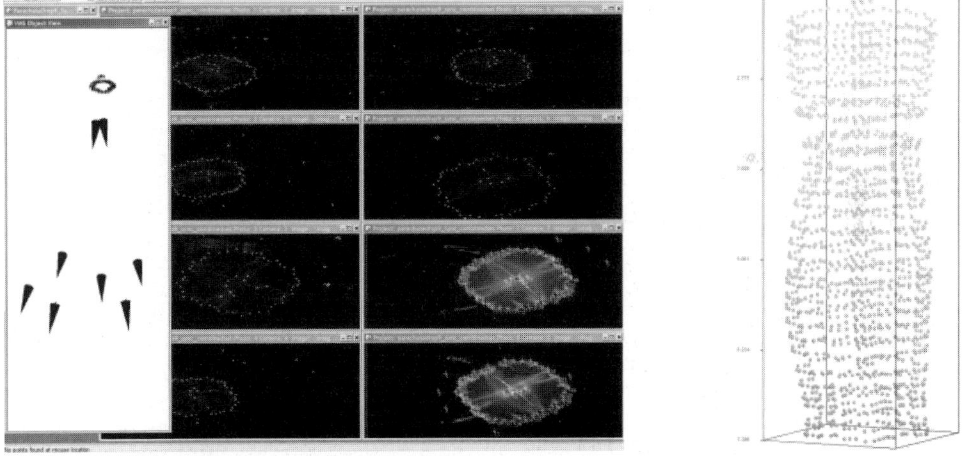

Figure 30.10. Example images taken with a network of eight cameras used to simultaneously capture image sequences of a parachute in flight (Left). Off-line post processed output demonstrating the movement of the parachute. As processing capability and more notably tracking algorithms improve, such systems will become on-line. Images Courtesy of NASA Langley (Right) (Shortis et al. 2007). (see colour plate page 514)

of three reference targets that can be observed by the cameras. The cameras are exposed by a switch located on the probe. Spatial intersection is used to determine the 3-D coordinates of the reference points. Consequently, the position of the calibrated probing tip can be calculated.

The typical measuring process can be described as follows:

Pre-calibration

Due to a limited imaging configuration, the simultaneous calibration of the cameras is not possible. Hence,

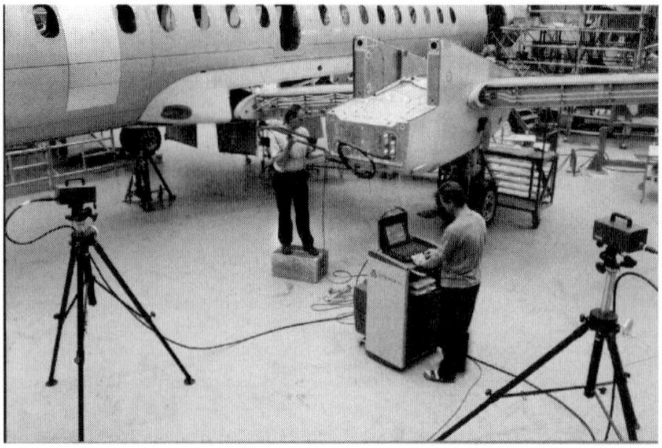

a) Example of a Metronor system being used to measure an aircraft enginemounting.

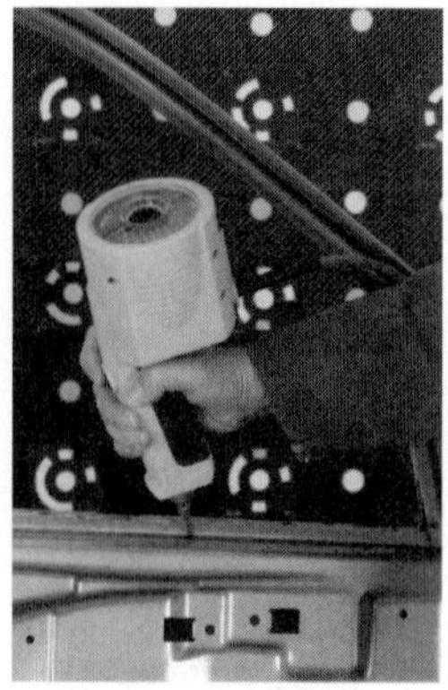

b) An AICON ProCam system being used to coordiate 3-D locations on a car door.

Figure 30.11. Tactile on-line metrology systems.

cameras are usually calibrated prior to measurement, e.g. by a separate test field. Therefore the parameters of interior orientation are assumed to be constant for the subsequent measurements.

Camera orientation
After setting up the (mobile) cameras, a relative orientation is determined by means of a field of targeted points. This step provides a local 3-D coordinate system.

Tactile measurement
Similar to a tactile coordinate measuring machine (CMM), the object is measured point by point in this case using a targeted touch probe.

Post-calibration
Some system suppliers allow the permanent orientation of the cameras using a target field (see step 2) in order to compensate for possible changes of the camera positions and orientations.

Output of results
A tactile on-line system can be regarded as an optical CMM. Measuring results can be transferred directly into another processing system for analysis or directly into the manufacturing process.

The procedure described above can be simplified significantly when the cameras are mounted in a common housing in a stable mechanical device, so that the relative orientation is constant. Such a design is well proven from older metric stereo cameras. Such systems are commonly used in medical environments (navigation for computer assisted surgery), but increasingly in industrial applications as well (e.g. NDI Optotrak, AXIOS 3-D CamBar).

The on-line measuring principle can be extended by integrating the camera and probe into a single unit. The AICON ProCam system (Fig. 30.11b) provides a touch probe solution whereby a closely mounted panel equipped with calibrated coded targets is used to automatically determine the exterior camera orientation. This automatic orientation allows continuous computation of the position of the calibrated probing tip in 3-D space. Details on optically determining the six degrees of freedom of a mechanical system in space are given in Loser & Kyle (2003).

30.3 PHOTOGRAMMETRIC SYSTEMS FOR SURFACE MEASUREMENT

For the measurement of free form industrial surfaces, the use of targets is often not possible or permitted. In practice, fringe projection systems are mostly used for this task. They use a sequential projection of shifted periodic fringe patterns in different wavelengths that are observed by one or more cameras. Height information can be extracted from a phase measurement of the recorded fringe pattern for each single pixel within a few seconds (Breuckmann 1993). Commercial industrial systems are available from a wide range of manufacturers including GOM (Fig. 30.12), Steinbichler, Breuckmann and Minolta.

Larger objects or more complex surfaces can be measured by fringe projection only by combining several partial measurements. Usually the measuring system is oriented in space by means of targets that are previously measured photogrammetrically (Winter & Reich 1997). Alternatively, the sensor device can

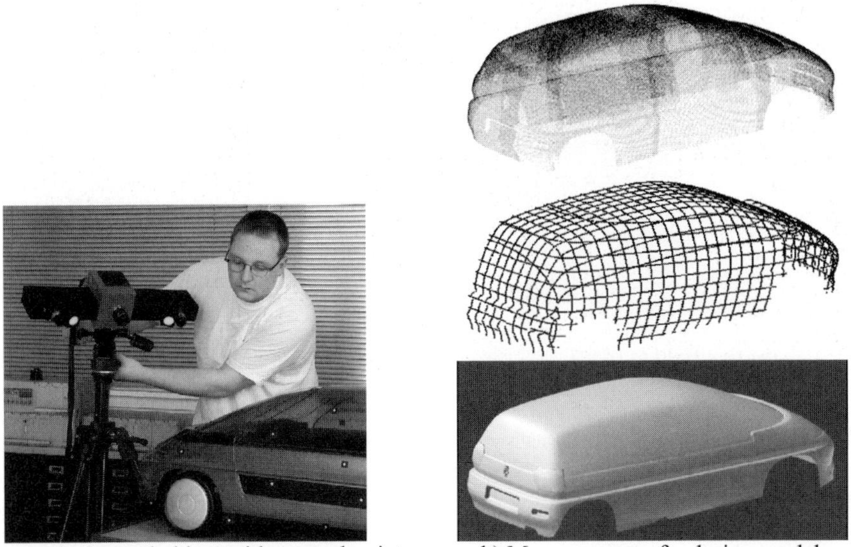

a) Sensor device and object with control points. b) Measurement of a design model.

Figure 30.12. Photogrammetric fringe projection system (GOM ATOS).

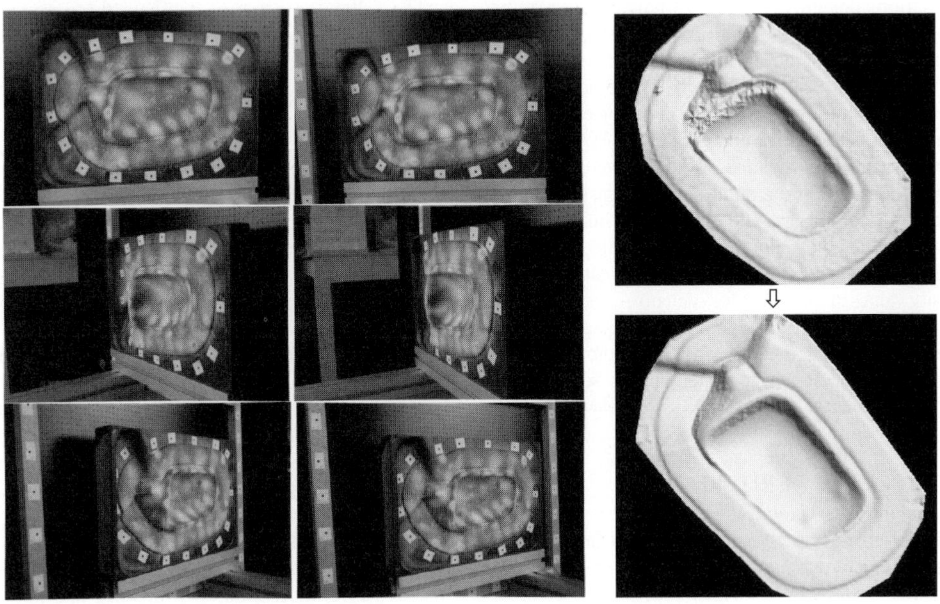

a) Multiple image acquisition. b) Iterative calculation of the 3D model.

Figure 30.13. Photogrammetric surface measurement. (see colour plate page 515)

be navigated by another measuring system, such as a laser tracker, a single camera, a stereo system or mechanical device (CMM, robot). If all sensor positions have their corresponding absolute orientation parameters, the single 3-D point clouds can be transformed into one global coordinate system.

Surface measurement by photogrammetric image matching is based either on stereo or multi-image

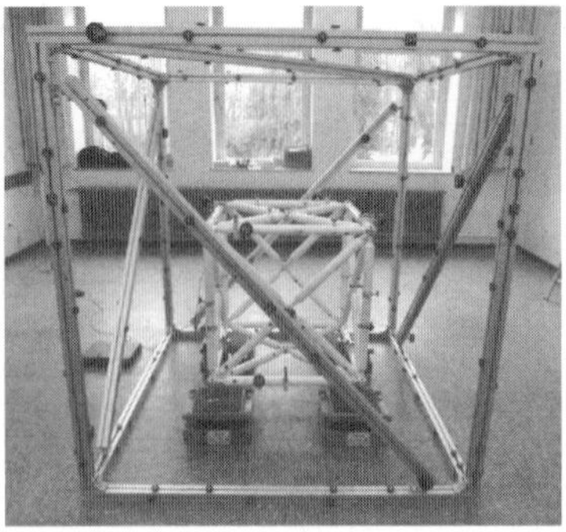

(a) Test field at IAPG.

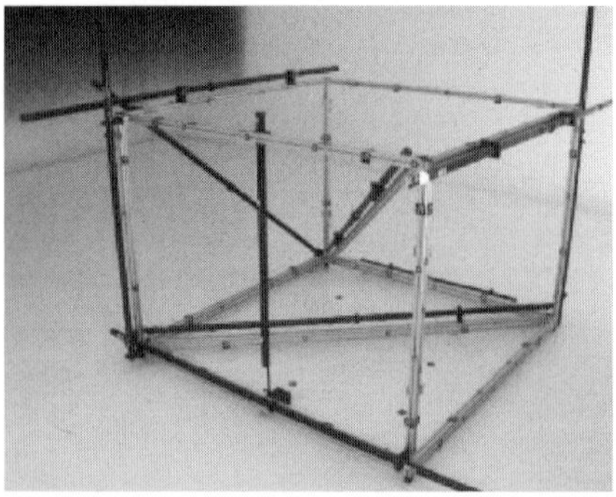

(b) Test field with scale bars according to VDI 2634 (AICON).

Figure 30.14. Examples of photogrammetric length reference bodies.

matching. In both cases the surface must provide a sufficient texture, either by its natural surface structure or by projection of an artificial pattern. Corresponding image patches are detected by conventional correlation or least squares matching under the geometric conditions of epipolar lines. Digital stereo workstations designed for aerial imagery can also be used for close range applications when object coordinate systems and camera parameters can be handled.

A more sophisticated approach can be achieved by object-based multi-image matching. The object is imaged by a multi-image set (example in Fig. 30.13). The object is reconstructed by simultaneous iterative calculation of a geometric model and a radiometric (intensity) model. If the complete surface and the optical imaging path are known, including illumination and reflectance, this (ideal) object model can be used to calculate artificial intensity values in the image for each pixel. Using the differences between calculated and real image grey values as observations, both the geometric and radiometric surface model are iteratively corrected by least squares estimation (Wrobel 1987, Heipke 1991, Schneider 1991).

30.4 ACCURACY CRITERIA

The achievable accuracy is a key criteria for usability and acceptance of a photogrammetric or optical 3-D measuring system in industrial practice. In contrast to other system characteristics like technical design, measuring times, user friendliness, flexibility etc, measuring accuracy provides a single quantitative parameter for the evaluation of the complete system.

Conventional photogrammetric or geodetic assessment of accuracy is often based on inner numerical or statistical parameters such as standard deviation of unit weight or RMS values of estimated point coordinates. These terms characterize the statistical precision of how the mathematical model describes the modelling between observations and estimated unknowns. This approach is useful for the evaluation of internal precision and reliability; however, the achieved measuring uncertainty in object space with respect to reference values of higher accuracy cannot be derived.

The following equation can be used to provide an *a priori* estimation of the resulting point accuracy in photogrammetry:

$$s_{XYZ} = q \cdot m \cdot s_{x'y'}$$

The mean error of image measurement $s_{x'y'}$ is transferred by image scale number m into object measuring error S_{XYZ}. The design factor q includes the impact of image configuration (number, position and orientation of images) and other, not exactly defined influences, on the final result. Multi-image sets with sufficient configurations (minimum of 4 to 6 convergent images per point) allow for a design factor between 0.4 and 0.8, valid for all three coordinate axes. For object points with fewer or weak image rays, or for conventional stereo set-ups, the design factor can easily reach values of 2 and more, e.g. expressed by the base-to-height ratio in stereo cases.

Verification of the real measuring uncertainty in object space is only possible with calibrated and certified reference data. In principle, this could be provided by measured reference points with given 3-D coordinates that are used as independent control points. However, it is often not possible to provide such points with reliable and high-accurate coordinates under economical side conditions.

A more practical solution is given by calibrated scale bars that can be used to determine ISO conforming length measuring errors (examples in Fig. 30.14). Since 2000, the German guideline VDI/VDE 2634 "Optical 3-D Measuring Systems" (VDI 2000) offers

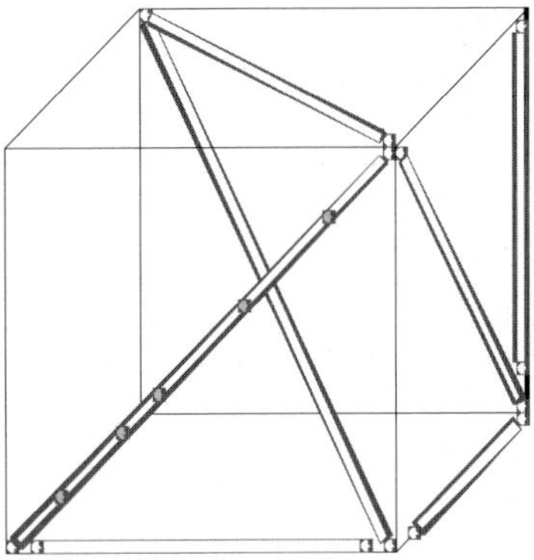

Figure 30.15. Configuration of measuring line in object space.

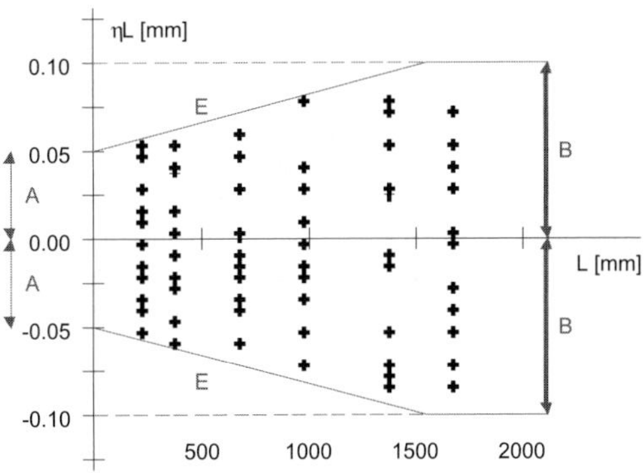

Figure 30.16. Diagram of length measuring error.

recommendations for the determination of suitable parameters for point-wise and area-based measuring systems. The recommended procedures also guarantee traceability to the metre length standard.

For point-wise probing optical systems (e.g. off-line photogrammetry systems), a diagram of length measuring deviations between calibrated nominal lengths and measured lengths can be derived (Fig. 30.16). It simply shows the upper and lower limits of measured length deviations as a function of the length itself. The length measuring error is defined as the maximum positive or negative deviation from the nominal length.

The measured scale bars should represent the object space as recommended in Figure 30.15. Each scale bar consists of a number of sub-distances so that a high number of reference lengths in all principal object directions are available.

The derived length measuring error conforms to international standards, and is comparable to well-accepted guidelines in tactile coordinate measuring technology (CMM). For surface measuring systems, a sphere distance error is defined in addition to plane measuring error and probing error (Luhmann & Wendt 2000).

In practical cases, photogrammetric off-line systems with high-resolution digital cameras and quality targets achieve length measuring errors of the order of 50 μm in a measuring volume of about $2 \times 2 \times 2$ m. Methods and strategies for camera calibration and achievable measuring accuracy are directly related to each other (Luhmann & Godding 2004).

30.5 SUMMARY

Industrial photogrammetry has undergone significant growth over the past ten years. The key drivers for this are high-resolution digital cameras, powerful algorithms for image processing and object reconstruction and high computing power. Systems with a high degree of automation and easy-to-use interfaces are available and accepted in practice by end users.

Most applications are related to point-wise probing of object points, either in off-line mode with sequential image acquisition and subsequent processing, or in on-line mode with simultaneous multi-image recording and tactile probes. Photogrammetric and optical surface measurement becomes more and more important, usually based on fringe projection methods.

Since standardized guidelines for acceptance and re-verification of optical 3-D measuring systems are available, systems and solutions can be evaluated and compared by objective means. ISO-conform accuracy parameters and traceability to standard units are inevitable preconditions for the worldwide acceptance of close range photogrammetry in industry.

Future developments can be identified in the field of dynamic measurements, e.g. for fast processes or high-dynamic deformation analysis. An additional trend is given by the combination of different sensors with photogrammetric components (e.g. laser trackers with cameras) to build hybrid systems. Last but not least, photogrammetric systems are being integrated into the manufacturing processes in order to perform non-contact 3-D measurements that are fully automated for process and quality control purposes.

REFERENCES

Atkinson, K.B. (ed.). 1996. *Close Range Photogrammetry and Machine Vision*. Caithness, UK: Whittles Publishing.

Bösemann, W. 1996. The optical tube measurement system OLM—photogrammetric methods used for industrial automation and process control. *International Archives of Photogrammetry and Remote Sensing* 31, B5:55–58.

Breuckmann, B. 1993. *Bildverarbeitung und optische Meßtechnik in der industriellen Praxis*. München: Franzis Verlag.

Cronk, S., Fraser, C.S. & Hanley, H. 2006. Automatic metric calibration of colour digital cameras. *The Photogrammetric Record* 21, 116: 355–372.

Dold, J., 1996. Influence of large targets on the results of photogrammetric bundle adjustment. *International Archives of Photogrammetry and Remote Sensing* 31(B5): 119–123.

Dold, J. 1997. *Ein hybrides photogrammetrisches Industriemeßsystem höchster Genauigkeit und seine Überprüfung.* Dissertation, Heft 54, Schriftenreihe Studiengang Vermessungswesen, Universität der Bundeswehr, München.

Dold, J. 1999. Stand der Technik in der Industriephotogrammetrie. *Photogrammetrie-Fernerkundung-Geoinformation* Heft 2: 113–126.

Fraser, C.S. 1984. Network Design for Close Range Photogrammetry. *Photogrammetric Engineering and Remote Sensing* 50(8): 1115–1126.

Fraser, C.S. & Brown, D.C. 1986. Industrial photogrammetry—new developments and recent applications. *The Photogrammetric Record* 12(68): 197–217.

Fraser, C.S. & Shortis, M. 1990. A Correction Model for Variation of Distortion within the Photographic Field. *International Archives of Photogrammetry and Remote Sensing* 28(5): 244–251.

Fryer, J.G., Mitchell, H.L. & Chandler, J.H. 2007. Applications of 3-D measurement from images. Caithness, Scotland, UK: Whittles Publishing.

Granshaw, S.I. 1980. Bundle adjustment methods in engineering photogrammetry. *The Photogrammetric Record* 10(56): 181–207.

Grün, A. 1985. Adaptive least squares correlation—a powerful image matching technique. *South African Journal of Photogrammetry, Remote Sensing and Cartography* 14(3): 175–187.

Harding, J.E. 1974. The behaviour of high strength friction grip bolted joints in box girder construction. *CESLIC Report BC2.*

Heipke, C. 1991. *Integration von Bildzuordnung, Punktbestimmung, Oberflächen-rekonstruktion und Orthoprojektion innerhalb der digitalen Photogrammetrie.* Dissertation, Deutsche Geodätische Kommission, Reihe C, Nr. 366.

Loser, R. & Kyle, S. (2003). Concepts and components of a novel 6DOF tracking system for 3-D metrology. In A. Grün & H. Kahmen (eds.), *Optical 3-D Measurement Techniques VI, vol. 2*: 55–62.

Luhmann, T. 2003. *Nahbereichsphotogrammetrie. 2. Auflage,* Heidelberg: Wichmann Verlag.

Luhmann,T. & Wester-Ebbinghaus,W. 1988. Digital Image Processing by Means of Reseau-Scanning. *Allgemeine Vermessungsnachrichten* 5/1988.

Luhmann, T. & Wendt, K. 2000. Recommendations for an Acceptance and Verification Test of Optical 3-D Measurement Systems. *International Archives for Photogrammetry and Remote Sensing* 33/5: 493–499.

Luhmann, T. & Godding, R. 2004. Messgenauigkeit und Kameramodellierung—Kernfragen der Industriephotogrammetrie. *Photogrammetrie-Fernerkundung-Geoinformation* Heft 1/2004: 13–21.

Luhmann, T. 2005. On the determination of objects rotation and translation in 3-D space (6 DOF) by a single camera. *Optical 3-D Measurement Techniques, Vienna, 2005.*

Luhmann, T., Robson, S., Kyle, S. & Harley, I. 2006. *Close-range photogrammetry*. Caithness, UK: Whittles Publishing.

Luhmann, T., Hastedt, H. & Tecklenburg, W. 2006. Modelling of chromatic aberration for high-precision photogrammetry. *International Archives for Photogrammetry and Remote Sensing* 36(5): 173–178.

Peipe, J. 1997. High-resolution CCD area array sensors in digital close range photogrammetry. *Videometrics V, SPIE Vol. 3174*: 153–156.

Newton, I. 1974. Dimensional quality control of large ship structures by photogrammetry. *The Photogrammetric Record* 8(44): 139–153.

Robson, S. & Shortis, M.R. 1998. Practical influences of geometric and radiometric image quality provided by different digital camera systems. *The Photogrammetric Record* 16(92): 225–248.

Schneider, C.-T. 1991. *Objektgestützte Mehrbildzuordnung.* Dissertation, Deutsche Geodätische Kommission, Reihe C, Nr. 375.

Shortis, M.R., Clarke, T.A. & Robson, S. 1995. Practical testing of the precision and accuracy of target image centring algorithms. *Videometrics IV, SPIE Vol. 2598*: 65–76.

Shortis, M.R., Robson, S., Jones, T.W. & Lunsford, C.B. 2007. Parachute model validation using image sequences. In A. Grun & H. Kahmen (eds.), *Eighth Conference on Optical 3-D Measurement Techniques, ETH Zurich, Switzerland*, ISBN 3-906467-67-8, Volume I: 72–79.

Staiger, R. 1992. *Automatische und dynamische Koordinatenmessung mit mobilen Sensorsystemen. In Welsch et al. (eds.), Geodätische Meßverfahren im Maschinenbau. Schriftenreihe DVW*, 1/1992: 81–96. Stuttgart: Wittwer Verlag.

Tecklenburg, W., Luhmann, T. & Hastedt, H. 2001. Camera Modelling with Image-variant Parameters and Finite Elements. In *Optical 3-D Measurement Techniques V.* Heidelberg: Wichmann Verlag.

VDI. 2000. Optische 3-D-Messtechnik. *VDI/VDE-Richtlinie 2634*: 1–2. Berlin: Beuth Verlag.

Wester-Ebbinghaus, W. 1985. Bündeltriangulation mit gemeinsamer Ausgleichung photogrammetrischer und geodätischer Beobachtungen. *Zeitschrift für Vermessungswesen* 110(3):101–111.

Winter, D. & Reich, C. 1997. Video-3-D-Digitalisierung komplexer Objekte mit frei beweglichen, topometrischen Sensoren. *DGZfP—VDI/VDA-GMA Fachtagung "Optische Formerfassung", Langen, GMA-Bericht 30*: 119–127.

Wrobel, B. 1987. Digitale Bildzuordnung durch Facetten mit Hilfe von Objektraummodellen. *Bildmessung und Luftbildwesen* Heft 3/1987: 93–101.

Advances in Photogrammetry, Remote Sensing and Spatial Information Sciences: 2008 ISPRS Congress Book – Li, Chen & Baltsavias (eds)

CHAPTER 31

Medical applications

Nicola D'Apuzzo & Harvey Mitchell

ABSTRACT: The wide variety of uses of measurements from imagery to provide spatial information for medicine and human sciences is categorized into three distinct groups. The first category of applications involves surface shape measurement of any part of the body, whether very small or large. It is now commonly undertaken by laser scanning and structured light measurement, and finds uses in plastic surgery, reconstructive medicine, orthopaedics, prosthetics orthodontics and other dentistry, dermatology and cosmetics. The second group involves the recording of the human body in motion, which is widely used predominantly but not solely for studies of walking people. The third group represents the most novel application: the surgical use of real-time measurement to precisely position surgical instruments and prostheses in the operating theatre. The latter two fields are distinguished from the first by being well suited to conventional photogrammetric triangulation using camera imagery in conjunction with the matching of features in the images. This chapter identifies the main challenges in these three groups, specifies the technical features which characterize each, and records some of the applications for purposes involving human beings.

Keywords: Medical photogrammetry, body motion tracking, body surface scanning, surgical procedures

31.1 INTRODUCTION

Photogrammetrists who are interested in applications in human measurement will find that there are many papers in this field, and they are found in both biomedical and photogrammetric journals. They will also find that the authors have a variety of backgrounds, including medical, engineering and photogrammetry. The papers show that some of the photogrammetric systems are intended for short term medical use, others are used over longer terms; some systems emphasize simplicity and low cost while others are more sophisticated. To understand the distinctive characteristics and challenges of medical photogrammetric measurement, it is useful to first recognize some of the more common uses. Motion tracking, typically of the whole body but sometimes of just a part of the body, seems to dominate current applications. There is also a demand for static measurement of certain body regions, the back, the entire head, face, female breast, the hand, arm and foot, undertaken for disease detection, treatment, prosthesis fitting, anatomical research, ergonomic studies and so on. Occasionally, there is a requirement for measuring smaller and very specialized body areas such as the eye, ear, and teeth, or part of the skin in the location of melanoma or wound. Finally, it is important to recognize the current major challenges faced in the dynamics of surgical monitoring in which the goal is to guide the surgeon during surgical interventions.

The aim of medical photogrammetry is to assist in health matters, usually related to a disease and its detection, treatment or monitoring, but perhaps also related to disease avoidance as in the case of sports and ergonomic studies. Medical measurement necessarily involves human beings as subjects for imaging, and humans have certain characteristics that require consideration. Unlike inanimate objects, humans need physical comfort, and do not want to be left in one fixed or unnatural position, especially for an unspecified length of time. Humans also have a range of requirements for mental comfort, including at times the need for privacy and confidentiality.

Although each type of medical measurement faces its own challenges, there are certain distinctive and common photogrammetric characteristics. Body parts are never simple geometric shapes, and the measurement task is generally to define surface shape, and not to detect edges or to locate characteristic points as may be required in industrial measurement, for example.

Also, body surfaces are notoriously free from good natural targets, so that for imaging purposes the surfaces of interest must somehow be given texture. At the photographic stage, the imaging is invariably synchronous, the various cameras being held fixed, and so multiple cameras are needed. Fortunately for the photogrammetrist, accuracy requirements are generally not as high as in other areas of photogrammetry such as industrial measurement.

31.2 SENSORS AND DATA ACQUISITION

31.2.1 *Human motion tracking*

Human motion recording by photogrammetry is used primarily to observe the movement of the entire human body, mostly for gait analysis, but it is also used to record people involved in sporting activity. Less commonly, motion recording by photogrammetry concerns movements of regions less than the entire body. Recording of movement creates severe photographic challenges. In the most common cases, a person, often shrouded with targets, is imaged moving at a relatively high speed while using actions that cause the occlusion of the targets. A stream of confusing imagery may be collected from a number of cameras, so that the subsequent analysis requires photogrammetric systems that are sophisticated, and hence may also be relatively expensive. However, the demand for this type of measurement is high, and a number of advanced photogrammetric systems are available commercially. (The commercial systems also find use in health related, e.g. ergonomic, studies and even in the entertainment industry.)

The extensive commercial involvement in motion recording has meant that reports of applications are widespread but technical details not common. Nor do commercial websites provide the underlying photogrammetric theory and practice, which might interest the photogrammetric reader. In a number of cases, however, technical detail has been reported during a photogrammetric motion recording system's development stage. Consequently, papers reporting the photogrammetric facts can seem to be dated, but they are nevertheless important reports of the developments in that era; see e.g. Ferrigno & Pedotti (1985) and Walton (1990).

Movement recording systems require a number of cameras in order to both ensure patient coverage and to cope with occluded targets. High frame rates are necessary, often rates higher than the 30 images per second typically encountered with common cameras. Targets are normally placed on the subjects, both passive and active targeting being used. The technical features of the many available motion recording systems have been compared by, e.g. Ehara et al. (1995, 1997) and more recently by Richards (1999), who refers to eight systems, which use up to about 30 cameras observing up to 500 targets. The relative accuracies of point measurement are typically 0.1 to 1 mm. Information about the current versions of the systems referred to by Richards are provided by commercial websites: Ariel Dynamics (2008), Charnwood Dynamics Ltd. (2008), BTS SpA (2008), Motion Analysis Corporation (2008), Northern Digital Inc. (2008), Peak Performance (2008), Qualisys AB (2008) and Vicon Motion Systems (2008), while references can be found to yet other systems, e.g. Biogesta (2008) and Phoenix Technologies (2008). It is now relatively rare to find reports of the use of motion analysis systems other than those of commercial origin, but some developments can be noted. A development by Garrido Castro et al. (2006) contrasts with commercial systems referred to above, because the system is intended to be low-cost, with off-line processing involving percentages of manual digitizing. It also aims to be used without wired targets to provide greater versatility, such as outdoors.

Calibration of motion recording systems is a crucial matter, and techniques are included in the comparison by Richards (1999).

31.2.2 *Human body surface scanning*

Laser scanning and structured light projection represent the optical measurement technologies mostly employed for the three-dimensional digitizing of the surface of the human body.

They are both based on the same rule, namely triangulation, in the way that light structures (normally in the form of stripes) are projected onto the human body, with light sensors acquiring the scene; by known geometry of the set-up, 3-D information can then be drawn from the imaged data.

The difference between the two methods resides in the way the light structure is projected and imaged. By laser scanning, laser light sources are used to project on the human body one or multiple thin and sharp stripes (see Fig. 31.1 on the left). To assure the inoffensiveness of the light beam, only eye-safe lasers are employed. A laser scanning unit is composed of the laser light source, the optical system and the light sensor (variations are possible also in the number of employed laser sources and light sensors). The unit is moved across the human body to digitize the surface. The type of movement and the number of employed units can vary depending on the human body parts to be measured. For example, full body scanners are usually composed of three or four scanner units that move vertically synchronously along three or four pillars placed around the person.

The high costs of producing hardware components for laser scanning technology have to be considered as a disadvantage. Additionally to the laser, the light sensor and the optical system, precise electric motors also have to be used for the displacement of the scanner units. Moreover, the complete scanner system has to be calibrated so that the geometrical disposition of all the elements can be determined exactly. A second disadvantage of this method is the time required for the digitizing of large surfaces, as movements and displacements of the person occurring during the scanning process lead to noise and measurement errors in the resulting data. There is no problem for the measurement of extremities such as feet and hands, since these body parts can be kept immobile for some seconds. But in the case of the measurements of head or full body, it is practically impossible to stay immobile for several seconds. Additionally, uncontrolled movements such as breathing or muscle contraction have also to be considered.

The second technology used extensively for human body measurement is based on the projection of light patterns. It comes closer to the solution of the problems described above. Instead of moving the scanning unit, a light pattern (usually in the form of a bundle of stripes) is projected onto the entire interested area of the human body (see Fig. 31.1 right), where a light sensor (usually a digital camera) acquires the scene. In this case, the scanning device is usually composed of a white light projector and a camera. More complex systems employ multiple light sensors and/or multiple projectors. The measurement process is similar to the method of laser scanning: stripes on the surface are measured singularly by triangulation. The difference is that this happens in a single step allowing the entire surface to be digitized by a single acquisition. For the increment of the resolution, the projected stripes may be shifted and multiple scenes acquired.

The short time of the scanning process (mostly under one second) reduces the problems caused by uncontrolled movements of the person during the measurement process. However, to measure large parts of the human body (e.g. entire head or full body), multiple scanning devices are required. This procedure has the disadvantage that multiple units cannot be used simultaneously since they interfere with each other. In practice, multiple equipment has to be used serially. This implies again an extension of the acquisition time.

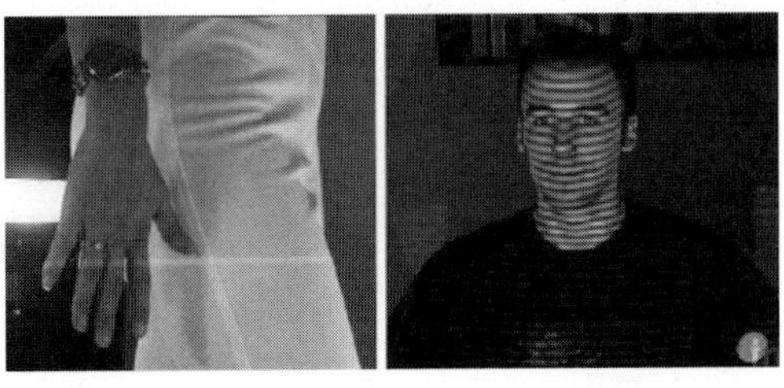

Figure 31.1. Left: laser stripe projected onto the human body. Right: structured light in the form of multiple white stripes projected onto the human body. (see colour plate page 515)

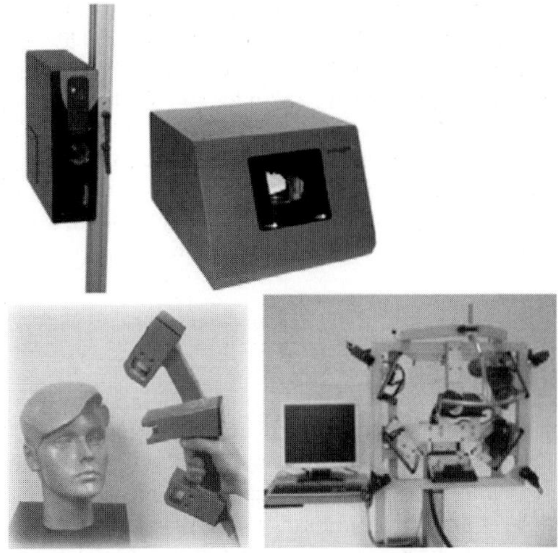

Figure 31.2. Example of surface measurement systems employed in medicine (from left to right, top to bottom): InSpeck surface digitizer Mega Capturor II (mounted on a vertical bar), 3Shape dental scanner DentalSCAN, Polhemus hand-held surface digitizer FastSCAN Scorpion, IVB-Jena face scanner gscan.

Which of the two technologies—laser scanning or structured light projection—should be more adequately employed, depends on the measurement task to be performed. This fact is also demonstrated by the multitude of commercial measurement systems based on different methods and technologies available nowadays. Among them, four groups can be identified as the main systems used in the medical field: (1) modular structured light projection systems (multiple units can be combined into a complete system) - for a fast digitizing of body areas; (2) desktop solutions with multi-axial moving platform—for the automatic measurement of small objects (e.g. dental impressions, ear impressions); (3) hand-operated or hand-held laser profilers—for complete freedom during the acquisition; and (4) dedicated systems designed for a specific measurement task—as for example face scanner, foot scanner, full body scanner. Figure 31.2 shows examples of the four groups.

31.3 APPLICATIONS IN MEDICINE

The medical field represents an established group of users of both human body surface scanning and human body motion tracking techniques. In this regard, various applications can be found in biomechanics, plastic surgery, orthopaedics, prosthetics, orthodontics, forensic medicine, dermatology, etc. In the following sections, some examples of applications are briefly described. They represent uniquely a small part of possible exploitations. Nevertheless, they serve to illustrate the potentials of 3-D measurement techniques applied in medicine.

31.3.1 *Dynamic 3-D measurement*

31.3.1.1 *Human motion tracking*

The end uses of human motion tracking are very diverse, and extend beyond anticipated application areas in anatomical and physiological research, sporting studies, rehabilitation and ergonomic planning to applications in the neurological, psychological and anthropological areas. Even a fairly cursory examination of the literature will show the full range of commercial systems being used for the analysis of walking (Whittle & Levine 1999, Troje 2002), running (Krosshaug et al. 2007), in sport (Ford et al. 2003) and also for studies involving the effects of cold exposure on leg muscles (Oksa et al. 1997), cerebral palsy (Meinecke et al. 2006), obesity (Achard de Leluardière et al. 2006) and idiopathic arthritis (Broström et al. 2007), and so on.

Not all studies concern the full body. Puce et al. (2003) and Baroni et al. (1998) observed movements of the face using commercial systems intended for studies of the movement by the entire human. Liu et al. (2007) investigated the measurement of teeth movement, again using a standard commercial system. "Optical motion analysis techniques have been widely used in biomechanics for measuring large-scale motions such as gait, but have not yet been significantly explored for measuring smaller movements such as the tooth displacements under load. In principle, very accurate measurements could be possible and this could provide a valuable tool in many engineering applications", according to the authors, who, in their evaluation of the "accuracy and repeatability of the Qualisys ProReflex-MCU120 system when measuring small displacements", noted that system repeatability was less than 0.005 mm, demonstrating "that the system suffices accuracy for measuring tooth displacements and could potentially be useful in many other applications."

Cases of measurement when the trajectory itself is not relevant but the movement causes an effect, such as a deformation, which is of interest are not strictly motion recording, but we will include them here especially because they often face the same challenges and use similar solutions to pure motion recording. For example, Gao et al. (2000) used a pair of cameras to record the shape of a bioprosthetic heart valve at critical stages during a laboratory experiment to observe the valve's deformation in an action induced by a mechanical apparatus. The valve surfaces were marked to provide photogrammetric targets. The only medical part of the experiment were the valves, but the issue is still seen here as medical photogrammetry. The experiment faced the interesting challenge that the valve opening and closing could last less than 0.04 seconds, while affordable cameras recorded with a period of 0.03 seconds. The interline transfer cameras recorded two events per frame by stroboscopic illumination, the two images being segregated subsequently.

Photogrammetry is also finding applications in the task of positioning patients for irradiation. While this may not strictly be motion recording, it can be noted that Baroni et al. (2003) used a commercial motion recording system for this purpose, and the challenges of the task can put it into the motion recording category. The benefit of precise patient positioning lies in reducing the radiation that must be allowed to overflow to nearby healthy cells in cases of poor positioning. The obvious tactic is to use targets fixed to the body, but another approach is to mathematical match measured surface shapes to identify whether the relevant part of the patient is in the required location. This is especially suited to positioning distinctive shapes such as the female breasts (e.g. Baroni et al. 2003), but it has also been used for neurosurgery, e.g. Raabe et al. (2002). Bert et al. (2006) refer to the use of a commercial patient positioning system; see Vision RT (2008). Accuracy is obviously a primary concern; Rogus et al. (1999) consider accuracy issues in patient positioning, but high speed measurement and processing are also crucial.

Although recording is of primary interest to the photogrammetrist, the spatial data manipulation and motion analysis is an important adjunct to motion recording to provide information of the sort shown in Figure 31.3. Most commercial photogrammetric motion recording systems are available with linked

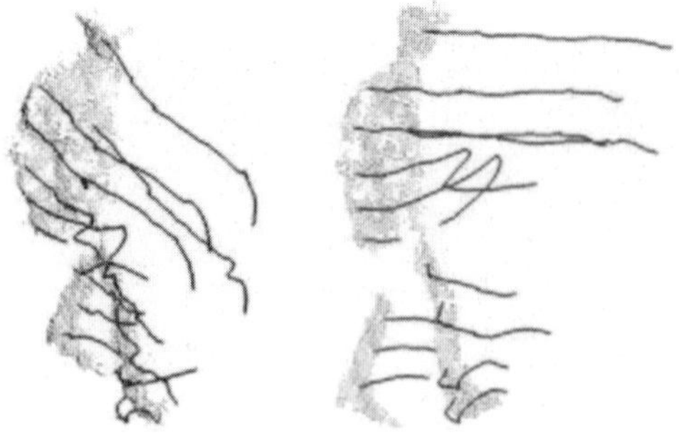

Figure 31.3. Three dimensional trajectories representing human body motion, from D'Apuzzo (2000).

motion analysis software. Occasionally analysis software is available independent of the recording system (e.g. see C-Motion Inc. (2008)).

31.3.1.2 *Surgical navigation systems*

Optical measurement systems are an established facet of surgical procedures using computer aided assistance. The importance of these systems has grown with the increasing popularity of minimally invasive procedures, which are associated with a reduction in operative field visibility. These measurement systems are mostly based on stereo photogrammetry and are usually composed of stereo camera acquisition units and surgical instrumentations marked with signalized target points. Figure 31.4 shows a typical example: the infra-red photogrammetric measurement system Cambar of the company Axios 3D Services (2008). Recent studies (Broers & Jansing 2007) evaluated positively the reliability and precision of such stereo measurement systems for the overall success rate of surgical navigation.

The stereo photogrammetric measurement system is uniquely a component of the surgical navigation system. In fact, complete solutions for computer assisted surgery also include real-time navigation and real-time computer assisting software solutions. Moreover, medical imaging data acquired pre-operatively (as for example by computer tomography or by magnetic resonance imaging) has also to be integrated into the navigation system and has to be aligned with the photogrammetrically acquired data. This is required both for planning and for executing the surgical intervention.

Various solutions are available commercially and dedicated solutions have been implemented for typical interventions, as for example knee replacement or spinal surgery. Figure 31.5 shows snapshots of different parts of a complete navigation system for total knee replacement: VectorVision Knee of BrainLAB AG (2008). The complete navigation solution enables knee replacement surgery with great precision and control. The system helps the surgeon, in real-time and with on-screen information, in the different phases during the intervention, from the bone cuts execution to the precise insertion of the knee replacement components in the bone. The surgeon also has the option to make individual modifications of the treatment plan, including the size, position and orientation of the implant. Surgical instruments are continuously tracked by the navigation system to judge their precise position and alignment relative to the operated parts and in accordance with the planned and

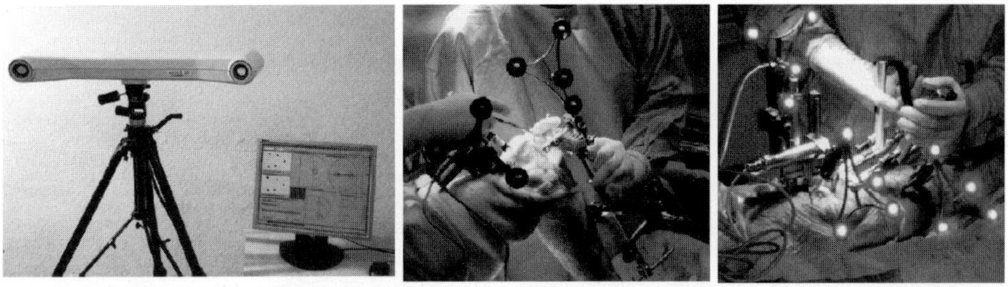

Figure 31.4. Axios3D Cambar photogrammetric measurement system and surgical instrumentation with signalized targets. (see colour plate page 516)

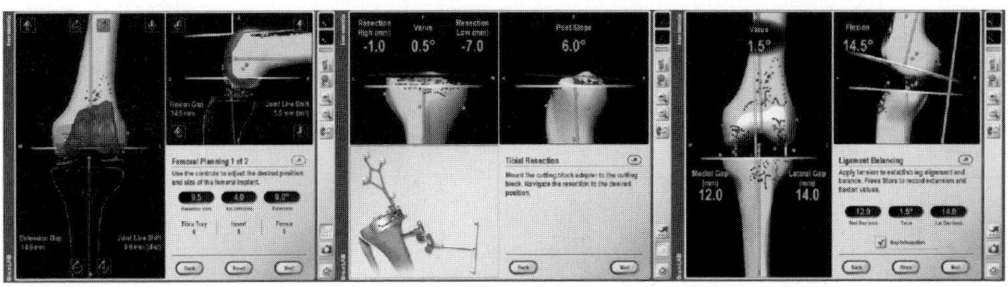

Figure 31.5. Snapshots of VectoVision Knee of BrainLAB (2008). Initial planning and alignment of the implant position (left), real-time bone cut navigation and verification (centre), kinematical analysis for optimal leg alignment balance (right). (see colour plate page 516)

executed treatment. This provides the surgeon with comprehensive real-time information, which allows the surgeon to make decisions for greater control over the surgical outcomes.

31.3.2 *Human body surface measurement*

31.3.2.1 *Plastic surgery, reconstructive medicine*

Even though, at first sight, plastic surgery will be thought as an ideal field of application of human surface measurement techniques, in practice this is not the case. Commercial systems dedicated to the measurement of various parts of the human body and especially developed for surgical applications are available. Two examples are shown in Figures 31.6 and 31.7: the breast and face digitizer of InSpeck (2008) and the portable face and body scanner by GFMesstechnik (Benderoth 2005).

Advantages of using such systems in combination with analysis and measurement software could be manifold, e.g. for the planning of surgical interventions, for the comparison between pre- and post-operative shape, or for a better interface between patient and surgeon. Other software solutions go even further by simulating the effects of surgical interventions in a virtual environment. Regardless of all these available solutions, resistance is still present from the surgeons for their use in actual practice. Nevertheless, interest begins to arise especially if advantages can be demonstrated, as in the case of the two examples described in the following paragraphs.

The first regards a study on the simulation of surgical intervention done by Buskirk et al. (2005). In their work, they describe a project aimed at collecting and comparing standard body contouring data from 3-D scanners and from manual anthropometric measurements. Additionally, forecasted models of prospective body contouring procedures were mathematically implemented for the purpose of generating 3-D views of multiple predictive outcomes of surgical interventions. The special case of breast implant surgery was considered.

The scanning system employed for the study, Novaptus Systems Scanner Suite (Novaptus 2008), is based on structured light pattern projection and is able to rapidly produce a three-dimensional surface model of the entire body of a patient. From the scan data, the system extracts automatically relevant measures, such as suprasternal-notch-to-nipple and nipple-to-nipple linear lengths and bust, contoured-bust, under-bust, waist, hip, abdomen, thigh and knee circumferences. These findings were compared to manual anthropometric measurements taken on more than 70 patients. In this way, it could be demonstrated that digital scanning was as accurate as classical anthropometric measurement techniques.

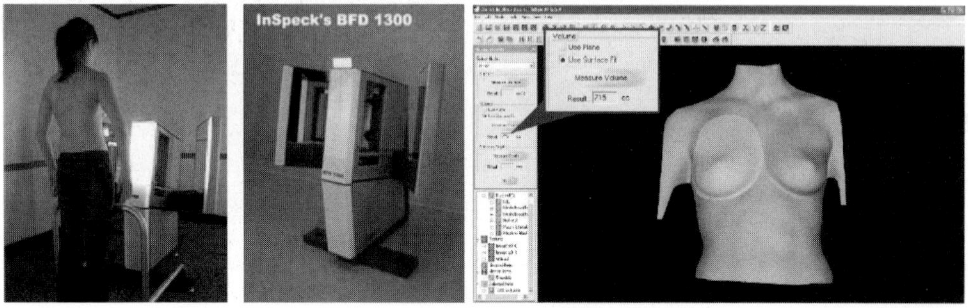

Figure 31.6. InSpeck Breast and Facial Digitizer BFD 1300 and measurement software EM-Measurement. (see colour plate page 516)

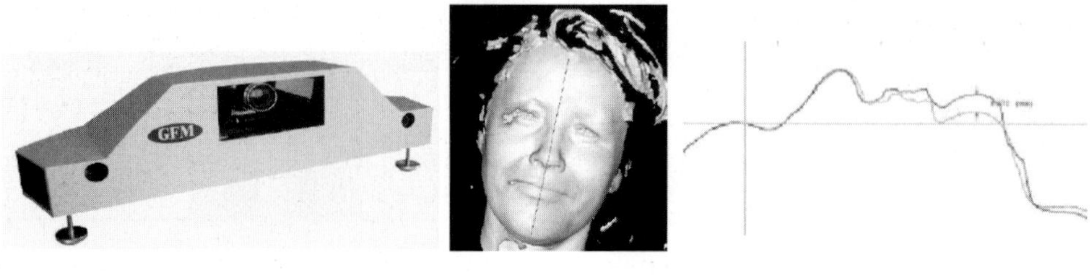

Figure 31.7. PRIMOS body II—portable 3D face and body scanner, 3-D scan data and comparison (colours code the difference ranges), 2-D cross sections showing difference between pre- and post-operative 3-D scan data. (see colour plate page 517)

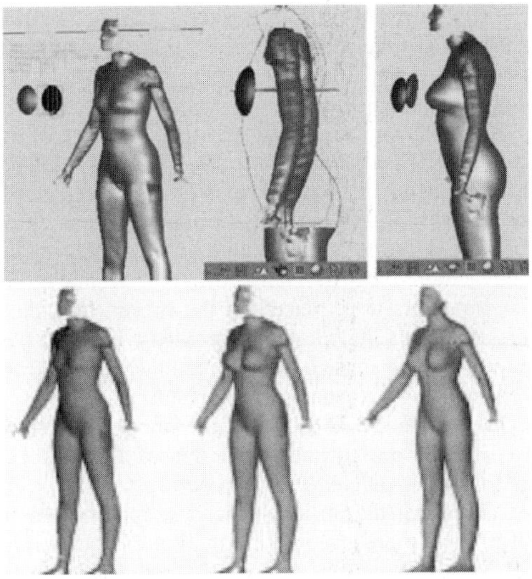

Figure 31.8. Top: placing the virtual breast implants on the pre-operative scan and the resulting forecasted 3-D model. Bottom: comparison of breast augmentation pre-operative subject's 3-D scan image, forecasted 3-D model and seven month post-operative subject scan.

Virtual predictive modelling, to forecast post-operative results, was also performed during the study. In this case, the scanner was employed in conjunction with a predictive software tool, Novaptus Virtual Surgery Suite (Novaptus 2008), allowing the surgeon to volumetrically manipulate the subject's pre-operative contour to virtually model it into the desired shape (see Fig. 31.8). Real surgical interventions such as breast augmentation, breast reduction and abdominoplasty, performed during the study, demonstrated that both pre- and post-operative measurements were concordant with the simulated results.

In the cases where a patient has lost part of their faces (as eyes, ears or nose), plastic surgery commonly uses the method of generating an artificial replacement (called "epithesis") for the missing part. 3-D scanning technologies can be very useful in these situations.

A project at the Institute of Production Engineering of the Technical University of Dresden aims at the use of face scanning and rapid prototyping technologies for the design and production of epitheses (Stelzer et al. 2006).

A face scanner based on white light projection, gscan of IVB-Jena (2008), is used to measure three-dimensionally the face of patients. Modelling software is then employed to virtually design and model in 3-D the replacement of the missing part. In the cases where the missing parts are still present on the face of the patient as mirrored parts (as eye or ear), the missing part can be modelled by mirroring the existing part. In the case of a missing nose, the artificial replacement can be chosen from an extensive 3-D database of virtual noses and visualized in 3-D over the digitized face. The accurate 3-D modelling of the artificial replacement is completely performed in a virtual environment using commercial modelling software. The obtained CAD model is then converted into a physical model with rapid prototyping techniques. The model is first generated on 3-D printers in a polymer material and then used as master for the production of the final epithesis from silicone. Figure 31.9 shows an example of 3-D scan data and the modelled epithesis (to protect the privacy of the patient, a 3-D point cloud of the face is shown; in the real situations fully textured surface models are visualized).

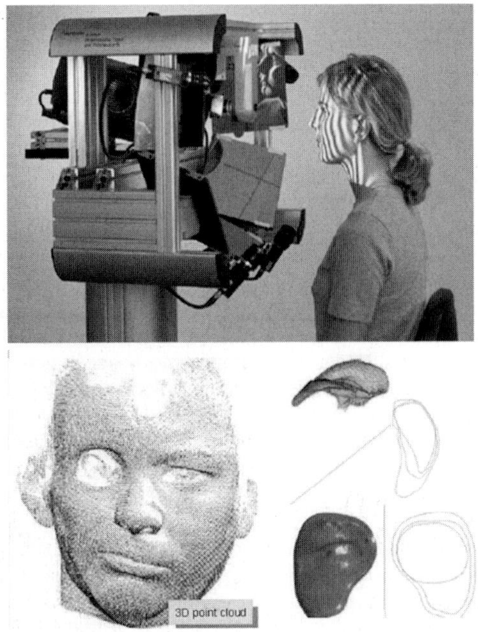

Figure 31.9. Human face scanner gscan (left), 3-D point cloud of patient with missing eye (centre), digital design of the missing part (right).

31.3.2.2 *Orthopaedics, prosthetics*

The peculiarity of orthopaedics and prosthetics is that measurements of the interested body part are always required. In a large proportion of the cases, in order to build best fit ortheses and/or prostheses, negatives of the human body part are required; these are usually obtained by casting. For these reasons, 3-D scanning technologies have the best presuppositions to become replacements of classical methods. In the next paragraphs, two typical cases are described where optical

3-D measurement techniques are already successfully employed instead of classic measurement methods (negative of the human body part by casting).

A good example of how 3-D surface scanning can be successfully employed in the field of prosthetics is well described by the work of Bonacini et al. (2007) and regards the design process of lower limb prostheses. An important aspect concerns the reconstruction of the 3-D digital model of the stump, which replaces a plaster cast. The work of Bonacini et al. (2007) goes even further by combining optical 3-D measurement techniques for the external surface of the limb and computer tomography (CT) and magnetic resonance imaging (MRI) for the inner structure (bone structure, muscle tissue, soft tissue and dermis). With the integration of external surface and internal components, it is possible to obtain a complete digital model of the residual limb, which permits the simulation of the behaviour and interaction of a designed socket with the stump.

For the acquisition of the external surface of the stump, a laser scanner, Minolta Vivid 9i, was used. This system is normally employed for industrial applications and it is not particularly suited for the measurement of human body parts. Nevertheless, it was suitable for the preliminary studies of the project. The principal problems occurring during acquisition were caused by noise generated by downy skin on the stump, and by uncontrolled muscular contractions. The first problem could be solved with a preparation of the skin of the stump whereas for the second problem a more suitable measurement system should be adopted.

After the acquisition of the inner structure of the stump by CT and MRI and after alignment of the different datasets, a complete 3-D digital integrated stump was obtained using commercial modelling software. In order to guarantee an alignment of the data acquired by the three different technologies, fixed markers were placed on the stump (lead shots for laser and CT, vitamin E for MRI). Figure 31.10 shows the 3-D surface model, the inner bones structure and the obtained combined result.

The obtained 3-D model of the stump can then be used for the design and production of a customized socket in a CAD/CAM environment.

A proof of the utility of the described process is given by the presence in the market of commercial CAD/CAM solutions especially dedicated to the design of lower limb sockets, based on 3-D scan models of the stump.

An example is shown in Figure 31.11, the CANFIT-PLUS P&O System from the Canadian company Vorum Research Corporation (2008). In this specific case, a hand-help laser scanner is employed for the 3-D acquisition of the external surface. Dedicated software tools are then employed for the accurate modelling of prostheses on the base of the scanned data.

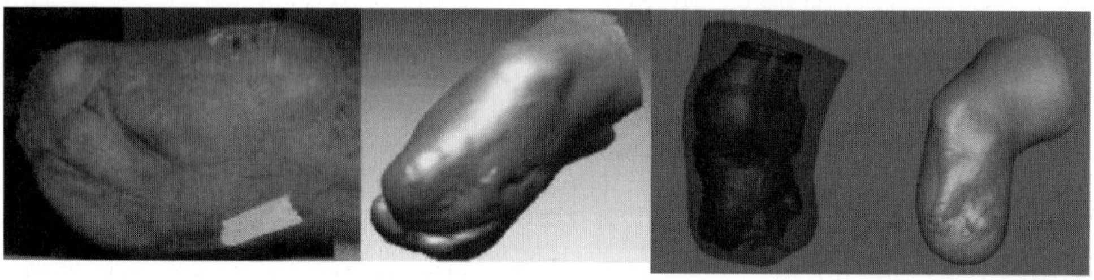

Figure 31.10. From left to right: stump image (note the signalized landmarks on the top), 3-D laser scanning model, CT skin-bone model, integrated laser/bone model (right). (see colour plate page 517)

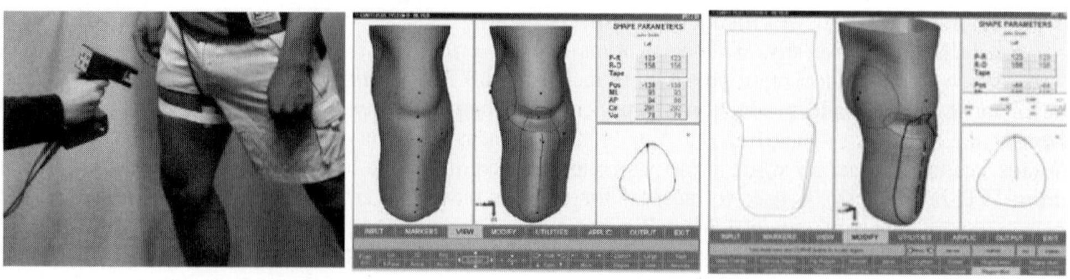

Figure 31.11. Lower limb socket design system CANFIT PLUS P&O: hand-held laser scanner and snapshots of the CAD/CAM solution. (see colour plate page 517)

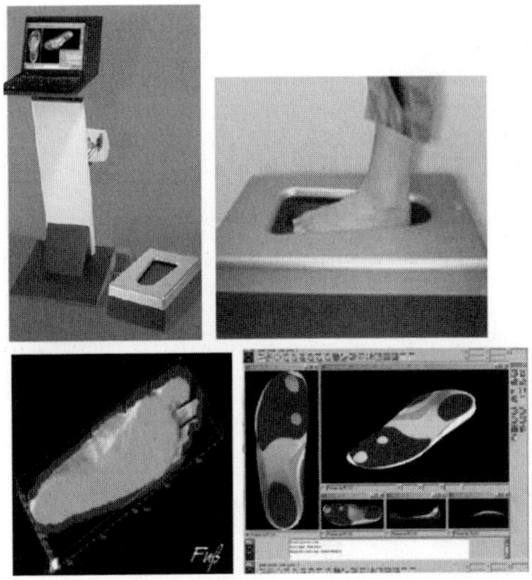

Figure 31.12. 3-D scanning system of pedcad (2008), acquired 3-D data and designed 3-D sole.

The second example of application of 3-D scanning technologies in the orthopaedic field regards the design and production of customized orthopaedic soles. The classical method consists also in this case of first getting an imprint of the foot (i.e. a negative of the foot plantar) and then building the shoe sole according to it. The advantages of using modern 3-D scanning technologies to acquire the shape of the foot are manifold compared to the classic foot imprint method. The design of customized shoe soles can in fact be completely performed in CAD/CAM solutions starting from 3-D scan data of the plantar of the feet of a person.

Various commercial systems have been developed especially for this purpose. Inexpensive scanning solutions are available, as the measurement task is relatively simple. For these reasons, this application became commercially more interesting.

Figure 31.12 shows an example of a complete solution from the German manufacturer pedcad Orthopaedieprodukte (2008). The scanning system is based on the structured light projection method and acquires the shape of the plantar of the foot in a very short time. The integrated software solution allows for the analysis of the scan data and for the complete design process of customized orthopaedic shoe soles in a CAD/CAM environment. The generated data can be sent directly to milling machines for the production. Also in this case, the integration of 3-D scanning systems with CAD/CAM solutions serves for a smoother and simplified design and production.

31.3.2.3 *Orthodontics, dentistry*

In dentistry and orthodontics, the advantages of employing optical 3-D digitizing techniques can be as straightforward as for orthopaedics and prosthetics. In fact, already available for many years are CAD/CAM solutions for the 3-D scanning of dental cast impressions for archiving and for a simplified design of orthodontical restoration, as well as dedicated intra-oral 3-D scanners for a direct 3-D measurement of teeth without the need for taking dental impressions. Clinical studies and basic reviews were published assessing positively the employment of CAD/CAM technologies both in dental restoration laboratories (Williams et al. 2006), as well as in dental practices (Christensen 2006). Moreover, manufacturers of CAD/CAM dental solutions constantly improve their products in order to increase the acceptance from potential users (i.e. dentists and dental laboratories) and in order to increase the penetration in the specific market. However, dentists' acceptance of 3-D technologies and CAD/CAM solutions still varies from excellent to poor. The cost of the investments involved and the technical know-how required to operate the systems are the significant deterrents. Dentists have to examine various factors in order to determine if the purchase of CAD/CAM technology is financially and clinically advantageous for their practice (Trost et al. 2006). Nevertheless, CAD/CAM dental solutions have been a reality for several years and are successfully exploited by thousands of dental laboratories and dental practices (Christensen 2006).

A typical example of solutions for modern design and production of dental restorations starting from 3-D scans of dental cast impressions is the given by the products of the company LaserDenta (Blumenschein 2006), which cover all the process steps: 3-D scanning, 3-D processing, 3-D visualization, 3-D virtual planning of restorations (see Fig. 31.13). For this purpose, a desktop laser scanning solution was especially designed and developed to automatically scan dental impressions with complete automation and with high precision (absolute accuracy of 0.02 mm, value given by the manufacturer). The scanning system employs a 5-axis (2 rotations and 3 translations) moving platform in combination with a laser profiler in order to scan entirely a dental cast impression. A complete jaw dental cast impression can be scanned fully automatically in about 5–8 minutes. The scanning equipment (scanner and control software) is designed to be easily used by non experts, so that the scanning process can be performed by unqualified (inexpensive) staff.

The data resulting from the scanning process will then be used by the dentist and/or dental laboratories to directly plan and design dental restorations (e.g. bridges, ceramic inlays, crowns, etc.) in a virtual environment. The virtually designed components of a restoration can then be produced in a fast and accurate

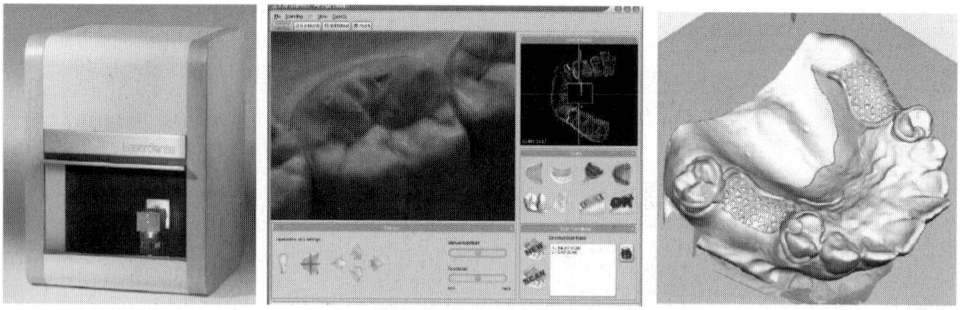

Figure 31.13. Dental solution of LaserDenta (Blumenschein 2006): desktop dental cast laser scanner (left), snapshot of the scanning control software (centre), virtual planning of a complex dental restoration (right). (see colour plate page 517)

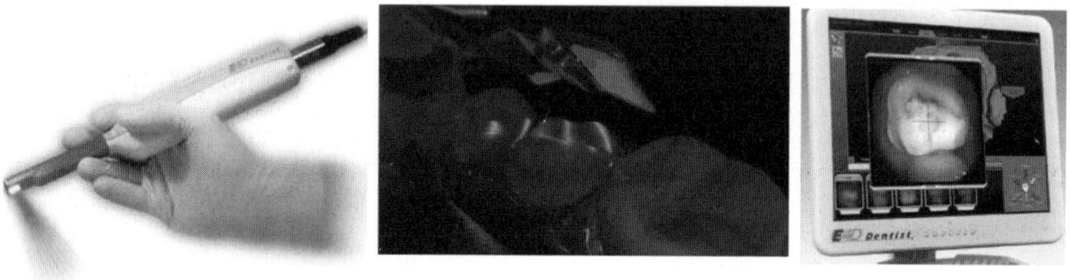

Figure 31.14. The intra-oral scanner E4D Dentist of D4D Technologies (2008): the hand held miniaturized scanner (left), surface scanning by multiple lines projection (centre), real-time scanning control visualization (right). (see colour plate page 518)

way by modern rapid prototyping technologies: metal and ceramic parts can be produced automatically by CAM solutions starting from CAD data.

The employment of 3-D scanning technologies and modern CAD/CAM solutions in the dental domain can increase the efficiency and at the same time reduce the costs of the design and production of the different parts (in ceramic and/or metal) required by a dental restoration. For this reason, these technologies have a great potential for expansion in the dental domain.

A step further to simplify the entire process for the design and production of dental restorations is given by the intra-oral scanners. Various companies offer solutions for a direct scanning of teeth inside the oral cavity, i.e. without the need to first take an impression of the teeth and after to digitize it with a desktop scanner (as described above). Figure 31.14 shows an example of such a device from the company D4D Technologies (2008). The hand-held 3-D scanning system is so miniaturized that it can be placed inside the mouth of a patient to accurately record the 3-D shape of the teeth.

The main advantage resulting from the use of intra-oral scanners is the immediate generation of a 3-D surface model of teeth, thus allowing an immediate and direct design and planning of dental restorations in a virtual environment. In some simple cases (e.g. ceramic inlays), the entire procedure can be performed during the same session, i.e. without the need of have the patient leave. In fact, 3-D intra-oral scanning systems are usually accompanied by virtual restoration planning software, as well as with small milling machines for the production of ceramic parts directly from the generated 3-D data.

31.3.2.4 *Forensic medicine*

Documentation of morphological findings on and in the living and deceased is essential in forensic medicine. Nowadays, most of the documentation of forensic relevant medical findings is limited mainly to traditional 2-D photography, 2-D radiographs, sketches and verbal descriptions. More and more forensic institutes begin however to employ modern 3-D acquisition techniques, such as photogrammetry, laser scanning and surface scanning for the documentation and analysis processes. Among them, the Institute of Forensic Medicine at the University of Bern (Switzerland) represents a pioneer. Within the research project Virtopsy (Naether et al. 2007, Thali et al. 2005), 3-D optical surface digitizing and cross-sectional radiological modalities were introduced for documentation and reconstruction of the internal and

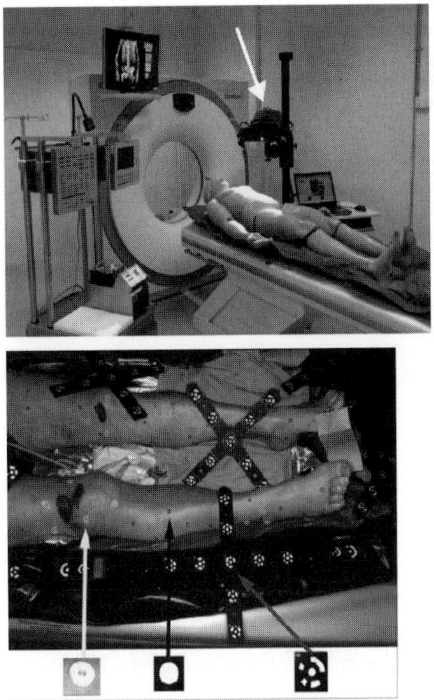

Figure 31.15. Left: Computer tomograph Siemens Emotion 6 and 3-D surface scanner GOM ATOS III (arrow). Right: different markers used for the measurement with GOM TRITOP. (see colour plate page 518)

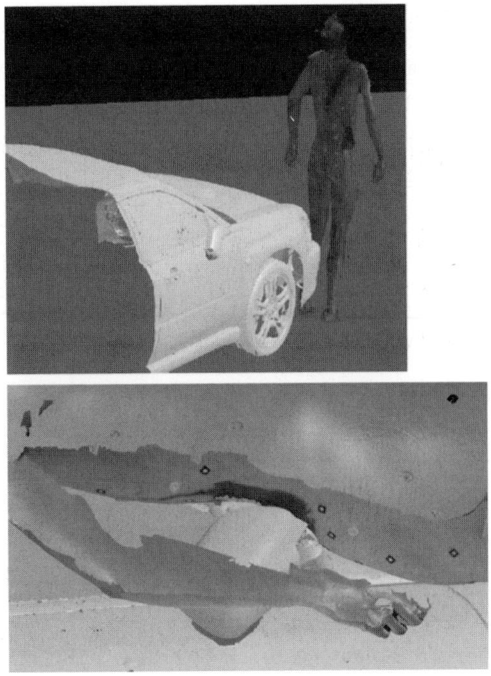

Figure 31.16. Top: 3-D model of involved vehicle and 3-D model of the injured pedestrian. Bottom: 3-D models of the injured corpse and the injury-causing instrument, the broken mirror of the car.

external body morphology of living or deceased persons, as well as of objects involved in the case (e.g. vehicles, weapons).

For the digitizing of the external body morphology, the forensic institute is using the structured light based surface scanning system ATOS III and the photogrammetric measurement system TRITOP, both of GOM (2008). The combined TRITOP/ATOS III system has the advantage that it can be used for surface documentation ranging from fine detailed structures, as skin lesion or fine instrument structures, to overview documentation, as whole body or entire vehicles. Additionally, the obtained 3-D models are fully coloured. Such a complete and detailed documentation could not previously be efficiently performed, in terms of time and quality, using classical approaches. Moreover, with the precise 3-D measurement of markers placed on the corpse, 3-D data acquired optically can be aligned and merged with the internal body morphology digitally documented using medical imaging systems such as multi-slice computed tomography (MSCT) and magnetic resonance imaging (MRI). Figure 31.15 shows the employed computer tomograph and the optical acquisition system.

The integration of the radiological data and the 3-D body surface data allow for a complete and detailed analysis of injuries. Figure 31.16 shows for example the 3-D surface model of the whole body of an injured pedestrian and the 3-D model of the involved vehicle. The 3-D human body model is completed with the internal body structure (bones, muscles, soft tissues, etc.) and the comprehensive injured damages. Both the 3-D surface model of the vehicle and the 3-D complete model of the corpse are very useful to determine the causes of the injury. In fact, the documented geometric findings (on and in the body, and involved objects) can be animated step by step or even by movie clips. Using data merging methods and animation, it is possible to answer reconstructive questions of the dynamic development of patterned injuries (i.e. morphologic imprints). In Figure 31.16, for example, the results achieved by the analysis of the case can be seen: the broken mirror of the involved vehicle was the cause of the fatal injury.

This example serves to illustrate how modern technologies of three-dimensional documentation can open new possibilities for forensic reconstruction by bringing added values and real quality improvements into forensic science. Moreover, they enable real data based 3-D reconstructions at any time, without the need for storage of the bodies, vehicles and instruments. By bringing the body into a virtual 3-D accident or crime scene, further information about the course of events can be delivered than in the conventional way. This is path-breaking for accident and crime scene analyses in the future.

31.3.2.5 *Dermatology, cosmetics*

Dermatology and cosmetics represent medical fields where optical 3-D measurement techniques can be successfully employed. For this reason, manufacturers of 3-D scanning equipment have developed products especially dedicated to the 3-D measurement of the human skin, as for example the handheld 3-D skin scanner PRIMOS compact portable of GFMesstechnik GmbH (Benderoth 2005). This optical measuring system, shown in Figure 31.17, is based on digital fringe projection technology and is designed to measure with extremely high resolution (about 60 microns in the plane and 6 microns in the vertical direction) in order to acquire the fine structures of the human skin. The main fields of application of such devices are, in fact, the 3-D measurement of wounds, scars, wrinkles, melanoma, naevi, birthmarks, etc. By means of analysis software, the received skin profile can then be evaluated in terms of all parameters of interest such as roughness, volumes of wrinkles or scars or, respectively, their geometrical dimensions.

An interesting application of 3-D skin measurement in dermatology is presented by the work of Smalls (2005). In this case, cooperation between private companies and medical universities has resulted in the implementation of a laser surface scanning system of Cyberware (2008) along with analysis

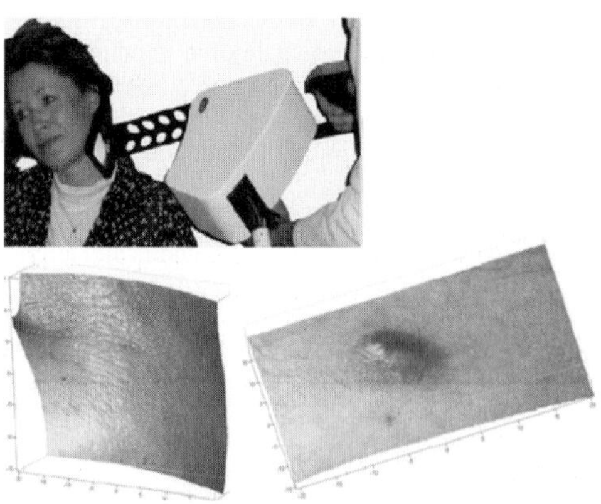

Figure 31.17. 3-D data acquisition by PRIMOS compact portable and 3-D data examples: wrinkle around the eye and wound measurements.

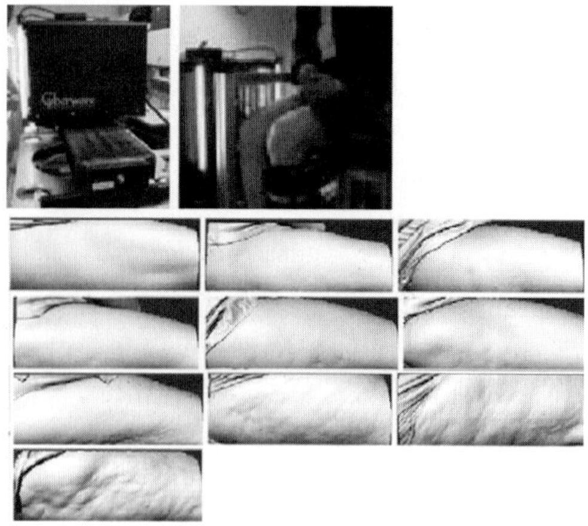

Figure 31.18. Top: the 3-D laser scanning system. Bottom: 3-D scan images arranged with increasing cellulite severity grades.

software in order to quantify gynoid lipodystrophy, more commonly known as cellulite. Cellulite affects 85% of post-adolescent women and is for this reason of chief interest to the pharmaceutical and cosmetic industry. The distinctive surface morphology is believed to result when subcutaneous adipose tissue protrudes into the lower reticular dermis, thereby creating irregularities at the surface.

The three-dimensional laser scanning system was employed in order to objectively characterize and quantify the three-dimensional features of the skin surface for this specific work. The scanner was used to capture surface data of cellulite-affected thigh sites in women with varying degrees of cellulite, as well as subjects exhibiting non cellulite. Three-dimensional skin surface data were captured by using a customized laser scanning solution of Cyberware (2008): a laser scanning head situated on a linear platform, with user defined platform speed and scanner resolution. The outer aspects of both thighs were scanned while subjects were seated on a level surface, with knees bent at a 90° angle, as shown in Figure 31.18.

The acquired surface data were analysed to quantify the skin surface morphology and to determine specific roughness values. Customized software solutions were employed to firstly remove features not associated with cellulite, including thigh curvature, noise, hair, varicose veins. The edited 3-D data were finally processed in order to calculate the surface roughness parameters for the region of interest. A comparison of the roughness parameters extracted from the 3-D scan data, with the grades (on a scale from 0 to 9) established by an expert by visual analysis, proved the correctness of the method.

In Figure 31.18, on the bottom are shSown details of the acquired 3-D surface of the thighs skin arranged by increasing grade of cellulite.

31.4 FINAL REMARKS AND CONCLUSIONS

This study of the applications of image measurement for purposes involving human beings shows that they are not restricted to simple static measurement of body parts. In fact, the more common usage and the more exciting applications involve the relatively complex challenges caused by movement, whether for studies of the human body in motion or in the exciting area of surgical navigation. It can also be discerned from this review of medical measurement that among the various application areas there is a range of levels of acceptance of the use of spatial data from image measurement. It is more noticeable that many commercial organisations are involved in developing and marketing imaging systems, which have medical or other related human applications.

Overall, medical measurement is not a field that is solely the field of study of conventional measurement specialists as it is a field that involves a variety of scientists and engineers who presumably wish to contribute to bettering the lives of people. For the future, the surgical field in particular would seem to offer opportunities for further development, as it is clearly work that has not reached its ultimate objective.

REFERENCES

Achard de Leluardière, F., Hajri, L., Lacouture, P., Duboy, J., Frelut, M. & Peres, G. 2006. Validation and influence of anthropometric and kinematic models of obese teenagers in vertical jump performance and mechanical internal energy expenditure. *Gait and Posture* 23(2): 149–158.

Ariel Dynamics Inc. 2008. http://www.arielnet.com. Accessed 21.02.2008.

Axios 3D Services GmbH. 2008. Complete solution for 3D surgical navigation systems. http://www.axios3d.de. Accessed 21.02.2008.

Baroni, G., Ferrigno, G. & Pedotti, A. 1998. Implementation and application of real-time motion analysis based on passive markers. *Medical and Biological Enginering and Computing* 36: 693–703.

Baroni, G., Troia, A., Riboldi, M., Orecchia, R., Ferrigno, G. & Pedotti, A. 2003. Evaluation of methods for opto-electronic body surface sensing applied to patient position control in breast radiation therapy. *Medical and Biological Engineering and Computing* 41: 679–688.

Benderoth, C. 2005. PRIMOS handheld and PRIMOS body II: New products from GFM—Germany and applications in maxillofacial surgery. In N. D'Apuzzo (ed.), *Human Body Measurement Newsletter, Vol.1 No.2.* Zurich, Switzerland. http://www.hometrica.ch/en/newsletter.html. Accessed 21.02.2008.

Bert, C., Metheany, K.G., Doppke, K.P., Taghian, A.G., Powell, S.N. & Chen, G.T.Y. 2006. Clinical experience with a 3-D surface patient setup system for alignment of partial-breast irradiation patients. *International Journal of Radiation Oncology, Biology and Physics* 64: 1265–1274.

Biogesta. 2008. http://www.biogesta.fr. Accessed 21.02.2008.

Blumenschein, P. 2006. State of the art of the 3-D scanning devices used for dental applications: open or closed systems. *3-D Modelling, Proceedings of Conference, June 13–14, 2006, Paris, France.*

Bonacini, D., Corradini, C. & Magrassi, G. 2007. 3-D Digital Model Reconstruction: Residual Limb Analysis to Improve Prosthesis Design. In A. Gruen & H. Kahmen (eds.), *Optical 3-D Measurement Techniques VIII.* Zurich, Switzerland.

BrainLAB AG. 2008. *VectorVision Knee*. http://www.brainlab.com Accessed 21.02.2008.

Broers, H. & Jansing, N. 2007. How precise is navigation for minimally invasive surgery? *International Orthopaedics* 31(1).

Broström, E., Örtqvist, M., Haglund-Åkerlind, Y., Stefan Hagelberg, S. & Gutierrez-Farewik, E. 2007. Trunk and center of mass movements during gait in children with juvenile idiopathic arthritis. *Human Movement Science* 26(2): 296–305.

BTS SpA. 2008. http://www.bts.it Accessed 21.02.2008.

Buskirk, T.D. et al. 2005. 3D Body Scanner: an aid in planning body contouring surgery. *Proc. of Plastic Surgery 2005, September 24–28, Chicago, USA.*

Charnwood Dynamics Ltd. 2008. http://www.charndyn.com Accessed 21.02.2008.

Christensen, G.J. 2006. Is now the time to purchase an in-office CAD/CAM device? *Journal of the American Dental Association* 137: 235–238.

C-Motion Inc. 2008. http://www.c-motion.com Accessed 21.02.2008.

Cyberware Inc. 2008. http://www.cyberware.com Accessed 21.02.2008.

D'Apuzzo, N. 2000. Motion capture by least squares matching tracking algorithm. In *AVATARS'2000 Workshop, November 30–December 1, 2000, Lausanne, Switzerland. IFIP Conference Proceedings*: 196. Kluwer. ISBN 0-7923-7446-0.

D4D Technologies L.P. 2008. *E4D intraolar 3D dental surface digitizer.* http://www.d4dtech.com Accessed 21.02.2008.

Ehara, Y., Fujimoto, H., Miyazaki, S., Tanaka, S. & Yamamoto, S. 1995. Comparison of the performance of 3-D camera systems. *Gait and Posture* 3: 166–169.

Ehara, Y. Ehara, Y., Fujimoto, H., Miyazaki, S., Mochimaru, M., Tanaka, S. & Yamamoto, S. 1997. Comparison of the performance of 3-D camera systems II. *Gait and Posture* 5: 251–255.

Ferrigno, G. & Pedotti, A. 1985. ELITE: a digital dedicated hardware system for movement analysis via TV signal processing. *IEEE Transaction Biomedical Engineering* BME-32: 943–950.

Ford, K.R., Myer, G.D. & Hewett, T.E. 2003. Valgus Knee Motion during Landing in High School Female and Male Basketball Players. *Medicine and Science in Sports and Exercise* 35(10): 1745–1750.

Gao, Z., Pandya, S., Arana, C. & Hwang, N.H.C. 2000. Bioprosthetic heart valve leaflet motion monitored by dual camera stereo photogrammetry. *Journal of Biomechanics* 33(2): 199–207.

Garrido Castro, J.L., Medina-Carnicer, R. & Galisteo, A.M. 2006. Design and evaluation of a new three-dimensional motion capture system based on video. *Gait and Posture* 24(1): 126–129.

GOM mbH. 2008. http://www.gom.com Accessed 21.02.2008.

Hoyer, K., Holzner, M., Guala, M. & Kinzelbach,W. 2006. Simultaneous 3D-flow field and compliant wall measurements in a abdominal aortic aneurysm flow using scanning-PTV. ISPRS *Commission V Symposium, September 25–27. International Archives of the Photogrammetry, Remote Sensing and Spatial Information Sciences* XXXVI, Part 5.

InSpeck Inc. 2008. *Breast and Facial Digitizer BFD 1300*. http://www.inspeck.com/products/PlasticSurgery/PlasticSurfery.asp Accessed 21.02.2008.

IVB Jena GmbH. 2008. *GSCAN 3D Body Scanner*. http://www.ivb-jena.de Accessed 21.02.2008.

Krosshaug, T., Nakamae, A., Boden, A., Engebretsen, L., Smith, G., Slauterbeck, J., Hewett, T.E. & Bahr, R. 2007. Estimating 3D joint kinematics from video sequences of running and cutting maneuvers—assessing the accuracy of simple visual inspection. *Gait and Posture* 26(3): 378–385.

Liu, H., Holt, C. & Evans, S. 2007. Accuracy and repeatability of an optical motion analysis system for measuring small deformations of biological tissues. *Journal of Biomechanics* 40(1): 210–214.

Meinecke, L., Breitbach-Faller, N., Bartz, C., Damen, R., Rau G. & Disselhorst-Klug, C. 2006. Movement analysis in the early detection of newborns at risk for developing spasticity due to infantile cerebral palsy. *Human Movement Science* 25(2): 125–144.

Motion Analysis Corporation. 2008. http://www.motionanalysis.com Accessed 21.02.2008.

Naether S., Buck U. & Thali J.M. 2007. Virtopsy—Scientific 3-D Documentation of Human Bodies in Forensic Medicine: with High Resolution Surface Scanning Radiological MSCT/MRI Scanning. In A. Gruen & H. Kahmen (eds.), *Optical 3-D Measurement Techniques VIII*. Zurich, Switzerland.

Northern Digital Inc. 2008. http://www.ndigital.com Accessed 21.02.2008.

Novaptus Systems, Inc. 2008, *Novaptus Systems Scanner Suite and Novaptus Virtual Surgery Suite*. http://www.navaptus.com Accessed 21.02.2008.

Oksa, J., Rintama, H. & Rissanen, S. 1997. Muscle performance and electromyogram activity of the lower leg muscles with different levels of cold exposure. *European Journal of Applied Physiology* 75: 484–490.

Peak Performance Technologies, Inc. 2008. http://www.vicon.com/company Accessed 21.02.2008.

pedcad Orthopaedieprodukte GmbH. 2008. *PedCad scanning system*. http://www.pedcad.de Accessed 21.02.2008.

Phoenix Technologies Inc. 2008. http://www.ptiphoenix.com Accessed 21.02.2008.

Puce, A., Syngeniotis, A., Thompson, J.C., Abbott, D.F., Wheaton, K.J. & Castielloc, U. 2003. The human temporal lobe integrates facial form and motion: evidence from fMRI and ERP studies. *NeuroImage* 19: 861–869.

Qualisys AB. 2008. http://www.qualisys.se Accessed 21.02.2008.

Richards, J.G. 1999. The measurement of human motion: A comparison of commercially available systems. *Human Movement Science* 18(5): 589–602.

Rogus, R.D., Stern, R.L. & Kubo, H.D. 1999. Accuracy of a photogrammetry-based patient positioning and monitoring system for radiation therapy. *Medical Physics* 26(5): 721–728.

Raabe, A., Krishnan, R., Wolff, R., Hermann, E., Zimmermann, M. & Seifert, V. 2002. Laser Surface Scanning for Patient Registration in Intracranial Image-guided Surgery. *Neurosurgery* 50(4):797–803.

Stelzer, S., Schreiber, S. & Schoene, C. 2006. Utilisation of 3D Face Scanner in Medical Faculty. *3D Modelling, Proceedings of Conference, June 13–14, 2006, Paris, France.*

Smalls, L.R.K. 2005. Development of quantitative models for the investigation of gynoid lipodystrophy (cellulite). PhD Thesis, University of Cincinnati, USA, April 2005.

Thali, J.M. & Aghayev, E. 2005.Virtopsy research project, documentation of the internal and external body morphology in forensic medicine. In N. D'Apuzzo (ed.), *Human Body Measurement Newsletter, Vol.1 No.2*. Zurich, Switzerland. http://www.hometrica.ch/en/newsletter.html Accessed 21.02.2008.

Troje, N.F. 2002. Decomposing biological motion: A framework for analysis and synthesis of human gait patterns. *Journal of Vision* 2(5)2: 371–387.

Trost, L., Stines, S. & Burt, L. 2006. Making informed decisions about incorporating a CAD/CAM system into dental practice. *Journal of the American Dental Association* 137: 32s–36s.

Vicon Motion Systems. 2008. http://www.vicon.com Accessed 21.02.2008.

Vision RT Ltd. 2008. http://www.visionrt.com Accessed 21.02.2008.

Vorum Research Corporation. 2008. *CANFIT-PLUS P&O System*. http://www.vorum.com Accessed 21.02.2008.

Walton, J.S. (ed.). 1990. *Image-based motion measurement, Proc. of Symposium, University of California, San Diego*. Bellingham, USA: International Society for Optical Engineering, SPIE.

Whittle, M.W. & Levine, D. 1999. Three-dimensional relationships between the movements of the pelvis and lumbar spine during normal gait. *Human Movement Science* 18(5): 681–692.

Williams, R.J., Bibb, R., Eggbeer, D. & Collis, J. 2006. Use of CAD/CAM technology to fabricate a removable partial denture framework. *Journal of Prosthetic Dentistry* 96: 96–99.

Advances in Photogrammetry, Remote Sensing and Spatial Information Sciences: 2008 ISPRS Congress Book – Li, Chen & Baltsavias (eds)

CHAPTER 32

Forestry applications

Barbara Koch & Matthias Dees
(with other contributors listed in the end of paper)

ABSTRACT: Starting with an in depth description of the user requirements that highlight an increased demand for up to date information on forests and of cost-efficient monitoring systems, applications using different data sources and techniques at various geographic levels are presented. The status of the established operational use of airborne and spaceborne optical data is shown together with recent developments. New opportunities arising out of the option to use LiDAR data on the forest enterprise and regional inventory level are presented. The opportunity of height measurements and information on the vertical forest structure opens a wide set of new applications in the forest domain, especially in combination with optical data. Radar data from spaceborne system are offering new possibilities for regional to large scale forest monitoring due to a new generation of sensors that are being implemented in this decade. These will allow monitoring systems that are not restricted by cloud coverage. Furthermore, options to combine different data sources and techniques for the challenge to assess changes over time with remote sensing are shown. Two themes of high importance in the forestry sector are highlighted in specific sub-sections: forest fire monitoring and the use of remote sensing in combination with field sampling data.

Keywords: Optical sensors, SAR, LIDAR, forest management, national forest inventories, global forest monitoring, forest fire, biodiversity, UNFCCC, energy.

32.1 INTRODUCTION

Forest applications that are based solely, mainly or partly on remote sensing are manifold. They could be structured by data type (e.g. optical, SAR, LIDAR or combinations of these), geographic scope of the application (local, sub-national, national, supra-national, global), by the information need addressed (local forest planning, legal forest register, forest policy, UNFCCC, forest fire, storm damage assessment, afforestation control) or by information retrieval and data utilization technique. Forestry applications cover a wide variety of applications, some having been operational for decades, while others have only appeared recently and are undergoing fast development.

To structure this great variety of forestry applications, the section starts with a brief description of the information needs, followed by the presentation of the status quo and recent developments of forestry applications per remote sensing data type used: passive optical, LiDAR, radar and multi-sensoral. Then specific attention is given in sub-chapters to the detection of change (an important cross cutting issue), forest fire (an important thread to forests that needs specific remote sensing techniques) and on utilizing remote sensing data in combination with sample based *in situ* data (an important approach to using remote sensing data in the forest sector).

32.2 INFORMATION NEEDS IN FORESTRY

Forest information requirements have grown over time and will continue to do so in the future. Prior to the 1960s, most of the focus was on the timber resource for commercial use. Since then society has moved from an economic focus on wood resources towards multi-functional forestry, supported by multi-resource inventories. This reflects the development of a more inclusive attention by society and policy for the environment, thereby recognizing the important role of forests in the functioning of the global ecosystem (see Table 32.1).

The information and reporting requirements on forests from international environmental conventions and agreements and their down-stream policy implementation often create a potential role for Earth Observation implicitly. Policy has acknowledged as

Table 32.1. Growing forest information needs over time at the global level. Based on the development of thematic elements in Global and Regional Forest Resources Assessments of the Food and Agricultural Organisation.

Pre-1960s	1960s	1970s	1980s	1990s	2000+
Extent of forest resources (area of forest, by species groups, by ownership and productivity categories)					
Productive functions of forest (growing stock on forest, by species groups)					
			Forest Health and vitality		
				Biological diversity	
				Protective functions of forest	
				Social services, NWFP	
				Carbon balance	
					LULUCF[1]
					Biomass for energy
					Forests and water

[1]LULUCF: Land Use, Land-Use Change and Forestry such as defined in the IPCC Good Practice Guidelines on greenhouse gas reporting to the United Nations Framework Conference on Climate Change.

early as at the United Nations Conference on Environment and Development (1992) that Earth Observation should be used in data collection as a cost-effective and efficient means for the collection of large amounts of data with high levels of detail and accuracy.

Several over-arching high level forest-related reporting obligations and interrelated information needs emerge from international and national policy drivers as well as market-based drivers for geographical information on forests. International policy was reviewed and included in the United Nations Environmental Conferences, the Framework Convention on Climate Change and its Kyoto Protocol, the Convention on Biological Diversity and the Convention to Combat Desertification; also the United Nations Forum on Forests and processes based on reporting by Criteria and Indicators (such as the Ministerial Conference on the Protection of Forests in Europe), national forest programmes and market-related issues such as biomass for the production of renewable energy, forest law enforcement, governance and trade (FLEGT) and forest certification were covered (Van Brusselen & Schuck 2005). While these policy and market drivers differ in their main objectives, they often require identical or similar information.

The identification of these information needs allows reaction to them in a coordinated way, thus taking into account different requirements simultaneously.

Three major categories of information needs on forests emerged from policy reviews. These are: (1) Land Cover Area (surface area and specification of location and boundaries of an area and changes per land cover class); (2) Biomass (biomass as such, volume, changes of biomass and volume, growth and increment); and (3) Disturbances (biotical damage and abiotical damage, including atmospheric pollution effects on forests, flooding, drought, forest fires and effects of climate change).

On forest management level, the information needs are driven by the need to manage forests as economic, ecological and recreational resources. Typically, both a separation into forest stands and management units is required together with a forest description covering a wide set of attributes including forest types, species, assortments and ecological parameters.

Reviews of demand for remote sensing products generally identified a strong potential for increases in geo-spatial monitoring capacity. Several potential users could benefit from remote sensing support to monitor activities; however this would depend on external financial assistance to achieve their objectives. The same applies for increasing GIS penetration of forest administrations and also downstream implementation of GIS systems to assist *in situ* forest management (Van Brusselen 2006a).

User-specific definitions of observed parameters, application of best practice, minimum mapping unit and geographical extent are often considered to be crucial when considering the success of remote sensing implementation.

Both the EO and the forestry sector are working towards intensifying the cooperation in combining *in situ* and *ex situ* data in order to develop highly accurate, flexible and timely products and services. This requires as a basis the establishment of flexible and transparent data access policies.

An assessment of future user needs and requirements amongst a group of existing users of remote sensing in the forest sector in Europe revealed the following important needs for the period 2008–2012: monitoring of forest area by forest types and by detailed natural habitat categories; tree species composition; forest fragmentation; landscape pattern; storm damage assessment; flooding areas and soil moisture; stand density maps and stand density change maps (for thinning assessments and planning);

and tree height layer. Issues as 'likely places for forest conversion and forest expansion' have been also identified as emerging needs both in Europe and North America (Van Brusselen 2006b).

32.3 PASSIVE OPTICAL DATA

32.3.1 *Airborne optical data*

Historically, aerial photography represents the most popular application of remote sensing in forestry (Spurr 1960, Aldrich 1979, Gillis & Leckie 1996, Lund 1997). It was especially successfully incorporated in operational procedures of forest practice at local level in forest management inventories and mapping and as a traditional source of national forest inventories. Well accepted applications for aerial photographs comprise the measurement and analysis in ortho-rectified products as well as in stereoscopic mode. For many years, airborne optical data have been successfully incorporated to support large area mapping projects, in the acquisition of training areas, or as a source of verification data (White et al. 2005). Aerial photographs have been mainly used for forest inventory parameters such as forest species mapping, stand volume estimation, tree height estimation (Magnusson et al. 2007) and to a smaller extent for the estimation of forest biophysical parameters (Hoffmann 2008). In addition, forests have been divided into management units with the help of aerial photographs. A considerable part of the literature is related to methods developed to recognize individual tree crowns of forest species (Wulder 1998, Bolduc et al. 1999, Gross 2000).

New perspectives and opportunities for forest applications are given by digital airborne data obtained from Ultracam or ADS40, which provide stereo-overlap of up to 90% or entire image strips with higher radiometric resolution. Considering analogue and digital airborne optical data, much of the literature is related to methods developed to detect single trees (Leckie et al. 2005, Hirschmugl et al. 2007), individual tree crowns of forest species (Wulder 1998, Bolduc et al. 1999, Gross 2000, Wang et al. 2004). Combining airborne optical data with 3-D information from others sources, e.g. LIDAR data, offers a wider range of information retrieval on forest attributes (St- Onge & Achaichia 2001, St-Onge et al. 2004), (see also Chapter 6 Multisource data).

Future technical development also aims at the (semi-)automatic analysis with the help of segmentation and classification involving approaches such as fuzzy logic or artificial neural networks (Pillai et al. 2005, Binner et al. 2006, Mallinis et al. 2008).

Using shape and colour characteristics existing in CIR photographs, it is possible to assess forest health by identifying damage levels of single tree crowns or stands. Photo-interpretation offers an alternative to the practical monitoring systems dominating terrestrial tree damage assessment. An example is the interpretation of aerial CIR images within the assessment of forest health status in the Czech Republic (Zidek 2001). Here, during the 20th century, mountain forests have been heavily damaged by air pollution (SO_2). Levels of the damage of mature spruce stands were estimated by aerial photographs interpretation based on large scale colour infrared aerial photographs (1:7000) using a European wide standardized interpretation key (Gross 2000) that has been developed for detailed single tree crown analysis.

Aerial photography can also be integrated in a process for complex mapping tasks such as the mapping of habitat types as required for EU Natura 2000 protection sites. Such an approach has been developed and is currently implemented by the Bavarian forest administration for mapping and monitoring of Alpine biotopes, where using field work on 100% of the large habitat areas for that purpose would be a financially unaffordable solution. The approach combines the results of complex modelling of existing GIS data and of a semi-automated analysis of CIR images (Binner et al. 2006). Ortho-rectified CIR aerial photographs are merged, segmented (see Fig. 32.1) and classified with a nearest-neighbourhood classifier based on visually selected reference areas into classes "forest", "open land" and "miscellaneous". Using decision matrices on all segments qualified as "forest", potential habitats and habitat types are then distilled out of a set of relevant geo-parameters like exposition, slope and sun exposure time. This method has been

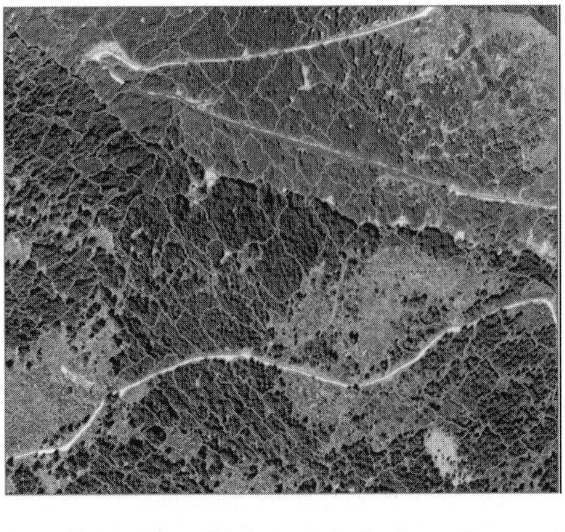

Figure 32.1. CIR aerial photograph after the semi-automatic object-based segmentation used for habitat mapping by the forest administration of the German federal state of Bavaria for monitoring of Alpine habitats within EU Natura 2000 protection sites.

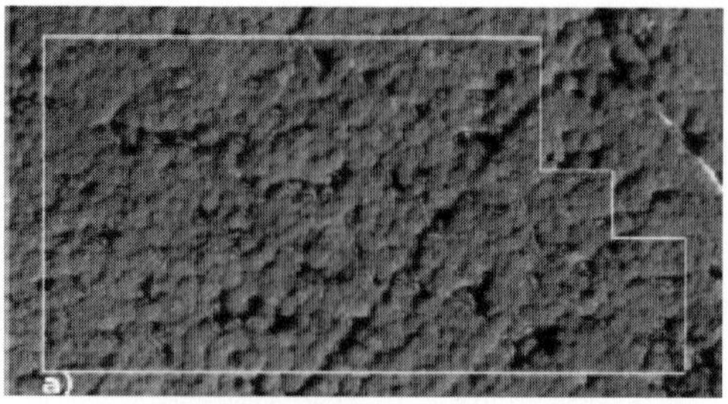

Figure 32.2. Full automatic assessment of gaps (Nuske 2006). (see colour plate page 518)

tested in an alpine Natura 2000 forest protection site of the German federal state of Bavaria and proved to be entirely adequate for the majority (about 85%) of the Natura 2000 habitat types. The approach allows high quality mapping at moderate cost by avoiding cost intensive fieldwork (Binner et al. 2006).

The future application of digital camera data will further improve the capabilities of landscape analysis for the (semi-) automatic assessment of structural parameters for habitat validation. An example is shown in Figure 32.2. Details of future options based on this technology are given in Waser & Ginsler (2008).

First approaches concerning the automatic detection of single tree crowns allow the optimistic view that the human interpreter will be assisted decisively by technical automations in the future (Hájek 2007, Waser & Ginsler 2008).

32.3.2 *Spaceborne optical data*

32.3.2.1 *Introduction*

The capacity of optical satellite systems to record data on the Earth have proved to be useful for forest monitoring, to study landscape patterns or climate change impacts. They serve our understanding of the Earth system to predict environmental changes and based on this knowledge help to mitigate negative effects of global change. Satellites also help scientists to better understand the carbon cycle by providing measurements of some of the variables required as inputs to carbon-cycle models. In the frame of the Global Monitoring for Environment and Security Programme (GMES), forestry related issues are considered especially important. A new generation of spaceborne optical systems is going to be launched from commercial industry in the very near future, providing better temporal resolutions and, therefore, offers of better opportunities for a wide range of forestry related applications.

32.3.2.2 *Spaceborne optical systems*

Currently available low spatial resolution satellite systems, i.e. WiFS, MODIS or SPOT VEGETATION, have proven to be a big advantage in the ascertainment of forest areas over huge regions, although the low spatial resolutions can only deliver small scale mapping. For example, in the Alpine region this kind of data was used to calculate the spatial pattern of forest cover in terms of forest area percentage or length of forest border in order to obtain trans-national homogenous data over the whole Alpine bow. The potential of these low resolution optical data could be used for forest topics related to climate change issues on a global scale.

For instance, operational applications on the derivation of the forest area have been reported for over a decade (Schardt et al. 2000). The most commonly used methods for forest delineation are thresholding, maximum likelihood classification, segmentation followed by nearest neighbour classification (Schardt et al. 2004; Granica et al. 2000; Gallaun et al. 2001, Ramminger & Koch 2007), visual interpretation (Ramminger & Koch 2007), but also unsupervised classification. A wall to wall map of the forest cover of the tropical belt was produced in the Tropical Ecosystem Environment Observation by Satellite (TREES) project (Malingreau et al. 1995, Achard et al. 2002). Häme et al. (2001) developed a method for a pan-European forest proportion map. The image data were first classified using an unsupervised clustering method. The method was applied over a mosaic of 49 Advanced Very High Resolution Radiometer (AVHRR) images. Also the medium resolution satellites, e.g. Landsat TM, SPOT or ASTER, are frequently used to provide forest cover map for large regions, e.g. for the European Community (Pekkarinen et al. 2007) or e.g. in Quebec, Canada (Valeria et al. 2007).

These parameters have been integrated on an operational basis into forest information systems for forest protection planning or forest management tasks during several projects in European countries (Granica et al. 2006, Hirschmugl et al. 2007, Gallaun et al. 2007). Within the project "GMES [Global Monitoring for Environment & Security] Service Element Forest Monitoring" several pre-operational forest information services and real applications have been initiated in Europe (Häusler et al. 2004, Ramminger & Koch 2007) (see Fig. 32.3). For the determination of forest habitat types in a pre-alpine area in Bavaria, ASTER and SPOT5 data were successfully used (Förster et al. 2005).

The five-satellite based optical system from the company RapidEye, although targeted on agricultural monitoring, expected to start in 2008/9, will have a high repetition rate, which will also be useful for forestry applications needing rapid and frequent observations, as for example the monitoring of windfall areas.

Sentinel-2 will be a multi-spectral imaging mission, jointly implemented by ESA and the EC (European Commission), at high resolution with high revisit capability to provide enhanced continuity of

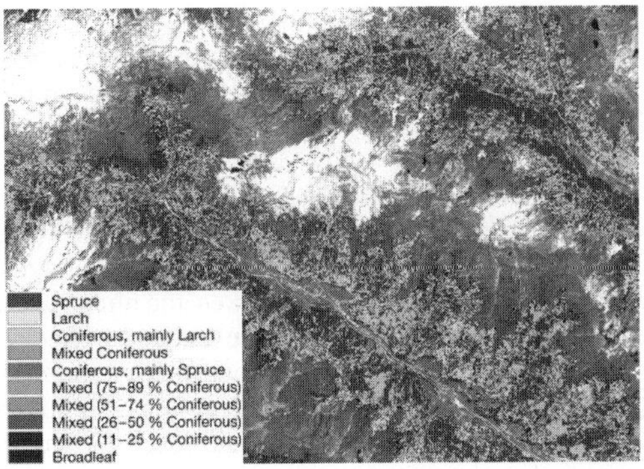

Figure 32.3. Classification result for tree species distribution using SPOT5 imagery in Austria (Schardt et al. 2004). (see colour plate page 519)

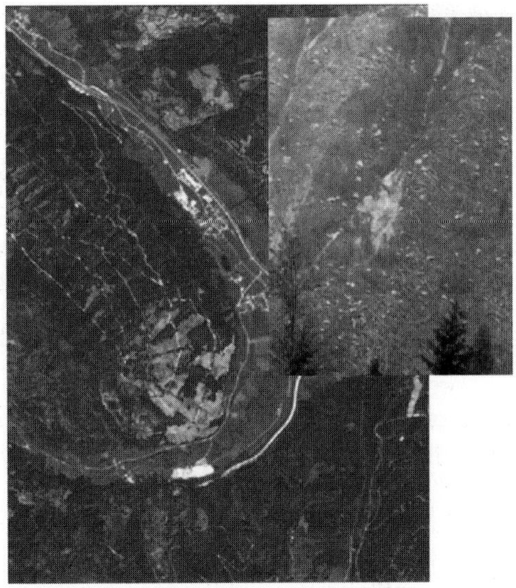

Figure 32.4. Automated derivation of windfall areas after a heavy storm in November 2002 in Salzburg using SPOT Imagery. (see colour plate page 519)

data so far provided by SPOT-5 and Landsat-7. The mission is dedicated to the full and systematic coverage of land surface globally with the objective of providing cloud-free products typically every 15 to 30 days over Europe and Africa. These systems, further complemented by the PLEIADES systems, are important to prolong the very well established derivation of forestry parameters from satellite systems at resolutions of about 5 m—30 m, thus enabling a continuous observation of forest areas over larger areas that are used within the global forest resource assessment of the FAO and in many national forest inventories (see Section 9).

The currently running VHR (very high resolution) optical systems, i.e. IKONOS, QUICKBIRD, KOMPSAT-2, EROS, ORBVIEW-3, CARTOSAT-1, have proved that their imagery can be used for the recognition of even single trees (Hirschmugl et al. 2005, 2007, Larsen 2007).

Several forthcoming VHR optical systems are designed to record with spatial resolutions up to 40 cm in the panchromatic mode and up to 1.65 m for the multispectral channels (GEOEYE1, RAPIDEYE, Worldview1, Worldview2, Worldview3, Pleiades). If this type of data is recorded in a stereo mode, it can deliver information on biomass and timber volume, with multitemporal data changes, such as growth or increment, also being measured. The derivation of disturbances in forestry, as biotic or abiotic damage, can be considered as a further capacity from these systems. An example is shown in Figure 32.4.

32.3.2.3 *Conclusions*

In many applications it could be demonstrated that the currently running systems are successfully used in different forestry related disciplines, like forest monitoring, disaster management or climate change monitoring. Specifically, the derivation of the forest parameters of forest area assessment, tree species distribution, crown coverage or growth classes can be performed on an operational basis from medium resolution optical satellites. Sustainability of such applications will be enabled through the forthcoming systems that are promoted by ESA and the EC GMES programme, as well as by the Landsat data continuity mission. The currently planned optical VHR spaceborne systems, which include systems with stereo mode, show a high potential for deriving more detailed information of forests, e.g. biomass or disturbances. The combination of the various spatial resolutions from the different systems can satisfy manifold needs for forestry related issues and will be used operationally for a broad spectrum of forest topics.

32.4 LiDAR DATA

32.4.1 *Introduction*

Especially during the last 10 years, the use of LiDAR data for local forestry applications has become of increasing interest. This is due to the fact that LiDAR data seem to provide very good options of assessing not only height or crown surface structure but also very detailed information about the vertical and horizontal structure of forests, allowing estimation of volume, crown density and crown layers. The need for forest information is no longer only focused on requirements related to timber production; biological diversity, carbon sequestration or general environmental protection are becoming equally important. Therefore, a system that provides both manifold information on forests and also on the terrain covered by forests is of highest interest. Today small footprint airborne laser scanner instruments (ALS) are generally used for precise measurements. The ALS instruments store reflected energy as geo-referenced point clouds. This geometric information is mainly used to model forest objects and extract relevant parameters with regard to the information needs in forestry. Besides the geometric information extracted from ALS data, physically-related information like intensity is of increasing interest for object characterization. A summary of laser scanning basics is given in Baltsavias (1999a,b) and Wehr & Lohr (1999). Today, the majority of the commercial laser systems are able to collect between 20,000 and 100,000 points per second. There are two airborne systems on the market, the discrete point laser systems, which in the past normally registered first and last echoes and, since 2004, also the full waveform systems. A description of the benefits of full waveform LiDAR is provided by Wagner et al. (2007) who also describe waveform analysis techniques.

32.4.2 *DSM and DTM modelling*

The use of ALS data in forestry can be divided into different applications. Initially, the development focus was on Digital Terrain Models (DTMs) and Digital Surface Models (DSMs), which are used to describe terrain and forest surfaces. Different algorithms are available for the calculation of DTMs. Kraus & Pfeiffer (1998) were among the first to develop an algorithm for DTMs under forests by distinguishing terrain and non-terrain points. General methods are described in Chapter III-5 of this publication. Elmquist et al. (2001) and Bretar & Chehata (2007) describe alternative approaches to generate DTMs under forest cover or vegetated areas (Fig. 32.5).

A comparison of different approaches can be found in Sithole & Vosselmann (2004). The quality of DTMs as well as DSMs is not only influenced by the algorithms used but also by the data taken and object characteristics. These influences are described by Hyyppä et al. (2000), Raber et al. (2002) and Hodgson et al. (2003). As Hyyppä et al. (2005) reports, under boreal forest conditions, with gentle terrain and a pulse density of more than 2 points per sqm, random errors of less than 20 cm are normal. Nevertheless, investigations in very difficult terrain and under dense forest cover indicate that larger errors, up to several metres are possible, especially along break lines.

32.4.3 *Discrete laser systems*

For the extraction of forest-related parameters, either a Crown Height Model (CHM) is calculated or raw data are used. The quality of the CHM and the quality of information directly extracted from raw data depends on the operational properties, the model algorithms, as well as on the object characteristics. Since the LiDAR datasets are gathered using spatial sampling by a scanning device, it is expected that the operational characteristics of the instrument as well as the sampling process might have an impact on the data acquired for a forest canopy. The issue is addressed using two different methodologies: (i) LiDAR surveys for forest areas using different operational conditions (i.e. flight altitude, scan angle etc.) (Maltamo et al. 2004, Yu et al. 2004, Hirata 2004, Næsset 2004, Goodwin et al.

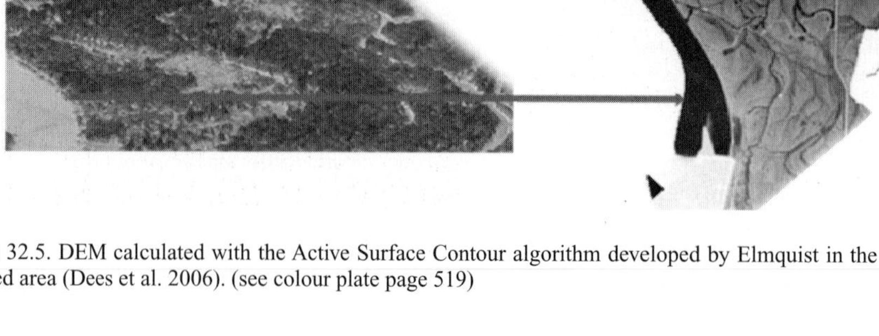

Figure 32.5. DEM calculated with the Active Surface Contour algorithm developed by Elmquist in the Rhine valley under forested area (Dees et al. 2006). (see colour plate page 519)

2006), and (ii) 3-D simulation models (Kalogirou 2007, Kalogirou et al. 2007, Lovell et al. 2005, Holmgren et al. 2003). It is noteworthy that such studies will not only contribute to more accurate LiDAR datasets but will also enable optimal and cost-effective planning of LiDAR surveys for forest environments. Results from studies based on both of the aforementioned methodologies suggest that increasing the frequency of pulses (high-pulse rate) increases the ability to detect individual trees and therefore improves the extracted CHM (Hyyppä & Inkinen 1999, Hirata 2004). Moreover, for higher flight altitudes, discrete return devices tend to record a single return, possibly due to the fact that the laser power is spread over a larger area and, thus, a weaker signal is returned from the understorey, which may be below the established threshold (Goodwin et al. 2006). Therefore, last-return pulse measurements tend to be affected by canopy conditions (Næsset 2005). There are also results describing the influence of the footprint diameter and scan angle, which seem to affect the penetration ability of the LiDAR signal in the forest canopy (Kalogirou 2007). Yu et al. (2004) and Gobakken & Naesset (2007) investigated the influence of system parameters on tree height estimation, which they found to be especially influenced by increasing flight altitude that is associated with decreasing point density. Figure 32.6 shows a simulation of a decreased point density from 2 points per sqm to 0.06 points per sqm that visualizes the possible difference in positions of reflected points within a conifer crown, which has a direct effect on height measurements.

LiDAR data can be used to retrieve a number of useful parameters that relate to the information needs for forests. There are two main approaches to assess this information: The geometrical information extracted from laser data is either used to derive estimations of forest parameters at stand level (Naesset et al. 2004, Holmgren 2004, Dees et al. 2006, Breidenbach et al. 2007, Koch et al. 2007b) or it is used to derive this information on an individual tree level (Wang et al. 2007a, Koch et al. 2006, Hyyppae et al. 2005, Maltamo et al. 2004). Irrespective of whether raw data or model data are used to derive forest-related parameters, the important information derived from LIDAR data reflected echoes is the estimation of height, crown cover or tree density. As reported by a number of authors (Hyyppae & Inkinen 1999, Persson et al. 2002, Lecki et al. 2003, Lefsky et al. 1999, Naesset et al. 2004, Chasmer et al. 2006, Dees et al. 2006), the height is typically underestimated by LiDAR data. In general, this underestimation is greater for cone-shaped trees like spruce or Douglas fir than for spherical-shaped trees like many broadleaf trees. However, Rönnholm et al. (2004) demonstrated that height estimations are reliable, even in leaf off conditions, if using a high point density. Different approaches are described in the literature for the detection of individual trees. In the past the separation of single trees was mainly based on CHM using the detection of local maxima for tree location and watershed or pouring algorithms (as well as derivations of these algorithms) for the delineation of single tree crowns (Gougeon 1995, Brandtberg et al. 2003, Popescu et al. 2003, Weinacker et al. 2004, Koch et al. 2006, Maltamo et al. 2006). However latterly, approaches are becoming increasingly based on raw data using clustering or blob detection methods (Morsdorf et al. 2003, Wack et al. 2003, Tiede et al. 2005 Wang et al. 2007a) (Fig. 32.7).

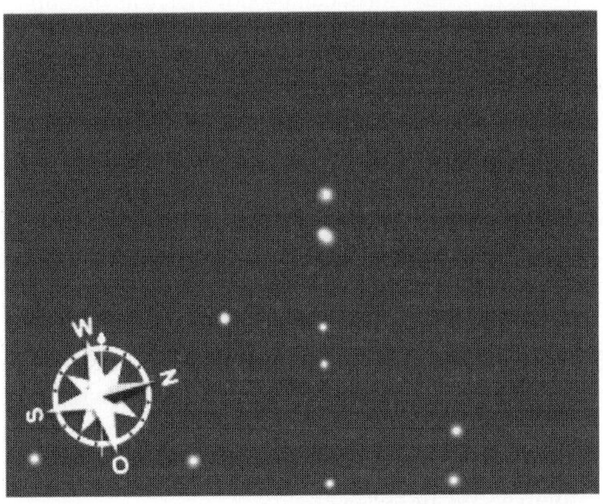

Figure 32.6. Vertical cut through a conifer crown with the reflected raw point positions for a laser point density of 2 per sqm and with simulated positions of reflected points (yellow) for a laser point density of 0.06 per sqm (Koch & Weinacker 2007). (see colour plate page 520)

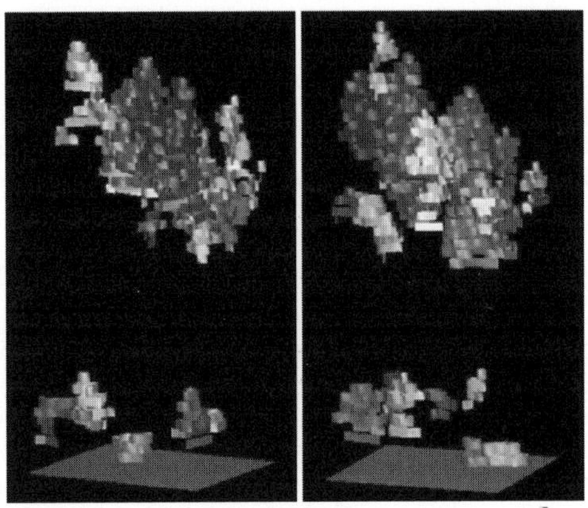

Figure 32.7. Single crown delineation based on blob detection method (Wang et al. 2007a). (see colour plate page 520)

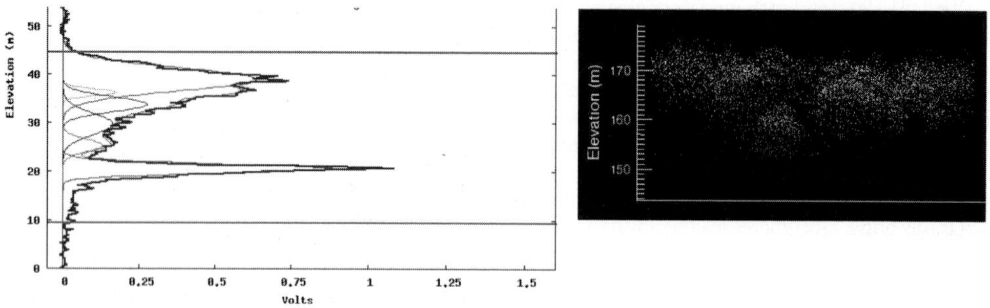

Figure 32.8. Bimodal full-waveform over vegetated area from airborne LiDAR (Rosette et al. 2007). (see colour plate page 520)

The single tree approach allows single tree counts, crown volume calculations, canopy closure or single tree height estimations. These are important inputs in order to derive estimations of diameter at breast height (DBH) distributions, volume or biomass. Published results show good results with 65% to 90% of correct tree counts within conifer stands (Hyyppä et al. 2001, Persson et al. 2002, Koch et al. 2006), while for broadleaf trees the results are less accurate (Hyyppä et al. 2001, Koch et al. 2006, Heurich 2006). An underestimation of volume is normally reported (Persson et al. 2002, van Aardt et al. 2006, Korpela 2007). For the area-based approach, the fundamental ideas have been described by Naesset (1997, 2002), Means et al. (2000) and Holmgren (2004). Investigations that analyse LiDAR data in combination with field sampling data, such as Anderson & Breidenbach (2007) and Breidenbach et al. (2007), show that with the stand-based approach, relatively good volume and biomass estimations are possible. As indicated by many authors, single tree-based approaches as well as area-based approaches show high potential for forest inventory tasks (Suárez et al. 2007, Maltamo et al. 2007). Further information extracted are forest/non-forest mapping, stand delineation, coniferous/deciduous tree identification, species detection, growth assessment, understorey and regeneration mapping, mapping of gaps and standards (Yu et al. 2004, Maltamo et al. 2006, Dees et al. 2006, Koch et al. 2007b).

32.4.4 *Full-wave laser systems*

Recently, investigations have been carried out using full-waveform laser systems (Reitberger et al. 2006, Reitberger et al. 2007, Wagner et al. 2007, GRL 2005, Gupta et al. 2007, Litkey et al. 2007). Most of the investigations have been based on airborne systems but a few have also used GLAS/ICESat spaceborne data. The validation of airborne full waveform data is at its initial stages and most of the research is still focused on data processing. Nevertheless, the first results strongly suggest that the use of full waveform data will improve the forest-related information content such as vegetation layer detection, tree counts, tree species identification and crown volume estimations.

The ability for the laser pulse to penetrate gaps between vegetation foliage means that the full waveform represents both the canopy profile and surface topography. The waveform is recorded within 1 nanosecond height bins, equivalent to 15 cm vertical resolution (Brenner et al. 2005, GRL 2005, Hese et al. 2005 and Zwaly et al. 2002) (See also Chapter III-5 of this publication). For vegetated footprints, a bimodal waveform is produced consisting of a broad, complex return from the canopy and a narrow ground peak (Fig. 32.8).

The varying amplitude within the waveform is a function of both the surface area of returns and the reflectivity of the intercepted surface. The airborne LiDAR point cloud illustrates how features for the same area are being represented within the waveform plot. The raw waveform can be simplified as the sum of six Gaussian peaks in order to facilitate interpretation.

The ICESat/GLAS spaceborne system produces footprints with an average equivalent circular area diameter of 64 m. The potential of satellite LiDAR data for forest applications has been recently investigated. Waveform indices such as estimated heights, percentiles of cumulative energy or waveform area can be used to infer important forest biophysical parameters such as vegetation height profile (Harding & Carabajal 2005), volume (Lefsky et al. 2005) and canopy cover (Rosette et al. 2007). This can be achieved using regression analysis with field or model predictions. Potential is shown for satellite LiDAR to provide sampling-based stemwood volume estimates over large areas.

32.4.5 *Outlook*

In the future there will be more focus on the combination of LiDAR data with other data types (Rönnholm et al. 2007, Tiede et al. 2006 and Korpela et al. 2007). Tree species classification in particular may be improved using combined datasets of LiDAR and spectral data. In addition, more attention has recently been given to the use of intensity values from laser data (Wagner et al. 2007, Orka et al. 2007). Although the research is at its early stages, first investigations indicate that intensity values could be helpful for species classification.

32.5 RADAR DATA

32.5.1 *Introduction*

With radar it is possible to acquire data independent of cloud cover and light conditions, which is an important capability in cases where frequent updates of forest conditions are required. The most commonly used radar technique for mapping and monitoring is SAR (Synthetic Aperture Radar). The SAR systems operate at different frequency bands and the radar signatures from forest vary with the used wavelength. This is in part due to the fact that the longer the wavelength is, the larger is the penetration of the forest canopy. The polarization of the radar signal is also important because different polarizations are sensitive to different orientations of the tree and forest structures, consisting of tree and forest specific spatial arrangements of stems, branches, twigs and leaves or needles.

Compared to passive optical sensors, SAR is a relatively new technique. There have been several airborne SAR systems, but it was not until the launch of the European Remote Sensing Satellite-1 (ERS-1) in 1991 that SAR became more widely available for forestry applications. As described in Chapter II-2, the number of satellites with SAR sensors has increased rapidly over the last few years. With these satellites, spaceborne SAR data will now be available at three different frequencies. In addition they bring increased resolutions and multiple polarization channels, which altogether will improve the possibilities to retrieve information about forest status and forest parameters.

Many of these new systems are designed for operational use rather than for research. For example, COSMO-SkyMed will consist of a constellation of four identical satellites that will be put in orbits that give significantly shorter revisit times. This could, for example, allow rapid detection of storm damage or illegal clear-cutting. It is also foreseen that the number of SAR satellites will continue to increase and that there will be a stronger commitment to ensuring data continuity. This will lead to a larger availability of SAR data, which will facilitate a broader use in forestry applications.

32.5.2 *Intensity data*

32.5.2.1 *Change detection of clear-cuts*

The forestry application that is closest to becoming operational is the use of intensity SAR data for detection of clear-cuts. The research results for this application indicate that L-band SAR data is the most promising for discriminating clear-cuts, due to the larger contrast between forested and non-forested areas at longer wavelengths. The potential for clear-cut mapping has been demonstrated in studies where L-band SAR data from the JERS-1 satellite have been used (Thiel et al. 2006). In this study, about 90% of the clear-cuts were detected, although differences were observed between different forest areas depending on the landscape, topography, logging practices and weather conditions. At present, the Japanese PALSAR data are now being collected world-wide over forests through the global systematic observation strategy of JAXA (Japan Aerospace Exploration Agency) (Rosenqvist et al. 2007). Recent research has shown that PALSAR data potentially can be used for large-scale mapping of clear-cuts. A study of a boreal forest area showed that the radar backscatter decreased in the order of 2–3 dB when mature forest stands were compared with clear-cuts (Fransson et al. 2007). This drop in backscatter should be satisfactory for an operational application applying change detection techniques. Promising results have also been found for tropical forest. Figure 32.9 shows an area of extensive deforestation along the Xingu River in Brazil and demonstrates the possibility to map changes over a 10-year period when data from JERS-1 and ALOS PALSAR are combined. For many years, the Brazilian National Institute for Space Research (INPE) has had a programme to monitor deforestation with satellite images. The main information source has been Landsat images, but due to abundant cloud cover over the rain forest there are examples of areas where only one cloud free image has been acquired in 20 years. To get more frequent updates, INPE has now started to incorporate SAR data in their monitoring efforts.

32.5.2.2 *Estimation of biomass and stem volume*

Many studies have shown that backscatter from microwave SAR is partially correlated with stem volume and that the sensitivity of SAR data to forest stem volume (biomass) increase with longer wavelengths (e.g. Dobson et al. 1992, Le Toan et al. 1992). At shorter wavelengths (e.g. X- or C-band), the high attenuation of the forest canopy gives scattering dominated by small structures in the upper part of the canopy, and is thus only indirectly related to stem volume. This results in saturation of the backscatter at low stem volumes. With longer wavelengths (L- and P-band), the sensitivity to stem volume increases, as the scattering

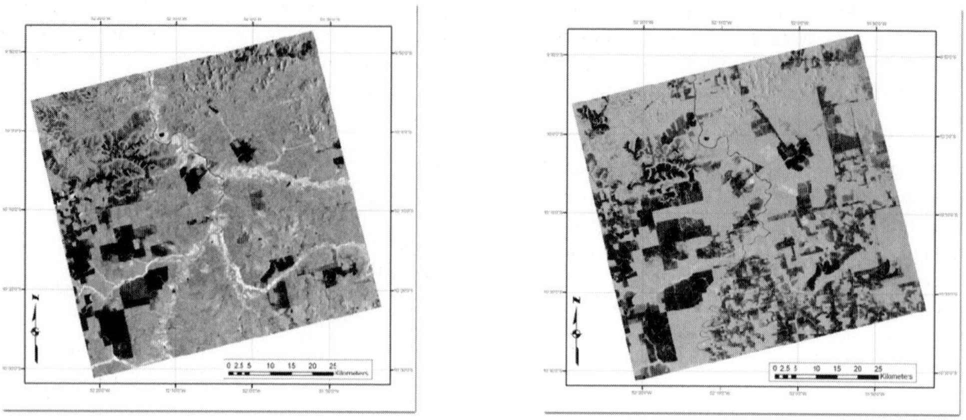

Figure 32.9. Soyfield expansion and smallholder deforestation in the Xingu basin in Brazil observed with L-band SAR. On the left side a JERS-1 image from 1996 and on the right side an ALOS PALSAR image from 2006 are shown. (Courtesy Josef Kellndorfer, Woods Hole Research Center. Images provided by JAXA/MITI.)

from trunks and large branches increases. However, even at P-band, a saturation effect is seen for stem volume, resulting in about 80% of the estimated total terrestrial phytomass being above the P-band saturation point (Imhoff 1995). For accurate measurements of stem volume in more dense forests, even longer wavelengths might be needed. One of the most promising SAR techniques for stem volume estimation is the Swedish airborne CARABAS SAR system that operates in the lower VHF-band (20–90 MHz corresponding to wavelengths between 3.3 and 15 m). Previous studies show that there is a strong correlation between forest stem volume and measured backscatter, with no sign of signal saturation for stem volumes up to about 1000 m^3 ha^{-1} (Fransson et al. 2000, Melon et al. 2001). The strong correlation can be explained physically as being due to the dominance of the ground-trunk dihedral scattering, the strength of which is directly related to the stem volume (Smith & Ulander 2000).

32.5.2.3 *Detection of wind-thrown forest*

SAR images from the Swedish airborne CARABAS (VHF-band) and LORA (wavelengths in the range of 0.5–1.5 m) systems have been visually analysed over artificially wind-thrown forest at both single tree and stand level (Fransson et al. 2007). In ideal conditions, where all trees have been wind-thrown, the results show that LORA is more accurate than CARABAS at detecting wind-thrown trees, regardless of tree size and direction of the fallen trees relative to the flight heading. Furthermore, the visible single trees in the LORA images appeared more distinct than in the CARABAS images, which could be explained by the higher resolution in the LORA images (about 0.5 m instead of 3 m). However, in natural storm damage, the attenuation of the radio waves from surrounding undamaged forest may have a larger effect on LORA than on CARABAS. Based on visual interpretation, it is likely that the detection of wind-thrown forests could be improved using VHF/UHF SAR images acquired both prior to and after a storm event.

32.5.2.4 *Interferometry*

In addition to the intensity of the backscattered radar signal, SAR interferometry can also be an important source of information for forestry applications. Interferometry requires that at least two SAR images with similar observation configuration (orbit, frequency, incidence angle, etc.) have been acquired over the same area. For forestry applications, a large part of the research has been focused on analysis of the magnitude of the interferometric coherence, in most cases referred to simply as coherence. Any changes that occur between the two image acquisitions will result in reduced coherence values, so to minimize the temporal decorrelation the time between the acquisitions should be relatively short. This temporal decorrelation is stronger for shorter wavelengths. For C-band, the tandem operation between ERS-1 and ERS-2 allowed observation of the same area with only one day between the acquisitions, which was favourable for forest studies.

It has been observed that in most cases the decorrelation increases with increasing stem volume. For estimation of stem volume in boreal forest, semi-empirical models have been developed that under ideal conditions give estimation errors at stand level as low as 10 m^3 ha^{-1} for coherence from ERS-1/2 tandem data (Santoro et al. 2002). In the project SIBERIA (SAR Imaging for Boreal Ecology and Radar

Interferometry Applications) coherence from ERS-1/2 and intensity data from JERS-1 were combined to create a forest map covering 900,000 km^2 of central Siberia (Wagner et al. 2003). Although the map only contained a few stem volume classes, it was used by Russian foresters to update old inventory maps.

Since ERS-1 stopped delivering data there is no SAR system in orbit that is well suited for interferometry for stem volume estimation. It is possible to use image pairs from the same satellite, but the time between two acquisitions is in these cases so long that temporal decorrelation becomes a significant limitation. However, for L-band the coherence levels can still be high enough for discrimination between mature forest and open areas, and thereby suitable for clear-cut detection (Eriksson et al. 2003). Recent studies funded by ESA indicate that a satellite with a P-band SAR could deliver even better results. Future SAR systems for operational services will in most cases consist of a constellation of satellites, which will reduce the repeat cycles considerably. This will likely increase the possibilities of using SAR interferometry for forestry applications.

One technique that has been developed during the last ten years and has shown great potential for tree height estimation is polarimetric SAR interferometry (PolInSAR) (Papathanassiou & Cloude 2001). In addition to the interferometric requirements mentioned above, PolInSAR requires a radar system where all polarizations (HH, HV, VH and VV) are acquired. Combining tree height estimations from PolInSAR with forest allometry, it is possible to make estimations of forest biomass. In a study where L-band PolInSAR data acquired with the German airborne E-SAR system over temperate forest, it has been shown that biomass levels up to 450 tons ha^{-1} could be extracted without saturation (Mette 2007). ALOS PALSAR is the first satellite system that allows the use of PolInSAR techniques, but with a repeat cycle of 46 days temporal decorrelation will restrict this kind of applications.

32.6 MULTISENSORAL DATA

32.6.1 *Introduction*

This sub-chapter gives a short overview of forestry applications by means of multisensoral remotely sensed data. In this chapter, we contrast image fusion that is based on optical data with data fusion that is based on multi source data (optical data and 3-D information from active sensors).

Optical remotely sensed imagery is well suited for capturing horizontally distributed conditions, structures and changes, while 3-D information as obtained by e.g. LiDAR are more appropriate for capturing vertically distributed elements of forest structure and change. Therefore, the main advantage of combining data of different sensors is that more forest parameters can be obtained in comparison to the use of single sensors. This enables an improvement of the accuracy of measurements (e.g. forest area, tree heights, crown diameter), the improvement of existing predictions (e.g. species, structural parameters) that are only derived from single sensors.

Several authors compared and contrasted the results obtained from optical data and from LiDAR, digital stereo-photogrammetry and SAR (Coops et al. 2004, Wulder et al. 2007) and reported that the combination produced forest area estimates of stem volume, age, and composition that were superior to those obtained using only a single sensor (Holström & Fransson 2003). They concluded that the combination of both 3-D information and optical data enables the derivation of forest parameters that were not available from either source independently. Using both data sources simultaneously enables the production of more accurate forest parameters and more reliable results (St-Onge et al. 2004, Popescu et al. 2004) and also facilitates a wide range of forest resource variables (McRoberts & Tomppo 2007). These variables are essential and will play a key role in future national forest inventory programmes tasked to produce timely and accurate estimations (Wulder & Seemann 2003, Leckie et al. 2003, Hudak et al. 2006).

Thus, an integration of spaceborne and airborne optical remotely sensed imagery and airborne digital surface models (DSMs) provides improved opportunities to fully characterize forest canopy attributes and dynamics. DSMs obtained from either airborne LiDAR or optical stereo aerial images are of sufficient quality and level of detail for a variety of derivations of forest parameters. However, the level of detail of these forest parameters strongly depends on the scale of the data acquired.

32.6.2 *Image fusion*

Definitions and terms of reference for image fusion, with special reference to mostly spaceborne remote sensing and GIS, were first presented to academia in the late 1990s. Image fusion techniques are generally used to combine multi-source imagery using advanced image processing techniques that generate an interpretation of the scene not obtainable from a single sensor. Alternatively, the technique may be used to reduce uncertainty associated with the data from individual sensors (Solberg et al. 1994). Pohl & van Genderen (1998) identified the following applications: image sharpening or pan-sharpening, improved geometric corrections (Strobl et al. 1990), stereo-photogrammetry (Bloom et al. 1988), feature enhancement (Leckie 1990), improved classification

(Fabre & Dherete 2003), change detection (Duguay et al. 1987), substitution of missing data (Aschbacher & Lichtenegger 1990) and finally replacement of defective data (Suits et al. 1998). Methods used to perform fusion include band selection, colour related techniques and statistical/numerical techniques, as well as Artificial Neural Networks (ANNs), Bayesian inference and Dempster-Shafer schemes, to name but a few.

32.6.3 *Data fusion*

Data fusion has been recently performed by using 3-D information as provided by DSMs (derived from LiDAR, optical stereo aerial images and SAR) combined with either spaceborne or airborne optical imagery. Particularly the increasing need of accurate estimation for a wide range of forest resource variables for a variety of users and applications explains the recent progress in three-dimensional remote sensing including digital stereophotogrammetry, radar interferometry (Hyyppä et al. 2000), LiDAR (Naesset 2002) and a combination of them (St-Onge & Achaichia 2001). Meanwhile, several airborne LiDAR systems are commercially available (e.g. Baltsavias 1999, Heurich et al. 2003), enabling the derivation of DSMs and DTMs from such data as well. These data are now commonly used for forest applications ranging from local to sub-national levels with respect to scale restrictions.

A number of studies reveal the successful use of combining LiDAR with optical data to mapping the 3-D canopy structure as canopy height models (CHM) and estimate tree and stand attributes such as tree height, crown diameter, basal area, stem volume, e.g. Lefsky et al. (2002), Persson et al. (2004), and Naesset & Gobakken (2005). Tanaka et al. (1998) and McCombs et al. (2003) point out in their studies that LiDAR data were suitable for predicting tree height, while optical data were used to obtain crown delineation or tree species. The 3-D information of LiDAR data is often combined with Landsat ETM images (Wulder & Seemann 2003, Wulder et al. 2007) or aerial images (St-Onge et al. 2004, Suarez et al. 2005, Baltsavias et al. 2008, Hirschmugl et al. 2007 and Wang et al. 2007b).

Some studies point out the modern regression approaches using optical and 3-D remote sensing data that have proven particularly useful for modelling of the spatial distribution of tree species and communities and also for changes. For example, Hudak et al. (2001, 2002) used spatial statistical models for predicting canopy height using an airborne LiDAR system (Aeroscan) and Landsat ETM+ data. Waser et al. (2007) and Waser et al. (2008) present methods to spatially predict shrub and tree occurrence and level of change within different time periods in sensitive mire forests by the use of derived parameters from high-quality DSM data and CIR aerial images.

All these studies clearly reveal the benefit of merging or fusing disparate data sources to produce timely and accurate forest parameters as needed by a variety of users.

32.7 CHANGE DETECTION

32.7.1 *Introduction*

Reliable information on changes in forest ecosystems are required for a wide range of market and policy driven needs, e.g. for forest management, planning of regional and national forest programmes and for monitoring of international agreements. This section reviews the current status as well as new developments of remote sensing based digital change detection methods for monitoring forests.

Change detection is the process of identifying differences in the state of an object or phenomenon by observing it at different times (Singh 1989). As forests are in a permanent flux at a variety of spatial and temporal scales, the selection of appropriate multi-temporal imagery is as crucial as is the choice of the sensors, change categories and change detection algorithms (Coppin et al. 2004).

To allow the direct overlay of multi-temporal imagery, a high geometric accuracy is required. Gallaun et al. (1999) performed a simulation to analyse displacement effects for forest monitoring with Landsat Thematic Mapper imagery. The simulation was carried out by shifting a forest type classification by 0.5 pixels to east and 0.5 pixels to north. The superposition error resulting from this image displacement amounted to an overall error of 13%. Olsson (1995) showed that errors in the geometry between multitemporal data sets results in calibration functions that underestimate light areas. High geometric and radiometric preprocessing accuracies are in general of utmost importance.

The change detection methods can be grouped into Bi-temporal change detection methods, e.g. change detection between two peak-green summer images, and temporal trajectory analysis, e.g. including time series of images over two seasons.

32.7.2 *Bi-temporal change detection*

A widely used approach is the post-classification comparison, where the classification results are directly compared at the pixel-by-pixel or at the segment-by-segment level (Halounová 2003). The main advantage of this approach is the fact that the two dates of imagery are separately classified, thereby minimizing the problem of radiometric calibration or phenological differences between the image acquisition dates. However, the final accuracy closely resembles that resulting from the multiplication of the accuracies of each individual classification and may be considered intrinsically low (Coppin et al. 2004).

Another group of bi-temporal approaches directly compares the spectral characteristics of the multi-temporal data. First, certain transformations are applied to the two images (e.g. subtraction) and yield a temporal difference image which contains the change information (Häme et al. 1998, Schardt et al. 1998). Then, the difference image is analysed to identify the areas that have changed. Afterwards, classification techniques can be employed to yield from-to change types. Image differencing, image rationing, change vector analysis (CVA), principal component analysis (PCA) are typical methods of employed. Usually, the separation of the changed areas from the unchanged areas is performed by application of thresholds. These can be determined empirically, or by employing an automatic thresholding strategy (Gallaun et al. 1999, Zhang et al. 2007).

32.7.3 *Temporal trajectory analysis*

In this approach, seasonal development curves, or time profiles for different seasons, are compared to monitor changes. In addition to the spectral information, information on the phenological development of the vegetation is incorporated in this approach. An operational example is the MODIS Vegetation Continuous Field (VCF) product on per cent tree cover. For deriving yearly estimates on per cent tree cover at a spatial resolution of 500 metres, annual phenological metrics are derived from the MODIS data, which are appropriate for mapping global forest tree cover as they are independent of the specific timing of forest dynamics. For the classification, an automatic regression tree algorithm with subsequent stepwise regression is implemented (Hansen et al. 2003). However, a serious disadvantage of this approach is,that currently only data with low spatial resolution sensors (e.g. MODIS, Spot-VEGETATION, MERIS) provide the required high temporal frequency that is necessary to establish appropriate time profiles.

32.7.4 *New developments*

Capabilities of satellite based forest monitoring will be enhanced by improvements in satellite sensor data that will become available over the next years. Specifically, larger swath-widths of new satellites such as Sentinel 2 associated with increased temporal resolution will enhance operational forest monitoring capabilities significantly and allow the application of new methods, e.g. temporal trajectory approaches also for monitoring at the local to regional level.

Main achievements can further be expected from approaches where field data and/or high resolution imagery are used for sampling, e.g. in the context of the global forest resource assessment 2010 (Mayaux et al. 2005, Achard et al. 2002, Ridder 2007).

Ground validation of change detection applications is often insufficient to produce accurate and well-accepted figures (Mayaux et al. 2005). Further development of cost-effective and operationally feasible methods for the accuracy assessment of change detection results is required.

32.8 FOREST FIRE

32.8.1 *Introduction*

Forest fires are a widespread phenomenon that affects ecosystems and humans from local to global scales. While at the local scale administrations aim at establishing operational systems for forest fire prediction, detection and control, at the regional to global scales, information systems on forest fires are mainly aimed at forest fire danger prediction and evaluation of forest fire impacts. In these systems, fire detection is not intended for operational forest fire fighting. Remote sensing can play an important role in these systems. In the first part of this section, we will present the use of remote sensing in pre-fire conditions monitoring. The second part of the section will deal with detecting fires and burn scar mapping.

32.8.2 *Pre-fire conditions*

Ignition and spread of forest fires depends on fuel moisture and weather conditions as well as on fuel types and topography. Several fire danger parameters can be monitored using remote sensing. The first one is the fuel type, which can be mapped, like classical vegetation mapping, from high spatial resolution optical or radar images (see the review of Leblon 2005). These maps can then be linked within a wildfire threat system, to other fire danger parameters, such as topography, proximity to roads and to urban areas (Leblon 2005). Another fire danger parameter, which can be estimated by remote sensing, is the fuel moisture conditions. They were first estimated using optical data, mainly NOAA-AVHRR NDVI images (Leblon 2005). These studies listed several problems related to the use of NDVI images in fuel moisture mapping, namely the saturation of relationships and the difficulty of using NDVI over forests, due to the spectral overstory/ understory mixture both being different in nature and in moisture content.

Fuel moisture is theoretically better related to another optical band, the shortwave infrared (1300–2500 nm), but the relationships were significant only when the water stress was already well developed (Leblon 2005). In addition, from the operational point of view, shortwave infrared data are acquired by only a few numbers of spaceborne sensors. By contrast, thermal infrared data are provided often and at the same time as the optical ones, by several existing spaceborne sensors, among others, NOAA-AVHRR, MODIS and LANDSAT-TM. Thermal infrared data were better correlated than NDVI to fire danger codes (Leblon 2005, Oldford et al. 2003, Oldford et al. 2006, Leblon et al. 2007) and to foliar moisture content.

Figure 32.10 shows that water stress boreal forest areas of high fire danger codes correspond to high surface temperature areas, but to high NDVI value areas indicating no drought conditions in the case of NOAA-AVHRR images (Oldford et al. 2003).

Most recent studies use both the NDVI and thermal infrared images, directly or through drought indices (Leblon 2005). However, both images have limited image availability during cloudy days. By contrast, active microwave (or radar) images can be acquired in any weather conditions. Good correlations were obtained between fuel moisture codes and backscatter images from ERS-1 (Leblon 2005) or RADARSAT-1 (Abbott et al. 2007) acquired over boreal forests. The operational use of radar images in fire danger monitoring is limited by the influence of factors other than moisture, like vegetation type, species, and structure, vegetation biomass, topography and surface roughness, as well as canopy height. For these reasons, the most promising use of remote sensing in fire danger monitoring is by combining optical, thermal infrared and radar images. Such a combination has been shown to be superior when multi-polarized ratios are used. Multipolarized radar images are provided by several existing or planned satellites, among others RADARSAT-2. This new satellite has the additional advantage of providing images at a fine spatial resolution (pixel of 3 m) and fully polarimetric radar images, which have more information than multipolarized SAR images because they have the phase information in addition to the four polarizations (HH, HV, VH and VV).

32.8.3 *Fire detection and burnt area mapping*

Mapping forest fires is perhaps one of the most widely used applications of satellite remote sensors. Fire detection from remote sensors is based on the discrimination of the burning area from the surroundings on the basis of differential temperature. The fire line of active wildfires is characterized by high temperatures that are distinct (above) from the temperatures of the surrounding areas. These high temperatures induce

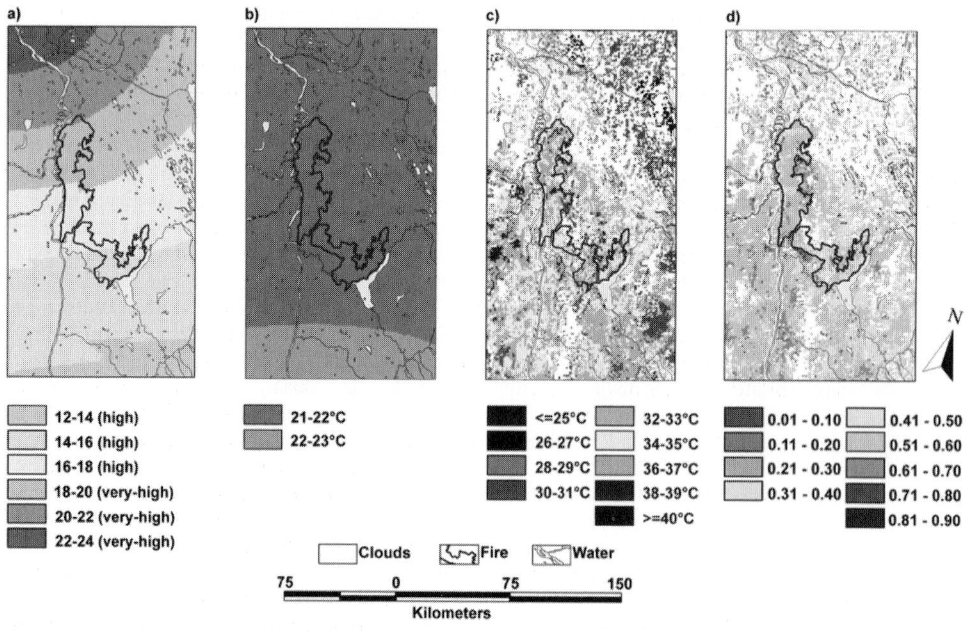

Figure 32.10. Map of (a) interpolated fire danger code (Fire Weather Index), (b) interpolated air temperatures (T_a), and corresponding (c) T_s NOAA-AVHRR image and (d) NDVI NOAA-AVHRR image for the area, where the fire "*vq0332*" starts. The image was acquired 3 days before fire ignition (after Oldford et al. 2003). (see colour plate page 521)

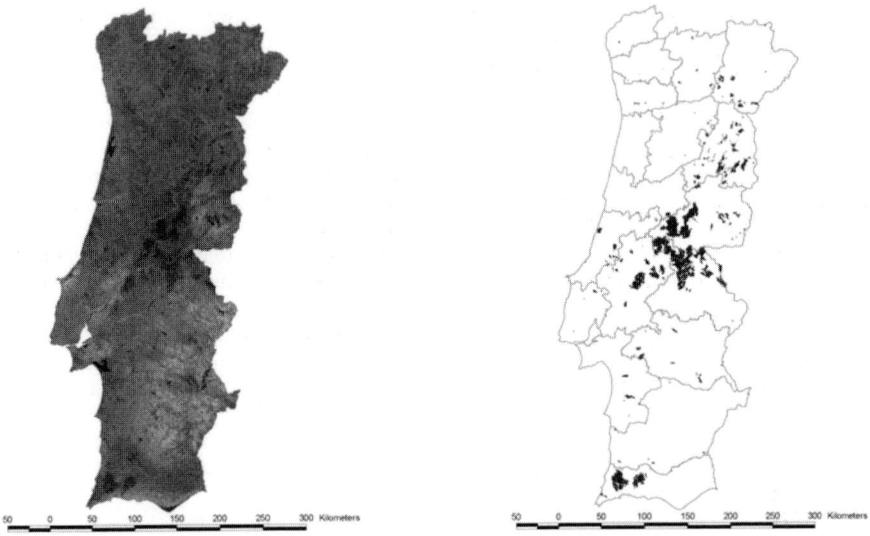

Figure 32.11. MODIS image (250 m spatial resolution) showing the burnt scars in black (left side), and mapped burnt areas in red on the administrative map of Portugal (right side) (Source EFFIS). (see colour plate page 521)

a high reflectance in both thermal and mid-infrared wavelengths that is detectable from satellite sensors. Contextual algorithms are subsequently used to process satellite images in order to discriminate these so-called hot spots from other surrounding areas (see review of San-Miguel-Ayanz et al. 2005, Kennedy et al. 1994).

The methods for the detection of fire affected areas are either based on the identification of spectrally homogeneous burnt areas that are distinct from other cover types, or based on change detection analysis. In this later approach, the identification of the burnt areas is done through the comparison of pre-fire and post-fire satellite imagery. When fires burn healthy vegetation, they induce a significant drop in the spectral reflectance of this vegetation in the near-infrared spectrum and additional changes in the optical and mid-infrared spectra These changes in reflectance are easily detected when comparison of the pre-fire and post-fire scenes is performed; changes in backscatter between pre-fire and post-fire imagery is also detectable when using Synthetic Aperture Radar imagery (Gimeno & San-Miguel-Ayanz 2004a, 2004b). Often, change detection methodologies are guided through the use of satellite detected hot-spots, which identify potential active fires that may have the consequence of burnt areas (Fraser et al. 2000).

Wildfires burn more than half a million ha of forested areas in Europe every year; over 95% of the burnt areas are located in the Mediterranean region in which critical fire events have taken place in recent years (e.g. 2003, 2005, 2006 and 2007) (Source EFFIS: http://effis.jrc.it). Mapping of burnt areas and analysis of damage in Europe is carried out by the European Forest Fire Information System (EFFIS). The EFIS is an operational regional information system on forest fires, which provides forest fire danger for the European region and evaluates the damage caused by forest fires in this region in terms of burned area, forest types damaged by fires and emissions into the atmosphere. The EFFIS currently includes twenty countries, from northern Europe (e.g. Finland, Sweden) to eastern (e.g. Cyprus and Turkey) and southern Europe (Italy, Greece). It provides operationally fire danger forecast using the Canadian Fire Weather Index (FWI) to all the participant countries and evaluates the forest fire damage through Europe using satellite imagery of medium spatial resolution (WiFS and MODIS). In this system, burnt area identification and mapping is performed following the above described methodology, i.e. hot-spot detection (for the identification of active fires) followed by change detection analysis. Figure 32.11 shows fire scars in Portugal in 2003, where more than 400,000 ha of forested areas were burnt (European Commission 2004). Foreseen enhancements of EFFIS include the use high spatial resolution imagery (AWiFs or DMC) for burnt area mapping and the developing of fire danger forecast up to 7 days, and the modelling of the impact of forest fire emissions in nearby populations. Additional efforts will be put on the analysis of forest fire causes, as nearly 99% of the fires in Europe are human-induced.

32.9 STATISTICAL APPROACHES FOR COMBINING FIELD DATA AND IMAGE DATA FOR FORESTRY

32.9.1 *Prediction of forest variables using remote sensing trained field data*

Forest inventories provide information on the structure, location and extent of forest resources. Typically, the area for which information is needed is described by estimates of means or totals of variables measured on field plots; for example, volume, biomass, tree height, diameter at breast height (DBH), density (number of trees per unit area), and tree species composition. The quality of such forest attribute estimates may be improved by combining the field data with remotely sensed data.

Empirical models of forest attributes are often constructed using statistical techniques whereby forest variables measured during fieldwork campaigns are related to remotely sensed imagery. These empirical models are then used to construct spatially explicit maps by predicting forest attributes such as volume and biomass for individual remote sensing units. The model of the relationship between the measured forest variables and the remotely sensed data is then expressed in a statistical form; for example:

$$y = \beta_0 + \beta_1 dn + \varepsilon \qquad (1)$$

where y is the forest variable being analysed, β_0 and β_1 are parameters to be estimated, dn is the digital number of the satellite image, and ε is a residual or error term. The idea is that if you can construct a model that explains most of the variance in the measured field data using remotely sensed imagery, then you can construct spatially explicit maps of the variable of interest (Lefsky et al. 1999, Cohen et al. 2003, Foody et al. 2003, Zheng et al. 2004, McRoberts 2006). The example given is fairly simple but introduces the basic underlying principles behind using remotely sensed data in forest resource management. In many cases, the regression models relate forest attribute variables, or transformations of the variables, to spectral bands, band products, band ratios and band transformations. However, the relationship between forest variables such as volume and age and satellite spectral data from sensors such as Landsat TM may be weaker in mature forest stands (Gemmell 1995, Steininger 2000, Lu 2005).

Statistical models are also frequently used to relate data from LiDAR (Means et al. 2000, Hyyppä et al. 2001; Næsset 2002) and radar sensors (Fransson & Israelsson 1999, Austin et al. 2003, Rauste 2005), or features extracted from such data, to forest variables. For example, airborne LiDAR data has been demonstrated to be an accurate predictor of height, basal area, and volume, even in mature forest stands (Næsset 2002, Holmgren 2004). For synthetic aperture radar (SAR), the relationship between volume and radar backscattering increases as the wavelength increases. For C-band SAR data from ERS-1, the backscattered signal shows a weak relationship with volumes greater than 60 m^3/ha, and for L-band data from JERS-1 the relationship becomes weak for volumes greater than 140 m^3/ha (Fransson & Israelsson 1999).

An alternative to regression techniques is the non-parametric k-NN method (Tomppo 1996, McRoberts et al. 2007). With this method forest variables are predicted as weighted averages of observed variable values for the k most similar plots in a feature space consisting of spectral data and other ancillary variables. Similarity is defined by the feature space distance d, and the weight $w_{i,p}$ for field plot i when predicting a forest variable for pixel p is defined as:

$$w_{i,p} = \frac{1}{d^t_{i,p}} \bigg/ \sum_{j=1}^{k} \frac{1}{d^t_{j,p}}, \qquad (2)$$

where $d_{1,p} \le d_{2,p} \le \ldots \le d_{k,p}$; $d_{i,p}$ = feature space distance from pixel p to plot i; and $t \in [0,2]$. A pixel-level prediction, $\hat{y}_p$, for a forest variable, Y, for pixel p in an image is defined as:

$$\hat{\boldsymbol{y}}_p = \sum_{i=1}^{k} w_{i,p} \cdot \boldsymbol{y}_{i,p} \qquad (3)$$

The k-NN technique has been used extensively with inventory data and Landsat TM imagery to map forest attributes: site quality, volume by species, age, height, basal area, mean diameter and other variables in Finland (Tomppo & Halme 2004, Tomppo 2006); age, height, and volume in Sweden (Reese et al. 2003); age and volume in Sweden (Holmstrom & Fransson 2003); forest type in Austria (Koukal et al. 2005); species groups in Norway (Gjertsen 2005); volume, stocking, diameter, height, and basal area in Ireland (McInerney et al. 2005); and forest type, proportion forest area, basal area, volume, and tree density in the USA (Franco-Lopez et al. 2001, McRoberts et al. 2007).

Traditionally, National Forest Inventories (NFIs) have used sample data to produce estimates of forest variables that respond to the user question "How much?" Increasingly, however, users are also asking "Where?" and requesting that NFIs construct maps depicting the spatial distribution of different forest variables. To satisfy user requirements, inventory mapping applications should be multivariate because consistency among the maps of related variables is essential. Unfortunately, most parametric multivariate modelling techniques require that variables have a multivariate Gaussian distribution. However, the multivariate k-NN technique has no such assumptions and, therefore, is particularly well-suited for NFI mapping applications.

Multiple other techniques have been used to predict forest variables from remote sensing data, including spatial statistics (Tuominen et al. 2003, Wallerman 2003), classification and regression trees (CART) (Breiman et al. 1984, Xian et al. 2002), and artificial neural networks (ANN) (Ingram et al. 2005).

Remotely sensed data and maps with estimated or classified forest information can also be used as an aid in large area sampling inventories, a topic for discussion in the following chapter.

32.9.2 *Use of remote sensing data as an aid in sampling inventories*

National forest inventories (NFI) are tasked to produce and report timely and accurate estimates of forest resources. Because enumeration and measurement of all trees on forest lands would be prohibitively expensive and time-consuming, NFIs rely on sampling procedures to obtain representative data which serve as the basis for probability-based forest resource estimates. The same applies for forest management inventories when accurate estimates of local forest resources are required.

In the context of sampling inventories, the major reasons for using remote sensing data are twofold: (i) to complement the statistical information with maps, and (ii) to increase the economic efficiency of sampling designs. Remotely sensed data increases economic efficiency by serving as an information source for the forest-nonforest decision necessary to reduce and avoid fieldwork on non-forest sampling plots in NFIs (e.g. McRoberts & Tompoo 2007, Keller 2001) and for reducing the sampling error of the probability-based estimates. The focus of this section is on the latter technically more challenging option.

Auxiliary data in form of remote sensing or GIS data (often retrieved from RS data) can be used without any influence on the sampling design for probability-based statistical estimation. In addition, it can also be used to optimize the design. Of the two major strategies for reducing sampling error, regression and stratification, the major strategy used in forest inventory praxis is stratification in the forms of pre-stratification, post-stratification or double sampling for stratification. The statistical estimation procedures that produce unbiased or approximately unbiased probability-based estimates when using stratification have been developed and described for various sampling designs by Cochran (1977). In addition, Särndal et al. (1992) described stratification more generally as a special case of regression in the form of model-assisted estimation. Starting from the general estimator defined by Särndal et al. (1992), Dees & Koch (1997) derived an estimator for cluster sampling designs that are widely used by NFIs and the post-stratification situation in which strata extend across clusters.

The use of aerial photography for stratification was initiated for NFIs in the 1940s and 1950s by the U.S. Forest Inventory and Analysis (FIA) programme (Osborne 1942, Bickford 1952) and for forest management inventories in the 1960s in the United States and Switzerland (Bickford 1961, Schmid 1969). This technique is still used by many national and local forest inventories (Dees 2006). The use of satellite remote sensing data rather than aerial photography for stratification started in the 1980s (e.g. Schade 1980, Dees 1988, Dees et al. 1998) and has been used operationally for several years in the redesigned U.S. FIA programme (Hansen & Wendt 2000, McRoberts 2005, Mc Roberts & Tompoo 2007).

When using stratification for estimation purposes, the effectiveness of the stratification can be simply quantified using relative efficiency, *RE*, defined as,

$$RE = \frac{\hat{Var}\left(\bar{Y}_{srs}\right)}{\hat{Var}\left(\bar{Y}_{str}\right)} \tag{4}$$

where $\hat{Var}\left(\bar{Y}_{srs}\right)$ is the variance of the estimated mean obtained under the assumption of simple random sampling (srs) and no stratification and $\hat{Var}\left(\bar{Y}_{str}\right)$ is the variance of the estimate using stratification. For more complex designs (e.g. a cluster sampling), *RE* is defined equivalently by replacing $\hat{Var}\left(\bar{Y}_{srs}\right)$ by the appropriate probability-based variance estimator. For single phase sampling designs, *RE* directly corresponds to *FSRP*, the potential to reduce the field sampling size using stratification compared to not using stratification with the same sampling design,

$$FSRP = 1 - RE^{-1} \tag{5}$$

The potential reduction in field sampling attributed to stratification has been estimated by the Swiss NFI to be approximately 50% when focusing on the major attributes and when using a double sampling approach (Köhl 2002, Köhl & Brassel 1998).

For single inventories within the U.S. FIA programme, McRoberts et al. (2002) obtained 2.00 < RE < 3.50 for forest areas and 1.25 < RE < 1.75 for volume when using a satellite based forest-nonforest map. When using a forest type classification that discriminated nine forest type classes and data of the German NFI in the Black Forest region, Dees et al. (1998) achieved RE values for various major attributes that ranged from 9.0 for forest area (*FSRP* = 88.9%) and 4.1 (*FSRP* = 75.8%) for forest volume to lower values such as 1.4 (*FSRP* = 35.7%) for volume per forest area or 1.6 for beech volume. Nilsson et al. (2004) reported RE = 3.0 for forest area for the Swedish NFI. In a later study, McRoberts et al. (2006) reported an approach to stratification in which a logistic regression model was used to predict proportion forest area, $\hat{p}$, for each pixel (another way to use the approaches described in Section 9.1). Strata were defined in terms of categories of predictions, (e.g. $0.0 \leq \hat{p} \leq 0.1$, $0.1 < \hat{p} \leq 0.5$, $0.5 < \hat{p} \leq 0.9$, and $0.9 < \hat{p} \leq 1.0$). For forest area and volume in the USA, they obtained RE ≥ 5.5 and RE ≥ 2.5, respectively. In summary, stratification has the potential to reduce sampling errors, although the magnitude of the reduction depends on the target attribute, the type of map strata used and regional forest properties.

Another motivation for using remote sensing-based techniques is to obtain estimates for small geographic sub-units. Small area estimates can be obtained by adding or averaging pixel predictions obtained using techniques such as regression and *k*-Nearest Neighbours (*k*-NN). In statistical terms, such estimates are characterized as model-based. Estimation of uncertainty for these techniques requires advanced model-based variance estimators. For forestry applications involving satellite imagery such model-based estimators that bridge the gap between *k*-NN techniques and probability-based estimators are now appearing in the literature (McRoberts 2006, McRoberts et al. 2007).

When evaluating economic efficiency, the additional cost of remote sensing phases must be balanced against the potential cost saving due to reduced field sampling. Examples of such cost saving calculations have shown them to be substantial (Köhl 2002, McRoberts et al. 2006, Dees 2006). In many parts of the world such as Europe, landuse mapping based on high resolution satellite data obtained from Landsat TM, SPOT XS or equivalent sensors is becoming routine as the result of efforts such as the European GMES initiative. In these situations, use of remote sensing-based data sets as aids to reducing sampling errors in NFIs is becoming economically more and more attractive as mapping services become available freely or at moderate costs to public end users.

When designing a new NFI, investigation of this potential may be quite valuable. For example, Turkey has begun preparations for an NFI and considerable economic efficiency is expected through use of available digital forest management maps based on CIR photography for public forests and satellite remote sensing data for the remaining private forests (mainly poplar plantations) (Dees et al. 2004, Asan et al. 2005).

32.10 CONCLUSIONS

While applications based on optical airborne and satellite data will still be developed further, it is shown that a new generation of remote sensing applications in the forest sector is currently under development. LiDAR based applications will allow more detailed information on quantitative forest attributes, such as wood volume or biomass that is not retrievable from optical data in that detail and accuracy. SAR based applications will allow applications for disaster monitoring and high frequent forest monitoring where optical data have restrictions due to cloud coverage and poor solar illumination during the winter season at middle to high latitudes.

ACKNOWLEDGEMENTS

The authors would like to thank Gyde Lund, Forest Information Services. Gainesville, United States of America for his valuable contributions and internal reviewing assistance.

The authors would also like to thank the following scientists for their valuable contribution to the chapter "Forestry applications": Ünal Asan, University of Istanbul., Dept. of Forest Management Planning, Istanbul, Turkey; Armin Troycke, Stefan Binner and Gudrun Faißt, Bavarian State Institute of Forestry—LWF, Department of GIS and Remote Sensing, Freising ,Germany; Ionnnis Gitas and George Mallinis, University of Thessaloniki., Lab of Forest Management and Remote Sensing, Aristotle Thessaloniki, Greece; Klaus Granica, Institute of Digital Image Processing, Joanneum Research. Graz, Austria; Vasileios Kalogirou, University College, London, UK; Jesús San-Miguel-Ayanz, European Commission, DG Joint Research Centre, Institute for Environment and Sustainability, Land Management & Natural Hazards Unit, Ispra, Italy; Wesley Roberts, Council for Scientific and Industrial Research, Forestry and Forest Products Research Centre, South Africa; Jackie Rosette, Department of Geography, Swansea University, Swansea, UK; Andreas Schuck, European Forest Institute, Joensuu, Finland, Juan Suárez, Forest Research Northern Research Station, Midlothian, UK; Dirk Tiede, University of Salzburg, Z_GIS Centre for Geoinformatics, Salzburg, Austria; Erkki Tomppo, The Finnish Forest Research Institute—Metla. Helsinki, Finland and Vladimir Zidek, Mendel University of Agriculture and Forestry, Brno, Czech Republic.

REFERENCES

Abbott, K., Leblon, B., Staples, G., Alexander, M. & MacLean, D. 2007. Fire danger monitoring in a northern boreal forest region using RADARSAT-1 imagery. *International Journal of Remote Sensing* 28(5–6): 1317–1338.

Achard, F., Eva, H.D., Stibig, H.J., Mayaux, P., Gallego, J., Richards, T. & Malingreau, J.P. 2002. Determination of deforestation rates of the world's humid tropical forests. *Science* 297(5583): 999–1002.

Andersen, H.E. & Breidenbach, J. 2007. Statistical Properties of Mean Stand Biomass Estimators in a Lidar-Based Double Sampling Forest Survey Design. *Proc. ISPRS III/3, III/4, V/3 and VIII/11. "LaserScanning 2007 and SilviLaser 2007". Vol.XXXVI, Part 3/W52, 12–14 September 2007, Espoo, Finland*: 8–14.

Asan, Ü., Yesil, A., Ozkan, U.Y. & Saglam, S. (eds.). 2005. *Proceedings of the International Symposium on "National Forest Inventory for Turkey", 24–28th September 2002, Istanbul, Turkey*. ISBN: 975-404-748-0 Istanbul, Turkey: Istanbul Üniv. Orman Fakültesi.

Aschbacher, J. & Lichtenegger, J. 1990. Complimentary nature of SAR and data: a case study in the Tropics. *Earth Observation Quarterly* 31: 4–8.

Austin, J.M., Mackey, B.G. & Van Niel, K.P. 2003. Estimating forest biomass using satellite Radar: an exploratory study in a temperate Australian Eucalyptus forest. *Forest Ecology and Management* 176: 575–583.

Baltsavias, E.P. 1999a. Airborne laser scanning: basic relation and formulas. *ISPRS J. of Photogrammetry and Remote Sensing* 54: 199–214.

Baltsavias, E.P. 1999b. Airborne laser scanning: existing systems and firms and other resources. *ISPRS J. of Photogrammetry and Remote Sensing* 54: 164–198.

Baltsavias, E., Gruen, A., Zhang, L. & Waser, L.T. 2008. High Quality Image Matching and Automated Generation of 3-D Tree Models. *International Journal of Remote Sensing* 29(5): 1243–1259.

Bickford, C.A. (1952): The sampling design used in the forest survey of the Northeast. *J. For* 50(4): 290–293.

Bickford, C.A. 1961, Stratification for Timercruising. *Journal of Forestry* 59, 10: 761–763.

Binner, S., Ewald, J., Faißt, G. & Seitz, R. 2006. Die Abgrenzung von FFH-Lebensraumtypen im Hochgebirge mit Hilfe von GIS und Fernerkundung. In: Strobl, Blaschke, Griesebner (Hrsg.): *Angewandte Geoinformatik 2006 Beiträge zum 18. AGIT-Symposium, Salzburg*: 53–59.

Bloom, A., Fielding, E. & Fu, X. 1988. A demonstration of stereophotography with combined SIR-B and Landsat-TM images. *International Journal of Remote Sensing*: 1023–1038.

Bolduc, P., Lowell, K. & Edwards, G. 1999. Automated estimation of localized forest volume from large-scale aerial photographs and ancillary cartographic information in a Boreal forest. International Journal of Remote Sensing 20: 3611–3624.

Brandtberg, T., Warner, T.A., Landenberger, R.E. & McGraw, J.B. 2003. Detection and analysis of individual leaf-off tree crowns in small footprint, high sampling density LIDAR data from the eastern deciduous forest in North America. *Remote Sensing of Environment* 85: 290–303.

Breidenbach, J., McGaughey, R., Andersen, H.E., Kändler, G. & Reutebuch, S. 2007. A Mixed Effects Model to Estimate Stand Volume by Means of Small Footprint Airborne Lidar Data for an American and German Study Site. In *Proc. ISPRS III/3, III/4, V/3 and VIII/11. "LaserScanning 2007 and SilviLaser 2007"*. Vol.XXXVI, Part 3/W52, 12–14 September 2007, Espoo, Finland: 77–84.

Breiman, L., Friedman, J.H., Olshen, R.A. & Stone, C.J. 1984. *Classification and Regression Trees*. Belmont: Wadsworth.

Brenner, A., Zwally, H., Bentley, C., Csatho, B., Harding, D., Hofton, M., Minster, J., Roberts, L. Saba, J., Thomas, R. & Yi, D. 2003. *Algorithm Theoretical Basis Document V4.1*. Available online at: http://www.csr.utexas.edu/glas/pdf/Atbd_20031224.pdf (accessed March 2005).

Bretar, F. & Chehata, N. 2007. Digital Terrain Model on Vegetated Areas: Joint Use of Airborne Lidar Data and Optical Images. In U. Stilla et al. (eds.), *PIA07. International Archives of Photogrammetry, Remote Sensing and Spatial Information Sciences* 36 (3/W49A): 19–24.

Chasmer, L., Hopkinson, C. & Treitz, P. 2006. Investigating laser pulse penetration through a conifer canopy by integrating airborne and terrestrial lidar. *Can. J Remote Sensing* 32(2): 116–125.

Cochran, W.G. 1977. *Sampling techniques, 3rd Ed.* New York: Wiley.

Cohen, W.B., Maiersperger, T.K., Gower, S.T. & Turner, D.P. 2003. An improved strategy for regression of biophysical variables and Landsat ETM+ data. *Remote Sensing of Environment* 84: 561–571.

Coops, N.C., Wulder, M.A., Culvenor, D.S. & St-Onge, B. 2004. Comparison of forest attributes extracted from fine spatial resolution multispectral and LiDAR data. *Canadian Journal of Remote Sensing* 30: 855–866.

Coppin, P., Jonckheere, I., Nackaerts, K., Muys, B. & Lambin, E. 2004. Digital change detection methods in ecosystem monitoring: a review. *Int. J. Remote Sensing* 25, 9,:1565–1596.

Dees, M. & Koch, B. 1997. Poststratifizierung bei Klumpenstichprobendesign zur Nutzung von Fernerkundung und GIS bei Waldinventuren. In Pelz, D.R. (Hrsg.): *Sammlung von Beiträgen aus der 10. Jahrestagung der Sektion Forstliche Biometrie und Angewandte Informatik des Deutschen Verbandes Forstlicher Forschungsanstalten, 24. 26. Sept. 1997 in Freiburg i. Br.* Universität Freiburg i. Br., S.: 109–118.

Dees, M. 1998. Regressionsschätzer bei forstlichen Trakt-Großrauminventuren. *Allgemeine Forst- und Jagd-Zeitung*, 169: 177–184.

Dees, M. 2006. Kombination von Fernerkundung und Stichprobeninventur bei betrieblichen und nationalen Waldinventuren. Freiburg: Fakultät für Forst und Umweltwissenschaften der Albert-Ludwigs-Universität Freiburg. *Schriftenreihe Freiburger Forstliche Forschung*, Bd. 37.

Dees, M., Asan, U. & Yesil, A. 2004. Ideas and options for a National Forest Inventory in Turkey. In P. Corona, M Köhl & M. Marchetti (eds.), *Advances in Forest Inventory for Sustainable Forest Management and Biodiversity Monitoring*: 375–394. Kluiver Academic Publishers.

Dees, M., Koch, B. & Pelz, D.R. 1998. Integrating Satellite Based Forest Mapping with Landsat TM in a Concept of a Large Scale Forest Information System. *PFG* 4/1998: 209–220.

Dees, M., Straub, C. Langar, P. & Koch, B. 2006. Remote sensing based concepts utilising SPOT 5 and LIDAR for forest habitat mapping and monitoring under the EU Habitat Directive. In A. Lamb, R. Hill, D. Wilson, M. Bock, E. Ivits, M. Dees, S. Hemphil & Koch, B. (eds.), *ONP 10 Test & Benchmarks* Report. Part of the geoland project reporting dossier: 53–78. www.geoland-gmes.info. http://portal.uni-freiburg.de/felis/Forschung/proj/lauf/geol/, (accessed 6 Dec. 2007).

Dobson, M.C., Ulaby, F.T., Le Toan, T., Beaudoin, A., Kasischke, E.S. & Christensen, N, 1992. Dependence of radar backscatter on coniferous forest biomass. *IEEE Transactions on Geoscience and Remote Sensing* 30, 2: 412–415.

Duguay, G., Holder, G., Howarth, P. & LeDrew, E. 1987. Integrating remotely sensed data from different sensors for change detection. *Proceedings of the IEEE International Geoscience and Remote Sensing Symposium.*

Elmquist, M., Jungert, E., Lantz, F., Persson, A. & Söderman, U. 2001. Terrain modelling and analysis using laser scanner data estimations of laser data using shape models. *International Archives of Photogrammetrie and Remote Sensing* 34–3/W4: 219–227.

Eriksson, L.E.B., Santoro, M., Wiesmann, A. & Schmullius, C. 2003. Multi-temporal JERS repeat-pass coherence for growing stock volume estimation of Siberian forest. *IEEE Transactions on Geoscience and Remote Sensing* 41, 7: 1561–1570.

European Commission. 2004. San-Miguel-Ayanz, J., Barbosa, P., Camia, A., Kucera, J., Liberta, G., Schmuck, G., E. Schulte, Bucella, P., Colletti, L. & Flies, R. *Forest Fires in Europe—2003 fire campaign, Official Publication of the European Communities, SPI.04.124 EN.*

Fabre, S. & Dherete, P. 2003. Data fusion applications: classification and mapping. *Proceedings of the IEEE International Geoscience and Remote Sensing Symposium* 2: 1053–1055.

Förster, M., Kleinschmit, B. & Walentowski, H. 2005. Monitoring Natura2000 Forest Habitats in Bavaria by the Use of ASTER, SPOT5 and GIS Data—an integrated Approach. *Session 3b: Classification studies; ForestSat 2005 Conference—May 31–June 3, 2005, Boras, Sweden: 21–25.*

Foody, G.M., Boyd, D.S. & Cutler, M.E.J. 2003. Predictive relations of tropical forest biomass from Landsat Tm data and their transferability between regions. *Remote Sensing of Environment* 85: 463–474.

Franco-Lopez, H., Ek, A.R. & Bauer, M.E. 2001. Estimation and mapping of forest stand density, volume, and cover type using the k-nearest neighbors method. *Remote Sensing of Environment* 77: 251–274.

Fransson, J.E.S. & Israelsson, H. 1999. Estimation of stem volume in boreal forests using ERS-C- and JERS-1 L-band SAR data. *International Journal of Remote Sensing* 20: 123–137.

Fransson, J.E.S., Magnusson, M., Folkesson, K., Hallberg, B., Sandberg, G., Smith-Jonforsen, G., Gustavsson, A. & Ulander, L.M.H. 2007. Mapping of wind-thrown forests using VHF/UHF SAR images. *Proceedings of IGARSS 2007 Symposium, Sensing and Understanding Our Planet, Barcelona, Spain, 23–27 July, 2007.*

Fransson, J.E.S., Magnusson, M., Olsson, H., Eriksson, L.E.B., Sandberg, G., Smith-Jonforsen, G. & Ulander, L.M.H. 2007. Detection of forest changes using ALOS PALSAR satellite images. In *Proceedings of IGARSS 2007 Symposium, Sensing and Understanding Our Planet, Barcelona, Spain, 23–27 July, 2007.*

Fransson, J.E.S., Walter, F. & Ulander, L.M.H. 2000. Estimation of forest parameters using CARABAS-II VHF SAR data. *IEEE Transactions on Geoscience and Remote Sensing* 38, 2: 720–727.

Fraser, R., Li, Z. & Cihlar, J. 2000. Hotspot and NDVI differencing synergy: A new method for mapping fire scars. *Remote Sensing of Environment* 74: 362–375.

Gallaun, H., Schardt, M. & Häusler, T. 1999. Pilot Study on Monitoring of European Forests. *Tagungsband der Jahrestagung des Arbeitskreises, Interpretation von Fernerkundungsdaten der DGPF, Halle* 28–29.4.1999: 157–161.

Gallaun, H., Schardt, M., Granica, K. & Flaschberger, G. 2001. Monitoring von Schutzwäldern mit Satelliten-Fernerkundung. VGI—Österr. Zeitschr. für Vermessung & Geoinformation, Heft3/01:157–161.

Gallaun, H., Schardt, M. & Linser, S. 2007. Remote sensing based forest map of Austria and derived environmental indicators; Plenary session: Large area forest monitoring using RS data. *ForestSat. 2007. Conference—November 5–7, 2007—Montpellier, France.* http://forestsat.teledetection.fr/index.php?option=com_docman&task=doc_details&gid=50&Itemid=31 (accessed 7 March, 2008).

Gemmell, F.M. 1995. Effects of Forest Cover, Terrain, and Scale on Timber Volume Estimation with Thematic Mapper Data in a Rocky mountain Site. *Remote Sensing of Environment* 51: 291–305.

Gillis, M. & D. Leckie, 1996. Forest inventory update in Canada. *The Forestry Chronicle* 72: 138–156.

Gimeno, M. & San-Miguel-Ayanz, J. 2004a. Evaluation of RADARSAT-1 data for identification of burnt areas in Southern Europe. *Remote Sensing of Environment* 92: 370–375.

Gimeno, M. & San-Miguel-Ayanz, J. 2004b. Identification of burnt areas in Mediterranean forest environments from ERS-2 SAR time series. *International Journal of Remote Sensing* 25(22): 4873–4888.

Gjertsen, A. 2005. Accuracy of forest mapping based on Landsat TM data and a kNN-based method. *Remote Sensing of Environment* 110: 420–430.

Gobakken T. & Naesset E. 2007. Assessing Effects of Laser Point Density on Biophysical Stand Properties Derived from Airborne Laser Scanner Data in Mature Forest. *Proc. ISPRS III/3, III/4, V/3 and VIII/11. "LaserScanning 2007 and SilviLaser 2007" XXXVI, 3/W52, 12–14 September 2007, Espoo, Finland.*

Goodwin, N.R., Coops, N.C. & Culvenor, D.S. 2006. Assessment of forest structure with airborne LiDAR and the effects of platform altitude. *Remote Sensing of Environment* 103(2): 140–152.

Gougeon, F. 1995. A crown-following approach to the automatic delineation of individual tree crowns in high spatial resolution aerial images. *Canadian Journal of Remote Sensing* 21(3): 274–284.

Granica, K., Schardt, M. & Gallaun, H. 2000. Monitoring of Protection Forests by Means of SPOT4 Satellite Data. *Prodeedings, 20th EARSeL Symposium, 14th—16th June, 2000, Dresden, Germany*: 59–66.

Granica, K., Senitza, E., Hochbichler, E., Panagiotis, B., Schardt, M., Zuba, C. & Flaschberger, G. 2006. Water & Forest—Management of Water Resources by Forest Cultivation Practices and Usage of Satellite Remote Sensing Imagery. *Proc. All About Karst & Water, Decision Making in a Sensitive Environment, Vienna, Austria*: 296–304.

GRL. 2005. *Geophysical Research Letters 32. ICESat Special Issue October–November 2005.*

Gross, C.P. (ed.). 2000. Remote Sensing Application for Forest Health Status Assessment (second edition). European Communities. ISBN 92-828-8144-x.

Gupta, S. Weinacker, H. & Koch, B. 2007. An approach for 3d single tree detection using airborne fullwave LiDAR raw point clouds. In M. Deshayes (ed.), *Proceedings of ForestSat 2007 Conference—November 5–7, 2007—Montpellier, France* (submitted, to be published soon under

Halounová, L. 2003. Textural classification of B&W aerial photos for the forest classification. *Symposium of European Association of Remote Sensing Laboratories, Gent, Belgium*: 173–179.

Häme, T., Heiler I. & San Miguel-Ayanz, J. 1998. An unsupervised change detection and recognition system for forestry. *International Journal of Remote Sensing* 19, 6: 1079–1099.

Häme, T., Stenberg, P., Andersson, K., Rauste, Y., Kennedy, P., Folving, S. & Sarkeala, J. 2001. AVHRR-based forest proportion map of the Pan-European area. *Remote Sensing of Environment* 77: 76–91.

Häusler, T., Dees, M., Fockelmann, R., Pryomko, R. & Strunz, G. (2004). Land Use, Land Use Change and Forestry Information for Kyoto Protocol Reporting and Sustainable Forest Management. In M. Keil & B. Mohaupt-Jahr (edss), *Tagungsband "CORINE LAND COVER 2000 in Germany and Europe and its use for environmental applications" 21–23. Jan 2004, Berlin*: 123–134. Umweltbundesamt, Berlin.

Hansen, M.C., DeFries, R.S., Townshend, J.R.G., Carroll, M., Dimiceli, C. & Sohlberg, R.A. 2003. Global percent tree cover at a spatial resolution of 500 meters: First results of the MODIS vegetation continuous fields algorithm. *Earth Interactions* 7, paper no. 10. http://landval.gsfc.nasa.gov/pdf/hansen_01.pdf, (accessed 4. Dec. 2007).

Hansen, M.H. & Wendt, G.D. 2000. Using Classified Landsat Thematic Mapper Data for Stratification in a Statewide Forest Inventory. *Proceedings of the 1st Annual Forest Inventory and Analysis Symposium. US Department of Agriculture—Forest Service St. Paul, Minnesota, USA, North Central Forest Experiment Station*, General Technical Report NC-GTR-213 : 20–27. http://nrs.fs.fed.us/pubs/jrnl/2000/nc_2000_Hansen_001.pdf (accessed 6. Dec. 2007).

Hájek, F. 2007. Process-based approach to automated classification of forest structures using medium format digital aerial photos and ancillary GIS information. *European Journal of Forest Research* 127(2): 115–124.

Harding, D. & Carabajal, C. 2005. ICESat waveform measurements of within-footprint topographic relief and vegetation vertical structure. *Geophysical Research Letters* 32: L21S10.

Hese, S., Lucht, W., Schmullius, C., Barnsley, M., Dubayah, R., Knorr, D., Neumann, K., Riedel, T. & Schröter, K. 2005. Global biomass mapping for an improved understanding of the CO2 balance. *Remote Sensing of Environment* 94: 94–104.

Heurich, M. 2006. Evaluierung und Entwicklung von Methoden zur automatisierten Erfassung von Waldstrukturen aus Daten flugzeuggetragener Fernerkundungssensoren. Ph.D. Thesis, Department of Ecosystem and Landscapemanagement, Technical University of Munich, Munich, Germany.

Heurich, M., Schneider, T. & Kennel, E. 2003. Laser scanning for identification of forest structure in the Bavarian forest national park. Proceedings of the ScandLaser Scientific Workshop on Airborne Laser Scanning of Forests, 3–4 September 2003, Umea, Sweden.

Hirata, Y. 2004. The effects of footprint size and sampling density in airborne laser scanning to extract individual trees in mountainous terrain. *Proc. ISPRS WG VIII/2 "Laser-Scanners for Forestry and Landscape Assessment"*, 3–6 October 2004, Freiburg, Germany. XXXVI, 8/W2.

Hirschmugl, M., Franke, M., Ofner, M., Schardt, M. & Raggam, H. 2005. Single tree detection in very high resolution remote sensing data. *Proceedings of ForestSat 2005, Borås*: 38–43.

Hirschmugl, M., Ofner, M., Raggam, J. & Schardt, M. 2007. Single tree detection in very high-resolution remote sensing data. *Remote Sensing of Environment* 110: 533–544.

Hodgson, M.E., Jensen, J.R., Schmidt, L., Schill, S. & Davis, B. 2003. An evaluation of LIDAR and IFSAR derived digital elevation models in leaf-on conditions with USGS Level 1 and Level 2 DEMs. *Rem. Sens. Environ* 84: pp. 295–308.

Hoffmann, K. 2008. Assessment of forest structure and forest health parameters from digital aerial photographs of the Saxonia aerial photography campaign 2005. Proceedings of the Symposium "The pixeled forest" Freising 2008. Forstliche Forschungsberichte (in print).

Holmgren, I., Nilsson, M. & Olsson, H. 2003. Simulating the effects of lidar scan angle for estimation of mean tree height and canopy closure. *Canadian Journal of Remote Sensing* 29(5), 623–632.

Holmgren, J. 2004. Prediction of Tree Height, Basal Area and Stem Volume in Forest Stands Using Airborne Laser Scanning. *Scand. Journal of Forest Research* 19: 543–553.

Holström, H. & Fransson, J.E.S. 2003. Combining remotely sensed optical and radar data in *K*NN-estimation of forest variables. *Forest Science* 49: 409–418.

Hudak, A.T., Crookston, N.L., Evans, J.S., Falkowski, M.J., Smith, A.M.S., Gesseler, P.E. & Morgan, P. 2006. Regression modelling and mapping of coniferous forest basal are and tree density from discrete-return lidar and multispectral satellite data. *Canadian Journal of Remote Sensing* 32: 126–138.

Hudak, A.T., Lefsky, M.A., Cohen, W.B. & Berterretche, M. 2002. Integration of LiDAR and Landsat ETM+ data for estimating and mapping forest canopy height. *Remote Sensing of Environment* 82: 397–416.

Hudak, A.T., Lefsky, M.A., Cohen, W.B. & Berterretche, M. 2001. Integration of lidar and landsat etm+ data. *International Archives of Photogrammetry and Remote Sensing* XXXIV-3/W4.

Hyyppä, J, Pyysalo, U., Hyyppä, H., Haggren, H. & Ruppert, G. 2000. *Proceedings of SPIE* 4035: 119–130.

Hyyppä, J, Schardt, M., Haggren, H., Koch, B., Lohr, B., Scherrer, H.U., Paananen, R., Luukkonen, H., Ziegler, M., Ziegler, M., Hyyppä, H., Pyysalo, U., Friedländer, H., Uuttera, J., Wagner, S., Inkinen, M., Wimmer, A., Kukko, A., Ahoka, E. & Karjalainen, M. 2001. High-Scan: The first European-wide attempt to derive single-tree information from laser scanner data. *The Photogrammetric Journal of Finland* 17: 58–68.

Hyyppä, J. & Inkinen, M. 1999. Detecting and estimating attributes for single trees using laser scanner. *The Photogrammetric Journal of Finland* 16: 27–42.

Hyyppä, J., Hyyppä, H., Inkinen, M., Engdahl, M., Linko, S. & Yi-hong, Z. 2000. Accuracy comparison of various remote sensing data sources in the retrieval of forest stand attributes. *Forest Ecology and Management* 128: 109–120.

Hyyppä, J., Kelle, O., Lehikoinen, M. & Inkinen, M. 2001. A segmentation-based method to retriev stem volume estimates from 3–D tree height models produced by laser scanners. *IEEE Transactions on Geosciences and Remote Sensing* 39(5): 969–975.

Hyyppä, J., Mielonen, T., Hyyppä, H., Maltamo, M., Yu, X., Honkavaara, E. & Kaartinen, H. 2005. Using individual tree crown appraoch for forest volume extraction with aerial images and laser point clouds. *Proceedings of ISPRS Workshop Laser Scanning 2005, September 12–14, 2005, Enschede, Netherlands* 36, 3/W19: 144–149.

Imhoff, M.L. 1995. Radar backscatter and biomass saturation—Ramifications for global biomass inventory. *IEEE Transactions on Geoscience and Remote Sensing* 33(2): 511–518.

Ingram, J.C., Dawson, T.P. & Whittaker, R.J. 2005. Mapping tropical forest structure in southeastern Madagascar using remote sensing and artificial neural networks. *Remote Sensing of Environment* 94:491–507.

Kalogirou, V. 2007. *Simulation of discrete-return LiDAR signal from conifer stands for forestry applications.* MSc Thesis, Department of Geomatic Engineering, University College London, UK.

Kalogirou, V., Disney, M.I. & Lewis, P. 2007. The impact of system characteristics on discrete-return LiDAR: a simulation study over a conifer forest. Manuscript submitted for publication.

Keller, M. 2001. Aerial Photography. In P. Brassel & H. Lischke (eds.), *Swiss National Forest Inventory: Methods and Models of the Second Assessment*: 45–64. Birmensdorf: Swiss Federal Research Institute.

Kennedy, P.J., Belward, A.S. & Grégoire, J.M. 1994. An improved approach to fire monitoring in West Africa using AVHRR data. *International Journal of Remote Sensing* 15: 2235–2255.

Koch, B., Heyder, U., Straub, Ch. & Weinacker, H. 2007a. 3D data for forest and environmental planning. *Proceedings Workshop on 3-D Remote Sensing in Forestry, 14–15. Feb. 2006, Vienna, Austria*: 1–15.

Koch, B., Heyder U. & Weinacker, H. 2006. Detection of individual tree crowns in airborne LIDAR data. An approach to delineate tree crowns in mixed and deciduous temperate forests. *Photogrammetric Engineering and Remote Sensing* 72(4): 357–363.

Koch, B., Straub, Ch., Dees, M., Wang, Y. & Weinacker, H. 2007b. Airborne laser data for stand delineation and information extraction. *ISPRS Journal of Photogrammetry & Remote Sensing* (accepted for publication).

Koch, B. & Weinacker, H. 2007. Simulation berechnet für die Landesanstalt für Wald und Holz. *Interner Bericht* an die LWF 2007.

Köhl, M. 2002. Inventory concept LFI II. In P. Brassel & H. Lischke (eds.), 2001: *Swiss National Forest Inventory: Methods and Models of the Second Assessment*: 19–44. Birmensdorf, Swiss Federal Research Institute WSL.

Köhl, M. & Brassel, P. 1998. Einsatz der Fernerkundung bei Nationalen Forstinventuren. *Allgemeine Forstzeitschrift für Waldwirtschaft und Umweltvorsorge/DerWald* 24: 1472–1474.

Korpela, I., Dahlin, B., Schäfer, H., Bruun, E., Haapaniemi, F., Honkasalo, J., Ilvesniemi, S., Kuutti, V., Linkosalmi, M., Mustonen, J., Salo, M., Suomi, O. & Virtanen, H. 2007. Single-Tree Forest Inventory Using Lidar and Aerial Images for 3D Treetop Positioning, Species Recognition, Height and Crown Width Estimation. In *Proc. ISPRS III/3, III/4, V/3 and VIII/11. "LaserScanning 2007 and SilviLaser 2007", 12–14 September 2007, Espoo, Finland,* XXXVI, Part 3/W52: 227–234.

Koukal, T., Suppan, F. & Schneider, W. 2007. The impact of radiometric calibration on the accuracy of kNN-predictions of forest attributes. *Remote Sensing of Environment* 110: 431–437.

Kramer, H.J. 2002. Observation of the Earth and Its Environment: survey of missions and sensors, 4th Edition. Springer.

Kraus, K. & Pfeiffer, N. 1998. Determination of terrain models in wooded areas with airborne laser scanner data. *ISPRS J. of Photogrammetry and Remote Sensing* 53: 193–203.

Larsen, M. 2007. Single tree species classification with a hypothetical multi-spectral satellite. *Remote Sensing of Environment* 110, *Issue 4*: 523–532. (ForestSAT Special Issue, ForestSAT 2005 Conference "Operational tools in forestry using remote sensing techniques").

Le Toan, T., Beaudoin, A., Riom, J. & Guyon, D. 1992. Relating forest biomass to SAR data. *IEEE Transactions on Geoscience and Remote Sensing* 30, 2: 403–411.

Leblon, B. 2005. Using remote sensing for fire danger monitoring. *Natural Hazards* 35(3): 343–359.

Leblon, B., Fernandez-Garcia, P.A., Oldford, S., MacLean, D. & Flannigan, M. 2007. Using NOAA-AVHRR cumulative indices for estimating fire danger codes in northern boreal forests. *International Journal of Applied Earth Observations and Geoinformation* 9: 335–342.

Lecki, D., Gougeon, F., Hill, D., Quinn, R., Armstrong, L. & Shreeman, R. 2003. Combined high-density lidar and multispectral imagery for individual tree crown analysis. *Can. J. Remote Sensing* 29(5): 633–649.

Leckie, D.G. 1990, Advances in remote sensing technologies for forest surveys and management. *Canadian Journal of Forestry Research* 20: 464–483.

Leckie, D.G., Gougeon, F.A., Hill, D., Quinn, R., Armstrong, L. & Shreenan, R. 2003. Combined high-density lidar and multispectral imagery for individual tree crown analysis. *Canadian Journal of Remote Sensing* 29: 633–649.

Leckie, D.G., Gougeon, F.A., Tinis, S., Nelson, T., Burnett, Ch.N. & Paradine, D. 2005. Automated tree recognition in old growth conifer stands with high resolution digital imagery. *Remote Sensing of Environment* 94: 311–326.

Lefsky, M.A., Cohen, W.B., Hudak, A., Acker, S.A. & Ohmann, J.A. 1999. Integration of Lidar, Landsat ETM+ and Forest Inventory data for Regional Forest Mapping. *ISPRS Workshop: Mapping Forest Structure and Topography by Airborne and Spacebourne Lasers, November 9–11, La Jolla CA.*

Lefsky, M.A., Cohen, W.B., Parker, G.G. & Harding, D.J. 2002. LiDAR remote sensing for ecosystem studies. *BioScience* 52: 19–30.

Lefsky, M., Harding, D., Keller, M., Cohen, W., Carabajal, C., Del Bom Espirito-Santo, F., Hunter, M., de Oliveira Jr., R. & de Camargo, P. 2005. Estimates of forest canopy height and aboveground biomass using ICESat. *Geophysical Research Letters* 32: L22S02.

Lefsky, M.A., Harding, D., Cohen, W.B., Parker, G. & Shugart, H.H. 1999. Surface LIDAR remote sensing of basal area and biomass in deciduous forests of Eastern Maryland, USA. *Remote Sensing of Environment* 67: 83–98.

Litkey, P., Rönnholm, P., Lumme, J. & Lian, X. 2007. Waveform Features for Tree Identification. *Proc. ISPRS III/3, III/4, V/3 and VIII/11 "LaserScanning 2007 and SilviLaser 2007" 12–14 September 2007, Espoo, Finland*: 258–264. XXXVI, Part 3/W52.

Lovell, J.L., Jupp, D.L.B., Newnham, G.J., Coops, N.C. & Culvenor, D.S. 2005. Simulation study for finding optimal lidar acquisition parameters for forest height retrieval. *Forest Ecology and Management* 214: 398–412.

Lu, D. 2005. Aboveground biomass estimation using Landsat TM data in the Brazilian Amazon. *International Journal of Remote Sensing* 26: 2509–2525.

Lund, H.G. 1997. Chapter 11—Forestry. In W.R. Philipson, (editor-in-chief), Manual of Photographic Interpretation, 2nd edition: 399–440. Bethesda, MD: American Society for Photogrammetry and Remote Sensing.

Magnusson, M., Fransson, J. & Olsson, H. 2007, Aerial photo-interpretation using Z/I DMC images for estimation of forest variables. Scandinavian Journal of Forest Research 22, 3: 25–266.

Malingreau, J.P., Achard, F., D'Souza, G., Stibig, H.J., D'Souza, J., Estregeul, C. & Eva, H. 1995. AVHRR for global tropical forest monitoring: The lessons of the TREES project. Remote Sensing Reviews, 12: 29–40.

Mallinis, G., Karamanolis, D., Karteris, M. & Gitas, I. 2008. An object based approach for the implementation of forest legislation in Greece using very high resolution satellite data. In T. Blaschke, S. Lang & G. Hay (eds.), *Lecture Notes in Geoinformation and Cartography. Object-Based Image Analysis Spatial concepts for knowledge-driven remote sensing applications*. Springer-Verlag (in press) ISBN: 978-3-540-77057-2.

Maltamo, M., Eerikäinen, K., Pitkänen, J., Hyyppä, J. & Vehmas, M. 2004. Estimation of timber volume and stem density based on scanning laser altimetry and expected tree size distribution functions. *Remote Sensing of Environment* 90(3): 319–330.

Maltamo, M., Malinen, J., Packalén, P., Suvanto, A. & Kangas, J. 2006. Nonparametric estimation of stem volume using airborne laser scanning, aerial photography, and stand-register data, *Canadian Journal of Forest Research* 36, Number 2: 426–436.

Maltamo, M., Packalén,P., Peuhkurinen, J., Suvanto, A., Pesonen, A. & Hyyppä, J. 2007. Experiences and Possibilities of ALS Based Forest Inventory in Finland. *Proc. ISPRS III/3, III/4, V/3 and VIII/11. "LaserScanning 2007 and SilviLaser 2007", 12–14 September 2007, Espoo, Finland* XXXVI, Part 3/W52: 270–280.

Mayaux, P., Holmgren, P., Achard, F., Eva, H., Stibig, H.-J. & Branthomme, A. 2005. Tropical forest cover change in the 1990s and options for future monitoring. *Phil. Trans. R. Soc. B.* 360: 373–384.

McCombs, J.W., Roberts, S.D. & Evans, D.L. 2003. Influence of fusing LiDAR and multispectral imagery on remotely sensed estimates of stand density and mean tree height in a managed loblolly pine plantation. *Forest Science* 49: 457–466.

McInerney, D., Pekkarinen, A. & Haakana, M. 2005. Combining Landsat ETM+ with field data for Ireland's National Forest Inventory—a pilot study for Co. Clare. In H. Olsson (ed.), *Proceedings of ForestSat 2005. Borås, Sweden, 31 May—3 June 2005*: 12–16.

McRoberts, R.E, 2006. A model-based approach to estimating forest area. *Remote Sensing of Environmen,* 103: 56–66

McRoberts, R.E. 2005. Remote sensing support for the national forest inventory for the United States of America. F*orest SAT 2005 Proceedings*: http://www.svo.se/episerver4/templates/SNormalPage.aspx?id=16101, (accessed at 6 Nov. 2007).

McRoberts, R.E. 2006. A model-based approach to estimating forest area. *Remote Sensing of Environment* 103: 56–55.

McRoberts, R.E., Holden, G.R., Nelson, M.D., Liknes, G.C. & Gormanson, D.D. 2006. Using satellite imagery as ancillary data for increasing the precision of estimates for the Forest Inventory and Analysis program of the USDA Forest service. *Canadian Journal of Forest Research* 36: 2968–2980.

McRoberts, R.E., Tomppo, E.O. 2007. Remote Sensing for national forest inventories. *Remote Sensing of Environment* 110: 412–419.

McRoberts, R.E., Tomppo, E.O., Finley, A.O. & Heikkinen, J. 2007. Estimating areal means and variances of forest attributes using the k-Nearest Neighbor technique and satellite imagery. *Remote Sensing of Environment* 111:466–480.

McRoberts, R.E., Wendt, D.G. & Nelson, M.D. 2002. Using a land cover classification based on satellite imagery to improve the precision of forest inventory area estimates. *Remote Sensing of Environment* 81: 36–44.

Means, J.E., Acker, S.A., Fitt, J.B., Renslow, M., Emerson, L. & Hendix, C.J. 2000. Predicting forest stand characteristics with airborne scanning LIDAR. *Photogrammetric Engineering & Remote Sensing* 66(11): 1367–1371.

Melon, P., Martinez, J.-M., Le Toan, T., Ulander, L.M.H. & Beaudoin, A. 2001. On the retrieving of forest stem volume from VHF SAR data: observation and modeling. *IEEE Transactions on Geoscience and Remote Sensing* 39, 11: 2364 –2372.

Mette, T. 2007. *Forest biomass estimation from polarimetric SAR interferometry.* Ph.D. thesis, Technische Universität München, Munich, Germany, 164 pp.

Morsdorf, F., Meier, E., Allgöwer, B. & Nuesch, D. 2003. Clustering in airborne laser scanning raw data for segmentation of single trees. *International Archives of the Photogrammetry, Remote Sensing and Spatial Information Sciences* 34, part 3/W13: 27–33.

Naesset E., Gobakken, T., Holmgren, J., Hyyppä, H., Hyyppä, J., Maltamo, M., Nilsson, M., Olsson, H., Persson, Å. & Söderman, U. 2004. Laser scanning of forest resources: the Nordic experience. *Scand. J. of For. Res.* 19(6): 482–499.

Naesset, E. & Gobakken, T. 2005. Estimating forest growth using canopy metrics derived from airborne laser scanner data. *Remote Sensing of Environment* 96(3–4): 453–465.

Naesset, E. 1997b. Estimating timber volume of forest stands using airborne laser scanner data. *Rem. of Environ.* 61: 246–253.

Naesset, E. 2002. Predicting forest stand characteristics with airborne scanning laser using practical two-stage procedure and field data. *Remote Sensing of Environment* 80, (1): 88–99.

Næsset, E. 2004. Effects of different flying altitudes on biophysical stand properties estimated from canopy height and density measured with a small-footprint airborne scanning laser. *Remote Sensing of Environment* 91(2), 243–255.

Næsset, E. 2005. Assessing sensor effects and effects of leaf-off and leaf-on canopy conditions on biophysical stand properties derived from small-footprint airborne laser data. *Remote Sensing of Environment* 98(2–3), 356–370.

Nilsson, M., Folving, S., Kennedy, P., Puumalainen, J., Chirici, G., Corona, P., Marchetti, M., Olsson, H., Ricotta, C., Ringvall, A., Ståhl, E. & Tomppo, E. 2004. Combining remote sensing and field data for deriving unbiased estimates of forest parameters over large regions. In P Corona, M. Köhl & M. Marchetti (eds.), *Advances in forest inventory for sustainable forest management and biodiversity monitoring. Forestry Sciences 76*: 19–32. Kluwer Academic Publishers.

Nuske, R. 2006. A retrospective study of canopy gap dynamics of an European beech stand. *Proceedings Workshop on 3-D Remote Sensing in Forestry, 14–15. Feb. 2006. Vienna, Austria*: 40–44.

Oldford, S., Leblon, B., Gallant, L. & Alexander, M. 2003. Mapping pre-fire conditions in the Northwest Territories, Canada, using NOAA-AVHRR images. *Geocarto International,*18(4): 21–32.

Oldford, S., Leblon, B., MacLean, D. & Flannigan, M. 2006. Predicting slow drying fuel moisture codes using NOAA-AVHRR images. *International Journal of Remote Sensing* 27(18): 3881–3902.

Olsson, H. 1995. Reflectance calibration of thematic mapper data for forest change detection. Int.J.Remote Sensing 16, No.1: 81–96.

Ørka, H.O., Næsset, E. & Bollandsås, O.M. 2007. Utilizing Airborne Laser Intensity for Tree Species Classification. *Proc. ISPRS III/3, III/4, V/3 and VIII/11. "LaserScanning 2007 and SilviLaser 2007", 12–14 September 2007, Espoo, Finland* : 300–305. Vol.XXXVI, Part 3/W52.

Osborne, J.G. 1942. Sampling errors for systematic and random surveys of cover type areas. *Journal of the American Statistical Association* 37: 256–264.

Papathanassiou, K.P. & Cloude, S.R. 2001. Single-baseline polarimetric SAR interferometry. *IEEE Transaction on Geoscience and Remote Sensing* 39, no. 11: 2352–2363.

Pekkarinen, A., Reithmaier, L. & Strobl, P. 2007. Pan-European Forest/Non-Forest Mapping Based on Landsat Data. *Proceedings of the 10th AGILE International Conference on Geographic Information Science, Aalborg University, Denmark*: http://www.svo.se/forlag/rapporter/1741.pdf (accessed 4. Dec. 2007).

Persson, Å., Holmgren, J. & Söderman, U. 2002. Detecting and measuring individual trees using an airborne laser scanner. *Photogrammetric Engineering and Remote Sensing* 68: 925–932.

Pillai, R., Weisberg, P. & Lingua, E. 2005. Object-oriented classification of repeat aerial photography for quantifying woodland expansion in central Nevada. 20 Biennial Workshop on Aerial Photography, Videography, and High Resolution Digital Imagery for Resource Assessment October 4–6, 2005 * Weslaco, Texas. http://www.cabnr.unr.edu/weisberg/publications/ (accessed 7 March 2008)

Pohl, C. & van Genderen, J.L. 1998. Multisensor image fusion in remote sensing: concepts, methods and applications. *International Journal of Remote Sensing* 19: 823–854.

Popescu, S.C., Wynne, R.H. & Scrivani, J.A. 2004. Fusion of small-footprint Lidar and multispectral data to estimate plot-level volume and biomass in deciduous and pine forests in Virginia, USA. *Forest Science* 50: 551–565.

Popescu, S., Wynne, R. & Nelson, R.F. 2003. Measuring individual tree crown diameter with LIDAR and assessing its influence on estimating forest volume and biomass. *Canadian Journal of Remote Sensing* 29(5): 564–577

Raber, G.T., Jensen, J.R., Schill, S.R. & Schuckman, K. 2002a. Creation of digital terrain models unsing an adaptive lidar vegetation point removal process. *Photogr. Eng. & Rem. Sens.* 68(12): 1307–1315.

Ramminger, G. & Koch, B. 2007. GSE-FM Sub-National Forest Information Update. *Forest SAT 2007, Montpellier, France,* http://forestsat.teledetection.fr/index.php?option=com_docman&task=cat_view&gid=24&Itemid=31 (accessed 4 Dec. 2007).

Rauste, R. 2005. Multi-temporal JERS SAR data in boreal forest biomass mapping. *Remote Sensing of Environment* 97: 263–275.

Reese, H., Nilsson, M., Pahlén, T.G., Hagner, O., Joyce, S., Tingelöf, U., Egberth, M. & Olsson, H. 2003. Country-wide estimates of forest variables using satellite data and field data from the National Forest Inventory. *Ambio* 32: 542–548.

Reitberger, J., Krzystek P. & Stilla, U. 2007. Combined Tree Segmentation and Stem Detection Using Full Waveform Lidar data. *Proc. ISPRS III/3, III/4, V/3 and VIII/11. "LaserScanning 2007 and SilviLaser 2007", 12–14 September 2007, Espoo, Finland* XXXVI, Part 3/W52: 332–338.

Reitberger, J., Krzystek, P. & Heurich, M. 2006. Full-waveform analysis of small footprint airborne laser scanning data in the Bavarian forest national park for tree species classification. *Proceedings Workshop on 3D Remote Sensing in Forestry, 14–15. Feb. 2006. Vienna, Austria*: 218–228.

Ridder, R.M. 2007. Global Forest Resources Assessment 2010. Options and Recommendations for a Global Remote Sensing Survey of Forests. *FAO working paper* 141, Rome, 2007, ftp://ftp.fao.org/docrep/fao/010/ai074e/ai074e00.pdf, (accessed 4. Dec. 2007)

Rönnholm, P., Honkavaara, E., Litkey, P., Hyyppä, H. & Hyyppä, J. 2007. Integration of Laser Scanning and Photogrammetry. *Proc. ISPRS III/3, III/4, V/3 and VIII/11. "LaserScanning 2007 and SilviLaser 2007", 12–14 September 2007, Espoo, Finland* XXXVI, Part 3/W52: 355–363.

Rönnholm, P., Hyyppä, J., Hyyppä, H., Haggren, H., Yu, X., Pyysalo, U., Pöntinen, P. & Kaartinen, H. 2004. Calibration of laser-derived tree height estimates by means of photogrammetric techniques. *Scand. J. of For. Res.*19(6): 524–528.

Rosenqvist, Å., Shimada, M., Ito, N. & Watanabe, M. 2007. ALOS PALSAR: A pathfinder mission for global-scale monitoring of the environment. *IEEE Transaction on Geoscience and Remote Sensing* 45: 3307–3316.

Rosette, J., North, P., Suárez, J. & Armston, J. 2007. A comparison of biophysical parameter retrieval for forestry using airborne and satellite lidar. *"ForestSat 2007" 5–7 November 2007, Montpellier, France*, http://forestsat.teledetection.fr/index.php?option=com_docman&task=doc_details&gid=50&Itemid=31 (accessed 7 March 2008).

San-Miguel-Ayanz, J., Ravail, N., Kelha, V. & Ollero, A. 2005. Active Fire Detection for Fire Emergency Management: Potential and Limitations for the Operational Use of Remote Sensing. *Natural Hazards* 35(3): 361–376.

Santoro, M., Askne, J., Smith, G. & Fransson, J.E.S. 2002. Stem volume retrieval in boreal forests with ERS-1/2 interferometry. *Remote Sensing of Environment* 81, 1: 19–35.

Särndal, C.E., Swenson, B. & Wretman, J. 1992. *Model Assisted Survey Sampling.* New York: Springer-Verlag.

Schade, J. 1980. Ein mehrphasiges Stichprobensystem für forstliche Großrauminventuren, gestützt auf Landsat MSS Daten, Luftbilder 1:50 000 und ergänzende terrestrische Messungen. Dissertation, Universität Freiburg i.Br, 130 p.

Schardt, M., Gallaun, H. & Häusler, T. 1998. Monitoring of Environmental Parameters in the Alpine Regions by Means of Satellite Remote Sensing. *Proceedings of the International ISPRS-Symposium on "Resource and Environmenal Monitoring, Local, Regional, Global", Commission VII, September 7–4, 1998, Budapest, Hungary.*

Schardt, M., Granica, K., Schmitt, U. & Gallaun, H. 2000. Monitoring of Protection Forests in alpine Regions. Proceedings of the 4.02.05 Group Session „Remote Sensing and World Forest Monitoring", IUFRO XXI World Congress, 7–12 August 2000, Kuala Lumpur, Malaysia, 2000: 117–118.

Schardt, M., Granica, K., Hirschmugl, M., Luckel, W. & Klaushofer, F. 2004. Mapping Protection Forests in the Province of Salzburg Using Remote Sensing. *Proc. Göttingen GIS & Remote Sensing Days 2004*: 204–213. Göttinger Geographische Schriften.

Schmid, P. 1969. Ergebnisse einer Zweitaufnahme mit Kontroll-Stichproben. *Berichte der Eidgenössischen Anstalt für das forstliche Versuchswesen* Nr. 15.

Singh, A. 1989. Digital change detection techniques using remotely-sensed data. *Int. J. Remote Sensing* 10, 6: 989–1003.

Sithole G. & Vosselman, G. 2004. Experimental comparison of filter algorithms for bare earth extraction from airborne laser scanning point clouds. *International Archives of Photogrammetrie and Remote Sensing*: 59: 85–101.

Smith, G. & Ulander, L.M.H. 2000. A model relating VHF-band backscatter to stem volume of coniferous boreal forest. *IEEE Transactions on Geoscience and Remote Sensing* 38, 2: 728–740.

Solberg, A.H., Jain, A.K. & Taxt, T. 1994. Multisource Classification of Remotely Sensed Data: Fusion of Landsat TM and SAR Images. *IEEE Transactions on Geoscience and Remote Sensing* 32: 768–778.

Spurr, S.H. 1960. *Photogrammetry and Photo-Interpretation (second ed.)*: 382–392. New York: Ronald Press.

Steininger, M.K. 2000. Satellite estimation of tropical secondary forest above-ground biomass: data from Brazil and Bolivia. *International Journal of Remote Sensing* 21: 1139–1157.

St-Onge, A. & Achaichia, N. 2001. Measuring forest canopy height using a combination of LIDAR and aerial photography data. *International Archives of Photogrammetry and Remote Sensing* XXXIV-3/W4: 131–137.

St-Onge, B., Jumelet, J., Cobello, M. & Vega, C. 2004. Measuring individual tree height using a combination of stereophotogrammetry and LiDAR. *Canadian Journal of Forest Research* 34: 2122–2130.

Strobl, D., Raggam, J. & Buchroithner, M.F. 1990. Terrain correction geocoding of a multisensor image data set. *Proceedings 10th EARseL Symposium, Toulouse, France :* 98–107. Paris: European Space Agency.

Suarez, J.C., Ontiveros, C., Smith, S. & Snape, S. 2005. Use of LiDAR and aerial photography in the estimation of individual tree heights in forestry. *Computers and Geosciences* 31: 253–262.

Suárez, J.C., García, R., Gardiner, B.A. & Patenaude, G. 2007. The estimation of wind risk in forest stands using ALS. *Journal of Forest Planning* (In press).

Suits, G., Malila, W. & Weller, T. 1998. Procedures for using signals from one sensor as substitutes for signals of another. *Remote Sensing of Environment* 25: 395–408.

Tanaka, S., Sugimura, T., Senoo, T. and Nishi, T. 1998. Forest stand structure analysis in Mt Fuji area with OPS and SAR Data. *Advances in Space Research* 22: 685–688.

Thiel, C., Drezet, P., Weise, C., Quegan, S. & Schmullius, C. 2006. Radar remote sensing for the delineation of forest cover maps and the detection of deforestation. *Forestry* 79, 5: 589–597.

Tiede, D., Hochleitner, G. & Blaschke, T. 2005. A full GIS-based workflow for tree identification and delineation using laser scanning. *The International Archives of Photogrammetry, Remote Sensing and Spatial Information Sciences*XXXVI, Part 3/W24: 9–14.

Tiede, D., Lang, S. & Hoffmann, C. 2006. Supervised and forest type-specific multi-scale segmentation for a one-level-representation of single trees. *International Archives of Photogrammetry, Remote Sensing and Spatial Information Sciences*,XXXVI-4/C42.

Tomppo, E. 1996. Multi-source National Forest Inventory of Finland. In R. Päivinen, J. Vanclay & S. Miina (eds.), New thrusts in forest inventory, proceedings of the Subject Group S4.02-00, Forest Resource Inventory and Monitoring, and Subject Group S4.12-00, Remote Sensing Technology, volume I. IUFRO XX World Congress, Tampere, Finland, 6–12 August 1995: 27–41.

Tomppo, E. 2006. The Finnish multi-source national forest inventory—small area estimation and map production. Forest inventory—methodology and applications: 195–224. Dordrecht, The Netherlands: Springer.

Tomppo, E. & Halme, M. 2004. Using coarse scale forest variables as ancillary information and weighting of variable sin k-NN estimation: a genetic algorithm approach. *Remote Sensing of Environment* 92: 1–20.

Tuominen, S., Fish, S. & Poso, S. 2003. Combining Remote Sensing data from earlier inventories, and geostatistical interpolation in multisource forest inventory. *Canadian Journal of Forest Research* 33: 624–634.

Valeria, O., Laarman, A., Beaudoin, A., Cote, S. & Simard, G. 2007. Use of multi-temporal satellite imagery over large territories for monitoring the state of the boreal forest of north-western Quebec, Canada; Plenary Session: Global Forest Monitoring and International reporting. *ForestSat 2007 Conference—November 5–7, 2007*. Montpellier, France. http://forestsat.teledetection.fr/index.php?option=com_docman&task=doc_details&gid=8&Itemid=31 (accessed 7 March 2008).

Van Aardt, J., Wynne, R.H. & Oderwald, R.G. 2006. Forest Volume and Biomass Estimation Using Small-Footprint Lidar-Distributional Parameters on a Per-Segment Basis. *Forest Science* 52, Number 6, December 2006: 636–649.

Van Brusselen, J. (ed.). 2006a. Core User Needs and User Standards Dossier. *GMES Service Element for Forest Monitoring project report. Doc. No. GSE-FM-T1-U1-Ph1*. GAF AG, München, ESA ESRIN. Contract number 17063/03/I-LG.

Van Brusselen. J. (ed.). 2006b. Service Utility Reports. *GMES Service Element for Forest Monitoring project report. Doc. No. GSE-FM-T5-U7-Ph1*. GAF AG, München, ESA ESRIN. Contract number 17063/03/I-LG.

Van Brusselen, J. & Schuck, A. 2005. Policy Foundations Review for Geographically Explicit Information Needs on Forests. *Technical Report* 18, *European Forest Institute*.

Wack, R., Schardt, M., Lohr, U., Barrucho, L. & Oliveira, T. 2003. Forest inventory for Eucalyptus plantations based on airborne laser scanner data. *International Archives of the Photogrammetry, Remote Sensing and Spatial Information Sciences* 34, part 3/W13: 40–46.

Wagner, W., Luckman, A., Vietmeier, J., Tansey, K., Balzter, H., Schmullius, C., Davidson, M., Gaveau, D., Gluck, M., Le Toan, T., Quegan, S., Shvidenko, A., Wiesmann, A. & Yu, J.J. 2003. Large-scale mapping of boreal forest in SIBERIA using ERS tandem coherence and JERS backscatter data. *Remote Sensing of Environment* 85, 2: 125–144.

Wagner, W., Roncat,A., Melzer, T. & Ullrich A. 2007. Waveform Analysis Techniques in Airborne Laser Scanning. *Proc. ISPRS III/3, III/4, V/3 and VIII/11. "LaserScanning 2007 and SilviLaser 2007"*, 12–14 September 2007, Espoo, Finland. XXXVI, Part 3/W52 :413–419.

Wallerman, J. 2003. Remote sensing aided spatial prediction of forest stem volume. *Acta Universitatis agriculturae Sueciae. Silvestria nr* 27.

Wang, L., Gong, P. & Biging, G.S. 2004. Individual Tree-Crown Delineation and Treetop Detection in High-Spatial-Resolution Aerial Imagery. *Photogrammetric Engineering & Remote Sensing* 70: 351–357.

Wang, Y., Weinacker, H. & Koch, B. 2007a. Development of a Procedure for Vertical Structure Analysis and 3D Single Tree Extraction within Forests Based on Lidar Point Cloud. *Proc. ISPRS III/3, III/4, V/3 and VIII/11. "Laser-Scanning 2007 and SilviLaser 2007", 12–14 September 2007, Espoo, Finland* XXXVI, Part 3/W52: 419–424.

Wang, Z,. Boesch, R. & Ginzler, C. 2007b. Color and LiDAR data fusion: Application to automatic forest boundary delineation in aerial images. *International Archives of Photogrammetry and Remote Sensing* XXXVI-1/W51.

Waser, L.T., Baltsavias, E., Ecker, K., Eisenbeiss, H., Feldmeyer-Christe, E., Ginzler, C., Küchler, M., Thee, P. & Zhang, L. 2007. Assessing changes of forest area and shrub encroachment in a mire ecosystem using digital surface models and CIR-aerial images. *Remote Sensing of Environment* doi:10.1016/j.rse.2007.09.015.

Waser, L.T., Baltsavias, E., Ecker, K., Eisenbeiss, H., Ginzler, C., Küchler, M., Thee, P. & Zhang, L. 2008. High-resolution digital surface models (DSM) for modelling fractional shrub/tree cover in a mire environment. *International Journal of Remote Sensing* 29(5): 1261–1276.

Waser, L. & Ginzler, C. 2008. Semi-automatiic tree species differenciation on the basis of ADS 40 digital aerial photographs. Proceedings of the Symposium "The pixeled forest" Freising 25.1.2008 Forstliche Forschungsberichte (in print).

Wehr, A. & Lohr, U. 1999. Airborne laser scanning—an introduction and overview. *ISPRS J. of Photogrammetry and Remote Sensing* 54 (2–3): 68–82.

Weinacker, H., Koch, B., Heyder, U. & Weinacker, R. 2004. Development of filtering, segementation and modelling modules for LIDAR and multispectral data as a fundamental of an automatic forest inventory system. *Laser-Scanners for Forest and Landscape Assessment, International Archives of Photogrammetry, Remote Sensing and Spatial Information Sciences* XXXVI, 8/W2.

White, J.C., Wulder, M.A., Brooks, D., Reich, R. & Wheate, R.D. 2005. Detection of red attack stage mountain pine beetle infestation with high spatial resolution satellite imagery. *Remote Sensing of Environment* 96: 340–351.

Wulder, M.A. & Seemann, D. 2003. Forest inventory height update through the integration of LiDAR data with segmented Landsat imagery. *Canadian Journal of Remote Sensing* 29: 536–543.

Wulder, M. 1998. Optical remote-sensing techniques for the assessment of forest inventory and biophysical parameters *Progress in Physical Geography* 22: 449–476.

Wulder, M.A., Han, T., White, J.C., Sweda, T. &Tsuzuki, H. 2007. Integrating profiling LIDAR with Landsat data for regional boreal forest canopy attribute estimation and change characterization. *Remote Sensing of Environment* 110: 123–137.

Xian, G., Zhu, Z., Hoppus, M. & Fleming, M. 2002. Application of Decision-Tree Techniques to forest group and basal area mapping using satellite imagery and forest inventory data. *Integrated Remote Sensing at the conjunction with Pecora 15/Land Satellite Information IV Conference, 10–15 November 2002, Denver, Colorado, USA*

Yu, X., Hyyppä, J., Hyyppä, H. & Maltamo, M. 2004. Effects of flight altitude on tree height estimation using airborne laser scanning. *Proc. ISPRS WG VIII/2 "Laser-Scanners for Forestry and Landscape Assessment" 3–6 October 2004, Freiburg, Germany* XXXVI, Part 8/W2.

Zhang, L., Liao, M., Yang L. & Lin, H. 2007. Remote Sensing Change Detection Based on Canonical Correlation Analysis and Contextual Bayes Decision. *Photogrammetric Engineering & Remote Sensing* 73,No.3: 311–318.

Zheng, D., Rademacher, J., Chen, J., Crow, T., Breese, M., Le Moine, J. & Ryu, S. 2004. Estimating aboveground biomass using Landsat 7 ETM+ data across a managed landscape in northern Wisconsin, USA. *Remote Sensing of Environment* 93: 402–411.

Zidek, V. 2001. Nature reserve Knehyne—spatial database and tree health status assessment. *1st Workshop of the EARSeL Special Interest Group on Remote Sensing for Developing Countries* : 183–195. EARSeL, Paris.

Zwally, H J., Schutz, B. & Abdalati, W. 2002. ICESat's Laser Measurements of Polar Ice, Atmosphere, Ocean and Land. *Journal of Geodynamics* 34: 405–445.

OTHER CONTRIBUTORS

Jo v. Brusselen[1], Leif Eriksson[2], Johan Fransson[3], Heinz Gallaun[4], Brigitte Leblon[5], Ronald E. McRoberts[6], Mats Nilsson[3], Mathias Schardt[4], Rudolf Seitz[7], Lars Waser[8]

[1]European Forest Institute, Joensuu, Finland
[2]Chalmers University of Technology, Göteborg, Sweden
[3]Swedish University of Agricultural Sciences, Umea, Sweden
[4]Institute of Digital Image Processing, Joanneum Research, Graz, Austria
[5]University of New Brunswick, Fredericton, Canada
[6]USDA Forest Service, Northern Central Research Station, St. Paul, Minnesota, USA
[7]Bavarian State Institute of Forestry—LWF, Department of GIS and Remote Sensing, Freising, Germany
[8]Swiss Federal Research Institute—WSL, Birmensdorf, Switzerland

Part VI
Education and cooperation

Advances in Photogrammetry, Remote Sensing and Spatial Information Sciences: 2008 ISPRS Congress Book – Li, Chen & Baltsavias (eds)

CHAPTER 33

Educational developments and outreach

Kohei Cho, Gerhard König & Joachim Höhle

ABSTRACT: The ISPRS Commission VI on "Education and Outreach" mainly deals with educational matters related to photogrammetry, remote sensing and spatial information science. The many changes in technology and in social and economic conditions are leading to innovation in education and learning in these three fields. This chapter first covers the situation at universities. Examples of various curricula are given where these three disciplines are either integrated or these subjects are specializations. The core knowledge in each of these three fields is outlined. The second part of the paper deals with education outside the universities. Some of the existing courses in distance learning are presented. Special emphasis is given to e-learning, including its state of the art. The important subject of accreditation and evaluation of e-learning is discussed thereafter. The summary of the achievements of Commission VI on education in the last four years is also a part of this chapter. Finally, the possible future of education and learning is outlined. Learning is becoming a lifelong process and e-learning will have an important place in complying with the needs of the workforce in these three fields. ISPRS has an important role in supporting lifelong education and continuing professional development.

Keywords: Education, teaching, learning, e-learning, internet, cooperation

33.1 INTRODUCTION

The technological advancements in photogrammetry, remote sensing and spatial information science are numerous. They have created new production methods in mapping and also many new applications. Amateur photography, Internet-based information services and car navigation are widely used by the public today. Changed social and economic conditions have come with globalization. Mapping tasks are carried out around the globe in cooperation with many organizations. Mobility, flexibility and readiness for learning new knowledge are requirements for the workforce in mapping and the production of geo-products. Google Earth and other net-based services eased the public access to satellite/aerial images and other geographic information of almost any place around the world. The interests and needs in education in these technical fields are increasing.

In general, education and training are no longer restricted to a few years of study at universities; it is a lifelong task. The Internet and e-learning are playing significant roles and will therefore be dealt with in detail in this chapter. The methodologies in teaching and training have to be adapted to the new conditions. Many new e-learning courses are available: good courses as well as bad ones. There is a need for accreditation and evaluation of curricula and e-learning courses. Education and training are then no longer a matter of a single university or a single country, but more inter-organizational, national and international cooperation has to take place (Dittler et al 2005). This outreach is important to support and extend teaching and learning. ISPRS Commission VI on "Education and Outreach" deals mainly with educational matters related to photogrammetry, remote sensing and spatial information science. The major achievements of Commission VI on educational developments and outreach during 2004 to 2008, as well as a presentation of the state of the art in education in the three fields connected with the activities of Commission VI, are the goals of this chapter.

33.2 EDUCATION AND TRAINING AT UNIVERSITIES

33.2.1 *International situation*

Photogrammetry (P) and remote sensing (RS) are usually not study programmes in themselves at universities. Education and training in these subjects is embedded in various curricula. In Central Europe and Canada, it has been part of surveying engineering or geodesy. In recent years, with the development of global positioning and spatial information science (SIS), the study programme is often called geomatics or geo-informatics. The contents, however, can be very different, even when the study programme has the same name. Many European countries recently structured their five-year diploma programmes into three-year Bachelor and a two-year Master Degree programmes after the Anglo-American example and the recommendations of the European Union. The change to a two-level study programme is connected with the wish for more mobility and comparability. Some of the European degree programmes are now taught in English in order to attract more foreign students. A recent FIG symposium on the development of curricula revealed that North American universities are planning to open campuses in other parts of the world (FIG 2007). Globalization of education in photogrammetry, remote sensing and spatial information science is rapidly advancing. Table 33.1 gives some examples of curricula that contain photogrammetry, remote sensing and spatial information science. A more complete list of all academic institutions and their programmes in geomatics is published by the Geodetic Institutes in Munich and is available on a Web page (TU München 2007).

The list does not reveal how many hours (or credit points) the subjects of photogrammetry, remote sensing and spatial information science contain. Such numbers will also depend on the form of the studies and the educational system of each country.

33.2.2 *Equipment and personnel resources*

Beside the changes in the curriculum, available equipment and personnel influence the quality of education. The hardware and software of photogrammetric workstations and the specialized software packages in photogrammetry, remote sensing and spatial information science have to be used in exercises and project work. Such equipment changes constantly. In addition, a variety of data needs to be acquired and updated for practical work. It is therefore relatively expensive to educate and train students in these subjects. Not all of the educational institutions have the resources to invest in the latest equipment and data and to maintain appropriate levels. Furthermore, the number of qualified personnel is another prerequisite for laying the foundations for good education. Good education at universities should be based on research activities. Photogrammetry, remote sensing and spatial information science are mostly connected to certain application fields. Active participation of the teachers in research projects is therefore necessary. Supervision of Master and PhD students is an important task for a teacher at universities. In some countries, special research schools have been created. They organize courses, visits to conferences and short stays at other universities.

Table 33.1. Overview of study programmes in photogrammetry, remote sensing and geomatics.

Study Programme	University	Country	Degree
Geomatics Engineering	Ohio State University	USA	MSc
	The University of Melbourne	Australia	MSc
	Stuttgart University	Germany	MSc (E)
	University of New Brunswick	Canada	BSc, MSc
Geodesy & Geoinformatics	University of Hannover	Germany	BSc, MSc
	Technical University Vienna	Austria	BSc, MSc
Geoinformatics	ITC	The Netherlands	M (E) MSc (E)
Geomatics	Aalborg University	Denmark	BSc, MSc
	Helsinki University of Technology	Finland	BSc, MSc
	University of Cape Town	South Africa	BSc, MSc
	The Hong Kong Polytechnic University	P.R. China	BSc, MSc (E)
Civil Engineering	Aristotle University of Thessaloniki	Greece	BSc
Geodesy and Cartography, Geoinformatics	CTU Prague	Czech Republic	BSc, MSc
Geographical Information Systems	Lund University	Sweden	MSc (E)
Photogrammetry and Remote sensing	Wuhan University	P.R.China	BSc, MSc

BSc…Bachelor of Science Programme, MSc…Master of Science Programme, M…Master Programme, (E)…taught in English.

33.2.3 *Pedagogical models*

With all the changes in technology and social-economic conditions, the form of education has to be adapted to these new conditions. Various pedagogical models are used at universities today.

The problem-based and project-organized learning is one approach. Other approaches are the 'progressive inquiry' and 'integrated micro learning'. The traditional model, which is based on lectures and small exercises, needs no further explanation.

Problem-based and project-organized learning (PBL) means that a group of students carries out projects. A problem has to be defined, analysed, discussed and solved, and the search for a solution and the result have to be documented, presented and defended. Teachers have a different role in this type of learning; they are advisors and facilitators for the group. The students have a very active role in their studies. The Figure 33.1 illustrates the difference between the traditional and a PBL model. The ability to learn new things and to go deeper into a few fields of knowledge will be trained in PBL in a unique way. Detailed information on PBL is published in Kolmos et al (2004) and Queens University (2006).

In progressive inquiry, the students take part in a research process. The students learn to inquire and investigate by means of relevant methods. Information that is more detailed can be found in Muukkonen et al (1999).

Another approach is Integrated Micro Learning (IML). IML is based on subdividing the learning process into small activities. Learning takes place along with professional activities and as such, it is in strong contrast to "artificial" learning found in conventional course settings. More information on this approach can be found in Hug (2005).

Which model can be applied depends on the attitude of the teachers, the available resources, the level of education and the maturity of the students.

33.2.4 *Core knowledge*

Due to the significant changes in technology, the number of subjects has substantially increased. This is also the case in many other fields of the curriculum. It is, therefore, necessary to reduce obsolete knowledge and to concentrate on core knowledge. The core knowledge will depend on the application areas of photogrammetry, remote sensing and spatial information science. The requirements may be different in each country. What should students know and master after completion of a course or after graduation with a bachelor degree? Accreditation of courses and curricula should give the answer (see Section 7). Quoting the credit points may help to compare educations. In the European Credit Transfer System (ECTS), for example, one credit point is equivalent to 30 hours of student's work. A semester then corresponds to 900 hours of student work. Some basic principles regarding the definition of core knowledge in the three fields will be mentioned in the following.

33.2.4.1 *Photogrammetry*

It is obvious that photogrammetry will completely change from analogue to digital in the near future. Digital systems of acquisition and processing should therefore be in focus. Theoretical and practical knowledge of how to produce maps, digital elevation models (DEMs) and orthoimages are important for future employees in the mapping organizations. The use of professional stereo workstations with their overwhelming number of functions can make the training of students difficult. Only a few main functions should

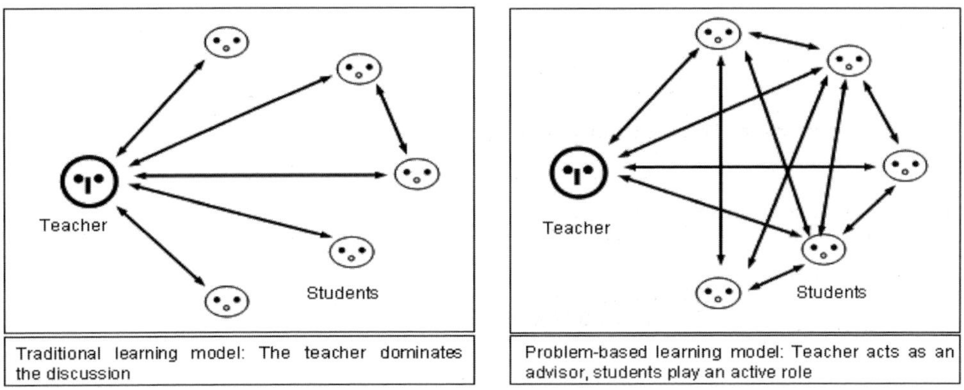

Figure 33.1. Comparison of pedagogical models.

be introduced and be well described in a compact manual. The students should be able to evaluate the quality of different photogrammetric products.

How a simplified spatial data acquisition system is designed and used in education is described in a recent publication (Takeda 2006). It is an example of how the production companies cooperate with universities and have some of their needs for qualified personnel realized.

A more advanced course in photogrammetry should contain analytical photogrammetry, aero triangulation techniques, terrestrial photogrammetry, calibration of non-metric cameras and automation of photogrammetric processes. Some practical work with digital cameras, for example point determination or camera calibration, should also be included.

Other advanced courses in the Master programme could contain:

- integration of positional and imaging sensors
- production of true orthoimages
- advanced automated mapping procedures.

33.2.4.2 *Remote sensing*

Nowadays, various kinds of sensors on board satellites and/or airplanes are available in remote sensing. Understanding geometric and radiometric characteristics of sensors as well as targets is essential for extracting proper information from remote sensing data. Image processing techniques are also important in remote sensing. Derivation of DEMs from laser scanning and interferometric SAR data is becoming important. Basic courses on remote sensing may cover the following subjects:

- Fundamentals of Remote Sensing
- Sensors (optical sensors/ microwave sensors, laser scanners)
- Platforms for sensors
- Image interpretation
- Image processing (pre-processing, transformation, classification, etc.)
- Modelling of point clouds
- Integration of sensors
- Applications of remote sensing.

More advanced courses may include analysis of hyper spectral imagery, analysis of multi-temporal images, extraction of objects and object oriented classification. It will again depend on the application where the focus on the advanced course is placed.

33.2.4.3 *Spatial information science*

Spatial information science is one of the fastest growing technology sectors of the last decade. Graduates in this field can take up employment in a broad field of applications. A curriculum in SIS has to provide the skills and knowledge required to be competent in one or more of a variety of different market segments.

Based on foundations in mathematics, physics and computer science, knowledge in acquisition, storage, analysis, processing and visualization of spatial information have to be the core of a degree programme.

The curriculum for spatial information science may include the following subjects:

- Geodetic reference systems
- Methods of data collection
- Data and topological modelling
- Addresses and geo-coding
- Metadata
- Specifications and standards
- Geospatial databases
- Geographical Information Systems
- Management of spatial data
- Spatial analysis
- Geo-Statistics
- Data representation
- Visualization of geo-data
- Web mapping and web services
- Data quality

Students of advanced courses should have the choice between a more theoretical and a more practice oriented education. Hardware and software development, 3-D and 4-D modelling, mobile GIS and location based services are examples for advanced topics (Balstrøm et al 2006).

The three fields are no longer strongly separated. Photogrammetry and remote sensing cover similar or overlapping topics. Spatial information science uses data of imaging sensors to a great extent. Many other disciplines will support education in photogrammetry, remote sensing and spatial information science. Subjects like surveying, adjustment theory, map projections and programming are essential for a good education in these three fields. Only good teamwork between the teachers of all disciplines will guarantee a successful education. Students should practise interdisciplinary group work during their studies.

33.2.5 *Specialization in Master of Science programmes*

Another form of MSc programme has been created where students with a Bachelor degree in various subjects can enrol to deepen their knowledge in photogrammetry, remote sensing and spatial information science. Table 33.2 shows some examples.

It is very likely that such specialization will increase in the future. The accreditation of such courses by national and international professional organizations will be an important task. ISPRS should play a role in these tasks.

33.3 EDUCATION OUTSIDE UNIVERSITIES

Other education and training in photogrammetry, remote sensing and spatial information science is carried out outside universities, for example by means of short courses, summer schools and master degree programmes by distance learning (see Table 33.3). Mid-career professionals usually attend the last one of these.

A mixture between distance learning and a few seminars at a university is also practised; it is named blended learning. The European Organization for Spatial Data Research (EuroSDR) carries out blended learning on the topics of its research projects.

These so-called EduServ modules start with a two-day introductory seminar at different places and they are supplemented by a two-week learning course by means of the Internet. Six EduServ modules and 12 courses (Table 33.4) have taken place so far. In this way, continuing professional development (CPD) is realized.

Table 33.2. Master of Science programmes with specialization in photogrammetry, remote sensing and spatial information science.

Specialization	University, country, (language in education)
Geodesy & Geoinformation Science	TU Berlin, Germany, (E)
Photogrammetry & Geoinformatics	Stuttgart University of Applied Sciences, Germany, (E)
Airborne Photogrammetry & Remote Sensing	Institute of Geomatics, Barcelona, Spain, (E)
Remote Sensing	University College London, UK & Chiba University, Japan

33.4 THE ROLE OF E-LEARNING

There are many reasons for using computers and the Internet in education and training. Today many courses are offered on the Internet. The participants at remote places will study and communicate with their teachers and fellow students. At the universities, e-learning is also used by students in their group rooms or at their dormitories (see Figure 33.2). Project work and work for a thesis can then also be done outside the university, for example in a mapping organization. Even MSc programmes are offered as a mixture of e-learning in the student's home town and a few weekend seminars at the university. Companies instruct their clients by means of e-learning on how to properly use sophisticated hardware or software. The learning material uses text, images, videos and sound, and a lot of interactivity. Other features for successful e-learning are quick feedback from the teacher or a learning system, and collaborative exercises.

Another approach is the use of video conferencing. Lectures and discussions between teacher and students are transferred via the Internet in real-time and in both directions. The teacher can monitor the reactions of the student and the student can observe the "body language" of the teacher. The video and audio equipment (cameras, microphones, projectors, loudspeakers) and other auxiliary equipment (white board, disk systems, etc.) are rather expensive. Simultaneous video conferencing among three or more remote points is also possible. Web plug-ins and software (for example NetMeeting, Skype) also enable cheaper communication between two persons at a lower quality.

Table 33.3. Examples of courses and programmes in photogrammetry, remote sensing and spatial information science, which take place outside of universities.

Organization	Subjects	Type of education
European Organization for Spatial Data Research (EuroSDR)	Photogrammetry, Remote Sensing, SIS	Short courses with introductory seminar
E-Learning Academic Network (ELAN) Niedersachsen, Germany	Photogrammetry, Image processing	Short courses
ITC, The Netherlands	Remote Sensing, SIS	Short courses
GITTA, Switzerland	SIS	Short courses
Institute for Advanced Education in Geospatial Sciences, USA	Geospatial Information Technology	Short courses, curriculum
JICA / RESTEC, Japan	Remote Sensing	Short courses

Table 33.4. Blended learning courses of the European Organization of Spatial Data Research.

Name of the course	University/Research Institute
Integrated Sensor Orientation	University of Hanover, Germany
Automatic Orientation of Aerial Images on Databases	Aalborg University, Denmark
Laser scanning & Airborne Interferometric Synthetic Aperture Radar	ITC Enschede, The Netherlands
Digital Cameras/Sensors	Ohio State University, USA
Co-ordinate Systems & Transformations for Spatial Data Position	Dublin Institute of Technology, Ireland
Positional Accuracy Improvements in Geographic Information Databases	Ordnance Survey, England & TU Berlin, Germany
Quality of Geospatial Data & Related Statistical Concepts	ITC Enschede, The Netherlands
Methods for Checking & Improving of Digital Terrain Models	University of Aalborg, Denmark & Charles University Prague, Czech Rep.
Mapping with Synthetic Aperture Radar	TU Berlin, Germany
Laser Scanning for 3D City Models	Finnish Geodetic Institute
Geometric Performance of Digital Airborne Cameras	University of Stuttgart, Germany
City GML	TU Berlin, Germany & University of Gävle, Sweden

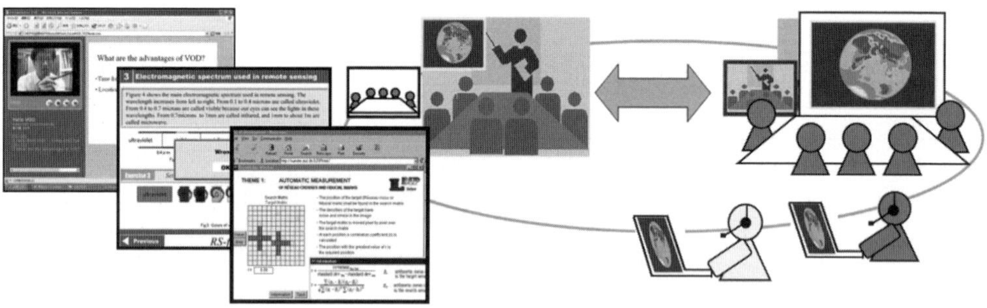

Figure 33.2. Various ways of e-learning.

Many universities have now established research and/or service groups for e-learning. The production of good courseware requires many resources and good teamwork from different specialists. The carrying out of the courses, including marketing, requires substantial efforts.

33.5 E-LEARNING TECHNOLOGIES

Developing e-learning courses requires much effort from the original idea to the final product. Innovative approaches to learner interactions have complicated the process. For producing interactions for e-learning, appropriate tools such as Adobe Flash, Java, JavaScript, Active Server Pages (ASP), XML (Extensible Markup Language), etc. are necessary.

To be successful, experimental learning, project learning, action-oriented learning and experience based learning approaches should be considered during the course development.

Learning management systems (LMSs) are now widely used for constructing and operating e-learning courses. A LMS is a software package designed to handle administrative functions, deliver online content, provide tracking and reporting, offer assessments and report testing scores, measuring learner competencies, and supporting portability and standards. Moreover LMSs give e-learning modules a unique, area-wide visibility and corporate design. Various kinds of LMS software are now available on the market. Most of them are web-based to facilitate "anytime, any place, any pace" access to learning content and administration. Beyond high priced but in general excellently maintained commercial systems, a variety of powerful open-source products exist. It is not easy to decide which LMS fits to one's needs, but there are several studies and benchmarks available that may help a purchaser make the right selection. As software changes rapidly, valuable information on different LMSs are mainly to be found on the Internet. The most recent and comprehensive, but also most expensive study, is published by Brandon-Hall

Research (http://www.brandon-hall.com/publications/lmskb/lmskb.shtml). Results of an evaluation of Open Source LMSs conducted by order of the Swiss Virtual Campus project appear on http://www.edutech.ch/lms/ev3/. Also of interest can be the comparison of a commercial and an Open Source LMS (Blackboard and Moodle), considering the rating of system administrators, instructors and students (http://www.naccq.co.nz/conference05.html?page=51-previous links accessed 7. Aug. 2007).

The need for international cooperation and partnerships in education is increasing and involves integration or exchange of learning material for sharing and re-use. In order to transfer e-learning contents to another software environment or to other LMSs, interoperability and standardization are necessary. In the educational community, standards are often associated with restricting creativity and flexibility. On the other hand, standards reduce development costs, and allow users to be independent from proprietary software and allow re-use of well-known e-learning content after software changes. The following sections will focus on the most important e-learning standards.

33.5.1 *IEEE Learning Object Metadata (IEEE LOM)*

Learning Object Metadata (LOM) IEEE 1484.12.1—2002 is an open standard published by the Institute of Electrical and Electronics Engineers Standards Association (IEEE). The main aim of LOM is to facilitate search, retrieval, acquisition and the use of learning objects, either by humans or by automated query. For that purpose, the well structured LOM data model was specified. The corresponding XML-binding—the way LOM records should be represented in XML—is defined in IEEE 1484.12.3.

LOM consists of nine categories with a set of hierarchically structured elements as shown in Figure 33.3. The guidelines for each category's elements can be downloaded in PDF format from the CanCore Website (http://www.cancore.ca—Accessed 7.Aug. 2007). CanCore is the most thorough guide to the LOM.

When implementing the LOM, it is not necessary to support all the elements shown in the data model. The report on Learning Object Metadata Editors by Norm Friesen (2005) can be found on the webpage http://www.cancore.ca/editors.html, which refers to various helpful tools.

33.5.2 *IMS Content Packaging (IMS CP)*

Standardized content packages for easy exchange and reuse of learning products between LMS authoring tools are specified by the IMS (Instructional Management System) Global Learning Consortium, a worldwide non-profit organization. IMS specifications form internationally accepted *de facto* standards for delivering learning products and services. IMS specifications and related publications can be found on the web-page www.imsglobal.org.

A content package is used in e-learning to define learning content or an assessment that can be delivered, for example by a LMS. It is a standard way of describing learning content that can be interpreted by other IMS compliant programs, guaranteeing a straightforward import. The IMS Content Packaging specification consists of two components:

- the Information Model, which describes the data structure, and
- the Content packing XML binding, which defines its representation using XML.

Figure 33.4 schematically sketches the Package Interchange File, a zip file which includes the XML manifest file (imsmanifest.xml) containing the information needed for transparent content exchange. Metadata used for description of the package correspond to the LOM-standard previously described. The organization section as the key element contains the structure of the learning resource.

There are at least three possibilities to form an IMS Content Package. First manually, which means collecting the files needed, creating the imsmanifest.xml-file using a standard text editor and finally generating the zip archive. A more comfortable solution requires Adobe's Dreamweaver software. A plug-in (Content Packager for IMS) is ready for download from the Adobe website. Another easy to handle procedure is provided by the RELOAD Content Package and Metadata Editor. This free downloadable Java tool includes support for IEEE LOM, IMS MD 1.2.4, IMS CP 1.1.4 and also SCORM 2004 (3rd edition). Once the zip-archive is created, the e-learning material can be re-used in different systems, without the need to convert the material into new formats.

33.5.3 *Sharable Content Object Reference Model (SCORM)*

Harmonizing the variety of e-learning specifications has led to SCORM, a widely adopted standard accomplished by the US Secretary of Defence Initiative 'Advanced Distributed Learning (ADL)'. SCORM combines the specifications of the Aviation Industry CBT Committee (AICC), IMS Global Learning Consortium, the IEEE standards and the work initiated by the Alliance for Remote Instructional Authoring and Distribution Networks for Europe (ARIADNE). SCORM conform content will work with any SCORM conform LMS.

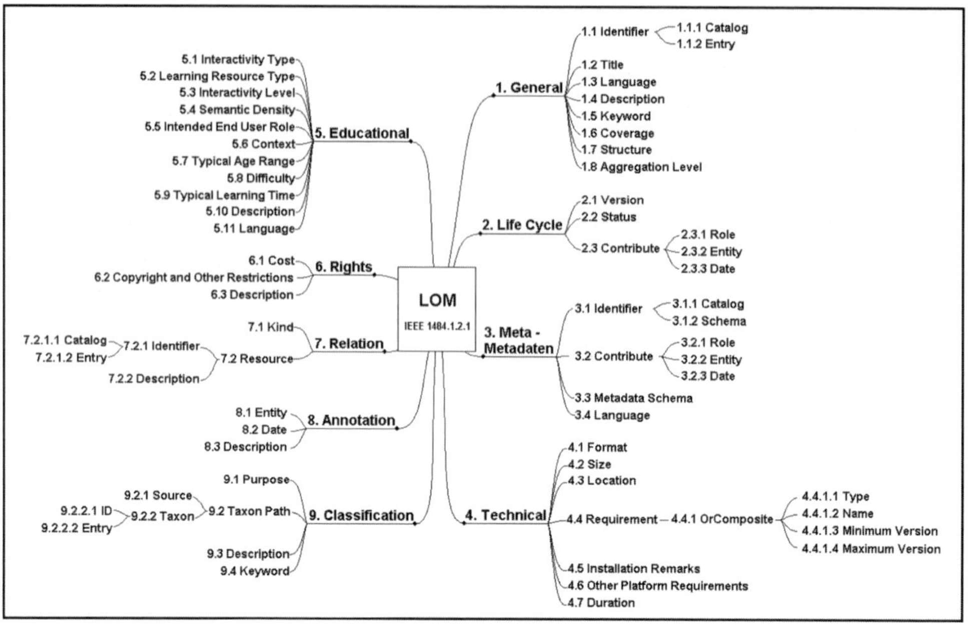

Figure 33.3. The elements and structure of the LOM conceptual data schema. Adapted from IMS Meta-data Best Practice Guide for IEEE 1484.12.1-2002 Standard for Learning Object Metadata, Version 1.3 Public Draft. Adapted from: http://www.imsglobal.org/metadata/mdv1p3pd/imsmd_bestv1p3pd.html (Fig. 2.2).

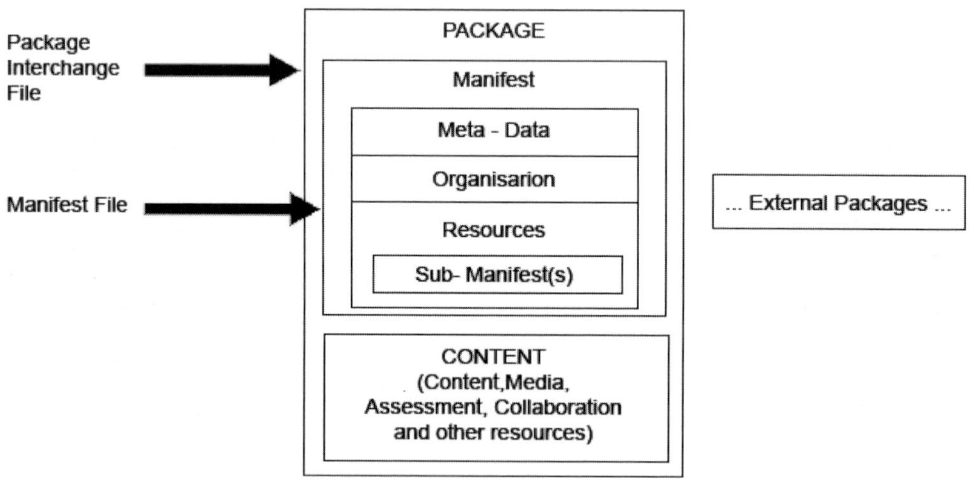

Figure 33.4. IMS Content framework. Adapted from IMS Content Packaging Best Practice and Implementation Guide Version 1.1.4 Final Specification—4. October 2004 (excerpt).

In principle, SCORM follows the LOM metadata concept and the IMS content packaging. Comparable to IMS CP, a SCORM package contains a zip-archive described by a XML-formatted manifest-file. The package consists of so-called Sharable Content Objects (SCO), which could be a module, a chapter, or a page including all the necessary files (assets) required to function properly. Additionally, it contains a SCORM Application Program Interface (API)-adapter client software, which allows communication

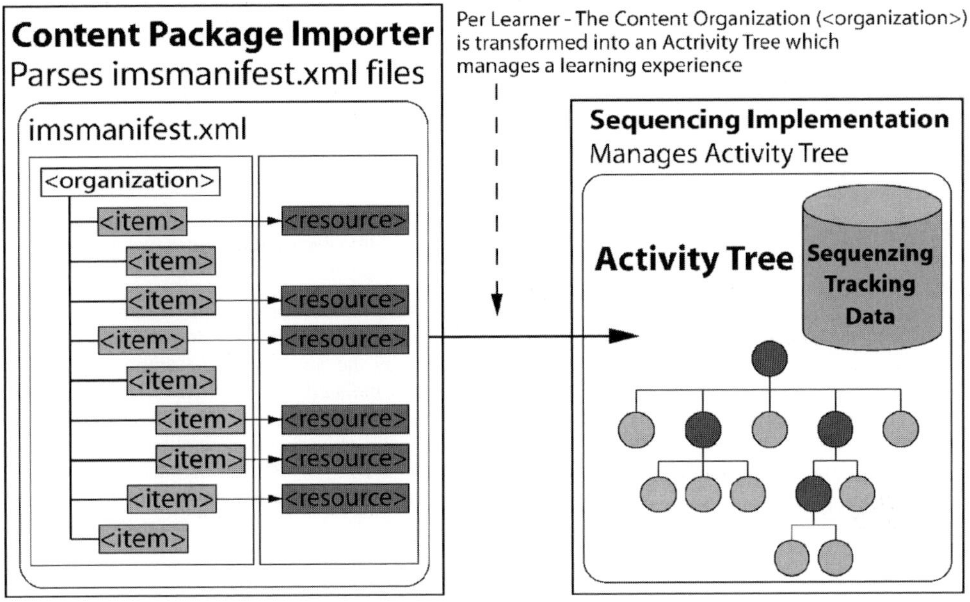

Figure 33.5. Relationships between a Content Organization and an Activity Tree.
Adapted from: SCORM® 2004 3rd Edition Sequencing and Navigation, Version 1.0, p SN-2-4.

with a SCORM compatible LMS. A SCO has to be independent from other SCOs to be reused in other learning modules.

Learning resources are usually organized hierarchically. This is in the form of a Content Organization that is represented in a SCORM content package as a single <organization> element as seen in Figure 33.5. Each item in the structured hierarchy represents an instructionally relevant learning unit consisting of SCO and assets.

In the SCORM 2004 3rd edition, the idea of sequencing is introduced in order to present a specific learning experience as intended by an author or content developer. "This learning experience may be free play, in which the learner can choose any item in the content organization in any order, or it may be guided by a flow through the structure of the content organization. The learning experience can be adaptive, with different behaviours that depend on the learner's performance or other variables that can be tracked by the LMS. … In those cases, the content developer can define specific sequencing information that prescribes how an LMS will manage the learning experience" (SCORM® 2004 3rd Content Aggregation Model, Version 1.0, p CAM-3-13—see ref.). For that purpose, so-called Activity Trees are introduced, defining the order in which a learner is supposed to experience content objects. A reference implementation presenting SCORM conformal Photoshop course examples can be downloaded from ADLs website: http://www.adlnet.gov/downloads/DownloadPage.aspx?ID=143 (Accessed 7. August 2007). The examples demonstrate sequencing the same content in nine different instructional strategies.

33.5.4 *Further developments*

Up to now many e-learning projects were focused on the production of single learning modules. With the increasing reorganization of Europe's higher education system and the introduction of the bachelor and master courses, an increasing demand for self-learning services has occurred. New techniques supporting the change from teacher to student centred lessons are necessary. In this context, Web 2.0/e-learning2.0 is discussed. Web 2.0 is not a redesign of web-technology but a merge of (often long existing) web-based services and features. File repositories, wikis, podcasts, folksonomic tagging, blogs, syndication, etc. are techniques to set up social networking sites, facilitating collaboration and sharing between users.

33.6 EXAMPLES OF E-LEARNING COURSES

Around the world a variety of well-designed e-learning products exist. An overview presented at the ISPRS symposium in Tokyo on activities in Germany can be found in Koenig & Schiewe (2005). Links to the authors known websites and resources are listed on the web-pages of Commission VI Working Groups 2 and 4 (http://www.commission6.isprs.org/wg2 and ~wg4—Accessed 7.Aug. 2007). In the following, a range of best practice examples are introduced.

33.6.1 *Canadian Centre of Remote Sensing Tutorials*

The e-learning products of the Canadian Center of Remote Sensing (CCRS) are classics among the tutorials. CCRS provides several interactive training modules mainly focused on remote sensing. They can be found on CCRS homepage (URL:http://ccrs.nrcan.gc.ca/resource/index_e.ph). Each tutorial is characterized by rich images, graphics and animations. A print version is also available. Quiz tests at the end of most sections allow good self-control. CCRS tutorials are of high value for students interested in remote sensing technology and applications.

33.6.2 *Curtin University of Technology, Virtual Online Learning*

At Curtin University, Australia, authentic e-learning projects have been developed. One of them, the Virtual Online Learning (VOL) project, covers topics concerning geographic information science (GISc), global positioning systems (GPS), cartography and surveying. The aim is to provide stimulating and interactive learning environments, which adequately prepare a student for real world conditions.

For preparations for a field trip, where students have to apply their theoretical knowledge about principles and methods of spatial sampling, a virtual training scenario was set up (URL: http://www.cage.curtin.edu.au/volearning).

Evaluation on the basis of an assessment proved that students who participated in the virtual field trip achieved better grades compared to students who had no access. Overall, student attitudes to field trips were very positive. Real trips were found to be more stimulating, whereas virtual field trips were found to be more useful.

33.6.3 *Indian Institute of Technology, E-Tutor*

E-Tutor is a development by the Centre of Studies in Resources Engineering, Indian Institute of Technology, Bombay, India. The multimedia based tutor was developed with Macromedia Authorware. For hands-on exercises, GRAM++, an indigenous GIS package, must be installed. The tutor was funded by the United Nations Development Programme (UNDP) in the context of the project "GIS Based Technologies for Local Level Development Planning". It is expected to be a useful teaching aid for spreading GIS education across different levels of users in the developing countries. Each topic is explained by text, graphical illustrations and animations. After each topic, the trainee has to answer quizzes and perform hands-on exercises. E-Tutor received the Gold award during CATCON 2006. The software may be downloaded from the following website: http://www.csre.iitb.ac.in/gram++/

33.6.4 *Swiss Virtual Campus, Geographic Information Technology Training Alliance*

The Geographic Information Technology Training Alliance project (GITTA) started in 2001. It was funded by the Swiss Virtual Campus (SVC), a programme initiated by the Swiss Confederation. Ten partners from universities, federal institutes of technology and universities of applied sciences throughout the country, cover a wide variety of disciplines. The consortium offers over 80% of the capacity in academic Geographic Information System Technology (GIST) education in Switzerland.

Easily accessible teaching modules for basic and intermediate study programmes have been created. GITTA learning objects are no longer static-, reactive- or text-based, but highly multimedia-enriched. Great importance is attached to flexibility and self-exploration. GITTA is distributed in a variety of languages (German, English, French and Italian).

SVGs funding ended in 2006, but GITTA preserved sustainability because it was opened to the general public and is now available as open content under the creative commons licence. Among all tutorials discussed, GITTA is the state-of-the-art. It follows recent software concepts and fits to previously mentioned e-learning standards, which means that GITTA's learning objects can easily be imported by IMS CP or SCORM compatible LMS (URL: http://www.gitta.info).

33.7 ACCREDITATION AND EVALUATION OF E-LEARNING COURSES

Quality is the most important factor determining acceptance of e-learning products. But developing high-quality material is a real challenge. It is nearly impossible to meet the demands of all parties involved who have different perspectives on quality:

- from a learners' view the structure of the course should facilitate long-term learning results
- teachers' perspectives focus on the quality of content and emphasize that materials meet their requirements; openness and interoperability are also main topics
- developers are interested in the quality of code, which is a prerequisite for reuse, making the e-learning modules profitable

Table 33.5. The SEEQUEL core quality framework—excerpt.

Object	Criteria	Rating
1. Learning sources		
1.2 Learning materials		
Content reliability and updated		
Integration property (it can be used in different contexts/org)		
Technical quality		
Aesthetic quality		
Conformance to standards		
Interoperability with other systems		
Portability with other systems		
Reusability		
Security		
Metadata attribution		
Relevant background of the authors of material		
Brand recognition		
Suitability of the language and contents to the targets		
Clear objectives		

- governments' interests aim at making education more effective with constant or even decreasing investment
- businesses value a praxis oriented modern curriculum meeting the emerging demands of the employers.

Moreover, the term 'quality' greatly depends on social background, regional provenance and cultural imprint. Although quality is a very diverse term, different attempts have been made to classify and evaluate e-learning material. Ehlers & Pawlowski (2005, pp. 6–7) specify methodologies and instruments for quality control. They list:

- quality management approaches, which are directed to creation, implementation and performance processes
- quality assessment on the basis of criteria lists and checklists, which must contain not only technical but also pedagogical and didactic criteria
- evaluation as a quality approach, which does not focus on the product itself, but on the learning process and thus draws the trainees to the focus of attention
- further attempts such as
  - benchmarking, comparing different products based on specified criteria
  - accreditations and certification procedures, in which providers must submit to one-time or regular audits
  - quality marks, often self-developed by organizations that determine the release of the award.

A very detailed criteria list was compiled during the SEEQUEL project (Sustainable Environment for the Evaluation of Quality in e-learning), which was co-funded by the European Commission, Directorate General of Education and Culture as part of the eLearning Initiative. The project originates from the joint initiative of the eLearning Industry Group (eLIG) and a number of European expert organizations and associations. In Table 33.5, only the part covering learning materials is listed. The detailed criteria catalogue is published on the SEEQUEL homepage.

Up to now only a few attempts have been made to evaluate learning materials in photogrammetry, remote sensing and GIS. This makes a potential user feel uncertain as to which modules may be worth studying. It should be of great importance that e-learning products are peer-reviewed or are shown to comply with a quality assurance process. This will provide confidence and strengthen the acceptance of e-learning. For future education and training packages, certification should be obtained. However, ISPRS may clear the way by providing a more transparent process for evaluating existing education and training offers. Further, certification should help interested students to find suitable applications. WG VI/2 members presently are preparing a first study enumerating necessary steps for introducing quality control. First results will be presented at the Beijing Conference.

33.8 COMMISSION VI CONTRIBUTIONS

The ISPRS Commission VI on "Education and Outreach" deals mainly with various educational matters related to photogrammetry, remote sensing and spatial information sciences. As described above, the highly integrated multimedia technologies of e-learning are rapidly changing the style of education, and technical matters being discussed within Commission VI are increasing. The advancements in the topics of the Commission VI during 2004 to 2008 are introduced in this section including pedagogy, e-learning, educational resources and student activities. Commission VI also deals with capacity building and international cooperation. However, those are discussed in the next chapter.

33.8.1 *Structure of Commission VI*

The Commission VI consists of five working groups (WG) and one special interest group (SIG) as shown in Table 33.6. Each WG and SIG has its own clear target to optimize the activities of Commission VI for education.

As described in previous sections, the importance of pedagogy is increasing. Without certain educational theory, good e-learning materials or systems cannot exist. WG VI/1 is collecting information on new curricula, and searching for ways and tools for evaluating individual courses of academic education.

The WG VI/2 on e-learning is mainly working on development, promotion and dissemination of e-learning software in the three fields related to ISPRS. Organizing the educational software contest CATCON is one of the important activities of the WG (see also Section 8.3)

The main theme of WG VI/3 is international cooperation and capacity building. Various kinds of international and regional organizations are now involved in capacity building, including technology transfer and training. WG VI/3 is working on developing connections with organizations, such as ESCAP, FAO, and CEOS, to optimize the role of ISPRS for capacity building. In capacity building, each region or continent has its own needs or problems. In order to harmonize the regional needs with ISPRS activities, WG VI/3 has set up a new position, Regional Coordinator (RC). The RC is assigned for each continent or region to oversee, coordinate and organize regional and international activities related to education. The Special Interest Group (SIG) on Technology Transfer will be organizing small workshops or seminars once or twice every year mainly in developing countries for technology transfer. The SIG is organized by the highly skilled and well experienced professionals of ISPRS.

Table 33.6. Structure of Commission VI.

WG VI/1	Educational Frameworks and Methodologies
WG VI/2	e-Learning
WG VI/3	International Cooperation and Capacity Building
WG VI/4	Internet Resources and Datasets
WG VI/5	Promotion of the Profession to Students
Special Interest Group	Technology Transfer Caravan

The number of home pages providing educational materials is rapidly increasing. However, more than a few teachers and students have been frustrated when they cannot find proper educational materials from the information flood on the internet. Certain categorization and reviewing processes are necessary for assisting students to find suitable educational materials from the internet to meet their requirement. WG VI/4 is constructing a procedure for categorizing and reviewing homepages for education in the related fields of ISPRS.

In order to promote student and young scholars' activity within ISPRS, WG VI/5 on Promotion of the Profession to Students and the Student Consortium were set up in 2004. The WG VI/5 and the Student Consortium are working together to involve more students and young scientists in ISPRS.

33.8.2 *Mid term symposium*

One of the highlights of Commission VI activity during 2004 to 2008 was the Commission VI Symposium, which was organized in Tokyo, Japan from 27 to 30 of June 2006 (Fig. 33.6). The main theme of the symposium was "e-Learning and the Next Steps for Education". A total of 150 participants from 27 countries attended, and 53 papers were presented in 16 technical sessions. The educational software contest CATCON4, JICA NET distance learning demonstration and a technical exhibition were successfully organized during the Symposium.

33.8.3 *CATCON*

In order to promote the development and dissemination of good/user-friendly software as well as data sets for computer assisted teaching (CAT), the educational software contest CATCON was set up by WG VI/2 in 1996 at the Vienna Congress. CATCON2 was organized at the Amsterdam Congress in 2000, and CATCON3 at the Istanbul Congress in 2004 (Fig. 33.7). CATCON has become one of the most popular official events organized at the ISPRS Congresses.

In order to reflect the quick technology advancement of e-Learning, Commission VI decided to organize CATCON4 at the Commission VI Symposium in Tokyo in 2006, the first time CATCON was organized at the mid-term symposium. A total of 14 software packages were nominated for the contest. More than half of them were from Asia, and the quality level was much higher than expected. The following are the winners of CATCON4.

No.1. E-Tutor for GIS (version 1.0)
Author: P. Venkatachalam from the Indian Institute of Technology

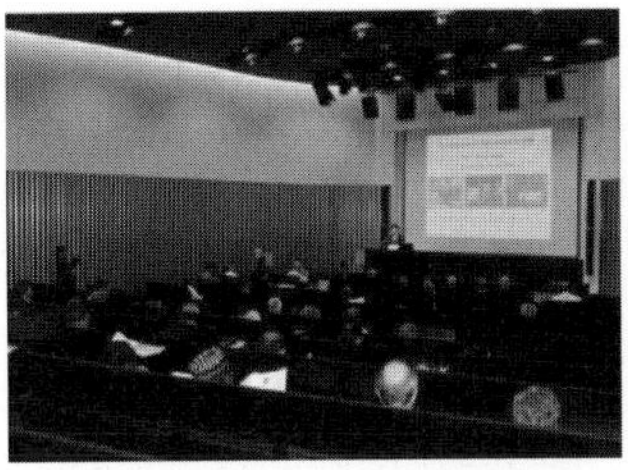

(a) The opening session.

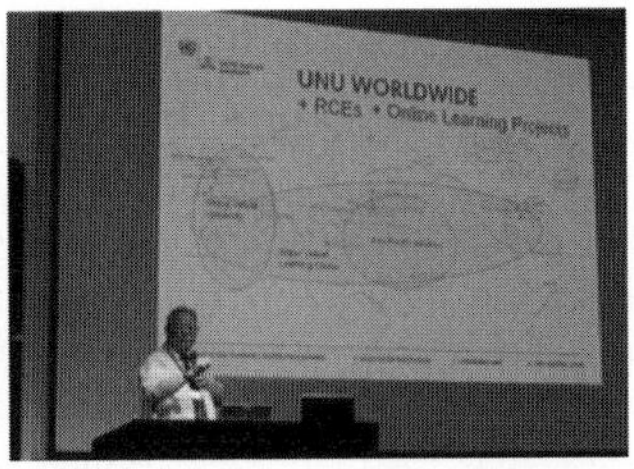

(b) Keynote speech.

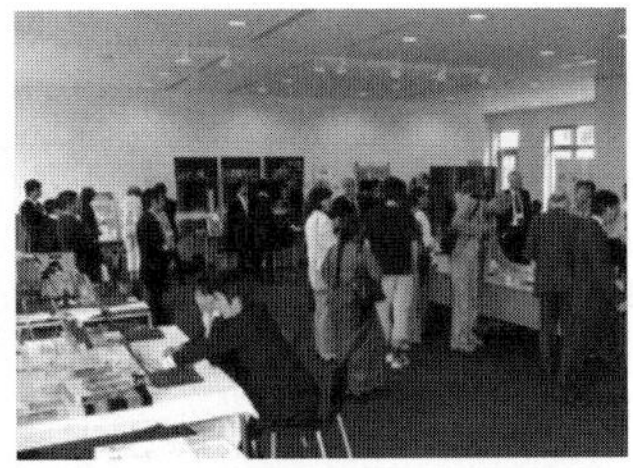

(c) Technical exhibition.

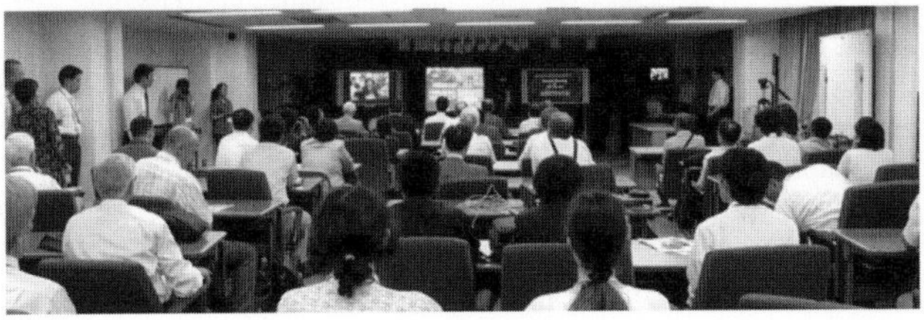

(d) JICA distance learning demo: Palestine, Kenya, Bangladesh and Sri Lanka were connected to Tokyo.

Figure 33.6. Snapshots at the Commission VI Symposium in Tokyo in June, 2006. (see colour plate page 522)

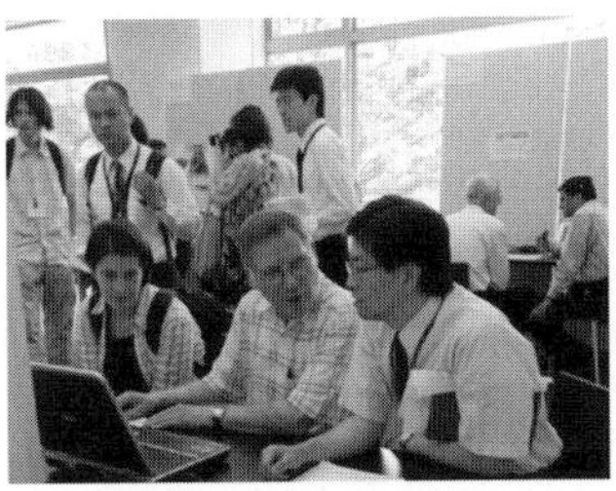

(a) Software demonstration.

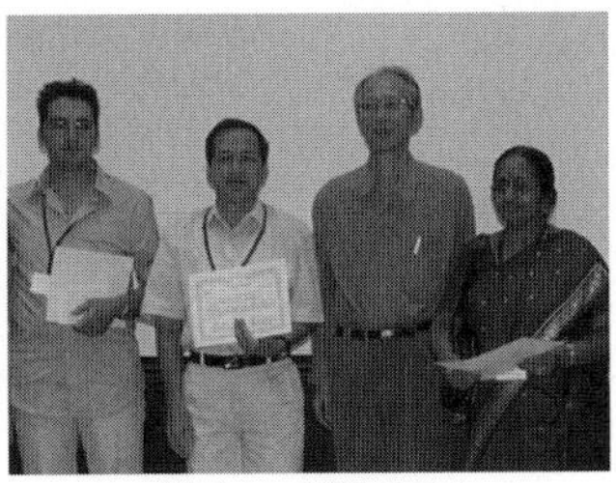

(b) Award Winners.

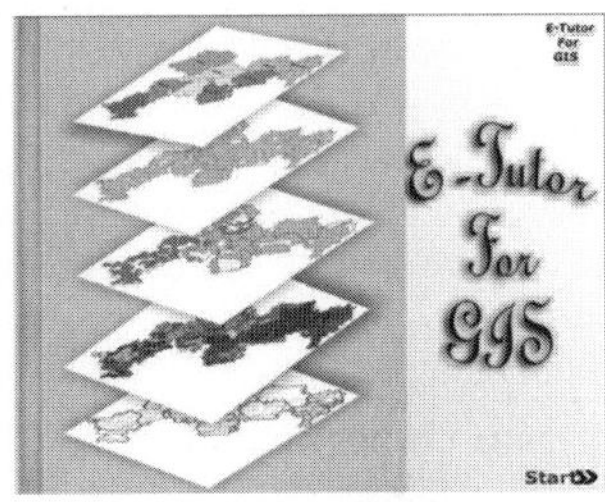

(c) E-Tutor for GIS.

Figure 33.7. Snapshots at CATCON4. (see colour plate page 522)

No.2. CARST 1.0: Computer Assisted Remote Sensing Training package version 1.0
Author: Nguyen Dinh Duong from the Vietnamese Academy of Sciences and Technology

No.3. SV3DVISION: Didactical Photogrammetric Software for Single Image-Based Modelling
Author: D. G. Aguilera from the University of Salamanca

The awards—US$1000, US$700 and US$500, sponsored by the ISPRS Foundation—were given to the winners at the Closing Ceremony. Most of the software presented to CATCON is open to public free of charge and are accessible at the address http://www.isprs.org/catcon /index.html

Following the success of CATCON4, WG VI/2 plans to organize CATCON5 at the ISPRS Congress in Beijing in 2008.

33.8.4 *JICA NET*

JICA NET Distance Education on Remote Sensing and GIS Course has been implemented by the Japan International Development Agency (JICA) in cooperation between the Japan Society of Photogrammetry and Remote Sensing (JSPRS), ISPRS Commission VI and the Special Interest Group on Technology Transfer Caravan, since August 2004. Eight courses have been conducted up to November 2006 with 445 participants for the Remote Sensing Course and 471 participants for the GIS Course from 13 countries; Turkey, Thailand, Philippines, Vietnam, Malaysia, Laos, Kenya, Sri Lanka, Jordan, Cambodia, Bangladesh, Palestine and Mongolia.

Each time the JICA Tokyo Office is connected to several countries via the JICA network, lectures are given from Tokyo (see Fig. 33.8 (a), (b)). Each course is divided into nine half days including an examination. A certificate is conferred on those who successfully complete the course. A certificate of participation is given to those who fail the examination but participated in more than 75% of the classes. 344 participants of the Remote Sensing Course and 472 of GIS Course have been given the Certificate of Successful Completion.

The power Point teaching materials and texts in English for 12 modules in total in each course have been developed by JSPRS. Video conferencing for Q&A sessions after each module is conducted with the support of a resource person (Prof. Shunji Murai) and a site facilitator from each participating country. The summary of Q&A sessions is also published. In order to let more people in the international community access this educational content, the President of JICA kindly agreed to allow JSPRS & ISPRS to put the JICA NET teaching materials on the web page of ISPRS Commission VI.

33.8.5 *Educational gateway*

In order to overcome the frustration in searching for proper educational materials from the information flood of the internet, Commission WG VI/2 has started the short term project "Analysis of E-Learning Software and Guidelines for Quality Assurance in Photogrammetry, Remote Sensing and GIS", which is funded by ISPRS. Their work aims not only to inform about aspects such as content, level, didactic approach and accessibility—as standard metadata could do—but also to examine quality aspects.

Moreover WG VI/4 has been collecting and disseminating information on Internet resources on education and is taking the following approach to reinforce this activity.

a. Maintain a comprehensive list of educational material on the web, eliminate outdated links, complement by sending proposals to URLs on recent education, e-learning and distance training material, promote this information source in order to achieve a wider recognition for common use.
b. Certain categorization and reviewing processes are necessary to assist students find suitable educational materials meeting their requirement. WG VI/4 is constructing a procedure for categorizing and reviewing homepages on education in the related fields of ISPRS. The categorization includes the following items.

- Category: remote sensing, GIS, photogrammetry, image processing, etc.
- Type of homepage: Tutorials, text book, power point materials, etc.
- Target user: University students, high school students, children, etc.
- Level: Basic, medium, high.

33.8.6 *Student activities*

During the ISPRS Congress in Istanbul, the youth forum was successfully organized mainly by young researchers and students. At the panel discussion of the youth forum, Council members and the Commission VI members discussed with students the future activities of students within ISPRS. Through the discussion it became clear that students not only need support from ISPRS but also freedom to develop their activities in ISPRS. Following the discussion, it was decided to set up the Student Consortium (SC) as the first ISPRS official framework for students. WG VI/5 was also set up to promote the profession to students and act as the interface of SC to ISPRS. Until now, around 100 persons from over 30 countries have registered as members of SC.

One of the main events of WG VI/5 and SC was an ISPRS Summer School, the purpose of which was to provide an intensive and educational week with training activities for the students and young researchers the fields of activity of ISPRS. The first ISPRS Summer School took place at Istanbul Technical University's Maslak campus during 19–26 June 2006

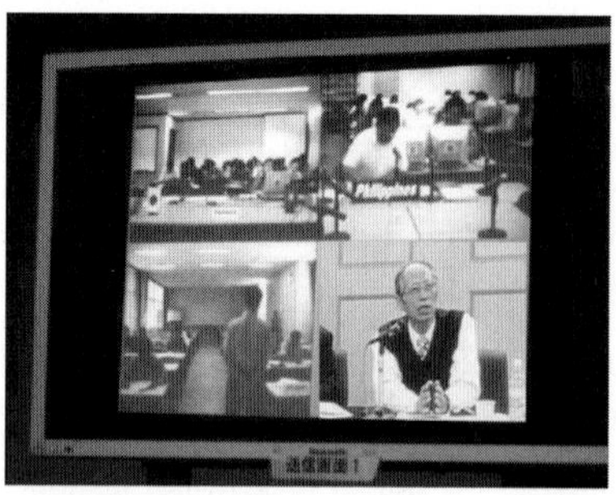

(a) A lecture to three countries at a time.

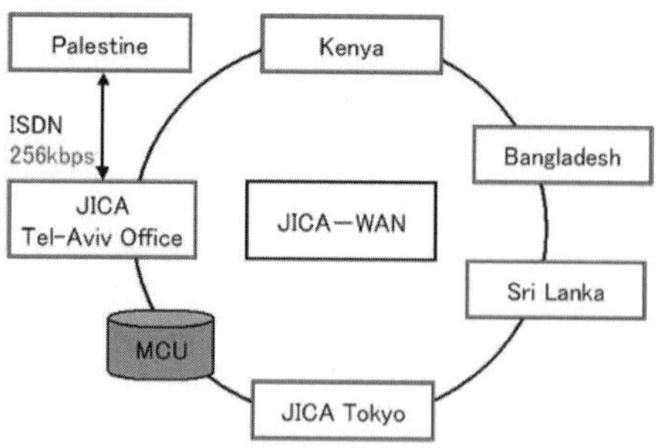

(b) Example of network connection of JICA NET.

Figure 33.8. JICA NET.

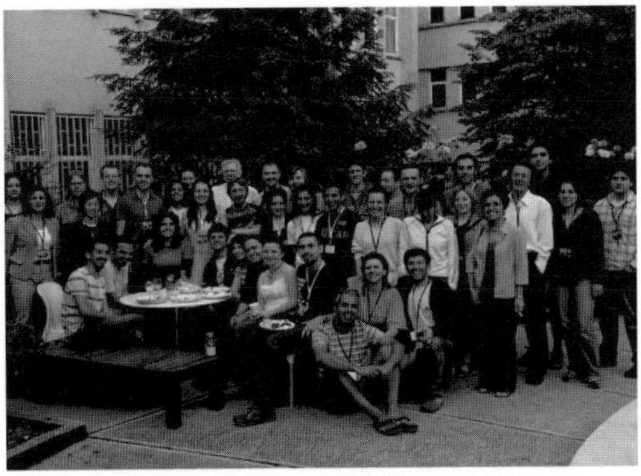

Istanbul, June 2005.

Figure 33.9. The first Summer School. (see colour plate page 522)

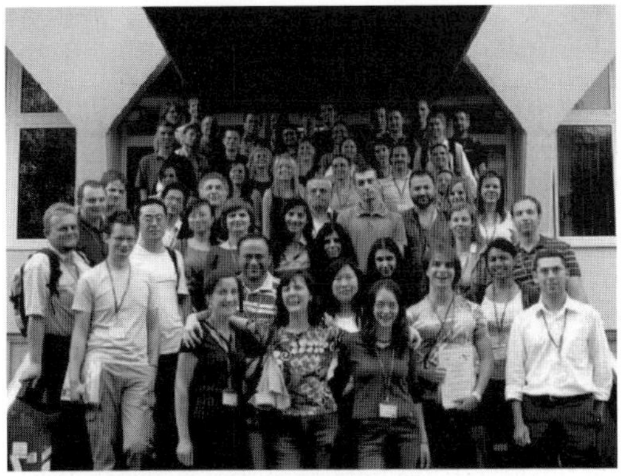

Ljubljana, July 2007.

Figure 33.10. The second Summer School. (see colour plate page 522)

(Fig. 33.9). The general topic of the Summer School was "Satellite Data Processing and Spatio-Temporal Analysis (for Resource and Disaster Mapping, Monitoring and Management)". Altogether 39 advanced diploma students and young researchers from 11 countries participated. Every evening after the lectures the Local Organizing Committee arranged various social events, which provided valuable chances for building personal relationships between the participants and for getting to know the Turkish culture. Students were also able to meet and discuss with the lecturers in person. The participants were very satisfied with the first Summer School. It was a unanimous decision that this kind of events should also be arranged in future.

The 2nd Summer School (SS) on Laser Scanning was organized in Ljubljana, Slovenia, 1–7 July 2007 (Fig. 33.10). The topic in Ljubljana was selected because laser scanning technology is inevitably entering the everyday surveying practice. The needs of practices are demanding and enterprises employing young professionals expect that they are well acquainted with the latest developments when finishing their studies. The summer school was hosted by the University of Ljubljana—Faculty of Civil and Geodetic Engineering, and locally organized by the Slovenian Geodetic Student Association (more than 20 students of geodesy were involved in the organization activities) and the Association of Slovenian Surveyors—Section of Photogrammetry and Remote Sensing. A total of 52 foreign participants from 20 countries (China, Malaysia, India, Iran, Egypt, Turkey and many European countries), and around 30 domestic participants (students, faculty teaching staff, representatives from sponsors) attended.

The 3rd Summer School will take place in Nanjing, China, 27 June to 1 July 2008, just before the ISPRS Congress in Beijing, and the proposed topic is "Acquisition, processing and representation of 3D geospatial information".

33.9 EDUCATION AND LEARNING IN THE FUTURE

In closing this chapter, we should like to look towards the future of education and learning in the field of photogrammetry, remote sensing, and spatial information sciences. Due to the rapid advancements in technology, the importance of lifelong education is increasing in our fields. For example, airborne and terrestrial laser scanning is one of the important technologies of the last few years. These subjects were possibly not part of the curricula a few years ago. Now, many companies are recognizing the importance of lifelong education including on-the-job training to familiarize their employees with the latest technologies. In various countries, professional organizations including academic societies have started Continuing Professional Development (CPD). The idea of CPD is to systematically improve and broaden professional knowledge and skills of the members of the organizations in order to optimize their career opportunities. ISPRS as an international academic society should play a leading role in CPD.

Globalization and e-learning are leading to the borderless education. Distance learning courses of some universities or organizations are now open to students in other countries. Once good e-learning material is developed, it could be used at any place and at any time. E-learning is expanding the opportunity of education on a global basis. However, to realize true borderless education, there are still various problems to be solved. One is the infrastructure problem. A poor Internet infrastructure generates a serious disparity in opportunities of those students affected. This means that the advancement of e-learning is somehow increasing the gap between equal and unequal opportunities. Students in developed countries are getting great benefits from e-learning, but many students in less developed countries are handicapped because they do not have easy access to the Internet. This infrastructure problem cannot be solved by an academic society. However, the need for capacity building in photogrammetry, remote sensing and spatial information sciences is increasing in these countries. It is quite important for ISPRS to discuss this problem and to assist in providing better facilities for education to those countries.

Accreditation is another problem. Technically, it is not difficult to take e-learning courses or attend e-learning classes of universities in other countries. However, the transfer of the credits is often not accepted because a consistent quality assessment does not exist. In order to promote borderless education an international accreditation system has to be established. Moreover, an evaluation of e-learning materials and the development of guidelines on developing high quality e-learning materials are required. Quality remains an utmost priority! Continuous optimization is essential and an ongoing pedagogical challenge. Although e-learning is an important factor in educational curricula, the importance of face-to-face education remains. To optimize education, we may have to seek for a blending of e-learning with face to face education.

Furthermore, knowledge in photogrammetry, remote sensing and spatial information science is used in many other fields. Cooperation with other professional organizations is necessary in order to find new applications and new knowledge. Dissemination of knowledge in our fields should begin in elementary and high schools.

The importance of education is increasing as it is accepted as one of the most important goals of any society. Setting up of rules and procedures on education in our field, promoting generation and distribution of free educational software and data sets, supporting summer camps for students and encouraging cooperation and collaboration between scientists and organizations continue to be among the main goals of ISPRS's work in the future.

REFERENCES

API. URL: http://en.wikipedia.org/wiki/API (Accessed 6 November 2007).

ADL. 2006. *Advanced Distributed Learning. Sharable Content Object Reference Model (SCORM) 2004 3rd Edition Documentation Suite*. http://www.adlnet.gov/downloads/DownloadPage.aspx?ID=237 (Accessed 6 November 2007).

Balstrøm, T., Jacobi, O. & Bodum, L. 2006. *Bogen om GIS og geodata* (Book on GIS and geo-data). ISBN87-991446-0-3.

Dittler, U., Kahler, H. & Kindt, M. (eds.). 2005. *E-Learning in Europe—Learning Europe*. Medien in der Wissenschaft; volume 35, Gesellschaft für Medien in der Wissenschaft.

Ehlers, U.-D. & Pawlowski, J.M. (eds.). 2006. *Handbook on Quality and Standardisation in E-Learning*. Springer.

FIG. 2007. *FIG Commission 2 Symposium, June 7–9, 2007, Prague, Czech Republic*. URL:http://gama.fsv.cvut.cz/data/geoinformatics/2007/06/07-09/scientia-est-potentia.pdf.

Hug, T. 2005. *Micro Learning and Narration. Fourth Media in Transition conference, May 6–8, 2005, MIT Cambridge, MA, USA*. URL:http://web.mit.edu/comm-forum/mit4/papers/hug.pdf (Accessed 8 Sept. 2007).

IEEE Learning Technology Standards Committee. 2002. *Learning Object Metadata Standard, IEEE 1484.12.1-2002*. URL:http://ltsc.ieee.org/wg12/files/LOM_1484_12_1_v1_Final_Draft.pdf

IMS Global Learning Consortium. URL: http://www.imsglobal.org/ (Accessed 6 November 2007).

IMS Global. 2003. *IMS Content Packaging Information Model*. URL:http://www.imsglobal.org/content/packaging/cpv1p1p4/imscp_infov1p1p4.html(Accessed8Sept. 2007).

Koenig, G., Schiewe, J., 2006, E-Learning courses for GIS and Remote Sensing in Germany—Status and perspectives. *Proceedings ISPRS-Symposium,Commission VI, June, 27–30, 2006, Tokyo, Japan.*

Kolmos, A., Fink, F. & Krogh, L. (eds.). 2004. The AAU model—Problem-based and Project-organized Learning. In *The Aalborg PBL Model*: 9–18. Aalborg University Press. ISBN 87-7307-700-3.

Muukkonen, H., Hakkarainen, K. & Lakkala, M. 1999. Collaborative Technology for Facilitating Progressive Inquiry: the Future Learning Environment Tools. In C. Hoadley & J. Roschelle (eds.), *CSCL '99 conference, December, 12–15, 1999, Palo Alto*: 406–415. Mahwah, NJ: Lawrence Erlbaum and Associates. URL:http://www.helsinki.fi/science/networkedlearning/texts/muukkonenetal1999.pdf (Accessed 8 Sept. 2007).

Queens University. 2006. *Handbook on PBL*. Kingston, Canada. URL: http://meds.queensu.ca/pbl/assets/pblhndbk2006.pdf (Accessed 8 Sept. 2007)

SEEQUEL. 2005. *Sustainable Environment for the Evaluation of Quality in e-Learning, "the e-learner Bill of Rights"*. URL: http://www.education-observatories.net/seequel/SEEQUEL%20_elearners_Bill_of_rights.pdf (Accessed 8 Sept. 2007)

Takeda, H., Isobe, K., Akamatsu, Y., Matsuoka, R. & Cho, K. 2006. Application of the simplified spatial data acquisition system K-Scope to education and research. *International Archives of the Photogrammetry, Remote Sensing and Spatial Information Sciences* XXXVI-Part 6: 223–228.

The CanCore Metadata Initiative. 2005. *CanCore: Dynamic Guidelines 2.0* URL: http://www.cancore.ca/en/dynamic (Accessed 8 Sept. 2007)

The Geographical Resource Portal. 2007. URL:http://www.gisdevelopment.net/education/inst/index.htm (Accessed 8 Sept. 2007)

TU München. 2007. *Academic Sites for Geomatic Engineering (Geodesy, Surveying, Photogrammetry, Cartography, GIS)* URL: http://www.lrz-muenchen.de/%7Et5831aa/WWW/Links.html (Accessed 12 July 2007).

Advances in Photogrammetry, Remote Sensing and Spatial Information Sciences: 2008 ISPRS Congress Book – Li, Chen & Baltsavias (eds)
 ISBN 978-0-415-47805-2

CHAPTER 34

International cooperation and capacity building

Ian Dowman and Shunji Murai

ABSTRACT: This chapter outlines the objectives and activities of international organizations that are active in the areas of interest of The International Society for Photogrammetry and Remote Sensing (ISPRS). The relationship between these organizations and ISPRS is also discussed. The organizations covered fall into two main groups: those concerned with collaboration between international societies, such as the Joint Board of Geo-spatial Information Societies (JBGIS) and The International Council for Science (ICSU); and Inter Government Agencies such as the United Nations and GEO, the Group on Earth Observations. The chapter shows that there is very healthy collaboration in Geomatics, and particularly in the use of Earth Observation data. ISPRS is involved in many of these activities and is benefiting from this interaction.

Keywords: ISPRS, NGOs, GEO, UN, societies, capacity building

34.1 OVERVIEW OF ISPRS ACTIVITES 2004–2008

International collaboration in the geo-information sciences has been vigorous during the past four years. ISPRS has been active in many areas and has developed and forged important links with a number of international organizations to promote cooperation and capacity building. This chapter reviews the relevant activities of the Joint Board of Geo-spatial Information Societies (JBGIS), the Group on Earth Observation (GEO), the International Council of Science (ICSU) and the United Nations, as well as other organizations.

The most important role of ISPRS is to develop the science of using images within geo-spatial information technology. The work of the Society is carried out by the Technical Commissions, which bring together workers from many countries to form working groups that hold workshops and other meetings. Such meetings are often the most productive when held with other organizations and ISPRS has collaborated with, for example, the conference on Optical 3-D Measurement Techniques, sponsored by ISPRS, International Federation of Surveyors (FIG) and the International Association of Geodesy (IAG), and the conference on Geo-information for Disaster Management (GI4DM) organized by ISPRS, The United Nations Office for Outer Space Affairs (UNOOSA), FIG, The European Commission (EC), AGILE (Association Geographic Information Laboratories Europe), International Cartographic Association (ICA) and TU Delft. There are only a few conferences that bring together all the various ofrganizations involved with geo-information. Map World Forum, held in India in January 2007, was one such meeting, which concentrated on commercial and political aspects of geo-information. Many of the international societies were represented there. The International ESRI User Conference is a meeting that brings together all components of the geo-spatial industry, including the geo-spatial societies. The societies also met at the Cambridge Conference, organized by Ordnance Survey of Great Britain in July 2007. This meeting was primarily for national mapping agencies but provided a very valuable overview of developments and issues within the mapping profession. The International Society for Digital Earth (ISDE) was formally established during the last four years and has organized a number of conferences bringing together organizations that support this concept, and this inevitably includes the use of imagery.

34.2 ORGANIZATIONS THAT PROMOTE INTERNATIONAL COLLABORATION

34.2.1 *Joint Board of Geo-spatial Information Societies (JBGIS)*

The Joint Board of Geo-spatial Information Societies (JBGIS)—[http://www.fig.net/jbgis/]—is a coalition of leading international geo-spatial societies, having the primary goal of speaking on behalf of the geo-spatial profession at international level, especially to the United Nations and other global stakeholders. Its second goal is to coordinate activities within the geo-spatial society and organizations. The current members of the JBGIS are: The Global Spatial Data Infrastructure Association (GSDI), International Association of Geodesy (IAG), International Cartographic Association (ICA), International Federation of Surveyors (FIG), The International Hydrographic Bureau (IHB), The International Map Trade Association (IMTA), International Steering Committee for Global Mapping (ISCGM) and the International Society of Photogrammetry and Remote Sensing (ISPRS).

The JBGIS can appoint *ad hoc* committees for certain projects and important topics. Normally these committees have specific objectives and are normally appointed for a limited time period. Currently there are committees on Capacity Building in Africa; the committee on Risk and Disaster Management operated for two years and was then disbanded after making recommendations to the Board.

The Board has identified capacity building in Africa as a priority activity for societies involved in Geo-spatial Information (GI). A major activity has been the development of a knowledge portal for Africa. In the final plenary session of CC: The Exchange, an event hosted by Ordnance Survey in July 2005 on behalf of National Mapping Agencies (NMAs) from across the world, the notion of transferring information from NMAs in developed countries to those in developing countries was discussed. GSDI agreed to scope the requirements to set up such a portal, which will have three primary components:

- A knowledge database;
- A commitment database;
- An equipment database.

Much information already exists and work needs to be done to integrate this into the portal. An existing portal, AGIRN (African Geo-Information Research Network) run by EIS Africa, has been chosen as a pilot study, which will give experience of the required structure based on an existing portal and allow feedback from providers and users. The AGIRN portal is now operating and can be viewed at [http://www.AGIRN.org] and the Joint Board is supporting the development of this portal and the GSDI portal.

The Risk and Disaster Management Committee had the role of coordinating the many initiatives in these areas and has recommended that each organization within JBGIS organizes sessions on disaster management topics during their main conferences. The societies should also provide experts to UN conferences and other UN activities and also nominate a contact person or working group for disaster management. The GI4DM conference series is recognized as the main international event in this area.

Contributions are also made to other projects initiated by particular societies, such as the African Reference Frame (AFREF), supported by the Economic Commission for Africa (ECA), Committee on Development Information (CODI), (see below) and Mapping Africa for Africa (MAFA) organized by ICA and UN SPIDER.

34.2.2 *Group on Earth Observation (GEO)*

A major initiative, formally established in 2006, is the Group on Earth Observation (GEO) [http://www.earthobservations.org/], which is working towards a Global Earth Observing System of Systems (GEOSS). The GEO vision for GEOSS is to realize a future wherein decisions and actions for the benefit of humankind are informed via coordinated, comprehensive and sustained Earth observations and information. GEOSS will build on and add value to existing Earth observation systems by coordinating their efforts, addressing critical gaps, supporting their interoperability, sharing information, reaching a common understanding of user requirements and improving delivery of information to users. The work is centred on nine areas of societal benefit, shown in Table 34.1.

In order to meet the objectives of GEOSS, a number of tasks have been defined and many of these are relevant to the activities of ISPRS. There are also cross cutting themes overseen by the GEO Committees.

The Architecture and Data Committee supports GEO in all architecture and data management aspects of the design, coordination and implementation of the Global Earth Observation System of Systems (GEOSS) for comprehensive, coordinated and sustained Earth observations. The Capacity Building Committee supports GEO in strengthening the capability of all countries, in particular developing countries, to use Earth observation data and products in a sustainable manner and to contribute observations and systems to GEOSS. The Science and Technology Committee engages the scientific and technological communities in the development, implementation and use of a sustained GEOSS in order to ensure that GEO has access to sound scientific and technological advice. The User Interface Committee engages users in the nine societal benefit areas in the development, implementation and use of a sustained GEOSS that provides the data and information required by user groups on national, regional and global scales. The User Interface Committee has a specific goal to address cross-cutting issues by coordinating user communities of practice, ensuring continuity and avoiding duplication. The Working Group on Tsunami Activities supports the coordinating activities of the UNESCO Intergovernmental

Table 34.1. The GEO societal benefit areas.

Disasters	Reducing loss of life and property from natural and human induced disasters
Health	Understanding environmental factors affecting human health and well being
Energy	Improving management of energy resources
Climate	Understanding, predicting, mitigating and adapting to climate variability and change
Water	Improving water resource management through better understanding of the water cycle
Weather	Improving weather information, forecasting and warning
Ecosystems	Improving the management and protection of terrestrial, coastal and marine resources
Agriculture	Supporting sustainable agriculture and combating desertification
Biodiversity	Understanding, monitoring and conserving biodiversity

Oceanographic Commission (IOC) and other national, regional and international initiatives to realize effective warning and mitigation systems for natural hazards, as an integral part of a multi-hazard approach supported by GEOSS.

ISPRS is represented on these committees.

ISPRS has collaborated with the Institution of Electrical and Electronic Engineers (IEEE) and the Open Geo-spatial Consortium (OGC) under the umbrella of GEO to organize a series of workshops to provide a forum for discussing the architecture and user applications of the GEOSS. These include on-line demonstrations by OGC on the web services that will be part of GEOSS and feedback sessions for user comment. ISPRS has taken the lead in several of these including 'Applications in Public Health' in Goa, India, during the ISPRS Commission IV Symposium; Agriculture, Sustainable development through Earth Observation at San Jose, Costa Rica, during the International Symposium on Remote Sensing of Environment; Climate and Biodiversity at Yellowknife, Canada, during IPY GeoNorth; Water and Health at Burkina Faso before Africa GIS 2007; and Disaster Preparedness and Management Environment Disaster/Tsunami at Kuala Lumpur, Malaysia, during the Asian Conference on Remote Sensing.

CEOS, The Committee on Earth Observation Satellites, with its Working Groups on Information Systems and Services, Calibration and Validation and Education has continued to work in close collaboration with GEO and is leading the key task of virtual constellations.

34.2.3 *ICSU – The International Council for Science*

To strengthen international science for the benefit of society, ICSU [http://www.icsu.org/index.php] mobilizes the knowledge and resources of the international science community to:

- Identify and address major issues of importance to science and society.
- Facilitate interaction amongst scientists across all disciplines and from all countries.
- Promote the participation of all scientists—regardless of race, citizenship, language, political stance, or gender—in the international scientific endeavour.
- Provide independent, authoritative advice to stimulate constructive dialogue between the scientific community and governments, civil society, and the private sector.

ICSU is a valuable forum for interdisciplinary activity. ICSU has two major initiatives at the moment, both of which involve geo-spatial data: these are Health and Wellbeing in the Changing Urban Environment, and Natural and Human-induced Environmental Hazards and Disasters. A major agent for scientific collaboration has been the grouping of the GeoUnions within ICSU. This group aims to be a focus within ICSU for issues relating to the science of the Earth and space and the collection, processing and management of geo-spatial data and particularly to coordinate activities and develop joint interdisciplinary projects. The members of the GeoUnions are: International Astronomical Union (IAU), International Geographical Union (IGU), International Union for Quaternary Research (INQUA), International Society for Photogrammetry and Remote Sensing (ISPRS), International Union of Geodesy and Geophysics (IUGG), International Union of Geological Sciences (IUGS), International Union of Soil Sciences (IUSS) and International Union of Radio Science (URSI). The group has identified six themes that are suitable for interdisciplinary collaboration: Cities and Megacities, Desertification, Groundwater, Hazards, Health and Polar.

ICSU has also established regional offices in Africa, South America and the Caribbean, and Asia Pacific. These will work with local organizations on local problems.

34.2.4 *International science years*

In the period 2007–2009, a number of science years have been declared, which are of interest to the photogrammetry and remote sensing community. These are briefly described with ISPRS involvement.

- In recognizing that Earth science knowledge is greatly under-utilized, the International Year of Planet Earth (IPYE) aims to raise awareness among governments, decision makers and the general public of the largely hidden contribution Earth science know-how contributes to everyday life, and the considerable potential Earth science has to offer in the quest for a safer, healthier and wealthier global society. A number of themes have been identified, these include hazards, Earth and health and megacities, which overlap areas defined by GEO, for example, and indicate the perceived importance of these topics.
- The electronic Geophysical Year (eGY) is an IUGG initiative that sets out to help provide ready and open access to geo-scientific data and services worldwide using the capabilities offered by modern digital communications and information management technologies. Key features are the promotion of virtual observatories to complement in cyberspace the contribution from physical observatories, and the strengthening of data stewardship practices.
- The primary objectives of the International Heliophsyical Year (IHY) are to advance understanding of the heliophysical processes that govern the Sun, Earth and Heliosphere and, in so doing, continue the tradition of international research and advance the scientific legacy of the International Geophysical Year on its 50th anniversary, so demonstrating the beauty, relevance and significance of Space and Earth Science to the World.
- The International Polar Year (IPY) is an interdisciplinary and internationally coordinated campaign of research and observation, designed to deepen understanding of polar processes and their global linkages. IPY knowledge and the observations upon which it is built will be managed so as to ensure greatest future benefit. IPY, timed for 2007–2008, also coincides with the 50th anniversary of the IGY, as well as with a former Polar Year. ISPRS is leading a project on 'Assessment of Cryospheric Changes from Historical Photographs'.

34.2.5 *United Nations*

The United Nations is concerned with geospatial information in many aspects of its work. The Committee on Peaceful Uses of Outer Space (COPUOS) [http://www.unoosa.org/oosa/] discusses, amongst other topics, remote sensing and the Office of Outer Space Affairs (OOSA), which provides the secretariat for COPUOS, runs the programme on Space applications, which includes the United Nations Platform for Space-based Information for Disaster Management and Emergency Response (UN SPIDER) with the following mission statement: "Ensure that all countries have access to and develop the capacity to use all types of space-based information to support the full disaster management cycle."

Whereas there have been some initiatives in recent years that have contributed to making space technologies available for humanitarian and emergency response, such as the International Charter Space and Major Disasters, UN SPIDER is the first to focus on the need to ensure access to and use of such solutions during all phases of the disaster, including the risk reduction phase, which will significantly contribute to an increasing reduction in loss of lives and property. The new programme will achieve this by focusing on being a gateway to space information for disaster management support, serving as a bridge to connect the disaster management and space communities and being a facilitator of capacity building and institutional strengthening, in particular for developing countries.

The Regional Cartographic Conferences (RCCs) provide a forum for discussion on the use of geo-information within the regions. During the past four years these have been held in New York (RCC for The Americas in July 2005), and Bangkok (RCC for Asia Pacific, September 2006). In Africa a similar forum is provided by the Geo-information Sub Committee of CODI (Committee on Data and Information), which is run by the UN Economic Commission for Africa (ECA). CODI GEO is more proactive than the RCCs and supports AFREF and MAFA (see above) as well as capacity building. All of these fora link upwards to ministers and pass resolutions, which go before the UN General Assembly.

34.2.6 *University Network for Disaster Risk Reduction in Africa*

The University Network for Disaster Risk Reduction in Africa (UNEDRA) [http://www.itc.nl/unu/dgim/unedra/default.asp] was formed during a workshop held at Makerere University, Kampala, Uganda from 12 to 13 September 2005. The Workshop was organized by the International Institute for Geo-Information Science and Earth Observation (ITC), The Netherlands and the Department of Geography, Makerere University and it was organized in the framework of the United Nations University Programme on Disaster Risk Information Management. The workshop was attended by relevant participants from universities of eastern and southern Africa. The main aim was to forge interaction amongst universities in Africa with interests in teaching disaster risk reduction, through information sharing, capacity building and collaborative research. The objective of the network is to share experiences on course activities that have already been initiated and identify universities that have interest in developing disaster reduction courses and to develop joint education and training activities where staff and students from different African universities collaborate. There is also the intention to develop joint research activities and to support the setting up of educational programmes on Geo-information for Disaster Management.

34.3 INTERNATIONAL GEODETIC STUDENT ORGANIZATION (ISGO)

IGSO [http://www.igso.de/] is an international, non-governmental organization, whose purpose is to support international collaboration for the progress of geodesy in all fields and applications. It should be noted that the definition of geodesy is that of central Europe, and includes all branches of what is now commonly called Geomatics. IGSO was founded on the 10 May 1991, during the 4th meeting at the Technical University of Graz (Austria). The organization realizes the following aims:

- grouping the associations or organizations of geodetic students from all countries in order to exchange views on matters of their general interest
- representing the students of geodesy in the public field
- organizing liaisons between different member associations
- gathering information about the social conditions amongst geodetic students in individual countries, so that each could take advantage of the improvement attained
- improving and supporting good relations with the proper authorities and the exchange of geodetic students between different countries.

IGSO enables students of subjects related to geodesy, cartography, photogrammetry and remote sensing not only to exchange information, views and experiences, but also ensures contacts on professional matters with known agenda. During the past four years, meetings have been held in Espoo (Finland), Istanbul (Turkey), Cracow (Poland) and Sofia (Bulgaria).

34.4 SUMMARY AND CONCLUSIONS

Active collaboration in the area of photogrammetry and remote sensing exists between societies with similar interests and between these societies and non-governmental organizations (NGOs) and inter-governmental organizations (IGOs). The first group are well organized and groups like the Joint Board of Geo-spatial Information Societies and the ICSU GeoUnions meet regularly and foster collaborative activities, such as the science years, and particularly from the point of ISPRS, the International Polar Year. The United Nations acts as a focal point for thematic activities such as disaster management and global navigation satellite systems and enables NGOs to interact with the international agencies and with national organizations such as space agencies. A newcomer on the scene is the Group on Earth Observation (GEO), which has great potential for encouraging collaboration, with very specific objectives, aimed at benefiting society. More time is needed to assess the success of this group.

A key interest of many international organizations is education and youth. This chapter has mentioned the International Geodetic Student Organization, but other issues relating to students are covered elsewhere in this book.

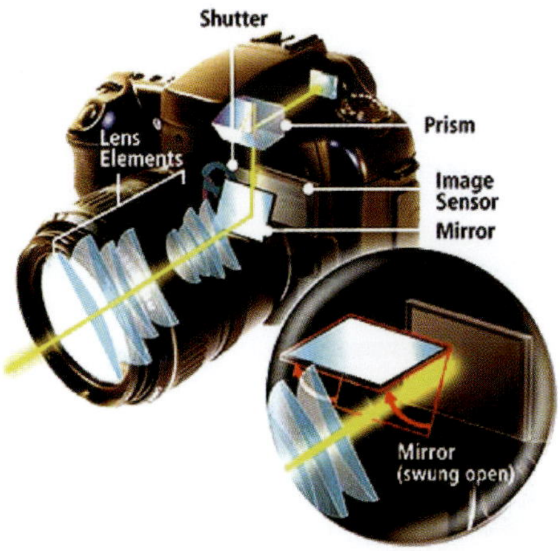

Figure 5.1. Digital SLR-camera principle (Long, 2005). (H.-G. Maas, see page 64)

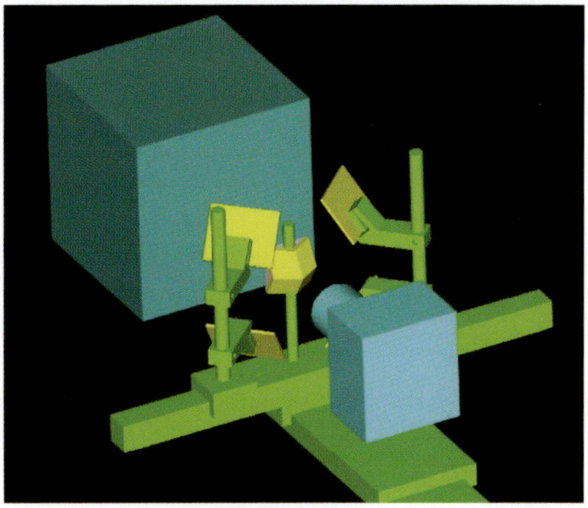

Figure 5.4. Multi-ocular high-speed camera system. (H.-G. Maas, see page 66)

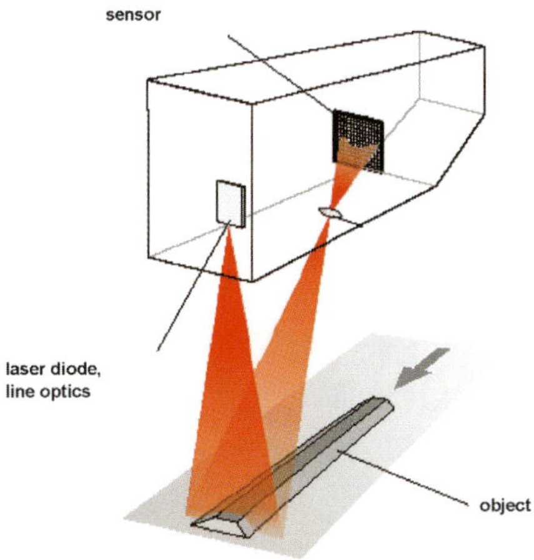

Figure 5.5. Laser light-sheet triangulation principle. (H.-G. Maas, see page 66)

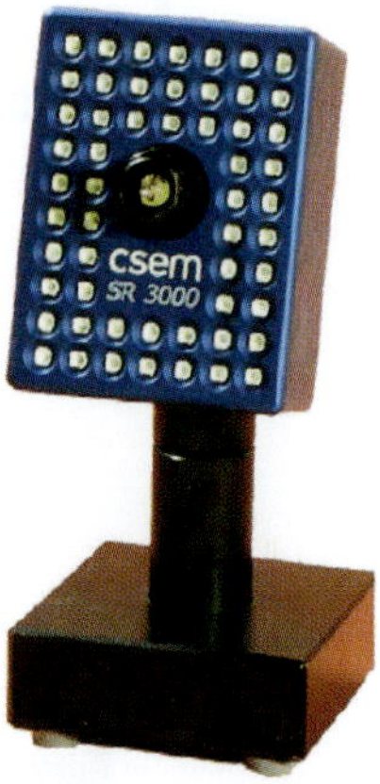

Figure 5.8. 3D-camera Swissranger-3000 (MESA Imaging). (H.-G. Maas, see page 68)

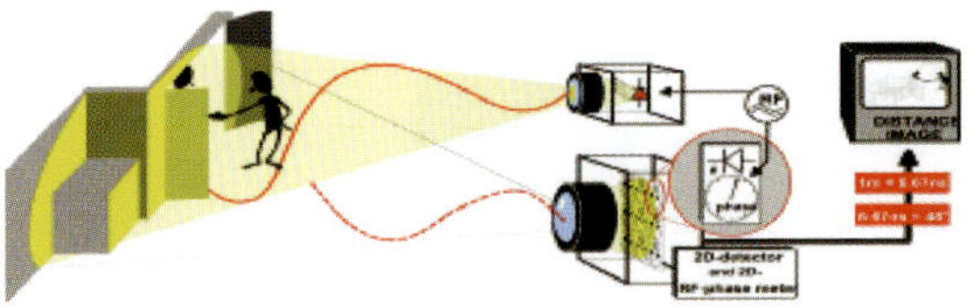

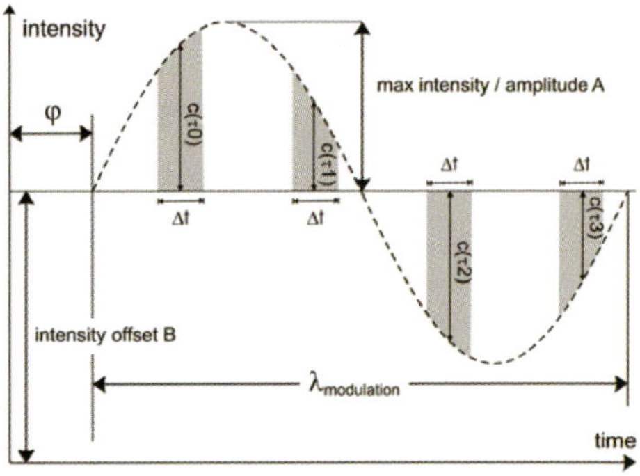

Figure 5.9. Range imaging configuration (Lange, 2000) and pixel phase demodulation (Kahlmann and Ingensand, 2005). (H.-G. Maas, see page 68)

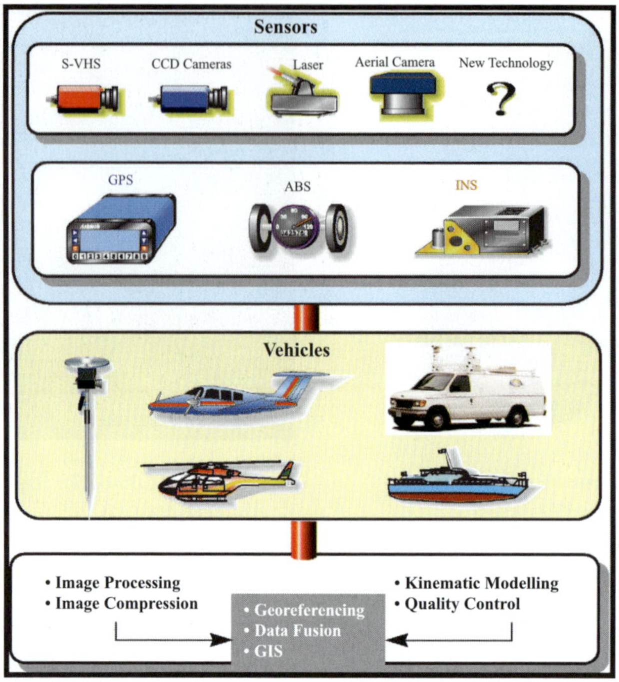

Figure 7.1. Multi-sensor integration for mapping application. (N. El-Sheimy, see page 87)

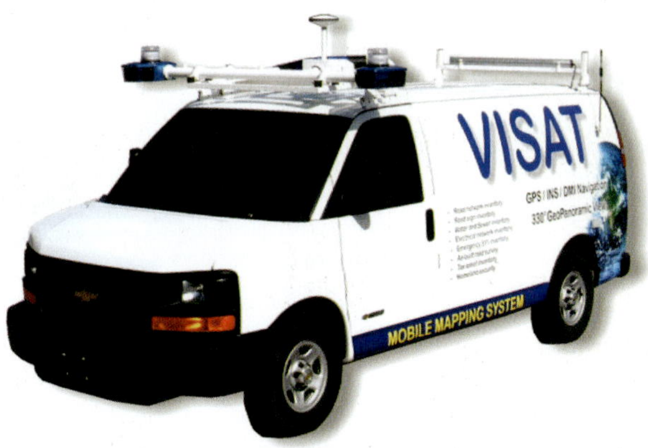

Figure 7.3. VISAT™ Van Mobile Mapping System. (N. El-Sheimy, see page 94)

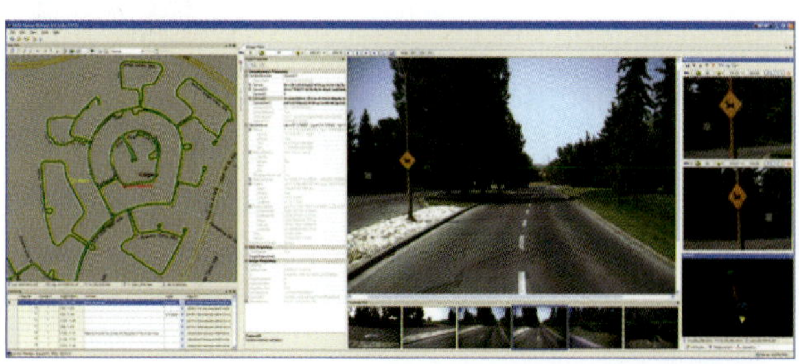

Figure 7.4. VISAT™ Station Measurement. (N. El-Sheimy, see page 96)

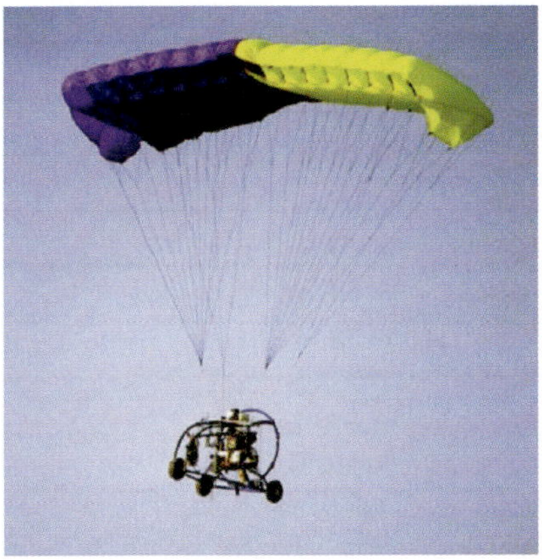

Figure 9.2. A powered paraglider UAV. (J. Everaerts, see page 121)

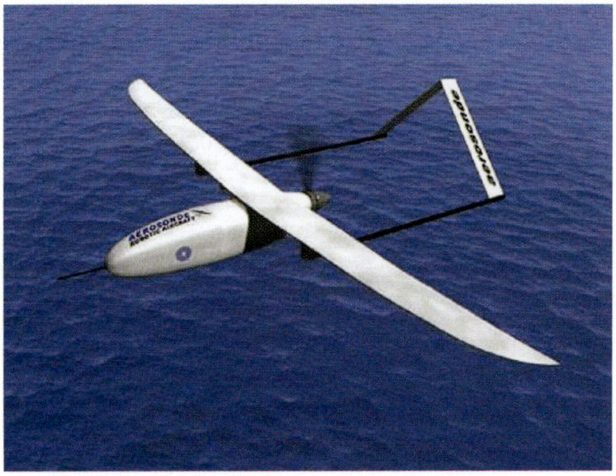

Figure 9.3. The Aerosonde UAV (©Jon Becker, Aerosonde Pty Ltd). (J. Everaerts, see page 122)

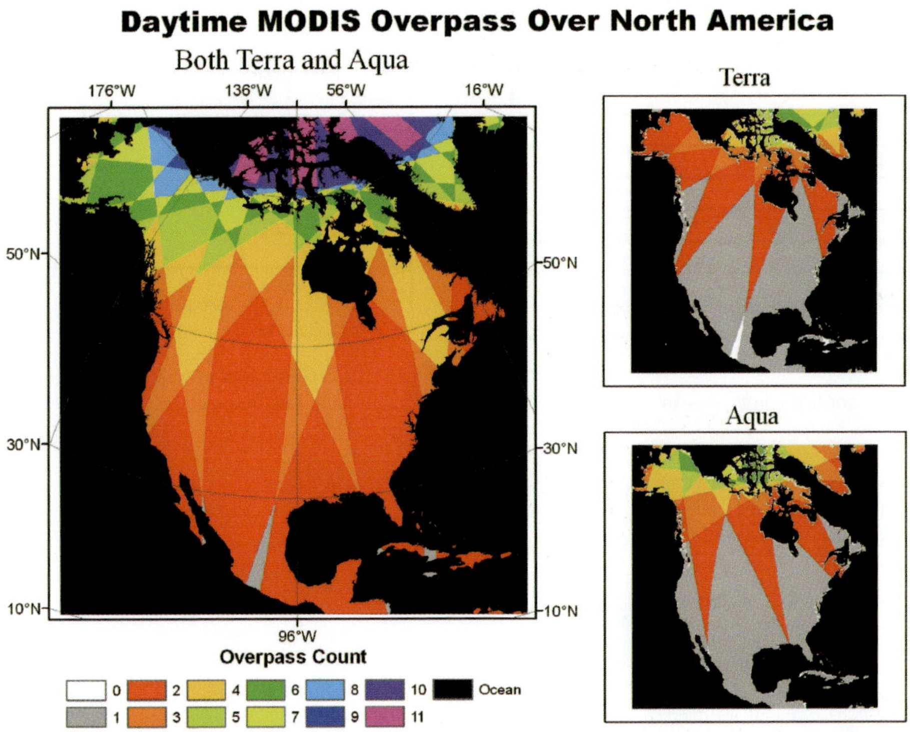

Figure 10.1. Number of MODIS overpass counts over North America based on the Terra and Aqua orbital simulation from GMT 12:00 to 24:00, June 16, 2006. (Wang et al., 2008). (S. Liang et al., see page 134)

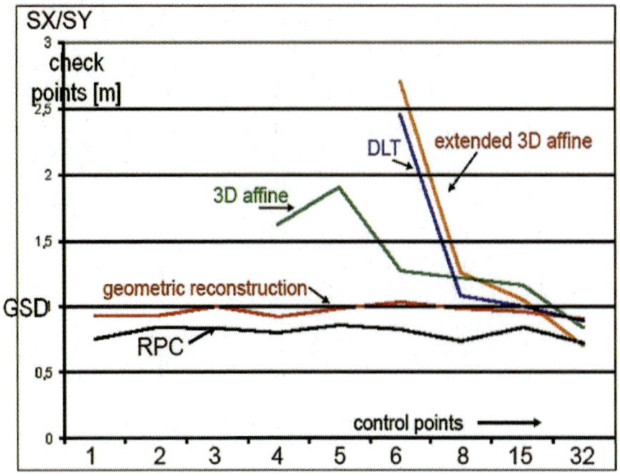

Figure 11.5. Orientation of IKONOS images, Zonguldak (mountainous), 1.0 m GSD, root mean square discrepancies at independent check points as function of number of used control points. (K. Jacobsen, see page 149)

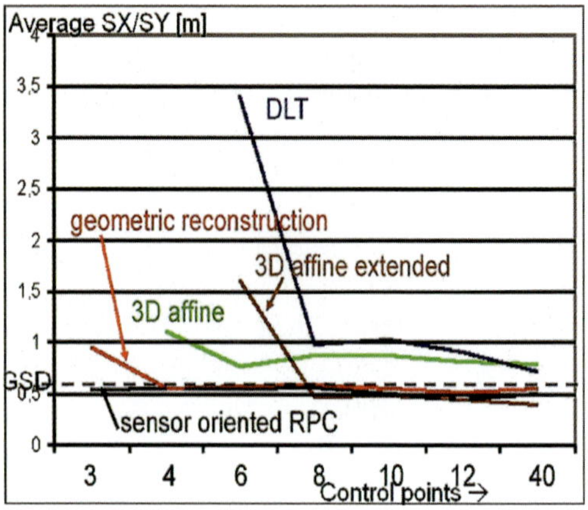

Figure 11.6. Orientation of QuickBird, Zonguldak, 0.62 m GSD, root mean square discrepancies at independent check points as a function of control point number. Exception: 40 GCPs = discrepancies at control points. (K. Jacobsen, see page 149)

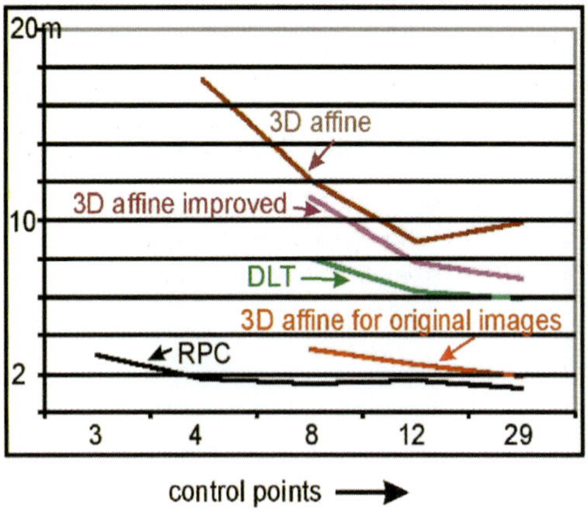

Figure 11.7. Orientation of OrbView-3 Basic, Zonguldak—root mean square differences at independent check points (vertical direction), depending upon number of used control points. (K. Jacobsen, see page 149)

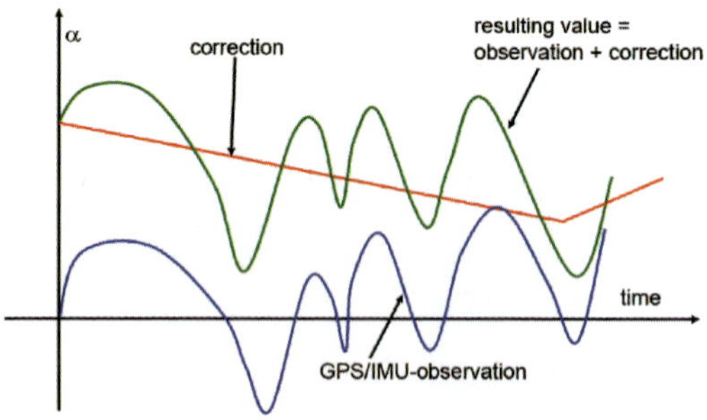

Figure 11.12. Correction of trajectory—high frequency observation with low frequency correction. (K. Jacobsen, see page 151)

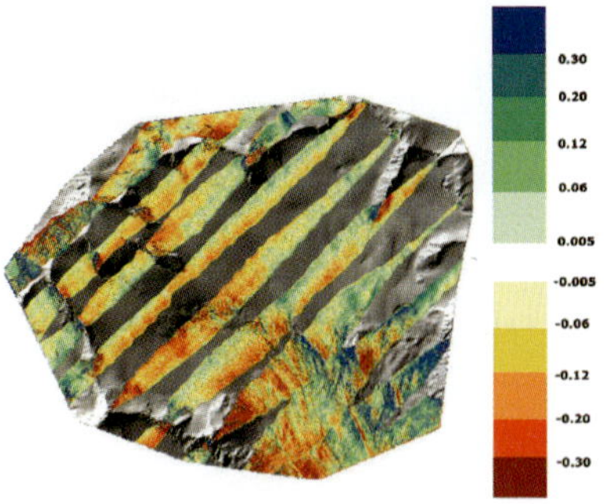

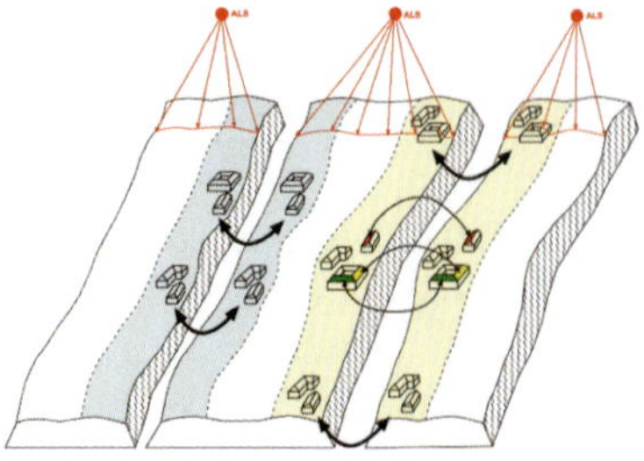

Figure 13.5. Left: The background shows airborne laser data acquired over a glacier. In the overlaps of adjacent strips the height differences of the strips are shown. Dark red and blue tones refer to differences larger than 30 cm. Right: Principle of airborne laser scanning strip adjustment based on tie patches (Kager 2004). (N. Pfeifer et al., see page 176)

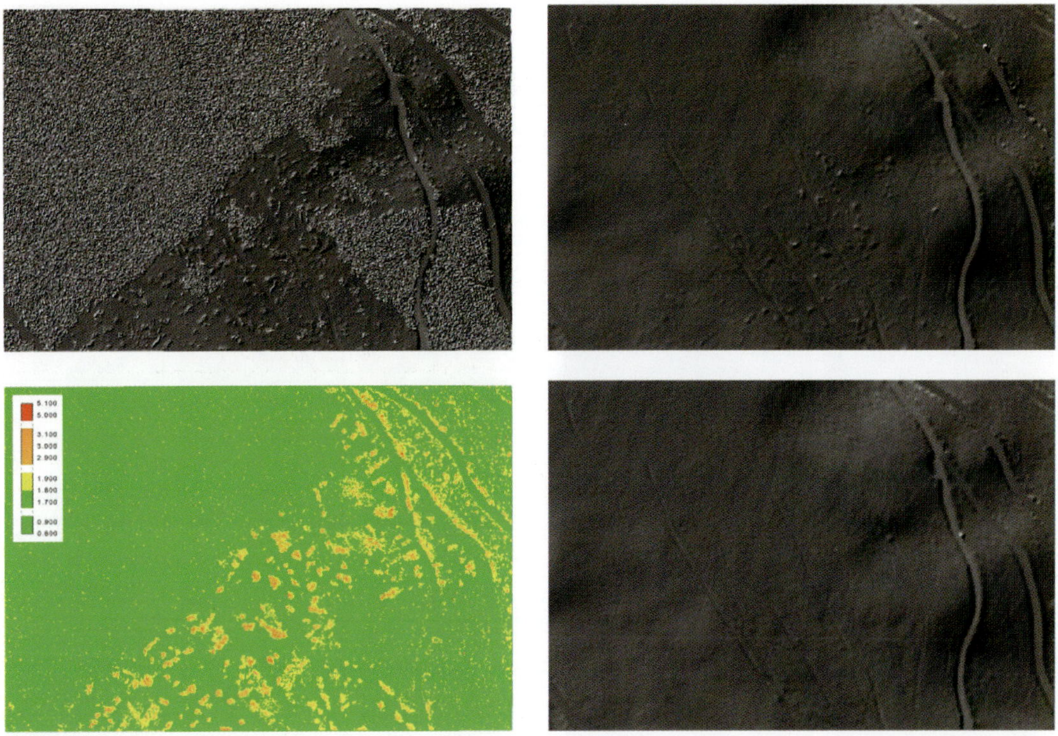

Figure 13.6. From top left to lower right: The digital surface model based on the ALS data; automatically computed terrain model with hierarchic robust interpolation (remaining vegetation artifacts); echo width shown, with red tones referring to wider echoes; terrain model computed with hierarchic robust interpolation after first removing points with large echo width. From Doneus and Briese (2006). (N. Pfeifer et al., see page 181)

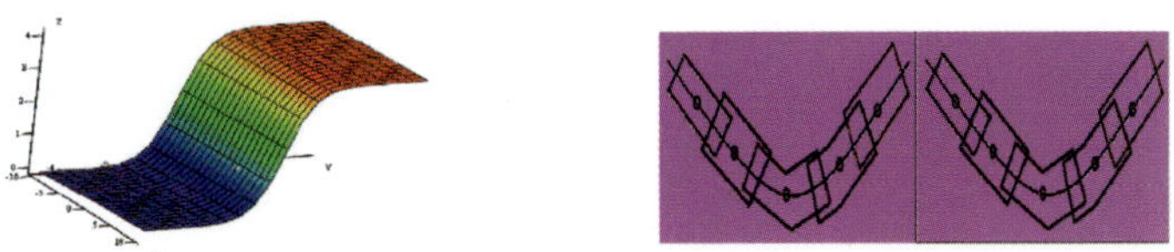

Figure 13.7. Left: Hyperbolic tangent function as model of an edge pair. Right: Overlapping break line patches (both from Brzank et al. (2005)). (N. Pfeifer et al., see page 181)

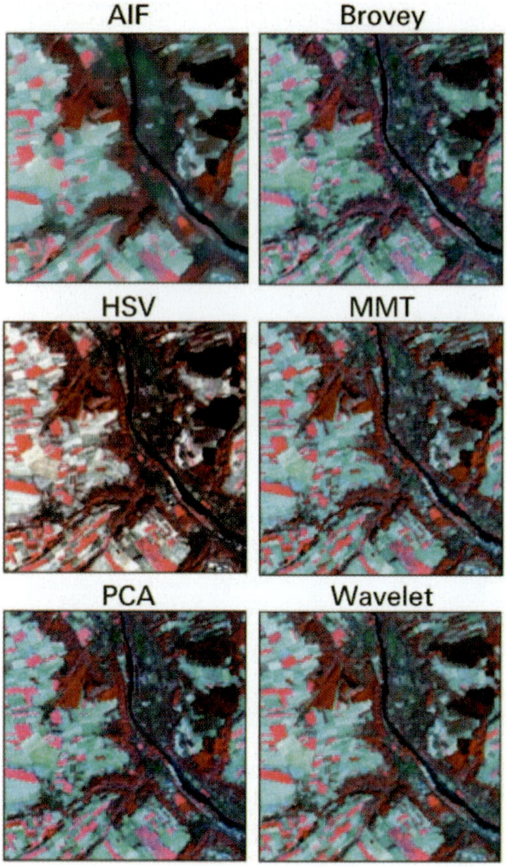

Figure 14.1. Landsat ETM+ Pan-MS fusion results of different fusion approaches. (Adopted from Colditz et al., 2006, rotated) (Y. Zhang, see page 189)

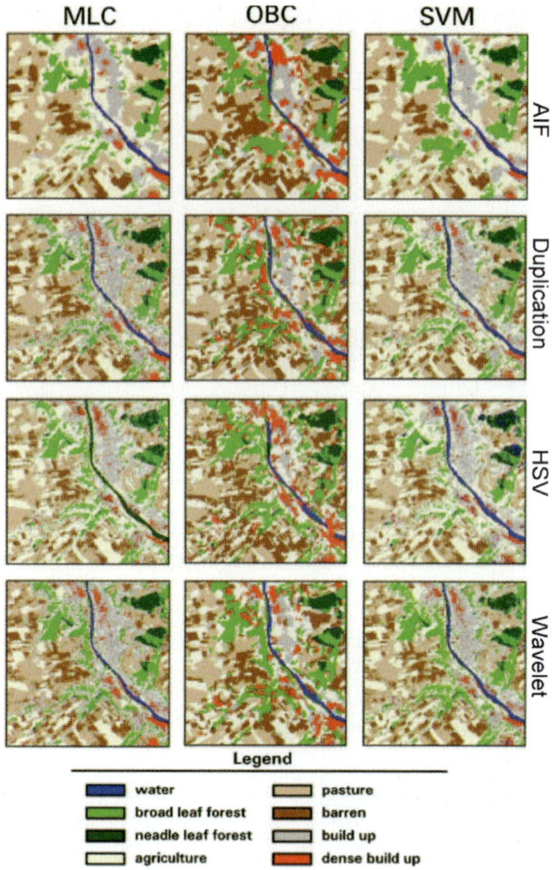

Figure 14.2. Selected classification results from Landsat ETM+ Pan-MS fusions for maximum likelihood classifier (MLC), object based approach (OBC), and support vector machine (SVM) classifications. Brovey, MMT and PCA were not shown due to similarity to other approaches. (Adopted from Colditz et al. 2006, rotated) (Note: "Duplication" was the original MS image but with duplicated pixels resampled from 30 m to 15 m using a nearest neighbour method) (Y. Zhang, see page 190)

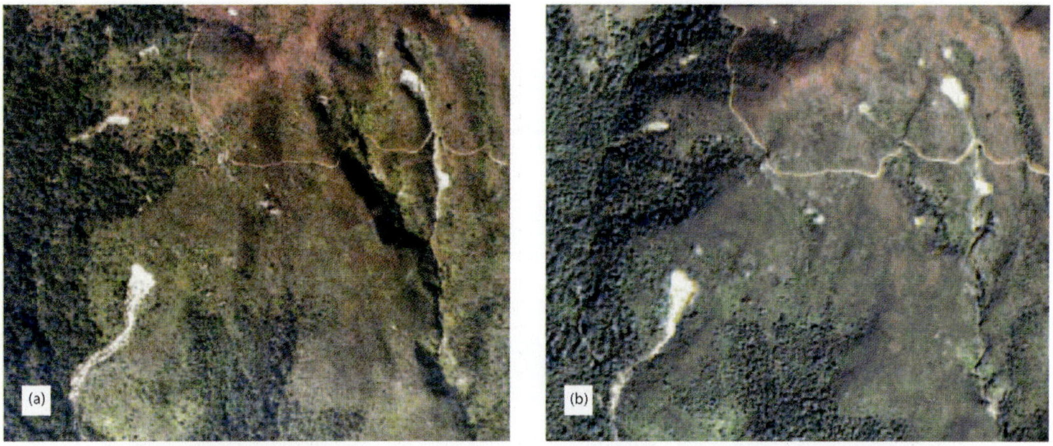

Figure 14.3. Comparison between aerial photograph and pan-sharpened Ikonos MS image showing landslide scars and patches of disturbed ground. (a) Aerial ortho-photograph (1:10 000); (b) Pan-sharpened Ikonos image (PCI Pansharp). (Adopted from Nichol and Wong 2005, subset) (Y. Zhang, see page 191)

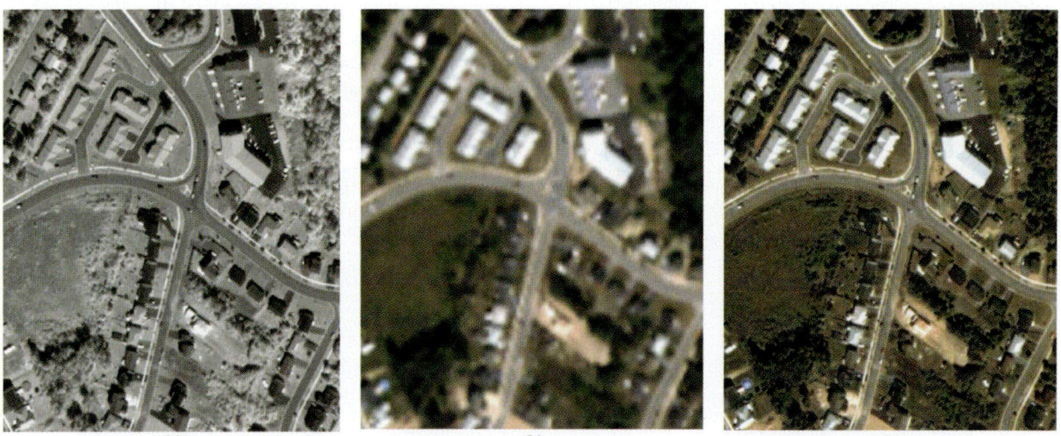

(a) (b) (c)

Figure 14.4. Fusion of QuickBird Pan and MS images for road extraction. (a) Original Pan image; (b) Original MS image; (c) Pan-sharpened MS image using UNB Pansharp. (Adopted from Zhang et al., 2004, subsets) (Y. Zhang, see page 191)

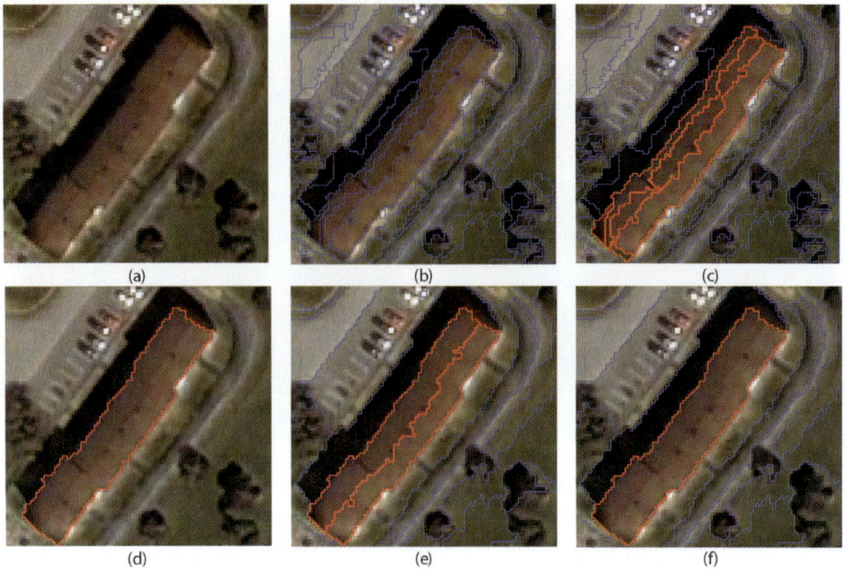

Figure 14.6. Process of the proposed FSS approach and the corresponding segmentation results in pan-sharpened QuickBird image (PCI Pansharp). (a) Object of interest in a training area; (b) Initial segmentation with sub-segments of an object; (c) Sub-segments selected for training the Fuzzy System; (d) Desired segment model formed by the selected sub-segments; (e) Segmentation result using the estimated segmentation parameters from the first iteration; (f) Segmentation result using the estimated, segmentation parameters from the final iteration (the 3rd iteration). (Adopted from Maxwell and Zhang 2005) (Y. Zhang, see page 193)

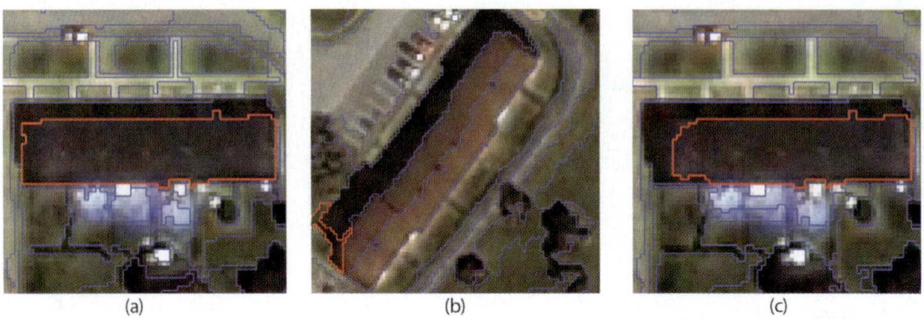

Figure 14.7. Comparison of segmentation results from the proposed FSS approach (a) and eCognition's trial-and-error approach (b) and (c) in pan-sharpened QuickBird image (the segmentation result of the FSS approach for the building in (b) can be found in Figure 14.6f). (Adopted from Maxwell & Zhang 2005) (Y. Zhang, see page 193)

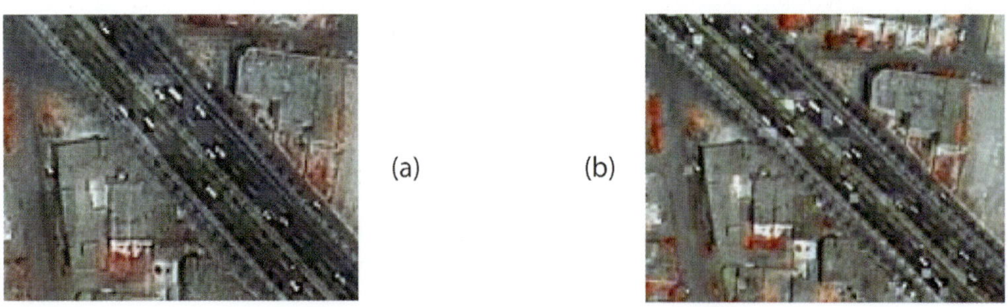

Figure 14.8. The wavelet fused Ikonos image using ERDAS Imagine 8.7 software. (a) Low-pass filtered fused data to remove artifacts; (b) Fused data without low-pass filtering. (Adopted from Boloorani et al., 2006) (Y. Zhang, see page 194)

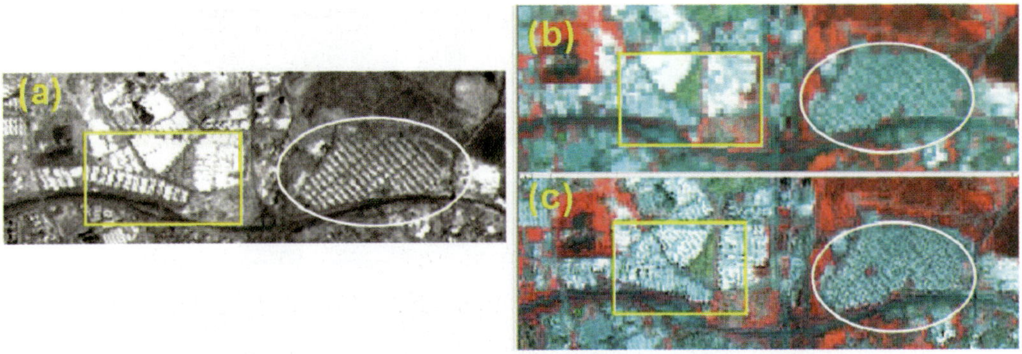

Figure 14.9. Landsat ETM+ Pan and MS fusion using the Wavelet Resolution Merge tool of ERDAS Imagine. (a) Original Pan image; (b) Original MS image; (c) Pan and MS fused image. (Adopted from Balcik and Sertel 2007) (Y. Zhang, see page 194)

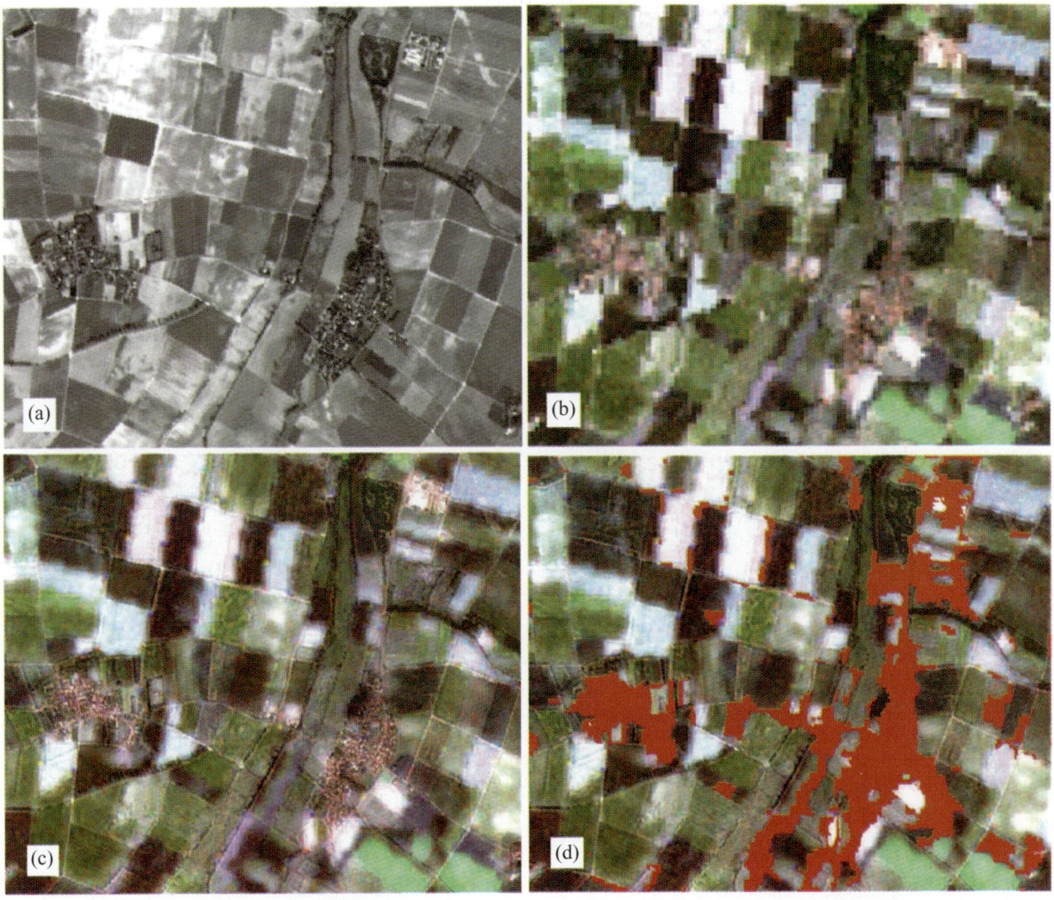

Figure 14.10. SPOT-5 Pan and Landsat ETM+ MS image fusion and object-oriented classification. (a) Original SPOT-5 Pan image, 5 m pixel size; (b) Landsat ETM+ image (bands 1, 2, and 3) co-registered to the SPOT-5 image and resampled to 5 m pixel size; (c) SPOT-ETM fused image using IHS-FFT integrated (Ehlers) fusion approach; (d) Result after a subsequent segmentation and classification using eCognition (settlement in red color). (Adopted from Ehlers et al., 2006, subsets) (Y. Zhang, see page 195)

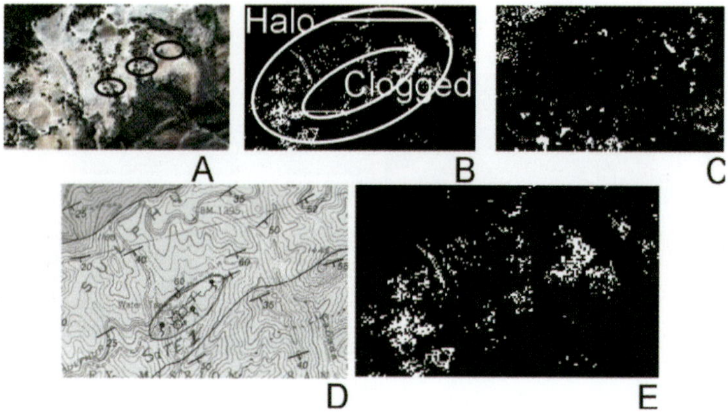

Figure 15.4. Results and interpretation. False color image of probe bands 16, 9, 4 in RGB with seeps (A), SDI probability minerals with interpretation of the anomaly (B) SDI probability vegetation (C), geological map (D) and SDI probability minerals with annotation. All probability maps display P>0.8. (F. van der Meer et al., see page 210)

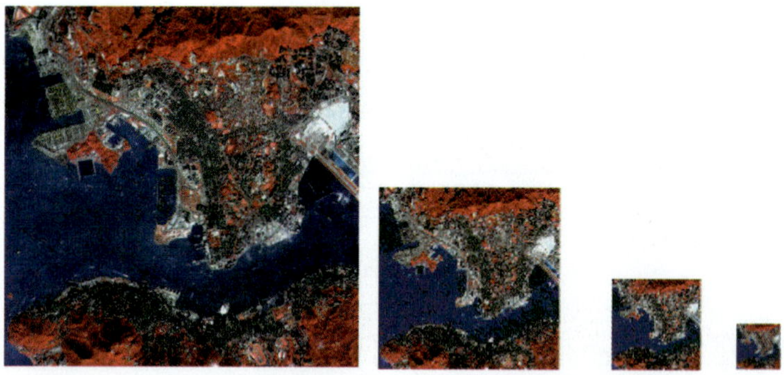

Figure 19.7. Scaling-up: Aggregation of digital images. (Z. Li, see page 272)

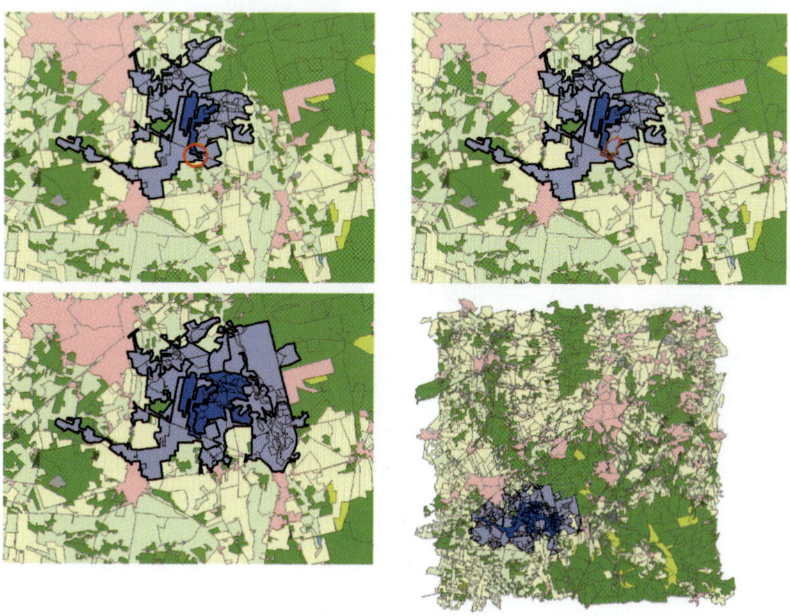

Figure 20.2. Incremental expansion of boundary of maximally affected objects (from top left to bottom right); whole area with possibly affected objects (in blue). (M. Sester, see page 283)

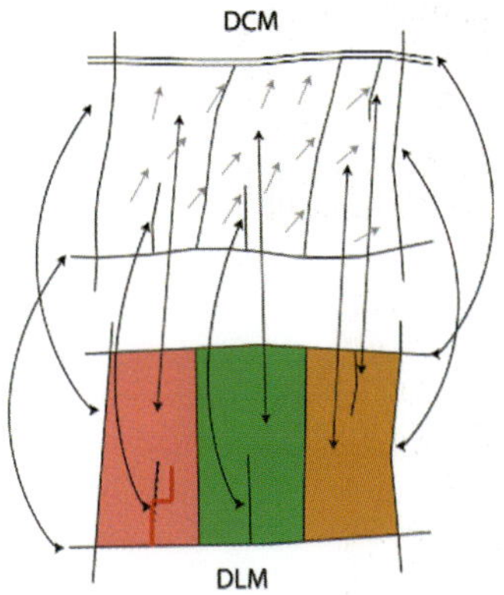

Figure 20.4. Update situations: modification of red object (left), insertion of red object (middle), deletion of object (right). (M. Sester, see page 284)

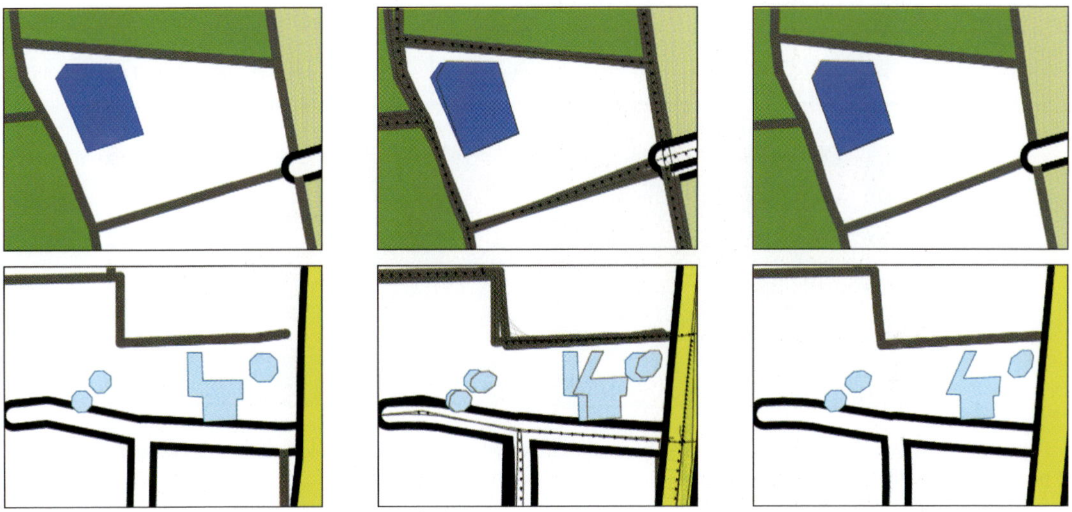

Figure 20.5. Inserting lake: situation before (left) and after applying rubber-sheeting (right); middle: both situations and displacement vectors. (M. Sester, see page 285)

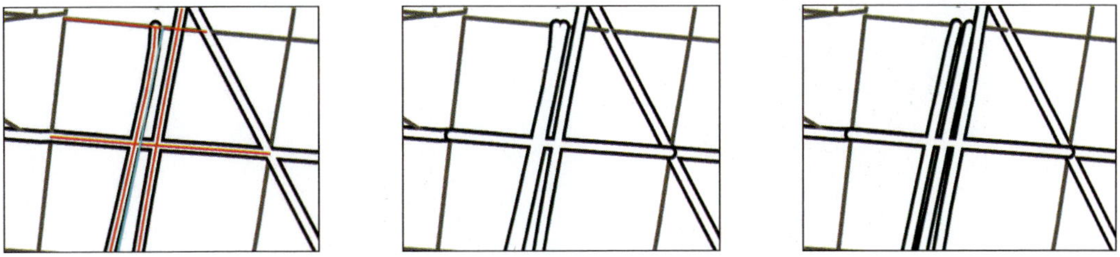

Figure 20.6. Insertion of new road (blue) (left); conflicts after applying signature (middle); situation after conflict resolution with displacement (right). (M. Sester, see page 285)

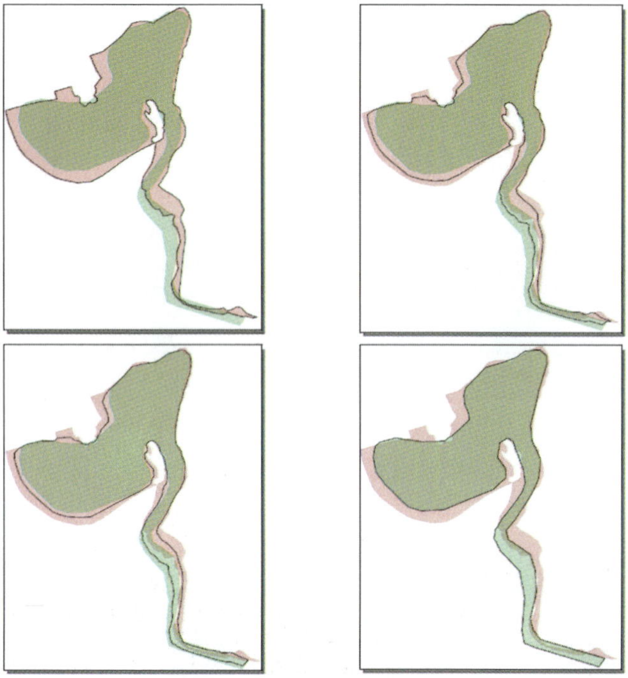

Figure 20.7. Incremental morphing of green and red geometry. The weighting scheme between red and green changes from top left to lower right: 1:0, 0.6:0.4, 0.4:0.6, 0:1. (M. Sester, see page 286)

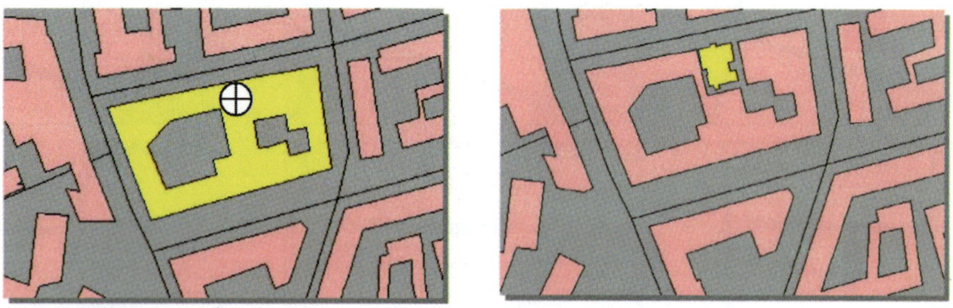

Figure 20.8. Visualization of a landmark in a small scale view: left: marking whole building block; right: delineating precise, detailed object geometry of landmark. (M. Sester, see page 286)

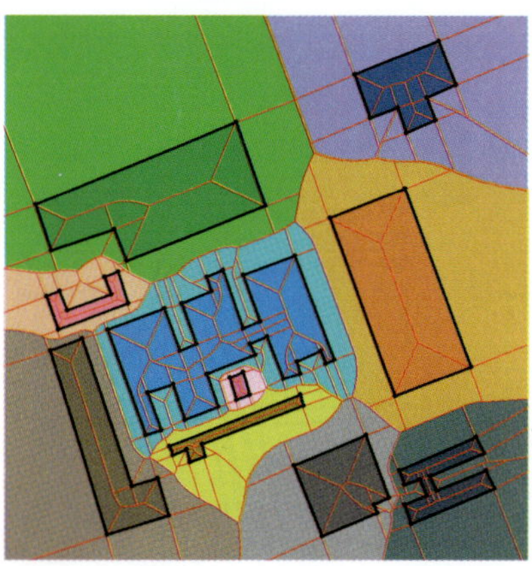

Figure 21.12. Line segment Voronoi diagram. (C.M. Gold et al., see page 298)

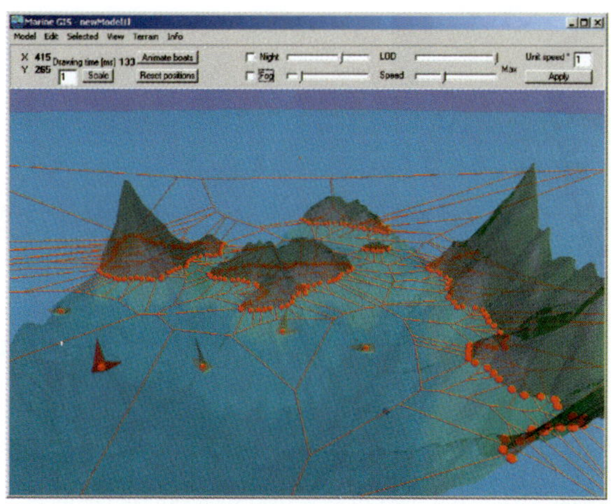

Figure 21.14. Dynamic Voronoi diagrams for Marine GIS. (C.M. Gold et al., see page 299)

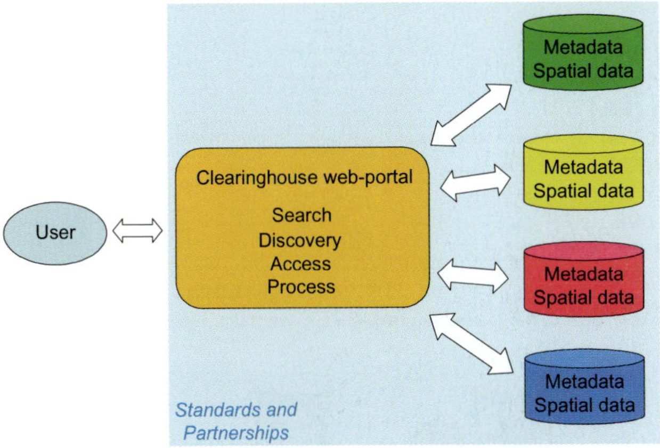

Figure 24.4. Main components of a SDI. (C. Armenakis, see page 328)

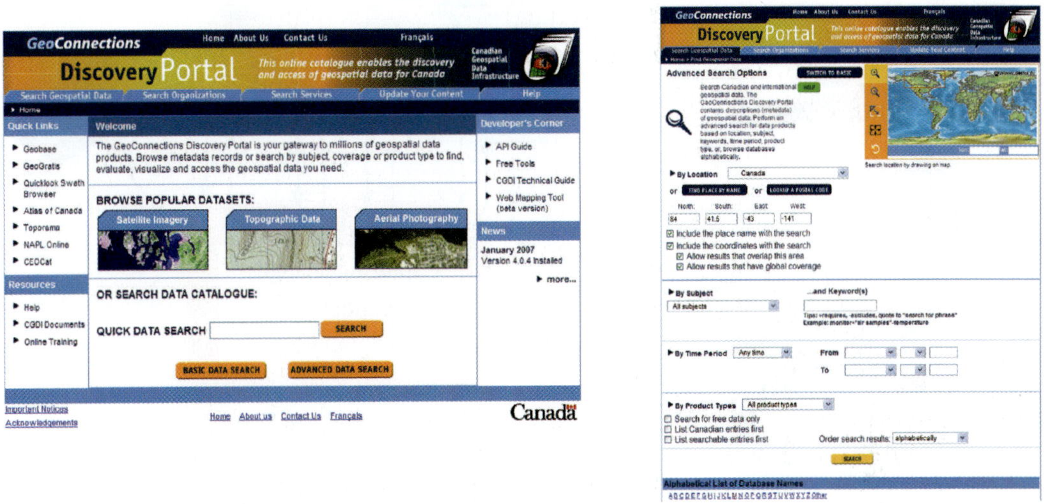

Figure 24.5. Examples of data discovery portals—Canadian GeoConnections (from: http://geodiscover.cgdi.ca/, accessed Feb. 21, 2008). (C. Armenakis, see page 329)

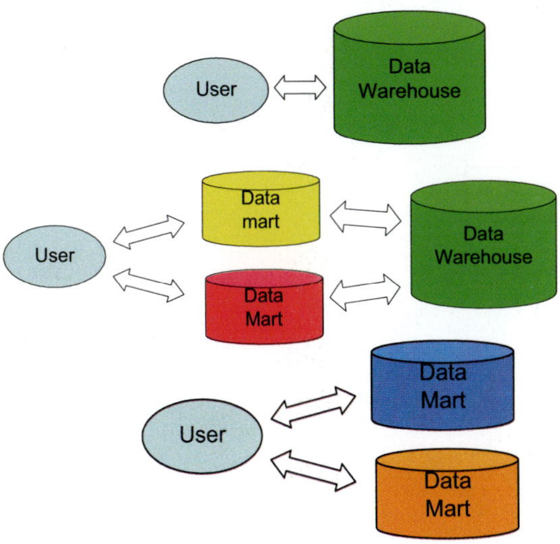

Figure 24.6. Typical architectures for accessing data warehouses and data marts. (C. Armenakis, see page 331)

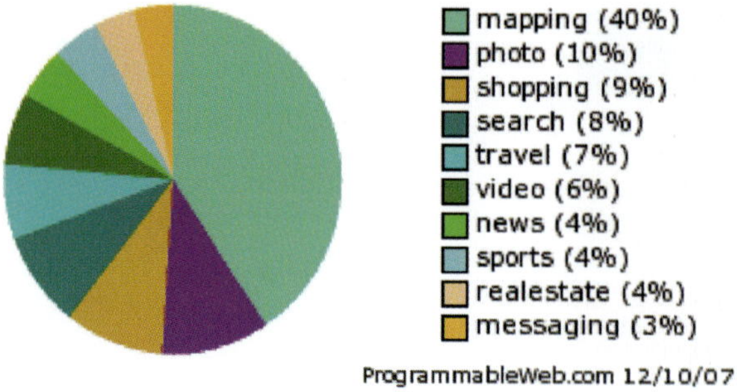

Figure 25.9. Mashup statistics by the Programmable.com. (S. Li, see page 346)

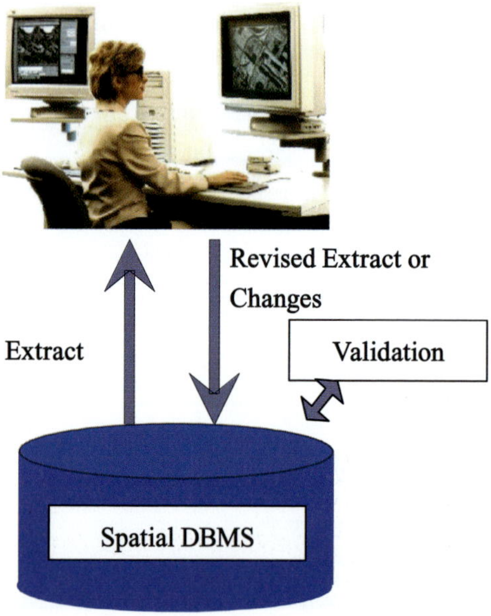

Figure 26.1. File-based data exchange: several cycles are needed, resulting in a very low efficiency. (C. Heipke et al., see page 357)

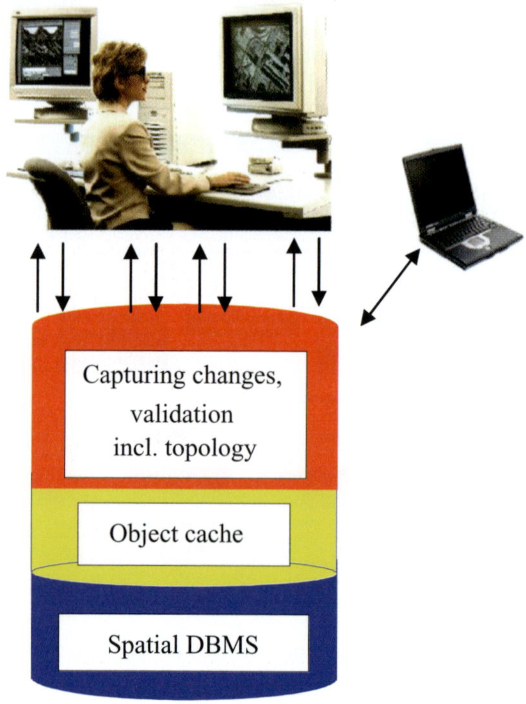

Figure 26.2. Direct link with multi-tier architecture. (C. Heipke et al., see page 357)

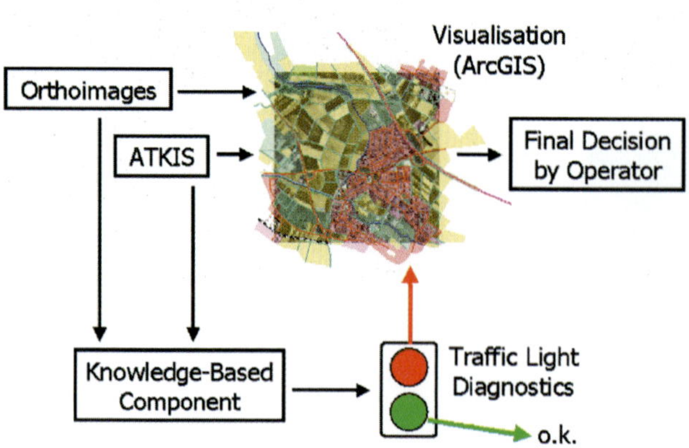

Figure 26.4. Workflow of quality control. (C. Heipke et al., see page 359)

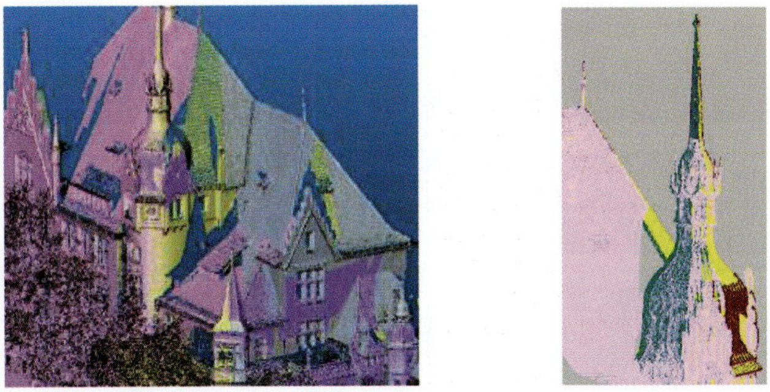

Figure 27.4. Different scans (colors) registered onto the same co-ordinate system. (P. Patias et al., see page 367)

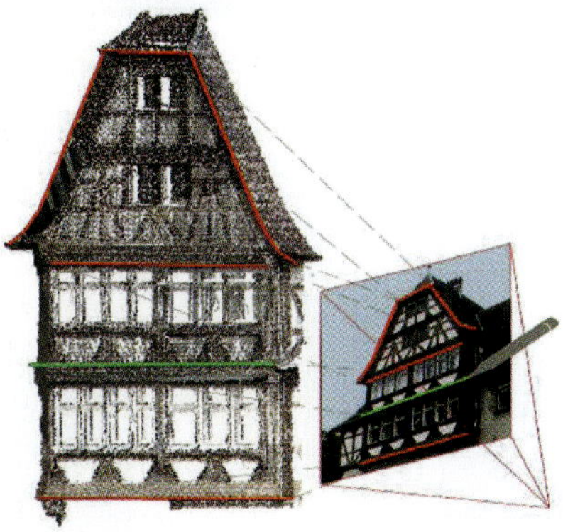

Figure 27.5. Monoplotting 3D coordinates and lines onto the image in relation with the point cloud mesh. (P. Patias et al., see page 367)

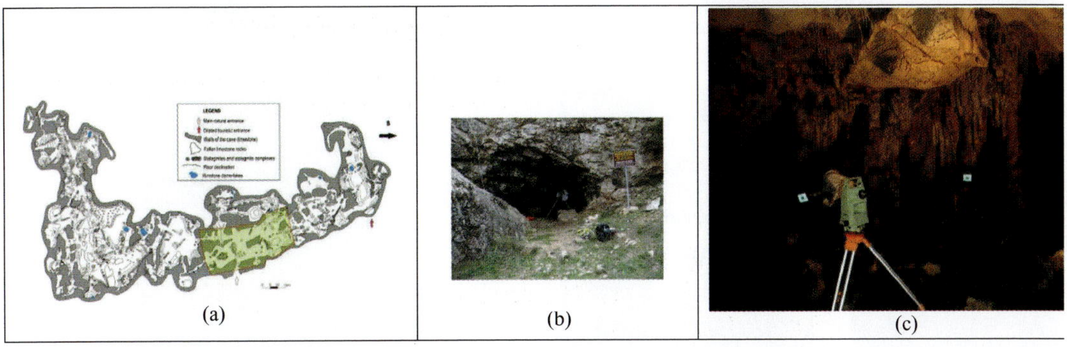

Figure 27.6. From the whole cave only the denoted part has been documented (a); The cave entrance (b); The cave interior (c). (P. Patias et al., see page 368)

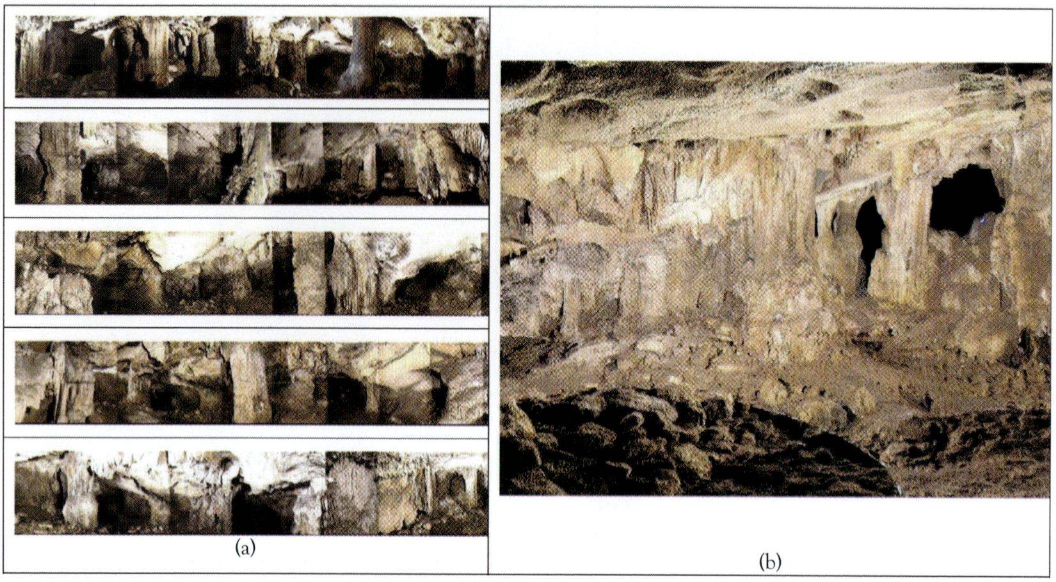

Figure 27.7. Panoramic views of the individual scans (a); Part of the final point cloud (b). (P. Patias et al., see page 368)

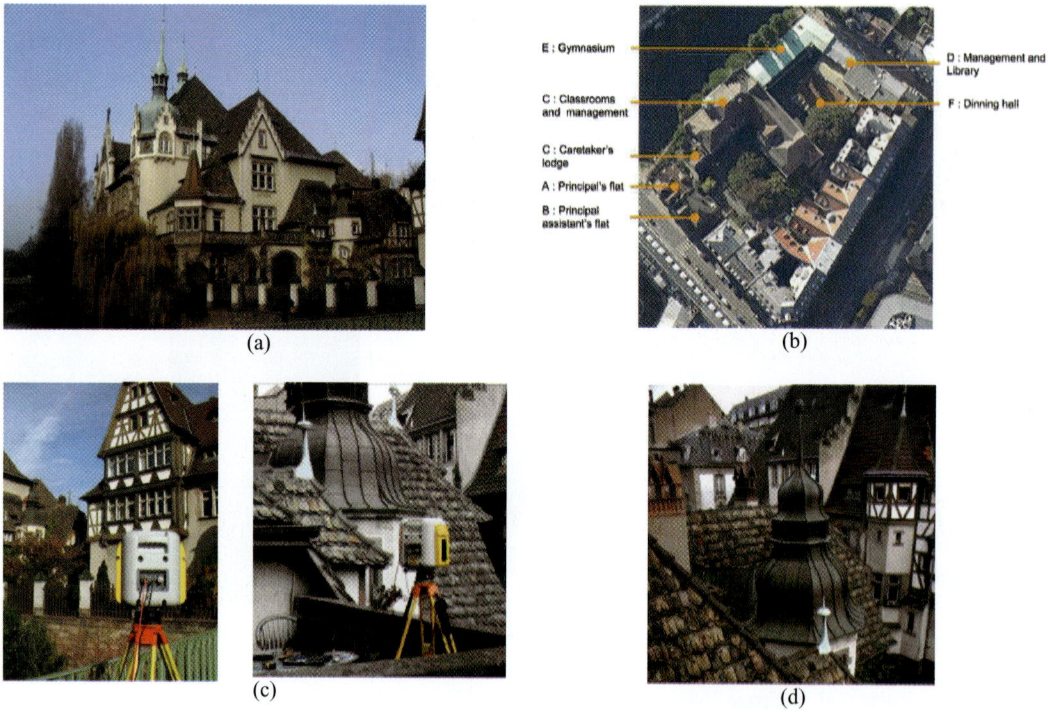

Figure 27.14. Overall view of the main building (a); Aerial view of the site (b); 3D scanning system (Trimble GX) (c); Roof details (d). (P. Patias et al., see page 372)

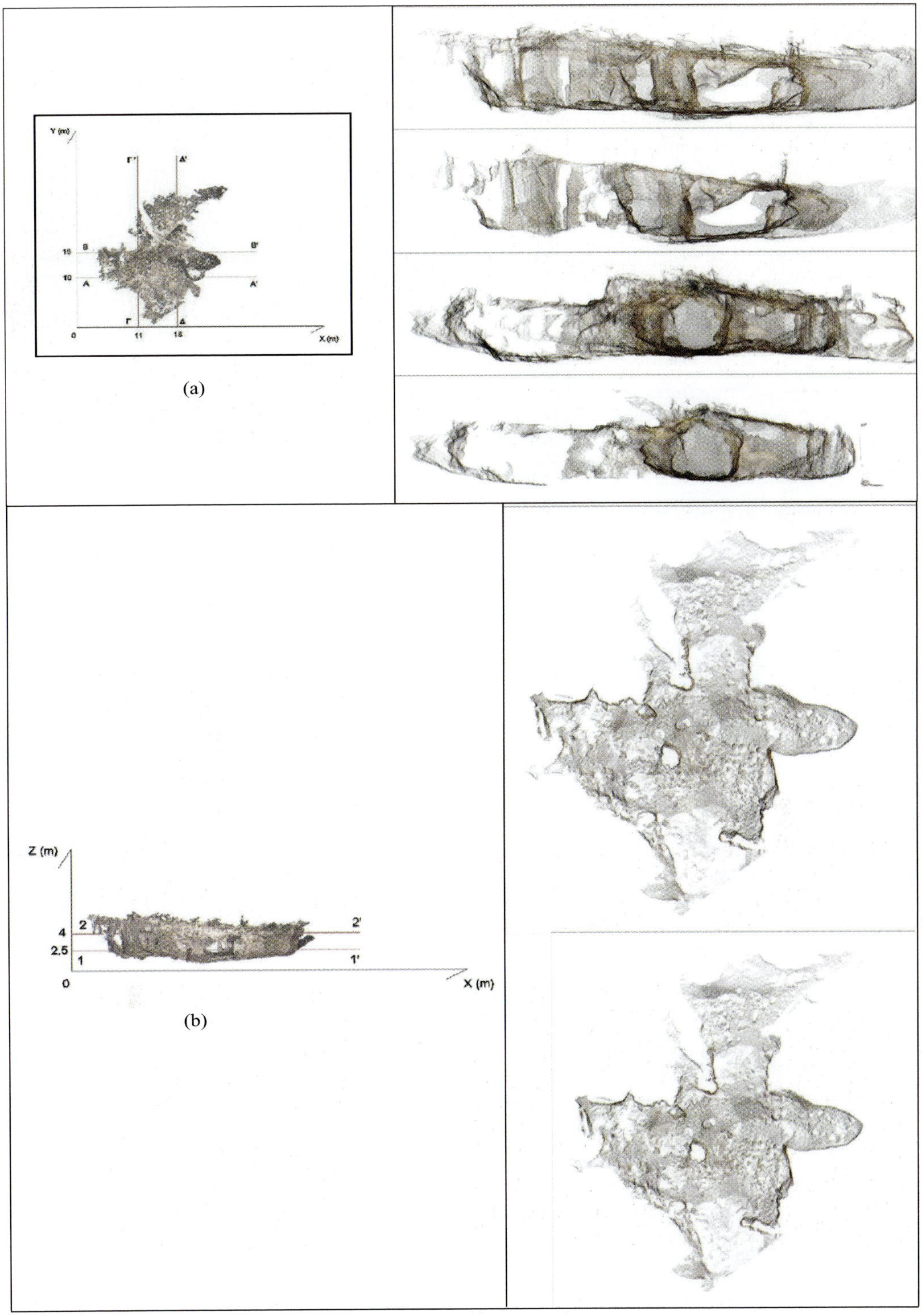

Figure 27.8. Four cross sections (a); two plan sections (b). (P. Patias et al., see page 369)

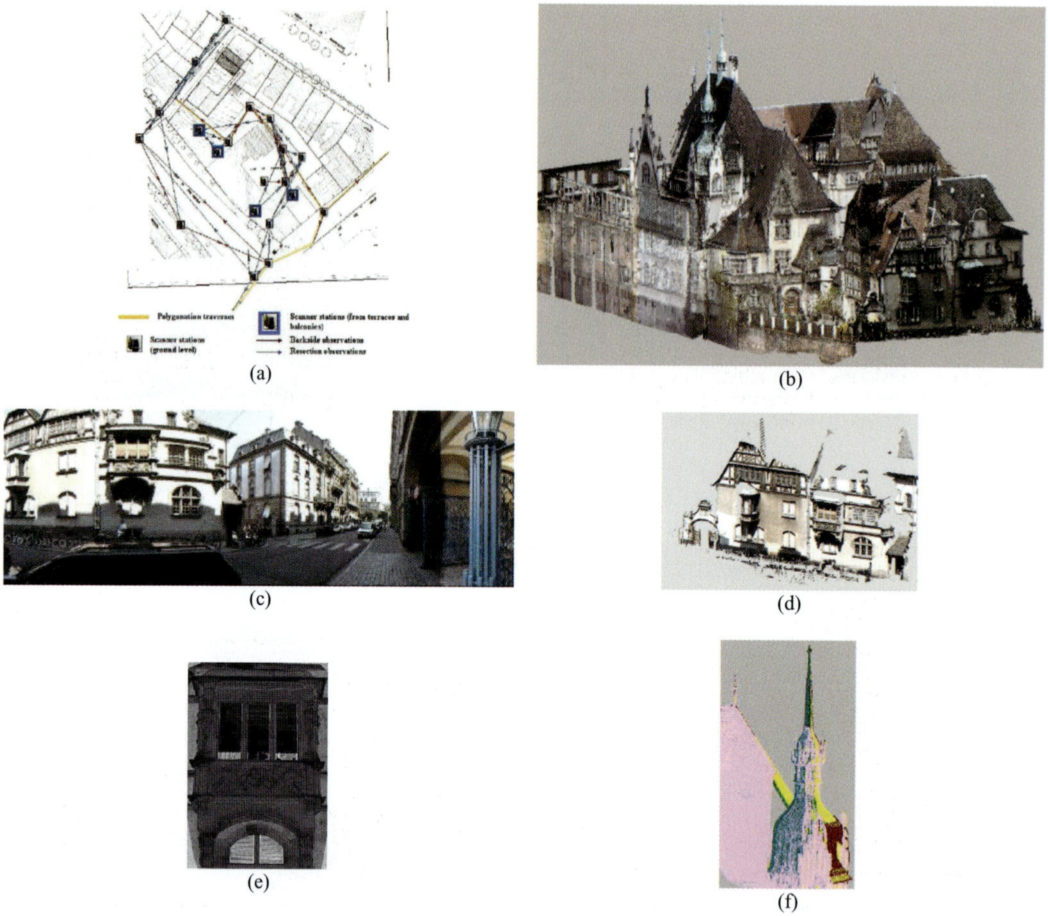

Figure 27.15. Ground control network used for the laser scanning survey (a); Final point cloud (with colors from the built-in video camera of the scanner) (b); Panoramic image computed from the video of the scanner (c); Point cloud (raw data) displayed with true color pixels (d); Detail of the façade (grey scale intensity image), 1 point / 1 cm (e); Control of the model by merging objects from different stations (f). (P. Patias et al., see page 373)

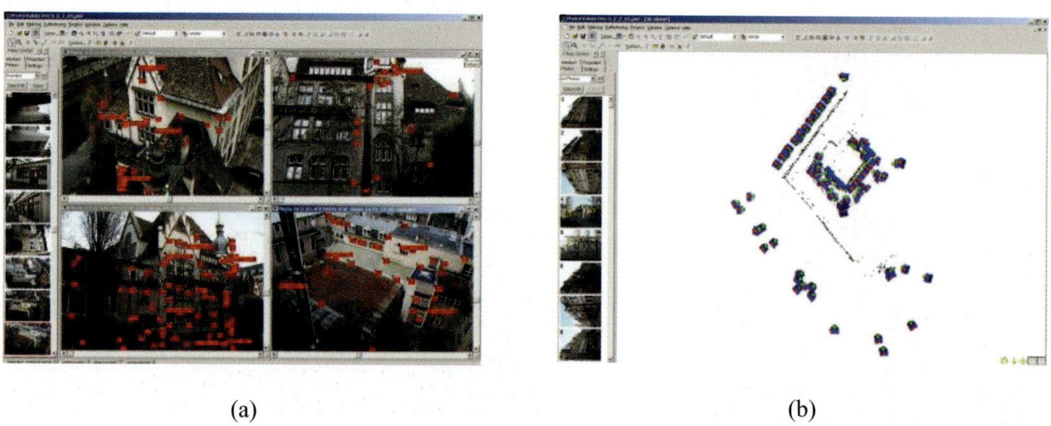

Figure 27.16. Set of photos in PhotoModeler's interface (a); Overall view of the photo stand points around the site (b). (P. Patias et al., see page 374)

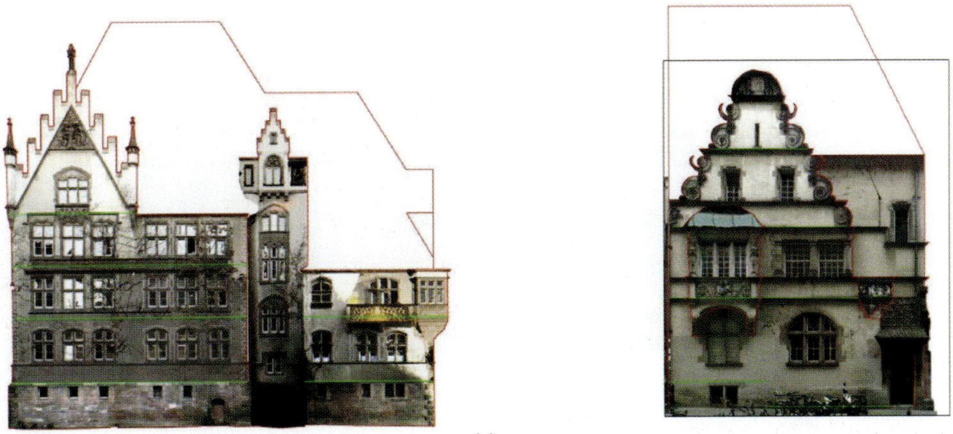

(a) (b)

Figure 27.17. Examples of orthophotos imported into the AutoCAD© software to be used as reference documentation for all providers in charge of the diagnosis. (P. Patias et al., see page 375)

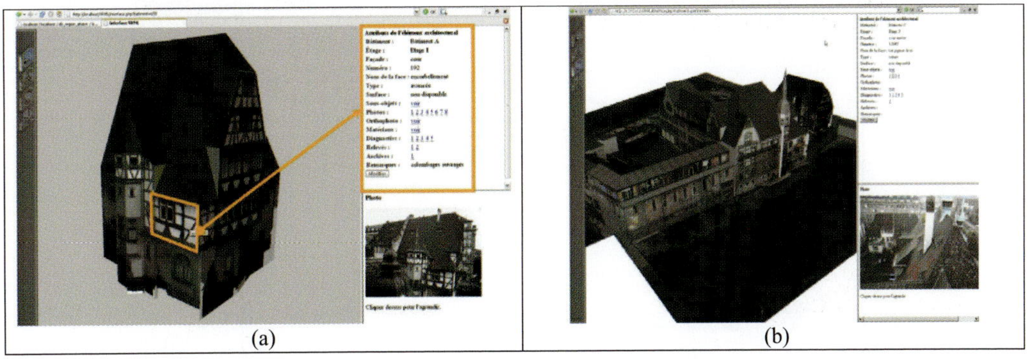

Figure 27.18. Interface allowing access to the documents of a building (a); View of the global 3D model showing the whole site in the web interface. The access to the information of each building is possible as shown in (a) (b). (P. Patias et al., see page 375)

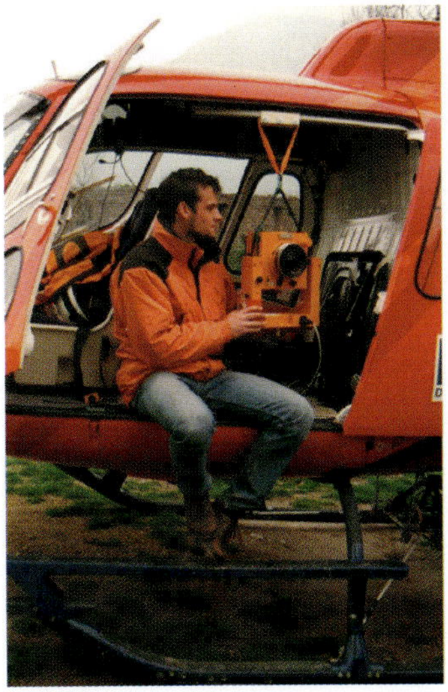

Figure 27.20. Zeiss UMK mounted in helicopter. (P. Patias et al., see page 376)

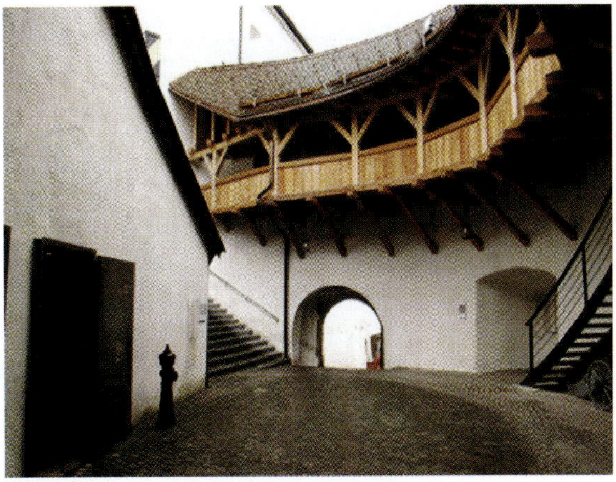

Figure 27.21. Digital photo of interior yard. (P. Patias et al., see page 376)

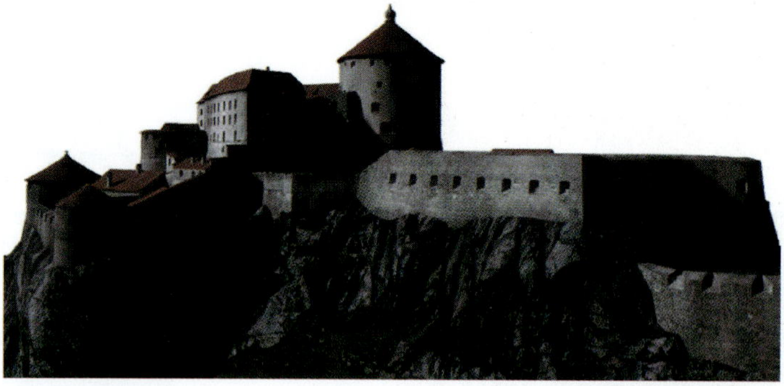

Figure 27.23. Partial view of the textured 3D model. (P. Patias et al., see page 377)

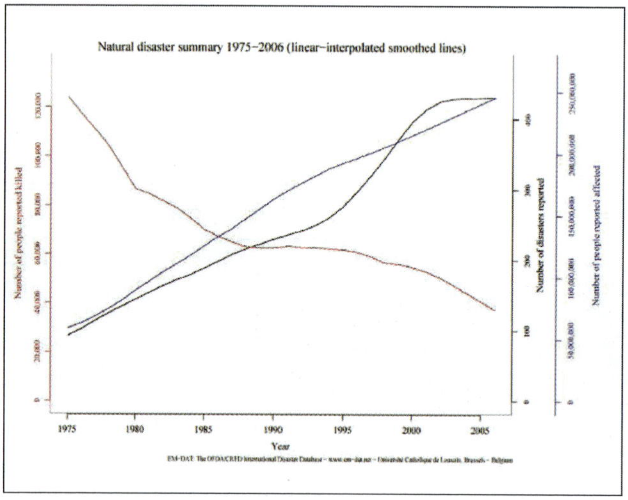

Figure 28.1. Natural Disaster summary 1975–2006 (CRED 2007). (P. Boccardo et al., see page 385)

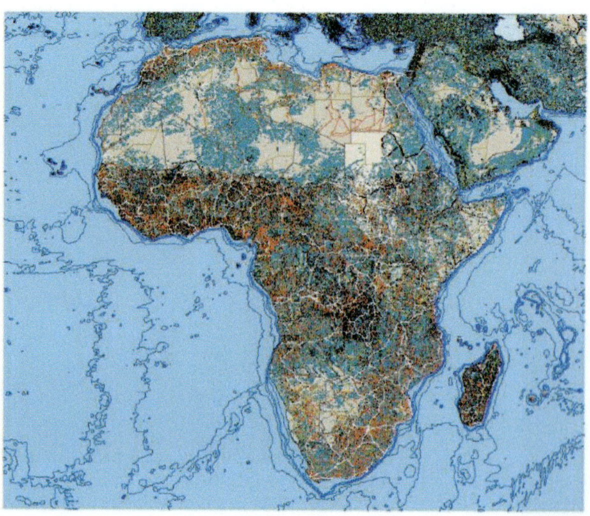

Figure 28.3. Test dataset of the African continent. (P. Boccardo et al., see page 389)

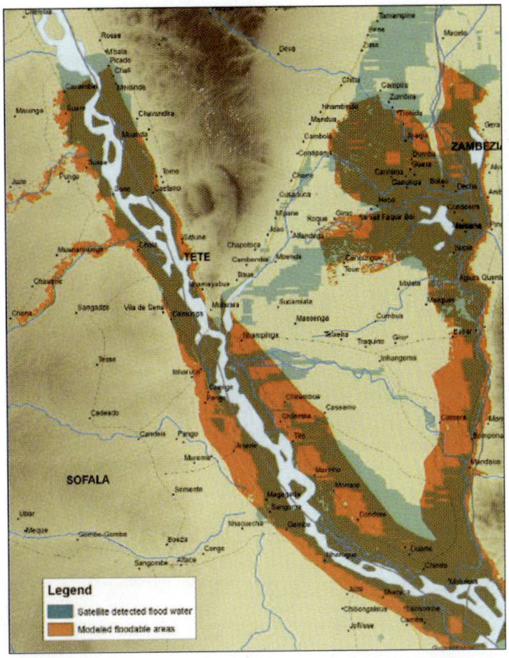

Figure 28.6. Modeled floodable areas overlapped to the real extent of the flood (Mozambique, 2007). (P. Boccardo et al., see page 392)

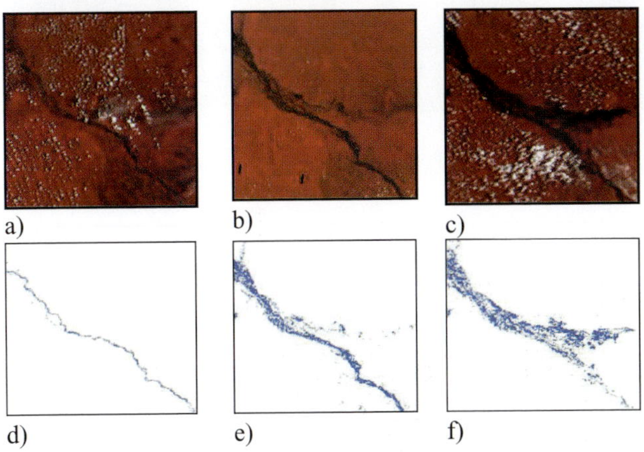

Figure 28.8. Mozambique 2007 flood event evolution. Original MODIS data (false colour): January 16-day composite (a), early February 16-day composite (b), late February 16-day composite (c) and corresponding identified water bodies and flooded areas - (d), (e) and (f). (P. Boccardo et al., see page 393)

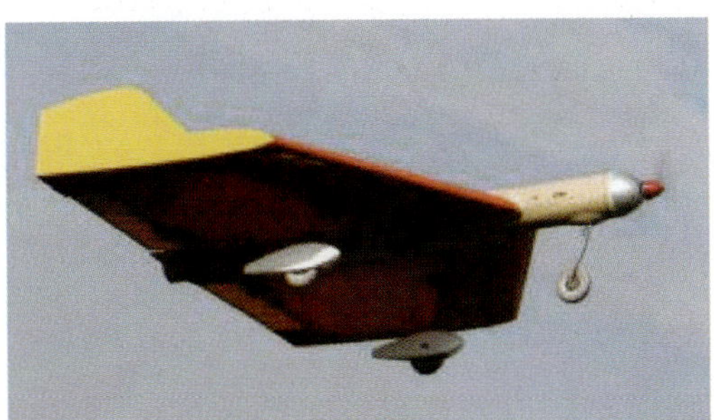

Figure 28.9. The UAV prototype ITHACA 01 during the first test flight. (P. Boccardo et al., see page 394)

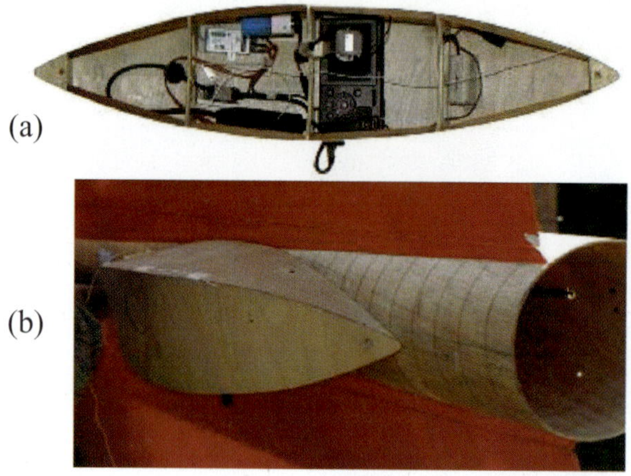

Figure 28.10. "Pod" with Ricoh GR digital camara (a), and "pod" position under the UAV (b). (P. Boccardo et al., see page 394)

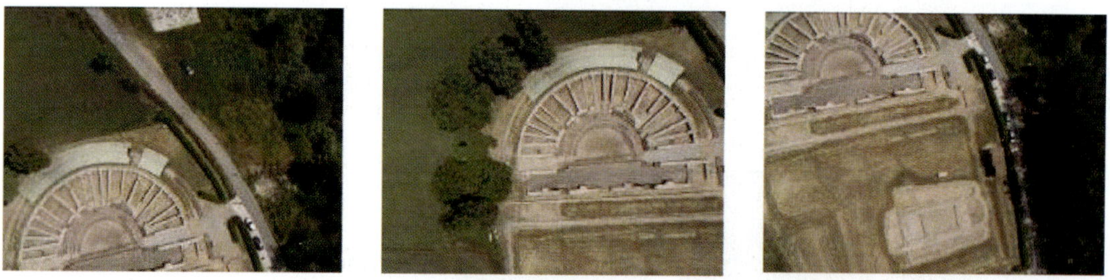

Figure 28.12. Images processed for the first orientation test. (P. Boccardo et al., see page 395)

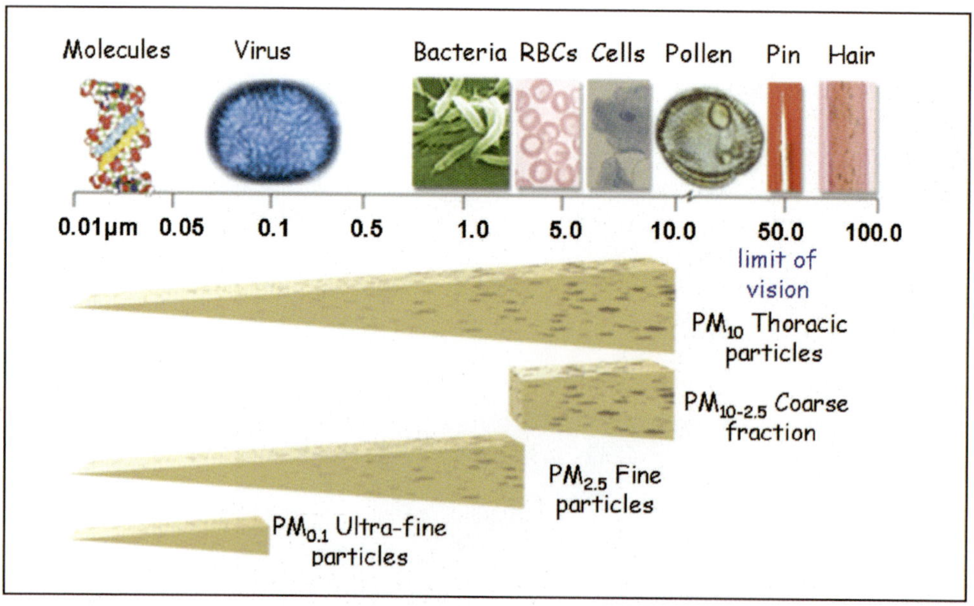

Figure 29.1. Particulate size distribution and related biophysical impacts. Modified from Kaiser (2005). (S.A. Morain et al., see page 398)

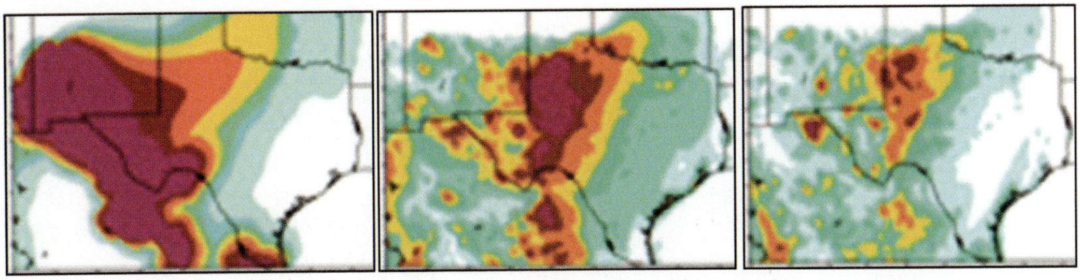

Figure 29.5. Dust storm in southwestern USA in January 2003: (left image) baseline model performance in 2005 before satellite data assimilation; (middle image) 2006 model performance after assimilation of satellite bare ground observations recorded by Terra/MODIS, and processed as the MOD12Q1 data product; (right image) 2007 model performance after the same EO data were assimilated into a higher resolution atmospheric model. (S.A. Morain et al., see page 406)

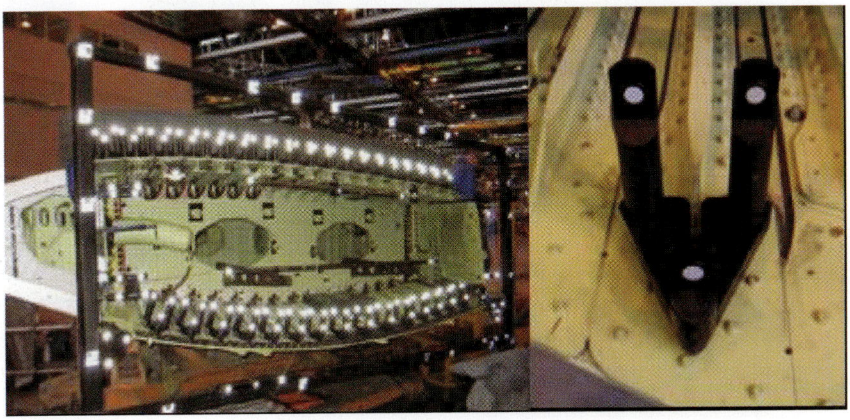

Figure 30.7. (Left) Airbus wing root fitted with targets for photogrammetric measurement. (Right) detailed view showing target design. (Courtesy of Airbus UK). (T. Luhmann et al., see page 418)

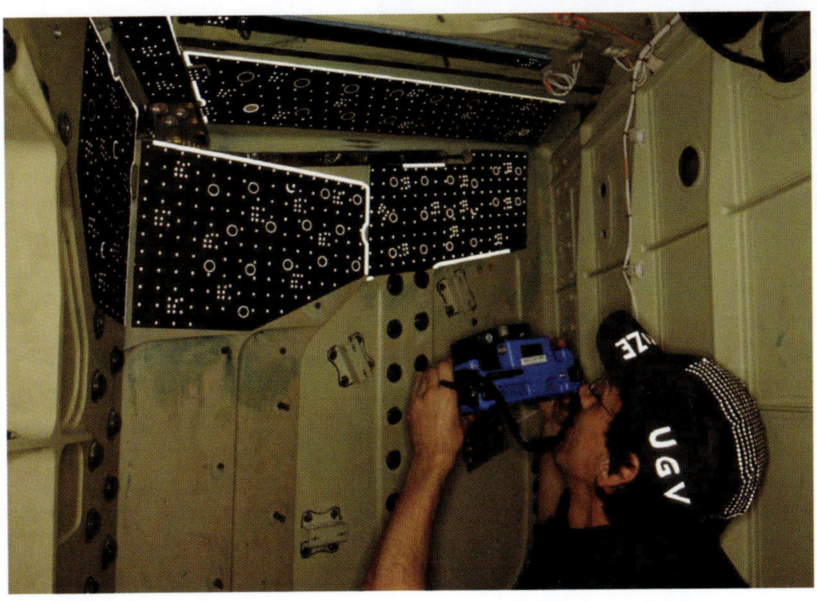

Figure 30.8. Photogrammetric measurement of the junction of the wing and fuselage in an Airbus A380 aircraft using a GSI V-STARS system. The photogrammetric system triggers an image processing function to automatically generate a 3-D surface and a 5 axis machining tool path. This process produces a titanium part that will fit in the corners. (Copyright: Airbus). (T. Luhmann et al., see page 418)

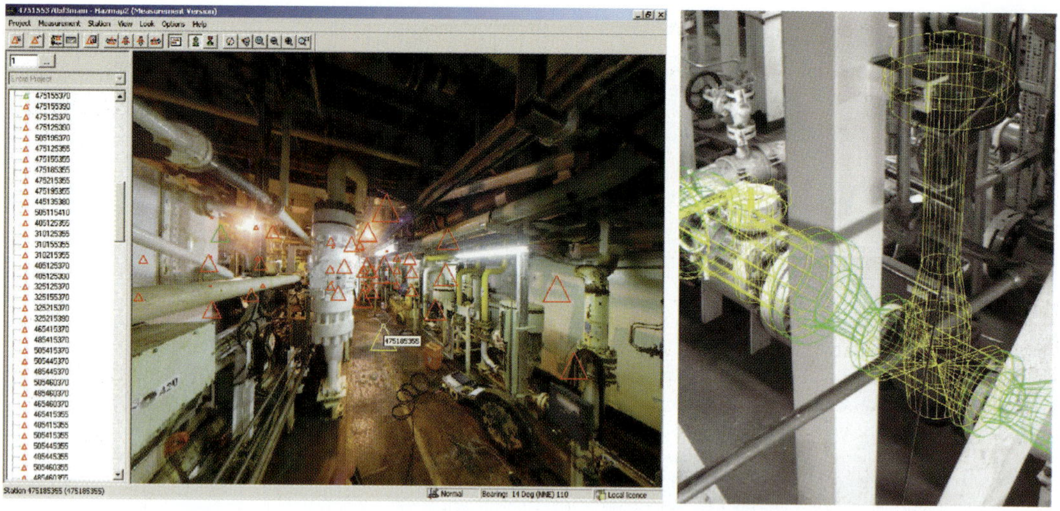

Figure 30.9. Example sub-image from the panoramic "Hazmap" site measurement system showing further panoramic image locations as triangles (Left). Example integration with CAD where a new pipe design has been integrated into the Hazmap environment for clash detection (Right). "Images Courtesy of Hazmap". (T. Luhmann et al., see page 419)

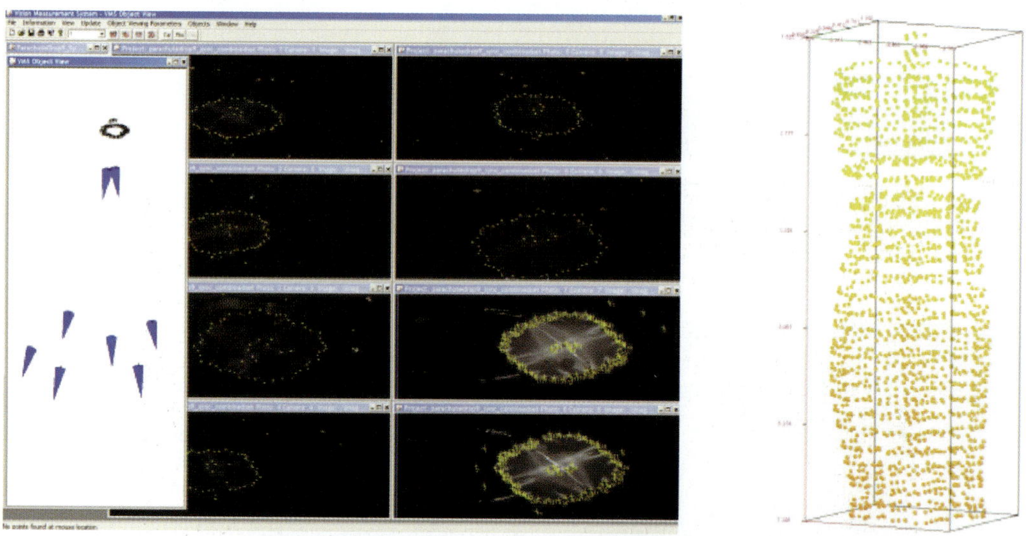

Figure 30.10. Left: Example images taken with a network of eight cameras used to simultaneously capture image sequences of a parachute in flight. Right: Off-line post processed output demonstrating the movement of the parachute. As processing capability and more notably tracking algorithms improve, such systems will become on-line. (Images Courtesy of NASA Langley, Shortis et al. 2007). (T. Luhmann et al., see page 419)

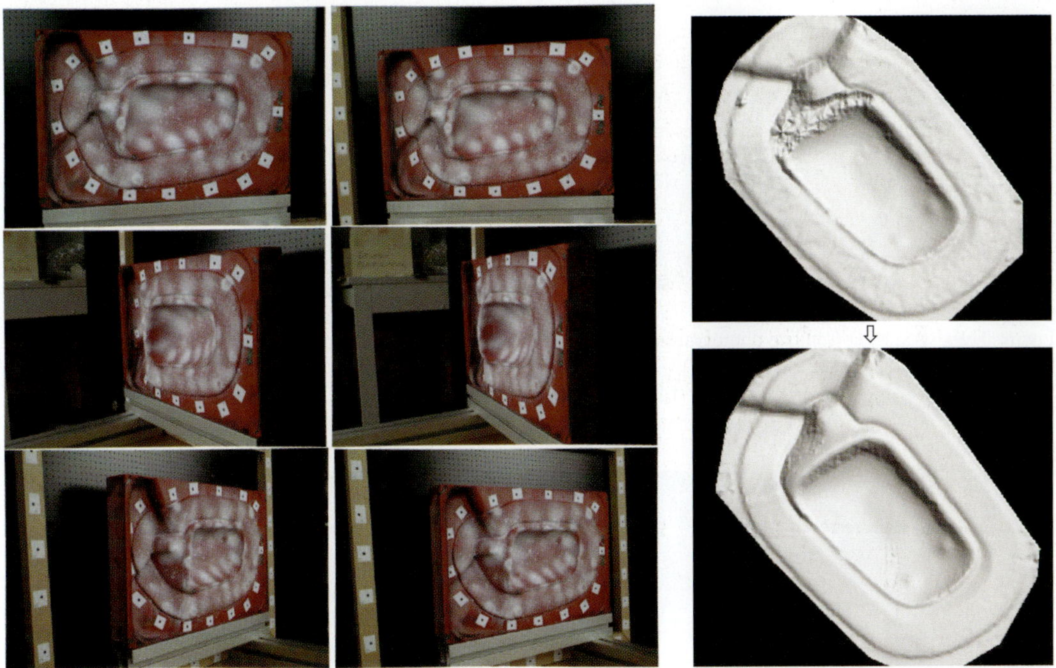

a) Multiple image acquisition

b) Iterative calculation of the 3D model

Figure 30.13. Photogrammetric surface measurement. (T. Luhmann et al., see page 421)

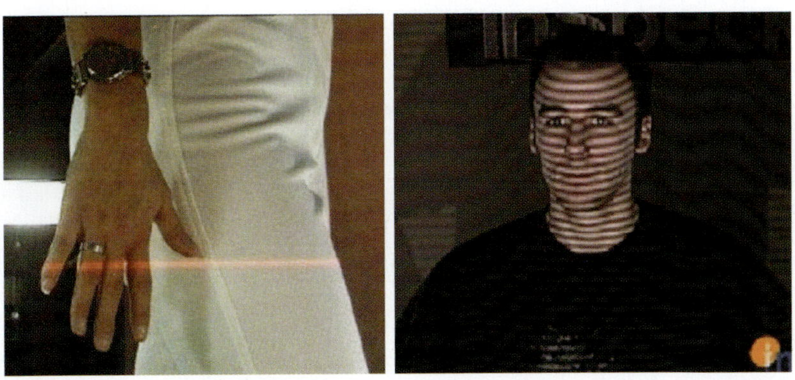

Figure 31.1. Left: laser stripe projected onto the human body. Right: structured light in form of multiple white stripes projected onto the human body. (N. D'Apuzzo et al., see page 427)

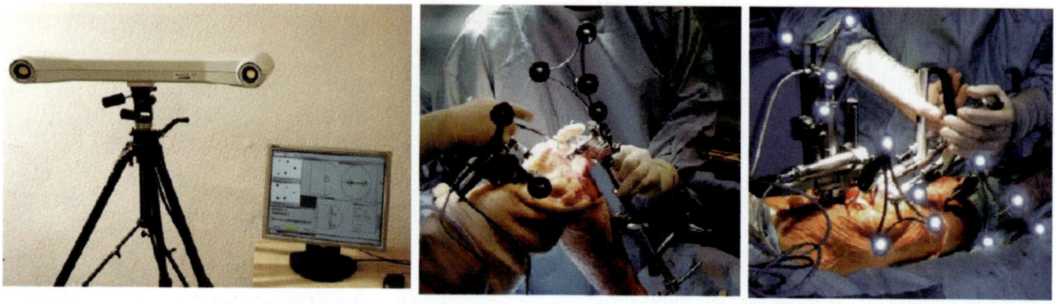

Figure 31.4. Axios3D Cambar photogrammetric measurement system and surgical instrumentation with signalized targets. (N. D'Apuzzo et al., see page 429)

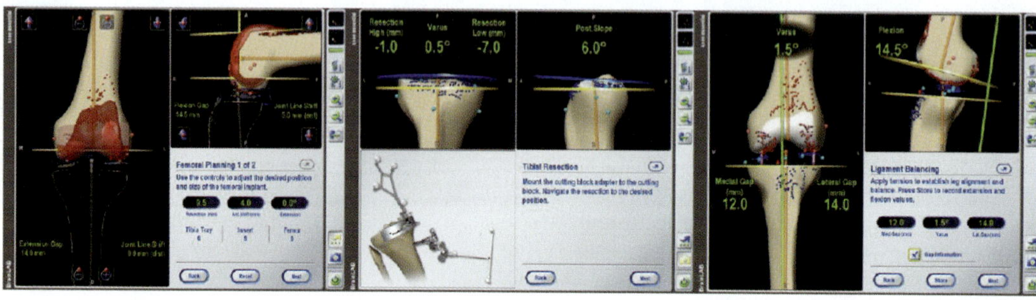

Figure 31.5. Snapshots of VectoVision Knee of BrainLAB (2008). Initial planning and alignment of the implant position (left), real-time bone cut navigation and verification (centre), kinematical analysis for optimal leg alignment balance (right). (N. D'Apuzzo et al., see page 429)

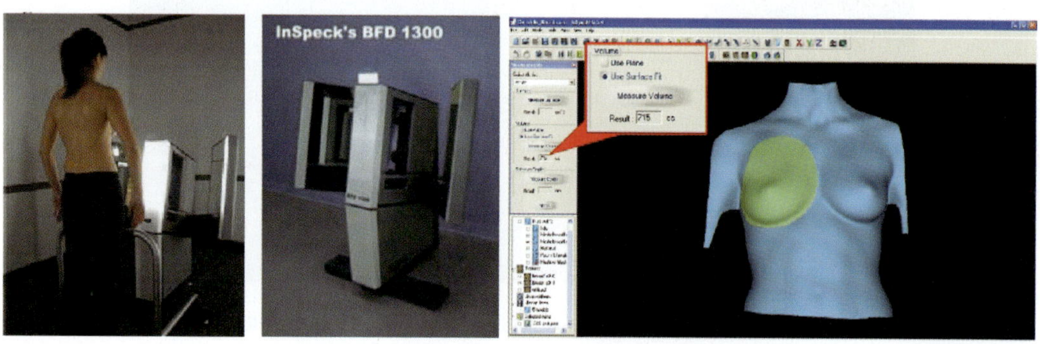

Figure 31.6. InSpeck Breast and Facial Digitizer BFD 1300 and measurement software EM-Measurement. (N. D'Apuzzo et al., see page 430)

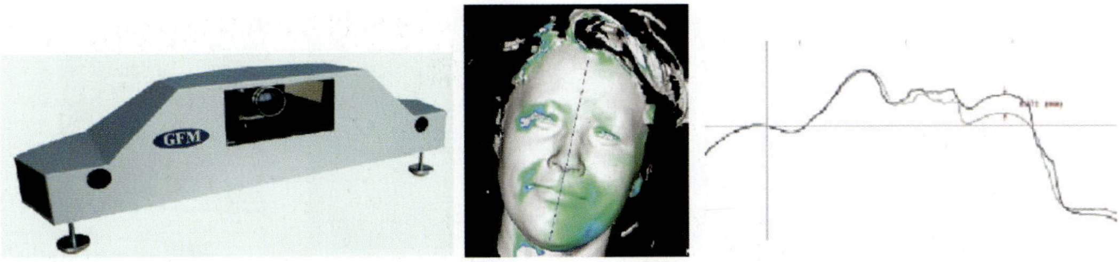

Figure 31.7. PRIMOS body II-portable 3D face and body scanner, 3D scan data and comparison (colours code the difference ranges), 2D cross sections showing difference between pre- and post-operative 3D scan data. (N. D'Apuzzo et al., see page 430)

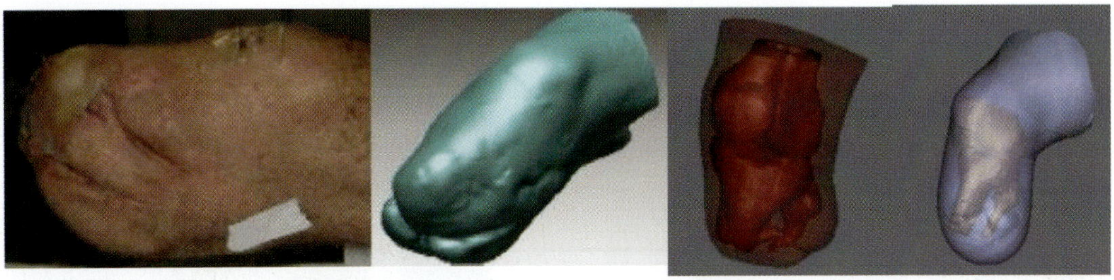

Figure 31.10. From left to right: stump image (note the signalized landmarks on the top), 3D laser scanning model, CT skin-bone model, integrated laser/bone model (right). (N. D'Apuzzo et al., see page 432)

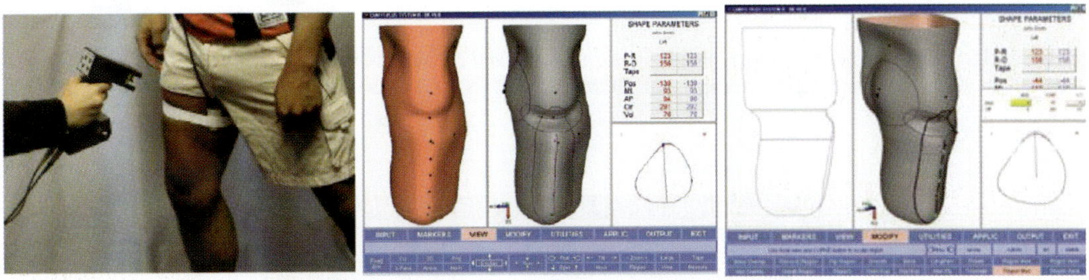

Figure 31.11. Lower limb socket design system CANFIT PLUS P&O: hand-held laser scanner and snapshots of the CAD/CAM solution. (N. D'Apuzzo et al., see page 432)

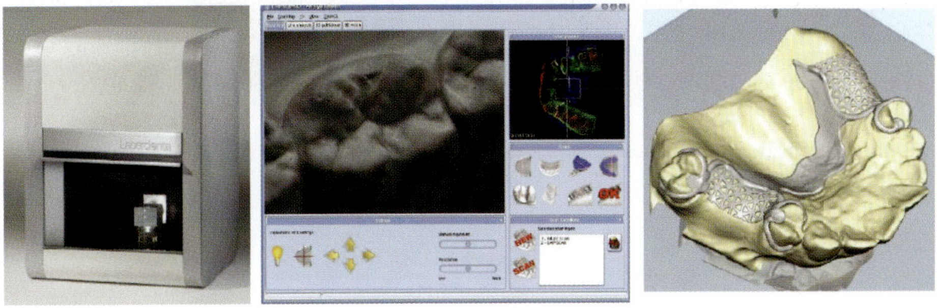

Figure 31.13. Dental solution of LaserDenta (Blumenschein 2006): desktop dental cast laser scanner (left), snapshot of the scanning control software (centre), virtual planning of a complex dental restoration (right). (N. D'Apuzzo et al., see page 434)

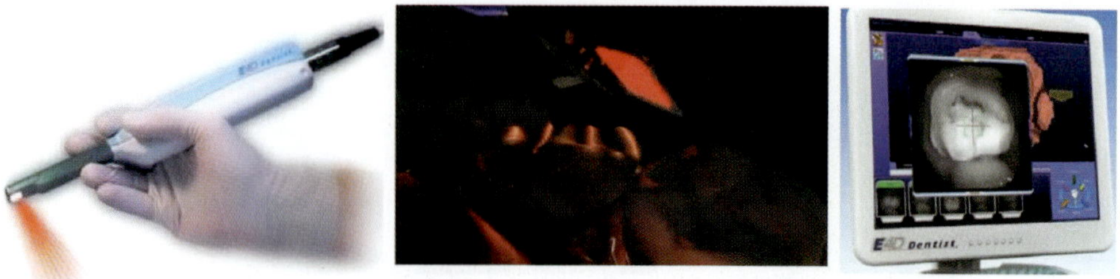

Figure 31.14. The intra-oral scanner E4D Dentist of D4D Technologies (2008): the hand-held miniaturized scanner (left), surface scanning by multiple lines projection (centre), real-time scanning control visualisation (right). (N. D'Apuzzo et al., see page 434)

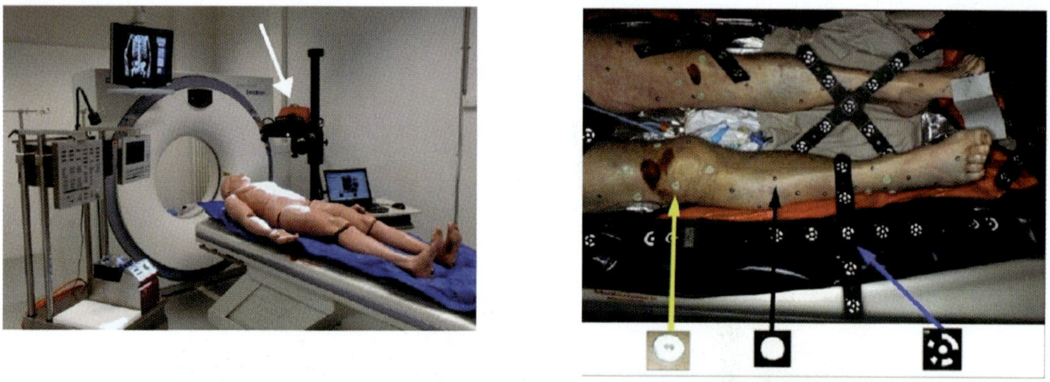

Figure 31.15. Left: Computer tomograph Siemens Emotion 6 and 3D surface scanner GOM ATOS III (arrow). Right: Different markers used for the measurement with GOM TRITOP. (N. D'Apuzzo et al., see page 435)

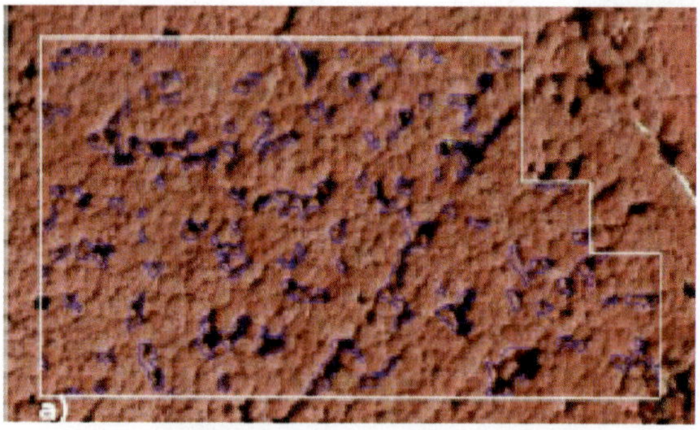

Figure 32.2. Full automatic assessment of gaps (Nuske 2006). (B. Koch et al., see page 442)

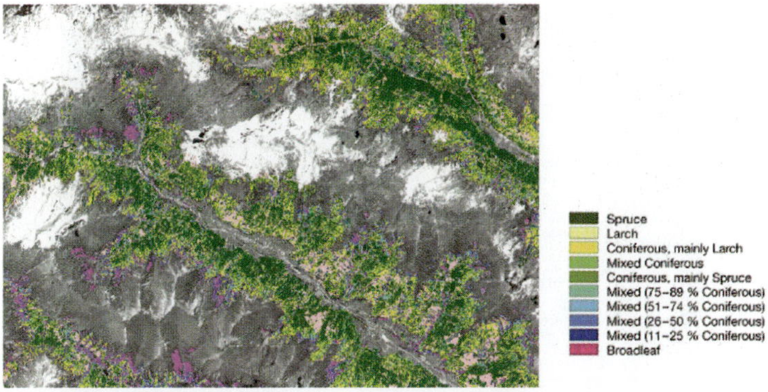

Figure 32.3. Classification result for tree species distribution using SPOT5 imagery in Austria (Schardt et al., 2005). (B. Koch et al., see page 443)

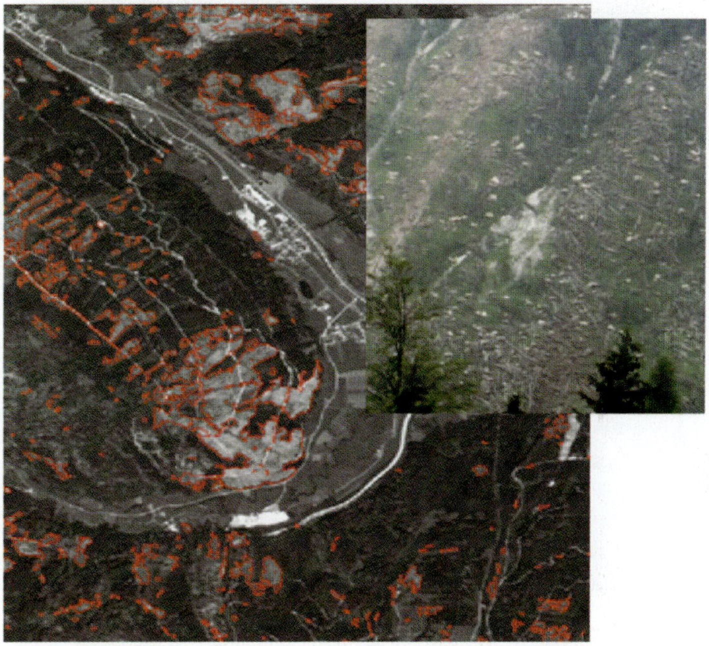

Figure 32.4. Automated derivation of windfall areas (red line) after a heavy Storm in November 2002 in Salzburg using SPOT Imagery. (B. Koch et al., see page 443)

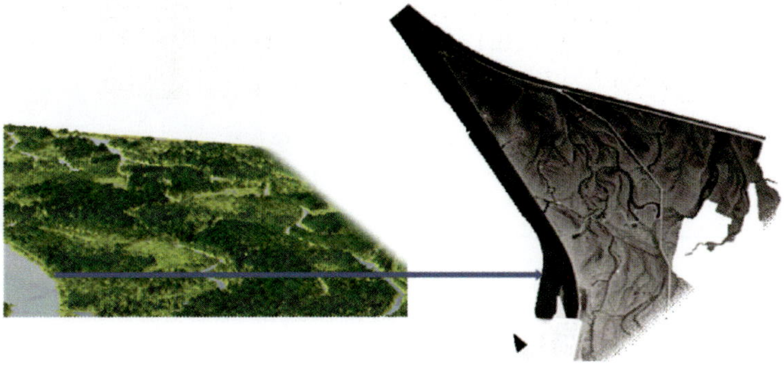

Figure 32.5. DEM calculated with the Active Surface Contour algorithm developed by Elmquist in the Rhine valley under forested area (Dees et al 2006). (B. Koch et al., see page 444)

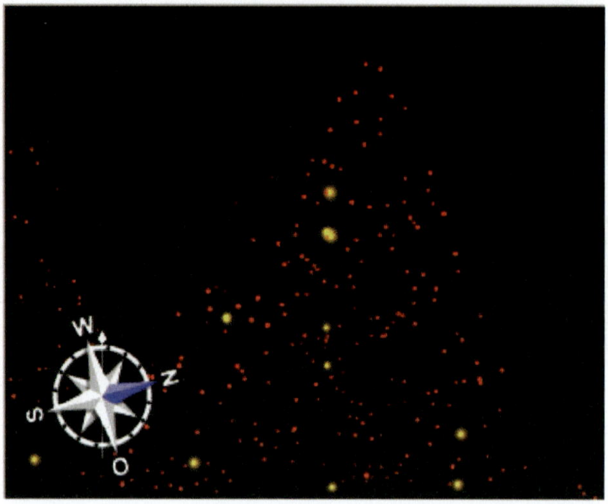

Figure 32.6. Vertical cut through a conifer crown with the reflected raw point (red) positions for a laser point density of 2 per sqm and with simulated positions of reflected points (yellow) for a laser point density of 0.06 per sqm (Koch & Weinacker 2007). (B. Koch et al., see page 445)

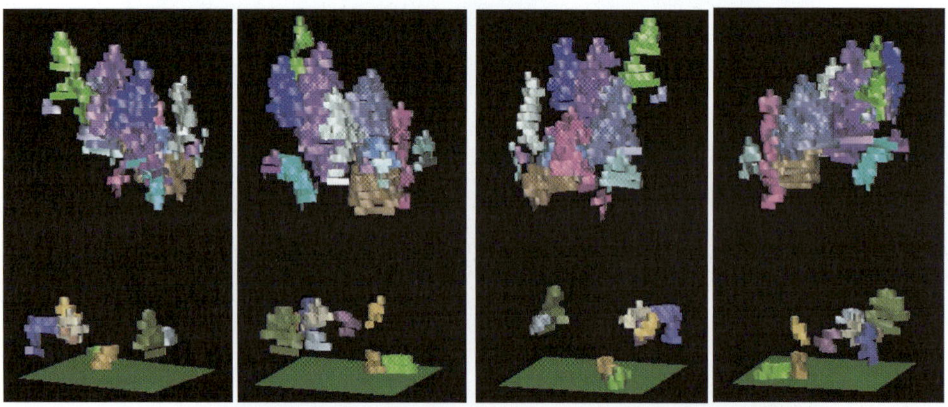

Figure 32.7. Single crown delineation based on blob detection method (Wang et al. 2007). (B. Koch et al., see page 445)

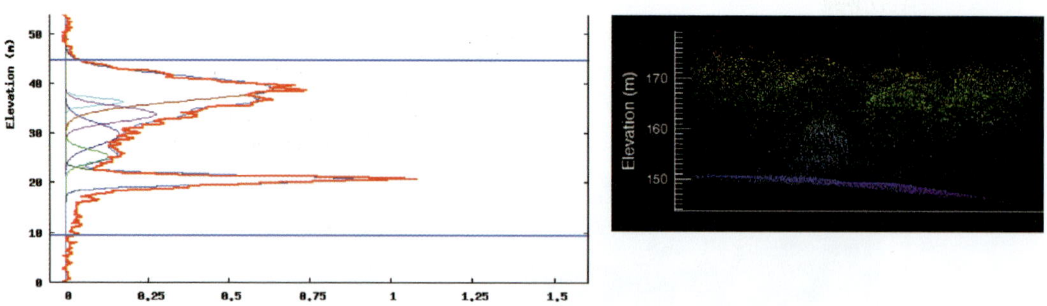

Figure 32.8. Bimodal full-wave form over vegetated area from airborne LIDAR (Rosette et al. 2007). (B. Koch et al., see page 446)

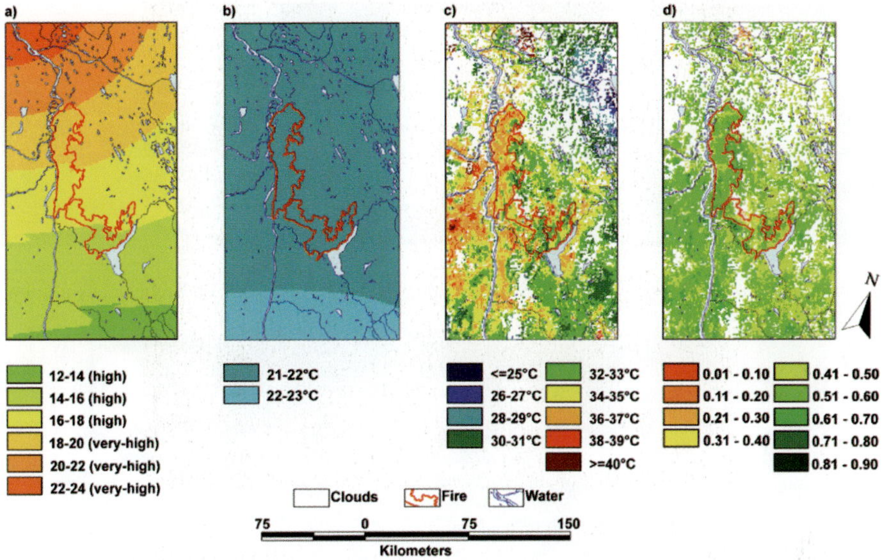

Figure 32.10. Map of (a) interpolated fire danger code (Fire Weather Index), (b) interpolated air temperatures (T_a), and corresponding T_s (c) NOAA-AVHRR image and (d) NDVI NOAA-AVHRR image for the area, where the fire "*vq0332*" starts. The image was acquired 3 days before fire ignition (after Oldford et al., 2003). (B. Koch et al., see page 452)

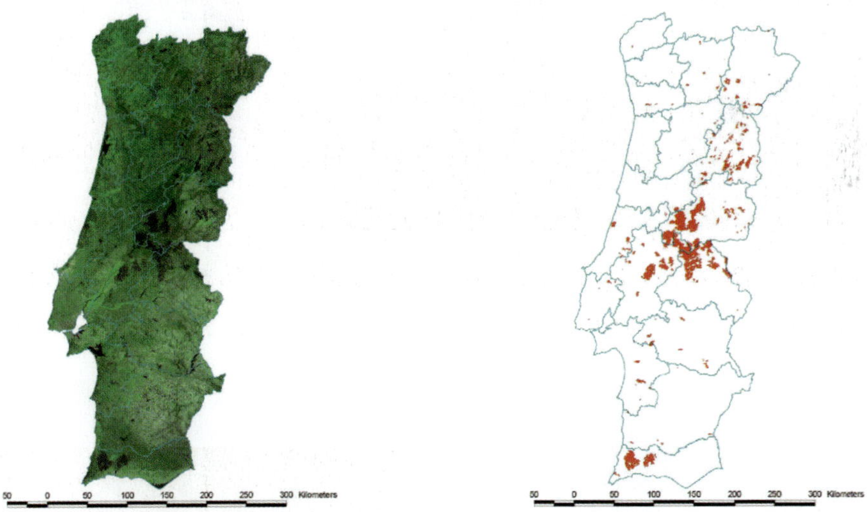

Figure 32.11. MODIS image (250 m spatial resolution) showing the burnt scars in black (left side), and mapped burnt areas in red on the administrative map of Portugal (right side) (Source EFFIS). (B. Koch et al., see page 453)

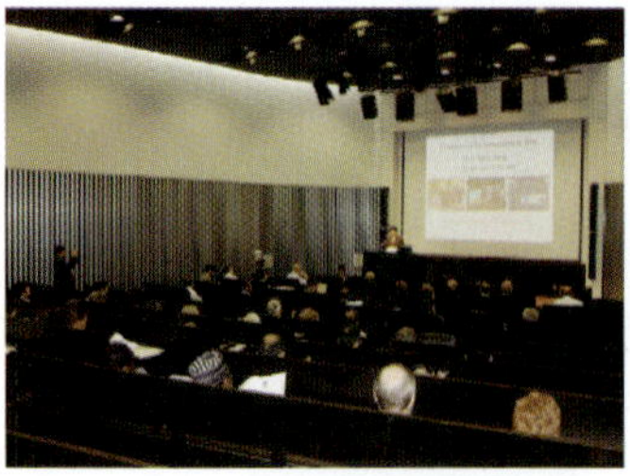

(a) The opening session

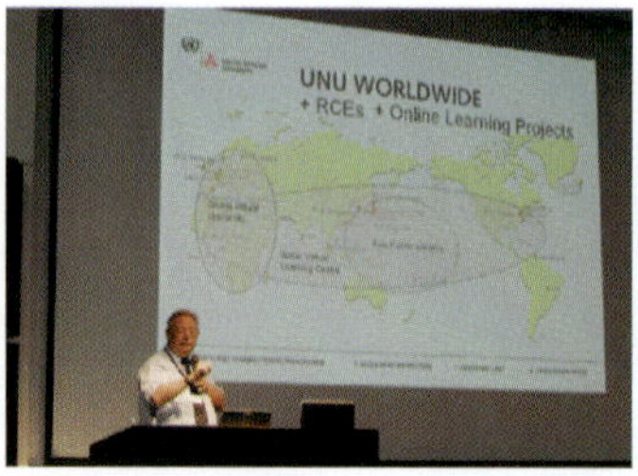

(b) Keynote speech

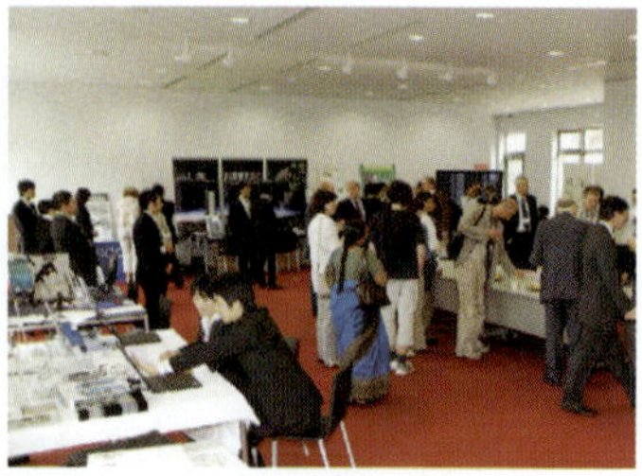

(c) Technical exhibition

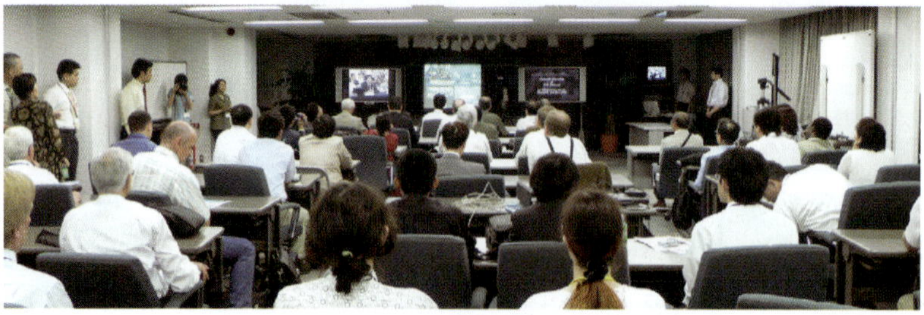

(d) JICA distance learning demo: Palestine, Kenya, Bangladesh and Sri Lanka were connected to Tokyo

Figure 33.6. Snapshots at the Commission VI Symposium in Tokyo in June, 2006. (K. Cho et al., see page 481)

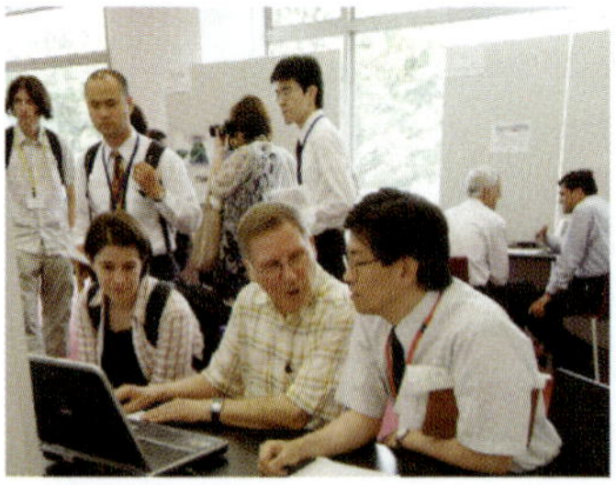

(a) Software demonstration

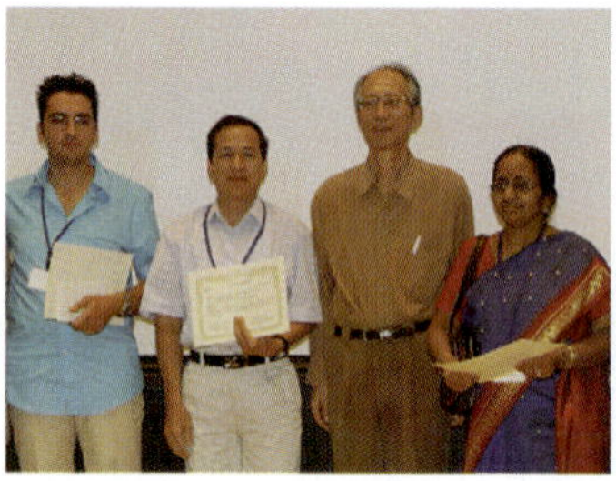

(b) Award Winners

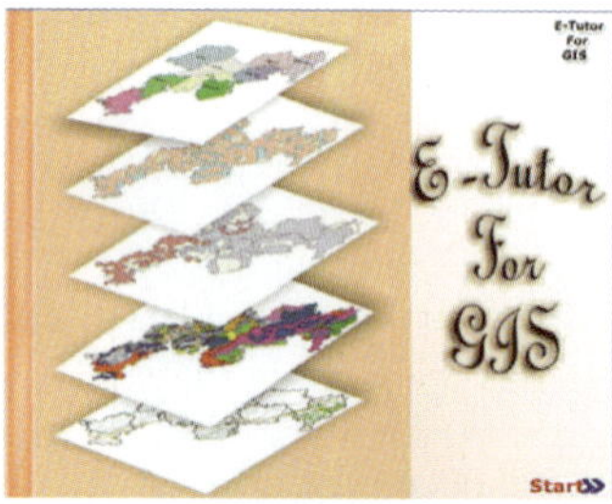

(c) E-Tutor for GIS

Figure 33.7. Snapshots at CATCON4. (K. Cho et al., see page 481)

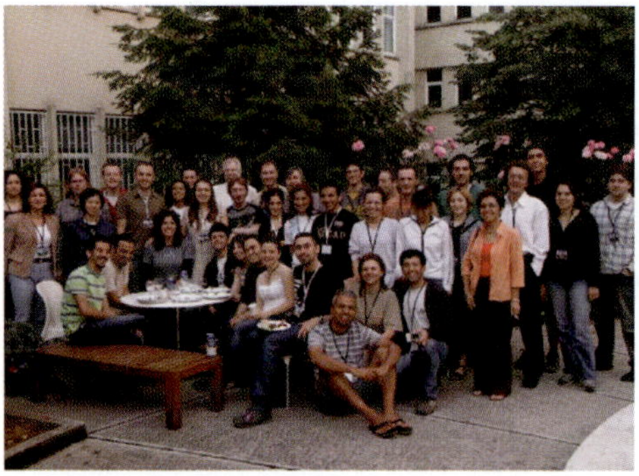

Istanbul, June, 2005

Figure 33.9. The first Summer School. (K. Cho et al., see page 483)

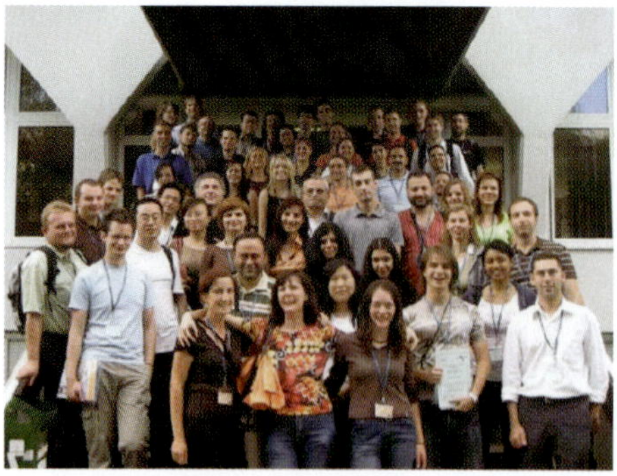

Ljubljana, July, 2007

Figure 33.10. The second Summer School. (K. Cho et al., see page 483)

Advances in Photogrammetry, Remote Sensing and Spatial Information Sciences: 2008 ISPRS Congress Book – Li, Chen & Baltsavias (eds)
© 2008 Taylor & Francis Group, London, ISBN 978-0-415-47805-2

Author index

Advances in Photogrammetry, Remote Sensing and Spatial Information Sciences: 2008 ISPRS Congress Book – Li, Chen & Baltsavias (eds)
© 2008 Taylor & Francis Group, London, ISBN 978-0-415-47805-2

Keyword index

ISPRS Book Series

1. Advances in Spatial Analysis and Decision Making (2004)
   Edited by Z. Li, Q. Zhou & W. Kainz
   ISBN: 978-90-5809-652-4 (HB)

2. Post-Launch Calibration of Satellite Sensors (2004)
   Stanley A. Morain & Amelia M. Budge
   ISBN: 978-90-5809-693-7 (HB)

3. Next Generation Geospatial Information: From Digital Image Analysis to Spatiotemporal Databases (2005)
   Peggy Agouris & Arie Croituru
   ISBN: 978-0-415-38049-2 (HB)

4. Advances in Mobile Mapping Technology (2007)
   Edited by C. Vincent Tao & Jonathan Li
   ISBN: 978-0-415-42723-4 (HB)
   ISBN: 978-0-203-96187-2 (E-book)

5. Advances in Spatio-Temporal Analysis (2007)
   Edited by Xinming Tang, Yaolin Liu, Jixian Zhang & Wolfgang Kainz
   ISBN: 978-0-415-40630-7 (HB)
   ISBN: 978-0-203-93755-6 (E-book)

6. Geospatial Information Technology for Emergency Response (2008)
   Edited by Sisi Zlatanova & Jonathan Li
   ISBN: 978-0-415-42247-5 (HB)
   ISBN: 978-0-203-92881-3 (E-book)

7. Advances in Photogrammetry, Remote Sensing and Spatial Information Sciences. 2008 ISPRS Congress Book
   Edited by Zhilin Li, Jun Chen & Emmanuel Baltsavias
   ISBN: 978-0-415-47805-2 (HB)
   ISBN: 978-0-203-88844-5 (E-book)